Further Progress in Analysis

Further Progress in Analysis

Proceedings of the 6th International ISAAC Congress

Middle East Technical University,
Ankara, Turkey

13 – 18 August 2007

Editors

H. G. W. Begehr
Freie Universität Berlin, Germany

A. O. Çelebi
Yeditepe University, Turkey

R. P. Gilbert
University of Delaware, USA

Editorial Assistant

H. T. Kaptanoğlu
Bilkent University, Turkey

World Scientific

NEW JERSEY • LONDON • SINGAPORE • BEIJING • SHANGHAI • HONG KONG • TAIPEI • CHENNAI

Published by

World Scientific Publishing Co. Pte. Ltd.
5 Toh Tuck Link, Singapore 596224
USA office: 27 Warren Street, Suite 401-402, Hackensack, NJ 07601
UK office: 57 Shelton Street, Covent Garden, London WC2H 9HE

British Library Cataloguing-in-Publication Data
A catalogue record for this book is available from the British Library.

FURTHER PROGRESS IN ANALYSIS
Proceedings of the 6th International ISAAC Congress

ISBN-13 978-981-283-732-5
ISBN-10 981-283-732-9

Printed in Singapore by World Scientific Printers

PREFACE

The *6th International ISAAC Congress* took place on the beautiful campus of Middle East Technical University (METU), Ankara, Turkey, during the period 13–18 August 2007. The Congress was carried out by the Ankara Branch of the Turkish Mathematical Society with the generous support of METU and particularly of the Department of Mathematics.

There were 303 registered participants from 49 countries of whom more than 250 were from abroad. These numbers made the Congress the largest mathematical scientific meeting ever held in Turkey. The governmental support obtained enabled the organizers to introduce 25 well-known mathematicians from Central Asia to the western world.

The ISAAC Board acted as the Scientific Committee of the Congress. The Local Organizing Committee consisted of Öner Çakar (Ankara University), A. Okay Çelebi (then METU, now Yeditepe University), H. Turgay Kaptanoğlu (Bilkent University) and Ağacık Zafer (METU). They had a long and exhausting preparation period. The illness and hospitalization of Prof. Çelebi for an extended time scared but did not slow them. Luckily, he was back to his normal health and activity well before the Congress.

The program of the Congress was a combination of plenary talks and special sessions. The 7 plenary speakers, who are international experts in their fields of interest, were selected by the ISAAC Board. The plenary speakers and the titles of their presentations were the following:

G. A. Barsegian (Armenia) : Some Interrelated Results in Different Branches of Geometry and Analysis

A. Bourgeat (France) : Spatial and Random Scaling Up of the Source Terms in a Diffusion Convection Model

P. L. Butzer (Germany) : Prediction in Terms of Samples from the Past: Error Estimates by a Modulus Covering Discontinuous Signals

W. D. Evans (UK) : Recent Results on Hardy and Rellich Inequalities

B. Gramsch (Germany) : Spectral Invariance for Pseudodifferential Operators and Fredholm Theory

J. Sjöstrand (France) : Some Results on Nonselfadjoint Operators: A Survey

C. Y. Yıldırım (Turkey) : The Distribution of Primes: Conjectures vs. Hitherto Provables

In addition, there were 281 contributed talks in 15 special sessions. The final form of the Special Sessions and their organizers were as follows:

Analytic Function Spaces and Their Operators
R. Aulaskari (Finland), H. T. Kaptanoğlu (Turkey)

Clifford and Quaternion Analysis
I. M. Sabadini (Italy), M. Shapiro (Mexico), F. Sommen (Belgium)

Complex Analysis and Potential Theory
T. A. Aliyev (Turkey), M. Lanza de Cristoforis (Italy), P. M. Tamrazov (Ukraine)

Complex Analytic Methods for Applied Sciences
S. V. Rogosin (Belarus), V. V. Mityushev (Poland)

Complex and Functional Analytic Methods in Partial Differential Equations
H. G. W. Begehr (Germany), D.-Q. Dai (China), J. Y. Du (China)

Dispersive Equations
V. Georgiev (Italy), M. Reissig (Germany)

Inverse and Ill-Posed Problems
A. Hasanov (Turkey), M. Yamamoto (Japan), S. I. Kabanikhin (Russia)

Modern Aspects of the Theory of Integral Transforms and Their Applications
A. Kilbas (Belarus), S. Saitoh (Japan), J. J. Trujillo (Spain)

Oscillation of Functional-Differential and Difference Equations
L. Berezansky (Israel), A. Zafer (Turkey)

Pseudodifferential Operators
L. Rodino (Italy), M. W. Wong (Canada)

Reproducing Kernels and Related Topics
D. Alpay (Israel), A. Berlinet (France), S. Saitoh (Japan), D.-X. Zhou (China)

Spaces of Differentiable Functions of Several Real Variables and Applications
V. I. Burenkov (UK), S. G. Samko (Portugal)

Numerical Functional Analysis
P. E. Sobolevski (Brazil), A. Ashyralyev (Turkey)

Integrable Systems

M. Gürses (Turkey), Ismagil T. Habibullin (Turkey)

General Session (topics not suitable for any of the sessions listed above)

A. O. Çelebi (Turkey)

The contributions to the Proceedings of the Congress are organized so as to follow the special sessions. Thanks are due to the organizers for accepting to be the editors of their sessions and for their extraordinary efforts to keep the scientific quality of the Proceedings at a high level. The texts of the plenary talks were edited by the editorial board of this volume. The organizers of two sessions, of Inverse and Ill-Posed Problems and of Pseudodifferential Operators, decided to publish their contributed papers elsewhere. Thus this volume contains 75 articles from 13 sessions along with the texts of 4 plenary talks, all refereed. A few of the manuscripts were not in the style of the publisher or even TEX. Dr. Uğur Yüksel converted them into expected forms. But the main job of preparing the Proceedings for publication was done by Prof. H. Turgay Kaptanoğlu who acted as editorial assistant. He did major work in compiling the individual papers into book form and correcting them in every respect, from mathematics to language to TEX. His efficient and meticulous work is evident in this volume. He was also the editor of the book of abstracts and the schedule of the Congress.

Two important events occurred during the Congress. The first was the presentation of the ISAAC award for researchers under the age of 40 to Dr. Michael Ruzhansky during the opening ceremony. The second was the meeting of the ISAAC Board. The Board reelected Prof. M. W. Wong as the president of ISAAC, and decided that the next Congress would take place in the United Kingdom at the host institution of the ISAAC awardee. So the 7th International ISAAC Congress will be held at Imperial College London, July 13–18, 2009.

The high scientific quality of the Congress can be seen from the contributions in this Proceedings volume. The presentations covered a wide range topics varying from analytic number theory to computational aspects of applied analysis. There were plenary talks directed to a more general audience as well as to specialists. The sessions of the Congress were well-attended despite the lack of air conditioning and the heat wave. The closing ceremony extraordinarily attracted a majority of the participants.

The support of the institutions listed below made the Congress possible. More than 25% of the participants were supported one way or another.

Republic of Turkey Prime Ministry Promotion Fund
Turkish International Cooperation and Development Agency (TİKA)
Scientific and Technological Research Council of Turkey (TÜBİTAK)
European Mathematical Society
Middle East Technical University
Ankara University
Bilkent University
Gama Trade and Tourism
Yüksel Construction
Kavaklıdere Anatolian Wines
Ertem Printing House
Birkhäuser

Also, volunteer work of colleagues and friends are gratefully acknowledged. The last, but not the least, the students in orange T-shirts should be thanked for their ever presence to help and joviality.

Congress services were successfully and efficiently handled by Asterya Congress and Event Management. They and the local organizers prepared an excellent social program and excursions with wining and dining, and music and dancing, sometimes in surprise times and locations. These all made the whole event a memorable one.

The Editors
15 October 2008

SCIENTIFIC COMMITTEE

of the Congress

(The ISAAC Board of 2007)

H. G. W. Begehr	Freie Universität, Berlin, Germany
A. Berlinet	Université des Sciences et Techniques de Languedoc, Montpellier, France
E. Brüning	University of KwaZulu-Natal, Durban, South Africa
V. I. Burenkov	Cardiff University, Cardiff, Wales, UK
A. O. Çelebi	Yeditepe University, İstanbul, Turkey
R. P. Gilbert	University of Delaware, Newark, DE, USA
A. Kilbas	Belarusian State University, Minsk, Belarus
M. Lanza de Cristoforis	Università di Padova, Padova, Italy
W. Lin	Zhongshan University, Guangzhou, China
M. Reissig	Technische Universität, Freiberg, Germany
L. Rodino	Università di Torino, Torino, Italy
J. Ryan	University of Arkansas, Fayetteville, AR, USA
S. Saitoh	Gunma University, Kiryu, Japan
B.-W. Schulze	Universität Potsdam, Potsdam, Germany
J. Toft	Växjö University, Växjö, Sweden
M. W. Wong	York University, Toronto, Canada
M. Yamamoto	University of Tokyo, Tokyo, Japan
S. Zhang	University of Delaware, Newark, DE, USA

LOCAL ORGANIZING COMMITTEE

of the Congress

Ö. Çakar	Ankara University, Ankara, Turkey
A. O. Çelebi	Yeditepe University, İstanbul, Turkey
H. T. Kaptanoğlu	Bilkent University, Ankara, Turkey
A. Zafer	Middle East Technical University, Ankara, Turkey

CONTENTS

Session 06 Dispersive Equations 413

Session 11 Oscillation of Functional Differential and Difference Equations 505

Plenary Talks

EDITORS

H. G. W. Begehr Freie Universität, Berlin, Germany
A. O. Çelebi Middle East Technical University, Ankara, Turkey
R. P. Gilbert University of Delaware, Newark, DE, USA

SOME INTERRELATED RESULTS IN DIFFERENT BRANCHES OF GEOMETRY AND ANALYSIS

G. A. BARSEGIAN

Institute of Mathematics of NAS, Armenia
E-mail: barseg@instmath.sci.am

This paper presents some new identities and inequalities in integral and differential geometry, in real and complex analysis in ODE. Ideas, methods and problems come from some preceding studies related to Gamma-lines. There are also certain bridges between the presented results and Nevanlinna theory of meromorphic functions.

Keywords: Differential Geometry, Integral Geometry, Real Analysis, Complex Analysis, ODE, Nevanlinna Theory, Gamma Lines.

Introduction

In the sections 1-4 of this paper we give some new interrelated inequalities for the broken lines, plane curves, real and complex functions of one variable. Derivation of these inequalities is based on a new *principle of angles and lengths* for curves (Section 1). Then we obtain some modifications that refer to the broken lines (Section 2) and to the real functions of one variable (Section 3). In the Section 4 we apply the principle of angles and lengths to the complex functions and obtain, particularly we obtain an inequality, *principle of derivatives*, valid for arbitrary analytic function in a given domain. In the Section 5 we apply the principle of angles and lengths to describe the windings of the solutions of some broad classes of ODE.

Some of these results resemble the second fundamental theorem of Nevanlinna theory and its deficiency relation. Thus we see that Nevanlinna type results (Nevanlinna theory [16], Ahlfors theory [1], Gamma-lines [7,8]) are valid far beyond complex analysis since we have now their corresponding analogues in differential geometry, real analysis and ODE.

Section 6 presents some generalized identities and inequalities in integral geometry. The last results are applied to the real functions of two variables in Section 7. We obtain an inequality, *principle of zeros*, which deals with the zeros of these functions and their derivatives.

The titles of the sections.

1. Some Nevanlinna type inequalities for the plane curves.
2. Consequences for the broken lines.
3. Consequences for the real functions of one variable.
4. Applications in complex analysis, particularly principle of derivatives of analytic functions.
5. Applications in ODE: the windings of solutions.
6. Some identities and inequalities in integral geometry.
7. Principles of zeros for the real functions of two variables.

The author thanks Professors R. Ambartzumian, H. Begehr, A. Fernandez, and D. T. Le for valuable discussions of results. Also, I should like to thank the ICTP at Trieste and the Departments of Mathematics of UNED at Madrid for their hospitality as well as INTAS (for two grants) and ANSEF (for two grants).

1. Some Nevanlinna type inequalities for the plane curves

1.1. *The principle of angles and lengths and consequent deficiency relation for curves*

Let γ be a curve in the plane (x, y). We can consider it as a curve in the complex plane $x+iy$ respectively $\gamma := f(t) := f_1(t)+if_2(t)$, $t \in [0,1]$, where $f(t)$ is a complex function of the real argument t. Denoting by $f_1^{(j)}(t)$ and $f_2^{(j)}(t)$ the derivatives we say $\gamma := f(t) \in F(k)$, k is an integer ≥ 1, if $\gamma^{(j)} := f^{(j)}(t) := f_1^{(j)}(t) + if_2^{(j)}(t)$ is continuous in $[0,1]$ for any j, $1 \leq j \leq k+1$, and if for any j, $0 \leq j \leq k+1$ and any $t \in [0,1]$ we have $f^{(j)}(t) \neq 0$.

Notice that $\arg f(t_0)$ means the angle between the x-axis and the vector connecting 0 and $f(t_0)$, while $\arg f'(t)$ means the angle between tangent to γ at the point $f(t_0)$ and the x-axis. Thus, if a is a point on the plane (x, y) then

$$R(a,\gamma) := \int_0^1 \left|(\arg(f(t) - a))'\right| dt$$

is the total rotation of γ around this point a. Then we consider the curve $\gamma^{(k)} := f_1^{(k)}(t) + if_2^{(k)}(t)$ and denote by

$$T(\gamma^{(k)}) := R(0,\gamma^{(k)}) := \int_0^1 \left|\left(\arg f^{(k)}(t)\right)'\right| dt$$

the total integral curvature of $\gamma^{(k-1)}$ (or the total rotation of the curve $\gamma^{(k)}$ around $a = 0$).

Denote by $l(\gamma)$ the length of γ.

Theorem 1.1 (principle of angles and lengths). *For any $\gamma := f(t) \in F(k)$, any integer $k \geq 1$, any point a*

$$R(a, \gamma) \leq T\left(\gamma^{(k)}\right) + k\pi. \tag{1.1}$$

For any collection of pairwise different points $a_\nu, \nu = 1, 2, ..., q$

$$\sum_{\nu=1}^{q} R(a_\nu, \gamma) \leq T\left(\gamma^{(k)}\right) + \frac{2k\pi q}{\rho} l(\gamma) + k\pi, \tag{1.2}$$

where ρ is the minimal distance between the points a_ν.

The inequalities have simple geometric meaning: the total rotation of γ around a does not exceed total integral curvature of γ plus π. For $k > 1$, $T(\gamma^{(k)})$ equals total rotation of the curve $\gamma^{(k)}$ so that both (1.1) and (1.2) admit corresponding interpretations.

The reader familiar with Nevanlinna's value distribution theory and (or) with Ahlfors theory of covering surfaces will see an analogy between (1.2) and the second fundamental theorem in Nevanlinna's theory; we will discuss this below.

Sharpness. We consider the case where $k = 1$. Let $\gamma := f(t)$, $t \in [0, 1]$ be the segment connecting the points $(-1, \varepsilon)$ and $(1, \varepsilon)$ in the plane. Then $R(0, \gamma)$ is as close to π as we please when we take ε sufficiently small, meantime $T(\gamma^{(1)})$ is equal to zero. Thus, (1.1) can not be improved.

Assume that our curve γ approaches to a circumference by a spiral. Then the part of γ having N "coils" contributes to both the first and the second integrals asymptotically as N when N tends to infinity. Thus the ratio of the left and the right magnitudes in (1.1) tends to 1 when N tends to infinity.

The inequality (1.2) is sharp as well. Let γ be the graph of the function $f_\varepsilon := \sqrt{\varepsilon} \sin \frac{1}{t+\varepsilon}$, $t \in [0, 1]$, where $0 < \varepsilon < \frac{1}{2}$ and take $a_\nu = 2(\nu - 1)$, $\nu = 1, 2, ..., q$. When ε tends to zero then both the left and the right sides of (1.2) tend to infinity but their ratio tends to 1.

Let $\gamma_i \in F(1)$ be a sequence of curves, each satisfying the conditions of Theorem 1.1, $\gamma_i \subset \gamma_{i+1}$, for which $T\left(\gamma^{(k)}\right) \to \infty$ when $i \to \infty$ and

$$\frac{l(\gamma_i)}{T\left(\gamma^{(k)}\right)} \to 0, \quad i \to \infty. \tag{1.3}$$

This is a *bee sequence* (frequent excursions on small distances in different directions) respectively T is comparatively large and l is comparatively small. The rotations of γ_i around a_ν determines the following magnitude

$$\Delta(a_\nu) := \liminf_{i\to\infty} \frac{R(a_\nu, \gamma_i)}{T\left(\gamma^{(k)}\right)}$$

which we refer as *deficiency.* Inequality (1.2) implies the following **Deficiency relation (for the curves).** For any bee sequence of curves and an arbitrary collection of pairwise different points a_ν, $\nu = 1, 2, ..., q$,

$$\sum_{\nu=1}^{q} \Delta(a_\nu) \leq 1. \tag{1.4}$$

1.2. *Bridges between inequalities (1.2), (1.4) and Nevanlinna theory*

We would like to stress the similarity between (1.2) and the second fundamental theorems in Nevanlinna theory [16], Ahlfors theory [1]. In Nevanlinna version it looks as follows: for any meromorphic in the complex plane function $w(z)$ and any collection of pairwise different points a_ν, $\nu = 1, 2, ..., q$ we have

$$\sum_{\nu=1}^{q} m(r, a_\nu, w) \leq 2T(r, w) + C(q)\lambda(r)T(r, w), \quad r = r_n \to \infty, \tag{1.5}$$

where $m(r, a_\nu, w)$ is the approximation function, $T(r, w)$ is Nevanlinna characteristic function, $C(q)$ is a constant depending on q and $\lambda(r) \to 0$ when $r \to \infty$, see [16]. The fact that the second term on the right side of (1.5) is essentially less than the first term plays a crucial role. Particularly it leads to the well known Nevanlinna deficiency relation:

$$\sum_{\nu=1}^{q} \delta(a_\nu) \leq 2, \tag{1.6}$$

where $\delta(a_\nu) := \liminf_{r\to\infty} \{m(r, a_\nu, w)/T(r, w)\}$.

Thus we see that inequality (1.2) is similar to the second fundamental theorem (1.5) and (1.4) is similar to the Nevanlinna deficiency relation (1.6).

1.3. *The proof Theorem 1.1. and some comments*

An inequality similar to (1.2) was proved in 1978 (see Lemma 2 in [3]). In this paper we studied meromorphic functions w in the disks $D(r) := \{z | \, |z| < r\}$ and in Lemma 2 we obtained an inequality of type (1.2) but for the particular class of curves γ which are the images $w(\partial D(r))$ and with the spherical length instead of ordinary length $l(\gamma)$ in (1.2). However, the proof was valid in fact for arbitrary smooth curves. Much later, passing to study ordinary differential equations [6] we again used this proof, this time in full generality (for arbitrary smooth curves). In an excellent book by Sheill-Small [15] (2002) the same magnitudes (R and T) were studied but for closed curves only and for only one value a. He proved (pp. 387-389) that in this case

$$R(a, \gamma) \leq T(\gamma'). \tag{1.7}$$

Unfortunately in [15] we did not find any hint as regards the origin of this inequality. Seemingly it should be Sheill-Small.

Below we prove this inequality making use in fact just a part of the proof of Lemma 2 in [3].

Observe that $\arg f(t_0)$ means the angle between x-axis and vector connecting 0 and $f(t_0)$ respectively $\arg f'(t)$ means the angle between tangent to γ at the point $f(t_0)$ and x-axis. Thus, if a is a point on the plane (x, y) then the magnitude $R(a,\gamma) := \int_0^1 \left|(\arg(f(t) - a))'\right| dt$ means total rotation of γ around this point a.

First we assume that on the curve γ do not involve some sub curves consisting only of points, where $(\arg(f(t) - a))' = 0$. This assumption does not restrict generality since we can consider the result for a new a^* (very close to the point a) such that the magnitudes occurring in (1.7) for a^* and a are as close as we please. Then proving the inequality for a^* we obtain the inequality for a.

Now we assume that our closed curve do not have points, where $(\arg(f(t) - a))'$ change its sign. This means that the curve rotate around a only clockwise (or anti clockwise). Since the curve is closed we have $\int_0^1 \left|(\arg(f(t) - a))'\right| dt = \int_0^1 (\arg(f(t) - a))' \, dt = 2\pi k$, where k is an integer. But then the tangential angle $\arg f'(t)$ also rotates at least k time so that inequality (1.7) is true in this simplest case.

Now we consider the case when on the curve we have some points, where$(\arg(f(t) - a))'$ change its sign. Denote corresponding points by $t_1 < t_2 < \ldots < t_I$. Since the curve is closed we have $f(0) = f(1)$ so that we can assume that $t_1 = 0$ take as t_{I+1} the point 1.

Consider three type of intervals:

(type 1) those intervals (t_i, t_{+1}) at whose both endpoints function $|f(t)|$ increases (or decreases);

(type 2) those intervals (t_i, t_{+1}) for which $|f(t)|$ increases at t_i and decreases at t_{i+1};

(type 3) those intervals (t_i, t_{+1}) for which $|f(t)|$ decreases at t_i and increases t_{i+1}.

Since $\left|(\arg(f(t)-a))'\right| > 0$ and $|\arg(f(t_{i+1})-a) - \arg(f(t_i)-a)| = |\arg f'(t_{i+1}) - \arg f'(t_i)|$ we have for the intervals (t_i, t_{i+1}) of type 1

$$\begin{aligned}&\textstyle\int_{t_i}^{t+1} \left|(\arg(f(t)-a))'\right| dt \\ &= |\arg(f(t_{i+1})-a) - \arg(f(t_i)-a)| = |\arg f'(t_{i+1}) - \arg f'(t_i)| \\ &\le \textstyle\int_{t_i}^{t+1} \left|(\arg f'(t))'\right| dt.\end{aligned}$$

For a given interval (t_i, t_{+1}) of type 2 we observe that there is a point $t^* \in (t_i, t_{+1})$ where $|f(t)|$ is maximal; if we have more than one similar points we take arbitrary one. Denote by α, $\alpha_1$and α_2 the angles $\int (\arg(f(t)-a))'\,dt$ taken for interval (t_i, t_{+1}), (t_i, t^*) and (t^*, t_{+1}) respectively and by η_1, η^* and η_2 the straight lines passing trough zero and the points having $f(t_i)$, $f(t^*)$ and $f(t_{i+1})$ respectively. Since the tangent of our curve at the point t_i coincides with η_1 and direction of $\arg f'(t^*)$ is perpendicular to the direction of η^* and since $|f(t)|$ increases at t_i we have $\alpha_1 + \pi/2 \le |\arg f'(t^*) - \arg f'(t_i)| \le \int_{t_i}^{t^*} \left|(\arg f'(t))'\right| dt$. Quite similarly we have $\alpha_2 + \pi/2 \le |\arg f'(t_{i+1}) - \arg f'(t^*)| \le \int_{t^*}^{t_{i+1}} \left|(\arg f'(t))'\right| dt$ so that we obtain for any interval of type 2

$$\int_{t_i}^{t+1} \left|(\arg(f(t)-a))'\right| dt \le \int_{t_i}^{t+1} \left|(\arg f'(t))'\right| dt - \pi.$$

Consider now intervals of type 3 and denote by $t^* \in (t_i, t_{+1})$ the point where $|f(t)|$ is minimal. Making use the above notations for angles and the straight lines and taking into account that the tangent of our curve at the point t_i coincides with η_1, that the direction of $\arg f'(t^*)$ is perpendicular to the direction of η^* and $|f(t)|$ decreases at the point we observe that for $\alpha_1 > \pi/2$ we have $\alpha_1 - \pi/2 \le |\arg f'(t^*) - \arg f'(t_i)| \le \int_{t_i}^{t^*} \left|(\arg f'(t))'\right| dt$. The same inequality is true for any α_1 since for $\alpha_1 \le \pi/2$ this inequality is obvious. Quite similarly we have $\alpha_2 - \pi/2 \le |\arg f'(t_{i+1}) - \arg f'(t^*)| \le \int_{t^*}^{t_{i+1}} \left|(\arg f'(t))'\right| dt$ so that we obtain for any interval of type 3

$$\int_{t_i}^{t+1} \left|(\arg(f(t)-a))'\right| dt \le \int_{t_i}^{t+1} \left|(\arg f'(t))'\right| dt + \pi.$$

Now we observe that when our curve is closed then the intervals of type 2 and 3 occur in pair since if the module $|f(t)|$ increases (decreases) at a given point t_i then at another point t_{i+k} this module $|f(t)|$ should decrease (increase) due to this closeness. From here and the above three inequalities for the intervals of type 1-3 we obtain inequality (1.1) for any closed curve γ which do with pass trough zero.

Now we consider a given non closed curves γ_1. Denote the coordinate of the plane by (x, y) and make use complex coordinates $z = x + iy$, so that $|z| = r$. Assume that $r^* > |f_1(0)|, |f_1(1)|$.

Let $\mathcal{T}(0)$ ($\mathcal{T}(1)$) be the tangent of γ_1 at the terminal points $f_1(0)$ ($f_1(1)$). Let us continue the curve γ_1 by adding to γ_1 the segment γ_2 lying on $\mathcal{T}(1)$ and connecting $f_1(1)$ with a point $t(1)$ lying on intersection of $\mathcal{T}(1)$ with the circumference $\{z|\ |z| = r^*\}$. Similarly we define the segment γ_4 lying on $\mathcal{T}(0)$ and connecting $f_1(0)$ with a point $t(0)$ lying on intersection of $\mathcal{T}(0)$ with the circumference $\{z|\ |z| = r^*\}$. The points $t(0)$ and $t(1)$ divide the circumference $\{z|\ |z| = r^*\}$ onto two parts. Denote by γ_3 that part of $\{z|\ |z| = r^*\}$ which corresponds to the smaller central angle.

Notice that the curve $\gamma = \gamma_1 \cup \gamma_2 \cup \gamma_3 \cup \gamma_4$ is a closed curve.

Also notice that if r^* tends to infinity then the angle formed by γ_2 and γ_3 at the point $t(1)$ tends to $\pi/2$ as well as the angle formed γ_3 and γ_4 at the point $t(0)$. In other words, for a given "small" $\varepsilon > 0$ we can chose such a big r^* that the differences between these angles and $\pi/2$ are less than $\varepsilon/4$.

After choosing similar r^* we cut from γ_3 two ends of γ_3 with the central angles less than $\varepsilon/4$; the rest part we denote by $\tilde{\gamma}_3$. Now we construct a new curve $\gamma_{2,3}$ which connects the point $f_1(1)$ with the initial point of $\tilde{\gamma}_3$ in such a way that the union $\gamma_1 \cup \gamma_{2,3} \cup \tilde{\gamma}_3$ is a smooth curve. Similarly we will construct a new curve $\gamma_{3,4}$ connecting the terminal point of $\tilde{\gamma}_3$ with the point $f_1(0)$ and again such that $\tilde{\gamma}_3 \cup \gamma_{3,4} \cup \gamma_4$ is a smooth curve. We clearly can construct $\gamma_{2,3}$ and $\gamma_{3,4}$ such that:

(a) the curve

$$\tilde{\gamma} = \gamma_1 \cup \gamma_2 \cup \gamma_{2,3} \cup \tilde{\gamma}_3 \cup \gamma_{3,4} \cup \gamma_4 \tag{1.8}$$

is a closed, smooth curve;

(b)

$$\left|T(\gamma_{2,3}) - \frac{\pi}{2}\right| \le \frac{\varepsilon}{2} \quad \text{and} \quad \left|T(\gamma_{3,4}) - \frac{\pi}{2}\right| \le \frac{\varepsilon}{2}.$$

Taking into account (a), (b), the inequality $R(0, \tilde{\gamma}_3) = T(\tilde{\gamma}_3')$ (since $\tilde{\gamma}_3$ is the part of the circumference centered at zero) and $|T(\gamma_2')| = |T(\gamma_4')| = 0$

(since γ_2 and γ_4 are straightforward) we obtain from (1.7) and (1.8)

$$\begin{aligned}&R(0,\gamma_1) \leq R(0,\tilde{\gamma}) \leq T(\tilde{\gamma}')\\&:= T(\gamma_1') + T(\gamma_2') + T(\gamma_{2,3}') + T(\gamma_3') + T(\gamma_{3,4}') + T(\gamma_4')\\&\leq T(\gamma_1') + \pi + \varepsilon.\end{aligned}$$

and since ε is arbitrary we obtain (1.1) when $k = 1$.
Taking into account that $T(\gamma_1') := R(0,\gamma_1')$ we can repeat the obtained inequality for $\gamma_1', \gamma_1'', ..., \gamma_1^{(k)}$ and come to (1.1) for arbitrary integer k.

1.4. *The proof of Theorem 1.2*

Denote $D(r,a) := \{z : |z-a| < r\}$, $\gamma(D(\rho,a)) := \gamma \cap D(\rho,a)$, $\rho = \min\left[\min_{i\neq,j} \frac{|a_i - a_j|}{2}; 1\right]$. For any a_ν we can consider our curve γ as a collection of the following type of curves. Type 1: curves $\gamma_1(i,a_\nu)$, $i = 1,2,...,I(\nu)$, which lie in $D(\rho,a_\nu)$ and have the length $\geq \rho/2$. Type 2: curves $\gamma_2(j,a_\nu)$, $j = 1,2,...,J(\nu)$, which lie in $D(\rho,a)\backslash D(\frac{\rho}{2},a)$ and have the length $< \rho/2$. Type 3: curves $\gamma_3(s,a_\nu)$, $s = 1,2,...,S(\nu)$, which lie out of $\cup_\nu D(\rho,a_\nu)$. Type 4: curve γ_4, which have common points with $D(\frac{\rho}{2},a)$ and have the length $< \rho/2$; obviously this is the case when our curve γ coincides with γ_4.

We will proceed now assuming that γ is not of view γ_4.

Obviously we have

$$\begin{aligned}&\textstyle\sum_{\nu=1}^{q} R(a_\nu,\gamma)\\&= \textstyle\sum_{\nu=1}^{q}\sum_{i=1}^{I(\nu)} R(a_\nu,\gamma_1(i,a_\nu)) + \sum_{\nu=1}^{q}\sum_{j=1}^{J(\nu)} R(a_\nu,\gamma_2(j,a_\nu))\\&\quad + \textstyle\sum_{\nu=1}^{q}\sum_{s=1}^{S(\nu)} R(a_\nu,\gamma_3(s,a_\nu))\end{aligned} \tag{1.9}$$

Denoting by $l(X)$ Euclidean length of a curve X and taking into account the definition of curves of type 1 we have

$$\sum_{\nu=1}^{q}\sum_{i=1}^{I(\nu)} k\pi \leq \frac{2k\pi}{\rho}\sum_{\nu=1}^{q}\sum_{i=1}^{I(\nu)} l(\gamma_1(i,a_\nu))$$

so that (1.1) yields

$$\begin{aligned}&\textstyle\sum_{\nu=1}^{q}\sum_{i=1}^{I(\nu)} R(a_\nu,\gamma_1(i,a_\nu))\\&\leq \textstyle\sum_{\nu=1}^{q}\sum_{i=1}^{I(\nu)} \int_{z\in\gamma_1(i,a_\nu)} \left|\frac{d}{dt}\arg f^{(k)}\right| dt + \frac{2k\pi}{\rho}\sum_{\nu=1}^{q}\sum_{i=1}^{I(\nu)} l(\gamma_1(i,a_\nu))\\&\leq R\left(0,\gamma^{(k)}\right) + \textstyle\frac{2k\pi}{\rho}\sum_{\nu=1}^{q}\sum_{i=1}^{I(\nu)} l(\gamma_1(i,a_\nu)).\end{aligned} \tag{1.10}$$

Observe that since $\gamma_2(j, a_\nu)$ lie in $D(\rho, a) \backslash D(\frac{\rho}{2}, a)$ we have that the increment of $\arg(z - a_\nu)$ on $\gamma_2(j, a_\nu)$ is comparable with the length $l\left(\gamma_2(j, a_\nu)\right)$ so that follows

$$R\left(a_\nu, \gamma_2(j, a_\nu)\right) \leq \frac{2}{\rho} l\left(\gamma_2(j, a_\nu)\right)$$

consequently

$$\sum_{\nu=1}^{q} \sum_{j=1}^{J(\nu)} R\left(a_\nu, \gamma_2(j, a_\nu)\right) \leq \frac{2}{\rho} \sum_{\nu=1}^{q} \sum_{j=1}^{J(\nu)} l\left(\gamma_2(j, a_\nu)\right). \tag{1.11}$$

Further we have

$$\begin{aligned}
&\left|\frac{d}{dt} \arg\left(z - a\right)\right| dt = \left|\frac{d}{dt} \arctan \frac{y - \mathrm{Im} a}{x - \mathrm{Re} a}\right| dt \\
&= \left|\frac{\left(\frac{y - \mathrm{Im} a}{x - \mathrm{Re} a}\right)'_t}{1 + \left(\frac{y - \mathrm{Im} a}{x - \mathrm{Re} a}\right)^2}\right| dt = \left|\frac{y'\left(x - \mathrm{Re} a\right) - x'\left(y - \mathrm{Im} a\right)}{\left(x - \mathrm{Re} a\right)^2 + \left(y - \mathrm{Im} a\right)^2}\right| dt \\
&\leq \frac{|y'|}{\sqrt{\left(x - \mathrm{Re} a\right)^2 + \left(y - \mathrm{Im} a\right)^2}} \frac{|x - \mathrm{Re} a|}{\sqrt{\left(x - \mathrm{Re} a\right)^2 + \left(y - \mathrm{Im} a\right)^2}} dt \\
&+ \frac{|x'|}{\sqrt{\left(x - \mathrm{Re} a\right)^2 + \left(y - \mathrm{Im} a\right)^2}} \frac{|y - \mathrm{Im} a|}{\sqrt{\left(x - \mathrm{Re} a\right)^2 + \left(y - \mathrm{Im} a\right)^2}} dt \\
&\leq \frac{\sqrt{|y'|^2 + |x'|^2}}{\sqrt{\left(x - \mathrm{Re} a\right)^2 + \left(y - \mathrm{Im} a\right)^2}}
\end{aligned}$$

so that taking into account that for any $z \in \gamma_3(s, a_\nu)$ is valid

$$\sqrt{\left(x - \mathrm{Re} a_\nu\right)^2 + \left(y - \mathrm{Im} a_\nu\right)^2} \geq \rho$$

we obtain

$$\begin{aligned}
\sum_{\nu=1}^{q} \sum_{s=1}^{S(\nu)} R\left(a_\nu, \gamma_3(s, a_\nu)\right) &\leq \frac{1}{\rho} \sum_{\nu=1}^{q} \sum_{s=1}^{S(\nu)} \int_{z \in \gamma_3(s, a_\nu)} \sqrt{|y'|^2 + |x'|^2} dt \\
&\leq \frac{1}{\rho} \sum_{\nu=1}^{q} \sum_{s=1}^{S(\nu)} l\left(\gamma_3(s, a_\nu)\right).
\end{aligned} \tag{1.12}$$

Summing up (1.9)-(1.12) we have

$$\sum_{\nu=1}^{q} R\left(a_\nu,\gamma\right) \le R\left(0,\gamma^{(k)}\right)$$
$$+\frac{2k\pi}{\rho}\left[\sum_{\nu=1}^{q}\sum_{i=1}^{I(\nu)} l\left(\gamma_1(i,a_\nu)\right)+\sum_{\nu=1}^{q}\sum_{j=1}^{J(\nu)} l\left(\gamma_2(j,a_\nu)\right)\right.$$
$$\left.+\sum_{\nu=1}^{q}\sum_{j=1}^{J(\nu)} l\left(\gamma_2(j,a_\nu)\right)\right]$$

and since the magnitude in the brackets $\le ql(\gamma)$ we obtain

$$\sum_{\nu=1}^{q} R\left(a_\nu,\gamma\right) \le R\left(0,\gamma^{(k)}\right)+\frac{2k\pi q}{\rho}l(\gamma). \tag{1.13}$$

Remember that we proved (1.13) provided that γ is not of view γ_4. Notice that any curve γ is of view γ_4 should lie in one of the disks $D(\rho,a_\nu)$, say in $D(\rho,a_1)$, and then inequality (1.1) applied for a_1 yields

$$R\left(a_1,\gamma\right) \le R\left(0,\gamma^{(k)}\right)+k\pi$$

meantime repeating for $a_\nu \ne a_1$ the above proofs of (1.11) and (1.12) we obtain

$$R\left(a_\nu,\gamma\right) \le \frac{2}{\rho}l(\gamma)$$

so that summing up we have

$$\sum_{\nu=1}^{q} R\left(a_\nu,\gamma\right) \le R\left(0,\gamma^{(k)}\right)+\frac{2\{q-1\}}{\rho}l(\gamma)+k\pi. \tag{1.14}$$

Inequality (1.2) of Theorem 1.1 follow now from (1.13) and (1.14).

2. Consequences for the broken lines

Let Γ be an arbitrary broken line (closed or open) in the plane (x,y) composed of n successive segments γ_i, $i=1,2,...,n$. Denote by $\alpha(a,\gamma_i)$ the angle under which γ_i is seen from the point a in the plane and by $L(X)$ the length of X.

We state first a very simple inequality: for an arbitrary broken line Γ (closed or open) and an arbitrary collection of pairwise different points a_ν, $\nu=1,2,...,q$, in the plane

$$\sum_{\nu=1}^{q} A(a_\nu,\Gamma) \le \pi n+\frac{2\pi q}{\rho}L(\Gamma), \tag{2.1}$$

where $A(a_\nu,\Gamma)=\sum_i^n \alpha(a_\nu,\gamma_i)$.

The inequality is a very rough corollary of our Theorem 2.2 below. Meantime, surprisingly, (2.1) appears to be rather sharp as shows the first example given after Theorem 1.1.

We may prescribe a direction to the segment γ_i (that is consider them as some successive vectors) and denote by $\beta(\gamma_i)$ the absolute value of the angle between vectors γ_i and γ_{i+1}. Further we write $B(\Gamma) := \sum_i^{n^*} \beta(\gamma_i)$, where $n^* = n$ if Γ is closed and $n^* = n-1$ if Γ is an open broken line: for an open Γ clearly $\beta(\gamma_n)$ is not defined.

Theorem 2.1. *For an arbitrary broken line Γ (closed or open) in the plane and an arbitrary point a in the plane*

$$A(a,\Gamma) \leq B(\Gamma) + \pi. \tag{2.2}$$

For any collection of pairwise different points a_ν, $\nu = 1, 2, ..., q$,

$$\sum_{\nu=1}^{q} A(a_\nu,\Gamma) \leq B(\Gamma) + \frac{2k\pi q}{\rho} L(\Gamma) + \pi, \tag{2.3}$$

where ρ is the minimal distance between a_ν.

Sharpness, inequalities (2.2) and (2.3) can be verified by considering broken lines "very close" to the curves that demonstrated sharpness in the previous section.

Let now Γ be a broken line consisting of infinitely many successive segments γ_i and $\Gamma_n := \cup_{\nu=1}^{n} \gamma_i$. We say that Γ is a bee broken line if

$$\frac{L(\Gamma_n)}{B(\Gamma_n)} \to 0, \quad n \to \infty. \tag{2.4}$$

Defining the deficiency of Γ as

$$\Delta(a_\nu) := \liminf_{n\to\infty} \frac{A(a_\nu,\Gamma_n)}{B(\Gamma_n)}$$

from the inequality (2.3) we obtain

Deficiency relation for arbitrary bee broken line. For an arbitrary bee broken line in the plane and an arbitrary collection of pairwise different points a_ν, $\nu = 1, 2, ..., q$ in the plane

$$\sum_{\nu=1}^{q} \Delta(a_\nu) \leq 1. \tag{2.5}$$

Clearly the inequalities (2.3) and (2.5) also can be considered as analogues of the second fundamental theorem in Nevanlinna-Ahlfors' theories.

Theorem 2.1 is an immediate consequence of Theorem 1.1. Indeed, rounding up Γ at the vertices (the ends of γ_i and the initial points of γ_{i+1}) we can obtain a curve $\gamma \in F(1)$ as close to Γ as we please. Then $A(a_\nu, \Gamma)$ and $B(\Gamma)$ will be as close as we please to $R(a_\nu, \gamma_i)$ and $T\left(\gamma^{(k)}\right)\Gamma$ correspondingly. Thus Theorem 1.1 implies Theorem 2.1.

3. Consequences for the real functions of one variable

Inequalities (1.1) and (1.2) for the curves imply corresponding corollaries for the real smooth functions of one variable.

The graph of a real function of one variable $\varphi(x) \in C^1[0,1]$ is the curve $\gamma := f(t) := x + i\varphi(x) \in F(1)$, $x \in [0,1]$. With the notation $S(u) := u' / \left(1+u^2\right)$ (the spherical derivative of u) and $a_\nu = (x_\nu, y_\nu)$ we have

Theorem 3.1. *For any $\varphi(x) \in C^1[0,1]$,*

$$\int_0^1 S\left(\frac{\varphi(x)}{x}\right) dx \leq \int_0^1 S\left(\varphi'(x)\right) dx + \pi. \tag{3.1}$$

For any collection of pairwise different points $a_\nu = (x_\nu, y_\nu)$, $\nu = 1, 2, ..., q$,

$$\sum_{\nu=1}^{q} \int_0^1 S\left(\frac{\varphi(x) - y_\nu}{x - x_\nu}\right) dx \leq \int_0^1 S\left(\varphi'(x)\right) dx + \frac{2\pi q}{\rho} \int_0^1 \sqrt{1 + \left(\varphi'(x)\right)^2} dx + \pi, \tag{3.2}$$

where ρ is the minimal distance between a_ν.

Thus we again obtain the Nevanlinna type inequality for real functions.

Proof. To derive these inequalities from Theorem 1.1 we need just to observe that

$$\left(\arg(f(t) - a_\nu)\right)' = \arctan\frac{\varphi(x) - y_\nu}{x - x_\nu} = S\left(\frac{\varphi(x) - y_\nu}{x - x_\nu}\right),$$

$$\left(\arg f'(t)\right)' = S\left(\varphi'(x)\right),$$

so that the inequality (1.1) of Theorem 1.1 applied to the curve γ and $a_1 = 0$ immediately yields (3.1) and inequality (1.2) yields (3.2).

Sharpness of Theorem 3.1. To show that (3.1) is sharp we consider the function $\varphi(x) = h(2x-1)$, $x \in [0,1]$, for which

$$\int_0^1 S\left(\frac{\varphi(x)}{x}\right) dx \to \pi, \quad \text{as} \quad h \to \infty.$$

Clearly, the integral means the angle under which we see the line segment connecting the points $(-h, 0)$ and $(1, h)$. On the other hand $\int_0^1 S\left(\varphi'(x)\right) dx = 0$ so that the difference between the left and the right sides of (3.1) will be as small as we please provided we take h sufficiently large.

Sharpness of (3.2) can be checked using the functions $f_\varepsilon := \sqrt{\varepsilon}\sin\frac{1}{x+\varepsilon}$, $x \in [0,1]$ by the same arguments as in Section 1.

4. Applications in complex analysis, particularly principle of derivatives of analytic functions

We feel pertinent to mention that the last principle is a "pure" type assertion in the sense that we deal only with the functions w and its derivatives like Cauchy formula and its consequences, Carleman formula for analytic functions and Cauchy-Pompeiu formula for smooth functions. One can remember that the majority of results valid for arbitrary analytic functions w in arbitrary domains are not of pure type since they deal either with a-points of w or their generalizations such as Ahlfors' islands (Nevanlinna-Ahlfors theories [1,16]) or they deal with Gamma-lines ([7,8]).

In what follows we denote by D a bounded domain with piecewise smooth boundary whose intersection with any line consists of finite number of intervals. Let $l(\partial D)$ be the length of the boundary ∂D.

Theorem 4.1 (principle of derivatives). *For any meromorphic function f in the closure of a given domain D and any integer $k \geq 1$,*

$$\int\int_D \left|\frac{f'(z)}{f(z)}\right| d\sigma < \int\int_D \left|\frac{f^{(k+1)}(z)}{f^{(k)}(z)}\right| d\sigma + \frac{k\pi}{2} l(\partial D). \tag{4.1}$$

For any collection of pairwise different points a_ν, $\nu = 1, 2, \ldots, q$,

$$\sum_{\nu=1}^{q} \iint_D \left|\frac{f'}{f - a_\nu}\right| dydx$$
$$\leq \iint_D \left|\frac{f^{(k+1)}}{f^{(k)}}\right| dydx + \frac{k\pi^2 q}{\rho} \iint_D |f'|\, dydx + \frac{k\pi}{2} l(\partial D), \tag{4.2}$$

where ρ is the minimal distance between the a_ν's.

Sharpness. For function $f(z) = \exp z$ in the disk $|z| < r$ we have $\int\int_D \left|\frac{f'(z)}{f(z)}\right| d\sigma = 2\pi r^2$, $\int\int_D \left|\frac{f^{(k+1)}(z)}{f^{(k)}(z)}\right| d\sigma = 2\pi r^2$ and $l(\partial D) = 2\pi r$ so that the ratio of the left and the right sides in (4.1) tends to 1 when $r \to \infty$. This means that (4.1) is asymptotically sharp.

Clearly (4.2) is an analogue of Nevanlinna theorem. The case $k = 1$ was considered in [4].

Now we pass to analogous results for "reasonably smooth" complex functions. The results can be proved for more general classes of functions but for simplicity we put some rather common restrictions. We write $f(z) := u(x,y) + iv(x,y) \in C(D,k)$ if for any integer j, $1 \leq j \leq k+1$, u and v have continuous j-th derivatives in x and y in the closure $\bar{D}$ of D and if for any j there is at most finite number of points in $\bar{D}$, where $u_x^{(j)}(x,y) = v_x^{(j)}(x,y) = 0$ or $u_y^{(j)}(x,y) = v_y^{(j)}(x,y) = 0$.

Also we need the following

Definition (projections with multiplicity). Let Γ be a given smooth plane curve Γ and γ be a given straight line in the plane P. Consider Γ as a union of subsets $\{\omega_m\} \cup \{\bar{\omega}_p\}$ ($\omega_m := \omega_m(\gamma)$, $\bar{\omega} = \bar{\omega}_p(\gamma)$) such that for any m the ortogonal projection $\pi(\omega_m)$ of ω_m onto γ is a homeomorphic map and for any p the length $l(\pi(\bar{\omega}_p))$ of the ortogonal projections $\pi(\bar{\omega}_p)$ onto γ is equal to zero. Denote by $\Gamma \perp \gamma$ the union $\cup_m \pi(\omega_m)$ and by $l(\Gamma \perp \gamma)$ the sum $\sum_m l(\pi(\omega_m)$. Thus the mentioned union counts the projections of Γ onto γ with multiplicities. We call $l(\Gamma \perp \gamma)$ the *projection length* of Γ on γ.

Denote by $l(\partial D \perp x)$ the projection length of ∂D on the axis x. Similarly we define $l(\partial D \perp x)$ substituting x by y.

Theorem 4.2. *For any $f(z) \in C(D,k)$ and any integer $k \geq 1$,*

$$\iint_D \left|(\arg f)'_y\right| dydx \leq \iint_D \left|\left(\arg f^{(k)}_{y\dots y}\right)'_y\right| dydx + \frac{k\pi}{2} l(\partial D \perp x). \quad (4.3)$$

For any collection of pairwise different points a_ν, $\nu = 1, 2, ..., q$,

$$\begin{aligned} &\textstyle\sum_{\nu=1}^{q} \left|(\arg(f - a_\nu)'_y\right| dydx \leq \\ &int_D \left|\left(\arg f^{(k)}_{y\dots y}\right)'_y\right| dydx + \tfrac{2k\pi q}{\rho} \textstyle\int_D \left|f'_y\right| dydx + \tfrac{k\pi}{2} l(\partial D \perp x). \end{aligned} \quad (4.4)$$

Thus we have again an Nevanlinna type inequality, this time for broad classes of (non analytic) functions belonging to $C(D,k)$.

Clearly these inequalities remain true if we substitute y by x. Summing up and taking into account that $l(\partial D \perp x) + l(\partial D \perp y) \leq \sqrt{2} l(D)$ we obtain

$$\begin{aligned} &\textstyle\int_D \left[\left|(\arg f)'_x\right| + \left|(\arg f)'_y\right|\right] dydx \leq \\ &\textstyle\int_D \left[\left|\left(\arg f^{(k)}_{x\dots x}\right)'_y\right| + \left|\left(\arg f^{(k)}_{y\dots y}\right)'_y\right|\right] dydx + \frac{\sqrt{2}k\pi}{2} l(D) \end{aligned}$$

and

$$\sum_{\nu=1}^{q}\int_D\left[\left|(\arg(f-a_\nu)'_x\right|+\left|(\arg(f-a_\nu)'_y\right|\right]dydx\le$$
$$\int_D\left[\left|\left(\arg f^{(k)}_{x\ldots x}\right)'_y\right|+\left|\left(\arg f^{(k)}_{y\ldots y}\right)'_y\right|\right]dydx+$$
$$\frac{2k\pi q}{\rho}\int_D\left(|f'_x|+|f'_y|\right)dydx+\frac{\sqrt{2}k\pi}{2}l(D).$$

Proof of Theorem 4.2. Denote by I_x the straight line passing trough point $(x,0)$ and perpendicular to the axes x. Similarly we define I_y as the straight line passing trough point $(0,y)$ and perpendicular to the axes y. Since the domain D is proper we have only finite number of intervals on $D\cap I_x$; denote these intervals by $m_x^{(p)}$, where p is the counting index. Consider the following function $g(y):=f(x,y):=u(x,y)+iv(x,y)\in\mathbb{C}(k,D)$ (of one variable) on $m_x^{(p)}$. Function $g(y)$ on $m_x^{(p)}$ determines a curve. Denote by $(x,y^*(p))$ and $(x,y^{**}(p))$ the endpoints of $m_x^{(p)}$. The determined curve we can consider on $[0,1]$ after linear transformation $[y^*(p),y^{**}(p)]$ to $[0,1]$. Since $f\in\mathbb{C}(D,k)$ we conclude that this curve on $[0,1]$ belongs to the class $F(k)$. So that we can apply inequality (1.1) to this curve. This yields

$$\int_{m_x^{(p)}}\left|(\arg g)'_y\right|dy\le\int_{m_x^{(p)}}\left|\left(\arg g^{(k)}_{y\ldots y}\right)'_y\right|dy+k\pi \tag{4.5}$$

so that summing up by all $m_x^{(p)}$ we have

$$\int_{I_x}\left|(\arg g)'_y\right|dy\le\int_{I_x}\left|\left(\arg g^{(k)}_{y\ldots y}\right)'_y\right|dy+\sum_{\{m_x^{(k)}\}}k\pi.$$

Now we integrate this inequality by x and obtain

$$\begin{aligned}&\int_{D|_x}\int_{I_x}\left|(\arg g)'_y\right|dydx\le\\&\int_{D|_x}\int_{I_x}\left|\left(\arg g^{(k)}_{y\ldots y}\right)'_y\right|dydx+k\pi\int_{D|_x}\left(\sum_{\{m_x^{(k)}\}}1\right)dx,\end{aligned} \tag{4.6}$$

where $D|_x$ is the interval on x-axis which is the ortogonal projection of D on x-axis.

Suppose that the part $D(I_{x'},I_{x''})$ of the domain D contained in the strip between $I_{x'}$ and $I_{x''}$ satisfies the following conditions:

(A) the part is decomposed into n connected components $D_j(I_{x'},I_{x''})$, $j=1,2,\ldots,n$, each with boundaries having common points both with $I_{x'}$ and $I_{x''}$;

(B) intersection of each of these components $D_j(I_{x'}, I_{x''})$ with any I_x, $x' \leq x \leq x''$ consists of only one interval on I_x.

Then for every $x \in (x', x'')$ we have

$$\sum_{\{m_x^{(k)}\}} 1 = n,$$

where the constant n is independent of x and consequently

$$\int_{x'}^{x''} \left(\sum_{\{m_x^{(k)}\}} 1 \right) dx = n(x'' - x').$$

But the quantity $n(x'' - x')$ is half of the sum of the lengths of total projections on axis x of all boundary components of $D_j(I_{x'}, I_{x''})$ occurring in the mentioned strip; here the multiplier 2 (half) arise since each $D_j(I_{x'}, I_{x''})$ has two boundary components projected on x. Hence, since the domain D is assumed to have a piecewise smooth boundary, we can split the interval (x_1, x_2) onto appropriate parts and obtain

$$\int_{D|_x} \left(\sum_{\{m_x^{(k)}\}} 1 \right) dx = \frac{l(\partial D \perp x)}{2}. \tag{4.7}$$

Now (4.6) and (4.7) yield the inequality

$$\int_{D|_x} \int_{I_x} \left| (\arg g)'_y \right| dy dx \leq \int_{D|_x} \int_{I_x} \left| \left(\arg g^{(k)}_{y \dots y} \right)'_y \right| dy dx + \frac{k\pi}{2} l(\partial D \perp x) \tag{4.8}$$

which implies inequality (4.3) of Theorem 4.2.

To prove inequality (4.4) we need to apply inequality (1.2) similarly as we have applied above inequality (1.1). This yields

$$\sum_{\nu=1}^{q} \int_{D|_x} \int_{I_x} \left| (\arg(f - a_\nu)'_y \right| dy dx \leq$$

$$\int_{D|_x} \int_{I_x} \left| \left(\arg g^{(k)}_{y \dots y} \right)'_y \right| dy dx +$$

$$\tfrac{2k\pi q}{\rho} \int_{D|_x} \int_{I_x} \left| f'_y \right| dy dx + k\pi \int_{D|_x} \left(\sum_{\{m_x^{(k)}\}} 1 \right) dx$$

so that taking into account (4.7) we obtain (4.4).

Proof of Theorem 4.1. Now we deal with the meromorphic functions f in $\bar{D}$. First we assume that f, f', $f^{(2)}$,...,$f^{(k+1)} \neq 0, \infty$ in $\bar{D}$. From the definition of $g(y)$ we have

$$(\arg g)'_y = \left(\arctan \frac{v}{u}\right)'_y = \frac{v'_y u - v u'_y}{u^2+v^2}$$

$$= \frac{\sqrt{(u'_y)^2+(v'_y)^2}}{\sqrt{u^2+v^2}}\left[\frac{v'_y}{\sqrt{(u'_y)^2+(v'_y)^2}}\frac{u}{\sqrt{u^2+v^2}} - \frac{v}{\sqrt{u^2+v^2}}\frac{u'_y}{\sqrt{(u'_y)^2+(v'_y)^2}}\right]$$

$$= -\frac{\sqrt{\left(u'_y\right)^2+\left(v'_y\right)^2}}{\sqrt{u^2+v^2}}\sin\left(\arg f - \arg f'_y\right) = -\left|\frac{f'}{f}\right|\sin\left(\arg f - \arg f'_y\right)$$

and similarly we have

$$\left(\arg g^{(k)}_{y\ldots y}\right)'_y = -\left|\frac{f^{(k+1)}_{y\ldots y}}{f^{(k)}_{y\ldots y}}\right|\sin\left(\arg f^{(k)}_{y\ldots y} - \arg f^{(k+1)}_{y\ldots y}\right)$$

so that (4.3) yields

$$\int\int_D \left|\frac{f'_y}{f}\right|\left|\sin\left(\arg f - \arg f'_y\right)\right| d\sigma \le$$

$$\int\int_D \left|\frac{f^{(k+1)}_{y\ldots y}}{f^{(k)}_{y\ldots y}}\right|\left|\sin\left(\arg f^{(k)}_{y\ldots y} - \arg f^{(k+1)}_{y\ldots y}\right)\right| d\sigma + \tfrac{k\pi}{2} l(\partial D \perp x),$$

where $d\sigma$ is the area element.

Let $\bar{X}(\theta)$ be the straight line $\{(x,y)|\ 0 \le \theta := \arctan(y/x) < 2\pi\}$ on the plane (x,y), $\bar{Y}(\theta)$ is the straight line perpendicular to $\bar{X}(\theta)$.

Denoting by f'_η the partial derivative of f in direction η perpendicular to $\bar{X}(\theta)$ we rewrite the last inequality in the new coordinates $(\bar{X}(\theta), Y(\theta))$ as follows:

$$\begin{aligned}&\int\int_D \left|\frac{f'_\eta}{f}\right|\left|\sin\left(\arg f - \arg f'_\eta\right)\right| d\sigma \le \\ &\int\int_D \left|\frac{f^{(k+1)}_{\eta\ldots\eta}}{f^{(k)}_{\eta\ldots\eta}}\right|\left|\sin\left(\arg f^{(k)}_{\eta\ldots\eta} - \arg f^{(k+1)}_{\eta\ldots\eta}\right)\right| d\sigma + \tfrac{k\pi}{2} l(\partial D \perp \bar{X}(\theta)).\end{aligned} \tag{4.9}$$

Now we observe that due to conformity of f, f', $f^{(2)}$,...,$f^{(k+1)}$ we have

$$\left|\frac{f'_\eta}{f}\right| = \left|\frac{f'(z)}{f(z)}\right|$$

and

$$\left|\frac{f^{(k+1)}_{\eta\ldots\eta}}{f^{(k)}_{\eta\ldots\eta}}\right| = \left|\frac{f^{(k+1)}(z)}{f^{(k)}(z)}\right|.$$

Substituting the above two equalities in (4.9), observing that $\left|\frac{f'(z)}{f(z)}\right|$ and $\left|\frac{f^{(k+1)}(z)}{f^{(k)}(z)}\right|$ do not depend on θ and integrating the obtained inequality in θ we have

$$\begin{aligned}
&\int_0^{2\pi}\left[\int\int_D\left|\frac{f'}{f}\right|\left|\sin\left(\arg f-\arg f'_\eta\right)\right|d\sigma\right]d\theta\\
&=\int\int_D\left|\frac{f'}{f}\right|\left[\int_0^{2\pi}\left|\sin\left(\arg f-\arg f'_\eta\right)\right|d\theta\right]d\sigma\\
&\le\int_0^{2\pi}\left[\int\int_D\left|\frac{f^{(k+1)}(z)}{f^{(k)}(z)}\right|\left|\sin\left(\arg f^{(k)}_{\eta\ldots\eta}-\arg f^{(k+1)}_{\eta\ldots\eta}\right)\right|d\sigma\right]d\theta \qquad (4.10)\\
&\quad+\int_0^{2\pi}\frac{k\pi}{2}l(\partial D\perp\bar X(\theta))d\theta\\
&=\int\int_D\left|\frac{f^{(k+1)}(z)}{f^{(k)}(z)}\right|\left[\int_0^{2\pi}\left|\sin\left(\arg f^{(k)}_{\eta\ldots\eta}-\arg f^{(k+1)}_{\eta\ldots\eta}\right)\right|d\theta\right]d\sigma\\
&\quad+\int_0^{2\pi}\frac{k\pi}{2}l(\partial D\perp\bar X(\theta))d\theta.
\end{aligned}$$

Since $\arg f'_\eta=\arg f'_x+\theta+\frac{\pi}{2}=\arg f'+\theta+\frac{\pi}{2}$ (conformity of f) and the magnitudes $\arg f$, $\arg f'$ do not depend on θ we have

$$\begin{aligned}
\int_0^{2\pi}\left|\sin\left(\arg f-\arg f'_{X(\theta)}\right)\right|d\theta&=\int_0^{2\pi}\left|\sin\left(\arg f-\arg f'-\theta-\frac{\pi}{2}\right)\right|d\theta\\
&=\int_0^{2\pi}|\sin t|\,dt=4.
\end{aligned}$$

On the other hand we have $f'_\zeta=f'e^{i\theta}$, where ζ is the direction of the straight line $\bar X(\theta)$ composing angle θ with x-axis. Observing that $e^{i\theta}$ is a constant when we move along $\bar X(\theta)$ we obtain $f^{(k)}_{\zeta\ldots\zeta}=f^{(k)}e^{ik\theta}$. Applied to the direction η this yields $f^{(k)}_{\eta\ldots\eta}=f^{(k)}e^{ik(\theta+\frac{\pi}{2})}$. Therefore

$$\begin{aligned}
&\int_0^{2\pi}\left|\sin\left(\arg f^{(k)}_{\eta\ldots\eta}-\arg f^{(k+1)}_{\eta\ldots\eta}\right)\right|d\theta\\
&=\int_0^{2\pi}\left|\sin\left(\arg f^{(k)}+\arg e^{ik(\theta+\frac{\pi}{2})}-\arg f^{(k+1)}-\arg e^{i(k+1)(\theta+\frac{\pi}{2})}\right)\right|d\theta\\
&=\int_0^{2\pi}\left|\sin\left(\arg f^{(k)}-\arg f^{(k+1)}-\theta-\frac{\pi}{2}\right)\right|d\theta=\int_0^{2\pi}|\sin t|\,dt=4.
\end{aligned}$$

The above two equalities and (4.10) yield

$$4\int\int_D\left|\frac{f'(z)}{f(z)}\right|d\sigma\le4\int\int_D\left|\frac{f^{(k+1)}(z)}{f^{(k)}(z)}\right|d\sigma+\int_0^{2\pi}\frac{k\pi}{2}l(\partial D\perp\bar X(\theta))d\theta. \qquad (4.11)$$

Due to the main identity of integral geometry ([9], [14], see also Section 6 of the present paper) we have

$$\int_0^{2\pi} l(\partial D \perp \bar{X}(\theta))d\theta = 4l(D). \tag{4.12}$$

The last two assertions imply inequality (4.1) in Theorem 4.1 in the case when f, f', $f^{(2)}$,...,$f^{(k+1)} \neq 0, \infty$ in $\bar{D}$. The general case can be proved making use quite standard arguments. Namely, we exclude from $\bar{D}$ some small neighborhoods s_c of all the zeros and poles of all these functions. Then we apply in the remained domain the proved inequality and then shrink the neighborhoods to the corresponding points (zeros or poles). When s_c tends to the corresponding point we have

$$\int\int_{s_c} \left|\frac{f'(z)}{f(z)}\right| d\sigma \to 0, \int\int_{s_c} \left|\frac{f^{(k+1)}(z)}{f^{(k)}(z)}\right| d\sigma \to 0, \ l(\partial s_c) \to 0$$

what completes the proof of (4.1).

In the proof of (4.1) we have utilized inequality (4.3) to obtain (4.11). Arguing similarly and making use (4.4) we obtain instead of (4.11) the following inequality

$$4\sum_{\nu=1}^{q} \iint_D \left|\frac{f'}{f-a_\nu}\right| dydx \leq$$

$$4\iint_D \left|\frac{f^{(k+1)}}{f^{(k)}}\right| dydx + \frac{2k\pi q}{\rho}\int_0^{2\pi}\iint_D |f'_\eta|\, dydxd\theta + \int_0^{2\pi} \frac{k\pi}{2} l(\partial D \perp \bar{X}(\theta))d\theta.$$

Inequality (4.2) follows now from the last inequality, (4.12) and identity $|f'_\eta| = |f'|$.

5. Applications in ODE: the windings of the solutions

5.1. *The problem, its connections with Poincaré theory and applicability*

Many phenomena in physics, technics, biology, economics have a cyclic nature. Often these phenomena are described by the differential equations $y' = F_1(t, x, y)$, $x' = F_2(t, x, y)$. In similar cases the solutions are the curves $\gamma := (x(t), y(t))$, $t \in (t_1, t_2)$, that rotate (wind up) around a given center a on the plane (x, y).

There are numerous studies (in pure mathematics and in all the mentioned sciences) which utilize Poincaré theory to describe the windings of the solutions $(x(t), y(t))$ of these equations. The differential geometric principle of the angles and lengths in Section 1 leads to another approach for studying these windings.

Below we discuss some parallels and differences between the Poincaré theory and the new approach.

Poincaré theory deals with the concepts (the spiral and periodic solutions, the limit cycles) that determine the asymptotic shape or asymptotic windings of the solutions. Respectively the theory considers the solutions on infinite intervals $t \in (-\infty, +\infty)$, $(-\infty, 0)$, $(0, +\infty)$. This is an essential restriction in applications since in the practice we very often need to consider namely finite time intervals $t \in (t_1, t_2)$. Also the Poincaré theory deals in fact with only one center. The new approach permits to study the solutions on finite time intervals. In this case the windings are determined by ordinary total rotation of γ defined in Section 1. Also we able to study an interplay between the windings around different centers what leads to some Nevanlinna type consequences for the solutions.

The approach leads to some new problems related to the interplay between the solutions on finite and infinite intervals. We state qualitatively one problem connected with Hilbert Problem 16 (part b) [10] which asks about the number of limit cycles for the solutions with polynomial F_1 and F_2 of the degree n. Following this problem we can consider (instead of the limit cycles) the solutions with rather strong rotations and we may ask whether the number of all possible similar solutions also depend on n.

The approach first arose in the study [6] by K. Barseghyan and the author, where we dealt with some particular classes of equations in biomathematics (Lotka-Volterra's [11], [17] and Kolmogorov's [12] equations).

Below we study the windings of the solutions of much larger classes of autonomous equations.

5.2. *The windings of solutions of autonomous equations*

Consider the following autonomous system of equations

$$\begin{cases} y' = F_1(x, y) \\ x' = F_2(x, y) \end{cases} \tag{5.1}$$

with continuously differentiable F_1 and F_2 in the closure $\bar{D}$ of a given domain D. Assume that the coefficient satisfy the usual restrictions

$$\left|(F_1(x, y))'_x\right|, \left|(F_1(x, y))'_y\right|, \left|(F_2(x, y))'_x\right|, \left|(F_2(x, y))'_y\right| \le C(D) = const. \tag{5.2}$$

Let $\gamma := (x(t), y(t))$, $t \in (t_1, t_2)$, be a part of an integral curve of (5.1) lying in $\bar{D}$. We will refer simply γ as a is a solution of (5.1), [13].

Theorem 5.1. *For any smooth solution* $\gamma := (x(t), y(t))$, $t \in (t_1, t_2)$, *of equation (5.1) satisfying (5.2)*

$$R(a, \gamma) \leq T(\gamma') + \pi \leq 3C(D)|t_2 - t_1| + \pi. \tag{5.3}$$

Notice that (5.3) gives upper bounds both for the rotations R of the solutions around a and for the integral curvature T of γ. Another advantage is that the upper bounds can be given simply by making use $|t_2 - t_1|$ and the derivatives of F_1 and F_2.

As in section 1 we consider again a collection of pairwise different points a_ν, $\nu = 1, 2, ..., q$, $q \geq 2$, in the plane (x, y) and denote $\rho = \min\left[\min_{i\neq,j} \frac{|a_i - a_j|}{2}; 1\right]$. Then we consider a collection of the bounded non intersecting domains $D_\nu \ni a_\nu$, $\nu = 1, 2, ..., q$, in the plane (x, y). Denote $d := \cup_{\nu=1}^{q} D_\nu$ and assume that for $(x, y) \in \bar{d}$ functions $F_1(x, y)$ and $F_2(x, y)$ in (5.1) satisfy

$$|(F_1(x, y))|, |(F_2(x, y))| \leq K(d) = const. \tag{5.4}$$

The set $\gamma \cap D$ consists of some (one or more) curves $\gamma_i(D_\nu)$, $i = 1, 2, ..., I_\nu$. Denote by $A(a_\nu, D_\nu, \gamma)$ the total rotations of all similar curves around a, that is $A(a_\nu, D_\nu, \gamma) := \sum_{i=1}^{I_\nu} R(a_\nu, \gamma_i(D_\nu))$.

For the similar total rotations we prove the following Nevanlinna type result.

Theorem 5.2. *Assume that for a given above type set d the coefficients of the equation (5.1) satisfy (5.2) and (5.4) in d.* Then for *any smooth solution* $(x(t), y(t))$, $t \subset (t_1, t_2)$, *we have*

$$\sum_{\nu=1}^{q} A(a_\nu, D_\nu, \gamma) \leq \left\{3C(d) + \frac{4\sqrt{2}\pi}{\rho} K(d)\right\} |t_2 - t_1| + \pi. \tag{5.6}$$

Remark 5.1. It is interesting that ρ is the only magnitude depending on the geometry of points a_ν, $\nu = 1, 2, ..., q$, and that the geometry of domains D_ν does not affect to inequality (5.6).

Proof of Theorem 5.1. Consider a solution $\gamma := (x(t), y(t))$, $t \in (t_1, t_2)$, of (5.1) lying in D. Upper bounds for $T(\gamma')$ can be given easily. Indeed, according to the definitions we have

$$T(\gamma') := \int_{t_1}^{t_2} \left| \frac{d}{dt} \arg(x'(t) + iy'(t)) \right| dt = \int_{t_1}^{t_2} \left| \frac{d}{dt} \arctan \frac{y'(t)}{x'(t)} \right| dt.$$

Further

$$\left|\frac{d}{dt}\arctan\frac{y'(t)}{x'(t)}\right| := \left|\frac{y''x' - x''y'}{(x')^2 + (y')^2}\right|.$$

Since

$$y''(t) = (F_1(x,y))'_x\, x' + (F_1(x,y))'_y\, y' =$$
$$(F_1(x,y))'_x\, F_2(x,y) + (F_1(x,y))'_y\, F_1(x,y)$$

and

$$x''(t) = (F_2(x,y))'_x\, x' + (F_2(x,y))'_y\, y' =$$
$$(F_2(x,y))'_x\, F_2(x,y) + (F_2(x,y))'_y\, F_1(x,y)$$

we have

$$\frac{y''x' - x''y'}{(x')^2 + (y')^2} =$$
$$\frac{\left[(F_1(x,y))'_x\, F_2(x,y) + (F_1(x,y))'_y\, F_1(x,y)\right] F_2(x,y)}{F_2^2(x,y) + F_1^2(x,y)} -$$
$$\frac{\left[(F_2(x,y))'_x\, F_2(x,y) + (F_2(x,y))'_y\, F_1(x,y)\right] F_1(x,y)}{F_2^2(x,y) + F_1^2(x,y)} =$$
$$\frac{(F_1(x,y))'_x\, F_2^2(x,y) - (F_2(x,y))'_y\, F_1^2(x,y)}{F_2^2(x,y) + F_1^2(x,y)} +$$
$$\frac{\left[(F_1(x,y))'_y - (F_2(x,y))'_x\right] F_1(x,y)\, F_2(x,y)}{F_2^2(x,y) + F_1^2(x,y)}$$

so that

$$\left|\frac{y''x' - x''y'}{(x')^2 + (y')^2}\right| \le \left|(F_1(x,y))'_x\right| +$$
$$\frac{\left|(F_1(x,y))'_y\right| + \left|(F_2(x,y))'_x\right|}{2} + \left|(F_2(x,y))'_y\right|$$

and applying (5.2) we obtain the inequality

$$T(\gamma') \le 3C(D)\,|t_2 - t_1| \tag{5.7}$$

which being applied to inequality (1.1) of Theorem 1.1 yields Theorem 4.1.

Proof of Theorem 5.2. Arguing similarly as in the proof of inequality (1.2) we can obtain the following quite similar inequality

$$\sum_{\nu=1}^{q} A(a_\nu, D_\nu, \gamma) := \sum_{\nu=1}^{q}\sum_{i=1}^{I_\nu} R\left(a_\nu, \gamma_i(D_\nu)\right)$$
$$\leq \sum_{\nu=1}^{q}\sum_{i=1}^{I_\nu} T(\gamma_i'(D_\nu)) + \frac{4\pi}{\rho}\sum_{\nu=1}^{q}\sum_{i=1}^{I_\nu} l(\gamma_i(D_\nu)) + \pi. \quad (5.8)$$

Thus to prove Theorem 5.2 we need to obtain upper bounds for the sums in the right side of (5.8). Assume that $\gamma_i(D_\nu)$ is the image of $(t'_{i,\nu}, t''_{i,\nu}) \subset (t_1, t_2)$. Applying (5.7) to this curve and obtain $T\left(\gamma_i'(D_\nu)\right) \leq 3C(D_\nu)\left|t''_{i,\nu} - t'_{i,\nu}\right|$ so that

$$\sum_{\nu=1}^{q}\sum_{i=1}^{I_\nu} T(\gamma_i'(D_\nu)) \leq 3\sum_{\nu=1}^{q}\sum_{i=1}^{I_\nu} C(D_\nu)\left|t''_{i,\nu} - t'_{i,\nu}\right| \leq 3C(d)\left|t_2 - t_1\right|.$$

For the length we have $l(\gamma_i(D_\nu)) := \int_{\gamma_i(D_\nu)} \sqrt{(x'(t))^2 + (y'(t))^2}dt$ and from the equation we have $x'(t) = F_1(x,y)$, $y'(t) = F_2(x,y)$ so that (5.4) yields

$$\sum_{\nu=1}^{q}\sum_{i=1}^{I_\nu} l(\gamma_i(D_\nu)) \leq \sqrt{2}K(d)\sum_{\nu=1}^{q}\sum_{i=1}^{I_\nu}\left|t''_{i,\nu} - t'_{i,\nu}\right| \leq \sqrt{2}K(d)\left|t_2 - t_1\right|.$$

Substituting the last two inequalities in (5.8) we obtain Theorem 5.2.

6. Some identities and inequalities in integral geometry

6.1. *Introduction*

In this section we present some new identities and inequalities concerning integrals of function along the curves, among them those generalizing some classical formulae of integral geometry.

The study has arisen as follows. The theory of Gamma-lines [7] (complex analysis) studies particularly the level sets of harmonic functions. The level sets occur in numerous branches of pure and applied mathematics (isoterm, isobar, potential line, stream line etc.). Naturally we tried to find some ways to apply and extend Gamma-lines technic for studying level sets of much larger classes of functions $u(x, y)$ that can be of interest in different applied situations. This have led us to an extensive list of open problems [8] in nearly thirty fields.

Our attempts to develop these ideas and solve some of the problems posed in [8] have led to some new inequalities [5] for curves, real and complex functions. Then we have applied the Gamma-lines' methods to obtain some other identities and inequalities which we present in this paper. It is interesting that the obtained results turned out to be rather close to those obtained earlier in Integral Geometry [9], [14] and Combinatorial integral Geometry [2]. I got idea about the mentioned interrelation with Integral geometry thanks to R. Ambartzumyan and V. Oganian who kindly presented me the key ideas and results in [2], [9], [14]. I express my deep thanks to them for that as well as for their valuable comments.

In this section we present some of the obtained results in this direction. In full generality they will be presented elsewhere.

6.2. *The main identity in integral geometry*

We say that a given oriented plane curve $\Gamma := (x(t)), y(t))$, $t \in [0,1]$, is *proper* if Γ is a smooth curve with bounded length $l(\Gamma)$ and uniformly continuous on Γ curvature k. Let f be a function given on Γ.

In this paper we give an identity for $\int f ds$. In the particular case when $f \equiv 1$ this identity coincides with the classical Identity 1 (see below) in integral geometry.

To present the identity for $\int f ds$ we need to modify the classical concepts and to give an interpretation.

Definition (projections with multiplicity). Let Γ be a given smooth plane curve Γ and γ be a given straight line in the plane P. Consider Γ as a union of subsets $\{\omega_m\} \cup \{\bar{\omega}_p\}$ $(\omega_m := \omega_m(\gamma),\ \bar{\omega} = \bar{\omega}_p(\gamma))$ such that for any m the orthogonal projection $\pi(\omega_m)$ of ω_m onto γ is a homeomorphic map and for any p the length $l(\pi(\bar{\omega}_p))$ of the orthogonal projections $\pi(\bar{\omega}_p)$ onto γ is equal to zero. Denote by $\Gamma \perp \gamma$ the union $\cup_m \pi(\omega_m)$ and by $l(\Gamma \perp \gamma)$ the sum $\sum_m l(\pi(\omega_m)$. Thus the mentioned union counts the projections of Γ onto γ with multiplicities. We call $l(\Gamma \perp \gamma)$ the *projection length* of Γ on γ.

Let $\bar{X}(\theta)$ be the straight line $\{(x,y)|\ 0 \leq \theta := \arctan(y/x) < 2\pi\}$ on the plane (x,y). We will use notation $X(\theta)$ for the coordinate on $\bar{X}(\theta)$. Denote by $J_{X(\theta)}$ the straight line perpendicular to $\bar{X}(\theta)$ and intersecting $\bar{X}(\theta)$ at the point $X(\theta)$.

Notice that $\Gamma \cap J_{X(\theta)}$ may consist of some points and some intervals on the straight lines $J_{X(\theta)}$: these intervals clearly should be the intervals of type $\bar{\omega}_p(\bar{X}(\theta))$ belonging to Γ. Ignoring similar intervals we denote by $N(\Gamma, J_{X(\theta)})$ the total number of similar points.

We make use notations $l_\theta(\Gamma)$ for $l(\Gamma \perp \bar{X}(\theta))$ when γ is the straight line $\bar{X}(\theta)$.Notations $\Gamma|_{\bar{X}(\theta)}$ stands for the set of all those points $X(\theta)$ on $\bar{X}(\theta)$ for which $J_{X(\theta)}$ intersects Γ in at least one point.

With the above notations and definitions we able now to present

Identity 6.1. *For any proper curve* Γ *we have*

$$l(\Gamma) = \frac{1}{4}\int_0^{2\pi}\int_{\Gamma|\bar{X}(\theta)} N(\Gamma, J_{X(\theta)})\,|dX(\theta)|\,d\theta = \frac{1}{4}\int_0^{2\pi} l_\theta(\Gamma)d\theta, \qquad (6.1)$$

(here we make use Lebesgue integration).

Comment 6.1. Notice that $l_\theta(\Gamma) = l_{\theta+\pi}(\Gamma)$ and $N(\Gamma, J_{X(\theta)}) = N(\Gamma, J_{X(\theta+\pi)})$ for any θ so that the above inequality we may rewrite as

$$l(\Gamma) = \frac{1}{2}\int_0^{\pi}\int_{\Gamma|\bar{X}(\theta)} N(\Gamma, J_{X(\theta)})\,|dX(\theta)|\,d\theta = \frac{1}{2}\int_0^{\pi} l_\theta(\Gamma)d\theta. \qquad (6.1')$$

Similar comments are true for other identities and inequalities below.

Comment 6.2. Identity (6.1) is a bit simplified and complemented version of the classical identity whose modifications and generalizations constitute in fact an essential part of integral geometry: here contributed Barbier, Poincaré, Blaschke, Santalo (see [14]). The difference between the first identity in (6.1) and its classical counterparts is that we make use more simple pointwise intersections $N(\Gamma, J_{X(\theta)})$ (instead of forms and densities). Also we have the second part in (6.1) which despite its triviality we did not meet elsewhere.

Notice that in (6.1) we deal with all length projections $l_\theta(\Gamma)$ for all θ, $0 \le \theta \le 2\pi$. Now we give another identity for the length of Γ which make use only two projections of Γ on perpendicular axes $J_{X(\theta)}$ and $J_{X(\eta)}$, where $\eta := \theta + \pi/2$. Since our curve Γ is oriented we can fix starting point of Γ and define the angle $\alpha(s)$ between the tangent to Γ at the point $s \in \Gamma$ and $\bar{X}(0)$. Notice that the last straight line is x-axis.

Let us consider identity (6.1) from a bit different point of view. Moving along the curve Γ we fix the elements $s_{i(\theta)}$ of $\Gamma \cap J_{X(\theta)}$, $s_{i(\theta)} = 1, 2, ..., N(\Gamma, J_{X(\theta)})$: here i is, clearly, the index of these elements (points or curves). Thus we can rewrite (6.1′) as

$$l(\Gamma) = \frac{1}{2}\int_0^{\pi} l_\theta(\Gamma)d\theta = \frac{1}{2}\int_0^{\pi}\int_{\Gamma|\bar{X}(\theta)} \left\{\sum_{i(\theta)=1}^{N(\Gamma, J_{X(\theta)})} 1\right\} |dX(\theta)|\,d\theta.$$

6.3. *Identities for the integrals along curves*

In this section we give some formulas of the above types for the integrals

$$\int_\Gamma f(s)ds,$$

where s is the natural parameter of Γ. Particularly, when $f(s) \equiv 1$ this formulae coincide with identities (6.1) and (6.2).

Instead of the projection length of Γ on $\bar{X}(\theta)$, we consider now the following *weighted projection length* of Γ on $\bar{X}(\theta)$

$$\mathbb{L}_\theta(\Gamma, f) := \int_\Gamma f(s) \left|\cos\left(\alpha(s) - \theta\right)\right| ds = \sum_m \int_{\omega_m(\bar{X}(\theta))} f(s) \left|\cos\left(\alpha(s) - \theta\right)\right| ds$$

and, instead of the length $\mathbb{D}_\theta(\Gamma)$ of double projection, we consider the following *weighted double projection length* $\int_{\pi(\omega_m(\bar{X}(\theta))}$

$$\mathbb{I}_\theta(\Gamma, f) := \int_\Gamma f(s) \cos^2\left(\alpha(s) - \theta\right) ds = \sum_m \int_{\omega_m(\bar{X}(\theta)} f \cos^2\left(\alpha(s) - \theta\right) ds.$$

Theorem 6.1. *For any proper curve Γ and any continuous function $f(s) > 0$ on Γ we have*

$$\begin{aligned} \int_\Gamma f(s) ds &= \frac{1}{4} \int_0^{2\pi} \mathbb{L}_\theta(\Gamma, f) d\theta \\ &= \frac{1}{4} \int_0^{2\pi} \int_{\Gamma|_{X(\theta)}} \sum_{i(\theta)=1}^{N(\Gamma, J_{X(\theta)})} f(s_{i(\theta)}) \left|dX(\theta)\right| d\theta. \end{aligned} \tag{6.2}$$

6.4. *Inequalities for the lengths of curves and inequalities for the integrals along curves*

From the above identities we derive some inequalities giving upper bounds for the integrals $\int_\Gamma f(s)ds$ in terms of the integral curvature $C(\Gamma) := \int_\Gamma k(s)ds$, where $k(s)$ is the ordinary curvature of Γ at the point $s \in \Gamma$.

Observe that for the proper curves $C(\Gamma) = \mathrm{Var}_\Gamma \alpha(s)$ so that we can consider also $C(\Gamma)$ as the variation of the tangential angle.

We give first some simple inequalities related to the particular case when $f(s) \equiv 1$; respectively $\int_\Gamma f(s)ds$ become ordinary length $l(\Gamma)$ of Γ.

Denote by $\Delta_\theta(\Gamma)$ the length of $\Gamma|_{X(\theta)}$.

Inequality 6.1. *For any proper curve Γ and any θ we have*

$$\begin{aligned} l(\Gamma) \le &\left(\frac{1}{2} \int_\Gamma \left|k(s) \sin\alpha(s)\right| ds + 1\right) \Delta_0(\Gamma) \\ &+ \left(\frac{1}{2} \int_\Gamma \left|k(s) \cos\alpha(s)\right| ds + 1\right) \Delta_{\pi/2}(\Gamma). \end{aligned} \tag{6.3}$$

If Γ is closed we have

$$l(\Gamma) \le \left(\frac{1}{2} \int_\Gamma \left|k(s) \sin\alpha(s)\right| ds\right) \Delta_0(\Gamma) + \left(\frac{1}{2} \int_\Gamma \left|k(s) \cos\alpha(s)\right| ds\right) \Delta_{\pi/2}(\Gamma).$$

Notice that (6.3) implies the following more simple (and more rough) inequality

$$l(\Gamma) \leq \left(\frac{1}{2}\int_{\Gamma} |k(s)|\, ds + 1\right)\left(\Delta_0(\Gamma) + \Delta_{\pi/2}(\Gamma)\right). \tag{6.3'}$$

Notice also that when Γ is a segment of a straight line then $k(s) \equiv 0$ and inequality (6.3′) become a usual Pythagorean (or triangular) inequality in the form: $l(\Gamma) \leq \Delta_0(\Gamma) + \Delta_{\pi/2}(\Gamma)$.
Sharpness. We have equality in (6.3) and (6.3′) and for any segment lying on $\bar{X}(0)$ and $\bar{X}(\pi/2)$.

For the proper closed curves Γ Santalo proved the following interesting inequality, see [14], p. 37:

$$l(\Gamma) \leq \frac{1}{2}\int_{\Gamma} |k(s)|\, ds\Delta_{\max}(\Gamma).$$

We establish similar result for arbitrary (non closed) proper curves.
Inequality 6.2. *For any proper curve Γ we have*

$$\begin{aligned} l(\Gamma) &\leq \frac{1}{2}\int_{\Gamma} |k(s)|\, ds\Delta_{\max}(\Gamma) + \frac{1}{2}\int_0^{\pi} \Delta_\theta(\Gamma)d\theta \\ &\leq \frac{1}{2}\int_{\Gamma} |k(s)|\, ds\Delta_{\max}(\Gamma) + \frac{\pi}{2}\Delta_{\max}(\Gamma). \end{aligned} \tag{6.4}$$

Sharpness The ratio of the left and the right sides in (6.4) tends to 1 for any sequence of spirals $(1 - \frac{1}{\varphi(r)})e^{i\omega(r)}$, $r_n \in (0, 1)$, where $\varphi(r)$ and $\omega(r)$ are some monotone differentiable functions tending to infinity when $r_n \to 1$. Thus (6.4) is asymptotically sharp

6.5. *Inequalities for the integrals along curves*

Theorem 6.2. *For any proper curve Γ and any continuously differentiable function $f(s) > 0$ on Γ and any θ* we have

$$\begin{aligned} &\textstyle\int_\Gamma f(s)ds \\ &\leq \tfrac{1}{2}\sup_\Gamma |f| \left\{\textstyle\int_\Gamma |k(s)\sin\alpha(s)|\, ds\Delta_0(\Gamma) + \int_\Gamma |k(s)\cos\alpha(s)|\, ds\Delta_{\pi/2}(\Gamma)\right\} \\ &+ \left\{\tfrac{1}{2}\sup_\Gamma |f'_\lambda|\, l(\Gamma) + \sup_\Gamma |f|\right\}\left(\Delta_0(\Gamma) + \Delta_{\pi/2}(\Gamma)\right) \\ &\leq \left\{\tfrac{1}{2}\sup_\Gamma |f| \textstyle\int_\Gamma k(s)ds + \tfrac{1}{2}\sup_\Gamma |f'_\lambda|\, l(\Gamma) + \sup_\Gamma |f|\right\}\left(\Delta_0(\Gamma) + \Delta_{\pi/2}(\Gamma)\right) \end{aligned} \tag{6.5}$$

and

$$\int_\Gamma f(s)ds \leq \frac{1}{2\pi}\left[\sup_\Gamma |f| \int_\Gamma |k(s)|\, ds + \sup_\Gamma |f'_\lambda|\, l(\Gamma) + 2\sup_\Gamma |f|\right]\int_0^{2\pi} \Delta_\theta(\Gamma)d\theta. \tag{6.6}$$

Notice that (6.5) implies (6.3) when $f(s) \equiv 1$. In turn the sharpness of (6.5) follows from the sharpness of (6.3).

7. Principle of zeros for real functions of two variables

In this section we present an inequality related to the zeros of real functions of two variables and the zeros of derivatives of this function. We show that these inequalities can be considered as some analogues of Rolle's theorem for functions of two variables.

In what follows we denote by D domains with piecewise smooth boundary ∂D of length $l(\partial D)$.

The results are valid for arbitrary functions $u(x, y) \in C^2(\bar{D})$, $\bar{D} = D \cup \partial D$, but below we present only the case of the *proper functions* in $\bar{D}$, that is functions $u \in C^2(\bar{D})$ which admit only finite number of points in $\bar{D}$, where $\text{grad}u = 0$ or $\text{grad}u' = 0$. Observe that the zeros of $u(x, y)$ (that is the solutions of $u(x, y) = 0$ or level sets of $u(x, y)$) are piecewise smooth curves $\gamma_i(u)$ for the proper functions. The same is true for zeros $u'_\theta(x, y) = 0$ (for any θ, $0 \leq \theta < \pi$). Corresponding curves we denote by $\gamma_j(u'_\theta)$.

Denote by $L(D, 0, u)$ (by $L(D, 0, u'_\theta)$ the total length of the curves $\gamma_i(u)$ ($\gamma_j(u'_\theta)$) in $\bar{D}$.

Theorem 1 (Principle of zeros for proper functions). *For any proper function $u(x, y)$ in $\bar{D}$*

$$\begin{aligned} L(D, 0, u) &\leq \frac{1}{2} \int_0^\pi L(D, 0, u'_\theta) d\theta + \frac{1}{2} l(\partial D) \\ &\leq \frac{\pi}{2} \sup_{0 \leq \theta < \pi} L(D, 0, u'_\theta) + \frac{1}{2} l(D). \end{aligned} \tag{7.1}$$

Sharpness. The following simple example shows that this inequality is sharp. Consider $u = \sqrt{x^2 + y^2 - 1}$ in the unit disk. The set of solutions of $u = 0$ coincides with the boundary of the disk and the set of solutions of $u'_\theta = 0$ coincides with the diameter of the disk having direction θ. We have therefore $L(D, 0, u) = 2\pi$, $L(D, 0, u'_\theta) = 2$ for any θ (consequently $\sup_{0 \leq \theta < 2\pi} L(D, 0, u'_\theta) = 2$) and $l(D) = 2\pi$. Thus we have equality in the double inequality (7.1) for this function.

Comment 1 (Theorem 7.1 as a Rolle type results). One of the key results in real analysis, the Rolle's theorem, asserts: between arbitrary two zeros of a real differentiable function of one variable $f(x)$ in $[a, b]$ there is a zero of $f'(x)$.

At the early stage of mathematical education we learn that this theorem has no any analogue for functions of several variables. We ask how can look like possible results for functions of several variables that can serve as some analogues of the Rolle's theorem? We meet essential difference: zeros of functions of several variables are curves, surfaces etc. and these functions have directional derivatives (unlike functions of one variable whose zeros are points in general case and we deal with only one derivative).

Let us consider the Rolle's theorem in a bit enlarged version: if f has n zeros on $[x_1, x_2]$ then f' should have at least $n-1$ zeros on $[x_1, x_2]$. The last statement we can express qualitatively as follows: if f has "powerful" set of zeros of f on $[x_1, x_2]$ then f' should also have "powerful" set of zeros.

The power of zeros we can define also in multivariate case. A natural measure for the "power" of zeros of functions $u(x,y)$ $(u'_\theta(x,y))$ of two variables in D is the length $L(D,0,u)$ $(L(D,0,u'_\theta))$.

Inequality 1 has exactly the same meaning: if $u(x,y)$ has "powerful" set of zeros of u in $\bar{D}$ (that is if $L(D,0,u)$ is large) we should have also "powerful" set of zeros of u'_θ for at least one value θ (that is $L(D,0,u'_\theta)$ should be large as well). Respectively this inequality can be considered as an analogue of Rolle's theorem for functions of two variables.

Comment 2. There were many attempts to obtain Rolle's type results for analytic functions: the Internet shows nearly five hundreds papers. This goes back to Gauss who has proved the following astonishing results: any disk involving all zeros of a given complex polynomial P involves also all zeros of derivative P'. This was likely the first and till present the the most simple and applicable result.

We are not aware whether someone tried to get an analog of the Rolle's theorem in real case, say for function of two real variables. As we see in this case we deal with essentially different type of objects (curves instead of points) and one need to consider the problem from a different point of view, for instance as we do it in Theorem 7.1.

Hydrodynamic and electromagnetic interpretations of inequality (7.1). Let $u(x,y)$ be the velocity at the point (x,y) in a plane flow of an ideal liquid. This implies $u'_y = -v'_x$ since the corresponding complex potential w is an analytic function. Then solutions of $u(x,y) = A$ are those lines where the velocity is equal to A. The magnitudes u'_x and u'_y mean component of the accelerations in directions x and y; again $u'_y = -v'_x$. Thus all magnitudes in this inequality have hydrodynamic interpretations and inequality (7.1) can be considered as a result in hydrodynamics. Quite similar interpretation we have for the case when $u(x,y)$ is an electric potential. Then solutions of $u(x,y) = A$ equipotential lines, where the potential is equal to A. The magnitudes u'_x and u'_y mean components of the electric field in directions x and y. Clearly, inequality (7.1) can now be considered as a result in electromagnetism.

References

1. L. Ahlfors, *Zur Theorie der Überlagerungsflächen*, Acta Soc. Sci. Fenn. **9** (1), 1-40 (1930).
2. R. V. Ambatzumian, *Combinatorial integral geometry*, John Wiley, 1982.
3. G. A. Barsegian, *On geometric structure of image of disks under mappings by meromorphic functions*, Math. Sbornik **106(148)** (1), 35-43 (1978).
4. G. A. Barsegian, *Exceptional values associated with logarithmic derivatives of meromorphic functions*, Izvestia Acad. Nauk ArmSSR, Matematika **16** (5), 408-423 (1981).
5. G. Barsegian, *Some new inequalities in geometry and analysis*, Izvestia Academii Nauk Armenii, vol. 42, no. 2, 2007, 17-28.
6. K. G. Barseghyan and G. A. Barsegian, *Some Nevanlinna type results in differential equations with application to the population dynamics*, Preprint.
7. G. A. Barsegian, *Gamma-lines: on the Geometry of Real and Complex Functions*, Taylor and Francis, London, New York, 2002.
8. G. A. Barsegian, *A new program of investigations in analysis: Gamma-lines approaches,* pp. 1-73, in *Value Distribution and Related Topics*, Eds. by G. Barsegian, I. Laine and C. Yang (Kluwer, Series: Advances in complex analysis and its applications, 2004.
9. W. Blaschke, *Vorlezungen über Integralgeometrie*, Chelsea Publ. Company, New York, 1949.
10. D. Hilbert, *Mathematical problems*, Archif fur Math. und Phys., series III, v. 11, 1901, 44-63, 213-237.
11. A. J. Lotka, *Elements of physical biology*, Williams and Wilkins, Baltimore, 1924, also Dover Publ., New York, 1956.
12. A. N. Kolmogorov, *Sulla theoria do Volterra della Lotka per l'Esistttenza,* Giorn Ist. Iyal. Attuary 7, 1936, 74-80.
13. L. S. Pontryagin, *Ordinary differential equations* (in Russian), Nauka, 1965.
14. L. Santalo, *Integral geometry and geometric probability*, Addison-Wesley, London, 1979.
15. R. Sheill-Small, *Complex Polynomials*, Ser. Cambridge Studies in Advanced Mathematics, No. 75, 2002.
16. R. Nevanlinna, *Eindeutige analytische Funktionen*, Springer, Berlin, 1936.
17. V. Volterra, *Lecons sur la theorie mathematique de la lutte pour la vie*, Marcel Brelot, Paris, 1931.

RECENT RESULTS ON HARDY AND RELLICH INEQUALITIES

W. D. EVANS

School of Mathematics, Cardiff University
Cardiff, CF24 4AG, UK
E-mail: EvansWD@cardiff.ac.uk

1. Introduction

In the 1920s, G. H. Hardy proved the following one-dimensional inequality:

$$\int_0^\infty \left(\frac{F(x)}{x}\right)^p dx \le \left(\frac{p}{p-1}\right)^p \int_0^\infty f(x)^p dx, \tag{1}$$

where $1 \le p < \infty, f(x) \ge 0$ and $F(x) := \int_0^x f(t)dt$. The multi-dimensional form of the inequality is

$$\int_{\mathbb{R}^n} |\nabla f(\mathbf{x})|^p d\mathbf{x} \ge \left|\frac{n-p}{p}\right|^p \int_{\mathbb{R}^n} \frac{|f(\mathbf{x})|^p}{|\mathbf{x}|^p} d\mathbf{x}, \quad f \in C_0^\infty(\mathbb{R}^n \setminus \{0\}). \tag{2}$$

The constants exhibited in (1) and (2) are sharp and there is equality if and only if $f = 0$. These inequalities play a crucial role throughout Analysis and its applications and continue to be the subject of extensive study and research. They have been generalised in many directions: weights have been introduced and scales of spaces other than L^p spaces have been considered; analogues have been established on manifolds and also ones involving fractional derivatives and derivatives of higher order; the analogue of (2) on a bounded open subset Ω of $\mathbb{R}^n$, with $|\mathbf{x}|$ replaced by the distance $\delta(\mathbf{x})$ of $\mathbf{x}$ from the boundary of Ω, and also other distance functions, has been studied. The last category of problem will be of particular concern to us in this article.

The inequality (2) is intimately related to the Sobolev inequality

$$\left(\int_{\mathbb{R}^n} |f(\mathbf{x})|^{p^*} d\mathbf{x}\right)^{1/p^*} \le C(n,p) \left(\int_{\mathbb{R}^n} |\nabla f(\mathbf{x})|^p d\mathbf{x}\right)^{1/p}, \quad f \in C_0^\infty(\mathbb{R}^n), \tag{3}$$

where $p^* := np/(n-p), 1 \le p < n$, is the Sobolev conjugate of p : the optimal constant is

$$C(n,p) = \pi^{-1/2} n^{-1/p} \left(\frac{p-1}{n-p}\right)^{(p-1)/p} \left\{\frac{\Gamma(1+n/2)\Gamma(n)}{\Gamma(n/p)\Gamma(1+n-n/p)}\right\}^{1/n},$$

for $1 < p < n$, and

$$C(n,1) = \pi^{-1/2} n^{-1} \left(\Gamma(1+n/2)\right)^{1/n}.$$

On the Sobolev space defined as the completion of $C_0^\infty(\mathbb{R}^n)$ with respect to the norm $\|\nabla f\|_{L^p(\mathbb{R}^n)}$, there is equality in the case $1 < p < n$ if and only if $f(\mathbf{x})$ is a multiple of $u(\mathbf{x}) := [a + b|\mathbf{x}|^{p/(p-1)}]^{1-n/p}$, with a, b positive constants. When $p = n$, (2) is trivially true, while (3) does not hold for any constant C in this case (with $p^* = \infty$). The case $p = n = 2$ of (2) is in fact of particular interest. There is no non-trivial inequality (2) but Laptev and Weidl showed, *inter alia*, in [21] that

$$\int_{\mathbb{R}^2} |\nabla_{\mathbf{A}} f(\mathbf{x})|^2 d\mathbf{x} \ge \left(\operatorname{dist}(\widetilde{\Psi}, \mathbb{Z})\right)^2 \int_{\mathbb{R}^2} \frac{|f(\mathbf{x})|^2}{|\mathbf{x}|^2} d\mathbf{x}, \quad f \in C_0^\infty(\mathbb{R}^n \setminus \{0\}) \quad (4)$$

where $\nabla_{\mathbf{A}} := \nabla - i\mathbf{A}$ is the magnetic gradient associated with the magnetic potential $\mathbf{A}$ which, in polar co-ordinates, is of the form

$$\mathbf{A}(\mathbf{r}, \theta) = \frac{\boldsymbol{\Psi}(\theta)}{\mathbf{r}} \left(-\sin\theta, \cos\theta\right), \quad \boldsymbol{\Psi} \in \mathbf{L}^\infty(\mathbf{0}, \mathbf{2}\pi), \quad (5)$$

with magnetic flux $\widetilde{\Psi} = (1/2\pi) \int_0^{2\pi} \Psi(\theta) d\theta \notin \mathbb{Z}$. In (5), the magnetic field $\operatorname{curl}\mathbf{A} = \mathbf{0}$ in $\mathbb{R}^2 \setminus \{0\}$ and is of so-called Aharonov-Bohm type. If $\widetilde{\Psi} \in \mathbb{Z}$, the problem is equivalent to that with no magnetic field (by a gauge transformation) when there is no non-trivial inequality.

In section 2 we discuss inequalities of Hardy-Sobolev type established in [2] which combines features of (2) and (3). They are consequences of an improved Sobolev inequality derived by Ledoux in [23].

For a bounded open set Ω, the form of Hardy's inequality we consider is

$$\int_\Omega |\nabla f(\mathbf{x})|^p d\mathbf{x} \ge C \int_\Omega \frac{|f(\mathbf{x})|^p}{\delta(\mathbf{x})^p} d\mathbf{x}, \quad f \in C_0^\infty(\Omega), \quad (6)$$

where the positive constant C depends on p, n and Ω, and $\delta(\mathbf{x}) = \operatorname{dist}(\mathbf{x}, \partial\Omega)$ denotes the distance from $\mathbf{x}$ to the boundary $\partial\Omega$ of Ω. It is well-known that (6) requires some restrictions on Ω. For a convex Ω it is valid, with best constant $c_p := [(p-1)/p]^p$, although the convexity is not necessary for this result (see [12]). If Ω is a simply connected open subset of $\mathbb{R}^2$ and $p = 2$, $1/16$ is the best constant (see [1] and [12], Theorem 1.5.10).

For a general domain Ω, what can be said is that there is such an inequality when δ is replaced by the *mean distance* δ_M introduced by Davies (see [13]) and defined by

$$\frac{1}{\delta_M^2(\mathbf{x})} := \int_{\mathbb{S}^{n-1}} \frac{1}{\delta_\nu^2(\mathbf{x})} d\omega(\nu), \tag{7}$$

where $d\,\omega(\nu)$ is the normalised measure on the unit sphere $\mathbb{S}^{n-1}$ in $\mathbb{R}^n$ and $\delta_\nu(\mathbf{x})$ is the distance from $\mathbf{x}$ to $\partial\Omega$ in the direction ν. The inequality (6) and its various extensions are the subject of section 3. A fairly comprehensive treatment of (6) in a general setting may be found in [15].

In section 4, we survey some recent results on the Rellich inequality. The original, announced by Rellich in 1953 (see [27] and [28]), is

$$\int_{\mathbb{R}^n} |\Delta f(\mathbf{x})|^2 d\mathbf{x} \geq \left[\frac{n(n-4)}{4}\right]^2 \int_{\mathbb{R}^n} \frac{|f(\mathbf{x})|^2}{|\mathbf{x}|^4} d\mathbf{x}, \quad f \in C_0^\infty(\mathbb{R}^n \setminus \{0\}) \tag{8}$$

when $n \neq 2$; when $n = 2$, the inequality continues to hold for functions $f \in C_0^\infty(\mathbb{R}^n \setminus \{0\})$ which satisfy

$$\int_0^{2\pi} f(r,\theta)\cos\theta d\theta = \int_0^{2\pi} f(r,\theta)\sin\theta d\theta = 0.$$

The inequality (8) is trivially true when $n = 4$, but an analogue of the Laptev-Weidl inequality (4) was derived in [16]: for $\tilde{\Psi} \notin \mathbb{Z}$,

$$\int_{\mathbb{R}^4} |\Delta_{\mathbf{A}} f(\mathbf{x})|^2 d\mathbf{x} \geq C\left(\mathrm{dist}(\tilde{\Psi}, \mathbb{Z})\right) \int_{\mathbb{R}^4} \frac{|f(\mathbf{x})|^2}{|\mathbf{x}|^4} d\mathbf{x}, \quad f \in C_0^\infty(\mathbb{R}^4 \setminus \mathcal{L}_4), \tag{9}$$

where $\Delta_{\mathbf{A}} = \nabla_{\mathbf{A}} \cdot \nabla_{\mathbf{A}}, \mathbf{A}$ is a magnetic potential which is an analogue of (5) in $\mathbb{R}^4$, being now singular on a co-ordinate plane $\mathcal{L}_4$ of co-dimension 2, and $\tilde{\Psi}$ is an associated flux; see [16] for details. In (9) it is indicated that the constant depends on the distance from $\tilde{\Psi}$ to the set of integers $\mathbb{Z}$; specifically, from [16], Corollary 4, if $\tilde{\Psi} \in (0,1)$,

$$C(\mathrm{dist}(\tilde{\Psi}, \mathbb{Z}) = \min\left\{[(1+\tilde{\Psi})^2 - 1]^2, [(-2+\tilde{\Psi})^2 - 1]^2\right\}.$$

Weighted versions and higher-order analogues of (8) are given in [5], [14], [22] and [30]; see [30] for a comprehensive analysis of exceptional cases. Weighted versions and higher-order analogues of (9) are also given in [30].

The bounded domain version of Rellich's inequality takes the form

$$\int_\Omega |\Delta f(\mathbf{x})|^2 d\mathbf{x} \geq C \int_\Omega \frac{|f(\mathbf{x})|^2}{\delta(\mathbf{x})^4} d\mathbf{x}, \quad f \in C_0^\infty(\Omega). \tag{10}$$

This and some recent extensions are discussed in section 4.

2. Hardy's inequality in $\mathbb{R}^n$

In [2] it is shown that for all $n \geq 1$ and $1 \leq p < \infty$,

$$\int_{\mathbb{R}^n} |(\mathbf{x}\cdot\nabla)f|^p d\mathbf{x} \geq \left(\frac{n}{p}\right)^p \int_{\mathbb{R}^n} |f(\mathbf{x})|^p d\mathbf{x}, \quad f \in C_0^\infty(\mathbb{R}^n). \tag{11}$$

This follows on integration by parts and the application of Hölder's inequality:

$$n\int_{\mathbb{R}^n} |f(\mathbf{x})|^p d\mathbf{x} = \int_{\mathbb{R}^n} \mathrm{div}(\mathbf{x})|\mathrm{f}(\mathbf{x})|^{\mathrm{p}} \mathrm{d}\mathbf{x} \tag{12}$$

$$= -p\mathrm{Re}\int_{\mathbb{R}^n} (\mathbf{x}\cdot\nabla)\mathrm{f}(\mathbf{x})|\mathrm{f}(\mathbf{x})|^{\mathrm{p}-2}\overline{\mathrm{f}}(\mathbf{x})\mathrm{d}\mathbf{x} \tag{13}$$

$$\leq p\left(\int_{\mathbb{R}^n} |(\mathbf{x}\cdot\nabla)f(\mathbf{x})|^p d\mathbf{x}\right)^{1/p}\left(\int_{\mathbb{R}^n} |f(\mathbf{x})|^p d\mathbf{x}\right)^{(p-1)/p} \tag{14}$$

$$\tag{15}$$

which yields (11).

The inequality (11) implies (1) for $1 \leq p \leq n$. For we have from

$$\nabla(|\mathbf{x}|f) = \frac{\mathbf{x}}{|\mathbf{x}|}f + |\mathbf{x}|\nabla f$$

that

$$\begin{aligned}\|\nabla(|\mathbf{x}|f)\|_{L^p(\mathbb{R}^n)} &\geq \||\mathbf{x}||\nabla f|\|_{L^p(\mathbb{R}^n)} - \|f\|_{L^p(\mathbb{R}^n)}\\ &\geq \|(\mathbf{x}\cdot\nabla)f\|_{L^p(\mathbb{R}^n)} - \|f\|_{L^p(\mathbb{R}^n)}\\ &\geq \left(\frac{n-p}{p}\right)\|f\|_{L^p(\mathbb{R}^n)}\end{aligned}$$

whence (2) on replacing $f(\mathbf{x})$ by $f(\mathbf{x})/|\mathbf{x}|$.

The inequality (11), which holds without restriction on n, is the basis of a Hardy-Sobolev inequality discussed in [2] on the lines of the improved Sobolev inequality of Ledoux in [23]; see also [8], [9] and [10]. The result of Ledoux is the following: for every $1 \leq p < q < \infty$ and every function f in the Sobolev space $W^{1,p}(\mathbb{R}^n)$,

$$\|f\|_{L^q(\mathbb{R}^n)} \leq C\|\nabla f\|^\theta_{L^p(\mathbb{R}^n)}\|f\|^{1-\theta}_{B^{\theta/(\theta-1)}_{\infty,\infty}}, \tag{16}$$

where $\theta = p/q, C$ is a positive constant which depends only on p, q and n, and $B^\alpha_{\infty,\infty}$ is the homogenous Besov space of indices (α,∞,∞). The latter is the space of tempered distributions for which the norm

$$\|f\|_{B^{\alpha}_{\infty,\infty}} := \sup_{t>0}\{t^{-\alpha/2}\|P_t f\|_{L^\infty(\mathbb{R}^n)}\}$$

is finite, where $P_t = e^{t\Delta}, t \geq 0$, is the heat semigroup on $\mathbb{R}^n$: recall that $\{P_t\}_{t\geq 0}$ is defined by $P_0 f = f$ and

$$P_t f(\mathbf{x}) = \frac{1}{(4\pi t)^{n/2}}\int_{\mathbb{R}^n} f(\mathbf{y})\mathbf{e}^{-|\mathbf{x}-\mathbf{y}|^2/4\mathbf{t}}\mathbf{dy}$$

for $t > 0, \mathbf{x} \in \mathbb{R}^n$. Cases of (16) were earlier established in [8], [9] and [10], the proofs being based on wavelet decompositions together with weak-ℓ^1 type estimates and interpolation theory. Ledoux's approach relies on the use of pseudo-Poincaré inequalities and the argument easily extends to more general frameworks, including manifolds or graphs which are out of reach of wavelet methods. The inequality (16) is easily seen to include the classical Sobolev inequality (3) on using the readily established estimate

$$\|P_t f\|_{L^\infty(\mathbb{R}^n)} \leq C t^{-n/2r}\|f\|_{L^r(\mathbb{R}^n)}$$

for any $r \geq 1$. For then with $r = q = p^* = np/(n-p)$, we have $\theta/(\theta-1) + n/r = 0$ and so

$$\|f\|_{B^{\theta/(\theta-1)}_{\infty,\infty}} \leq C\|f\|_{L^q(\mathbb{R}^n)}.$$

Ledoux's technique requires specific information on the heat semi-group $e^{t\Delta}$ in $L^2(\mathbb{R}^n)$. The operator semigroup associated with the inequality in [2] is e^{-tL^*L}, where $L = \mathbf{x}\cdot\nabla$. In [2], Theorem 2, it is proved that

$$(e^{-tL^*L}\psi)(\mathbf{x}) = \frac{e^{-tn^2/4}}{\sqrt{4\pi t}}r^{-n/2}\int_0^\infty e^{-\frac{(\ln r - \ln s)^2}{4t}}s^{-n/2}\psi(s,\omega)\,s^{n-1}ds. \quad (17)$$

A significant fact is that after a co-ordinate change Φ, L^*L is related to the Laplacian on $\mathbb{R}$. To be specific, Φ is an isometry of $L^2(\mathbb{R}^n)$ onto $L^2(\mathbb{R}\times\mathbb{S}^{n-1})$ defined by

$$(\Phi\psi)(s,\omega) := e^{sn/2}\psi(e^s,\omega), \quad \omega \in \mathbb{S}^{n-1}, \quad s \in \mathbb{R}.$$

It is readily seen that $L = iA - n/2$, where A is the non-negative self-adjoint generator of dilations in $L^2(\mathbb{R}^n)$ which in fact is the momentum operator on $\mathbb{R}$. It follows that

$$\Phi A\Phi^{-1} = -i\partial_r \otimes I_{\mathbb{S}^{n-1}} \quad (18)$$

where $I_{\mathbb{S}^{n-1}}$ denotes the identity on the unit sphere $\mathbb{S}^{n-1}$ and, consistent with (11), $L^*L = A^2 + n^2/4 \geq n^2/4$. This leads to the representation

$$\Phi e^{-tA^2}\Phi^{-1}g(r,\omega) = \frac{1}{\sqrt{4\pi t}}\int_{\mathbb{R}}\exp\{-\frac{1}{4t}(r-s)^2\}g(s,\omega)ds \quad (19)$$

and $e^{-tL^*L} = e^{-tA^2}e^{-tn^2/4}$. The fact that $\Phi e^{-tA^2}\Phi^{-1}$ in (19) is essentially radial means that the analogue of (16) derived by Ledoux's technique is a consequence of the one-dimensional case of (16). Defining B^α to be the space of all tempered distributions g on $\mathbb{R} \times \mathbb{S}^{n-1}$ for which the norm

$$\|g\|_{B^\alpha} := \sup_{t>0}\{t^{-\alpha/2}\|\Phi e^{-tA^2}\Phi^{-1}g\|_{L^\infty(\mathbb{R}\times\mathbb{S}^{n-1})}\} < \infty, \tag{20}$$

one obtains from the $n = 1$ case of (16), that for any $\omega \in \mathbb{S}^{n-1}$,

$$\begin{aligned}
&\int_{\mathbb{R}} |g(r,\omega)|^q dr \le C^q \int_{\mathbb{R}} \left|\frac{\partial g(r,\omega)}{\partial r}\right|^p dr \\
&\times \left(\sup_{t>0, r\in\mathbb{R}} t^{\theta/2(1-\theta)} \left|\frac{1}{\sqrt{4\pi t}}\int_{\mathbb{R}} e^{-(r-s)^2/4t} g(s,\omega)ds\right|\right)^{q(1-\theta)} \\
&= C^q \int_{\mathbb{R}} \left|\frac{\partial g(r,\omega)}{\partial r}\right|^p dr \left(\sup_{t>0, r\in\mathbb{R}} t^{\theta/2(1-\theta)} \left|\Phi e^{-tA^2}\Phi^{-1}g(r,\omega)\right|\right)^{q(1-\theta)} \\
&\le C^q \int_{\mathbb{R}} \left|\frac{\partial g(r,\omega)}{\partial r}\right|^p dr \left(\sup_{t>0} t^{\theta/2(1-\theta)} \left\|\Phi e^{-tA^2}\Phi^{-1}g\right\|_{L^\infty(\mathbb{R}\times\mathbb{S}^{n-1})}\right)^{q(1-\theta)} \\
&\le C^q \int_{\mathbb{R}} \left|\frac{\partial g(r,\omega)}{\partial r}\right|^p dr \|g\|^{q(1-\theta)}_{B^{\theta/(\theta-1)}}.
\end{aligned}$$

On integrating with respect to ω over $\mathbb{S}^{n-1}$ we obtain the following result.

Theorem 2.1. *Let $1 \le p < q < \infty$ and suppose that g is such that $\Phi A\Phi^{-1}g \equiv -i(\partial/\partial r)g \in L^p(\mathbb{R} \times \mathbb{S}^{n-1})$ and $g \in B^{\theta/(\theta-1)}, \theta = p/q$. Then there exists a positive constant C, depending on p and q, such that*

$$\|g\|_{L^q(\mathbb{R}\times\mathbb{S}^{n-1})} \le C\|(\partial/\partial r)g\|^{\theta}_{L^p(\mathbb{R}\times\mathbb{S}^{n-1})}\|g\|^{1-\theta}_{B^{\theta/(\theta-1)}}. \tag{21}$$

Various corollaries of Hardy-Sobolev type featuring the inequality (2.1) are derived in [2].

Corollary 2.1. *Let $p^* := np/(n-p), 1 \le p \le n-1$, and suppose $(\partial/\partial r)g \in L^p(\mathbb{R} \times \mathbb{S}^{n-1})$ and $\sup_{\omega\in\mathbb{S}^{n-1}} \|g(\cdot,\omega)\|_{L^p(\mathbb{R})} < \infty$. Then*

$$\|g\|_{L^{p^*}(\mathbb{R}\times\mathbb{S}^{n-1})} \le C\|(\partial/\partial r)g\|^{1/n}_{L^p(\mathbb{R}\times\mathbb{S}^{n-1})} \sup_{\omega\in\mathbb{S}^{n-1}} \|g(\cdot,\omega)\|^{(n-1)/n}_{L^p(\mathbb{R})}. \tag{22}$$

If $G = \mathcal{M}(g)$ denotes the integral mean of g, namely,

$$G(r) = \mathcal{M}(g)(r) := \frac{1}{|\mathbb{S}^{n-1}|}\int_{\mathbb{S}^{n-1}} g(r,\omega)d\omega,$$

then if $g, (\partial/\partial r)g \in L^p(\mathbb{R} \times \mathbb{S}^{n-1})$,

$$\|G\|_{L^{p^*}(\mathbb{R})} \le C\|(\partial/\partial r)g\|^{1/n}_{L^p(\mathbb{R}\times\mathbb{S}^{n-1})}\|g\|^{(n-1)/n}_{L^p(\mathbb{R}\times\mathbb{S}^{n-1})}. \tag{23}$$

If g is supported in $[-\Lambda,\Lambda]\times\mathbb{S}^{n-1}$, then

$$\|g\|_{L^{p^*}(\mathbb{R})} \le C\Lambda^{(n-1)/n^2}\|(\partial/\partial r)g\|_{L^p(\mathbb{R}\times\mathbb{S}^{n-1})}\sup_{\omega\in\mathbb{S}^{n-1}}\|g(\cdot,\omega)\|^{(n-1)/n}_{L^{p^*}(\mathbb{R})}; \quad (24)$$

also

$$\|G\|_{L^{p^*}(\mathbb{R})} \le C\Lambda^{(n-1)/n}\|(\partial/\partial r)g\|_{L^p(\mathbb{R}\times\mathbb{S}^{n-1})}. \quad (25)$$

The case $p=2$ is of special interest; for instance one gets from (22) and (23) the following.

Corollary 2.2. *Let f be such that $Lf \in L^2(\mathbb{R}^n)$ and $\sup_{\omega\in\mathbb{S}^{n-1}}\|f\|_{L^2(\mathbb{R}^+;d\mu)}<\infty$, where $L=\mathbf{x}\cdot\nabla$. Then for $n\ge 3$,*

$$\|rf(r,\omega)\|^2_{L^{2^*}(\mathbb{R}^n)} \le C\left\{\|Lf\|^2_{L^2(\mathbb{R}^n)} - \frac{n^2}{4}\|f\|^2_{L^2(\mathbb{R}^n)}\right\}^{1/n} \times \sup_{\omega\in\mathbb{S}^{n-1}}\|f\|^{2(1-1/n)}_{L^2(\mathbb{R}^+;d\mu))}, \quad (26)$$

where $2^*=2n/(n-2)$ and $d\mu=r^{n-1}dr$. If $f, Lf\in L^2(\mathbb{R}^n)$ then with $F=\mathcal{M}(f)$,

$$\|rF(r)\|^2_{L^{2^*}(\mathbb{R}^+;d\mu)} \le C\left\{\|Lf\|^2_{L^2(\mathbb{R}^n)} - \frac{n^2}{4}\|f\|^2_{L^2(\mathbb{R}^n)}\right\}^{1/n} \times \|f\|^{2(1-1/n)}_{L^2(\mathbb{R}^n)}. \quad (27)$$

On setting $f=h/|\cdot|, n\ge 3$, in (26) and (27) it follows that there exist positive constants C depending only on n such that

$$\|h\|^2_{L^{2^*}(\mathbb{R}^n)} \le C\{\|\nabla h\|^2_{L^2(\mathbb{R}^n)} - (\frac{n-2}{2})^2\|h/|\cdot|\|^2_{L^2(\mathbb{R}^n)}\}^{1/n} \times \sup_{\omega\in\mathbb{S}^{n-1}}\{\|h(\cdot,\omega)/|\cdot|\|^2_{L^2(\mathbb{R}^+;d\mu)}\}^{1-1/n} \quad (28)$$

and

$$\|\mathcal{M}(h)\|^2_{L^{2^*}(\mathbb{R}^+;d\mu)} \le C\{\|\nabla h\|^2_{L^2(\mathbb{R}^n)} - (\frac{n-2}{2})^2\|h/|\cdot|\|^2_{L^2(\mathbb{R}^n)}\}^{1/n} \times \{\|h/|\cdot|\|^2_{L^2(\mathbb{R}^n)}\}^{1-1/n}. \quad (29)$$

From (29) and Young's inequality, we have that, for any $\varepsilon>0$,

$$\varepsilon^{1-1/n}\|\mathcal{M}(h)\|^2_{L^{2^*}(\mathbb{R}^+;d\mu)} \le C\{\|\nabla h\|^2_{L^2(\mathbb{R}^n)} - [(\frac{n-2}{2})^2-\varepsilon]\|h/|\cdot|\|^2_{L^2(\mathbb{R}^n)}\},$$

which, on taking $\delta = \left(\frac{n-2}{2}\right)^2 - \varepsilon > 0$, yields

$$\|\mathcal{M}(h)\|^2_{L^{2^*}(\mathbb{R}^+;d\mu)} \le C[\frac{(n-2)^2}{4} - \delta]^{-\frac{(n-1)}{n}} \left\{\|\nabla h\|^2_{L^2(\mathbb{R}^n)} - \delta\|\frac{h}{|\cdot|}\|^2_{L^2(\mathbb{R}^n)}\right\}. \tag{30}$$

Since

$$\|\mathcal{M}(h)\|^{2^*}_{L^{2^*}(\mathbb{R}^+;d\mu)} \le \frac{1}{|\mathbb{S}^{n-1}|}\|h\|^{2^*}_{L^{2^*}(\mathbb{R}^n)}$$

by Hölder's inequality, it follows that (30) is a consequence of Stubbe's inequality in [29], Theorem 1, namely, for $\delta < (n-2)^2/4$,

$$\|h\|^2_{L^{2^*}(\mathbb{R}^n)} \le C^2[1 - \frac{4\delta}{(n-2)^2}]^{-\frac{(n-1)}{n}}\{\|\nabla h\|^2_{L^2(\mathbb{R}^n)} - \delta\|\frac{h}{|\cdot|}\|^2_{L^2(\mathbb{R}^n)}\}, \tag{31}$$

where the optimal constant C is that for (3) in the case $p = 2$, that is,

$$C = [\pi n(n-2)]^{-1/2} \left(\Gamma(n)/\Gamma(n/2)\right)^{1/n}. \tag{32}$$

The fact that the integral mean, $\mathcal{M}h$, rather than h appears on the left-hand-side of (30) is due to the fact that (30) is derived from (23), where the presence of the integral mean is a consequence of the fact that e^{-tA^2} does not map any $L^p(\mathbb{R} \times \mathbb{S}^{n-1}), 1 < p < \infty$, into $L^\infty(\mathbb{R} \times \mathbb{S}^{n-1})$. This is in contrast to $e^{t\Delta}$ which maps $L^p(\mathbb{R}^n)$ into $L^\infty(\mathbb{R}^n)$.

If in (25) we put $g = \Phi f$, where f is supported in the annulus $A(1/R, R) := \{\mathbf{x} \in \mathbb{R}^n : 1/R \le |\mathbf{x}| \le R\}$, then G is supported in the interval $[-\ln R, \ln R]$ and we obtain the following local form of a Hardy-Sobolev inequality.

Corollary 2.3. *Let $f \in C_0^\infty(A(1/R, R))$. Then*

$$\|\mathcal{M}(f)\|^2_{L^{2^*}(\mathbb{R}^+;d\mu)} \le C(\ln R)^{\frac{2(n-1)}{n}} \left\{\|\nabla f\|^2_{L^2(\mathbb{R}^n)} - \frac{(n-2)^2}{4}\|\frac{f}{|\cdot|}\|^2_{L^2(\mathbb{R}^n)}\right\}. \tag{33}$$

The following related result is proved in [7] and [19] (as the case $s = 1$ of (2.6) whose proof is given in section 6.4): if $f \in C_0^\infty(\Omega)$ and $2 \le q < 2^*$,

$$\|f\|^2_{L^q(\mathbb{R}^n)} \le C|\Omega|^{2(1/q-1/2^*)} \left\{\|\nabla f\|^2_{L^2(\mathbb{R}^n)} - \frac{(n-2)^2}{4}\left\|\frac{f}{|\cdot|}\right\|^2_{L^2(\mathbb{R}^n)}\right\}, \tag{34}$$

where $|\Omega|$ denotes the volume of Ω. It is noted in [19], Remark 2.4, that, in contrast to (33), the q in (34) must be strictly less than the critical Sobolev exponent $2^* = 2n/(n-2)$ if Ω includes the origin. A similar inequality to (34) in $L^p(\Omega), 1 < p < n$, when Ω is convex and $f(\mathbf{x})/|\mathbf{x}|$ is replaced by $f(\mathbf{x})/\text{dist}(\text{x}, \partial\Omega)$, was established in [18]; see the discussion after (41) in the next section.

3. Hardy's inequality on bounded domains

It is known (see [24] and [25]) that if Ω is a convex domain in $\mathbb{R}^n$, the best constant in (6) is $C = c_p := [(p-1)/p]^p$, but there are smooth domains for which $C < c_p$. In [6], Brézis and Markus proved that for any convex Ω, the largest possible constant $\lambda(\Omega)$ such that

$$\int_\Omega |\nabla f(\mathbf{x})|^2 d\mathbf{x} \geq (1/4) \int_\Omega \frac{|f(\mathbf{x})|^2}{\delta(\mathbf{x})^2} d\mathbf{x} + \lambda(\Omega) \int_\Omega |f(\mathbf{x})|^2 d\mathbf{x}, \quad f \in C_0^\infty(\Omega), \tag{35}$$

satisfies

$$\lambda(\Omega) \geq (4 \operatorname{diam}(\Omega)^2)^{-1}. \tag{36}$$

This was improved by M. Hoffmann-Ostenhof, T. Hoffmann-Ostenhof and A. Laptev, to

$$\lambda(\Omega) \geq \mu_n |\Omega|^{-2/n}, \quad \mu_n = n^{(n-2)/n} |\mathbb{S}^{n-1}|^{2/n}/4 \tag{37}$$

in [20] and subsequently by Evans and Lewis in [16] to

$$\lambda(\Omega) \geq 6\mu_n |\Omega|^{-2/n}. \tag{38}$$

Filippas, Maz'ya and Tertikas [18] obtained the following bound as a special case of L^p Hardy inequalities:

$$\lambda(\Omega) \geq 3D_{int}(\Omega)^{-2} \tag{39}$$

where $D_{int}(\Omega) := 2\sup_{\mathbf{x}\in\Omega} \delta(\mathbf{x})$, the interior diameter of Ω. Since we have $3D_{int}(\Omega)^{-2} \geq (3/n)\mu_n|\Omega|^{-2/n}$, this is an improvement of (37) when $n = 2, 3$, but not of (38); for $n > 3$, the results are not comparable. The analogue of (35) in L^p for $p > 1$ has been studied in [17], [18] and [31]; in [17] weighted versions of the inequality and also arbitrary domains were considered. As illustrations, we quote the following results for a convex domain Ω from [31] and [18] respectively: Tidblom proves an extension of (35) and (37) to the following in $L^p(\Omega), 1 < p < \infty$:

$$\int_\Omega |\nabla f|^p d\mathbf{x} \geq c_p \int_\Omega \frac{|f(\mathbf{x})|^p}{\delta(\mathbf{x})^p} d\mathbf{x} + \lambda_p(\Omega) \int_\Omega |f(\mathbf{x})|^p d\mathbf{x}, \quad f \in C_0^\infty(\Omega), \tag{40}$$

where $\lambda_p(\Omega) \geq \mu_{n,p}|\Omega|^{-p/n}$ and

$$\mu_{n,p} = \frac{(p-1)^{p+1}}{p^p} \left(\frac{|\mathbb{S}^{n-1}|}{n}\right)^{p/n} \frac{\sqrt{\pi}\,\Gamma(\frac{n+p}{2})}{\Gamma(\frac{p+1}{2})\Gamma(\frac{n}{2})}.$$

In [18] the result is a corollary of the following Hardy-Sobolev type inequality: for $1 < p < n$ and $p \le q < p^* := np/(n-p)$, we have

$$\int_\Omega |\nabla f|^p d\,\mathbf{x} \ge c_p \int_\Omega \frac{|f(\mathbf{x})|^p}{\delta(\mathbf{x})^p} d\,\mathbf{x} + \lambda_{p,q}(\Omega) \left(\int_\Omega |f(\mathbf{x})|^q d\,\mathbf{x} \right)^{p/q}, \quad (41)$$

for $f \in C_0^\infty(\Omega)$, where, for some positive constants $k_i = k_i(p,q,n), i = 1,2,$ independent of Ω,

$$k_1(p,q,n) D_{int}(\Omega)^{n-p-\frac{np}{q}} \ge \lambda_{p,q}(\Omega) \ge k_2(p,q,n) D_{int}(\Omega)^{n-p-\frac{np}{q}}.$$

As we have already observed, $D_{int}(\Omega) \le (n/\mu_n)^{1/2} |\Omega|^{1/n}$. Hence (41) yields the following analogue of (34) in $L^p(\Omega)$ when Ω is convex with finite internal diameter $D_{int}(\Omega)$: for $1 < p < n, p \le q < p^*$, and $f \in C_0^\infty(\Omega)$,

$$\|f\|^p_{L^q(\Omega)} \le C(n,p,q) |\Omega|^{p(1/q - 1/p^*)} \left\{ \|\nabla f\|^p_{L^p(\Omega)} - c_p \|f/\delta\|^p_{L^p(\Omega)} \right\}. \quad (42)$$

A comprehensive and unified treatment of improvements to Hardy's inequality is given in [3] in which distance functions other than δ are also considered and optimal weights and best constants established.

The technique used in [17], [20] and [31] is to first derive an appropriate one-dimensional Hardy-type inequality on an interval and then use this to give an inequality of type (35) in which δ is replaced by the mean distance δ_M of (7). This leads to (35) when Ω is convex, on using the following estimate given in [13] for $p = 2$ and [31] for general p:

$$\int_{\mathbb{S}^{n-1}} \frac{1}{\delta_M(\mathbf{x})^p} d\omega(\nu) \ge B(n,p)/\delta_M(\mathbf{x})^p,$$

where, with $\mathbf{e}$ a fixed unit vector in $\mathbb{R}^n$,

$$B(n,p) := \int_{\mathbb{S}^{n-1}} |\mathbf{e} \cdot \nu|^p d\omega(\nu) = \frac{\Gamma(\frac{p+1}{2}) \cdot \Gamma(\frac{n}{2})}{\sqrt{\pi} \cdot \Gamma(\frac{n+p}{2})}.$$

4. Rellich's inequality

The same technique as that described at the end of the last section has been used in [4] and [17] to prove extensions of the Rellich inequality(10) of the form

$$\int_\Omega |\Delta f(\mathbf{x})|^2 \ge \mu(\Omega) \int_\Omega \frac{|f(\mathbf{x})|^2}{\delta(\mathbf{x})^4} d\,\mathbf{x} + \lambda(\Omega) \int_\Omega |f(\mathbf{x})|^2 d\,\mathbf{x}$$

for a convex Ω : weighted versions are also given in [17]. To incorporate the weights, the following is established in [17], Proposition 1, in which

$\partial_\nu := \nu \cdot \nabla$ and Ω is any domain in $\mathbb{R}^n$: for all $u \in C^2(\mathbb{R}^n)$

$$\int_{\mathbb{S}^{n-1}} |\partial_\nu^2 u(\mathbf{x})|^2 d\omega(\nu) = \frac{1}{n(n+2)} \left[|\Delta u(\mathbf{x})|^2 + 2 \sum_{i,j=1}^{n} \left| \frac{\partial^2 u(\mathbf{x})}{\partial x_i \partial x_j} \right|^2 \right]. \qquad (43)$$

A consequence of this is that for all $u \in C_0^2(\Omega)$

$$\int_\Omega \int_{\mathbb{S}^{n-1}} |\partial_\nu^2 u(\mathbf{x})|^2 d\omega(\nu) d\mathbf{x} = \frac{3}{n(n+2)} \int_\Omega |\Delta u(\mathbf{x})|^2 d\mathbf{x}, \qquad (44)$$

an identity first proved by Owen in [26]. When Ω is convex, the best constant in (10) is 9/16 (see [26]), and in [4] it is proved that

$$\int_\Omega |\Delta u(\mathbf{x})|^2 d\mathbf{x} \geq \frac{9}{16} \int_\Omega \frac{|u(\mathbf{x})|^2}{\delta(\mathbf{x})^4} d\mathbf{x} + cn(n+2) \left[\frac{\mathbb{S}^{n-1}}{n|\Omega|} \right]^{4/n} \int_\Omega |u(\mathbf{x})|^2 d\mathbf{x}, \qquad (45)$$

with $c = 11/48$; the corresponding result in [17]) has the improvement $c \approx 1.25$.

References

1. A. Ancona, On strong barriers and an inequality of Hardy for domains in $\mathbb{R}^n$, *J. London Math. Soc.* **34** (1986), 274-290.
2. A. Balinsky, W. D. Evans, D. Hundertmark and R. T. Lewis, On inequalities of Hardy-Sobolev type, *preprint.*
3. G. Barbatis, S. Filippas and A. Tertikas, A unified approach to improved L^p Hardy inequalities with best constants, *Trans. Amer. Math. Soc.* **356** (6) (2004), 2169-2196.
4. G. Barbatis, Improved Rellich inequalities for the polyharmonic operator, *Indiana Univ. Math. J.* **55**(4) (2006), 1401-1422.
5. D. M. Bennett, An extension of Rellich's inequality, *Proc. Amer. Math. Soc.* **106** (1989), 987-993.
6. H. Brezis and M. Marcus, Hardy's inequalities revisited, Dedicated to Ennio De Giorgi, *Ann. Scuola Norm. Sup. Pisa Cl. Sci.* **25** (1997, 1998), 217-237.
7. H. Brezis and J.-L. Vázquez, Blow-up solutions of some non-linear elliptic problems, *Rev. Mat. Univ. Comp. Madrid* **10** (1997), 443-469.
8. A. Cohen, W. Dahmen, I. Daubechies and R. DeVore, Harmonic analysis of the space BV, *Rev. Mat. Iberoamericana* **19** (2003), 235-263.
9. A. Cohen, R. DeVore, P. Petrushev and H. Xu, Non-linear approximation and the space $BV(\mathbb{R}^2)$, *Amer. J. Math.* **121** (1999), 587-628.
10. A. Cohen, Y. Meyer and F. Oru, Improved Sobolev inequalities, *Séminaires X-EDP, École Polytechnique* (1998).
11. E. B. Davies, *Heat Kernels and Spectral Theory*, Cambridge Univ. Press, Cambridge, 1989.
12. E. B. Davies, The Hardy constant, *Quart. J. Math. Oxford (2),* **46** (1995), 417-431.

13. E. B. Davies, *Spectral Theory and Differential Operators*, Cambridge Studies in Advanced Mathematics, Vol. 42, Cambridge Univ. Press, Cambridge, 1995.
14. E. B. Davies and A. Hinz, Explicit constants for Rellich inequalities in $L_p(\Omega)$, *Math. Z.* **227** (1998), 511-523.
15. D. E. Edmunds and W.D. Evans, *Hardy Operators, Function Spaces and Embeddings*, Springer, Berlin, Heidelberg, New York, 2004.
16. W. D. Evans and R. T. Lewis, On the Rellich inequality with magnetic potentials, *Math. Z.* **251** (2005), 267-284.
17. W. D. Evans and R. T. Lewis, Hardy and Rellich inequalities with remainders, to appear in *J. Math. Inequal.*
18. S. Filippas, V. Maz'ya, and A. Tertikas, On a question of Brezis and Marcus, *Calc. Var. Partial Differential Equations* **25**(4) (2006), 491-501.
19. R. L. Frank, E. H. Lieb and R. Seiringer, Hardy-Lieb-Thirring inequalities for fractional Schrödinger operators, *preprint* (2006). `http://xxx.lanl.gov/math.SP/0610593`
20. M. Hoffmann-Ostenhof, T. Hoffmann-Ostenhof, and A. Laptev, A geometrical version of Hardy's inequality, *J. Funct. Anal.* **189** (2002), 539-548.
21. A. Laptev and T. Weidl, Hardy inequalities for magnetic Dirichlet forms, in *Mathematical results in quantum mechanics (Prague, 1998)*, 299-305, Oper. Theory Adv. Appl., 108, Birkhäuser, Basel, 1999.
22. R. T. Lewis, Singular elliptic operators of second-order with purely discrete spectrum, *Trans. Amer. Math. Soc.* **271** (1982), 653-666.
23. M. Ledoux, On improved Sobolev embedding theorems, *Math. Res. Lett.* **10** (2003), 659-669.
24. M. Marcus, V. J. Mizel, and Y. Pinchover, On the best constant for Hardy's inequality in $\mathbb{R}^n$, *Trans. Amer. Math. Soc.* **350** (1998), 3237-3255.
25. T. Matskewich and P. E. Sobolevskii, The best possible constant in a generalized Hardy's inequality for convex domains in $\mathbb{R}^n$, *Nonlinear Analysis TMA*, **28** (1997), 1601-1610.
26. M. P. Owen, The Hardy-Rellich inequality for polyharmonic operators, *Proc. Royal Society of Edinburgh A* **129** (1999), 825-839.
27. F. Rellich, Halbbeschränkte Differentialoperatoren höherer Ordnung, in *Proceedings of the International Congress of Mathematicians 1954* (J. C. H. Gerretsen, J. de Groot, eds.), Vol. III (pp. 243-250), Groningen, Noordhoff, 1956.
28. F. Rellich and J. Berkowitz, *Perturbation Theory of Eigenvalue Problems,* Gordon and Breach, New York, 1969.
29. J. Stubbe, Bounds on the number of bound states for potentials with critical decay at infinity, *J. Math. Phys.* **31**, no. 5, (1990), 1177-1180.
30. J. C. Thomas, Some problems associated with sum and integral inequalities, *Ph.D. thesis, Cardiff University*, 2007.
31. J. Tidblom, A geometrical version of Hardy's inequality for $W_0^{1,p}(\Omega)$, *Proc. Amer. Math. Soc.* **132**(8) (2004), 2265-2271.

SOME RESULTS ON NONSELFADJOINT OPERATORS: A SURVEY

JOHANNES SJÖSTRAND

Centre de Mathématiques Laurent Schwartz, Ecole Polytechnique,
FR-91128 Palaiseau Cédex, France
and UMR 7640, CNRS
E-mail: johannes@math.polytechnique

This text is a survey of recent results obtained by the author and collaborators on different problems for non-self-adjoint operators. The topics are: Kramers-Fokker-Planck type operators, spectral asymptotics in two dimensions and Weyl asymptotics for the eigenvalues of non-self-adjoint operators with small random perturbations. In the introduction we also review the notion of pseudo-spectrum and its relation to non-self-adjoint spectral problems.

Keywords: Non-self-adjoint, spectrum, resolvent, random, Fokker-Planck.

1. Introduction

1.1. *Some background*

For self-adjoint and more generally normal operators on some complex Hilbert space $\mathcal{H}$ we have a nice theory, including the spectral theorem and a nice estimate on the norm of the resolvent:

$$\|(z-P)^{-1}\| \le (\mathrm{dist}\,(z,\sigma(P)))^{-1}, \quad \sigma(P) = \text{ the spectrum of } P. \tag{1}$$

This has a consequence for the corresponding evolution problem: If $\sigma(P) \subset \{z \in \mathbf{C}; \Re z \ge \lambda_0\}$, then

$$\|e^{-tP}\| \le e^{-\lambda_0 t},\ t \ge 0. \tag{2}$$

However, non-normal operators appear frequently: Scattering poles, convection-diffusion problems, Kramers-Fokker-Planck equation, damped wave equations, linearized operators in fluid dynamics. Then typically, $\|(z-P)^{-1}\|$ may be very large even when z is far from the spectrum. This implies mathematical difficulties:
– When studying the distribution of eigenvalues,

– When studying functions of the operator, like e^{-tP} and its norm.

It also implies numerical difficulties like:

– Eigenvalue instability.

There are (in the author's opinion) two ways out:

– Change the Hilbert space norm to make the operators look more normal. (Complex scaling methods.)

– Recognize that the region of the z-plane where $\|(z-P)^{-1}\|$ is large, has its own interest. (Pseudospectrum.)

The option to choose depends on the problem.

- In some problems, like those related to scattering poles, there is no obvious choice of Hilbert space and we are free to make the most natural one. This option is particularly natural when considering a differential operator with analytic coefficients.
- In other problems the canonical Hilbert space is L^2 and we are at most allowed to change the norm into an equivalent one. Here the notion of pseudospectrum is likely to be important.

Let $P:\mathcal{H}\to\mathcal{H}$ be closed, densely defined, $\mathcal{H}$ a complex Hilbert space and let $\rho(P)=\mathbf{C}\backslash\sigma(P)$ denote the resolvent set. The notion of pseudospectrum is important in numerical analysis and we refer to L.N. Trefethen [52], Trefethen–M. Embree [54] and further references given there. Thanks to works of E.B. Davies [5], [7], M. Zworski [57] and others it has become popular in the non-self-adjoint spectral theory of differential operators.

Definition 1.1. Let $\epsilon>0$. The ϵ-pseudospectrum of P is

$$\sigma_\epsilon(P):=\sigma(P)\cup\{z\in\rho(P);\ \|(z-P)^{-1}\|>1/\epsilon\}.$$

Unlike the spectrum, the pseudospectrum will in general change when we change the norm on $\mathcal{H}$. Moreover, it can be characterized as a set of spectral instability as follows from the following version of a theorem of Roch-Silberman [43]:

Theorem 1.1. $\sigma_\epsilon(P)=\bigcup\{\sigma(P+Q):Q\in\mathcal{L}(\mathcal{H},\mathcal{H}),\ \|Q\|<\epsilon\}.$

In his survey [52] L.N. Trefethen discusses some linearized operators from fluid dynamics:

- Orr-Sommerfeld equation (Orzag, Reddy, Schmid, Hennigson).
- Plane Poiseuille flow (L.N and A.N Trefethen, Schmid).
- Pipe Poiseuille flow (L.N and A.N Trefethen, Reddy, Driscoll),

and to what extent stability can be predicted from the study of the spectrum of these non-self-adjoint operators: Eigenvalue analysis alone leads in some cases to the prediction of stability for Reynolds numbers $R < 5772$. Experimentally however, we have stability only for $R < 1000$.

The rough explanation of this is that the ϵ-pseudospectrum (for a suitable ϵ) crosses the imaginary axis before the spectrum does, when R increases. Then $\|e^{-tP}\|$ will grow fast for a limited time even though the growth for very large times is determined by the spectrum. However, since P appears as a linearization of a non-linear problem, that suffices to cause instability.

In the case of differential operators the pseudospectral phenomenon is very general and related to classical works in PDE on local solvability and non-hypoellipticity. E.B. Davies [5] studied the non-self-adjoint semiclasscial Schrödinger operator with a smooth (complex-valued) potential in dimension 1 and showed under "generic" assumptions that one can construct quasimodes with the spectral parameter varying in an open complex set, containing points that are possibly very far from the spectrum (as can be verified in the case of the complex harmonic oscillator). M. Zworski [57] observed that this is essentially a rediscovery of an old result of Hörmander [32,33], and was able to generalize considerably Davies' result by adapting the one of Hörmander to the semi-classical case. With N. Dencker and M.Zworski [9] we also gave a direct proof and a corresponding adaptation of old results of Sato-Kawai-Kashiwara [44] to the analytic case:

Theorem 1.2. ([9,57]) *Let*

$$P(x, hD_x) = \sum_{|\alpha| \le m} a_\alpha(x)(hD_x)^\alpha, \ D_x = \frac{\partial}{\partial x} \tag{3}$$

have smooth coefficients in the open set $\Omega \subset \mathbf{R}^n$. *Put* $p(x,\xi) = \sum_{|\alpha|\le m} a_\alpha(x)\xi^\alpha$. *Assume* $z = p(x_0, \xi_0)$ *with the Poisson bracket* $\frac{1}{i}\{p, \overline{p}\}(x_0, \xi_0) > 0$. *Then there exists a* $u = u_h \in C_0^\infty(\Omega)$ *with* $\|u\| = 1$ *such that* $\|(P - z)u\| = \mathcal{O}(h^\infty)$ *when* $h \to 0$.
Analytic case: We can replace "h^∞*" by "*$e^{-1/Ch}$*".*

Here, we have used standard multi-index notation: $\xi^\alpha = \xi_1^{\alpha_1} \cdot ... \cdot \xi_n^{\alpha_n}$, $|\alpha| = \alpha_1 + ... + \alpha_n$, and the norms $\|\cdot\|$ are the ones of L^2 or ℓ^2 if nothing else is indicated.

This result was subsequently generalized by K. Pravda-Starov [42]. Notice that this implies that when the theorem applies and if the resolvent

$(P-z)^{-1}$ exists then its norm is greater than any negative power of h when $h \to 0$.

Example 1.1. Let $P = -h^2\Delta + V(x)$, $p(x,\xi) = \xi^2 + V(x)$, $\xi^2 = \xi_1^2 + \ldots + \xi_n^2$. Then $\frac{1}{i}\{p,\overline{p}\} = -4\xi \cdot \Im V'(x)$.

"Generically", if $z = p(x,\xi)$, then $\{p,\overline{p}\}(x,\xi) \neq 0$ and one can show quite generally that if this happens then there is also another point (y,η) with $p(y,\eta) = z$ such that $\{p,\overline{p}\}(x,\xi)$ takes the opposite sign. This justifies the following simplified terminology in the semi-classical limit:

The semi-classical pseudospectrum of P is the range $\mathcal{R}(p)$ of p.

In [9] we also showed under suitable assumptions (inspired from scattering theory and from the theory of sub-elliptic operators), that there is no spectrum near the boundary of the semi-classical pseudospectrum and that we may have quite a good control of the norm of the resolvent there. Generalizations to the case of systems were given by Dencker [8].

1.2. *The topics of this survey*

We will discuss three subjects involving non-self-adjoint differential and pseudodifferential operators, we will always work in the semi-classical limit, which means that our operators are of the form $P(x, hD_x; h)$, where P is a suitable symbol and $0 < h \ll 1$. It is quite clear however that some of our results will also apply to non-semiclassical situations in the limit of large eigenvalues.

The subjects are:

- The Kramers-Fokker-Planck operator,
- Bohr-Sommerfeld rules in dimension 2,
- Weyl asymptotics for non-self-adjoint operators with small random perturbations.

and most of the works discussed are the results of collaborations with A. Melin, M. Hitrik, F. Hérau, C. Stolk, S. Vũ Ngọc and M. Hager.

In the first two topics we exploit the possibility of changing the Hilbert space norm by introducing exponential weights on phase space.

In the case of the Kramers-Fokker-Planck operator, we make no analyticity assumptions and the phase-space weights are correspondingly quite weak. In this case however our operator is a differential one, so we are allowed to apply strong exponential weights depending only on the base variables, and this is important when studying small exponential corrections of the eigenvalues via the so called tunnel effect.

For the Bohr-Sommerfeld rules, we make analyticity assumptions that allow stronger phase space weights. In both cases the effect of the exponential weights is to make the operator under consideration more normal.

In the third topic, we do not use any deformations of the given Hilbert space, but exponential weights play an important role at another level, namely to count zeros of holomorphic functions with exponential growth.

The pseudospectrum will not be discussed explicitly below. In the Kramers-Fokker-Planck case, the problems are located near the boundary of the semi-classical pseudospectrum, and it turns out that we have a very nice control of the resolvent there. In the 2 dimensional Bohr-Sommerfeld rules, we have stronger exponential weights, reflecting stronger pseudospectral phenomena. Finally in the subject of Weyl asymptotics, we often have strong pseudospectral behaviour for the unperturbed operator. From the proofs it appears that the random perturbations will weaken the pseudospectral behaviour and this might have very interesting consequences for the associated evolution problems. This is still very much an open problem.

2. Kramers-Fokker-Planck type operators, spectrum and return to equilibrium

2.1. *Introduction*

There has been a renewed interest in the problem of "return to equilibrium" for various 2nd order operators. One example is the Kramers-Fokker-Planck operator:

$$P = y \cdot h\partial_x - V'(x) \cdot h\partial_y + \frac{\gamma}{2}(-h\partial_y + y) \cdot (h\partial_y + y), \tag{4}$$

where $x, y \in \mathbf{R}^n$ correspond respectively to position and speed of the particles and $h > 0$ corresponds to temperature. The constant $\gamma > 0$ is the friction. (Since we will only discuss L^2 aspects we here present right away an adapted version of the operator, obtained after conjugation by a Maxwellian factor.)

The associated evolution equation is:

$$(h\partial_t + P)u(t, x, y) = 0.$$

Problem of return to equilibrium: Study the rate of convergence of $u(t, x, y)$ to a multiple of the "ground state" $u_0(x, y) = e^{-(y^2/2+V(x))/h}$ when $t \to +\infty$, assuming that $V(x) \to +\infty$ sufficiently fast when $x \to \infty$ so that $u_0 \in L^2(\mathbf{R}^{2n})$. Notice here that $P(u_0) = 0$ and that the vector field part of P is h times the Hamilton field of $y^2/2 + V(x)$, when we identify $\mathbf{R}^{2n}_{x,y}$ with the cotangent space of $\mathbf{R}^n_x$.

A closely related problem is to study the difference between the first eigenvalue (0) and the next one, $\mu(h)$. (Since our operator is non-self-adjoint, this is only a very approximate formulation however.)

Some contributions: L. Desvillettes–C. Villani [10], J.P. Eckmann–M. Hairer [12], F. Hérau–F. Nier [27], B. Helffer–F. Nier [22], Villani [55]. In the work [27] precise estimates on the exponential rates of return to equilibrium were obtained with methods close to those used in hypoellipticity studies and this work was our starting point. With Hérau and C. Stolk [28] we made a study in the semi-classical limit and studied small eigenvalues modulo $\mathcal{O}(h^\infty)$. More recently with Hérau and M. Hitrik [25] we have made a precise study of the exponential decay of $\mu(h)$ when V has two local minima (and in that case $\mu(h)$ turns out to be real). This involves tunneling, i.e. the study of the exponential decay of eigenfunctions. As an application we have a precise result on the return to equilibrium [26]. This has many similarities with older work on the tunnel effect for Schrödinger operators in the semi-classical limit by B. Helffer–Sjöstrand [23,24] and B. Simon [45] but for the Kramers-Fokker-Planck operator the problem is richer and more difficult since P is neither elliptic nor self-adjoint. We have used a supersymmetry observation of J.M. Bismut [1] and J. Tailleur–S. Tanase-Nicola–J. Kurchan [51] that allow arguments similar to those for the standard Witten complex [24].

2.2. *Statement of the main results*

Let P be given by (4) where $V \in C^\infty(\mathbf{R}^n; \mathbf{R})$, and

$$\partial^\alpha V(x) = \mathcal{O}(1),\ |\alpha| \geq 2, \tag{5}$$

$$|\nabla V(x)| \geq 1/C,\ |x| \geq C, \tag{6}$$

$$V \text{ is a Morse function.} \tag{7}$$

We also let P denote the graph closure of P from $\mathcal{S}(\mathbf{R}^{2n})$ which coincides with the maximal extension of P in L^2 (see [22,26,27]). We have $\Re P \geq 0$ and the spectrum of P is contained in the right half plane. In [28] the spectrum in any strip $0 \leq \Re z \leq Ch$ (and actually in a larger parabolic neighborhood of the imaginary axis, in the spirit of [27]) was determined asymptotically mod $(\mathcal{O}(h^\infty))$. It is discrete and contained in a sector $|\Im z| \leq C\Re z + \mathcal{O}(h^\infty)$:

Theorem 2.1. *The eigenvalues in the strip* $0 \leq \Re z \leq Ch$ *are of the form*

$$\lambda_{j,k}(h) \sim h(\mu_{j,k} + h^{1/N_{j,k}}\mu_{j,k,1} + h^{2/N_{j,k}}\mu_{j,k,2} + ..) \tag{8}$$

where $\mu_{j,k}$ are the eigenvalues of the quadratic approximation ("non-selfadjoint oscillator")

$$y \cdot \partial_x - V''(x_j)x \cdot \partial_y + \frac{\gamma}{2}(-\partial_y + y) \cdot (\partial_y + y),$$

at the points $(x_j, 0)$, where x_j are the critical points of V.

The $\mu_{j,k}$ are known explicitly and it follows that when x_j is not a local minimum, then $\Re\lambda_{j,k} \geq h/C$ for some $C > 0$. When x_j is a local minimum, then precisely one of the $\lambda_{j,k}$ is $\mathcal{O}(h^\infty)$ while the others have real part $\geq h/C$. Furthermore, when $V \to +\infty$ as $x \to \infty$, then 0 is a simple eigenvalue. *In particular, if V has only one local minimum, then*

$$\inf \Re(\sigma(P) \setminus \{0\}) \sim h(\mu_1 + h\mu_2 + \dots), \quad \mu_1 > 0$$

(or possibly an expansion in fractional powers) and we obtained a corresponding result for the problem of return to equilibrium. It should be added that when $\mu_{j,k}$ is a simple eigenvalue of the quadratic approximation then $N_{j,k} = 1$ so there are no fractional powers of h in (8).

The following is the main new result that we obtained with F. Hérau and M. Hitrik in [25]:

Theorem 2.2. *Assume that V has precisely 3 critical points; 2 local minima, $x_{\pm 1}$ and one "saddle point", x_0 of index 1. Then for $C > 0$ sufficiently large and h sufficiently small, P has precisely 2 eigenvalues in the strip $0 \leq \Re z \leq h/C$, namely 0 and $\mu(h)$, where $\mu(h)$ is real and of the form*

$$\mu(h) = h(a_1(h)e^{-2S_1/h} + a_{-1}(h)e^{-2S_{-1}/h}), \tag{9}$$

where a_j are real,

$$a_j(h) \sim a_{j,0} + ha_{j,1} + \dots, \; h \to 0, \quad a_{j,0} > 0,$$

$$S_j = V(x_0) - V(x_j).$$

As for the problem of return to equilibrium, we obtained the following result with F. Hérau and M. Hitrik in [26]:

Theorem 2.3. *We make the same assumptions as in Theorem 2.2 and let Π_j be the spectral projection associated with the eigenvalue μ_j, $j = 0, 1$, where $\mu_0 = 0$, $\mu_1 = \mu(h)$. Then we have*

$$\Pi_j = \mathcal{O}(1) : L^2 \to L^2, \quad h \to 0. \tag{10}$$

We have furthermore, uniformly as $t \geq 0$ and $h \to 0$,

$$e^{-tP/h} = \Pi_0 + e^{-t\mu_1/h}\Pi_1 + \mathcal{O}(1)e^{-t/C}, \; in \; \mathcal{L}(L^2, L^2), \tag{11}$$

where $C > 0$ is a constant.

Actually, as we shall see in the outline of the proofs, these results (as well as (8)) hold for more general classes of supersymmetric operators.

2.3. *A partial generalization of [28]*

Consider on $\mathbf{R}^n$ ($2n$ is now replaced by n):

$$P = \sum_{j,k} hD_{x_j} b_{j,k}(x) hD_{x_k} + \frac{1}{2}\sum_j (c_j(x)h\partial_{x_j} + h\partial_{x_j} \circ c_j(x)) + p_0(x)$$
$$= P_2 + iP_1 + P_0,$$

where $b_{j,k}, c_j, p_0$ are real and smooth. The associated symbols are:

$$p(x,\xi) = p_2(x,\xi) + ip_1(x,\xi) + p_0(x),$$
$$p_2 = \sum b_{j,k}\xi_j\xi_k,\ p_1 = \sum c_j\xi_j.$$

Assume $p_2 \geq 0$, $p_0 \geq 0$,

$$\partial_x^\alpha b_{j,k} = \mathcal{O}(1),\ |\alpha| \geq 0,$$
$$\partial_x^\alpha c_j = \mathcal{O}(1),\ |\alpha| \geq 1,$$
$$\partial_x^\alpha p_0 = \mathcal{O}(1),\ |\alpha| \geq 2.$$

Assume that

$$\{x;\ p_0(x) = c_1(x) = .. = c_n(x) = 0\}$$

is finite $= \{x_1, ..., x_N\}$ and put $\mathcal{C} = \{\rho_1, ..., \rho_n\}$, $\rho_j = (x_j, 0)$. Put

$$\widetilde{p}(x,\xi) = \langle\xi\rangle^{-2} p_2(x,\xi) + p_0(x),\ \langle\xi\rangle = \sqrt{1+|\xi|^2}$$

$$\langle\widetilde{p}\rangle_{T_0} = \frac{1}{T_0}\int_{-T_0/2}^{T_0/2} \widetilde{p}\circ\exp(tH_{p_1})dt,\ T_0 > 0 \text{ fixed.}$$

Here in general we let $H_a = a'_\xi \cdot \frac{\partial}{\partial x} - a'_x \cdot \frac{\partial}{\partial \xi}$ denote the Hamilton field of the C^1-function $a = a(x,\xi)$.

Dynamical assumptions: Near each ρ_j we have $\langle\widetilde{p}\rangle_{T_0} \sim |\rho - \rho_j|^2$ and in any compact set disjoint from $\mathcal{C}$ we have $\langle\widetilde{p}\rangle_{T_0} \geq 1/C$. (Near infinity this last assumption has to be modified slightly and we refer to [25] for the details.) The following result from [25] is very close to the main result of [28] and generalizes Theorem 2.1:

Theorem 2.4. *Under the above assumptions, the spectrum of P is discrete in any band $0 \leq \Re z \leq Ch$ and the eigenvalues have asymptotic expansions as in (8).*

Put $q(x,\xi) = -p(x,i\xi) = p_2(x,\xi) + p_1(x,\xi) - p_0(x)$.

The linearization of the Hamilton field H_q at ρ_j (for any fixed j) has eigenvalues $\pm\alpha_k$, $k = 1,..,n$ with real part $\neq 0$. Let $\Lambda_+ = \Lambda_{+,j}$ be the unstable manifold through ρ_j for the H_q-flow. Then Λ_+ is Lagrangian and of the form $\xi = \phi_+'(x)$ near x_j ($\phi_+ = \phi_{+,j}$), where

$$\phi_+(x_j) = 0,\ \phi_+'(x_j) = 0,\ \phi_+''(x_j) > 0.$$

The next result is from [25]:

Theorem 2.5. *Let $\lambda_{j,k}(h)$ be a simple eigenvalue as in (8) and assume there is no other eigenvalue in a disc $D(\lambda_{j,k}, h/C)$ for some $C > 0$. Then, in the L^2 sense, the corresponding eigenfunction is of the form $e^{-\phi_+(x)/h}(a(x;h) + \mathcal{O}(h^\infty))$ near x_j, where $a(x;h)$ is smooth in x with an asymptotic expansion in powers of h. Away from a small neighborhood of x_j it is exponentially decreasing.*

The proof of the first theorem uses microlocal weak exponential estimates, while the one of the last theorem also uses local exponential estimates.

2.4. *Averaging and exponential weights.*

The basic idea of the proof of Theorem 2.4 is taken from [28], but we reworked it in order to allow for non-hypoelliptic operators. We will introduce a weight on $T^*\mathbf{R}^n$ of the form

$$\psi_\epsilon = -\int J(\frac{t}{T_0})\widetilde{p}_\epsilon \circ \exp(tH_{p_1})dt, \tag{12}$$

for $0 < \epsilon \ll 1$. Here $J(t)$ is the odd function given by

$$J(t) = \begin{cases} 0, & |t| \geq \frac{1}{2}, \\ \frac{1}{2} - t, & 0 < t \leq \frac{1}{2}, \end{cases} \tag{13}$$

and we choose $\widetilde{p}_\epsilon(\rho)$ to be equal to $\widetilde{p}(\rho)$ when $\mathrm{dist}\,(\rho, \mathcal{C}) \leq \epsilon$, and flatten out to $\epsilon\widetilde{p}$ away from a fixed neighborhood of $\mathcal{C}$ in such a way that $\widetilde{p}_\epsilon = \mathcal{O}(\epsilon)$. Then

$$H_{p_1}\psi_\epsilon = \langle\widetilde{p}_\epsilon\rangle_{T_0} - \widetilde{p}_\epsilon. \tag{14}$$

We let $\epsilon = Ah$ where $A \gg 1$ is independent of h. Then the weight $\exp(\psi_\epsilon/h)$ is uniformly bounded when $h \to 0$. Indeed, $\psi_\epsilon = \mathcal{O}(h)$.

Using Fourier integral operators with complex phase, we can define a Hilbert space of functions that are "microlocally $\mathcal{O}(\exp(\psi_\epsilon/h))$ in the L^2

sense". The norm is uniformly equivalent to the one of L^2, but the natural leading symbol of P, acting in the new space, becomes

$$p(\exp(iH_{\psi_\epsilon})(\rho)),\ \rho \in T^*\mathbf{R}^n \tag{15}$$

which by Taylor expansion has real part $\approx p_2(\rho) + p_0(\rho) + \langle \widetilde{p}_\epsilon \rangle - \widetilde{p}_\epsilon$.

Very roughly, the real part of the new symbol is $\geq \epsilon$ away from $\mathcal{C}$ and behaves like dist $(\rho, \mathcal{C})^2$ in a $\sqrt{\epsilon}$-neighborhood of $\mathcal{C}$. This can be used to show that the spectrum of P (viewed as an operator on the weighted space) in a band $0 \leq \Re z < \epsilon/C$ comes from an $\sqrt{\epsilon}$-neighborhood of $\mathcal{C}$. In such a neighborhood, we can treat P as an elliptic operator and the spectrum is to leading order determined by the quadratic approximation of the dilated symbol (15). This gives Theorem 2.4.

We next turn to the proof of Theorem 2.5, and we work near a point $\rho_j = (x_j, \xi_j) \in \mathcal{C}$. Recall that $\Lambda_+ : \xi = \phi'_+(x)$ is the unstable manifold for the H_q-flow, where $q(x,\xi) = -p(x, i\xi)$. We have $q(x, \phi'_+(x)) = 0$.

In general, if $\psi \in C^\infty$ is real, then $P_\psi := e^{\psi/h} \circ P \circ e^{-\psi/h}$ has the symbol

$$p_\psi(x,\xi) = p_2(x,\xi) - q(x, \psi'(x)) + i(q'_\xi(x, \psi'(x)) \cdot \xi \tag{16}$$

- As long as $q(x, \psi'(x)) \leq 0$, we have $\Re p_\psi \geq 0$ and we may hope to establish good apriori estimates for P_ψ.
- This is the case for $\psi = 0$ and for $\psi = \phi_+$. Using the convexity of $q(x, \cdot)$, we get suitable weights ψ with $q(x, \psi'(x)) \leq 0$, equal to $\phi_+(x)$ near x_j, strictly positive away from x_j and constant outside a neighborhood of that point.
- It follows that the eigenfunction in Theorem 2.5 is (roughly) $\mathcal{O}(e^{-\phi_+(x)/h})$ near x_j in the L^2 sense.
- On the other hand, we have quasi-modes of the form $a(x;h)e^{-\phi_+(x)/h}$ as in [23].
- Applying the exponentially weighted estimates, indicated above, to the difference of the eigenfunction and the quasi-mode, we then get Theorem 2.5.

2.5. *Supersymmetry and the proof of Theorem 2.2*

We review the supersymmetry from [1], [51], see also G. Lebeau [36]. Let $A(x) : T_x^*\mathbf{R}^n \to T_x\mathbf{R}^n$ be linear, invertible and smooth in x. Then we have the nondegenerate bilinear form

$$\langle u|v\rangle_{A(x)} = \langle \wedge^k A(x) u|v\rangle,\ u, v \in \wedge^k T_x^*\mathbf{R}^n,$$

and we also write $(u|v)_{A(x)} = \langle u|\overline{v}\rangle_{A(x)}$.

If u, v are smooth k-forms with compact support, put

$$(u|v)_A = \int (u(x)|v(x))_{A(x)} dx.$$

The formal "adjoint" $Q^{A,*}$ of an operator Q is then given by

$$(Qu|v)_A = (u|Q^{A,*}v)_A.$$

Let $\phi : \mathbf{R}^n \to \mathbf{R}$ be a smooth Morse function with $\partial^\alpha \phi$ bounded for $|\alpha| \geq 2$ and with $|\nabla \phi| \geq 1/C$ for $|x| \geq C$. Introduce the Witten-De Rham complex:

$$d_\phi = e^{-\frac{\phi}{h}} \circ h d \circ e^{\frac{\phi}{h}} = \sum_j (h\partial_{x_j} + \partial_{x_j}\phi) \circ dx_j^\wedge,$$

where d denotes exterior differentiation and $dx_j^\wedge$ left exterior multiplication with dx_j. The corresponding Laplacian is then: $-\Delta_A = d_\phi^{A,*} d_\phi + d_\phi d_\phi^{A,*}$. Its restriction to q-forms will be denoted by $-\Delta_A^{(q)}$. Notice that:

$$-\Delta_A^{(0)}(e^{-\phi/h}) = 0.$$

Write $A = B + C$ with $B^t = B$, $C^t = -C$. $-\Delta_A$ is a second order differential operator with scalar principal symbol in the semi-classical sense $(\frac{h}{i}\frac{\partial}{\partial x_j} \mapsto \xi_j)$ of the form:

$$p(x,\xi) = \sum_{j,k} b_{j,k}(\xi_j \xi_k + \partial_{x_j}\phi \partial_{x_k}\phi) + 2i \sum_{j,k} c_{j,k} \partial_{x_k}\phi\, \xi_j.$$

Example 2.1. Replace n by $2n$, x by (x, y), let

$$A = \frac{1}{2}\begin{pmatrix} 0 & I \\ -I & \gamma \end{pmatrix}.$$

Then

$$\begin{aligned} -\Delta_A^{(0)} &= h(\phi'_y \cdot \partial_x - \phi'_x \cdot \partial_y) \\ &\quad + \frac{\gamma}{2} \sum_j (-h\partial_{y_j} + \partial_{y_j}\phi)(h\partial_{y_j} + \partial_{y_j}\phi). \end{aligned}$$

When $\phi = y^2/2 + V(x)$ we recover the KFP operator (4).

The results of Subsection 2.3 apply, if we make the additional dynamical assumptions there; $-\Delta_A^{(q)}$ has an asymptotic eigenvalue $= o(h)$ associated to the critical point x_j precisely when the index of x_j is equal to q (as for the Witten complex and analogous complexes in several complex variables). In order to cover the cases $q > 0$ we also assume that

$$A = \text{Const.} \tag{17}$$

The Double well case. Keep the assumption (17). Assume that ϕ is a Morse function with $|\nabla\phi| \geq 1/C$ for $|x| \geq C$ such that $-\Delta_A$ satisfies the extra dynamical conditions of Subsection 2.3 and having precisely three critical points, two local minima $U_{\pm 1}$ and a saddle point U_0 of index 1.

Then $-\Delta_A^{(0)}$ has precisely 2 eigenvalues: 0, μ that are $o(h)$ while $-\Delta_A^{(1)}$ has precisely one such eigenvalue: μ. (Here we use as in the study of the Witten complex, that d_ϕ and $d_\phi^{A,*}$ intertwine our Laplacians in degeree 0 and 1. The detailed justification is more complicated however.) $e^{-\phi/h}$ is the eigenfunction of $\Delta_A^{(0)}$ corresponding to the eigenvalue 0. Let $S_j = \phi(U_0) - \phi(U_j)$, $j = \pm 1$, and let D_j be the connected component of $\{x \in \mathbf{R}^n;\ \phi(x) < \phi(U_0)\}$ containing U_j in its interior.

Let $E^{(q)}$ be the corresponding spectral subspaces so that $\dim E^{(0)} = 2$, $\dim E^{(1)} = 1$. Truncated versions of the function $e^{-\phi(x)/h}$ can be used as approximate eigenfunctions, and we can show the following.

Proposition 2.1. *$E^{(0)}$ has a basis e_1, e_{-1}, where*

$$e_j = \chi_j(x) e^{-\frac{1}{h}(\phi(x)-\phi(U_j))} + \mathcal{O}(e^{-\frac{1}{h}(S_j-\epsilon)}),\ \textit{in the } L^2\textit{-sense.}$$

Here, we let $\chi_j \in C_0^\infty(D_j)$ be equal to 1 on $\{x \in D_j;\ \phi(x) \leq \phi(U_0) - \epsilon\}$.

Theorems 2.4, 2.5 can be adapted to $-\Delta_A^{(1)}$ and lead to the following.

Proposition 2.2. *$E^{(1)} = \mathbf{C}e_0$, where*

$$e_0(x) = \chi_0(x) a_0(x;h) e^{-\frac{1}{h}\phi_+(x)} + \mathcal{O}(e^{-\epsilon_0/h}),$$

$\phi_+(x) \sim (x - U_0)^2$, $\epsilon_0 > 0$ is small enough, a_0 is an elliptic symbol, $\chi_0 \in C_0^\infty(\mathbf{R}^n)$, $\chi_0 = 1$ near U_0.

Let the matrices of $d_\phi : E^{(0)} \to E^{(1)}$ and $d_\phi^{A,*} : E^{(1)} \to E^{(0)}$ with respect to the bases $\{e_{-1}, e_1\}$ and $\{e_0\}$ be

$$\begin{pmatrix} \lambda_{-1} & \lambda_1 \end{pmatrix} \text{ and } \begin{pmatrix} \lambda_{-1}^* \\ \lambda_1^* \end{pmatrix} \text{ respectively.}$$

Using the preceding two results in the spirit of tunneling estimates and computations of Helffer–Sjöstrand ([23,24]) we can show the following.

Proposition 2.3. *Put $S_j = \phi(U_0) - \phi(U_j)$, $j = \pm 1$. Then we have*

$$\begin{pmatrix} \lambda_{-1} \\ \lambda_1 \end{pmatrix} = h^{\frac{1}{2}}(I + \mathcal{O}(e^{-\frac{1}{Ch}})) \begin{pmatrix} \ell_{-1}(h) e^{-S_{-1}/h} \\ \ell_1(h) e^{-S_1/h} \end{pmatrix},$$

$$\begin{pmatrix} \lambda_{-1}^* \\ \lambda_1^* \end{pmatrix} = h^{\frac{1}{2}}(I + \mathcal{O}(e^{-\frac{1}{Ch}})) \begin{pmatrix} \ell_{-1}^*(h) e^{-S_{-1}/h} \\ \ell_1^*(h) e^{-S_1/h} \end{pmatrix},$$

where $\ell_{\pm 1}$, $\ell^*_{\pm 1}$ *are real elliptic symbols of order* 0 *such that* $\ell_j \ell^*_j > 0$, $j = \pm 1$.

From this we get Theorem 6, since $\mu = \lambda^*_{-1}\lambda_{-1} + \lambda^*_1\lambda_1$. □

Thanks to the fact that we have only two local minima, certain simplifications were possible in the proof. In particular it was sufficent to control the exponential decay of general eigenfunctions in some small neighborhood of the critical points. For more general configurations, it might be necessary to get such a control also further away and this seems to lead to interesting questions, involving degenerate and non-symmetric Finsler distances.

2.6. *Return to equilibrium, ideas of the proof of Theorem 2.3*

Keeping the same assumptions, let Π_0, Π_1 be the rank 1 spectral projections corresponding to the eigenvalues $\mu_0 := 0$, $\mu_1 := \mu$ of $-\Delta_A^{(0)}$ and put $\Pi = \Pi_0 + \Pi_1$. Then $e_{-1,}, e_1$ is a basis for $\mathcal{R}(\Pi)$ and the restriction of P to this range, has the matrix

$$\begin{pmatrix} \lambda^*_{-1} \\ \lambda^*_1 \end{pmatrix} \begin{pmatrix} \lambda_{-1} & \lambda_1 \end{pmatrix} = \begin{pmatrix} \lambda^*_{-1}\lambda_{-1} & \lambda^*_{-1}\lambda_1 \\ \lambda^*_1\lambda_{-1} & \lambda^*_1\lambda_1 \end{pmatrix} \tag{18}$$

with the eigenvalues 0 and $\mu = \lambda^*_{-1}\lambda_{-1} + \lambda^*_1\lambda_1$. A corresponding basis of eigenvectors is given by

$$\begin{aligned} v_0 &= \frac{1}{\sqrt{\mu_1}}(\lambda_1 e_{-1} - \lambda_{-1} e_{-1}) \\ v_1 &= \frac{1}{\sqrt{\mu_1}}(\lambda^*_{-1} e_{-1} + \lambda^*_1 e_{-1}). \end{aligned} \tag{19}$$

The corresponding dual basis of eigenfunctions of P^* is given by

$$\begin{aligned} v^*_0 &= \frac{1}{\sqrt{\mu}}(\lambda^*_1 e^*_{-1} - \lambda^*_{-1} e^*_{-1}) \\ v^*_1 &= \frac{1}{\sqrt{\mu}}(\lambda_{-1} e^*_{-1} + \lambda_1 e^*_1), \end{aligned} \tag{20}$$

where $e^*_{-1}, e^*_1 \in \mathcal{R}(\Pi^*)$ is the basis that is dual to e_{-1}, e_1. It follows that $v_j, v^*_j = \mathcal{O}(1)$ in L^2, when $h \to 0$.

From this discussion we conclude that $\Pi_j = (\cdot|v^*_j)v_j$, *are uniformly bounded when* $h \to 0$. A non-trivial fact, based on the analysis described in Subsections 2.3, 2.4, is that after replacing the standard norm and scalar product on L^2 by certain uniformly equivalent ones, we have

$$\Re(Pu|u) \geq \frac{h}{C}\|u\|^2, \quad \forall u \in \mathcal{R}(1 - \widetilde{\Pi}), \tag{21}$$

where $\widetilde{\Pi}$ is the spectral projection corresponding to the spectrum of P in $D(0, Bh)$ for some $B \gg 1$.

This can be applied to the study of $u(t) := e^{-tP/h}u(0)$, where the initial state $u(0) \in L^2$ is arbitrary: Write

$$u(0) = \Pi_0 u(0) + \Pi_1 u(0) + (1 - \Pi)u(0) =: u^0 + u^1 + u^\perp. \tag{22}$$

Then

$$\|u^0\|, \|u^1\|, \|u^\perp\| \leq \mathcal{O}(1)\|u(0)\| \tag{23}$$

$$\|e^{-tP/h}u^\perp\| \leq Ce^{-t/C}\|u(0)\| \tag{24}$$

$$e^{-tP/h}u_j = e^{-t\mu_j/h}u_j, \; j = 0, 1. \tag{25}$$

Here (24) follows if we write $u^\perp = (1 - \widetilde{\Pi})u + (\Pi - \widetilde{\Pi})u$, apply (21) to the evolution of the first term, and use that the last term is the (bounded) spectral projection of u to a finite dimensional spectral subspace of P, for which the corresponding eigenvalues all have real part $\geq h/C$. □

3. Spectral asymptotics in 2 dimensions

3.1. *Introduction*

This section is mainly based on recent joint works with S. Vũ Ngọc and M. Hitrik [29] [30], but we shall start by recalling some earlier results that we obtained with A. Melin [40] where we discovered that in the two dimensional case one often can have Bohr-Sommerfeld conditions to determine all the individual eigenvalues in some region of the spectral plane, provided that we have analyticity. This was first a surprise for us since in the self-adjoint case such results are known only in 1 dimension and in very special cases for higher dimensions.

Subsequently, with M. Hitrik we have studied small perturbations of self-adjoint operators. First we studied the case when the classical flow of the unperturbed operator is periodic, then also with S. Vũ Ngọc we looked at the more general case when it is completely integrable, or just when the energy surface contains some invariant diophantine Lagrangian tori.

3.2. *Bohr-Sommerfeld rules in two dimensions*

For (pseudo-)differential operators in dimension 1, we often have a Bohr-Sommerfeld rule to determine the asymptotic behaviour of the eigenvalues. Consider for instance the semi-classical Schrödinger operator

$$P = -h^2 \frac{d^2}{dx^2} + V(x), \text{ with symbol } p(x, \xi) = \xi^2 + V(x),$$

where we assume that $V \in C^\infty(\mathbf{R};\mathbf{R})$ and $V(x) \to +\infty$, $|x| \to \infty$. Let $E_0 \in \mathbf{R}$ be a non-critical value of V such that (for simplicity) $\{x \in \mathbf{R}; V(x) \le E_0\}$ is an interval. Then in some small fixed neighborhood of E_0 and for $h > 0$ small enough, the eigenvalues of P are of the form $E = E_k$, $k \in \mathbf{Z}$, where

$$\frac{I(E)}{2\pi h} = k - \theta(E;h), \quad I(E) = \int_{p^{-1}(E)} \xi \cdot dx, \ \theta(E;h) \sim \theta_0(E) + \theta_1(E)h + \dots$$

In the non-self-adjoint case we get the same results, provided that $\Im V$ is small and V *is analytic.* The eigenvalues will then be on a curve close to the real axis.

For self-adjoint operators in dimension ≥ 2 it is generally admitted that Bohr-Sommerfeld rules do not give all eigenvalues in any fixed domain except in certain (completely integrable) cases. Using the KAM theorem one can sometimes describe some fraction of the eigenvalues.

With A. Melin [40]: we considered an h-pseudodifferential operator with leading symbol $p(x,\xi)$ that is bounded and holomorphic in a tubular neighborhood of $\mathbf{R}^4$ in $\mathbf{C}^4 = \mathbf{C}^2_x \times \mathbf{C}^2_\xi$. Assume that

$$\mathbf{R}^4 \cap p^{-1}(0) \neq \emptyset \text{ is connected.} \tag{26}$$

$$\text{On } \mathbf{R}^4 \text{ we have } |p(x,\xi)| \ge 1/C, \text{ for } |(x,\xi)| \ge C, \tag{27}$$

for some $C > 0$,

$$d\Re p(x,\xi), d\Im p(x,\xi) \text{ are linearly independent for all } (x,\xi) \in p^{-1}(0) \cap \mathbf{R}^4. \tag{28}$$

(Here the boundedness assumption near ∞ and (27) can be replaced by a suitable ellipticity assumption.) It follows that $p^{-1}(0) \cap \mathbf{R}^4$ is a compact (2-dimensional) surface.

Also assume that

$$|\{\Re p, \Im p\}| \text{ is sufficiently small on } p^{-1}(0) \cap \mathbf{R}^4. \tag{29}$$

Here "sufficiently small" refers to some positive bound that can be defined whenever the the other conditions are satisfied uniformly.

When the Poisson bracket vanishes on $p^{-1}(0)$, this set becomes a Lagrangian torus, and more generally it is a torus. The following is a complex version of the KAM theorem without small divisors (cf T.W. Cherry [3](1928), J. Moser [41](1958)).

Theorem 3.1. ([40]) *There exists a smooth 2-dimensional torus* $\Gamma \subset p^{-1}(0) \cap \mathbf{C}^4$, *close to* $p^{-1}(0) \cap \mathbf{R}^4$ *such that* $\sigma_{|\Gamma} = 0$ *and* $I_j(\Gamma) \in \mathbf{R}$, $j = 1, 2$.

Here $I_j(\Gamma) := \int_{\gamma_j} \xi \cdot dx$ *are the actions along the two fundamental cycles* $\gamma_1, \gamma_2 \subset \Gamma$, *and* $\sigma = \sum_1^2 d\xi_j \wedge dx_j$ *is the complex symplectic (2,0)-form.*

Replacing p by $p - z$ for z in a neighborhood of $0 \in \mathbf{C}$, we get tori $\Gamma(z)$ depending smoothly on z and a corresponding smooth action function $I(z) = (I_1(\Gamma(z)), I_2(\Gamma(z)))$, which are important in the Bohr-Sommerfeld rule for the eigen-values near 0 in the semi-classical limit $h \to 0$:

Theorem 3.2. ([40]) *Under the above assumptions, there exists* $\theta_0 \in (\frac{1}{2}\mathbf{Z})^2$ *and* $\theta(z; h) \sim \theta_0 + \theta_1(z)h + \theta_2(z)h^2 + ..$ *in* $C^\infty(\text{neigh}\,(0, \mathbf{C}))$, *such that for* z *in an* h*-independent neighborhood of 0 and for* $h > 0$ *sufficiently small, we have that* z *is an eigenvalue of* $P = p(x, hD_x)$ *iff*

$$\frac{I(z)}{2\pi h} = k - \theta(z; h), \text{ for some } k \in \mathbf{Z}^2. \quad (BS)$$

Recently, a similar result was obtained by S. Graffi, C. Villegas Bas [13].

An application of this result is that we get all resonances (scattering poles) in a fixed neighborhood of $0 \in \mathbf{C}$ for $-h^2\Delta + V(x)$ if V is an analytic real potential on $\mathbf{R}^2$ with a nondegenerate saddle point at $x = 0$, satisfying $V(0) = 0$ and having $\{(x, \xi) = (0, 0)\}$ as its classically trapped set in the energy surface $\{p(x, \xi) = 0\}$.

3.3. *Diophantine case*

In this and the next subsection we describe a result from [30] and the main result of [29] about individual eigenvalues for small perturbations of a self-adjoint operator with a completely integrable leading symbol. We start with the case when only Diophantine tori play a role.

Let $P_\epsilon(x, hD; h)$ on $\mathbf{R}^2$ have the leading symbol $p_\epsilon(x, \xi) = p(x, \xi) + i\epsilon q(x, \xi)$ where p, q are real and extend to bounded holomorphic functions on a tubular neighborhood of $\mathbf{R}^4$. Assume that p fulfills the ellipticity condition (27) near infinity and that

$$P_{\epsilon=0} = P(x, hD) \tag{30}$$

is self-adjoint. (The conditions near infinity can be modified and we can also replace $\mathbf{R}_x^2$ by a compact 2-dimensional analytic manifold.)

Also, assume that $P_\epsilon(x, \xi; h)$ depends smoothly on $0 \le \epsilon \le \epsilon_0$ with values in the space of bounded holomorphic functions in a tubular neighborhood of $\mathbf{R}^4$, and $P_\epsilon \sim p_\epsilon + hp_{1,\epsilon} + h^2 p_{2,\epsilon} + ...$, when $h \to 0$.

Assume

$$p^{-1}(0) \text{ is connected and } dp \neq 0 \text{ on that set.} \tag{31}$$

Assume complete integrability for p: There exists an analytic real valued function f on $T^*\mathbf{R}^2$ such that $H_p f = 0$, with the differentials df and dp being linearly independent almost everywhere on $p^{-1}(0)$. ($H_p = p'_\xi \cdot \frac{\partial}{\partial x} - p'_x \cdot \frac{\partial}{\partial \xi}$ is the Hamilton field.)

Then we have a disjoint union decomposition

$$p^{-1}(0) \cap T^*\mathbf{R}^2 = \bigcup_{\Lambda \in J} \Lambda, \tag{32}$$

where Λ are compact connected sets, invariant under the H_p flow. We assume (for simplicity) that J has a natural structure of a graph whose edges correspond to families of regular leaves; Lagrangian tori (by the Arnold-Mineur-Liouville theorem). The union of edges $J \setminus S$ possesses a natural real analytic structure.

Each torus $\Lambda \in J \setminus S$ carries real analytic coordinates x_1, x_2 identifying Λ with $\mathbf{T}^2 = \mathbf{R}^2 / 2\pi\mathbf{Z}^2$, so that along Λ, we have

$$H_p = a_1 \frac{\partial}{\partial x_1} + a_2 \frac{\partial}{\partial x_2}, \tag{33}$$

where $a_1, a_2 \in \mathbf{R}$. The rotation number is defined as the ratio $\omega(\Lambda) = [a_1 : a_2] \in \mathbf{RP}^1$, and it depends analytically on $\Lambda \in J \setminus S$. We assume that $\omega(\Lambda)$ is not identically constant on any open edge.

We say that $\Lambda \in J \setminus S$ is respectively rational, irrational, diophantine if a_1/a_2 has the corresponding property. Diophantine means that there exist $\alpha > 0$, $d > 0$ such that

$$|(a_1, a_2) \cdot k| \geq \frac{\alpha}{|k|^{2+d}}, \quad 0 \neq k \in \mathbf{Z}^2. \tag{34}$$

We introduce

$$\langle q \rangle_T = \frac{1}{T} \int_{-T/2}^{T/2} q \circ \exp(tH_p) dt, \quad T > 0, \tag{35}$$

and consider the compact intervals $Q_\infty(\Lambda) \subset \mathbf{R}$, $\Lambda \in J$, defined by,

$$Q_\infty(\Lambda) = [\lim_{T \to \infty} \inf_\Lambda \langle q \rangle_T, \lim_{T \to \infty} \sup_\Lambda \langle q \rangle_T]. \tag{36}$$

A first localization of the spectrum $\sigma(P_\epsilon(x, hD_x; h))$ ([30]) is given by

$$\Im(\sigma(P_\epsilon) \cap \{z; |\Re z| \leq \delta\}) \subset \epsilon[\inf \bigcup_{\Lambda \in J} Q_\infty(\Lambda) - o(1), \sup \bigcup_{\Lambda \in J} Q_\infty(\Lambda) + o(1)], \tag{37}$$

when $\delta, \epsilon, h \to 0$.

For each torus $\Lambda \in J\setminus S$, we let $\langle q\rangle(\Lambda)$ be the average of $q_{|_\Lambda}$ with respect to the natural smooth measure on Λ, and assume that the analytic function $J\setminus S \ni \Lambda \mapsto \langle q\rangle(\Lambda)$ is not identically constant on any open edge.

By combining (33) with the Fourier series representation of q, we see that when Λ is irrational then $Q_\infty(\Lambda) = \{\langle q\rangle(\Lambda)\}$, while in the rational case,

$$Q_\infty(\Lambda) \subset \langle q\rangle(\Lambda) + \mathcal{O}(\frac{1}{(|n|+|m|)^\infty})[-1,1], \tag{38}$$

when $\omega(\Lambda) = \frac{m}{n}$ and $m \in \mathbf{Z}$, $n \in \mathbf{N}$ are relatively prime.

Let $F_0 \in \cup_{\Lambda\in J} Q_\infty(\Lambda)$ and assume that there exists a Diophantine torus Λ_d (or finitely many), such that

$$\langle q\rangle(\Lambda_d) = F_0, \quad d_\Lambda\langle q\rangle(\Lambda_d) \neq 0. \tag{39}$$

With M. Hitrik and S. Vũ Ngọc we obtained:

Theorem 3.3. ([30]) *Assume also that F_0 does not belong to $Q_\infty(\Lambda)$ for any other $\Lambda \in J$. Let $0 < \delta < K < \infty$. Then $\exists C > 0$ such that for $h > 0$ small enough, and $k^K \leq \epsilon \leq h^\delta$, the eigenvalues of P_ϵ in the rectangle $|\Re z| < h^\delta/C$, $|\Im z - \epsilon\Re F_0| < \epsilon h^\delta/C$ are given by*

$$P^{(\infty)}(h(k-\frac{k_0}{4}) - \frac{S}{2\pi}, \epsilon; h) + \mathcal{O}(h^\infty), \;\; k \in \mathbf{Z}^2.$$

Here $P^{(\infty)}(\xi,\epsilon;h)$ is smooth, real-valued for $\epsilon = 0$ and when $h \to 0$ we have

$$P^{(\infty)}(\xi,\epsilon;h) \sim \sum_{\ell=0}^{\infty} h^\ell p_\ell^{(\infty)}(\xi,\epsilon), \;\; p_0^{(\infty)} = p_\infty(\xi) + i\epsilon\langle q\rangle(\xi) + \mathcal{O}(\epsilon^2), \tag{40}$$

corresponding to action angle coordinates.

In [30] we also considered applications to small non-self-adjoint perturbations of the Laplacian on a surface of revolution. Thanks to (38) the total measure of the union of all $Q_\infty(\Lambda)$ over the rational tori is finite and sometimes small, and we could then show that there are plenty of values F_0, fulfilling the assumptions in the theorem.

3.4. *The case with rational tori*

Let F_0 be as in (39) but now also allow for the possibility that there is a rational torus (or finitely many) Λ_r, such that

$$F_0 \in Q_\infty(\Lambda_r), \quad F_0 \neq \langle q\rangle(\Lambda_r), \tag{41}$$

$$d_\Lambda(\langle q\rangle)(\Lambda_r) \neq 0,\ d_\Lambda(\omega)(\Lambda_r) \neq 0. \tag{42}$$

Assume also that

$$F_0 \notin Q_\infty(\Lambda),\ \text{for all } \Lambda \in J \setminus \{\Lambda_d, \Lambda_r\}. \tag{43}$$

With M. Hitrik we showed the following result:

Theorem 3.4. ([29]) *Let $\delta > 0$ be small and assume that $h \ll \epsilon \leq h^{\frac{2}{3}+\delta}$, or that the subprincipal symbol of P vanishes and that $h^2 \ll \epsilon \leq h^{\frac{2}{3}+\delta}$. Then the spectrum of P_ϵ in the rectangle*

$$[-\frac{\epsilon}{C}, \frac{\epsilon}{C}] + i\epsilon[F_0 - \frac{\epsilon^\delta}{C}, F_0 + \frac{\epsilon^\delta}{C}]$$

is the union of two sets: $E_d \cup E_r$, where the elements of E_d form a distorted lattice, given by the Bohr-Sommerfeld rule (40), with horizontal spacing $\asymp h$ and vertical spacing $\asymp \epsilon h$. The number of elements $\#(E_r)$ of E_r is $\mathcal{O}(\epsilon^{3/2}/h^2)$.

NB that $\#(E_d) \asymp \epsilon^{1+\delta}/h^2$.

This result can be applied to the damped wave equation on surfaces of revolution.

3.5. *Outline of the proofs of Theorem 3.3 and 3.4*

The principal symbol of P_ϵ is $p_\epsilon = p + i\epsilon q + \mathcal{O}(\epsilon^2)$. Put

$$\langle q\rangle_T = \frac{1}{T}\int_{-T/2}^{T/2} q \circ \exp(tH_p)dt.$$

As in Section 2 we will use an averaging of the imaginary part of the symbol. Let $J(t)$ be the piecewise affine function with support in $[-\frac{1}{2}, \frac{1}{2}]$, solving

$$J'(t) = \delta(t) - 1_{[-\frac{1}{2},\frac{1}{2}]}(t),$$

and introduce the weight

$$G_T(t) = \int J(-\frac{t}{T})q \circ \exp(tH_p)dt.$$

Then $H_p G_T = q - \langle q\rangle_T$, implying

$$p_\epsilon \circ \exp(i\epsilon H_{G_T}) = p + i\epsilon\langle q\rangle_T + \mathcal{O}_T(\epsilon^2). \tag{44}$$

The left hand side of (44) is the principal symbol of the isospectral operator $e^{-\frac{\epsilon}{h}G_T(x,hD_x)} \circ P_\epsilon \circ e^{\frac{\epsilon}{h}G_T(x,hD_x)}$ and under the assumptions of Theorem 3.3 resp. 3.4 its imaginary part will not take the value $i\epsilon F_0$ on

$p^{-1}(0)$ away from Λ_d resp. $\Lambda_d \cup \Lambda_r$. This means that we have localized the spectral problem to a neighborhood of Λ_d resp. $\Lambda_d \cup \Lambda_r$.

Near Λ_d we choose action-angle coordinates so that Λ_d becomes the zero section in the cotangent space of the 2-torus, and

$$p_\epsilon(x,\xi) = p(\xi) + i\epsilon q(x,\xi) + \mathcal{O}(\epsilon^2). \tag{45}$$

We follow the quantized Birkhoff normal form procedure in the spirit of V.F. Lazutkin and Y. Colin de Verdière [4,35]: solve first

$$H_p G = q(x,\xi) - \langle q(\cdot,\xi)\rangle, \tag{46}$$

where the bracket indicates that we take the average over the torus with respect to x. Composing with the corresponding complex canonical transformation, we get the new conjugated symbol

$$p(\xi) + i\epsilon\langle q(\cdot,\xi)\rangle + \mathcal{O}(\epsilon^2 + \xi^\infty).$$

Here the Diophanticity condition is of course important.

Iterating the procedure we get for every N,

$$p_\epsilon \circ \exp(H_{G^{(N)}}) = \underbrace{p(\xi) + i\epsilon(\langle q\rangle(\xi) + \mathcal{O}(\epsilon,\xi))}_{\text{independent of } x} + \mathcal{O}((\xi,\epsilon)^{N+1})$$

This procedure can be continued on the operator level, and up to a small error we see that P_ϵ is microlocally equivalent to an operator $P_\epsilon(hD_\xi, \epsilon; h)$. At least formally, Theorem 3.3 then follows by considering Fourier series expansions, but in order to get a full proof we also have take into account that we have constructed complex canonical transformations that are quantized by Fourier integral operators with complex phase and study the action of these operators on suitable exponentially weighted spaces.

Near Λ_r we can still use action-angle coordinates as in (45) but the homological equation (46) is no longer solvable. Instead, we use secular perturbation theory (cf the book [37]), which amounts to making a partial Birkhoff reduction.

After a linear change of x-variables, we may assume that $p(\xi) = \xi_2 + \mathcal{O}(\xi^2)$ and in order to fix the ideas $= \xi_2 + \xi_1^2$. Then we can make the averaging procedure only in the x_2-direction and reduce p_ϵ in (45) to

$$\widetilde{p}_\epsilon(x,\xi) = \underbrace{\xi_2 + \xi_1^2 + \mathcal{O}(\epsilon)}_{\substack{\text{independent of } x_2,\\ \approx \xi_2 + \xi_1^2 + i\epsilon\langle q\rangle_2(x_1,\xi)}} + \mathcal{O}((\epsilon,\xi)^\infty),$$

where $\langle q\rangle_2(x_1,\xi)$ denotes the average with respect to x_2.

Carrying out the reduction on the operator level, we obtain up to small errors an operator $\widetilde{P}_\epsilon(x_1, hD_{x_1}, hD_{x_2}; h)$ and after passing to Fourier series in x_2, a family of non-self-adjoint operators on $S^1_{x_1}$: $\widetilde{P}_\epsilon(x_1, hD_{x_1}, hk; h)$, $k \in \mathbf{Z}$.

The non-self-adjointness and the corresponding possible wild growth of the resolvent makes it hard to go all the way to study individual eigenvalues. However, it can be shown that in the region $|\xi_1| \gg \epsilon^{1/2}$ (inside the energy surface $p = 0$) we can go further and (as near Λ_d) get a sufficiently good elimination of the x-dependence. This leads to the conclusion that the contributions from a vicinity of Λ_r to the spectrum of P_ϵ in the rectangle

$$|\Re z| \leq \frac{\epsilon}{C}, \; |\Im z - \epsilon F_0| \leq \frac{\epsilon^{1+\delta}}{C},$$

come from a neighborhood of Λ_r of phase space volume $\mathcal{O}(\epsilon^{3/2})$.

This explains heuristically why the rational torus will contribute with $\mathcal{O}(\epsilon^{3/2}/h^2)$ eigenvalues in the rectangle.

The actual proof is more complicated. We use a Grushin problem reduction in order to reduce the study near Λ_r to that of a square matrix of size $\mathcal{O}(\epsilon^{3/2}/h^2)$. However, even if we avoid the eigenvalues of such a matrix, the inverse can only be bounded by

$$\exp \mathcal{O}(\epsilon^{3/2}/h^2). \tag{47}$$

What saves us is that away from $\Lambda_r \cup \Lambda_d$, we can conjugate the operator with exponential weights and show that the resolvent has an "off-diagonal decay" like $\exp(-1/(Ch))$. This implies that we can confine the growth in (47) to a small neighborhood of Λ_r, if

$$\frac{1}{Ch} \gg \frac{\epsilon^{\frac{3}{2}}}{h^2},$$

leading to the assumption $\epsilon \ll h^{2/3}$ in Theorem 3.4.

4. Weyl asymptotics for non-self-adjoint operators

4.1. *Introduction*

For self-adjoint differential (pseudo)differential operators we have (under suitable assumptions) the Weyl law for the asymptotic distribution of eigenvalues, established in higher dimensions by H. Weyl [56] in 1912 in the case of second order elliptic boundary value problems.

In the semiclassical setting such results were obtained by J. Chazarain, B. Helffer–D. Robert, V. Ivrii and many others (see [11] and further references there). Under suitable additional assumptions it states that if $P = P^w(x, hD_x; h)$ is a self-adjoint h-pseudodifferential operator with leading (real) symbol $p(x,\xi)$, then if $\Omega \subset \mathbf{C}$ is a domain intersecting $\mathbf{R}$ along a bounded interval, the number of eigenvalues of P in Ω (intersected with $\mathbf{R}$) satisfies

$$\#(\sigma(P)\cap\Omega) = \frac{1}{(2\pi h)^n}(\mathrm{vol}\,(p^{-1}(\Omega)) + o(1)),\ h\to 0.$$

A simple example is provided by the semiclassical harmonic oscillator $\frac{1}{2}((hD_x)^2 + x^2)$ on the real line which has the eigenvalues $(k+\frac{1}{2})h$, $k = 0, 1, \ldots$.

In the non-self-adjoint case Weyl-asymptotics is known to hold in some cases close to the self-adjoint case or for normal operators.

We do not always have Weyl-asymptotics: Following Davies and Boulton (see [7]), we can consider the non-self-adjoint harmonic operator: $P = \frac{1}{2}((hD_x)^2 + ix^2)$ whose eigenvalues are given by $e^{i\pi/4}(k+\frac{1}{2})h$, $k\in\mathbf{N}$ ([46]):

The set of values of $p = \frac{1}{2}(\xi^2 + ix^2)$ is the closed first quadrant and if we choose the open bounded set Ω to intersect the 1st quadrant but not the line $\Im z = \Re z$, we get $\mathrm{vol}\,(p^{-1}(\Omega)) > 0$, while there are no eigenvalues in Ω.

More generally, h-differential operators with analytic coefficients often have their spectrum determined by complex-geometric quantities, and are likely not to obey the Weyl law. In particular, in the one dimensional case it often happens that the eigenvalues are concentrated to certain curves with branch points.

As we have seen in Theorem 1.2 we are often confronted with the pseudospectral phenomenon: On the image of p the resolvent may be very large even far from the spectrum. This causes the eigenvalues to be very sensitive to small perturbations of the operator (by Theorem 1.1).

In her thesis M. Hager (see [18]) considered a class of perturbed h-pseudodifferential operators on the real line of the form $P_\delta = P(x, hD; h) + \delta q_\omega(x)$, where P is analytic and q_ω is a random linear combination of the C/h first eigen-functions of an auxiliary operator. She showed that with probability very close to 1 when $h\to 0$, P_δ obeys Weyl asymptotics.

Here, we shall discuss a generalization to the multidimensional case obtained with Hager [21]. The results will be much more general in many ways, but the class of perturbations will be slightly different.

4.2. *The result*

a) *The unperturbed operator.* Let $m(\rho) \geq 1$, $\rho = (x, \xi)$ be an order function on $\mathbf{R}^{2n}$ so that $0 < m(\rho) \leq C_0 \langle \rho - \mu \rangle^{N_0} m(\mu)$. We may assume that $m \in S(m) = \{u \in C^\infty(\mathbf{R}^{2n});\ \partial_\rho^\alpha u = \mathcal{O}_\alpha(m),\ \forall \alpha \in \mathbf{N}^{2n}\}$. Assume $m \geq 1$ and let

$$P(\rho; h) \sim p(\rho) + hp_1(\rho) + ... \text{ in } S(m).$$

Assume $\exists z_0 \in \mathbf{C}$, $C_0 > 0$ such that

$$|p(\rho) - z_0| \geq m(\rho)/C_0 \text{ (ellipticity)}.$$

Let

$$\Sigma = \overline{p(\mathbf{R}^{2n})} = p(\mathbf{R}^{2n}) \cup \Sigma_\infty,$$

$$\Sigma_\infty = \{\lim_{j \to \infty} p(\rho_j);\ \mathbf{R}^{2n} \ni \rho_j \to \infty\}$$

We write $P = P^w(x, hD; h)$.

Let $\Omega \subset\subset \mathbf{C} \setminus \Sigma_\infty$ be open and simply connected containing z_0. Then using the pseudodifferential calculus, it is easy to show:

1) $\sigma(P) \cap \Omega$ is discrete when $h > 0$ is small enough.

2) $\forall \epsilon > 0$, $\exists h(\epsilon) > 0$, such that $\sigma(P) \cap \Omega \subset \Sigma + D(0, \epsilon)$, $0 < h \leq h(\epsilon)$.

b) *The random pertubation.* Let $0 < \widetilde{m}, \widehat{m} \leq 1$ be square integrable order functions, one of which is integrable. Let $\widetilde{S} \in S(\widetilde{m})$, $\widehat{S} \in S(\widehat{m})$ be elliptic symbols. The corresponding operators are Hilbert-Schmidt with $\|\widetilde{S}\|_{\mathrm{HS}}, \|\widehat{S}\|_{\mathrm{HS}} = \mathcal{O}(h^{-n/2})$. Let

$$Q_\omega = \widehat{S} \circ \sum_{j,k} \alpha_{j,k}(\omega) \widehat{e}_j \widetilde{e}_k^* \circ \widetilde{S},$$

where $\alpha_{j,k}$, $j, k \in \mathbf{N}$, are independent complex Gaussian random variables with expectation value 0 and variance 1, and $(\widehat{e}_j)_1^\infty$ and $(\widetilde{e}_j)_1^\infty$ are orthonormal bases in $L^2(\mathbf{R}^n)$, $\widehat{e}_j \widetilde{e}_k^* u = (u|\widetilde{e}_k)\widehat{e}_j$.

Let $M = C_1 h^{-n}$ with $C_1 \gg 1$. Then with probability $\geq 1 - Ce^{-h^{-2n}/C}$ the Hilbert-Schmidt and trace class norms of Q fulfil

$$\|Q\|_{\mathrm{HS}} \leq M,\ \|Q\|_{\mathrm{tr}} \leq M^{\frac{3}{2}} \tag{48}$$

Let $\Gamma \subset\subset \Omega$ be open with smooth boundary.

Theorem 4.1. ([21]) *Assume*

$$p(\rho) \in \partial\Gamma \Rightarrow dp(\rho),\ d\overline{p}(\rho) \text{ are linearly independent.} \tag{49}$$

Let ϵ, δ depend on h with $0 < \epsilon \ll 1$,

$$e^{-\frac{\epsilon}{Ch}} \le \delta \ll h^{3n+\frac{1}{2}},\ C \gg 1$$

(implying that $\epsilon \ge$ Const.$h \ln \frac{1}{h}$). Then with probability $\ge 1 - \frac{C}{\sqrt{\epsilon}} e^{-\frac{\epsilon}{2(2\pi h)^n}}$, we have

$$|\#(\sigma(P_\delta) \cap \Gamma) - \frac{1}{(2\pi h)^n} \mathrm{vol}\,(p^{-1}(\Gamma))| \le C \frac{\sqrt{\epsilon}}{h^n}. \tag{50}$$

There is a similar result giving (50) simultaneously for all Γ in a suitable family.

The assumption (49) implies that $\partial\Gamma \cap \partial\Sigma = \emptyset$, so if we want to count the eigenvalues near $\partial\Sigma$, we need to weaken that assumption. For $z \in \mathrm{neigh}\,(\partial\Gamma)$, we put

$$V_z(t) = \mathrm{vol}\,\{\rho \in \mathbf{R}^{2n};\ |p(\rho) - z|^2 \le t\}. \tag{51}$$

Introduce the assumption

$$\exists \kappa \in]0,1],\ \text{s. t. } V_z(t) = \mathcal{O}(t^\kappa) \text{ uniformly for } z \in \mathrm{neigh}\,(\partial\Gamma),\ 0 \le t \ll 1. \tag{52}$$

Example. (49) $\Rightarrow$ (52) with $\kappa = 1$.

Example. The best that can happen when $\partial\Gamma \cap \partial\Sigma \ne \emptyset$ is that

$$p(\rho) \in \partial\Gamma \Rightarrow \{p, \overline{p}\}(\rho) \ne 0 \text{ or } \{p, \{p, \overline{p}\}\}(\rho) \ne 0. \tag{53}$$

It is easy to see that (53) implies (52) with $\kappa = 3/4$. (53) holds for the non-self-adjoint harmonic oscillator when $0 \notin \partial\Gamma$.

Theorem 4.2. ([21]) *We assume (52). Let ϵ, δ depend on h with $0 < \epsilon \ll 1$,*

$$e^{-\frac{\epsilon}{Ch^\kappa}} \le \delta \ll h^{3n+\frac{1}{2}},\ C \gg 1,$$

implying that $\epsilon \ge$ Const.$h^\kappa \ln \frac{1}{h}$. Then for $0 < r \ll 1$ we have with probability $\ge 1 - \frac{C}{r} e^{-\frac{\epsilon}{2}(2\pi h)^{-n}}$, that

$$|\#(\sigma(P_\delta) \cap \Gamma) - \frac{1}{(2\pi h)^n} \mathrm{vol}\,(p^{-1}(\Gamma))| \le \tag{54}$$
$$\frac{C}{h^n}\left(\frac{\epsilon}{r} + C_N\left(r^N + \ln(\frac{1}{r})\mathrm{vol}\,\left(p^{-1}(\partial\Gamma + D(0,r))\right)\right)\right),$$

for every fixed $N \in \mathbf{N}$.

If $\kappa > \frac{1}{2}$, we have $\mathrm{vol}\,(p^{-1}(\partial\Gamma + D(0,r))) = \mathcal{O}(r^{2\kappa-1})$ with $2\kappa - 1 > 0$ and in all cases we may assume that $\ln(\frac{1}{r})\mathrm{vol}\,(p^{-1}(\partial\Gamma + D(0,r))) = \mathcal{O}(r^{\alpha_0})$, where $\alpha_0 > 0$. Then we choose $N \gg 1$, $r = \epsilon^{1/(1+\alpha_0)}$ and the right hand side of (54) becomes $\mathcal{O}(1)\epsilon^{\alpha_0/(1+\alpha_0)}h^{-n}$.

Again we have a similar theorem where the conclusion (54) is valid simultaneously for all Γ in a suitable family.

Recently, the author obtained similar results when Q_ω is an operator of multiplication, see [49].

4.3. *Outline of the proofs*

We can construct $\widetilde{P} \in S(m)$ such that $\widetilde{P}(\rho;h) = P(\rho;h)$ for $|\rho| \gg 1$ and $|\widetilde{P}(\rho;h) - z| \geq m(\rho)/C$ for $\rho \in \mathbf{R}^{2n}$, $z \in \overline{\Omega}$. The eigenvalues of P in Ω coincide with the zeros of the holomorphic function

$$F(z;h) = \det P_z, \qquad (55)$$
$$P_z = (\widetilde{P}(x,hD;h) - z)^{-1}(P(x,hD;h) - z)$$

The same remark holds for P_δ and F_δ defined as in (55) with P replaced by P_δ, provided that (48) holds.

For $z \in \operatorname{neigh}(\partial\Gamma)$, put $Q = P_z^* P_z$. Let $1_\alpha(E) = \max(E, \alpha)$, where $\alpha = Ch$, $C \gg 1$. Using semiclassical analysis, we can show that under the assumption (52) (cf [38])

$$\ln\det Q \leq \ln\det 1_\alpha(Q) = \frac{1}{(2\pi h)^n}\Big(\iint \ln q\, dx d\xi + \mathcal{O}(1) h^\kappa \ln\frac{1}{h}\Big), \qquad (56)$$

where $q = |p_z|^2$ is the leading symbol of Q.

Since

$$\ln\det Q = \ln\det P_z^* P_z = \ln|F(z;h)|^2$$

we have

$$\ln|F(z;h)| \leq \frac{1}{(2\pi h)^n}\Big(\iint \ln|p_z| dx d\xi + \mathcal{O}(1) h^\kappa \ln\frac{1}{h}\Big). \qquad (57)$$

For $\delta > 0$ small enough, we get the same upper bound for $\ln|F_\delta(z;h)|$ (provided that (48) holds).

The main step in the proof is to get a corresponding lower bound for each fixed z with a probability close to 1. In the multidimensional case this boils down to a question about random determinants. Let $z \in \operatorname{neigh}(\partial\Gamma)$. Let $e_1, e_2, \ldots$ be the first eigen-functions of $Q = P_z^* P_z$ and let $f_1, f_2, \ldots$ be the first eigen-functions for $P_z P_z^*$. The two operators have the same eigenvalues $0 \leq \lambda_1 \leq \lambda_2 \leq \ldots$

We can arrange so that

$$P_z e_j = \sqrt{\lambda_j} f_j,\ P_z^* f_j = \sqrt{\lambda_j} e_j.$$

Let $N = N(\alpha) = \#\{j; \lambda_j \leq \alpha\}$ ($\alpha = Ch$, $C \gg 1$). Semiclassical analysis gives that $N = \mathcal{O}(h^{\kappa-n})$.

Consider

$$\mathcal{P}^0 = \begin{pmatrix} P_z & R_- \\ R_+ & 0 \end{pmatrix} : L^2 \times \mathbf{C}^N \to L^2 \times \mathbf{C}^N,$$

$$R_+ : L^2 \to \mathbf{C}^N,\ R_+ u(j) = \sqrt{\alpha}(u|e_j),$$

$$R_- : \mathbf{C}^N \to L^2,\ R_- = \sqrt{\alpha}\sum_1^N u_-(j) f_j.$$

$\mathcal{P}^0$ has an inverse

$$\mathcal{E}^0 = \begin{pmatrix} E^0 & E_+^0 \\ E_-^0 & E_{-+}^0 \end{pmatrix} = \mathcal{O}(\frac{1}{\sqrt{\alpha}}),$$

with E_+^0, E_-^0, E_{-+}^0 "explicit", and

$$\ln|\det \mathcal{P}^0|^2 = N \ln\alpha + \det 1_\alpha(P_z^* P_z).$$

For $P_\delta = P + \delta Q_\omega$, we form

$$P_z^\delta = (\widetilde{P} - z)^{-1}(P - z + \delta Q_\omega) = P_z + \delta\widetilde{Q}_\omega,$$

and since $\delta\|Q_\omega\| \leq \delta C_0 h^{-n} \ll 1$,

$$\mathcal{P}^\delta := \begin{pmatrix} P_z^\delta & R_- \\ R_+ & 0 \end{pmatrix}$$

is invertible with inverse

$$\mathcal{E}^\delta = \begin{pmatrix} E^\delta & E_+^\delta \\ E_-^\delta & E_{-+}^\delta \end{pmatrix} \approx \mathcal{E}^0.$$

Here

$$E_{-+}^\delta = E_{-+}^0 + \delta E_-^0 \widetilde{Q}_\omega E_+^0 + \text{"small"}, \tag{58}$$

and we can show by perturbative arguments that

$$\ln\det\mathcal{P}^\delta = \ln\det\mathcal{P}^0 + \mathcal{O}(\frac{\delta}{\sqrt{\alpha}} M^{3/2}),$$

leading to

$$\ln|\det\mathcal{P}^\delta| = \frac{1}{(2\pi h)^n}(\iint \ln|p_z| dx d\xi + \mathcal{O}(h^\kappa \ln\frac{1}{h})). \tag{59}$$

On the other hand, computations in [50] can be used to get

$$\ln|\det P_z^\delta| = \ln|\det\mathcal{P}^\delta| + \ln|\det E_{-+}^\delta| \tag{60}$$

Using (58), we can view E_{-+}^{δ} as a random matrix of size $\mathcal{O}(h^{\kappa-n})$, close in a suitable sense to one with independent Gaussian random variables as its entries. This can be used to show:

For every $z \in \mathrm{neigh}\,(\partial\Gamma)$, we have a nice lower bound on $\ln|\det E_{-+}^{\delta}|$ with probability close to 1. (60) then gives a corresponding lower bound on $\ln|\det P_z^{\delta}|$.

To complete the proof of Theorem 4.1 we can apply the following result of M. Hager [17,18] with $\widetilde{h} = h^n$:

Proposition 4.1. ([17,18]) *Let Γ and Ω be as above. Let $\phi \in C(\Omega;\mathbf{R})$ be smooth near $\partial\Gamma$. Let $f = f(z;\widetilde{h})$ be holomorphic in Ω with*

$$|f(z;\widetilde{h})| \le e^{\phi(z)/\widetilde{h}}, \; z \in \mathrm{neigh}\,(\partial\Gamma), \; 0 < \widetilde{h} \ll 1.$$

Assume there exist $\epsilon = \epsilon(\widetilde{h}) \ll 1$, $z_k = z_k(\widetilde{h}) \in \Omega$, $k \in J = J(h)$, such that

$$\partial\Gamma \subset \bigcup_{k\in J} D(z_k,\sqrt{\epsilon}), \; \#J = \mathcal{O}(\frac{1}{\sqrt{\epsilon}}),$$

$$|f(z_k;\widetilde{h})| \ge e^{(\phi(z_k)-\epsilon)/\widetilde{h}}, \; k \in J.$$

Then,

$$\#(f^{-1}(0)\cap\Gamma) = \frac{1}{2\pi\widetilde{h}}\iint_{\Gamma}(\Delta\phi)d(\Re z)d(\Im z) + \mathcal{O}(\frac{\sqrt{\epsilon}}{\widetilde{h}}).$$

For the proof of Theorem 1.2 we use an improved version of this result, see [21].

4.4. *Comparison with Theorem 3.2*

From the example with the non-self-adjoint harmonic oscillator in dimension 1, we have seen that Weyl asymptotics does not always hold for differential operators in one dimension, when the coefficients are analytic. If we add a small random perturbation to the non-self-adjoint harmonic oscillator, the theorems above and the main result in [18] show that with probability close to 1 the eigenvalues will no longer be confined to a half-line but will tend to fill up the range of the principal symbol p with a density that is given by $(2\pi h)^{-n}p_*(dv)$, where dv denotes the symplectic volume element on $\mathbf{R}^{2n}$ and $p_*(dv)$ is the direct image under p.

From this simple one-dimensional example it is easy to build examples in higher dimension when Weyl asymptotics does not hold. In the 2-dimensional case, we can also consider the situation when the unperturbed operator P satisfies the assumptions of Theorem 3.2. It is then natural to compare the distribution law given by Theorem 3.2 for P and the one given by the theorems 4.1, 4.2 for the random perturbations. To leading order in h, we get Weyl asymptotics already for P in the (close to normal) case when $\{p, \overline{p}\}$ vanishes identically. In general however, we get different asymptotic distributions already to leading order ([48]).

References

1. J. M. Bismut, *The hypoelliptic Laplacian on the cotangent bundle,* J. Amer. Math. Soc. 18 (2005), 379–476.
2. W. Bordeaux-Montrieux, in preparation
3. T. W. Cherry, *On the solution of Hamiltonian systems of differential equations in the neighboorhood of a singular point,* Proc. London. Math. Soc. 27 (1928), 151–170.
4. Y. Colin de Verdière, *Quasi-modes sur les variétés Riemanniennes,* Inv. Math. 43 (1977), 15–52.
5. E. B. Davies, *Semi-classical states for non-self-adjoint Schrödinger operators,* Comm. Math. Phys. 200 (1999), 35–41.
6. E. B. Davies, *Non-self-adjoint differential operators,* Bull. London Math. Soc. 34(5) (2002), 513–532.
7. E. B. Davies, *Semi-classical analysis and pseudospectra,* J. Diff. Eq. 216(1) (2005), 153–187.
8. N. Dencker, *The pseudospectrum of systems of semiclassical operators.* `http://arxiv.org/abs/0705.4561`
9. N. Dencker, J. Sjöstrand, *Pseudospectra of semiclassical (pseudo-) differential operators,* M.Zworski, Comm. Pure Appl. Math. 57(3) (2004), 384–415.
10. L. Desvillettes, C. Villani, *On the trend to global equilibrium in spatially inhomogeneous entropy-dissipating systems: the linear Fokker-Planck equation,* Comm. Pure Appl. Math. 54(1) (2001), 1–42.
11. M. Dimassi, J. Sjöstrand, *Spectral asymptotics in the semi-classical limit,* London Math. Soc. Lecture Notes Ser., 268, Cambridge Univ. Press, 1999.
12. J. P. Eckmann, M. Hairer, *Spectral properties of hypoelliptic operators,* Comm. Math. Phys. 235(2) (2003), 233–253.
13. S. Graffi, C. Villegas Blas, *A uniform quantum version of the Cherry theorem.* `http://arxiv.org/abs/math-ph/0702021`
14. V. I. Girko, *Theory of random determinants,* Mathematics and its applications, Kluwer academic publishers, Dordrecht, 1990.
15. I. C. Gohberg, M. G. Krein, *Introduction to the theory of linear nonselfadjoint operators*, Translations of mathematical monographs, Vol. 18, AMS, Providence, R.I., 1969.
16. A. Grigis, J. Sjöstrand, *Microlocal analysis for differential operators,* London Math. Soc. Lecture Notes Ser., 196, Cambridge Univ. Press, 1994.
17. M. Hager, *Instabilité spectrale semiclassique pour des opérateurs nonautoadjoints. I. Un modèle,* Ann. Fac. Sci. Toulouse Math. (6)15(2) (2006), 243–280.

18. M. Hager, *Instabilité spectrale semiclassique d'opérateurs non-autoadjoints. II.* Ann. Henri Poincaré 7(6) (2006), 1035–1064.
19. M. Hager, *Bound on the number of eigenvalues near the boundary of the pseudospectrum,* Proc. Amer. Math. Soc. 135(12) (2007), 3867–3873.
20. M. Hager, E. B. Davies, *Perturbations of Jordan matrices.* `http://arxiv.org/abs/math/0612158`
21. M. Hager, J. Sjöstrand, *Eigenvalue asymptotics for randomly perturbed non-self-adjoint operators,* Math. Ann., to appear. `http://arxiv.org/abs/math.SP/0601381`
22. B. Helffer, F. Nier, *Hypoelliptic estimates and spectral theory for Fokker-Planck operators and Witten Laplacians,* Lecture Notes in Mathematics, 1862, Springer-Verlag, Berlin, 2005.
23. B. Helffer, J. Sjöstrand, *Multiple wells in the semiclassical limit. I,* Comm. Partial Differential Equations 9(4) (1984), 337–408.
24. B. Helffer, J. Sjöstrand, Puits multiples en mécanique semi-classique. IV. Étude du complexe de Witten, Comm. Partial Differential Equations 10(3) (1985), 245–340.
25. F. Hérau, M. Hitrik, J. Sjöstrand, *Tunnel effect for Kramers-Fokker-Planck operators,* Annales H. Poincaré, to appear. `http://arxiv.org/abs/math.SP/0703684`
26. F. Hérau, M. Hitrik, J. Sjöstrand, *Tunnel effect for Kramers-Fokker-Planck operators: return to equilibrium and applications.* `http://arxiv.org/abs/0801.3615`
27. F. Hérau, F. Nier, *Isotropic hypoellipticity and trend to equilibrium for the Fokker-Planck equation with a high-degree potential,* Arch. Ration. Mech. Anal. 171(2) (2004), 151–218.
28. F. Hérau, J. Sjöstrand, C. Stolk, *Semiclassical analysis for the Kramers-Fokker-Planck equation,* Comm. Partial Differential Equations 30(4-6) (2005), 689–760.
29. M. Hitrik, J. Sjöstrand, *Rational invariant tori, phase space tunneling, and spectra for non-selfadjoint operators in dimension 2,* Ann. Sci. E.N.S., to appear. `http://arxiv.org/abs/math/0703394`
30. M. Hitrik, J. Sjöstrand, S. Vũ Ngọc, *Diophantine tori and spectral asymptotics for non-selfadjoint operators* Amer. J. Math. 129 (2007), 105–182.
31. L. Hörmander, *Fourier integral operators I,* Acta Math. 127 (1971), 79–183.
32. L. Hörmander, *Differential operators of principal type,* Math. Ann. 140 (1960), 124–146.
33. L. Hörmander, *Differential equations without solutions,* Math. Ann. 140 (1960), 169–173.
34. V. N. Kolokoltsov, *Semiclassical analysis for diffusions and stochastic processes,* Lecture Notes in Mathematics, 1724, Springer-Verlag, Berlin, 2000.
35. V. F. Lazutkin, *KAM theory and semiclassical approximations to eigenfunctions.* With an addendum by A.I. Shnirelman. Ergebnisse der Mathematik und ihrer Grenzgebiete, 24, Springer-Verlag, Berlin, 1993.
36. G. Lebeau, *Le bismutien,* Sém. é.d.p., Ecole Pol. 2004–05, I.1–I.15.
37. A. J. Lichtenberg, M. A. Lieberman, *Regular and chaotic dynamics,* Second

ed., Springer-Verlag, New York, 1992.

38. A. Melin, J. Sjöstrand, *Fourier integral operators with complex-valued phase functions,* Fourier integral operators and partial differential equations (Colloq. Internat., Univ. Nice, Nice, 1974), pp. 120–223, Lecture Notes in Math., Vol. 459, Springer, Berlin, 1975.
39. A. Melin, J. Sjöstrand, *Determinants of pseudodifferential operators and complex deformations of phase space,* Methods Appl. Anal. 9(2) (2002), 177–237.
40. A. Melin, J. Sjöstrand, *Bohr-Sommerfeld quantization condition for non-selfadjoint operators in dimension 2,* Astérisque 284 (2003), 181–244.
41. J. Moser, *On the generalization of a theorem of A. Liapounoff,* Comm. Pure Appl. Math. 11 (1958), 257–271.
42. K. Pravda Starov, *Etude du pseudo-spectre d'operateurs non auto-adjoints,* Thesis Rennes 2006. `http://tel.archives-ouvertes.fr/tel-00109895`
43. S. Roch, B. Silbermann, *C^*-algebra techniques in numerical analysis,* J. Oper. Theory 35 (1996), 241-280.
44. M. Sato, T. Kawai, M. Kashiwara, *Microfunctions and pseudo-differential equations. Hyperfunctions and pseudo-differential equations*, pp. 265–529, Lecture Notes in Math., Vol. 287, Springer, Berlin, 1973.
45. B. Simon, *Semiclassical analysis of low lying eigenvalues. II. Tunneling,* Ann. of Math. (2) 120(1) (1984), 89–118.
46. J. Sjöstrand, *Parametrices for pseudodifferential operators with multiple characteristics,* Ark. Mat. 12(1) (1974), 85–130.
47. J. Sjöstrand, *Resonances for bottles and trace formulae,* Math. Nachr. 221 (2001), 95–149.
48. J. Sjöstrand, *Eigenvalue distributions and Weyl laws for semi-classical non-self-adjoint operators in 2 dimensions*, in preparation.
49. J. Sjöstrand, *Eigenvalue distribution for non-self-adjoint operators with small multiplicative random perturbations.* `http://arxiv.org/abs/0802.3584`
50. J. Sjöstrand, M. Zworski, *Elementary linear algebra for advanced spectral problems*, Ann. Inst. Fourier, to appear. `http://arxiv.org/abs/math/0312166`
51. J. Tailleur, S. Tanase-Nicola, J. Kurchan, *Kramers equation and supersymmetry,* J. Stat. Phys. 122(4) (2006), 557–595.
52. L. N. Trefethen, *Pseudospectra of linear operators,* SIAM Review 39(3) (1997), 383–406.
53. L. N. Trefethen, *Nonhermitian systems and pseudospectra,* sém. é.d.p. Ecole Polytechnique, 2005–06. `http://www.math.polytechnique.fr/seminaires/seminaires-edp/2005-2006/sommaire2005-2006.html`
54. L. N. Trefethen, M. Embree, *Spectra and pseudospectra, the behaviour of nonnormal matrices and operators,* Princeton University Press, 2005.
55. C. Villani, Hypocoercivity, preprint.
56. H. Weyl, *Das asymptotische Verteilungsgesetz der Eigenwerte linearer partieller Differentialgleichungen (mit einer Anwendung auf die Theorie der Hohlraumstrahlung),* Math. Ann. 71(4) (1912), 441–479.
57. M. Zworski, *A remark on a paper of E. B. Davies,* Proc. Amer. Math. Soc. 129 (2001), 2955–2957.

THE DISTRIBUTION OF PRIMES: CONJECTURES vs. HITHERTO PROVABLES

C. Y. YILDIRIM

Department of Mathematics, Boğaziçi University
& Feza Gürsey Institute for Fundamental Sciences,
İstanbul, Turkey
E-mail: yalciny@boun.edu.tr

We present the main results, conjectures and ideas concerning the distribution of primes. We recount only the most important milestones and we do not go into technical details. Instead, suggestions for further reading are provided.

Keywords: Distribution of prime numbers.

1. Introduction

Here I endeavour to give an exposition of what are 'known' about the distribution of prime numbers. Of course, there is more than one meaning of 'known'. In the strictest sense a statement is known if there is a proof agreed upon by all mathematicians; in a lenient sense instead of a proof there could be numerical data or heuristic arguments or evidence of other sorts which indicate the truth of a statement and these may be sufficient for some people to accept the assertion. Number theory is rife with long-standing questions, ranging from completely solved claims to problems about which not even conjectures can be formulated. In what follows we will dwell upon the main results and conjectures concerning the distribution of prime numbers, skipping most of the intermediate developments which along with other relevant questions not mentioned here can be tracked down from the references.

Notation

For two natural numbers a and b we denote by (a, b) their greatest common divisor. For a natural number n, Euler's totient $\phi(n)$ is the number of positive integers less than or equal to n and relatively prime to n. The letter p will always stand for a prime number. By p_n, the n-th largest prime is meant (so that $p_1 = 2$, $p_2 = 3, \ldots$). On some occasions γ will denote Euler's constant, and on some other occasions it will denote the imaginary part of a zero of the Riemann zeta-function. In general ϵ will mean an arbitrarily small positive real number, the letters $A, C, c, \ldots$ will be used for constants, and these need not have the same value at each occurrence.

For two functions f, g we write $f(x) = O(g(x))$, or equivalently $f(x) \ll g(x)$, if there is a constant C such that $|f(x)| \leq Cg(x)$ on the common domain where both f and g are defined. If C depends on a parameter α, we write $f = O_\alpha(g)$ or $f \ll_\alpha g$. When x tends to a limit, $f(x) \sim g(x)$ means $\lim f(x)/g(x) = 1$, and $f(x) = o(g(x))$ means $\lim f(x)/g(x) = 0$. We write $f(x) = \Omega(g(x))$ in place of $\limsup |f(x)|/g(x) > 0$, and $f(x) = \Omega_\pm(g(x))$ if $\limsup f(x)/g(x) > 0$ and $\liminf f(x)/g(x) < 0$.

2. The dawn: Up until the 19th century

We can't pinpoint at which epoch and where some person(s) first hit upon the property that some of the counting numbers are prime. The most ancient records we are aware of are the almost 4000 years old clay tablets by mathematicians of the Sumer-Akkad civilization who compiled tables of factors of integers and also studied equations and extraction of square roots involving primes. One of the oldest known mathematical proofs, the proof of irrationality of $\sqrt{2}$ attributed to Pythagoras's school (6th century B.C.), relies on the concepts of divisibility and primes. Euclid (4th century B.C.), upon giving the fundamental theorem of arithmetic, proved the existence of infinitely many primes which we regard as the oldest result on the distribution of primes. Eratosthenes (3rd century B.C.) described the basic sieve method for determining the primes up to a given bound. Eratosthenes's method evolved to powerful sieve methods which played key roles in almost all of the strongest results in number theory obtained in the 20th century.

A most influential mathematician of antiquity Diophantus (circa 3rd century A.D.) was mostly occupied with finding integer or rational solutions to various equations, but his books were lost in the fire that destroyed the library of Alexandria. Some of Diophantus's books were found in the 15th century, and then translated into Latin by Bachet (1621). Fermat and Mersenne, upon studying Bachet's publication, announced new results on divisibility, primes and Diophantine equations. From their works the branches of (as they are now called) elementary number theory, algebraic number theory, Diophantine equations, elliptic curves ... developed. The 20th century witnessed several culminations of these developments, the completion of the proof of Fermat's Last Theorem by Wiles, and applications of primes and factorization to cryptography to name two.

As far as we know after Euclid and Eratosthenes the next mathematician to produce results on the distribution of primes was Euler (1737). The only still problematic statement from the times in between is that every even number is a sum of one, two or three primes due to Descartes. *Euler's product identity*

$$\sum_{n=1}^{\infty} \frac{1}{n^s} = \prod_{p:\,\text{prime}} (1 - \frac{1}{p^s})^{-1}, \tag{1}$$

valid for $s > 1$, is an analytic way of expressing the unique factorization of natural numbers into a product of primes. Euler had no scruples about using this identity with $s = 1$, whereupon he deduced that the sum of the reciprocals of the primes diverges. He also corresponded with Goldbach, who conjectured that every number > 2 can be written as a sum of three primes and every even number > 2 is a sum of two primes. (These statements are believed to be true, with the obvious modification to > 5 in the first statement, according to the present understanding that 1 is not a prime number contrary to Goldbach and many mathematicians before him who at times took 1 as a prime. We can immediately notice some properties which distinguish 1 from the primes: 1 is the multiplicative unit of the ring of integers and divisibility by 1 is not a special feature of any integer. Moreover, the value of any multiplicative arithmetic function at 1 is 1, whereas $\phi(p) = p - 1$ for example, so that unnecessary inconvenience in the definitions of arithmetic functions is avoided by excluding 1 from the set of primes. Some information about the status of Goldbach's still unproved statements will be given in §9).

Gauss mentioned much later in his life that around 1792, upon examining tables of primes, he conjectured that a good approximation to

$$\pi(x) := \sum_{p \le x} 1 \tag{2}$$

is given by the logarithmic integral

$$\mathrm{li}(x) := \int_2^x \frac{dt}{\log t}, \tag{3}$$

and that

$$\lim_{x\to\infty} \frac{\pi(x)}{x/\log x} = 1 \tag{4}$$

which is equivalent to $p_n \sim n \log n$. A similar but slightly wrong conjecture was in fact first published in 1798 and 1808 by Legendre, also on empirical grounds. Gauss also stated that the number of integers $\leq x$ which are products of two distinct primes (we call these E_2-*numbers*) is approximately $x \log \log x / \log x$.

3. The launch of rigorous analytic methods for primes

Euler stated that an arithmetic progression with the first term 1 contains infinitely many primes. Elementary number theory textbooks contain examples of generalizations of Euclid's theorem to some arithmetic progressions, i.e. sequences of the form $a + kq$ where a and q are fixed relatively prime positive integers and k runs through all natural numbers, for example, that the progressions $3 + 4k, 1 + 4k, 1 + 3k, 5 + 6k$... each contain an infinitude of primes. Some of these are proved by simple modifications in Euclid's proof, and some require other elementary ingredients. M.R. Murty [88], defining a Euclidean proof for the arithmetic progression $a + kq$ via the existence of an irreducible polynomial $f(x)$ with integer coefficients such that all but finitely many prime divisors of the values $f(n)$, $n \in \mathbb{Z}$ are either $\equiv 1(\text{mod}\, q)$ or $\equiv a(\text{mod}\, q)$, proved that a Euclidean proof exists if and only if $a^2 \equiv 1(\text{mod}\, q)$. (The 'if' part of this theorem had been proved by I. Schur).

The full generalization of Euclid's theorem that if a and q are relatively prime natural numbers, then there are infinitely many primes of the form $a + kq$ was conjectured by Legendre (1788) and proved by Dirichlet (1837-1840) in some memoirs which are regarded as the origin of analytic number theory. Dirichlet's idea was to adapt Euler's method to the case when the primes are restricted to the residue class $a(\text{mod}\, q)$. The proof uses the so-called *Dirichlet's L-functions*, defined by

$$L(s, \chi) := \sum_{n=1}^{\infty} \frac{\chi(n)}{n^s} = \prod_p (1 - \frac{\chi(p)}{p^s})^{-1} \tag{5}$$

in $s > 1$ where the series and the product are absolutely convergent. Here χ is a *Dirichlet's character to the modulus* q, a function of an integer variable n which is multiplicative and periodic with period q. It follows that if $(n, q) = 1$, then $\chi(n)$ is a root of unity. For $(n, q) > 1$, it is apt to define $\chi(n) = 0$. The character χ_0 which assumes the value 1 at all n coprime to q is called the *principal character* in which case (5) differs from (1) by the finite factor $\prod_{p|q}(1 - \frac{1}{p^s})$. It could be that for values of n coprime to q, the least period of a nonprincipal $\chi(n)$ is not q but a divisor of q, in which case χ is called an *imprimitive* character, and otherwise *primitive*. There are $\phi(q)$ characters in all to the modulus q, which form an abelian group (defining $\chi_1\chi_2(n) =$

$\chi_1(n)\chi_2(n)$) isomorphic to the group of those residue classes which are relatively prime to the modulus q. The characters to the modulus q satisfy

$$\sum_{n(\bmod q)} \chi(n) = \begin{cases} \phi(q) & \text{if} \quad \chi = \chi_0, \\ 0 & \text{otherwise,} \end{cases}$$

and

$$\sum_{\chi(\bmod q)} \chi(n) = \begin{cases} \phi(q) & \text{if} \quad n \equiv 1 \pmod q, \\ 0 & \text{otherwise.} \end{cases}$$

From a given set of integers those belonging to a particular residue class $a(\bmod q)$ can be selected by invoking

$$\frac{1}{\phi(q)} \sum_{\chi(\bmod q)} \overline{\chi}(a)\chi(n) = \begin{cases} 1 & \text{if} \quad n \equiv a \pmod q \quad \text{and } (a,q) = 1, \\ 0 & \text{otherwise.} \end{cases} \qquad (6)$$

For nonprincipal χ the series in (5) is conditionally convergent for $0 < s \leq 1$ so that $L(s,\chi)$ is regular at $s = 1$. Dirichlet's proof of the infinitude of primes of the form $a + kq$ hinges on the fact that for χ nonprincipal, $\log L(s,\chi)$ is bounded as $s \to 1^+$.

Dirichlet's innovations were wonderful, but the challenge of estimating the number of primes up to x, as $x \to \infty$, was not overcome yet. Around 1849 Chebyshev made advances in this problem. He showed that if $\lim_{x\to\infty} \pi(x)/x \log x$ exists it should be 1, and found lower and upper bounds to this limit as 0.92.. and 1.10.. by an argument based on equating the Stirling formula estimate for $\log n!$ with the expression of $\log n!$ as a sum over the logarithms of primes up to n. But Chebyshev's method would not pin down the limit to 1.

Mertens (1874) went further along Chebyshev's lines and established the following asymptotic formulae:

$$\sum_{p \leq x} \frac{\log p}{p} = \log x + O(1), \qquad (7)$$

$$\sum_{p \leq x} \frac{1}{p} = \log\log x + A + O(\frac{1}{\log x}), \qquad (8)$$

$$\prod_{p \leq x} (1 - \frac{1}{p})^{-1} = e^{\gamma} \log x + O(1). \qquad (9)$$

Mertens also showed that if χ is a nonprincipal character then $\sum_p \frac{\chi(p)}{p}$ converges (Dirichlet had used $\lim_{\sigma \to 1^+} \sum_p \frac{\chi(p)}{p^\sigma}$ is finite), wherefrom he obtained the generalization of the second of his formulas given above, a quantified form of Dirichlet's theorem:

$$\sum_{\substack{p\le x\\ p\equiv a(\bmod q)}} \frac{1}{p} = \frac{\log\log x}{\phi(q)} + A(q,a) + O(\frac{1}{\log x}). \tag{10}$$

4. Riemann's revolution

It is indeed interesting that partial summation applied to (7) or (8) is of no avail for obtaining an asymptotic formula for $\pi(x)$. In between Chebyshev and Mertens there was a development which turned out to be of utmost importance, not only for the goal of settling (4), but also for the impact on the theory of functions of a complex variable. In an article published in 1859, Riemann [98] took the quantity given by either of the two expressions in the Euler product identity (1) as a function of a complex variable $s = \sigma + it \in \mathbb{C}$, $\sigma, t \in \mathbb{R}$. In this way the *Riemann zeta-function* $\zeta(s)$ was defined in the half-plane $\sigma > 1$ where both sides of (1) are absolutely convergent and therefore make sense. He started from the well-known equation $\int_0^\infty e^{-nx}x^{s-1}\,dx = \frac{\Gamma(s)}{n^s}$, which upon summing over n gives $\zeta(s) = \frac{1}{\Gamma(s)}\int_0^\infty \frac{x^{s-1}}{e^x-1}\,dx$ for $\sigma > 1$. Riemann's remarkable insight to take $s\in\mathbb{C}$ allowed him to apply the methods of complex integration, and upon using the functional equation for a Jacobi theta function he obtained the representation

$$\pi^{-\frac{s}{2}}\Gamma(\frac{s}{2})\zeta(s) = \frac{1}{s(s-1)} + \int_1^\infty (x^{\frac{s}{2}-1} + x^{-\frac{s+1}{2}})(\sum_{n=1}^{\infty} e^{-n^2\pi x})\,dx\,, \tag{11}$$

where the integral converges absolutely for any s, and uniformly in compact subsets of the s-plane. Thus Riemann showed that from its original domain of definition $\zeta(s)$ can be continued analytically over the whole complex plane as a single-valued and meromorphic function with its only pole at $s = 1$, a simple pole with residue 1. Since the right-hand side of (11) is unchanged when $1-s$ is used in place of s, this also revealed the *functional equation of* $\zeta(s)$:

$$\pi^{-\frac{s}{2}}\Gamma(\frac{s}{2})\zeta(s) = \pi^{-\frac{1-s}{2}}\Gamma(\frac{1-s}{2})\zeta(1-s). \tag{12}$$

The value of $\zeta(s)$ can be calculated at any s with $\sigma > 1$ to any desired accuracy from the expressions in (1). Then, by the functional equation, $\zeta(s)$ can also be calculated for any s with $\sigma < 0$. It is clear from the product in (1) that $\zeta(s) \neq 0$ in $\sigma > 1$, and it is immediate from (12) that at the negative even integers where $\Gamma(\frac{s}{2})$ has simple poles, $\zeta(s)$ vanishes - these are called the *trivial zeros*. In the rather mysterious region $0 \leq \sigma \leq 1$, the so-called *critical strip*, one may use (11) or

$$\zeta(s) = \frac{s}{s-1} - s\int_1^\infty (x)x^{-s-1}\,dx \quad (\sigma > 0), \tag{13}$$

where (x) is the fractional part of x. In this formula which is obtained by applying partial summation to the series in (1), the integral converges absolutely for $\sigma > 0$ and uniformly for $\sigma \geq \delta > 0$, so that we have an analytic continuation of $\zeta(s)$ to $\sigma > 0$.

Riemann also made several assertions on the zeros of $\zeta(s)$, and provided sketches of the proofs for some. The matter of justifying Riemann's statements inspired Hadamard's work on general results for entire functions. Hadamard (1893) showed that the entire function $\xi(s) := \frac{1}{2}s(s-1)\pi^{-\frac{s}{2}}\Gamma(\frac{s}{2})\zeta(s)$ is bounded in size by $\exp(C|s|\log|s|)$ as $|s| \to \infty$, from which he deduced that $\zeta(s)$ has infinitely many *nontrivial zeros* in the critical strip. The nontrivial zeros must be situated symmetrically with respect to the real axis, and with respect to the point $\frac{1}{2}$. Applying the argument principle to $\xi(s)$, von Mangoldt (1895) gave the proof of Riemann's assertion that the number of nontrivial zeros $\rho = \beta + i\gamma$ with $0 < \gamma \leq T$ is $\frac{T}{2\pi}\log T - \frac{T}{2\pi} + O(\log T)$, as $T \to \infty$.

Riemann's expression for $\pi(x)$ in terms of the zeros of $\zeta(s)$, and a related formula we will now dwell upon, were also fully proved by von Mangoldt. Logarithmic differentiation of the product expression of $\zeta(s)$ gives

$$\frac{\zeta'}{\zeta}(s) = -\sum_{n=2}^{\infty}\frac{\Lambda(n)}{n^s} \quad (\sigma > 1), \tag{14}$$

where $\Lambda(n)$ is *von Mangoldt's function*

$$\Lambda(n) := \begin{cases} \log p & \text{if} \quad n = p^a, \quad p\text{: prime, } a \in \mathbb{N}\,, \\ 0 & \text{otherwise.} \end{cases} \tag{15}$$

Most researchers find it more convenient to work with

$$\psi(x) := \sum_{n \leq x}\Lambda(n) = \sum_{p^m \leq x}\log p \tag{16}$$

and then convert a result involving $\psi(x)$ to a result for $\pi(x)$ than working straightforwardly with $\pi(x)$. For instance, (4) is equivalent to $\psi(x) \sim x$.

From (14) one has

$$\psi_0(x) = \frac{1}{2\pi i}\int_{c-i\infty}^{c+i\infty}[-\frac{\zeta'}{\zeta}(s)]\frac{x^s}{s}\,ds \quad (c>1), \tag{17}$$

where $\psi_0(x) = \psi(x) - \dfrac{\Lambda(x)}{2}$. Considering the integral from $c-iT$ to $c+iT$, and moving the line of integration all the way to the left in the complex plane one obtains, by the residue theorem, for any $x \geq 2$,

$$\begin{aligned}\psi_0(x) = x - \sum_{|\gamma|<T}\frac{x^\rho}{\rho} - \frac{\zeta'}{\zeta}(0) - \frac{1}{2}\log(1-x^{-2}) \\ + O(\frac{x\log^2(xT)}{T}) + O(\log x \min(1, \frac{x}{T\langle x\rangle}))\end{aligned} \tag{18}$$

(here $\langle x\rangle$ denotes the distance from x to the nearest prime power - other than x itself in case x is a prime power). This is known as the *Riemann-von Mangoldt formula* or the *explicit formula*: it provides an explicit link between a (weighted) count of the primes and a sum over the nontrivial zeros of $\zeta(s)$. The more elegant form obtained by taking the limit $T \to \infty$ in (18),

$$\psi_0(x) = x - (\lim_{T\to\infty}\sum_{|\gamma|<T}\frac{x^\rho}{\rho}) - \frac{\zeta'}{\zeta}(0) - \frac{1}{2}\log(1-x^{-2}) \tag{19}$$

which is valid for $x > 1$, is less useful in applications.

The time was ripe, in 1896 Hadamard and de la Vallée-Poussin proved independently that $\zeta(s)$ has no zeros on the line $\sigma = 1$, from which they reached (4), the *prime number theorem*. (In fact, the non-vanishing of $\zeta(s)$ on the line $\sigma = 1$ is equivalent to the prime number theorem).

It is clear from the explicit formula that the more we know about the location of the zeta zeros, the more we can say about $\psi(x) - x$. So de la Vallée-Poussin's (1899) result $\zeta(s) \neq 0$ for $\sigma > 1 - c/\log(|t|+2)$ for the *zero-free region of* $\zeta(s)$ yielded the prime number theorem with the error term $\psi(x) = x + O(x\exp[-c(\log x)^{\frac{1}{2}}])$. This can be seen by an argument which combines the fact that $3\frac{\zeta'}{\zeta}(\sigma) + \Re[4\frac{\zeta'}{\zeta}(\sigma+it) + \frac{\zeta'}{\zeta}(\sigma+2it)] \leq 0$ (a consequence of $3 + 4\cos\theta + \cos 2\theta \geq 0$, $\forall\theta \in \mathbb{R}$) with simple inequalities obtained from the partial fraction expansion of $\frac{\zeta'}{\zeta}(s)$.

The hope for an elementary, i.e. without recourse to the theory of functions of a complex variable and in fact to any infinite summation, though not necessarily easy, proof of the prime number theorem seemed to fade.

Hardy [54] said in 1921 "No elementary proof is known, and one may ask whether it is reasonable to expect one. Now we know that the theorem is roughly equivalent to a theorem about an analytic function, the theorem that Riemann's zeta-function has no zeros on a certain line. A proof of such a theorem, not fundamentally dependent upon the ideas of the theory of functions, seems to me extraordinarily unlikely. It is rash to assert that a mathematical theorem cannot be proved in a particular way; but one thing seems quite clear. We have certain views about the logic of the theory; we think that some theorems, as we say, 'lie deep', and others nearer to the surface. If anyone produces an elementary proof of the prime number theorem, he will show that these views are wrong, that the subject does not hang together in the way we have supposed, and that it is time for the books to be cast aside and for the theory to be rewritten". A century after Chebyshev, Selberg [103] succeeded in giving an elementary proof based on *Selberg's identity*

$$\sum_{n\le x}\Lambda(n)\log n+\sum_{mn\le x}\Lambda(m)\Lambda(n)=2x\log x+O(x) \tag{20}$$

(and Erdös [20] too; Goldfeld [32] has told the story of the dispute over their roles in the proof). Being a kind of Stirling's formula of higher degree, (22) leads to achieving more than Chebyshev's results. There is some folklore to the effect that in fact Selberg might have been inspired by zeta function theory in arriving at his identity (Ingham, in his review [68] of Selberg's and Erdos's articles pointed out that it is quicker to obtain the identity starting out from equating coefficients in the Dirichlet series expression for $(\frac{\zeta'}{\zeta}(s))'+(\frac{\zeta'}{\zeta}(s))^2=\frac{\zeta''}{\zeta}(s))$, but Selberg presented an argument which avoids these. Selberg [104] also gave an elementary proof of Dirichlet's theorem on the infinitude of primes in an arithmetic progression.

Using I.M. Vinogradov's method for strong majorizations of *exponential sums* (i.e. sums of the kind $\sum_{a<n\le b} g(n)e^{2\pi i f(n)}$), in 1958 Korobov and Vinogradov independently established that the region

$$\sigma>1-\frac{c}{(\log(3+|t|))^{2/3}(\log\log(3+|t|))^{1/3}} \tag{21}$$

is free of zeta zeros. This implies the prime number theorem with the smallest proved error term

$$\pi(x)=\mathrm{li}\,x+O(x\exp[-c\frac{(\log x)^{\frac{3}{5}}}{(\log\log x)^{\frac{1}{5}}}]). \tag{22}$$

5. The Riemann Hypothesis

In view of the fact that zeta zeros do exist in the critical strip and they are situated symmetrically with respect to the critical line, the strongest improvement on the zero-free region (21) one can ever hope for is $\sigma > \frac{1}{2}$. Of all the assertions in Riemann's memoir the guess that *'all of the nontrivial zeros lie on the critical line* $\sigma = \frac{1}{2}$*'* is the only one that still awaits a proof. This statement which has turned out to be very profound is known as the *Riemann Hypothesis* (RH). Riemann wrote that after some fleeting futile attempts he put it aside, correctly assessing that it wasn't necessary for his immediate aim of reaching the prime number theorem.

Other than the point of view that it is esthetically pleasing and ideally desirable to have all of the zeros of a function which is deeply related to the primes and which has a simple Dirichlet series in the sense that all of the coefficients are 1, lie on one line (and are simple), there is considerable evidence in favour of RH.

Beginning with Riemann himself, computations have now reached to cover the first 10^{13} zeros, as well as millions of zeros in certain intervals at ordinates up to 10^{24}, and they are all simple zeros on the critical line ([46]). By the theoretical studies of Hardy and Littlewood, Selberg, Levinson, and finally Conrey [10], it is now known that more than 40 % of the zeros of $\zeta(s)$ lie on the critical line $\sigma = \frac{1}{2}$ and are simple. This result was attained by the *mollifier method*, in which a suitable *Dirichlet polynomial* (an expression of the kind $\sum_{n \leq y} \frac{a_n}{n^s}$), producing the effect of smoothing out the irregularities, is introduced in the integrands involving $\zeta(s)$ of the integrals needed for the application of the argument principle. There are also *zero-density estimates* which are upper bounds for $N(\sigma, T) := \{\rho : \zeta(\rho) = 0, \sigma < \beta, 0 < \gamma \leq T\}$. Zero-density estimates are useful in tackling many problems of analytic number theory unconditionally (i.e. not assuming the truth of RH or any other unproved hypotheses). It is known that $N(\sigma, T) \ll T^{3(1-\sigma)/(2-\sigma)} \log^5 T$ uniformly for $\sigma > \frac{1}{2}$. (This result due to Ingham (1940) has been improved for some σ by later researchers). The density estimates say that exceptions to RH, if ever they exist, are rare and become rarer as σ moves away from $\frac{1}{2}$.

The importance of $\zeta(s)$ is not limited by its connections to the distribution of primes and other applications in analytic number theory by virtue of its Euler product being a key tool in dealing with similar products arising from multiplicative functions. The Riemann zeta-function is the protoype of a class of functions, *global L-functions*, associated with various algebraic, arithmetic or geometric objects (as well as some which are not directly associated as such). These functions are Dirichlet series possessing a product representation of the Euler type, they can be analytically continued from their original domain of definition in keeping with a functional equation, and a version of RH can be stated for them. The existence of these functions and the properties satisfied by them have enormous implications in many areas. In many cases some parts of these conditions have been shown to hold, the rest remain conjectural. There are numerical data concerning some cases which confirm the hypotheses. For no global L-function the truth or falsity of RH has been established. However, for many problems the implications of the relevant form of RH have been proved by independent means. Thus global L-functions can be regarded to provide additional belief in the truth of RH by dint of the (mostly hypothetical) coherence in the big picture which includes many theories. Furthermore, for the zeta functions of algebraic varieties over finite fields, the analogue of RH has been proved.

Other mostly hypothetical support comes from additional conjectures which predict results that are confirmed by numerical computations. There are even problems for which predictions were not possible before. The predictions from these conjectures on already proved results coincide with them. We will say a little more about these in §7.

There are diverse statements, in complex analysis and functional analysis, formulated as tests or equivalent conditions for the truth of RH, and nothing that points to the falsity of RH has come up.

The following probabilistic argument supports RH. Consider the Möbius function $\mu(n)$ defined on the natural numbers as follows. First of all, as all multiplicative arithmetic functions, $\mu(1) = 1$. If $n = p_{\alpha_1} \cdots p_{\alpha_k}$ is a product of k distinct primes, then $\mu(n) = (-1)^k$. If n is divisible by the square of a prime, then $\mu(n) = 0$. It is easy to see that

$$\frac{1}{\zeta(s)} = \sum_{n=1}^{\infty} \frac{\mu(n)}{n^s}, \quad (\sigma > 1), \tag{23}$$

and that

$$M(x) := \sum_{n \leq x} \mu(n) = O(x^{\frac{1}{2}+\epsilon}) \Longleftrightarrow \text{RH is true.} \tag{24}$$

The sequence $\{\mu(n)\}_{n=1}^{\infty}$ seems to be quite random in the long run, and for almost all random sequences of entries 0 or ± 1, the summatory function is bounded as $O(x^{\frac{1}{2}+\epsilon})$. (We note that Mertens' original conjecture that $|M(x)| \leq x^{\frac{1}{2}}$ was refuted by Odlyzko and te Riele [91] who showed that $\limsup_{x\to\infty} M(x)x^{-\frac{1}{2}} > 1.06$ and $\liminf_{x\to\infty} M(x)x^{-\frac{1}{2}} < -1.009$. Ingham [67] proved that if at most a finite number of sums of the type $\sum_{n\le N} c_n\gamma_n$, with integers c_i having greatest common divisor 1, are 0, then the just mentioned lim inf and lim sup are ∞ and $-\infty$ respectively. Here $\gamma_1, \gamma_2, \ldots$ denote the imaginary parts of the distinct nontrivial zeros of $\zeta(s)$. Bateman et. al. [2] weakened the condition demanded by Ingham's theorem to having $|c_n| \le 2$ with at most one of the $|c_n| = 2$ and not all of them 0).

If RH is false, then the distribution of primes would have to exhibit irregularities much wilder than expected, and the first zeta zero off the critical line would be a very significant mathematical constant.

If RH is true, then all the ρ have real part $\frac{1}{2}$, and knowing the count of zeta zeros it is easy to deduce from the explicit formula that

$$\psi(x) = x + O(x^{\frac{1}{2}}\log^2 x) \quad \text{and} \quad \pi(x) = \mathrm{li}\, x + O(x^{\frac{1}{2}}\log x) \tag{25}$$

with a much smaller error term than (22). The exponent $\frac{1}{2}$ in the error term is the best possible because there are zeta zeros on the critical line.

6. More about the error term in the prime number theorem

Littlewood (1914) used Dirichlet's theorem on Diophantine approximation to show that the sum over ρ cannot be very small at all x, and proved that

$$\psi(x) - x = \Omega_\pm(x^{\frac{1}{2}}\log\log\log x) \quad \text{and} \quad \pi(x) - \mathrm{li}\, x = \Omega_\pm\Big(\frac{x^{\frac{1}{2}}\log\log\log x}{\log x}\Big). \tag{26}$$

This also answers Riemann's [98] thoughts on whether it could always be that $\pi(x) < \mathrm{Li}\, x := \int_0^\infty \frac{dt}{\log t}$ for $x > 2$. However, Pintz [93] proved that $\int_2^X (\pi(t) - \mathrm{Li}\, t)dt < 0$ for all sufficiently large X if and only if RH holds, and that $\pi(x) - \mathrm{Li}\, x$ is negative in a particular average sense. Furthermore, considerations based on some basic beliefs about the zeta zeros suggest that the probability that $\pi(x) > \mathrm{li}\, x$ is about $2.6 \cdot 10^{-7}$ (see [99]). Montgomery [84] suggested on a probabilistic argument, assuming RH and the linear independence over rational numbers of the imaginary parts of the nontrivial zeros above the real axis so that linear forms in the γ don't take on very small values, that the sharpest form of the prime number theorem can be

$$\overline{\underline{\lim}} \frac{\psi(x) - x}{x^{\frac{1}{2}}(\log\log\log x)^2} = \pm\frac{1}{2\pi}, \quad (x \to \infty). \tag{27}$$

Note that even under (27), Legendre's conjecture that there is always a prime between n^2 and $(n+1)^2$ for any positive integer n is inaccessible.

Averages of the error term in the prime number theorem are also of great relevance in studying the distribution of primes. For brevity call

$$R(x) := \psi(x) - x\,. \tag{28}$$

By (26) it is known that, as $x \to \infty$, $R(x)$ changes sign infinitely many times. Cramér [14] showed that on RH

$$\int_1^X \frac{(R(u))^2}{u}\, du = O(X), \tag{29}$$

and

$$\int_1^X \left(\frac{R(u)}{u}\right)^2 du \sim C \log X\,. \tag{30}$$

Gallagher's article [31] contains compact proofs of such results. Pintz [92] has shown that for all sufficiently large X

$$\frac{X^{\frac{3}{2}}}{400} \le \int_1^X |R(u)|\, du \le X^{\frac{3}{2}}\,, \tag{31}$$

where the lower bound is unconditional and the upper bound depends essentially on RH. Jurkat [73], by developing concepts on almost-periodic functions proved upon RH that,

$$\int_x^{x+\frac{x}{\log\log x}} \frac{R(u)}{u^{\frac{3}{2}}}\, du = \Omega_{\pm}\left(\frac{\log\log\log x}{\log\log x}\right) \tag{32}$$

(with the implied constants $\pm\frac{1}{2}$), and that this cannot be improved upon much for he also showed that the left-hand side of (32) is $O((\log\log\log x)^2/\log\log x)$. This result contains Littlewood's result (26). From (31) and (25) we see that as $x \to \infty$, $|R(x)|$ spends most of its time roughly around the value $x^{\frac{1}{2}}$ (instead of much smaller values), and (32) reveals the existence of quite long intervals throughout which $|R(x)|$ is almost as large as possible.

7. Pair correlation of zeta zeros and primes

In 1972 H. L. Montgomery [83], assuming RH, found a way of looking into the distribution of the nontrivial zeros on the line $\sigma = \frac{1}{2}$. Upon developing a new version of the explicit formula, Montgomery was led to define

$$F(\alpha, T) = \left(\frac{T}{2\pi}\log T\right)^{-1} \sum_{0<\gamma,\gamma'\le T} \frac{4}{4+(\gamma-\gamma')^2} T^{i\alpha(\gamma-\gamma')}, \tag{33}$$

where $\frac{1}{2}+i\gamma$ and $\frac{1}{2}+i\gamma'$ run through the zeros of $\zeta(s)$. It is easily seen that $F(\alpha,T)=F(-\alpha,T)$, and from the count of zeta zeros that $F(\alpha,T)\ll\log T$ for any $\alpha\in\mathbb{R}$. By using the mean-value result for Dirichlet series (see [86])

$$\int_0^T |\sum_n a_n n^{-it}|^2\,dt = \sum_n |a_n|^2(T+O(n)) \tag{34}$$

(with a_n's composed of the von Mangoldt function $\Lambda(n)$ in this application), Montgomery showed that upon RH one can obtain

$$F(\alpha,T)=(1+o(1))T^{-2\alpha}\log T+\alpha+o(1), \quad (0\le\alpha\le 1), \tag{35}$$

as $T\to\infty$. What is in question here is the size of

$$\int_0^T |\frac{1}{T^\alpha}\sum_{n\le T^\alpha}\Lambda(n)n^{\frac{1}{2}-it}+T^\alpha\sum_{n>T^\alpha}\Lambda(n)n^{-\frac{3}{2}-it}-\frac{2T^{\alpha(\frac{1}{2}-it)}}{(\frac{1}{2}+it)(\frac{3}{2}-it)}|^2\,dt.$$

For $\alpha>1$ the estimate for the contribution of the nondiagonal terms become large enough to prevent an asymptotic estimate. Montgomery, drawing upon the conjectured size of expressions of the kind $\sum_{n\le x}\Lambda(n)\Lambda(n+h)$ (see §9), conjectured that

$$F(\alpha,T)=1+o(1), \qquad (1\le\alpha\le A). \tag{36}$$

In (35) and (36) the estimates are uniform in the respective domains of α, and $A>1$ is arbitrary but fixed. Here the condition that A is fixed cannot be relaxed, since from Dirichlet's theorem on Diophantine approximation it can be seen that there exist large $\alpha=\alpha(T)$ for which $F(\alpha,T)$ is close to $F(0,T)\sim\log T$. We expect that $F(\alpha,T)\ll 1$ holds on average, for Goldston [34] proved, assuming RH, that for sufficiently large T,

$$\int_c^{c+1}F(\alpha,T)<\frac{8}{3}+\epsilon, \qquad \int_{c-1}^{c+1}F(\alpha,T)>\frac{2}{3}-\epsilon,$$

uniformly for any $c\in\mathbb{R}$ (c may even be a function of T).

Using convolutions of $F(\alpha,T)$ with appropriate kernels $\hat{r}(\alpha)$ so that

$$\sum_{0<\gamma,\gamma'\le T} r((\gamma-\gamma')\frac{\log T}{2\pi})w(\gamma-\gamma')=(\frac{T}{2\pi}\log T)\int_{-\infty}^{\infty}F(\alpha,T)\hat{r}(\alpha)\,d\alpha$$

where r and $\hat{r}$ are Fourier transforms of each other, from (35) and (36) Montgomery obtained

$$\sum_{\substack{0<\gamma,\gamma'\le T\\ \frac{2\pi\alpha}{\log T}<\gamma-\gamma'\le\frac{2\pi\beta}{\log T}}} 1\sim(\frac{T\log T}{2\pi})\{\int_\alpha^\beta 1-(\frac{\sin\pi u}{\pi u})^2\,du+\delta(\alpha,\beta)\}, \tag{37}$$

as $T \to \infty$ for fixed $\alpha < \beta$ (here $\delta(\alpha,\beta) = 1$ if $0 \in [\alpha,\beta]$ and $\delta(\alpha,\beta) = 0$ if $0 \notin [\alpha,\beta]$). The assertions (36) or (37) are essentially equivalent, they are known as the *pair correlation conjecture* (PC), and from either it is easy to deduce that almost all zeta zeros are simple. Since on RH, $F(\alpha,T)$ can be calculated for $|\alpha| \le 1$ as in (35), one can use an $\hat{r}(\alpha)$ supported in $[-1,1]$ to see the implications of RH. By taking $r(u) = (\frac{\sin \pi\alpha u}{\pi\alpha u})^2$, Montgomery deduced that on RH at least $\frac{2}{3}$ of the zeros of $\zeta(s)$ are simple. We also note that for $M(x)$ defined in (24), the *Mertens hypothesis* in the weaker form

$$\int_1^X \left(\frac{M(x)}{x}\right)^2 dx = O(\log X), \tag{38}$$

implies that all zeros of $\zeta(s)$ which are on the critical line are simple (see [110, §14.29]).

The emergence of $1-(\frac{\sin \pi u}{\pi u})^2$ as the pair correlation function of the zeta zeros has turned out to be very fruitful for the unleashing of new methods for studying $\zeta(s)$ and functions allied to it. F. J. Dyson told Montgomery that this is precisely the same pair correlation function for the eigenvalues of large order random Hermitian matrices, specifically the Gaussian Unitary Ensemble (GUE). From the random matrix model many details about $\zeta(s)$ (such as the moments of $\zeta(s)$ on the critical line, the maximal size of $\arg\zeta(\frac{1}{2}+it)$, the maximal and minimal gaps between zeta zeros) which were formerly inaccessible even by heuristics can now be predicted. Furthermore, random matrix models for general L-functions are also being thoroughly investigated in tandem with the philosophy of studying ensembles of such functions rather than regarding each one, and in particular their prototype $\zeta(s)$, as an isolated object. Over the last three decades considerable numerical and theoretical evidence pointing to the validity of random matrix models have accumulated. All these add to our belief in the existence of a linear operator whose eigenvalues characterize the zeros of $\zeta(s)$ (and similarly for other L-functions), an idea that legend dates back to Hilbert and Pólya (see [90]). We refer the reader to Conrey's surveys [12] and [13] for a detailed exposition of this research area.

Goldston and Montgomery [39] proved an equivalence between the pair correlation conjecture (36) and the asymptotic estimate for the second moment for primes (note that the first moment for primes is easily obtained from the prime number theorem). The result is, setting $x = T^\alpha$ and

$F_T(x) = \dfrac{T\log T}{2\pi}F(\alpha, T)$ and assuming RH, as follows:

$$\int_1^x \{\psi((1+\delta)y) - \psi(y) - \delta y\}^2\,dy \sim \frac{1}{2}\delta x^2(\log 1/\delta) \tag{39}$$

holds uniformly for $x^{-B_2} \le \delta \le x^{-B_1}$ where $0 < B_1 \le B_2 \le 1$, provided that

$$F_T(x) \sim \frac{T\log T}{2\pi} \tag{40}$$

is true uniformly for $x^{B_1}(\log x)^{-3} \le T \le x^{B_2}(\log x)^3$; conversely, if (39) holds uniformly in $x^{-1/A_1}(\log x)^{-3} \le \delta \le x^{-1/A_2}(\log x)^3$ where $1 \le A_1 \le A_2 < \infty$, then (40) is true uniformly for $T^{A_1} \le x \le T^{A_2}$.

Heath-Brown [57] showed that from RH and (36) in the weaker form $F_T(x) = o(T\log^2 T)$ for $1 \le \alpha \le A$ it follows that

$$\psi(x) = x + o(x^{\frac{1}{2}}\log^2 x). \tag{41}$$

The effect of assuming PC on top of RH can be assessed by comparing (41) with (25)-(27). With the support of PC one would expect to obtain nontrivial estimates on sums such as $\sum_\rho \dfrac{x^\rho}{\rho}$. For instance, Heath-Brown derived from the representation

$$\int_{-\infty}^{\infty} e^{-2|u|}\Big|\sum_{0<\gamma\le T} x^{i\gamma}e^{i\gamma u}\Big|^2\,du = F_T(x) \tag{42}$$

(which incidentally reveals that $F_T(x) \ge 0$ for all x), that

$$\sum_{0<\gamma\le T} x^{i\gamma} \ll T^{\frac{1}{2}}\{\max_{t\le T} F_t(x)\}^{\frac{1}{2}} \tag{43}$$

whereas the trivial bound is

$$\sum_{0<\gamma\le T} x^{i\gamma} \ll T\log T. \tag{44}$$

Since $\psi_0(x)$ is discontinuous at the prime powers, so is $\sum_\rho \dfrac{x^\rho}{\rho}$ $\big(= \lim_{T\to\infty}\sum_{|\gamma|<T}\dfrac{x^\rho}{\rho}\big)$ by the explicit formula (19); the series is boundedly convergent in fixed intervals $1 < a \le x \le b$. The sum $\sum_{0<\gamma\le T} x^\rho$ too is discontinuous at the prime powers. By considering $\int \dfrac{\zeta'(s)}{\zeta(s)}x^s\,ds$ taken around an appropriate rectangular contour, Gonek [45] proved a uniform (in both x and T) version of an old result of Landau, that for $x, T > 1$,

$$\sum_{0<\gamma\le T} x^{\rho} = -\frac{T}{2\pi}\Lambda(x) + O(x\log 2xT\,\log\log 3x) \tag{45}$$
$$+O(\log x\min(T, \frac{x}{\langle x\rangle})) + O(\log 2T\min(T, \frac{1}{\log x})).$$

If one assumes RH, then (45) implies for $x, T > 1$ that

$$\sum_{0<\gamma\le T} x^{i\gamma} \ll Tx^{-\frac{1}{2}}\log 2x + x^{\frac{1}{2}}\log 2x\log\log 3x. \tag{46}$$

Comparison with (44) shows that (46) is nontrivial for $2 \le x \le T^{2-\epsilon}$. If one assumes further that the $x^{i\gamma}$ behave like independent random variables, then one may expect that for almost all $x > 1$,

$$\sum_{0<\gamma\le T} x^{i\gamma} \ll T^{\frac{1}{2}+\epsilon}. \tag{47}$$

However, by Dirichlet's theorem on Diophantine approximation there exist arbitrarily large x for which

$$\sum_{0<\gamma\le T} x^{i\gamma} \gg T\log T, \tag{48}$$

so (47) does not hold for all $x > 1$. The observation (47) along with the heuristics of using $n/\log n$ in place of γ_n in $\sum x^{i\gamma}$ led Gonek to conjecture that

$$\sum_{0<\gamma\le T} x^{i\gamma} \ll (Tx^{-\frac{1}{2}+\epsilon} + T^{\frac{1}{2}}x^{\epsilon}), \quad (x, T \ge 2). \tag{49}$$

This would imply that, for $1 \le h \le x$,

$$\psi(x+h) - \psi(x) = h + O(h^{\frac{1}{2}}x^{\epsilon}), \tag{50}$$

which in turn yields

$$p_{n+1} - p_n \ll p^{\epsilon}. \tag{51}$$

For comparison we note the results on difference between consecutive primes under various conditions:

$$\begin{aligned} p_{n+1} - p_n &\ll p_n^{0.525} && \text{(unconditional)} \quad [1] \\ p_{n+1} - p_n &\ll p_n^{\frac{1}{2}}\log p_n && \text{(on RH)} \quad [14] \\ p_{n+1} - p_n &= o\big((p_n\log p_n)^{\frac{1}{2}}\big) && \text{(on RH \& PC)} \quad [38] \end{aligned} \tag{52}$$

8. Primes in arithmetic progressions

The theory of Dirichlet L-functions parallels that of $\zeta(s)$ for the most part although $L(s,\chi)$ does not have a pole at $s=1$ for nonprincipal χ. But a major trouble is encountered in the calculation of the zero-free region. The analogue of the inequality used for $\zeta(s)$ is $3\frac{L'}{L}(\sigma,\chi_0)+\Re[4\frac{L'}{L}(\sigma+it,\chi)+\frac{L'}{L}(\sigma+2it,\chi^2)]\le 0$. If χ is a real character, so that $\chi^2=\chi_0$, the argument does not work completely, and one cannot rule out the possibility of a real *exceptional zero* $\beta\in(1-\frac{c}{\log q},1)$ where c is an absolute constant, which can occur for at most one of the real nonprincipal characters modulo q. Of course, this often creates complications and limitations in dealing with many problems. It is believed that exceptional zeros do not exist, and the *Generalized Riemann Hypothesis* (GRH) states that all zeros of Dirichlet's L-functions lie on the line $\sigma=\frac{1}{2}$.

Analogous to (16), let

$$\psi(x;q,a):=\sum_{\substack{n\le x\\ n\equiv a\ (\bmod\ q)}}\Lambda(n). \tag{53}$$

We will take $(a,q)=1$ all throughout this section. Writing

$$\psi(x,\chi):=\sum_{n\le x}\chi(n)\Lambda(n), \tag{54}$$

we have by (6),

$$\psi(x;q,a)=\frac{1}{\phi(q)}\sum_{\chi\ (\bmod\ q)}\overline{\chi}(a)\psi(x,\chi). \tag{55}$$

Here one aims for a *prime number theorem for arithmetic progressions* of the type $\psi(x;q,a)=\frac{x}{\phi(q)}(1+o(1))$. Proceeding as in the proof of the prime number theorem, an explicit formula for $\psi(x,\chi)$ is obtained leading to

$$\psi(x;q,a)=\frac{x}{\phi(q)}-\frac{\overline{\chi_1}(a)x^{\beta_1}}{\phi(q)\beta_1}+O(xe^{-C'\sqrt{\log x}}) \tag{56}$$

uniformly for $q\le\exp[C\sqrt{\log x}]$, where C' is a positive constant depending only on C, and χ_1 is the only real character to the modulus q, if it exists, for which $L(s,\chi_1)$ has an exceptional zero β_1. If the exceptional zero β (for characters to the modulus q) exists, then it is known to satisfy $1-\beta\gg\frac{1}{\sqrt{q}(\log q)^2}$ with a specifiable constant, and this leads to

$$\psi(x;q,a)=\frac{x}{\phi(q)}+O(xe^{-c\sqrt{\log x}}) \tag{57}$$

valid for $q \leq (\log x)^{1-\delta}$ for some fixed $\delta > 0$. This result is effective in the sense that the numerical value of both of the constants in (57) can be determined if the value of δ is given. Siegel showed that for any $\epsilon > 0$ there exists a positive number $C = C(\epsilon)$ such that $L(s, \chi) \neq 0$ for $s > 1 - Cq^{-\epsilon}$. The upper bound for β is thus greatly improved in a sense but by the nature of its proof this result is noneffective (i.e. it is not known how to assign a value to C corresponding to a given value of ϵ). It leads to the *Siegel-Walfisz theorem* (1936), the best unconditionally established result that (57) holds uniformly for $q \leq (\log x)^N$ for any fixed $N > 0$ with a constant $c = c(N)$ which is unspecifiable. Assuming GRH, one easily gets

$$\psi(x; q, a) = \frac{x}{\phi(q)} + O(x^{\frac{1}{2}} \log^2 x), \quad (q \leq x), \tag{58}$$

which is an asymptotic relation for q almost up to $x^{\frac{1}{2}}$.

The main inquiry is: For which ranges of the relevant variables are the primes evenly distributed in the permissible congruence classes modulo q, and to what extent and in which sense is this distribution regular? When various averages over q and a are taken, results that hold in greater ranges of the parameters can be obtained. The *Bombieri-Vinogradov theorem* (1965) says that given any constant $A > 0$, we have

$$\sum_{q \leq Q} \max_{y \leq x} \max_{\substack{a \\ (a,q)=1}} |\psi(y; q, a) - \frac{y}{\phi(q)}| \ll \frac{x}{(\log x)^A} \tag{59}$$

with $Q = \dfrac{x^{\frac{1}{2}}}{(\log x)^B}$ where $B = B(A)$. The meaning of the Bombieri-Vinogradov theorem is that the asymptotic formula for $\psi(x; q, a)$ usually holds for q roughly as large as $x^{\frac{1}{2}}$, almost the same extent GRH delivers for an individual $\psi(y; q, a)$.

The important ingredients in the now standard version of the proof of the Bombieri-Vinogradov theorem include the Siegel-Walfisz theroem, a so-called *large-sieve* result

$$\sum_{q \leq Q} \frac{q}{\phi(q)} \sum_{\chi}{}^{*} |\sum_{M+1}^{M+N} a_n \chi(n)|^2 \ll (N + Q^2) \sum_{M+1}^{M+N} |a_n|^2$$

(here $\sum_{\chi}^{*}$ denotes a sum over all primitive characters $\chi(\text{mod} q)$), the *Pólya-Vinogradov inequality*

$$\sum_{M+1}^{M+N} \chi(n) \ll q^{\frac{1}{2}} \log q$$

for a nonprincipal character to the modulus q, and a simple identity due to Vaughan which simplifies Vinogradov's method for evaluating sums of the sort $\sum_{p\leq N} f(p)$ when f is oscillatory but not multiplicative. (For monotonic f the prime number theorem and partial summation usually suffices; for multiplicative f arguments involving Perron's summation formula such as (4) and the knowledge about the zero-free regions for $\zeta(s)$ and $L(s,\chi)$ are employed). The original proof of the Bombieri-Vinogradov theorem also used zero-density estimates for Dirichlet L-functions, it was Gallagher who first essentially avoided arguments that involve the critical strip, and then Vaughan's method provided further simplification.

The Bombieri-Vinogradov theorem is a very deep and useful result in number theory. So it was natural when Elliott and Halberstam [19] wondered whether it is possible that this theorem is valid for larger values of Q, in particular up to $x/(\log x)^B$. Friedlander and Granville [27], upon proving that this is not possible, cast the present form of the *Elliott-Halberstam conjecture* as the hope that the Bombieri-Vinogradov theorem is valid for Q up to $x^{1-\epsilon}$. They also conjectured the uniform estimates (with any fixed $\epsilon > 0$)

$$\psi(x;q,a) = \frac{x}{\phi(q)} + O((\frac{x}{q})^{\frac{1}{2}} x^{\epsilon}), \quad (q < x), \tag{60}$$

and

$$\psi(x;q,a) \ll_{\epsilon} \frac{x}{\phi(q)}, \quad (q < \frac{x}{(\log x)^{2+\epsilon}}). \tag{61}$$

With further averaging over the permissible residue classes a for each q, one has the *Barban-Davenport-Halberstam theorem* which reads in its asymptotic form (proved by Montgomery)

$$\sum_{q\leq Q} \sum_{\substack{1\leq a\leq q \\ (a,q)=1}} |\psi(x;q,a) - \frac{x}{\phi(q)}|^2 \sim Qx\log x, \quad (\frac{x}{(\log x)^A} \leq Q \leq x), \tag{62}$$

where $A > 0$ is any fixed number ([82, Thm. 17.2]). Upon GRH this holds for $x^{\frac{1}{2}} \log^2 x \leq Q \leq x$ ([25], [44]). Here the range of q is much longer than that for the Bombieri-Vinogradov theorem, but the mean-square error in the prime number theorem for arithmetic progressions is considered instead of the maxima in (59). For the average taken only over a there is Hooley's result [64], depending on GRH, that

$$\sum_{\substack{1\leq a\leq q \\ (a,q)=1}} \max_{u\leq x} |\psi(u;q,a) - \frac{u}{\phi(q)}|^2 \ll x(\log x)^4, \quad (q \leq x). \tag{63}$$

Hooley conjectured that this sum without the max behaves asymptotically as $x \log q$ in some ranges of q, and provided [63] evidence for his guess. The results of Friedlander and Goldston [25] also support this conjecture. Hooley [65] also did some work for the third moment

$$\sum_{q \leq Q} \phi(q) \sum_{\substack{1 \leq a \leq q \\ (a,q)=1}} (\psi(x;q,a) - \frac{x}{\phi(q)})^3,$$

and showed that as $x \to \infty$ this quantity is

$$\ll (Qx \log x)^{\frac{3}{2}} + \frac{x^3}{(\log^A x)} \quad \text{if } Q = o(\frac{x}{\log x}),$$

and

$$= \frac{3}{\pi^2} Q^2 x \log^2 x + O(Q^2 x \log x \log^2(\frac{2x}{Q})) \quad \text{if } \frac{x}{\log x} \leq Q \leq x.$$

These results are consistent with the belief that

$$(\psi(x;q,a) - \frac{x}{\phi(q)}) \ll (\frac{x \log x}{\phi(q)})^{\frac{1}{2}}$$

in most cases for q up to $x/\log x$, and that $\psi(x;q,a) - \frac{x}{\phi(q)}$ is symmetrically distributed about its zero mean.

Although the asymptotic values of $\psi(x;q,a_1)$ and $\psi(x;q,a_2)$ are the same, estimates for their difference (or with $\pi(x;q,a)$ which is defined by the same summation as in (53) but with the summand 1) are also of interest. Chebyshev observed that $\pi(x;4,3) - \pi(x;4,1)$ is more often positive than negative. One simple reason for this is that 1 is a square modulo 4, and 3 is not. Studies in this direction were taken up by Hardy and Littlewood, Landau, Pólya and continued by Turán, Knapowski, Staś and Wiertelak. In their works q is mostly taken to be either fixed or very small compared to x. For these we refer the reader to Ingham's tract [66], Turán's Collected Works Vol. 3 [111] and Pintz's review articles therein. Rubinstein and Sarnak [99], assuming GRH and that for each q the collection of non-trivial zeros of all $L(s,\chi)$ for primitive $\chi(\bmod q)$ are linearly independent over the rational numbers, characterized the moduli and residue classes for which Chebyshev type bias occurs and calculated the logarithmic densities for the set of x giving $\pi(x;q,a) - \pi(x;q,b) > 0$ for $q \geq 3$, $(ab,q) = 1$. As a continuation of these results and methods, Bays, Ford, Hudson and Rubinstein [3] examined the connections between small zeros of L-functions, class numbers of imaginary quadratic fields and Chebyshev's bias.

A very important result concerning the distribution of primes in arithmetic progressions is the unconditional *Brun-Titchmarsh theorem*

$$\pi(x;q,a) \le \frac{cx}{\phi(q)\log(x/q)}, \quad (1 \le q < x). \tag{64}$$

Here c is a constant (the theorem is known to hold with $c = 2$), and x is assumed to be sufficiently large. For $1 \le q \le x^{1-\epsilon}$ the upper-bound is at the expected order of magnitude. For $q > \sqrt{x}$, (64) says more than (58) which depends upon GRH. There has been considerable effort for proving (64) with smaller values of c (the theorem should be true with any $c > 1$). The critical barrier of $c = 2$ has been overcome for $x^\alpha < q < x^\beta$ for any fixed $0 < \alpha < \beta < 1$ (see Friedlander-Iwaniec [28]). From a similar improvement for small q, the nonexistence of exceptional zeros would follow as can be seen from (56).

Another famous problem in this topic is determining the size of the least prime in an arithmetic progression, denoted as $p_{\min}(a,q) := \min\{p : p \equiv a (\bmod q)\}$. The Siegel-Walfisz theorem gives a very weak estimate. Upon discounting the small contribution of proper prime powers to $\psi(x;q,a)$ in (58), we see that GRH implies $p_{\min}(a,q) \ll q^2 \log^5 q$. In a pioneering deep work Linnik obtained $p_{\min}(a,q) \ll q^C$ with effectively computable constant. The best known value for the Linnik constant is $C = 5.5$ due to Heath-Brown [60]. It is conjectured that $C = 1+\epsilon$ would work; it is easy to see that $p_{\min}(a,q) \ll o(q\log q)$ is false.

9. Additive questions about the primes

It has long been believed but not yet proved that there exist infinitely many twin primes, i.e. p and $p+2$ both prime. Brun (1919) showed that the series formed by the reciprocals of twin primes is convergent. His method involved the truncation of certain sums which arise in Legendre's formulation of Eratosthenes's sieve. This not only rendered the Eratosthenes sieve usable for his purpose, but also opened the way for the development of various sieve methods to be applied to many formerly unapproachable problems. For example, it is now known that $\pi_{(0,2)}(x)$, the number of prime $p \le x$ such that $p+2$ is also prime, is $\lesssim 6.84 \prod_{p>2}(1 - \frac{1}{(p-1)^2})\frac{x}{(\log x)^2}$. (Of course for the twin primes problem one needs a lower bound which tends to infinity with x. A curious result due to Heath-Brown [58] is that the existence (in an appropriate sense) of exceptional zeros implies the infinitude of twin primes).

Hardy and Littlewood's [55] *prime r-tuples conjecture* for the number $\pi_{\mathbf{d}}(N)$ of positive integers $n \le N$ for which $n+d_1, \ldots, n+d_r$ are all prime (here r is fixed, $d_1, \ldots, d_r$ are distinct integers and $\mathbf{d} = (d_1, \ldots, d_r)$) in the form of an asymptotic formula is

$$\pi_{\mathbf{d}}(N) \sim \mathfrak{S}(\mathbf{d}) \frac{N}{\log^r N}, \quad (N \to \infty) \tag{65}$$

when $\mathfrak{S}(\mathbf{d}) \neq 0$, where

$$\mathfrak{S}(\mathbf{d}) = \prod_p \frac{p^{r-1}(p - \nu_{\mathbf{d}}(p))}{(p-1)^r}, \tag{66}$$

and $\nu_{\mathbf{d}}(p)$ is the number of distinct residue classes modulo p occupied by $d_1, \ldots, d_r$. A strong form of the conjecture with an error term is

$$\sum_{n=1}^{N} \prod_{i=1}^{r} \Lambda(n+d_i) = \mathfrak{S}(\mathbf{d})N + E_r(N, \mathbf{d}), \quad E_r(N, \mathbf{d}) \ll_{\epsilon, r} N^{\frac{1}{2}+\epsilon}, \tag{67}$$

uniformly for $|d_i| \le N$. The $r = 1$ case of (65) is the prime number theorem. For $r \geq 2$ the conjecture remains unproved for any $\mathbf{d}$. The heuristics for arriving at the r-tuples conjecture in the case $r = 2$, $d_1 = 0$, $d_2 = 2$ (see Hardy and Wright [56, §22.20]), is based on counting the number of positive integers $\le x$ which are relatively prime to $\prod_{p \le \sqrt{x}} p$ in two ways, using Mertens's result (9) and using the prime number theorem, that leads to a discrepancy by a factor of $2e^{-\gamma}$. The error lies in the calculation that uses (9), because it involves the assumption that the number of such integers in an interval is always proportional to the length of the interval, even if the interval is very short as in this application. The conjectural step comes in when this reasoning is adopted for pairs of such integers, and the square of the correction factor $e^\gamma/2$ is introduced. Let us also mention in this context, that Hensley and Richards [62] proved that the conjecture $\pi(x+y) \le \pi(x) + \pi(y)$ is incompatible with the prime r-tuples conjecture.

Assuming that for each r, (65) holds uniformly for $1 \le d_1, \ldots, d_r \le h$, Gallagher [30] showed that if $P_k(h, N)$ is the number of integers $n \le N$ for which the interval $(n, n+h]$ contains exactly k primes, then $P_k(\lambda \log N, N) \sim N \frac{e^{-\lambda}\lambda^k}{k!}$ as $N \to \infty$, i.e. the distribution tends to the Poisson distribution with parameter λ. Numerical studies [109] indicate that small primes seem to obey Gaussian Orthogonal Ensemble statistics and as more primes are included there is a transition towards Poisson statistics.

Assuming the conjecture (67) for $1 \le r \le K$, Montgomery and Soundararajan [85] calculated the K-th moment for the primes in short intervals as

$$\sum_{n=1}^{N}(\psi(n+h)-\psi(n)-h)^K = \mu_K h^{\frac{K}{2}} \int_1^N (\log(\frac{x}{h}+b)^{\frac{K}{2}}\, dx \tag{68}$$

$$+O(N(\log N)^{\frac{K}{2}} h^{\frac{K}{2}} (\frac{h}{\log N})^{-\frac{1}{8K}} + h^K N^{\frac{1}{2}+\epsilon}),$$

uniformly for $\log N \le h \le N^{\frac{1}{K}}$, where $\mu_K = 1\cdot 3 \cdots (K-1)$ if K is even, and $\mu_K = 0$ if K is odd, and $b = 1-\gamma-\log 2\pi$. We notice that in the case $K=1$ the assumption is equivalent to RH, and for $K=2$ the assumption is equivalent to a strong form of PC. Later on Chan [8] obtained the same result under the weaker assumption that for $N \ge h$,

$$\sum_{\substack{1\le d_i \le h \\ i=1,\dots,r}} E_r(N,\mathbf{d})^2 \ll_r N^{1+\epsilon} h^r.$$

Montgomery and Soundararajan conjectured that the right-hand side of (68) continues to be $(\mu_K + o(1))N(h\log N/h)^{K/2}$ uniformly for $(\log N)^{1+\delta} \le h \le N^{1-\delta}$, and that this conjecture is equivalent to

$$\int_1^X (\sum_{0<\gamma\le T} \cos(\gamma \log x))^k\, dx = (\mu_k + o(1))X(\frac{T}{4\pi}\log T)^{\frac{k}{2}},$$

where the γ run through the ordinates of the zeta zeros.

One of the oldest unsolved problems concerning primes is the *Goldbach conjecture* (1742) that every even number > 2 is the sum of two prime numbers. Beginning with the work of Brun on sieve methods the furthest achievement in this direction is the theorem of Chen from 1966 [9], that every sufficiently large even number can be expressed as the sum of a prime and a number which has at most two prime factors - counted with multiplicity. Pintz [94], making a great improvement over former results, has shown that the number of even integers $\le x$ which are not expressible as a sum of two primes is $O(x^{\frac{2}{3}})$. In this context we note that Schnirelman (1930) proved by developing a quite simple method that there exists a constant C such that every natural number > 1 can be expressed as a sum of at most C primes. Ramaré's work [96] shows that $C=7$ suffices. Vinogradov (1937), using his method of estimating exponential sums, proved that every sufficiently large odd number can be expressed as a sum of three primes (see [16]), from which it immediately follows that every sufficiently large integer can be expressed as a sum of at most 4 primes. As for conditional results, it is known that RH implies $C=6$ is fine, while GRH allows Vinogradov's result for all odd numbers > 5 and this makes $C=4$ sufficient.

The root of the difficulty in such conjectures is that they involve constraints of an additive nature on the primes which are defined in terms of the operation of multiplication. So what needs to be understood deeply is the relation between the operations of addition and multiplication, which are linked only by the distributive law. Some mathematicians, e.g. Knuth [75], think that Goldbach's conjecture may well be true but there is no rigourous way to prove it. It might be one of the unprovable assertions that Gödel showed exist (and it is known that in some sense almost all correct statements about mathematics are unprovable).

Schinzel formulated two grand conjectures which embody the Goldbach and twin-prime conjectures as well as many other questions. Let $P_i(n)$ ($i = 1, \ldots, r$) be distinct irreducible polynomials in $\mathbb{Z}[x]$ having positive leading coefficients such that the product $P = P_1 \cdots P_r$ has no fixed prime divisor. The first conjecture is: There exist infinitely many $n \in \mathbb{Z}$ such that each $P_i(n)$ is prime. The second conjecture is: Let $N \in \mathbb{N}$ and $G \in \mathbb{Z}[x]$ be another polynomial with positive leading coefficient such that $N - G$ is irreducible, and also such that $(N - G)P$ has no fixed prime divisor. For all sufficiently large N, there exists $n \in \mathbb{Z}^+$ such that $N - G(n) > 0$ and each of $N - G(n), P_1(n), \ldots, P_r(n)$ is prime. Iwaniec [70] showed that any inhomogeneous irreducible quadratic polynomial in two variables which depends essentially on both of the variables, and whose coefficients are integers without a common factor takes infinitely many prime values. Iwaniec also proved an asymptotic formula for the number of these primes $\leq z$ as $z \to \infty$. In this case, as well as in the previously dealt cases of the prime number theorem and binary quadratic forms, the number $N(z)$ of the values of the argument of the polynomial for which the value of the polynomial is $\leq z$ satisfies $N(z) \gg z$. Fouvry and Iwaniec [21] proved an asymptotic formula for the number of primes represented by $x^2 + p^2$ (where p is a prime), in which case $N(z) \ll z/\log z$. In a later development, Friedlander and Iwaniec [29] settled the problem for $x^2 + y^4$, which is a quite sparse sequence with $N(z) = O(z^{\frac{3}{4}})$. More recently Heath-Brown [61] worked out the problem for the even sparser sequence $x^3 + 2y^3$ with $N(z) = O(z^{\frac{2}{3}})$ in which case the homogeneity of the polynomial is helpful. In the case of $x^2 + 1$, Iwaniec [71] showed that it is infinitely often a number having at most two (not necessarily distinct) prime factors.

A recent major development in a question of additive nature took place when Green and Tao [51] showed that the primes contain arbitrarily long arithmetic progressions. Their proof makes use of combinatorial methods, harmonic analysis and ergodic theory in addition to methods of analytic number theory.

10. Primes in short intervals

After the prime number theorem it is natural to ask for which functions $\Phi(x)$, as $x \to \infty$,

$$\pi(x+\Phi(x)) - \pi(x) \sim \frac{\Phi(x)}{\log x}. \tag{69}$$

Here one would try to find $\Phi(x)$ as slowly increasing as possible. Heath-Brown [59] proved that one can take $\Phi(x) = x^{\frac{7}{12}-\epsilon(x)}$ ($\epsilon(x) \to 0$, as $x \to \infty$), and $\Phi(x) = x^{\frac{1}{2}+\epsilon}$ is allowed if RH is assumed. The gap between these upper-bounds and the known lower-bounds is huge. We know due to Rankin [97] that (69) doesn't hold with

$$\Phi(x) = c\frac{\log x \log\log x \log\log\log\log x}{(\log\log\log x)^2}, \tag{70}$$

because Rankin showed that there exists a sequence of values of x tending to ∞ with intervals around x of length given in (70) which don't contain a prime. This is the best proved order of magnitude for the largest gaps between consecutive primes. Maier [77] generalized this to the existence, for any fixed k, of k consecutive gaps between primes each of size at the order of magnitude (70). (This in turn was generalized to the case of arithmetic progressions by Shiu [105]). On the other hand Selberg [102] showed assuming RH that, (69) holds for almost all x if $\dfrac{\Phi(x)}{(\log x)^2} \to \infty$ as $x \to \infty$. Here what is meant by 'almost all x' is that, while $X \to \infty$ the measure of the set of $x \in [0, X]$ for which (69) doesn't hold is $o(X)$. Without assuming RH, this almost-all result is known to hold with $\Phi(x) = x^{\frac{1}{6}-\epsilon}$ (Zaccagnini [114]). Whether Selberg's result is true without exceptions or not was answered by Maier [78] who showed that exceptions to (69) exist with $\Phi(x)$ as large as $(\log x)^\lambda$ with any fixed $\lambda > 1$, i.e.

$$\limsup_{x\to\infty} \frac{\pi(x+(\log x)^\lambda) - \pi(x)}{(\log x)^{\lambda-1}} > 1, \ \liminf_{x\to\infty} \frac{\pi(x+(\log x)^\lambda) - \pi(x)}{(\log x)^{\lambda-1}} < 1. \tag{71}$$

By the prime number theorem the difference between consectuive primes, $p_{n+1} - p_n$ is on average $\sim \log p_n$. Cramér [15] conjectured upon probabilistic reasoning that the largest possible gap between consecutive primes satisfies $\limsup_{n\to\infty} \dfrac{p_{n+1}-p_n}{(\log p_n)^2} = 1$, but now it is widely thought that

$$p_{n+1} - p_n = O(\log^2 p_n) \tag{72}$$

is a safer conjecture. Cramér's reasoning was based on the simple model that the probability of n being a prime is approximately $1/\log n$, and that this can be taken into account independently for different integers (which is the basic flaw in Cramér's model; for example, n and $n+2$ both being primes are not independent events: if n is even we automatically know that $n+2$ can not be prime). Since Rankin's estimate (70), only the value of the constant c in (70) has been improved (see [80]). The estimates in (51) and (52), even under very strong conjectures for the zeta zeros, fall dismally short of (72). Naturally the more that is known or assumed about the zeta zeros, the stronger results are deduced on the distribution of primes, notwithstanding the zeta zeros by themselves do not pertain steadfastly to primes as

$$\zeta^*(s) = \prod_{n=1}^{\infty}(1 - \frac{1}{q_n^s})^{-1} \quad (p_n \leq q_n \leq p_{n+1}),$$

considered by Grosswald and Schnitzer [52], reveals. This product converges absolutely for $\sigma > 1$ where it does not vanish, it can be analytically continued to $\sigma > 0$ and then it is seen that $\zeta^*(s)$ has the same zeros as $\zeta(s)$ in $\sigma > 0$. But some significant properties of $\zeta(s)$ are not valid for $\zeta^*(s)$, namely the analytic continuation is not possible beyond $\sigma > 0$ so that for $\zeta^*(s)$ there is no functional equation, and it has a simple pole at $s = 1$ with residue r, $\frac{1}{2} \leq r \leq 1$.

As for small gaps between primes, there has been considerable progress recently. In the 1990's Goldston [35] developed a method which gives lower bounds of the correct order of magnitude in many problems about the distribution of primes. This method rested upon using short divisor sum approximations to the von Mangoldt function such as

$$\Lambda_R(n) := \sum_{d|n,\, d\leq R} \mu(d) \log(\frac{R}{d}),$$

with a parameter R to be chosen as large as possible in an application because without the condition $d \leq R$ this sum is equal to $\Lambda(n)$ for $n > 1$. After a sequence of works by Goldston et. al. either concerning various applications or for improving the results by employing better approximants, Goldston, Pintz and the author in [40] and [42] have reached

$$\liminf_{n\to\infty} \frac{p_{n+r} - p_n}{\log p_n} \leq e^{-\gamma}(\sqrt{r} - 1)^2 \tag{73}$$

for any fixed positive integer r. With $r = 1$ this shows that for any fixed $\epsilon > 0$, there are infinitely many consecutive primes differing by less than ϵ

times the average gap. Through a more refined analysis in [41] the sharper result

$$\liminf_{n\to\infty} \frac{p_{n+1} - p_n}{\sqrt{\log p_n}(\log\log p_n)^2} \leq \infty \tag{74}$$

was obtained. The method also allows a generalization of (73) to primes in arithmetic progressions where the modulus q can grow with the size N of the primes subject to $q \ll_A (\log\log N)^A$. Furthermore, it turns out somewhat surprisingly, the assumption that (59) holds with $Q = x^{\theta-\epsilon}$ for any fixed $\theta \in (\frac{1}{2}, 1]$ implies that there exist bounded gaps between primes infinitely often, the size of the gaps being a function of θ. This is an instance revealing the importance and strength of the Bombieri-Vinogradov theorem, and that showing its result continues to hold beyond $\theta = \frac{1}{2}$ will be an utmost breakthrough.

Together with Graham the same authors applied their methods to the problem of small gaps between E_2-numbers q_n. Corresponding to (73), it was shown in [36] that

$$\liminf_{n\to\infty} q_{n+r} - q_n \leq C(r) \tag{75}$$

for a constant $C(r)$, in particular $C(1) = 6$. The ideas involved in this work also yielded in [37] stronger variants of the Erdös-Mirsky conjecture: There are infinitely many integers n which simultaneously satisfy $d(n) = d(n+1)$, $\Omega(n) = \Omega(n+1)$, $\omega(n) = \omega(n+1)$, even by specifying the value of these functions along with generalizations to shifts $n + b$ with an arbitrary positive integer b (here $d(n)$, $\Omega(n)$, $\omega(n)$ denote respectively the number of positive integer divisors of n, the number of prime divisors of n counted with multiplicity, and the number of distinct prime divisors of n).

Suggestions for further reading

In addition to the articles and books cited in the text we suggest the following works for a presentation of the topics and the proofs.

Section 2: Dickson's three volume set [17] is a compendium of results in number theory up to approximately 1920. The book by Hardy and Wright [56] is a standard reference for most topics of elementary number theory.

Sections 3, 4 & 8: For detailed introductions to the topics of these sections we refer the reader to the books by Davenport [16], and by Montgomery and Vaughan [87].

Sections 4 & 5: Titchmarsh's book (with notes added by Heath-Brown for the 2nd edition) [110] is a classic treatise on the Riemann zeta-function.

There are other treatises devoted to the theory of $\zeta(s)$, notably those of Edwards [18], Ivic [69], Karatsuba and Voronin [74]. The last two books contain a proof of (21) and (22). Elementary proofs of the prime number theorem are given in Hardy and Wright [56], and in the introductory level book by Tenenbaum and Mendes-France [108]. The articles by Bombieri [6], Sarnak [101], and Conrey [11] are recent essays on the Riemann Hypothesis by three of the leading experts in the field.

Section 6: A classical book on the subject which includes the proof of (26) is Ingham's tract [66].

Section 7: For further details on the pair correlation conjecture and prime numbers we refer the reader to Goldston's survey article [33]. The n-level correlations of zeros of $\zeta(s)$ were studied by Bogomolny and Keating [4] using heuristic arguments and based on the Hardy-Littlewood conjecture (equations (65)-(66) above with $r = 2$), and rigorously by Rudnick and Sarnak [100]. For the theory of random matrices, correlation functions and relations of zeta zeros to eigenvalues of a Hermitian operator the reader may consult the book by Mehta [81].

Section 8: The survey article by Iwaniec [72] gives an overview of the current directions in prime number theory. In a series of three papers, the last being [7], Bombieri, Friedlander and Iwaniec obtained improvements in some directions (not including the version stated in §8) on the Bombieri-Vinogradov theorem. Concerning the Barban-Davenport-Halberstam theorem there are the works of Hooley titled 'On the Barban - Davenport - Halberstam theorem I – XVIII'), and of Friedlander and Goldston [26]. The books by Montgomery [82], and Bombieri [5] contain further treatment of most of the topics of this section. For the Brun-Titchmarsh theorem see Friedlander's survey article [22].

Section 9: The methods of attacking problems having an additive nature, sieve methods, and the conjectures on the distribution of primes mentioned in this section can be found at an introductory level in Friedlander [24], Greaves [49] and Nathanson [89]. At more advanced levels there are the books of Greaves [50], Halberstam and Richert [53], and Vaughan [112]. The article by Kra [76] presents an introduction to the ideas in Green and Tao's work.

Section 10: Maier's papers listed in the references brought forth important developments in prime number theory. For comments on the error terms in the prime number theorem for arithmetic progressions, and for the consequences of Maier's work we refer the reader to Friedlander's survey article [23]. For Cramér's model, Maier's method and related matters on the distribution of primes we refer the reader to the articles by Granville [47], [48], Pintz [95] and Soundararajan [106], [107]. The last of these articles is mainly concerned with the problem of small gaps between primes. The developments leading to the results of Goldston, Pintz and Yıldırım are recounted in [43].

It is much easier to find the articles of Riemann, Hardy, Littlewood and Selberg from their collected works. Some of the original articles which have shaped the development of the theory in the 20th century have been collected in a book edited by Wang [113]. And most of the recent articles can be found in the websites of the authors.

References

1. R. C. Baker, G. Harman and J. Pintz, *Proc. London Math. Soc. (3)* **83**, 532-562 (2001).
2. P. T. Bateman, J. W. Brown, R. S. Hall, K. E. Kloss and R. M. Stemmler, Linear relations connecting the imaginary parts of the zeros of the zeta function, in *Computers in Number Theory* (Proc. Sci. Res. Council Atlas Sympos. No. 2, Oxford, 1969), Academic Press, London, 1971, pp. 1-19.
3. C. Bays, K. Ford, R. H. Hudson and M. Rubinstein, *J. Number Theory* **87**, 54-76 (2001).
4. E. B. Bogomolny and J. P. Keating, *Nonlinearity* **9**, 911-935 (1996).
5. E. Bombieri, *Le grand crible dans la théorie analytique des nombres*; *Astérisque* **18** (1987/1974), (Société Mathématique de France, 1987).
6. E. Bombieri, (2000), `http://www.claymath.org/library/MPP.pdf`, pp. 99-111.
7. E. Bombieri, J. B. Friedlander and H. Iwaniec, *J. Amer. Math. Soc.* **2**, no. 2, 215-224 (1989).
8. T. H. Chan, *Int. J. Number Theory* **2**, no.1, 105-110 (2006).
9. J. Chen, *Sci. Sinica* **16**, 157-176 (1973).
10. J. B. Conrey, *J. Reine Angew. Math.* **399**, 1-26 (1989).
11. J. B. Conrey, *Notices Amer. Math. Soc.* **50**, no.3, 341-353 (2003).
12. J. B. Conrey, L-functions and random matrices, in *Mathematics unlimited - 2001 and beyond*, Springer-Verlag, Berlin, 2001, pp. 331-352.
13. J. B. Conrey, Notes on L-functions and random matrix theory, in *Frontiers in Number Theory, Physics and Geometry I*, Springer-Verlag, Berlin, 2006, pp. 107-162.
14. H. Cramér, *Ark. Mat. Astronom. Fys.* **15**, 1-32 (1920).
15. H. Cramér, *Acta Arith.* **2**, 23-46 (1936).
16. H. Davenport, *Multiplicative number theory*, 3rd edn., revised and with a preface by H. L. Montgomery, Springer-Verlag, New York, 2000.
17. L. E. Dickson, *History of the theory of numbers* (3 vol.s), 1923, reprinted, Amer. Math. Soc., Providence, 2002.
18. H. M. Edwards, *Riemann's zeta function*, 1974, reprinted, Dover, New York, 2001.

19. P. D. T. A. Elliott and H. Halberstam, *Symposia Mathematica* **4** (INDAM, Rome, 1968/69), Academic Press, London, 1970, pp. 59-72.
20. P. Erdös, *Proc. Nat. Acad. Sci.* **35**, 374-384 (1949).
21. E. Fouvry and H. Iwaniec, *Acta Arith.* **79**, no. 3, 249-287 (1997).
22. J. B. Friedlander, On the Brun-Titchmarsh theorem, in *Number Theory, Trace Formulas and Discrete Groups* (Selberg Symposium, Oslo, 1987) (eds. Aubert, Bombieri and Goldfeld), Academic Press, Boston, 1989, pp 219-228.
23. J. B. Friedlander, Irregularities in the distribution of primes, in *Advances in number theory* (Kingston, ON, 1991), (eds. Gouvêa and Yui) Oxford Univ. Press, New York, 1993, pp. 17-30.
24. J. B. Friedlander, Topics in analytic number theory, in *Number Theory and its Applications* (Ankara, 1996), (eds. Yıldırım and Stepanov), Lecture Notes in Pure and Applied Math. **204**, Marcel Dekker, New York, 1999, pp. 47-64.
25. J. B. Friedlander and D. A. Goldston, *Quart. J. Math. Oxford (2)* **47**, 313-336 (1995).
26. J. B. Friedlander and D. A. Goldston, Note on a variance in the distribution of primes, in *Number theory in progress, Vol. 2* (Zakopane, 1997), (eds. Györy, Iwaniec and Urbanowicz), de Gruyter, Berlin, 1999, pp. 841-848.
27. J. B. Friedlander and A. Granville, *Ann. of Math.(2)* **129**, no. 2, 363-382 (1989).
28. J. B. Friedlander and H. Iwaniec, The Brun-Titchmarsh theorem, in *Analytic number theory (Kyoto 1996)*, London Math. Soc. Lect. Note Ser. **247**, Cambridge Univ. Press, 1997, pp. 85-93.
29. J. B. Friedlander and H. Iwaniec, *Ann. of Math. (2)* **148**, no. 3, 945-1040 (1998).
30. P. X. Gallagher, *Mathematika* **23**, 4-9 (1976).
31. P. X. Gallagher, *Acta Arithmetica* **37**, 339-343 (1980).
32. D. Goldfeld, `http://www.math.columbia.edu/~goldfeld/ErdosSelbergDispute.pdf`
33. D. A. Goldston, `http://aimath.org/preprints.html`, Preprint 2004-28.
34. D. A. Goldston, *J. Number Theory* **27**, no. 2, 149-177 (1987).
35. D. A. Goldston, *Expo. Math.* **13**, 366-376 (1995).
36. D. A. Goldston, S. W. Graham, J. Pintz and C. Y. Yıldırım, `http://arxiv.org/PS_cache/arxiv/pdf/0609/0609615v1.pdf`
37. D. A. Goldston, S. W. Graham, J. Pintz and C. Y. Yıldırım, `http://arxiv.org/PS_cache/arxiv/pdf/0803/0803.2636v1.pdf`
38. D. A. Goldston and D. R. Heath-Brown, *Math. Ann.* **266**, 317-320 (1984).
39. D. A. Goldston and H. L. Montgomery, Pair correlation of zeros and primes in short intervals, in *Analytic Number Theory and Diophantine Problems, Proc. of a conference at Oklahoma State Univ. 1984*, (eds. Adolphson, Conrey, Ghosh and Yager), Birkhäuser Boston 1987, pp. 183-203.
40. D. A. Goldston, J. Pintz and C. Y. Yıldırım, to appear in *Ann. of Math.* `http://arxiv.org/PS_cache/arxiv/pdf/0508/0508185v1.pdf`
41. D. A. Goldston, J. Pintz and C. Y. Yıldırım, `http://arxiv.org/PS_cache/arxiv/pdf/0710/0710.2728v1.pdf`
42. D. A. Goldston, J. Pintz and C. Y. Yıldırım, *Funct. Approx. Comment.*

Math. **35**, 79-89 (2006).
43. D. A. Goldston, J. Pintz and C. Y. Yıldırım, The path to recent progress on small gaps between primes, in *Analytic Number Theory, Clay Math. Proc.* **7**, Amer. Math. Soc., Providence, 2007, pp. 129-139.
44. D. A. Goldston and R. C. Vaughan, On the Montgomery-Hooley asymptotic formula, in *Sieve methods, exponential sums, and their applications to number theory (Cardiff, 1995)* London Math. Soc. Lecture Note Ser. **237**, Cambridge Univ. Press, 1997, pp. 117-142.
45. S. M. Gonek, An explicit formula of Landau and its applications to the theory of the zeta-function, in *A tribute to Emil Grosswald: Number theory and related analysis* (eds. Knopp and Sheingorn) Contemporary Math. **143**, Amer. Math. Soc., Providence, 1993, pp. 395-413.
46. X. Gourdon, (2004), `http://numbers.computation.free.fr/Constants/Miscellaneous/zetazeros1e13-1e24.pdf`
47. A. Granville, *Scand. Actuarial J.*, no.1, 12-28 (1995).
48. A. Granville, Unexpected irregularities in the distribution of prime numbers, in *Proc. International Congress of Mathematicians, Zurich 1994*, Birkäuser, Basel, 1995, pp. 388-399.
49. G. Greaves, Sieve methods, in *Number Theory and its Applications* (Ankara, 1996), (eds. Yıldırım and Stepanov), Lecture Notes in Pure and Applied Math. **204**, Marcel Dekker, NY, 1999, 65-107.
50. G. Greaves, *Sieves in Number Theory*, Springer-Verlag, Berlin, 2001.
51. B. Green and T. Tao, *Ann. of Math. (2)* **167**, no. 2, 481-547 (2008).
52. E. Grosswald and F. J. Schnitzer, *Pacific J. Math.* **74**, no.2, 357-364 (1978).
53. H. Halberstam and H.-E. Richert, *Sieve Methods*, Academic Press, London, 1974.
54. G. H. Hardy, *Matematisk Tidsskrift B*, 1-16 (1922).
55. G. H. Hardy and J. E. Littlewood, *Acta Mathematica* **44**, 1-70 (1922).
56. G. H. Hardy and E. M. Wright, *An introduction to the theory of numbers*, 5th ed., Oxford Univ. Press, 1985.
57. D. R. Heath-Brown, *Acta Arithmetica* **XLI**, 85-99 (1982).
58. D. R. Heath-Brown, *Proc. London Math. Soc. (3)* **47**(2), 193-224 (1983).
59. D. R. Heath-Brown, *J. Reine Angew. Math.* **389**, 22-63 (1988).
60. D. R. Heath-Brown, *Proc. London Math. Soc. (3)* **64**(2), 265-338 (1992).
61. D. R. Heath-Brown, *Acta Math.* **186** no. 1, 1-84 (2001).
62. D. Hensley and I. Richards, On the incompatibility of two conjectures concerning primes, in *Proc. Symp. Pure Math.* **24**, Amer. Math. Soc., Providence, 1973, pp. 181-193.
63. C. Hooley, *J. London Math. Soc. (2)* **11**, 399-407 (1975).
64. C. Hooley, *J. London Math. Soc. (2)* **13**, 57-64 (1976).
65. C. Hooley, *J. Reine Angew. Math.* **499**, 1-46 (1998).
66. A. E. Ingham, *The distribution of prime numbers*, 1932, reprinted with a foreword by R. C. Vaughan, Cambridge Univ. Press, 1992.
67. A. E. Ingham, *Amer. J. Math.* **64**, 313-319 (1942).
68. A. E. Ingham, (1949), `http://www.ams.org/mathscinet/pdf/29411.pdf`
69. A. Ivic, *The Riemann zeta-function*, 1985, reprinted, Dover, New York, 2003.

70. H. Iwaniec, *Acta Arith.* **24**, 435-459, (1973/74).
71. H. Iwaniec, *Invent. Math.* **47** no. , 171-188, (1973/74).
72. H. Iwaniec, Prime numbers and *L*-functions, in *Proc. International Congress of Mathematicians, Madrid, 2006*, Eur. Math. Soc., Zurich, 2007, pp. 279-306.
73. W. B. Jurkat, The Mertens conjecture and related general Ω-theorems, in *Proc. Symp. Pure Math.* **24** (Amer. Math. Soc., Providence, 1973), pp. 147-158.
74. A. A. Karatsuba and S. M. Voronin, *The Riemann zeta-function* (translated from the Russian by N. Koblitz), de Gruyter, Berlin, 1992.
75. D. Knuth, *Notices Amer. Math. Soc.* **49**, no. 3, 318-324 (2002).
76. B. Kra, *Bull. Amer. Math. Soc. (N.S.)* **43**, no. 1, 3-23 (2006).
77. H. Maier, *Adv. in Math.* **39**, no. 3, 257-269 (1981).
78. H. Maier, *Michigan Math. J.* **32**, no. 2, 221-225 (1985).
79. H. Maier, *Michigan Math. J.* **35**, (1988), no. 3, 323-344.
80. H. Maier and C. Pomerance, *Trans. Amer. Math. Soc.* **322**, no. 1, 201-237 (1990).
81. M. L. Mehta, *Random Matrices*, 3rd ed., Elsevier/Academic Press, 2004.
82. H. L. Montgomery, *Topics in multiplicative number theory*, Lecture Notes in Math. **227**, Springer, Berlin, 1971.
83. H. L. Montgomery, The pair correlation of zeros of the zeta function, in *Proc. Symp. Pure Math.* **24**, Amer. Math. Soc., Providence, 1973, pp. 181-193.
84. H. L. Montgomery, The zeta function and prime numbers in *Proc. Queen's Number Theory Conference, 1979*, Queen's Papers in Pure and Appl. Math. **54**, Queen's University, Kingston, Ont., 1980, pp. 1-31.
85. H. L. Montgomery and K. Soundararajan, *Commun. Math. Phys.* **252**, 589-617 (2004).
86. H. L. Montgomery and R. C. Vaughan, *J. London Math. Soc. (2)* **8**, 73-82, (1974).
87. H. L. Montgomery and R. C. Vaughan, *Multiplicative Number Theory I. Classical Theory*, Cambridge University Press, 2007.
88. M. R. Murty, *J. Madras Univ.*, 161–169 (1988); or `http://www.math.mast.queensu.ca/~murty/index2.html`
89. M. B. Nathanson, *Additive Number Theory, The Classical Bases*, Springer-Verlag, New York, 1996.
90. A. M. Odlyzko, `http://www.dtc.umn.edu/~odlyzko/`
91. A. M. Odlyzko and H. J. J. te Riele, *J. Reine Angew. Math.* **357**, 138-160 (1985).
92. J. Pintz, On the remainder term of the prime number formula and the zeros of Riemann's zeta-function, in *Lec. Notes in Math.* **1068** (ed. H. Jager), Springer-Verlag, Berlin, 1984, pp. 186-197.
93. J. Pintz, *Acta Math. Hungar.* **58** no. 3-4, 383-387 (1991).
94. J. Pintz, Approximations to the Goldbach and twin prime problem and gaps between consecutive primes, in *Adv. Studies in Pure Math.* **43**, Int. Conference on Probability and Number Theory, Kanazawa, 2005, (2006), pp. 1-40.

95. J. Pintz, *Functiones et Approximatio* **XXXVII.2**, 361-376 (2007).
96. O. Ramaré, *Ann. Scuola Norm. Sup. Pisa Cl. Sci.* **22**, 645-706, (1995).
97. R. A. Rankin, *J. London Math. Soc.* **13**, 242-247 (1938).
98. B. Riemann, *Monatsberichte d. Preuss. Akad. d. Wissens.*, 671–680 (1859).
99. M. Rubinstein and P. Sarnak, *Experimental Math.* **3**, 173-197 (1994).
100. Z. Rudnick and P. Sarnak, *C.R. Acad. Sci. Paris* **319**, Série I, 1027-1032 (1994).
101. P. Sarnak, (2004), `http://www.claymath.org/library/04report_sarnak.pdf`
102. A. Selberg, *Arch. Math. Naturvid.* **47**, no. 6, 87-105 (1943).
103. A. Selberg, *Ann. of Math.* **50**, 305-313 (1949).
104. A. Selberg, *Ann. of Math.* **50**, 297-304 (1949).
105. D. K. L. Shiu, *J. London Math. Soc. (2)* **61**, no. 2, 359-373 (2000).
106. K. Soundararajan, The distribution of prime numbers, in *Equidistribution in number theory, an introduction*, NATO Sci. Ser. II. Math. Phys. Chem. **237**, Springer, Dordrecht, 2007, pp. 59-83.
107. K. Soundararajan, *Bull. Amer. Math. Soc. (N.S.)* **44**, no. 1, 1-18 (2007).
108. G. Tenenbaum and M. Mendes-France, *The prime numbers and their distribution* (translated from the French), Amer. Math. Soc, Providence, 2000.
109. T. K. Timberlake and J. M. Tucker, (2008), `http://arxiv.org/PS_cache/arxiv/pdf/0708/0708.2567v2.pdf`
110. E. C. Titchmarsh, *The theory of the Riemann zeta-function*, 2nd ed. revised by D.R. Heath-Brown, Oxford Univ. Press, 1986.
111. P. Turán, *Collected papers of Paul Turán, Vol. 3*, ed. P. Erdös Akademiai Kiado, Budapest, 1990.
112. R. C. Vaughan, *The Hardy-Littlewood method*, 2nd ed., Cambridge Univ. Press, 1997.
113. Y. Wang (editor), *The Goldbach Conjecture*, 2nd ed., World Scientific, Singapore, 2002.
114. A. Zaccagnini, *Acta Arith.* **84**, no. 3, 225-244, (1998).

Session 01

Analytic Function Spaces and Their Operators

SESSION EDITORS

R. Aulaskari	University of Joensuu, Joensuu, Finland
H. T. Kaptanoğlu	Bilkent University, Ankara, Turkey

A C^*-ALGEBRA OF FUNCTIONAL OPERATORS WITH SHIFTS HAVING A NONEMPTY SET OF PERIODIC POINTS

M. A. BASTOS

Departamento de Matemática, Instituto Superior Técnico
Av. Rovisco Pais, 1049-001 Lisboa, Portugal
E-mail: abastos@math.ist.utl.pt

C. A. FERNANDES

Departamento de Matemática, Faculdade de Ciências e Tecnologia
Universidade Nova de Lisboa
Quinta da Torre, 2825 Monte de Caparica, Portugal
E-mail: caf@fct.unl.pt

Y. I. KARLOVICH

Facultad de Ciencias, Universidad Autónoma del Estado de Morelos
Av. Universidad 1001, Col. Chamilpa, C. P. 62209
Cuernavaca, Morelos, México
E-mail: karlovich@buzon.uaem.mx

Let $\mathcal{A}$ be a C^*-algebra of functional operators on the space $L^2(\mathbb{T})$ generated by multiplication operators by piecewise slowly oscillating functions and by a group of unitary operators associated to an amenable group G of orientation-preserving homeomorphisms of the unit circle $\mathbb{T}$ onto itself such that all $g \in G \setminus \{e\}$ have piecewise continuous derivatives and the same nonempty set Λ of periodic points, and there is a point $t \in \Lambda$ with a finite G-orbit. An invertibility criterion for the operators $A \in \mathcal{A}$ is established.

Keywords: Representations of C^*-algebras; Amenable group; Local-trajectory method; Functional operator; Invertibility.

1. Introduction

Let $\mathcal{B}(L^2(\mathbb{T}))$ be the C^*-algebra of all bounded linear operators acting on the Lebesgue space $L^2(\mathbb{T})$ where $\mathbb{T}$ is the unit circle in $\mathbb{C}$ with the length measure and the usual anticlockwise orientation. The aim of this paper is to study the invertibility in the C^*-algebra of functional operators

$$\mathcal{A} := \mathrm{alg}\,(PSO(\mathbb{T}), U_G) \subset \mathcal{B}(L^2(\mathbb{T})) \tag{1}$$

generated by all multiplication operators by piecewise slowly oscillating ($PSO(\mathbb{T})$) functions and by a group $U_G = \{U_g : g \in G\}$ of unitary operators where $(U_g\varphi)(t) := |g'(t)|^{1/2}\varphi(g(t))$ for $t \in \mathbb{T}$, and G is an amenable discrete group (see, e.g., [2,5,7]) of orientation-preserving homeomorphisms of $\mathbb{T}$ onto itself such that all $g \in G\setminus\{e\}$ have piecewise continuous derivatives g' and the same nonempty set Λ of periodic points, and there is a point $t \in \Lambda$ with the finite G-orbit $G(t) := \{g(t) : g \in G\}$.

Functional operators with shifts and continuous or piecewise continuous coefficients are studied for a long time (see, e.g., [1,2,6,8]). The C^*-algebra $\mathcal{A} = \mathrm{alg}\,(PSO(\mathbb{T}), U_G)$ was recently investigated in [3,4] under the condition that all $g \in G\setminus\{e\}$ have the same (finite, resp. arbitrary) nonempty set Λ of fixed points. Using the local-trajectory method and its generalizations [3,5, 7] (also see [1,2]), we got in [3,4] invertibility criteria for the operators $A \in \mathcal{A}$. In the present paper, Λ is an arbitrary nonempty set of common periodic points for all $g \in G\setminus\{e\}$. Appearance of periodic points of multiplicity $n > 1$ leads to the necessity to combine local-trajectory methods and the scheme of studying functional operators with Carleman shifts exposed in [8].

The paper is organized as follows. In Section 2 we describe the maximal ideal space of the commutative algebra $PSO(\mathbb{T})$. In Section 3 we study the structure of the group of homeomorphisms G. In Section 4 we describe a reduction of functional operators with shifts having periodic points to matrix functional operators with shifts having fixed points (see [8]). Section 5 contains the main results of the paper. First we reduce the study of invertibility of functional operators $A \in \mathcal{A}$ to studying the invertibility of their restrictions A_{arc}, A° and A_* on the spaces $L^2(\mathbb{T}^*_{arc})$, $L^2(\Lambda^\circ)$ and $L^2(\Lambda_*)$, respectively, where $\mathbb{T}^*_{arc} := \mathbb{T} \setminus \overline{\Lambda^\circ}$, Λ° is the interior of Λ, and $\Lambda_* = \mathbb{T} \setminus (\mathbb{T}^*_{arc} \cup \Lambda^\circ)$. Then we establish invertibility criteria for A_{arc} and A°, by applying the local-trajectory method and the scheme in Section 4, respectively, and show that the invertibility of A_{arc} and A° implies the invertibility of A_*. As a result, we obtain the invertibility criterion for $A \in \mathcal{A}$.

2. Preliminaries

Let $C(\mathbb{T})$ and $PC(\mathbb{T})$ denote the C^*-subalgebras of $L^\infty(\mathbb{T})$ consisting, respectively, of the functions continuous on $\mathbb{T}$ and the functions having finite one-sided limits at each point of $\mathbb{T}$.

A function $f \in L^\infty(\mathbb{T})$ is called *slowly oscillating at a point* $\lambda \in \mathbb{T}$ if

$$\lim_{\varepsilon \to 0} \operatorname{ess\,sup} \Big\{ |f(z_1) - f(z_2)| : z_1, z_2 \in \mathbb{T}_\varepsilon(\lambda) \Big\} = 0,$$

where $\mathbb{T}_\varepsilon(\lambda) := \big\{ z \in \mathbb{T} : \varepsilon/2 \le |z - \lambda| \le \varepsilon \big\}$. We denote by $SO(\mathbb{T})$ the C^*-subalgebra of $L^\infty(\mathbb{T})$ consisting of the functions slowly oscillating at each point of $\mathbb{T}$. Let $PSO(\mathbb{T}) := \operatorname{alg}(SO(\mathbb{T}), PC(\mathbb{T}))$ be the C^*-subalgebra of $L^\infty(\mathbb{T})$ generated by the C^*-algebras $PC(\mathbb{T})$ and $SO(\mathbb{T})$.

Let $M(\mathcal{A})$ be the maximal ideal space of a commutative unital C^*-algebra $\mathcal{A}$. We have (cf. [3]):

$$M(SO(\mathbb{T})) = \bigcup_{t \in \mathbb{T}} M_t(SO(\mathbb{T})), \quad M(PSO(\mathbb{T})) = \bigcup_{\xi \in M(SO(\mathbb{T}))} M_\xi(PSO(\mathbb{T})),$$

where the fibers $M_t(SO(\mathbb{T}))$ for $t \in \mathbb{T}$ are given by

$$M_t(SO(\mathbb{T})) = \big\{ \xi \in M(SO(\mathbb{T})) : \xi|_{C(\mathbb{T})} = t \big\},$$

and, if $\xi \in M_t(SO(\mathbb{T}))$ with $t \in \mathbb{T}$, then $M_\xi(PSO(\mathbb{T})) = \{(\xi, 0), (\xi, 1)\}$, where for $\mu \in \{0, 1\}$, $(\xi, \mu)|_{SO(\mathbb{T})} = \xi$, $(\xi, \mu)|_{C(\mathbb{T})} = t$, $(\xi, \mu)|_{PC(\mathbb{T})} = (t, \mu)$.

The Gelfand topology on $M(PSO(\mathbb{T})) = M(SO(\mathbb{T})) \times \{0, 1\}$ can be described as follows. If $\xi \in M_t(SO(\mathbb{T}))$ $(t \in \mathbb{T})$, a base of neighborhoods for $(\xi, \mu) \in M(PSO(\mathbb{T}))$ consists of all open sets of the form

$$U_{(\xi,\mu)} = \begin{cases} \big(U_{\xi,t} \times \{0\}\big) \cup \big(U^-_{\xi,t} \times \{0,1\}\big) & \text{if } \mu = 0, \\ \big(U_{\xi,t} \times \{1\}\big) \cup \big(U^+_{\xi,t} \times \{0,1\}\big) & \text{if } \mu = 1, \end{cases}$$

where $U_{\xi,t} = U_\xi \cap M_t(SO(\mathbb{T}))$, U_ξ is an open neighborhood of ξ in $M(SO(\mathbb{T}))$, and $U^-_{\xi,t}$, $U^+_{\xi,t}$ consist of all $\zeta \in U_\xi$ such that $\tau = \zeta|_{C(\mathbb{T})}$ belong, respectively, to the sets $(-t, t) := \{z \in \mathbb{T} : -\pi < \arg(z/t) < 0\}$ and $(t, -t) := \{z \in \mathbb{T} : 0 < \arg(z/t) < \pi\}$.

3. Structure of the group of homeomorphisms G

Let G be a group of orientation-preserving homeomorphisms of $\mathbb{T}$ onto itself such that all $g \in G \setminus \{e\}$ have the same nonempty set Λ of periodic points. The group operation in G is given by $(gh)(t) = h(g(t))$ for $g, h \in G$, $t \in \mathbb{T}$.

Lemma 3.1. *If there is $t_0 \in \Lambda$ with finite G-orbit then, for all $t \in \Lambda$, the cardinalities* $\operatorname{card} G(t)$ *of the G-orbits $G(t) = \{g(t) : g \in G\}$ coincide.*

Proof. Let $n := \operatorname{card} G(t_0) < \infty$ for a point $t_0 \in \Lambda$. If $n = 1$, then t_0 and hence [8] all $t \in \Lambda$ are fixed points for each $g \in G$, that is, $\operatorname{card} G(t) = 1$ for all $t \in \Lambda$. Let $n > 1$ and let $t_0 \prec t_1 \prec t_2 \prec \ldots \prec t_{n-1} (\prec t_0)$ be the points in $G(t_0)$ enumerated anticlockwise. Fix $\alpha \in G$ such that $\alpha(t_0) = t_1$. Then

$$\alpha^k(t_0) = t_k \quad \text{for} \quad k = 0, 1, \ldots, n \quad \text{where} \quad t_n := t_0. \tag{2}$$

Indeed, if $\alpha^2(t_0) \neq t_2$, then $\alpha^2(t_0) \in \{t_3, t_4, \ldots, t_{n-1}, t_0\}$. In that case, since α preserves the orientation of $\mathbb{T}$ and $\alpha(t_0) = t_1 \prec t_2 \prec \alpha^2(t_0) = \alpha(t_1)$, we conclude that $\alpha^{-1}(t_2)$ belongs to the open arc (t_0, t_1). But there are no points of $G(t_0)$ between t_0 and t_1. Thus, we arrive at a contradiction, and hence $\alpha^2(t_0) = t_2$. Analogously we obtain all the other relations in (2). Consequently, t_0 is a periodic point of α of multiplicity $n = \operatorname{card} G(t_0)$, and then every point $t \in \Lambda$ is a periodic point of α of the same multiplicity (see Chapter 1 in [8]): for all $t \in \Lambda$, $\alpha^k(t) \neq t$ $(k = 1, 2, \ldots, n-1)$ and $\alpha^n(t) = t$. Fix $g \in G \setminus \{e\}$. Clearly $g(t_0) = t_j$ for some $j = 0, 1, \ldots, n-1$. Then we put

$$\widehat{g} := \alpha^j g^{-1} (\in G). \tag{3}$$

By (2) and (3), we obtain $\widehat{g}(t_0) = g^{-1}(\alpha^j(t_0)) = g^{-1}(t_j) = t_0$. Hence, $\widehat{g}(t) = t$ for all $t \in \Lambda$ (see again [8]), and from (3) it follows that $g(t) = \alpha^j(t)$ for all $t \in \Lambda$. Thus, for every $g \in G$ there is a number $k(g) \in \{0, 1, \ldots, n-1\}$ such that $g(t) = \alpha^{k(g)}(t)$ for all $t \in \Lambda$, which implies that the G-orbit of each point $t \in \Lambda$ is given by $G(t) = \{\alpha^j(t) : j = 0, 1, \ldots, n-1\}$. □

Setting $\widetilde{g} := \widehat{g}^{-1}$ in (3), we get by the proof of Lemma 3.1 the following.

Proposition 3.1. *If G is the group of orientation-preserving homeomorphism of $\mathbb{T}$ onto itself such that all $g \in G \setminus \{e\}$ have the same nonempty set Λ of periodic points and* $\operatorname{card} G(t_0) < \infty$ *for some $t_0 \in \Lambda$, then there exists a shift $\alpha \in G$ for which all points $t \in \Lambda$ have the multiplicity* $n = \operatorname{card} G(t_0)$, *and for any shift $g \in G$ there are a number $k(g) \in \{0, 1, \ldots, n-1\}$ and a shift $\widetilde{g} \in G$ with the set Λ of fixed points such that $g = \widetilde{g}\,\alpha^{k(g)}$.*

For $j = 0, 1, \ldots, n-1$, we consider the sets

$$G_j := \{g \in G : k(g) = j\} \subset G. \tag{4}$$

It is easily seen that G_0 is a normal subgroup of G. Further, $g = \widetilde{g}\,\alpha^{k(g)}$ and (4) imply that $G_j = G_0 \alpha^j$ for $j = 0, 1, \ldots, n-1$, and hence the sets G_j are exactly the cosets of the quotient group G/G_0.

4. Reduction to the case of fixed points

Consider the following reduction procedure (see Chapter 3 in [8]). Let $\widehat{\mathcal{A}}$ be the dense subalgebra of $\mathcal{A}$ consisting of all operators $\sum_{g \in F} a_g U_g$ with finite sets $F \subset G$. By (4) and $g = \widetilde{g}\,\alpha^{k(g)}$, any $A \in \widehat{\mathcal{A}}$ is represented in the form

$$A = \sum_{g\in F} a_g U_g = \sum_{k=0}^{n-1} B_k U_\alpha^k, \quad \text{with} \quad B_k = \sum_{g\in G_k\cap F} a_g U_{\widetilde{g}}, \tag{5}$$

where $\widetilde{g}, \alpha^n \in G_0$. By direct computation, we get the following (cf. [8]).

Lemma 4.1. *If $B_0, B_1, \dots, B_{n-1}$ are given by (5) and $\varepsilon = e^{2\pi i/n}$, then*

$$C^{-1} B_A C = D_A := \operatorname{diag}(A_0, A_1, \dots, A_{n-1}), \tag{6}$$

$$C := \left[\varepsilon^{kj} U_\alpha^k\right]_{k,j=0}^{n-1}, \quad C^{-1} := n^{-1}\left[\varepsilon^{-kj} U_\alpha^{-j}\right]_{k,j=0}^{n-1},$$

$$A_j := \sum_{k=0}^{n-1} \varepsilon^{kj} B_k U_\alpha^k \quad (j = 0, 1, \dots, n-1), \quad B_A := \left[H_{k,j}\right]_{k,j=0}^{n-1}, \tag{7}$$

$$H_{k,j} := U_\alpha^k B_{j-k} U_\alpha^{-k} \quad (k \le j), \quad H_{k,j} := U_\alpha^k B_{n+j-k} U_\alpha^{n-k} \quad (k > j). \tag{8}$$

Fix $t_0 \in \Lambda$ and suppose that the points $t_s := \alpha^s(t_0)$ for $s = 0, 1, \dots, n-1$ are ordered anticlockwise: $t_0 \prec t_1 \prec t_2 \prec \dots \prec t_{n-1} \prec t_n := t_0$. Define the invertible piecewise constant function b on $\mathbb{T}$ by $b(t) = \varepsilon^{-s}$ if $t \in (t_s, t_{s+1})$ and $s = 0, 1, \dots, n-1$. Since $b = b \circ \widetilde{g}$ for all $\widetilde{g} \in G_0$, we infer that $b^j(t) b^{-j}[(\widetilde{g}\alpha^k)(t)] = \varepsilon^{jk}$ for all $t \in \mathbb{T}$ and all $j, k = 0, 1, \dots, n-1$. Hence, $A_j = b^j A b^{-j} I$, these operators are invertible only simultaneously, and

$$D_A = D \operatorname{diag}(A, \dots, A) D^{-1}, \quad \text{with} \quad D := \operatorname{diag}(I, bI, \dots, b^{n-1} I). \tag{9}$$

In view of (6) and (9), the map defined on $\widehat{\mathcal{A}}$ by

$$\Theta : A = \sum_{g\in F} a_g U_g \mapsto B_A = CD \operatorname{diag}(A, \dots, A) D^{-1} C^{-1} \tag{10}$$

extends by continuity to a C^*-algebra homomorphism $\Theta : \mathcal{A} \to \mathcal{B}(L_n^2(\mathbb{T}))$, where $L_n^2(\mathbb{T})$ is the Hilbert space of vector functions $\widetilde{u} = [u_k]_{k=1}^n$ with $u_k \in L^2(\mathbb{T})$. Applying (10), we immediately obtain the following criterion.

Theorem 4.1. *Any operator $A \in \mathcal{A}$ with shifts $g \in G$ is invertible on the space $L^2(\mathbb{T})$ if and only if the operator $B_A := \Theta(A)$ with shifts $g \in G_0$ is invertible on the space $L_n^2(\mathbb{T})$.*

5. Invertibility in the C^*-algebra $\mathcal{A}$

5.1. *Decomposition*

Using a G-invariant decomposition of $\mathbb{T}$, we will decompose the C^*-algebra $\mathcal{A}$ defined by (1) into an orthogonal sum of three operator C^*-algebras $\mathcal{A}_{arc}$, $\mathcal{A}^\circ$ and $\mathcal{A}_*$, and study separately the invertibility in these new algebras.

Let $\Lambda^\circ := \operatorname{Int}\Lambda$ be the interior of Λ, $\partial\Lambda$ the boundary of Λ, and let $\Lambda_\pm$ be the set of all $t \in \partial\Lambda$ that are limit points of the sets $\gamma_t^\pm \cap \Lambda^\circ$, respectively, where γ_t^+ (γ_t^-) is a right (left) semi-neighborhood of t on $\mathbb{T}$. Then the closure $\overline{\Lambda^\circ}$ of Λ° is given by $\overline{\Lambda^\circ} = \Lambda^\circ \cup \Lambda_+ \cup \Lambda_-$, and $\Lambda_r^\circ \subset \Lambda_-$, $\Lambda_l^\circ \subset \Lambda_+$, where Λ_l° and Λ_r° denotes, respectively, the at most countable set of the initial and final points of all open arcs which compose the set Λ°.

Let χ_{arc}, χ° and χ_* be the characteristic functions of the sets $\mathbb{T}_{arc}^* := \mathbb{T} \setminus \overline{\Lambda^\circ}$, Λ° and $\Lambda_* := \Lambda_+ \cup \Lambda_-$, respectively. Since $\mathbb{T}_{arc}^* \cup \Lambda^\circ \cup \Lambda_*$ is a G-invariant partition of $\mathbb{T}$, we obtain the following decomposition result.

Lemma 5.1. *An operator $A \in \mathcal{A}$ is invertible on $L^2(\mathbb{T})$ if and only if:*

(i) *the operator $A_{arc} := \chi_{arc} A$ is invertible on $L^2(\mathbb{T}_{arc}^*)$,*

(ii) *the operator $A^\circ := \chi^\circ A$ is invertible on $L^2(\Lambda^\circ)$,*

(iii) *the operator $A_* := \chi_* A$ is invertible on $L^2(\Lambda_*)$ if $\operatorname{mes}\Lambda_* > 0$.*

For every $A \in \mathcal{A}$, the operators A_{arc}, A° and A_* belong to the C^*-algebras $\mathcal{A}_{arc} := \chi_{arc}\mathcal{A} \subset \mathcal{B}(L^2(\mathbb{T}_{arc}^*))$, $\mathcal{A}^\circ := \chi^\circ\mathcal{A} \subset \mathcal{B}(L^2(\Lambda^\circ))$ and $\mathcal{A}_* := \chi_*\mathcal{A} \subset \mathcal{B}(L^2(\Lambda_*))$, respectively.

5.2. *Invertibility in the C^*-algebra $\mathcal{A}_{arc}$*

Consider $\mathcal{A}_{arc}$ as the C^*-algebra $\operatorname{alg}(\chi_{arc}PSO(\mathbb{T}), \chi_{arc}U_G) \subset \mathcal{B}(L^2(\mathbb{T}_{arc}^*))$. Since the set $\mathbb{T}_{arc}^*$ is open, the commutative C^*-algebra $\mathcal{Z}_{arc} := \{\chi_{arc}aI : a \in PSO(\mathbb{T})\}$ is isometrically isomorphic to the C^*-algebra $C(\widetilde{\mathcal{M}}_{arc}^*)$ of all continuous functions on the set $\widetilde{\mathcal{M}}_{arc}^*$ given by

$$\begin{aligned}\widetilde{\mathcal{M}}_{arc}^* := \overline{\mathcal{M}_{arc}^*} = &\Big(\bigcup\nolimits_{t\in\mathbb{T}_{arc}^*} M_t(SO(\mathbb{T})) \times \{0,1\}\Big)\\ &\cup\Big(\bigcup\nolimits_{t\in\Lambda_+\setminus\Lambda_-} M_t(SO(\mathbb{T}))\times\{0\}\Big) \cup \Big(\bigcup\nolimits_{t\in\Lambda_-\setminus\Lambda_+} M_t(SO(\mathbb{T}))\times\{1\}\Big)\\ &\cup\Big(\bigcup\nolimits_{t\in\Lambda_+\setminus\Lambda_l^\circ} M_{t,+}^{arc} \times \{1\}\Big) \cup \Big(\bigcup\nolimits_{t\in\Lambda_-\setminus\Lambda_r^\circ} M_{t,-}^{arc} \times \{0\}\Big),\end{aligned} \tag{11}$$

where $M_{t,+}^{arc}$ for $t \in \Lambda_+ \setminus \Lambda_l^\circ$ denotes the closed set of all $\xi \in M(SO(\mathbb{T}))$ which are limits of nets δ_{t_α} where $t_\alpha \to t$ and $t_\alpha \in \gamma_t^+ \cap \mathbb{T}_{arc}$; $M_{t,-}^{arc}$ for $t \in \Lambda_- \setminus \Lambda_r^\circ$ is the closed set of all $\xi \in M(SO(\mathbb{T}))$ which are limits of nets δ_{t_α} where $t_\alpha \to t$ and $t_\alpha \in \gamma_t^- \cap \mathbb{T}_{arc}$ (see [4]). For every $g \in G$, the *-automorphism $\alpha_g : a \mapsto U_g a U_g^{-1}$ of $\mathcal{Z}_{arc}$ induces the homeomorphisms $\beta_g : (\xi, \mu) \mapsto (g(\xi), \mu)$ on $M(PSO(\mathbb{T}))$ where $\xi \mapsto g(\xi)$ are homeomorphisms on $M(SO(\mathbb{T}))$ given by $a(g(\xi)) = (a \circ g)(\xi)$ for all $a \in SO(\mathbb{T})$ and all $\xi \in M(SO(\mathbb{T}))$ (as usual, $a(\xi) := \xi(a)$). Hence, in view of the topologically free action (see [1–4]) of the group G on $\mathbb{T} \setminus \Lambda^\circ$ and the group $\{\beta_g : g \in G\}$

on $\widetilde{\mathcal{M}}^*_{arc}$ because $g(\xi) \in M_{g(t)}(SO(\mathbb{T}))$ for $\xi \in M_t(SO(\mathbb{T}))$, we obtain an invertibility criterion for the operators in the C^*-algebra $\mathcal{A}_{arc}$ by analogy with [4], by applying the local-trajectory method and the techniques of spectral measures [3,5,7].

Let $\mathcal{O}_{arc}$ be a subset of $\mathbb{T}_{arc} = \mathbb{T} \setminus \Lambda$ containing exactly one point in each G-orbit defined by the action of the group G on $\mathbb{T}_{arc}$, and let

$$\mathfrak{R}_{arc} := \{(\xi, \mu) \in M(PSO(\mathbb{T})) : \ \xi \in M_\tau(SO(\mathbb{T})), \ \tau \in \mathcal{O}_{arc}, \ \mu = 0, 1\}.$$

With each ideal $(\xi, \mu) \in \widetilde{\mathcal{M}}^*_{arc}$ we associate the representation

$$\Pi_{(\xi,\mu)} : \mathcal{A} \to \mathcal{B}(l^2(G)), \ \ A \mapsto A_{(\xi,\mu)}, \tag{12}$$

given for the operators $A = \sum_{g\in F} a_g U_g \in \widehat{\mathcal{A}}$, with finite sets $F \subset G$, by

$$(A_{(\xi,\mu)} f)(h) = \sum\nolimits_{g\in F} [(a_g \circ h)(\xi, \mu)] f(hg) \quad (h, g \in G, \ f \in l^2(G)), \tag{13}$$

where $(a_g \circ h)(\xi, \mu)$ means the value of the Gelfand transform of the function $a_g \circ h \in PSO(\mathbb{T})$ at the point $(\xi, \mu) \in M(PSO(\mathbb{T}))$.

Theorem 5.1. *For each functional operator $A \in \mathcal{A}$, the operator A_{arc} is invertible on the space $L^2(\mathbb{T}^*_{arc})$ if and only if for all $(\xi, \mu) \in \mathfrak{R}_{arc}$ the operators $A_{(\xi,\mu)}$ given by* (12)–(13) *are invertible on the space $l^2(G)$ and*

$$\sup_{(\xi,\mu)\in\mathfrak{R}_{arc}} \left\| (A_{(\xi,\mu)})^{-1} \right\| < \infty. \tag{14}$$

5.3. *Invertibility in the C^*-algebra $\mathcal{A}^\circ$*

Consider $\mathcal{A}^\circ$ as the C^*-subalgebra $\operatorname{alg}(\chi^\circ PSO(\mathbb{T}), \chi^\circ U_G)$ of $\mathcal{B}(L^2(\Lambda^\circ))$. Let $g \in G$. By Proposition 3.1, $g = \widetilde{g}\alpha^{k(g)}$, where $\alpha \in G$, $\widetilde{g} \in G_0$ and $k(g) \in \{0, 1, \ldots, n-1\}$. Consequently, $\chi^\circ U_g = \chi^\circ U_{\widetilde{g}} U_{\alpha^{k(g)}} = \chi^\circ U_\alpha^{k(g)}$ for each $g = \widetilde{g}\alpha^{k(g)} \in G$. Then for every $A = \sum_{g\in F} a_g U_g \in \widehat{\mathcal{A}}$ we get

$$\chi^\circ A = \chi^\circ \sum_{g\in F} a_g U_g = \sum_{k=0}^{n-1} \chi^\circ b_k^A U_\alpha^k,$$

where the functions $b_k^A \in PSO(\mathbb{T})$ are given by

$$b_k^A = \sum_{g\in G_k \cap F} a_g \quad (k = 0, 1, \ldots, n-1). \tag{15}$$

According to (7)–(8), with every operator $A^\circ = \chi^\circ \sum_{g\in F} a_g U_g \in \widehat{\mathcal{A}}$ we associate the operator matrix $B_A^\circ := \chi^\circ B_A$ of the form

$$B_A^\circ = \begin{bmatrix} \chi^\circ b_0^A I & \chi^\circ b_1^A I & \chi^\circ b_2^A I & \dots & \chi^\circ b_{n-1}^A I \\ \chi^\circ (b_{n-1}^A \circ \alpha) I & \chi^\circ (b_0^A \circ \alpha) I & \chi^\circ (b_1^A \circ \alpha) I & \dots & \chi^\circ (b_{n-2}^A \circ \alpha) I \\ \chi^\circ (b_{n-2}^A \circ \alpha^2) I & \chi^\circ (b_{n-1}^A \circ \alpha^2) I & \chi^\circ (b_0^A \circ \alpha^2) I & \dots & \chi^\circ (b_{n-3}^A \circ \alpha^2) I \\ \vdots & \vdots & \vdots & \ddots & \vdots \\ \chi^\circ (b_1^A \circ \alpha^{n-1}) I & \chi^\circ (b_2^A \circ \alpha^{n-1}) I & \chi^\circ (b_3^A \circ \alpha^{n-1}) I & \dots & \chi^\circ (b_0^A \circ \alpha^{n-1}) I \end{bmatrix}, \tag{16}$$

where the functions b_k^A are given by (15).

For $A \in \widehat{\mathcal{A}}$, we deduce from (10) that $B_A^\circ := \chi^\circ \Theta(A) \in \mathcal{B}(L_n^2(\Lambda^\circ))$. Fix $A \in \mathcal{A}$. Hence, $B_A^\circ = \lim_{m\to\infty} B_{A_m}^\circ$ for every sequence $\{A_m\} \subset \widehat{\mathcal{A}}$ convergent to the operator $A \in \mathcal{A}$, which defines $B_A^\circ \in \mathcal{B}(L_n^2(\Lambda^\circ))$ for all $A \in \mathcal{A}$. Since $B_A^\circ := \chi^\circ B_A$ for all $A \in \mathcal{A}$, Theorem 4.1 immediately implies the following.

Theorem 5.2. *For $A \in \mathcal{A}$, the operator $A^\circ := \chi^\circ A$ is invertible on the space $L^2(\Lambda^\circ)$ if and only if the operator B_A° is invertible on the space $L_n^2(\Lambda^\circ)$.*

Consider the commutative C^*-subalgebra $\mathcal{Z}^\circ := \{\chi^\circ a I : a \in PSO(\mathbb{T})\}$ of $\mathcal{A}^\circ$. By [4], the maximal ideal space of $\mathcal{Z}^\circ$ is given by

$$\begin{aligned} \widetilde{\mathcal{M}}^\circ := \overline{\mathcal{M}^\circ} &= \Big(\bigcup\nolimits_{t\in\Lambda^\circ} M_t(SO(\mathbb{T})) \times \{0,1\}\Big) \\ &\cup \Big(\bigcup\nolimits_{t\in\Lambda_l^\circ} M_t(SO(\mathbb{T})) \times \{1\}\Big) \cup \Big(\bigcup\nolimits_{t\in\Lambda_r^\circ} M_t(SO(\mathbb{T})) \times \{0\}\Big) \\ &\cup \Big(\bigcup\nolimits_{t\in\Lambda_+\setminus\Lambda_l^\circ} M_{t,+}^\circ \times \{1\}\Big) \cup \Big(\bigcup\nolimits_{t\in\Lambda_-\setminus\Lambda_r^\circ} M_{t,-}^\circ \times \{0\}\Big), \end{aligned} \tag{17}$$

where $M_{t,+}^\circ$ for $t \in \Lambda_+ \setminus \Lambda_l^\circ$ denotes the closed set of all $\xi \in M(SO(\mathbb{T}))$ which are limits of nets δ_{t_α} where $t_\alpha \to t$ and $t_\alpha \in \gamma_t^+ \cap \Lambda^\circ$; $M_{t,-}^\circ$ for $t \in \Lambda_- \setminus \Lambda_r^\circ$ is the closed set of all $\xi \in M(SO(\mathbb{T}))$ which are limits of nets δ_{t_α} where $t_\alpha \to t$ and $t_\alpha \in \gamma_t^- \cap \Lambda^\circ$.

Since the C^*-algebras $\mathcal{Z}^\circ$ and $C(\widetilde{\mathcal{M}}^\circ)$ are isometrically isomorphic, we get the following invertibility criterion for the operators $A^\circ \in \mathcal{A}^\circ$.

Theorem 5.3. *For any $A \in \mathcal{A}$, the operator $A^\circ := \chi^\circ A$ has the form*

$$A^\circ = \sum_{k=0}^{n-1} \chi^\circ b_k^A U_\alpha^k \tag{18}$$

where $b_k^A \in PSO(\mathbb{T})$ for all $k = 0, 1, \dots, n-1$. The operator A° is invertible on the space $L^2(\Lambda^\circ)$ if and only if

$$\det B_A(\xi,\mu) \neq 0 \quad \text{for all} \quad (\xi,\mu) \in \widetilde{\mathcal{M}}^\circ, \tag{19}$$

where $B_A(\xi,\mu) :=$

$$\begin{bmatrix} b_0^A(\xi,\mu) & b_1^A(\xi,\mu) & b_2^A(\xi,\mu) & \dots & b_{n-1}^A(\xi,\mu) \\ (b_{n-1}^A\circ\alpha)(\xi,\mu) & (b_0^A\circ\alpha)(\xi,\mu) & (b_1^A\circ\alpha)(\xi,\mu) & \dots & (b_{n-2}^A\circ\alpha)(\xi,\mu) \\ (b_{n-2}^A\circ\alpha^2)(\xi,\mu) & (b_{n-1}^A\circ\alpha^2)(\xi,\mu) & (b_0^A\circ\alpha^2)(\xi,\mu) & \dots & (b_{n-3}^A\circ\alpha^2)(\xi,\mu) \\ \vdots & \vdots & \vdots & \ddots & \vdots \\ (b_1^A\circ\alpha^{n-1})(\xi,\mu) & (b_2^A\circ\alpha^{n-1})(\xi,\mu) & (b_3^A\circ\alpha^{n-1})(\xi,\mu) & \dots & (b_0^A\circ\alpha^{n-1})(\xi,\mu) \end{bmatrix}. \tag{20}$$

Proof. Fix $A\in\mathcal{A}$. Let $\{A_m\}$ be a sequence of operators in $\widehat{\mathcal{A}}$ such that $A=\lim\limits_{m\to\infty}A_m$. Then $\chi^\circ A=\lim\limits_{m\to\infty}\sum_{k=0}^{n-1}\chi^\circ b_k^{A_m}U_\alpha^k$ where, for $k=0,1,\dots,n-1$, $b_k^{A_m}\in PSO(\mathbb{T})$. By (10) and (16), there are constants $C_k\in(0,\infty)$ such that

$$\|b_k^{A_m}\|_{L^\infty(\Lambda^\circ)}\le\|B_{A_m}\|_{\mathcal{B}(L_n^2(\Lambda^\circ))}\le C_k\|A_m\|_{\mathcal{B}(L^2(\Lambda^\circ))}\quad(m\in\mathbb{N}).$$

Hence, the restrictions $\{\chi^\circ b_k^{A_m}\}$ can be considered as convergent sequences in $C(\widetilde{\mathcal{M}}^\circ)$. Since $\widetilde{\mathcal{M}}^\circ$ is a closed subset of $M(PSO(\mathbb{T}))$, there exist functions $b_0^A, b_1^A,\dots,b_{n-1}^A\in PSO(\mathbb{T})$ such that $\chi^\circ b_k^A=\lim\limits_{m\to\infty}\chi^\circ b_k^{A_m}$, which gives (18). Because $A\in\mathcal{A}$ is given by (18) with $b_k^A\in PSO(\mathbb{T})$, we conclude that $B_A^\circ:=\Theta(A^\circ)$ is given by (16) with entries identified with their Gelfand transforms in $C(\widetilde{\mathcal{M}}^\circ)$. Hence, the operator B_A° is invertible on the space $L_n^2(\Lambda^\circ)$ if and only if (19) holds. It remains to apply Theorem 5.2. □

5.4. *Invertibility in the C^*-algebra $\mathcal{A}_*$*

Let us study the invertibility of the operator $A_*:=\chi_*A$ on the space $L^2(\Lambda_*)$. Consider the C^*-algebra $\mathcal{A}_0:=\mathrm{alg}\,(PSO(\mathbb{T}),U_{G_0})\subset\mathcal{B}(L^2(\mathbb{T}))$. Let

$$\mathcal{M}_*:=\bigcup_{t\in\Lambda_*}M_t(SO(\mathbb{T}))\times\{0,1\}\subset M(PSO(\mathbb{T})).$$

To every $(\xi,\mu)\in\mathcal{M}_*$ we assign the representation $\widetilde{\Pi}_{\xi,\mu}:\mathcal{A}_0\to\mathcal{B}(l^2(G_0))$, $A\mapsto A_{\xi,\mu}$, given on the dense set of operators $A=\sum_{g\in F}a_gU_g\in\mathcal{A}_0$ with finite sets $F\subset G_0$ by

$$(A_{\xi,\mu}f)(h)=\sum_{g\in F}a_g(\xi,\mu)f(h)\quad(h,g\in G_0,\ f\in l^2(G_0)). \tag{21}$$

Since $\widetilde{\Pi}_{\xi,\mu}(\mathcal{A}_0)\subset\mathcal{B}(l^2(G_0))$ is a C^*-algebra, applying the homomorphism

$$\Pi:\ \widetilde{\Pi}_{\xi,\mu}(\mathcal{A}_0)\to\mathbb{C},\quad A_{\xi,\mu}=\sum_{g\in F}a_g(\xi,\mu)I\mapsto\sum_{g\in F}a_g(\xi,\mu),$$

we infer that for every $(\xi, \mu) \in \mathcal{M}_*$ the composition map

$$\Pi^*_{\xi,\mu} = \Pi \circ \widetilde{\Pi}_{\xi,\mu} : \sum_{g \in F} a_g U_g \mapsto \sum_{g \in F} a_g(\xi, \mu)$$

extends to a C^*-algebra homomorphism of $\mathcal{A}_0$ onto $\mathbb{C}$. Since $\widetilde{g}, \alpha^n \in G_0$, we deduce from (7)–(8) that each entry of B_A belongs to $\mathcal{A}_0$. Applying the map $\Pi^*_{\xi,\mu}$ to B_A entry-wise, we obtain the representations $\nu^*_{\xi,\mu} : B_A \mapsto B_A(\xi, \mu)$ for all $(\xi, \mu) \in \mathcal{M}_*$, where $B_A(\xi, \mu)$ for these (ξ, μ) is also defined by (20).

Since $B^*_A := \chi_* B_A = \Theta(\chi_* A)$ is given by (16) with χ° replaced by χ_*, and therefore B^*_A is the operator of multiplication by the restriction of a $PSO(\mathbb{T})$ matrix function to Λ_*, it follows that B^*_A is invertible on the space $L^2_n(\Lambda_*)$ if $\det B_A(\xi, \mu) \neq 0$ for all $(\xi, \mu) \in \mathcal{M}_*$. Thus, we get the following.

Lemma 5.2. *For any operator $A \in \mathcal{A}$, if $\det B_A(\xi, \mu) \neq 0$ for all $(\xi, \mu) \in \mathcal{M}_*$, then the operator $A_* := \chi_* A$ is invertible on the space $L^2(\Lambda_*)$.*

By [7], with every $(\xi, \mu) \in \widetilde{\mathcal{M}}^*_{arc} \cap \mathcal{M}_*$ we associate the representation

$$\Pi^{arc}_{\xi,\mu} : \chi_{arc} \mathcal{A}_0 \to \widetilde{\Pi}_{\xi,\mu}(\mathcal{A}_0) \ (\subset \mathcal{B}(l^2(G_0))), \quad A_{arc} \mapsto A_{\xi,\mu},$$

given for $A_{arc} = \chi_{arc} \sum_{g \in F} a_g U_g \in \mathcal{A}_0$ with finite sets $F \subset G_0$ by (21). Applying now $\Pi \circ \Pi^{arc}_{\xi,\mu}$ to $\chi_{arc} B_A$ entry-wise, we obtain the representation $\nu^{arc}_{\xi,\mu} : \chi_{arc} B_A \mapsto B_A(\xi, \mu)$ for every $(\xi, \mu) \in \widetilde{\mathcal{M}}^*_{arc} \cap \mathcal{M}_*$ along with $\nu^*_{\xi,\mu}$.

Theorem 5.4. *If $A \in \mathcal{A}$ and the operators A_{arc} and A° are invertible, then the operator A_* is also invertible.*

Proof. If the operator A_{arc} is invertible on the space $L^2(\mathbb{T}^*_{arc})$ then, by Theorem 4.1, the operator $\chi_{arc} B_A = \Theta(\chi_{arc} A)$ is invertible on the space $L^2_n(\mathbb{T}^*_{arc})$, and hence the operator $\nu^{arc}_{\xi,\mu}(\chi_{arc} B_A) I$ is also invertible. Therefore, $\det B_A(\xi, \mu) \neq 0$ for all $(\xi, \mu) \in \widetilde{\mathcal{M}}^*_{arc} \cap \mathcal{M}_*$. On the other hand, by (19), the invertibility of A° implies that $\det B_A(\xi, \mu) \neq 0$ for all $(\xi, \mu) \in \widetilde{\mathcal{M}}^\circ \cap \mathcal{M}_*$. But $(\widetilde{\mathcal{M}}^*_{arc} \cup \widetilde{\mathcal{M}}^\circ) \cap \mathcal{M}_* = \mathcal{M}_*$ due to (11) and (17). Hence $\det B_A(\xi, \mu) \neq 0$ for all $(\xi, \mu) \in \mathcal{M}_*$, and then, by Lemma 5.2, the operator A_* is invertible on the space $L^2(\Lambda_*)$. □

5.5. *Main result*

Finally, combining Lemma 5.1 and Theorems 5.1, 5.3 and 5.4, we get the following invertibility criterion in the C^*-algebra $\mathcal{A}$.

Theorem 5.5. *An operator $A \in \mathcal{A}$ is invertible on $L^2(\mathbb{T})$ if and only if*

(i) *for all* $(\xi, \mu) \in \widetilde{\mathcal{M}}^\circ$ *and* $B_A(\xi, \mu)$ *given by* (20), $\det B_A(\xi, \mu) \neq 0$;
(ii) *for all* $(\xi, \mu) \in \mathfrak{R}_{arc}$, *the operators* $A_{(\xi,\mu)}$ *given by* (13) *are invertible on the space* $l^2(G)$ *and* (14) *holds.*

References

1. A. B. Antonevich, *Linear Functional Equations. Operator Approach*, Oper. Theory Adv. Appl., vol. 83, Birkhäuser, Basel, 1996. Russian original: University Press, Minsk, 1988.
2. A. Antonevich, A. Lebedev, *Functional Differential Equations: I. C*-Theory*, Pitman Monographs and Surveys in Pure and Applied Mathematics, vol. 70, Longman Scientific & Technical, Harlow, 1994.
3. M. A. Bastos, C. A. Fernandes, and Y. I. Karlovich, *J. Funct. Anal.* **242**, 86–126 (2007).
4. M. A. Bastos, C. A. Fernandes, and Y. I. Karlovich, C^*-algebras of singular integral operators with shifts having the same nonempty set of fixed points, *Complex Anal. Oper. Theory* **2**, 241–272 (2008).
5. Y. I. Karlovich, *Soviet Math. Dokl.* **37**, 407–411 (1988).
6. Y. I. Karlovich, *Soviet Math. Dokl.* **38**, 301–307 (1989).
7. Y. I. Karlovich, A local-trajectory method and isomorphism theorems for nonlocal C^*-algebras, In: Oper. Theory Adv. Appl., vol. 170, Birkhäuser, Basel, 2006, pp. 137–166.
8. V. G. Kravchenko, G. S. Litvinchuk, *Introduction to the Theory of Singular Integral Operators with Shift*, Kluwer, Dordrecht, 1994.

ON BLOCH AND NORMAL FUNCTIONS ON COMPLEX BANACH MANIFOLDS

P. V. DOVBUSH

Institute of Mathematics and Computer Science of Academy of Sciences of Moldova,
5 Academy Street, Kishinev, MD-2028, Republic of Moldova
E-mail: peter.dovbush@gmail.com

Let X be a complex Banach manifold. A holomorphic function $f : X \to \mathbb{C}$ is called a Bloch function (resp., a normal function) if the family $\mathcal{F}_f = \{f \circ \varphi - f(\varphi(0)) : \varphi \in \mathcal{O}(\Delta, X)\}$ (resp., $\mathcal{F} = \{f \circ \varphi : \varphi \in \mathcal{O}(\Delta, X)\}$) forms a normal family in the sense of Montel, where $\mathcal{O}(\Delta, X)$ denotes the set of holomorphic maps from the complex unit disc to X. Characterizations of normal and Bloch functions are presented. A sufficient condition for the sum of a normal function and a nonnormal function to be nonnormal is given. Criteria for a holomorphic function to be non-normal are obtained.

Keywords: Kobayashi pseudometric, infinitesimal Kobayashi pseudometric, Bloch function, normal function, Banach space, complex Banach manifold.

1. Introduction

The concept of a Bloch function was introduced in 1952 by Hayman [1] and has been widely studied. (See e.g. [2], p. 269 et seq.) The idea of associating to a meromorphic function f on the complex unit disc $\Delta = \{z \in \mathbb{C} : |z| < 1\}$ a family $\mathcal{F} = \{f \circ g : g \in \text{Aut}(\Delta)\}$, where $\text{Aut}(\Delta)$ is the group of biholomorphic automorphisms of Δ, and of ascribing to the function f properties of the family $\mathcal{F}$ apparently arose in work of Yosida [3] in 1934 and was considered by Noshiro [4] in 1937. In 1957, Lehto and Virtanen [5] defined "normal functions" to be those meromorphic functions f whose associated families $\mathcal{F}$ were normal in the sense of Montel.

Since that time, the subject of normal functions and Bloch functions has been studied extensively, resulting in substantial development in the single complex variable context and in generalizations to the setting of several complex variables (see lists of references in [6–8]).

In this paper, we generalize the theory of normal holomorphic functions to the case of holomorphic functions on a complex Banach manifold.

2. Basic Definitions

We refer the reader to the paper [9] and the books [10,11] for background on complex analysis in infinite dimension.

Let X be a complex Banach manifold modelled on a complex Banach space of positive, possibly infinite, dimension; X assumed to be a connected Hausdorff space. For each x in X the tangent space to X at point x will be denoted by $T_x(X)$. The tangent bundle $T(X)$ of X consist of the ordered pairs (y, v) such that $x \in X$ and $v \in T_x(X)$. We shall denote the space of all holomorphic maps from the unit disk Δ into X by $\mathcal{O}(\Delta, X)$.

The infinitesimal Kobayashi pseudometric on the complex Banach manifold X is the function k_X on $T(X)$ defined by the formula

$$k_X(x, v) = \inf\{|a| : \exists\, \varphi \in \mathcal{O}(\Delta, X),\ \varphi(0) = x,\ \varphi_*(0)a = v\}$$

where $\varphi_*(0)$ is the linear map induced by φ from $T_0(\Delta)$ to $T_{\varphi(0)}(X)$.

We say that k_X is a metric if $k_X(x, v) > 0$ for all $(x, v) \in T_x(X)$, $v \neq 0$.

The Kobayashi length of a piecewise C^1 curve $\gamma : [0, 1] \to X$ in X is to be the upper Riemann integral

$$L_k(\gamma) = \overline{\int_0^1} k_X(\gamma(t), \gamma'(t))\, dt$$

and the pseudometric $\widetilde{K}_X(x, y)$ is the infimum of the lengths of all piecewise C^1 curves joining x to y in X.

The Kobayashi pseudometric between two points x, y in X is defined as follows. Consider all finite sequences of points $p_0 = x, p_1, \ldots, p_{k-1}, p_k = y$ of X such that there exist points $z_1, \ldots, z_k$, $z_1, \ldots, z_k$ of Δ and maps $\varphi_1, \ldots, \varphi_k \in \mathcal{O}(\Delta, X)$ satisfying $\varphi_j(z_j) = p_{j-1}$ and $\varphi_j(\widetilde{z}_j) = p_j$, $j = 1, \ldots, k$. The Kobayashi pseudometric $K_X(x, y)$ is, by definition,

$$K_X(x, y) = \inf \sum_{j=1}^{k} d_\Delta(z_j, \widetilde{z}_j) \text{ where } d_\Delta(z_j, \widetilde{z}_j) = \tanh^{-1}\left|\frac{z_j - \widetilde{z}_j}{1 - z_j \overline{\widetilde{z}}_j}\right|$$

is the Poincaré metric on Δ and the infimum is taken over all possible choices of points and maps.

The *spherical arc length element* ds on the Riemann sphere $\overline{\mathbb{C}}$ is given by

$$ds(z, dz) = \frac{|dz|}{1 + |z|^2}.$$

The spherical length

$$s(\gamma) = \int_\gamma ds(z, dz)$$

of a curve γ in $\overline{\mathbb{C}}$ induces a metric in the following manner. Given distinct points a, b on the Riemann sphere, define $s(a,b) = \inf_\gamma\{s(\gamma)\}$ where the infimum is taken over all piecewise C^1 curves on $\overline{\mathbb{C}}$ which join a with b. Then $s(a,b)$ defines a metric on the sphere known as the *spherical metric.*

Denote by $\mathcal{O}(X)$ the set of all holomorphic functions on X. A family $F \subseteq \mathcal{O}(\Delta)$ is said to be *normal* in Δ if every sequence $\{f_n\} \subset F$ has a subsequence which converges uniformly (with respect to the Euclidean metric) on compacta in Δ or diverges uniformly to ∞ on compacta in Δ.

Definition 2.1. A holomorphic function $f : X \to \mathbb{C}$ is called a Bloch function (resp., a normal function) if the family $\mathcal{F}_f = \{f \circ \varphi - f(\varphi(0)) : \varphi \in \mathcal{O}(\Delta, X)\}$, (resp., $\mathcal{F} = \{f \circ \varphi : \varphi : \Delta \to X$ is holomorphic $\}$) is normal.

A family $F \subseteq \mathcal{O}(\Delta)$ is said to be spherically equicontinuous at a point $z_0 \in \Delta$ if for each positive number ϵ there is a positive number δ such that $s(f(z), f(z_0)) < \epsilon$ for $K_\Delta(z, z_0) < \delta$ and f in F.

Following Gauthier [12] we shall define a sequence $\{x_n\}$ of points in X to be a P-sequence of $f \in \mathcal{O}(X)$ if there is a sequence $\{y_n\}$ of points in X such that $K_X(x_n, y_n) \to 0$ as $n \to \infty$ but $s(f(x_n), f(y_n)) \geq \epsilon$ for some $\epsilon > 0$ for each positive integer n.

3. Background Results

Lemma 3.1 (Zalcman's Lemma [13]). *Let $\mathcal{F}$ be a family of analytic functions in Δ. Then $\mathcal{F}$ is not normal in Δ if and only if there exist (i) a number r with $0 < r < 1$; (ii) points z_n satisfying $|z_n| < r$; (iii) functions $f_n \in \mathcal{F}$; (iv) positive numbers $\rho_n \to 0$ as $n \to \infty$; such that*

$$f_n(z_n + \rho_n \xi) \to g(\xi) \quad \text{as } n \to \infty, \tag{1}$$

uniformly on compact subsets of $\mathbb{C}$, where g is a nonconstant entire function in $\mathbb{C}$. The function g may be taken to satisfy the normalization $g^\sharp(z) < g^\sharp(0) = 1 \quad (z \in \mathbb{C})$.

Here $g^\sharp(z)$ denotes the spherical derivative

$$g^\sharp(z) = \frac{|g'(z)|}{1 + |g(z)|^2}.$$

Lemma 3.2. *Let f be a normal function on a complex Banach manifold X and suppose k_X is a metric. There exists a constant $c > 1$ such that*

$$\log(c\mu(f,x)) \leq \log(c\mu(f,y)) \exp^{2K_X(x,y)} \quad \text{for all } x, y \in X.$$

Here $\mu(f,x) := \max\{1, |f(x)|\}$.

Remark 3.1. This result was originally proved by Zaidenberg [8] in the case of complex manifolds. The proof given in [14], p. 39, extends immediately to the complex Banach manifold case.

4. Preliminary Properties of Normal Functions

Theorem 4.1. *Let X be a complex Banach manifold and suppose k_X is a metric. The following statements are equivalent for $f \in \mathcal{O}(X)$:*
(a) f is normal;
(b) there exists a constant $Q > 0$ such that

$$Q_f(x) := \sup_{v \in T_x(X)\setminus\{0\}} \frac{ds(f(x), f_*(x)v)}{k_X(x,v)} < Q \text{ for all } x \in X; \tag{2}$$

(c) there exists a constant $L > 0$ such that

$$s(f(x), f(y)) \leq L \cdot K_X(x,y) \quad \text{for all} \quad x, y \in X; \tag{3}$$

(d) there is no P-sequence $\{x_n\}$ of points in X possessed by f.

Proof. $(a) \Rightarrow (b)$: Assume that the family $\mathcal{F} = \{f \circ \varphi : \varphi \in \mathcal{O}(\Delta, X)\}$ is a normal family. By Marty's theorem [13], p. 75, there exists a constant $L > 0$ such that

$$\frac{|(f \circ \varphi)'(0)|}{1 + |f \circ \varphi(0)|^2} < L \quad \text{for all } \varphi \in \mathcal{O}(\Delta, X) \tag{4}$$

By the definition of k_X there exists $\psi \in \mathcal{O}(\Delta, X)$ such that $\psi(0) = x$, $\psi_*(0)a = v$ for $a > 0$ and $a/2 < k_X(x,v) \leq a$. Therefore, from (4),

$$ds(f(x), f_*(x)v) < 2L \cdot k_X(x,v) \quad \text{for all } (x,v) \in T(X).$$

Namely, $Q_f \leq 2L$.

$(b) \Rightarrow (c)$: Let x and y be distinct points of X. It follows readily from (2) that $ds(f(x), f_*(x)v) < Q \cdot k_X(x,v)$. By integrating both sides of the above inequality along the piecewise C^1 curves joining x to y in X and by the definitions we have $s(f(x), f(y)) \leq L \cdot \widetilde{K}_X(x,y)$. Since K_X is the integrated form of the infinitesimal metric k_X (see [9], Corollary 3) we have (3).

$(c) \Rightarrow (d)$: This follows immediately.

$(d) \Rightarrow (a)$: If (d) holds, then the family $\mathcal{F} = \{f \circ \varphi : \varphi \in \mathcal{O}(\Delta, X)\}$ is equicontinuous at each point of Δ, since otherwise there is a point $z_0 \in \Delta$, some $\epsilon > 0$, a sequence $\{z_n\}$ of points in Δ with $z_n \to z_0$, and a sequence $\{f \circ \varphi_n\} \subseteq \mathcal{F}$ satisfying

$$s(f \circ \varphi_n(z_n), f \circ \varphi_n(z_0)) \geq \epsilon, \quad n = 1, 2, \dots \quad . \tag{5}$$

Since $K_\Delta = d_\Delta$, and $z_n \to z_0$ then an application of Proposition 3.2 in [10] gives $K_\Delta(z_n, z_0) \to 0$, as $z_n \to z_0$. By the contracting property of the Kobayashi metric,

$$K_X(\varphi(z_n), \varphi(z_0)) \leq K_\Delta(z_n, z_0) \to 0, \text{ as } z_n \to z_0. \tag{6}$$

From (5), (6) follows that $\{\varphi_n(z_0)\}$ is a P-sequence for f in X which contradicts (d).

Therefore, $\mathcal{F} = \{f \circ \varphi : \varphi \in \mathcal{O}(\Delta, X)\}$ is spherically equicontinuous family at each point of Δ, and hence, by Montel's theorem [13], p. 74, $\mathcal{F}$ is normal. This proves (a). □

Remark 4.1. Hahn [15] has also published a similar characterization of a normal function on a complex hyperbolic manifold M of finite dimension. Unfortunately, Hahn's argument is incorrect, because the assumption that $\mathcal{F} = \{f \circ \varphi : \varphi \in \mathcal{O}(\Delta, M)\}$ is not a equicontinuous family does not entail that "*there exist an* $\epsilon > 0$ *such that for all* $n \in N$ *there exist sequences* $\{z_n\}$ *and* $\{w_n\}$ *in* Δ *with* $K_\Delta(z_n, w_n) < 1/n$ *but* $s(f(\psi(z_n)), f(\psi(w_n))) \geq \epsilon$ *for* ***some*** $\psi \in \mathcal{O}(\Delta, M)$" (see [15], p. 60).

5. Main Results

A sufficient condition for the non-normality of a holomorphic function in Δ is given by Lappan [16], Lemma 3. For holomorphic functions on a complex Banach manifold we obtain the following criterion.

Theorem 5.1. *Let X and k_X be given as in Theorem 4.1. The following statements are equivalent for $f \in \mathcal{O}(X)$:*

(a) *f is not a normal function;*

(b) *There exist sequences $\{y_m\}$, $\{x_m\}$ in X, and a constant $M > 0$ such that $K_X(y_m, x_m) < M$ for all $m \geq 1$, $\lim_{m\to\infty} f(x_m) = \infty$, and $\lim_{m\to\infty} f(y_m) = a \in \mathbb{C}$.*

Proof. (a) $\Rightarrow$ (b) : Since h is not normal it follows that the family $\mathcal{F} = \{f \circ \varphi : \varphi \in \mathcal{O}(\Delta, X)\}$ is not normal in Δ. Apply the Zalcman Lemma to $\mathcal{F}$ with r, z_n, ρ_n, $f \circ \varphi_n$, and g as given therein. Since g is nonconstant entire function in $\mathbb{C}$ it follows that there exist a sequence $\{\xi_m\} \subset \mathbb{C}$ such that $|g(\xi_m)| > m$. In fact, if this were not the case, g would have been bounded in $\mathbb{C}$. By Liouville's theorem $g \equiv$ constant, a contradiction with the assumption that g is nonconstant entire function.

For fixed ξ_m chose n_m large enough so that

$$\text{(i)}\ |z_{n_m} + \rho_{n_m}\xi_m| < (1+r)/2;$$
$$\text{(ii)}\ |f \circ \varphi_{n_m}(z_{n_m} + \rho_{n_m}\xi_m)| > m/2.$$

Put $w_{n_m} = z_{n_m} + \rho_{n_m}\xi_m$, $x_m = \varphi_{n_m}(w_{n_m})$, and $y_m = \varphi_{n_m}(z_{n_m})$.

By the contracting property of the Kobayashi metric,

$$K_X(y_m, x_m) = K_X(\varphi_{n_m}(w_{n_m}), \varphi_{n_m}(z_{n_m})) \leq d_\Delta(w_{n_m}, z_{n_m}).$$

By the triangle inequality $d_\Delta(w_{n_m}, z_{n_m}) \leq d_\Delta(0, z_{n_m}) + d_\Delta(0, w_{n_m})$. For any $z \in \Delta$ the Poincaré distance $d_\Delta(0,z) = \log((1+|z|)/(1-|z|))$. Since $|z_{n_m}| < r$, and $|w_{n_m}| < (1+r)/2$, we have

$$d_\Delta(w_{n_m}, z_{n_m}) \leq \log((1+r)/(1-r)) + \log((3+r)/(1-r)).$$

Putting the above together we get $K_X(y_m, x_m) \leq M$ where

$$M = \log((1+r)/(1-r)) + \log((3+r)/(1-r)) < \infty.$$

By (ii), we have that

$$\lim_{m\to\infty} f(x_m) = \lim_{m\to\infty} f \circ \varphi_{n_m}(z_{n_m} + \rho_{n_m}\xi_m) = \infty.$$

By (1), we have that

$$\lim_{m\to\infty} f(y_m) = \lim_{m\to\infty} f \circ \varphi_{n_m}(z_{n_m}) = g(0) \in \mathbb{C}.$$

$(b) \Rightarrow (a)$: Assume, to get a contradiction, that f is a normal function. By Lemma 3.2,

$$\log(c\mu(f, x_m)) \leq \log(c\mu(f, y_m)) \exp^{2K_X(x_m, y_m)} \text{ for all } m \geq 1.$$

The left hand side of the above inequality tends to infinity as $m \to \infty$ while the right-hand side tends to a number which is less than $\log(c \max\{1, |a|\}) \exp^{2r}$ and we have a contradiction. This contradiction proves the theorem. □

Corollary 5.1. *If under the conditions of the theorem, $f_1, \ldots, f_l$ are a finite number of normal holomorphic function on X such that each sequence $\{x_n\}$ of points in X contains a subsequence $\{x_{n_m}\}$ on which at most one of f_j $(1 \leq j \leq l)$ is unbounded, then $h := \Sigma_{j=1}^l f_j$ is a normal function.*

Proof. Suppose, on the contrary, that h is not a normal function. By Theorem 5.1, we can find two sequences $\{x_n\}$ and $\{y_n\}$ of points in X, and a positive constant M such that $K_X(x_m, y_m) < M$ for all $m \geq 1$, $\lim_{m\to\infty} h(x_m) = \infty$, and $\lim_{m\to\infty} h(y_m) = a \in \mathbb{C}$. Since $\lim_{m\to\infty} h(x_m) = \infty$ then $\{x_m\}$ contains a subsequence again denoted by $\{x_m\}$ such that at most one of f_j, say f_1, is unbounded on $\{x_m\}$. Since $\lim_{m\to\infty} h(y_m) = a \in \mathbb{C}$ then $\{y_m\}$ contains a subsequence $\{y_{m_k}\}$ such that either:

(i) at least two of f_j $(1 \leq j \leq l)$ is unbounded on $\{y_{m_k}\}$;
(ii) or $\lim_{k\to\infty} f_j(y_{m_k}) = \alpha_j \in \mathbb{C}$ $(1 \leq j \leq l)$.

The case (i) is excluded by the assumption of the corollary. Hence $\lim_{k\to\infty} f_1(x_{m_k}) = \infty$, $\lim_{k\to\infty} f_1(y_{m_k}) = \alpha_1 \in \mathbb{C}$, and $K_X(x_{m_k}, y_{m_k}) < M$ for all $k \geq 1$. By Theorem 5.1 f_1 is not a normal function, a contradiction which proves the corollary. □

Theorem 5.2. *Let X and k_X be given as in Theorem 4.1. If $f \in \mathcal{O}(X)$ is a normal function and if h is not a normal holomorphic function on X such that each sequence $\{x_m\}$ of points in X contains a subsequence $\{x_{m_k}\}$ on which at most one of f or h is unbounded, then $f + h$ is not a normal function.*

Proof. Since h is not normal on X it follows that the family $\mathcal{H} = \{h \circ \varphi : \varphi \in \mathcal{O}(\Delta, X)\}$ is not normal in Δ. By the Zalcman Lemma, we can find $h \circ \varphi_n \in \mathcal{H}$, $z_n \subseteq \Delta_r$, and $\rho_n \to 0^+$ such that $h \circ \varphi_n(z_n + \rho_n\xi) \to g(\xi)$ uniformly on compacta, where g is a nonconstant entire function on $\mathbb{C}$. But then

$$\rho_n \frac{|h_*(\varphi_n(z_n + \rho_n\xi))\varphi_n'(z_n + \rho_n\xi)|}{1 + |h(\varphi_n(z_n + \rho_n\xi))|^2} \to g^\#(\xi).$$

We claim that there exist a sequence $\{\xi_m\} \subset \mathbb{C}$ such that $|g(\xi_m)| > m$ and $g'(\xi_m) \neq 0$. Let us suppose the contrary.

(1) If $g'(\xi) \neq 0$ and g is bounded in $\mathbb{C}$ by Liouville's theorem $g \equiv$ constant which contradicts the assumption that g is nonconstant entire function.
(2) Let $\{a_m\}$ be the zeroes of $g'(\xi)$. Suppose that $|g(\xi)| < L < \infty$ on $\mathbb{C}\backslash\{\xi \in \mathbb{C} : g'(\xi) = 0\}$. The zeros of $g'(\xi)$ are isolated. Since nonconstant holomorphic function has no local maxima, we conclude that $|g(a_m)| < L$. Therefore g is bounded in $\mathbb{C}$, hence constant by Liouville's theorem. But this contradicts with the assumption that g is nonconstant entire function, and the assertion is proven.

For fixed ξ_m choose n_m large enough so that

(i) $|\lambda_{n_m}| < (1+r)/2$;
(ii) $g^\sharp(\xi_m)/\rho_{n_m} > m/2$;
(iii) $\rho_{n_m}|h_*(x_m)v_m|/(1+|h(x_m)|^2) > g^\sharp(\xi_m)/2$;
(iv) $|h(x_m)| > m/2$.

Here $\lambda_{n_m} = z_{n_m} + \rho_{n_m}\xi_m$, $x_m = \varphi_{n_m}(\lambda_{n_m})$, and $v_m = \varphi'_{n_m}(\lambda_{n_m})$. If (i), (ii), and (iii) hold, it is easy to see that

$$\frac{|h_*(x_m)v_m|}{1+|h(x_m)|^2} > \frac{g^\sharp(\xi_m)}{2\rho_{n_m}} > \frac{(4-(1+r)^2)m}{16}\cdot\frac{1}{1-|\lambda_{n_m}|^2}.$$

By the distance decreasing property of the Kobayashi metric

$$\frac{1}{1-|\lambda_{n_m}|^2} = k_\Delta(\lambda_{n_m}, 1) \geq k_X(\varphi_{n_m}(\lambda_{n_m}), \varphi'_{n_m}(\lambda_{n_m})\cdot 1).$$

Hence

$$\frac{|h_*(x_m)v_m|}{1+|h(x_m)|^2} > \frac{(4-(1+r)^2)m}{16}\cdot k_X(x_m, v_m).$$

Since $|h(x_m)| \to \infty$, as $m \to \infty$, by considering a subsequence if necessary, we may assume from the hypothesis of the theorem that f is bounded on $\{x_m\}$, namely $|f(x_m)| < M < \infty$ for all $m \geq 1$, and hence, for m sufficiently large, we have

$$\frac{|(f+h)_*(x_m)v_m|}{1+|f(x_m)+h(x_m)|^2} \geq \frac{|h_*(x_m)v_m| - |f_*(x_m)v_m|}{2(1+|f(x_m)|^2)(1+|h(x_m)|^2)} \geq$$

$$\frac{1}{2}\left[\frac{1}{1+|M|^2}\cdot\frac{|h_*(x_m)v_m|}{1+|h(x_m)|^2} - \frac{|f_*(x_m)v_m|}{1+|f(x_m)|^2}\right].$$

By hypothesis, f is a normal function on X hence, by Theorem 4.1, there exists $Q > 0$ such that $Q_f(x_m) < Q$ for all $m \geq 1$. Thus

$$Q_{f+h}(x_m) \geq \frac{1}{2}\left[\frac{(4-(1+r)^2)m}{16(1+|M|^2)} - Q\right] \to \infty \text{ as } m \to \infty.$$

By Theorem 4.1, $f+h$ is not a normal function on X. Thus the proof is complete. □

Theorem 5.3. *Let X and k_X be given as in Theorem 4.1. If $Q_f(x_m) \to \infty$ as $m \to \infty$, then $\{x_m\}$ contains a subsequence which is a P-sequence of f.*

Proof. Since $Q_f(x_m) \to \infty$ as $m \to \infty$, then by (2) there exist sequences $\{v_m\}$, $v_m \in T_{x_m}(X)$, and $\{r_m\} \subset R$, $r_m \to \infty$ as $m \to \infty$, with

$$ds(f(x_m), f_*(x_m)v_m) > r_m \cdot k_X(x_m, v_m). \tag{7}$$

By the definition of k_X there exists $\psi_m \in \mathcal{O}(\Delta, X)$ such that $\psi_m(0) = x_m$, $\psi_{m*}(0)a_m = v_m$ for $a_m > 0$ and $a_m/2 < k_X(x_m, v_m) \leq a_m$. Hence, inequality (7) gives,

$$\frac{|(f \circ \psi_m)'(0)|}{1 + |f \circ \psi_m(0)|^2} > r_m \cdot \frac{k_X(x_m, v_m)}{a_m} > \frac{r_m}{2} \to \infty \text{ as } m \to \infty.$$

By Marty's theorem [13], p. 75, the family $\{f \circ \psi_m\} \subseteq \mathcal{O}(\Delta)$ is not normal in any disc $\Delta_{\frac{1}{n}} = \{z \in \mathbb{C} : |z| < \frac{1}{n}\}$. Then by a local adaptation of the Zalcman Lemma [13], p. 152, there is a subsequence $\{f \circ \varphi_n\}$ of $\{f \circ \psi_m\}$, $z_n \to 0$, $\rho_n \to 0^+$ and a nonconstant entire function g with $f \circ \varphi_n(z_n + \rho_n \xi) \to g(\xi)$ normally in $\mathbb{C}$. Passing to a subsequence if necessary, we can assume that $\{f \circ \varphi_n(0)\}$ converges to a point $\beta \in \overline{\mathbb{C}}$. Set $g_n(\xi) = f \circ \varphi_n(z_n + \rho_n \xi)$. The sequence of functions $\{g_n\}$ converges locally uniformly to g. Let α be any complex number, $\alpha \neq \beta$, for which the equation $g(\xi) = \alpha$ has a solution ξ_0 which is not a multiple solution, that is, $g^\sharp(\xi_0) \neq 0$. By a theorem of Hurwitz [13], p. 9, in each neighborhood of ξ_0 all but a finite number of the functions $\{g_n\}$ assume the value α. Thus there exists a sequence of points $\{\xi_n\} \subseteq \mathbb{C}$ such that $\xi_n \to \xi_0$ and $g_n(\xi_n) = \alpha$ for n sufficiently large. It follows $s(f \circ \varphi_n(z_n + \rho_n \xi_n), f \circ \varphi_n(0)) \geq s(\alpha, \beta)/2 > 0$ for $n \geq N_0$. The Poincaré metric d_Δ is very closed to the Euclidian metric near 0 so $d_\Delta(0, z_n + \rho_n \xi_n) \to 0$ as $n \to \infty$. Hence $K_X(\varphi_n(z_n + \rho_n \xi_n), \varphi_n(0)) \leq d_\Delta(z_n + \rho_n \xi_n, 0) \to 0$ as $n \to \infty$. Which proves that a subsequence $\{\varphi_n(0)\}_{n=N_0}^{\infty}$ of $\{x_m\}$ is a P-sequence of f. The proof is complete. □

Theorem 5.4. *Let X and k_X be given as in Theorem 4.1. Then the following statements are mutually equivalent for $f \in \mathcal{O}(X)$.*

(a) f is a Bloch function.

(b) The quantity $\sup\{|(f \circ \varphi)'(0)| : \varphi \in \mathcal{O}(\Delta, X)\}$ *is finite.*

(c) There exists a constant $L > 0$ such that

$$|f_*(x)v| \leq L \cdot k_X(x, v) \quad \textit{for all } (x, v) \textit{ in } T(X),$$

where $f_*(x)$ is the linear map induced by f from $T_x(X)$ to $T_{f(x)}(C)$.

(d) There exists a constant $L > 0$ such that

$$|f(x) - f(y)| \leq L \cdot K_X(x, y) \quad \textit{for all } x \textit{ and } y \textit{ in } X.$$

Remark 5.1. See [17] for the proof of Theorem 5.4.

References

1. W. K. Hayman, *Functions with values in a given domain*, Proc. Amer. Math. Soc. 3 (1952), pp. 428–432.
2. C. Pommerenke, *Univalent functions*, Göttingen, Vandenhoek and Ruprecht, 1975.
3. K. Yosida, *On a class of meromorphic functions*, Proc. Phys.-Math. Soc. Japan, 3 (1934), ser. 1, pp. 227–235.
4. K. Noshiro, *Contributions to the theory of meromorphic functions in the unit circle*, J. Fac. Sci. Hokkaido Imp. Univ., Ser. I (1938), pp. 149–159.
5. O. Lehto and K. I. Virtanen, *Boundary behaviour and normal meromorphic functions*, Acta Math., 97(1957), pp. 47–65.
6. A. J. Lohwater, *The boundary behavior of analytic functions*, in *Current problems in mathematics, Fundamental directions, vol. 10, Itogi Nauki i Tekhniki, Akad. Nauk SSSR, Vsesoyuz. Inst. Nauchn. i Tekhn. Inform.*, Moscow, 1973, pp. 99–259 (in Russian).
7. P. Lappan, *Normal families and normal functions: results and techniques*, in: Function Spaces and Complex Analysis, Joensuu 1997, Univ. Joensuu, Department of Mathematics Rep. Ser. 2 (1997), pp. 63–78.
8. M. G. Zaidenberg, *Schottky–Landau growth estimates for s-normal families of holomorphic mappings*, Math. Ann. 293 (1992), pp. 123–141.
9. C. J. Earle, L. A. Harris, J. H. Hubbard, and S. Mitra, Schwarz's lemma and the Kobayashi and Carathéodory metrics on complex Banach manifolds, in: *Kleinian Groups and Hyperbolic 3-Manifolds*, (Cambridge Univ. Press, Cambridge, Lond. Math. Soc. Lec. Notes 299, 2003), pp. 363–384. `http://www.ms.uky.edu/~larry/paper.dir/minsky.ps`
10. S. Dineen, *The Schwarz Lemma*, Oxford Mathematical Monographs, Oxford, 1989.
11. T. Franzoni and E. Vesentini, *Holomorphic Maps and Invariant Distances*, North-Holland Mathematical Studies 40, North-Holland Publishing, Amsterdam, 1980.
12. P. Gauthier, *A criterion of normalcy*, Nagoya Math. J. 32 (1968), pp. 272–282.
13. J. L. Shchiff, *Normal families*, (Springer, New York, 1993).
14. M. H. Kwack, *Families of normal maps in several variables and classical theorems in complex analysis* Lecture Notes Series 33, Res. Inst. Math., Global Analysis Res. Center, Seoul, Korea, 1996.
15. K. T. Hahn, *Non-tangential limit theorems for normal mappings*, Pacific J. Math. 135 (1988), pp. 57–64.
16. P. Lappan, *Non-normal sums and prodacts of unbounded normal functions*, Mich. Math. J. 8 (1961), pp. 187–192.
17. P. V. Dovbush, *Bloch functions on complex Banach manifolds*, Math. Proc. R. Ir. Acad., to appear.

STRONG-TYPE ESTIMATES AND CARLESON MEASURES FOR WEIGHTED BESOV-LIPSCHITZ SPACES

VAGIF S. GULIYEV

Baku State University, Baku, Azerbaijan
E-mail: vagif@guliyev.com

ZHIJIAN WU

University of Alabama, USA
E-mail: zwu@gp.as.ua.edu

We establish a capacitory strong-type estimate for weighted inhomogeneous (homogeneous) Besov-Lipschitz space $B_{pq}^{\alpha,\varphi}$ ($\dot{B}_{pq}^{\alpha,\varphi}$). As an application, we characterize the related Carleson measures.

Keywords: Weighted Besov-Lipschitz spaces, Carleson measures, Poisson kernel, Hardy-Littlewood maximal function, capacity.

1. Introduction

Let $\mathbb{R}^n$ be the n-dimensional Euclidean space of points $x = (x_1, ..., x_n)$ with norm $|x| = \left(\sum_{i=1}^n x_i^2\right)^{1/2}$. For $x \in \mathbb{R}^n$ and $r > 0$, denote by $B(x,r)$ the open ball centered at x of radius r.

Let $E \subset \mathbb{R}^n$ be a measurable set. By a *weight function* we mean any nonnegative function φ that is Lebesgue integrable on any bounded and measurable subset of E. If f is a measurable function on E, we denote

$$\|f\|_{L_p(E,\varphi)} \equiv \|f\|_{p,\varphi;E} = \left(\int_E |f(x)|^p \varphi(x)dx\right)^{1/p}, \quad 1 \le p < \infty,$$

$$\|f\|_{L_\infty(E,\varphi)} \equiv \|f\|_{\infty,\varphi;E} = esssup_{x\in E}|f(x)|\,\varphi(x), \quad p = \infty.$$

If $\varphi \equiv 1$ we write $\|f\|_{L_p(E)}$, and if $E = \mathbb{R}^n$ we write $\|f\|_{L_{p,\varphi}} \equiv \|f\|_{p,\varphi}$.

We say that a weight function φ *satisfies condition* A_p, $1 < p < \infty$, if there exists a positive number C_p such that for every cube $Q \subset \mathbb{R}^n$

$$\frac{1}{|Q|}\int_Q \varphi(x)dx \left(\frac{1}{|Q|}\int_Q (\varphi(x))^{-p'/p}dx\right)^{p/p'} \le C_p,$$

where $|Q|$ is the Lebesgue measure of Q and $1/p+1/p'=1$. Condition A_p, in this form, first appeared in Muckenhoupt's fundamental paper [8].

For $p, q \geq 1$ and $0<\alpha<1$, the weighted Besov-Lipschitz space $B_{pq}^{\alpha,\varphi}(\mathbb{R}^n)$ consist of all functions f in $L_{p,\varphi}$ so that

$$\|f\|_{B_{pq}^{\alpha,\varphi}} = \|f\|_{p,\varphi} + \left(\int_{\mathbb{R}^n} \frac{\|f(\cdot+h)-f(\cdot)\|_{p,\varphi}^q}{|h|^{n+\alpha q}} dh\right). \tag{1}$$

For any open set $\mathcal{O} \subset \mathbb{R}^n$, the $B_{pq}^{\alpha,\varphi}(\mathbb{R}^n)$-capacity of $\mathcal{O}$ is defined by

$$cap_{B_{pq}^{\alpha,\varphi}}(\mathcal{O}) = cap\left(\mathcal{O}; B_{pq}^{\alpha,\varphi}(\mathbb{R}^n)\right) = \inf\left\{\|f\|_{B_{pq}^{\alpha,\varphi}};\ f\geq 1 \text{ on } \mathcal{O}\right\}.$$

Let $\mathbb{R}^{n+1} = \{(x,t) : x\in\mathbb{R}^n,\ t>0\}$ be the upper half space. For any open set $\mathcal{O}\subset\mathbb{R}^n$, the $T(\mathcal{O})$ the tent of $\mathcal{O}$ in $\mathbb{R}^{n+1}$, which is $T(\mathcal{O}) = \{(x,t)\in\mathbb{R}^{n+1}$: the open ball centered at x with radius t lies in $\mathcal{O}\}$.

Let

$$P_t(x) = c_n \frac{t}{(|x|^2+t^2)^{\frac{n+1}{2}}}$$

be the Poisson kernel, where the constant c_n is chosen so that $\int_{\mathbb{R}^n} P_t(x)dx = 1$. For any function $f\in L_{p,\varphi}$, the harmonic extension of f onto $\mathbb{R}^{n+1}$ is the convolution between P_t and f, i.e.

$$F(x,t) = P_t * f(x) = \int_{\mathbb{R}^n} P_t(x-y)f(y)dy.$$

The main results of this note are the following:

Theorem 1.1. *For $1\leq p\leq q\leq\infty$, $0<\alpha<1$ and for weighted function φ, there is a constant $C>0$ such that the following strong type estimate*

$$\int_0^\infty cap_{B_{pq}^{\alpha,\varphi}}(|f|>t)\ dt^q \leq C\|f\|_{B_{pq}^{\alpha,\varphi}}^q$$

holds for all $f\in B_{pq}^{\alpha,\varphi}$.

Theorem 1.2. *Suppose $1<p\leq q\leq\infty$, $0<\alpha<1$, $\varphi\in A_p$. A nonnegative measure μ on $\mathbb{R}_+^{n+1}$ satisfies*

$$\int_{\mathbb{R}_+^{n+1}} |P_t * f(x)|^q\ d\mu(x,t) \leq C\|f\|_{B_{pq}^{\alpha,\varphi}}^q, \qquad \forall f\in B_{pq}^{\alpha,\varphi} \tag{2}$$

if and only if

$$\mu(T(\mathcal{O})) \leq C\ cap\left(\mathcal{O}; B_{pq}^{\alpha,\varphi}\right), \quad \textit{for any bounded open set } \mathcal{O}\subset\mathbb{R}^n. \tag{3}$$

For $\varphi \equiv 1$, both Theorems 1.1 and 1.2 are established in [10]. Strong type estimates (Theorem 1.1) for L_p and Sobolev type spaces have been studied systematically in the past. They play an important role in the imbedding theory of function spaces and related areas. We refer the reader to [1,5,7,14] and the references therein. Estimate (2) defines the usual Carleson measure if the space $B_{pq}^{\alpha,\varphi}$ is replaced by L_q. Therefore, Theorem 1.2 gives a characterization for Carleson measures for weighted Besov-Lipschitz space $B_{pq}^{\alpha,\varphi}$ with $1 < p \le q \le \infty$, $0 < \alpha < 1$ and $\varphi \in A_p$. In general, the bounded open set in characterization (3) cannot be replaced by an open ball as it can in the characterization for usual Carleson measures. Carleson measures play an important role in harmonic analysis and operator theory on function spaces. For example, it is well-known that the measure $\left|\frac{\partial}{\partial t} P_t * f(x)\right|^2 t dx dt$ is a usual Carleson measure if and only if $f \in BMO$. Carleson measures for Dirichlet space (analytic) on the unit disk of complex plane is characterized in [9] (see also [6] for a "single-box condition" characterization). Applications of Carleson measures for Besov space on commutators and related bilinear forms are studied in [11,12].

For the sake of simplicity, the letter C always denotes a positive constant which may change from one step to the next. For two positive functions a and b, we write $a \asymp b$, if there is a constant $C > 0$ such that both $a \le Cb$ and $b \le Ca$ hold.

2. Proof of Theorem 1.1

Let ψ be a nondecreasing function in $C_0^\infty(0,\infty)$ which satisfies

$$\psi(t) = \begin{cases} 0, & \text{if} \quad t \le \frac{1}{2}; \\ 1, & \text{if} \quad t \ge 1. \end{cases}$$

Consider the smooth truncation $\{F_j\}_{-\infty}^{\infty}$ given by

$$F_j(f) = 2^j \psi\left(\frac{|f|}{2^j}\right), \qquad j = 0, \pm 1, \pm 2, \ldots.$$

D. R. Adams first used this smooth truncation in [1] to prove a strong type estimate for Sobolev space on $\mathbb{R}^n$ (see also [9]). The key properties of this smooth truncation are the following.

$$0 \le F_j(f) \le 2^j, \qquad \{F_j(f) > 0\} \subset \left\{|f| > 2^{j-1}\right\},$$

and if $2^{j-1} \le |f| < 2^j$, then

$$F_k(f) = \begin{cases} 2^k, & \text{if} \quad k < j; \\ 0, & \text{if} \quad k > j. \end{cases}$$

Lemma 2.1. *For $1 \le p \le q \le \infty$, $0 < \alpha < 1$ and for weighted function φ the smooth truncation $\{F_j\}_{-\infty}^{\infty}$ maps $B_{pq}^{\alpha,\varphi}$ to itself, and there is a constant $C > 0$ such that*

$$\sum_{j=-\infty}^{\infty} \|F_j(f)\|_{B_{pq}^{\alpha,\varphi}}^q \le C \|f\|_{B_{pq}^{\alpha,\varphi}}^q$$

holds for all $f \in B_{pq}^{\alpha,\varphi}$.

Proof. It is easy to check that

$$\|F_j(f)\|_{L_{p,\varphi}}^p \le 2^{jp} \int_{\{|f|>2^{j-1}\}} \varphi(x)\, dx \equiv 2^{jp}\varphi\left(|f| > 2^{j-1}\right).$$

Since $q/p \ge 1$, we have therefore

$$\begin{aligned}\sum_{j=-\infty}^{\infty} \|F_j(f)\|_{L_{p,\varphi}}^q &\le \sum_{j=-\infty}^{\infty} \left(2^{jp}\varphi\left(|f| > 2^{j-1}\right)\right)^{q/p} \\ &\le \left(\sum_{j=-\infty}^{\infty} 2^{jp}\varphi\left(|f| > 2^{j-1}\right)\right)^{q/p}.\end{aligned}$$

It is standard that

$$\|f\|_{L_{p,\varphi}}^p = \int_0^\infty \varphi\left(|f| > t\right) dt^p \asymp \sum_{j=-\infty}^{\infty} 2^{jp}\varphi\left(|f| > 2^j\right).$$

Combining the above two estimate, we obtain

$$\sum_{j=-\infty}^{\infty} \|F_j(f)\|_{L_{p,\varphi}}^q \le C\left(\int_0^\infty \varphi\left(|f| > t\right)\ dt^p\right)^{q/p} = C\|f\|_{L_{p,\varphi}}^q.$$

On the other hand, it is clear that

$$\begin{aligned}\sum_{l=-\infty}^{\infty} \|F_l(f(\cdot+h)) - F_l(f(\cdot))\|_{L_{p,\varphi}}^q &\le \left(\sum_{l=-\infty}^{\infty} \|F_l(f(\cdot+h)) - F_l(f(\cdot))\|_{L_{p,\varphi}}^p\right)^{q/p} \\ &\le \left\|\sum_{l=-\infty}^{\infty} |F_l(f(\cdot+h)) - F_l(f(\cdot))|^p\right\|_{L_{1,\varphi}}^{q/p}\end{aligned}$$

By the definition of the weighted Besov space $B_{pq}^{\alpha,\varphi}$, it suffices to show that for any x, $h \in \mathbb{R}^n$

$$\sum_{l=-\infty}^{\infty} |F_l(f(x+h)) - F_l(f(x))|^p \le C\,|f(x+h) - f(x)|^p.$$

This estimate is proved in [10]. The proof is complete. □

Proof of Theorem 1.1. It is standard that

$$\int_0^\infty cap_{B_{pq}^{\alpha,\varphi}}(|f|>t)\,dt^q \asymp \sum_{k=-\infty}^{\infty} 2^{kq} cap_{B_{pq}^{\alpha,\varphi}}(|f|>2^k).$$

Since $2^{-k}F_k(f) \geq 1$ on the set $\{|f|>2^k\}$, and hence

$$cap_{B_{pq}^{\alpha,\varphi}}(|f|>2^k) \leq 2^{-kq}\,\|F_k(f)\|^q_{B_{pq}^{\alpha,\varphi}}$$

we have therefore

$$\sum_{k=-\infty}^{\infty} 2^{kq} cap_{B_{pq}^{\alpha,\varphi}}(|f|>2^k) \leq \sum_{k=-\infty}^{\infty} \|F_k(f)\|^q_{B_{pq}^{\alpha,\varphi}} \leq C\|f\|^q_{B_{pq}^{\alpha,\varphi}}.$$

The last inequality follows from Lemma 2.1. □

Consider the weighted homogeneous Besov-Lipschitz space $\dot{B}_{pq}^{\alpha,\varphi}$ which is the completion of C_0^∞ under the norm

$$\|f\|_{\dot{B}_{pq}^{\alpha,\varphi}} = \left(\int_{\mathbb{R}^n} \frac{\|f(\cdot+h)+f(\cdot)\|_{p,\varphi}{}^q}{|h|^{n+\alpha q}}\,dh\right).$$

We can define $\dot{B}_{pq}^{\alpha,\varphi}$-capacity similarly. The above discussion proves also the following result.

Theorem 2.1. *For $1\leq p\leq q\leq\infty$, $0<\alpha<1$ and for weighted function φ there is a constant $C>0$ such that the following strong type estimate*

$$\int_0^\infty cap_{\dot{B}_{pq}^{\alpha,\varphi}}(|f|>t)\,dt^q \leq C\|f\|^q_{\dot{B}_{pq}^{\alpha,\varphi}(\mathbb{R}^n)}$$

holds for all $f\in \dot{B}_{pq}^{\alpha,\varphi}$.

3. Proof of Theorem 1.2

For a function f defined on $\mathbb{R}^n$, the Hardy-Littlewood maximal function $M(f)$ is defined by

$$M(f)(x) = \sup_{r>0}\frac{1}{|B(x,r)|}\int_{B(x,r)} |f(y)|\,dy.$$

The proof of next lemma has its root in [10] (which is for the case $\varphi\equiv 1$).

Lemma 3.1. *For $1<p<\infty$, $0<\alpha<1$ and $\varphi\in A_p$ the Hardy-Littlewood maximal operator is bounded on $B_{pq}^{\alpha,\varphi}$. More precisely, there is a constant $C>0$ such that*

$$\|M(f)\|_{B_{pq}^{\alpha,\varphi}} \leq C\|f\|_{B_{pq}^{\alpha,\varphi}}$$

hold for all $f \in B_{pq}^{\alpha,\varphi}$.

Proof. It is well-known (see [8]) that M is bounded in $L_{p,\varphi}$, i.e. $\|M(f)\|_{p,\varphi} \le C\|f\|_{p,\varphi}$. For $h \in \mathbb{R}^n$, let T_h be the translation by h, i.e. $T_h f(x) = f(x+h)$. By the definition of the weighted Besov-Lipschitz space it suffices to show that

$$\|T_h M(f) - M(f)\|_{p,\varphi} \le C\|T_h f - f\|_{p,\varphi}.$$

It is easy to see that T_h commutes with M, i.e. $T_h M(f) = M(T_h f)$. Hence we have

$$|T_h M(f) - M(f)| = |M(T_h f) - M(f)| \le M(|T_h f - f|).$$

Taking $L_{p,\varphi}$ norm on both ends of the above inequality, by the boundedness of M on $L_{p,\varphi}$, we obtain the desired result. Lemma 3.1 is proved. □

For $x \in \mathbb{R}^n$ denote by $\Gamma(x) = \{(u,t) \in \mathbb{R}_+^{n+1} : |u - x| < t\}$, the cone with vertex at x. For a function g defined on $\mathbb{R}_+^{n+1}$, the nontangential maximal function of g is a function on $\mathbb{R}^n$ defined by

$$\mathcal{N}(g)(x) = \sup_{(u,t)\in\Gamma(x)} |g(u,\tau)|.$$

The following lemma is proved in [10].

Lemma 3.2. *Let μ be a nonnegative Borel measure on $\mathbb{R}_+^{n+1}$. Then for any measurable function g on $\mathbb{R}_+^{n+1}$ and any $t > 0$*

$$\mu(\{|g(x,\tau)| > t\}) \le \mu(T(\{\mathcal{N}(g)(x) > t\})).$$

Proof of Theorem 1.2. The approach in the following has its root in [9] (see also [10]). There Stegenga proved a result for Carleson measures for the (analytic) Dirichlet space on the unit disk of the complex plane which is similar to the case $p = q = 2$ and $\varphi = 1$ here.

We prove the "only if" part of Theorem 1.2 first. Given a bounded open set $\mathcal{O} \subset \mathbb{R}^n$, by the definition of $B_{pq}^{\alpha,\varphi}$-capacity, there is a function $f \in B_{pq}^{\alpha,\varphi}$ such that $f \ge 1$ on the set $\mathcal{O}$ and $\|f\|_{B_{pq}^{\alpha,\varphi}}^q \le 2cap\left(\mathcal{O}; B_{pq}^{\alpha,\varphi}\right)$. We can assume further that $f \ge 0$ on $\mathbb{R}^n$, because it is easy to show (similar to the proof of Lemma 3.1) that $\||f|\|_{B_{pq}^{\alpha,\varphi}} \le \|f\|_{B_{pq}^{\alpha,\varphi}}$. Since $(x,\tau) \in T(\mathcal{O})$ is equivalent to $B(x,\tau) \subset \mathcal{O}$, we have for any $(x,\tau) \in T(\mathcal{O})$ (see [10]),

$$P_t * f(x) \ge \int_{B(0,1)} P_1(y)\, dy = \delta.$$

Therefore, we obtain

$$\mu\left(T(\mathcal{O})\right) \leq \delta^q \int_{\mathbb{R}^{n+1}_+} |P_\tau * f(y)|^q \, d\mu(y,\tau) \leq C\|f\|^q_{B^{\alpha,\varphi}_{pq}} \leq C cap\left(\mathcal{O}; B^{\alpha,\varphi}_{pq}\right).$$

To prove the "if" part of Theorem 1.2, we consider the measure

$$\mu_k = \mu|_{T(B(0,k))}, \qquad k > 0.$$

For any $f \in B^{\alpha,\varphi}_{pq}$, let $F(y,\tau) = P_\tau * f(y)$. By Lemma 3.2 and the standard result that $\mathcal{N}(F)(y) \leq CM(f)(y)$, we have

$$\mu_k\left(\{|F(y,\tau)| > t\}\right) \leq \mu_k\left(T\left(\{\mathcal{N}(F)(y) > t\}\right)\right) \leq \mu_k\left(T\left(\{M(f) > t/C\}\right)\right).$$

It is clear that for any pair of open sets $\mathcal{O}_1$, $\mathcal{O}_2 \subset \mathbb{R}^n$, we have $T\left(\mathcal{O}_1 \cap \mathcal{O}_2\right) = T\left(\mathcal{O}_1\right) \cap T\left(\mathcal{O}_2\right)$ and if $\mathcal{O} \subset \mathcal{O}^*$ then $cap_{B^{\alpha,\varphi}_{pq}}(\mathcal{O}) \leq cap_{B^{\alpha,\varphi}_{pq}}(\mathcal{O}^*)$. We can continue the above estimate by

$$\begin{aligned}\mu_k\left(\{|F(y,\tau)| > t\}\right) &\leq \mu_k\left(T\left(\{M(f) > t/C\} \cap B(0,k)\right)\right) \\ &\leq cap_{B^{\alpha,\varphi}_{pq}}\left(\{M(f) > t/C\} \cap B(0,k)\right).\end{aligned}$$

Hence, by Theorem 1.1 and Lemma 3.1, we obtain

$$\begin{aligned}\int_{T(B(0,k))} |P_\tau * f(y)|^q \, d\mu(y,\tau) &= \int_0^\infty \mu_k\left(\{|P_\tau * f(y)| > u\}\right) \, du^q \\ &\leq \int_0^\infty cap_{B^{\alpha,\varphi}_{pq}}\left(\{M(f) > u/C\}\right) \, du^q \\ &\leq C\|M(f)\|^q_{B^{\alpha,\varphi}_{pq}} \leq C\|f\|^q_{B^{\alpha,\varphi}_{pq}}.\end{aligned}$$

Theorem 1.2 is proved. □

4. Similar Results on Unit Disk

Let $\mathbb{U}$ be the unit disk of the complex plane, $\mathbb{T}$ be the boundary of $\mathbb{U}$. For $z \in \mathbb{U}$, we can write $z = r\zeta$ with $r = |z|$ and $\zeta \in \mathbb{T}$. Let $\phi \geq 0$ be a measurable function on $\mathbb{T}$. If f is a measurable function defined on $\mathbb{T}$, then

$$\|f\|_{L_p(\mathbb{T},\phi)} \equiv \|f\|_{p,\phi;\mathbb{T}} = \left(\int_{\mathbb{T}} |f(\zeta)|^p \phi(\zeta) d\zeta\right)^{1/p}, \qquad for\ 1 \leq p < \infty,$$

$$\|f\|_{L_\infty(\mathbb{T},\phi)} \equiv \|f\|_{\infty,\phi;\mathbb{T}} = esssup_{\zeta\in\mathbb{T}} |f(\zeta)|\phi(\zeta), \qquad for\ p = \infty.$$

If $\phi \equiv 1$ we write $\|f\|_{L_p(\mathbb{T})} \equiv \|f\|_{p,\mathbb{T}}$.

We say that a weight function ϕ on $\mathbb{T}$ satisfies condition $A_p(\mathbb{T})$, $1 < p < \infty$, if there exists a positive number C_p such that for every interval $I \subset \mathbb{R}$

$$\frac{1}{|I|} \int_I \phi(x) \, dx \left(\frac{1}{|I|} \int_I (\phi(x))^{-p'/p} \, dx\right)^{p/p'} \leq C_p,$$

where $|I|$ is the Lebesgue measure of I.

For $p,\ q \geq 1$ and $0 < \alpha < 1$, the weighted Besov-Lipschitz space $B_{pq}^{\alpha,\phi}(\mathbb{T})$ consist of all functions f in $L_p(\mathbb{T},\phi)$ so that

$$\|f\|_{B_{pq}^{\alpha,\phi}(\mathbb{T})} = \|f\|_{p,\phi;\mathbb{T}} + \left(\int_{\mathbb{T}} \frac{\|f(e^{i(\cdot+h)}) - f(e^{i\cdot})\|_{p,\phi;\mathbb{T}}^q}{|h|^{1+\alpha q}}\, dh\right).$$

For any open set $\mathcal{O} \subset \mathbb{T}$, the $B_{pq}^{\alpha,\phi}(\mathbb{T})$-capacity of $\mathcal{O}$ is defined by

$$cap_{B_{pq}^{\alpha,\phi}(\mathbb{T})}(\mathcal{O}) = cap\left(\mathcal{O}; B_{pq}^{\alpha,\phi}(\mathbb{T})\right) = \inf\left\{\|f\|_{B_{pq}^{\alpha,\phi}(\mathbb{T})};\ f \geq 1 \text{ on } \mathcal{O}\right\}.$$

Denote by $\mathcal{D}$ and $\mathcal{D}'$ the spaces of test functions and distributions on $\mathbb{T}$, respectively. For $g \in \mathcal{D}'$, let the Fourier series of g be

$$g(\zeta) \sim \sum_{k=-\infty}^{\infty} g_k \zeta^k.$$

We define accordingly $(Q_0 g)(\zeta) = g_0$ and

$$(Q_m g)(\zeta) = \sum_{k=2^{m-1}}^{2^m-1} \left(g_k \zeta^k + g_{-k}\zeta^{-k}\right), \qquad m = 1, 2, \ldots.$$

It is clear that g can be represented by the weakly convergent series

$$g(\zeta) \overset{\mathcal{D}'}{=} \sum_{m=0}^{\infty} (Q_m g)(\zeta).$$

In [4] we study the weighted holomorphic Besov space on the unit disk $\mathbb{U}$ and their boundary values on $\mathbb{T}$. There the norm of f in $B_{pq}^{\alpha,\phi}(\mathbb{T})$ is defined as

$$b_{p,q}^{\alpha,\phi}(g,\mathbb{T}) = \left\|\left\{2^{m\alpha}\|Q_m g\|_{L_p(\mathbb{T},\phi)}\right\}\right\|_{l_q},$$

which is equivalent to the norm defined in (4) (see, for example, [2,3] for the case $\phi \equiv 1$. For general $\phi \in A_p(\mathbb{T})$ the proof of equivalences of these norms is the same.). Under certain restrictions on the weighted function and parameters, we establish in [4] the equivalent norms for holomorphic functions in terms of their boundary functions.

For any open set $\mathcal{O} \subset \mathbb{T}$, the tent of $\mathcal{O}$ in $\mathbb{U}$ is defined as $T(\mathcal{O}) = \{z \in \mathbb{U}:$ the open interval centered at ζ with radius $1-r$ is contained in $\mathcal{O}\}$.

Let

$$P_r(e^{it}) = \frac{1}{\pi}\frac{1-r^2}{|e^{it}-r|^2}$$

be the Poisson kernel. For any function $f \in L_p(\mathbb{T}, \phi)$, the harmonic extension of f onto $\mathbb{U}$ is the convolution between P_r and f, i.e.

$$F(z) = P_r * f(e^{it}) = \int_{\mathbb{T}} P_r(e^{i(t-\tau)}) f(e^{i\tau})\, d\tau, \qquad z = re^{it}.$$

Similarly we can prove the following results.

Theorem 4.1. *For $1 \le p \le q \le \infty$, $0 < \alpha < 1$ and for weighted function ϕ there is a constant $C > 0$ such that the following strong type estimate*

$$\int_0^\infty cap_{B^{\alpha,\phi}_{pq}(\mathbb{T})}(|f| > t)\, dt^q \le C \|f\|^q_{B^{\alpha,\phi}_{pq}(\mathbb{T})}$$

holds for all $f \in B^{\alpha,\phi}_{pq}(\mathbb{T})$.

Theorem 4.2. *Suppose $1 < p \le q \le \infty$, $0 < \alpha < 1$, $\phi \in A_p(\mathbb{T})$. A nonnegative measure μ on $\mathbb{U}$ satisfies ($z = re^{it}$)*

$$\int_{\mathbb{U}} |P_r * f(e^{it})|^q\, d\mu(z) \le F \|f\|^q_{B^{\alpha,\phi}_{pq}(\mathbb{T})}, \qquad \forall f \in B^{\alpha,\phi}_{pq}(\mathbb{T}),$$

if and only if

$$\mu(T(\mathcal{O})) \le C\ cap\left(\mathcal{O}; B^{\alpha,\phi}_{pq}(\mathbb{T})\right), \qquad \textit{for any bounded open set} \quad \mathcal{O} \subset \mathbb{T}.$$

Remark 4.1. In Theorem 4.2 the open set $\mathcal{O}$ can be replaced by finite union of open intervals.

Acknowledgments

The authors are thankful to the referee for his/her valuable comments.

References

1. D. R. Adams, *On the existence of capacitory strong type estimates in* $\mathbb{R}^n$. Ark. Mat. 14 (1966), 125-140.
2. V. S. Guliev and P. I. Lizorkin, *$\mathcal{B}-$ and $\mathcal{L}-$spaces of harmonic and holomorphic functions in the disc, and classes of boundary values.* (Russian) Dokl. Akad. Nauk SSSR 319 (1991), no. 4, 806-810. English transl. in Soviet Math. Dokl. 44 (1992), no. 1, 215-219.
3. V. S. Guliev and P. I. Lizorkin, *Classes of holomorphic and harmonic functions in a polydisk in connection with their boundary values.* Trudy Math. Inst. Steklov, 204 (1993), no. 16, 137-159. English trans. in Proc. Steklov Inst. 204 (1994), no. 3, 117-135.
4. V. S. Guliyev and Z. Wu, *Weighted holomorphic Besov spaces and their boundary values.* Anal. Theory Appl. 21 (2005), 143-156.

5. K. Hansson, *Imbedding theorems of Sobolev type in potential theory.* Math. Scand. 45 (1979), 67-109.
6. R. Kerman and E. Sawyer, *Carleson measures and multipliers of Dirichlet spaces.* Trans. Amer. Math. Soc. 329 (1988), 87-99.
7. V. G. Mazya and T. O. Shaposhnikova, *Theory of Multipliers in Spaces of Differentiable Functions.* Pitman Advanced Publishing Program, Boston and London.
8. B. Muckenhoupt, *Weighted norm inequalities for the Hardy maximal functions.* Trans. Amer. Math. Soc. 165 (1972), 207-226.
9. D. Stegenga, *Multipliers of the Dirichlet space.* Illinois J. Math. 24 (1980), 113-139.
10. Z. Wu, *Strong type estimate and Carleson measures for Lipschitz spaces.* Proc. Amer. Math. Soc. 137 (1999), no. 11, 3243-3249.
11. Z. Wu, *Clifford analysis and commutators on the Besov spaces.* J. Funct. Anal. 169 (1999), no. 1, 121-143.
12. Z. Wu, *Carleson measures and multipliers for Dirichlet spaces.* J. Funct. Anal. 169 (1999), no. 1, 148-164.
13. E. M. Stein, *Harmonic Analysis.* Princeton University Press, 1993.
14. W. P. Ziemer, *Weakly Differentiable Functions.* Springer-Verlag, New York, 1989.

COMPOSITION THEOREMS IN BESOV SPACES, THE VECTOR-VALUED CASE

MADANI MOUSSAI

Department of Mathematics, Lab. Math. Pure and Applied,
University of M'Sila, P.O. Box 166, M'Sila 28000, Algeria
E-mail: mmoussai@yahoo.fr

Let U^d be the set of the functions $g : \mathbb{R} \to \mathbb{R}^d$, $g = (g_1, \dots, g_d)$, such that $g_i' = g_j'$ for $i \neq j$. For a Lipschitz continuous function $f : \mathbb{R}^d \to \mathbb{R}$ vanishing at the origin, we will show that the composition operator $g \mapsto T_f(g) = f \circ g$ takes the Besov space $B^s_{p,q}(\mathrm{U}^d)$ into $B^s_{p,q}(\mathbb{R})$ under some conditions on $\partial_j f$ and on the parameters s, p, q.

Keywords: Composition operators, Besov spaces, Lipschitz spaces.

1. Introduction

Let $T_f : g \mapsto f \circ g$ be a composition operator defined from a function $f : \mathbb{R} \to \mathbb{R}$. In the Besov space $B^s_{p,q}(\mathbb{R}^n)$ it is well know the following result (see e.g. [7]):

Theorem 1.1. *Let $1 + (1/p) < s < (n/p)$ and $f : \mathbb{R} \to \mathbb{R}$. Then T_f takes $B^s_{p,q}(\mathbb{R}^n)$ into itself if and only if there exists some constant c such that $f(t) = ct$. The same result holds in the limit case $1 + (1/p) = s < (n/p)$ as soon as $q > 1$.*

In the space $B^s_{p,q}(\mathbb{R}^n) \cap L_\infty(\mathbb{R}^n)$, the composition operator problem becomes nontrivial, in the sense that $f : \mathbb{R} \to \mathbb{R}$ is not automatically linear. Then the extension of Theorem 1.1 to a nonlinear function f needed some conditions on f and on the parameters s, p, q; for example $f(0) = 0$, $f \in B^{s,\ell oc}_{p,q}(\mathbb{R})$ and $(1/p) < s - [s]$, cf. [2,4,5].

In this paper we consider the set

$$\mathrm{U}^d = \{g : \mathbb{R} \to \mathbb{R}^d,\ g = (g_1, \dots, g_d)\,,\ \text{such that } g_i' = g_j' \ \text{for } i \neq j\}$$

and we will study the composition operator problem in Besov space of vector-valued case in the context

$$f : \mathbb{R}^d \to \mathbb{R} \quad \text{and} \quad g = (g_1, \ldots, g_d) \in \mathrm{U}^d, \quad g_i \in B^s_{p,q}(\mathbb{R}), \quad i = 1, \ldots, d.$$

Notation

All functions are assumed to be real-valued. We define the difference operator for an arbitrary function f, by

$$\Delta_h f(x) := f(x+h) - f(x), \qquad \forall h, x \in \mathbb{R}^d.$$

For a space E, of functions defined on $\mathbb{R}^d$, the associated local space is defined as

$$E^{\ell oc} = \{f \in \mathcal{S}'(\mathbb{R}^d) : \varphi f \in E, \forall \varphi \in \mathcal{D}(\mathbb{R}^d)\}.$$

If s is a real number, then $[s]$ denotes the integer part of s, i.e. the largest integer less than or equal to s. The constants $c, c_1, \ldots$ are strictly positive and depend only on the fixed parameters d, s, p and q; their values may vary from line to line.

Some Definitions

In this section we will define some spaces considered in this work.

(i) *The Lipschitz and Besov spaces*:

Definition 1.1. If $f \in \mathrm{Lip}_1(\mathbb{R}^d)$, then

$$\|f\|_{\mathrm{Lip}_1(\mathbb{R}^d)} = \sup_{x \neq y} \frac{|f(x) - f(y)|}{|x-y|} < +\infty.$$

Definition 1.2. Let $1 \leq p, q \leq \infty$ and $0 < s < 1$. If $f \in B^s_{p,q}(\mathbb{R})$, then

$$\|f\|_{B^s_{p,q}(\mathbb{R})} = \|f\|_{L_p(\mathbb{R})} + \left(\int_{|h|\leq 1} |h|^{-sq} \|\Delta_h f\|^q_{L_p(\mathbb{R})} \frac{dh}{|h|}\right)^{1/q} < +\infty.$$

Remark 1.1. The following expressions define equivalent norms in Besov spaces cf. [8, pp. 58-59]. For an integer $m \geq 1$,

$$\|f\|_{B^s_{p,q}(\mathbb{R})} = \|f\|_{L_p(\mathbb{R})} + \|f^{(m)}\|_{B^{s-m}_{p,q}(\mathbb{R})}, \quad (m < s < m+1), \tag{1}$$

and

$$\|f\|_{B^m_{p,q}(\mathbb{R})} = \|f\|_{L_p(\mathbb{R})} + \left(\int_{|h|\leq 1} |h|^{-q} \|\Delta^2_h f^{(m-1)}\|^q_{L_p(\mathbb{R})} \frac{dh}{|h|}\right)^{1/q}. \tag{2}$$

Remark 1.2. If $f \in E(\mathrm{U}^d)$, then we will identify $\|f\|_{E(\mathrm{U}^d)}$ with $\|f\|_{E(\mathbb{R},\mathbb{R}^d)}$.

(ii) *The space* $M^s_{p,q}(\mathbb{R}^d)$: For $f : \mathbb{R}^d \to \mathbb{R}$, $x = (x_1, \ldots, x_{d-1})$, $d \in \{2, 3, \ldots\}$ and $j \in \{1, \ldots, d\}$ we put

$$f_{j,x}(y) = f(x_1, \ldots, x_{j-1}, y, x_j, \ldots x_{d-1}), \quad (\forall y \in \mathbb{R}).$$

Definition 1.3. Let $1 \leq p, q \leq \infty$, $0 < s < 1$ and $d \in \{2, 3, \ldots\}$. If $f \in M^s_{p,q}(\mathbb{R}^d)$, then

$$\|f\|_{M^s_{p,q}(\mathbb{R}^d)} = \sum_{j=1}^{k} \left(\int_0^\infty t^{-sq} \left(\int_{\mathbb{R}} \sup_{|h| \leq t} \|\Delta_h f_{j,(\cdot)}(y)\|^p_{L_\infty(\mathbb{R}^{d-1})} \mathrm{d}y \right)^{q/p} \frac{\mathrm{d}t}{t} \right)^{1/q} < +\infty .$$

Remark 1.3. The extension of Definition 1.3 to the case $[s] \geq 2$ requires to replace Δ_h by $\Delta_h(\Delta_h^{L-1})$, where L is an integer $\geq 1 + [s]$. This is known as well for Besov spaces, cf. [8,9].

Remark 1.4. We obtain equivalent norm in $M^s_{p,q}(\mathbb{R}^d)$, by replacing the integration for $t \in]0, +\infty[$ by the integration for $t \in]0, 1[$, and by replacing the part of the integral for which $t \in]1, +\infty[$ by

$$\sum_{j=1}^{k} \left(\int_{\mathbb{R}} \|f_{j,(\cdot)}(y)\|^p_{L_\infty(\mathbb{R}^{d-1})} \mathrm{d}y \right)^{1/p} .$$

(iii) *Functions of bounded variation*: Let $h \in]0, +\infty]$ and $1 \leq p < \infty$. For a function $g : \mathbb{R} \to \mathbb{R}$, we denote by $\nu_p(g, h)$ the positive number defined by the supremum of

$$\left(\sum_{k=1}^{N} |g(t_k) - g(t_{k-1})|^p \right)^{1/p}$$

taken on all reals $\{t_k\}_{k=0}^N$ such that $t_0 < t_1 < \cdots < t_N$ and $|t_k - t_{k-1}| \leq h$; and if $h = +\infty$, we simply put $\nu_p(g, \infty) = \nu_p(g)$.

Definition 1.4. Let $1 \leq p < \infty$.
1- A function g is said to be of bounded p-variation if $\nu_p(g) < +\infty$.
2- The space $BV^1_p(\mathbb{R})$ is the set of the primitives of functions of bounded p-variation, and endowed with the seminorm

$$\|f\|_{BV^1_p(\mathbb{R})} = \inf \nu_p(g) ,$$

where the infimum is taken on all functions g whose primitive is f.

Remark 1.5. We have at our disposal the Peeter embedding type (cf. [3, Thm. 5])

$$B_{p,1}^{1+(1/p)}(\mathbb{R}) \hookrightarrow BV_p^1(\mathbb{R}), \quad (1 \le p < \infty). \tag{3}$$

2. Statement of the Result

For brevity, we will limit our study to the case $[s] = 1$.

Theorem 2.1. *Let $1 < p < \infty$, $1 \le q \le \infty$, and a real number s such that*

$$\frac{p}{\min(p,q)} + \frac{1}{p} < s < 2. \tag{4}$$

If a Borel-measurable function $f : \mathbb{R}^d \to \mathbb{R}$ satisfying

- *f is Lipschitz continuous,*
- *$f(0) = 0$,*
- *$\partial_j f \in M_{p,q}^{s-1,\ell oc}(\mathbb{R}^d)$, $j = 1, \ldots, d$,*

then the superposition operator T_f takes the space $B_{p,q}^s(\mathrm{U}^d)$ to $B_{p,q}^s(\mathbb{R})$.

Remark 2.1. 1- In the case $[s] \ge 2$: the condition (4) remains

$$(1/p) < s - [s] \quad \text{and} \quad \frac{p}{\min(p,q)} + \frac{1}{p} - 1 < s - [s].$$

By induction on $[s]$ we can prove this case, that we omit the details, cf. [5].
2- We obtain [5, Thm. 2] by taking $d = 1$ in Theorem 2.1.
3 It is possible to extend Theorem 2.1 to the case $s - [s] \le (1/p)$, cf. [5].

Examples. We will discuss two examples.
1- For $\beta > 1$ we consider the function $f_\beta(x) = |x|^\beta$, $(x \in \mathbb{R}^d)$. Let us accept for the moment that $\partial_j f_\beta \in M_{p,q}^{s-1,\ell oc}(\mathbb{R}^d)$ if, either $0 < s < \beta + (1/p)$ and $1 \le q < \infty$, or $s = \beta + (1/p)$ and $q = \infty$. By Theorem 2.1 we have

$$g \in B_{p,q}^s(\mathrm{U}^d) \Rightarrow |g|^\beta \in B_{p,q}^s(\mathbb{R}),$$

as soon as s, p and q satisfy (4) and, either $0 < s < \beta + (1/p)$ and $1 \le q < \infty$, or $s = \beta + (1/p)$ and $q = \infty$. Furthermore, the following elementary inequality

$$(a^\mu + b^\mu)^{1/\mu} \le a + b \le 2^{1-(1/\mu)} (a^\mu + b^\mu)^{1/\mu}, \qquad (a, b > 0, \mu > 1),$$

and the embedding $B_{p,q}^s(\mathbb{R}) \hookrightarrow B_{\beta p, \beta q}^{s/\beta}(\mathbb{R})$ yield

$$\| \, |g|^\beta \, \|_{B_{p,q}^s(\mathbb{R})} \le c \, \|g\|_{B_{p,q}^s(\mathrm{U}^d)}^\beta, \qquad \forall g \in B_{p,q}^s(\mathrm{U}^d)$$

with some constant $c > 0$ independent of g.

We now turn to the proof of the following: The function $\Phi(x) = |x|^{\beta}\psi(x)$ defined on $\mathbb{R}^d$, such that $\psi \in \mathcal{D}(\mathbb{R}^d)$ and $\beta \in \mathbb{R}$, belongs to $M^s_{p,q}(\mathbb{R}^d)$ if, either $0 < s < \beta + (1/p)$ and $1 \le q < \infty$, or $s = \beta + (1/p)$ and $q = \infty$. Indeed, we put $L = 1 + [\beta + (1/p)]$, then we have

$$\int_{|y|\le 2t} \sup_{|h|\le t} \left\|\Delta_h^L \Phi_{j,(\cdot)}(y)\right\|^p_{L_\infty(\mathbb{R}^{d-1})} dy$$
$$\le (L+1)^p \int_{|y|\le (L+2)t} \left\|\Phi_{j,(\cdot)}(y)\right\|^p_{L_\infty(\mathbb{R}^{d-1})} dy \le c\, t^{\beta p+1};$$

and by the mean value theorem we have

$$\int_{|y|\ge 2t} \sup_{|h|\le t} \left\|\Delta_h^L \Phi_{j,(\cdot)}(y)\right\|^p_{L_\infty(\mathbb{R}^{d-1})} dy \le c_1\, t^{Lp} \int_{|y|\ge 2t} |y|^{p(\beta-L)}\, dy$$
$$\le c_2\, t^{\beta p+1}.$$

2- For $\beta > 1$ and $\delta > 0$ we define

$$f_{\beta,\delta}(x) = \psi(x)\,|x|^{\beta}\,(-\log|x|)^{-\delta}, \qquad (x \in \mathbb{R}^d),$$

where $\psi \in \mathcal{D}(\mathbb{R}^d)$ with $\psi(0) \ne 0$. As in the first example, we have $\partial_j f_{\beta,\delta} \in M^{s-1,\ell oc}_{p,q}(\mathbb{R}^d)$ if, either $s < \beta + (1/p)$, or $s = \beta + (1/p)$ and $\delta > 1/q$, cf. [7, 2.3.1]. Furthermore if the parameters s, p and q satisfy (4), then $f_{\beta,\delta} \circ g \in B^s_{p,q}(\mathbb{R})$ for all $g \in B^s_{p,q}(\mathrm{U}^d)$.

3. Proof of Theorem 2.1

We propose the following formulation more precise than Theorem 2.1.

Proposition 3.1. *Let p, q, s be a real numbers as in Theorem 2.1. There exists a constant $c = c(s,p,q,d) > 0$ such that the estimate*

$$\|f \circ g\|_{B^s_{p,q}(\mathbb{R})}$$
$$\le c\Big(\|f\|_{\mathrm{Lip}_1(\mathbb{R}^d)} + \|\nabla f\|_{M^{s-1}_{p,q}(\mathbb{R}^d)}\Big)\Big(1 + \|g\|_{B^s_{p,q}(\mathrm{U}^d)}\Big)^{s-(1/p)} \tag{5}$$

holds, for all $g \in B^s_{p,q}(\mathrm{U}^d)$ and all Lipschitz continuous function $f : \mathbb{R}^d \to \mathbb{R}$ such that $f(0) = 0$ and $\partial_j f \in M^{s-1}_{p,q}(\mathbb{R}^d)$, $(j = 1, \ldots, d)$.

If $d = 1$, Proposition 3.1 is given by [4,5]. Also its proof is based on the following lemma, which is proved in [4, Prop. 2]:

Lemma 3.1. *For $1 < p < \infty$, $(1/p) < s < 1$, and $1 \le q \le \infty$, there exists a constant $c = c(s,p,q) > 0$ such that*

$$\left(\int_0^\infty \left(\frac{\nu_p(g,h)}{h^{s-(1/p)}}\right)^q \frac{dh}{h}\right)^{1/q} \le c\,\|g\|_{B^s_{p,q}(\mathbb{R})}, \quad \forall g \in B^s_{p,q}(\mathbb{R}). \tag{6}$$

Remark 3.1. The exponent $s-(1/p)$ in (5) is optimal for all functions f except the polynomials. Then as in [3, Prop., 11] we can easily obtain the following assertion: Let $s \ge 1$ and $0 < \alpha < s-(1/p)$, let $f : \mathbb{R}^d \to \mathbb{R}$ be a continuous function. If there exists a constant $c = c(f) > 0$ such that the inequality

$$\|f \circ g\|_{B^s_{p,q}(\mathbb{R})} \le c\left(1 + \|g\|_{B^s_{p,q}(\mathrm{U}^d)}\right)^\alpha$$

holds for all $g \in \mathcal{D}(\mathbb{R},\mathbb{R}^d)$, then f is a polynomial of degree less than $[s]$.

Proof of Proposition 3.1. The proof will be done on two steps. We will use, in the first, the real analytic functions in $B^s_{p,q}(\mathrm{U}^d)$, cf. [3, proof of Thm. 7]. In the second, the general case will be obtained by the property of Fatou (cf. [5,6]).

Step 1. We take $d = 2$. Let $g = (g_1, g_2) \in B^s_{p,q}(\mathrm{U}^2)$ real analytic function. By Remark 1.1 (1) we have to to estimate

$$\left(\int_0^1 h^{-(s-1)q}\left(\int_{\mathbb{R}} |\Delta_h(f\circ g)'(x)|^p\,\mathrm{d}x\right)^{q/p}\frac{\mathrm{d}h}{h}\right)^{p/q},$$

where $(f\circ g)' = (\partial_1 f\circ g + \partial_2 f \circ g)\, g_1'$. The Lipschitz continuity of f and the assumption $f(0) = 0$ yield

$$\|f \circ g\|_{L_p(\mathbb{R})} \le \|f\|_{\mathrm{Lip}_1(\mathbb{R}^2)}\|g\|_{L_p(\mathrm{U}^2)}\,. \tag{7}$$

Now from the following decomposition

$$\Delta_h(f\circ g)'(x) = \sum_{j=1}^2 \Big((\partial_j f\circ g)(x+h)\,\Delta_h g_1'(x) + g_1'(x)\,\Delta_h(\partial_j f\circ g)(x)\Big),$$

and (7), and $\|\partial g\|_{B^{s-1}_{p,q}(\mathrm{U}^2)} \le c_1\,\|g\|_{B^s_{p,q}(\mathrm{U}^2)}$, we have

$$\|f\circ g\|_{B^s_{p,q}(\mathbb{R})} \le c_2\left(\|f\|_{\mathrm{Lip}_1(\mathbb{R}^2)} + \|\nabla f\|_{L_\infty(\mathbb{R}^2)}\right)\|g\|_{B^s_{p,q}(\mathrm{U}^2)} + V(f;g)$$

where $V = V_1 + V_2$, with

$$V_j(f;g) = \left(\int_0^\infty h^{-(s-1)q}\left(\int_{\mathbb{R}} |\Delta_h(\partial_j f\circ g)(x)|^p\,|g_1'(x)|^p\mathrm{d}x\right)^{q/p}\frac{\mathrm{d}h}{h}\right)^{1/q},$$

for $j = 1, 2$. We note that the estimate of V_1 and of V_2 are similar. Again using the following decomposition

$$\Big(\int_{\mathbb{R}} |\Delta_h(\partial_1 f\circ g)(x)|^p\, |g_1'(x)|^p\,\mathrm{d}x\Big)^{1/p}$$
$$\leq \Big(\int_{\mathbb{R}} |\partial_1 f(g_1(x), g_2(x+h)) - \partial_1 f(g_1(x), g_2(x))|^p\, |g_1'(x)|^p\,\mathrm{d}x\Big)^{1/p}$$
$$+\Big(\int_{\mathbb{R}} |\partial_1 f(g_1(x+h), g_2(x+h)) - \partial_1 f(g_1(x), g_2(x+h))|^p\, |g_1'(x)|^p\,\mathrm{d}x\Big)^{1/p},$$

then we have $V_1 \leq V_3 + V_4$, where

$$V_{j+2}(f;g) =$$
$$\Big(\int_0^\infty h^{-(s-1)q}\Big(\int_{\mathbb{R}} \|\Delta_h(\partial_1 f_{j,(\cdot)}\circ g_j)(x)\|^p_{L_\infty(\mathbb{R})}\, |g_j'(x)|^p\,\mathrm{d}x\Big)^{q/p}\frac{\mathrm{d}h}{h}\Big)^{1/q}$$

for $j = 1, 2$, and also their estimates are similar.

Estimate of V_3. Let $\{I_l\}_l$ a family of nonempty open disjoint intervals defined such that the complement of $\bigcup_l I_l$ in $\mathbb{R}$ is the discrete set $\{x \in \mathbb{R} : g_1'(x) = 0\}$. For any $h > 0$ we set

$$I'_{l,h} = \{x \in I_l : \mathrm{dist}(x, \text{the right endpoint of } I_l) \geq h\} \text{ and } I''_{l,h} = I_l \setminus I'_{l,h}.$$

($I'_{l,h}$ is possibly empty). Then we write $V_3 \leq V_5 + V_6$, where

$$V_5(f;g) = \Big(\int_0^\infty h^{-(s-1)q}\Big(\sum_l \int_{I'_{l,h}} \cdots\Big)^{q/p}\frac{\mathrm{d}h}{h}\Big)^{1/q},$$

and $V_6(f;g)$ is the part corresponding to $I''_{l,h}$ in the integral with respect to h.

Estimate of V_5. We put $\alpha = \min(p,q)$, $a_l = \sup_{t\in I_l} |g_1'(t)|$ and

$$\Omega_{p,j}(f,t) = \Big(\int_{\mathbb{R}} \sup_{|v|\leq t} \|\Delta_v f_{j,(\cdot)}(y)\|^p_{L_\infty(\mathbb{R})}\,\mathrm{d}y\Big)^{1/p}, \qquad j = 1, 2.$$

The fact that the function $g_{1|_{I_l}}$ is a diffeomorphism of I_l into $g_1(I_l)$, then we have

$$|g_1(g^{-1}_{1|_{I_l}}(y) + h) - y| \leq a_l\, h \quad \text{for} \quad y \in g_{1|_{I_l}}(I'_{l,h}), \tag{8}$$

where $g^{-1}_{1|_{I_l}}$ denotes the inverse function of $g_{1|_{I_l}}$. By (8), and by a change of variable $g_{1|_{I_l}}(x) = y$ on each $I'_{l,h}$, and by the embedding $\ell_{\alpha/p} \hookrightarrow \ell_p$, and by Minkowski's inequality with respect to $L_{q/\alpha}$, successively, we find

$$\begin{aligned} V_5(f;g)^q &\leq \int_0^\infty h^{-(s-1)q}\Big(\sum_l a_l^{p-1}\,\Omega_{p,1}^p(\partial_1 f, a_l h)\Big)^{q/p}\frac{\mathrm{d}h}{h} \\ &\leq \int_0^\infty h^{-(s-1)q}\Big(\sum_l a_l^{(p-1)\alpha/p}\,\Omega_{p,1}^\alpha(\partial_1 f, a_l h)\Big)^{q/\alpha}\frac{\mathrm{d}h}{h} \\ &\leq \Big(\sum_l a_l^{(p-1)\alpha/p}\Big(\int_0^\infty h^{-(s-1)q}\,\Omega_{p,1}^q(\partial_1 f, a_l h)\,\frac{\mathrm{d}h}{h}\Big)^{\alpha/q}\Big)^{q/\alpha} \\ &= \|\partial_1 f\|^q_{M^{s-1}_{p,q}(\mathbb{R}^2)}\Big(\sum_l a_l^{\alpha(s-(1/p))}\Big)^{q/\alpha}. \end{aligned}$$

We put $I_l =]\beta_{l-1}, \beta_l[$, with $g_1'(\beta_{l-1}) = g_1'(\beta_l) = 0$. In addition, there is at least one point $\xi_l \in I_l$ such that

$$|g_1'(\xi_l)| = \sup_{t\in I_l} |g_1'(t)|.$$

Hence

$$\begin{aligned} \sum_l \sup_{t\in I_l} |g_1'(t)|^u &= \sum_l |g_1'(\xi_l) - g_1'(\beta_l)|^u \\ &\leq \nu_u(g_1')^u, \quad \text{with} \quad u = \alpha(s-(1/p)). \end{aligned} \tag{9}$$

Now by (3) we have

$$B^s_{p,q}(\mathbb{R}) \hookrightarrow B^{1+1/u}_{u,1}(\mathbb{R}) \hookrightarrow BV^1_u(\mathbb{R}).$$

We combine this chain of embeddings with (9) then we obtain

$$\begin{aligned} V_5(f;g) &\leq c_1\|\partial_1 f\|_{M^{s-1}_{p,q}(\mathbb{R}^2)}\|g_1\|^{s-(1/p)}_{BV^1_u(\mathbb{R})} \\ &\leq c_2\,\|\partial_1 f\|_{M^{s-1}_{p,q}(\mathbb{R}^2)}\|g_1\|^{s-(1/p)}_{B^s_{p,q}(\mathbb{R})}. \end{aligned}$$

Estimate of V_6. Since the nonempty intervals $I''_{l,h}$ have length equal to h, we have

$$\begin{aligned} \int_{I''_{l,h}} \|\Delta_h(\partial_1 f_{1,(\cdot)} \circ g_1)(x)\|^p_{L_\infty(\mathbb{R})}\,|g_1'(x)|^p\,\mathrm{d}x \\ \leq\ h\,(2\,\|\partial_1 f\|_\infty)^p \sup_{t\in I''_{l,h}} |g_1'(t)|^p. \end{aligned} \tag{10}$$

As above, the right endpoint of $I''_{l,h}$ is β_l, also there exists $\xi_l \in I''_{l,h}$ such that

$$\xi_l < \beta_l \quad \text{and} \quad |g_1'(\xi_l)| = \sup_{t\in I''_{l,h}} |g_1'(t)|.$$

Then we have

$$\sum_l \sup_{I''_{l,h}} |g'_1|^p = \sum_l |g'_1(\xi_l) - g'_1(\beta_l)|^p \leq \nu_p^p(g'_1, h).$$

Now by (10), and by Lemma 3.1, we conclude that

$$V_6(f;g) \leq c \|\partial_1 f\|_\infty \|g_1\|_{B^s_{p,q}(\mathbb{R})}.$$

Step 2. Let g be any function in $B^s_{p,q}(\mathrm{U}^d)$. We consider a sequence $\{\phi_j\}_{j\in\mathbb{N}}$ of the functions defined as $\phi_j(x) = j\phi(jx)$, $\forall x \in \mathbb{R}$, where $\widehat{\phi} \in \mathcal{D}(\mathbb{R})$ and $\widehat{\phi}(0) = 1$. We put

$$\widetilde{g}_j = (g_1 * \phi_j, \ldots, g_d * \phi_j), \quad (\forall j \in \mathbb{N}).$$

Clearly $\widetilde{g}_j \in \mathrm{U}^d$, and it is easy to obtain

$$\|g_\ell * \phi_j\|_{B^s_{p,q}(\mathbb{R})} \leq \|\phi\|_{L_1(\mathbb{R})} \|g_\ell\|_{B^s_{p,q}(\mathbb{R})}, \quad \forall \ell \in \{1, \ldots, d\}.$$

Then the sequence $\{\widetilde{g}_j\}_{j\in\mathbb{N}}$ is of real analytic functions, and satisfies

$$\sup_{j\geq 0} \|\widetilde{g}_j\|_{B^s_{p,q}(\mathrm{U}^d)} < +\infty \qquad \text{and} \qquad \lim_{j\to\infty} \|\widetilde{g}_j - g\|_{L_p(\mathbb{R}^d)} = 0.$$

Now, the Lipschitz continuity of f yields

$$\lim_{j\to\infty} \|f \circ \widetilde{g}_j - f \circ g\|_{L_p(\mathbb{R})} = 0.$$

Then by Step 1 and by Fatou's property, we have both $f \circ g \in B^s_{p,q}(\mathbb{R})$ and (5). □

Proof of Theorem 2.1. Let ρ be a real-valued function in $\mathcal{D}(\mathbb{R}^2)$ such that $\rho(x) = 1$ if $|x| \leq 1$ and $\rho(x) = 0$ if $|x| \geq 2$. We put $\rho_t(x) = \rho(x/t)$ for $t > 0$. Since

$$f \circ g = (f\rho_t) \circ g \quad \text{with} \quad t = \|g\|_{L_\infty(\mathbb{R})},$$

the result follows from Proposition 3.1. □

The condition "*f is Lipschitz continuous*" in Theorem 2.1 must be necessary at least locally. This is proved recently by S. E. Allaoui, cf. [1]:

Theorem 3.1. *If T_f takes $B^s_{p,q}(\mathbb{R}, \mathbb{R}^d) \cap L_\infty(\mathbb{R}^d)$ to $B^s_{p,\infty}(\mathbb{R}^d)$, then f is locally Lipschitz continuous.*

Remark 3.2. The necessity of condition $f(0) = 0$ is obvious. However the necessity of $\partial_j f \in M^{s-1,\ell oc}_{p,q}(\mathbb{R}^d)$ still open (for $d = 1$, it has been known before cf. [7, Thm. 1, p. 292]). This remark was reported by the referee in the first version of this paper. I would like to thank him.

References

1. S. E. Allaoui. *Remarques sur le calcul symbolique dans certains espaces de Besov à valeurs vectorielles* (submitted).
2. G. Bourdaud. *Une propriété de composition dans l'espace H^s (II).* C. R. Acad. Sci. Paris, Ser. I **342** (2006), 243–246.
3. G. Bourdaud, M. Lanza de Cristoforis, W. Sickel. *Superposition operators and functions of bounded p-variation.* Rev. Mat. Iberoamer. **22** (2006), 455–485.
4. G. Bourdaud, M. Moussai, W. Sickel. *An optimal symbolic calculus on Besov algebras.* Ann. Inst. H. Poincaré (C) Anal. Non Linéarire **23** (2006), 949–956.
5. G. Bourdaud, M. Moussai, W. Sickel, *Towards sharp superposition theorems in Besov and Lizorkin-Triebel spaces.* Nonlinear Anal. **68** (2008), 2889–2912.
6. J. Franke. *On the spaces $F^s_{p,q}(\mathbb{R})$ of Triebel-Lizorkin type: Pointwise multipliers and spaces on domains.* Math. Nachr. **125** (1986), 29–68.
7. T. Runst, W. Sickel. *Sobolev spaces of fractional order, Nemytskij operators, and nonlinear partial differential equations.* De Gruyter, Berlin, 1996.
8. H. Triebel. *Theory of Function Spaces.* Birkhäuser, Basel, 1983.
9. H. Triebel. *Theory of Function Spaces II.* Birkhäuser, Basel, 1992.

POINTWISE APPROXIMATION FOR CERTAIN MIXED SZÁSZ-BETA OPERATORS

QIULAN QI* and YUPING ZHANG†

College of Mathematics and Information Science
Hebei Normal University
Shijiazhuang 050016, People's Republic of China
**E-mail: qiqiulan@hebtu.edu.cn*
†E-mail: yuping.zh@163.com.

We study the mixed summation-integral type operators having Szász and beta basis functions. We obtain the pointwise approximation equivalence theorems using the unified modulus of smoothness.

Keywords: Szász-Beta operators, Approximation, Modulus of smoothness, K-functional.

1. Introduction

In recent years, there are many results for summation-integral type operators (see [3, 4, 7–11]) and Szász type operators (see [5]). Recently, Z. Finta [3] and V. Gupta [9] introduced a mixed sequence of summation-integral type operators defined for $x \in [0, \infty)$ by

$$S_n(f,x) = \sum_{v=1}^{\infty} s_{n,v}(x) \int_0^{\infty} b_{n,v}(t) f(t) dt + s_{n,0}(x) f(0), \tag{1}$$

where $\delta(t)$ is the Dirac delta function, and

$$s_{n,v}(x) = e^{-nx} \frac{(nx)^v}{v!},$$

$$b_{n,v}(t) = \frac{1}{B(n+1,v)} \frac{t^{v-1}}{(1+t)^{n+v+1}} = \frac{(n+v)!}{(v-1)!n!} \frac{t^{v-1}}{(1+t)^{n+v+1}}$$

are respectively Szász and beta basis functions. We define

$$s_{n,v}(x) = 0 \ (v < 0), \qquad b_{n,v}(t) = 0 \ (v < 1).$$

It is easily verified that the operators (1) are linear positive.

As far as approximation properties of the operators S_n are concerned, in [4] the authors extended the study of [9] and obtained some direct results and global direct approximation theorems using $\omega_\varphi^2(f,t)_\infty$. Some other operators of type S_n were recently considered in [7,8].

Ditzian [1] introduced $\omega_{\varphi^\lambda}^2(f,t)$ $(0 \le \lambda \le 1)$ and obtained a direct theorem with $\omega_{\varphi^\lambda}^2(f,t)$ for the Bernstein operator. In this paper, we introduce the idea of Ditzian to the operators S_n with the unified modulus of smoothness $\omega_{\varphi^\lambda}^2(f,t)$, and obtain direct, inverse and equivalence theorems. To state our results, we give some notation: If $0 \le \lambda \le 1$, let

$$\omega_{\varphi^\lambda}^2(f,t) = \sup_{0<h\le t} \sup_{x\pm h\varphi^\lambda(x)\in[0,\infty)} |\Delta_{h\varphi^\lambda}^2 f|,$$

$$D_\lambda^2 = \left\{f \in C_B[0,\infty) : f' \in A.C.loc_{(0,\infty)},\ \|\varphi^{2\lambda} f''\| < +\infty\right\},$$

$$K_{\varphi^\lambda}^2(f,t^2) = \inf\left\{\|f-g\| + t^2\|\varphi^{2\lambda}g''\| : g \in D_\lambda^2\right\},$$

$$\overline{D}_\lambda^2 = \{f \in D_\lambda^2 : \|f''\| < +\infty\},$$

$$\overline{K}_{\varphi^\lambda}^2(f,t^2) = \inf\left\{\|f-g\| + t^2\|\varphi^{2\lambda}g''\| + t^{\frac{4}{2-\lambda}}\|g''\| : g \in \overline{D}_\lambda^2\right\}.$$

It is well known that [2] $\omega_{\varphi\lambda}^2(f,t) \sim K_{\varphi^\lambda}^2(f,t^2) \sim \overline{K}_{\varphi^\lambda}^2(f,t^2)$. Our main results can be stated as follows:

Theorem 1.1. *For $f \in C_B[0,\infty)$, $0 \le \lambda \le 1$, $\delta_n^2(x) = \varphi^2(x) + \frac{1}{n}$, $\varphi(x) = \sqrt{x(2+x)}$, $n \ge 4$, then $|S_n(f,x) - f(x)| \le C\omega_{\varphi^\lambda}^2(f, n^{-\frac{1}{2}}\delta_n^{1-\lambda}(x))$.*

Theorem 1.2. *For $f \in C_B[0,\infty)$, $0 < \alpha < 2$, $0 \le \lambda \le 1$, $\delta_n^2(x) = \varphi^2(x) + \frac{1}{n}$, $\varphi(x) = \sqrt{x(2+x)}$, $n \ge 3$, then $|S_n(f,x) - f(x)| = O\big((n^{-\frac{1}{2}}\delta_n^{1-\lambda}(x))^\alpha\big)$ implies $\omega_{\varphi^\lambda}^2(f,t) = O(t^\alpha)$.*

From Theorems 1.1 and 1.2, we can easily get the equivalence theorem.

Theorem 1.3. *For $f \in C_B[0,\infty)$, $0 < \alpha < 2$, $0 \le \lambda \le 1$, $\delta_n^2(x) = \varphi^2(x) + \frac{1}{n}$, $\varphi(x) = \sqrt{x(2+x)}$, $n \ge 4$, then $|S_n(f,x) - f(x)| = O\big((n^{-\frac{1}{2}}\delta_n^{1-\lambda}(x))^\alpha\big)$ if and only if $\omega_{\varphi^\lambda}^2(f,t) = O(t^\alpha)$.*

Remark 1.1. $f(x) \sim g(x)$ means that there exists a positive constant C such that $C^{-1}f(x) \le g(x) \le Cf(x)$, and $f(x) = O(g(x))$ means that there exists a positive constant C such that $f(x) \le Cg(x)$.

Remark 1.2. $C_B[0,\infty)$ is the space of all real-valued continuous bounded functions on $[0,\infty)$. The norm $\|\cdot\|$ on it is $\|f\| = \sup\{|f(x)| : x \in [0,\infty)\}$.

Throughout this paper, C denotes a constant independent of n and x which is not necessarily the same in different cases.

2. The Direct Theorem

Lemma 2.1. *If $f \in C_B[0,\infty)$, then $\|S_n f\| \le \|f\|$.*

Proof. Noting that $\int_0^\infty b_{n,v}(t)dt = 1$ and $\sum_{v=0}^\infty s_{n,v}(x) = 1$, we have

$$|S_n(f,x)| \le \sum_{v=1}^{\infty} s_{n,v}(x) \int_0^\infty b_{n,v}(t)|f(t)|dt + e^{-nx}|f(0)| \le \|f\|. \qquad \square$$

Lemma 2.2. *Let $\delta_n^2(x) = \varphi^2(x) + 1/n$. Then $S_n(1,x) = 1$, $S_n(t-x,x) = 0$, $S_n((t-x)^2,x) \le 2\varphi^2(x)/n$ for $n \ge 2$, $S_n((t-x)^4,x) \le C\delta_n^4(x)/n^2$ for $n \ge 3$.*

Proof. From the recursion relation in [9, Lemma 2], by simple calculations, we can obtain these results about moments. $\square$

Lemma 2.3. *(1) For u between t and x, we have [2]*

$$\frac{|t-u|}{\varphi^{2\lambda}(u)} \le \frac{|t-x|}{x^\lambda}\left(\frac{1}{(2+x)^\lambda} + \frac{1}{(2+t)^\lambda}\right).$$

(2) For u between t and x, $x \in E_n^c = [0, n^{-1})$, we have $\dfrac{|t-u|}{\delta_n^{2\lambda}(u)} \le \dfrac{|t-x|}{\delta_n^{2\lambda}(x)}$.

Proof. For $x < t$, the proof is obvious. For $x > t$, observe that the function $g(x) = \left(x/\delta_n^{2\lambda}(x)\right)^\lambda$ is monotone increasing in E_n^c and combine it with $|t-u|x \le |t-x|u$. $\square$

Lemma 2.4. *$S_n((2+t)^{-2},x) \le 20(2+x)^{-2}$ for $n \ge 2$.*

Proof. Since $1 + x \sim 2 + x$, it is sufficient to show that $(1+x)^2 S_n((1+t)^{-2},x) \le 20$.

$$\begin{aligned}&(1+x)^2 S_n((1+t)^{-2},x)\\ &= (1+x)^2\left(\sum_{v=1}^{\infty} s_{n,v}(x)\int_0^\infty \frac{(n+1)(n+2)}{(n+v+1)(n+v+2)} b_{n+2,v}(t)dt + e^{-nx}\right) =: I.\end{aligned}$$

Case I: $x \le 1$. By $(1+x)^2 \le 4$, $\frac{(n+1)(n+2)}{(n+v+1)(n+v+2)} \le 1$, we have

$$I \le 4\left(\sum_{v=1}^{\infty} s_{n,v}(x)\int_0^\infty b_{n+2,v}(t)dt + e^{-nx}\right) = 4. \qquad (2)$$

Case II: $x > 1$. Using

$$(1+x)^2 \le 4x^2, \qquad \frac{(n+1)(n+2)}{(n+v+1)(n+v+2)}\,\frac{(v+1)(v+2)}{n^2} \le 4,$$

for $n \geq 2$, we have

$$I \leq 4\left(\sum_{v=1}^{\infty} s_{n,v+2}(x)\int_0^{\infty}\frac{(n+1)(n+2)(v+1)(v+2)}{(n+v+1)(n+v+2)n^2}b_{n+2,v}(t)dt + x^2e^{-nx}\right)$$
$$\leq 20. \tag{3}$$

The proof of Lemma 2.4 is completed from (2) and (3). □

Proof of Theorem 1.1. By the definition of $\overline{K}_{\varphi^\lambda}(f,t^2)$ and the equivalence of $\omega^2_{\varphi^\lambda}(f,t)$ and $\overline{K}^2_{\varphi^\lambda}(f,t^2)$, we can choose $g \in \overline{D}^2_\lambda$ such that

$$\|f-g\| + \left(n^{-\frac{1}{2}}\delta_n^{1-\lambda}(x)\right)^2\|\varphi^{2\lambda}g''\| + \left(n^{-\frac{1}{2}}\delta_n^{1-\lambda}(x)\right)^{\frac{4}{2-\lambda}}\|g''\|$$
$$\leq C\omega^2_{\varphi^\lambda}(f, n^{-\frac{1}{2}}\delta_n^{1-\lambda}(x)).$$

Using the Taylor expansion of $g(t)$, we will consider $x \in E_n$ and $x \in E_n^c$ respectively.

For $x \in E_n$, $\delta_n^2(x) \sim \varphi^2(x)$, by Lemmas 2.2–2.4 and the Hölder inequality, we have

$$\left|S_n\left(\int_x^t (t-u)g''(u)du, x\right)\right|$$
$$\leq S_n\left(\frac{(t-x)^2}{\varphi^{2\lambda}(x)}, x\right)\|\varphi^{2\lambda}g''\| + S_n\left(\frac{(t-x)^2}{x^\lambda(2+t)^\lambda}, x\right)\|\varphi^{2\lambda}g''\|$$
$$\leq C\frac{\varphi^{2(1-\lambda)}(x)}{n}\|\varphi^{2\lambda}g''\| + \left[S_n\left(\frac{1}{(2+t)^2}, x\right)\right]^{\lambda/2}\left[S_n\left(\frac{(t-x)^4}{x^{2\lambda}}, x\right)\right]^{1/2}\|\varphi^{2\lambda}g''\|$$
$$\leq C\frac{\varphi^{2(1-\lambda)}(x)}{n}\|\varphi^{2\lambda}g''\| + ((2+x)^{-2})^{\lambda/2}n^{-1}\varphi^2(x)x^{-\lambda}\|\varphi^{2\lambda}g''\|$$
$$\leq C\left(n^{-\frac{1}{2}}\varphi^{1-\lambda}(x)\right)^2\|\varphi^{2\lambda}g''\| \leq C\omega^2_{\varphi\lambda}(f, n^{-1/2}\delta_n^{1-\lambda}(x)).$$

Hence, we get for $x \in E_n$,

$$|S_n(g,x) - g(x)| \leq C\omega^2_{\varphi\lambda}(f, n^{-\frac{1}{2}}\delta_n^{1-\lambda}(x)). \tag{4}$$

For $x \in E_n^c$, $\delta_n^2(x) \sim 1/n$, by Lemmas 2.2 and 2.3, we get

$$\left|S_n\left(\int_x^t (t-u)g''(u)du, x\right)\right| \leq S_n\left(\frac{(t-x)^2}{\delta_n^{2\lambda}(x)}, x\right)\|\delta_n^{2\lambda}g''\|$$
$$\leq C\left(n^{-\frac{1}{2}}\delta_n^{1-\lambda}(x)\right)^2\|\delta_n^{2\lambda}g''\|$$
$$\leq C\left(n^{-\frac{1}{2}}\delta_n^{1-\lambda}(x)\right)^2\|\varphi^{2\lambda}g''\| + C\left(n^{-\frac{1}{2}}\delta_n^{1-\lambda}(x)\right)^{\frac{4}{2-\lambda}}\|g''\|$$
$$\leq C\omega^2_{\varphi\lambda}(f, n^{-\frac{1}{2}}\delta_n^{1-\lambda}(x)).$$

Therefore

$$|S_n(g,x)-g(x)| \le C\omega^2_{\varphi\lambda}(f, n^{-\frac{1}{2}}\delta_n^{1-\lambda}(x)). \tag{5}$$

From (4), (5) and Lemma 2.1, we can write

$$\begin{aligned}|S_n(f,x)-f(x)| &\le C\left(\|f-g\| + |S_n(g,x)-g(x)|\right)\\ &\le C\omega^2_{\varphi^\lambda}(f, n^{-\frac{1}{2}}\delta_n^{1-\lambda}(x)).\end{aligned}$$

□

3. The Inverse Theorem

To prove Theorem 1.2, we need some new notation. Let $C^0 = \{f \in C_B[0,\infty), f(0) = 0\}$, $C^2 = \{f \in C^0, f'' \in C_B[0,\infty)\}$, $\|f\|_0 = \sup\limits_{x\in[0,\infty)} \{|\delta_n^{\alpha(\lambda-1)}(x)f(x)|\}$, $\|f\|_2 = \sup\limits_{x\in[0,\infty)} \{|\delta_n^{2+\alpha(\lambda-1)}(x)f''(x)|\}$, $C^0_{\alpha,\lambda} = \{f \in C^0, \|f\|_0 < +\infty\}$, $C^2_{\alpha,\lambda} = \{f \in C^2, f' \in A.C.loc, \|f\|_2 < +\infty\}$, and $K^2_{\alpha,\lambda}(f,t^2) = \inf\limits_{g\in C^2_{\alpha,\lambda}} \{\|f-g\|_0 + t^2\|g\|_2\}$.

Lemma 3.1. *Let* $\varphi_1^2(t) = t(1+t)$, $t \in E_n = [1/n, \infty)$. *Then*

$$\left|\varphi_1^2(t)b''_{n,v}(t)\right| \le C\sum_{i=0}^{2} n^{1+\frac{i}{2}} b_{n,v}(t)\left|\frac{v-1-2t}{n} - t\right|^i \varphi_1^{-i}(t).$$

Proof. We use the relations

$$b'_{n,v}(t) = \frac{n}{t(1+t)}b_{n,v}(t)\left(\frac{v-1}{n} - \frac{n+2}{n}t\right),$$

$$|b''_{n,v}(t)| \le \frac{1}{(t(1+t))^2}\sum_{i=0}^{2} Q_i(n,t)n^i b_{n,v}(t)\left|\frac{v-1}{n} - \frac{n+2}{n}t\right|^i,$$

where $Q_0(n,t) = nt(1+t)$, $Q_1(n,t) = 2t+1$, $Q_2(n,t) = 1$. On the other hand

$$\left|(t(1+t))^{-2}Q_i(n,t)n^i\right| \le C\left(\frac{n}{t(1+t)}\right)^{1+\frac{i}{2}},$$

which implies the lemma. □

Lemma 3.2. *For* $\varphi_1^2(x) = x(1+x)$, *one has*

$$\int_{E_n} b_{n,v}(x)\varphi_1^{-2i}(x)\left(\frac{v}{n} - x\right)^{2i} dx \le Cn^{-i}, \quad i \in N_0 = N \cup \{0\}.$$

Proof. We use induction since $\int_{E_n} b_{n,v}(x)dx \le 1$ is our lemma for the case $i=0$. We assume that

$$\int_{E_n} b_{n,v}(x)\varphi_1^{-2l}(x)\left(\frac{v}{n}-x\right)^{2l} dx \le Cn^{-l}.$$

Now we only need to show that

$$\int_{E_n} b_{n,v}(x)\varphi_1^{-2l-2}(x)\left(\frac{v}{n}-x\right)^{2l+2} dx \le Cn^{-l-1}. \tag{6}$$

For $v \ge 10l+10$, using the integration by parts, we write

$$\int_{E_n} \frac{b_{n,v}(x)}{x^l(1+x)^l}\left(\frac{v}{n}-x\right)^{2l} dx = \frac{1}{2l+1}\frac{b_{n,v}(\frac{1}{n})}{(\frac{1}{n})^l(1+\frac{1}{n})^l}\left(\frac{v}{n}-\frac{1}{n}\right)^{2l+1}$$
$$+\frac{2l+n+2}{2l+1}\int_{E_n} \frac{b_{n,v}(x)}{x^{l+1}(1+x)^{l+1}}\left(\frac{v}{n}-x\right)^{2l+1}\left(\frac{v-l-1}{2l+n+2}-x\right)dx,$$

or

$$\left|\int_{E_n} \frac{b_{n,v}(x)}{x^{l+1}(1+x)^{l+1}}\left(\frac{v}{n}-x\right)^{2l+1}\left(\frac{v-l-1}{2l+n+2}-x\right)dx\right|$$
$$\le Cn^{-l-1}+\frac{1}{n}\frac{b_{n,v}(\frac{1}{n})}{(\frac{1}{n})^l(1+\frac{1}{n})^l}\left(\frac{v}{n}-\frac{1}{n}\right)^{2l+1}.$$

Since

$$\frac{b_{n,v}(\frac{1}{n})}{(\frac{1}{n})^l(1+\frac{1}{n})^l}\left(\frac{v}{n}-\frac{1}{n}\right)^{2l+1} \le Cn^{-l},$$

we can get

$$\left|\int_{E_n} \frac{b_{n,v}(x)}{x^{l+1}(1+x)^{l+1}}\left(\frac{v}{n}-x\right)^{2l+1}\left(\frac{v-l-1}{2l+n+2}-x\right)dx\right| = O(n^{-l-1}).$$

Let [2, pp. 125–129]

$$G_{n,v}=\left\{x : \frac{nv-2(l+1)(n+v)}{n(n+2l+2)} < x \le \frac{v+1}{n}\right\} \subset E_n=[1/n,\infty).$$

On $G_{n,v}^c$, we have the relations

$$\left(\frac{v}{n}-x\right)\left(\frac{v-l-1}{n+2l+2}-x\right) > 0, \quad \left|\frac{v}{n}-x\right| \le C\left|\frac{v-l-1}{n+2l+2}-x\right|.$$

We get

$$\int_{E_n} \frac{b_{n,v}(x)}{x^{l+1}(1+x)^{l+1}}\left|\frac{v}{n}-x\right|^{2l+2} dx$$
$$=\int_{E_n\setminus G_{n,v}} \frac{b_{n,v}(x)}{x^{l+1}(1+x)^{l+1}}\left|\frac{v}{n}-x\right|^{2l+2} dx+\int_{G_{n,v}} \frac{b_{n,v}(x)}{x^{l+1}(1+x)^{l+1}}\left|\frac{v}{n}-x\right|^{2l+2} dx$$
$$=: T_1+T_2, \tag{7}$$

$$\int_{E_n\setminus G_{n,v}} \frac{b_{n,v}(x)}{x^{l+1}(1+x)^{l+1}}\left(\frac{v}{n}-x\right)^{2l+1}\left(\frac{v-l-1}{n+2l+2}-x\right)dx$$
$$\le C\left|\int_{E_n} \frac{b_{n,v}(x)}{x^{l+1}(1+x)^{l+1}}\left(\frac{v}{n}-x\right)^{2l+1}\left(\frac{v-l-1}{n+2l+2}-x\right)dx\right|$$
$$+C\int_{G_{n,v}} \frac{b_{n,v}(x)}{x^{l+1}(1+x)^{l+1}}\left|\frac{v}{n}-x\right|^{2l+1}\left|\frac{v-l-1}{n+2l+2}-x\right|dx = O(n^{-l-1})+T_3.$$

One obtains

$$T_1 = O\left(n^{-l-1}\right) + T_3. \tag{8}$$

The integrals T_2 and T_3 are defined on $G_{n,v}$. For $x \in G_{n,v}$, one has

$$\left|\frac{v}{n}-x\right| = O\left(\frac{1}{n}\left(1+\frac{v}{n}\right)\right), \quad \left|\frac{v-l-1}{n+2l+2}-x\right| = O\left(\frac{1}{n}\left(1+\frac{v}{n}\right)\right),$$
$$\varphi_1^2(x) \ge \frac{1}{2}\left(\frac{v}{n}\left(1+\frac{v}{n}\right)\right).$$

We can deduce

$$T_2 = O(n^{-l-1}), \qquad T_3 = O(n^{-l-1}). \tag{9}$$

From (7)–(9), we get the relation (6) for $v \ge 10l+10$.

For $v < 10l+10$, we have $|\frac{v}{n}-x| \le C(l)|x|$ and

$$\int_{E_n} b_{n,v}(x)\varphi_1^{-2l-2}(x)\left(\frac{v}{n}-x\right)^{2l+2}dx \le C(l)\int_{E_n} b_{n,v}(x)\frac{x^{l+1}}{(1+x)^{l+1}}dx$$
$$\le C(l)\int_{E_n} b_{n,v+l+1}(x)\frac{(v+l)!(n+v)!}{(n+v+l+1)!(v-1)!}dx \le Cn^{-l-1}. \qquad \square$$

Lemma 3.3. *If $c_1, c_2, c_3, c_4 \in Z, k=2$ or $k=-2$, we have*

$$\sum_{v=1}^{\infty} s_{n+c_1,v+c_2}(x)\int_0^{\infty} b_{n+c_3,v+c_4}(t)\delta_n^k(t)dt \le C\delta_n^k(x).$$

Proof. For $k=2$, by direct calculation, we have

$$\sum_{v=1}^{\infty} s_{n+c_1,v+c_2}(x)\int_0^{\infty} b_{n+c_3,v+c_4}(t)\delta_n^2(t)dt$$
$$= \sum_{v=1}^{\infty} s_{n+c_1,v+c_2}(x)\int_0^{\infty} b_{n+c_3,v+c_4}(t)\varphi^2(t)dt + \frac{1}{n}$$
$$= \sum_{v=1}^{\infty} s_{n+c_1,v+c_2}(x)\int_0^{\infty} \frac{(v+c_4)(v+c_4+1)}{(n+c_3)(n+c_3-1)}b_{n+c_3-2,v+c_4+2}(t)dt$$
$$+2\sum_{v=1}^{\infty} s_{n+c_1,v+c_2}(x)\int_0^{\infty} \frac{v+c_4}{n+c_3}b_{n+c_3-1,v+c_4+1}(t)dt + \frac{1}{n}$$

$$= \sum_{v=1}^{\infty} s_{n+c_1,v+c_2}(x) \frac{(v+c_4)(v+c_4+1)}{(n+c_3)(n+c_3-1)} + 2\sum_{v=1}^{\infty} s_{n+c_1,v+c_2}(x) \frac{v+c_4}{n+c_3} + \frac{1}{n}$$
$$\leq C(x^2+2x) + \frac{1}{n} \leq C\delta_n^2(x).$$

For $k=-2$, by a similar calculation, we complete our proof. □

Lemma 3.4. *For* $f \in C^2_{\alpha,\lambda}$, $0 < \alpha < 2$, $0 \leq \lambda \leq 1$, $n \geq 2$, *we have* $\|S_n f\|_2 \leq C\|f\|_2$.

Proof. Using the expressions $s''_{n,v}(x) = n^2 \sum\limits_{j=0}^{2} \binom{2}{j} s_{n,v-j}(x)(-1)^{2-j}$ and $b''_{n,v}(x) = (n+1)(n+2) \sum\limits_{j=0}^{2} \binom{2}{j} b_{n+2,v-j}(x)(-1)^{2-j}$, and also noting that $f \in C^2_{\alpha,\lambda}$, $f(0)=0$, we have

$$S''_n(f,x) = \frac{n}{n-1} \sum_{v=1}^{\infty} s_{n,v}(x) \int_0^{\infty} b''_{n-2,v+2}(t) f(t)dt. \tag{10}$$

By integration by parts, Jensen's inequality and Lemma 3.3, we write

$$\left|\delta_n^{2+\alpha(\lambda-1)}(x) S''_n(f,x)\right|$$
$$\leq C\|f\|_2 \delta_n^{2+\alpha(\lambda-1)}(x) \sum_{v=1}^{\infty} s_{n,v}(x) \int_0^{\infty} b_{n-2,v+2}(t) \delta_n^{-2+\alpha(1-\lambda)}(t)dt$$
$$\leq C\|f\|_2 \delta_n^{2+\alpha(\lambda-1)}(x) \left[\sum_{v=1}^{\infty} s_{n,v}(x) \int_0^{\infty} b_{n-2,v+2}(t) \delta_n^{-2}(t)dt\right]^{\frac{2+\alpha(\lambda-1)}{2}}$$
$$\leq C\|f\|_2.$$
□

Lemma 3.5. *For* $f \in C^0_{\alpha,\lambda}$, $0 < \alpha < 2$, $0 \leq \lambda \leq 1$, $n \geq 3$, *we have* $\|S_n f\|_2 \leq Cn\|f\|_0$.

Proof. For $x \in E_n$, $m = 0, 1, 2, \cdots$, one has [2]

$$\sum_{v=0}^{\infty} s_{n,v}(x) \left(\frac{v}{n} - x\right)^{2m} \leq C\frac{x^m}{n^m}, \tag{11}$$

$$\left|x s''_{n,v}(x)\right| \leq C \sum_{i=0}^{2} \frac{n^{1+\frac{i}{2}} s_{n,v}(x)}{x^{i/2}} \left|\frac{v}{n} - x\right|^i. \tag{12}$$

Next we will estimate the lemma in the two cases $x \in E_n^c$ and $x \in E_n$.

Case I: For $x \in E_n^c$, noting that $\delta_n^2(x) \sim 1/n$, using (10), Jensen's inequality and Lemma 3.3, we have

$$\left|\delta_n^{2+\alpha(\lambda-1)}(x)\sum_{v=1}^{\infty} s_{n,v}(x)\int_0^{\infty} b''_{n-2,v+2}(t)f(t)dt\right|$$

$$\leq C\|f\|_0\delta_n^{2+\alpha(\lambda-1)}(x)n^2\sum_{v=1}^{\infty} s_{n,v}(x)\int_0^{\infty}[b_{n,v+2}+2b_{n,v+1}+b_{n,v}]\delta_n^{\alpha(1-\lambda)}dt$$

$$\leq C\|f\|_0\delta_n^{2+\alpha(\lambda-1)}(x)n^2\left[\sum_{v=1}^{\infty} s_{n,v}(x)\int_0^{\infty}(b_{n,v+2}+2b_{n,v+1}+b_{n,v})\delta_n^2 dt\right]^{\frac{\alpha(1-\lambda)}{2}}$$

$$\leq Cn\|f\|_0.$$

Therefore, for $x \in E_n^c$, we get

$$\left|\delta_n^{2+\alpha(\lambda-1)}(x)S''_n(f,x)\right| \leq Cn\|f\|_0. \tag{13}$$

Case II: For $x \in E_n$, $\delta_n^2(x) \sim \varphi^2(x)$, we write

$$\left|\delta_n^{2+\alpha(\lambda-1)}(x)S''_n(f,x)\right| \leq \left|\delta_n^{2+\alpha(\lambda-1)}(x)\frac{2}{2+x}\sum_{v=1}^{\infty} s''_{n,v}(x)\int_0^{\infty} b_{n,v}(t)f(t)dt\right|$$
$$+\left|\delta_n^{2+\alpha(\lambda-1)}(x)\frac{x}{2+x}\sum_{v=1}^{\infty} s''_{n,v}(x)\int_0^{\infty} b_{n,v}(t)f(t)dt\right| =: I_1 + I_2. \tag{14}$$

In order to complete our proof, we need estimate I_1 and I_2. Choosing $m \in N$ such that $\frac{1}{2m} + \frac{\alpha(1-\lambda)}{2} < 1$, using (11), (12), Lemma 3.3, Hölder's and Jensen's inequalities, we have

$$\left|\delta_n^{2+\alpha(\lambda-1)}(x)\frac{2}{2+x}\sum_{v=1}^{\infty} s''_{n,v}(x)\int_0^{\infty} b_{n,v}(t)dt\right|$$

$$\leq C\varphi^{\alpha(\lambda-1)}(x)\|f\|_0\sum_{i=0}^{2} n^{1+\frac{i}{2}}\sum_{v=1}^{\infty} s_{n,v}(x)x^{-\frac{i}{2}}\left|\frac{v}{n}-x\right|^i\int_0^{\infty} b_{n,v}f\delta_n^{\alpha(1-\lambda)}dt$$

$$\leq C\varphi^{\alpha(\lambda-1)}(x)\|f\|_0\sum_{i=0}^{2} n^{1+\frac{i}{2}}x^{-\frac{i}{2}}\left\{\sum_{v=1}^{\infty} s_{n,v}(x)\left(\frac{v}{n}-x\right)^{2mi}\right\}^{\frac{1}{2m}}$$
$$\cdot\left\{\sum_{v=1}^{\infty} s_{n,v}(x)\int_0^{\infty} b_{n,v}(t)\delta_n^2(t)dt\right\}^{\frac{\alpha(1-\lambda)}{2}}$$

$$\leq C\varphi^{\alpha(\lambda-1)}(x)\|f\|_0\sum_{i=0}^{2} n^{1+\frac{i}{2}}n^{-\frac{i}{2}}\delta_n^{\alpha(1-\lambda)}(x) \leq Cn\|f\|_0.$$

Then

$$I_1 \leq Cn\|f\|_0. \tag{15}$$

Noting that when $x \leq 1$, $I_2 \leq I_1$, from (13)–(15), we can get our result. For $x > 1$, we can split I_2 into two parts

$$\begin{aligned} I_2 \leq & \left|\delta_n^{2+\alpha(\lambda-1)}(x)\frac{x}{2+x}\sum_{v=1}^{\infty} s''_{n,v}(x)\int_0^1 b_{n,v}(t)f(t)dt\right| \\ & + \left|\delta_n^{2+\alpha(\lambda-1)}(x)\frac{x}{2+x}\sum_{v=1}^{\infty} s''_{n,v}(x)\int_1^{\infty} b_{n,v}(t)f(t)dt\right| =: J_1 + J_2. \end{aligned} \tag{16}$$

Using (11), (12), Lemma 3.3, Hölder's and Jensen's inequalities, we have

$$\begin{aligned} & \left|\delta_n^{2+\alpha(\lambda-1)}(x)\frac{1}{2+x}\sum_{i=0}^{2} n^{1+\frac{i}{2}}\sum_{v=1}^{\infty} s_{n,v}(x)x^{-\frac{i}{2}}\left|\frac{v}{n}-x\right|^i\int_0^1 b_{n,v}(t)f(t)dt\right| \\ & \leq C\|f\|_0\varphi^{\alpha(\lambda-1)}(x)\sum_{i=0}^{2} n^{1+\frac{i}{2}}\sum_{v=1}^{\infty}\frac{s_{n,v+1}(x)}{x^{i/2}}\left|\frac{v}{n}-x\right|^i\int_0^1\frac{v+1}{n}b_{n,v}\delta_n^{\alpha(1-\lambda)}dt \\ & \leq C\|f\|_0\varphi^{\alpha(\lambda-1)}(x)\sum_{i=0}^{2} n^{1+\frac{i}{2}}\sum_{v=1}^{\infty}\frac{s_{n,v+1}(x)}{x^{i/2}}\left|\frac{v+1}{n}-x\right|^i \\ & \quad\cdot\int_0^1[b_{n+1,v-1}(t)+b_{n,v}(t)]\delta_n^{\alpha(1-\lambda)}(t)dt \\ & \quad + C\|f\|_0\varphi^{\alpha(\lambda-1)}(x)\sum_{i=0}^{2} n^{1+\frac{i}{2}}\sum_{v=1}^{\infty}\frac{s_{n,v+1}(x)}{x^{i/2}n^i}\int_0^1[b_{n+1,v-1}+b_{n,v}]\delta_n^{\alpha(1-\lambda)}dt \\ & \leq C\|f\|_0\varphi^{\alpha(\lambda-1)}(x)\sum_{i=0}^{2} n^{1+\frac{i}{2}}x^{-\frac{i}{2}}\left\{\sum_{v=1}^{\infty} s_{n,v+1}(x)\left(\frac{v+1}{n}-x\right)^{2mi}\right\}^{\frac{1}{2m}} \\ & \quad\cdot\left\{\sum_{v=1}^{\infty} s_{n,v+1}(x)\int_0^1[b_{n+1,v-1}(t)+b_{n,v}(t)]\delta_n^2(t)dt\right\}^{\frac{\alpha(1-\lambda)}{2}} \\ & \quad + C\|f\|_0\varphi^{\alpha(\lambda-1)}(x)\sum_{i=0}^{2}\frac{n^{1+\frac{i}{2}}}{x^{i/2}n^i}\left[\sum_{v=1}^{\infty} s_{n,v+1}(x)\int_0^1[b_{n+1,v-1}+b_{n,v}]\delta_n^2 dt\right]^{\frac{\alpha(1-\lambda)}{2}} \\ & \leq C\|f\|_0\varphi^{\alpha(\lambda-1)}(x)\sum_{i=0}^{2} n^{1+\frac{i}{2}}x^{-\frac{i}{2}}\left(\frac{x^{mi}}{n^{mi}}\right)^{\frac{1}{2m}}\delta_n^{\alpha(1-\lambda)}(x) \\ & \quad + C\|f\|_0\varphi^{\alpha(\lambda-1)}(x)\sum_{i=0}^{2} n^{1+\frac{i}{2}}n^{\frac{i}{2}}n^{-i}\delta_n^{\alpha(1-\lambda)}(x) \leq Cn\|f\|_0. \end{aligned}$$

We now get

$$J_1 \le Cn\|f\|_0. \tag{17}$$

Using (11), (12) and Lemma 3.1, we write

$$\begin{aligned}
J_2 \le& \left|\delta_n^{2+\alpha(\lambda-1)}(x)\frac{x}{2+x}\sum_{v=1}^{\infty} s_{n,v}(x)\int_1^{\infty} b''_{n-2,v+2}(t)f(t)dt\right| \\
\le& C\|f\|_0\varphi^{\alpha(\lambda-1)}(x)x^2\sum_{v=1}^{\infty} s_{n,v}(x)\int_1^{\infty}\sum_{i=0}^{2}\frac{n^{1+\frac{i}{2}}b_{n-2,v+2}}{\varphi_1^i(t)}\left|\frac{v+1-2t}{n-2}-t\right|^i\frac{\delta_n^{\alpha(1-\lambda)}}{t(1+t)}dt \\
\le& C\|f\|_0\varphi^{\alpha(\lambda-1)}(x)\sum_{v=1}^{\infty} s_{n,v+2}(x)\int_1^{\infty}\sum_{i=0}^{2}\frac{n^{1+\frac{i}{2}}b_{n,v}(t)}{\varphi_1^i(t)}\left|\frac{v+1-2t}{n-2}-t\right|^i\delta_n^{\alpha(1-\lambda)}(t)dt \\
\le& C\|f\|_0\varphi^{\alpha(\lambda-1)}(x)\sum_{v=1}^{\infty} s_{n,v+2}(x)\int_1^{\infty}\sum_{i=0}^{2}\frac{n^{1+\frac{i}{2}}b_{n,v}(t)}{\varphi_1^i(t)}\left|\frac{v}{n}-t\right|^i\delta_n^{\alpha(1-\lambda)}(t)dt \\
&+ C\|f\|_0\varphi^{\alpha(\lambda-1)}(x)\sum_{v=1}^{\infty} s_{n,v+2}(x)\int_1^{\infty}\sum_{i=0}^{2}\frac{n^{1+\frac{i}{2}}b_{n,v}(t)}{\varphi_1^i(t)}\left(\frac{2v}{n(n-2)}\right)^i\delta_n^{\alpha(1-\lambda)}(t)dt \\
&+ C\|f\|_0\varphi^{\alpha(\lambda-1)}(x)\sum_{v=1}^{\infty} s_{n,v+2}(x)\int_1^{\infty}\sum_{i=0}^{2}\frac{n^{1+\frac{i}{2}}b_{n,v}(t)}{\varphi_1^i(t)}\left(\frac{2t-1}{n-2}\right)^i\delta_n^{\alpha(1-\lambda)}(t)dt \\
&=: L_1 + L_2 + L_3.
\end{aligned}$$

Similar to I_2 and Lemma 3.2, we can get

$$\begin{aligned}
&\|f\|_0\varphi^{\alpha(\lambda-1)}(x)\sum_{i=0}^{2} n^{1+\frac{i}{2}}\left\{\sum_{v=1}^{\infty} s_{n,v+2}(x)\int_1^{\infty}\frac{b_{n,v}(t)}{\varphi_1^{2m}(t)}\left(\frac{v}{n}-t\right)^{2m}dt\right\}^{\frac{1}{2m}} \\
&\quad\cdot\left\{\sum_{v=1}^{\infty} s_{n,v+2}(x)\int_1^{\infty} b_{n,v}(t)\delta_n^2(t)dt\right\}^{\frac{\alpha(1-\lambda)}{2}} \\
&\le C\|f\|_0\varphi^{\alpha(\lambda-1)}(x)\sum_{i=0}^{2} n^{1+\frac{i}{2}}\left(n^{-mi}\right)^{\frac{1}{2m}}\delta_n^{\alpha(1-\lambda)}(x) \le Cn\|f\|_0.
\end{aligned}$$

Therefore, one has

$$L_1 \le Cn\|f\|_0. \tag{18}$$

Similarly we can get the estimates for L_2, L_3. Combining (16)–(18), we obtain $I_2 \le Cn\|f\|_0$ for $x > 1$. □

Lemma 3.6. [6, Theorem 3.2] *If* $f \in C_B[0,\infty)$, $0 < \alpha < 2$, $0 \le \lambda \le 1$, *we have* $|\varphi^{\alpha(\lambda-1)}(x)\Delta^2_{h\varphi^\lambda} f(x)| \le CK^2_{\alpha,\lambda}(f, h^2\varphi^{2(\lambda-1)}(x))$, *where* $\Delta^2_{h\varphi^\lambda} f(x) = f(x+h\varphi^\lambda(x)) - 2f(x) + f(x - h\varphi^\lambda(x))$.

Proof of Theorem 1.2. By the definition of $K^2_{\alpha,\lambda}(f,t^2)$, we can choose $g \in C^2_{\alpha,\lambda}$ such that $\|f-g\|_0 + n^{-1}\|g\|_2 \le 2K^2_{\alpha,\lambda}(f,n^{-1})$. The condition $|S_n(f,x) - f(x)| = O((n^{-\frac{1}{2}}\delta_n(x)^{1-\lambda}))^\alpha$ implies that $\|S_n(f,x) - f(x)\|_0 = O(n^{-\frac{\alpha}{2}})$. Using Lemmas 3.4 and 3.5, we have

$$\begin{aligned} K^2_{\alpha,\lambda}(f,t^2) &\le \|f - S_n f\|_0 + t^2\|S_n f\|_2 \le Cn^{-\frac{\alpha}{2}} + t^2(\|S_n(f-g)\|_2 + \|S_n g\|_2) \\ &\le Cn^{-\frac{\alpha}{2}} + t^2(Cn\|f-g\|_0 + C\|g\|_2) \\ &\le C\left(n^{-\frac{\alpha}{2}} + \left(\frac{t}{1/\sqrt{n}}\right)^2 K^2_{\alpha,\lambda}(f,n^{-1})\right). \end{aligned}$$

Using the Berens-Lorentz Lemma [2], we get $K^2_{\alpha,\lambda}(f,t^2) \le Ct^\alpha$. From Lemma 3.6, one has $\omega^2_{\varphi^\lambda}(f,t) = O(t^\alpha)$. □

References

1. Z. Ditzian, Direct estimate for Bernstein polynomials, *J. Approx. Theory* **79** (1994), 165–166.
2. Z. Ditzian and V. Totik, *Moduli of Smoothness*, Springer, New York, 1987.
3. Z. Finta, On converse approximation theorems, *J. Math. Anal. Appl.* **312** (2005), 159–180.
4. Z. Finta, N. K. Govil and V. Gupta, Some results on modified Szász-Mirakjan operators, *J. Math. Anal. Appl.* **327** (2007), 1284–1296.
5. S. Guo and Q. Qi, Simultaneous approximation for Szász-Mirakin quasi-interpolants, *Czechoslovak Math. J.* **56** (2006), 789–803.
6. S. Guo and Q. Qi, Strong converse inequalities for Baskakov operators, *J. Approx. Theory* **124** (2003), 219–231.
7. V. Gupta, Error estimate for mixed summation-integral type operators, *J. Math. Anal. Appl.* **313** (2006), 632–641.
8. V. Gupta, R. N. Mohapatra and Z. Finta, A certain family of mixed summation-integral type operators, *Math. Comput. Modelling* **42** (2005), 181–191.
9. V. Gupta and M. A. Noor, Convergence of derivatives for certain mixed Szász-beta operators, *J. Math. Anal. Appl.* **321** (2006), 1–9.
10. M. K. Gupta and V. Vasishtha, The iterative combinations of a new sequence of linear positive operators, *Math. Comput. Modelling* **39** (2004), 521–527.
11. H. M. Srivastava and V. Gupta, A certain family of summation-integral type operators, *Math. Comput. Modelling* **37** (2003), 1307–1315.

Session 02

Clifford and Quaternion Analysis

SESSION EDITORS

I. Sabadini Politecnico di Milano, Milano, Italy
M. Shapiro Instituto Politécnico Nacional, México, Mexico
F. Sommen Ghent University, Ghent, Belgium

CRITERIA FOR MONOGENICITY OF CLIFFORD-ALGEBRA-VALUED FUNCTIONS

RICARDO ABREU BLAYA

Facultad de Informática y Matemática, Universidad de Holguín, Holguín 80100, Cuba
E-mail: rabreu@facinf.uho.edu.cu

JUAN BORY REYES

Department of Mathematics, Universidad de Oriente, Santiago de Cuba 90500, Cuba
E-mail: jbory@rect.uo.edu.cu

DIXAN PEÑA PEÑA* and FRANK SOMMEN†

Department of Mathematical Analysis, Ghent University, 9000 Gent, Belgium
**E-mail: dixan@cage.UGent.be*
†E-mail: fs@cage.UGent.be

A differential and integral criterion for monogenicity is presented within the framework of Clifford analysis. The results obtained provide characterizations for the holomorphic and the two-sided biregular functions.

Keywords: Clifford analysis, Dirac operator, Cauchy transform.

1. Preliminaries

Let $(e_1, \ldots, e_m)$ be an orthonormal basis of the Euclidean space $\mathbb{R}^m$. Let $\mathbb{C}_m$ be the complex Clifford algebra constructed over $\mathbb{R}^m$. The non-commutative multiplication in $\mathbb{C}_m$ is governed by the rules

$$e_j^2 = -1, \quad j = 1, 2, \ldots, m \quad \text{and} \quad e_j e_k + e_k e_j = 0, \quad 1 \leq j \neq k \leq m.$$

The Clifford algebra $\mathbb{C}_m$ is generated additively by elements of the form

$$e_A = e_{j_1} \ldots e_{j_k},$$

where $A = \{j_1, \ldots, j_k\} \subset \{1, \ldots, m\}$ is such that $j_1 < \cdots < j_k$. For the empty set $\emptyset$, we put $e_\emptyset = 1$, the latter being the identity element.

Any Clifford number $a \in \mathbb{C}_m$ may thus be written as

$$a = \sum_A a_A e_A, \quad a_A \in \mathbb{C},$$

or still as $a = \sum_{k=0}^m [a]_k$, where $[a]_k = \sum_{|A|=k} a_A e_A$ is a so-called k-vector ($k = 0, 1, \ldots, m$). Denoting the space of k-vectors by $\mathbb{C}_m^{(k)}$, we have that

$$\mathbb{C}_m = \bigoplus_{k=0}^m \mathbb{C}_m^{(k)}.$$

The 0-vectors and 1-vectors are simply called scalars and vectors respectively.

The product of two Clifford vectors splits up into a scalar part and a 2-vector or a so-called bivector part

$$\underline{a}\,\underline{b} = -\langle \underline{a}, \underline{b} \rangle + \underline{a} \wedge \underline{b}$$

where

$$\langle \underline{a}, \underline{b} \rangle = \sum_{j=1}^m a_j b_j$$

and

$$\underline{a} \wedge \underline{b} = \sum_{j<k} e_j e_k (a_j b_k - a_k b_j).$$

If we identify the $\mathbb{R}^m$-vector $(x_1, \ldots, x_m)$ with the real Clifford vector $\underline{x} = \sum_{j=1}^m e_j x_j$, then $\mathbb{R}^m$ may be considered as a subspace of $\mathbb{C}_m$.

The conjugation and the main involution are defined respectively for $a \in \mathbb{C}_m$ as

$$\overline{a} = \sum_A \bar{a}_A \bar{e}_A, \quad \bar{e}_A = (-1)^{\frac{k(k+1)}{2}} e_A, \quad |A| = k,$$

$$\tilde{a} = \sum_A a_A \tilde{e}_A, \quad \tilde{e}_A = (-1)^k e_A, \qquad |A| = k.$$

It is easy to check that

$$\overline{ab} = \bar{b}\bar{a}, \quad \widetilde{ab} = \tilde{a}\tilde{b}, \quad a, b \in \mathbb{C}_m.$$

By means of the conjugation, a norm $|a|$ may be defined for each $a \in \mathbb{C}_m$ by putting

$$|a|^2 = [a\bar{a}]_0 = \sum_A |a_A|^2.$$

The Dirac operator $\partial_{\underline{x}}$ is defined as

$$\partial_{\underline{x}} = \sum_{j=1}^{m} e_j \partial_{x_j},$$

its fundamental solution being given by

$$E_m(\underline{x}) = \frac{1}{\omega_m} \frac{\overline{\underline{x}}}{|\underline{x}|^m}, \quad \underline{x} \in \mathbb{R}^m \setminus \{0\},$$

where ω_m stands for the area of the unit sphere S^{m-1} in $\mathbb{R}^m$.

The function $f : \mathbb{R}^m \to \mathbb{C}_m$ is said to be left monogenic in the open set Ω of $\mathbb{R}^m$ iff f is continuously differentiable in Ω and moreover satisfies the equation $\partial_{\underline{x}} f = 0$.

We notice that $\Delta_m = \sum_{j=1}^{m} \partial_{x_j}^2 = -\partial_{\underline{x}}^2$. Thus any monogenic function in Ω is harmonic in Ω.

As a natural generalization of the holomorphic functions in the complex plane to higher dimensional spaces, monogenic functions are the main objects in Clifford analysis (see [5] and [11] for more details).

A special case of monogenic function arises when the dimension of $\mathbb{R}^m$ is even. Let $m = 2n$ be even and denote by $\mathbb{C}_n$ the complex Clifford algebra generated by $(e_1, \ldots, e_n)$. Next, we introduce the primitive idempotent

$$I = \prod_{j=1}^{n} I_j,$$

with

$$I_j = \frac{1}{2}(1 + i e_j e_{n+j}), \quad j = 1, \ldots, n.$$

We have the following conversion relations

$$e_{n+j} a I = i \tilde{a} e_j I, \tag{1}$$

with $a \in \mathbb{C}_n$. Note that for $a \in \mathbb{C}_n$ one also has that

$$aI = 0 \quad \text{iff} \quad a = 0.$$

If we now take $f : \mathbb{R}^{2n} \to \mathbb{C}_n$, then by (1) we obtain

$$\partial_{\underline{x}}(fI) = (\partial_{\underline{x}_1} f + i \tilde{f} \partial_{\underline{x}_2}) I,$$

with

$$\partial_{\underline{x}_1} = \sum_{j=1}^{n} e_j \partial_{x_j} \quad \text{and} \quad \partial_{\underline{x}_2} = \sum_{j=1}^{n} e_j \partial_{x_{n+j}}.$$

Hence the function fI is left monogenic iff f satisfies the equation

$$\partial_{\underline{x}_1} f + i\tilde{f}\partial_{\underline{x}_2} = 0.$$

Solutions of the latter equation are called isotonic functions (see [6,16,18]). It is worth pointing out that if in particlular f takes values in the space of scalars $\mathbb{C}$, then

$$(\partial_{x_j} + i\partial_{x_{n+j}})f = 0, \quad j = 1, \ldots, n,$$

which means that f is a holomorphic function of several complex variables. On the other hand, if f takes values in the real Clifford algebra $\mathbb{R}_{0,n}$, then

$$\begin{aligned} \partial_{\underline{x}_1} f &= 0 \\ \tilde{f}\partial_{\underline{x}_2} &= 0 \end{aligned}$$

or, equivalently, by the action of the main involution on the second equation:

$$\begin{aligned} \partial_{\underline{x}_1} f &= 0 \\ f\partial_{\underline{x}_2} &= 0 \end{aligned}$$

i.e. f is a two-sided biregular function (see [7–9,15]).

Throughout the paper Ω will stand for an open bounded and connected set in $\mathbb{R}^m$ with boundary Σ.

The aim of this paper is the following. Let F be a $\mathbb{C}_m$-valued function harmonic in Ω and continuous up to the boundary Σ. We wish to investigate necessary and sufficient conditions for the monogenicity of F in terms of its boundary value $f = F|_{\Sigma}$.

The basic idea is to apply a differential or integral criterion for monogenicity. Some applications are also indicated.

2. A differential criterion for monogenicity

In this section we require Σ to be a C^1-smooth surface. Then for $\underline{x}$ sufficiently close to Σ we may assume that the orthogonal projection of $\underline{x}$ onto Σ is unique and it is denoted by $\underline{x}_{\perp}$. Let us denote by $\underline{\nu} = \sum_{j=1}^{m} e_j \nu_j$ the unit normal vector on Σ at the point $\underline{x}_{\perp}$.

In a neighborhood of Σ we have the decomposition of $\partial_{\underline{x}}$ in the normal and the tangential parts (see [17])

$$\partial_{\underline{x}} = -\underline{\nu}(\underline{\nu}\,\partial_{\underline{x}}) = \underline{\nu}\partial_{\nu} + \partial_{\|\underline{x}}, \tag{2}$$

where

$$\partial_{\nu} = \langle \underline{\nu}, \partial_{\underline{x}} \rangle \quad \text{and} \quad \partial_{\|\underline{x}} = -\underline{\nu}(\underline{\nu} \wedge \partial_{\underline{x}}).$$

Theorem 2.1. *Let $F : \overline{\Omega} \to \mathbb{C}_m$ be a function of class C^1. Then F is left monogenic in Ω iff F is harmonic in Ω and*

$$\partial_\nu F - \underline{\nu}\partial_{\|\underline{x}} F = 0 \quad on \quad \Sigma. \tag{3}$$

Proof. If F is left monogenic in Ω, then from the decomposition (2) of the Dirac operator we get that F satisfies the equation

$$\partial_\nu F - \underline{\nu}\partial_{\|\underline{x}} F = 0$$

in a neighbourhood of Σ intersected with Ω. By continuity, we get on Σ the relation (3).

Conversely, assume that F is harmonic in Ω and $\partial_\nu F - \underline{\nu}\partial_{\|\underline{x}} F = 0$ on Σ. Define $G = \partial_{\underline{x}} F$. Therefore in a neighbourhood of Σ intersected with Ω we obtain

$$G = \underline{\nu}\partial_\nu F + \partial_{\|\underline{x}} F.$$

We thus get $G|_\Sigma = 0$. As F is harmonic in Ω it follows that G is left monogenic in Ω. By the maximum modulus theorem for monogenic functions, $G \equiv 0$ in Ω. □

3. An integral criterion for monogenicity

Let us denote by $\mathcal{H}^{m-1}$ the Hausdorff $(m-1)$-dimensional measure (see [12–14]). In the remainder of this paper we assume Σ to be an Ahlfors-David regular surface (see [10]), i.e., there exists $c > 0$ such that for all $\underline{x} \in \Sigma$ and all $0 < r \leq \operatorname{diam} \Sigma$,

$$c^{-1} r^{m-1} \leq \mathcal{H}^{m-1}(\Sigma \cap \{|\underline{y} - \underline{x}| \leq r\}) \leq c r^{m-1}.$$

We also assume that f belongs to the Hölder space $C^{0,\alpha}(\Sigma)$, $0 < \alpha < 1$. The Cauchy transform C_Σ and the principal value Hilbert transform H_Σ of f are defined respectively by

$$C_\Sigma f(\underline{x}) = \int_\Sigma E_m(\underline{y} - \underline{x})\underline{\nu}(\underline{y}) f(\underline{y}) d\mathcal{H}^{m-1}(\underline{y}), \quad \underline{x} \in \mathbb{R}^m \setminus \Sigma,$$

$$H_\Sigma f(\underline{x}) = \int_\Sigma E_m(\underline{y} - \underline{x})\underline{\nu}(\underline{y})(f(\underline{y}) - f(\underline{x})) d\mathcal{H}^{m-1}(\underline{y}), \quad \underline{x} \in \Sigma.$$

Here and subsequently, $\underline{\nu}(\underline{y})$ stands for the unit normal vector on Σ at the point $\underline{y}$ introduced by Federer (see [13]).

We notice that $C_\Sigma f$ is left monogenic in $\mathbb{R}^m \setminus \Sigma$. Let us now formulate some important properties of $C_\Sigma f$ and $H_\Sigma f$. For their proofs we refer the reader to [1] and [3].

(A) $H_\Sigma f \in C^{0,\alpha}(\Sigma)$.
(B) (Sokhotski-Plemelj formula) For $\underline{z} \in \Sigma$,

$$\lim_{\Omega \ni \underline{x} \to \underline{z}} C_\Sigma f(\underline{x}) = H_\Sigma f(\underline{z}) + f(\underline{z}).$$

Theorem 3.1. *Let $F : \overline{\Omega} \to \mathbb{C}_m$ be a continuous function such that $f = F|_\Sigma$ belongs to $C^{0,\alpha}(\Sigma)$ where Σ is a Ahlfors-David regular surface. Then F is left monogenic in Ω iff F is harmonic in Ω and $H_\Sigma f = 0$.*

Proof. Suppose that F is left monogenic in Ω. By Cauchy's integral formula (see [5]), we have $F(\underline{x}) = C_\Sigma f(\underline{x})$ for $\underline{x} \in \Omega$. Now (B) yields

$$f(\underline{x}) = H_\Sigma f(\underline{x}) + f(\underline{x}), \quad \underline{x} \in \Sigma.$$

Consequently, $H_\Sigma f(\underline{x}) = 0$ for all $\underline{x} \in \Sigma$.

Conversely, assume that F is harmonic in Ω and $H_\Sigma f = 0$. Let us define

$$G(\underline{x}) = \begin{cases} C_\Sigma f(\underline{x}), & \underline{x} \in \Omega \\ f(\underline{x}), & \underline{x} \in \Sigma. \end{cases}$$

The function G is left monogenic in Ω and hence harmonic in Ω. By (A) and (B), G is also continuous on $\overline{\Omega}$. As $F-G$ is harmonic in Ω and $(F-G)|_\Sigma = 0$ it follows that $F(\underline{x}) = C_\Sigma f(\underline{x})$ for $\underline{x} \in \Omega$. □

Let $m = 2n$ be even and consider in particular F taking values in $\mathbb{C}_n$. As mentioned above F is isotonic in Ω iff FI is left monogenic in Ω. By Theorem 3.1, FI is left monogenic in Ω iff $H_\Sigma(fI) = 0$ with $f = F|_\Sigma$, or, equivalently, applying (1):

$$H_\Sigma^{\text{isot}} f = 0. \tag{4}$$

Here H_Σ^{isot} stands for the isotonic Hilbert transform defined for $\underline{x} \in \Sigma$ as

$$\begin{aligned} &H_\Sigma^{\text{isot}} f(\underline{x}) \\ &= -\frac{1}{\omega_{2n}} \int_\Sigma \left[\frac{(\underline{y}_1 - \underline{x}_1)(\underline{\nu}_1(\underline{y})(f(\underline{y}) - f(\underline{x})) + i(\tilde{f}(\underline{y}) - \tilde{f}(\underline{x}))\underline{\nu}_2(\underline{y}))}{|\underline{y} - \underline{x}|^{2n}} \right. \\ &\left. + \frac{((f(\underline{y}) - f(\underline{x}))\underline{\nu}_2(\underline{y}) - i\underline{\nu}_1(\underline{y})(\tilde{f}(\underline{y}) - \tilde{f}(\underline{x})))(\underline{y}_2 - \underline{x}_2)}{|\underline{y} - \underline{x}|^{2n}} \right] d\mathcal{H}^{2n-1}(\underline{y}), \end{aligned}$$

where $\underline{\nu}_1(\underline{x}) = \sum_{j=1}^n e_j \nu_j(\underline{x})$ and $\underline{\nu}_2(\underline{x}) = \sum_{j=1}^n e_j \nu_{n+j}(\underline{x})$ (see [2] and [4]).

Using Theorem 3.1 we can state two important consequences related to holomorphic and two-sided biregular functions.

Corollary 3.1. *Let $m = 2n$ and suppose that F takes values in the space of scalars $\mathbb{C}$. Then F is holomorphic in Ω iff F is harmonic in Ω and for all $\underline{x} \in \Sigma$ the following equalities hold*

$$\int_\Sigma \frac{\langle \underline{y}_1 - \underline{x}_1, \underline{\nu}_1 + i\underline{\nu}_2\rangle + \langle \underline{y}_2 - \underline{x}_2, \underline{\nu}_2 - i\underline{\nu}_1\rangle}{|\underline{y} - \underline{x}|^{2n}} (f(\underline{y}) - f(\underline{x}))d\mathcal{H}^{2n-1}(\underline{y})$$
$$= \int_\Sigma \frac{(\underline{y}_1 - \underline{x}_1) \wedge (\underline{\nu}_1 + i\underline{\nu}_2) - (\underline{y}_2 - \underline{x}_2) \wedge (\underline{\nu}_2 - i\underline{\nu}_1)}{|\underline{y} - \underline{x}|^{2n}} (f(\underline{y}) - f(\underline{x}))d\mathcal{H}^{2n-1}(\underline{y})$$
$$= 0.$$

Proof. The proof follows easily by taking the scalar and the bivector part of (4). □

Corollary 3.2. *Let $m = 2n$ and suppose that F takes values in the real Clifford alegra $\mathbb{R}_{0,n}$. Then F is a two-sided biregular function in Ω iff F is harmonic in Ω and for all $\underline{x} \in \Sigma$ the following equalities hold*

$$\int_\Sigma \frac{(\underline{y}_1 - \underline{x}_1)\underline{\nu}_1(\underline{y})(f(\underline{y}) - f(\underline{x})) + (f(\underline{y}) - f(\underline{x}))\underline{\nu}_2(\underline{y})(\underline{y}_2 - \underline{x}_2)}{|\underline{y} - \underline{x}|^{2n}} d\mathcal{H}^{2n-1}(\underline{y})$$
$$= \int_\Sigma \frac{(\underline{y}_1 - \underline{x}_1)(f(\underline{y}) - f(\underline{x}))\underline{\nu}_2(\underline{y}) - \underline{\nu}_1(\underline{y})(f(\underline{y}) - f(\underline{x}))(\underline{y}_2 - \underline{x}_2)}{|\underline{y} - \underline{x}|^{2n}} d\mathcal{H}^{2n-1}(\underline{y})$$
$$= 0.$$

Proof. It is sufficient to take the real and the imaginary part of (4). □

Acknowledgments

This paper was written while the second author was visiting the Department of Mathematical Analysis of Ghent University. He was supported by the Special Research Fund No. 01T13804, obtained for collaboration between the Clifford Research Group in Ghent and the Cuban Research Group in Clifford analysis, on the subject *Boundary values theory in Clifford Analysis*. J. Bory Reyes wishes to thank all members of this Department for their kind hospitality.

D. Peña Peña was supported by a Doctoral Grant of the Special Research Fund of Ghent University. He would like to express his sincere gratitude.

References

1. R. Abreu Blaya, J. Bory Reyes and D. Peña Peña, Clifford Cauchy type integrals on Ahlfors-David regular surfaces in $\mathbb{R}^{m+1}$. *Adv. Appl. Clifford Algebras* 13 (2003), no. 2, 133–156.
2. R. Abreu Blaya, J. Bory Reyes, D. Peña Peña and F. Sommen, The isotonic Cauchy transform. *Adv. Appl. Clifford Algebras* 17 (2007), no. 2, 145–152.
3. J. Bory Reyes and R. Abreu Blaya, Cauchy transform and rectifiability in Clifford analysis. *Z. Anal. Anwendungen* 24 (2005), no. 1, 167–178.
4. J. Bory Reyes, D. Peña Peña and F. Sommen, A Davydov theorem for the isotonic Cauchy transform. *J. Anal. Appl.* 5 (2007), no. 2, 109–121.
5. F. Brackx, R. Delanghe and F. Sommen, Clifford analysis. Research Notes in Mathematics, 76. *Pitman Advanced Publishing Program, Boston, MA*, 1982.
6. F. Brackx, H. De Schepper and F. Sommen, *The Hermitean Clifford analysis toolbox*, accepted for publication in Proceedings of ICCA7.
7. F. Brackx and W. Pincket, A Bochner-Martinelli formula for the biregular functions of Clifford analysis. *Complex Variables Theory Appl.* 4 (1984), no. 1, 39–48.
8. F. Brackx and W. Pincket, *Two Hartogs theorems for nullsolutions of overdetermined systems in Euclidean space.* Complex Variables Theory Appl. 4 (1985), no. 3, 205–222.
9. F. Brackx and W. Pincket, *Series expansions for the biregular functions of Clifford analysis.* Simon Stevin 60 (1986), no. 1, 41–55.
10. G. David and S. Semmes, Analysis of and on uniformly rectifiable sets. Mathematical Surveys and Monographs, 38. *American Mathematical Society, Providence, RI*, 1993.
11. R. Delanghe, F. Sommen and V. Souček, Clifford algebra and spinor-valued functions. Mathematics and its Applications, 53. *Kluwer Academic Publishers Group, Dordrecht*, 1992.
12. K. J. Falconer, The geometry of fractal sets. Cambridge Tracts in Mathematics, 85. *Cambridge University Press, Cambridge*, 1986.
13. H. Federer, Geometric measure theory. Die Grundlehren der mathematischen Wissenschaften, Band 153. *Springer-Verlag New York Inc., New York* 1969.
14. P. Mattila, Geometry of sets and measures in Euclidean spaces. Fractals and rectifiability. Cambridge Studies in Advanced Mathematics, 44. *Cambridge University Press, Cambridge*, 1995.
15. W. Pincket, *Biregular functions.* Ph.D. Thesis, Ghent University, 1984.
16. R. Rocha-Chávez, M. Shapiro and F. Sommen, Integral theorems for functions and differential forms in $\mathbb{C}^m$. Research Notes in Mathematics, 428. *Chapman & Hall/CRC, Boca Raton, FL*, 2002.
17. F. Sommen, Tangential Cauchy-Riemann operators in $\mathbb{C}^m$ arising in Clifford analysis. *Simon Stevin* 61 (1987), no. 1, 67–89.
18. F. Sommen and D. Peña Peña, Martinelli-Bochner formula using Clifford analysis. *Archiv der Mathematik*, 88 (2007), no. 4, 358–363.

HYPERMONOGENIC FUNCTIONS AND BIHYPERMONOGENIC FUNCTIONS IN REAL CLIFFORD ANALYSIS

YUYING QIAO[‡] and HEJU YANG*
College of Mathematics and Information,
Hebei Normal University Shijiazhuang , 050016, P. R. China
[‡]*E-mail: yuyingqiao@163.com*

XIAOLI BIAN[†]
Department of Mathematics and Physics,
Tianjin University of Technology and Education,
Tianjin 300222, P. R. China
E-mail: bianxiaoli@yeah.net

We discuss the properties of bihypermonogenic functions in Clifford analysis. Firstly, we obtain equivalent conditions for bihypermonogenic functions. Secondly, we give a Cauchy integral formula and a Plemelj formula for bihypermonogenic functions.

Keywords: Bihypermonogenic; Plemelj Formula; Cauchy Integral Formula; Bihypermononic Function.

1. Introduction

Clifford algebra is an associative and noncommutative algebraic structure that was introduced in the middle of the 1800s. Clifford analysis is an important branch of modern analysis that studies functions defined on $\mathbb{R}^{n+1}$ with the values in a Clifford algebra [1]. Clifford analysis possesses important theoretical and applicable value in many fields. Since 1987, Wen, Huang, Qiao, J. Ryan, etc. have done a lot of work on boundary value problems and the properties of monogenic functions in real Clifford analysis [2-5]. In Clifford analysis, let $D = \sum_{i=0}^{n} e_i \frac{\partial}{\partial x_i}$. A solution of the equation

*Research supported by National Natural Science Fundamental of China (No.10771049,10671207) and National Natural Science Fundamental of Hebei (No.A2007000225)
[†]Corresponding author.

$Df = 0$ is called a monogenic function (or regular function). Monogenic functions are a generalization of analytic functions in complex analysis and their properties have been studied [1] . W. Hengartner and H. Leutwiler have researched hypermonogenic functions of $\mathbb{R}^3$ [7-8]. S.-L. Eriksson and H. Leutwiler introduced hypermonogenic functions in Clifford analysis and research some of its properties, discuss the integral representation for hypermonogenic functions [9-10]. Qiao discuss the boundary value problems of hypermonogenic functions [11]. In recent years, Z. Zhang and J. Du research Riemann boundary value problems and singular integral equations in Clifford analysis [12]. They get Laurent expansion and residue theorems in universal Clifford analysis [13]. In real and complex Clifford analysis [14], S. Huang introduced biregular functions and researched some of their properties. In this paper, we discuss the extension theorem of hypermonogenic functions. Using the idea of quasi-permutation created by Sha Huang [6], we give an equivalent definition of bihymonogenic functions.

2. Preliminaries

2.1. *Clifford Algebra and Clifford Analysis*

Let $A_{n+1}(R)$ be a real Clifford algebra over an $n+1$-dimensional real vector space R^{n+1} with orthogonal basis $e := \{e_0, e_1, \cdots, e_n\}$, where $e_0 = 1$ is a unit element in R^{n+1}. Then $A_{n+1}(R)$ has its basis $e_0 = 1, e_1, \cdots, e_n; e_1e_2, \cdots, e_{n-1}e_n; \cdots; e_1 \cdots e_n$. Hence an arbitrary element of the basis may be written as $e_A = e_{i_1,\cdots,i_k} = e_{i_1}e_{i_2}\cdots e_{i_k}$, here $A = \{i_1, i_2, \cdots, i_k \mid 1 \le i_1 < i_2 < \cdots < i_k \le n\}$ and when $A = \emptyset$ (empty set), $e_A = e_0$. So real Clifford algebra is composed of the elements having the type $x = \sum_A x_A e_A$, in which $x_A (\in R)$ are real numbers. In general, one has $e_i^2 = -1$, $i = 1, \cdots, n$, and $e_ie_j + e_je_i = 0$, $i, j = 1, \cdots, n, i \ne j$. Note that the real vector space R^{n+1} consists of the elements

$$x := x_0e_0 + \cdots + x_ne_n.$$

2.2. *Dirac Operator*

Let $\Omega \subset R^{n+1}$ be an open connected set. A function f defined in Ω with values in $A_{n+1}(R)$ is a C^1-function. The Dirac operator is defined as

$$D_l f = \sum_{i=0}^{n} e_i \frac{\partial f}{\partial x_i}, \quad D_r f = \sum_{i=0}^{n} \frac{\partial f}{\partial x_i} e_i,$$

$$\overline{D}_l f = \sum_{i=0}^{n} \overline{e}_i \frac{\partial f}{\partial x_i}, \quad \overline{D}_r f = \sum_{i=0}^{n} \frac{\partial f}{\partial x_i} \overline{e}_i.$$

Then f is left regular in Ω, if $D_l f(x) = 0$ in Ω; and f is right regular in Ω, if $D_r f(x) = 0$ in Ω. By calculation we can get $D\overline{D} = \overline{D}D = \Delta$.

2.3. *An Involution and a Decomposition in $A_{n+1}(R)$*

The involution $' : A_{n+1}(R) \longrightarrow A_{n+1}(R)$ is defined by

$$x' = \sum_A x_A (-1)^{|A|} e_A$$

where usually $|A|$ is the number of elements in the set A. Clearly we have $e_0' = e_0 = 1$ and $e_i' = -e_i$, if $i = 1, \cdots, n$. Moreover the mapping $'$ is an isomorphism since $(ab)' = a'b'$.

Recall that any element $x \in A_{n+1}(R)$ may be uniquely decomposed as $x = b + ce_n$, for $b, c \in A_n(R)$.Using this decomposition we define the mappings $P : A_{n+1}(R) \longrightarrow A_n(R)$ and $Q : A_{n+1}(R) \longrightarrow A_n(R)$ by $Px = b$ and $Qx = c$. Note that if $x = \sum_A x_A e_A \in A_{n+1}(R)$ then

$$Px = \sum_{n \notin A} x_A e_A, \quad Qx = \sum_{n \in A} x_A e_{A\backslash\{n\}}.$$

We call Px and Qx be the P part and Q part of x respectively. P part and Q part are similarly with the real part and imaginary part of complex number.

2.4. *The First and the Third Class Quasipermutation*

Let $A = \{h_1, h_2, \cdots, h_k\}$, $1 \leq h_1 < h_2 < \cdots < h_k \leq n$, where $h_i \in N$, $i = 1, 2, \cdots, k$. $1 \leq m \leq n$, m be natural numbers. For the arrangement

$$\overline{mA} = \begin{cases} A\backslash\{m\}, & m \in A; \\ \{g_1, g_2, \cdots, g_{k+1}\}, & m \notin A, \end{cases}$$

where $g_i \in A \cup \{m\}, 1 \leq g_1 < g_2 < \cdots < g_{k+1}$. We call $\overline{mA}$ the first class quasi-permutation for arrangement mA, and define $\overline{mm} = \emptyset, \overline{m0} = m$. If there are p natural numbers $h_i \in [1, m]$, then the sign $\delta_{\overline{mA}} = \delta_{\overline{Am}} = (-1)^p$ is called the sign of the first class quasi-permutation $\overline{mA}$.

The properties of the first class quasi-permutation:

1. If $\overline{mA} = B$, then $\overline{mB} = A$;
2. If $m = 0$, then $\delta_{\overline{mA}} = 1$;

3. If $\overline{mA} = B$, then $\delta_{\overline{mA}} = -\delta_{\overline{mB}}$;
4. $e_m e_A = \delta_{\overline{mA}} e_{\overline{mA}}$.

Let j be a natural number and $D = \{h_1, h_2, \cdots, h_s\}, 1 < h_1 < h_2 < \cdots < h_s \leq n$. $h_p\{1 \leq p \leq s\}$ be natural numbers, the arrangement D_j is called the third class quasi-permutation of arrangement jD, and the sign

$$\varepsilon_{jD} = \begin{cases} (-1)^s, & \forall p \in [1, s], \quad h_p \neq j; \\ (-1)^{s-1}, & \exists p \in [1, s], \quad h_p = j, \end{cases}$$

$\varepsilon_{0D} = 1$, is called the sign of the third class quasi-permutation. If j, D are as stated above, then $e_j e_D = \varepsilon_{jD} e_D e_j$.

3. Some Special Functions

3.1. *Hypermonogenic Functions*

We consider the modified Dirac operators M^l, M^r, $\overline{M}^l$ and $\overline{M}^r$ in $A_{n+1}(R)$ defined by

$$M^l f(x) = D_l f(x) + (n-1)\frac{Q'f}{x_n},$$
$$M^r f(x) = D_r f(x) + (n-1)\frac{Qf}{x_n},$$
$$\overline{M}^l f(x) = \overline{D}_l f(x) - (n-1)\frac{Q'f}{x_n},$$
$$\overline{M}^r f(x) = \overline{D}_r f(x) - (n-1)\frac{Qf}{x_n},$$

where $Q'f = (Qf)'$. Let $\Omega \subset \mathbb{R}^{n+1}_+ = \{x = (x_0, x_1, \cdots x_n) \mid x_n > 0\}$. A mapping $f : \Omega \to A_{n+1}(R)$ is called *left hypermonogenic* (briefly *hypermonogenic*) if $f \in C^1(\Omega)$ and $M^l f = 0$ for any $x \in \Omega$. A right hypermonogenic function is defined similarly. Hypermonogenic functions were introduced in [9].

3.2. *Bihypermonogenic Functions*

Suppose Ω_1 and Ω_2 be open subset of $R_{m+1} \setminus \{x_m \leq 0\}$ and $R_{k+1} \setminus \{y_k \leq 0\}$ respectively. If

$$\begin{cases} M^l_x f(x,y) = 0, \\ M^r_y f(x,y) = 0, \end{cases}$$

for any $(x, y) \in \Omega_1 \times \Omega_2$, then a function$f(x, y)$ is called bihypermonogenic in $\Omega_1 \times \Omega_2$.

4. An Equivalent Condition for Bihypermonogenic Functions

Lemma 4.1. *Let the set* $\beta = \{e_A|\ A = \{h_1, h_2, \cdots, h_r\}, 1 \le h_1 < h_2 < \cdots < h_r \le n\}, r = \{e_{\overline{jA}}|\ A = \{h_1, h_2, \cdots, h_r\}, 1 \le h_1 < h_2 < \cdots < h_r \le n\}$*, where* $1 \le j \le n$ *is a fixed integer, then* $\beta = r$.

Proof. For j, $1 \le j \le n$, and any $e_A \in \beta$,
(1) If $j \in A$, we can find e_B from r, $B = A\backslash\{j\}$, then $e_{\overline{jB}} = e_A$. (2) If $j \notin A$, we can find e_B from r, $B = A\bigcup\{j\}$, then $e_{\overline{jB}} = e_A$.
Summarizing the above discussion, we can get $e_A \in r$, so $\beta \subset r$, similarly $r \subset \beta$ can be obtained, hence $\beta = r$. □

Theorem 4.1. *If*

$$f(x,y) = \sum_{A\subset\{1,\ldots,m\}} \sum_{B\subset\{m+1,\ldots,m+k\}} f_{A,B}(x,y)e_Ae_B,$$

where $f_{A,B} \in C^r(\Omega)$ *and* $x \in \mathbb{R}^{m+1}, y \in \mathbb{R}^{k+1}$*, then the* $f(x,y)$ *is bihypermonogenic functions if and only if it satisfying the system*

$$\frac{\partial f_{A,B}}{\partial x_0}(x,y) - \sum_{j=0}^{m} \delta_{\overline{jA}} \frac{\partial f_{\overline{jA},B}}{\partial x_j}(x,y) = 0, \quad \text{if } m \in A.$$

$$\frac{\partial f_{A,B}(x,y)}{\partial x_0} - \sum_{j=0}^{m} \delta_{\overline{jA}} \frac{\partial f_{\overline{jA},B}}{\partial x_j}(x,y) + \frac{m-1}{x_m}(-1)^{|A|} f_{\overline{mA},B}(x,y) = 0,$$

if $m \notin A$.

$$\frac{\partial f_{A,B}(x,y)}{\partial y_0} - \sum_{i=0}^{k} \varepsilon_{B(i+k)} \delta_{\overline{(i+k)B}} \frac{\partial f_{A\overline{(i+k)B}}}{\partial y_i}(x,y) + \frac{k-1}{y_k} f_{A,\overline{Bk}}(x,y) = 0,$$

if $m+k \notin B$.

$$\frac{\partial f_{A,B}(x,y)}{\partial y_0} - \sum_{i=0}^{k} \varepsilon_{B(i+k)} \delta_{\overline{(i+k)B}} \frac{\partial f_{A,\overline{iB}}(x,y)}{\partial y_i} = 0, \quad \text{if } m+k \in B.$$

Proof. Assume that $f(x,y)$ is bihypermonogenic. We obtain

$$\begin{aligned} D_x^l f(x,y) &= \sum_{A,B} e_Ae_B \frac{\partial f_{A,B}(x,y)}{\partial x_0} + \sum_{A,B}\sum_{j=1}^{m} e_je_Ae_B \frac{\partial f_{A,B}(x,y)}{\partial x_j} \\ &= \sum_{A,B} e_Ae_B \frac{\partial f_{A,B}(x,y)}{\partial x_0} + \sum_{B}\sum_{A}\sum_{j=1}^{m} \delta_{\overline{jA}} e_{\overline{jA}} e_B \frac{\partial f_{A,B}(x,y)}{\partial x_j}. \end{aligned}$$

Denoting $\overline{jA}=C$, we infer $A=\overline{jC}$ and $\delta_{\overline{jA}}=-\delta_{\overline{jC}}$, which implies

$$D^l_x f(x,y)=\sum_{A,B} e_A e_B \frac{\partial f_{A,B}(x,y)}{\partial x_0}-\sum_B\sum_C\sum_{j=1}^m \delta_{\overline{jC}} e_C e_B \frac{\partial f_{\overline{jC},B}(x,y)}{\partial x_j}. \quad (1)$$

Denoting Q-operator in $A_{n+1}(R)$ by Q_x we obtain

$$\begin{aligned} Q'_x f(x,y) &= \sum_{m\in A}\sum_B f_{A,B}(x,y) e'_{A\backslash\{m\}} e_B \\ &= \sum_{m\in A}\sum_B (-1)^{|A\backslash\{m\}|} f_{A,B}(x,y) e_{A\backslash\{m\}} e_B \end{aligned}$$

Denoting $D=A\backslash\{m\}$, we infer $A=\overline{mD}$ and

$$Q'_x f(x,y)=\sum_{m\notin D}\sum_B (-1)^{|D|} f_{\overline{mD},B}(x,y) e_D e_B. \quad (2)$$

Combining (1) and (2) we deduce

$$\begin{aligned} M^l_x f(x,y) &= D^l_x f(x,y)+\frac{m-1}{x_m} Q'_x f(x,y) \\ &= \sum_{A,B} e_A e_B \Big(\frac{\partial f_{A,B}(x,y)}{\partial x_0}-\sum_{j=1}^m \delta_{\overline{jA}} \frac{\partial f_{\overline{jC},B}(x,y)}{\partial x_j}\Big) \\ &\quad + \frac{m-1}{x_m}\sum_{m\notin A}\sum_B (-1)^{|A|} e_A e_B f_{\overline{mA},B}(x,y). \end{aligned}$$

Collecting the coefficients of $e_A e_B$ we obtain that $M^l_x f(x,y)=0$ is equivalent to the the system

$$\begin{aligned} &\frac{\partial f_{A,B}(x,y)}{\partial x_0}-\textstyle\sum_{j=1}^m \delta_{\overline{jA}} \frac{\partial f_{\overline{jA},B}(x,y)}{\partial x_j}=0, \text{ if } m\in A, \\ &\frac{\partial f_{A,B}(x,y)}{\partial x_0}-\textstyle\sum_{j=1}^m \delta_{\overline{jA}} \frac{\partial f_{\overline{jA},B}(x,y)}{\partial x_j}+\frac{m-1}{x_m}(-1)^{|A|} f_{\overline{mA},B}(x,y)=0, \text{ if } m\notin A. \end{aligned}$$

Similarly, we obtain the equivalent equations for $M^r_y f(x,y)=0$. □

The preceeding theorem gives the relation between the system

$$\begin{cases} M^l_x f(x,y)=0 \\ M^r_y f(x,y)=0 \end{cases}$$

in Clifford analysis and the system that has 2^n real partial differential equations.

5. Cauchy-Riemann-Type Condition

In this part we give a kind of Cauchy-Riemann-type condition for hypermonogenic function like holomorphic function in complex variables.

Lemma 5.1. *[10] Let Ω be stated as above, $f \epsilon C^1(\Omega, \mathcal{A}_n(R))$, then*

$$D_l\left(\frac{f}{x_n^{n-1}}\right) = \frac{M^l f}{x_n^{n-1}} - (n-1)\frac{P'f}{x_n^n}e_n$$
$$D_r\left(\frac{f}{x_n^{n-1}}\right) = \frac{M^r f}{x_n^{n-1}} - (n-1)\frac{Pf}{x_n^n}e_n$$
$$P(M^l f) = x_n^{n-1} P D_l\left(\frac{f}{x_n^{n-1}}\right)$$
$$Q(M^l f) = Q(D_l f) \qquad D_l\left(\frac{Qfe_n}{x_n^{n-1}}\right) = \frac{M^l(Qfe_n)}{x_n^{n-1}}$$

Theorem 5.1. *Let Ω, f be stated as above, a function f is a hypermonogenic function if and only if*

$$\begin{cases} \dfrac{\partial P'f(x)}{\partial x_n} + \displaystyle\sum_{i=0}^{n-1} e_i \frac{\partial Qf(x)}{\partial x_i} = 0, \\ \displaystyle\sum_{i=0}^{n-1} e_i \frac{\partial Pf(x)}{\partial x_i} - \frac{\partial Q'f(x)}{\partial x_n} + (n-1)\frac{Q'f(x)}{x_n} = 0. \end{cases}$$

Proof. Applying Lemma 5.1, since

$$\begin{aligned} M^l f(x) &= P\left(M^l f(x)\right) + Q\left(M^l f(x)\right) e_n \\ &= x_n^{n-1} P\left(D_l\left(\frac{f(x)}{x_n^{n-1}}\right)\right) + Q\left(D_l f(x)\right) e_n \end{aligned}$$

$$\begin{aligned} Q(D_l f(x)) &= Q D_l(Pf(x) + Qf(x)e_n) = Q\sum_{i=0}^{n} e_i \frac{\partial(Pf(x) + Qf(x)e_n)}{\partial x_i} \\ &= Q\left(\sum_{i=0}^{n-1} e_i \frac{\partial Pf(x)}{\partial x_i} + e_n \frac{\partial Pf(x)}{\partial x_n}\right. \\ &\quad \left. + \sum_{i=0}^{n-1} e_i \frac{\partial Qf(x)}{\partial x_i} e_n + e_n \frac{\partial Qf(x)}{\partial x_n} e_n\right) \\ &= \frac{\partial P'f(x)}{\partial x_n} + \sum_{i=0}^{n-1} e_i \frac{\partial Qf(x)}{\partial x_i}. \end{aligned}$$

$$x_n^{n-1}PD\left(\frac{f}{x_n^{n-1}}\right) = x_n^{n-1}P\sum_{i=0}^{n} e_i \frac{\partial\left(\frac{f(x)}{x_n^{n-1}}\right)}{\partial x_i}$$

$$= x_n^{n-1}P\sum_{i=0}^{n} e_i \frac{\partial \frac{Pf(x)+Qf(x)e_n}{x_n^{n-1}}}{\partial x_i}$$

$$= x_n^{n-1}P\left[\sum_{i=0}^{n-1} e_i \frac{\partial \frac{Pf(x)}{x_n^{n-1}}}{\partial x_i} + e_n \frac{\partial \frac{Pf(x)}{x_n^{n-1}}}{\partial x_n} + \sum_{i=0}^{n-1} e_i \frac{\partial \frac{Qf(x)}{x_n^{n-1}}}{\partial x_i} e_n + e_n \frac{\partial \frac{Qf(x)}{x_n^{n-1}}}{\partial x_n} e_n\right]$$

$$= \sum_{i=0}^{n-1} e_i \frac{\partial Pf(x)}{\partial x_i} - x_n^{n-1} \frac{\partial \frac{Q'f(x)}{x_n^{n-1}}}{\partial x_n}$$

$$= \sum_{i=0}^{n-1} e_i \frac{\partial Pf(x)}{\partial x_i} - \frac{\partial Q'f(x)}{\partial x_n} + \frac{n-1}{x_n} Q'f(x). \qquad \square$$

6. The Extension Theorem of Hypermonogenic Functions

Lemma 6.1. *[10] A function $f \epsilon C^1(\Omega, \mathcal{A}_n(R))$ is hypermonogenic in an open connected subset $\Omega \subset R^{n+1} \setminus \{x_n = 0\}$ if and only if the property*

$$\int_{\partial K} P(d\sigma_{n-1} f(x)) + \int_{\partial K} Q(d\sigma_0 f(x))e_n = 0,$$

or equivalently

$$\int_{\partial K} \left((d\sigma_{n-1} + d\sigma_0)f(x) + (\widehat{d\sigma}_{n-1} - \widehat{d\sigma}_0)\widehat{f(x)}\right) = 0.$$

holds for any n+1-chain K satisfying $\overline{K} \subset \Omega$.

We call this lemma a Cauchy-Morera theorem for hypermonogenic functions.

Theorem 6.1. *Let Ω be as above and $\Omega' = \Omega \bigcap \{(x_0, x_1, \dots, x_{n-1}, a_n)\}$, where $a_n > 0$ is a constant. If f is a hypermonogenic function in $\Omega \backslash \Omega'$ and $f \epsilon C^1(\Omega, \mathcal{A}_n(R))$, then f is a hypermonogenic function in Ω.*

Proof. For any $\widetilde{\Omega}$ contained in Ω, if $\widetilde{\Omega} \bigcap \Omega' = \emptyset$, we obtain that f is a hypermonogenic function, since f is a hypermonogenic function in $\Omega \backslash \Omega'$. If $\widetilde{\Omega} \bigcap \Omega' \neq \emptyset$, we put $K_{1\varepsilon} = \{x = (x_0, x_1, \dots, x_{n-1}, x_n) \mid x \in \widetilde{\Omega}, x_n \geq a_n + \varepsilon\}$, $K_{2\varepsilon} = \{x =$

$(x_0, x_1, \ldots, x_{n-1}, x_n) \mid x \in \widetilde{\Omega}, x_n \leq a_n - \varepsilon\}$, where $a_n - \varepsilon > 0$. Using Lemma 6.1, we obtain

$$\int_{\partial K_{1\varepsilon}} \Big((d\sigma_{n-1} + d\sigma_0) f(x) + (\widehat{d\sigma}_{n-1} - \widehat{d\sigma}_0)\widehat{f(x)}\Big) = 0,$$

$$\int_{\partial K_{2\varepsilon}} \Big((d\sigma_{n-1} + d\sigma_0) f(x)(\widehat{d\sigma}_{n-1} - \widehat{d\sigma}_0)\widehat{f(x)}\Big) = 0.$$

Since f is a continuous function in Ω, we have

$$\begin{aligned}
&\int_{\partial\widetilde{\Omega}} \Big((d\sigma_{n-1} + d\sigma_0) f(x) + (\widehat{d\sigma}_{n-1} - \widehat{d\sigma}_0)\widehat{f(x)}\Big) \\
&= \lim_{\varepsilon\to 0^+} \int_{\partial K_1} \Big((d\sigma_k + d\sigma_0) f(x) + (\widehat{d\sigma}_{n-1} - \widehat{d\sigma}_0)\widehat{f(x)}\Big) \\
&\quad + \lim_{\varepsilon\to 0^+} \int_{\partial K_2} \Big((d\sigma_{n-1} + d\sigma_0) f(x) + (\widehat{d\sigma}_{n-1} - \widehat{d\sigma}_0)\widehat{f(x)})\Big) \\
&= 0.
\end{aligned}$$

By Lemma 6.1 again, the result follows. □

References

1. R. Delanghe, F. Sommen, V. Soucek, Clifford Algebra and Spinor-Valued Functions, Dordrecht, Kluwer, 1992.
2. G. C. Wen, Clifford Analysis and Elliptic Systems, Hyperbolic Systems of First Order Equations, World Scientific, Singapore, 1991, 230-237.
3. H. Sha, Nonlinear Boundary Value Problems for Biregular Functions in Clifford Analysis, Sci. China Ser. A, 39 (1996), 1152-1164.
4. Y. Qiao, Nonlinear Boundary Value Problems of Generalized Biregular Functions with Haseman Shift in Real Clifford Analysis, J. Sys. Sci. Math. Sci., 19 (1999), 484-489.
5. E. Franks and J. Ryan, Bounded Monogenic Functions on Unbounded Domains, Contemp. Math., 1998, American Mathematical Society.
6. S. Huang, Quasi-Permutations and Generalized Regular Functions in Real Clifford Analysis, J. Sys. Sci. Math. Sci., 18 (1998), 380-384.
7. W. Hengartner and H. Leutwiler, Hyperholomorphic Functions in $\mathbb{R}^3$, Adv. Appl. Clifford Algebras, 11 (2001), S1, 247-259.
8. H. Leutwiler, Modified Clifford Analysis, Complex Variables, 17 (1992), 153-171.
9. H. Leutwiler, Modified Quaternionic Analysis in $\mathbb{R}^3$, Complex Variables, 20 (1992), 19-51.
10. S.-L. Eriksson and H. Leutwiler, Hypermonogenic Functions, In Clifford Algebras and Their Applications in Mathematical Physics, 2000, Vol. 2, Clifford Analysis, 286-301.
11. S.-L. Eriksson and H. Leutwiler, Hypermonogenic Functions and Their Cauchy-Type Theorems, in Trends in Mathematics, Advances in Analysis and Geometry, Birkhäuser, 2004, 97-112.

12. S.-L. Eriksson, Integral Formulas for Hypermonogenic Functions, Bull. Belgian Math. Soc., 11 (2004), 705-717.
13. Y. Y. Qiao, A Boundary Value Problem for Hypermonogenic Functions in Clifford Analysis, Sci. China Ser. A, 48 (2005), Supp., 324-332.
14. Z. Zhang, J. Du, Laurent Expansions and Residue Theorems in Universal Clifford Analysis, Acta Math. Sci., 23A (2003), 692-703.
15. Z. Zhang, J. Du, On Certain Riemann Boundary Value Problems and Singular Integral Equations in Clifford Analysis, Chinese Ann. Math., 23A (2001), 421-426.
16. S. Huang, Y. Y. Qiao, G. C. Wen, Real and Complex Clifford Analysis, Springer, 2005.
17. R. P. Gilbert, J. L. Buchanan, First Order Elliptic Systems: A Function Theoretic Approach, New York, Academic Press, 1983.

INITIAL VALUE PROBLEMS FOR REGULAR QUATERNION-VALUED INITIAL FUNCTIONS

LE HUNG SON*, NGUYEN CANH LUONG† and NGUYEN QUOC HUNG‡

Faculty of Applied Mathematics, Hanoi University of Technology,
Hanoi, Vietnam
** E-mail: sonlh_fami@mail.hut.edu.vn*
† E-mail: ncluong@mail.hut.edu.vn
‡ E-mail: nguyenquochung7337@gmail.com

In the talk at 15th ICFIDCAA, Osaka, Japan, we constructed a method to solve the initial value problem in Clifford and quaternion analysis. On the other hand, in [9] we also constructed first-order operator of functions that are associated to a space of regular functions in quaternionic analysis. We will now find out conditions under which the initial value problem is uniquely solvable in this associated space.

Keywords: Initial value problems of Cauchy-Kovalevskaya type; Associated differential operators; Quaternionic analysis; Fixed-point methods.

1. Introduction

An initial value problem (IVP) of Cauchy-Kovalevskaya type has the form

$$\frac{\partial w}{\partial t} = L\left(t, x, w, \frac{\partial w}{\partial x_j}\right) \tag{1}$$

$$w(0, x) = \varphi(x) \tag{2}$$

where the desired function $w = w(t, x)$ depends on the time t and the spacelike variable x, and L is a first order differential operator depending on t, x, w and the derivatives of w with respect to the spacelike variables. The classical approach to IVP (1)-(2) is based on power-series representations. In this case, one assumes that both the right-hand side L and the initial function φ as well have power-series representations in their variables. On the other hand, the IVP (1)-(2) can be formulated within the framework of continuously differentiable functions. Therefore it is natural to ask whether or not there exist continuously differentiable solutions in case the right-hand side of (1) is only a continuously differentiable function. In [8] H. Lewy gives a famous example which shows that there exist such linear first order differential equations with infinitely differentiable coefficients not having any solutions. However, the technique of *associated* differential operators allows to construct operators L such that the IVP (1)-(2) is solvable for each initial function φ satisfying $l\varphi = 0$, where l is an operator which satisfies the *interior estimate* property. The first step of this technique is M. Nagumo's integral rewriting of the IVP (1)-(2):

$$w(t,x) = \varphi(x) + \int_0^t L\left(\tau, x, w(\tau,x), \frac{\partial w(\tau,x)}{\partial x_j}\right) d\tau \tag{3a}$$

Consequently, the solution of the IVP (1)-(2) is a fixed point of the operator

$$Fw(t,x) = \varphi(x) + \int_0^t L\left(\tau, x, w(\tau,x), \frac{\partial w(\tau,x)}{\partial x_j}\right) d\tau \tag{3b}$$

W. Walter (see [7]) uses M. Nagumo's lemma which is such an *interior estimate* for a weighted supremum norm to prove the classical Cauchy-Kovalevskaya theorem. Later, W. Tutschke ([3,4]) also points out the method of weighted function spaces for solving IVP which is based on the technique of associated differential operators. In [9], we consider a linear differential operator of first order of the form

$$Lw = \sum_{j=1}^{3} A^{(j)} \frac{\partial w}{\partial x_j} + Bw + C, \tag{3}$$

where $w(t,x)$ are quaternion-valued functions, the initial funtion $\varphi(x)$ is a regular function (satisfying $D\varphi = 0$, D being the Dirac operator) and the coefficients $A^{(j)} = A^{(j)}(t,x)$, $B = B(t,x)$, $C = C(t,x)$ are also quaternion-valued functions and $x \in \Omega \subset \mathbb{R}^3$ and $t \in [0,T]$ is the time variable. We formulated conditions for the coefficients of L under which L is associated to the Dirac operator D.

In the present paper, we will prove an *interior estimate* for the Dirac operator and use the results of associated pair in [9] to formulate conditions under which the IVP (1)-(2) in quaternionic analysis is uniquely solvable.

2. Preliminaries and notation

We denote by $e_0 = 1$, $e_1 = i$, $e_2 = j$, $e_3 = k$, where i, j, k are the units of the (real) quaternionic algebra $\mathbb{H}$. A function f defined in the bounded domain $\Omega \subset \mathbb{R}^3$ and takes value in $\mathbb{H}$ is a mapping

$$f : \Omega \to \mathbb{H}$$

and thus f can be represented in the form

$$f = \sum_{j=0}^{3} e_j f_j(x)$$

where the $f_j(x)$ are real valued functions of $x = (x_1, x_2, x_3) \in \Omega$. Properties such as continuity, differentiability, integrability and so on, which are ascribed to the function f have to be fulfiled by all components f_j.

We denote by $C_{\mathbb{H}}(\Omega)$, $C_{\mathbb{H}}^k(\Omega)$,... the corresponding spaces of continuous or k-times continuously differentiable functions.

Now let us introduce the Dirac differential operator as

$$D = \sum_{k=1}^{3} e_k \frac{\partial}{\partial x_k}$$

For other definitions concerning quaternions we refer the reader to [1,2].

Lemma 2.1. *([2]) Let $u, v \in C_{\mathbb{H}}^1(\Omega)$. Then*

$$D(u, v) = D(u)v + \bar{u}(Dv) - 2\sum_{j=1}^{3} u_j \frac{\partial v}{\partial x_j}.$$

Definition 2.1. The function $u \in C_{\mathbb{H}}^1(\Omega)$ is called a regular function if $Du = 0$, and we denote by $A_{\mathbb{H}}(\Omega)$ the space of regular functions on Ω.

Definition 2.2. ([4]) Let L be a first order differential opeator depending on $t, x, w = w(t, x)$ and on the spacelike first order derivatives $\frac{\partial w}{\partial x_j}$, while l is any differential operator with respect to the space variables x_j whose coefficients do not depend on the time t. Then L is associated to l if L transforms solutions of $lw = 0$ into solutions of the same equation for fixedly chosen t, i.e, $lw = 0$ implies $l[Lw] = 0$.

Theorem 2.1. *([9]) Suppose that $A^{(j)}(t, x) \in C_{\mathbb{H}}^2(\Omega)$ for each $t \in [0, T]$ and $B(t, x), C(t, x) \in C_{\mathbb{H}}^1(\Omega)$ for each $t \in [0, T]$. The operator L (right-hand side of (1)) is associated to the Dirac operator if the following conditions are satisfied:*

(i) $A_0^{(1)} = A_3^{(2)} = -A_2^{(3)}, \quad A_1^{(1)} = A_2^{(2)} = A_3^{(3)},$
$A_2^{(1)} = -A_1^{(2)} = A_0^{(3)}, \quad A_3^{(1)} = -A_0^{(2)} = -A_1^{(3)},$

(ii) $(DA^{(1)} - 2B_1 e_0)e_1 = (DA^{(2)} - 2B_2 e_0)e_2 = (DA^{(3)} - 2B_3 e_0)e_3,$

(iii) $B, C \in A_{\mathbb{H}}(\Omega)$ *for each* $t \in [0, T]$.

With the condition (i) of Theorem 2.1, if $A^{(1)} = \sum_{k=0}^{3} A_k^{(1)} e_k$ is given, then we can get $A^{(2)}$ and $A^{(3)}$. Furthermore (ii) of Theorem 2.1 gives us one system with 8 equations for 12 unknowns $\frac{\partial A_k^{(1)}}{\partial x_j}$, $(k = 0,1,2,3; j = 1,2,3)$. In [9] we also illustrated this result by giving a general example.

3. Solving the initial value problem

3.1. *Exhaustion*

First, we introduce an exhaustion of Ω by a family of subdomains Ω_s, $0 < s < s_0$, satisfying the following conditions:

- To each point $x \neq x_0$ of Ω where $x_0 \in \Omega$ is fixedly chosen there exists uniquely determined $s(x)$ with $0 < s(x) < s_0$ such that $x \in \partial\Omega_{s(x)}$.
- If $0 < s' < s'' < s_0$, then $\overline{\Omega_{s'}}$ is a compact subset of $\Omega_{s''}$.
- There exists a positive constant C_1 such that for any s', s'' with $0 < s' < s'' < s_0$ the distance of $\Omega_{s'}$ from $\partial\Omega_{s''}$ can be estimated by:

$$\text{dist}\,(\Omega_{s'}, \partial\Omega_{s''}) \geq C_1(s'' - s'),$$

  where C_1 does not depend on the choice of s' and s''.

Define, finally $s(x_0) = 0$. Then $s_0 - s(x)$ is a meansure of the distance of a point x of Ω from the boundary $\partial\Omega_{s(x)}$.

Now we consider the conical set

$$M = \{(t,x) : x \in \Omega, 0 \leq t < \eta(s_0 - s(x))\}$$

in the (t,s)-space. The parameter η will describes the height of the conical set and will be fixed later (Theorem 3.1). The base of M is the given domain Ω, whereas its lateral surface is defined by

$$t = \eta(s_0 - s(x)) \tag{4}$$

The nearer a point x to the boundary $\partial\Omega$, the shorter the coresponding time interval (4). The expression

$$d(t,x) = s_0 - s(x) - \frac{t}{\eta} \tag{5}$$

is positive in M, while it vanishes identically on the lateral surface of M. Thus (5) can be interpreted as some pseudo-distance of a point (t,x) of M from the lateral surface of M.

Again let $\Omega_s, 0 < s < s_0$, be an exhaution of a given (bounded) domain in $\mathbb{R}^3$. Let B_s be the space of all $w(x) \in A_{\mathbb{H}}(\Omega_s)$ equipped with the norm

$$\|w\|_s = \|w\|_{\Omega_s} = \max_{i=0,1,2,3} (\sup_{\Omega_s} |w_j|).$$

The limit function of a Cauchy sequence of functions belonging to B_s is continuous since convergence with respect to the supremum norm means uniform convergence. In view of Weierstrass convergence theorem, moreover, the limit function of an uniformly convergent sequence of regular functions is regular, too. Thus the space B_s is proved to be complete, i.e., B_s is a Banach space.

For a chosen fixed $\tilde{t} < \eta s_0$ the intersection of M with the plane $t = \tilde{t}$ in the (t, x)-space is given by

$$\{(t,x) : t = \tilde{t}; s(x) < \tilde{s}\} \quad \text{where} \quad \tilde{s} = s_0 - \frac{\tilde{t}}{\eta} \tag{6}$$

Let $B_*(M)$ be the set of all function $w = w(t, x)$ which are continuous in M. We see $w(\tilde{t}, x)$ belongs to $B_{s(x)}$ for fixed $\tilde{t}$ if only $s(x) < \tilde{s}$ where $\tilde{s}$ is given by (6). And $B_*(M)$ is equipped with the norm

$$\|w\|_* = \sup_{(t,x)\in M} \|w(t,.)\|_{s(x)} d(t,x). \tag{7}$$

The definition (7) of the norm $\|.\|_*$ implies the estimate

$$\|w(t,.)\|_{s(x)} \leq \frac{\|w\|_*}{d(t,x)} \tag{8}$$

for any point $(t, x) \in M$.

Proposition 3.1. *$B_*(M)$ is a Banach space.*

Proof. Note that the inequality $d(t,x) \geq \delta > 0$ defines a closed subset M_δ of the conical domain M. Each point of M is contained in such a subset M_δ provided δ is suitably chosen. For points (t, x) in M_δ, the definition (7) implies the estimate

$$\|w(t,.)\|_{s)(x)} \leq \frac{1}{\delta}\|w\|_*$$

Now consider a fundamental sequence $w_1, w_2, ...$ with respect to the norm $\|.\|_*$. Then one has

$$\|w_n(t,.) - w_m(t,.)\|_{s(x)} \leq \frac{1}{\delta}\epsilon \tag{9}$$

for points in M_δ provided n and m are sufficiently large. This implies also

$$|w_n - w_m| \leq \frac{1}{\delta}\epsilon$$

for points in M_δ. Consequently, a fundamental sequence converges uniformly in each M_δ, i.e., the w_n have a continuous limit function $w_*(t,x)$ in M. Similarly, estimate (9) shows that for $t = \tilde{t}$ and $s(x) < \tilde{s}$ the limits function belongs to $B_{s(x)}$ because of the completeness of this space. Carrying out the limiting process $m \to \infty$ in the inequality $\|w_n - w_m\|_* < \epsilon$, it follows, finally, $\|w_n - w_*\|_* \leq \epsilon$ and, therefore, $\|w_*\|_*$ is finite. □

3.2. *Interior estimate*

Proposition 3.2. *If $0 < s' < s'' < s_0$ and $w \in A_{\mathbb{H}}(\Omega)$, then*

$$\left\| \frac{\partial w}{\partial x_i} \right\|_{s'} \leq \frac{3}{dist\,(\Omega_{s'}, \Omega_{s''})} \|w\|_{s''}$$

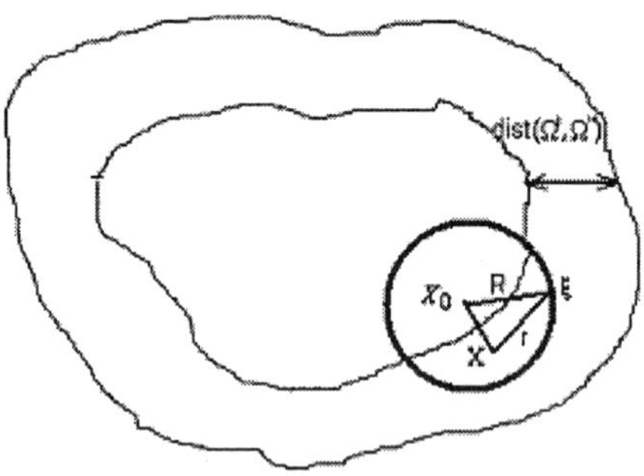

Proof. Let x_0 be an arbitrary point of $\Omega_{s'}$, then $B_R(x_0)$ is a ball with centre x_0 and radius R will belong to $\Omega_{s''}$ if $R < \text{dist}\,(\Omega_{s'}, \partial\Omega_{s''})$. Since

$$w(x) \in A_{\mathbb{H}}(\Omega) \Rightarrow \Delta w = 0,$$

we have $\Delta w_k = 0$ $(k = 0, 1, 2, 3)$.

By using Poisson integral formula for harmonic functions, we get

$$w_k(x) = \frac{1}{4\pi R} \int_{|\xi - x_0| = R} w_k(\xi) \frac{R^2 - |x - x_0|^2}{|\xi - x|^3} d\mu \tag{10}$$

We denote $r = |\xi - x|$. Then $r = \left(\sum_{i=1}^{3}(\xi_i - x_i)^2\right)^{1/2}$ and $\frac{\partial r}{\partial x_0} = \frac{x_i - \xi_i}{r}$. Formula (10) implies

$$\frac{\partial w_k(x)}{\partial x_i} = \frac{1}{4\pi R}\int_{|\xi - x_0|=R} w_k(\xi)\left[\frac{-2(x_i - x_{0i})}{r^3} - 3(R^2 - |x - x_0|^2)\frac{x_i - \xi_i}{r^5}\right] d\mu \tag{11}$$

where $x_0 = (x_{01}, x_{02}, x_{03})$. Substitute $x = x_0$ into (11), one gets $|x - x_0| = 0$; $|x_i - x_{0i}| = 0$. Thus

$$\frac{\partial w_k(x)}{\partial x_i} = \frac{1}{4\pi R}\int_{|\xi - x_0|=R} w_k(\xi)(-3)R^2\frac{(x_{0i} - \xi_i)}{R^5} d\mu$$

On the other hand, $\left|\frac{x_{0i} - \xi_i}{R}\right| \le 1$; thus

$$\left|\frac{\partial w_k}{\partial x_i}(x_0)\right| \le \frac{1}{4\pi R}\int |w_k(\xi)|\frac{3}{R^2} d\mu \qquad \text{implies}$$

$$\left|\frac{\partial w_k}{\partial x_i}(x_0)\right| \le \frac{3}{4\pi R}\sup_{|\xi - x_0|=R} |w_k(\xi)|\frac{1}{R^2}4\pi R^3.$$

By maximum principle (see [2]). we get $\sup_{|x - x_0|\le R} |w_k| = \sup_{|\xi - x_0|=R} |w_k|$. Hence

$$\left|\frac{\partial w_k}{\partial x_i}(x_0)\right| \le \frac{3}{R}\sup_{|x - x_0|\le R} |w_k|. \tag{12}$$

In view of the definition of the norm in $B(\Omega)$, (12) leads to

$$\left\|\frac{\partial w}{\partial x_i}\right\|_{s'} \le \frac{3}{R}\|w\|_{s''}.$$

Let $R \to \operatorname{dist}(\Omega_{s'}, \partial\Omega_{s''})$; we get

$$\left\|\frac{\partial w}{\partial x_i}\right\|_{s'} \le \frac{3}{\operatorname{dist}(\Omega_{s'}, \partial\Omega_{s''})}\|w\|_{s''}.$$

But $\operatorname{dist}(\Omega_{s'}, \partial\Omega_{s''}) \ge C_1(s'' - s')$ provided $s' < s''$. Thus

$$\left\|\frac{\partial w}{\partial x_i}\right\|_{s'} \le \frac{3}{C_1(s'' - s')}\|w\|_{s''}. \tag{13}$$

(13) is called an *interior estimate*. □

3.3. *The main theorem*

Theorem 3.1. *The initial value problem (1)-(2) is uniquely solvable, and the solution $w = w(t,x)$ satisfies the side condition $Dw = 0$ for each t if*

- *hypotheses (i), (ii), (iii) of Theorem 2.1 are satisfied;*
- *there exists positive constants K_0, K_1 and K_2 such that*

$$\max_{i=0,1,2,3}\Big(\sup_{\Omega\times[0;+\infty)}|B_i|\Big) \le K_0$$

$$\max_{i=0,1,2,3}\Big(\sup_{\Omega\times[0;+\infty)}|A_i^{(1)}|\Big) \le K_1$$

$$\max_{i=0,1,2,3}\Big(\sup_{\Omega\times[0;+\infty)}|C_i|\Big) \le K_2;$$

- *the parameter η (describing the height of the conical set M) satisfies*

$$\eta < \frac{1}{s_0K_0 + 3\frac{12}{C_1}K_1}.$$

Proof. Let (t,x) be an arbitrary point of M. Then

$$d(t,x) = s_0 - s(x) - \frac{t}{\eta} > 0.$$

Define $\tilde{s} = s(x) + \frac{1}{2}d(t,x)$ implying

$$\tilde{s} \le s(x) + \frac{1}{2}(s_0 - s(x)) = \frac{1}{2}s(x) + \frac{1}{2}s_0 < s_0.$$

Thus there exists a point $\tilde{x}$ with $s(\tilde{x}) = \tilde{s}$, i.e., $\tilde{x} \in \partial\Omega_{\tilde{s}}$. One has

$$d(t,\tilde{x}) = s_0 - s(\tilde{x}) - \frac{t}{\eta} = \frac{1}{2}d(t,x).$$

Taking into account the estimate (8), the last relation gives

$$\|w(t,.)\|_{\tilde{s}} \le \frac{\|w\|_*}{d(t,\tilde{x})} \le \frac{2\|w\|_*}{d(t,x)}.$$

In view of (13) one gets, therefore,

$$\left\|\frac{\partial w}{\partial x_i}\right\|_{s(x)} \le \frac{3}{C_1(\tilde{s}-s)}\|w\|_{\tilde{s}} \le \frac{3}{C_1}\frac{4}{d^2(t,x)}\|w\|_*. \tag{14}$$

To be brief, denote $L\Big(t,x,w,\dfrac{\partial w}{\partial x_j}\Big)$ by Lw. In particular, one has $L\theta = L(t,x,0,0)$. Now, we have to estimate the norm of Lw. We get

$$\begin{aligned} \|Lw - Lv\|_s &= \|\sum_{j=1}^{3} A^{(j)}\left(\frac{\partial w}{\partial x_j} - \frac{\partial v}{\partial x_j}\right) + B(w-v)\|_s \\ &\leq K_0\|w-v\|_s + K_1 \sum_{j=1}^{3} \|\frac{\partial w}{\partial x_j} - \frac{\partial v}{\partial x_j}\|_s \end{aligned} \tag{15}$$

Using (8) and (14), condition (15) implies

$$\|Lw\|_{s(x)} - \|L\theta\|_{s(x)} \leq \|Lw - L\theta\|_{s(x)} \leq K_0\|w\|_{s(x)} + K_1\sum_{j=1}^{3}\|\frac{\partial w}{\partial x_j}\|_{s(x)}$$

and

$$\begin{aligned} \|Lw\|_{s(x)} &\leq \|L\theta\|_{s(x)} + K_0\|w\|_{s(x)} + K_1\sum_{j=1}^{3}\|\frac{\partial w}{\partial x_j}\|_{s(x)} \\ &\leq \|C(t,x)\|_* \frac{s_0}{d^2(t,x)} + K_0\|w\|_* \frac{s_0}{d^2(t,x)} + 3\frac{12}{C_1}K_1\|w\|_* \frac{1}{d^2(t,x)}, \end{aligned}$$

because $d(t,x) < s_0$. The definition (5) of the weight function $d(t,x)$ implies $\displaystyle\int_0^t \frac{1}{d^2(\tau,x)}d\tau < \frac{\eta}{d(t,x)}$, and thus it follows that

$$\left\|\int_0^t L(\tau,x,w(\tau,x),\frac{\partial w(\tau,x)}{\partial x_i})d\tau\right\|_{s(x)} \leq \frac{\eta}{d(t,x)}(K_2.s_0 + C_2\|w\|_*) \tag{16}$$

where $C_2 = s_0K_0 + 3\frac{12}{C_1}K_1$.

The estimate (16) of the $s(x)$-norm yields

$$\left\|\int_0^t L\left(\tau,x,w(\tau,x),\frac{\partial w(\tau,x)}{\partial x_i}\right)d\tau\right\|_* \leq \eta\left(K_2s_0 + C_2\|w\|_*\right).$$

From (3), (3b), it follows that

$$\|F\|_* \leq \|\varphi\|_* + \eta(K_2s_0 + C_2\|w\|_*). \tag{17}$$

The conditions (17) and (15) imply

$$\|Fw - Fv\|_* \leq \eta C_2\|w-v\|_* = \eta\left(s_0K_0 + 3\frac{12}{C_1}K_1\right)\|w-v\|_*. \tag{18}$$

Now we consider the space

$$B_*^D(M) = \{\, w \in B_*(M) : Dw(t,\cdot) = 0 \text{ for each fixed } t\,\}.$$

We see that $B_*^D(M)$ is closed subset of $B_*(M)$ in view of Weierstrass convergence theorem. Because of Theorem 2.1, if $w \in B_*^D(M)$, that means $Dw(t, \cdot) = 0$ for each fixed t which implies $D[Fw(t, \cdot)] = 0$, because $D\varphi = 0$ and $D(Lw) = 0$. So the operator F maps the subspace $B_*^D(M)$ into itself.

From the hypothesis of Theorem 3.1, we have $\eta < \left(s_0 K_0 + 3(12)K_1/C_1\right)^{-1}$, which together with (18) imply that F is a contractive mapping. Theorem 3.1 is now proved. □

References

1. F. Brackx, R. Delanghe and F. Sommen, *Clifford analysis*, Pitman Advanced Publishing Program, Boston-London-Melbourne, 1982.
2. K. Gürlebeck and W. Sprössig, *Quaternionic analysis and elliptic boundary value problems*, Akademie-Verlag, Berlin, 1989.
3. W. Tutschke, *Solution of initial value problems in classes of generalized analytic functions*, Teubner, Leipzig and Springer Verlag, 1989.
4. H. Florian, N. Ortner, F. J. Schnitzer, and W. Tutschke, *Functional analytic and complex methods, their interactions, and applications to partial differential equations*, World Scientific, 2001, 75–90.
5. L. H. Son and W. Tutschke (2003), First order differential operators associated to the Cauchy-Riemann operator in the plane, *Complex Variables*, 48, 797–801.
6. L. H. Son and N. Q. Hung, *The inital value problems in Clifford and quaternion analysis*, Proceedings of the 15th ICFIDCAA, Osaka Municipal Universities Press 3, 2008, 317–323.
7. W. Walter (1985), An elementary proof of the Cauchy-Kovalevskaya theorem, *Amer. Math. Monthly*, 92, 115–125.
8. H. Lewy (1957), An example of a smooth linear partial differential equaiton without solution, *Ann. of Math.*, 66, 155–158.
9. N. Q. Hung and N. C. Luong, *First order differential operators associated to the Dirac operator in quaternionic analysis*, Proceedings of the 2004 International Conference on Applied Mathematics, SAS International Publications, Delhi, 369–378.

Session 03

Complex Analysis and Potential Theory

SESSION EDITORS

T. Aliyev Azeroğlu	Gebze Institute of Technology, Kocaeli, Turkey
M. Lanza de Cristoforis	Università di Padova, Padova, Italy
P. M. Tamrazov	National Academy of Sciences, Kiev, Ukraine

THE GEOMETRY OF BLASCHKE PRODUCTS MAPPINGS

ILIE BARZA

Karlstad University, Department of Mathematics,
S-651 88 Karlstad, Sweden
E-mail: ilie.barza@kau.se

DORIN GHISA

York University, Glendon College, Department of Mathematics,
2275-Bayview Av., Toronto, Canada, M4N3M6
E-mail: dghisa@yorku.ca

A Blaschke product B generates a covering Riemann surface (W, B) of the complex plane. The study of such surfaces is undertaken here in a very general context, where the cluster points of the zeros of B form a (generalized) Cantor set. Explicit forms and fundamental domains for the covering transformations are revealed for a particular case of Blaschke products.

Keywords: Blaschke product, covering surface, covering transformation, fundamental domain, Cantor set, conformal module.

1. Blaschke Products

A Blaschke product is an expression of the form

$$w = B(z) = \prod_{k=1}^{n\leq\infty} b_k(z), \tag{1}$$

where

$$b_k(z) = \frac{\bar{a}_k}{|a_k|}\frac{a_k - z}{1 - \bar{a}_k z} \tag{2}$$

where $a_k \in D := \{z \in \mathbb{C}|\ |z| < 1\}$.

The Moebius transformations (2) are called the Blaschke factors of B and the complex numbers a_k are called the zeros of B. The product B can have a finite number of factors and then it is called finite Blaschke product, or infinitely many factors and then it is called infinite Blaschke product. If B has m Blaschke factors (counted with their multiplicities), we say that

B is a Blaschke product of order m. In the infinite case, we understand by B(z) the limit (if it exists) of the sequence $(B_n(z))$, where

$$B_n(z) = \prod_{k=1}^{n} b_k(z).$$

It is known (see [6]) that under the following Blaschke condition:

$$\sum_{n=1}^{\infty}(1 - |a_n|) < \infty \tag{3}$$

the sequence (B_n) converges uniformly on compact subsets of $W = \mathbb{C} \setminus (E \cup A)$, where E is the set of cluster points of zeros of B (necessarily a subset of ∂D) and

$$A = \{z \in \mathbb{C} \mid \ z = \frac{1}{\bar{a}_k}, \ k = 1, 2, ...\}.$$

While $B_n(z)$ are meromorphic functions in $\overline{\mathbb{C}}$, every point of E is an essential singularity of B. Moreover,these points are non isolated essential singularities, since they are also cluster points of the poles $\frac{1}{\bar{a}_k}$. There are known examples (see [3]) where $E = \partial D$, and therefore the domain of B is not a connected subset of $\mathbb{C}$. However, if $\partial D \setminus E$ contains an arc of the unit circle, then W is connected and (W, B), where W is endowed with the conformal structure induced by the identity mapping, represents a (branched) Riemann covering surface of the complex plane (see [2], page 119).

The most general setting in which we will treat the infinite Blaschke products in this paper is that where E is the union of an usual Cantor set (see [5], page 41) and a discrete subset of the unit circle ∂D. We will call such a set generalized Cantor set, since the last subset can also be obtained by arc removal, namely when the removed open arcs are contiguous. This kind of generalized Cantor set might not be any more perfect. We consider that E is the empty set when B is finite.

2. The Behavior of a Blaschke Product on the Unit Circle

We would like to know what is the group of covering transformations of (W, B) over $\mathbb{C}$, and determine the fundamental domains of that group. Therefore, we are looking for conformal mappings $U : W \longrightarrow W$ such that

$$B \circ U(z) = B(z), \text{for every } z \in W. \tag{4}$$

Functions U defined in a neighborhood of the unit circle and fulfilling (4) have been found in [4] in the case of finite Blaschke products and this result has been generalized to some infinite Blaschke products in [3].

The main fact proven in [3] is the following:

If the Blaschke sequence (a_n) has a unique cluster point $e^{i\theta_0}$, then there are infinitely many disjoint half-open arcs

$$\Gamma_n = \{z = e^{i\theta} \mid \alpha_n \leq \theta < \alpha_{n+1}\},\ n \in \mathbb{Z},\ lim_{n->\pm\infty}\alpha_n = \theta_0 \quad (5)$$

such that $\bigcup \Gamma_n = \partial D \setminus \{e^{i\theta_0}\}$ and B maps every Γ_n continuously and bijectively on the unit circle in the w- plane. There are continuous functions

$$U_k : \partial D \setminus \{e^{i\theta_0}\} \longrightarrow \partial D \setminus \{e^{i\theta_0}\}$$

such that $U_k(\Gamma_n) = \Gamma_m$, $k, m, n \in \mathbb{Z}$ and $B \circ U_k(\zeta) = B(\zeta)$ for every $\zeta \in \partial D \setminus \{e^{i\theta_0}\}$. The set of these functions form an infinite cyclic group. Moreover, the functions U_k can be extended to analytic functions in a neighborhood (in $\mathbb{C}$) of every compact set included in $\partial D \setminus \{e^{i\theta_0}\}$.

Theorem 2.1. *If E is a generalized Cantor subset of ∂D, then every removed open arc $I = \{z = e^{i\theta} \mid \alpha < \theta < \beta\}$ contains infinitely many disjoint half-open sub-arcs $\Gamma_n = \{z = e^{i\theta} \mid \alpha_n \leq \theta < \alpha_{n+1}\}$, $n \in \mathbb{Z}$, $\lim_{n\to+\infty} \alpha_n = \alpha$, $\lim_{n\to-\infty} \alpha_n = \beta$, and $I = \bigcup_{n=-\infty}^{+\infty} \Gamma_n$, such that B maps every subarc Γ_n continuously and bijectively on the unit circle in the w-plane.*

Proof. We know (see [3]) that, for every $n \in \mathbb{N}^*$, the equation $B_n(z) = 1$ has exactly n distinct solutions, no matter if the zeros a_k of $B_n(z)$ are distinct or not. Let us notice that

$$\lim_{n\to\infty} \frac{B(z)}{B_n(z)} = 1 \quad (6)$$

uniformly on compact subsets of $\mathbb{C} \setminus E \cup A$.

This means that, for every $\varepsilon > 0$ and $\delta > 0$, there is $n_0 \in \mathbb{N}^*$ such that $n \geq n_0$ and $\mid z - e^{i\alpha} \mid \geq \delta$, where $e^{i\alpha} \in E$ imply

$$\mid \frac{B(z)}{B_n(z)} - 1 \mid < \varepsilon.$$

Therefore, if ζ is a solution of the equation $B_n(z) = 1$, $n \geq n_0$ and $\mid \zeta - e^{i\alpha} \mid \geq \delta$, then

$$\mid B(\zeta) - 1 \mid < \varepsilon.$$

Due to the continuity of the local inverse of B at $B(\zeta)$, given $\eta > 0$, we can choose $\epsilon > 0$ such that $\mid B(\zeta) - 1 \mid < \epsilon$ implies $\mid \zeta - B^{-1}(1) \mid < \eta$. Therefore, there is ζ' such that $B(\zeta') = 1$ and $|\zeta - \zeta'| < \eta$. In other words, the solutions of the equation $B(z) = 1$ are as close as we want of the solutions of $B_n(z) = 1$, if n is big enough.

This means that the equation $B(z) = 1$ has infinitely many solutions. They cannot cluster to any point $\zeta_0 \in I$ since then the function $B(z) - 1$ which is analytic in a neighborhood V of ζ_0, would be identically zero in V, which is impossible.

Consequently, they accumulate to the ends of those arcs. Let us show that the end of every removed arc is a cluster point of the roots of the equation $B(z) = 1$.

Suppose that the contrary happens, i.e. there is a point $e^{i\alpha}$ such that a subsequence $(a_{n_k})_{k\in\mathbb{N}}$ of $(a_n)_{n\in\mathbb{N}}$ converges to $e^{i\alpha}$ and there is a neighborhood V of $e^{i\alpha}$ such that $B(z) \neq 1$ in V. Then $\varphi(z) = 1/[B(z) - 1]$ is analytic in V, and $\varphi(1/\overline{a_{n_k}}) = 0$, $k = 1, 2, ...$ This would imply that $\varphi(z)$is identically equal to 0 in V since $\lim\limits_{k\to\infty} 1/\overline{a_{n_k}} = e^{i\alpha} \in V$, which is a contradiction. Now the conclusion of the theorem is obvious. □

Remark 2.1. Actually, even more can be said about the behavior of B in any neighborhood of an $e^{i\alpha} \in E$. We can replace 1 at the denominator of φ by any complex number $w \neq 0$. Then again φ would be analytic in V and $\varphi(1/\overline{a_{n_k}}) = 0$, $k = 1, 2, ...$, which contradicts the fact that φ is not identically 0 in V. Consequently, in any neighborhood V of $e^{i\alpha} \in E$, the Blaschke product B takes every complex value (zero included, for an obvious reason!) infinitely many times. We will show in the following that this happens in a very orderly way, namely there are infinitely many regions in V which are mapped by B bijectively on $\widehat{\mathbb{C}}$, the mapping being conformal in the interior of each one of these regions.

3. Simultaneous Continuation

In order to describe the geometry of the mappings realized by a Blaschke product B, we need to perform a partition of the domain of B in subsets in which B is injective.

The following example illustrates this idea in a very simple case, namely when the Blaschke product is of the form

$$w = B(z) = \left[\frac{\bar{a}}{|a|}\frac{a - z}{1 - \bar{a}z}\right]^n, \tag{7}$$

where $a = re^{i\theta}$, $0 \leq r < 1$ and n is a natural number greater than 1. Let $0 \leq \lambda \leq 1$ and let us solve the equation $B(z) = \lambda^n$ for the unknown z. We get the solutions of the form:

$$z_k(\lambda) = \frac{\omega_k\lambda - r}{\omega_k\lambda r - 1}e^{i\theta},\ k = 0, 1, 2, ..., n-1,$$

where ω_k are the roots of order n of the unity. It can be easily checked that, for $\lambda \neq 0$, $z_k(\lambda)$ are distinct numbers, while $z_k(0) = a$ for every k.

Moreover, as λ varies from o to $1, z_k(\lambda)$ describes a Jordan arc γ_k connecting the point a with

$$\zeta_k = \frac{\omega_k - r}{\omega_k r - 1}e^{i\theta}.$$

The points ζ_k are all distinct points on the unit circle and consequently the arcs γ_k have no other common point except for a. We notice that in this particular example the arcs γ_k , as well as their symmetric arcs with respect to the unit circle should be arcs of circle or line segments. Indeed, $z_k(\lambda)$ are Moebius transformations and λ varies on an interval.

Let us denote by Γ_k the arc of the unit circle between ζ_k and ζ_{k+1}. Then

$$C_k = \gamma_k + \Gamma_k - \gamma_{k+1}$$

is a Jordan curve bounding a domain D_k. The image of C_k in the $w-$ plane is a curve obtained by adding to the unit circle the superior border and subtracting the inferior border of the slit alongside the interval $[0, 1]$ of the real axis. Since $B'(z) \neq 0$ in D_k, the function B represents conformally D_k on the slit unit disc. This map can be extended by symmetry to the domain

$$\Omega_k - D_k \bigcup \widehat{D_k} \bigcup \Gamma_k^0,$$

where $\widehat{D_k}$ is the symmetric of D_k with respect to the unit circle and $\Gamma_k^0 = \Gamma_k \setminus \{\zeta_k, \zeta_{k+1}\}$.

The function $B\ |_{\Omega_k}$ realizes a conformal mapping of Ω_k on the domain obtained from the complex w-plane by removing the positive real half-axis. For $k \neq n-1$, there is a continuous passage from $B\ |_{\Omega_k}$ to $B\ |_{\Omega_{k+1}}$, when crossing γ_{k+1} as well as from $B\ |_{\Omega_{n-1}}$ to $B\ |_{\Omega_0}$ when crossing γ_0. In fact the function B is locally injective in $W = \mathbb{C} \setminus \{a; 1/\bar{a}\}$ and (W, B) is a covering Riemann surface of the complex $w-$ plane.

The example (7) is useful also from another point of view, namely because we can find explicitly in that case the relation between the points having the same image by B. Indeed, suppose that $B(z') = B(z)$, i.e.

$$\left[\frac{\bar{a}}{|a|}\frac{a - z'}{1 - \bar{a}z'}\right]^n = \left[\frac{\bar{a}}{|a|}\frac{a - z}{1 - \bar{a}z}\right]^n.$$

Then, $\frac{\bar{a}}{|a|}\frac{a-z'}{1-\bar{a}z'} = \frac{\bar{a}}{|a|}\frac{a-z}{1-\bar{a}z}\omega_k$, where ω_k, $k = 0, 1, 2, ..., n-1$ are the roots of order n of the unity.This gives

$$z' = \frac{a(1-\omega_k) - (|\,a\,|^2 - \omega_k)z}{1-|\,a\,|^2\,\omega_k - \bar{a}\,(1-\omega_k)z}.$$

If we denote $z' = T_k(z)$, we can easily see that $\{T_k\}$ is a group of order n with respect to the usual composition of functions.

Indeed, $T_0(z) = z$ and for arbitrary k and k' we have $T_k \circ T_{k'} = T_{k+k'(modn)}$. Particularly,$T_k \circ T_{n-k} = T_0$. These formulas imply that $T_k(\Omega_{k'}) = \Omega_{k+k'(modn)}$. We notice also that if $z \neq a$, and $z \neq 1/\bar{a}$, then $T_k(z) \neq z$ and $T_k(a) = a$, $T_k(1/\bar{a}) = 1/\bar{a}$, for every k. Therefore, the Moebius transformations T_k are covering transformations of (W, B) over $\mathbb{C}$.

We can also think of $(\widehat{\mathbb{C}}, B)$ as being a branched covering surface of $\widehat{\mathbb{C}}$, with the branch points a and $1/\bar{a}$, both of them of multiplicity n. Obviously, the group of covering transformations of $(\widehat{\mathbb{C}}, B)$ over $\widehat{\mathbb{C}}$ is the same, if we consider T_k as being this time self-mappings of $\widehat{\mathbb{C}}$.

The purpose of this section is to show that similar results can be obtained in a much more general context.

The partition of the corresponding W will have infinitely many subsets when B is an infinite product and it can be realized by a technique that we call *simultaneous continuation.*

Theorem 3.1. *For every Blaschke product B, with E a generalized Cantor subset of ∂D, there is a partition of W into connected sets, such that the interior of each one of these sets is mapped by B conformally onto a slit w-plane. There is a continuous passage from the mapping of every set to that of a contiguous set in the partition.*

Proof. Let us deal first, for simplicity, with the case of a finite Blaschke product B. Let $b_1, b_2, ..., b_p$ be the branch points of B situated in the unit disc. If $q_1, q_2, ..., q_p$ are their order of multiplicity, then an easy computation shows that $q_1 + q_2 + ... + q_p = n$. We might have $w_j = w_{j'}$, for $b_j \neq b_{j'}$; particularly $w_j = 0$, for all multiple zeros of B, which are obviously brunch points. Let $w_j = B(b_j), j = 1, 2, ..., p$. We connect all the points w_j by a polygonal line with no self-intersection, in such a way that one of its ends can be connected with $w = 1$ by a segment not intersecting the polygonal line. Let η be the new polygonal line which is a Jordan arc.

Now we do continuations over η from every root $e^{i\theta_k}$ of the equation $B(z) = 1$. There are exactly n such arcs and an easy topological argument shows that they must all end in points b_j. Moreover, every b_j should be reached at some moment by at least two consecutive arcs γ_j and γ_{j+1}. If the point b_j is a branch point of multiplicity q_j, then exactly q_j consecutive arcs γ_k will end in b_j. Indeed, a corollary of the open mapping theorem (see [9]) guarantees that there is a local parameter at b_j, $z = z(u), z(0) = b_j$ such that $w = B(z(u)) = w_j + u^{q_j}$ and there is a disc centered at 0 in the u-plane such that $B(z(u))$ maps p rays of that disc on η with $B(z(0)) = w_j$. The images by $z(u)$ of these rays are sub-arcs of γ_k converging to b_j, and they are all mapped by B on η.

Let us denote $C_k = \Gamma_k + \gamma_{k+1} - \gamma_k$, which is a closed Jordan curve delimiting a domain D_k. The image of C_k by B is the closed curve obtained by adding to the unit circle the lower border and subtracting the upper border of a cut alongside η ending in $w_k = B(b_k)$, where b_k is the common end of γ_k and γ_{k+1}. Let Δ_k be the domain in the w-plane bounded by that curve.

The conformal correspondence theorem (see [8], p. 154) shows that $B\,|_{D_k}$ is a conformal map of D_k on Δ_k with homeomorphic correspondence of their boundaries. Moreover, if $A_k = D_k \bigcup L_k \bigcup l_k$, where L_k and l_k represent the set of points of Γ_k, respectively γ_k, then $B\,|_{A_k}$ is a bijective and continuous map of A_k on the closed unit disc.

We can extend analytically $B\,|_{A_k}$ by symmetry to $\widehat{A_k}$, the symmetric of A_k with respect to the unit circle. Obviously, the image of $\Omega_k = A_k \bigcup \widehat{A_k}$ is the whole Riemann sphere. In fact B realizes a conformal map of the interior of Ω_k on the Riemann sphere with a slit alongside η from w_k to $w = 1$ and alongside the symmetric of η with respect to the unit circle from $w = 1$ to $1/\overline{w_k}$.

The sets Ω_k form the partition we were referring to in the theorem.

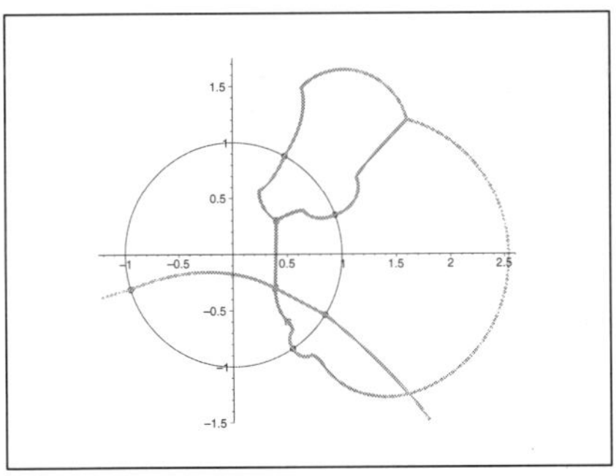

Fig. 1.

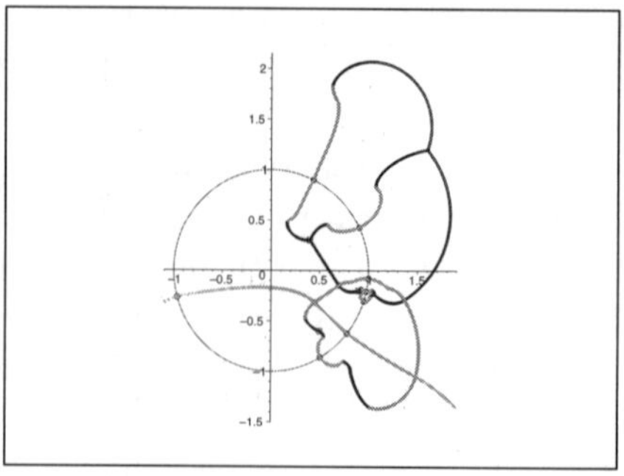

Fig. 2.

The figures 1 and 2 exhibit computer generated fundamental domains for two Blaschke products.The first is of the form $B_1(z) = b_1^3(z)b_2^2(z)$ where b_k are of the form (2) with $a_1 = 0.4 + i\,0.3$ and $a_2 = 0.5 - i\,0.6$. The second is of the form $B_2(z) = B_1(z)b_3^3(z)$, where $a_3 = 0.9 - i\,0.2$.We notice that the addition of $b_3^3(z)$ changed in an obvious way only the fundamental domain containing a_3. Also, since a_3 is closed to the unit circle, some of the corresponding domains are very small.

We can reduce the infinite case to the finite one by using an exhaustion sequence (U_m) of $\widehat{\mathbb{C}} \setminus E$. Let us denote by F_m and by H_m the finite sets of branch points of B, respectively of roots of the equation $B(z) = 1$, situated in (U_m). Let us show that (U_m) can be chosen in such a way that every continuation over η_m (the polygonal line constructed with F_m as in the finite case) starting from a point of H_m lands in a point of F_m. Indeed, let η_m stretch from $w = 1$ to $w = 0$ and have other vertices exactly at the branch points $b_j \in F_m$. Let $\varepsilon > 0$ be small enough such that every circle C_j of radius ε centered at b_j intersects only the sides of η_m ending in b_j. Let h_j be the point of intersection of C_j and η_m situated between b_j and 1. For ε small enough, the equation $B(z) = h_j$ has exactly q_j distinct solutions z_k where q_j is the multiplicity of b_j as solution of the equation $B'(z) = 0$. The continuation over η_m from every z_k towards b_j are all arcs connecting z_k to b_j. We complete these continuations up to $w = 1$. They are Jordan arcs $\gamma_{n,j}$ connecting b_j with the points H_m on the unit circle. The arcs $\gamma_{n,j}$ cannot intersect each other, since then the image of the domain they are bounding would have no interior point, which is absurd. An easy analysis shows that we can arrange the exhaustion (U_m) in such a way that F_m and H_m are for every m the vertices of a connected bilateral graph with edges $\gamma_{n,j}$.

Switching from (U_m) to (U_{m+1}) implies just adding a finite number of vertices and edges to that graph, and hence obtaining a new one of the same type. Finally, every element of $\cup_{k=1}^{\infty} F_k$ and $\cup_{k=1}^{\infty} H_k$ will appear in these graphs starting with an m. As in the finite case, the curves $C_{n,j} = \Gamma_{n,j} - \gamma_{n,j+1} + \gamma_{n,j}$ contain domains which are mapped by B conformally on slit unit discs. The partition of $\widehat{\mathbb{C}} \setminus E$ into fundamental domains follow the same steps as in the finite case. Later on in this section we will find more about such a partition in the infinite case. □

Corollary 3.1. *If B is a Blaschke product of order m, then*

$$\frac{1}{\pi}\iint_{|z|<1} |B'(z)|^2 dxdy = \frac{1}{\pi}\iint_{|z|>1} |B'(z)|^2 dxdy = m, \tag{8}$$

$$\frac{1}{2\pi}\int_{|z|=1} |B'(z)| \mid dz \mid \geq m. \tag{9}$$

Proof. Indeed, it is known (see [8], page 155) that if f represents conformally a domain D on D^*, then $area(D^*) = \iint_D |f'(z)|^2 dxdy$ and that if f is regular on the closed contour Γ and in the domain enclosed by Γ, then the length of $f(\Gamma)$ is $\int_\Gamma |f'(z)| \mid dz \mid$.

With D_k, previously defined, instead of D and $B(z)$ instead of $f(z)$, we have that D^* is the slit unit disc, therefore $\frac{1}{\pi}\iint_{D_k} |B'(z)|^2 dxdy = 1$ for every k, consequently

$$m = \sum_{k=1}^{m} \frac{1}{\pi}\iint_{D_k} |B'(z)|^2 dxdy = \frac{1}{\pi}\iint_{|z|<1} |B'(z)|^2 dxdy.$$

The change of variable $\zeta = 1/\bar{z}$, where $\zeta = \xi + i\eta$, shows that we have also $\frac{1}{\pi}\iint_{|\zeta|>1} |B'(\zeta)|^2 d\xi d\eta = m$, and therefore the formula (8) is true. A trivial consequence is that, for every infinite Blaschke product B, $\iint_{|z|<1} |B'(z)|^2 dxdy = \iint_{|z|>1} |B'(z)|^2 dxdy = \infty$.

One might be tempted to replace this integral by an integral in the Lebesgue sense in the infinite case, if the Lebesgue measure of E is zero, as it happens frequently in harmonic analysis. However, it is obvious from (8) that such an integral is infinite and there is no gain in adopting that approach. On the other hand, if we take $\Gamma = \Gamma_k + \gamma_k - \gamma_{k+1}$, then $B(\Gamma)$ is the unit circle plus the two borders of the slit and therefore the length of $B(\Gamma)$ is $\geq 2\pi$, i.e. $1 \leq (1/2\pi)\int_{\Gamma_k+\gamma_k-\gamma_{k+1}} |B'(z)||dz|$. When we take the sum with respect to k in both terms, the integrals on γ_k cancel and we obtain $m \leq \frac{1}{2\pi}\int_{|z|=1} |B'(z)||dz|$. □

The formulas (8) and (9) allow us to describe more exactly the behavior of B in the neighborhood of a cluster point of its zeros. Indeed, the domains bounded by the curves C_k can be made quadrilaterals Q_k (see [7], page 14) by choosing the "vertices" $e^{i\alpha_k}$,$e^{i\beta_k}$, $e^{i\alpha_{k+1}}$ and b_k, where $B(e^{i\beta_k}) = -1$ and b_k is a branch point of (W, B) situated on C_k. The conformal image of Q_k by B is the slit unit disc made quadrilateral Q'_k with vertices $1, -1, 1, B(b_k)$. Let us denote by Γ'_k the side of Q_k between $e^{i\alpha_k}$ and $e^{i\beta_k}$. The conformal modules of Q_k and Q'_k are

$$M(Q_k) = M(Q'_k) \geq \frac{\iint_{Q_k} |B'(z)|^2 dxdy}{(\int_{\Gamma'_k} |B'(z)| \mid dz \mid)^2} = \frac{\pi}{\pi^2} = \frac{1}{\pi}.$$

The quadrilateral Q_k has then as canonical image a rectangle whose ratio of the side lengths is $\frac{a_k}{b_k} \geq \frac{1}{\pi}$. If we let a_k be proportional with $l(\Gamma'_k)$, then b_k must be proportional with $l(\gamma_k)$. As $\lim_{k\to\infty} l(\Gamma'_k) = 0$, we must also have that $\lim_{k\to\infty} l(\gamma_k) = 0$. Consequently, not only the arcs Γ_k accumulate to the respective cluster point of zeros of B, but also the whole domains Q_k contract themselves towards that point. Obviously, the same is true for the symmetric image $\hat{Q}_k$ of Q_k with respect to the unit circle, hence for Ω_k. In other words, for every point $\zeta \in E$,every neighborhood V_ζ of ζ contains infinitely many domains Ω_k. Therefore, every value $w \in \widehat{\mathbb{C}}$ is taken by B infinitely many times in V_ζ. On the other hand, if $K \subset \widehat{\mathbb{C}} \setminus E$ is a compact set, there is a neighborhood V_K of K and a positive integer n_K such that every value $w \in \widehat{\mathbb{C}}$ is taken by B exactly n_K times in V_K. Here the brunch points of (W, B) are counted with their multiplicities. Indeed, K is covered by a finite number n_K of domains Ω_j, etc. A movie of the mapping would be the following. Suppose that a rubber thread lays over $\gamma_k + \hat{\gamma}_k$ with the ends fixed at the branch points b_k and $\hat{b}_k$. There is an image rubber thread laying over $\eta + \hat{\eta}$ with the ends fixed at $B(b_k)$, respectively $B(\hat{b}_k)$. When the first rubber thread stretches scanning the domain Ω_k, its image stretches correspondingly scanning the whole $w-$sphere. It passes over the North Pole of the sphere when the first thread passes over $1/\bar{a}_k$. The angles on the two threads are changed with the same quantities at corresponding points except at their ends. The scanning continues from Ω_k to Ω_{k+1}, $k \in \mathbb{Z}$ with possible change of the fixed ends of the threads.

Consequently (W, B) is a branched covering surface of $\widehat{\mathbb{C}}$, i.e. B is locally injective in a neighborhood of every point $z \in W$, except for a sequence of points accumulating to E, where its behavior expressed in terms of local parameters is as that of $z \to z^n$. In a subsequent paper, we will characterize the covering transformations of (W, B). This will give an even more detailed picture of the geometry of Blaschke product mappings in a very general context.

References

1. L. V. Ahlfors, *Complex Analysis*, International Series in Pure and Applied Mathematics, 1979.
2. L. V. Ahlfors, L. Sario, *Riemann Surfaces*, Princeton University Press, 1960.
3. T. Cao-Huu, D. Ghisa, *Invariants of Infinite Blaschke Products*, Matematica, Tome 45(68), No. 1, pp. 1-8, 2007.
4. G. Cassier, I. Chalendar, *The Group of Invariants of a Finite Blaschke Product*, Complex Variables, Vol. 42, pp. 193-206, 2000.
5. C. Constantinescu et al., *Integration Theory*, Vol. 1, John Wiley & Sons, 1985.
6. J. B. Garnett, *Bounded Analytic Functions*, Academic Press, 1981.
7. O. Lehto, K. I. Virtanen, *Quasiconformal Mappings in the Plane*, Springer Verlag, 1973.
8. Z. Nehari, *Conformal Mappings*, International Series in Pure and Applied Mathematics, 1951.
9. W. Rudin, *Real and Complex Analysis*, McGraw-Hill, 1987.

THE LAYER POTENTIALS OF SOME PARTIAL DIFFERENTIAL OPERATORS: REAL ANALYTIC DEPENDENCE UPON PERTURBATIONS

M. DALLA RIVA

Dipartimento di Matematica Pura ed Applicata,
Università degli Studi di Padova,
Padova 35131, Italy
E-mail: mdallari@math.unipd.it

We introduce a particular single layer potential corresponding to the fundamental solution of a given elliptic partial differential operator of order $2k$ with constant coefficients. If the operator can be factorized with operators of order two, we show a real analyticity result in the frame of Schauder spaces for the dependence of the single layer potential and its derivatives up to order $2k-1$ upon suitable perturbations of the density, the support and the coefficients of the operator. Exploiting such a result, we investigate the dependence upon perturbations of the density, the support and the coefficients of the single and double layer potentials which arise in certain boundary value problems, such as the Dirichlet and Neumann problems for the Lamé equations and the Stokes system. We show also in this case that the dependence is real analytic. The results are in part obtained in collaboration with Prof. M. Lanza de Cristoforis.

Keywords: Layer potentials; Domain perturbation; Coefficient perturbation.

1. Introduction

Let $\mathbf{L}$ be an elliptic partial differential operator in $\mathbb{R}^n$ with constant coefficients. Let S be a a fundamental solution of $\mathbf{L}$. Let Ω be an open and bounded subset of $\mathbb{R}^n$. Let $\partial\Omega$ denote the boundary of Ω. Let μ be a real valued function defined on $\partial\Omega$. Under reasonable regularity assumptions on Ω and μ, the expression

$$v(x) \equiv \int_{\partial\Omega} S(x-y)\mu(y)\, d\sigma_y, \quad \forall\, x \in \mathbb{R}^n, \tag{1}$$

defines a function of $\mathbb{R}^n$ to $\mathbb{R}$. We say that v is the *single layer potential* of the operator $\mathbf{L}$ with support $\partial\Omega$, and with density μ, and with respect to the fundamental solution S. Our aim is to investigate the dependence of v

and of its derivatives upon perturbation of the coefficients of the operator $\mathbf{L}$, of the support $\partial\Omega$ and of the density μ. In particular, we focus our attention to the restriction of v and its derivatives to the boundary of Ω.

Our main result is stated in the following Theorem 5.1 and it is a real analyticity theorem in the frame of Schauder spaces, which can be considered as a generalization of the results for the Cauchy integral and for the layer potentials corresponding to the Laplace and Helmholtz operators due to Lanza de Cristoforis and Preciso [1] and to Lanza de Cristoforis and Rossi [2,3]. Here we consider constant coefficient elliptic operators of order $2k$, with $k \in \mathbb{N} \setminus \{0\}$, which can be factorized with operators of order 2.

The paper is organized as follows. In Section 2, we introduce a particular fundamental solution for a given constant coefficient elliptic partial differential operator and we describe the dependence of such a fundamental solution upon the coefficients of the operator. Then, in Section 3, we introduce the support of our single layer potential. This will be a compact submanifold of $\mathbb{R}^n$ parametrized by an *admissible* diffeomorphism defined on the boundary of a fixed domain Ω. In Section 4 we define a particular single layer potential for a given constant coefficient elliptic operator by exploiting the notation introduced in the previous Sections 2 and 3. In Section 5, we state our main real analyticity result under the assumption that the operator can be factorized with operators of order two. The last Section 6 is devoted to the presentation of some applications. In Subsection 6.1, we consider a single layer potential for the bi-Helmholtz operator. In Subsections 6.2 and 6.3, we consider particular single and double layer potentials for the Lamé equations and for the Stokes system, respectively.

2. A Particular Fundamental Solution

As announced, we begin by introducing a particular fundamental solution for a given elliptic partial differential operator with constant coefficients. We also describe the dependence of such a fundamental solution upon the coefficients of the operator (cf. Theorem 2.1.) To do so we need some notation.

Let $n, l \in \mathbb{N}$, $n > 0$. We denote by $N(n,l)$ the set of all multi-indexes $\alpha \equiv (\alpha_1, \dots, \alpha_n) \in \mathbb{N}^n$ with $|\alpha| \equiv \alpha_1 + \cdots + \alpha_n \le l$. We say that $\mathbf{a}$ is a *vector of coefficients of order* l if $\mathbf{a} \equiv (a_\alpha)_{\alpha \in N(n,l)}$ is a real valued function on $N(n,l)$. We denote by $R(n,l)$ space of all vectors of coefficients of order l. Then, by ordering $N(n,l)$ on arbitrary way, we identify $R(n,l)$ with a finite dimension real vector space and we endow $R(n,l)$ with the corresponding Euclidean norm $|\cdot|$. For each $\mathbf{a} \in R(n,l)$ we denote by

$P[\mathbf{a}](\xi) = P[\mathbf{a}](\xi_1, \dots, \xi_n)$ the polynomial $\sum_{\alpha \in N(n,l)} a_\alpha \xi^\alpha$ and we denote by $P_j[\mathbf{a}](\xi)$ the homogeneous polynomial $\sum_{|\alpha|=j} a_\alpha \xi^\alpha$, for all $j = 0, \dots, l$. Then, we set $\mathbf{L}[\mathbf{a}] \equiv P[\mathbf{a}](\partial_{x_1}, \dots, \partial_{x_n})$, so that $\mathbf{L}[\mathbf{a}]$ is a partial differential operator with constant coefficients. We say that $\mathbf{L}[\mathbf{a}]$ is an elliptic operator of order l if $P_l[\mathbf{a}](\xi) \neq 0$ for all $\xi \in \partial\mathbb{B}_n$, where $\mathbb{B}_n \equiv \{\xi \in \mathbb{R}^n : |\xi| < 1\}$ is the unit ball in $\mathbb{R}^n$. We have the following.

Theorem 2.1. *Let $n, k \in \mathbb{N}$, $n \geq 2$, $k \geq 1$. Let $\mathcal{U}$ be a bounded open subset of $R(n, 2k)$ such that $\mathbf{L}[\mathbf{a}]$ is an elliptic operator of order $2k$ for all vector of coefficients $\mathbf{a}$ in the closure of $\mathcal{U}$. Then, there exist a real analytic function $A(\cdot,\cdot,\cdot)$ of $\mathcal{U} \times \partial\mathbb{B}_n \times \mathbb{R}$ to $\mathbb{R}$ and real analytic functions $B(\cdot,\cdot)$, $C(\cdot,\cdot)$ of $\mathcal{U} \times \mathbb{R}^n$ to $\mathbb{R}$, such that the following statements hold.*

(i) There exists a sequence $\{f_j(\cdot,\cdot)\}_{j\in\mathbb{N}}$ of real analytic functions of $\mathcal{U} \times \partial\mathbb{B}_n$ to $\mathbb{R}$ such that

$$f_j(\mathbf{a}, -\theta) = (-1)^j f_j(\mathbf{a}, \theta), \quad \forall\ (\mathbf{a}, \theta) \in \mathcal{U} \times \partial\mathbb{B}_n$$

and

$$A(\mathbf{a}, \theta, r) = \sum_{j=0}^{\infty} f_j(\mathbf{a}, \theta) r^j, \quad \forall\ (\mathbf{a}, \theta, r) \in \mathcal{U} \times \partial\mathbb{B}_n \times \mathbb{R},$$

where the series converges absolutely and uniformly in all compact subsets of $\mathcal{U} \times \partial\mathbb{B}_n \times \mathbb{R}$.

(ii) There exists a family $\{b_\alpha(\cdot) : \alpha \in \mathbb{N}^n, |\alpha| \geq 2k - n\}$ of real analytic functions of $\mathcal{U}$ to $\mathbb{R}$ such that

$$B(\mathbf{a}, z) = \sum_{|\alpha| \geq 2k-n} b_\alpha(\mathbf{a}) z^\alpha, \ \forall\ (\mathbf{a}, z) \in \mathcal{U} \times \mathbb{R}^n,$$

where the series converges absolutely and uniformly in all compact subsets of $\mathcal{U} \times \mathbb{R}^n$. Furthermore, the function B can be chosen to be identically 0 if n is odd.

(iii) The function C can be chosen to be identically 0 if n is odd.

(iv) For all $\mathbf{a} \in \mathcal{U}$ the function $S(\mathbf{a}, \cdot)$ of $\mathbb{R}^n \setminus \{0\}$ to $\mathbb{R}$ defined by

$$S(\mathbf{a}, z) \equiv |z|^{2k-n} A(\mathbf{a}, z/|z|, |z|) + B(\mathbf{a}, z) \log|z| + C(\mathbf{a}, z)$$

for $z \in \mathbb{R}^n \setminus \{0\}$ is a fundamental solution of $\mathbf{L}[\mathbf{a}]$.

For the proof of Theorem 2.1 we refer to [4]. Here we only remark that the argument of the proof heavily exploits the construction of a fundamental solution given by John in [5]. We also observe that Theorem 2.1 resembles the results obtained by Mantlik [6,7] with much more general assumptions on the operator (see also Trèves [8].) Nevertheless it is not a corollary. Indeed, by Theorem 2.1 we obtain a suitably detailed expression for the fundamental solution, which cannot be deduced by Mantlik's results.

3. The Support of the Single Layer Potential

Our next step is to introduce the support of the single layer potential To do so we recall some technical facts of Lanza de Cristoforis and Rossi [2,3].

Let $n, m \in \mathbb{N}$, $n \geq 2$, $m \geq 1$. Let $0 < \lambda < 1$. We fix an open connected and bounded subset Ω of the euclidean space $\mathbb{R}^n$, whose exterior $\mathbb{R}^n \setminus \mathrm{cl}\Omega$ is also connected. Moreover we assume that the boundary $\partial\Omega$ of Ω is a compact sub-manifold of $\mathbb{R}^n$ of Hölder class $C^{m,\lambda}$. Then we introduce the class $\mathcal{A}_{\partial\Omega}$ of the admissible functions on $\partial\Omega$ by the following definition.

Definition 3.1. $\mathcal{A}_{\partial\Omega}$ is the set of all functions $\phi \in C^1(\partial\Omega, \mathbb{R}^n)$ which are injective and whose differential $d\phi(x)$ is injective for all $x \in \partial\Omega$.

One can verify that $\mathcal{A}_{\partial\Omega} \cap C^{m,\lambda}(\partial\Omega, \mathbb{R}^n)$ is an open subset of the Banach space $C^{m,\lambda}(\partial\Omega, \mathbb{R}^n)$. Moreover, if $\phi \in \mathcal{A}_{\partial\Omega} \cap C^{m,\lambda}(\partial\Omega, \mathbb{R}^n)$, then $\phi(\partial\Omega)$ splits $\mathbb{R}^n$ into two connected components. We denote by $\mathbb{I}[\phi]$ the bounded one. Then $\mathbb{I}[\phi]$ is an open and bounded subset of $\mathbb{R}^n$, connected and with connected exterior, and whose boundary $\partial\mathbb{I}[\phi] = \phi(\partial\Omega)$ is a compact sub-manifold of $\mathbb{R}^n$ of Hölder class $C^{m,\lambda}$ parametrized by the function ϕ. The set $\phi(\partial\Omega)$ will be the support of our single layer potential.

4. The Single Layer Potential

Now we introduce our particular single layer potential. Let $n, k \in \mathbb{N}$, $n \geq 2$, $k \geq 1$. Let $\mathcal{U}$ be an open bounded subset of $R(n, 2k)$ such that $\mathbf{L}[\mathbf{a}]$ is an elliptic operator of order $2k$ for all $\mathbf{a}$ in the closure of $\mathcal{U}$. Then $\mathcal{U}$ satisfies the assumptions of Theorem 2.1 and we can consider the corresponding function $S(\mathbf{a}, \cdot)$ for all $\mathbf{a} \in \mathcal{U}$. Let $m \in \mathbb{N}$, $m \geq 1$, and $\lambda \in]0, 1[$. We fix an open bounded and connected subset Ω of $\mathbb{R}^n$ with connected exterior and with boundary of class $C^{m,\lambda}$. If $\mathbf{a} \in \mathcal{U}$, $\phi \in \mathcal{A}_{\partial\Omega} \cap C^{m,\lambda}(\partial\Omega, \mathbb{R}^n)$, $\mu \in C^{m-1,\lambda}(\partial\Omega)$ and $\beta \in N(n, 2k-1)$, we denote by $v_\beta[\mathbf{a}, \phi, \mu]$ the function of $\mathbb{R}^n$ to $\mathbb{R}$ defined by

$$v_\beta[\mathbf{a}, \phi, \mu](x) \equiv \int_{\phi(\partial\Omega)} (\partial_z^\beta S)(\mathbf{a}, x - y)\, \mu \circ \phi^{(-1)}(y)\, d\sigma_y, \quad \forall x \in \mathbb{R}^n, \quad (2)$$

where the integral is understood in the sense of singular integrals if $|\beta| = 2k-1$ and $x \in \phi(\partial\Omega)$, namely

$$v_\beta[\mathbf{a},\phi,\mu](x) \equiv \lim_{\epsilon\to 0^+} \int_{\phi(\partial\Omega)\setminus(x+\epsilon\mathbb{B}_n)} (\partial_z^\beta S)(\mathbf{a}, x-y)\, \mu\circ\phi^{(-1)}(y)\, d\sigma_y.$$

Clearly, $v_{(0,\ldots,0)}[\mathbf{a},\phi,\mu]$ is the single layer potential with support $\phi(\partial\Omega)$ and with density $\mu\circ\phi^{(-1)}$ of the operator $\mathbf{L}[\mathbf{a}]$ with respect to the fundamental solution $S(\mathbf{a},\cdot)$. For $|\beta| \le 2k-2$, $v_\beta[\mathbf{a},\phi,\mu]$ coincides with the β-derivative of the single layer potential $v_{(0,\ldots,0)}[\mathbf{a},\phi,\mu]$, namely $v_\beta[\mathbf{a},\phi,\mu](x) = \partial_x^\beta v_{(0,\ldots,0)}[\mathbf{a},\phi,\mu](x)$ for all $x \in \mathbb{R}^n$. For $|\beta| = 2k-1$, $v_\beta[\mathbf{a},\phi,\mu]$ coincides with the β-derivative of $v_{(0,\ldots,0)}[\mathbf{a},\phi,\mu]$ on $\mathbb{R}^n \setminus \phi(\partial\Omega)$, but not on $\phi(\partial\Omega)$. Indeed, for $|\beta| = 2k-1$, the β derivative of $v_{(0,\ldots,0)}[\mathbf{a},\phi,\mu]$ displays a jump across the boundary $\phi(\partial\Omega)$ and its value at $x\in\phi(\partial\Omega)$ is not defined. Nevertheless, one can verify that the restriction of $\partial_x^\beta v_{(0,\ldots,0)}[\mathbf{a},\phi,\mu]$ to the domain $\mathbb{I}[\phi]$ admits a continuous extension to the closure of $\mathbb{I}[\phi]$. If we denote by $\bar{v}_\beta[\mathbf{a},\phi,\mu]$ such an extension, then

$$\bar{v}_\beta[\mathbf{a},\phi,\mu](x) = J(\mathbf{a},\nu_\phi)\mu\circ\phi^{(-1)}(x) + v_\beta[\mathbf{a},\phi,\mu](x), \quad \forall\, x\in\phi(\partial\Omega),$$

where $J(\cdot,\cdot)$ is a real analytic function of $\mathcal{U}\times(\mathbb{R}^n\setminus\{0\})$ to $\mathbb{R}$ and ν_ϕ denotes the outer unit normal of $\mathbb{I}[\phi]$ (cf. [4].)

5. A Real Analyticity Theorem

We state in this section our main Theorem 5.1, which is in some sense a natural extension of Theorem 3.23 of Lanza de Cristoforis and Preciso [1], where the Cauchy integral has been considered, of Theorem 3.25 of Lanza de Cristoforis and Rossi [2], where the Laplace operator Δ has been considered, and of Theorem 3.45 of Lanza de Cristoforis and Rossi [3], where the Helmholtz operator has been considered. Here we consider more general operators, the constant coefficient elliptic operators of order $2k$, with $k\in\mathbb{N}\setminus\{0\}$, which can be factorized with operators of order 2.

Let $n,k\in\mathbb{N}$, $n\ge 2$, $k\ge 1$. Let $\mathcal{U}_1,\ldots,\mathcal{U}_k$ be bounded open subsets of $R(n,2)$ such that $\mathbf{L}[\mathbf{a}_i]$ is an elliptic operator of order 2 for all $\mathbf{a}_i$ in the closure of $\mathcal{U}_i$ and for all $i=1,\ldots,k$. One easily verifies that, for each k-tuple $(\mathbf{a}_1,\ldots,\mathbf{a}_k)$ of $\mathcal{U}_1\times\cdots\times\mathcal{U}_k$, there exists a unique element $\bar{\mathbf{a}}$ of $R(n,2k)$ such that

$$P[\bar{\mathbf{a}}](\xi) = P[\mathbf{a}_1](\xi)\cdot P[\mathbf{a}_2](\xi)\cdot\cdots\cdot P[\mathbf{a}_k](\xi), \quad \forall\,\xi\in\mathbb{R}^n. \tag{3}$$

Moreover, the map of $\mathcal{U}_1\times\cdots\times\mathcal{U}_k$ to $R(n,2k)$ which takes $(\mathbf{a}_1,\ldots,\mathbf{a}_k)$ to $\bar{\mathbf{a}}$ is real analytic. So, the set $\mathcal{V}$ of all the vectors of coefficients $\bar{\mathbf{a}}\in R(n,2k)$

which satisfy Eq. (3), for some $(\mathbf{a}_1, \ldots, \mathbf{a}_k) \in \mathcal{U}_1 \times \cdots \times \mathcal{U}_k$, is a bounded subset of $R(n, 2k)$ and $\mathbf{L}[\bar{\mathbf{a}}]$ is an elliptic operator of order $2k$ for all $\bar{\mathbf{a}}$ in $\mathcal{V}$. By means of a simple topological argument, it follows that there exists a bounded open neighborhood $\mathcal{U}$ of $\mathcal{V}$ in $R(n, 2k)$ such that $\mathbf{L}[\mathbf{a}]$ is an elliptic operator of order $2k$ for all $\mathbf{a}$ in the closure of $\mathcal{U}$. Thus, $\mathcal{U}$ satisfies the assumption of Theorem 2.1 and we can introduce the corresponding function $S(\mathbf{a}, \cdot)$ of $\mathbb{R}^n \setminus \{0\}$ to $\mathbb{R}$ for all $\mathbf{a} \in \mathcal{U}$.

Now, let $m \in \mathbb{N} \setminus \{0\}$, $\lambda \in]0, 1[$. Let Ω be a bounded open connected subset of $\mathbb{R}^n$ with connected exterior and with boundary of class $C^{m,\lambda}$. Let $v_\beta[\mathbf{a}, \phi, \mu]$ be the function of $\mathbb{R}^n$ to $\mathbb{R}$ defined by Eq. (3), for all $\mathbf{a} \in \mathcal{U}$, $\phi \in \mathcal{A}_{\partial\Omega} \cap C^{m,\lambda}(\partial\Omega, \mathbb{R}^n)$, $\mu \in C^{m-1,\lambda}(\partial\Omega)$ and $\beta \in N(n, 2k-1)$. We denote by $V_\beta[\mathbf{a}, \phi, \mu]$ the function defined on the boundary of the fixed domain Ω by the following equation,

$$V_\beta[\mathbf{a}, \phi, \mu](x) \equiv v_\beta[\mathbf{a}, \phi, \mu] \circ \phi(x), \quad \forall\, x \in \partial\Omega, \tag{4}$$

for all $(\mathbf{a}, \phi, \mu) \in \mathcal{U} \times (\mathcal{A}_{\partial\Omega} \cap C^{m,\lambda}(\partial\Omega, \mathbb{R}^n)) \times C^{m-1,\lambda}(\partial\Omega)$ and for all $\beta \in N(n, 2k-1)$. Then we have the following real analyticity theorem.

Theorem 5.1. *The following statements hold.*

(i) If $|\beta| \leq 2k-2$, then the map of $(\mathcal{U}_1 \times \cdots \times \mathcal{U}_k) \times (\mathcal{A}_{\partial\Omega} \cap C^{m,\lambda}(\partial\Omega, \mathbb{R}^n)) \times C^{m-1,\lambda}(\partial\Omega)$ to $C^{m,\lambda}(\partial\Omega)$, which takes $(\mathbf{a}_1, \ldots, \mathbf{a}_k, \phi, \mu)$ to $V_\beta[\bar{\mathbf{a}}, \phi, \mu]$, is real analytic.

(ii) If $|\beta| = 2k-1$, then the map of $(\mathcal{U}_1 \times \cdots \times \mathcal{U}_k) \times (\mathcal{A}_{\partial\Omega} \cap C^{m,\lambda}(\partial\Omega, \mathbb{R}^n)) \times C^{m-1,\lambda}(\partial\Omega)$ to $C^{m-1,\lambda}(\partial\Omega)$, which takes $(\mathbf{a}_1, \ldots, \mathbf{a}_k, \phi, \mu)$ to $V_\beta[\bar{\mathbf{a}}, \phi, \mu]$, is real analytic.

For a proof of Theorem 5.1 we refer to [4]. The main idea stems from the papers of Lanza de Cristoforis and Preciso [1] and Lanza de Cristoforis and Rossi [2,3] and exploits the Implicit Mapping Theorem for analytic functions. Let β be a multi-index with $|\beta| \leq 2k-2$. Let $(\mathbf{a}_1, \ldots, \mathbf{a}_k, \phi, \mu)$ belong to $(\mathcal{U}_1 \times \cdots \times \mathcal{U}_k) \times (\mathcal{A}_{\partial\Omega} \cap C^{m,\lambda}(\partial\Omega, \mathbb{R}^n)) \times C^{m-1,\lambda}(\partial\Omega)$. We introduce a system of coupled boundary value problems defined in a suitable neighborhood of $\partial\Omega$. Such a system has a unique solution which uniquely identifies the function $V_\beta[\bar{\mathbf{a}}, \phi, \mu]$. By means of the Implicit Mapping Theorem, we show that the solution of the system depends real analytically on $(\mathbf{a}_1, \ldots, \mathbf{a}_k, \phi, \mu)$. We deduce that $V_\beta[\bar{\mathbf{a}}, \phi, \mu]$ depends real analytically on $(\mathbf{a}_1, \ldots, \mathbf{a}_k, \phi, \mu)$. To prove the statement for $|\beta| = 2k-1$, we exploit the jumping properties of the single layer potential.

6. Some Applications

6.1. *The Bi-Helmholtz Operator*

We denote by $\mathbf{H}^2[b_1, b_2]$ the operator $\mathbf{H}^2[b_1, b_2] \equiv (\Delta + b_1)(\Delta + b_2)$, where b_1 and b_2 are real coefficients. We fix two open and bounded subsets $\mathcal{B}_1$ and $\mathcal{B}_2$ of $\mathbb{R}$. By a straightforward corollary of Theorem 2.1, we introduce a particular fundamental solution $S_{H^2}(b_1, b_2, \cdot)$ of $\mathbf{H}^2[b_1, b_2]$, for all $(b_1, b_2) \in \mathcal{B}_1 \times \mathcal{B}_2$. Then, we fix two constants $m \in \mathbb{N} \setminus \{0\}$ and $\lambda \in]0, 1[$. We fix a bounded open connected subset Ω of $\mathbb{R}^n$ with connected exterior and with boundary of class $C^{m,\lambda}$. We denote by $(v_{H^2})_\beta[b_1, b_2, \phi, \mu]$ the function of $\mathbb{R}^n$ to $\mathbb{R}$ defined by

$$(v_{H^2})_\beta[b_1, b_2, \phi, \mu](x) \equiv \int_{\phi(\partial\Omega)} (\partial_z^\beta S_{H^2})(b_1, b_2, x - y)\, \mu \circ \phi^{(-1)}(y)\, d\sigma_y, \quad \forall\, x \in \mathbb{R}^n,$$

for all $(b_1, b_2, \phi, \mu) \in \mathcal{B}_1 \times \mathcal{B}_2 \times (\mathcal{A}_{\partial\Omega} \cap C^{m,\lambda}(\partial\Omega, \mathbb{R}^n)) \times C^{m-1,\lambda}(\partial\Omega)$ and for all $\beta \in N(n, 3)$, where the integral is understood in the sense of singular integrals if $|\beta| = 3$ and $x \in \phi(\partial\Omega)$. Then, we set

$$(V_{H^2})_\beta[b_1, b_2, \phi, \mu](x) \equiv (v_{H^2})_\beta[b_1, b_2, \phi, \mu] \circ \phi(x), \quad \forall\, x \in \partial\Omega,$$

for all $(b_1, b_2, \phi, \mu) \in \mathcal{B}_1 \times \mathcal{B}_2 \times (\mathcal{A}_{\partial\Omega} \cap C^{m,\lambda}(\partial\Omega, \mathbb{R}^n)) \times C^{m-1,\lambda}(\partial\Omega)$ and for all $\beta \in N(n, 3)$. By Theorem 5.1, we immediately deduce the following.

Proposition 6.1. *If $|\beta| \le 2$, the map $(V_{H^2})_\beta[\cdot, \cdot, \cdot, \cdot]$ is real analytic from $\mathcal{B}_1 \times \mathcal{B}_2 \times (\mathcal{A}_{\partial\Omega} \cap C^{m,\lambda}(\partial\Omega, \mathbb{R}^n)) \times C^{m-1,\lambda}(\partial\Omega)$ to $C^{m,\lambda}(\partial\Omega)$. If $|\beta| = 3$, the map $(V_{H^2})_\beta[\cdot, \cdot, \cdot, \cdot]$ is real analytic from $\mathcal{B}_1 \times \mathcal{B}_2 \times (\mathcal{A}_{\partial\Omega} \cap C^{m,\lambda}(\partial\Omega, \mathbb{R}^n)) \times C^{m-1,\lambda}(\partial\Omega)$ to $C^{m-1,\lambda}(\partial\Omega)$.*

6.2. *The Lamé Equations*

We denote by $\mathbf{L}[b_1, b_2]$ the vector valued operator

$$\mathbf{L}[b_1, b_2] \equiv \Delta + b_1 \nabla \mathrm{div} + b_2,$$

where b_1 and b_2 are real coefficients. We fix two open and bounded subsets $\mathcal{B}_1$ and $\mathcal{B}_2$ of $\mathbb{R}$ and we assume that $\mathcal{B}_1$ does not contain the point -1. A fundamental solution $S_L(b_1, b_2, \cdot)$ of the operator $\mathbf{L}[b_1, b_2]$ is given by the matrix-valued function defined by

$$(S_L(b_1, b_2, z))_{ij} \equiv \left(\delta_{ij}\left(\Delta_z + \frac{b_2}{b_1 + 1}\right) - \frac{b_1}{b_1 + 1}\frac{\partial^2}{\partial z_i \partial z_j}\right) S_{H^2}\Big(b_2, b_2/(b_1 + 1), z\Big), \tag{5}$$

for all $z \in \mathbb{R}^n \setminus \{0\}$, and for all $i, j = 1, \dots, n$, and for all $(b_1, b_2) \in \mathcal{B}_1 \times \mathcal{B}_2$, where δ_{ij} denotes the Kronecker delta symbol and S_{H^2} is the fundamental solution of the bi-Helmholtz operator $\mathbf{H}^2$ introduced in the previous Subsection 6.1 (cf. Kupradze, Gegelia, Basheleĭshvili and Burchuladze [9].) Now, let $m \in \mathbb{N} \setminus \{0\}$ and $\lambda \in]0,1[$. Let Ω be a bounded open connected subset of $\mathbb{R}^n$ with connected exterior and with boundary of class $C^{m,\lambda}$. We set

$$(v_L)_\beta[b_1, b_2, \phi, \mu](x) \equiv \int_{\phi(\partial\Omega)} (\partial_z^\beta S_L)(b_1, b_2, x-y)\, \mu \circ \phi^{(-1)}(y)\, d\sigma_y, \quad \forall\, x \in \mathbb{R}^n,$$

where the integral is understood in the sense of singular integrals if $|\beta| = 1$ and $x \in \phi(\partial\Omega)$, and

$$(V_L)_\beta[b_1, b_2, \phi, \mu](x) \equiv (v_L)_\beta[b_1, b_2, \phi, \mu] \circ \phi(x), \quad \forall\, x \in \partial\Omega,$$

for all $(b_1, b_2, \phi, \mu) \in \mathcal{B}_1 \times \mathcal{B}_2 \times (\mathcal{A}_{\partial\Omega} \cap C^{m,\lambda}(\partial\Omega, \mathbb{R}^n)) \times C^{m-1,\lambda}(\partial\Omega, \mathbb{R}^n)$ and for all $\beta \in N(n, 1)$. Then, by Theorem 5.1, we immediately have the following.

Proposition 6.2. *If $|\beta| = 0$, the map $(V_L)_\beta[\cdot,\cdot,\cdot,\cdot]$ is real analytic from $\mathcal{B}_1 \times \mathcal{B}_2 \times (\mathcal{A}_{\partial\Omega} \cap C^{m,\lambda}(\partial\Omega, \mathbb{R}^n)) \times C^{m-1,\lambda}(\partial\Omega, \mathbb{R}^n)$ to $C^{m,\lambda}(\partial\Omega, \mathbb{R}^n)$. If $|\beta| = 1$, the map $(V_L)_\beta[\cdot,\cdot,\cdot,\cdot]$ is real analytic from $\mathcal{B}_1 \times \mathcal{B}_2 \times (\mathcal{A}_{\partial\Omega} \cap C^{m,\lambda}(\partial\Omega, \mathbb{R}^n)) \times C^{m-1,\lambda}(\partial\Omega, \mathbb{R}^n)$ to $C^{m-1,\lambda}(\partial\Omega, \mathbb{R}^n)$.*

We now consider the double layer potential $w_L[b_1, b_2, \phi, \mu]$ of $\mathbf{L}[b_1, b_2]$. We denote by $S_L^{(i)}$ the vector valued function given by the i-th column of S_L, for all $i = 1, \dots, n$. We denote by $T(b, A)$ the matrix $(b-1)(\mathrm{tr}A)\mathbf{1}_n + (A + A^t)$, for all $n \times n$ real matrix A and all scalar $b \in \mathbb{R}$, where $\mathbf{1}_n$ is the $n \times n$ unit matrix. We denote by ν_ϕ the outward unit normal to $\phi(\partial\Omega)$, for all $\phi \in \mathcal{A}_{\partial\Omega} \cap C^{m,\lambda}(\partial\Omega, \mathbb{R}^n)$. Then we set

$$w_L[b_1, b_2, \phi, \mu](x) \equiv -\int_{\phi(\partial\Omega)} \left[\Big(T(b_1, D_z S_L^{(i)}(b_1, b_2, x-y))\nu_\phi(y)\Big) \cdot \mu \circ \phi^{(-1)}(y)\right]_{i=1,\dots,n} d\sigma_y$$

for all $x \in \mathbb{R}^n$ and all $(b_1, b_2, \phi, \mu) \in \mathcal{B}_1 \times \mathcal{B}_2 \times (\mathcal{A}_{\partial\Omega} \cap C^{m,\lambda}(\partial\Omega, \mathbb{R}^n)) \times C^{m,\lambda}(\partial\Omega, \mathbb{R}^n)$, where the integral is understood in the sense of singular integrals if $x \in \phi(\partial\Omega)$. We denote by $W_L[b_1, b_2, \phi, \mu]$ the composition $w_L[b_1, b_2, \phi, \mu] \circ \phi$. Then we have the following real analyticity Proposition 6.3 for $W_L[b_1, b_2, \phi, \mu]$. We refer to [4] for a proof.

Proposition 6.3. *The map* $W_L[\cdot,\cdot,\cdot,\cdot]$ *of* $\mathcal{B}_1\times\mathcal{B}_2\times(\mathcal{A}_{\partial\Omega}\cap C^{m,\lambda}(\partial\Omega,\mathbb{R}^n))\times C^{m,\lambda}(\partial\Omega,\mathbb{R}^n)$ *to* $C^{m,\lambda}(\partial\Omega,\mathbb{R}^n)$ *is real analytic.*

6.3. *The Stokes System*

We say that $S_S \equiv (S_V, S_P)$ is a fundamental solution for the Stokes system in $\mathbb{R}^n$ if S_V is a real analytic matrix-valued function of $\mathbb{R}^n \setminus \{0\}$, S_P is a real analytic vector valued function of $\mathbb{R}^n \setminus \{0\}$ and

$$\Delta S_V(z) - \nabla S_P(z) = \delta(z)\mathbf{1}_n, \quad \operatorname{div} S_V(z) = 0, \quad \forall\, z \in \mathbb{R}^n \setminus \{0\},$$

(cf. Ladyzhenskaya [10].) One can verify that a suitable choice of the functions S_V and S_P is given by the following equalities,

$$(S_V(z))_{ij} \equiv \left(\delta_{ij}\Delta - \frac{\partial^2}{\partial z_i \partial z_j}\right) S_{\Delta^2}(z), \quad (S_P)_i(z) \equiv -\frac{\partial}{\partial z_i} S_\Delta(z), \qquad (6)$$

for all $z \in \mathbb{R}^n\setminus\{0\}$, where we understand $S_{\Delta^2}(z) \equiv S_{H^2}(0,0,z)$ and $S_\Delta(z) \equiv \Delta S_{\Delta^2}(z)$, in accordance with the notation of subsection 6.1. Now, let $m \in \mathbb{N} \setminus \{0\}$ and $\lambda \in]0,1[$. Let Ω be a bounded open connected subset of $\mathbb{R}^n$ with connected exterior and with boundary of class $C^{m,\lambda}$. We introduce the single layer potentials $v_V[\phi,\mu]$ and $v_P[\phi,\mu]$ by the equalities

$$v_V[\phi,\mu](x) \equiv \int_{\phi(\partial\Omega)} S_V(x-y)\,\mu\circ\phi^{(-1)}(y)\, d\sigma_y, \qquad \forall\, x \in \mathbb{R}^n, \qquad (7)$$

$$v_P[\phi,\mu](x) \equiv \int_{\phi(\partial\Omega)} S_P(x-y)\cdot\mu\circ\phi^{(-1)}(y)\, d\sigma_y, \quad \forall\, x \in \mathbb{R}^n, \qquad (8)$$

for all $(\phi,\mu) \in (\mathcal{A}_{\partial\Omega}\cap C^{m,\lambda}(\partial\Omega,\mathbb{R}^n)) \times C^{m-1,\lambda}(\partial\Omega,\mathbb{R}^n)$, where the integral in equation (8) is understood in the sense of singular integrals if $x \in \phi(\partial\Omega)$. As usual, we set $V_V[\phi,\mu] \equiv v_V[\phi,\mu]\circ\phi$ and $V_P[\phi,\mu] \equiv v_P[\phi,\mu]\circ\phi$. Then, by equation (6) and by Proposition 6.1, we have the following.

Proposition 6.4. *The maps* $V_V[\cdot,\cdot]$ *and* $V_P[\cdot,\cdot]$ *of* $(\mathcal{A}_{\partial\Omega}\cap C^{m,\lambda}(\partial\Omega,\mathbb{R}^n))\times C^{m-1,\lambda}(\partial\Omega,\mathbb{R}^n)$ *to* $C^{m,\lambda}(\partial\Omega,\mathbb{R}^n)$ *and to* $C^{m-1,\lambda}(\partial\Omega,\mathbb{R}^n)$*, respectively, are real analytic.*

Now for each scalar $b \in \mathbb{R}$ and each $n\times n$ real matrix A we set $\mathcal{T}(b,A) \equiv -b\mathbf{1}_n + (A + A^t)$. Then we denote by $S_V^{(i)}$ the vector valued function given by the i-th column of S_V for each $i = 1,\dots,n$. We define the double layer potential $w_V[\phi,\mu]$ by

$$w_V[\phi,\mu](x) \equiv -\int_{\phi(\partial\Omega)} \Big[\Big(\mathcal{T}\Big((S_P)_i(x-y), DS_V^{(i)}(x-y)\Big) \cdot\nu_\phi(y)\Big)\cdot\mu\circ\phi^{(-1)}(y)\Big]_{i=1,\dots,n} d\sigma_y, \quad \forall\, x \in \mathbb{R}^n,$$

for all $(\phi, \mu) \in (\mathcal{A}_{\partial\Omega} \cap C^{m,\lambda}(\partial\Omega, \mathbb{R}^n)) \times C^{m-1,\lambda}(\partial\Omega, \mathbb{R}^n)$. As usual, we set $W_V[\phi, \mu] \equiv w_V[\phi, \mu] \circ \phi$ and we have the following (see [4] for the proof.)

Proposition 6.5. *The map* $W_V[\cdot, \cdot]$ *defined from* $(\mathcal{A}_{\partial\Omega} \cap C^{m,\lambda}(\partial\Omega, \mathbb{R}^n)) \times C^{m,\lambda}(\partial\Omega, \mathbb{R}^n)$ *to* $C^{m,\lambda}(\partial\Omega, \mathbb{R}^n)$ *is real analytic.*

We can also define the double layer potential $w_P[\phi, \mu]$ for the pressure and we can prove a real analyticity proposition for it similar to Proposition 6.5. To do so, we first need to obtain a suitable expression for $w_P[\phi, \mu]$. For the sake of brevity, we omit to present here more details. We refer to [4] for a complete analysis.

References

1. M. Lanza de Cristoforis and L. Preciso, On the analyticity of the Cauchy integral in Schauder spaces, *J. Integral Equations Appl.* **11**, 363 (1999).
2. M. Lanza de Cristoforis and L. Rossi, Real analytic dependence of simple and double layer potentials upon perturbation of the support and of the density, *J. Integral Equations Appl.* **16**, 137 (2004).
3. M. Lanza de Cristoforis and L. Rossi, Real analytic dependence of simple and double layer potentials for the Helmholtz equation upon perturbation of the support and of the density, to appear (2005).
4. M. Dalla Riva, *Potential theoretic methods for the analysis of singularly perturbed problems in linearized elasticity*, Doctoral dissertation, Advisor M. Lanza de Cristoforis, Università degli Studi di Padova, 2008.
5. F. John, *Plane waves and spherical means applied to partial differential equations*, Interscience Publishers, New York-London, 1955
6. F. Mantlik, Partial differential operators depending analytically on a parameter, *Ann. Inst. Fourier (Grenoble)* **41**, 577 (1991).
7. F. Mantlik, Fundamental solutions for hypoelliptic differential operators depending analytically on a parameter, *Trans. Amer. Math. Soc.* **334**, 245 (1992).
8. F. Trèves, Fundamental solutions of linear partial differential equations with constant coefficients depending on parameters, *Amer. J. Math.* **84**, 561 (1962).
9. V. D. Kupradze, T. G. Gegelia, M. O. Basheleĭshvili and T. V. Burchuladze, *Three-dimensional problems of the mathematical theory of elasticity and thermoelasticity*, North-Holland Series in Applied Mathematics and Mechanics, Vol. 25, Russian ed., North-Holland Publishing Co., Amsterdam, 1979, Edited by V. D. Kupradze.
10. O. A. Ladyzhenskaya, *The mathematical theory of viscous incompressible flow*, Gordon and Breach Science Publishers, New York, 1963. Revised English edition, translated from the Russian by R. A. Silverman.

DIFFERENTIAL PROPERTIES OF (α, Q)-HOMEOMORPHISMS

ANATOLY GOLBERG

Department of Applied Mathematics, Holon Institute of Technology,
52 Golomb St., P.O.Box 305, Holon 58102, Israel
E-mail: golberga@hit.ac.il

Dedicated to Professor Promarz M. Tamrazov
on the occasion of his 75th birthday.

Quasi-invariance of an n-module is a characteristic property of quasiconformal mappings. We give inequalities that generalize quasi-invariance. Topological mappings satisfying these inequalities are called (α, Q)-homeomorphisms. We establish the ACL property and differentiability a.e. of (α, Q)-homeomorphisms, distortions of α-modules and other properties.

Keywords: Quasiconformal mappings, Q-homeomorphisms, inner and outer dilatations, α-moduli of families of k-dimensional surfaces.

1. Introduction

In this paper, we continue to investigate the classes of mappings with finite integral dilatations, which play an important role in geometric function theory in multi-dimensional spaces. They include the well-known classes of mappings that are quasiconformal, quasiconformal in mean, etc.

One of the characteristic properties of quasiconformal mappings is the *quasi-invariance* of an n-module of families of joining curves (or, equivalently, conformal capacity). It states that the module of a curve family can be changed by a K-quasiconformal mapping only up to a factor at most K. All other properties of K-quasiconformal mappings can be derived from this condition (see, e.g., [9]). In this paper, we give inequalities that generalize quasi-invariance. Our approach involves p-moduli of k-dimensional surfaces.

This type of inequalities have been recently considered in [10,11]. The corresponding mappings are called Q-homeomorphisms. The authors studied their various properties, like distortion, removability, boundary behavior under some conditions provided that Q belongs to BMO, FMO, etc.

In this paper, we introduce more general (α, Q)-homeomorphisms that include all the above classes and investigate their differential and geometric properties: absolute continuity on lines (ACL), belonging to Sobolev's class $W^{1,1}_{loc}$, differentiability almost everywhere, distortion of moduli, etc.

2. Quasiconformality in $\mathbb{R}^n$

Let $A: \mathbb{R}^n \to \mathbb{R}^n$ be a linear bijection. The numbers

$$H_I(A) = \frac{|detA|}{l^n(A)}, \quad H_O(A) = \frac{L^n(A)}{|detA|}, \quad H(A) = \frac{L(A)}{l(A)},$$

are called the *inner*, *outer* and *linear* dilatations of A, respectively. Here $l(A) = \min\limits_{|h|=1} |Ah|$, $L(A) = \max\limits_{|h|=1} |Ah|$, and $detA$ is the determinant of A (see, e.g., [20]).

Obviously, all three dilatations are not less than 1. They have the following geometric interpretation. The image of the unit ball B^n under A is an ellipsoid $E(A)$. Let $B_I(A)$ and $B_O(A)$ be the inscribed and the circumscribed balls of $E(A)$, respectively. Then

$$H_I(A) = \frac{mE(A)}{mB_I(A)}, \quad H_O(A) = \frac{mB_O(A)}{mE(A)},$$

and $H(A)$ is the ratio of the greatest and the smallest semi-axis of $E(A)$. Here $mA = m_n A$ denotes the n-dimensional Lebesgue measure of a set A.

Let $\lambda_1 \geq \lambda_2 \geq \ldots \geq \lambda_n$ be the semi-axes of $E(A)$. More precisely, the numbers λ_i are positive square root of the proper values of A^*A where A^* is the adjoint of A. Then

$$L(A) = \lambda_1, \quad l(A) = \lambda_n, \quad |detA| = \lambda_1 \cdot \ldots \cdot \lambda_n,$$

and we can also write

$$H_I(A) = \frac{\lambda_1 \cdot \ldots \cdot \lambda_{n-1}}{\lambda_n^{n-1}}, \quad H_O(A) = \frac{\lambda_1^{n-1}}{\lambda_2 \cdot \ldots \cdot \lambda_n}, \quad H(A) = \frac{\lambda_1}{\lambda_n}.$$

If $n = 2$ then $H_I(A) = H_O(A) = H(A)$. In the general case, we have the relations:

$$\begin{aligned} H(A) &\leq \min(H_I(A), H_O(A)) \leq H^{n/2}(A) \\ &\leq \max(H_I(A), H_O(A)) \leq H^{n-1}(A). \end{aligned} \tag{1}$$

Let G and G^* be two bounded domains in $\mathbb{R}^n$, $n \geq 2$, and let a mapping $f: G \to G^*$ be differentiable at a point $x \in G$. This means there exists a linear mapping $f'(x) : \mathbb{R}^n \to \mathbb{R}^n$, called the (strong) derivative of the mapping f at x, such that

$$f(x+h) = f(x) + f'(x)h + \omega(x,h)|h|,$$

where $\omega(x,h) \to 0$ as $h \to 0$.

We denote

$$H_I(x,f) = H_I(f'(x)), \quad H_O(x,f) = H_O(f'(x)),$$

and

$$L(x,f) = L(f'(x)), \quad l(x,f) = l(f'(x)), \quad J(x,f) = J(f'(x)).$$

Proposition 2.1. *Let $f : G \to G^*$ be a K-quasiconformal homeomorphism, $1 \le K < \infty$. Then*
(i) f is ACL (absolutely continuous on lines);
(ii) $f \in W^{1,n}_{loc}(G)$ (Sobolev class);
(iii) for almost every $x \in G$,

$$H_I(x,f) \le K, \quad H_O(x,f) \le K.$$

3. α-module of the families of k-dimensional surfaces

Now we define the quasiconformality of a homeomorphism in other terms (geometric or modular). Let $\mathcal{S}_k$ be a family of k-dimensional surfaces $\mathcal{S}$ in $\mathbb{R}^n$, $1 \le k \le n-1$ (curves for $k=1$). $\mathcal{S}$ is a k-dimensional surface if $\mathcal{S} : D_s \to \mathbb{R}^n$ is a homeomorphic image of the closed domain $D_s \subset \mathbb{R}^k$.

The p-module of $\mathcal{S}_k$ is defined as

$$M_p(\mathcal{S}_k) = \inf \int_{\mathbb{R}^n} \rho^p \, dx, \quad p \ge k,$$

where the infimum is taken over all Borel measurable functions $\rho \ge 0$ and such that

$$\int_S \rho^k \, d\sigma_k \ge 1$$

for every $\mathcal{S} \in \mathcal{S}_k$. We call each such ρ an *admissible function* for $\mathcal{S}_k$.

The following proposition characterizes a K-quasiconformality in the terms of n-moduli of $\mathcal{S}_k$, (see, e. g., [17], cf. [1]).

Proposition 3.1. *A homeomorphism f of a domain $G \subset \overline{\mathbb{R}^n}$ is K-quasiconformal, $1 \le K < \infty$, if for any family $\mathcal{S}_k$, $1 \le k \le n-1$, of k-dimensional surfaces in G the double inequality*

$$K^{\frac{k-n}{n-1}} M_n(\mathcal{S}_k) \le M_n(f(\mathcal{S}_k)) \le K^{\frac{n-k}{n-1}} M_n(\mathcal{S}_k), \tag{2}$$

holds.

It suffices to use even one of these double inequalities to get the basic properties of quasiconformal mappings.

Now we will give an extension of (2) to a class of mappings which is essentially wider than the quasiconformal mappings. For linear bijections $A: \mathbb{R}^n \to \mathbb{R}^n$, we consider the following quantities

$$H_{I,\alpha}(A) = \frac{|J(A)|}{l^\alpha(A)}, \quad H_{O,\alpha}(A) = \frac{L^\alpha(A)}{|J(A)|}$$

assuming that $\alpha \geq 1$.

Note that these dilatations range between 0 and ∞, while the classical dilatations are greater than or equal to 1. In the case $\alpha = n$, the quantities $H_{I,\alpha}$ and $H_{O,\alpha}$ coincide with the inner and outer dilatations, respectively.

Now we consider the homeomorphisms f which are differentiable almost everywhere in G, and fix the real numbers α, β satisfying $1 \leq \alpha < \beta < \infty$. Define

$$HI_{\alpha,\beta}(f) = \int_G H_{I,\alpha}^{\frac{\beta}{\beta-\alpha}}(x,f)\,dx, \quad HO_{\alpha,\beta}(f) = \int_G H_{O,\beta}^{\frac{\alpha}{\beta-\alpha}}(x,f)\,dx,$$

where $H_{I,\alpha}(x,f) = H_{I,\alpha}(f'(x))$, $H_{O,\beta}(x,f) = H_{O,\beta}(f'(x))$. We call these quantities the *inner* and *outer mean dilatations* of a mapping $f: G \to \mathbb{R}^n$.

Define for the fixed real numbers $\alpha, \beta, \gamma, \delta$ such that $1 \leq \alpha < \beta < \infty$, $1 \leq \gamma < \delta < \infty$, the *class of mappings with finite mean dilatations* of such homeomorphisms $f: G \to G^*$ which satisfy

(iv) f and f^{-1} are ACL-homeomorphisms,

(v) f and f^{-1} are differentiable, and their Jacobians $J(x,f) \neq 0$ and $J(y, f^{-1}) \neq 0$ a.e. in G and G^*, respectively,

(vi) the inner and the outer mean dilatations $HI_{\alpha,\beta}(f)$ and $HO_{\gamma,\delta}(f)$ are finite.

On properties of mappings with finite mean dilatations, we refer to [4].

The relations (1) show that in the classical case of quasiconformal mappings, their inner and outer dilatations are finite or infinity simultaneously. However, this does not be true for the mean dilatations.

The following example shows that the unboundedness of one from these dilatations does not depend on the value of another mean dilatation.

Example 3.1. Consider the unit cube

$$G = \{x = (x_1, \dots, x_n) : 0 < x_k < 1, k = 1, \dots, n\}$$

and let

$$f(x) = \left(x_1, \dots, x_{n-1}, \frac{x_n^{1-c}}{1-c}\right), \quad 0 < c < 1.$$

An easy computation shows that f belongs to the class of mappings with finite mean dilatations if and only if the following inequalities

$$0 < c < 1 - \alpha/\beta \quad \text{and} \quad 0 < c < 1 - (\gamma - 1)\delta/(\delta - 1)\gamma$$

are fulfilled. The cases $1 - \alpha/\beta \leq c < 1$ and $1 - (\gamma - 1)\delta/(\delta - 1)\gamma \leq c < 1$ imply $HI_{\alpha,\beta}(f) = \infty$ and $HO_{\gamma,\delta}(f) = \infty$, respectively. Thus, by suitable choice of the parameters $c, \alpha, \beta, \gamma, \delta$, one obtains the desired relations between $HI_{\alpha,\beta}(f)$ and $HO_{\gamma,\delta}(f)$.

The class of mappings with finite mean dilatations can be generalized, in turn, so that the conditions (iv)-(v) are preserved, but instead of (vi), one has a more general inequality.

Theorem 3.1. *Let $f : G \to G^*$ be a homeomorphism satisfying (iv)-(v) with $H_{I,\alpha} \in L^1_{loc}$. Then for any family $\mathcal{S}_k$, $1 \leq k \leq n-1$, of k-dimensional surfaces in G, and for each admissible ρ for $\mathcal{S}_k$, we have the inequality*

$$M_\alpha(f(\mathcal{S}_k)) \leq \int_G \rho^\alpha H_{I,\alpha}(x, f)dx. \tag{3}$$

The proof goes in the line to the proof of Theorem 2 [4], cf. [13].

4. (α, Q)-homeomorphisms

Let $Q : G \to [1, \infty]$ be a measurable function. Due to [10], [11], a homeomorphism $f : G \to \mathbb{R}^n$ is a *Q-homeomorphism* if

$$M(f(\Gamma)) \leq \int_G Q(x)\rho^n(x)\, dx$$

for every family Γ of curves in G and for every admissible function ρ for Γ (see also [15]).

Given a function $Q : G \to [1, \infty]$, we say that a sense preserving homeomorphism $f : G \to \mathbb{R}^n$ is *$Q(x)$-quasiconformal* if $f \in W^{1,n}_{loc}(G)$ and

$$\max\{H_I(x, f), H_O(x, f)\} \leq Q(x) \quad \text{a.e.}$$

Proposition 4.1. *([11]) Let $f : G \to \mathbb{R}^n$ be a $Q(x)$-quasiconformal mapping. Then*
(a) f is differentiable a.e.,
(b) f satisfies N-property,
(c) $J(x, f) \geq 0$ a.e.

If, in addition, $Q \in L^{n-1}_{loc}$, then $f^{-1} \in W^{1,n}_{loc}(G^*)$, and
(d) f^{-1} is differentiable a.e.,

(e) f^{-1} *satisfies* N*-property,*
(f) $J(x, f) > 0$ *a.e.*

The proof of Proposition 4.1 is based on the results of [14], [5] and [12].

Let $Q: G \to [0, \infty]$ be a measurable function. For a given integer k, $1 \leq k \leq n-1$, a homeomorphism $f: G \to \mathbb{R}^n$ is called *(α, Q)-homeomorphism*, if

$$M_\alpha(f(\mathcal{S}_k)) \leq \int_G Q(x)\rho^\alpha(x)\, dx \tag{4}$$

for every family of k-dimensional surfaces $\mathcal{S}_k$ in G and for every admissible function ρ for $\mathcal{S}_k$.

This inequality is a natural generalization of the right-hand side in (2). Note also that the integral in (4) can be interpreted as a weighted module (cf. [1], [19]).

We now restrict ourselves by two cases: $k = 1$ and $k = n - 1$, which correspond to curve families and $(n-1)$-dimensional surface families, respectively, and for similarity adapt them to ring domains.

A ring domain $D \subset \mathbb{R}^n$ is a finite domain whose complement consists of two components C_0 and C_1. Setting $F_0 = \partial C_0$ and $F_1 = \partial C_1$. We obtain two boundary components of D. For definiteness, let us assume that $\infty \in C_1$.

We say that a curve γ *joins the boundary components in* D if γ lies in D, except for its endpoints, one of which lies on F_0 and the second on F_1. A compact set Σ is said to *separate the boundary components of* D if $\Sigma \subset D$ and if C_0 and C_1 are located in different components of $C\Sigma$. Denote by Γ_D the family of all locally rectifiable curves γ which join the boundary components of D and by Σ_D the family of all compact piecewise smooth $(n-1)$-dimensional surfaces Σ which separate the boundary components of D. For each quantity V associated with D such as subset of D or a family of sets contained in D, we let V^* denote its image under f.

We consider also certain set functions. Let Φ be a finite nonnegative function in G defined for open subsets E of G satisfying

$$\sum_{k=1}^{m} \Phi(E_k) \leqslant \Phi(E)$$

for any finite collection $\{E_k\}_{k=1}^m$ of nonintersecting open sets $E_k \subset E$. We denote the class of all such set functions Φ by $\mathcal{F}$.

The *upper* and *lower* derivatives of a set function $\Phi \in \mathcal{F}$ at a point

$x \in G$ are defined by

$$\overline{\Phi'}(x) = \lim_{h\to 0} \sup_{d(Q)<h} \frac{\Phi(Q)}{mQ}, \qquad \underline{\Phi}'(x) = \lim_{h\to 0} \inf_{d(Q)<h} \frac{\Phi(Q)}{mQ},$$

where Q ranges over all open cubes and open balls such that $x \in Q \subset G$. Due to [12], these derivatives have the following properties:
(vii) $\overline{\Phi'}(x)$ and $\underline{\Phi}'(x)$ are Borel's functions;
(viii) $\overline{\Phi'}(x) = \underline{\Phi}'(x) < \infty$ a.e. in G;
(ix) for each open set $V \subset G$, $\int_V \overline{\Phi'}(x)\,dx \le \Phi(V)$.

Theorem 4.1. *Let $f : G \to G^*$ be an (α, Q)-homeomorphism with $Q \in L^1_{loc}(G)$ and $\alpha > n-1$. Then f is ACL-homeomorphism which is differentiable a.e. in G.*

Proof. We begin to prove the theorem estimating the α-module $M_p(\Gamma_D)$. Let $B^n(x,r)$ denote the open ball $\{y \in \mathbb{R}^n : |y-x| < r\}$, and $S^{n-1}(x,r)$ be the sphere $\partial B^n(x,r)$. Furthermore, let $d(V)$ be the diameter of V. Denote $A = D \cup C_0$. For $\alpha > n-1$, we have the low estimate

$$M_\alpha^{n-1}(\Gamma_D) \ge \left(\frac{\alpha-1}{\alpha-n+1}\right)^{\frac{n-\alpha-1}{n-1}} \omega_{n-2}^{\frac{\alpha}{n-1}} \frac{d^\alpha(C_0)}{(mA)^{\alpha-n+1}}, \tag{5}$$

which is given in [7] for the α-capacities of condensers; here ω_{n-2} denotes the $(n-2)$-dimensional Lebesgue measure of the unit sphere S^{n-2}. The equivalence between the α-module of Γ_D and the α-capacity of the condenser (C_0, A) was obtained in [6].

Now the proof of Theorem 4.1 will be accomplished in two steps following the construction applied in [9] (see, also [7]).

First, let us show that f is ACL. Denote

$$\Theta(V) = mV^* \quad \text{and} \quad \Psi(V) = \int_V Q(x)\,dx. \tag{6}$$

It is easy to verify that $\Theta, \Psi \in \mathcal{F}$. Let Q be an open n-dimensional interval (parallelepiped) in G. Write $Q = Q_0 \times J$, where Q_0 is $(n-1)$-dimensional interval in $\mathbb{R}^{n-1}$ and J is an open segment of x_n-axis.

Using the notation of [12], we have $\Theta(T,Q) = \Theta(T \times J)$ and $\Psi(T,Q) = \Psi(T \times J)$. The functions $\Theta(T,Q)$ and $\Psi(T,Q)$ also belong to the class $\mathcal{F}$ for Borel set $T \subset Q_0$.

Fix $z \in Q_0$ such that $\overline{\Theta}'(z,Q) < \infty$ and $\overline{\Psi}'(z,Q) < \infty$. Let $\Delta_1, \dots, \Delta_k$ be the disjoint closed subintervals of the segment $J_z = \{z\} \times J$. Set $C_{0,i} = \Delta_i + r\overline{B}^n$ and $A_i = \Delta_i + 2rB^n$, where $\overline{B}^n$ is the closure of $B^n = B^n(x,1)$.

The positive number r is chosen so that the domains D_i are disjoint and $D_i \subset Q$.

Now from (4) and the admissibility of $\rho = 1/\mathrm{dist}(F_{0,i}, F_{1,i})$ for Γ_{D_i} one obtains

$$M_\alpha(\Gamma^*_{D_i}) \le \frac{1}{r^\alpha} \int_{A_i} Q(x)dx, \tag{7}$$

where $\mathrm{dist}(F_{0,i}, F_{1,i})$ denotes the distance between $F_{0,i}$ and $F_{1,i}$. Using (5), we have that $M_\alpha(\Gamma^*_{D_i})$ is estimated by

$$M_\alpha(\Gamma^*_{D_i}) \ge C^{\frac{1}{n-1}} \frac{\left(d(C^*_{0,i})\right)^{\frac{\alpha}{n-1}}}{\left(mA^*_i\right)^{\frac{\alpha-n+1}{n-1}}}. \tag{8}$$

Substituting (8) into (7) together with the notations from (6) gives

$$d(C^*_{0,i}) \le C_1 \left(\Theta(A_i)\right)^{\frac{\alpha-n+1}{\alpha}} \left(\Psi(A_i)\right)^{\frac{n-1}{\alpha}} r^{1-n},$$

where C_1 depends only on α and n. We summarize such inequalities for $1 \le i \le k$ and apply Hölder's inequality. Then we obtain

$$\sum_{i=1}^k d(C^*_{0,i}) \le C_1 \left(\sum_{i=1}^k \Theta(A_i)\right)^{\frac{1}{n}} \left(\sum_{i=1}^k \Psi(A_i)\right)^{\frac{n-1}{n}} r^{1-n}.$$

Applying the inequalities

$$\sum_{i=1}^k \Theta(A_i) \le \Theta(B^{n-1}(z,r), Q), \quad \sum_{i=1}^k \Psi(A_i) \le \Psi(B^{n-1}(z,r), Q),$$

and letting $r \to 0$, we get the estimate

$$\sum_{i=1}^k d(C^*_{0,i}) \le C_2 \left(\overline{\Theta}'(z,Q)\right)^{\frac{\alpha-n+1}{\alpha}} \left(\overline{\Psi}'(z,Q)\right)^{\frac{n-1}{\alpha}},$$

where the constant C_2 depends only on q and n. Thus f is ACL.

Let us now show that a given homeomorphism f belongs to $W^{1,1}_{loc}$. For $\tilde{x}, x \in G, \tilde{x} \ne x$, we get

$$k(x) = \limsup_{\tilde{x} \to x} \frac{|f(\tilde{x}) - f(x)|}{|\tilde{x} - x|}.$$

Consider for a point $x \in G$ the spherical ring $D_r(x) = \{y : r < |x-y| < 2r\}$, choosing $r > 0$ so that $D_r(x) \subset G$. Estimating the integral in the right-hand side of (4) we have

$$M_\alpha(\Gamma^*_{D_r}) \le \frac{1}{r^\alpha} \int_{A_i} Q(x)dx. \tag{9}$$

Substituting (5) and (6) into (9) yields

$$\left(d(C_{0,r}^*)\right)^n \le C_3\left(\Theta(A_r)\right)^{\frac{\alpha-n+1}{\alpha}}\left(\Psi(A_r)\right)^{\frac{n-1}{\alpha}} r^{1-n},$$

where C_3 depends only on q, n and K. Since

$$k(x) \le \limsup_{r\to 0}\frac{d(C_{0,r}^*)}{r} \le C_4\left(\overline{\Theta}'(x)\right)^{\frac{\alpha-n+1}{\alpha}}\left(\overline{\Psi}'(x)\right)^{\frac{n-1}{\alpha}},$$

it follows that $k(x) < \infty$ for a.e. $x \in G$. Here C_4 also depends only on n and α. Applying now the Stepanov theorem [18], we conclude that f is differentiable a.e. in G. Moreover, for an arbitrary Borel set $V \subset G$, using Hölder's inequality we have

$$\int_V k(x)\,dx \le C_4\left(\Theta(V)\right)^{\frac{\alpha-n+1}{\alpha}}\left(\Psi(V)\right)^{\frac{n-1}{\alpha}} < \infty. \qquad \square$$

In the case of Q-homeomorphisms a similar result was obtained in [16].

The following result shows that dilatation $H_{I,\alpha}$ is dominated in G by the upper derivative of the set function $\Psi(V) = \int_V Q(x)\,dx$; here V is an open subset of G.

Theorem 4.2. *Let $f: G \to G^*$ be an (α, Q)-homeomorphism with $Q \in L^1_{loc}(G)$ and $\alpha > n-1$. Then $H_{I,\alpha} \in L^1_{loc}(G)$.*

Sketch of the proof. Let $a \in G$ be an arbitrary point at which f is differentiable, and with $J(a,f) \neq 0$. Preceding f, if necessary, by a rotation and a reflection, one reduces the proof to the case when $f(a) = a = 0$ and $|f'(0)e_i| = \lambda_i$, $i = 1, \ldots, n$; here e_ν denotes the νth unit basis vector.

Let D be a ring domain obtained from n-dimensional interval

$$\{x : |x_i| < r(t\lambda_i + 1), i = 1, \ldots, n-1, |x_n| < rt\lambda_n\}, t > 0,$$

by deleting the points of $(n-1)$-dimensional interval

$$Q_{n-1}(0,r) = \{x : |x_i| \le r, i = 1, \ldots, n-1, x_n = 0\}.$$

We choose $r > 0$ so that $\overline{D} \subset G$ and will show that

$$\frac{\lambda_1 \cdot \ldots \cdot \lambda_n}{\lambda_n^\alpha} \le \overline{\Psi}'(0).$$

Using that a is arbitrary, one can conclude that $H_{I,\alpha}(x,f) \le \overline{\Psi}'(x)$ for almost all $x \in G$. Applying (ix) completes the proof of Theorem 4.2. $\square$

There are the well-known distortion estimates of the module of planar ring domains under quasiconformal mappings obtained by Grötzsch and

Belinskii (see, [2], [8]). One result of such type in multidimensional case is due to [3].

The following theorem presents the inequalities of Grötzsch and Belinskii types for homeomorphisms with locally integrable dilatation $H_{I,\alpha}$ and more general for (α, Q)-homeomorphisms.

Theorem 4.3. *Let $f : \mathbb{R}^n \to \mathbb{R}^n$, $f(0) = 0$, $n \geq 2$, be a homeomorphism with $H_{I,\alpha} \in L^1_{loc}$. Then for every spherical ring $R(a,b) = \{x \in \mathbb{R}^n : 0 < a < |x| < b < \infty\}$ the double inequality*

$$\left(\frac{n-\alpha}{b^{n-\alpha} - a^{n-\alpha}}\right)^{\frac{\alpha}{n-1}} \left(\int_{R(a,b)} \frac{H_{I,\frac{\alpha}{\alpha-n+1}}(x,f)\, dx}{|x|^\alpha}\right)^{\frac{n-1-\alpha}{n-1}} \leq M_\alpha(f(\Sigma_R)) \leq \omega_{n-1}^{\frac{\alpha}{1-n}} \int_{R(a,b)} \frac{H_{I,\alpha}(x,f)\, dx}{|x|^\alpha}$$

holds.

The proof of Theorem 4.3 is based on applying Theorem 3.1 for suitable admissible functions ρ and on using the relationship between $M_p(f(\Gamma_{\mathcal{R}}))$ and $M_\alpha(f(\Sigma_R))$

$$M_p(\Gamma_R) = M^{1-p}_{\frac{p(n-1)}{p-1}}(\Sigma_R)$$

found in [21].

References

1. C. Andreian Cazacu, *Some formulae on the extremal length in n-dimensional case*, Proceedings of the Romanian-Finnish Seminar on Teichmüller Spaces and Quasiconformal Mappings (Braşov, 1969), 87–102. Publ. House of the Acad. of the Socialist Republic of Romania, Bucharest, 1971.
2. P. P. Belinskii, *General properties of quasiconformal mappings*, Nauka, Novosibirsk, 1974 (Russian).
3. C. Bishop, V. Y. Gutlyanskiĭ, O. Martio, M. Vuorinen, *On conformal dilatation in space*, Int. J. Math. Math. Sci. **22** (2003), 1397–1420.
4. A. Golberg, *Homeomorphism with finite mean dilatations*, Contemp. Math. **382** (2005), 177–186.
5. J. Heinonen, P. Koskela, *Sobolev mappings with integrable dilatation*, Arch. Rational Mech. Anal. **125** (1993), 81–97.
6. J. Hesse, *A p-extremal length and p-capacity equality*, Ark. Mat. **13** (1975), 131–144.
7. V. I. Kruglikov, *Capacities of condensors and quasiconformal in the mean mappings in space*, Mat. Sb. **130** (1986), no. 2, 185–206.

8. S. L. Krushkal, *Quasiconformal mappings and Riemann surfaces*, V. H. Winston & Sons, Washington, D.C.; John Wiley & Sons, New York-Toronto, Ont.-London, 1979.
9. O. Martio, S. Rickman, and J. Väisälä, *Definitions for quasiregular mappings*, Ann. Acad. Sci. Fenn. Ser. A I **448** (1969), 1–40.
10. O. Martio, V. Ryazanov, U. Srebro, E. Yakubov, *Q-homeomorphisms*, Contemp. Math., **364** (2004), 193–203.
11. O. Martio, V. Ryazanov, U. Srebro, E. Yakubov, *On Q-homeomorphisms*, Ann. Acad. Sci. Fenn. Math. **30** (2005), no. 1, 49–69.
12. T. Rado, P. Reichelderfer, *Continuous Transformations in Analysis*, Springer-Verlag, 1955.
13. K. Rajala, *Surface families and boundary behavior of quasiregular mappings*, Illinois J. Math. **49** (2005), no. 4, 1145–1153.
14. Y. G. Reshetnyak, *Space Mappings with Bounded Distortion*, Transl. Math. Monogr., Amer. Math. Soc., vol. 73, 1989.
15. V. Ryazanov, R. Salimov, *Weakly flat spaces and boundaries in the mapping theory*, Ukrainian Math. Bull., **4** (2007), no. 2, 199–234.
16. R. Salimov, *Absolute continuity and differentiability of Q-homeomorphisms*, Ann. Acad. Sci. Fenn. Math., 2007, to appear.
17. B. V. Shabat, *The modulus method in space*, Soviet Math. Dokl. **130** (1960), no. 6, 1210–1213.
18. V. V. Stepanov, *Sur les conditions de l'existence de la differentielle totale*, Mat. Sb. **30** (1924), 487–489.
19. P. M. Tamrazov, *Moduli and extremal metrics in nonorientable and twisted Riemannian manifolds*, Ukrainian Math. J. **50** (1998), no. 10, 1586–1598.
20. J. Väisälä, *Lectures on n-dimensional Quasiconformal Mappings*, Springer-Verlag, 1971.
21. W. P. Ziemer, *Extremal length and p-capacity*, Michigan Math. J. **16** (1969), 43–51.

ON CHARACTERIZATION OF THE EXTENSION PROPERTY

A. P. GONCHAROV

Department of Mathematics, Bilkent University,
Ankara, 06800, Turkey
E-mail: goncha@fen.bilkent.edu.tr

The geometric characterization of the extension property for Cantor-type sets, found in [3], is related to the rate of growth of the values of the discrete logarithmic energies of compact sets that locally form the set.

Keywords: Extension property, Cantor-type sets, discrete logarithmic energies.

1. Extension Problem

Let K be a compact set in $\mathbb{R}^d$. Then $\mathcal{E}(K)$ is the space of Whitney jets on K, that is the space of traces on K of C^∞ functions. Topology in the space $\mathcal{E}(K)$ can be given by the system of seminorms

$$|| f ||_q = \inf\ | F |_q, \quad q \in \mathbb{N},$$

where the infimum is taken for all possible extensions of f to F and $| F |_q$ denotes the qth norm of F in $C^\infty(\mathbb{R}^d)$.

The Extension Problem is to characterize when there exists a linear continuous extension operator $L : \mathcal{E}(K) \longrightarrow C^\infty(\mathbb{R}^d)$. We say that the compact set K has the extension property if there exists a such operator.

Tidten in [6] applied Vogt's condition for a splitting of exact sequences of Fréchet spaces and gave a topological characterization of the extension property: a compact set K has the extension property if and only if the space $\mathcal{E}(K)$ has a dominating norm (see for instance [2] for the definition of a dominating norm and for a recent account of the theory).

Nevertheless, the problem of a geometric characterization of the extension property that goes back to the work [5] of Mityagin, is still open even for the one-dimensional case, in spite of the presence of numerous particular results. Here we consider the geometric characterization of the extension property for Cantor-type sets found in [3].

2. Cantor-type Sets

Given l_1 with $0 < l_1 < 1/2$ and a sequence $(\alpha_s)_{s=2}^{\infty}$ with $\alpha_s > 1$ let us define the sequence $(l_s)_{s=0}^{\infty}$ in the following way: $l_0 = 1, l_1, l_2 = l_1^{\alpha_2}, \cdots, l_s = l_1^{\alpha_2\alpha_3\cdots\alpha_s}, \cdots$. Then by $K^{(\alpha_n)}$ we denote the symmetric Cantor-type set $\bigcap_{s=0}^{\infty}\bigcup_{j=1}^{2^s} I_{j,s}$, where $|I_{j,s}| = l_s$ for all j. Here the closed intervals $I_{j,s}$ we call *basic* intervals. Let x be an endpoint of some basic interval. Then there exists the minimal number q (the *type* of x) such that x is the endpoint of some $I_{j,m}$ for every $m \geq q$.

As in [3] we suppose that $\alpha_s \geq 1 + \varepsilon_0, s \geq s_0$ for some positive ε_0 and $l_s \geq 4\,l_{s+1}$ for all s. Let $h_s = l_s - 2l_{s+1}$ be the gap between two adjacent intervals.

We follow the notations used in [3]: $\pi_{n,0} = 1$, and for $n \geq 1, s \geq 1$, we let

$$\pi_{n,s} = 2^{-s}\alpha_{n+1}\alpha_{n+2}\cdots\alpha_{n+s}, \qquad \sigma_{n,s} = \sum_{k=0}^{s}\pi_{n,k}.$$

Theorem [3]. The following are equivalent:
(i) The set $K^{(\alpha_n)}$ has the extension property.
(ii) $\forall M > 0 \quad \exists\, s_M : \quad l_{n+s}^M > l_n^{2^s} l_{n+1}^{2^{s-1}} \cdots l_{n+s}, \quad \forall n \ \forall s \geq s_M.$
(iii) $\sigma_{n,s+1} \,/\, \sigma_{n,s} \rightrightarrows 1, \quad$ as $\quad s \to \infty$ uniformly with respect to n.

We see that the condition (ii) is purely geometrical, whereas the condition (iii) is related to the theory of logarithmic potential. In what follows log denotes the natural logarithm.

3. Discrete Logarithmic Energies

Let K be a compact set in $\mathbb{C}$ and for given N points $z_1, \cdots, z_N \subset K$ let $\mu_N = \mu_N(z_1, \cdots, z_N)$ denote the discrete measure that associates the mass $1/N$ to any point z_k, $1 \leq k \leq N$. The logarithmic potential of measure μ_N is given by

$$U^{\mu_N}(z) = \frac{1}{N}\sum_{k=1}^{N}\log\frac{1}{|z - z_k|}.$$

Any discrete measure has infinite energy, but if we use the truncated kernels (see e.g.[1]), then we can define the corresponding logarithmic energy as in [4]:

$$I(\mu_N) = \frac{1}{N^2}\sum_{k,j=1,k\neq j}^{N}\log\frac{1}{|z_j - z_k|}.$$

Clearly,

$$I(\mu_N) = \frac{-2}{N^2} \log |V(z_1, \cdots, z_N)|,$$

where for the corresponding Vandermonde determinant we have

$$V(z_1, \cdots, z_N) = \prod_{1 \le j < k \le N} (z_k - z_j).$$

Points $(z_k)_{k=1}^N \subset K$ that maximize the determinant $V(z_1, \cdots, z_N)$ (or minimize the corresponding discrete logarithmic energy) are known as Fekete points of order N for K.

Given compact set $K^{(\alpha_n)}$ let us fix $n \in \mathbb{N}$ and $N = 2^{s+1}$ for some $s \in \mathbb{N}_0 := \{0, 1, \cdots\}$.

Let $(z_k)_{k=1}^N$ consist of all endpoints of the type $\le s + n$ on the first basic interval $[0, l_n]$ ordered increasingly, that is $z_1 = 0, z_2 = l_{s+n}, z_3 = l_{s+n-1} - l_{s+n}, \cdots, z_N = l_n$.

Lemma 3.1.

$$e^{-2^{2s+3} l_n^{\epsilon_0}} \; l_n^{4^s \sigma_{n,s}} < |V(z_1, \cdots, z_N)| < l_n^{4^s \sigma_{n,s}}.$$

Proof. Temporarily we denote $l_{n+s} \, l_{n+s-1}^2 \cdots l_n^{2^s}$ by λ. We see at once that $\prod_{k=2}^N |z_k - z_1| < \lambda$. The product $\prod_{k=3}^N |z_k - z_2|$ with the upper bound $l_{n+s-1}^2 \cdots l_n^{2^s}$ we join with $|z_N - z_{N-1}| = l_{n+s}$. We continue in this fashion to join the product $\prod_{k=m+1}^N |z_k - z_m|$ with the product $\prod_{k=N-m+2}^N |z_k - z_{N-m+1}|$ for $m = 3, \cdots, 2^s$. For each m we can estimate from above the joint product through λ. Since there are 2^s pairs of products, we get the bound

$$|V(z_1, \cdots, z_N)| < \lambda^{2^s}.$$

By definition, λ has the form $l_n^{\varkappa}$, where $\varkappa = \alpha_{n+1} \alpha_{n+2} \cdots \alpha_{n+s} + 2\alpha_{n+1} \cdots \alpha_{n+s-1} + \cdots + 2^{s-1} \alpha_{n+1} + 2^s = 2^s [1 + 2^{-1}\alpha_{n+1} + \cdots + 2^{-s}\alpha_{n+1} \alpha_{n+2} \cdots \alpha_{n+s}] = 2^s \sigma_{n,s}$, which gives the desired upper bound of the lemma.

To estimate $|V(z_1, \cdots, z_N)|$ from below we replace all l_k in λ by h_k for $k = n, n+1, \cdots, n+s-1$, since the distance between any two points z_i, z_j that belong to the same basic interval of the length l_k, but do not belong to an interval of the length l_{k+1}, is not smaller than h_k.

Therefore,

$$|V(z_1, \cdots, z_N)| > (l_{n+s} \, h_{n+s-1}^2 \cdots h_n^{2^s})^{2^s} = \lambda^{2^s} \cdot a$$

with $\log a = 2^s\,(2\log\frac{h_{n+s-1}}{l_{n+s-1}} + \cdots + 2^s\log\frac{h_n}{l_n})$. By condition, $\frac{h_k}{l_k} = 1 - 2\frac{l_{k+1}}{l_k} > \frac{1}{2}$. From this $\log\frac{h_k}{l_k} > -4\,\frac{l_{k+1}}{l_k} \geq -4\,l_k^{\varepsilon_0}$ and $\log a > -2^s\sum_{k=n}^{n+s-1} 2^{n+s-k+2}\,l_k^{\varepsilon_0} > -2^{2s+3}\,l_n^{\varepsilon_0}$, which completes the proof. □

Corollary 3.1.

$$\frac{1}{2}\,\sigma_{n,\,s}\,\log\frac{1}{l_n} < I(\mu_N) < \frac{1}{2}\,\sigma_{n,\,s}\,\log\frac{1}{l_n} + 4\,l_n^{\varepsilon_0}. \tag{1}$$

Theorem 3.1. *If the set $K^{(\alpha_n)}$ has the extension property then*

$$I_n(\mu_{2^{s+1}})/I_n(\mu_{2^s}) \to 1, \quad as \quad s\to\infty, n\to\infty.$$

Here $I_n(\mu_{2^{s+1}})$ stands for the discrete logarithmic energy defined by all endpoints of the type $\leq s+n$ on any basic interval of the length l_n.

Proof. Write $\gamma_n = 8\,l_n^{\varepsilon_0}\,\log^{-1}\,l_n^{-1}$. From (1) we have

$$\frac{I_n(\mu_{2^{s+1}})}{I_n(\mu_{2^s})} < \frac{\sigma_{n,\,s}+\gamma_n}{\sigma_{n,\,s-1}} < \frac{\sigma_{n,\,s}}{\sigma_{n,\,s-1}} + \gamma_n,$$

since $\sigma_{n,\,s-1} > 1$. Now the result follows on the condition (iii) and decrease of the sequence γ_n. □

One can conjecture that the existence of a linear continuous extension operator for the space $\mathcal{E}(K)$ (at least for Cantor-type sets) is characterized by a regularity of growth of the minimal discrete logarithmic energies corresponding to compact sets that locally form the set K. The points $(z_k)_{k=1}^N$ considered in Lemma give rather rough approximation of the minimal energy for the set $K^{(\alpha_s)}\cap[0,l_n]$. The exact position of the Fekete points is not known even for rarefied Cantor-type sets.

References

1. V. V. Andrievskii, H. P. Blatt, *Discrepancy of Signed Measures and Polynomial Approximation*, Springer, 2002.
2. L. Frerick, *Extension operators for spaces of infinite differentiable Whitney jets*, J. Reine Angew. Math. **602** (2007), 123–154.
3. A. Goncharov, *On the geometric characterization of the extension property*, Bull. Belg. Math. Soc. Simon Stevin **14** (2007), 513–520.
4. J. Korevaar, *Fekete extreme points and related problems*, in Approximation Theory and Function Series, Budapest, 1996, 35–62.
5. B. S. Mitiagin, *Approximative dimension and bases in nuclear spaces*, Russian Math. Surveys **16** (1961), 59–127.
6. M. Tidten, *Fortsetzungen von C^∞-Funktionen, welche auf einer abgeschlossenen Menge in $\mathbb{R}^n$ definiert sind*, Manuscripta Math. **27** (1979), 291–312.

ON SOME BOUNDARY PROPERTIES OF CONFORMAL MAPPING

OLENA W. KARUPU

Department of Higher and Numerical Mathematics,
National Aviation University, 1 Komarov ave,
Kyiv, Ukraine
E-mail: karupu@ukr.net

We consider some new estimates for the uniform curvilinear moduli of smoothness of arbitrary order of the function realizing a conformal mapping of the unit disk and a domain bounded by a smooth Jordan curve.

Keywords: Conformal mapping; modulus of smoothness; finite difference smoothnesses.

1. Introduction

Let G be a simply connected domain in the complex plane bounded by a smooth Jordan curve Γ, $\tau = \tau(s)$ be the angle between the tangent to Γ and the positive real axis, $s = s(w)$ be the arc length on Γ. Let $w = \varphi(z)$ be a homeomorphism of the closed unit disk $\overline{D} = \{z : |z| \leq 1\}$ onto the closure $\overline{G}$ of the domain G, conformal in the open unit disk D. Let $z = \psi(w)$ be the function inverse to the function $w = \varphi(z)$.

The aim of this paper is to receive information about connection between the function $\tau = \tau(s)$ and the functions $w = \varphi(z)$ and $z = \psi(w)$ in terms of the uniform curvilinear moduli of smoothness of arbitrary order.

Kellogg in 1912 proved the theorem in which it had been established that if $\tau = \tau(s)$ satisfies Holder condition with index α, $0 < \alpha < 1$, then the derivative $\varphi'(e^{i\theta})$ of the function $\varphi(z)$ on ∂D satisfies Holder condition with the same index α. Afterwards this result was generalized in works by several authors: S. E. Warshawski, J. L. Geronimus, S. J. Alper, R. N. Kovalchuk, L. I. Kolesnik.

P. M. Tamrazov [1] obtained solid reinforcement for the modulus of continuity for the derivative $\varphi'(z)$ of the function $\varphi(z)$ on $\overline{D}$. E. P. Dolzenko [2–4] received estimates for the modulus of continuity for the function realizing conformal mapping of Jordan domains in the case when boundaries of domains were characterized by moduli of oscilation or moduli of rectifiability.

P. M. Tamrazov [5] solved the problem of estimating of finite difference smoothnesses for composite function. These results gave posibility to receive generalizations and inversations of Kellogg's type theorems for general moduli of smoothness of arbitrary order.

In particular, results in the terms of the arithmetic and uniform curvilinear moduli of smoothness of arbitrary order were received by author in [6–8] (for more details, see [1,6,9]).

2. Estimates for moduli of smoothness for the derivative of a function realizing a conformal mapping of the unit disk onto a Jordan domain

Let the finite function $w = f(z)$ be defined on a curve $\gamma \subset C$. Let $(z_0, ..., z_k)$ be the collection of the points on the curve γ and $[z_0, ..., z_k; f, z_0]$ be the finite difference of order k for the function $w = f(z)$. P. M. Tamrazov [1] defined on rectifiable curves the uniform curvilinear moduli of smoothness of order k for the function $w = f(z)$ as $\widetilde{\omega}_{k,N,\gamma}((z), \delta)_p = \sup_{w\in\gamma} \sup_{(z_0,...,z_k)\in\gamma_{w,\delta}(N)} |[z_0, ..., z_k; f, z_0]|$, where $\gamma_{w,\delta}(N)$ is the set of collections $(z_0, ..., z_k)$ such that curvilinear (with respect to the curve γ) distances between points $z_0, ..., z_k \in \gamma$ satisfy the condition ${\rho(z_i, z_{i+1})}/{\rho(z_j, z_{j+1})} \leq N$, $(N \in [1, \infty))$, and $\rho(z_i, w) \leq \delta$ $(i, j = 1, ..., k)$.

In partial case when the function $w = f(z)$ is defined on the real axis and points $x_0, ..., x_k$ form the arithmetic progression we receive the real arithmetic moduli of smoothness $\omega_k(f(x), \delta)$.

The following results for the arithmetic and for the uniform curvilinear moduli of smoothness of arbitrary order for the the derivative $\varphi'(z)$ of the function $\varphi(z)$ on ∂D generalizing Kellogg's theorem were earlier obtained by the author [6].

Theorem 2.1. *([6]) Let modulus of smoothness $\omega_k(\tau(s), \delta)$ of order k $(k \in N)$ for the function $\tau(s)$ satisfy the condition $\omega_k(\tau(s), \delta) = O[\omega(\delta)]$ $(\delta \to 0)$, where $\omega(\delta)$ is normal majorant satisfying the condition*

$$\int_0^l \frac{\omega(t)}{t} dt < +\infty \tag{1}$$

Then the nonzero continuous on $\overline{D}$ derivative $\varphi'(z)$ of the function $\varphi(z)$

exists satisfying on ∂D the conditions

$$\omega_k(\arg\varphi'(e^{i\theta}),\delta) = O\left(\mu\left(\delta\right)\right)\ \ (\delta\to 0)\,,$$
$$\omega_k(\log\varphi'(e^{i\theta}),\delta) = O\left(\mu^*\left(\delta\right)\right)\ \ (\delta\to 0)\,,$$
$$\omega_k(\varphi'(e^{i\theta}),\delta) = O\left(\mu^*\left(\delta\right)\right)\ \ (\delta\to 0)\,,$$

where

$$\mu\left(\delta\right) = \omega_k(\tau(s),\delta) + \sum_{j=1}^{k-1}\sum_{r_1=1}^{j-1}\cdots\sum_{r_j=1}^{r_{j-1}-1}\left[\omega_k\left(\tau\left(s\right),\delta\right)\right]^{\frac{k-r_1}{k}}$$
$$\times\int_0^l\cdots\int_0^l \frac{\left[\omega_k\left(\tau\left(s\right),x_j\right)\right]^{\frac{r_j}{k}}\prod_{i=2}^{j}\left[\omega_k\left(\tau\left(s\right),x_{i-1}\right)\right]^{\frac{r_{i-1}-r_i}{k}}}{\prod_{p=1}^{j}x_p\left(1+\left({}^{x_p}\!/\!{}_{x_{p-1}}\right)^{r_p}\right)}dx_1\ldots dx_j,$$

$$\mu^*\left(\delta\right) = \int_0^l\frac{\omega\left(x_1\right)}{x_1\left(1+\left({}^{x_1}\!/\!{}_{\delta}\right)^k\right)}dx_1$$
$$+\sum_{j=1}^{k-1}\sum_{r_1=1}^{j-1}\cdots\sum_{r_j=1}^{r_{j-1}-1}\delta^{k-r_1}\int_0^l\cdots\int_0^l x_{j+1}^{r_{j-1}}\left(1+\int_{x_{j+1}}^l\frac{\omega(y)}{y_j^{r_j+1}}dy\right)$$
$$\times\left(1+\left(\frac{x_{j+1}}{x_j}\right)^{r_{j+11}}\right)\prod_{p=1}^{j}\left(1+\int_{x_p}^l\frac{\omega\left(t_p\right)}{t_p^{r_{p-1}-r_p+1}}dt_p\right)$$
$$\times\left(1+\left(\frac{x_p}{x_{p-1}}\right)^{r_{p-1}}\right)^{-1}x_p^{r_{r-1}-r_r-1}dx_1\ldots dx_{j+1}.$$

Corollary 2.1. *If modulus of smoothness $\omega_k(\tau(s),\delta)$ of order k for the function $\tau(s)$ satisfies Holder condition $\omega_k(\tau(s),\delta) = O\left(\delta^\alpha\right)(\delta\to 0)$, $0<\alpha<k$, then the modulus of smoothness $\omega_k(\varphi'(e^{i\theta}),\delta)$ of the same order k for the derivative $\varphi'(e^{i\theta})$ of the function $\varphi(z)$ on ∂D satisfies the condition $\omega_k(\varphi'(e^{i\theta}),\delta) = O\left(\delta^\alpha\right)(\delta\to 0)$ with the same index α.*

Theorem 2.2. *([6]) Let modulus of smoothness $\omega_k(\tau(s),\delta)$ of order k $(k\in N)$ for the function $\tau(s)$ satisfy the condition $\omega_k(\tau(s),\delta) = O\left[\omega\left(\delta\right)\right](\delta\to 0)$, where $\omega(\delta)$ is normal majorant satisfying the condition (1). Then the uniform curvilinear modulus of smoothness $\widetilde{\omega}_{k,1,\partial D}(\varphi',\delta)$ of the same order k for the derivative $\varphi'(z)$ of the function $\varphi(z)$ on ∂D satisfies the conditions*

$$\widetilde{\omega}_{k,1,\partial D}(\arg\varphi',\delta) = O\left(\widetilde{\mu}\left(\delta\right)\right)\ \ (\delta\to 0)\,,$$
$$\widetilde{\omega}_{k,1,\partial D}(\log\varphi',\delta) = O\left(\widetilde{\mu}^*\left(\delta\right)\right)\ \ (\delta\to 0)\,,$$
$$\widetilde{\omega}_{k,1,\partial D}(\varphi',\delta) = O\left(\widetilde{\mu}^*\left(\delta\right)\right)\ \ (\delta\to 0)\,,$$

where

$$\widetilde{\mu}(\delta) = \mu(\delta) + \delta^k \int_\delta^l \frac{\mu(t)}{t^{k+1}} dt, \qquad \widetilde{\mu}^*(\delta) = \mu^*(\delta) + \delta^k \int_\delta^l \frac{\mu^*(t)}{t^{k+1}} dt.$$

Corollary 2.2. *If modulus of smoothness $\omega_k(\tau(s),\delta)$ of order k for the function $\tau(s)$ satisfies Holder condition $\omega_k(\tau(s),\delta) = O(\delta^\alpha)(\delta \to 0)$, $0 < \alpha < k$, then the modulus of smoothness $\widetilde{\omega}_{k,1,\partial D}(\varphi',\delta)$ of the same order k for the derivative $\varphi'(z)$ of the function $\varphi(z)$ on ∂D satisfies the condition $\widetilde{\omega}_{k,1,\partial D}(\varphi',\delta) = O(\delta^\alpha)(\delta \to 0)$ with the same index α.*

3. Estimates for moduli of smoothness for a function realizing a conformal mapping of the unit disk onto a Jordan domain

Let us consider estimates for the modulus of smoothness for the function $\varphi(z)$.

Theorem 3.1. *Let modulus of smoothness $\omega_k(\tau(s),\delta)$ of order k for the function $\tau(s)$ satisfy the condition $\omega_k(\tau(s),\delta) = O[\omega(\delta)](\delta \to 0)$, where $\omega(\delta)$ is normal majorant satisfying the condition [1].*

Then the uniform curvilinear modulus of smoothness $\widetilde{\omega}_{k,1,\partial D}(\varphi,\delta)$ of the function $\varphi(z)$ on ∂D satisfies the condition

$$\widetilde{\omega}_{k,1,\partial D}(\varphi,\delta) = O\left(\delta^k \int_\delta^{2\pi} \frac{\mu*(t)}{t} dt\right)(\delta \to 0). \tag{2}$$

Proof. Let us introduce the function χ of real variable θ as $\chi(\theta) = \varphi\left(e^{i\theta}\right)$. Then on ∂D the identity $\varphi(z) = \chi \circ \theta(z)$ takes place. Applying inequalities for finite difference smoothnesses of composite function obtained by P. M. Tamrazov ([5], p. 17), we receive the estimates of the modulus of smoothness $\widetilde{\omega}_{k,1,\partial D}(\varphi,\delta) = \widetilde{\omega}_{k,1,\partial D}(\varphi \circ \vartheta,\delta)$ via the moduli of smoothness $\omega_j(\chi,\delta), j = 1,...,k$. We have

$$\widetilde{\omega}_{k,1,\partial D}(\varphi,\delta) = \widetilde{\omega}_{k,1,\partial D}(\varphi \circ \vartheta,\delta) \le \omega_k(\chi,\delta) + c_1 \sum_{j=1}^{k-1} \omega_j(\chi,\delta)\delta^{k-j},$$

where constant c_1 does not depend on δ.

We will also use the Marchaud inequalities:

$$\omega_j(\chi,\delta) \le \delta^j \left(c_2 + c_3 \int_\delta^{2\pi} \frac{\omega_k(\chi,t)}{t^{j+1}} dt\right), j = 1,...,k-1, \tag{3}$$

where the constants c_2 and c_3 do not depend on δ. So, as the estimates

$$\int_{\delta}^{2\pi} \frac{\omega_k(\chi, t)}{t^{j+1}} dt \leq a_j \int_{\delta}^{2\pi} \frac{\omega_k(\chi, t)}{t^k} dt, j = 1, ..., k-1,$$

where constants $a_1, ..., a_{k-1}$ do not depend on δ, are evident, then we obtain the estimates

$$\begin{aligned} \widetilde{\omega}_{k,1,\partial D}(\varphi, \delta) &\leq \omega_k(\chi, \delta) + c_4 \delta^k + c_5 \delta^k \textstyle\int_{\delta}^{2\pi} \frac{\omega_k(\chi,t)}{t^k} dt \\ &\leq c_6 \omega_k(\chi, \delta) + c_7 \delta^k \textstyle\int_{\delta}^{2\pi} \frac{\omega_k(\chi,t)}{t^k} dt, \end{aligned} \tag{4}$$

where constants c_4, c_5, c_6 and c_7 do not depend on δ.

But $\chi'(\theta) = \frac{d\varphi(e^{i\theta})}{d\theta} = ie^{i\theta}\varphi\left(e^{i\theta}\right)$, so

$$\omega_n(\chi', \delta) = \omega_n(ie^{i\theta}\varphi'(e^{i\theta}), \delta), n = 1, 2, ...$$

We have to estimate the modulus of smoothness for the the product of functions $ie^{i\theta}$ and $\varphi'(e^{i\theta})$. As

$$\Delta_h^k \left\{ie^{i\theta} \cdot \varphi'\left(e^{i\theta}\right)\right\} = \sum_{j=1}^{k} \binom{k}{j} \Delta_h^j \left\{ie^{i\theta}\right\} \Delta_h^{k-j} \left\{\varphi'\left(e^{i\theta}\right)\right\}$$

and

$$\omega_r(ie^{i\theta}), \delta) \leq b_r \delta^r, r = 0, ..., k,$$

where constants $b_0, ..., b_k$ do not depend on δ, then

$$\omega_n(\chi', \delta) \leq c_8 \sum_{j=0}^{n} \omega_j(\varphi'\left(e^{i\theta}\right), \delta)\delta^{k-j}, n = 1, 2, ...,$$

where constant c_8 does not depend on δ.

For the real arithmetic moduli of smoothness of the function $\chi(\theta)$ the Marchaud inequalities take the form

$$\omega_n(\chi, \delta) \leq \delta\omega_{n-1}(\chi', \delta), n = 1, 2, ...$$

So

$$\omega_k(\chi, \delta) \leq c_9 \delta \sum_{j=0}^{k-1} \omega_j(\varphi'\left(e^{i\theta}\right), \delta)\delta^{k-1-j} = c_9 \sum_{j=0}^{k-1} \omega_j(\varphi'\left(e^{i\theta}\right), \delta)\delta^{k-j},$$

where constant c_9 does not depend on δ. Taking into account the Marchaud inequalities (3) we will rewrite last estimate as

$$\omega_k(\chi, \delta) \leq c_{10} \delta^k \int_{\delta}^{2\pi} \frac{\omega_k(\varphi'\left(e^{i\theta}\right), t)}{t^k} dt, \tag{5}$$

where constant c_{10} does not depend on δ. But by Theorem 2.1 the estimate

$$\omega_k(\varphi'(e^{i\theta}), \delta) \le c_{11}\mu * (\delta)\,, \tag{6}$$

where constant c_{11} does not depend on δ, is true. So, from (5) and (6) we have

$$\omega_k(\varphi\left(e^{i\theta}\right), \delta) = \omega_k(\chi, \delta) \le c_{12}\delta^k \int_\delta^{2\pi} \frac{\mu * (t)}{t^k} dt, \tag{7}$$

where constant c_{12} does not depend on δ. Then we obtain

$$\begin{aligned} \int_\delta^{2\pi} \frac{\omega_k(\chi, t)}{t^k} dt &\le c_{13} \int_\delta^{2\pi} dt \int_t^{2\pi} \frac{\mu^*(y)}{y^k} dy \\ &\le c_{14} \int_t^{2\pi} \frac{\mu^*(y)}{y^{k-1}} dy \le c_{15} \int_t^{2\pi} \frac{\mu^*(y)}{y^k} dy, \end{aligned} \tag{8}$$

where constants c_{13}, c_{14} and c_{15} do not depend on δ.

Let us substitute the estimates (7) and (8) to the estimate (4) and obtain

$$\widetilde{\omega}_{k,1,\partial D}(\varphi, \delta) \le c_{16}\delta^k \int_\delta^{2\pi} \frac{\mu^*(t)}{t} dt,$$

where constant c_{16} does not depend on δ. Theorem 3.1 is proved. □

Corollary 3.1. *If modulus of smoothness $\omega_k(\tau(s), \delta)$ of order k for the function $\tau(s)$ satisfies Holder condition*

$$\omega_k(\tau(s), \delta) = O\left(\delta^\alpha\right)(\delta \to 0)\,, 0 < \alpha < k,$$

then the uniform curvilinear modulus of smoothness $\widetilde{\omega}_{k,1,\partial D}(\varphi, \delta)$ of the function $\varphi(z)$ on ∂D satisfies the condition

$$\widetilde{\omega}_{k,1,\partial D}(\varphi, \delta) = \begin{cases} O\left(\delta^{\alpha+1}\right)(\delta \to 0) & if \quad 0 < \alpha < k-1, \\ O\left(\delta^k \log \frac{1}{\delta}\right)(\delta \to 0) & if \quad \alpha = k-1, \\ O\left(\delta^k\right)(\delta \to 0) & if \quad k-1 < \alpha < k. \end{cases}$$

4. Estimates for moduli of smoothness for a function realizing a conformal mapping of a Jordan domain onto the unit disk

The following result for the uniform curvilinear modulus of smoothness of arbitrary order for the the derivative $\psi'(w)$ of the function $\psi(w)$ on the curve Γ was earlier obtained by author [6].

Theorem 4.1. *([6]) Let modulus of smoothness $\omega_k(\tau(s), \delta)$ of order k for the function $\tau(s)$ satisfy the condition $\omega_k(\tau(s), \delta) = O\left[\omega\left(\delta\right)\right](\delta \to 0)$, where $\omega(\delta)$ is normal majorant satisfying the condition (1).*

Then the nonzero continuous on $\overline{G}$ *derivative* $\psi'(w)$ *of the function* $\psi(w)$ *exists satisfying on* Γ *the conditions*

$$\widetilde{\omega}_{k,1,\Gamma}(\arg \psi', \delta) = O\left(\widetilde{\eta}\left(\delta\right)\right) \ (\delta \to 0),$$
$$\widetilde{\omega}_{k,1,\Gamma}(\log \psi', \delta) = O\left(\widetilde{\eta}^*\left(\delta\right)\right) \ (\delta \to 0),$$
$$\widetilde{\omega}_{k,1,\Gamma}(\psi', \delta) = O\left(\widetilde{\eta}^*\left(\delta\right)\right) \ (\delta \to 0),$$

where

$$\widetilde{\eta}\left(\delta\right) = \widetilde{\mu}\left(\delta\right) + \delta^{1-k(k-1)/2} \int_\delta^l \frac{\widetilde{\mu}\left(y\right)}{y^{k+1}} dy \left(\delta^k \int_\delta^l \frac{\mu\left(t\right)}{t^k} dt\right)^{\frac{k(k+1)/2-1}{k}},$$

$$\widetilde{\eta}^*\left(\delta\right) = \widetilde{\mu}^*\left(\delta\right) + \delta^{1-k(k-1)/2} \int_\delta^l \frac{\widetilde{\mu}^*\left(y\right)}{y^{k+1}} dy \left(\delta^k \int_\delta^l \frac{\widetilde{\mu}\left(t\right)}{t^k} dt\right)^{\frac{k(k+1)/2-1}{k}}.$$

Proof of this theorem is similar to the proof of Theorem 3.1.

Corollary 4.1. *If modulus of smoothness* $\omega_k(\tau(s), \delta)$ *of order* k *for the function* $\tau(s)$ *satisfies Holder condition*

$$\omega_k(\tau(s), \delta) = O\left(\delta^\alpha\right)(\delta \to 0), 0 < \alpha < k,$$

then the uniform curvilinear modulus of smoothness $\widetilde{\omega}_{k,1,\Gamma}(\psi', \delta)$ *of the derivative* $\psi'(w)$ *of the function* $\psi(w)$ *satisfies the condition* $\widetilde{\omega}_{k,1,\Gamma}(\psi', \delta) = O(\delta^\alpha)(\delta \to 0)$ *with the same index* α.

Let us consider estimates for the modulus of smoothness for the function $\psi(w)$.

Theorem 4.2. *Let modulus of smoothness* $\omega_k(\tau(s), \delta)$ *of order* $k(k \in N)$ *for the function* $\tau(s)$ *satisfy Holder condition* $\omega_k(\tau(s), \delta) = O\left(\delta^\alpha\right)(\delta \to 0)$, $0 < \alpha < k$, *then the uniform curvilinear modulus of smoothness* $\widetilde{\omega}_{k,1,\Gamma}(\psi, \delta)$ *of the function* $\psi(w)$ *satisfies the condition*

$$\widetilde{\omega}_{k,1,\Gamma}(\psi, \delta) = \begin{cases} O\left(\delta^{\alpha+1}\right)(\delta \to 0) & if \quad 0 < \alpha < k-1, \\ O\left(\delta^k \log \frac{1}{\delta}\right)(\delta \to 0) & if \quad \alpha = k-1, \\ O\left(\delta^k\right)(\delta \to 0) & if \quad k-1 < \alpha < k. \end{cases}$$

Proof of this theorem is based on Theorems 2.2 and 4.1 and is similar to the proof of Theorem 3.1.

Theorem 4.3. *Let* G_1 *and* G_2 *be the simply connected domains in the complex plane bounded by the smooth Jordan curves* Γ_1 *and* Γ_2. *Let* $\tau_1(s_1)$ *be the angle between the tangent to* Γ_1 *and the positive real axis,* $s_1(\zeta)$ *be* *the arc length on* Γ_1. *Let* $\tau_2(s_2)$ *be the angle between the tangent to* Γ_2 *and the positive real axis,* $s_2(w)$ *be the arc length on* Γ_2. *Let* $w = f(\zeta)$ *be a homeomorphism of the closure* $\overline{G_1}$ *of the domain* G_1 *onto the closure* $\overline{G_2}$ *of the domain* G_2, *conformal in the domain* G_1.

Let moduli of smoothness $\omega_k(\tau_1(s_1),\delta)$ *and* $\omega_k(\tau_2(s_2),\delta)$ *of order* k $(k \in N)$ *for the functions* $\tau_1(s_1)$ *and* $\tau_2(s_2)$ *satisfy Holder condition with the same index* α, $0 < \alpha < k$.

Then the uniform curvilinear modulus of smoothness $\widetilde{\omega}_{k,1,\Gamma_1}(f,\delta)$ *of the function* $f(\zeta)$ *on* Γ_1 *satisfies the condition*

$$\widetilde{\omega}_{k,1,\Gamma_1}(f,\delta) = \begin{cases} O\left(\delta^{\alpha+1}\right)(\delta \to 0) & if \quad 0 < \alpha < k-1, \\ O\left(\delta^k \log \frac{1}{\delta}\right)(\delta \to 0) & if \quad \alpha = k-1, \\ O\left(\delta^k\right)(\delta \to 0) & if \quad k-1 < \alpha < k. \end{cases}$$

Proof of this theorem is based on Theorems 3.1 and 4.2 and on estimates of finite difference smoothnesses for composite function.

References

1. P. M. Tamrazov, *Smoothnesses and polynomial appoximations*, Kiev: Naukova dumka, 1975 [in Russian].
2. E. P. Dolzenko, Smoothness of harmonic and analytic functions at boundary points of domain, *Dokl., Acad. Nauk SSSR*, **29** (1965), no 5, 1069-1084 [in Russian].
3. E. P. Dolzenko, Notes on modulus of continuity of the conformal mapping of the unit disk onto the Jordan domain, *Mat. Zametki*, **60** (1996), no 2, 176-184 [in Russian].
4. E. P. Dolzenko, On boundary smoothness of conformal mappings for domains with nonsmooth boundary, *Russian Acad. Nauk Dokl.*, **415** (2007), no 2, 155-159 [in Russian].
5. P. M. Tamrazov, *Finite difference identities and estimates for moduli of smoothness of composite functions*, Kiev: Institute of mathematics of Ukrainian Academy of sciences Publishers, 1977 [in Russian].
6. O. W. Karupu, On moduli of smoothness of conformal mappings, *Ukr. Math. J.*, **30** (1978), no 4, 540-545 [in Russian].
7. O. W. Karupu, On some chracteristics of conformal mapping, in: Aviation in the XXI century, *Proceedings of the second world congress*, Ukraine, Kiev, 19-21 September, 2005, **5**, (2005), NAU Publishers, 586-591.
8. O. W. Karupu, Finite difference properties of conformal mappings, *Acad. Nauk Ukrain. Inst. Mat. Works*, **3** (2006), no 4, 175-180 [in Ukranian].
9. O. W. Karupu, On properties of moduli of smoothness of conformal mappings, in: Complex Analysis and Potential Theory, *Proceedings of the Conference Satellite to ICM 2006*, 231-238, World Scientific, Singapore, 2007.

THE JUMP PROBLEM ON NONRECTIFIABLE CURVES AND METRIC DIMENSION

B. A. KATS*

Chair of Mathematics, Kazan State Architecture and Building University,
Zelenaya Street, 1, Kazan, Tatarstan, 430043, Russian Federation
E-mail: architec@mi.ru

We introduce a new version of metric dimension for plane compact sets and apply it to solving the so-called jump problem for holomorphic functions on open nonrectifiable arcs.

Keywords: Jump problem, Nonrectifiable curve, Metric dimension.

Introduction

The jump problem is a well-known boundary value problem for holomorphic functions. We cite its formulation for the case of an open arc. Let Γ be an arc on the complex plane $\mathbb{C}$ with beginning at a point a_1 and end point a_2. Let a function $f(t)$ be defined on this arc. We seek a holomorphic in $\overline{\mathbb{C}} \setminus \Gamma$ function $\Phi(z)$, which has limit values $\Phi^+(t)$ and $\Phi^-(t)$ from the left and and from the right correspondingly at any point $t \in \Gamma^\circ := \Gamma \setminus \{a_1, a_2\}$ and satisfies relation

$$\Phi^+(t) - \Phi^-(t) = f(t), t \in \Gamma^\circ; \tag{1}$$

in addition, we assume that $\Phi(\infty) = 0$.

Let function f satisfy the Hölder condition

$$\sup\{\frac{|f(t') - f(t'')|}{|t' - t''|^\nu} : t', t'' \in \Gamma, t' \neq t''\} := h_\nu(f, \Gamma) < \infty \tag{2}$$

with certain exponent $\nu \in (0, 1]$. We denote by $H_\nu(\Gamma)$ the set of all defined on Γ functions satisfying (2).

*The author was supported in part by RFBR Grant #07-01-00166-a

It is well known (see [1,2]), that the jump problem on a piecewise-smooth arc Γ has solution

$$\Phi(z) = \frac{1}{2\pi i}\int_{\Gamma} \frac{f(\zeta)d\zeta}{\zeta - z}. \tag{3}$$

If the jump $f(t)$ satisfies the Hölder condition (2), then the Cauchy integral (3) over piecewise - smooth arc has continuous boundary values on Γ° from both sides, and difference of these values equals to f.

Then the restrictions on the arc Γ were weakened. Thus, E. M. Dyn'kin [3] and T. Salimov [4] proved independently that for closed non-smooth rectifiable arc the boundary values of the Cauchy integral (3) exist and are continuous if $f \in H_\nu(\Gamma)$ for $\nu > 1/2$. E. M. Dyn'kin [3] proved also that this result cannot be improved. For any $\nu \in (0, 1/2]$ he constructs a rectifiable curve Γ and a function $f(t) \in H_\nu(\Gamma)$ such that the Cauchy integral (3) has not boundary values at a point of integration path. Thus, the jump problem is resolvable for rectifiable arcs under restriction $\nu > 1/2$, and this restriction cannot be weakened on the whole class of rectifiable curves. R. K. Seifullaev [5] obtained important results concerning the jump problem on open non-smooth rectifiable arcs.

Then the author [6] proved that for any (in general, nonrectifiable) simple closed curve Γ and for any function $f \in H_\nu(\Gamma)$ such that

$$\nu > \frac{1}{2}\operatorname{Dm}\Gamma \tag{4}$$

there exists a holomorphic in $\overline{\mathbb{C}} \setminus \Gamma$ function $\Phi(z)$ satisfying relation (1) on the whole curve Γ. Generally speaking, it cannot be written as the Cauchy integral (3). Here $\operatorname{Dm}\Gamma$ is well known [7] upper metric dimension, i.e.

$$\operatorname{Dm}\Gamma = \limsup_{\varepsilon \to 0} \frac{\log N(\varepsilon;\Gamma)}{-\log \varepsilon},$$

where $N(\varepsilon;\Gamma)$ is the least number of disks of diameter ε covering Γ. The upper metric dimension of rectifiable curve equals to 1, i.e. for rectifiable curves the condition (4) turns into the Dyn'kin - Salimov condition $\nu > 1/2$. Then the author [8] obtained analogs of these results for open arcs.

The restriction (4) cannot be weakened, too. If $0 < \nu \leq d/2 < 1$, then there exists a curve Γ a function $f \in H_\nu(\Gamma)$ such that $\operatorname{Dm}\Gamma = d$ and the jump problem (1) has not solutions [6]. But the conditions for solvability of the jump problem can be improved by means of another metric characteristics of Γ.

The author [9] has introduced new characteristics (so called refined metric dimension) for closed curves and has proved that this dimension allows us to weaken the solvability conditions for the jump problem on that curves.

In the present paper we define the refined metric dimension of arbitrary compact set on the complex plane. Then we use this dimension for improvement of conditions of solvability of the jump problem on open arcs.

1. Refined metric dimension of plane compact sets

We use a concept of decomposition from the paper [9]. Let $\delta_0, \delta_1, \delta_2, \dots$ be finite simply connected domains, and let $s_0, s_1, s_2, \dots$ be integer numbers such that $s_0 = +1$ and s_j for $j \geq 1$ equals either to $+1$ or to -1. We denote by Δ a sequence of pairs $\{\delta_j, s_j\}, j = 0, 1, 2, \dots,$ and define partial sums Δ_n of this sequence as follows: $\Delta_0 = \delta_0$; if $n \geq 1$, then we put Δ_n equaling to interior of union $\overline{\Delta_{n-1}} \cup \overline{\delta_n}$ for $s_n = +1$ and to difference $\Delta_{n-1} \setminus \overline{\delta_n}$ for $s_n = -1$. Everywhere below we suppose that $\delta_n \cap \Delta_{n-1} = \emptyset$ for $s_n = +1$ and $\delta_n \subset \Delta_{n-1}$ for $s_n = -1$. If the sequence Δ is infinite, then we denote by $\Delta_\infty = \sum_{j=0}^{\infty} s_j \delta_j$ a set of all points $z \in \mathbf{C}$ such that $z \in \Delta_j$ for all j beginning from certain $j = j(z)$.

We say that a sequence Δ is decomposition of a domain D if $D = \Delta_m$, where m is number of pairs in this sequence (it is either finite integer or infinity). We call a decomposition rectifiable if all domains δ_j have rectifiable boundaries. Let $\lambda(\delta)$ stand for a length of rectifiable boundary of domain δ, and $w(\delta)$ is the most diameter of open disk which can be embedded in δ.

Definition 1.1. Let Γ be a compact set on the complex plane. Let Q be a domain with rectifiable boundary (for instance, a square) such that $\Gamma \subset \overline{Q}$ and the set $Q \setminus \Gamma$ is not connected. Let $Q_1, Q_2, \dots$ be connected components of $Q \setminus \Gamma$. The set $e(\Gamma)$ consists of all real numbers $p \geq 1$ such that each of components $Q_1, Q_2, \dots$ has rectifiable decomposition $\Delta = \{\{\delta_j, s_j\}, 0 \leq j \leq m\}$ such that sum $\sigma(\Delta) := \sum_{j=0}^{m} \lambda(\delta_j) w^{p-1}(\delta_j)$ is finite.

Then the value $\operatorname{rdm} \Gamma := \inf e(\Gamma)$ is called refined metric dimension of the compact Γ.

Theorem 1.1. *If Γ is a continuum, then*

$$1 \leq \operatorname{rdm} \Gamma \leq \operatorname{Dm} \Gamma \leq 2. \tag{5}$$

In addition, for any $d \in (1, 2)$ there exists an open arc $\mathfrak{G}$ such that $\operatorname{rdm} \mathfrak{G} < d = \operatorname{Dm} \mathfrak{G}$.

If Γ is closed curve, then definition 1.1 coincides with Definition 1 from the paper [9]. The theorem 1.1 is a generalization of concerning closed curves

Theorem 1 from the same paper. The proof of relation (5) is analogous. Here we have to construct an arc $\mathfrak{G}$, because the corresponding curve in [9] is closed.

Let a value β be defined by equality $\frac{2\beta}{\beta+1} = d$. We put $K_n = 2^{[n\beta]}$, where square brackets mean entire part, and divide the segment $[2^{-n}, 2^{-n+1}]$ of real axis into K_n equal parts of length $\alpha_n = 2^{-n}/K_n$ each one. We denote by $x_{n,j}$ the ends of these parts, i.e. $x_{n,j} = 2^{-n} + j\alpha_n, j = 0, 1, \dots, 2^{[n\beta]} - 1$, and consider vertical segments $I_{n,j} = [x_{n,j}, x_{n,j} + i2^{-n}]$. As shown in [6], the upper metric dimension of the union $A = \cup_{n=1}^{\infty} \cup_{j=0}^{2^{[n\beta]}-1} I_{n,j}$ equals to $\frac{2\beta}{\beta+1}$, i.e. $\operatorname{Dm} A = d$.

Then we put $\varepsilon_n = \alpha_n^{\mu}$, where $\mu > 1$, and consider mutually disjoint rectangles $\delta_{n,j} = \{z = x+iy : x_{n,j} < x < x_{n,j}+\varepsilon_n, 0 < y < 2^{-n}\}$. Let δ_0 be square $\{z = x+iy : 0 < x < 1, 0 < y < 1\}$ and $Q_1 \equiv \delta_0 \setminus (\cup_{n=1}^{\infty} \cup_{j=0}^{2^{[n\beta]}-1} \delta_{n,j})$, i.e. Q_1 is unit square with countable set of rectangular cuts condensing to origin. The boundary of domain Q_1 consists of three unit segments $[0, i]$, $[i, 1+i]$, $[1, 1+i]$, and broken line connecting the points 0 and 1. We denote this line by $\mathfrak{G}$. Obviously, $A \subset \mathfrak{G}$. Easy estimation shows that $\operatorname{Dm} \mathfrak{G} = \operatorname{Dm} A = d$.

Let us obtain an upper bound for $\operatorname{rdm} \mathfrak{G}$. We put $Q = \{z = x + iy; 0 < x < 1, -1 < y < 1\}$. Then the set $Q \setminus \mathfrak{G}$ consists of two components Q_1 and $Q_2 = Q \setminus Q_1$. The pairs $\{\delta_0, +1\}, \{\delta_{n,j}, -1\}, n = 1, 2, \dots, j = 0, 1, \dots, 2^{[n\beta]} - 1$ compose rectifiable decomposition Δ of domain Q_1. We have $\lambda(\delta_{n,j}) \sim 2^{-n}$ and $w(\delta_{n,j}) = \varepsilon_n$, hence, the series $\sigma_p(\Delta)$ converges simultaneously with the series $\sum_{n=1}^{\infty} 2^{[n\beta]-n} \varepsilon_n^{p-1} = \sum_{n=1}^{\infty} 2^{[n\beta]-n-\mu(p-1)([n\beta]+n)}$, i.e. for $p > \frac{(\mu+1)\beta+(\mu-1)}{\mu(\beta+1)}$. The domain Q_2 has decomposition Δ' consisting of pairs $\{Q, +1\}, \{\delta_0, -1\}, \{\delta_{n,j}, +1\}, n = 1, 2, \dots, j = 0, 1, \dots, 2^{[n\beta]} - 1$ with the same parameters. Consequently,

$$\operatorname{rdm} \mathfrak{G} \le \frac{(\mu+1)\beta + (\mu-1)}{\mu(\beta+1)}. \tag{6}$$

The right side of (6) is less than $d = \operatorname{Dm} \mathfrak{G}$ for $\beta > 1$ and $\mu > 1$. Thus, the arc $\mathfrak{G}$ proves the last proposition of the theorem.

2. Solvability of the jump problem on open arcs

We consider first a case where end points of an arc Γ belongs to piecewise-smooth boundary of a domain Q such that $\Gamma \subset \overline{Q}$. Then Γ divides Q into two connected components Q_1 and Q_2. If that domain Q exists, then we say that Γ is separating arc.

Let $f^w(\zeta)$ be the Whitney continuation (see, for instance, [10]) of a jump $f \in H_\nu(\Gamma)$ on the whole complex plane. The boundary of component Q_1 consists of nonrectifiable arc Γ and piecewise-smooth arc γ, which is a part of ∂Q. Let $\Delta = \{\{\delta_j, s_j\}, 0 \le j < \infty\}$ be rectifiable decomposition of Q_1. We shall seek a solution of the jump problem (1) in the form

$$\Phi(z) = \sum_{j=0}^{\infty} \frac{s_j}{2\pi i} \int_{\partial\delta_j} \frac{f^w(\zeta)d\zeta}{\zeta - z} - \frac{1}{2\pi i} \int_\gamma \frac{f^w(\zeta)d\zeta}{\zeta - z} \tag{7}$$

We represent the Cauchy integrals over curves $\partial\delta_j$ in the form

$$\frac{1}{2\pi i} \int_{\partial\delta_j} \frac{f^w(\zeta)d\zeta}{\zeta - z} = f^w(z)\chi_j(z) - \frac{1}{2\pi i} \int\int_{\delta_j} \frac{\partial f^w}{\partial\bar\zeta} \frac{d\zeta \wedge d\bar\zeta}{\zeta - z}.$$

Here $\chi_j(z)$ is characteristic function of the domain δ_j. By means of the Whitney decomposition [10], we can prove the following.

Lemma 2.1. *If δ is finite domain with rectifiable boundary, $f \in H_\nu(\gamma)$ and $p < \frac{1}{1-\nu}$, then*

$$\int\int_\delta |\nabla f^w|^p \, dxdy \le C h_\nu^p(f, \gamma) \lambda(\delta) w^{1-p(1-\nu)}(\delta),$$

where C is absolute constant.

By virtue of the lemma the series (7) converges and gives a solution of the boundary value problem (1) if

$$\nu > \frac{1}{2} \operatorname{rdm} \Gamma. \tag{8}$$

As the arc γ is piecewise-smooth, since near the points $a_j, j = 1, 2$, we have $\Phi(z) = O(\log|z - a_j|), j = 1, 2$. Thus, the following is valid.

Lemma 2.2. *If Γ is a separating arc and $f \in H_\nu(\Gamma)$, then under restriction (8) the jump problem (1) has a solution such that its singularities at the end points of the arc are logarithmic.*

Now we consider arbitrary nonrectifiable arc. In general, it is not separating. Let $\Gamma_1 \subset \Gamma$ be an arc with beginning and end at points b_1 and b_2, $b_1 \ne b_2$. Boundary of its closed convex hull contains at least one interior point of the arc Γ_1. It is end point of a small segment of a straight line without another common points with Γ. We say that this point is attainable. As we has seen, any non-trivial arc of Γ contains at least one attainable point. Thus, the set of attainable points is everywhere dense on Γ. Hence, Γ is representable as union $\bigcup \Gamma_j$ of countable family of arcs with attainable end points. All these arcs are separating. We denote by $\Phi_j(z)$ a solution of the jump problem on the separating arc Γ_j; it exists by virtue of lemma 2.2. One can show easily, that if Γ_m and Γ_n are two neighboring arcs with common end point, then singularities of functions Φ_m and Φ_n at this point annihilate, i.e. the sum $\Phi_m + \Phi_n$ is solution of the jump problem on arc $\Gamma_m \cup \Gamma_n$. Hence, if series $\sum_{j=1}^{\infty} \Phi_j(z)$ converges, then its sum is solution of the jump problem on the whole arc Γ. In general this series diverges, but we can regularize it. Indeed, there exists a sequence of rational functions $R_j(z)$ with poles at points a_1 and a_2 such that the series $\sum_{j=1}^{\infty}(\Phi_j(z) - R_j(z))$ uniformly converges in domain $\overline{\mathbb{C}}\setminus\Gamma$. These these rational functions exist by virtue of considerations analogous to the proof of Mittag-Leffler theorem. Thus, the following is valid.

Theorem 2.1. *The jump problem (1) on nonrectifiable arc* Γ *under assumption (8) has a solution.*

Unlike lemma 2.2, here we do not obtain any information on behavior of solution near the end points. That behavior is of importance for applications of this boundary value problem (see [1]). In this connection we imply certain estimates for solution near the end points.

We consider a problem on unit jump

$$\Phi^+(t) - \Phi^-(t) = 1, t \in \Gamma^\circ;$$

its solution is function

$$k_\Gamma(z) := \frac{1}{2\pi i} \ln \frac{z - a_2}{z - a_1},$$

where the branch of logarithm is selected by means of cut along Γ and condition $k_\Gamma(\infty) = 0$. At the ends of arc Γ imaginary part of this function has logarithmic singularities. Its real part can be bounded (for instance, if the arc Γ is smooth), but it can have singularities of arbitrary high order if this arc curls into spiral. We shall show that under certain restrictions the jump problem (1) has a solution with the same order of singularities at the ends as k_Γ.

We fix $r > 0$ such that the arc Γ is situated in the disk $|z| < r$, and denote by $\omega(z)$ a smooth function equaling unit for $|z| \le r$ and zero for $|z| \ge 2r$. Let $f^w(z)$ stand for the Whitney extension (see, for instance, [10]) of the function f onto the whole complex plane. Then the product $\psi(z) := k_\Gamma(z)\omega(z)f^w(z)$ is continuous in $\overline{\mathbb{C}}\setminus\Gamma$ and has there partial derivatives with respect to $\mathrm{Re}\, z$ and $\mathrm{Im}\, z$ of all orders. The function ψ has jump f on the arc Γ. In addition, this function has compact support. We seek solution of the problem (1) in the form

$$\Psi(z) := \psi(z) - \frac{1}{2\pi i}\iint_{\mathbb{C}} \frac{\partial\psi}{\partial\overline{\zeta}}\frac{d\zeta\wedge d\overline{\zeta}}{\zeta - z}. \tag{9}$$

According to lemma 2.1, the derivative $\frac{\partial f^w}{\partial\overline{\zeta}}$ is locally integrable with any degree lesser than $(2 - \mathrm{rdm}\,\Gamma)/(1-\nu)$ Thus, under restriction (8) it is integrable with certain degree $p > 2$. Hence, under assumption

$$k_\Gamma(z) = O(|z - a_j|^{-\alpha}), \alpha < 1, j = 1, 2, \tag{10}$$

the function with compact support $\psi_{\overline{\zeta}} := \frac{\partial\psi}{\partial\overline{\zeta}}$ is integrable, what make our definition of $\Psi(z)$ correct. Moreover, under the same assumptions the function $\psi_{\overline{\zeta}}$ is locally integrable with certain degree $p > 2$ outside of arbitrary neighborhoods of ends of the arc Γ. Consequently, (see, for instance, [11]) the integral term of (9) is continuous in the whole plane with possible exception of end points of the arc Γ. In addition, the well known properties of integral operator

$$\varphi(z) \mapsto \frac{1}{2\pi i}\iint_{\mathbb{C}} \frac{\varphi(\zeta)d\zeta\wedge d\overline{\zeta}}{\zeta - z}$$

(see, for instance, [11]) imply that the function $\Psi(z)$ is holomorphic in $\overline{\mathbb{C}}\setminus\Gamma$ and vanishes at infinity point. Thus, under assumptions (8) and (10) the function (9) is a solution of the jump problem (1). It remains to estimate its singularities at the end points of Γ.

The first term $\psi(z)$ has at the points $a_j, j = 1, 2$ singularities of the same order as $k_\Gamma(z)$, i.e., $\psi(z) = O(|z - a_j|^{-\alpha})$. Let p stand for degree of local integrability of $\frac{\partial f^w}{\partial\overline{\zeta}}$. As above, $2 < p < (2 - \mathrm{rdm}\,\Gamma)/(1-\nu)$. Then near the point a_j we have

$$\left|\frac{1}{2\pi i}\iint_{\mathbb{C}} \frac{\partial\psi}{\partial\overline{\zeta}}\frac{d\zeta\wedge d\overline{\zeta}}{\zeta - z}\right|$$

$$< C\left(\iint_{|z|<2r}\left|\frac{\partial f^w}{\partial\overline{\zeta}}\right|^p dxdy\right)^{1/p}\left(\iint_{|z|<2r}\frac{dxdy}{|\zeta - a_j|^{q\alpha}|\zeta - z|^q}\right)^{1/q},$$

where $p^{-1} + q^{-1} = 1$, $\zeta = x + iy$, and C is a positive constant. The bounds for the last integral are well known (see, for instance, [11]): it is bounded for $q\alpha + q < 2$, has logarithmic singularity at the point a_j for $q\alpha + q = 2$, and singularity of order $O(|z - a_j|^{-(q\alpha+q-2)})$ for $q\alpha + q > 2$. As $q < 2$, since maximal order of singularity of the integral term of (9) is $(q\alpha + q - 2)/q = \alpha + 1 - 2/q < \alpha$. Thus, the following is valid.

Theorem 2.2. *If $f \in H_\nu(\Gamma)$, then under restrictions (8) and (10) the problem (1) has a solution Φ such that $\Phi(z) = O(|z - a_j|^{-\alpha}), j = 1, 2$.*

By virtue of theorem 1.1 the restriction (8) is weaker than the known condition (4).

References

1. F. D. Gakhov, *Boundary value problems*, Nauka publishers, Moscow, 1977.
2. N. I. Muskhelishvili, *Singular integral equations*, Nauka publishers, Moscow, 1962.
3. E. M. Dynkin, Smoothness of the Cauchy type integral, *Zapiski nauchn. sem. Leningr. dep. mathem. inst. AN USSR* **92** (1979), 115–133.
4. T. Salimov, A direct bound for the singular Cauchy integral along a closed curve, *Nauchn. Trudy Min. vyssh. i sr. spec. obraz. Azerb. SSR* **5** (1979), 59–75.
5. R. K. Seifullaev, The Riemann boundary value problem on a non-smooth open curve, *Math USSR Sb.* **40** (1981), 135–148.
6. B. A. Kats, The Riemann boundary value problem on closed Jordan curve, *Izv. VUZov, Math.* **4** (1983), 68–80.
7. I. Feder, *Fractals*, Mir Publishers, Moscow, 1991.
8. B. A. Kats, The Riemann boundary value problem on non-smooth arcs and fractal dimensions, *Algebra i Analiz* **6** (1994), 147–171.
9. B. A. Kats, The refined metric dimension with applications, *Comput. Methods Funct. Theory* **7** (2007), 77–89.
10. E. M. Stein, *Singular integrals and differentiability properties of functions*, Mir Publishers, Moscow, 1973.
11. I. N. Vekua, *Generalized analytical functions*, Nauka publishers, Moscow, 1988.

A FUNCTIONAL ANALYTIC APPROACH TO THE ANALYSIS OF THE ASYMPTOTIC BEHAVIOR OF THE ENERGY INTEGRAL OF A NONLINEAR TRANSMISSION PROBLEM

M. LANZA DE CRISTOFORIS

Dipartimento di Matematica Pura ed Applicata,
Università degli Studi di Padova,
Via Trieste 63, Padova 35121, Italy
E-mail: mldc@math.unipd.it

Keywords: Nonlinear transmission problem; singularly perturbed domain; Laplace operator; real analytic continuation.

1. Introduction

In this paper, we shall consider an application of an approach based on potential theory, on functional analysis, and on harmonic analysis, to analyze a nonlinear transmission problem on a domain with a small hole.

We fix $\alpha \in]0,1[$. We consider a subset $\mathbb{I}^o$ of $\mathbb{R}^n$ satisfying the following assumption.

$$\mathbb{I}^o \textit{ is a bounded open connected subset of } \mathbb{R}^n \textit{ of class } C^{1,\alpha} \tag{1}$$
$$\textit{containing } 0\,, \textit{ and its exterior is connected.}$$

For the definition of functions and sets of the usual Schauder class $C^{0,\alpha}$ or $C^{1,\alpha}$, we refer for example to Gilbarg and Trudinger [3, § 6.2]. Then we assign

$$f^o \in C^{1,\alpha}(\partial\mathbb{I}^o)\,. \tag{2}$$

As is well known, the Dirichlet problem

$$\begin{cases} \Delta u = 0 \text{ in } \mathbb{I}^o, \\ u = f^o \quad \text{on } \partial\mathbb{I}^o\,, \end{cases} \tag{3}$$

has one and only one solution $\tilde{u}^o \in C^{1,\alpha}(\mathrm{cl}\mathbb{I}^o)$. Next we perturb singularly problem (3) by making a hole in $\mathbb{I}^o$ around 0. To do so, we consider another subset $\mathbb{I}^i$ of $\mathbb{R}^n$ as in (1) with unit outward normal ν^i, and we consider the annular domain

$$\mathbb{A}(\epsilon) \equiv \mathbb{I}^o \setminus \epsilon \mathrm{cl}\mathbb{I}^i$$

for all $|\epsilon| \leq \epsilon_0$, with

$$\epsilon_0 \equiv \sup\{\theta \in]0, +\infty[: \epsilon \mathrm{cl}\mathbb{I}^i \subseteq \mathbb{I}^o, \ \forall \epsilon \in] - \theta, \theta[\} . \tag{4}$$

Obviously,

$$\partial \mathbb{A}(\epsilon) = (\epsilon \partial \mathbb{I}^i) \cup \partial \mathbb{I}^o .$$

Next we consider a nonlinear transmission problem in the pair of domains $\epsilon\mathbb{I}^i$, $\mathbb{A}(\epsilon)$ for $\epsilon \in]0, \epsilon_0[$. To do so, we assign the following data.

$$\text{An increasing real analytic diffeomorphism } F \text{ of } \mathbb{R} \text{ onto itself}, \tag{5}$$

$$\gamma \in]0, +\infty[, \qquad g \in C^{0,\alpha}(\partial \mathbb{I}^i) . \tag{6}$$

Also, we denote by $F^{(-1)}$ the inverse function of F. Next we consider $\epsilon \in]0, \epsilon_0[$ and the following nonlinear transmission problem

$$\begin{cases} \Delta u^i = 0 & \text{in } \epsilon\mathbb{I}^i \\ \Delta u^o = 0 & \text{in } \mathbb{A}(\epsilon), \\ u^o = F(u^i) & \text{on } \epsilon\partial\mathbb{I}^i , \\ \frac{\partial u^o}{\partial \nu_{\epsilon\mathbb{I}^i}}(x) = \gamma \frac{\partial u^i}{\partial \nu_{\epsilon\mathbb{I}^i}}(x) + g(x/\epsilon) & \forall x \in \ \epsilon\partial\mathbb{I}^i , \\ u^o = f^o & \text{on } \partial\mathbb{I}^o , \end{cases} \tag{7}$$

where $\nu_{\epsilon\mathbb{I}^i}$ denotes the outward unit normal to $\epsilon\partial\mathbb{I}^i$. Problems as (7) arise in the study of heat conduction and of composite materials (see Mityushev and Rogosin [13, Ch. 5]). Under reasonable conditions on the data, one can prove existence of a solution for a problem as (7) (see Wegert, Khimshiashvili, and Spitkovsky [23], Pflüger [16].) Here, we do not aim at investigating the optimal conditions of existence of solutions for (7). Instead, by invoking [8, § 6, 7], one knows that possibly shrinking ϵ_0, there is a unique family $\{(u^i(\epsilon, \cdot), u^o(\epsilon, \cdot))\}_{\epsilon \in]0,\epsilon_0[}$ of solutions of (7), close to the pair $(F^{(-1)}(\tilde{u}^o(0)), \tilde{u}^o)$ in a sense which we do not clarify here, and we are interested in analyzing the behavior of the energy integrals of $u^i(\epsilon, \cdot)$ and $u^o(\epsilon, \cdot)$ as ϵ degenerates to 0. Linear singular perturbation problems in a domain with a small hole have long been investigated with the methods of asymptotic analysis, which aims at giving complete asymptotic expansions of the solutions in terms of the parameter ϵ. Thus in the case of our problem, one may try to resort to asymptotic analysis and show that there exist suitable real coefficients a_j, $j = 1, \ldots, r$ such that

$$\int_{\epsilon\mathbb{I}^i} |Du^i(\epsilon, x)|^2 \, dx = \sum_{j=0}^{r} a_j \epsilon^j + o(\epsilon^r) \qquad \text{as } \epsilon \to 0^+ \,, \tag{8}$$

and as we shall see, one may succeed to do so for $n \geq 3$. It is perhaps difficult to provide a complete list of the contributions in this direction. Here, we mention the work of Kozlov, Maz'ya and Movchan [4], Maz'ya, Nazarov and Plamenewskii [12], Movchan [14], Ozawa [15], Vogelius and Volkov [19], Ward and Keller [22]. For nonlinear problems on domains with small holes far less seems to be known. We mention the results which concern the existence of a limiting value of the solutions or of their energy integral as the holes degenerate to points, as those of Ball [1], Sivaloganathan, Spector and Tilakraj [17], and the vast literature of homogenization theory (cf. Dal Maso and Murat [2]). We also mention the computation of the expansions in the case of quasilinear equations of Titcombe and Ward [18], Ward, Henshaw and Keller [20], Ward and Keller [21]. Instead, in order to analyze the behavior of the energy integral $\int_{\epsilon\mathbb{I}^i} |Du^i(\epsilon, x)|^2 \, dx$, we would try to show that the real valued function $\int_{\epsilon\mathbb{I}^i} |Du^i(\epsilon, x)|^2 \, dx$ of ϵ, which has been defined only for ϵ in a right neighborhood of 0 can be continued real analytically also for ϵ negative and small, or in other words that the energy integral of $u^i(\epsilon, \cdot)$ coincides for ϵ positive and small with a real analytic function defined in a whole neighborhood of 0. Thus we can say that we would try to represent the energy integral of $u^i(\epsilon, \cdot)$ by means of a real analytic map. We observe that our approach does have its advantages. Indeed, if we succeed (and we do for $n \geq 3$), then we can assert that $\int_{\epsilon\mathbb{I}^i} |Du^i(\epsilon, x)|^2 \, dx = \sum_{j=0}^{\infty} a_j \epsilon^j$ for suitable coefficients a_j and ϵ small enough, as a convergent series. Now in general, and in particular for our problem with $n = 2$, one cannot expect that the energy integral can be expanded in a power series of ϵ. Thus more generally, our goal will be to represent the energy integral in terms of real analytic functions, and of possibly singular but known functions of ϵ such as $\log \epsilon$, ϵ^{-1}, *etc.*

Such a project has been carried out for suitable nonlinear operators associated to the dependence of the conformal representation of $\mathbb{A}(\epsilon)$ in the planar case (see [5], [6]), and for the linear Dirichlet problem for the Laplace and for the Poisson equation (see [9], [10], [11]), where the dependence has been considered upon the complex of variables determined by ϵ, and by global charts of $\partial\mathbb{I}^i$, $\partial\mathbb{I}^o$, and for other nonlinear problems (see [7].)

2. Existence of a family of solutions and corresponding functional representation around $\epsilon = 0$

We first introduce the following theorem, which asserts the existence of a family of solutions.

Theorem 2.1. *Let* $\alpha \in]0,1[$. *Let* $\mathbb{I}^i$, $\mathbb{I}^o$, f^o, ϵ_0, F, γ, g *be as in (1), (2), (4)–(6). Then there exist* $\epsilon'' \in]0, \inf\{\epsilon_0, 1\}[$, *and a family* $\{(u^i(\epsilon,\cdot), u^o(\epsilon,\cdot))\}_{\epsilon\in]0,\epsilon''[}$ *such that* $(u^i(\epsilon,\cdot), u^o(\epsilon,\cdot))$ *belongs to* $C^{1,\alpha}(\epsilon \mathrm{cl}\mathbb{I}^i) \times C^{1,\alpha}(\mathrm{cl}\mathbb{A}(\epsilon))$ *and solves (7) for all* $\epsilon \in]0, \epsilon''[$.

For a proof of Theorem 2.1, we refer to [8, § 5]. Then we can show that for ϵ sufficiently small, the pair $(u^i(\epsilon,\cdot), u^o(\epsilon,\cdot))$ is the unique solution of (7) which is close to the pair $(F^{(-1)}(\tilde{u}^o(0)), \tilde{u}^o)$ in a sense which we do not clarify here (see [8, § 7].)

In order to illustrate the limiting behavior of our family of solutions, we need to introduce the following auxiliary '*limiting boundary value problem*'.

$$
\begin{cases}
\Delta \tilde{u}_1^{i,r} = 0 & \text{in } \mathbb{I}^i\,,\\
\Delta \tilde{u}_1^{o,r} = 0 & \text{in } \mathbb{R}^n \setminus \mathrm{cl}\mathbb{I}^i\,,\\
\Delta \tilde{u}_1^{o} = 0 & \text{in } \mathbb{I}^o\,,\\
F'(F^{(-1)}(\tilde{u}^o(0)))\tilde{u}_1^{i,r}(t) = \tilde{u}_1^{o,r}(t) & \forall t \in \partial\mathbb{I}^i\,,\\
g(t) + \gamma \frac{\partial \tilde{u}_1^{i,r}}{\partial \nu^i}(t) = \frac{\partial \tilde{u}_1^{o,r}}{\partial \nu^i}(t) & \forall t \in \partial\mathbb{I}^i\,,\\
\tilde{u}_1^{o} = 0 & \text{on } \partial\mathbb{I}^o\,,\\
\sup_{x\in\mathbb{R}^n\setminus\mathbb{I}^i} |\tilde{u}_1^{o,r\sharp}(x)|\,|x|^{n-2} < \infty & \text{if } n \geq 3\,,\\
\lim_{x\to\infty} |\tilde{u}_1^{o,r\sharp}(x)|\,|x|^{n-2} = 0 & \text{if } n = 2\,,
\end{cases}
\tag{9}
$$

where we have set

$$
\tilde{u}_1^{o,r\sharp}(x) \equiv \tilde{u}_1^{o,r}(x) - D\tilde{u}^o(0)\cdot x
$$
$$
-\delta_{2,n}\int_{\partial\mathbb{I}^i} g\,d\sigma\left[\int_{\partial\mathbb{I}^o} S_n(s)\rho(s)\,d\sigma_s + \frac{1}{2\pi}\log|x|\right] \qquad \forall x \in \mathbb{R}^n \setminus \mathbb{I}^i\,,
$$

where $\rho \in C^{0,\alpha}(\partial\mathbb{I}^o)$ is the unique solution of the integral equation

$$
\frac{1}{2}\rho(t) + \int_{\partial\mathbb{I}^o} \frac{\partial}{\partial\nu^o(t)}\left(S_n(t-s)\right)\rho(s)\,d\sigma_s = -\nu^o(t)\cdot DS_n(t) \qquad \forall t \in \partial\mathbb{I}^o\,,
$$

and where

$$
S_n(\xi) \equiv \begin{cases}
\frac{1}{s_n}\log|\xi| & \forall \xi \in \mathbb{R}^n\setminus\{0\},\ \text{if } n = 2\,,\\
\frac{1}{(2-n)s_n}|\xi|^{2-n} & \forall \xi \in \mathbb{R}^n\setminus\{0\},\ \text{if } n > 2\,,
\end{cases}
\tag{10}
$$

where s_n denotes the $(n-1)$ dimensional measure of the unit sphere in $\mathbb{R}^n$, and where $\delta_{2,n} = 1$ if $n = 2$, $\delta_{2,n} = 0$ if $n \geq 3$. Problem (9) can be shown to have one and only one solution $(\tilde{u}_1^{i,r}, \tilde{u}_1^{o,r}, \tilde{u}_1^o)$ in $C^{1,\alpha}(\mathrm{cl}\mathbb{I}^i) \times C^{1,\alpha}(\mathbb{R}^n \setminus \mathbb{I}^i) \times C^{1,\alpha}(\mathrm{cl}\mathbb{I}^o)$ (see [8, § 5].) Moreover, $\tilde{u}_1^o = 0$. Then we have the following statement, which clarifies the behavior of our family of solutions for ϵ small and positive.

Theorem 2.2. *Let $n \geq 3$. Let the assumptions of Theorem 2.1 hold. Then the following statements hold.*

(i) There exist a real analytic map $U_{n,1}^{i,r}[\cdot]$ of $]-\epsilon'', \epsilon''[$ to $C^{1,\alpha}(\mathrm{cl}\mathbb{I}^i)$ such that

$$u^i(\epsilon, \epsilon t) = F^{(-1)}(\tilde{u}^o(0)) + \epsilon U_{n,1}^{i,r}[\epsilon](t) \qquad \forall t \in \mathrm{cl}\mathbb{I}^i\,,$$

for all $\epsilon \in]0, \epsilon''[$. Moreover, $U_{n,1}^{i,r}[0] = \tilde{u}_1^{i,r}$.

(ii) Let $\tilde{\Omega}$ be a bounded open subset of $\mathbb{R}^n \setminus \mathrm{cl}\mathbb{I}^i$. Then there exist $\epsilon_{\tilde{\Omega},1} \in]0, \epsilon''[$ and a real analytic map $U_{n,\tilde{\Omega},1}^{o,r}[\cdot]$ of $]-\epsilon_{\tilde{\Omega},1}, \epsilon_{\tilde{\Omega},1}[$ to $C^{1,\alpha}(\mathrm{cl}\tilde{\Omega})$ such that $\mathrm{cl}\tilde{\Omega} \subseteq \frac{1}{\epsilon}\mathrm{cl}\mathbb{A}(\epsilon)$ for all $\epsilon \in]-\epsilon_{\tilde{\Omega},1}, \epsilon_{\tilde{\Omega},1}[\setminus\{0\}$ and such that

$$u^o(\epsilon, \epsilon t) = \tilde{u}^o(0) + \epsilon U_{n,\tilde{\Omega},1}^{o,r}[\epsilon](t) \qquad \forall t \in \mathrm{cl}\tilde{\Omega}\,,$$

for all $\epsilon \in]0, \epsilon_{\tilde{\Omega},1}[$. Moreover, $U_{n,\tilde{\Omega},1}^{o,r}[0]$ equals the restrictions of $\tilde{u}_1^{o,r}$ to $\mathrm{cl}\tilde{\Omega}$.

(iii) Let $\tilde{\Omega}$ be a bounded open subset of $\mathbb{I}^o \setminus \{0\}$ such that $0 \notin \mathrm{cl}\tilde{\Omega}$. Then there exist $\epsilon_{\tilde{\Omega}} \in]0, \epsilon''[$ and a real analytic map $U_{n,\tilde{\Omega},1}^{o}[\cdot]$ of $]-\epsilon_{\tilde{\Omega}}, \epsilon_{\tilde{\Omega}}[$ to $C^{1,\alpha}(\mathrm{cl}\tilde{\Omega})$ such that $\mathrm{cl}\tilde{\Omega} \subseteq \mathrm{cl}\mathbb{A}(\epsilon)$ for all $\epsilon \in]-\epsilon_{\tilde{\Omega}}, \epsilon_{\tilde{\Omega}}[$ and such that

$$u^o(\epsilon, \cdot)_{|\mathrm{cl}\tilde{\Omega}} = \tilde{u}^o_{|\mathrm{cl}\tilde{\Omega}}(\cdot) + \epsilon U_{n,\tilde{\Omega},1}^{o}[\epsilon](\cdot) \qquad \text{if } \epsilon \in]0, \epsilon_{\tilde{\Omega}}[\,.$$

Moreover, $U_{n,\tilde{\Omega},1}^{o}[0]$ equals the restriction of $\tilde{u}_1^o$ to $\mathrm{cl}\tilde{\Omega}$, i.e, it equals 0.

For the proof of Theorem 2.2, we refer to [8, § 6]. In the two-dimensional case instead, we have the following result.

Theorem 2.3. *Let $n = 2$. Let the assumptions of Theorem 2.1 hold. Then the following statements hold.*

(i) There exist ϵ_1'', $\epsilon_2'' \in]0, +\infty[$ and a real analytic map $U_{2,1}^{i,r}[\cdot,\cdot,\cdot]$ of the set $]-\epsilon'', \epsilon''[\times]-\epsilon_1'', \epsilon_1''[\times\,]-\epsilon_2'', \epsilon_2''[$ to $C^{1,\alpha}(\mathrm{cl}\mathbb{I}^i)$ such that

$$\log^{-1}\epsilon \in]-\epsilon_1'', \epsilon_1''[, \qquad \epsilon\log^2\epsilon \in]-\epsilon_2'', \epsilon_2''[\qquad \forall \epsilon \in]0, \epsilon''[\,,$$

and

$$u^i(\epsilon, \epsilon t) = F^{(-1)}(\tilde{u}^o(0)) + \frac{(\epsilon\log\epsilon)\int_{\partial\mathbb{I}^i} g\, d\sigma}{2\pi F'(F^{(-1)}(\tilde{u}^o(0)))}$$
$$+ \epsilon U_{2,1}^{i,r}[\epsilon, \log^{-1}\epsilon, \epsilon\log^2\epsilon](t)$$

for all $t \in \mathrm{cl}\mathbb{I}^i$ *and for all* $\epsilon \in]0, \epsilon''[$. *Moreover,* $U^{i,r}_{2,1}[0,0,0] = \tilde{u}^{i,r}_1$.

(ii) *Let* $\tilde{\Omega}$ *be a bounded open subset of* $\mathbb{R}^n \setminus \mathrm{cl}\mathbb{I}^i$. *Then there exist* $\epsilon_{\tilde{\Omega},1} \in]0, \epsilon''[$ *and a real analytic map* $U^{o,r}_{2,\tilde{\Omega},1}[\cdot,\cdot,\cdot]$ *of* $]-\epsilon_{\tilde{\Omega},1}, \epsilon_{\tilde{\Omega},1}[\times]-\epsilon''_1, \epsilon''_1[\times]-\epsilon''_2, \epsilon''_2[$ *to* $C^{1,\alpha}(\mathrm{cl}\tilde{\Omega})$ *such that* $\mathrm{cl}\tilde{\Omega} \subseteq \frac{1}{\epsilon}\mathrm{cl}\mathbb{A}(\epsilon)$ *for all* $\epsilon \in]-\epsilon_{\tilde{\Omega},1}, \epsilon_{\tilde{\Omega},1}[\setminus\{0\}$ *and such that*

$$u^o(\epsilon, \epsilon t) = \tilde{u}^o(0) + \frac{(\epsilon \log \epsilon)}{2\pi} \int_{\partial \mathbb{I}^i} g \, d\sigma + \epsilon U^{o,r}_{2,\tilde{\Omega},1}[\epsilon, \log^{-1}\epsilon, \epsilon \log^2 \epsilon](t)$$

forall $t \in \mathrm{cl}\tilde{\Omega}$ *and for all* $\epsilon \in]0, \epsilon_{\tilde{\Omega},1}[$. *Moreover,* $U^{o,r}_{2,\tilde{\Omega},1}[0,0,0]$ *equals the restriction of* $\tilde{u}^{o,r}_1$ *to* $\mathrm{cl}\tilde{\Omega}$.

(iii) *Let* $\tilde{\Omega}$ *be a bounded open subset of* $\mathbb{I}^o \setminus \{0\}$ *such that* $0 \notin \mathrm{cl}\tilde{\Omega}$. *Then there exist* $\epsilon_{\tilde{\Omega}} \in]0, \epsilon''[$ *and a real analytic map* $U^o_{2,\tilde{\Omega},1}[\cdot,\cdot,\cdot]$ *of the set* $]-\epsilon_{\tilde{\Omega}}, \epsilon_{\tilde{\Omega}}[\times]-\epsilon''_1, \epsilon''_1[\times]-\epsilon''_2, \epsilon''_2[$ *to* $C^{1,\alpha}(\mathrm{cl}\tilde{\Omega})$ *such that* $\mathrm{cl}\tilde{\Omega} \subseteq \mathrm{cl}\mathbb{A}(\epsilon)$ *for all* $\epsilon \in]-\epsilon_{\tilde{\Omega}}, \epsilon_{\tilde{\Omega}}[$ *and such that*

$$u^o(\epsilon, \cdot)_{|\mathrm{cl}\tilde{\Omega}} = \tilde{u}^o_{|\mathrm{cl}\tilde{\Omega}}(\cdot) + \epsilon U^o_{2,\tilde{\Omega},1}[\epsilon, \log^{-1}\epsilon, \epsilon \log^2 \epsilon](\cdot) \qquad \text{if } \epsilon \in]0, \epsilon_{\tilde{\Omega}}[\,.$$

Moreover, $U^o_{2,\tilde{\Omega},1}[0,0,0]$ *equals the restriction of* $\tilde{u}^o_1$ *to* $\mathrm{cl}\tilde{\Omega}$, *i.e,* 0.

For the proof of Theorem 2.3, we refer to [8, § 6].

3. Functional analytic representation for energy integrals

We now turn to analyze the behavior of the energy integrals of $u^i(\epsilon, \cdot)$, $u^o(\epsilon, \cdot)$ as ϵ approaches 0. We first introduce the following elementary Lemma, which can be proved by an immediate application of the Divergence Theorem and of the chain rule for ordinary derivatives.

Lemma 3.1. *Let the assumptions of Theorem 2.1 hold. Let* $\epsilon \in]0, \epsilon''[$. *Then the following identities hold.*

$$\int_{\epsilon \mathbb{I}^i} |Du^i(\epsilon, x)|^2 \, dx = \epsilon^{n-2} \int_{\partial \mathbb{I}^i} u^i(\epsilon, \epsilon s) \frac{\partial}{\partial \nu^i(s)} (u^i(\epsilon, \epsilon s)) \, d\sigma_s \,, \tag{11}$$

and

$$\int_{\mathbb{A}(\epsilon)} |Du^o(\epsilon, x)|^2 \, dx = -\epsilon^{n-2} \int_{\partial \mathbb{I}^i} u^o(\epsilon, \epsilon s) \frac{\partial}{\partial \nu^i(s)} (u^o(\epsilon, \epsilon s)) \, d\sigma_s \tag{12}$$
$$+ \int_{\partial \mathbb{I}^o} u^o(\epsilon, s) \frac{\partial}{\partial \nu^o(s)} (u^o(\epsilon, s)) \, d\sigma_s \,.$$

Now we split our analysis into case $n \geq 3$ and case $n = 2$.

Theorem 3.1. *Let $n \geq 3$. Let the assumptions of Theorem 2.1 hold. Then there exist $\epsilon''' \in]0, \epsilon''[$ and real analytic maps $\mathcal{E}^i_{n,1}[\cdot]$ and $\mathcal{E}^o_{n,1}[\cdot]$ of $]-\epsilon''', \epsilon'''[$ to $\mathbb{R}$ such that*

$$\int_{\epsilon\mathbb{I}^i} |Du^i(\epsilon, x)|^2 \, dx = \epsilon^n \mathcal{E}^i_{n,1}[\epsilon] \qquad \forall \epsilon \in]0, \epsilon'''[, \tag{13}$$

$$\int_{\mathbb{A}(\epsilon)} |Du^o(\epsilon, x)|^2 \, dx = \int_{\mathbb{I}^o} |D\tilde{u}^o(x)|^2 \, dx + \epsilon \mathcal{E}^o_{n,1}[\epsilon] \quad \forall \epsilon \in]0, \epsilon'''[. \tag{14}$$

Furthermore, $\mathcal{E}^i_{n,1}[0] = \int_{\mathbb{I}^i} |D\tilde{u}^{i,r}_1(x)|^2 \, dx$, $\mathcal{E}^o_{n,1}[0] = 0$.

Proof. By Theorem 2.2 (i) and by equality (11) and by the obvious equality $\int_{\partial\mathbb{I}^i} \frac{\partial}{\partial\nu^i} w \, d\sigma = 0$ which holds for all functions which are continuously differentiable in $\mathrm{cl}\mathbb{I}^i$ and harmonic in $\mathbb{I}^i$, we have

$$\int_{\epsilon\mathbb{I}^i} |Du^i(\epsilon, x)|^2 \, dx = \epsilon^n \int_{\mathbb{I}^i} |DU^{i,r}_{n,1}[\epsilon]|^2 \, dx \qquad \forall \epsilon \in]0, \epsilon''[.$$

Thus it is natural to define $\mathcal{E}^i_{n,1}[\epsilon] \equiv \int_{\mathbb{I}^i} |DU^{i,r}_{n,1}[\epsilon]|^2 \, dx$ for all $\epsilon \in]-\epsilon'', \epsilon''[$. Since the gradient is linear and continuous from $C^{1,\alpha}(\mathrm{cl}\mathbb{I}^i)$ to $(C^{0,\alpha}(\mathrm{cl}\mathbb{I}^i))^n$, and the pointwise product is continuous in $C^{0,\alpha}(\mathrm{cl}\mathbb{I}^i)$, and the operation of integration is linear and continuous from $C^{0,\alpha}(\mathrm{cl}\mathbb{I}^i)$ to $\mathbb{R}$, the real analyticity of $U^{i,r}_{n,1}[\cdot]$ and equality $U^{i,r}_{n,1}[0] = u^{i,r}_1$ implies that $\mathcal{E}^i_{n,1}$ satisfies the required properties.

We now turn to consider the energy integral of $u^o(\epsilon, \cdot)$. Let $R > 0$ be such that $\{x \in \mathbb{R}^n : |x| < R\} \equiv \mathbb{B}(0, R) \supseteq \mathrm{cl}\mathbb{I}^i$. By Theorem 2.2 (ii), (iii), there exist $\epsilon''' \in]0, \epsilon''[$ and a real analytic map $U_1[\cdot] \equiv U^{o,r}_{n,\mathbb{B}(0,R)\setminus\mathrm{cl}\mathbb{I}^i,1}[\cdot]$ of $]-\epsilon''', \epsilon'''[$ to $C^{1,\alpha}(\mathrm{cl}\mathbb{B}(0,R) \setminus \mathbb{I}^i)$ and a real analytic map $U_2[\cdot] \equiv U^o_{n,\mathbb{A}(\epsilon''),1}[\cdot]$ of $]-\epsilon''', \epsilon'''[$ to $C^{1,\alpha}(\mathrm{cl}\mathbb{A}(\epsilon''))$ as in Theorem 2.2 (ii), (iii). Then by equality (12), we have

$$\begin{aligned}
\int_{\mathbb{A}(\epsilon)} |Du^o(\epsilon, x)|^2 \, dx &= -\epsilon^{n-2}\Big\{\tilde{u}^o(0) \int_{\partial\mathbb{I}^i} \frac{\partial}{\partial\nu^i(s)}(u^o(\epsilon, \epsilon s)) \, d\sigma_s \\
&\quad + \epsilon^2 \int_{\partial\mathbb{I}^i} U_1[\epsilon] \frac{\partial}{\partial\nu^i} U_1[\epsilon] \, d\sigma\Big\} + \int_{\partial\mathbb{I}^o} \tilde{u}^o \frac{\partial\tilde{u}^o}{\partial\nu^o} \, d\sigma \\
&\quad + \epsilon\Big\{\int_{\partial\mathbb{I}^o} \tilde{u}^o \frac{\partial}{\partial\nu^o} U_2[\epsilon] + U_2[\epsilon] \frac{\partial}{\partial\nu^o} \tilde{u}^o \, d\sigma\Big\} \\
&\quad + \epsilon^2 \int_{\partial\mathbb{I}^o} U_2[\epsilon] \frac{\partial}{\partial\nu^o} U_2[\epsilon] \, d\sigma \qquad \forall \epsilon \in]0, \epsilon'''[.
\end{aligned}$$

Next we note that

$$\epsilon^{n-2}\int_{\partial\mathbb{I}^i}\frac{\partial(u^o(\epsilon,\epsilon s))}{\partial\nu^i(s)}\,d\sigma_s=\int_{\epsilon\partial\mathbb{I}^i}\gamma\frac{\partial u^i(\epsilon,x)}{\partial\nu_{\epsilon\mathbb{I}^i}}+g(x/\epsilon)\,d\sigma_x$$
$$=\epsilon^{n-1}\int_{\partial\mathbb{I}^i}g\,d\sigma\,, \tag{15}$$

and that accordingly, equality (14) holds if we set

$$\mathcal{E}^o_{n,1}[\epsilon]\equiv-\tilde{u}^o(0)\epsilon^{n-2}\int_{\partial\mathbb{I}^i}g\,d\sigma-\epsilon^{n-1}\int_{\partial\mathbb{I}^i}U_1[\epsilon]\frac{\partial}{\partial\nu^i}U_1[\epsilon]\,d\sigma$$
$$+\left[\int_{\partial\mathbb{I}^o}\tilde{u}^o\frac{\partial}{\partial\nu^o}U_2[\epsilon]+U_2[\epsilon]\frac{\partial}{\partial\nu^o}\tilde{u}^o\,d\sigma\right]$$
$$+\epsilon\int_{\partial\mathbb{I}^o}U_2[\epsilon]\frac{\partial}{\partial\nu^o}U_2[\epsilon]\,d\sigma\qquad\forall\epsilon\in]-\epsilon''',\epsilon'''[\,.$$

Since $U_2[0]=\tilde{u}^o_1=0$ in $\mathrm{cl}\mathbb{A}(\epsilon'')$, we have $\mathcal{E}^o_{n,1}[0]=0$. As for $\mathcal{E}^i_{n,1}$, one can easily prove that $\mathcal{E}^o_{n,1}$ is real analytic. □

For case $n=2$, we can prove the following.

Theorem 3.2. *Let $n=2$. Let the assumptions of Theorem 2.1 hold. Then there exist $\epsilon'''\in]0,\epsilon''[$ and real analytic maps $\mathcal{E}^i_{2,1}[\cdot,\cdot,\cdot]$ and $\mathcal{E}^o_{2,1}[\cdot,\cdot,\cdot]$ of $]-\epsilon''',\epsilon'''[\times]-\epsilon''_1,\epsilon''_1[\times\,]-\epsilon''_2,\epsilon''_2[$ to $\mathbb{R}$ such that*

$$\int_{\epsilon\mathbb{I}^i}|Du^i(\epsilon,x)|^2\,dx=\epsilon^2\mathcal{E}^i_{2,1}[\epsilon,\log^{-1}\epsilon,\epsilon\log^2\epsilon]\,, \tag{16}$$

$$\int_{\mathbb{A}(\epsilon)}|Du^o(\epsilon,x)|^2\,dx=\int_{\mathbb{I}^o}|D\tilde{u}^o(x)|^2\,dx+\epsilon\mathcal{E}^o_{2,1}[\epsilon,\log^{-1}\epsilon,\epsilon\log^2\epsilon]\,, \tag{17}$$

for all $\epsilon\in]0,\epsilon'''[$. Furthermore, $\mathcal{E}^i_{2,1}[0,0,0]=\int_{\mathbb{I}^i}|D\tilde{u}^{i,r}_1(x)|^2\,dx$ and $\mathcal{E}^o_{2,1}[0,0,0]=-\tilde{u}^o(0)\int_{\partial\mathbb{I}^i}g\,d\sigma$.

Proof. By arguing as in the proof of Theorem 3.1, and by replacing the use of Theorem 2.2 (i) with that of Theorem 2.3 (i), one can see that we can define

$$\mathcal{E}^i_{2,1}[\epsilon,\epsilon_1,\epsilon_2]\equiv\int_{\mathbb{I}^i}|DU^{i,r}_{2,1}[\epsilon,\epsilon_1,\epsilon_2]|^2\,dx$$

for $\epsilon\in]-\epsilon'',\epsilon''[\times]-\epsilon''_1,\epsilon''_1[\times]-\epsilon''_2,\epsilon''_2[$. We now consider the energy integral of $u^o(\epsilon,\cdot)$. Let R be as in the proof of Theorem 3.1. By Theorem 2.3 (ii), (iii), there exist $\epsilon'''\in]0,\epsilon''[$ and a real analytic map $U_1[\cdot,\cdot,\cdot]\equiv U^{o,r}_{2,\mathbb{B}(0,R)\setminus\mathrm{cl}\mathbb{I}^i,1}[\cdot,\cdot,\cdot]$ of $]-\epsilon''',\epsilon'''[\times]-\epsilon''_1,\epsilon''_1[\times]-\epsilon''_2,\epsilon''_2[$ to $C^{1,\alpha}(\mathrm{cl}\mathbb{B}(0,R)\setminus\mathbb{I}^i)$ and a real analytic map $U_2[\cdot,\cdot,\cdot]\equiv U^o_{2,\mathbb{A}(\epsilon''),1}[\cdot,\cdot,\cdot]$ of

$]-\epsilon''',\epsilon'''[\times]-\epsilon_1'',\epsilon_1''[\times]-\epsilon_2'',\epsilon_2''[$ to $C^{1,\alpha}(\mathrm{cl}\mathbb{A}(\epsilon''))$ as in Theorem 2.3 (ii), (iii). Then by equality (12), we have

$$\int_{\mathbb{A}(\epsilon)} |Du^o(\epsilon,x)|^2\,dx =$$
$$-\Big\{\Big[\tilde{u}^o(0)+\frac{\epsilon\log\epsilon}{2\pi}\int_{\partial\mathbb{I}^i} g\,d\sigma\Big]\int_{\partial\mathbb{I}^i}\frac{\partial}{\partial\nu^i(s)}(u^o(\epsilon,\epsilon s))\,d\sigma_s$$
$$+\epsilon^2\int_{\partial\mathbb{I}^i} U_1[\epsilon,\log^{-1}\epsilon,\epsilon\log^2\epsilon]\frac{\partial}{\partial\nu^i}U_1[\epsilon,\log^{-1}\epsilon,\epsilon\log^2\epsilon]\,d\sigma\Big\}$$
$$+\int_{\partial\mathbb{I}^o}\tilde{u}^o\frac{\partial\tilde{u}^o}{\partial\nu^o}\,d\sigma$$
$$+\epsilon\Big\{\int_{\partial\mathbb{I}^o}\tilde{u}^o\frac{\partial}{\partial\nu^o}U_2[\epsilon,\log^{-1}\epsilon,\epsilon\log^2\epsilon]+U_2[\epsilon,\log^{-1}\epsilon,\epsilon\log^2\epsilon]\frac{\partial}{\partial\nu^o}\tilde{u}^o\,d\sigma\Big\}$$
$$+\epsilon^2\int_{\partial\mathbb{I}^o} U_2[\epsilon,\log^{-1}\epsilon,\epsilon\log^2\epsilon]\frac{\partial}{\partial\nu^o}U_2[\epsilon,\log^{-1}\epsilon,\epsilon\log^2\epsilon]\,d\sigma\quad\forall\epsilon\in]0,\epsilon'''[.$$

By equation (15), we can see that equality (17) holds if we set

$$\mathcal{E}^o_{2,1}[\epsilon,\epsilon_1,\epsilon_2]\equiv-\Big[\tilde{u}^o(0)+\frac{\epsilon_1\epsilon_2}{2\pi}\int_{\partial\mathbb{I}^i} g\,d\sigma\Big]\int_{\partial\mathbb{I}^i} g\,d\sigma$$
$$-\epsilon\int_{\partial\mathbb{I}^i} U_1[\epsilon,\epsilon_1,\epsilon_2]\frac{\partial}{\partial\nu^i}U_1[\epsilon,\epsilon_1,\epsilon_2]\,d\sigma$$
$$+\Big[\int_{\partial\mathbb{I}^o}\tilde{u}^o\frac{\partial}{\partial\nu^o}U_2[\epsilon,\epsilon_1,\epsilon_2]+U_2[\epsilon,\epsilon_1,\epsilon_2]\frac{\partial}{\partial\nu^o}\tilde{u}^o\,d\sigma\Big]$$
$$+\epsilon\int_{\partial\mathbb{I}^o} U_2[\epsilon,\epsilon_1,\epsilon_2]\frac{\partial}{\partial\nu^o}U_2[\epsilon,\epsilon_1,\epsilon_2]\,d\sigma\qquad\forall\epsilon\in]-\epsilon''',\epsilon'''[.$$

Since $U_2[0,0,0]=\tilde{u}^o_1=0$ in $\mathrm{cl}\mathbb{A}(\epsilon'')$, we have $\mathcal{E}^o_{2,1}[0,0,0]=-\tilde{u}^o(0)\int_{\partial\mathbb{I}^i} g\,d\sigma$. As for $\mathcal{E}^i_{2,1}$, one can easily prove that $\mathcal{E}^o_{2,1}$ is real analytic. □

Acknowledgement. The author is indebted to V. V. Mityushev and S. V. Rogosin for several discussion on transmission problems in the frame of the applications, and for pointing out to the author problem (7).

References

1. J. M. Ball, *Discontinuous equilibrium solutions and cavitation in nonlinear elasticity*, Philos. Trans. Roy. Soc. London Ser. A, **306**, (1982), 557–611.
2. G. Dal Maso and F. Murat, *Asymptotic behaviour and correctors for linear Dirichlet problems with simultaneously varying operators and domains* Ann. Inst. H. Poincaré Anal. Non Linéaire, **21**, (2004), 445–486.
3. D. Gilbarg and N. S. Trudinger, *Elliptic Partial Differential Equations of Second Order*, Springer Verlag, 1983.

4. V. A. Kozlov, V. G. Maz'ya and A. B. Movchan, *Asymptotic analysis of fields in multistructures*, Oxford Mathematical Monographs, The Clarendon Press, Oxford university Press, New York, 1999.
5. M. Lanza de Cristoforis, *Asymptotic behavior of the conformal representation of a Jordan domain with a small hole, and relative capacity*, in 'Complex Analysis and Dynamical Systems', Proc. Conf. Karmiel, June 19-22, 2001, edited by M. Agranovsky, L. Karp, D. Shoikhet, and L. Zalcman, Contemp. Math., **364**, (2004) 155-167.
6. M. Lanza de Cristoforis, *Asymptotic behavior of the conformal representation of a Jordan domain with a small hole in Schauder spaces*, Comput. Methods Funct. Theory, **2**, 2002, 1–27.
7. M. Lanza de Cristoforis, *Asymptotic behaviour of the solutions of a nonlinear Robin problem for the Laplace operator in a domain with a small hole: a functional analytic approach*, Complex Var. Elliptic Equ., **52**, (2007), 945–977.
8. M. Lanza de Cristoforis, *Asymptotic behavior of the solutions of a nonlinear transmission problem for the Laplace operator in a domain with a small hole. A functional analytic approach*, typewritten manuscript, (2007).
9. M. Lanza de Cristoforis, *Asymptotic behavior of the solutions of the Dirichlet problem for the Laplace operator in a domain with a small hole. A functional analytic approach*, submitted, (2004).
10. M. Lanza de Cristoforis, *A singular domain perturbation problem for the Poisson equation*, submitted, (2005).
11. M. Lanza de Cristoforis, *A singular perturbation Dirichlet boundary value problem for harmonic functions on a domain with a small hole*, Proceedings of the 12th International Conference on Finite or Infinite Dimensional Complex Analysis and Applications, Tokyo July 27–31, 2004, edited by H. Kazama, M. Morimoto, C. Yang, Kyushu University Press, (2005), 205–212.
12. V. G. Mazya, S. A. Nazarov and B. A. Plamenewskii, *Asymptotic theory of elliptic boundary value problems in singularly perturbed domains*, I, II, translation of the original in German published by Akademie Verlag 1991, Oper. Theory Adv. Appl., 111, 112, Birkhäuser Verlag, Basel, 2000.
13. V. V. Mityushev and S. V. Rogosin, *Constructive methods for linear and nonlinear boundary value problems for analytic functions, theory and applications*, Monographs and surveys in pure and applied mathematics, Chapman & Hall/CRC, London, *etc.*, 2000.
14. A. B. Movchan, *Contributions of V. G. Maz'ya to analysis of singularly perturbed boundary value problems*, The Maz'ya anniversary collection, 1 (Rostock, 1998), Oper. Theory Adv. Appl., **109**, Birkhäuser, Basel, 1999, 201–212.
15. S. Ozawa, *Electrostatic capacity and eigenvalues of the Laplacian*, J. Fac. Sci. Univ. Tokyo Sect. IA Math., **30**, (1983), 53–62 .
16. K. Pflüger, *Nonlinear transmission problems in bounded domains of R^n*, English summary, Appl. Anal. **62**, (1996), 391–403.
17. J. Sivaloganathan, S. J. Spector and V. Tilakraj, *The convergence of regularized minimizers for cavitation problems in nonlinear elasticity*, SIAM J. Appl. Math., **66**, (2006), 736–757.

18. M. S. Titcombe, M. J. Ward, *Summing logarithmic expansions for elliptic equations in multiply-connected domains with small holes*, Canad. Appl. Math. Quart., **7**, (1999), 313-343.
19. M. S. Vogelius and D. Volkov, *Asymptotic formulas for perturbations in the electromagnetic fields due to the presence of inhomogeneities of small diameter*, Math. Model. Numer. Anal., **34**, (2000), 723–748.
20. M. J. Ward, W. Henshaw, and J. Keller, *Summing logarithmic expansions for singularly perturbed eigenvalue problems*, SIAM J. Appl. Math., **53**, (1993), 799-828.
21. M. J. Ward and J. Keller, *Nonlinear eigenvalue problems under strong localized perturbations with applications to chemical reactors*, Stud. Appl. Math., **85**, (1991), 1-28.
22. M. J. Ward and J. B. Keller, *Strong localized perturbations of eigenvalue problems*, SIAM J. Appl. Math., **53**, (1993), 770–798.
23. E. Wegert, G. Khimshiashvili, and I. Spitkovsky, *Nonlinear transmission problems.International Symposium on Differential Equations and Mathematical Physics (Tbilisi, 1997)*, Mem. Differential Equations Math. Phys., **12**, (1997), 223–230.

ANALOGUE OF JACKSON-BERSTEIN THEOREM IN L_p ON CLOSED CURVES IN THE COMPLEX PLANE II

J. I. Mamedkhanov

Institute of Mathematics and Mechanics, Baku, Azerbaijan
E-mail: jamalmamedkhanov@rambler.ru

I. B. Dadashova

Baku State University, Azerbaijan
E-mail: irada-dadashova@rambler.ru

This paper is devoted to the analogue of Jackson-Bernstein classic theorem on curves in the complex plane in the integral metric. In the uniform metric this matter in the form of a problem was formulated by J. Walsh and was investigated in the works of S. Y. Alper, A. I. Markushevich, S. N. Mergelian, J. I. Mamedkhanov and others. In the uniform metric this problem is close to a terminal solution. The similar problem on the curves in a complex plane in the integral metric was not studied enough and in this formulation it is considered for the first time.

Keywords: Jackson-Bernstein theorem; closed curve; polynomial approximation; continuity modulus; complex plane.

In this paper we consider the analogue of a problem of J. Walsh[a] on closed curves in a complex plane in the metric L_p related with Jackson-Bernstein classic theorem for periodic functions from the class $f \in Lip_{[0,\,2\pi]}\alpha$ $(0 < \alpha < 1)$ that affirms that for $f \in Lip_{[0,\,2\pi]}\alpha$ $(0 < \alpha < 1)$, it is necessary and sufficient the fulfillment of the condition

$$E_n\,(f;\,[0,\,2\pi]\,) = \inf_{p_n} \|\, f - P_n \,\|_{C[0,\,2\pi]} \leq const \cdot n^{-\alpha}.$$

Direct theorems of polynomial approximation on closed curves in a complex plane in the metric L_p are considered.

[a] J. Walsh [1] formulated the following problem: What necessary and sufficient conditions the closed curve Γ must satisfy in order that Jackson-Bernstein theorem be valid on it?

We need some results from [2] where a survey of basic stages of investigations related with J. Walsh problem in the metric C is given.

Let the function $u(z)$ be given on a closed rectifiable curve Γ and $L_p(\Gamma, u)$ denote a weight space of functions f with weight u and norm

$$\|f\|_{L_p(\Gamma, u)} = \left(\int_\Gamma | f(z)\, u(z)|^p \, |dz|\right)^{1/p} < \infty \quad (p \geq 1).$$

Let $K(0, a]$ be a set of positive non-decreasing functions determined on $(0, a]$ (a is an arbitrary positive number) and $P(0, a]$ be a set of positive non-increasing functions on $(0, a]$. By $K_0(0, a]$ we denote a set of those functions $f \in K(0, a]$ for which $f(t) \to 0$ as $t \to 0$.

Let the function φ ($\tilde{\varphi}$) map the exterior (interior) of a curve Γ onto the exterior (interior) of a unit circle γ_0 and ψ ($\tilde{\psi}$) be a function inverse to φ ($\tilde{\varphi}$). Assuming Γ to be smooth or piecewise-smooth and $f \in L_p(\Gamma)$, we consider the generalized moduli of continuity

$$z_h = \psi\left(\varphi(z)\, e^{ih}\right), \quad \tilde{z}_h = \tilde{\psi}\left(\tilde{\varphi}(z)\, e^{ih}\right);$$

$$u_p(\delta) = u_p(f, \delta)_\Gamma = \sup_{|h| \leq \delta} \| f(z_h) - f(z) \|_{L_p(\Gamma)}$$

$$\tilde{u}_p(\delta) = \tilde{u}_p(f, \delta)_\Gamma = \sup_{|h| \leq \delta} \| f(\tilde{z}_h) - f(z) \|_{L_p(\Gamma)}.$$

Obviously, these quantities possess all the properties of the ordinary modulus of continuity:

1^0. $\tilde{u}_p(\delta), \ u_p(\delta) \in K_0(0, l]$, where l is the length of the curve Γ;

2^0. The functions u_p and $\tilde{u}_p$ are semiadditive, i.e., $u_p(\delta_1 + \delta_2) \leq u_p(\delta_1) + u_p(\delta_2)$ and $\tilde{u}_p(\delta_1 + \delta_2) \leq \tilde{u}_p(\delta_1) + \tilde{u}_p(\delta_2)$ for $\delta_1, \delta_2 \in (0, l]$;

3^0. $u_p, \tilde{u}_p \in C[0, l]$, i.e., they are continuous on $[0, l]$.

It is easy to show that, on the curves Γ, for which $|\psi'(\omega)| \asymp 1$ or $\psi(\omega) \in H'(\gamma)$ ($H'(\gamma)$ is a Hölder class) a modulus of continuity of the form $u_p(\delta) = u_p(f, \delta)_\Gamma$ is equivalent as $\delta \to 0$ to the modulus of continuity

$$\omega_p^*(f, \delta)_\Gamma = \omega_p(f_0, \delta) = \sup_{|h| \leq \delta} \left\| f_0\left(\omega\, e^{ih}\right) - f_0(\omega) \right\|_{L_p(\gamma_0)},$$

introduced by S. Y. Alper, where $f_0(\omega) = f(\psi(\omega))$, and also to the general modulus of continuity

$$\omega_p(\delta) = \omega_p(f, \delta) = \sup_{|h| \leq \delta} \| f(z(s+h)) - f(z(s)) \|_{L_p(\Gamma)},$$

where $z = z(s)$ is a parametric equation of the curve Γ.

In the case when the curve Γ belongs to a more wide class of curves, the fulfillment of condition 2^0 for the function $u_p\,(\tilde{u}_p)$ is not always possible, when condition 1^0 is fulfilled. In this connection, let us introduce the regularized integral moduli of continuity

$$\omega_p(\delta) = \omega_p(f,\,\delta)_\Gamma = \delta \sup_{t\geq\delta} t^{-1} u_p(f,\,t)_\Gamma$$

$$\tilde{\omega}_p(\delta) = \tilde{\omega}_p(f,\,\delta)_\Gamma = \delta \sup_{t\geq\delta} t^{-1} \tilde{u}_p(f,\,t)_\Gamma$$

whose construction belongs to S. B. Stechkin.

Obviously for these quantities the following statements hold:

a) $u_p(\delta) \leq \omega_p(\delta)$ and $\tilde{u}_p(\delta) \leq \tilde{\omega}_p(\delta)$;

b) $\omega_p(\delta) \in K_0(0,\,l]$ and $\tilde{\omega}_p(\delta) \in K_0(0,\,l]$;

c) $\delta^{-1}\omega_p(\delta) \in P(0,\,l]$ and $\delta^{-1}\tilde{\omega}_p(\delta) \in P(0,\,l]$;

d) if $\delta^{-1}g(\delta) \in P(0,\,l]$ and $u_p(\delta) \leq g(\delta)$ $(\tilde{u}_p(\delta) \leq g(\delta))$, then $\omega_p(\delta) \leq g(\delta)$ $(\tilde{\omega}_p(\delta) \leq g(\delta))$;

e) if $u_p\,(\tilde{u}_p)$ is semiadditive, then $u_p(\delta) \asymp \omega_p(\delta)$ $(\tilde{u}_p(\delta) \asymp \tilde{\omega}_p(\delta))$[b].

Notice that the statement d) shows that the function $\omega_p\,(\tilde{\omega}_p)$ is the best majorant for $u_p\,(\tilde{u}_p)$ on all the class of functions $g(\delta)$, for which $\delta^{-1}g(\delta) \in P(0,\,l]$.

Let us introduce the following regularized generalized integral smoothness modulus:

$$\omega_p^{(2)}(f,\delta)_\Gamma = \delta^2 \sup_{t\geq\delta} t^{-2}\, u_p^{(2)}(f,\,t)_\Gamma\,,$$

where $\quad u_p^{(2)}(f,t)_\Gamma = \sup\limits_{|h|\leq t} \|\, f(z_h) + f(z_{-h}) - 2f(z)\,\|_{L_p(\Gamma)},$

and also

$$\tilde{\omega}_p^{(2)}(f,\delta)_\Gamma = \delta^2 \sup_{t\geq\delta} t^{-2}\, \tilde{u}_p^{(2)}(f,\,t)_\Gamma\,,$$

where $\quad \tilde{u}_p^{(2)}(f,t)_\Gamma = \sup\limits_{|h|\leq t} \|\, f(\tilde{z}_h) + f(\tilde{z}_{-h}) - 2f(z)\,\|_{L_p(\Gamma)}.$

It is easy to see that $\omega_p^{(2)}(f,\delta)$ and $\tilde{\omega}_p^{(2)}(f,\delta)$ possess all the properties of ordinary smoothness modulus and they are the best majorants for the functions $u_p^{(2)}(f,\delta)_\Gamma$ and $\tilde{u}_p^{(2)}(f,\delta)_\Gamma$, respectively, among all the functions of smoothness modulus type.

[b]The relations $A \asymp B$ and $A \preccurlyeq B$ $(A > 0,\ B > 0)$ each time are determined with respect to some fixed collection of parameters and correspond to the inequalities $C_1 B \leq A \leq C_2 B$ and $A \leq C_3 B$, where $C_k > 0$ $(k = 1,2,3)$ are the constants independent of the mentioned collection of parameters.

Notice that for the cited generalized smoothness modulus the relations:

$$\omega_p^{(2)}(f,\eta) \le \delta^{-2}(\delta+\eta)^2\,\omega_p^{(2)}(f,\delta) \quad \forall\,\delta>0,\ \forall\,\eta>0,$$
$$\tilde\omega_p^{(2)}(f,\eta) \le \delta^{-2}(\delta+\eta)^2\,\tilde\omega_p^{(2)}(f,\delta) \quad \forall\,\delta>0,\ \forall\,\eta>0$$

and

$$\eta^{-2}\omega_p^{(2)}(f,\eta) \le 4\,\delta^{-2}\omega_p^{(2)}(f,\delta) \qquad \text{for} \quad 0<\delta<\eta\,,$$
$$\eta^{-2}\tilde\omega_p^{(2)}(f,\eta) \le 4\,\delta^{-2}\tilde\omega_p^{(2)}(f,\delta) \qquad \text{for} \quad 0<\delta<\eta$$

hold as well.

By $M(G)$ we denote a class of functions, analytical and bounded everywhere in the domain G and bounded almost everywhere (in linear sense) on ∂G.

Let $f \in M(G)$ and $F_h(z) = f(z_h) + f(z_{-h}) - 2f(z)$, where $h \in [0,\pi]$, $z_h = \psi\left(\varphi(z)\,e^{ih}\right)$ and

$$J\,F_h = (J\,F_h)_\Gamma(t) = \frac{1}{\pi\,i}\int_\Gamma \frac{F_h(z)}{z-t}dz\,, \qquad t\in\Gamma. \tag{1}$$

We say that a Jordan rectifiable curve Γ belongs to the class M (or $\Gamma \in M$) if for any $f \in M(\bar G)$ integral (1) exists almost everywhere on Γ and for each $p \ge 1$ it holds the estimate

$$\|J\,F_h\|_{L_p(\Gamma)} \le C(p)\,\|F_h\|_{L_p(\Gamma)}\,, \quad \forall h\in[0,\pi],$$

where $C(p)$ is a constant dependent only on the indicated parameter p.

Notice that such a definition of a class of curves M allows to extend a class of curves considered earlier and also to consider the case $p=1$.

We need the following auxiliary proposals.

Lemma A ([2], Lemma B). Let Γ be a closed Jordan rectifiable curve and f be a function determined almost everywhere on Γ and measurable. If the singular integral

$$Sf = (Sf)_\Gamma(t) = \frac{1}{\pi i}\int_\Gamma \frac{f(z)}{z-t}\,dz\,, \quad t\in\Gamma,$$

exists almost everywhere on Γ, then the Cauchy type integral

$$\Phi(\xi) = \frac{1}{\pi\,i}\int_\Gamma \frac{f(z)}{z-\xi}dz\,, \quad \xi\bar\in\Gamma,$$

almost everywhere on Γ has definite corner values equal

$$\Phi^{\pm}(t) = \pm\frac{1}{2}f(t) + \frac{1}{2}Sf(t)\,. \tag{2}$$

Conversely, if the Cauchy type integral Φ has almost everywhere on Γ corner boundary values from within and from outside of Γ then almost everywhere on Γ there exists a particular (singular) integral Sf and boundary values of Cauchy type integral Φ are expressed by the formula (2).

Lemma B ([2], Lemma Γ). For representability of the function f by the Cauchy integral by its boundary values, it is necessary and sufficient that $f \in E_1(G)$ ($E_p(G)$ is V. I. Smirnov's class).

Lemma C ([2], Lemma 2). Let $G \in S$, $f \in E_p(G)$ $(p > 0)$ and $\{f(t)\,;\ t \in \Gamma\}$ be boundary values of the function $f(z)$, $z \in G$. Then there will be found a sequence of functions $\{\Phi_k(t)\}$ being boundary values of bounded, analytic in G functions $\Phi_k(z)$, such that almost for all $t \in \Gamma$, $\lim\limits_{k\to\infty} \Phi_k(t) = f(t)$ and $|\Phi_k(t)| \le |f(t)|$, $(k = 1, 2, ...)$.

Theorem 0.3. *([2]). Under the hypotheses of lemma C there exists a sequence of functions $\{\Phi_k^*\}$ such that the following hold.*

a) $\lim\limits_{k\to\infty} \int_\Gamma |f(t) - \Phi_k^*(t)|^p |dz| = 0.$

b) The functions Φ_k^* are bounded on Γ and coincide almost everywhere on Γ with boundary values of analytic functions bounded in G.

c) $\lim\limits_{k\to\infty} \int_\Gamma |f(t_{\pm h}) - \Phi_k^*(t_{\pm h})|^p |dt| = 0$, *where* $t_{\pm h} = \psi\left(\varphi(t)e^{\pm ih}\right) \in \Gamma$ *and the functions ψ and φ conformally map the exterior (or interior) of a unit circle onto G^+ (G^-) and vice versa.*

d) $\lim\limits_{k\to\infty} u_p^{*(2)}(\Phi_k^*, h) \asymp u_p^{*(2)}(f, h)_\Gamma$ *and* $\lim\limits_{k\to\infty} \tilde{u}_p^{*(2)}(\Phi_k^*, h) \asymp u_p^{*(2)}(f, h)_\Gamma$, *where* $u_p^{*(2)}(g, h) = \|g(t_h) + g(t_{-h}) - 2g(t)\|_{L_p(\Gamma)}$.

By $E_p^*(G)$ $(p \ge 1)$ we denote a class of functions $f \in E_p(G)$ with a finite regularized smoothness modulus $\omega_p^{(2)}(\delta) = \omega_p^{(2)}(f, \delta)_\Gamma$.

Theorem 0.4. *If the curve $\Gamma \in M$ and $f \in E_p^*(G)$ $(p \ge 1)$, then for each natural n there exists a polynomial P_n such that*

$$\|f - P_n\|_{L_p(\Gamma)} \le const\, \omega_p^{(2)}\left(f, \frac{1}{n}\right)_\Gamma.$$

Proof. Let us consider a sequence of the functions $\{f_k(z)\}$ from the class $M(\bar{G})$ for which the following relations are fulfilled (see Theorem 0.3)

$$\lim_{k\to\infty} \int_\Gamma |f(t) - f_k(t)|^p |dt| = 0, \tag{3}$$

$$\lim_{k\to\infty} \int_\Gamma |f(t_{\pm h}) - f_k(t_{\pm h})|^p |dt| = 0, \tag{4}$$

$$\lim_{k\to\infty} \|f_k(t_h) + f_k(t_{-h}) - 2f_k(t)\|_{L_p(\Gamma)} = \|f(t_h) - f(t_{-h}) - 2f(t)\|_{L_p(\Gamma)} \tag{5}$$

Now, considering $\Gamma \in M$ and by the hypotheses of the theorem $f(z_{\pm h}) \in L_p(\Gamma)$, we can affirm, that the integral

$$B_k = \int_\Gamma \frac{f_k(z_h) - f_k(z_{-h})}{z-t} dt, \quad \forall t \in \Gamma,$$

exists on the sense of the chief value almost everywhere on Γ. This allows us to approximate the functions $f_k(z)$ by polynomials (see [3]), that for $t \in \Gamma$ are representable in the form

$$P_{n,k}(t) = \frac{1}{4\pi^2 i} \int_0^\pi K_n(h)\, dh \int_\Gamma \frac{f_k(z_h) + f_k(z_{-h})}{z-t} dz + \frac{1}{4\pi i} \int_0^\pi K_n(h) \left[f_k(t_h) + f_k(t_{-h})\right] dh, \tag{6}$$

where $K_n(h)$ is a positive kernel representing itself as a trigonometric polynomial of power not higher than n and satisfying the conditions

$$\frac{1}{\pi}\int_{-\pi}^{\pi} K_n(h)\, dh = 1 \qquad (n = 0, 1, 2, \ldots), \tag{7}$$

$$\int_{-\pi}^{\pi} |K_n(h)|\, dh \le C_1 \qquad (n = 0, 1, 2, \ldots), \tag{8}$$

$$\int_{-\pi}^{\pi} |t|^k |K_n(h)|\, dt \le C_2(k)\, (n+1)^{-k} \qquad (n = 0, 1, 2, \ldots), \tag{9}$$

where C_1 and C_2 are positive constants. Notice that to these conditions, in particular, satisfy the Jackson kernels (see [3]). Besides, from the fulfillment of conditions (7)-(9) we immediately get (see [3])

$$\int_{-\pi}^{\pi} \left(|t| + \frac{1}{n}\right)^k |K_n(h)|\, dt \le C_3(k)\, (n)^{-k} \qquad (n = 1, 2, \ldots). \tag{10}$$

Since the function $f_k \in M(\overline{G}) \subset E_1(G)$, by lemma B, we can represent them by the Cauchy integral

$$f_k(z) = \frac{1}{2\pi i} \int_\Gamma \frac{f_k(\xi)}{\xi - z} d\xi, \quad \forall z \in G.$$

This allows us, by Lemma A, to get the representation

$$f_k(t) = \frac{1}{2} f_k(t) + \frac{1}{2\pi i} \int_\Gamma \frac{f_k(\xi)}{\xi - t} d\xi, \quad t \in \Gamma,$$

almost everywhere on Γ. By relation (7), we can represent the last expression in the form

$$f_k(t) = \frac{1}{2\pi} \int_{-\pi}^{\pi} K_n(h)\, f_k(t)\, dh + \frac{1}{2\pi^2 i} \int_{-\pi}^{\pi} K_n(h)\, dh \int_\Gamma \frac{f_k(\xi)}{\xi - t} d\xi. \tag{11}$$

Now, having used relations (6) and (11) we estimate the norm of the difference $f_k(t) - P_{n,k}(t)$ in the metric $L_p(\Gamma)$ $(p > 1)$. We have

$$\begin{aligned}\|f_k - P_{n,k}(t)\|_{L_p(\Gamma)} &= \left\| \frac{1}{4\pi^2 i} \int_0^\pi K_n(h) dh \int_\Gamma \frac{f_k(z_h) + f_k(z_{-h}) - 2f_k(z)}{z-t} dz \right.\\ &\quad \left. + \frac{1}{4\pi} \int_0^\pi K_n(h)[f_k(t_h) + f_k(t_{-h}) - 2f_k(t)]\, dh \right\|_{L_p(\Gamma)} \\ &\preccurlyeq \left\| \int_0^\pi K_n(h) dh \int_\Gamma \frac{f_k(z_h) + f_k(z_{-h}) - 2f_k(z)}{z-t} dz \right\|_{L_p(\Gamma)} \\ &\quad + \left\| \int_0^\pi K_n(h)\,[f_k(t_h) + f_k(t_{-h}) - 2f_k(t)]\, dh \right\|_{L_p(\Gamma)}.\end{aligned}$$

Further, applying Minkovsky generalized inequality we get

$$\begin{aligned}\|f_k - P_{n,k}(t)\|_{L_p}(\Gamma) &\preccurlyeq \int_0^\pi |K_n(h)|\, dh \left\| \int_\Gamma \frac{f_k(z_h) + f_k(z_{-h}) - 2f_k(z)}{z-t} dz \right\|_{L_p(\Gamma)} \\ &\quad + \int_0^\pi |K_n(h)|\, \|f_k(t_h) + f_k(t_{-h}) - 2f_k(t)\|_{L_p(\Gamma)} \\ &\preccurlyeq \int_0^\pi |K_n(h)|\, A_{k,p}(h)\, dh + \int_0^\pi |K_n(h)|\, U^{*(2)}_{k,p}(h)\, dh, \end{aligned} \tag{12}$$

where

$$A_{k,p}(h) = \left(\int_\Gamma \left| \int_\Gamma \frac{f_k(z_h) + f_k(z_{-h}) - 2f_k(z)}{z-t} dz \right|^p |dt| \right)^{1/p},$$

$$U^{*(2)}_{k,p} = \|f_k(t_h) + f_k(t_{-h}) - 2f_k(t)\|_{L_p(\Gamma)}.$$

In view of the fact that $\Gamma \in M$, obviously, we have

$$A_{k,p}(h) \prec U_{k,p}(h). \tag{13}$$

Hence we get

$$\|f_k - P_{n,k}\|_{L_p(\Gamma)} \preccurlyeq \int_0^\pi |K_n(h)|\, U_{k,p}(h)\, dh. \tag{14}$$

Besides, it is easy to notice that the sequence $\{P_{n,k}\}$ for fixed n tends to some polynomial P_n as $k \to \infty$, i.e.

$$\lim_{k\to\infty} \|P_{n,k} - P_n\|_{L_p(\Gamma)} = 0. \tag{15}$$

Now, let us return to the function $f \in E_p^*(G)$ for which the sequence $\{f_k\}$ was constructed and that satisfies conditions (3) and (5). Obviously, we have

$$\|f - P_n\|_{L_p(\Gamma)} \le \|f - f_k\|_{L_p(\Gamma)} + \|f_k - P_{n,k}\|_{L_p(\Gamma)} + \|P_n - P_{n,k}\|_{L_p(\Gamma)}.$$

By (14) the last estimate is written in the form

$$\|f - P_n\|_{L_p(\Gamma)} \le \|f - f_k\|_{L_p(\Gamma)} + \|P_n - P_{n,k}\|_{L_p(\Gamma)} + \int_0^\pi |K_n(h)|\, U_{k,p}(h)\, dh.$$

Now, considering (3)-(5) and using Helly theorem (see [4], p.366) on passage to the limit under the Lebesgue-Stieltjes integral sign, we get

$$\|f - P_n\|_{L_p(\Gamma)} \le \int_0^\pi |K_n(h)|\, U_p^*(h)\, dh,$$

where $U_p^*(h) = \|f(z_h) + f(z_{-h}) - 2f(z)\|_{L_p(\Gamma)}$. Hence, the estimate

$$\|f - P_n\|_{L_p(\Gamma)} \le \int_0^\pi |K_n(h)|\, \omega_p^{(2)}(f, h)_\Gamma\, dh$$

immediately follows. Hence, by the known reasonings we get

$$\|f - P_n\|_{L_p(\Gamma)} \preccurlyeq \omega_p^{(2)}\left(f, \frac{1}{n}\right)_\Gamma .$$

In fact, we estimate the expression

$$\int_0^\pi |K_n(t)|\, \omega_p^{(2)}(f, t)_\Gamma\, dt = \int_0^\pi |K_n(t)|\, \vartheta(t)\, dt,$$

where $\vartheta(t) \stackrel{df}{=} \omega_p^{(2)}(f, t)_\Gamma$.

Using the properties of smoothness modulus $\omega_p^{(2)}(f, \delta)_\Gamma$, we can show the validity of the following relation

$$\vartheta(|t|) \le n^2 \left(|t| + \frac{1}{n}\right)^2 \vartheta\left(\frac{1}{n}\right).$$

Hence and from inequality (10) we get

$$\int_0^\pi |K_n(t)|\, \vartheta(t)\, dt \le n^2 \vartheta\left(\frac{1}{n}\right) \int_0^\pi \left(t + \frac{1}{n}\right)^2 |K_n(t)| dt$$
$$\le n^2 \vartheta\left(\frac{1}{n}\right) n^{-2} = \vartheta\left(\frac{1}{n}\right).$$

□

References

1. J. Walsh, Interpolation and approximation by rational functions in the complex domain, M., IL, 1961 (Russian).
2. J. I. Mamedkhanov and I. B. Dadashova, Analogue of Jackson-Bernstein theorem in L_p on closed curves in the complex plane. I, Vestnik BGU, ser. fiz. mat. nauk, **3** (2007), 32–40 (Russian).
3. V. K. Dzjadyk, Introduction to the theory of approximation of functions by polynomials, M., Nauka, 1977 (Russian).
4. A. N. Kolmogorov and S. V. Fomin, Elements of theory of functions and functional analysis, M., Nauka, 1976 (Russian).

AN INFINITE-DIMENSIONAL COMMUTATIVE BANACH ALGEBRA AND SPATIAL POTENTIAL FIELDS

S. A. PLAKSA

Institute of Mathematics of the National Academy of Sciences of Ukraine,
3 Tereshchenkivska Street, Kiev, Ukraine
E-mail: plaksa@imath.kiev.ua

In an infinite-dimensional commutative Banach algebra, we construct explicitly monogenic functions which have components satisfying the three-dimensional Laplace equation. We also describe a relation between these monogenic functions and harmonic vectors.

Keywords: Laplace equation; harmonic vector; harmonic commutative Banach algebra; monogenic function.

1. Introduction

Consider a spatial stationary vector field defined by means the vector-function $\mathbf{V} \equiv \mathbf{V}(x,y,z) := (v_1, v_2, v_3)$, where $v_k := v_k(x,y,z)$ for $k = 1,2,3$ are real scalar functions of the Cartesian coordinates x, y, z.

Defining a *potential solenoid* field in a simply connected domain Q of the three-dimensional real space $\mathbb{R}^3$, the vector-function $\mathbf{V}$ satisfies the system of equations

$$\operatorname{div} \mathbf{V} = 0\,, \qquad \operatorname{rot} \mathbf{V} = 0\,, \tag{1}$$

that we rewrite also in expanded form as

$$\begin{gathered} \frac{\partial v_1}{\partial x} + \frac{\partial v_2}{\partial y} + \frac{\partial v_3}{\partial z} = 0\,, \\ \frac{\partial v_3}{\partial y} - \frac{\partial v_2}{\partial z} = 0\,, \quad \frac{\partial v_1}{\partial z} - \frac{\partial v_3}{\partial x} = 0\,, \quad \frac{\partial v_2}{\partial x} - \frac{\partial v_1}{\partial y} = 0\,. \end{gathered} \tag{2}$$

Then there exists a scalar potential function $u(x,y,z)$ such that

$$\mathbf{V} = \operatorname{grad} u := \left(\frac{\partial u}{\partial x}, \frac{\partial u}{\partial y}, \frac{\partial u}{\partial z} \right),$$

and u satisfies the three-dimensional Laplace equation

$$\Delta u(x,y,z) \equiv \left(\frac{\partial^2}{\partial x^2} + \frac{\partial^2}{\partial y^2} + \frac{\partial^2}{\partial z^2}\right) u(x,y,z) = 0\,. \tag{3}$$

Doubly continuously differentiable functions satisfying the equation (3) are called *harmonic* functions, and solutions of the system (1) are called *harmonic* vectors.

Analytic function methods in the complex plane for plane potential fields inspire searching of analogous methods for spatial potential solenoid fields. The problem to construct such methods for spatial potential solenoid fields was posed by M. A. Lavrentyev [1, p. 205]). Besides, even independently of relation with applications in mathematical physics, for a long time a variety and an effectiveness of the analytic function methods stimulate developing of analogous methods for equations with partial derivatives.

Apparently, Hamilton made the first attempts to construct an algebra associated with the three-dimensional Laplace equation (3) in that sense that components of differentiable functions taking values in this algebra satisfy Eq. (3). However, after constructing quaternion algebra he had not studied a problem about constructing any other algebra (see [2]).

In the paper [3] I. Mel'nichenko considered the problem to construct commutative associative Banach algebra such that monogenic (i.e. differentiable in accordance with Gateaux) functions taking values in this algebra have components satisfying Eq. (3). It is obvious that the specified problem is appeared as an attempt to generalize the fundamental relation between the algebra of complex numbers and the two-dimensional Laplace equation. As it is well known, this relation means that, on the one hand, analytic functions of complex variable satisfy the two-dimensional Laplace equation, and, on the other hand, plane harmonic functions conjugate with Cauchy-Riemann conditions are components of certain analytic function of complex variable.

Because monogenic functions taking values in a commutative Banach algebra form a functional algebra, note that a relation between these functions and solutions of Eq. (3) is important for a development of effective methods for constructing mentioned solutions. It is quite natural that on such a way a quantity of fulfilled operations will be minimal in an algebra of third rank.

But in the paper [3] it is established that there does not exist commutative associative algebra of third rank with the main unit over the field of real numbers in which monogenic functions would satisfy Eq. (3). At the same time, for commutative associative algebras of third rank over the field of complex numbers in the papers [3,4] I. Mel'nichenko developed a method for extracting bases such that hypercomplex monogenic functions constructed in these bases have components satisfying Eq. (3).

However, it is impossible to obtain all solutions of Eq. (3) in the form of components of monogenic functions taking values in commutative algebras of third rank that were constructed in the papers [3,4]. In particular, for each mentioned algebra there exist spherical functions which are not components of specified hypercomplex monogenic functions.

Below, we consider an infinite-dimensional commutative Banach algebra $\mathbb{F}$ over the field of real numbers and establish that any spherical function is a component of some monogenic function taking values in this algebra. Thus, monogenic functions taking values in $\mathbb{F}$ form the widest of known functional algebras associated with Eq. (3).

2. A problem about extracting harmonic triad of vectors

Let $\mathbb{A}$ be a commutative associative Banach algebra over the field of real numbers $\mathbb{R}$ with the basis $\{e_k\}_{k=1}^{n}$, $3 \leq n \leq \infty$. Consider in $\mathbb{A}$ the linear subspace E_3 generated by the vectors e_1, e_2, e_3.

For a domain Q of the three-dimensional space $\mathbb{R}^3$ consider the domain $Q_\xi := \{\xi = xe_1 + ye_2 + ze_3 : (x, y, z) \in Q\} \subset E_3$ which are congruent to Q.

We say that a function $\Phi : Q_\xi \to \mathbb{A}$ is *monogenic* in the domain Q_ξ if Φ is differentiable in accordance with Gateaux in every point of Q_ξ, i.e. for every $\xi \in Q_\xi$ there exists an element $\Phi'(\xi) \in \mathbb{A}$ such that

$$\lim_{\varepsilon \to 0+0} [\Phi(\xi + \varepsilon h) - \Phi(\xi)]\, \varepsilon^{-1} = h\Phi'(\xi) \qquad \forall\, h \in E_3\,. \tag{4}$$

Note that if there exists a doubly differentiable in accordance with Gateaux function $\Phi : Q_\xi \to \mathbb{A}$ which satisfies Eq. (3) and the inequality $\Phi''(\xi) \neq 0$ at least in one point $\xi := xe_1 + ye_2 + ze_3 \in Q_\xi$, then in this case the basic elements e_1, e_2, e_3 satisfy the condition

$$e_1^2 + e_2^2 + e_3^2 = 0\,. \tag{5}$$

We say that an algebra $\mathbb{A}$ is *harmonic* if in $\mathbb{A}$ there exists a triad of linearly independent vectors $\{e_1, e_2, e_3\}$ satisfying the equality (5) provided that $e_j^2 \neq 0$ for $j = 1, 2, 3$. We say also that such a triad $\{e_1, e_2, e_3\}$ is *harmonic*.

In the paper [3] it is proved that there does not exist the three-dimensional harmonic algebra over the field of real numbers.

3. Monogenic functions in an infinite-dimensional harmonic algebra

Consider as harmonic algebra $\mathbb{A}$ an infinite-dimensional commutative associative Banach algebra

$$\mathbb{F} := \{g = \sum_{k=1}^{\infty} c_k e_k : c_k \in \mathbb{R}, \sum_{k=1}^{\infty} |c_k| < \infty\}$$

over the field $\mathbb{R}$ with the norm $\|g\|_{\mathbb{F}} := \sum_{k=1}^{\infty} |c_k|$ and the basis $\{e_k\}_{k=1}^{\infty}$, where the multiplication table for elements of basis is of the following form:

$$e_n e_1 = e_n, \qquad e_{2n+1}e_{2n} = \frac{1}{2}\, e_{4n} \quad \forall n \geq 1,$$

$$e_{2n+1}e_{2m} = \frac{1}{2}\left(e_{2n+2m} - (-1)^m e_{2n-2m}\right) \quad \forall n > m \geq 1,$$

$$e_{2n+1}e_{2m} = \frac{1}{2}\left(e_{2n+2m} + (-1)^n e_{2m-2n}\right) \quad \forall m > n \geq 1,$$

$$e_{2n+1}e_{2m+1} = \frac{1}{2}\left(e_{2n+2m+1} + (-1)^m e_{2n-2m+1}\right) \quad \forall n \geq m \geq 1,$$

$$e_{2n}e_{2m} = \frac{1}{2}\left(-e_{2n+2m+1} + (-1)^m e_{2n-2m+1}\right) \quad \forall n \geq m \geq 1.$$

It is evident that here e_1, e_2, e_3 form a harmonic triad of vectors.

Theorem 3.1. *In order that a function $\Phi : Q_\xi \longrightarrow \mathbb{F}$ of the form*

$$\Phi(xe_1 + ye_2 + ze_3) = \sum_{k=1}^{\infty} U_k(x, y, z)\; e_k, \tag{6}$$

where $U_k : Q \longrightarrow \mathbb{R}$, be monogenic in the domain $Q_\xi \subset E_3$, it is necessary and sufficient that the functions U_k, $k = 1, 2, \ldots$, be differentiable in the domain Q, that the conditions

$$\frac{\partial U_1(x,y,z)}{\partial y} = -\frac{1}{2}\,\frac{\partial U_2(x,y,z)}{\partial x},$$

$$\frac{\partial U_2(x,y,z)}{\partial y} = \frac{\partial U_1(x,y,z)}{\partial x} + \frac{1}{2}\,\frac{\partial U_5(x,y,z)}{\partial x},$$

$$\frac{\partial U_3(x,y,z)}{\partial y} = -\frac{1}{2}\,\frac{\partial U_4(x,y,z)}{\partial x},$$

$$\frac{\partial U_{2k}(x,y,z)}{\partial y} = \frac{1}{2}\,\frac{\partial U_{2k-1}(x,y,z)}{\partial x} + \frac{1}{2}\,\frac{\partial U_{2k+3}(x,y,z)}{\partial x}, \qquad k = 2, 3, \ldots,$$

$$\frac{\partial U_{2k+1}(x,y,z)}{\partial y} = -\frac{1}{2}\,\frac{\partial U_{2k-2}(x,y,z)}{\partial x} - \frac{1}{2}\,\frac{\partial U_{2k+2}(x,y,z)}{\partial x}, \quad k = 2, 3, \ldots,$$

$$\frac{\partial U_1(x,y,z)}{\partial z} = -\frac{1}{2}\,\frac{\partial U_3(x,y,z)}{\partial x}\,, \qquad \frac{\partial U_2(x,y,z)}{\partial z} = -\frac{1}{2}\,\frac{\partial U_4(x,y,z)}{\partial x}\,, \tag{7}$$

$$\frac{\partial U_3(x,y,z)}{\partial z} = \frac{\partial U_1(x,y,z)}{\partial x} - \frac{1}{2}\,\frac{\partial U_5(x,y,z)}{\partial x}\,,$$

$$\frac{\partial U_k(x,y,z)}{\partial z} = \frac{1}{2}\,\frac{\partial U_{k-2}(x,y,z)}{\partial x} - \frac{1}{2}\,\frac{\partial U_{k+2}(x,y,z)}{\partial x}\,, \qquad k = 4,5,\ldots\,,$$

be satisfied in Q, and that the following relations be fulfilled:

$$\sum_{k=1}^{\infty} \left| \frac{\partial U_k(x,y)}{\partial x} \right| < \infty, \tag{8}$$

$$\lim_{\varepsilon\to 0+0} \sum_{k=1}^{\infty} \Bigg| U_k(x+\varepsilon h_1, y+\varepsilon h_2, z+\varepsilon h_3) - U_k(x,y,z) - \frac{\partial U_k(x,y,z)}{\partial x}\,\varepsilon h_1$$
$$- \frac{\partial U_k(x,y,z)}{\partial y}\,\varepsilon h_2 - \frac{\partial U_k(x,y,z)}{\partial z}\,\varepsilon h_3 \Bigg|\, \varepsilon^{-1} = 0 \quad \forall\, h_1, h_2, h_3 \in \mathbb{R}\,. \tag{9}$$

Proof. *Necessity.* If the function (6) is monogenic in the domain Q_ζ, then for $h = e_1$ the equality (4) turns into the equality

$$\Phi'(\zeta) = \sum_{k=1}^{\infty} \frac{\partial U_k(x,y,z)}{\partial x}\, e_k\,,$$

and the relation (8) is fulfilled. Now, setting in series $h = e_2$ and $h = e_3$ in the equality (4), we obtain the conditions (7) for components of monogenic function (6).

At last, for $h_1, h_2, h_3 \in \mathbb{R}$ and $\varepsilon > 0$, denoting $h := h_1e_1 + h_2e_2 + h_3e_3$ and taking into account the equalities (7), we have

$$\big(\Phi(\zeta+\varepsilon h) - \Phi(\zeta)\big)\,\varepsilon^{-1} - h\,\Phi'(\zeta)$$
$$= \Bigg(\sum_{k=1}^{\infty} \Big(U_k(x+\varepsilon h_1, y+\varepsilon h_2, z+\varepsilon h_3) - U_k(x,y,z) \Big) e_k$$
$$-\varepsilon\,(h_1e_1 + h_2e_2 + h_3e_3) \sum_{k=1}^{\infty} \frac{\partial U_k(x,y,z)}{\partial x}\, e_k \Bigg)\varepsilon^{-1}$$
$$= \varepsilon^{-1} \sum_{k=1}^{\infty} \Bigg(U_k(x+\varepsilon h_1, y+\varepsilon h_2, z+\varepsilon h_3) - U_k(x,y,z)$$
$$- \frac{\partial U_k(x,y,z)}{\partial x}\,\varepsilon h_1 - \frac{\partial U_k(x,y,z)}{\partial y}\,\varepsilon h_2 - \frac{\partial U_k(x,y,z)}{\partial z}\,\varepsilon h_3 \Bigg) e_k\,. \tag{10}$$

Therefore, in as much as the function (6) is monogenic in the domain Q_ζ, the relation (9) is fulfilled in Q_ζ.

Sufficiency. Let $\varepsilon > 0$ and $h := h_1e_1+h_2e_2+h_3e_3$, where $h_1, h_2, h_3 \in \mathbb{R}$. Then under the conditions (7) and (8) the equality (10) is true. Now, it follows from the equality (10) and the relation (9) that the function (6) is monogenic in the domain Q_ζ. The theorem is proved. □

Note that conditions (7) are similar by nature to the Cauchy-Riemann conditions for monogenic functions of complex variables, and the relations (8), (9) are conditioned by the infinite dimensionality of the algebra $\mathbb{F}$.

It is clear that if the Gateaux derivative Φ' of monogenic function $\Phi : Q_\xi \longrightarrow \mathbb{F}$, in turn, is monogenic function in the domain Q_ξ, then all components U_k of expansion (6) satisfy Eq. (3) in Q in consequence of condition (5). At the same time, the following statement is true even independently of relation between solutions of the system of equations (7) and monogenic functions.

Theorem 3.2. *If the functions $U_k : Q \longrightarrow \mathbb{R}$ have continuous second-order partial derivatives in the domain Q and satisfy the conditions (7), then they satisfy Eq. (3) in Q.*

Proof. It is easy to show that if the functions U_k are doubly continuously differentiable in the domain Q, then the equalities $\Delta U_k(x, y, z) = 0$ for $k = 1, 2, \ldots$ are corollaries of the system (7). □

Note that the algebra $\mathbb{F}$ is isomorphic to the algebra $\mathbf{F}$ of absolutely convergent trigonometric Fourier series

$$g(\tau) = a_0 + \sum_{k=1}^{\infty}\Big(a_k\, i^k \cos k\tau + b_k\, i^k \sin k\tau\Big)$$

with real coefficients a_0, a_k, b_k and the norm $\|g\|_{\mathbf{F}} := |a_0| + \sum\limits_{k=1}^{\infty}\Big(|a_k|+|b_k|\Big)$. In this case, we have the isomorphism $e_{2k-1} \longleftrightarrow i^{k-1}\cos{(k-1)\tau}$, $e_{2k} \longleftrightarrow i^k \sin k\tau$ between basic elements.

Let us write the expansion of a power function of the variable $\xi = xe_1 + ye_2 + ze_3$ in the basis $\{e_k\}_{k=1}^{\infty}$, using spherical coordinates ρ, θ, ϕ which have the following relations with x, y, z:

$$x = \rho\cos\theta, \quad y = \rho\sin\theta\sin\phi, \quad z = \rho\sin\theta\cos\phi\,. \tag{11}$$

In view of the isomorphism of the algebras $\mathbb{F}$ and $\mathbf{F}$, the construction of expansions of this sort is reduced to the determination of relevant Fourier coefficients. So we have

$$\xi^n = \rho^n \Bigg(P_n(\cos\theta)\, e_1$$

$$+2 \sum_{m=1}^{n} \frac{n!}{(n+m)!}\, P_n^m(\cos\theta)\big(\sin m\phi\; e_{2m} + \cos m\phi\; e_{2m+1}\big)\Bigg), \tag{12}$$

where n is a positive integer, P_n and P_n^m are Legendre polynomials and associated Legendre polynomials, respectively, namely:

$$P_n(t) := \frac{1}{2^n\, n!}\, \frac{d^n}{dt^n}\, (t^2 - 1)^n, \qquad P_n^m(t) := (1-t^2)^{m/2}\, \frac{d^m}{dt^m}\, P_n(t)\,.$$

We obtain in exactly the same way the following expansion of the exponential function:

$$e^{\zeta} = e^{\rho\cos\theta}\Big(J_0(\rho\sin\theta)e_1 + 2\sum_{m=1}^{\infty} J_m(\rho\sin\theta)\big(\sin m\phi\, e_{2m} + \cos m\phi\, e_{2m+1}\big)\Big),$$

where J_m are Bessel functions, namely

$$J_m(t) := \frac{(-1)^m}{\pi} \int_0^{\pi} e^{it\cos\tau}\, \cos m\tau\; d\tau\,.$$

Thus, $2n+1$ linearly independent spherical functions of the n-th power are components of the expansion (12) of the function ξ^n. Using the expansion (12) and rules of multiplication for basic elements of the algebra $\mathbb{F}$, it is easy to prove the following statement.

Theorem 3.3. *Every spherical function*

$$\rho^n \Bigg(a_{n,0} P_n(\cos\theta) + \sum_{m=1}^{n} \Big(a_{n,m} \cos m\phi + b_{n,m} \sin m\phi \Big) P_n^m(\cos\theta) \Bigg)$$

where $a_{n,0}, a_{n,m}, b_{n,m} \in \mathbb{R}$, is the first component of expansion of the monogenic function

$$\Bigg(a_{n,0}\, e_1 + \sum_{m=1}^{n} (-1)^m\, \frac{(n+m)!}{n!}\, \Big(b_{n,m}\, e_{2m} + a_{n,m}\, e_{2m+1} \Big) \Bigg) \xi^n$$

in the basis $\{e_k\}_{k=1}^{\infty}$, where $\xi := xe_1 + ye_2 + ze_3$, and x, y, z have the relations (11) with the spherical coordinates ρ, θ, ϕ.

4. Relation between monogenic functions and harmonic vectors

Let us establish some important properties of solutions of the system (7).

Theorem 4.1. *Every solution of the system (7) in a domain $Q \subset \mathbb{R}^3$ generates a harmonic vector* $\mathbf{V} := (U_1, -\frac{1}{2}U_2, -\frac{1}{2}U_3)$ *in* Q.

Proof. The equalities (2) for a vector $\mathbf{V} := (U_1, -\frac{1}{2}U_2, -\frac{1}{2}U_3)$ are corollaries of the system (7). Really, the system (7) contains the third equation and the fourth equation from (2). The second equation from (2) is a corollary of the third condition and the seventh condition of the system (7). Finally, the first equation from (2) is a corollary of the second condition and the eighth condition of the system (7). □

Theorem 4.2. *For any function $U_1 : Q \longrightarrow \mathbb{R}$ harmonic in a simply connected domain $Q \subset \mathbb{R}^3$ there exist harmonic functions $U_k : Q \longrightarrow \mathbb{R}$, $k = 2, 3, \ldots$, such that the conditions (7) are fulfilled in Q.*

Proof. Let us to add and to subtract the second condition and the eighth condition as well as the third condition and the seventh condition of the system (7). Let us also to add and to subtract the fourth condition for $k = m$ and the ninth condition of the system (7) for $k = 2m + 1$, where $m = 2, 3, \ldots$. Let us else to add and to subtract the fifth condition for $k = m$ and the ninth condition of system (7) for $k = 2m$, where $m = 2, 3, \ldots$. Thus, we rewrite the system (7) in the following equivalent form:

$$\begin{gathered}
\frac{\partial U_1}{\partial x} - \frac{1}{2}\frac{\partial U_2}{\partial y} - \frac{1}{2}\frac{\partial U_3}{\partial z} = 0\,, \qquad \frac{\partial U_3}{\partial y} - \frac{\partial U_2}{\partial z} = 0\,, \\
\frac{\partial U_1}{\partial y} + \frac{1}{2}\frac{\partial U_2}{\partial x} = 0\,, \qquad \frac{\partial U_1}{\partial z} + \frac{1}{2}\frac{\partial U_3}{\partial x} = 0\,, \\
\frac{\partial U_{2k}}{\partial x} = -\frac{\partial U_{2k-2}}{\partial z} - \frac{\partial U_{2k-1}}{\partial y}\,, \qquad \frac{\partial U_{2k+1}}{\partial x} = \frac{\partial U_{2k-2}}{\partial y} - \frac{\partial U_{2k-1}}{\partial z} \\
\frac{\partial U_{2k}}{\partial z} - \frac{\partial U_{2k+1}}{\partial y} = \frac{\partial U_{2k-2}}{\partial x}\,, \qquad \frac{\partial U_{2k}}{\partial y} + \frac{\partial U_{2k+1}}{\partial z} = \frac{\partial U_{2k-1}}{\partial x}\,,
\end{gathered} \tag{13}$$

where $k = 2, 3, \ldots$.

First of all, note that there exists a harmonic vector $\mathbf{V}_0 := (U_1, v_2^0, v_3^0)$ in the domain Q. Moreover, for any vector $\mathbf{V} := (U_1, v_2, v_3)$ harmonic in Q, the components v_2, v_3 are determined accurate within the real part and the imaginary part of any function $f_1(t)$ holomorphic in the domain

$\{t = z + iy : (x, y, z) \in Q\}$ of the complex plane, i.e. the equalities

$$v_2(x,y,z) = v_2^0(x,y,z) + \mathrm{Re}\, f_1(z+iy), \quad v_3(x,y,z) = v_3^0(x,y,z) + \mathrm{Im}\, f_1(z+iy)$$

are true for all $(x, y, z) \in Q$. Then, using Theorem 4.1, we find the functions U_2 and U_3, namely: $U_2 := -2\, v_2, \quad U_3 := -2\, v_3$.

Now, let us show that the last four conditions of the system (13) allow to determine the functions U_{2k}, U_{2k+1}, if the functions $U_2, U_3, \ldots, U_{2k-1}$ are already determined. Really, integrating the fifth equation and the sixth equation of the system (13), we obtain the expressions

$$U_{2k}(x,y,z) = -\int_{x_0}^{x} \left(\frac{\partial U_{2k-2}(\tau,y,z)}{\partial z} + \frac{\partial U_{2k-1}(\tau,y,z)}{\partial y} \right) d\tau + \widetilde{u}_{2k}(z,y),$$

$$U_{2k+1}(x,y,z) = \int_{x_0}^{x} \left(\frac{\partial U_{2k-2}(\tau,y,z)}{\partial y} - \frac{\partial U_{2k-1}(\tau,y,z)}{\partial z} \right) d\tau + \widetilde{u}_{2k+1}(z,y)$$

for (x, y, z) belonging to a certain neighborhood $\mathcal{N} \subset Q$ of any point $(x_0, y_0, z_0) \in Q$.

Substituting these expressions into the seventh equation and eighth equation of the system (13) and taking into account that U_{2k-2}, U_{2k-1} are harmonic functions in the domain $\mathcal{N}$, we obtain the following inhomogeneous Cauchy–Riemann system (see, for example, [5]) for finding the functions $\widetilde{u}_{2k}$, $\widetilde{u}_{2k+1}$:

$$\begin{aligned} \frac{\partial \widetilde{u}_{2k}(z,y)}{\partial z} - \frac{\partial \widetilde{u}_{2k+1}(z,y)}{\partial y} &= \left.\frac{\partial U_{2k-2}(x,y,z)}{\partial x}\right|_{x=x_0}, \\ \frac{\partial \widetilde{u}_{2k}(z,y)}{\partial y} + \frac{\partial \widetilde{u}_{2k+1}(z,y)}{\partial z} &= \left.\frac{\partial U_{2k-1}(x,y,z)}{\partial x}\right|_{x=x_0}. \end{aligned} \tag{14}$$

Solutions of the system (14) are determined accurate within the real part and the imaginary part of any function holomorphic in the domain $\{t = z + iy : (x, y, z) \in \mathcal{N}\}$ of the complex plane. Therefore, in as much as the domain Q is simply connected, taking into account the unicity theorem for spatial harmonic functions, it is easy to continue the functions U_{2k}, U_{2k+1} defined in the neighborhood $\mathcal{N}$ into the domain Q. □

Let us note that the functions U_{2m}, U_{2m+1} satisfying the last four conditions of the system (13) for $k = m \geq 2$ are determined accurate within the real part and the imaginary part of any function $f_m(t)$ holomorphic in the domain $\{t = z + iy : (x, y, z) \in Q\}$ of the complex plane, i.e. the equalities

$$\begin{aligned} U_{2m}(x,y,z) &= U_{2m}^0(x,y,z) + \mathrm{Re}\, f_m(z+iy), \\ U_{2m+1}(x,y,z) &= U_{2m+1}^0(x,y,z) + \mathrm{Im}\, f_m(z+iy) \end{aligned}$$

are true for all $(x, y, z) \in Q$, where U_{2m}^0, U_{2m+1}^0 are functions forming together with the functions $U_1, U_2, \ldots, U_{2m-1}$ a particular solution of the system (13), in which $k = 2, 3, \ldots, m$.

References

1. M. A. Lavrentyev and B. V. Shabat, *Problems of hydrodynamics and their mathematical models*, Moscow: Nauka, 1977 [in Russian].
2. F. Klein, *Vorlesungen über die entwicklung der mathematik im 19 jahrhundert*, V.1, Berlin: Verlag von Julius Springer, 1926.
3. I. P. Mel'nichenko, On expression of harmonic mappings by monogenic functions, *Ukr. Math. J.*, **27** (1975), no. 5, 606–613.
4. I. P. Mel'nichenko, Algebras of functionally-invariant solutions of the three-dimensional Laplace equation, *Ukr. Math. J.*, **55** (2003), no. 9, 1284–1290.
5. I. N. Vekua, Generalized analytic functions, London: Pergamon Press, 1962.

Session 04

Complex Analytic Methods for Applied Sciences

SESSION EDITORS

V. V. Mityushev Pedagogical University, Kraków, Poland

S. V. Rogosin Belarus State University, Minsk, Belarus

ON THE CURVATURE OF AN INNER CURVE IN A SCHWARZ-CHRISTOFFEL MAPPING

A. ANDERSSON

International Centre for Mathematical Modelling,
Växjö University, Sweden
and
School of Engineering,
Jönköping University, Sweden

In the so-called outer polygon method, an approximate conformal mapping for a given simply connected region Ω is constructed using a Schwarz–Christoffel mapping for an outer polygon, a polygonal region of which Ω is a subset. The resulting region is then bounded by a C^∞-curve, which among other things means that its curvature is bounded.

In this work, we study the curvature of an inner curve in a polygon, i.e., the image under the Schwarz–Christoffel mapping of R, the unit disk or upper half-plane, to a polygonal region P of a curve inside R. From the Schwarz–Christoffel formula, explicit expressions for the curvature are derived, and for boundary curves appearing in the outer polygon method, asymptotic boundaries for the curvature are given.

Keywords: Curvature, Schwarz-Christoffel mapping, inner curve, outer polygon.

1. Introduction

The Schwarz–Christoffel mapping from the upper half-plane or the unit disk to a polygonal region in the complex plane is a well-known and often used conformal mapping for which also efficient numerical methods exist by the work of Trefethen and Driscoll [1–3]. In a recent paper [4] by the author, this mapping is used for different non-polygonal regions by seeing the region under consideration as an inner region in a polygon. The motivation for the method is that in some applications, see for example [5], it is necessary that the approximate conformal mapping maintains a good control of the boundary curve direction, especially towards infinity when concerning unbounded regions. Furthermore, if the boundary curve in the region has bounded curvature, this property should be maintained in the approximate region produced by the conformal mapping.

Hence, an important issue is to find bounds for the variation of the curvature. Curvature in the context of conformal mappings has been studied in for example [6] where mappings from the disk to convex domains is treated. When bounded regions are handled in [4], the approximate regions have boundaries that are level curves of Green's functions. Curvature for such curves is dealt with in for example [7–9].

The purpose of this paper is to investigate the curvature for some inner curves in polygons, both bounded and unbounded, convex and non-convex, curves that typically appear as boundary curves in the regions produced by the methods in [4]. From the Schwarz–Christoffel formula, explicit expressions for the curvature of inner curves are given, and furthermore, asymptotic bounds for the curvature are given.

The plan of the paper is as follows. In Sec. 2, a short description of the method and the ideas in [4] is given, and in Sec. 3 and Sec. 4, the curvature for inner curves in Schwarz–Christoffel polygons in general and in the special cases relevant for the methods in [4] is calculated. Finally, in Sec. 5, bounds for the curvature are given.

2. The outer polygon method

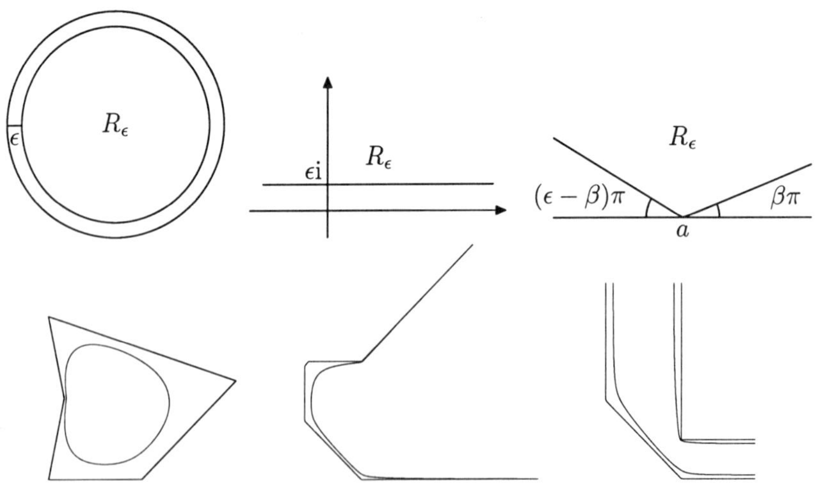

Fig. 1. Inner regions in polygons used in [4].

In [4], methods for constructing an outer polygon $P^\circ_{\epsilon,\Omega}$ to a region Ω

in the z-plane, such that a Schwarz–Christoffel mapping from R, the upper half-plane or unit disk in the w-plane, to $P^{\circ}_{\epsilon,\Omega}$ maps a region $R_\epsilon \in R$ approximately on Ω. In the examples in the paper, three different cases are treated:

(1) Ω is a bounded region, R is the unit disk and R_ϵ the disk $\{w : |w| \leq 1-\epsilon\}$.
(2) Ω is an unbounded region with one infinite boundary point, R is the upper half-plane and $R_\epsilon = \{w : \operatorname{Im} w \geq \epsilon\}$.
(3) Ω is an infinite channel with parallel straight line walls at both ends, R is the upper half-plane and R_ϵ the region above two rays in the upper half-plane going out from the point on the real axis which is mapped to infinity at one of the channel ends.

An algorithm with tangent polygons and a quasi-Newton method is used to determine an outer polygon and an approximate conformal mapping for Ω. The resulting region has a boundary with bounded curvature in all finite points. However, to make the algorithm work, it is often necessary to use a small ϵ, and as will be apparent, the curvature might in that case vary in a wide range.

3. Curvature of an inner curve in a polygon

Let s be a Schwarz–Christoffel mapping

$$s(w) = A\int_{w_0}^{w}\prod_{k=1}^{n}(\omega - w_k)^{\alpha_k - 1}d\omega + B \tag{1}$$

from R, the upper half-plane or the unit disk in the w-plane, to a polygonal region P in the z-plane with inner angles $\alpha_1\pi, \ldots, \alpha_n\pi$, which means that for each k, $\alpha_k \in [0,2]$. If s is a mapping from the unit disk, each w_k is a complex number on the unit circle; if s is a mapping from the upper half–plane, each w_k is a real number, and since there are three degrees of freedom in the Schwarz–Christoffel mapping, we can without loss of generality assume that they are all in the interval $[-1,1]$. (This restriction is used in SC toolbox [3], which is the numerical software that [4] uses. Note also that this type of restriction affects the value of A.) Let $w(t)$ with real parameter t be a curve in the interior of R. We will study the curvature of the curve $s(w(t))$, residing inside P.

The angular direction of the curve $s(w(t))$ is given by

$$\Phi(t) = \arg(s'(w(t))) = \arg(A) + \arg(w'(t)) + \sum_k (\alpha_k - 1)\arg(w(t) - w_k). \tag{2}$$

Setting $w(t) - w_k = x_k(t) + y_k(t)i$, $\arg(w(t) - w_k) = \arctan(y_k(t)/x_k(t))$, and we get

$$\Phi'(t) = \frac{d}{dt}\arg(w'(t)) + \sum_k \frac{(\alpha_k - 1)(x_k(t)y_k'(t) - x_k'(t)y_k(t))}{(x_k(t))^2 + (y_k(t))^2}. \tag{3}$$

Since the denominator $(x_k(t))^2+(y_k(t))^2 = |w(t) - w_k|^2$, and by setting

$$h_k(t) = \begin{vmatrix} x_k(t) & y_k(t) \\ x_k'(t) & y_k'(t) \end{vmatrix},$$

the curvature $\kappa(t)$ of $s(w(t))$ is

$$\kappa(t) = \frac{\Phi'(t)}{|s'(w(t))|} = \frac{\frac{d}{dt}\arg(w'(t)) + \sum_k (\alpha_k - 1)h_k(t)/|w(t) - w_k|^2}{|A|\,|w'(t)| \prod_k |w(t) - w_k|^{\alpha_k - 1}}. \tag{4}$$

If we separate the terms, Eq. (4) can be written

$$\kappa(t) = \frac{d}{dt}\arg(w'(t)) \frac{\prod_k |w(t) - w_k|^{1-\alpha_k}}{|A|\,|w'(t)|} + \sum_k \frac{(\alpha_k - 1)h_k(t) \prod_{j \neq k} |w(t) - w_j|^{1-\alpha_j}}{|A|\,|w'(t)|\,|w(t) - w_k|^{1+\alpha_k}}. \tag{5}$$

We will in the following mainly study curves passing close to a polygon vertex z_m and their curvature near this vertex. Let for some prevertex w_m, $\epsilon = \min\{|w(t) - w_m|\}$ and let $t = t_m$ be the point for which this minimum is reached. If furthermore

$$p_m = \prod_{k \neq m} |w(t_m) - w_k|^{1-\alpha_k}, \tag{6}$$

$$\kappa(t_m) = \frac{p_m}{|A|\,|w'(t_m)|\,\epsilon^{\alpha_k}} \left(\epsilon \frac{d}{dt}\arg(w'(t_m)) + \frac{(\alpha_m - 1)h_m(t_m)}{\epsilon} + \epsilon \sum_{k \neq m} \frac{(\alpha_k - 1)h_k(t_m)}{|w(t_m) - w_k|^2} \right). \tag{7}$$

4. Three special cases

For the boundary curve of the inner region $s(R_\epsilon)$ in the examples mentioned in Sec. 2 and treated in [4], we apply the result from the previous section.

4.1. *A bounded region*

Assume that P is a bounded region, i.e., with no infinite vertex, and that R is the unit disk. Let for some ϵ, $0 < \epsilon < 1$,

$$w(t) = (1-\epsilon)e^{it}, \quad 0 \le t < 2\pi. \tag{8}$$

Here, the prevertices $w_1, \ldots, w_n$ are all numbers on the unit circle; set $w_k = e^{it_k}$ for $k = 1, \ldots, n$. Furthermore, $\arg(w'(t)) = \pi/2 + t$, $x_k(t) = (1-\epsilon)\cos t - \cos t_k$ and $y_k(t) = (1-\epsilon)\sin t - \sin t_k$, and hence,

$$\begin{aligned} h_k(t) &= (1-\epsilon)\Big(\cos t\big((1-\epsilon)\cos t - \cos t_k\big) + \sin t\big((1-\epsilon)\sin t - \sin t_k\big)\Big) \\ &= (1-\epsilon)\big(1-\epsilon-\cos(t-t_k)\big). \end{aligned}$$

Therefore,

$$\begin{aligned} \kappa(t) = {} & \frac{\prod_k |w(t)-w_k|^{1-\alpha_k}}{|A|\,(1-\epsilon)} \\ & + \sum_k \frac{(\alpha_k-1)(1-\epsilon-\cos(t-t_k)) \prod_{j\neq k} |w(t)-w_j|^{1-\alpha_j}}{|A|\,|w(t)-w_k|^{1+\alpha_k}}. \end{aligned} \tag{9}$$

Since $\epsilon \le |w(t) - w_k| \le 2-\epsilon$, the first term in Eq. (9) dominates when ϵ is close to 1, and the curvature is almost constant. This means that the inner curve is nearly circular. Note that in this case, the images under the Schwarz–Christoffel mapping of origin-centered circles with radius less than one are level curves of Green's function for the polygonal region. The result harmonizes well with the results by Walsh [8], where the properties of Green's function level curves for general bounded regions are developed thoroughly.

On the other hand, for small ϵ, $\kappa(t)$ is near the mth vertex in P dominated by the corresponding term in Eq. (9). Since $w(t_m) = (1-\epsilon)w_m$, it follows that

$$\kappa(t_m) = \frac{p_m}{|A|\,\epsilon^{\alpha_m}}\left(\frac{\epsilon}{1-\epsilon} + 1 - \alpha_m + \epsilon \sum_{k\neq m} \frac{(\alpha_k-1)(1-\epsilon-\cos(t_m-t_k))}{(1-\epsilon)^2 - 2(1-\epsilon)\cos(t_m-t_k)+1}\right), \tag{10}$$

which, with an asymptotic expression, can be written

$$\kappa(t_m) = \frac{(1-\alpha_m)p_m}{|A|\,\epsilon^{\alpha_m}}\left(1 + \frac{\epsilon}{2} + \mathcal{O}\left(\epsilon^2\right)\right). \tag{11}$$

4.2. *Polygon with one infinite vertex*

Assume that P has one vertex at infinity and that R is the upper half-plane. Then,

$$s(w) = A \int_{w_0}^{w} \prod_{k=1}^{n-1} (\omega - w_k)^{\alpha_k - 1} d\omega + B, \tag{12}$$

with real prevertices w_k. Let for some $\epsilon > 0$, $w(t) = t + \epsilon i$. It follows that $x_k(t) = t - w_k$ and $y_k(t) = \epsilon$, so $h_k(t) = -\epsilon$ and the curvature

$$\kappa(t) = \sum_k \frac{\epsilon(1-\alpha_k) \prod_{j \neq k} |t + \epsilon i - w_j|^{1-\alpha_j}}{|A| \, |t + \epsilon i - w_k|^{1+\alpha_k}}. \tag{13}$$

If ϵ is small, the mth term dominates in Eq. (13) when $t \approx w_m$. Setting $t_m = w_m$, we have

$$\kappa(t_m) = \frac{p_m}{|A| \, \epsilon^{\alpha_m}} \left(1 - \alpha_m + \epsilon^2 \sum_{k \neq m} \frac{1 - \alpha_k}{|w_m - w_k + \epsilon i|^2} \right) \tag{14}$$

$$= \frac{(1-\alpha_m) p_m}{|A| \, \epsilon^{\alpha_m}} \left(1 + \mathcal{O}\left(\epsilon^2\right)\right). \tag{15}$$

4.3. *Channel with parallel walls at the ends*

Finally, assume that P is a polygonal channel with parallel walls at both ends, and that the mapping s from the upper half-plane to P maps infinity to one infinite end of the channel, and that the real number a is mapped on the other infinite channel end. Then,

$$s(w) = A \int_{w_0}^{w} (\omega - a)^{-1} \prod_{k=1}^{n-2} (\omega - w_k)^{\alpha_k - 1} d\omega + B. \tag{16}$$

Let for $\varphi \in (0, \pi)$, $w(t) = a + te^{i\varphi}$, $\quad 0 \leq t < \infty$, i.e., $w(t)$ is a ray from a in the upper half-plane making the angle φ with the positive real axis. It follows immediately that $\arg(w'(t)) = \varphi$, and furthermore, we have

$$x_k(t) = a - w_k + t\cos\varphi \qquad y_k(t) = t \sin\varphi$$
$$x_k'(t) = \cos\varphi \qquad y_k'(t) = \sin\varphi,$$

so $h_k(t) = (a - w_k)\sin\varphi$.

From Eq. (5), it follows that

$$\kappa(t) = \sum_k \frac{(\alpha_k - 1)(a - w_k)\sin\varphi \prod_{j \neq k} |a - w_j + te^{i\varphi}|^{1-\alpha_j}}{|A| \, |a - w_k + te^{i\varphi}|^{1+\alpha_k}}. \tag{17}$$

Consider the curvature near a vertex $s(w_m)$. A ray passing close to w_m must have a small φ if $w_m > a$ and a φ slightly less than π if $w_m < a$. In both cases, $\left|a - w_m + te^{i\varphi}\right| \geq |a - w_m| \sin\varphi$, with equality for $t = (w_m - a)\cos\varphi$. Hence, setting $t_m = (w_m - a)\cos\varphi$, $\epsilon = |a - w_m| \sin\varphi$ and $v_{k,m} = (w_k - a)/\left|w_m - a\right|$, it follows from Eq. (7) that

$$\kappa(t_m) = \frac{p_m}{|A|\,\epsilon^{\alpha_k}}\Big(\pm(1-\alpha_m) + \epsilon^2 \sum_{k\neq m} \frac{(1-\alpha_k)v_{k,m}}{(w_m - w_k)^2 - \epsilon^2\,(1 \pm 2v_{k,m})}\Big) \tag{18}$$

$$= \pm\frac{(1-\alpha_m)p_m}{|A|\,\epsilon^{\alpha_m}}\left(1 + \mathcal{O}\left(\epsilon^2\right)\right), \tag{19}$$

where $\pm$ is the sign of $w_m - a$.

5. Conclusions

In all the considered cases, an estimation of the curvature results in similar expressions, see Eqs. (11), (15) and (19). For small ϵ, a term

$$\pm\frac{1-\alpha_m}{|A|\,\epsilon^{\alpha_m}} \prod_{k\neq m} \left|w_m - w_k + \epsilon c\right|^{1-\alpha_k} \tag{20}$$

with $|c| = 1$ dominates near the mth vertex, and the curvature at that point grows for $\alpha_m \neq 1$ infinitely when $\epsilon \to 0$.

To give an asymptotic upper bound for the curvature, we start with a preliminary lemma.

Lemma 5.1. *Let f, g_0 and g_1 be functions, sufficiently smooth in a neighbourhood Ω of $\tilde{x}$, c a real and β a real non-zero constant, $0 < \epsilon < 1$, and let $x^*(\epsilon) \in \Omega$ be a local maximum of f. Assume that*

$$f(x) = (h(x))^\beta \left(g_0(x) + \epsilon g_1(x) + \mathcal{O}\left(\epsilon^2\right)\right), \tag{21}$$

where the function $h(x)$ is positive on Ω and defined by either of

$$h(x) = \epsilon^2 + (x - \tilde{x})^2, \tag{22}$$

$$h(x) = \epsilon^2 + (1-\epsilon)(x - \tilde{x})^2 + \mathcal{O}\left((x - \tilde{x})^4\right), \tag{23}$$

and furthermore that

$$\lim_{\epsilon\to 0} x^* = \tilde{x}. \tag{24}$$

Then, for $x^* \in \Omega$,

$$x^* - \tilde{x} = \mathcal{O}\left(\epsilon^2\right). \tag{25}$$

Proof.

$$f'(x) = (h(x))^{\beta-1}\Big(\beta h'(x)\left(g_0(x) + \epsilon g_1(x) + \mathcal{O}\left(\epsilon^2\right)\right) + \\ h(x)\left(g_0'(x) + \epsilon g_1'(x) + \mathcal{O}\left(\epsilon^2\right)\right)\Big),$$

From Eq. (24), $x^* - \tilde{x} = B\epsilon + \mathcal{O}\left(\epsilon^2\right)$. Since $h'(x^*) = 2B\epsilon + \mathcal{O}\left(\epsilon^2\right)$,

$$0 = f'(x^*) = (h(x^*))^{\beta-1}\left(2B\beta\epsilon g_0(x) + \mathcal{O}\left(\epsilon^2\right)\right).$$

Hence, $B = \mathcal{O}(\epsilon)$ and it follows that $x^* - \tilde{x} = \mathcal{O}\left(\epsilon^2\right)$. □

Let $\{w_k\}$ and $w(t)$ be defined as before, and let

$$q_m = \prod_{k\neq m} |w_m - w_k|^{1-\alpha_k}. \tag{26}$$

Then

$$p_m = \prod_{k\neq m} |w_m - w_k + \epsilon c|^{1-\alpha_k} = q_m\left(1 + \epsilon c\sum_{k\neq m}\frac{1-\alpha_k}{|w_m - w_k|} + \mathcal{O}\left(\epsilon^2\right)\right) \tag{27}$$

and with t_m defined as in any of Secs. 4.1, 4.2 or 4.3, it follows from Eqs. (11), (15) and (19) that

$$|\kappa(t_m)| = \frac{|1-\alpha_m|\, q_m}{|A|\,\epsilon^{\alpha_m}}\left(1 + r_m\epsilon + \mathcal{O}\left(\epsilon^2\right)\right), \tag{28}$$

where

$$r_m = \begin{cases} \displaystyle\sum_{k\neq m}\frac{|1-\alpha_k|}{|w_m - w_k|} + \frac{1}{2}, & \text{for bounded regions, Sec. 4.1} \\ \displaystyle\sum_{k\neq m}\frac{|1-\alpha_k|}{|w_m - w_k|}, & \text{for unbounded regions, Secs. 4.2, 4.3.} \end{cases} \tag{29}$$

Theorem 5.1. *Let t^* be a local optimum point for the boundary curvature in any of the three cases considered in Sec. 4, and let I_m be an interval containing t^* but no other optimum point and t_m, chosen such that $|\kappa(t)| \leq |\kappa(t^*)|$ for all $t \in I_m$. Then,*

$$|\kappa(t)| \leq \frac{|1-\alpha_m|\, q_m}{|A|\,\epsilon^{\alpha_m}}\left(1 + r_m\epsilon + \mathcal{O}\left(\epsilon^2\right)\right) \tag{30}$$

on I_m. Furthermore, $|\kappa(t^*) - \kappa(t_m)| = \mathcal{O}\left(\epsilon^4\right)$.

Proof. The first statement follows from the second together with Eq. (28).

From Eqs. (9), (13) and (17), it is seen that $\kappa(t)$ can be written in the form of Eq. (21) with $\tilde{x} = w_m$, $\beta = (1+\alpha_m)/2$ and $h(t) = |w(t) - w_m|^2$. This means that h takes the form of Eq. (23) for bounded regions and Eq. (22) for the two types of unbounded regions considered.

Since κ for $\epsilon > 0$ is an analytic function with a local optimum at t^*,

$$\kappa(t_m) = \kappa(t^*) + \frac{1}{2}\kappa''(t^*)(t_m - t^*)^2 + \mathcal{O}\left((t_m - t^*)^3\right),$$

and it follows from Lemma 5.1 that $\kappa(t^*) - \kappa(t_m) = \mathcal{O}\left(\epsilon^4\right)$. □

Assume that a curvature bounded by C is desired near the mth vertex. Then, if the vertex is convex, i.e., $\alpha_m < 1$, and if ϵ is small enough to leave out the higher order terms, Theorem 5.1 shows that ϵ should be chosen such that

$$\epsilon^{\alpha_m} > \frac{|1-\alpha_m|\, q_m}{|A|\, C}. \tag{31}$$

Similarly, if the vertex is concave and therefore $\alpha_m > 1$, ϵ should if possible be chosen such that

$$\frac{\epsilon^{\alpha_m}}{1 + r_m\epsilon} > \frac{|1-\alpha_m|\, q_m}{|A|\, C}. \tag{32}$$

This can of course be difficult to achieve.

However, in the method used in [4], tangent points can be put arbitrarily dense on the boundary curve of the considered region. This means that one can to some extent rule the size of α_m in the outer polygon. By using many tangent points to achieve obtuse angles with α_k close to 1, it is also possible to bring down the size of q_m.

Note also the different curvature in the curves we consider close to convex and concave vertices respectively. The parameter $\alpha_m \in [0, 2]$, and is less than one if the vertex is convex and greater than one if it is concave. For example, a right-angled convex vertex means $\alpha_m = 0.5$, but if it is concave, $\alpha_m = 1.5$. This results in a significantly greater curvature in the latter when ϵ is small. This difference is even greater for acute angles.

Acknowledgements

I would like to thank professor Börje Nilsson, Växjö University, Sweden, for many valuable comments, ideas and discussions on this work.

References

1. T. A. Driscoll and L. N. Trefethen, *Schwarz-Christoffel mapping*, Cambridge Monographs on Applied and Computational Mathematics, Vol. 8, Cambridge University Press, Cambridge, 2002.
2. L. N. Trefethen, *SIAM J. Sci. Statist. Comput.* **1**, 82 (1980).
3. T. A. Driscoll, Schwarz–Christoffel toolbox for Matlab available from `http://www.math.udel.edu/~driscoll/SC/`.
4. A. Andersson, *Schwarz–Christoffel mappings for non-polygonal regions*, Report 07104, MSI, Växjö University, Växjö, 2007; submitted to SIAM J. Sci. Comput.
5. B. Nilsson, *Proc. R. Soc. Lond. A* **458**, 1555 (2002).
6. V. Y. Gutlyanskiĭ and S. A. Kopanev, *Ukraïn. Mat. Zh.* **44**, 1330 (1992).
7. J. L. Walsh, *Amer. Math. Monthly* **44**, 202 (1937).
8. J. L. Walsh, *Amer. Math. Monthly* **60**, 671 (1953).
9. D. B. Shaffer, *Trans. Amer. Math. Soc.* **158**, 143 (1971).

AN $\mathbb{R}$-LINEAR PROBLEM WITH DERIVATIVES FOR DOUBLY PERIODIC FUNCTIONS AND ITS APPLICATION

PIOTR DRYGAŚ

Department of Mathematics, University of Rzeszow,
ul. Rejtana 16a, 35-959 Rzeszow, Poland
E-mail: drygaspi@univ.rzeszow.pl

We discuss the $\mathbb{R}$-linear conjugation problem for a disc in a class of doubly periodic functions with derivatives, i.e., the $\mathbb{R}$-linear problem with derivatives on the torus. We prove that this problem can be solved by the method of successive approximations in a Hardy-type space under some natural conditions.

Keywords: $\mathbb{R}$-Linear problem, functional equations, boundary value problem, Hardy space.

1. Introduction

Let D^+ be a simply connected domains in the complex plane $\mathbb{C}$ bounded by smooth curves ∂D^+, D be the complement of all closures of ∂D^+ to the extended complex plane $\mathbb{C} \cup \{\infty\}$. Let ∂D^+ be orientated in counter clockwise direction. Let $a(t)$, $b(t)$ and $c(t)$ be given Hölder continuous function on ∂D^+; $a(t) \neq 0$. The $\mathbb{R}$-linear conjugation problem on $\mathbb{C}$ is states as follows. To find a function $\varphi(z)$ analytic in D, D^+, continuous in the closures of the considered domains with the following conjugation condition

$$\varphi^+(t) = a(t)\varphi^-(t) + b(t)\overline{\varphi^-(t)} + c(t), \quad t \in \partial D^+. \tag{1}$$

The problem (1) in class of doubly periodic functions has application in composite. Mityushev in [13] shown that the condition (1) with approat coefficient a, b, c express the perfect contact between different materials. One can find constructive results for the effective conductivity tensor in [12]. The perfect contact condition is an idealization which is valid in a large variety of situations. However, imperfect contact is known to exist in composites. It is expressed by a discontinuity in the temperature distribution between matrix and inclusions. The effect of this interface phenomenon on the effective conductivity has been studied by Lipton ([9]), Torquato & Cheng ([2]), Benveniste & Miloh ([1]), Goncales & Kołodziej ([7]).

The present paper is devoted to $\mathbb{R}$-linear problem in a class of doubly periodic functions with derivatives

$$\varphi^-(t) = a(t)\varphi^+(t) - b(t)\overline{\varphi^+(t)} + c(t)(\varphi^+(t))' + d(t)\overline{(\varphi^+(t))'}, \quad |t| = r, \quad (2)$$

where $a(t)$, $b(t)$, $c(t)$, $d(t)$ be a given Hölder continuous function on ∂D^+. The condition (2) corresponds to the thermal resistance condition introduced by Kapitza ([1]). In contrast with the contributions, which were previously mentioned, we apply the method of functional equations ([10,12]). A simple analytical algorithm to obtain the coefficients is presented in Sec. 5.

This paper is organized as follows. The problem is presented in Sec. 2. Complex potential and functional equations which govern the local field are introduced in Sec. 3. Corresponding spaces and operators is introduced in Sec. 4. Solution to functional equations is given in Sec. 5.

2. Statement of the problem

It is convenient to use the complex variable $z = z + iy$, where i is the imaginary unit. Let $\mathbb{Z}[i]$ be a set of complex numbers with integer real and imaginary parts. Consider a square lattice $\mathcal{Q}$ which is defined by two fundamental translation vectors expressed in the form of the complex numbers 1 and i. Introduce the zero-th cell $Q_0 := \{z = t_1 + it_2 : -1/2 < t_j < 1/2 \ (j = 1, 2)\}$. The lattice $\mathcal{Q}$ consists of the cells $Q_m := \{z \in \mathbb{C} : z - m \in Q_0\}$, where $m \in \mathbb{Z}[i]$. Let the disc $D^+ = \{z \in \mathbb{C} : |z| < r\}$ lies in the cell Q_0, D^- be the complement of the closure of D^+ to Q_0.

Let $\mathbf{n} = (n_1, n_2)$ denote the outward unit normal vector to the circle ∂D^+, $\frac{\partial}{\partial \mathbf{n}}$ the normal derivative to ∂D^+. Then according to the direction of the normal vector the signs "+" and "−" are assigned to the domains D^+ and D^-. We have to find a function $u^+(z)$ harmonic in D^+ and $u^-(z)$ harmonic in D^-, continuously differentiable in the closures of the considered domains with the following conjugation conditions

$$\frac{\partial u^-}{\partial \mathbf{n}}(t) = \lambda \frac{\partial u^+}{\partial \mathbf{n}}(t), \quad (3)$$

$$\lambda \frac{\partial u^+}{\partial \mathbf{n}}(t) + \gamma(u^+(t) - u^-(t)) = 0, \quad t \in \partial D^+. \quad (4)$$

Hereafter, the letter z is used for a variable in domains, t for a variable on curves. The relation (3) and (4) model the imperfect thermal contact between inclusion and matrix [4].

Let the function $u^-(z)$ has the unit jump per a cell along the x–axis and is periodic along the y–axis

$$u^-(z+1) - u^-(z) = 1, \quad u^-(z+i) - u^-(z) = 0. \tag{5}$$

3. Complex potentials and boundary value problem

Introduce the complex potentials

$$\varphi^{\pm}(z) = u^{\pm}(z) + iv^{\pm}(z) \tag{6}$$

analytic in D^+, D^-, continuously differentiable in the closures of the considered domains. Two real relations (3) and (4) yield the complex relation (see for details [3,10])

$$\varphi^-(t) = \frac{\lambda+1}{2}\left[\varphi^+(t) - \rho\overline{\varphi^+(t)} + \mu t(\varphi^+(t))' + \mu\frac{r^2}{t}\overline{(\varphi^+(t))'}\right], \quad |t|=r, \tag{7}$$

where

$$\rho = \frac{\lambda-1}{\lambda+1}, \qquad \mu = \frac{1+\rho}{2r\gamma}. \tag{8}$$

For $\mu = 0$, (7) becomes the well-studied $\mathbb{R}$–linear conjugation problem ([10,13]).

The relations (5) in terms of the complex potentials become ([13])

$$\varphi^-(z+1) - \varphi^-(z) = 1 + id_1, \quad \varphi^-(z+i) - \varphi^-(z) = id_2, \tag{9}$$

where d_1 and d_2 are undetermined real constants.

Differentiating (7) along the curve ∂D^+ and using the relation ([10, p. 52])

$$\overline{\left[\varphi^+\left(\frac{r^2}{z}\right)\right]}' = -\left(\frac{r}{z}\right)^2 \overline{\psi^+\left(\frac{r^2}{z}\right)}, \tag{10}$$

we arrive at the following condition

$$\psi^-(t) = \frac{\lambda+1}{2}\left[(1+\mu)\psi^+(t) + (\rho-\mu)\left(\frac{r}{t}\right)^2\overline{\psi^+(t)} + \right.$$
$$\left. + \mu t[\psi^+(t)]' - \mu t\left(\frac{r^2}{t}\right)^2\overline{[\psi^+(t)]'}\right], \qquad |t| = r. \tag{11}$$

After differentiation the conditions (9) become

$$\psi^-(z+1) - \psi^-(z) = 0, \quad \psi^-(z+i) - \psi^-(z) = 0. \tag{12}$$

Lemma 3.1. *The function $\psi^+(z)$ satisfies the relations*

$$\psi^+(z) = \overline{\psi^+(\overline{z})} \quad \text{and} \quad \psi^+(z) = \psi^+(-z), \qquad |z| \le r.$$

The proof of Lemma 3.1 follows from the symmetry of the problem (11). It follows from lemma that the Taylor expansion of $\psi^+(z)$ has the form

$$\psi^+(z) = \sum_{k=0}^{\infty} \psi_{2k} z^{2k}, \tag{13}$$

where the coefficients ψ_{2k} are real.

4. Spaces and operators

We introduce the space C_A of function analytic on D^+ and continuous in the closure of on D^+, endowed with the norm $\|f\|_{C_A} = \sup\limits_{0\leq\theta\leq 2\pi} \left|f\left(re^{i\theta}\right)\right|$. Let C_A^1 be the space of functions analytic in D^+ and continuously differentiable in the closure of D^+, endowed with the norm $\|f\|_{C_A^1} = \|f\|_{C_A} + \|f'\|_{C_A}$.

We define the Hilbert spaces $\mathcal{L}^2(\partial D^+)$ of functions $f(t)$ which satisfy the condition $f\left(re^{i\theta}\right) \in \mathcal{L}^2(0, 2\pi)$ with respect to the variable $\theta \in (0, 2\pi)$, endowed with the norm $\|f\|^2_{\mathcal{L}^2(\partial D^+)} = \int_0^{2\pi} \left|f\left(re^{i\theta}\right)\right|^2 d\theta$.

The space $\mathcal{H}^2$ is introduced as the space of analytic function on D^+, satisfying the condition $\sup\limits_{0<r_0<r} \int_0^{2\pi} |f\left(r_0e^{i\theta}\right)|^2 d\theta < \infty$, endowed with the norm $\|f\|^2_{\mathcal{H}^2} = \sup\limits_{0<r_0<r} \int_0^{2\pi} |f\left(r_0e^{i\theta}\right)|^2 d\theta$. Note that for $r = 1$ we obtain the classical Hardy space (see [8]). The spaces are related by

$$C_A^1 \subset C_A \subset \mathcal{H}^2.$$

Let $\psi^+(z) = \sum\limits_{m=0}^{\infty} \alpha_m z^m$ be the Taylor series of $\psi^+(z)$ in D^+. We introduce the space l^2 of the sequences $\alpha = \{\alpha_m\}_{m=0}^{\infty}$, endowed with the norm $\|f\|_{l^2} = \sum\limits_{m=0}^{\infty} |\alpha_m|^2 r^{2m}$ corresponding to $\mathcal{H}^2$. It follows from [6] that $\mathcal{H}^2 \cong l^2$.

Introduce the operator $\mathrm{L}\colon C_A^1 \to \mathcal{H}^2$ defined as

$$\mathrm{L}f(z) = (1+\mu)\, f(z) + \mu z f'(z), \quad |z| < r. \tag{14}$$

Let us note that the direct operator L is not bounded. However, the iterative scheme constructed bellow is based on the inverse operator L^{-1} having the required properties described in Lemmas 4.1 and 4.2.

Lemma 4.1. [3] *If* $\mu > 0$, *then the inverse operator* L^{-1} *is bounded in* $\mathcal{H}^2$ *and* $||\mathrm{L}^{-1}|| \leq 1$.

Lemma 4.2. [3] *If* $f \in C_A^1$, *then* $\mathrm{L}^{-1} f \in C_A^1$.

Define the operator $\mathcal{A}$ in the space $\mathcal{H}^2$

$$\mathcal{A}f(z) = \sum_{m\neq 0} \mathcal{A}_m f(z), \tag{15}$$

where

$$\mathcal{A}_m f(z) = (\rho-\mu)\left[\frac{r}{z-m}\right]^2 \left[\overline{\psi^+\left[\frac{r^2}{\overline{\overline{z-m}}}\right]} - \overline{\psi^+(0)}\right] - \mu\frac{r^4}{(z-m)^3}\overline{\psi^{+\prime}\left[\frac{r^2}{\overline{\overline{z-m}}}\right]}. \tag{16}$$

At the beginning let us prove that the infinite series of (15) makes a sense. The series 15 convers encoding to Eisentein ([11,14]).

Lemma 4.3. *The series*

$$\sum_{m\neq 0} \mathcal{A}_m f(z) \tag{17}$$

converges absolutely and uniformly in $\mathrm{cl}Q_0$ *for each* $f \in C_A^1$.

Proof. Let us consider a finite set $T_N = \{m_1, m_2, \ldots, m_N\} \subset \mathbb{Z}[i]$, where the number N is chosen in such way that $|z-m| \geq h > 1$, for all $m \in T = \mathbb{Z}[i] \setminus T_N$ and $z \in \mathrm{cl}Q_0$. Its easy to see that $0 \in T_N$. Let $f(z) = \sum_{k=0}^{\infty} \alpha_k z^k$ be a Taylor expansion of $f(z)$. Then

$$\begin{aligned}\sum_{m\in T} \mathcal{A}_m f(z) = \sum_{m\in T} \Bigg((\rho-\mu) \sum_{k-1}^{\infty} r^{2k}\overline{\alpha_k}(z-m)^{-3}(z-m)^{-k+1} \\ - \mu \sum_{k=1}^{\infty} r^{2k+2}k\overline{\alpha_k}(z-m)^{-3}(z-m)^{-k+1}\Bigg).\end{aligned} \tag{18}$$

The series $S(z) = \sum_{m\in T} |z-m|^{-3}$ converges uniformly in $\mathrm{cl}Q_0$ ([5]). Moreover, $|z-m|^{-k+1} \leq h^{-k+1}$, $k = 1, 2, \ldots$. So we have

$$\begin{aligned}\sum_{m\in T} |\mathcal{A}_m f(z)| \leq |\rho-\mu| hM \sum_{k=1}^{\infty} |\alpha_k| r^{2k} h^{-k} \\ + |\mu| hMr^2 \sum_{k=1}^{\infty} |\alpha_k| k r^{2k} h^{-k},\end{aligned} \tag{19}$$

where $M := \max_{\mathrm{cl}Q_0} |S(z)|$. Since $f \in C_A^1$, $h > 1$ and $r < 1$, then the series in (19) are convergent. Therefore the series (18) and (17) converges absolutely and uniformly in $\mathrm{cl}Q_0$. □

Theorem 4.1. *Operator* $\mathcal{A}$ *is compact in* C_A^1.

Proof. Let us consider the sum (16) as a limit of a sequence of finite sums which are compact being finite sums of integral operators of the form

$$\mathcal{A}_m f(z) = (\rho-\mu)\left(\frac{r}{z-m}\right)^2 \left[\overline{\frac{1}{2\pi i}\int\limits_{\partial D+}\frac{f(t)}{t-\frac{r^2}{z-m}}dt} - \overline{f(0)}\right] \\ -\mu\frac{r^4}{(z-m)^3}\overline{\frac{1}{2\pi i}\int\limits_{\partial D+}\frac{f(t)}{\left(t-\frac{r^2}{z-m}\right)^2}dt} \tag{20}$$

This formula follows from Cauchy law in disk $|z|<r$. The function $\frac{r^2}{z-m}$ is bounded for any $t \in \partial D^+$. Then the integral operators are compact in $\mathcal{H}^2$ as integral operators with continuous and bounded on ∂D^+ kernels. The operator of multiplication by $(\rho-\mu)\left(\frac{r}{z-m}\right)^2$ and $\mu\frac{r^4}{(z-m)^3}$ is bounded in $\mathcal{H}^2$. Therefore, the operators of finite sums are compact in C_A^1 as a sum of compact operators with bounded multipliers. Moreover, it follows from Lemma 4.3 that sequence of finite sums converges uniformly to sum (16) for every $f \in C_A^1$. Therefore $\mathcal{A}$ is a compact operator in C_A^1. □

Lemma 4.4. *If* $f \in \mathcal{H}^2$, *then* $\mathrm{L}^{-1}\mathcal{A}f \in C_A^1$.

Proof. From the proof of Theorem 4.1 we obtain that operators in (20) transform $\mathcal{H}^2$ onto functions analytic in D^+ and continuously differentiable in their closure. Then the right hand side of (16) is a continuously differentiable function in the considered disc. Therefore $\mathcal{A}f \in C_A^1$ if $f \in \mathcal{H}^2$. Applying Lemma 4.2 we end the proof. □

5. Functional equation

Applying Eisenstein's method ([11,14]) it is convenient to use the series

$$\mathcal{W}f(z) = \sum_{m\neq 0} \mathcal{W}_m f(z), \tag{21}$$

where

$$\mathcal{W}_m f(z) = (\rho-\mu)\left(\frac{r}{z-m}\right)^2 \overline{\psi^+\left(\frac{r^2}{\overline{z-m}}\right)} - \mu\frac{r^4}{(z-m)^3}\overline{\psi^{+\prime}\left(\frac{r^2}{\overline{z-m}}\right)} \tag{22}$$

instead of (15)-(16). Here, according to Eisenstein's method ([11,14])

$$\sum_{m\neq 0}\frac{1}{(z-m)^2} := \wp(z) - \frac{1}{z^2} + S_2, \tag{23}$$

where $\wp(z)$ is Weierstras's function, S_2 is constans ([11]). Therefore, the operators $\mathcal{W}$ and $\mathcal{A}$ are related by identity

$$\mathcal{W}f(z) = \mathcal{A}f(z) + \left(\wp(z) - \frac{1}{z^2} + S_2\right) r^2\overline{f}(0). \tag{24}$$

This modification produces a functional-differential equation in simple form as follows. Introduce the function

$$\Phi(z) = \begin{cases} \frac{2}{\lambda+1}\left[(1+\mu)\psi^+(z) + \mu z\psi^{+\prime}(z) \right. \\ \quad - \sum\limits_{m\neq 0}\left\{(\rho-\mu)\left(\frac{r}{z-m}\right)^2\overline{\psi^+(\frac{r^2}{\overline{z-m}})}\right. & |z|<r, \\ \qquad \left.\left. -\mu\frac{r^4}{(z-m)^3}\overline{\psi^{+\prime}(\frac{r^2}{\overline{z-m}})}\right\}\right], \\ \psi^-(z) - \frac{2}{\lambda+1}\left[\sum\limits_{m}\left\{(\rho-\mu)\left(\frac{r}{z-m}\right)^2\overline{\psi^+(\frac{r^2}{\overline{z-m}})}\right.\right. & z\in D^-, \\ \qquad \left.\left. -\mu\frac{r^4}{(z-m)^3}\overline{\psi^{+\prime}(\frac{r^2}{\overline{z-m}})}\right\}\right]. \end{cases} \tag{25}$$

Let us calculate the jump of $\Phi(z)$ across the curve $|t|=r$

$$\Delta = \lim_{\substack{z\to t\\ z\in D^+}} \Phi(t) - \lim_{\substack{z\to t\\ z\in D^-}} \Phi(t).$$

Applying equation (11) we obtain that $\Delta = 0$.

The principle of analytic continuation and Liouville's theorem imply that $\Phi(z)$ is a constant, say c. Following [13] one can take $c=1$. The definition of $\Phi(z)$ in $|z|\le r$ yields the following functional-differential equation with respect to ψ^+ analytic in $|z|<r$ and continuously differentiable in $|z|\le r$:

$$\begin{aligned} &(1+\mu)\,\psi^+(z) + \mu z\psi^{+\prime}(z) - \frac{1}{2}(\lambda+1) \\ &= \sum_{m\neq 0}\left\{(\rho-\mu)\left(\frac{r}{z-m}\right)^2\overline{\psi^+\left(\frac{r^2}{\overline{z-m}}\right)} - \mu\frac{r^4}{(z-m)^3}\overline{\psi^{+\prime}\left(\frac{r^2}{\overline{z-m}}\right)}\right\}, \end{aligned} \tag{26}$$

where Eisenstein's summation is used.

Theorem 5.1. *For any $g\in\mathcal{H}^2$, and sufficiently small coefficients $|\rho|<1$ and $\mu\ge 0$, the method of successive approximations applied to equation $\psi^+ = \mathrm{L}^{-1}\mathcal{W}\psi^+ + \mathrm{L}^{-1}g$ converges in $\mathcal{H}^2$.*

Proof follows from the structure of the operator $\mathcal{A}$, (21) and Lemma 4.1. Eequation (26) can be considered as an equation

$$\psi^+ = \mathrm{L}^{-1}\mathcal{W}\psi^+ + \mathrm{L}^{-1}g \tag{27}$$

in the space C_A^1. Since it is hard to study this equation in C_A^1, we simplify the problem by using of the so-called "Pumping Principle" systematically applied to boundary value problems (cf. [10, p. 21]). This principle is based on the study of equation (27) first in a wider space $\mathcal{H}^2$, then it is assumed that g belongs to C_A^1. Let φ be a solution of equation (27) in the space $\mathcal{H}^2$ with $g \in C_A^1$. Using Lemma 4.2 and Lemma 4.4 we obtain from (27) that $\psi^+ \in C_A^1$.

It is convenient to apply the method of successive approximating in the following form. First, we have the zero approximation

$$\mathrm{L}\psi^+_{(0)}(z) = g(z). \tag{28}$$

Using the Taylor form of the function $g(z)$ and $\psi_0^+(z)$ in $|z| < r$, we write solution of (28) in the form

$$\psi^+_{(0)}(z) = \sum_{m=0}^{\infty} \frac{g_m}{1+\mu+\mu m} z^m. \tag{29}$$

Suppose we know the approximation of order $p-1$. Then the p-th approximation can be found from the system of differential equations with respect to $\psi^+_{(p)}(z)$

$$\mathrm{L}\psi^+_{(p)}(z) = g_{(p)}(z), \tag{30}$$

where

$$g_{(p)}(z) = \sum_{m\neq 0} \left\{ (\rho-\mu)\left(\frac{r}{z-m}\right)^2 \overline{\psi^+_{(p-1)}\left(\frac{r^2}{\overline{z-m}}\right)} - \mu \frac{r^4}{(z-m)^3} \overline{\psi^{+\,\prime}_{(p-1)}\left(\frac{r^2}{\overline{z-m}}\right)} \right\}, \quad |z| \le r. \tag{31}$$

Then according to (29) we obtain

$$\psi^+_{(p)}(z) = \sum_{m=0}^{\infty} \frac{g_m^{(p)}}{1+\mu+\mu m} z^m, \tag{32}$$

where $g_m^{(p)}$ are the Taylor coefficient of $g_{(p)}(z)$ in $|z| < r$.

We have proposed an efficient algorithm to solve functional equation (26) in the space $\mathcal{H}^2$. However, the right hand side of (26), the function $g(z)$ is continuously differentiable in $|z| \le r$, hence $\mathrm{L}^{-1}g \in C_A^1$ (see Lemma 4.2). Moreover, the operators $\mathrm{L}^{-1}\mathcal{W}$ transforms $\mathcal{H}^2$ to C_A^1 (see Lemma 4.4). Therefore, we actually construct $\psi^+(z)$ continuously differentiable in $|z| \le r$.

The function $u(z)$ satisfying the boundary problem (3)–(4) has the form

$$u(z) = \begin{cases} \mathrm{Re}\psi^+(z), & z \in D^+, \\ \mathrm{Re}\psi^-(z), & z \in D, \end{cases} \tag{33}$$

where $\psi^-(z)$ is calculated by (11). One can see that $u(z)$ depends on an arbitrary additive real constant $\mathrm{Re}\, c$.

References

1. Y. Benveniste, T. Miloh, *The effective conductivity of composites with imperfect contact at constituent interfaces*, Int. J. Eng. Sci., 24, 1986, 1537-1552.
2. H. Cheng, S. Torquato, *Effective conductivity of periodic arrays of spheres with interfacial resistance*, Proc. Royal Soc. London A, 453, 1997, 145-161.
3. P. Drygaś, *A functional-differential equation in a class of analytic functions and its application*, Aequationes Math., 73, 2007, 222-232.
4. P. Drygaś, *Steady heat conduction of material with coated incusion in the case of imperfect contact*, Math. Model. Anal., 12(3), 2007, 291-296.
5. A. Hurwitz, R. Courant, *Vorlesungen über Allgemeine Funktionentheorie und Elliptische Funktionen*, Springer-Verlag, Berlin, 1964.
6. A. N. Kolmogorov, S. V. Fomin, *Elements of the Theory of Functions and Functional Analysis*, Moscow, 1981 (in Russian).
7. L. C. Goncalves, J. A. Kołodziej, *Determination of effective thermal conductivity in fibrous composites with imperfect contact between constituents*, Int. Comm. Heat Mass Transfer, 20 (1993), 111-121.
8. P. Koosis, *Introduction to H_p Spaces*, Cambridge Univ. Press, Cambridge, 1980.
9. R. Lipton, D. R. S. Talbot, *Bounds for the effective conductivity of a composite with imperfect interface*, Proc. Royal Soc. London, 457, 2001, 1501 1517.
10. V. V. Mityushev, S. V. Rogosin, *Constructive Methods for Linear and Nonlinear Boundary Value Problems for Analytic Functions Theory*, Chapman&Hall/CRC, Boca Raton, 2000.
11. V. V. Mityushev, P. M. Adler, *Longitudial permeability of a doubly periodic rectangular array of circular cylinders*, I, ZAMM, 82, 2002, 335-345.
12. V. V. Mityushev, E. Pesetskaya, S. V. Rogosin, *Analytical Methods for Heat Conduction in Composites and Porous Media*, in print.
13. V. V. Mityushev, *$\mathbb{R}$-linear problem on torus and its application to composites*, Complex Variables, 50, 2005, 621-630.
14. A. Weil, *Elliptic Functions According to Eisenstein and Kronecker*, Springer-Verlag, Berlin etc., 1976.

Session 05

Complex and Functional Analytic Methods in Partial Differential Equations

SESSION EDITORS

H. G. W. Begehr	Freie Universität, Berlin, Germany
D.-Q. Dai	Zhongshan University, Guangzhou, China
J. Y. Du	Wuhan University, Wuhan, China

SOME PROBLEMS FOR ELLIPTIC SYSTEMS ON THE PLANE

G. AKHALAIA, G. MAKATSARIA, N. MANJAVIDZE *

Georgian Technical University, Tbilisi, Georgia
* *E-mail: ninomanjavidze@yahoo.com*

We study some problems for elliptic-type systems on the plane with regular as well as with irregular coefficients. Some models are investigated.

1. Introduction

For the first order elliptic systems on the plane, the boundary value problems were investigated extensively over the past years [4].

A vector $w(z) = (w_1, \dots, w_n)$ is called a generalized analytic vector in some domain D of the complex plane $\mathbb{C}$, if it is a solution of an elliptic system of the form

$$\partial_{\overline{z}} w - Q(z)\, \partial_z w + A(z)\, w + B(z)\, \overline{w} = 0, \tag{1}$$

where A, B are given square matrices of order n of the class $L_p(\overline{D})$, $p > 2$, $Q(z)$ is a given square matrix of order n of the following special quasi-diagonal form: Every block $Q^r = (q_{ik})^r$ is a lower (upper) triangular matrix satisfying the conditions

$$q_{11}^r = \cdots = q_{m_s,m_s}^r = q^r, \quad |q^r| \leq q_0 < 1,$$
$$q_{ik}^r = q_{i+s,k+s}^r \quad (i + s \leq n, \quad k + s \leq n);$$

moreover $Q(z) \in W_p^1(\mathbb{C})$, $p > 2$ and $Q(z) \equiv 0$ outside of some circle of the complex plane $\mathbb{C}$. By a solution of system (1), we mean a so-called regular solution, i.e., $w(z) \in L_2(\overline{D})$, $w_{\overline{z}}$, $w_z \in L_\lambda(D')$, $\lambda > 2$, $\overline{D}' \subset D$. System (1) is to be fulfilled almost everywhere on D.

The following equation

$$\partial_{\overline{z}} \Psi - \partial_z (Q' \, \Psi) - A'(z)\, \Psi - \overline{B'(z)\, \Psi} = 0 \tag{2}$$

---

*Corresponding author.

is called the conjugate equation of equation (1), where prime $'$ denotes the transposition of the matrices.

When $A = B = 0$, system (1) takes form

$$\partial_{\overline{z}} w - Q(z)\,\omega(z) = 0. \tag{1'}$$

The solutions of equation (1′) are called Q-holomorphic vectors.

The theory of generalized analytic vectors was constructed by B. Bojarski [1]. These results and some further development of this theory are presented in [4].

2. Boundary Value Problems for Elliptic System (1) on the Plane with Angular Points

We introduce suitable classes of generalized analytic vectors. Let Γ be a piecewise smooth curve. Denote by $E_p^{\pm}(\Gamma, Q, A, B, \rho)$ the class of solutions of the system (1) representable by generalized Cauchy type integrals

$$w(z) = \frac{1}{2\pi i}\int_\Gamma \left\{\Omega_1(z,t)\,d_Q t\varphi(t) - \Omega_2(r,t)\,\overline{d_Q t\varphi(t)}\right\} + \sum_{k=1}^{N} c_k\,W_k(z), \tag{3}$$

where c_k are arbitrary real constants, $W_k(z)$ are continuous vectors on the whole plane vanishing at infinity, $\{W_k\}$ $(k = \overline{1,N})$ forms a complete system of linearly independent solutions of the Fredholm equation

$$Kw \equiv w(z) - \frac{1}{\pi}\int_D v(t,z)\big[A(t)\,w(t) + B(t)\,\overline{W(t)}\,\big]\,d\sigma_t = 0, \tag{4}$$

the matrix $v(t,z)$ is generalized Cauchy kernel for the equation (1′), the weight function $\rho(t) = \prod\limits_{k=1}^{m} |t-t_k|^{\nu_k}$, $-1 < \nu_k < p-1$, $p > 1$, $\varphi(t) \in L_p(\Gamma, \rho)$ and satisfies the condition

$$\operatorname{Im}\int_\Gamma (d_Q\,t\,\varphi(t), \Psi_j) = 0 \quad (j = \overline{1,N}), \quad (d_Q\,t = I\,dt + Q\,\overline{dt}\,), \tag{5}$$

here Ψ_j form a similar system for the conjugate equation (2), Ω_1 and Ω_2 are the fundamental kernels of (1) representable by the resolvent Γ_1 and Γ_2

$$\begin{aligned} \Omega_1(z,t) &= v(t,z) + \int_D \Gamma_1(z,\tau)\,v(t,\tau)\,d\sigma_\tau\,, \\ \Omega_2(z,t) &= \int_D \Gamma_2(z,t)\,\overline{v(t,\tau)}\,d\sigma_\tau\,. \end{aligned} \tag{6}$$

Denote by $E_q^{\pm}(\Gamma, Q', -A', -B', \rho^{1-q})$, $q = p/p-1$ the class of solutions of the equation (2) representable in the form

$$\Psi(z) = \frac{1}{2\pi i}\int_\Gamma \left\{ \Omega_1'(t,z)\, d_{Q'}th(t) - \overline{\Omega_r'(t,z)}\,\overline{d_{Q'}th(t)} \right\} + \sum_{k=1}^{N} c_k\, \Psi_k(z), \quad (7)$$

where the density $h(t) \in L_q(\Gamma, \rho^{1-q})$ satisfies the conditions

$$\operatorname{Im} \int_\Gamma (d_{Q'}\, t\, h(t), W_j(t)) = 0 \quad (j = \overline{1,N}). \quad (8)$$

Our model problem is the following:

Find a vector $w(z) = (w_1, \ldots, w_n) \in E_p^{\pm}(\Gamma, Q, A, B, \rho)$ satisfying the boundary condition

$$w^+(t) = a(t)\, w^-(t) + b(t)\, \overline{w^-(t)} + c(t), \quad (9)$$

almost everywhere on Γ. Γ is piecewise smooth closed curve, the knot points of Γ (where Γ loosed the smoothness) are included in the set of $\{t_k\}$ points; $a(t)$ and $b(t)$ are given piecewise continuous $n \times n$-matrices on Γ, $\inf |\det a(t)| > 0$ and $c(t)$ is given vector of the class $L_p(\Gamma, \rho)$.

The boundary value problem (9) for holomorphic vectors was studied by A. Markushevich and is called the generalized Hilbert problem [8]. The boundary condition (9) contains the conjugate value of the desired vector. Therefore, the Noetherity condition and the index formula of this problem depend on the values of the angles at the knot points of the boundary curve Γ.

Substituting the integral representation formula (3) into the boundary condition (9) for the unknown vector $\varphi(t)$, we obtain the following singular integral equation

$$(M\,\varphi)(t) = f(t), \quad (10)$$

$$f(t) = 2c(t) + 2\sum_{k=1}^{N} c_k\, [a(t) - I]\, W_k(t) + b(t)\, \overline{W_k(t)}$$

and the solution is subject to conditions (5).

It is easy to see that problem (9) is Noetherian in the class $E_p^{\pm}(\Gamma, Q, A, B, \rho)$ if and only if the singular integral operator $(M\,\varphi)(t)$ is Noetherian in $L_p(\Gamma, \rho)$.

The necessary and sufficient Noetherity condition for the integral operator $M\,\varphi$ is [2]

$$\inf_{t\in\Gamma,\, \xi\in R^1} \left|\det M(t_k, \xi)\right| > 0 \quad (k = \overline{1,m}), \quad (11)$$

where the matrix $M(t_k, \xi)$ is defined in the following way: In the points different from the knot points

$$M(t_k, \xi) = (1 + \operatorname{sgn}\xi)\, I + (1 - \operatorname{sgn}\xi)\, a(t), \tag{12}$$

and in the knot points

$$M(t_k, \xi) = \begin{pmatrix} (1-s_0)I+(1+s_0)a(t-0) & s_1(I-a(t-0)) \\ s_2(I-a(t+0)) & (1+s_0)I+(1+s_0)a(t+0) \end{pmatrix}, \tag{13}$$

where $s_0 = \cos th\omega$, $s_1 = e^{(\alpha-1)\omega}/\sin h\omega$, $s_2 = e^{(1-\alpha)\omega}/\sin h\omega$, $\alpha = \eta/\pi$, and η is the angle between the tangents at the knot points, $0 < \eta < 2\pi$, $\omega_j = \pi(i\beta_j + \xi)$, $\beta_j = \frac{1+\nu_j}{p}$.

If the Noetherity condition (11) is fulfilled, then the necessary and sufficient solvability conditions are

$$\operatorname{Im} \int_\Gamma \big(f(t), d_Q\, f\, \Psi_k(t)\big) = 0, \tag{14}$$

where $\Psi_k(z)$ form a complete system of linearly independent solutions of the homogeneous problem

$$\Psi^+(t) = a'(t)\, \Psi^-(t) + b(t)[t'] + Q'(t)\, [\bar{t}\,']\, \Psi^-(t) \tag{15}$$

for the equation (2) in the class $E_q^{\pm}(\Gamma, Q', -A', -B', \rho^{1-q})$.

Starting from the properties of Mellin transform and the hyperbolic trigonometric functions, we define the index of singular integral operator and therefore the index of our problem (9) in $E_q^{\pm}(\Gamma, Q, A, B, \rho)$.

3. One Class of Two-Dimensional Third-Kind Fredholm Integral Equation

We have correct setting and complete analysis of boundary value problem for sufficiently wide class of equations. They are first order singular equations. In the works of L. Mikhailov and Z. Usmanov the integral equations with first order singularities are investigated. Further development of this theory is given in the works of A. Tungatarov, R. Akhmedov, M. Reissig, A. Timofeev, H. Begehr, D. Q. Dai and etc. Somewhat different equations were studied by L. Vinogradov and N. Bleiv. The equations of higher order singularities are undoubtedly of much more theoretical and practical interest. In the works of Kh. Najhmidinov and Z. Usmanov of such type equations are investigated, however the equations studied below cannot be reduced to them.

Let us consider the system

$$\sum_{k=0}^{m} z^{\nu_k} A_k \frac{\partial^k w}{\partial \bar{z}^{\,k}} = 0 \tag{16}$$

in some domain G of z plane, where m, ν are given natural numbers, A_k $(0 \le k \le m)$ are given complex $n \times n$-matrices; G is deleted neighborhood of $z = 0$. Assume that

$$\det \lambda_0 \neq 0, \quad \det A_m \neq 0, \quad A_k \cdot A_j = A_j \cdot A_k, \quad 0 \le j, \quad k \le m. \tag{17}$$

Construct all possible polynomials of the form

$$\tau_m \zeta^m + \tau_{m-1}\zeta^{m-1} + \cdots + \tau_1 \zeta + \tau_0 = 0, \tag{18}$$

where the coefficient τ_k is some eigen-value of the matrix A_k $(0 \le k \le n)$. Denote by Δ the set of all complex roots of these polynomials and introduce a number $\delta_0 = \max\limits_{\zeta \in \Delta} |\zeta|$, obviously $\delta_0 > 0$.

Along with the solution $w(z)$ of the system (16) construct its characteristic function

$$T_w(\rho) = \max_{0 \le \varphi \le 2\pi} \sum_{k=1}^{n} \sum_{p=0}^{m-1} \left| \frac{\partial^p \omega_k}{\partial \bar{z}^{\,p}} \left(\rho\, e^{iQ}\right) \right|, \quad \rho > 0. \tag{19}$$

Theorem 3.1. *Let $\nu \ge 2$ and $\psi(z)$ be some analytic function in G. Let the solution $w(z)$ of the system (16) satisfy the condition*

$$T_w(z) = O\left(|\psi(z)| \exp\left\{ \frac{\delta}{|z|^{\sigma}} \right\} \right), \quad z \to 0, \tag{20}$$

where δ is some number and $\sigma < \nu - 1$.

Then the solution $w(z)$ is identically the zero vector function. Moreover, when condition (20) is fulfilled then $w(n)$ is also trivial if

$$\sigma = \nu - 1, \quad \delta < \delta_0 \cos \pi\beta, \quad \beta = \max\left\{ \nu, \frac{\nu - 3}{2\nu - r} \right\}. \tag{21}$$

Note that in particular, when $\nu = 2$ for system (16) we are succeeded to state correct boundary value problem to take its complete analysis.

Example 3.1. Consider the integral equation

$$z^{\nu} w + p_n w = f, \tag{22}$$

where $\nu > 0$ is given integer, n is natural number, f is a given function on bounded domain G containing the origin,

$$\begin{aligned} p_1 w(z) &= -\frac{1}{\pi} \iint_G \frac{a_1(\zeta)\, w(\zeta)}{\zeta - z}\, dG(\zeta), \\ &\vdots \\ p_n w(z) &= -\frac{1}{\pi} \iint_G \frac{a_n(\zeta)\, p_{n-1}(w(\zeta))}{\zeta - z}\, dG(\zeta), \end{aligned} \tag{23}$$

is the sequence of singular integral operators, w is desired function; by $a_k = a_k(z)$ $(1 \le k \le n)$ are denoted the holomorphic in G and continuous in $\overline{G}$ functions. Here we mean that

$$m = \frac{\nu}{n} \ge 2. \tag{24}$$

Consider now the homogeneous integral equation

$$z^\nu w + p_n w = 0. \tag{25}$$

By a solution of this equation, we mean a function $w \in L_p(G)$, $p > 2$, satisfying (25) almost everywhere in $\overline{G}$. From the definition and on the basis of known properties of the operator $p_1 w$ [7] it follows that every solution of the equation (25) is continuous in every point of $\overline{G}$, except maybe the origin. Moreover, the function

$$w_0(z) \equiv z\, w(z), \quad z \in G, \tag{26}$$

has continuous derivatives of order $n-1$ in G and the function

$$w_1(z) \equiv z^\nu \frac{\partial^{n-1} w}{\partial \overline{z}^{\,n-1}}, \quad z \in G, \tag{27}$$

has generalized derivative with respect to $\overline{z}$ and

$$(w_1)_{\overline{z}} + R(z)\, w = 0, \tag{28}$$

where $R(z) = \prod_{k=1}^{n} a_k(z)$, $z \in G$.

Assume that there exists a holomorphic function $b(z)$ in domain G, such that

$$\big[b(z)\big]^n = -R(z), \quad z \in G \quad \text{and} \quad b(0) \ne 0. \tag{29}$$

Denote by

$$\alpha_k = \cos\frac{2\pi(k-1)}{n} + i \sin\frac{2\pi(k-1)}{n}, \quad 1 \le k \le n$$

all possible roots from 1 of order n. Taking into account (24), (27), (28) and (29) we obtain that every solution of the homogeneous equation (25) is representable in the form

$$w(z) = \sum_{k=1}^{m} Q_k(z), \quad z \in G \setminus \{0\}, \tag{30}$$

where $Q_k(z) = \Phi_k(z) \exp\{\alpha_k c(z) \overline{z}\}$, $c(z) = \frac{b(z)}{z^m}$ and $\Phi_k(z)$ $(k = \overline{1,n})$ are holomorphic in $G \setminus \{0\}$ functions.

It follows from (30) the validity of the representation

$$\frac{\partial^q w}{\partial \overline{z}^{\,q}} = \sum_{k=1}^{n} \alpha_k^q \, c(z)^q \, Q_k(z) \quad (1 \le q \le n-1), \tag{31}$$

which simultaneously with (30) admit the matrix form

$$A \cdot \Psi(z) = \Omega(z), \tag{32}$$

$$\Psi(z) = \text{column}\,(Q_1(z), Q_2(z), \dots, Q_n(z)),$$

$$\Omega(z) = \text{column}\left(w, \frac{1}{c(z)} \frac{\partial w}{\partial \overline{z}}, \dots, \frac{1}{c(z)^{n-1}} \frac{\partial^{n-1} w}{\partial \overline{z}^{\,n-1}}\right), \tag{33}$$

$$A = \begin{pmatrix} 1 & 1 & \cdots & 1 \\ \alpha_1 & \alpha_2 & \cdots & \alpha_n \\ \alpha_1^2 & \alpha_2^2 & \cdots & \alpha_n^2 \\ \dots & \dots & \dots & \dots \\ \alpha_1^{n-1} & \alpha_2^{n-1} & \cdots & \alpha_n^{n-1} \end{pmatrix}.$$

From (32) we have that every function Q_k $(k = \overline{1,n})$ is a linear combination of the components of the vector (33). Therefore, in the neighborhood of $z = 0$, the following estimation holds

$$Q(z) = O\left(\frac{1}{\overline{z}^{\,\nu}}\right), \quad z \to 0. \tag{34}$$

Consequently, we get the next result.

Theorem 3.2. *The homogeneous integral equation* (25) *has no nontrivial solutions in* $L_p(G)$, $p > 2$.

Let us consider the inhomogeneous case (22), when the right hand side function is s-analytic in G

$$f(z) = F_0(z) + \overline{z}\, F_1(z) + \cdots + \overline{z}^{\,s-1} F_{s-1}(z), \tag{35}$$

here s is a natural number, F_k $(0 \le k \le s-1)$ are arbitrary holomorphic functions in G. The following theorem is valid.

Theorem 3.3. *The inhomogeneous integral equation* (22) *is unsolvable in the class* $L_p(G)$, $p > 2$, *if the right hand side function has the form* (35), *where at least one function* F_k $(0 \leq k \leq s-1)$ *is not identically zero and* $s \leq n$.

Remark 3.1. We have established that the homogeneous integral equation (25) and the inhomogeneous equation (22) with sufficiently general right-hand sides have no solutions. Note that the exact description of the image of our integral operator for which equation (22) will have solutions is very complicated.

References

1. B. V. Bojarskiĭ, Theory of generalized analytic vectors. (Russian) *Ann. Polon. Math.* **17** (1966), 281–320.
2. R. Duduchava, On general singular integral operators of the plane theory of elasticity. *Rend. Sem. Mat. Univ. Politec. Torino* **42** (1984), no. 3, 15–41.
3. R. V. Duduchava, General singular integral equations and fundamental problems of the plane theory of elasticity. (Russian) *Trudy Tbiliss. Mat. Inst. Razmadze Akad. Nauk Gruzin. SSR* **82** (1986), 45–89.
4. R. P. Gilbert, J. L. Buchanan, First order elliptic systems: A function theoretic approach. *Mathematics in Science and Engineering,* 163. *Academic Press, Orlando, FL,* 1983.
5. G. Makatsaria, Correct boundary value problems for some classes of singular elliptic differential equations on a plane. *Mem. Differential Equations Math. Phys.* **34** (2005), 115–134.
6. G. Manjavidze, Boundary value problems of the theory of generalized analytic vectors. *Functional analytic methods in complex analysis and applications to partial differential equations* (Trieste, 1993), 13–51, *World Sci. Publ., River Edge, NJ,* 1995.
7. I. Vekua, Generalized analytic functions, *Pergamon, Oxford,* 1962.
8. N. P. Vekua, Systems of singular integral equations. *Noordhoff, Groningen* 1967.

NEUMANN PROBLEM FOR GENERALIZED POISSON AND BI-POISSON EQUATIONS

Ü. AKSOY

Department of Mathematics, Atılım University, Ankara, Turkey
E-mail: uaksoy@atilim.edu.tr

A. O. ÇELEBİ

Department of Mathematics, Yeditepe University, İstanbul, Turkey
E-mail: acelebi@yeditepe.edu.tr

Using the harmonic and biharmonic Neumann functions $N_1(z,\zeta)$ and $N_2(z,\zeta)$, and operators having these functions as kernels, the Neumann problem for second and fourth order complex partial differential equations with main parts as the harmonic and biharmonic operators, are discussed. Solvability conditions are obtained.

Keywords: Neumann Problem; Neumann Functions; Poisson Equation; Bi-Poisson Equation; Singular Integral Operators.

1. Introduction

The Neumann problem is investigated for several complex model equations. These equation include the inhomogeneous Cauchy-Riemann, Bitsadze, Poisson and bi-Poisson equations; see, e.g, [3–8].

The main subject of this paper is to investigate the solvability of the Neumann problem for linear general second and fourth order complex partial differential equations with harmonic and biharmonic operators as main parts, which will be called as generalized Poisson and generalized bi-Poisson equations, respectively. In Section 2 we give a review of the harmonic Neumann and bi-harmonic Neumann functions and the Neumann problems defined for Poisson and bi-Poisson equations. In Section 3, the operators related to these problems are introduced and some particular properties are discussed. In the last section, we consider the Neumann problems for generalized Poisson and bi-Poisson equations and we give the conditions for solvability of the problems.

2. Neumann Functions and Neumann Problems for Poisson and Bi-Poisson Equations

The harmonic Neumann function for the unit disc $\mathbb{D}$ is

$$N_1(z,\zeta) = \log|(\zeta - z)(1 - z\bar{\zeta})|^2 \ . \tag{1}$$

Its outward normal derivative is

$$\partial_{\nu_z} N_1(z,\zeta) = (z\partial_z + \bar{z}\partial_{\bar{z}})N_1(z,\zeta) = 2 \tag{2}$$

for $z \in \partial\mathbb{D}$, $\zeta \in \mathbb{D}$, and it satisfies

$$\frac{1}{2\pi i}\int_{\partial\mathbb{D}} N_1(z,\zeta)\frac{dz}{z} = 0$$

for $\zeta \in \mathbb{D}$ [3,5]. Neumann function is defined slightly different in [11].

Biharmonic Neumann function $N_2(z,\zeta)$ is attained by convolution of the harmonic Neumann function as

$$N_2(z,\zeta) = \frac{1}{\pi}\iint_{\mathbb{D}} N_1(z,\tilde{\zeta})N_1(\tilde{\zeta},\zeta)d\tilde{\xi}d\tilde{\eta} \tag{3}$$

and also called the Neumann function of second order [5]. For z and ζ in $\bar{\mathbb{D}}$ with $z \neq \zeta$

$$N_2(z,\zeta) = |\zeta - z|^2\,[\log|(\zeta - z)(1 - z\bar{\zeta})|^2 - 4] - (1-|z|^2)(1-|\zeta|^2)$$
$$+\, 2(2 - z\bar{\zeta} - \bar{z}\zeta) - \tilde{N}_2(z,\zeta)$$

where

$$\tilde{N}_2(z,\zeta) = -\frac{2}{2\pi i}\int_{\partial\mathbb{D}}[(2-z\bar{\tilde{\zeta}}-\bar{z}\tilde{\zeta})\log|\tilde{\zeta}-z|^2 - (1+|z|^2)]\log|\tilde{\zeta}-\zeta|^2\,\frac{d\tilde{\zeta}}{\tilde{\zeta}}$$
$$-\frac{1-|z|^2}{\pi}\iint_{\mathbb{D}}\Big(\frac{1}{1-z\bar{\tilde{\zeta}}} + \frac{1}{1-\bar{z}\tilde{\zeta}} - 1\Big)\log|(\tilde{\zeta}-\zeta)(1-\zeta\bar{\tilde{\zeta}})|^2\,d\tilde{\xi}d\tilde{\eta}\,. \tag{4}$$

Evaluating the integrals, another form of the biharmonic Neumann function is given by Begehr in [6,7] as

$$N_2(z,\zeta) = |\zeta - z|^2\,\left[4 - \log|(\zeta - z)(1 - z\bar{\zeta})|^2\right] - 4\sum_{k=2}^{\infty}\frac{1}{k^2}[(z\bar{\zeta})^k + (\bar{z}\zeta)^k]$$
$$-2(z\bar{\zeta} + \bar{z}\zeta)\log|1 - z\bar{\zeta}|^2 - (1+|z|^2)(1+|\zeta|^2)\left[\frac{\log(1-z\bar{\zeta})}{z\bar{\zeta}} + \frac{\log(1-\bar{z}\zeta)}{\bar{z}\zeta}\right].$$

The biharmonic Neumann function satisfies

$$\partial_z\partial_{\bar{z}}N_2(z,\zeta) = N_1(z,\zeta) \text{ in } \mathbb{D}, \quad \partial_{\nu_z}N_2(z,\zeta) = 2(|\zeta|^2 - 1) \text{ on } \partial\mathbb{D} \tag{5}$$

for any $\zeta \in \mathbb{D}$. Also, the normalization condition

$$\frac{1}{2\pi i}\int_{\partial\mathbb{D}} N_2(z,\zeta)\frac{dz}{z} = 0$$

holds for $\zeta \in \mathbb{D}$.

The following problems for the Poisson and bi-Poisson equations are posed and their unique solutions are obtained by Begehr [6,7].

Neumann problem. Find the solution to

$$\partial_z\partial_{\bar z}w = f \ \text{ in } \mathbb{D}, \quad \partial_\nu w = \gamma_1 \ \text{ on } \partial\mathbb{D}, \quad \frac{1}{2\pi i}\int_{\partial\mathbb{D}} w(z)\frac{dz}{z} = c,$$

for $f \in L^1(\mathbb{D}), \gamma_1 \in C(\partial\mathbb{D}), c \in \mathbb{C}$.

The solution to Neumann problem is

$$w(z) = c - \frac{1}{2\pi i}\int_{\partial\mathbb{D}} \gamma_1(\zeta)\log|\zeta - z|^2\frac{d\zeta}{\zeta} - \frac{1}{\pi}\iint_{\mathbb{D}} N_1(z,\zeta)f(\zeta)d\xi d\eta \tag{6}$$

if and only if

$$\frac{1}{2\pi i}\int_{\partial\mathbb{D}} \gamma_1(\zeta)\frac{d\zeta}{\zeta} = \frac{2}{\pi}\iint_{\mathbb{D}} f(\zeta)d\xi d\eta. \tag{7}$$

Neumann-2 problem. Find the solution to

$$(\partial_z\partial_{\bar z})^2 w = f \text{ in } \mathbb{D}\ , \ f \in L^p(\mathbb{D}), 1 < p < \infty\ ,$$

$$\partial_\nu w = \gamma_1\ ,\ \partial_\nu\partial_z\partial_{\bar z}w = \gamma_2 \text{ on } \partial\mathbb{D}\ ,\ \gamma_1,\gamma_2 \in C(\partial\mathbb{D}).$$

$$\frac{1}{2\pi i}\int_{\partial\mathbb{D}} w(\zeta)\frac{d\zeta}{\zeta} = c_1\ ,\ \frac{1}{2\pi i}\int_{\partial\mathbb{D}} w_{\zeta\bar\zeta}(\zeta)\frac{d\zeta}{\zeta} = c_2\ ,\ c_1,c_2 \in \mathbb{C}\ .$$

The unique solution is given as

$$w(z) = c_1 - (1-|z|^2)c_2 - \frac{1}{4\pi i}\int_{\partial\mathbb{D}}\{N_1(z,\zeta)\gamma_1(\zeta) + N_2(z,\zeta)\gamma_2(\zeta)\}\frac{d\zeta}{\zeta}$$
$$+\frac{1}{\pi}\iint_{\mathbb{D}} N_2(z,\zeta)f(\zeta)d\xi d\eta$$

if and only if

$$\frac{1}{2\pi i}\int_{\partial\mathbb{D}} \gamma_1(\zeta)\frac{d\zeta}{\zeta} = 2c_2 + \frac{2}{\pi}\iint_{\mathbb{D}}(|\zeta|^2 - 1)f(\zeta)d\xi d\eta\ ,$$
$$\frac{1}{2\pi i}\int_{\partial\mathbb{D}} \gamma_2(\zeta)\frac{d\zeta}{\zeta} = \frac{2}{\pi}\iint_{\mathbb{D}} f(\zeta)d\xi d\eta\ .$$

3. Integral Operators Related to Harmonic and Biharmonic Neumann Problems

In this section, we introduce the operators $P_{k,l}$ related to Poisson and bi-Poisson equations. Uniform boundedness and uniform continuity of them are obtained in case $k+l<4$. For the case $k+l=4$, L^p boundedness of $P_{k,l}$ is proved.

Using the second order Neumann functions defined in the previous section, we introduce the following operators.

Definition 3.1. For $k,l \in \mathbb{N}_0$ with $(k,l) \neq (2,2)$ and $k+l \leq 4$

$$P_{k,l}f(z) = \frac{1}{\pi}\iint_{\mathbb{D}} \partial_z^k \partial_{\bar z}^l N_2(z,\zeta) f(\zeta)d\xi d\eta$$

where $f \in L^1(\mathbb{D})$.

If $k+l<4$, then the operators $P_{k,l}$ are weakly singular linear operators and if $k+l=4$, then they are strongly singular operators.

Using this definition, we can write

$$P_{1,1}f(z) = \frac{1}{\pi}\iint_{\mathbb{D}} N_1(z,\zeta)f(\zeta)d\xi d\eta = \frac{1}{\pi}\iint_{\mathbb{D}} \log|(\zeta-z)(1-z\bar\zeta)|^2 f(\zeta)d\xi d\eta,$$

$$P_{2,1}f(z) = \frac{1}{\pi}\iint_{\mathbb{D}} \partial_z N_1 f(\zeta)d\xi d\eta = -\frac{1}{\pi}\iint_{\mathbb{D}}\left(\frac{1}{(\zeta-z)}+\frac{\bar\zeta}{(1-z\bar\zeta)}\right)f(\zeta)d\xi d\eta,$$

$$\begin{aligned}P_{3,1}f(z) &= \frac{1}{\pi}\iint_{\mathbb{D}} \partial_z^2 N_1 f(\zeta)d\xi d\eta\\ &= -\frac{1}{\pi}\iint_{\mathbb{D}}\left(\frac{1}{(\zeta-z)^2}+\frac{\bar\zeta^2}{(1-z\bar\zeta)^2}\right)f(\zeta)d\xi d\eta\,.\end{aligned}$$

$P_{1,1}$ and $P_{2,1}$ are modified forms of the operators $\hat\Pi_0$, $\hat\Pi_1$ and $P_{3,1}$ is the operator $\hat\Pi_2$ given by Vinogradov [12] which are also stated by Vekua [11] in his book.

Lemma 3.1. *Let $f \in L^p(\mathbb{D})$, $p>2$ and $k+l<4$. Then,*

$$|P_{k,l}f(z)| \leq C\|f\|_{L^p(\mathbb{D})}$$

for $z \in \mathbb{D}$.

Proof. Using the fact that $\left|\dfrac{\zeta-z}{1-z\bar\zeta}\right| \leq 1$ for $|z| \leq 1$, $|\zeta|<1$, it can be shown that

$$\|\partial_z^k \partial_{\bar z}^l N_2(z,\zeta)\|_{L^q(\mathbb{D})} < C$$

for $q < 2$ and for some constant C, where

$$\partial_z N_2 = -(\overline{\zeta - z})(\log|(\zeta - z)(1 - z\bar{\zeta})|^2 - 4)$$
$$- |\zeta - z|^2\left(\frac{1}{\zeta - z} + \frac{\bar{\zeta}}{1 - z\bar{\zeta}}\right) + \bar{z}(1 - |\zeta|^2) - 2\bar{\zeta} - \partial_z \tilde{N}_2 ,$$
$$\partial_z^2 N_2 = \frac{2(\overline{\zeta - z})}{(\zeta - z)} + \frac{2(\overline{\zeta - z})\bar{\zeta}}{1 - z\bar{\zeta}} - |\zeta - z|^2\left(\frac{1}{(\zeta - z)^2} + \frac{\bar{\zeta}^2}{(1 - z\bar{\zeta})^2}\right) - \partial_z^2 \tilde{N}_2,$$
$$\partial_z^3 N_2 = \frac{3(\overline{\zeta - z})}{(\zeta - z)^2} + \frac{3(\overline{\zeta - z})\bar{\zeta}^2}{(1 - z\bar{\zeta})^2} - 2|\zeta - z|^2\left(\frac{1}{(\zeta - z)^3} + \frac{\bar{\zeta}^3}{(1 - z\bar{\zeta})^3}\right) - \partial_z^3 \tilde{N}_2$$

and $\partial_z^k \partial_{\bar{z}}^l N_2 = \overline{\partial_{\bar{z}}^k \partial_z^l N_2}$ for $k = 1, 2, 3$ where $\tilde{N}_2$ is given by (4). The required result follows from the use of Hölder's inequality. □

The following lemma proves the uniform continuity of the operators $P_{k,l}$ for $k + l < 4$.

Lemma 3.2. *Let $f \in L^p(\mathbb{D})$, $p > 2$ and $k + l < 4$. Then for z_1, $z_2 \in \mathbb{D}$,*

$$|P_{k,l} f(z_1) - P_{k,l} f(z_2)| \leq C\|f\|_{L^p(\mathbb{D})} \begin{cases} |z_1 - z_2|^{(p-2)/p}, & \text{if } k + l = 3, \\ |z_1 - z_2|, & \text{otherwise.} \end{cases}$$

Proof. The result is achived for the case $k + l \neq 3$ by the use of the mean value theorem since we have

$$\partial_z P_{k,l} = P_{k+1,l} , \qquad \partial_{\bar{z}} P_{k,l} = P_{k,l+1}$$

which are known to be bounded. To prove the result for the remaining operators, we use the identity in [2, p. 680] and Hölder's inequality. □

The strongly singular operators $P_{k,l}$ for $k + l = 4$ are shown to be bounded on $L^p(\mathbb{D})$ in the next lemma.

Lemma 3.3. *For $f \in L^p(\mathbb{D})$ and $k + l = 4$, we have $P_{k,l} f \in L^p(\mathbb{D})$ and*

$$\|P_{k,l} f\|_{L^p(\mathbb{D})} \leq A_p \|f\|_{L^p(\mathbb{D})} \quad (p > 1)$$

with

$$\|P_{1,3}\|_{L^2(\mathbb{D})} = \|P_{3,1}\|_{L^2(\mathbb{D})} = 1 . \tag{8}$$

Proof. It is proved in [11] that $\hat{\Pi}_2$ is bounded in $L^p(\mathbb{D})$ and $\|\hat{\Pi}_2\|_{L^2(\mathbb{D})} = 1$. Thus the L^p boundedness of the operators $P_{3,1}$ and $P_{1,3}$ and equality (8) follows.

$P_{4,0}$ has the kernel $\partial_z^4 N_2$ given by

$$\partial_z^4 N_2 = \frac{8(\overline{\zeta - z})}{(\zeta - z)^3} + \frac{8(\overline{\zeta - z})\bar{\zeta}^3}{(1 - z\bar{\zeta})^3} - 6|\zeta - z|^2 \left(\frac{1}{(\zeta - z)^4} + \frac{\bar{\zeta}^4}{(1 - z\bar{\zeta})^4} \right) - \partial_z^4 \tilde{N}_2$$

and $P_{0,4}$ has the kernel $\partial_{\bar{z}}^4 N_2 = \overline{\partial_z^4 N_2}$. They can be decomposed as

$$P_{4.0} f = T_{-2,2} f + S_{4,0} f \text{ and } P_{0,4} f = T_{2,-2} f + S_{0,4} f$$

where the operators $T_{-2,2}$ and $T_{2,-2}$ are known to be bounded in L^p and unitary in L^2 space [2] and the operators $S_{4,0}$ and $S_{0,4}$ are shown to be bounded in L^p by the use of Schur's Theorem and Forelli-Rudin type estimates, see [9]. □

4. Neumann Problems for Generalized Poisson and Bi-Poisson Equations

We start with the problem for a generalized Poisson equation with the Neumann conditions.

Problem N1. Find $w \in W^{2,p}(\mathbb{D})$ as a solution to the second order complex differential equation

$$\frac{\partial^2 w}{\partial z \partial \bar{z}} + \sum_{\substack{k+l=2 \\ k \neq l}} \left(q_{kl}^{(1)}(z) \frac{\partial^2 w}{\partial \bar{z}^l \partial z^k} + q_{kl}^{(2)}(z) \frac{\partial^2 \overline{w}}{\partial z^l \partial \bar{z}^k} \right)$$
$$+ \sum_{k+l<2} \left[a_{kl}(z) \frac{\partial^{k+l} w}{\partial \bar{z}^l \partial z^k} + b_{kl}(z) \frac{\partial^{k+l} \overline{w}}{\partial z^l \partial \bar{z}^k} \right] = f(z), \qquad z \in \mathbb{D} \qquad (9)$$

with Neumann condition

$$\partial_\nu w = \gamma_0 \text{ on } \partial\mathbb{D} \qquad \gamma_0 \in C(\partial\mathbb{D}), \qquad (10)$$

satisfying

$$\frac{1}{2\pi i} \int_{\partial\mathbb{D}} w(\zeta) \frac{d\zeta}{\zeta} = c_0 \quad c_0 \in \mathbb{C}\,, \qquad (11)$$

where

$$a_{kl}, b_{kl}, f \in L^p(\mathbb{D}) \text{ and } \sum_{\substack{k+l=2 \\ k \neq l}} (|q_{kl}^{(1)}(z)| + |q_{kl}^{(2)}(z)|) \leq q_0 < 1\,. \qquad (12)$$

The above problem is introduced also in [5]. First we transform Problem N1 to a singular integral equation.

Lemma 4.1. *The Neumann problem (9), (10) and (11) is equivalent to the singular integral equation*

$$(I+\hat{N}_1+\hat{K}_1)g=\hat{f}\ , \tag{13}$$

if $w=\varphi_1+P_{1,1}g$, *where*

$$\varphi_1(z)=c-\frac{1}{2\pi i}\int_{\partial\mathbb{D}}\gamma_1(\zeta)\log|\zeta-z|^2\frac{d\zeta}{\zeta}$$

and

$$\hat{N}_1g=q_{20}^{(1)}P_{3,1}g+q_{02}^{(1)}P_{1,3}g+q_{20}^{(2)}\overline{P_{3,1}g}+q_{02}^{(2)}\overline{P_{1,3}g}\ ,$$

$$\hat{K}_1g=a_{10}P_{2,1}g+a_{01}P_{1,2}g+a_{00}P_{1,1}g+b_{10}\overline{P_{2,1}g}+b_{01}\overline{P_{1,2}g}+b_{00}\overline{P_{1,1}g}\ ,$$

$$\hat{f}=f-\sum_{\substack{k+l=2\\k\neq l}}\left(q_{kl}^{(1)}(z)\frac{\partial^2\varphi_1}{\partial\bar{z}^l\partial z^k}+q_{kl}^{(2)}(z)\frac{\partial^2\overline{\varphi_1}}{\partial z^l\partial\bar{z}^k}\right)$$
$$-\sum_{k+l<2}\left[a_{kl}(z)\frac{\partial^{k+l}\varphi_1}{\partial\bar{z}^l\partial z^k}+b_{kl}(z)\frac{\partial^{k+l}\overline{\varphi_1}}{\partial z^l\partial\bar{z}^k}\right]\ .$$

Consider the following problem for generalized bi-Poisson equation with the Neumann conditions.

Problem N2. Find $w\in W^{4,p}(\mathbb{D})$ as a solution to the fourth order complex differential equation

$$\frac{\partial^4w}{\partial z^2\partial\bar{z}^2}+\sum_{\substack{k+l=4\\k\neq l}}\left(q_{kl}^{(1)}(z)\frac{\partial^4w}{\partial\bar{z}^l\partial z^k}+q_{kl}^{(2)}(z)\frac{\partial^4\overline{w}}{\partial z^l\partial\bar{z}^k}\right)$$
$$+\sum_{k+l<4}\left[a_{kl}(z)\frac{\partial^{k+l}w}{\partial\bar{z}^l\partial z^k}+b_{kl}(z)\frac{\partial^{k+l}\overline{w}}{\partial z^l\partial\bar{z}^k}\right]=f(z),\qquad z\in\mathbb{D} \tag{14}$$

with Neumann conditions

$$\partial_\nu w=\gamma_0\quad\partial_\nu(\partial_z\partial_{\bar{z}}w)=\gamma_1\ \text{on}\ \partial\mathbb{D}\quad\ \gamma_0,\gamma_1\in C(\partial\mathbb{D}), \tag{15}$$

satisfying

$$\frac{1}{2\pi i}\int_{\partial\mathbb{D}}w(\zeta)\frac{d\zeta}{\zeta}=c_0\quad\frac{1}{2\pi i}\int_{\partial\mathbb{D}}\partial_\zeta\partial_{\bar{\zeta}}w(\zeta)\frac{d\zeta}{\zeta}=c_1\quad c_0,c_1\in\mathbb{C}\ , \tag{16}$$

where

$$a_{kl},b_{kl},f\in L^p(\mathbb{D})\quad\text{and}\quad\sum_{\substack{k+l=4\\k\neq l}}(|q_{kl}^{(1)}(z)|+|q_{kl}^{(2)}(z)|)\le q_0<1\ . \tag{17}$$

The next lemma transforms Problem N2 to a singular integral equation.

Lemma 4.2. *The Neumann problem (14), (15) and (16) is equivalent to the singular integral equation*

$$(I + B_1 + B_2 + \hat{K}_2)g = \hat{f}\ , \tag{18}$$

if $w = \varphi_2 + P_{0,0}g$, *where*

$$\varphi_2(z) = c_1 - (1- |z|^2)c_2 - \frac{1}{4\pi i}\int_{\partial\mathbb{D}} \{N_1(z,\zeta)\gamma_1(\zeta) + N_2(z,\zeta)\gamma_2(\zeta)\}\frac{d\zeta}{\zeta}$$

and

$$B_1 g = q_{31}^{(1)} P_{3,1}g + q_{31}^{(2)}\overline{P_{3,1}g} + q_{13}^{(1)} P_{1,3}g + q_{13}^{(2)}\overline{P_{1,3}g}$$

$$+ q_{40}^{(1)} T_{-2,2}g + q_{40}^{(2)}\overline{T_{-2,2}g} + q_{04}^{(1)} T_{2,-2}g + q_{04}^{(2)}\overline{T_{2,-2}g}$$

$$B_2 g = q_{40}^{(1)} S_{4,0}g + q_{40}^{(2)}\overline{S_{4,0}g} + q_{04}^{(1)} S_{0,4}g + q_{04}^{(2)}\overline{S_{0,4}g}$$

$$\hat{K}_2 g = \sum_{k+l<4} \left(a_{kl}P_{k,l}g + b_{kl}\overline{P_{k,l}g}\right)\ ,$$

$$\hat{f} = f - \sum_{\substack{k+l=4\\ k\neq l}} \left(q_{kl}^{(1)}(z)\frac{\partial^4\varphi_2}{\partial\bar{z}^l\partial z^k} + q_{kl}^{(2)}(z)\frac{\partial^4\overline{\varphi_2}}{\partial z^l\partial\bar{z}^k}\right)$$

$$- \sum_{k+l<4}\left[a_{kl}(z)\frac{\partial^{k+l}\varphi_2}{\partial\bar{z}^l\partial z^k} + b_{kl}(z)\frac{\partial^{k+l}\overline{\varphi_2}}{\partial z^l\partial\bar{z}^k}\right]\ .$$

4.1. *Solvability of the Problems*

4.1.1. *Solvability of Problem N1*

Theorem 4.1. *The equation (9) with the conditions (10) and (11) is solvable and a solution is of the form* $w(z) = \varphi_1(z) + P_{1,1}g(z)$ *where* $g \in L^p(\mathbb{D})$, $0 < p - 2 < \epsilon$, *is a solution of the singular integral equation (13) subject to the solvability condition*

$$\frac{1}{2\pi i}\int_{\partial\mathbb{D}} \gamma_0(\zeta)\frac{d\zeta}{\zeta} = \frac{2}{\pi}\iint_{\mathbb{D}} g(\zeta)d\xi d\eta. \tag{19}$$

Proof. Lemmas 3.1 and 3.2 imply that the operator $\hat{K}_1$ is a compact operator from $L^p(\mathbb{D})$ into itself and the operator $I + \hat{N}_1$ is an invertible operator from $L^p(\mathbb{D})$ into itself for p sufficiently close to 2, by Lemma 3.3 and Riesz-Thorin theorem. Thus the Fredholm alternative applies to the equation(13). If $g \in L^p(\mathbb{D})$ is a solution of the singular integral equation (13) satisfying (19), then $w(z) = \varphi_1(z) + P_{1,1}g(z)$ is a solution of Problem N1. □

4.1.2. *Solvability of Problem N2*

Theorem 4.2. *If the inequality*

$$q_0\|S_{4,0}\|_{L^p(\mathbb{D})}\|(I+B_1)^{-1}-K\|_{L^p(\mathbb{D})}<1 \tag{20}$$

holds for some $K\in K(L^p(\mathbb{D}))$, $0<p-2<\epsilon$, *then the equation (14) with the boundary conditions (15) has a solution of the form* $w(z)=P_{0,0}g(z)+\varphi_2(z)$, *where* $g\in L^p(\mathbb{D})$ *is a solution of the singular integral equation (18) subject to the solvability conditions*

$$\frac{1}{2\pi i}\int_{\partial\mathbb{D}}\gamma_1(\zeta)\frac{d\zeta}{\zeta}=2c_2+\frac{2}{\pi}\iint_{\mathbb{D}}(|\zeta|^2-1)g(\zeta)d\xi d\eta \tag{21}$$

and

$$\frac{1}{2\pi i}\int_{\partial\mathbb{D}}\gamma_2(\zeta)\frac{d\zeta}{\zeta}=\frac{2}{\pi}\iint_{\mathbb{D}}g(\zeta)d\xi d\eta\ . \tag{22}$$

Proof. $\hat{K}_2$ is a compact operator from $L^p(\mathbb{D})$ into itself by Lemma 3.1 and Lemma 3.2. Moreover Lemma 3.3 implies that B_1 is a linear combination of $P_{3,1}$, $P_{1,3}$, $T_{-2,2}$ and $T_{2,-2}$ which are unitary operators in $L^2(\mathbb{D})$ and B_2 is a linear combination of $S_{4,0}$ and $S_{0,4}$ which are bounded operators in $L^p(\mathbb{D})$. Thus $I+B_1+B_2+\hat{K}_2$ is the sum of an invertible operator, compact operator and a bounded operator. The result follows from the bounded index stability theorem given in [10]. □

The technique employed in the proof of Theorem 4.2 is used in the Schwarz problem for general higher order partial complex differential equations in [1].

References

1. Ü. Aksoy and A. O. Çelebi, Schwarz Problem for Higher-Order Complex Elliptic Partial Differential Equations, *Integral Transforms Spec. Funct.*, **19** (2008), 413–428.
2. H. Begehr and G. Hile, A Hierarchy of Integral Operators, *Rocky Mountain J. Math.*, **27** (1997), 669–706.
3. H. Begehr and C. J. Vanegas, Neumann Problem in Complex Analysis, in *Proc. 11th Int. Conf. Finite, Infinite Dimensional Complex Analysis Appl.*, Chiang Mai, Thailand, 2003.
4. H. Begehr, Boundary Value Problems in Complex Analysis II, *Bol. Asoc. Mat. Venez.*, **12** (2004), 223–241.
5. H. Begehr and C. J. Vanegas, Iterated Neumann Problem for Higher order Poisson Equation, *Math. Nachr.* **279** (2006), 38–57.

6. H. Begehr, Six Biharmonic Dirichlet Problems in Complex Analysis, in *Proc. 14th Int. Conf. Finite, Infinite Dimensional Complex Analysis Appl.*, Hue, Vietnam, 2006.
7. H. Begehr, Biharmonic Green Functions, *Mathematiche*, **61** (2006), 395–405.
8. H. Begehr, T. Vaitekhovich, Complex Partial Differential Equations in a Manner of I. N. Vekua, *Lecture Notes of TICMI*, **8** (2007), 15–26.
9. F. Forelli and W. Rudin, Projections on Spaces of Holomorphic Functions in Balls, *Indiana Univ. Math. J.*, **24** (1974), 593–602.
10. S. S. Kutateladze, *Fundamentals of Functional Analysis*, Kluwer Academic Publishers, Dordrecht-Boston-London, 1996.
11. I. N. Vekua, *Generalized Analytic Functions*, Pergamon Press, Oxford, 1962.
12. V. S. Vinogradov, On the Neumann Problem for Equations of Elliptic Type, *Dokl. Akad. Nauk*, **109** (1956), 13–16.

A NEW SOLUTION OF SOME WEIGHTED PROBLEMS FOR RIEMANN-LIOUVILLE AND WEYL OPERATORS

MUSTAFA AVCI *

Dicle University, Department of Mathematics, 21280 Diyarbakır, TURKEY
E-mail: mustafaxavci@hotmail.com

SEZAİ OĞRAŞ

Dicle University, Department of Mathematics, 21280 Diyarbakır, TURKEY
E-mail: sezaio@dicle.edu.tr

R. A. MASHIYEV

Dicle University, Department of Mathematics, 21280 Diyarbakır, TURKEY
E-mail: mrabil@dicle.edu.tr

Ee introduce necessary and sufficient conditions for the weight pairs (v, ω) that provide the boundedness of the Riemann-Liouville operator R_α and Weyl operator W_α from $L^p_\omega(0,\infty)$ to $L^q_v(0,\infty)$ when $1 < p \leq q < \infty$, and $1/q < \alpha < 1$ or $\alpha > 1$.

Keywords: Hardy-type operators, Riemann-Liouville operator, Weyl operator.

1. Introduction and preliminaries

A complete characterization of the weight pairs (v, ω) providing the boundedness of the operators R_α and W_α from $L^p_\omega(0,\infty)$ to $L^q_v(0,\infty)$ when $1 < p < q < \infty$ and $0 < \alpha < 1$ is given in [3]. In the case $1 < p \leq q < \infty$ and $\alpha > 1$ a similar problem was solved by many authors [4,5,8]; and also for the case $\alpha = 1$ the same work was done in [1,6,7,9,12]. Meanwhile, for the case $1 < q < p < \infty$ and $\alpha > 1$, a similar problem is solved in [10].

The necessary and sufficient conditions for the compactness of the above mentioned operators from $L^p_\omega(0,\infty)$ to $L^q_v(0,\infty)$ when $1 < p \leq q < \infty$ and $\alpha > 1$ was given in [11]; the case $\alpha = 1$ is in [2].

In this study, the necessary and sufficient conditions are found for the

*Corresponding Author, phone number: +905054502193.

weight functions υ and ω, which provide the boundedness of the Riemann-Liouville transform

$$R_\alpha f(x) = \int_0^x (x-t)^{\alpha-1} f(t)\, dt$$

and the Weyl transform

$$W_\alpha f(x) = \int_x^\infty (t-x)^{\alpha-1} f(t)\, dt$$

from L^p_ω to L^q_υ when $1 < p \le q < \infty$, $1/q < \alpha < 1$ or $\alpha > 1$.

As we mentioned above the characterization of the weight pairs (υ, ω) providing the boundedness of the operators R_α and W_α from $L^p_\omega(0,\infty)$ to $L^q_\upsilon(0,\infty)$ under the given conditions are not new, but there are many proof techniques which the authors used in previous studies. In this context our approach to the problem can be considered different.

We denote the weighted Lebesgue space by $L^p_\omega(0,\infty)$, where $1 < p < \infty$. It consists of all measurable functions $f(x)$ on $(0,\infty)$ such that

$$\|f\|_{p,\omega} = \|f\|_{L^p_\omega} = \left(\int_0^\infty |f(x)|^p \omega(x)\, dx\right)^{1/p} < \infty. \tag{1}$$

The measurable functions $\upsilon, \omega : (0,\infty) \to (0,\infty)$ are called weight functions.

We use the notation

$$W(x) = \int_0^x \omega^{1-p'}(t)\, dt \ , \ V(x) = \int_x^\infty \frac{\upsilon(t)}{t^{(1-\alpha)q}} dt \tag{2}$$

and assume that W and V satisfy

$$V \in D : \exists\, \eta > 0, \quad V\left(\frac{x}{2}\right) \le 2^\eta V(x) \qquad \text{for } x > 0, \tag{3}$$

$$W \in RD : \exists\, \delta > 0, \quad W\left(\frac{x}{2}\right) \le 2^{-\delta} W(x) \quad \text{for } x > 0. \tag{4}$$

Moreover, we use

$$\overline{W}(x) = \int_x^\infty \omega^{1-p'}(t)\, dt \ , \ \overline{V}(x) = \int_0^x \frac{\upsilon(t)}{t^{(1-\alpha)q}} dt \tag{5}$$

and assume that $\overline{W}$ and $\overline{V}$ satisfy

$$\overline{V} \in RD : \exists\delta > 0, \quad \overline{V}\left(\tfrac{x}{2}\right) \le 2^{-\delta}\overline{V}(x) \quad \text{for } x > 0, \tag{6}$$

$$\overline{W} \in D : \exists\eta > 0, \quad \overline{W}\left(\tfrac{x}{2}\right) \le 2^{\eta}\overline{W}(x) \qquad \text{for } x > 0. \tag{7}$$

2. Main results

Theorem 2.1. *Let $1 < p \le q < \infty$, $1/q < \alpha < 1$ or $\alpha > 1$. The inequality*

$$\|R_\alpha f\|_{L^q_v} \le A \|f\|_{L^p_\omega},$$

where $A > 0$ does not depend on f, is fulfilled if and only if

$$B := \sup_{t>0} \left(\int_t^\infty \frac{v(x)}{x^{(1-\alpha)q}} dx \right)^{1/q} \left(\int_0^t \omega^{1-p'}(x)\, dx \right)^{1/p'} < \infty, \quad \left(p' = \frac{p}{p-1} \right).$$

Morever, if A is the best constant, then $A \approx B$.

Proof. *Sufficiency.*

Let $f \in L^p$ and $\Omega_x^n = \left(2^{-n-1}x, 2^{-n}x\right]$, $n \in \mathbb{N}$. Let's first write

$$(R_\alpha f)(x) = \sum_{n=0}^{m} \int_{2^{-n-1}x}^{2^{-n}x} (x-t)^{\alpha-1} f(t)\, dt + \int_0^{2^{-m-1}x} (x-t)^{\alpha-1} f(t)\, dt.$$

Now, applying norm we obtain

$$\begin{aligned} \|R_\alpha f\|_{L^q_v} &\le \sum_{n=0}^{m} \left(\int_0^\infty v(x) \left(\int_{2^{-n-1}x}^{2^{-n}x} (x-t)^{\alpha-1} f(t)\, dt \right)^q dx \right)^{1/q} \\ &\quad + \left(\int_0^\infty v(x) \left(\int_0^{2^{-m-1}x} (x-t)^{\alpha-1} f(t)\, dt \right)^q dx \right)^{1/q} \\ &= J_n + J_m. \end{aligned}$$

First, let us show that J_n is finite. If we take $(x-t)^{\alpha-1}$ out of the inner integral, then we have

$$J_n^q \le \left(1 - \frac{1}{2^n}\right)^{q(\alpha-1)} \int_0^\infty \frac{v(x)}{x^{(1-\alpha)q}} \left(\int_{2^{-n-1}x}^{2^{-n}x} f(t)\, dt \right)^q dx.$$

Let $h(t) = \omega(t)^{p'-1} f(t)$ and apply Hölder's inequality to get

$$\begin{aligned} J_n^q &\le C_1 \int_0^\infty \frac{v(x)}{x^{(1-\alpha)q}} \left(\int_{2^{-n-1}x}^{2^{-n}x} |h(t)|^p\, dW(t) \right)^{q/p} \left(\int_{2^{-n-1}x}^{2^{-n}x} dW(t) \right)^{q/p'} dx \\ &= C_1 \int_0^\infty \frac{v(x)}{x^{(1-\alpha)q}} \left(\int_{\Omega_x^n} |h(t)|^p\, dW(t) \right)^{q/p} W(\Omega_x^n)^{q/p'}\, dx, \end{aligned}$$

where $C_1 = (1 - \frac{1}{2^n})^{q(\alpha-1)}$. But from (4) we have

$$W(\Omega_x^n)^{q/p'} \le (W(2^{-n}x))^{q/p'} \le 2^{-n\delta q/p'} W(x)^{q/p'} \le 2^{-n(\alpha-1)/q} W(x)^{q/p'}$$

since $\delta > 0$ is an arbitrary number and $\delta q/p' \geq (\alpha - 1)/q$. So

$$J_n^q \leq -C_1 2^{-n(\alpha-1)/q} \int_0^\infty V(x) W(x)^{q/p'} \left(\int_{\Omega_x^n} |h(t)|^p \, dW(t) \right)^{q/p} \frac{dV(x)}{V(x)}. \tag{8}$$

Now, let us assume that $\|f\|_{L^p_\omega} = \left(\int_0^\infty |f(t)|^p \omega(t) \, dt \right)^{1/p} \leq C$ and that $B_1 = V(x) W(x)^{q/p'} < \infty$ hold. Then $\left(\int_{\Omega_x^n} |f(t)|^p \omega(t) \, dt \right)^{1/p} \leq C$. Moreover, we can write $\left(\int_{\Omega_x^n} |f(t)|^p \omega(t) \, dt \right)^{q/p} \leq C^{q-p} \int_{\Omega_x^n} |f(t)|^p \omega(t) \, dt$.

Now, the inequality (8) together with the differentials of $V(x)$ and $W(x)$, imply that $J_n^q \leq -C_2 \int_0^\infty \int_{2^{-n-1}x}^{2^{-n}x} |f(t)|^p \omega(t) \, dt \frac{dV(x)}{V(x)}$, where $C_2 = C_1 C^{q-p} 2^{-n(\alpha-1)/q}$. Applying Fubini theorem, we have $J_n^q \leq -C_2 \int_0^\infty |f(t)|^p \omega(t) \int_{2^n t}^{2^{n+1}t} \frac{dV(x)}{V(x)} dt$. Since $V(x)$ satisfies (3), we can write $-\int_{2^n t}^{2^{n+1}t} \frac{dV(x)}{V(x)} = \log \frac{V(2^n t)}{V(2^{n+1}t)} \leq \eta \log 2, \quad t > 0, \quad n \in \mathbb{N}$. Eventually we obtain

$$J_n^q \leq \eta \log 2 C_2 \int_0^\infty |f(t)|^p \omega(t) \, dt, \quad J_n \leq (\eta \log 2)^{1/q} C_2^{1/q} C^{p/q} = C_3, \tag{9}$$

where $C_3 = C(1-2^{-n})^{(\alpha-1)} 2^{-n(\alpha-1)} (\eta \log 2)^{1/q}$.

Now let us show J_m is finite. From Lebesgue's convergence theorem, since we have $\int_0^{2^{-m-1}x} (x-t)^{\alpha-1} f(t) \, dt \leq \int_0^\infty (x-t)^{\alpha-1} f(t) \, dt \in L_1(0,\infty)$, as $m \to \infty$, we obtain immediately

$$J_m = \left(\int_0^\infty v(x) \left(\int_0^{2^{-m-1}x} (x-t)^{\alpha-1} f(t) \, dt \right)^q dx \right)^{1/q} \to 0. \tag{10}$$

As a result, (9) together with (10) imply that $\|R_\alpha f\|_{L^q_v} < \infty$.

Necessity.

Assume that $\|R_\alpha f\|_{L^q_v} < \infty$ holds. Let $B_2^{1/q} = B$ and choose test functions $f_t(x) = \chi_{(0,t)}(x) \left(\int_0^t \omega(x)^{1-p'} dx \right)^{1/p'} \in L^p$. Then the right-hand side will be smaller than $\|R_\alpha f\|_{L^q_v}^q$. So we have

$$\begin{aligned} \|R_\alpha f\|_{L^q_v}^q &\geq \int_t^\infty v(x) \left(\int_0^t (x-y)^{\alpha-1} \left(\int_0^t \omega(x)^{1-p'} dx \right)^{1/p'} dy \right)^q dx \\ &= \int_t^\infty v(x) \left(\int_0^t (x-y)^{\alpha-1} dy \right)^q \left(\int_0^t \omega(x)^{1-p'} dx \right)^{q/p'} dx \end{aligned}$$

$$= \frac{1}{\alpha^q} \left(\int_t^\infty v(x) \left(x^\alpha - (x-t)^\alpha \right)^q dx \right) \left(\int_0^t \omega(x)^{1-p'} dx \right)^{q/p'} .$$

For some positive k, it holds that $(x^\alpha - (x-t)^\alpha) \le k x^{\alpha-1}$. Therefore

$$\|R_\alpha f\|_{L^q_v} \ge C_4 \left(\int_t^\infty \frac{v(x)}{x^{(1-\alpha)q}} dx \right)^{1/q} \left(\int_0^t \omega(x)^{1-p'} dx \right)^{1/p'} = C_4 B.$$

The proof is complete. □

Remark 2.1. It is easy to see that, for the convergence of the series $\sum_{n=0}^{m} C_3$ as $m \to \infty$, the conditions $p > 1$ and $\alpha > 1$ are necessary.

Theorem 2.2. *Let $1 < p \le q < \infty$, and $1/q < \alpha < 1$ or $\alpha > 1$. The inequality*

$$\|W_\alpha f\|_{L^q_v} \le \overline{A} \, \|f\|_{L^p_\omega} ,$$

where $\overline{A} > 0$ does not depend on f, is fulfilled if and only if

$$\overline{B} := \sup_{t>0} \left(\int_t^\infty \omega^{1-p'}(x) \, dx \right)^{1/p'} \left(\int_0^t \frac{v(x)}{x^{(1-\alpha)q}} dx \right)^{1/q} < \infty.$$

Morever, if $\overline{A}$ is the best constant, then $\overline{A} \approx \overline{B}$.

Proof. The proof is similar to the proof of Theorem 2.1 and we omit it. □

References

1. J. S. Bradley, "Hardy inequality with mixed norms", *Canad. Math. Bull.*, **21** (1978), 405-408
2. D. E. Edmunds, W. D. Evans and D. J. Harris, "Approximation numbers of certain Volterra integral operators", *J. London Math. Soc.*, (2), **37** (1988), 471-489.
3. I. Genebashvili, A. Gogatishvili and V. Kokilashvili, "Solution of two weigth problems for integral transforms with positive kernels", *Georgian Math. J.*, **3** (1996), No. 1, 319-342.
4. H. P. Heinig, "Weigted inequalities in Fourier analysis", *Nonlinear Analysis, Function Spaces and Appl. 4, Proc. Spring School*, 1990, 42-85.
5. K. F. Andersen and H. P. Heinig, "Weighted norm inequalities for certain integral operators", *SIAM J. Math. Anal.*, **14** (1983), No. 4, 834-844.
6. V. M. Kokilashvili, "On Hardy's inequalities in weighted spaces" (in Russian), Soobshch. Akad. Nauk Gruzin. SSR, **96** (1979), 37-40.
7. V. G. Maz'ya, Sobolev spaces, Springer, 1985.
8. A. Meskhi, "Solution of some weighted problems for Riemann-Liouville and Weyl operators", *Georgian Math. J.*, **5** (1998), 319-342.

9. B. Muckhenhoupt, "Hardy's inequality with weights", *Studia Math.*, **44** (1972), 31-38.
10. V. Stepanov, "Two weighted estimates for Riemann-Liouville integrals", *Report* No. **39**, *Math. Inst., Czechoslovak Acad. Sci.*, 1988.
11. V. Stepanov, "Weighted inequalities of Hardy type for higher derivatives and their applications", *Soviet Math. Dokl.*, **38** (1989), 389-393.
12. G. Talenti, "Osservazioni spora una classe di disuguaglianze", *Rend. Sem. Mat. Fiz. Milano*, **39** (1996), 171-185.

A POLYHARMONIC DIRICHLET PROBLEM OF ARBITRARY ORDER FOR COMPLEX PLANE DOMAINS

H. BEGEHR

Math. Institut, Freie Universität Berlin
14195, Berlin, Germany
E-mail: begehr@math-fu.berlin.de

T. S. VAITEKHOVICH

Belarusian State University,
220030, Minsk, Belarus
E-mail: vaiteakhovich@mail.ru

Dedicated to Professor Dr. Ilppo Simo Louhivaara
on the occasion of his 80th *birthday.*

The consecutive convolution of the harmonic Green function with itself provides a polyharmonic Green function of any order. Convoluting such a polyharmonic Green function with a polyharmonic Green-Almansi function gives a polyharmonic Green function of new type. It leads to a certain Dirichlet problem for a Poisson equation of certain order. This problem is well-posed. Its solution is given here.

Keywords: Polyharmonic equation, Dirichlet boundary value problems, Green functions, Poisson equation.

1. Introduction

The harmonic Green function $G_1(z,\zeta)$ for a regular complex domain D satisfies for any $\zeta \in D$ the following properties.

- $G_1(\cdot,\zeta)$ is harmonic in $D \setminus \{\zeta\}$ and continuous in $\overline{D} \setminus \{\zeta\}$,
- $G_1(z,\zeta) + \log|\zeta - z|$ is harmonic in z in the neighborhood of ζ,
- $G_1(z,\zeta) = 0$ for $z \in \partial D$,
- $G_1(z,\zeta) = G_1(\zeta,z)$ for $z,\zeta \in D$, $z \neq \zeta$.

Any function $w \in C^2(D;\mathbb{C}) \cap C^1(\overline{D};\mathbb{C})$ can be represented through

$$w(z) = -\frac{1}{4\pi}\int_{\partial D} \partial_{\nu_\zeta} G_1(z,\zeta) w(\zeta) ds_\zeta - \frac{1}{\pi}\int_D G_1(z,\zeta) w_{\zeta\overline{\zeta}}(\zeta) d\xi d\eta. \quad (1)$$

This formula provides the unique solution to the Dirichlet problem for the Poisson equation

$$w_{z\overline{z}} = f \text{ in } D, \quad w = \gamma \text{ on } \partial D, \quad (2)$$

for $f \in L_p(D;\mathbb{C})$, $2 < p$, $\gamma \in C^\alpha(\partial D;\mathbb{C})$, $0 < \alpha < 1$.

The iteration process

$$G_n(z,\zeta) = -\frac{1}{\pi}\int_D G_{n-1}(z,\tilde{\zeta}) G_1(\tilde{\zeta},\zeta) d\tilde{\xi} d\tilde{\eta}, \ 2 \le n, \quad (3)$$

leads to a hierarchy of polyharmonic Green functions, which with respect to the Dirichlet problem and formula (1) satisfy for $\zeta \in D$

- $G_n(\,\cdot\,,\zeta)$ is polyharmonic of order n in $D \setminus \{\zeta\}$,
- $G_n(z,\zeta) + \dfrac{|\zeta - z|^{2(n-1)}}{(n-1)!^2} \log|\zeta - z|^2$ is polyharmonic of order n in z in the neighborhood of ζ,
- $(\partial_z\partial_{\overline{z}})^\mu G_n(z,\zeta) = 0$ for $z \in \partial D$, $0 \le \mu \le n-1$,
- $G_n(z,\zeta) = G_n(\zeta,z)$ for z, $\zeta \in D$, $z \ne \zeta$.

This polyharmonic Green function is related to the Dirichlet problem

$$(\partial_z\partial_{\overline{z}})^n w = f \text{ in } D, \quad (\partial_z\partial_{\overline{z}})^\mu w = \gamma_\mu \text{ on } \partial D, \ 0 \le \mu \le n-1, \quad (4)$$

for $f \in L_p(D;\mathbb{C})$, $2 < p$, $\gamma_\mu \in C^{0,\alpha}(\partial D;\mathbb{C})$, $0 < \alpha < 1$. Its unique solution is given via the representation formula for $w \in C^{2n}(D;\mathbb{C}) \cap C^{2n-1}(\overline{D};\mathbb{C})$, see [1],

$$\begin{aligned} w(z) = &-\sum_{\mu=1}^{n} \frac{1}{4\pi}\int_{\partial D} \partial_{\nu_\zeta} G_\mu(z,\zeta)(\partial_\zeta\partial_{\overline{\zeta}})^{\mu-1} w(\zeta) ds_\zeta \\ &- \frac{1}{\pi}\int_D G_n(z,\zeta)(\partial_\zeta\partial_{\overline{\zeta}})^n w(\zeta) d\xi d\eta. \end{aligned}$$

Theorem 1.1. *For $f \in L_p(D;\mathbb{C})$, $2 < p$, $\gamma_\mu \in C^{0,\alpha}(\partial D;\mathbb{C})$, $0 \le \mu \le n-1$, the unique solution to the Dirichlet problem (4) is given by*

$$w(z) = -\frac{1}{4\pi}\int_{\partial D} \sum_{\mu=0}^{n-1} \partial_{\nu_\zeta} G_{\mu+1}(z,\zeta)\gamma_\mu(\zeta) ds_\zeta - \frac{1}{\pi}\int_D G_n(z,\zeta) f(\zeta) d\xi d\eta.$$

Besides, this polyharmonic Green function $G_n(z,\zeta)$ there is the Green-Almansi function $\mathcal{G}_m(z,\zeta)$. It satisfies for $\zeta \in D$ the following properties (see [1,2]).

- $\mathcal{G}_m(\cdot\,,\zeta)$ is polyharmonic of order m in $D\setminus\{\zeta\}$,
- $\mathcal{G}_m(z,\zeta)+\dfrac{|\zeta-z|^{2(m-1)}}{(m-1)!^2}\log|\zeta-z|^2$ is polyharmonic of order m in z in the neighborhood of ζ,
- $(\partial_z\partial_{\overline{z}})^{\mu}\mathcal{G}_m(z,\zeta)=0,\ 0\le 2\mu\le m-1,$
  $\partial_{\nu_z}(\partial_z\partial_{\overline{z}})^{\mu}\mathcal{G}_m(z,\zeta)=0,\ 0\le 2\mu\le m-2,$ for $z\in\partial D$,
- $\mathcal{G}_m(z,\zeta)=\mathcal{G}_m(\zeta,z)$ for $z,\,\zeta\in D,\ z\neq\zeta$.

The related representation formula for $w\in C^{2m}(D;\mathbb{C})\cap C^{2m-1}(\overline{D};\mathbb{C})$ is

$$\begin{aligned}w(z)=&-\sum_{\mu=0}^{[\frac{m-1}{2}]}\frac{1}{4\pi}\int_{\partial D}\partial_{\nu_\zeta}(\partial_\zeta\partial_{\overline{\zeta}})^{m-\mu-1}\mathcal{G}_m(z,\zeta)(\partial_\zeta\partial_{\overline{\zeta}})^{\mu}w(\zeta)ds_\zeta\\&+\sum_{\mu=0}^{[\frac{m}{2}]-1}\frac{1}{4\pi}\int_{\partial D}(\partial_\zeta\partial_{\overline{\zeta}})^{m-\mu-1}\mathcal{G}_m(z,\zeta)\partial_{\nu_\zeta}(\partial_\zeta\partial_{\overline{\zeta}})^{\mu}w(\zeta)ds_\zeta\\&-\frac{1}{\pi}\int_D\mathcal{G}_m(z,\zeta)(\partial_\zeta\partial_{\overline{\zeta}})^{m}w(\zeta)d\xi d\eta,\end{aligned}$$

see [1]. Its provides the unique solution to the Dirichlet problem

$$(\partial_z\partial_{\overline{z}})^m w=f\ \text{ in }\ D,\quad(\partial_z\partial_{\overline{z}})^{\mu}w=\gamma_\mu,\ 0\le 2\mu\le m-1,\tag{5}$$

$$\partial_{\nu_\zeta}(\partial_z\partial_{\overline{z}})^{\mu}w=\widehat{\gamma}_\mu,\ 0\le 2\mu\le m-2,\ \text{ on }\ \partial D.$$

Theorem 1.2. *For $f\in L_p(D;\mathbb{C})$, $2<p$, $\gamma_\mu\in C^{m-2\mu,\alpha}(\partial D;\mathbb{C})$, $0\le 2\mu\le m-1$, $\widehat{\gamma}_\mu\in C^{m-2\mu-1,\alpha}(\partial D;\mathbb{C})$, $0\le 2\mu\le m-2$, $0<\alpha<1$, the unique solution to the Dirichlet problem (5) is given as*

$$\begin{aligned}w(z)=&-\frac{1}{4\pi}\int_{\partial D}\sum_{\mu=0}^{[\frac{m-1}{2}]}\partial_{\nu_\zeta}(\partial_\zeta\partial_{\overline{\zeta}})^{m-\mu-1}\mathcal{G}_m(z,\zeta)\gamma_\mu(\zeta)ds_\zeta\\&+\frac{1}{4\pi}\int_{\partial D}\sum_{\mu=0}^{[\frac{m}{2}]-1}(\partial_\zeta\partial_{\overline{\zeta}})^{m-\mu-1}\mathcal{G}_m(z,\zeta)\widehat{\gamma}_\mu(\zeta)ds_\zeta-\frac{1}{\pi}\int_D\mathcal{G}_m(z,\zeta)f(\zeta)d\xi d\eta.\end{aligned}$$

The explicit solution to this problem in the case of the unit disk is given in [3,4] and for the upper half plane in [5]. For the explicit expressions of the polyharmonic Green–Almansi functions for the unit disk see [1–4] and for the upper half plane see [5,7,8]. The function $G_n(z,\zeta)$ is evaluated only for $n = 2$ for the unit disk, half planes, quarter plane, circular ring, see [6,8–10].

In particular for the unit disk $\mathbb{D}$ and the upper half plane $\mathbb{H}_0$

$$(n-1)!^2\mathcal{G}_n(z,\zeta) = |\zeta - z|^{2(n-1)} \log\left|\frac{1-z\overline{\zeta}}{\zeta - z}\right|^2 - \sum_{\mu=1}^{n-1} \frac{1}{\mu} |\zeta - z|^{2(n-1-\mu)}(1-|z|^2)^\mu(1-|\zeta|^2)^\mu,$$

$$(n-1)!^2\mathcal{G}_n(z,\zeta) = |\zeta - z|^{2(n-1)} \log\left|\frac{\overline{\zeta}-z}{\zeta - z}\right|^2 + \sum_{\mu=1}^{n-1} \frac{1}{\mu} |\zeta - z|^{2(n-1-\mu)}(z-\overline{z})^\mu(\zeta-\overline{\zeta})^\mu,$$

respectively. Moreover, for $\mathbb{D}$, $\mathbb{H}_0$, the right upper quarter plane $\mathbb{Q}_1$ and the circular ring $R = \{0 < r < |z| < 1\}$

$$G_2(z,\zeta) = |\zeta - z|^2 \log\left|\frac{1-z\overline{\zeta}}{\zeta - z}\right|^2 + (1-|z|^2)(1-|\zeta|^2)\left[\frac{\log(1-z\overline{\zeta})}{z\overline{\zeta}} + \frac{\log(1-\overline{z}\zeta)}{\overline{z}\zeta}\right] \quad \text{for } \mathbb{D},$$

$$G_2(z,\zeta) = |\zeta - z|^2 \log\left|\frac{\overline{\zeta}-z}{\zeta - z}\right|^2 - (\zeta-\overline{\zeta})(z-\overline{z})\log|\overline{\zeta}-z|^2 \quad \text{for } \mathbb{H}_0,$$

$$G_2(z,\zeta) = |\zeta - z|^2 \log\left|\frac{\overline{\zeta}^2-z^2}{\zeta^2 - z^2}\right|^2 + (\zeta-\overline{\zeta})(z-\overline{z})\log\left|\frac{\zeta+z}{\overline{\zeta}-z}\right|^2 + (\zeta+\overline{\zeta})(z+\overline{z})\log\left|\frac{\overline{\zeta}+z}{\zeta+z}\right|^2 \quad \text{for } \mathbb{Q}_1,$$

and for R

$$G_2(z,\zeta) = |\zeta - z|^2\left[\frac{\log|z|^2\log|\zeta|^2}{\log r^2} - \log\left|\frac{\zeta-z}{1-z\overline{\zeta}}\prod_{k=1}^{\infty}\frac{(z-r^{2k}\zeta)(\zeta-r^{2k}z)}{(z\overline{\zeta}-r^{2k})(1-r^{2k}\overline{z}\zeta)}\right|^2\right]$$

$$+2\mathrm{Re}\left\{\frac{(1-|z|^2)\log|\zeta|^2}{\log r^2} + \frac{(1-|\zeta|^2)\log|z|^2}{\log r^2} - \frac{r^2(1-|z|^2)}{1-r^2}\frac{\zeta}{z}\log|\zeta|^2\right.$$

$$-\frac{r^2(1-|\zeta|^2)}{1-r^2}\frac{z}{\zeta}\log|z|^2 + \frac{z\overline{\zeta}\log|z|^2\log|\zeta|^2}{\log r^2} - \frac{(1-r^2)\log|z|^2\log|\zeta|^2}{(\log r^2)^2}$$

$$+\left(\frac{r^2}{1-r^2}\right)^2\frac{(1-|z|^2)(1-|\zeta|^2)\log r^2}{z\overline{\zeta}} + (1-|\zeta|^2)\left[\frac{1-|z|^2}{z\overline{\zeta}}\log(1-z\overline{\zeta})\right.$$

$$
\begin{aligned}
&- (1-r^2)\sum_{n=1}^{\infty}\Big(\frac{z}{r^{2n}\zeta}\log\Big(1-\frac{r^{2n}\zeta}{z}\Big) - \frac{1}{r^{2n}z\overline{\zeta}}\log(1-r^{2n}z\overline{\zeta})\Big)\Big] \\
&- \sum_{n=1}^{\infty}\frac{(r^{2n}|\zeta|^2-1)}{r^{2n}}\frac{(1-|z|^2)}{z\overline{\zeta}}\log(1-r^{2n}z\overline{\zeta}) \\
&- \sum_{n=1}^{\infty}(r^{2n}-|\zeta|^2)\Big[\frac{|z|^2-r^2}{z\overline{\zeta}}\log\Big(1-\frac{r^{2n}}{z\overline{\zeta}}\Big) - (1-r^2)\frac{z}{\zeta}\log\Big(1-\frac{r^{2n}z}{\zeta}\Big)\Big] \\
&- \sum_{n=1}^{\infty}(1-r^{2n})\Big[\frac{(|z|^2-r^2)\zeta}{z}\log\Big(1-\frac{r^{2n}\zeta}{z}\Big) - (1-r^2)z\overline{\zeta}\log(1-r^{2n}z\overline{\zeta})\Big] \\
&- \sum_{n=1}^{\infty}\frac{(1-r^{2n})}{r^{2n}}\frac{(1-|z|^2)\zeta}{z}\log\Big(1-\frac{r^{2n}z}{\zeta}\Big) + (1-r^2)\sum_{n=1}^{\infty}\sum_{k=1}^{\infty} \\
&\times\Big[\frac{r^{2n}|\zeta|^2-1}{r^{2n}}\Big[\frac{z}{r^{2k}\zeta}\log\Big(1-\frac{r^{2(n+k)}\zeta}{z}\Big) - \frac{1}{r^{2k}z\overline{\zeta}}\log(1-r^{2(n+k)}z\overline{\zeta})\Big] \\
&\quad - (r^{2n}-|\zeta|^2)\Big[\frac{r^{2k}}{z\overline{\zeta}}\log\Big(1-\frac{r^{2(n+k)}}{z\overline{\zeta}}\Big) - \frac{r^{2k}z}{\zeta}\log\Big(1-\frac{r^{2(n+k)}z}{\zeta}\Big)\Big] \\
&\quad - (1-r^{2n})\Big[\frac{r^{2k}\zeta}{z}\log\Big(1-\frac{r^{2(n+k)}\zeta}{z}\Big) - r^{2k}z\overline{\zeta}\log(1-r^{2(n+k)}z\overline{\zeta})\Big] \\
&\quad + \frac{1-r^{2n}}{r^{2n}}\Big[\frac{z\overline{\zeta}}{r^{2k}}\log\Big(1-\frac{r^{2(n+k)}}{z\overline{\zeta}}\Big) - \frac{\zeta}{r^{2k}z}\log\Big(1-\frac{r^{2(n+k)}z}{\zeta}\Big)\Big]\Big]\Big\}.
\end{aligned}
$$

2. A New Polyharmonic Dirichlet Problem

For a regular domain D on the complex plane the convolutions of the polyharmonic Green and Green–Almansi functions

$$
\begin{aligned}
H_{m,n}(z,\zeta) &= -\frac{1}{\pi}\int_D \mathcal{G}_m(z,\widetilde{\zeta})G_n(\widetilde{\zeta},\zeta)d\widetilde{\xi}d\widetilde{\eta}, \\
\widehat{H}_{n,m}(z,\zeta) &= -\frac{1}{\pi}\int_D G_n(z,\widetilde{\zeta})\mathcal{G}_m(\widetilde{\zeta},\zeta)d\widetilde{\xi}d\widetilde{\eta}
\end{aligned}
$$

provide two polyharmonic Green functions of order $m+n$. They differ from one another, but are related by

$$H_{m,n}(z,\zeta) = \widehat{H}_{n,m}(\zeta,z),$$

as $G_n(z,\zeta)$ and $\mathcal{G}_m(z,\zeta)$ are symmetric functions of their two variables. The properties of the hybrid Green function $H_{m,n}(z,\zeta)$ follow from the fact that as well $G_n(z,\zeta)$ as $\mathcal{G}_m(z,\zeta)$ are fundamental solutions to the operators $(\partial_z\partial_{\overline{z}})^n$ and $(\partial_z\partial_{\overline{z}})^m$, respectively, and from their related Dirichlet problems. As a function of z it satisfies for any $\zeta \in D$ the Dirichlet problem

$$(\partial_z\partial_{\overline{z}})^m H_{m,n}(z,\zeta) = G_n(z,\zeta) \ \text{ in } \ D,$$

$$(\partial_z\partial_{\overline{z}})^\mu H_{m,n}(z,\zeta) = 0, \ 0 \le 2\mu \le m-1,$$

$$\partial_{\nu_\zeta}(\partial_z\partial_{\overline{z}})^\mu H_{m,n}(z,\zeta) = 0, \ 0 \le 2\mu \le m-2, \ \text{ on } \partial D.$$

Thus $H_{m,n}(\cdot\,,\zeta)$ has for any $\zeta \in D$ the following properties.

- $H_{m,n}(\cdot\,,\zeta)$ is polyharmonic of order $m+n$ in $D\setminus\{\zeta\}$,
- $H_{m,n}(z,\zeta) + \dfrac{|\zeta - z|^{2(m+n-1)}}{(m+n-1)!^2}\log|\zeta - z|^2$ is polyharmonic of order $m+n$ in z in the neighborhood of ζ,
- $(\partial_z\partial_{\overline{z}})^\mu H_{m,n}(z,\zeta) = 0$, $0 \le 2\mu \le m-1$, $m \le \mu \le m+n-1$, $\partial_{\nu_z}(\partial_z\partial_{\overline{z}})^\mu H_{m,n}(z,\zeta) = 0$, $0 \le 2\mu \le m-2$, for $z \in \partial D$.

As a function of ζ it satisfies for any $z \in D$ the Dirichlet problem

$$(\partial_\zeta\partial_{\overline{\zeta}})^n H_{m,n}(z,\zeta) = \mathcal{G}_m(z,\zeta) \ \text{ in } \ D,$$

$$(\partial_\zeta\partial_{\overline{\zeta}})^\mu H_{m,n}(z,\zeta) = 0, \ 0 \le \mu \le n-1, \ \text{ on } \partial D.$$

Thus $H_{m,n}(z,\,\cdot)$ has for any $z \in D$ the following properties.

- $H_{m,n}(z,\,\cdot)$ is polyharmonic of order $m+n$ in $D\setminus\{z\}$,
- $H_{m,n}(z,\zeta) + \dfrac{|\zeta - z|^{2(m+n-1)}}{(m+n-1)!^2}\log|\zeta - z|^2$ is polyharmonic of order $m+n$ in ζ in the neighborhood of z,
- $(\partial_\zeta\partial_{\overline{\zeta}})^\mu H_{m,n}(z,\zeta) = 0$, $0 \le \mu \le n + \left[\frac{m-1}{2}\right]$, $\partial_{\nu_\zeta}(\partial_\zeta\partial_{\overline{\zeta}})^\mu H_{m,n}(z,\zeta) = 0$, $n \le \mu \le n + \left[\frac{m-2}{2}\right]$, for $\zeta \in \partial D$.

Theorem 2.1. *Any $w \in C^{2(m+n)}(D;\mathbb{C}) \cap C^{2(m+n)-1}(\overline{D};\mathbb{C})$ can be represented as*

$$\begin{aligned} w(z) = &\frac{-1}{4\pi}\int_{\partial D}\Big[\sum_{\mu=0}^{\left[\frac{m-1}{2}\right]} + \sum_{\mu=m}^{m+n-1}\Big]\partial_{\nu_\zeta}(\partial_\zeta\partial_{\overline{\zeta}})^{m+n-\mu-1}H_{m,n}(z,\zeta)(\partial_\zeta\partial_{\overline{\zeta}})^\mu w(\zeta)ds_\zeta \\ &+ \frac{1}{4\pi}\int_{\partial D}\sum_{\mu=0}^{\left[\frac{m}{2}\right]-1}(\partial_\zeta\partial_{\overline{\zeta}})^{m+n-\mu-1}H_{m,n}(z,\zeta)\partial_{\nu_\zeta}(\partial_\zeta\partial_{\overline{\zeta}})^\mu w(\zeta)ds_\zeta \\ &- \frac{1}{\pi}\int_D H_{m,n}(z,\zeta)(\partial_\zeta\partial_{\overline{\zeta}})^{m+n}w(\zeta)d\xi d\eta. \end{aligned}$$

Proof. From the Gauss theorem it follows that

$$\frac{1}{\pi}\int_D H_{m,n}(z,\zeta)(\partial_\zeta\partial_{\overline{\zeta}})^{m+n}w(\zeta)d\xi d\eta$$
$$= \frac{1}{2\pi}\int_D\Big(\partial_{\overline{\zeta}}\Big[H_{m,n}(z,\zeta)\partial_\zeta^{m+n}\partial_{\overline{\zeta}}^{m+n-1}w(\zeta)\Big]$$
$$+\partial_\zeta\Big[H_{m,n}(z,\zeta)\partial_\zeta^{m+n-1}\partial_{\overline{\zeta}}^{m+n}w(\zeta)\Big]-\partial_\zeta\Big[\partial_{\overline{\zeta}}H_{m,n}(z,\zeta)(\partial_\zeta\partial_{\overline{\zeta}})^{m+n-1}w(\zeta)\Big]$$
$$-\partial_{\overline{\zeta}}\Big[\partial_\zeta H_{m,n}(z,\zeta)(\partial_\zeta\partial_{\overline{\zeta}})^{m+n-1}w(\zeta)\Big]$$
$$+2\partial_\zeta\partial_{\overline{\zeta}}H_{m,n}(z,\zeta)(\partial_\zeta\partial_{\overline{\zeta}})^{m+n-1}w(\zeta)\Big)d\xi d\eta$$
$$=\frac{1}{4\pi i}\int_{\partial D}\Big(H_{m,n}(z,\zeta)\Big[\partial_\zeta^{m+n}\partial_{\overline{\zeta}}^{m+n-1}w(\zeta)d\zeta-\partial_\zeta^{m+n-1}\partial_{\overline{\zeta}}^{m+n}w(\zeta)d\overline{\zeta}\Big]$$
$$-(\partial_\zeta\partial_{\overline{\zeta}})^{m+n-1}w(\zeta)\Big[\partial_\zeta H_{m,n}(z,\zeta)d\zeta-\partial_{\overline{\zeta}}H_{m,n}(z,\zeta)d\overline{\zeta}\Big]\Big)$$
$$+\frac{1}{\pi}\int_D\partial_\zeta\partial_{\overline{\zeta}}H_{m,n}(z,\zeta)(\partial_\zeta\partial_{\overline{\zeta}})^{m+n-1}w(\zeta)d\xi d\eta$$
$$=\frac{1}{4\pi}\int_{\partial D}\Big(H_{m,n}(z,\zeta)\partial_{\nu_\zeta}(\partial_\zeta\partial_{\overline{\zeta}})^{m+n-1}w(\zeta)$$
$$-\partial_{\nu_\zeta}H_{m,n}(z,\zeta)(\partial_\zeta\partial_{\overline{\zeta}})^{m+n-1}w(\zeta)\Big)ds_\zeta$$
$$+\frac{1}{\pi}\int_D\partial_\zeta\partial_{\overline{\zeta}}H_{m,n}(z,\zeta)(\partial_\zeta\partial_{\overline{\zeta}})^{m+n-1}w(\zeta)d\xi d\eta.$$

Repeating this argumentation leads to

$$\frac{1}{\pi}\int_D H_{m,n}(z,\zeta)(\partial_\zeta\partial_{\overline{\zeta}})^{m+n}w(\zeta)d\xi d\eta$$
$$=\sum_{\mu=0}^{m+n-2}\frac{1}{4\pi}\int_{\partial D}\Big((\partial_\zeta\partial_{\overline{\zeta}})^\mu H_{m,n}(z,\zeta)\partial_{\nu_\zeta}(\partial_\zeta\partial_{\overline{\zeta}})^{m+n-\mu-1}w(\zeta)$$
$$-\partial_{\nu_\zeta}(\partial_\zeta\partial_{\overline{\zeta}})^\mu H_{m,n}(z,\zeta)(\partial_\zeta\partial_{\overline{\zeta}})^{m+n-\mu-1}w(\zeta)\Big)ds_\zeta$$
$$+\frac{1}{\pi}\int_D(\partial_\zeta\partial_{\overline{\zeta}})^{m+n-1}H_{m,n}(z,\zeta)\partial_\zeta\partial_{\overline{\zeta}}w(\zeta)d\xi d\eta.$$

Because

$$(\partial_\zeta\partial_{\overline{\zeta}})^{m+n-1}H_{m,n}(z,\zeta)=(\partial_\zeta\partial_{\overline{\zeta}})^{m-1}\mathcal{G}_m(z,\zeta)=G_1(z,\zeta),$$

and, see e.g. [1],

$$\frac{1}{\pi}\int_D G_1(z,\zeta)\partial_\zeta\partial_{\overline{\zeta}}w(\zeta)d\xi d\eta$$

$$= -\frac{1}{4\pi}\int_{\partial D}\Big(\partial_{\nu_\zeta}G_1(z,\zeta)w(\zeta) - G_1(z,\zeta)\partial_{\nu_\zeta}w(\zeta)\Big)ds_\zeta - w(z),$$

finally

$$w(z) = -\sum_{\mu=0}^{m+n-1}\frac{1}{4\pi}\int_{\partial D}\Big(\partial_{\nu_\zeta}(\partial_\zeta\partial_{\overline{\zeta}})^\mu H_{m,n}(z,\zeta)(\partial_\zeta\partial_{\overline{\zeta}})^{m+n-\mu-1}w(\zeta)$$
$$- (\partial_\zeta\partial_{\overline{\zeta}})^\mu H_{m,n}(z,\zeta)\partial_{\nu_\zeta}(\partial_\zeta\partial_{\overline{\zeta}})^{m+n-\mu-1}w(\zeta)\Big)ds_\zeta$$
$$-\frac{1}{\pi}\int_D H_{m,n}(z,\zeta)(\partial_\zeta\partial_{\overline{\zeta}})^{m+n}w(\zeta)d\xi d\eta$$
$$= -\sum_{\mu=0}^{m+n-1}\frac{1}{4\pi}\int_{\partial D}\Big(\partial_{\nu_\zeta}(\partial_\zeta\partial_{\overline{\zeta}})^{m+n-\mu-1}H_{m,n}(z,\zeta)(\partial_\zeta\partial_{\overline{\zeta}})^\mu w(\zeta)$$
$$- (\partial_\zeta\partial_{\overline{\zeta}})^{m+n-\mu-1}H_{m,n}(z,\zeta)\partial_{\nu_\zeta}(\partial_\zeta\partial_{\overline{\zeta}})^\mu w(\zeta)\Big)ds_\zeta$$
$$-\frac{1}{\pi}\int_D H_{m,n}(z,\zeta)(\partial_\zeta\partial_{\overline{\zeta}})^{m+n}w(\zeta)d\xi d\eta.$$

Observing the boundary behavior of $H_{m,n}(z,\zeta)$ for $z \in D$ gives the representation formula. □

Dirichlet problem. Find a solution to the $(m+n)$-Poisson equation

$$(\partial_z\partial_{\overline{z}})^{m+n}w = f$$

in the regular domain D, satisfying

$$(\partial_z\partial_{\overline{z}})^\mu w = \gamma_\mu,\ \ 0 \le 2\mu \le m-1,\ \ m \le \mu \le m+n-1,$$

$$\partial_{\nu_z}(\partial_z\partial_{\overline{z}})^\mu w = \widehat{\gamma}_\mu,\ \ 0 \le 2\mu \le m-2 \ \text{ on } \partial D.$$

Theorem 2.2. *For $f \in L_p(D;\mathbb{C})$, $2 < p$, $\gamma_\mu \in C^{0,\alpha}(\partial D;\mathbb{C})$, $0 < \alpha < 1$, for $m \le \mu \le m+n-1$, $\gamma_\mu \in C^{m-2\mu,\alpha}(\partial D;\mathbb{C})$ for $0 \le 2\mu \le m-1$, $\widehat{\gamma}_\mu \in C^{m-2\mu-1,\alpha}(\partial D;\mathbb{C})$ for $0 \le 2\mu \le m-2$, the unique weak solution to the Dirichlet problem is*

$$w(z) = -\frac{1}{4\pi}\int_{\partial D}\Big(\sum_{\mu=0}^{[\frac{m-1}{2}]} + \sum_{\mu=m}^{m+n-1}\Big)\partial_{\nu_\zeta}(\partial_\zeta\partial_{\overline{\zeta}})^{m+n-\mu-1}H_{m,n}(z,\zeta)\gamma_\mu(\zeta)ds_\zeta$$
$$+\frac{1}{4\pi}\int_{\partial D}\sum_{\mu=0}^{[\frac{m}{2}]-1}(\partial_\zeta\partial_{\overline{\zeta}})^{m+n-\mu-1}H_{m,n}(z,\zeta)\widehat{\gamma}_\mu(\zeta)ds_\zeta$$
$$-\frac{1}{\pi}\int_D H_{m,n}(z,\zeta)f(\zeta)d\xi d\eta.$$

Proof. To verify the boundary behavior one rewrites

$$
\begin{aligned}
w(z) = &-\frac{1}{4\pi}\int_{\partial D}\sum_{\mu=0}^{[\frac{m-1}{2}]}\partial_{\nu_\zeta}(\partial_\zeta\partial_{\overline{\zeta}})^{m-\mu-1}\mathcal{G}_m(z,\zeta)\gamma_\mu(\zeta)ds_\zeta \\
&-\frac{1}{4\pi}\int_{\partial D}\sum_{\mu=0}^{n-1}\partial_{\nu_\zeta}(\partial_\zeta\partial_{\overline{\zeta}})^{n-\mu-1}H_{m,n}(z,\zeta)\gamma_{\mu+m}(\zeta)ds_\zeta \\
&+\frac{1}{4\pi}\int_{\partial D}\sum_{\mu=0}^{[\frac{m}{2}]-1}(\partial_\zeta\partial_{\overline{\zeta}})^{m-\mu-1}\mathcal{G}_m(z,\zeta)\widehat{\gamma}_\mu ds_\zeta \\
&-\frac{1}{\pi}\int_D H_{m,n}(z,\zeta)f(\zeta)d\xi d\eta.
\end{aligned}
$$

Here the first three terms are relevant for the boundary behavior. According to Theorem 1.2 the first and the third terms together take care of the conditions

$$(\partial_z\partial_{\overline{z}})^\mu w = \gamma_\mu,\ \ 0\le 2\mu\le m-1,\ \ \partial_{\nu_z}(\partial_z\partial_{\overline{z}})^\mu w = \widehat{\gamma}_\mu,\ \ 0\le 2\mu\le m-2,$$

on ∂D because for $0\le 2\rho\le m-1$

$$
\begin{aligned}
&(\partial_z\partial_{\overline{z}})^\rho\int_{\partial D}\partial_{\nu_\zeta}(\partial_\zeta\partial_{\overline{\zeta}})^{n-\mu-1}H_{m,n}(z,\zeta)\gamma_\mu(\zeta)ds_\zeta \\
&= -\frac{1}{\pi}\int_{\partial D}\partial_{\nu_\zeta}(\partial_\zeta\partial_{\overline{\zeta}})^{n-\mu-1}\int_D(\partial_z\partial_{\overline{z}})^\rho\mathcal{G}_m(z,\widetilde{\zeta})G_n(\widetilde{\zeta},\zeta)d\widetilde{\xi}d\widetilde{\eta}\,\gamma_\mu(\zeta)ds_\zeta
\end{aligned}
$$

vanishes for $z\in\partial D$. This is similarly true for the $\partial_{\nu_z}(\partial_z\partial_{\overline{z}})^\rho$ derivative of these terms for $0\le 2\rho\le m-2$. According to Theorem 1.1 the second term gives for $m\le\rho\le m+n-1$

$$
\begin{aligned}
&(\partial_z\partial_{\overline{z}})^\rho\int_{\partial D}\partial_{\nu_\zeta}(\partial_\zeta\partial_{\overline{\zeta}})^{n-\mu-1}H_{m,n}(z,\zeta)\gamma_\mu(\zeta)ds_\zeta \\
&= (\partial_z\partial_{\overline{z}})^{\rho-m}\int_{\partial D}\partial_{\nu_\zeta}(\partial_\zeta\partial_{\overline{\zeta}})^{n-\mu-1}G_n(z,\zeta)\gamma_\mu(\zeta)ds_\zeta \\
&= (\partial_z\partial_{\overline{z}})^{\rho-m}\int_{\partial D}\partial_{\nu_\zeta}G_{\mu+1}(z,\zeta)\gamma_\mu(\zeta)ds_\zeta,
\end{aligned}
$$

which is $\gamma_{\rho-m}(z)$ on ∂D up to the factor -4π for $\mu=\rho-m$ and 0 for $\mu\ne\rho-m$.

As for $m\le\rho\le m+n-1$, $\zeta\in\partial D$,

$$(\partial_z\partial_{\overline{z}})^\rho\mathcal{G}_m(z,\zeta)=0,\ \ z\in D,$$

there are no contributions from the first and third terms. □

References

1. H. Begehr, T. Vaitekhovich. *Iterated Dirichlet problem for higher order Poisson equation.* Preprint, FU Berlin, 2006.
2. H. Begehr. *Orthogonal decomposition of the function space* $L_2(\overline{\mathbb{D}}; \mathbb{C})$. J. Reine Angew. Math., 549, 2002.
3. H. Begehr, T. N. H. Vu, Z.-X. Zhang. *Polyharmonic Dirichlet problems.* Proc. Steklov Inst. Math., 255, 2006.
4. H. Begehr. *A particular polyharmonic Dirichlet problem.* Complex Analysis and Potential Theory, eds. T. Aliyev Azeroğlu, P. M. Tamrazov, World Sci., Singapore, 2007.
5. H. Begehr, E. Gaertner. *A Dirichlet problem for the inhomogeneous polyharmonic equation in the upper half plane.* Georgian Math. J., 14, 2007.
6. H. Begehr, T. Vaitekhovich. *Green functions in complex plane domains.* Preprint, FU Berlin, 2007.
7. E. Gaertner. *Basic complex boundary value problems in the upper half plane.* Ph.D. thesis, FU Berlin, 2006. `http://www.diss.fu-berlin.de/2006/320`
8. H. Begehr, T. Vaitekhovich. *Polyharmonic Green functions for particular plane domains.* Preprint, FU Berlin, 2007.
9. T. S. Vaitekhovich. *Boundary value problems to second order complex partial differential equations in a ring domain.* Šiauliai Mathematical Seminar, 2(10), 2007.
10. T. S. Vaitekhovich. *Biharmonic Green function of a ring domain.* Preprint, FU Berlin, 2007.

THE FOURIER TRANSFORM METHOD IN CONTROLLABILITY PROBLEMS FOR THE FINITE STRING EQUATION WITH A BOUNDARY CONTROL BOUNDED BY A HARD CONSTANT

L. V. FARDIGOLA

Mathematical Division, Institute for Low Temperature Physics and Engineering of the National Academy of Sciences of Ukraine
Kharkiv, 61103, Ukraine
E-mail: fardigola@ukr.net

Necessary and sufficient conditions for (approximate) null-controllability are obtained for the control system $w_{tt} = w_{xx} - q^2 w$, $w_x(0,t) = u(t)$, $w_x(\pi,t) = 0$, $x \in (0,\pi)$, $t \in (0,T)$, where $q \geq 0$, $T \in (0,\pi]$, u is a control bounded by a hard constant. These problems are considered in Sobolev spaces. Controls solving them are found explicitly. Continuous controls solving the approximate null-controllability problem are constructed with the aid of the Cesàro means of a Fourier series generated by the data of the control system.

Keywords: Wave equation, half-axis, controllability problem, control bounded by a hard constant, Fourier transform, Sobolev space, Cesàro mean.

1. Introduction

Controllability problems for the wave equation on a segment were investigated in a number of papers [1–7] (and many others). We should note that most of the papers investigating controllability of the wave equation consider L^2-controllability or, more generally, L^p-controllability ($2 \leq p < +\infty$). But only L^∞-controls can be realized practically. Moreover, such controls should be bounded by a hard constant for practical purposes. Sometimes for a system arising in Nature and Technology we need to control the system letting it fluctuating when trying to drive it to a desired state without forcing it too much. Therefore we construct continuous controls solving the approximate null-controllability problem and bounded by a given constant.

Consider the control system

$$w_{tt} = w_{xx} - q^2 w, \quad w_x(0,t) = u(t), \quad w_x(\pi,t) = 0, \quad x \in (0,\pi), \quad t \in (0,T), \quad (1)$$

where $T \in (0, \pi]$, $q \geq 0$. We assume that the control u satisfies the restriction $u \in \mathcal{B}^U(0,T) = \{v \in L^2(0,T) \mid |v(t)| \leq U \text{ a. e. on } (0,T)\}$ where $U > 0$ is given.

The wave equation on a segment controlled by boundary conditions was studied for $q = 0$ in the Sobolev spaces [4,7]. In the present paper the most of these results are extended on the case of the Neumann control for $q \geq 0$. If $q = 0$, the cases of the Dirichlet control and the Neumann control are rather similar. If $q > 0$, investigation of the null-controllability problem for equation (1) is essentially more complicated. In the present paper the operator Φ describing the influence of a control on a target state is introduced and studied in the Sobolev spaces H_0^s, $s \in \mathbb{R}$. In fact, application of Φ is the most essential new point of the paper. In addition, continuous controls $u \in \mathcal{B}^U(0,T)$ solving the approximate null-controllability problem are constructed in the present paper. Continuous controls for a finite string were considered in the case $q = 0$ recently [2]. In the case $q > 0$ the wave equation controlled by the Dirichlet boundary conditions was investigated [6] in special classes of functions for the time $T \geq 2\pi$ (in contrast to the present paper where $T \leq \pi$). It is the principal distinction because in the case $T > \pi$ the considered boundary control conditions may be reduced to the ones $w_x(0,t) = u_1(t)$, $w_x(\pi,t) = u_2(t)$, $t \in (0,\pi)$, e.g., $u(t) = u_1(t) - u_2(t-\pi)$, if $T = 2\pi$.

If we replace q by iq in (1) then replacing q by iq throughout this paper we obtain the results analogous to ones obtained in Sections 3–5.

2. Notation

We define the spaces used in this work. Let $\mathcal{S}$ be the Schwartz space

$$\mathcal{S} = \left\{ \varphi \in C^\infty(\mathbb{R}) \mid \forall m \in \mathbb{N}, \forall l \in \mathbb{N}, \sup_{x \in \mathbb{R}} \left\{ \left|\varphi^{(m)}(x)\right| \left(1 + |x|^2\right)^l \right\} < +\infty \right\},$$

and let $\mathcal{S}'$ be the dual space [8], where $|\cdot|$ is the Euclidean norm.

Denote by H_l^s $(s, l \in \mathbb{R})$ the Sobolev spaces

$$H_l^s = \left\{ \varphi \in \mathcal{S}' \mid \left(1 + x^2\right)^{l/2} \left(1 + D^2\right)^{s/2} \varphi \in L^2(\mathbb{R}) \right\},$$

with $\|\varphi\|_l^s = \left(\int_{-\infty}^{\infty} \left|(1+x^2)^{l/2} (1+D^2)^{s/2} \varphi(x)\right|^2 dx \right)^{1/2}$ where $D = -i\partial/\partial x$. Let $\mathcal{F} : \mathcal{S}' \to \mathcal{S}'$ be the Fourier transform operator. For $\varphi \in \mathcal{S}$ we have $(\mathcal{F}\varphi)(\sigma) = (2\pi)^{-1/2} \int_{-\infty}^{\infty} e^{-ix\sigma} \varphi(x)\, dx$ and $(\mathcal{F}f, \psi) = (f, \mathcal{F}^{-1}\psi)$ for $f \in \mathcal{S}'$, $\psi \in \mathcal{S}$. It is well known [9, Chapter 1] that $\mathcal{F}H_0^s = H_s^0$ and $\|\varphi\|_0^s = \|\mathcal{F}\varphi\|_s^0$, if $\varphi \in H_0^s$. It is easy to see that for $f \in H_0^0$, $\operatorname{supp} f \subset [-\pi, \pi]$, we have

$$\frac{1}{\sqrt{1+\pi^2}} \|f\|_0^0 \leq \left\| \sum_{n=-\infty}^{\infty} f(x+2\pi n) \right\|_{-1}^{0} \leq 2 \|f\|_0^0 . \tag{2}$$

Let Ω be the odd extension operator and Ξ be the even extension operator.

Further, we use the spaces

$$\mathcal{H}_l^s = \left\{ \varphi \in \mathcal{S}' \times \mathcal{S}' \mid \operatorname{supp} \varphi \subset [0,\pi] \text{ and } \Xi\varphi \in H_l^s \times H_l^{s-1} \right\},$$
$$\widetilde{H}_l^s = \left\{ \varphi \in H_l^s \times H_l^{s-1} \mid \varphi \text{ is even } 2\pi\text{-periodic} \right\}$$

with $|||\varphi|||^s = \left((\|\varphi_0\|_0^s)^2 + (\|\varphi_1\|_l^{s-1})^2 \right)^{1/2}$.

3. Conditions for (approximate) null-controllability

For $w^0 = \begin{pmatrix} w_0^0 \\ w_1^0 \end{pmatrix} \in \mathcal{H}_0^1$ consider system (1) with the initial conditions

$$w(x,0) = w_0^0(x), \quad w_t(x,0) = w_1^0(x), \qquad x > 0. \tag{3}$$

Let W^0, $W(\cdot,t)$ be even 2π-periodic extensions for w^0, $\begin{pmatrix} w(\cdot,t) \\ \partial w(\cdot,t)/\partial t \end{pmatrix}$ respectively, $t \in [0,T]$. Evidently, $W^0 \in \widetilde{H}_{-1}^1$, $W(\cdot,t) \in \widetilde{H}_{-1}^1$, $t \in [0,T]$, since the norm $\|f\|_{-1}^1$ is equivalent [9] to the norm $\|f\|_{-1}^0 + \|f'\|_{-1}^0$. It is easy to see that control problem (1), (3) is equivalent to the following one

$$\frac{dW}{dt} = \begin{pmatrix} 0 & 1 \\ \left(\left(\frac{d}{dx}\right)^2 - q^2 \right) & 0 \end{pmatrix} W - \begin{pmatrix} 0 \\ 2\delta(x) \end{pmatrix} u, \quad t \in (0,T), \tag{4}$$

$$W(x,0) = W^0 \tag{5}$$

where $W(\cdot,t) \in \widetilde{H}_{-1}^1$, $t \in [0,T]$, $W^0 \in \widetilde{H}_{-1}^1$, $u \in \mathcal{B}^U(0,T)$. Here δ is the Dirac distribution, $\delta = H'$, H is the Heaviside function $H(\xi) = 1$, if $\xi > 0$, and $H(\xi) = 0$ otherwise. Consider for (4),(5) the steering condition

$$W(\cdot,T) = W^T \tag{6}$$

where $W^T \in \widetilde{H}_{-1}^1$.

For a given $T > 0$, $w^0 \in \mathcal{H}_0^1$ denote by $\mathcal{R}_T^U(w^0)$ the set of the states $W^T \in \widetilde{H}_{-1}^1$ for which there exists a control $u \in \mathcal{B}^U(0,T)$ such that problem (4)–(6) has a unique solution.

Definition 3.1. A state $w^0 \in \mathcal{H}_0^1$ is called null-controllable at a given time $T \in (0,\pi]$, if 0 belongs to $\mathcal{R}_T^U(w^0)$ and approximately null-controllable at a given time $T \in (0,\pi]$, if 0 belongs to the closure of $\mathcal{R}_T^U(w^0)$ in $\widetilde{H}_{-1}^1$.

Let $\Phi : \mathcal{S}' \to \mathcal{S}'$ with $D(\Phi) = \{g \in \mathcal{S}' \mid g \text{ is odd and } \operatorname{supp} g \subset [-T,T]\}$ such that $(\Phi g)(x) = \mathcal{F}^{-1}_{\sigma\to x}\left(\frac{-i}{\sqrt{\sigma^2+q^2}}\left(\mathcal{F}g\right)\left(\sqrt{\sigma^2+q^2}\right)\right)(x)$, $g \in D(\Phi)$. In the Appendix it is proved that Φ is invertible, $\Phi^{-1} : \mathcal{S}' \to \mathcal{S}'$, $D(\Phi^{-1}) = R(\Phi) = \{g \in \mathcal{S}' \mid g \text{ is even and } \operatorname{supp} g \subset [-T,T]\}$ (Lemma 5.3). Moreover, Φ is bounded from H_0^{p-1} to H_0^p (Lemma 5.1) and Φ^{-1} is bounded from H_0^p to H_0^{p-1} (Lemma 5.3), $p \in \mathbb{R}$.

Proposition 3.1. *Let $W^0 \in \mathcal{S}' \times \mathcal{S}'$, $u \in \mathcal{B}^U(0,T)$. Then*

$$W(x,T) = E(x,T) * \left[W^0(x) - \sum_{n=-\infty}^{\infty} \Phi\begin{pmatrix}\Omega\mathcal{U}\\ \Omega\mathcal{U}'\end{pmatrix}(x+2\pi n)\right], \quad t \in [0,T], \quad (7)$$

where $\mathcal{U}(t) = u(t)\,(H(t) - H(t-T))$, W is a unique solution of (4)–(5), $*$ is the convolution with respect to x,

$$E(x,t) \equiv \frac{1}{\sqrt{2\pi}}\begin{pmatrix}\partial/\partial t & 1\\ (\partial/\partial t)^2 & \partial/\partial t\end{pmatrix}\mathcal{F}^{-1}_{x\to\sigma}\frac{\sin\left(t\sqrt{\sigma^2+q^2}\right)}{\sqrt{\sigma^2+q^2}} \quad (8)$$

In particular, if $W^0 \in \widetilde{H}^p_{-1}$, $\Omega\mathcal{U} \in H_0^{p-1}$ and $\Xi\mathcal{U} \in H_0^{p-2}$ then $W(\cdot,T) \in \widetilde{H}^p_{-1}$, $p \in \mathbb{R}$.

Proof. It is easy to see that

$$\mathcal{F}^{-1}_{\sigma\to x}\left(\frac{\sin\left(t\sqrt{\sigma^2+q^2}\right)}{\sqrt{\sigma^2+q^2}}\right) = \sqrt{\frac{\pi}{2}}\,J_0\left(q\sqrt{t^2-x^2}\right)H\left(t^2-x^2\right)\operatorname{sgn} t \quad (9)$$

where $J_\nu(\xi)$ is the Bessel function ($\nu \in \mathbb{R}$).

Applying the Fourier transform with respect to x to problem (4)–(6) and using (9) we conclude that (7) and (8) are true.

With regard to Lemmas 5.1 and 5.7 we get that the last assertion of the proposition is also true. The proposition is proved. □

Theorem 3.1. *Let $w^0 \in \mathcal{H}_0^1$, $T \in (0,\pi]$ and*

$$\operatorname{supp} w_0^0 \subset [0,T]; \quad (10)$$

$$\left|\left(\Phi^{-1}\Xi w_0^0\right)(\xi)\right| \le U \text{ a.e. on } [0,T]; \quad (11)$$

$$\Xi w_1^0 = \Phi\left(\frac{d}{d\xi}\left(\operatorname{sgn}\xi\left(\Phi^{-1}\Xi w_0^0\right)\right)\right). \quad (12)$$

Then the state w^0 is null-controllable at the time T. Moreover, the solution of the null-controllability problem (the control u) is unique and

$$\begin{aligned} u(t) &= w_0^{0\prime}(t) + \int_t^T \frac{qtI_1\left(q\sqrt{x^2-t^2}\right)}{\sqrt{x^2-t^2}} w_0^{0\prime}(x)\,dx \\ &= w_1^0(t) + \int_t^T \frac{qxI_1\left(q\sqrt{x^2-t^2}\right)}{\sqrt{x^2-t^2}} w_1^0(x)\,dx \quad \text{a.e. on } (0,T). \end{aligned} \tag{13}$$

Here $I_\nu(\xi) = i^{-\nu} J_\nu(i\xi)$ is the modified Bessel function.

Proof. Put $\widehat{\mathcal{U}} = \Phi^{-1}\Xi w_0^0$. It follows from Lemma 5.3 and (11) that $\widehat{\mathcal{U}}$ is odd and $\widehat{\mathcal{U}} \in L^\infty(\mathbb{R})$. Put $\mathcal{U}(t) = H(t)\widehat{\mathcal{U}}(t)$ and denote by $u(t)$ its restriction on $[0,T]$. Using Lemma 5.4 we conclude that $u \in \mathcal{B}^U(0,T)$ and the first part of assertion (13) is true for it. Taking into account (12), we get

$$\Xi w_1^0 = \Phi\left(\frac{d}{dt}\left(\operatorname{sgn} t\left(\Phi^{-1}\Phi\Omega\mathcal{U}\right)\right)\right) = \Phi\left(\frac{d}{dt}\left(\operatorname{sgn} t\,\Omega\mathcal{U}\right)\right) = \Phi\Omega\mathcal{U}'. \tag{14}$$

With regard to (14) we obtain from (7) that w_0^0 is null-controllable.

It follows from (14) and Lemma 5.5 that $(\Xi\mathcal{U})' = \widetilde{w}_1^{0\prime}$ where $\widetilde{w}_1^0 \in H_0^0$ is even, $\operatorname{supp} \widetilde{w}_1^0 \subset [-T,T]$ and $\widetilde{w}_1^0(t) = \Xi w_1^0(t) + \int_{|t|}^T \frac{qxI_1\left(q\sqrt{x^2-t^2}\right)}{\sqrt{x^2-t^2}} w_1^0(x)\,dx$. Since $\Xi\mathcal{U}$ and $\widetilde{w}_1^0$ have compact supports then $\Xi\mathcal{U} = \widetilde{w}_1^0$. Therefore the second part of (13) is true. The theorem is proved. □

Theorem 3.2. *If a state $w^0 \in \mathcal{H}_0^1$ is approximately null controllable at a given time $T \in (0,\pi]$ then assertions (10)–(12) are valid.*

Proof. For each $n \in \mathbb{N}$ there exists a state $W^n \in \mathcal{R}_T^U(w^0)$ such that $|||W^n|||_{-1}^1 < 1/n$. According to Lemma 5.7 $|||E(\cdot,T) * W^n|||_{-1}^1 \to 0$ as $n \to \infty$. With regard to (7) we have $\sum_{n=-\infty}^\infty \Phi\begin{pmatrix}\Omega\mathcal{U}_n\\ \Omega\mathcal{U}_n'\end{pmatrix}(x+2\pi n) \to W^0(x)$ as $n \to \infty$ in $\widetilde{H}_{-1}^1$ for some $u_n \in \mathcal{B}^U(0,T)$, here $\mathcal{U}_n(t) = u_n(t)\left(H(t) - H(t-T)\right)$. Hence $\sum_{n=-\infty}^\infty \Phi\Omega\mathcal{U}_n^{(j)}(x+2\pi n) \to W_j^0(x)$ as $n \to \infty$ in H_{-1}^0. It follows from (2) that

$$\Phi\Omega\mathcal{U}_n^{(j)} \to \Xi w_j^0 \qquad \text{as } n \to \infty \text{ in } H_{-1}^0. \tag{15}$$

Put $\widehat{\mathcal{U}} = \Phi^{-1}\Xi w_0^0$. With regard to Lemma 5.3 we get $\operatorname{supp}\widehat{\mathcal{U}} \subset [-T,T]$, $\widehat{\mathcal{U}}$ is odd and $\Omega\mathcal{U}_n \to \widehat{\mathcal{U}}$ as $n \to \infty$ in H_0^{-1}. Then the sequence $\{\Omega\mathcal{U}_n\}_{n=1}^\infty$ converges to $\widehat{\mathcal{U}}$ as $n \to \infty$ in $\mathcal{S}'$ and consequently in $\left(L^2(\mathbb{R})\right)'$ because $\{\Omega\mathcal{U}_n\}_{n=1}^\infty$ is uniformly bounded on $\mathbb{R}$ and $\mathcal{S}$ is dense in $L^2(\mathbb{R})$. By the Riesz theorem $\widehat{\mathcal{U}} \in L^2(\mathbb{R})$. Since $u_n \in \mathcal{B}^U(0,T)$ then $|\widehat{\mathcal{U}}(t)| \le U$ a.e. on

$[-T,T]$. With regard to Lemma 5.1 we obtain from here that (10), (11) are true. Setting $\mathcal{U}(t) = H(t)\widehat{\mathcal{U}}(t)$ and taking into account (15) we obtain $\Omega\mathcal{U}_n^{(j)} \to \Omega\mathcal{U}^{(j)} = \Phi^{-1}\Xi w_j^0$ as $n \to \infty$ in H_{-1}^{-1}. Hence (12) holds. The theorem is proved. □

Remark 3.1. According to Lemma 5.6 condition (12) is equivalent to

$$w_1^0(x) = w_0^{0\prime}(x) + q\int_x^\infty w_0^0(\xi)\frac{I_1(q(\xi - x))}{\xi - x}\,d\xi, \quad x > 0. \qquad (12'')$$

Remark 3.2. If $q = 0$ then (12) is of the form $w_1^0 = w_0^{0\prime}$ and (13) is of the form $u(t) = w_0^{0\prime}(t) = w_1^0(t)$ a.e. on $[0,T]$.

Remark 3.3. According to Lemma 5.5 condition (11) is equivalent to

$$\left| w_1^0(t) + \int_t^T \frac{qxI_1\left(q\sqrt{x^2 - t^2}\right)}{\sqrt{x^2 - t^2}} w_1^0(x)\,dx \right| \le U \text{ a.e. on } [0,T], \qquad (11')$$

if (10) and (12) hold. With regard to Lemma 5.5 the condition $|w_1^0(x)| \le \frac{U}{I_0(qT)}$ a.e. on $[0,T]$ is sufficient for (11′) (and 11)) and the condition $|w_1^0(x)| \le U(1 + qT)$ a.e. on $[0,T]$ is necessary for (11′) (and 11)). If $q = 0$ we obtain from here that (11′) and (11) are equivalent to the condition $|w_1^0(x)| \le U$ a.e. on $[0,T]$.

4. Continuous controls

In this section we construct continuous controls that solve the approximate null-controllability problem and are bounded by U. They are obtained with the help of the Cesàro means for a Fourier series determined by the data of the considered system. These controls generate a continuous steering state, if an initial state is continuous.

Consider control system (4), (5) and assume that for $T = \pi$ and $w^0 \in \mathcal{H}_0^1$ conditions (10)–(12) hold. Set $\widetilde{\mathcal{U}}(t) = H(t)\left(\Phi^{-1}\Xi w_0^0\right)(t)$ and denote by $\widetilde{u}$ the restriction of $\widetilde{\mathcal{U}}$ on $[0,\pi]$. By Theorem 3.1 $\widetilde{u} \in \mathcal{B}^U(0,\pi)$. Put $\nu_n = \frac{2}{\pi}\int_0^\pi \widetilde{u}(t)\sin(nt)\,dt$, $n = \overline{0,\infty}$. According to Lemma 5.3 we get $\nu_n = \frac{2n}{\pi}\int_0^\pi \cos\left(x\sqrt{n^2 - q^2}\right) w_0^0(x)\,dx$, $n = \overline{0,\infty}$. Put $u_n(t) = \frac{1}{n+1}\sum_{k=0}^n \sum_{l=0}^k \nu_l \sin(lt) = \sum_{l=0}^n \frac{n+1-l}{n+1}\nu_l \sin(lt)$, $n = \overline{0,\infty}$. Evidently, u_n is a Cesàro mean for the Fourier series $\sum_{l=0}^\infty \nu_l \sin(lt)$, $n = \overline{0,\infty}$. Set $\nu(t) = \sum_{k=-\infty}^\infty (\Omega\widetilde{\mathcal{U}})(t - 2\pi k)$. It is well known [11, Section 3.3] that $u_n(t) = \frac{1}{\pi}\int_{-\pi}^\pi \nu(t+\xi)F_n(\xi)\,d\xi$ where $F_n(\xi) = \frac{1}{2(n+1)}\left(\frac{\sin((n+1)\xi)}{\sin(\xi)}\right)^2$ is the Fejér kernel and $\frac{1}{\pi}\int_{-\pi}^\pi F_n(\xi)\,d\xi = 1$. Therefore $u_n \in \mathcal{B}^U(0,\pi)$. It is easy to see that $\mathcal{U}_n \in H_0^p$, $p < 3/2$, $n = \overline{0,\infty}$, where $\mathcal{U}_n(t) = u_n(t)\left(H(t) - H(t-\pi)\right)$.

Since $\widetilde{u} \in \mathcal{B}^U(0,\pi) \subset L^2(0,\pi)$ then $\frac{\pi}{2}\sum_{l=0}^{\infty}\nu_l^2 = \left(\|\widetilde{u}\|_{L^2(0,\pi)}\right)^2$. Taking this into account we have from [11, Section 3.1] that the first and the second Cesàro means tend to $\|\widetilde{u}\|_{L^2(0,\pi)}$ as $n \to \infty$, i.e.

$$\sum_{l=0}^{n}\frac{n+1-l}{n+1}\nu_l^2 \to \frac{\pi}{2}\sum_{l=0}^{\infty}\nu_l^2, \quad \sum_{l=0}^{n}\frac{(n+1-l)(n+2-l)}{(n+1)(n+2)}\nu_l^2 \to \frac{\pi}{2}\sum_{l=0}^{\infty}\nu_l^2$$

as $n \to \infty$. Therefore $\sum_{l=0}^{n}\frac{l}{n+1}\nu_l^2 \to 0$ as $n \to \infty$ hence $\sum_{l=0}^{n}\frac{l^2}{(n+1)^2}\nu_l^2 \to 0$ as $n \to \infty$. Summarizing we obtain

$$\|\widetilde{u} - u_n\|_{L^2(0,\pi)} = \sqrt{\frac{\pi}{2}}\left(\sum_{l=n+1}^{\infty}\nu_l^2 + \sum_{l=0}^{n}\frac{l^2}{(n+1)^2}\nu_l^2\right)^{1/2} \to 0 \text{ as } n \to \infty.$$

Taking into account Lemmas 5.7, 5.1 and Proposition 3.1 we get the following result.

Theorem 4.1. *Let $T = \pi$, $w^0 \in \mathcal{H}_0^1$, $s \le 1$. Let also conditions (10)–(12) be fulfilled. Then $u_n \in \mathcal{B}^U(0,\pi)$, $n = \overline{0,\infty}$, and $|||W^n(\cdot,\pi)|||_{-1}^1 \le 2M_qL_q\|\widetilde{u} - u_n\|_{L^2(0,\pi)} \to 0$ as $n \to \infty$ where W^n is the solution of (4), (5) corresponding to the control u_n, L_q is the norm of Φ acting from H_0^0 to H_0^1. Moreover, $\mathcal{U}_n \in H_0^p$, $p < 3/2$, $n = \overline{0,\infty}$, where $\mathcal{U}_n(t) = u_n(t)\left(H(t) - H(t-\pi)\right)$. In addition, $W^n(\cdot,\pi) \in \widetilde{H}_{-1}^p \subset C^1(\mathbb{R}) \times C(\mathbb{R})$, if $w^0 \in \mathcal{H}_0^p$, $3/2 < p < 5/2$.*

Denote $\widetilde{\omega}_j^n = \sqrt{\frac{\pi}{2}}\left(\mathcal{F}\Xi w_0^0\right)(\lambda_n)$ where $\lambda_n = \sqrt{n^2 - q^2}$, $n \in \mathbb{N}$, $j = 1,2$. Then $\nu_n = i\sqrt{\frac{2}{\pi}}\left(\mathcal{F}\Omega\widetilde{\mathcal{U}}\right)(\pi n) = i\sqrt{\frac{2}{\pi}}\left(\mathcal{F}\left(\Phi^{-1}\Xi w_0^0\right)\right)(\lambda_n) - \frac{2n}{\pi}\widetilde{\omega}_0^n$. Condition (12) is equivalent to $\Omega w_1^0 = \Phi\Omega\widetilde{\mathcal{U}}'$. Therefore the following holds.

Remark 4.1. (12) is equivalent to $\widetilde{\omega}_1^n = \frac{2}{T}\sum_{l=0}^{\infty}\frac{\left((-1)^{l+n}-1\right)l^2\widetilde{\omega}_0^l}{l^2-n^2}$, $n = \overline{0,\infty}$.

5. Appendix

We assume throughout this section that $p \in \mathbb{R}$, $T > 0$.

Denote $\partial : \mathcal{S}' \to \mathcal{S}'$ with $D(\partial) = \{f \in \mathcal{S}' \mid f \text{ is even and } \operatorname{supp} f \subset [-T,T]\}$ such that $\partial f = f'$, $f \in D(\partial)$. Evidently, $R(\partial) = \{g \in \mathcal{S}' \mid f \text{ is odd and } \operatorname{supp} g \subset [-T,T]\}$. Then ∂ is bounded linear operator from H_0^p to H_0^{p-1}. Due to the Paley–Wiener theorem we obtain that ∂ is invertible. According to the inverse operator theorem we conclude that ∂^{-1} is also bounded.

Lemma 5.1. *Let $g \in D(\Phi)$. Then $\operatorname{supp}\Phi g \subset [-T,T]$, Φg is even, Φ is bounded from H_0^{p-1} to H_0^p. If $q = 0$ then $\Phi = \partial^{-1}$.*

Proof. Due to the generalized Paley–Wiener theorem [10, Chapter 3] we conclude that 1) $G = \mathcal{F}g$ is a regular functional; 2) G is of a polynomial growth on $\mathbb{R}$; 3) G can be extended to an entire function of the order ≤ 1 and the type $\le T$. Evidently, G is odd. Applying again the generalized Paley–Wiener theorem to $\frac{G(\sqrt{\sigma^2+q^2})}{\sqrt{\sigma^2+q^2}}$ we obtain that $\operatorname{supp}\Phi g \subset [-T,T]$. Obviously, Φg is even. We have

$$\|(\Phi g)'\|_0^{p-1} = \left(2\int_q^\infty \frac{|G(\mu)|^2}{(1-q^2+\mu^2)^{1-p}}\sqrt{1-\frac{q^2}{\mu^2}}\,d\mu\right)^{1/2} \le \widetilde{L}_q^p\,\|g\|_0^{p-1},$$

where $\widetilde{L}_q^p > 0$. Since ∂^{-1} is bounded then Φ is bounded from H_0^{p-1} to H_0^p. If $q = 0$ then $\Phi g = \mathcal{F}^{-1}\left(\frac{(\mathcal{F}g)(\sigma)}{(i\sigma)}\right) = \partial^{-1}g$. The lemma is proved. □

Lemma 5.2. *If $g \in D(\Phi) \cap H_0^0$, then*

$$(\Phi g)(x) = -\int_{|x|}^\infty J_0\left(q\sqrt{t^2-x^2}\right) g(t)\,dt.$$

Proof. With regard to the definition of Φ and (9) we conclude that the assertion of the lemma is true. □

Denote $\Psi : \mathcal{S}' \to \mathcal{S}'$ with $D(\Psi) = \{f \in \mathcal{S}' \mid f \text{ is even and } \operatorname{supp} g \subset [-T,T]\}$ such that $(\Psi f)(t) = \mathcal{F}^{-1}_{\mu\to t}\left(i\mu\left(\mathcal{F}f\right)\left(\sqrt{\mu^2-q^2}\right)\right)(t)$, $f \in D(\Psi)$.

Lemma 5.3. *Let $f \in D(\Psi)$. Then $\operatorname{supp}\Psi f \subset [-T,T]$, Ψf is odd, Ψ is bounded from H_0^p to H_0^{p-1}. If $q = 0$ then $\Psi = \partial$. In addition, $R(\Psi) = D(\Phi)$, $D(\Psi) = R(\Phi)$ and $\Psi = \Phi^{-1}$.*

Proof. Reasoning as in the proof of Lemma 5.1 we conclude that $\operatorname{supp}\Psi f \subset [-T,T]$ and Ψf is odd. Hence $R(\Psi) \subset D(\Phi)$. We have $R(\Phi) \subset D(\Psi)$ and $\Psi\Phi g = \mathcal{F}^{-1}_{\mu\to t}\left(i\mu\left(\mathcal{F}_{x\to\xi}\left(\Phi g\right)(x)\right)(\xi)\big|_{\xi=\sqrt{\mu^2-q^2}}\right) = g$, $g \in D(\Phi)$, i.e. $\Phi : D(\Phi) \to R(\Phi)$ is invertible and $\Phi^{-1} = \Psi$.

Denote by Φ_p the restriction of Φ on H_0^{p-1}. According to Lemma 5.1 Φ_p is bounded from H_0^{p-1} to H_0^p. Evidently, Φ_p is invertible and $\Phi_p^{-1} = \Psi_{p-1}$ where Ψ_{p-1} is the restriction of Ψ on H_0^p. Applying the inverse operator theorem we conclude that Ψ_{p-1} is bounded from H_0^p to H_0^{p-1}. If $q = 0$ then $\Phi f = \frac{d}{dt}\mathcal{F}^{-1}\left(\left(\mathcal{F}f\right)(\sigma)\right) = \partial f$. The lemma is proved. □

Lemma 5.4. *If $f \in D(\Phi^{-1}) \cap H_0^1$ then $(\Phi^{-1}f)(t) = f'(t) + qt\int_{|t|}^\infty \frac{I_1\left(q\sqrt{x^2-t^2}\right)}{\sqrt{x^2-t^2}} f'(x)\,dx$.*

Proof. We have $g(t) = \frac{d}{dt}\mathcal{F}^{-1}_{\mu\to t}(\frac{-i}{\sqrt{\mu^2-q^2}}(\mathcal{F}f')(\sqrt{\mu^2-q^2}))$. Since $\sin\left(t\sqrt{\sigma^2+q^2}\right)/\sqrt{\sigma^2+q^2}$ is an entire function with respect to (σ, q) then replacing q by iq and reasoning as in the proof of Lemma 5.2 we conclude that the assertion of the lemma is true. □

Lemma 5.5. *Let $f \in D(\Phi^{-1}) \cap H^0_0$. Then $\Phi^{-1}f = \widetilde{g}'$ where $\widetilde{g} \in H^0_0$ is even,* $\operatorname{supp}\widetilde{g} \subset [-T, T]$ *and*

$$\widetilde{g}(t) = -\int_{|t|}^{\infty} f'(x) I_0\left(q\sqrt{x^2-t^2}\right) dx, \tag{16}$$

$$f(x) = -\int_{|x|}^{\infty} \widetilde{g}'(t) J_0\left(q\sqrt{t^2-x^2}\right) dt. \tag{17}$$

Moreover, if $|f(x)| \le \mathcal{F}$ a.e. on $[-T, T]$ then $|\widetilde{g}(t)| \le \mathcal{F} I_0(qT)$ a.e. on $[-T, T]$, if $|\widetilde{g}(t)| \le \mathcal{G}$ a.e. on $[-T, T]$ then $|f(x)| \le \mathcal{G}(1 + qT)$ a.e. on $[-T, T]$.

Proof. It follows from Lemma 5.4 that $-\int_{|t|}^{\infty} I_0\left(q\sqrt{x^2-t^2}\right) f'(x)\, dx = \widetilde{g}(t) + C$. Evidently, $\operatorname{supp}\widetilde{g} \subset [-T, T]$ iff $C = 0$. Hence (16) is true. It follows from Lemma 5.2 that (17) is also true. If $|f(x)| \le \mathcal{F}$ a.e. on $[-T, T]$ then $|\widetilde{g}(t)| \le \mathcal{F}\int_{|t|}^{T}\left|\frac{\partial}{\partial x} I_0\left(q\sqrt{x^2-t^2}\right)\right| dx \le \mathcal{F} I_0(qT)$. If $|\widetilde{g}(t)| \le \mathcal{G}$ a.e. on $[-T, T]$ then $|f(x)| < \mathcal{G} + \mathcal{G}\int_{|x|}^{T} \frac{qt}{\sqrt{t^2-x^2}}\, dt < \mathcal{G}(1 + qT)$. □

Lemma 5.6. *Let $f \in H^1_0$ be even,* $\operatorname{supp} f \subset [-T, T]$, $h = \Phi\frac{d}{dt}\left(\operatorname{sgn} t\,\Phi^{-1}f\right)$. *Then* $h(x) = \operatorname{sgn} x\, f'(x) + q\int_{|x|}^{\infty} f(\xi)\frac{I_1(q(\xi-|x|))}{\xi-|x|}\, d\xi$.

Proof. Put $F = \mathcal{F}f$. Denote $G(\xi) = \frac{1}{i\xi}\mathcal{F}_{t\to\xi}\left(\frac{d}{dt}\left(\operatorname{sgn} t(\Phi^{-1}f)(t)\right)\right)(\xi)$. Then $G = \frac{2}{\pi}\mathrm{V.p.}\int_0^{\infty}\frac{1}{\xi^2-\mu^2}\mu^2 F\left(\sqrt{\mu^2-q^2}\right) d\mu$. Therefore $h = -\Delta_q\mathcal{F}^{-1}_{\sigma\to x}\frac{G\left(\sqrt{\sigma^2+q^2}\right)}{\sigma^2+q^2} = -\frac{1}{2}\Delta_q\left(\frac{e^{-q|x|}}{q} * \left(\mathcal{F}^{-1}_{\sigma\to x} G\left(\sqrt{\sigma^2+q^2}\right)\right)\right)$ where $\Delta_q = \left(\frac{d}{dx}\right)^2 - q^2$. Taking into account (9) where q is replaced by iq and $\frac{2}{\pi}\int_0^1\frac{\cosh(\xi t)}{\sqrt{1-t^2}}\, dt = I_0(\xi)$ we obtain

$$\begin{aligned} h(x) &= \frac{1}{2}\Delta_q\Bigg(\left(f'(x) * (I_0(qx)\operatorname{sgn} x)\right)\Big|_{x=0}\frac{e^{-q|x|}}{q} \\ &\qquad + \operatorname{sgn} x\left(f(x) * (I_0(qx)\operatorname{sgn} x)\right) - f(x) * I_0(qx)\Bigg) \\ &= \operatorname{sgn} x\, f'(x) - \int_0^{\infty} f(\xi + |x|)\left(\Delta_q I_0(q\xi)\right) d\xi. \end{aligned}$$

Since $\Delta_q I_0(q\xi) = (q^2/2)\,(I_2(q\xi) - I_0(q\xi)) = -qI_1(q\xi)/\xi$, we conclude that the assertion of the lemma os true. □

Lemma 5.7. *Let* $f = \begin{pmatrix} f_0 \\ f_1 \end{pmatrix} \in \widetilde{H}^p_{-1}$, $M_q = 40(2+\pi^2)\sqrt{(1+q^2)}$. *Then* $|||E(x,t) * f|||^p_{-1} \le M_q |||f|||^p_{-1}$, $t \in [-\pi, \pi]$.

Proof. Put $f_j^0 = (1+D^2)^{-\frac{p-j}{2}} \left[\left((1+D^2)^{\frac{p-j}{2}} f_j\right)(H(x+\pi) - H(x-\pi))\right]$, $j = 0,1$. Evidently, $f_j^0 \in H_0^{p-j}$, $\operatorname{supp}(1+D^2)^{\frac{p-j}{2}} f_j^0 \subset [-\pi,\pi]$. Let $E = \{e_{mj}\}_{m,j=0}^1$. Therefore for $t \in [-\pi,\pi]$ we have

$$|||E(x,t) * f|||^p_{-1} \le \sum_{m,j=0,1} \left\| \left(e_{mj}(\cdot,t) * (1+D^2)^{\frac{p-m}{2}} f_j^0\right)(x+2\pi n) \right\|^0_{-1}.$$

Taking into account (8) and applying the Paley–Wiener theorem we conclude that $\operatorname{supp}\left(e_{mj}(\cdot,t) * \left((1+D^2)^{\frac{p-m}{2}} f_j^0\right)\right) \subset [-2\pi, 2\pi]$. We have $\left\|\sum_{n=-\infty}^{\infty} h(x+2\pi n)\right\|^0_{-1} \le 20\,\|h\|_0^0$ for $h \in H_0^0$, $\operatorname{supp} h \subset [-2\pi, 2\pi]$. With regard to (2) and (8) it follows from here that

$$|||E(x,t) * f|||^p_{-1} \le 20 \sum_{m,j=0,1} \left\|e_{mj}(\cdot,t) * f_j^0\right\|_0^{p-m} \le M_q |||f|||^p_{-1}, \quad t \in [-\pi,\pi],$$

since $\left(1+|\sigma|^2\right)\left|\frac{\sin(t\sqrt{\sigma^2+q^2})}{\sqrt{\sigma^2+q^2}}\right|^2 \le t^2+2$. The lemma is proved. □

References

1. W. Kràbs and G. Leugering, *Math. Methods Appl. Sci.*, **17**, 77 (1994).
2. M. Gugat, *ZAMM*, **86**, 134 (2006).
3. M. Negreanu and E. Zuazua, *C. R. Math. Acad. Sci. Paris*, **338**, No. 5, 413 (2004).
4. M. Gugat, G. Leugering and G. M. Sklyar, *SIAM J. Control Optim.*, **44**, No. 1, 49 (2005).
5. H. O. Fattorini, *Infinite dimensional optimization and control theory,* Cambridge University Press, 1999.
6. V. A. Il'in and E. I. Moiseev, *Dokl. Akad. Nauk, Ross. Acad. Nauk,* **394**, No. 2, 154 (2004).
7. L. V. Fardigola and K. S. Khalina, *Ukr. Mat. Zh.*, **59**, No. 7, 939 (2007).
8. L. Schwartz, *Théorie des distributions, 1, 2,* Hermann, Paris, 1950–1951.
9. S. G. Gindikin and L. R. Volevich, *Distributions and convolution equations,* Gordon and Breach Sci. Publ., Philadelphia, 1992.
10. I. M. Gelfand and G. E. Shilov, *Generalized Functions, 3,* Fismatgiz, Moscow, 1958.
11. A. Zygmund, *Trigonometric series, 1,* Cambridge University Press, 1959.

DENSITY PROBLEM OF MONODROMY REPRESENTATION OF FUCHSIAN SYSTEMS

G. K. GIORGADZE

Department of Mathematics, Tbilisi State University,
Tbilisi, 0186, Georgia
E-mail: ggk@jinr.ru

This article is dedicated to the investigation of the density problem of monodromy groups of Fuchsian systems on complex manifolds in linear groups. We consider the so-called inverse problem. We apply the Riemann-Hilbert monodromy problem and we show that there exists a Fuchsian system with dense monodromy subgroup in the special unitary group.

Keywords: Fuchsian system; Monodromy representation; Dense subgroup; Fundamental group; Riemann surface.

1. Introduction and statement of the problem

In this paper we investigate some properties of monodromy represcnation of Fuchsian systems [12]. This problem arises in the theory of quantum computation in connection with the construction of a universal set of gates for a quantum computer [9].

Let X be an m-dimensional complex analytic manifold and suppose $D = \cup_{i=1}^{n} D_j$ is a divisor such that D_j are generic 1-codimensional submanifolds of X. Let

$$df = \omega f \tag{1}$$

be a completely integrable Pfaffian system on X, where ω is a $d \times d$ - matrix valued holomorphic 1-form on $X \setminus D$. Complete integrability condition means that ω satisfies $d\omega - \omega \wedge \omega = 0$. The system (1) defines the monodromy representation

$$\rho : \pi_1(X \setminus D) \to GL(n, \mathbb{C}). \tag{2}$$

The main problem is *to investigate the properties of the system* (1) *when the represenation* (2) *is a dense subset of* $GL(n, \mathbb{C})$, *i.e* $\overline{\rho(\pi_1(X \setminus d)} = GL(n, \mathbb{C})$.

In case, when $\pi_1(X\backslash D)$-not free, for example, braid group, then problem is difficult. M. Freedman, M. Larsen and Z. Wang [16] constructed Fuchsian system with density monodromy representation in $SU(n)$.When $\pi_1(X \setminus D)$ is free group and $\dim X = 1$, then existence of regular system (1) with density monodromy representation (2) follows from positivity solution of Riemann-Hilbert monodromy problem in class of regular systems [12]. In addition if system (1) is 2×2, then there exist the solution of problem in class of Fuchsian system [17].

2. Fuchsian system and entangled operator

Consider a system [17]

$$\frac{dy}{dz} = B(z)y \tag{3}$$

of p linear differential equations with rational coefficients on the Riemann sphere $\mathbb{CP}^1$. Let $D = \{a_1, ..., a_n\}$ be the set of singular points of (3), consisting of the poles of the matrix function $B(z)$.

Consider the matrix differential form $\omega = B(z)dz$. In what follows we will rather write (3) in its invariant form

$$dy = \omega y \tag{4}$$

in terms of which D is a singular divisor of ω.

Let Y denote a fundamental solution of (4), holomorphic in a neighbourhood of a given non-singular point $z_0 \in \mathbb{C}$. Analytic continuation of Y along a loop in $\mathbb{CP}^1 \setminus D$ yields a new fundamental solution $\tilde{Y} = YG$ for some matrix $G \in GL(p, \mathbb{C})$. This defines the monodromy representation

$$\chi : \pi_1(\mathbb{CP}^1 \setminus D; z_0) \to GL(p, \mathbb{C}) \tag{5}$$

of the system, with respect to Y. Since the fundamental group of $\mathbb{CP}^1 \setminus D$ is generated by the homotopy classes of all elementary loops γ_i, where $i, i = 1, ..., n$, encloses the only singular point a_i, the monodromy representation of (4) is defined by the local monodromy matrices G_i corresponding to these loops. These matrices satisfy a priori the only relation $G_1 \cdots G_n = I$.

The Fuchsian differential equation is a linear differential equation whose singularities are all regular [17]. It frequently appears in a range of problems in mathematics and physics. For example, the famous Gauss hypergeometric differential equation is a canonical form of the second-order Fuchsian differential equation with three singularities on the Riemann sphere $\mathbb{CP}^1$. Global properties of solutions, i.e., the monodromy, often play decisive roles in the applications of these equations in physics and other areas of mathematics.

For example consider a Schrödinger equation of the form

$$i\frac{\partial\Psi(t)}{\partial t} = H(t)\Psi(t) \tag{6}$$

with time dependent Hamiltonian $H(t) = (H_{ij}(t))$, $i, j = 1, ..., N$, where

$$H_{11} = \varepsilon(t), H_{12} = V_2, H_{13} = V_3, ..., H_{1N} = V_N,$$

$$H_{21} = V_2, H_{31} = V_3, ..., H_{N1} = V_N, \text{ and } H_{ij} = 0 \text{ otherwise},$$

$\Psi(t) = (\psi_1(t), ..., \psi_N(t))$ is a wave function, $V_j, j = 2, \ldots, n$ are constants, and the time dependent part ε has the form $\varepsilon(t) = E_1 \tanh(t/T)$ with constant E_1 and T.

Theorem 1. *([6,8]) Equation (6) is reducible to an N-dimensional Fuchsian system of Okubo type*

$$(zI_N - B)\frac{d\Phi(z)}{dz} = A\Phi(z)$$

with $B = \boldsymbol{diag}(i, -i, ..., -i)$.

In two-dimensional case (i.e., for $N = 2$) this result can be generalized by permitting Hamiltonians of more general form. Namely, consider a Schrödinger equation with two-component phase function

$$i\frac{\partial f(t)}{\partial t} = H(t)f(t), \tag{7}$$

where $f(t) = (f_1(t), f_2(t))$, and time dependent Hamiltonian $H(t)$ has the form

$$H(t) = \begin{pmatrix} \varepsilon(t) & V(t) \\ V(t) & -\varepsilon(t) \end{pmatrix} \tag{8}$$

where

$$\varepsilon(t) = \frac{E_0T + E_1Ty}{1 + y^2}\frac{dy}{dt}, \quad V(t) = \frac{V_0T}{\sqrt{1 + y^2}}\frac{dy}{dt}$$

with some monotonically increasing differentiable function $y : \mathbb{R} \to \mathbb{R}$ satisfying $y(t) \to \pm\infty$ as $t \to \pm\infty$.

Theorem 2. *([8]) Equation (7) is reducible to a system of two hypergeometric equations of the form*

$$z(z-1)\frac{d^2g}{dz^2} + (\gamma - (1 + \alpha + \beta))\frac{dg}{dz} - \alpha\beta g(z) = 0 \tag{9}$$

with appropriate constants α, β, γ.

For the proofs of Theorems 1 and 2, see [6–8].

3. Fuchsian system and entangled operator

One-qubit gates are elements of $SU(2)$ and act on $\mathbb{C}^2$. For realization of quantum algorithms one needs all operators from $SU(2^n)$. This problem can be reduced to finding such elements M_i of $SU(4)$ which together with $SU(2) \otimes SU(2)$ generate the whole $SU(4)$.

Definition 1. ([2]) A two-qubit gate A is universal for the quantum computation if A together with local unitary transformations (i.e transformation of one qubit states) generates all unitary transformations of the complex vector space of dimension 2^n to itself.

For example it is known that a controlled NOT [1] is a universal gate.

Definition 2. A gate A is a said to be entangling if there is a vector $|ab>=|a>\otimes|b>\in \mathbb{C}^2\otimes\mathbb{C}^2$ such that $A|ab>$ is not decomposable as a tensor product of two qubits.

Theorem 3. *([2]) A two-qubit gate A is universal if and only if it is entangling.*

Below we give, following Kauffman [3], a topological description of the entangling operator. Let $R: V\otimes V \to V\otimes V$ be a linear map. If R satisfies braided Yang-Baxter equation

$$(R\otimes \mathbb{I})(\mathbb{I}\otimes R)(R\otimes \mathbb{I}) = (\mathbb{I}\otimes R)(R\otimes \mathbb{I})(\mathbb{I}\otimes R),$$

then R is called *braiding.*

Theorem 4. *([4]) All 4×4 unitary solutions to the Yang-Baxter equation are similar to one of*

$$R1 = \begin{pmatrix} \frac{1}{\sqrt{2}} & 0 & 0 & \frac{1}{\sqrt{2}} \\ 0 & \frac{1}{\sqrt{2}} & -\frac{1}{\sqrt{2}} & 0 \\ 0 & \frac{1}{\sqrt{2}} & \frac{1}{\sqrt{2}} & 0 \\ -\frac{1}{\sqrt{2}} & 0 & 0 & \frac{1}{\sqrt{2}} \end{pmatrix}, R2 = \begin{pmatrix} a & 0 & 0 & 0 \\ 0 & 0 & b & 0 \\ 0 & c & 0 & 0 \\ 0 & 0 & 0 & d \end{pmatrix}, R3 = \begin{pmatrix} 0 & 0 & 0 & a \\ 0 & b & 0 & 0 \\ 0 & 0 & c & 0 \\ d & 0 & 0 & 0 \end{pmatrix},$$

where a, b, c, d are unit complex numbers.

Operator $R1$ acts on the standard basis of $\mathbb{C}^2\otimes\mathbb{C}^2$ and transforms it into the Bell basis [1]. As the operators $R2$ and $R3$ they are described in the following proposition.

Proposition 1. *([3]) The braidings R2 and R3 are universal gates when* $ad - bc \neq 0$ *for certain internal parameters* a, b, c *and* d.

From this proposition follows that nor every braiding is universal. For example, if $a = b = c = d = 1$, we obtain so called $SWAP$ operator, which does not universal. By the $SWAP$ operator, when by the definition act on the tensor product $u \otimes v \in V \otimes V$ the following manner $SWAP(u \otimes v) = v \otimes u$, we define the operator $R_{13} : V \otimes V \otimes V \to V \otimes V \otimes V$ as $R_{13} = (\mathbb{I} \otimes SWAP)(R \otimes \mathbb{I})(\mathbb{I} \otimes SWAP)$. Denote by $R_{12} = R \otimes \mathbb{I}$ and $R_{23} = \mathbb{I} \otimes R$.

Definition 3. The equation (with respect to R)

$$R_{12}R_{13}R_{23} = R_{23}R_{13}R_{12}$$

is called the algebraic Yang-Baxter equation.

Proposition 2. *R is a solution of algebraic Yang-Baxter equation if and only if $R \circ SWAP$ is solution of braided Yang-Baxter equation.*

In the paper [3], the relation between the topological and quantum entangling is investigated and it is proved that the *unitary solution of Yang-Baxter equation detect topological linking if and only if the gates corresponding to these solutions can entangle quantum states.*

We will apply results of this section to a special divisor. Namely, suppose $D = \cup_{i<j} D_{ij} \cup D_0$, where $D_{ij} = \{(z_0, ..., z_n) \in \mathbb{CP}^n | z_i = z_j, i, j \neq 0\}$, $D_0 = \{z \in \mathbb{CP}^n | z_0 = 0\}$. Let $X_n = \mathbb{CP}^n - D = \{(z_1, ..., z_n) \in \mathbb{C}^n | z_i \neq z_j, i \neq j\}$. The fundamental group of X_n is called the pure braid group on n strings, which we denote by P_n. The symmetric group S_n acts on X_n by $g.(z_1, ..., z_n) = (z_{g(1)}, ..., z_{g(n)})$, $g \in S_n$. The fundamental group of the quotient space X_n/S_n is called the braid group on n strings and denoted by B_n. The braid group has $n-1$ generators $\sigma_1, ..., \sigma_{n-1}$ with relations

$$\sigma_i\sigma_{i+1}\sigma_i = \sigma_{i+1}\sigma_i\sigma_{i+1}, 1 \leq i \leq n-2,$$

$$\sigma_i\sigma_j = \sigma_j\sigma_i, |i-j| \geq 2.$$

Choose a point $z_0 = (1, 2, ..., n) \in X_n$ and denote by $pr : X_n \to X_n/S_n$ the natural projection. We have exact sequence of groups:

$$1 \to P_n \to B_n \to S_n \to 1$$

and generators of the pure braid group P_n are τ_{ij}, $1 \leq i < j \leq n$, where

$$\tau_{ij} = \sigma_i\sigma_{i+1}...\sigma_{j-1}\sigma_j^2\sigma_{j-1}^{-1}...\sigma_i^{-1}, 1 \leq i < j \leq n,$$

satisfying the relations

$$\begin{aligned}
\tau_{rs}\tau_{ik} &= \tau_{ik}\tau_{rs}, \text{ for } s < i \text{or } k < s,\\
\tau_{ks}\tau_{ik}\tau_{ks}^{-1} &= \tau_{is}^{-1}\tau_{ik}\tau_{is} \text{ for } i < k < s,\\
\tau_{rk}\tau_{ik}\tau_{rk}^{-1} &= \tau_{ik}^{-1}\tau_{ir}^{-1}\tau_{ik}\tau_{ir}\tau_{ik} \text{ for } i < r < k,\\
\tau_{rs}\tau_{ik}\tau_{rs}^{-1} &= \tau_{is}^{-1}\tau_{is}\tau_{ir}\tau_{ik}\tau_{ir}^{-1}\tau_{is}^{-1}\tau_{ir}\tau_{is} \text{ for } i < r < k < s.
\end{aligned}$$

Consider the matrix valued 1-form

$$\Omega = \sum_{1 \le i < j \le n} \Omega_{ij} d\log(z_i - z_j) \tag{10}$$

holomorphic on X_n, where $\Omega_{ij} \in M_m(\mathbb{C}), 1 \le i < j \le n$. Let $E \to X_n$ be a trivial holomorphic vector bundle of rank m with connection ∇ for which Ω is connection form. Holomorphic sections $f = (f^1, ..., f^m)$ of this bundle are solutions of the Fuchs system

$$df = \Omega f, \tag{11}$$

where Ω has the form (10). The system (11) is integrable if and only if

$$[\Omega_{ij}, \Omega_{ik} + \Omega_{jk}] = [\Omega_{ij} + \Omega_{ik}, \Omega_{ji}], i < j < k, \tag{12}$$

$$[\Omega_{ij}, \Omega_{kl}] = 0, \text{ for distinct } i, j, k, l. \tag{13}$$

From the preceding results we obtain the solvability of Riemann-Hilbert problem [10,11]. Indeed, if $\rho : P_n \to \mathrm{GL}_m(\mathbb{C})$ is such representation that $||\rho(\tau_{ij}) - 1||$ is sufficiently small for each $1 \le i < j \le n$, then there exist matrices $\Omega_{ij} \in M_m(\mathbb{C})$, $1 \le i < j \le n$ close to 0, which satisfy the conditions (12)-(13) and monodromy representation of the Fuchs system

$$df = \sum_{1 \le i < j \le n} \Omega_{ij} \frac{d(z_i - z_j)}{z_i - z_j} f$$

coincides with ρ.

From this it follows, that for a given collection of unitary matrices $U_1, ..., U_n$ in general there *does not exist* a gauge potential $d - \sum_{i,j} \Omega_{ij} d\log(z_i - z_j)$ on the bundle $E \to X_n$ such that monodromy matrices coincide with those in the given collection. The reason of this lies in the *nonfreeness* of the braid group.

4. Construction of dense monodromy group

Let M_1, M_2, M_3, M_4 such matrices, such that $E_j = (1/2\pi i) ln M_j, j = 1, ..., 4$ and M_1, M_2, M_3 make a basis of the Lie algebra of $SU(2)$ and satisfy the condition $M_1M_2M_3M_4 = 1$. Let $s_1, s_2, s_3, s_4 \in \mathbb{CP}^1$ any points. Denote by $X_4 = \mathbb{CP}^1 - \{s_1, s_2, s_3, s_4\}$ and by $\gamma, \gamma_2, \gamma_3, \gamma_4$ the generators of $\pi(X_4, z_0)$ with relation $\gamma_1\gamma_2\gamma_3\gamma_4 = 1$. Consider the representation ρ of fundamental group $\pi(X_4, z_0)$ in $SU(2)$ defined by the correspondence

$$\rho : \pi(X_4, z_0) \to SU(2), \gamma_j \mapsto M_j.$$

Then for ρ there exits the system of differential equations of Fuchs type

$$dF(z) = (\frac{A_1}{z-s_1}dz + \frac{A_2}{z-s_2}dz + \frac{A_3}{z-s_3}dz + \frac{A_4}{z-s_4}dz)F(z)$$

where monodromy representation coincides with ρ. The representation ρ induces two-dimensional vector bundle $E \to \mathbb{CP}^1$ with Chern number $c_1(E) = \sum_{j=1}^{4} E_j$ and meromorphic connection form

$$\omega = \sum_{j=1}^{4} \frac{A_j}{z-s_j}dz.$$

The solution space Υ of the system is a two-dimensional vector space, $\pi(X_4, z_0)$ acts on Υ and any unitary operator is obtained this way.

Therefore we proved the following result (see [9,12]).

Theorem 5. *The connection form ω generates the basis for the local gates.*

Consider general case. Let X be a compact Riemann surface genus $g \geq 2$ and let $E \to X$ flat holomorphic vector bundle. Then by Weil theorem there exits the unitary representation ρ of the fundamental group $\pi_1(X)$ of surface X, where $\alpha_1, \beta_1, ..., \alpha_g, \beta_g$ generators of $\pi_1(X)$ with relations $\prod_{j=1}^{g}(\alpha_j\beta_j\alpha_j^{-1}\beta_j^{-1}) = 1$, that the bundle $E \to X$ is induced by ρ. Let $E \to X$ two-dimension flat holomorphic vector bundle, then there exits the holomorphic connection of $E \to X$, which holonomy coincides with ρ. Therefore we have proved the following theorem.

Theorem 6. *Let X be a compact Riemann surface genus $g \geq 2$ and let $E \to X$ stable flat holomorphic vector bundle with holomorphic connection ∇. Then ∇ gives the basis for the quantum computation.*

Let $\Gamma_A(\gamma) = P\exp\{i\int_0^\tau A(t)dt\}$ holonomy operator and $Hol(A)$ the holonomy group, where A is (nonintegrable) connection. If $Hol(A) = U(n)$,then A is called irreducible. It is known, that for $U(2)$ there exist irreducible connection. In the theory of holonomic quantum computing [18], is known, that for the controllability of the correspondence quantum system necessary conditions is irreducibility of the connection, in this case holonomy group of connections coincides of $SU(2)$.

The following theorem imposes on the hypergeometrical equation those conditions which ensure construction of local gates.

Theorem 7. *The physical system described by Schrödinger equation (7) with Hamiltonian (8) defines a universal set of gates iff satisfying the following conditions:*

$$(1)\ iT(-E_1+\sqrt{E_1^2+V_0^2})\overline{\in}\mathbb{Z}; \qquad (2)\ iT(-E_1-\sqrt{E_1^2+V_0^2})\overline{\in}\mathbb{Z};$$

$$(3)\ \frac{1}{2}+E_0T+iT\sqrt{E_1^2+V_0^2}\overline{\in}\mathbb{Z}; \qquad (4)\ \frac{1}{2}+E_0T-2iE_1T+iT\sqrt{E_1^2+V_0^2}\overline{\in}\mathbb{Z};$$

$$(5)\ \frac{1}{2}-E_0T+iT\sqrt{E_1^2+V_0^2}\overline{\in}\mathbb{Z}; \qquad (6)\ \frac{1}{2}-E_0T-iT\sqrt{E_1^2+V_0^2}\overline{\in}\mathbb{Z}.$$

Proof. It is known, that the equation (9) defines the two dimension holomorphic vector bundle on $\mathbb{CP}^1$ with holomorphic connection with logarithmic singularities [12]. The fibre of this bundle consider as space of qubits. The monodromy matrices of equation (9) acts on the fibre in natural way. It is know also, that associated connection gives universal set of gates, if and only if monodromy representation is irreducible [18]. For equation (9) irreducibility condition means, that in Riemann P function

$$P\begin{pmatrix} 0 & 1 & \infty \\ 0 & 0 & \alpha \\ 1-\gamma & \gamma-\alpha-\beta & \beta \end{pmatrix}$$

the constants α, β, γ satisfies the conditions

$$\alpha, \beta, \gamma-\alpha, \gamma-\beta, 1-\gamma+\alpha, 1-\gamma+\beta\overline{\in}\mathbb{Z}.$$

From the definition of universal set of gates and from irreducibility of the monodromy representation of the hypergeometric equation, it follows that the system is completely controllable if and only if the representation associated with equation (9) is irreducible. From the above condition of irreducibillity, it follows that the constants in (8) can satisfy the conditions (1)–(6) in the theorem. □

In [15] a certain equation is given on coefficients of the Heun equation which transforms the latter into a hypergeometric equation. The author calls this transformation a Heun to hypergeometric reduction. This property of reducibility of the Heun equation to the hypergeometric one we will call PHH and will use it without further explanations. In this direction let us give the following important statement: [15] a Heun equation which has four singular points and is nontrivial, can be transformed to a hypergeometric equation. From this it follows that if Heun equation satisfies the PHH properties [15], then it defines a completely controllable quantum system.

References

1. M. Nielsen and I. Chuang, *Quantum computation and quantum information.* Cambridge University Press, New York, 2000.
2. J.-L. Brylinski and R. Brylinski, Universal quantum gates. ArXiv: quant-ph/0108062.
3. L. Kauffman and S. Lomonaco, Braiding operators are universal quantum gates. LANL quant-ph/0401090.
4. H. Dye, Unitary solution to the Yang-Baxter equation in dimension four. LANL quant-ph/0211050.
5. G. Giorgadze, Gates for Quantum Computing Induced from Monodromy Operators *Particles and Nuclei, Lett.*, 4, 2(138), 295-299 (2007).
6. T. Kato, K. Nakamura and M. Lakshmanan, Analytic calculation of nonadiabatic transition probalities from monodromy of differetial equations. *J. Phys. A* 36, 5803-5815 (2003).
7. G. Giorgadze, On the Hamiltonians induced from a Fuchsian system. *Mem. Diff. Eq. Math. Phys.* 31, 69-82 (2004).
8. G. Giorgadze and G. Khimshiashvili, On Shrödinger equations of Okubo type. *J. Dyn. Control Syst.* 10, No. 2, 171-186 (2004).
9. G. Giorgadze, Monodromic approach to quantum computing. *Int. J. Modern Phys. B* 16, No. 30, 4593-4605 (2002).
10. T. Kohno, Hecke algebra representations of braid groups and classical Yang-Baxter equation. *Advanced Studies in Pure Mathematics.* 28, pp. 255-269 (1988).
11. V. P. Leksin, The Riemann-Hilbert problem for analytic families of representations, *Math. USSR-Zam.* 50, No.2, 89-97 (1991).
12. G. Giorgadze, Regular systems on Riemann surfaces. *Journal of Math. Sci.* 118 (5), 5347-5399 (2003).
13. V. A. Golubeva and V. P. Leksin, On two types of representations of braid group associated with the Knizhnik-Zamolodchikov equation of the B_n type. *J. Dyn. Control Syst.* 5, No. 4, 565-596 (1999).
14. A. Ronveaux (ed.), *Heun's differential equations.* Oxford Sci. Pub. Oxford University Press, Oxford, 1995.
15. R. Maier, Transforming the Heun equations to the hypergeometric equation: I. Polynomial transformations. ArXiv: math.CA/0203264.
16. M. H. Freedman, A. Kitaev and Z. Wang, Simulation of topological field theries by quantum computer. ArXiv: quant-ph/0001071.
17. D. V. Anosov and A. A. Bolibruch, *The Riemann-Hilbert problem*, Aspects of Mathematics, Vieweg, Braunschweig, Wiesbaden, 1994.
18. P. Zanardi and M. Rasetti, Holonomic quantum computation. *Phys. Lett.* A264, 94 (1999).

ABOUT ONE CLASS OF SECOND-ORDER LINEAR HYPERBOLIC EQUATIONS FOR WHICH ALL OF THE BOUNDARY CONSIST OF SINGULAR LINES

NUSRAT RAJABOV

Tajik State National University, Dushanbe, Tajikistan
E-mail: nusrat38@mail.ru

Let D denote the rectangle $D = \{(x, y) : \ a < x < b, \, c < y < d\}$. Accordingly, we introduce $\Gamma_1 = \{a < x < b, y = c\}$, $\Gamma_2 = \{x = b, c < y < d\}$, $\Gamma_3 = \{a < x < b, y = d\}$, $\Gamma_4 = \{x = a, c < y < d\}$. By $C^2_{xy}(D)$, we denote the class functions $u(x, y) \in C(D)$. Further introduce the following designation:

$$D_1 = \{(x,y) : a < x \leq x_0, c < y \leq y_0\}, \ \ D_2 = \{(x,y) : x_0 \leq x < b, c < y \leq y_0\},$$
$$D_3 = \{(x,y) : a < x \leq x_0, y_0 \leq y < d\}, \ \ D_4 = \{(x,y) : x_0 \leq x < b, y_0 \leq y < d\},$$

where $a < x_0 < b, \quad c < y_0 < d$.

In D, we consider the following hyperbolic equation

$$L(u) \equiv \frac{\partial^2 u}{\partial x \, \partial y} + \frac{\lambda}{(y-c)(d-y)} \frac{\partial u}{\partial x} + \frac{\mu}{(x-a)(b-x)} \frac{\partial u}{\partial y}$$
$$+ \frac{\delta u}{(x-a)(b-x)(y-c)(d-y)} = \frac{f(x,y)}{(x-a)(b-y)(y-c)(d-y)} \qquad (1)$$

where λ, μ, δ are given constants, $f(x, y)$ is a given function in D.

If we multiply both sides of equation (1) by $(x - a)(b - x)(y - c)(d - y)$, then we obtain a hyperbolic equation degenerate in all boundary domain D. Hyperbolic equations with singular and super-singular coefficients are investigated in [1,2,8].

Introduce the following designation:

$$\Pi^y_{c,\lambda}[\psi(y)] = \int_c^y \left(\frac{s-c}{d-s}\right)^{\frac{\lambda}{d-c}} \psi(s) ds, \quad \Pi^x_{a,\mu}[\varphi(x)] = \int_a^x \left(\frac{t-a}{b-t}\right)^{\frac{\mu}{b-a}} \varphi(t) dt$$

$$\Pi^x_{b,\mu}[\varphi(x)] = \int_b^x \left(\frac{t-a}{b-t}\right)^{\frac{\mu}{b-a}} \varphi(t) dt, \quad \Pi^y_{d,\lambda}[\psi(y)] = \int_y^d \left(\frac{s-c}{d-c}\right)^{\frac{\lambda}{d-c}} \psi(s) ds,$$

$$T^x_{a,\mu}[f(x,y)] = \int_a^x \left(\frac{t-a}{b-t}\right)^{\frac{\mu}{b-a}} \frac{f(t,y)dt}{(t-a)(b-t)},$$

$$T^x_{b,\mu}[f(x,y)] = \int_x^b \left(\frac{t-a}{b-t}\right)^{\frac{\mu}{b-a}} \frac{f(t,y)dt}{(t-a)(b-t)}$$

$$T^y_{a,\lambda}[f(x,y)] = \int_c^y \left(\frac{s-c}{d-s}\right)^{\frac{\lambda}{d-c}} \frac{f(x,s)ds}{(s-c)(d-s)},$$

$$T^y_{d,\lambda}[f(x,y)] = \int_y^d \left(\frac{s-c}{d-s}\right)^{\frac{\lambda}{d-c}} \frac{f(x,s)ds}{(s-c)(d-s)}.$$

We have the following confirmation:

Theorem 0.1. *Let in (1) $\lambda > 0$, $\mu > 0$, $\delta = \lambda\mu$, $f(x,y) \in C(\overline{D})$, $f(b,y) = 0$ with the asymptotic behavior $f(x,y) = O[(b-x)^{\gamma_1}]$, $\gamma_1 > \frac{\mu}{b-a}$, $f(x,b) = 0$ with the asymptotic behavior $f(x,y) = 0[(d-y)^{\gamma_2}]$, $\gamma_2 > \frac{\lambda}{d-c}$. Then solution of equation (1) of class $C^2_{xy}(D)$ is representable in the form*

$$\begin{aligned} u(x,y) &= K_1[\varphi_1(x),\ \psi_1(y),\ f(x,y)], \quad \textit{when} \quad (x,y) \in D_1,\\ u(x,y) &= K_2[\varphi_1(x),\ \psi_2(y),\ f(x,y)], \quad \textit{when} \quad (x,y) \in D_2,\\ u(x,y) &= K_3[\varphi_2(x),\ \psi_1(y),\ f(x,y)], \quad \textit{when} \quad (x,y) \in D_3,\\ u(x,y) &= K_4[\varphi_2(x),\ \psi_2(y),\ f(x,y)], \quad \textit{when} \quad (x,y) \in D_4, \end{aligned} \tag{2}$$

where $\varphi_1(x)$, $\psi_1(y)$ are arbitrary functions of points of Γ_1 and Γ_2, $\varphi_2(x)$, $\psi_2(y)$ are related to $\varphi_1(x)$, $\psi_1(y)$ by

$$\psi_2(y) + \int_a^b \left(\frac{t-a}{b-t}\right)^{\frac{\mu}{b-a}} \frac{f(t,y)}{(t-a)(b-t)(y-c)(d-y)}, \quad c < y < d \tag{3}$$

$$\varphi_2(x) - \varphi_1(x) - M_1|C^0_1 f(x,y)| \quad \textit{when} \quad a < x < x_0 \tag{4}$$

$$\varphi_2(x) - \varphi_1(x) = M_2|C^0_1 f(x,y)| \quad \textit{when} \quad x_0 < x < b \tag{5}$$

$$C_2 = C^0_1 + \int_a^b \left(\frac{t-a}{b-t}\right)^{\frac{\mu}{b-c}} \frac{dt}{(t-a)(b-t)} \int_c^d \left(\frac{s-c}{d-s}\right)^{\frac{\lambda}{d-c}} \frac{f(t,s)ds}{(s-c)(d-s)} \tag{6}$$

where C^0_1 is an arbitrary constant. In (2), the integral operators K_j $(1 \le j \le 4)$ are determined from

$$K_1[\varphi_1(x), \psi_1(y), f(x,y)] \equiv \left(\frac{d-y}{y-c}\right)^{\frac{\lambda}{(d-c)}} \varphi_1(x)$$

$$+ \left(\frac{d-y}{y-c}\right)^{\frac{\lambda}{(d-c)}} \left(\frac{b-x}{x-a}\right)^{\frac{\lambda}{(b-a)}} \left[\Pi^y_{c,\lambda}[\psi_1(y)] + T^x_{a,\mu} T^y_{c,\lambda}[f(x,y)]\right],$$

$$K_2[\varphi_1(x),\psi_2(y),f(x,y)] \equiv \left(\frac{d-y}{y-c}\right)^{\frac{\lambda}{(d-c)}} \varphi(x)$$
$$+\left(\frac{d-y}{y-c}\right)^{\frac{\lambda}{d-c}} \left(\frac{b-x}{x-a}\right)^{\frac{\lambda}{b-a}} \left[\Pi^y_{c,\lambda}[\psi_2(y)] - T^x_{b,\mu} T^y_{c,\lambda}[f(x,y)]\right],$$
$$K_3[\varphi_2(x),\psi_1(y),f(x,y)] \equiv \left(\frac{d-y}{y-c}\right)^{\frac{\lambda}{d-c}} \varphi_2(x)$$
$$+\left(\frac{d-y}{y-c}\right)^{\frac{\lambda}{d-c}} \left(\frac{d-y}{y-c}\right)^{\frac{\mu}{b-a}} \left[-\Pi^y_{d,\lambda}[\psi_1(y)] + T^x_{a,\mu} T^y_{d,\lambda}[f(x,y)]\right],$$
$$K_4[\varphi_2(x),\psi_2(y),f(x,y)] \equiv \left(\frac{d-y}{y-c}\right)^{\frac{\lambda}{d-c}} \varphi_2(x)$$
$$+\left(\frac{d-y}{y-c}\right)^{\frac{\lambda}{d-c}} \left(\frac{d-y}{y-c}\right)^{\frac{\mu}{b-a}} \left[-\Pi^y_{d,\lambda}[\psi_2(y)] + T^x_{b,\mu} T^y_{d,\lambda}[f(x,y)]\right],$$

and the integral operators M_1, M_2 *from*

$$M_1[C_1,\ f(x,y)] = \left(\frac{b-x}{x-a}\right)^{\frac{\mu}{b-a}} \left[C_1 + \int_a^x \left(\frac{t-a}{b-t}\right)^{\frac{\mu}{b-a}} \frac{dt}{(t-a)(b-t)}\right.$$
$$\left.\int_c^d \left(\frac{s-c}{d-s}\right)^{\frac{\lambda}{d-c}} \frac{f(t,s)ds}{(s-c)(d-s)}\right],$$
$$M_2[C_2,\ f(x,y)] = \left(\frac{b-x}{x-a}\right)^{\frac{\mu}{b-a}} \left[C_2 - \int_x^b \left(\frac{t-a}{b-t}\right)^{\frac{\mu}{b-a}} \frac{dt}{(t-a)(b-t)}\right.$$
$$\left.\int_c^d \left(\frac{s-c}{d-s}\right)^{\frac{\lambda}{d-c}} \frac{f(t,s)ds}{(s-c)(d-s)}\right].$$

Theorem 0.2. *Let in (1)* $\lambda < 0$, $\mu < 0$, $\delta = \lambda\mu$, $f(x,y) \in C(\overline{D})$, $f(a,y) = 0$ *with the asymptotic behavior* $f(x,y) = O[(x-a)^{\gamma_3}]$, $\gamma_3 > \frac{|\mu|}{b-a}$, $f(x,c) = 0$ *with the asymptotic behavior* $f(x,y) = O[(y-c)^{\gamma_4}]$, $\gamma_4 > \frac{|\lambda|}{d-c}$. *Then any solution of equation (1) of class* $C^2_{xy}(D)$ *is representable in the form (2), where* $\varphi_1(x), \psi_1(y)$ *are arbitrary functions of points of* Γ_1 *and* Γ_2, $\varphi_2(x), \psi_2(y)$ *and* $\varphi_1(x), \psi_1(y)$ *are related by (3), (4), (5),* C^0_1 *is an arbitrary constant,* C_2 *and* C^0_1 *are related by (6).*

Remark 0.1. Statements similar to Theorems 1 and 2 are obtained also in the cases $\lambda > 0$, $\mu < 0$ and $\lambda < 0$, $\mu > 0$.

Remark 0.2. By fulfillment of all conditions of Theorem 0.1, any solution of equation (1) of class $C^2_{xy}(D)$ as $x \to a$ and $y \to c$ is unbounded and its behavior is determined from the asymptotic formula $U(x,y) = O[(x-a)^{-\frac{\mu}{b-a}}]$ and $U(x,y) = O[(y-c)^{-\frac{\lambda}{d-c}}]$; as $x \to b$ and $y \to d$, it vanishes and its behavior is determined from the asymptotic formula $U(x,y) = o[(b-x)^{\frac{\mu}{b-a}}]$ and $U(x,y) = o[(d-x)^{\frac{\lambda}{d-c}}]$.

Remark 0.3. By fulfillment of conditions of Theorem 0.2, any solution of equation (1) of class $C^2_{xy}(D)$ as $x \to a$, $y \to c$ vanishes and its behavior is determined from $U(x,y) = o[(x-a)^{\frac{|\mu|}{b-a}}]$ and $U(x,y) = o[(y-c)^{\frac{|\lambda|}{d-c}}]$; as $x \to b$ and $y \to d$, it is unbounded and its behavior is determined from $U(x,y) = O[(x-a)^{\frac{|\mu|}{b-x}}]$ and $U(x,y) = O[(d-x)^{\frac{|\lambda|}{d-c}}]$.

In the case $\delta \neq \lambda\mu$ that which has been proved, the problem of determination of solutions of (1) in domain D is equivalent to the problem of determination of solutions of following equations in domains D_j $(1 \le j \le 4)$:

$$u(x,y) - \delta_1 \int_a^x \frac{dt}{(t-a)(b-t)} \int_c^y \left(\frac{\omega(x,y)}{\omega(t,s)}\right) \frac{u(t,s)ds}{(s-c)(d-s)}$$

$$= K_1[\varphi_1(x),\ \psi_1(y),\ f(x,y)],\ (x,y) \in D_1, \tag{7}$$

$$u(x,y) + \delta_1 \int_x^b \frac{dt}{(t-a)(b-t)} \int_c^y \left(\frac{\omega(x,y)}{\omega(t,s)}\right) \frac{u(t,s)ds}{(s-c)(d-s)}$$

$$= K_2[\varphi_1(x),\ \psi_2(y),\ f(x,y)],\ (x,y) \in D_2, \tag{8}$$

$$u(x,y) + \delta_1 \int_a^x \frac{dt}{(t-a)(b-t)} \int_y^d \left(\frac{\omega(x,y)}{\omega(t,s)}\right) \frac{u(t,s)ds}{(s-c)(d-s)}$$

$$= K_3[\varphi_2(x),\ \psi_1(y),\ f(x,y)],\ (x,y) \in D_3, \tag{9}$$

$$u(x,y) - \delta_1 \int_x^b \frac{dt}{(t-a)(b-t)} \int_y^d \left(\frac{\omega(x,y)}{\omega(t,s)}\right) \frac{u(t,s)ds}{(s-c)(d-s)}$$

$$= K_4[\varphi_2(x),\ \psi_2(y),\ f(x,y)],\ (x,y) \in D_4, \tag{10}$$

where $\delta_1 = \lambda\mu - \delta$, $\omega(x,y) = \left(\frac{b-x}{x-a}\right)^{\frac{\mu}{b-a}} \left(\frac{d-y}{y-c}\right)^{\frac{\lambda}{d-c}}$. The solution of equation (7) is found in the class of function $u(x,y)$ representable in D_1 in the form

$$u(x,y) = \omega(x,y) \sum_{n=0}^{\infty} (x-a)^{n+\gamma_1} (b-x) u_n^1(y) \tag{11}$$

or

$$u(x,y) = \omega(x,y) \sum_{n=0}^{\infty} (y-c)^{n+\gamma_2}(d-y)\nu_n^1(x) \tag{12}$$

where $U_n^1(y)$, $\nu_n^1(x)$ are unknown functions, $\gamma_1 > 0$, $\gamma_2 > 0$.

Correspondingly the solution of equation (8) is found in the class of functions $u(x,y)$ representable in D_2 in the form

$$u(x,y) = \omega(x,y) \sum_{n=0}^{\infty} (b-x)^{n+\gamma_3}(x-a)u_n^2(y) \tag{13}$$

or (12). The solution of equation (9) is found in the class of functions $u(x,y)$ representable in D_3 in the form (11) or

$$u(x,y) = \omega(x,y) \sum_{n=0}^{\infty} (d-y)^{n+\gamma_4}(y-c)\nu_n^2(x), \quad \gamma_4 > 0. \tag{14}$$

Similarly the solution of the integral equation (10) is found in the class of functions representable in D_4 in the form (13) or (14). In this connection assume that in (7)-(10) the right part, also is expandable in corresponding series. Then for determination of the unknown functions $u_n^1(y)$, $u_n^2(y)$, $\nu_n^1(x)$, $\nu_n^2(x)$ we obtained an infinite number of system of one-dimensional integral equations with boundary singular points, theory of which is developed in [3-8,9].

For example, when we seek the solution of (7) in the form (11), we assume that the function $f(x,y)$ is represented in the form

$$f(x,y) = \omega(x,y) \sum_{n=0}^{\infty} (x-a)^{n+\gamma}(b-x)f_n^1(y). \tag{f_1}$$

Then at $\varphi_1(x) = 0$, $\psi(y) = 0$, for determination of the unknown function $u_n^1(y)$, we obtain the following infinite number of one-dimensional systems of integral equations with singular left boundary points:

$$u_0^1(y) - \frac{\delta_1}{(b-a)\gamma} \int_c^y \frac{u_0^1(s)ds}{(s-c)(d-s)} = \frac{1}{(b-a)\gamma} \int_c^y \frac{f_0^1(s)ds}{(s-c)(d-s)}, \tag{u_0^1}$$

$$u_n^1(y) - \frac{\delta_1}{(n+\gamma)(b-a)} \int_c^y \frac{u_n^1(s)ds}{(s-c)(d-s)}$$

$$= \frac{(u_{n-1}^1(y))'}{(b-a)} + \frac{1}{(n+\gamma)(b-a)} \int_c^y \frac{f_n^1(s)ds}{(s-c)(d-s)}, \tag{u_n^1}$$

$n = 1, 2, 3, \ldots$, $c < y < y_0 < d$.

Similarly, when we seek the solution of (9) in the form (11) with unknown $u_n^3(y)$, and with condition that $f(x,y)$ repressented in the form (f_1), provided that $\varphi_2(x) = 0$, $\psi_1(y) = 0$, then for determining the unknown function $u_n^3(y)$, we obtain the following infinite number of one-dimensional system of integral equations with singular right boundary points:

$$u_0^3(y) + \frac{\delta_1}{(b-a)\gamma} \int_c^y \frac{u_0^3(s)ds}{(s-c)(d-s)} = \frac{1}{(b-a)\gamma} \int_c^y \frac{f_0^1(s)ds}{(s-c)(d-s)}, \quad (u_0^3)$$

$$u_0^3(y) + \frac{\delta_1}{(b-a)(b-a)} \int_y^d \frac{u_n^3(s)ds}{(s-c)(d-s)}$$

$$= \frac{(u_{n-1}^3(y))'}{(b-a)} - \frac{1}{(n+\gamma)(b-a)} \int_y^d \frac{f_n^1(s)ds}{(s-c)(d-s)}. \quad (u_n^3)$$

In according to [4] and [9], when the solution the system of the integral equation $(u_n^1)(n = 0, 1, 2, \ldots)$ in D_1 exist, then it is given by formulas

$$u_0^1(y) = \left(\frac{y-c}{d-y}\right)^{\frac{\delta_1}{\gamma(b-a)(d-c)}} \left[C_0^1\right.$$

$$\left. + \frac{1}{\gamma(b-a)} \int_c^y \left(\frac{d-c}{s-c}\right)^{\frac{\delta_1}{\gamma(b-a)(d-c)}} \frac{f_0^1(s)}{(s-c)(d-s)}\right] \equiv T_0^1[C_0^1, f_0^1(y)],$$

$$u_n^1(y) - \left(\frac{y-c}{d-y}\right)^{\frac{\delta_1}{(n+\gamma)\gamma(b-a)(d-c)}} \left[C_n^1 + \int_c^y \left(\frac{d-c}{s-c}\right)^{\frac{\delta_1}{(\gamma+n)(b-a)(d-c)}}\right.$$

$$\left.\left[\frac{(u_{n-1}^1(s))'}{b-a} + \frac{1}{(\gamma+n)(b-a)} \frac{f_n^1(s)}{(s-c)(d-s)}\right] ds\right] \equiv T_n^1[C_0^1, f_n^1(y)],$$

$n = 1, 2, 3, \cdots$. The solution of system integral equation $(u_n^3)(n = 0, 1, 2, \cdots)$ in D_3 is given by formulas

$$u_0^3(y) = \left(\frac{y-c}{d-y}\right)^{\frac{\delta_1}{\gamma(b-a)(d-c)}} \left[C_0^3 + \frac{1}{\gamma(b-a)} \int_y^d \left(\frac{d-c}{s-c}\right)^{\frac{\delta_1}{\gamma(b-a)(d-c)}}\right.$$

$$\left.\frac{f_0^1(s)}{(s-c)(d-s)}\right] \equiv T_0^1[C_0^1, f_0^1(y)],$$

$$u_n^3(y) = \left(\frac{y-c}{d-y}\right)^{\frac{\delta_1}{(n+\gamma)\gamma(b-a)(d-c)}} [C_n^3 - -\frac{1}{(b-a)} \int_y^d \left(\frac{d-c}{s-c}\right)^{\frac{\delta_1}{(\gamma+n)(b-a)(d-c)}}$$

$$\left[(u_{n-1}^3(s))' + \frac{f_n^1(s)}{(\gamma+n)(s-c)(d-s)} \right] ds \Big] \equiv T_n^3[C_0^3,\ f_n^1(y),\ u_{n-1}^3(s)].$$

Moreover from condition continuity solution equation (1) in lines $y = y_0(u_0^1(y_0) = u_n^3(y_0))$, follows that

$$C_0^3 - C_0^1 = \frac{1}{\gamma(b-a)} \int_c^d \left(\frac{d-s}{s-c}\right)^{\frac{\delta_1}{\gamma(b-a)(d-c)}} \frac{f_0^1(s)ds}{(s-c)(d-s)}, \qquad (A_0^1)$$

$$C_n^3 - C_n^1 = \frac{1}{(\gamma+n)(b-a)} \int_c^d \left(\frac{d-s}{s-c}\right)^{\frac{\delta_1}{(\gamma+n)(b-a)(d-c)}} \frac{f_n^1(s}{(s-c)(d-s)} ds + \frac{1}{b-a}$$

$$\left[\int_c^{y_0} \left(\frac{d-s}{s-c}\right)^{\frac{\delta_1}{(\gamma+n)(b-a)(d-c)}} (u_{n-1}^1(s))' ds + \int_{y_0}^d \left(\frac{d-s}{s-c}\right)^{\frac{\delta_1}{(\gamma+n)(b-a)(d-c)}} (u_{n-1}^3(s))' ds \right]. \qquad (A_n^1)$$

In the case, when we the solution of the Integral Equation (8) will be seek in the form (13) and will be assume that, function $f(x,y)$ representable in the form

$$f(x,y) = \omega(x,y) \sum_{n=0}^{\infty} (b-x)^{n+\gamma}(x-a) f_n^2(y), \qquad (f_2)$$

then at $\varphi_2(x) = 0$, $\psi_2(y) = 0$ for determined unknown functions $u_n^2(y)$, we obtained the following one dimensional Volterra type Integral Equation with left fixed singular point

$$u_0^2(y) + \frac{\delta_1}{\gamma(b-a)} \int_c^y \frac{u_0^2(s)ds}{(s-c)(d-c)} = \frac{1}{\lambda(b-a)} \int_c^y \frac{f_0^2(s)ds}{(s-c)(d-s)}, \qquad (u_0^2)$$

$$u_n^2(y) + \frac{\delta_1}{(\gamma+n)(b-a)} \int_c^y \frac{u_n^2(s)ds}{(s-c)(d-s)}$$

$$= \frac{1}{(b-a)} \left[(u_{n-1}^2(y))' + \frac{1}{(\gamma+n)} \int_c^y \frac{f_n^2(s)ds}{(s-c)(d-s)} \right]. \qquad (u_n^2)$$

Similarly, when the solution of the Integral Equation (10) we will be seek in the form (13) with unknown functions $u_n^4(y)$ and with condition that, function $f(x,y)$ repressing in the form (f_2), at provided that $\varphi_2(x) = 0$, $\psi_2(y) = 0$, then for determined unknown functions $u_n^4(y)$ we obtained the following infinity number the one - dimensional Volterra type system Integral Equation with right boundary singular point

$$u_0^4(y) + \frac{\delta_1}{\gamma(b-a)} \int_y^d \frac{u_0^4(s)ds}{(s-c)(d-c)} = \frac{1}{\gamma(b-a)} \int_y^d \frac{f_0^2(s)ds}{(s-c)(d-s)}, \qquad (u_0^4)$$

$$u_n^4(y)+\frac{\delta_1}{(\gamma+n)(b-a)}\int_y^d\frac{u_n^4(s)ds}{(s-c)(d-s)}$$

$$=\frac{1}{(b-a)}\left[(u_{n-1}^4(y))'+\frac{1}{(\gamma+n)}\int_y^d\frac{f_n^2(s)ds}{(s-c)(d-s)}\right]. \qquad (u_n^4)$$

The solution of the system integral Equation (u_n^2) in D_2 is given by formulas, $n=1,2,3,\cdots$,

$$u_0^2(y)=\left(\frac{d-y}{y-c}\right)^{\frac{\delta_1}{\gamma(b-a)(d-c)}}\left[C_0^2+\frac{1}{(b-a)}\int_c^y\left(\frac{s-c}{d-c}\right)^{\frac{\delta_1}{\gamma(b-a)(d-c)}}\right.$$

$$\left.\frac{f_0^2(s)}{(s-c)(d-s)}\right]\equiv T_0^2[C_0^2,\ f_0^2(y)]$$

$$u_n^2(y)=\left(\frac{d-y}{y-c}\right)^{\frac{\delta_1}{(n+\gamma)(b-a)(d-c)}}\left[C_n^2+\int_y^d\left(\frac{s-c}{d-s}\right)^{\frac{\delta_1}{(\gamma+n)(b-a)(d-c)}}\right.$$

$$\left.\left[\frac{(u_{n-1}^2(s))'}{b-a}+\frac{1}{(\gamma+n)(b-a)(s-c)(d-s)}\right]ds\right]\equiv T_n^2[C_n^2, f_n^2(y), u_{n-1}^2(y)].$$

The solution of system Integral Equation (u_n^4) in D_4 is given by formulas, $n=1,2,3,\cdots$

$$u_0^4(y)=\left(\frac{d-y}{y-c}\right)^{\frac{\delta_1}{\gamma(b-a)(d-c)}}\left[C_0^4+\frac{1}{(b-a)}\int_y^d\left(\frac{s-c}{d-c}\right)^{\frac{\delta_1}{\gamma(b-a)(d-c)}}\right.$$

$$\left.\frac{f_0^2(s)}{(s-c)(d-s)}\right]=T_0^4[C_0^4,\ f_0^2(y)]$$

$$u_n^4(y)=\left(\frac{d-y}{y-c}\right)^{\frac{\delta_1}{(n+\gamma)(b-a)(d-c)}}\left[C_n^4-\int_y^d\left(\frac{s-c}{d-s}\right)^{\frac{\delta_1}{(\gamma+n)(b-a)(d-c)}}\right.$$

$$\left.\left[\frac{(u_{n-1}^4(s))'}{b-a}+\frac{1}{(\gamma+n)(b-a)(s-c)(d-s)}\right]ds\right]\equiv T_n^4[C_n^4, f_n^2(y), u_{n-1}^4(y)].$$

Moreover from condition continuity the solution hyperbolic equation equation in lines $y=y_0$, follows, that $u_n^2(y_0)=u_n^4(y_0)$. From here it follows that the constants C_n^4 and $C_n^2(n=0,1,2,\dots)$ among themselves connected in the following way, $n=1,2,3,\cdots$

$$C_0^4-C_0^2=\frac{1}{\gamma(b-a)}\int_c^d\left(\frac{s-c}{d-s}\right)^{\frac{\delta_1}{\gamma(b-a)(d-c)}}\frac{f_0^2(s)ds}{(s-c)(d-s),} \qquad (B_0^1)$$

$$C_n^4-C_n^2=\frac{1}{(\gamma+n)(b-a)}\int_c^d\left(\frac{s-c}{d-s}\right)^{\frac{\delta_1}{(\gamma+n)(b-a)(d-c)}}\frac{f_n^2(s)}{(s-c)(d-s)}ds+\frac{1}{(b-a)}$$

$$\left[\int_c^{y_0}\left(\frac{s-c}{d-s}\right)^{\frac{\delta_1}{(\gamma+n)(b-a)(d-c)}}(\nu_{n-1}^2(s))'ds+\int_{y_0}^{d}\left(\frac{s-c}{d-s}\right)^{\frac{\delta_1}{(\gamma+n)(b-a)(d-c)}}(\nu_{n-1}^4(s))'ds\right]. \quad (B_n^1)$$

Thus we have proved the following.

Theorem 0.3. *Let in equation (1) the coefficients* λ, μ, δ *be such that* $\delta_1 = \lambda\mu - \delta > 0$, f *be a function in* D_1, D_3 *representable in the form* (f_1) *and in* D_2, D_4 *representable in the form* (f_2), *where* $f_n^1(c) = 0$ *with the asymptotic behavior* $f_n^1(y) = O[(y-c)^{\gamma_n^1}]$, $\gamma_n^1 > \frac{\delta_1}{(\gamma+n)(b-a)(d-c)}$, $n = 0, 1, 2, 3, \cdots$, *as* $y \to c$ *and* $f_n^2(d) = 0$, *with the asymptotic behavior* $f_n^2 = O[(d-c)^{\gamma_n^2}]$, $\gamma_n^2 > \frac{\delta_1}{(\gamma+n)(b-a)(d-c)}$, $n = 0, 1, 2, 3, \cdots$ *as* $y \to d$. *Then any solution of equation (1) is in the class of functions representable in domains* D_1, D_3 *in the form (11), and in domains* D_2, D_4 *in the form (13), that is,*

$$u(x,y) = \omega(x,y)\sum_{n=0}^{\infty}(x-a)^{n+\gamma}(b-x)T_n^1[C_n^1, f_n^1(y), u_{n-1}^1(y)], \quad (x,y) \in D_1,$$

$$u(x,y) = \omega(x,y)\sum_{n=0}^{\infty}(b-x)^{n+\gamma}(x-a)T_n^2[C_n^2, f_n^2(y), u_{n-1}^2(y)], \quad (x,y) \in D_2,$$

$$u(x,y) = \omega(x,y)\sum_{n=0}^{\infty}(x-a)^{n+\gamma}(b-x)T_n^3[C_n^3, f_n^1(y), u_{n-1}^3(y)], \quad (x,y) \in D_3,$$

$$u(x,y) = \omega(x,y)\sum_{n=0}^{\infty}(b-x)^{n+\gamma}(x-a)T_n^4[C_n^4, f_n^2(y), u_{n-1}^4(y)], \quad (x,y) \in D_4,$$

where constant C_n^1, C_n^3 *among themselves connected by formulas* (A_n^1) $(n = 0, 1, 2, 3, \cdots)$ *and constants* C_n^2, C_n^4 *among themselves connected by formulas* $(B_n^1)(n = 0, 1, 2, 3, \cdots)$.

Using the integral representation of the from (2), immediately testing, easily we can see, that if exist the solution of the equation (1), then it have the following characteristics

$$\left[\left(\frac{y-c}{d-c}\right)^{\frac{\lambda}{d-c}} u(x,y)\right]_{y=c} = \varphi_1(x), \quad \left[\left(\frac{x-a}{b-x}\right)^{\frac{\mu}{b-a}} P_\lambda^y(u)\right]_{x=a} = \psi_1(y),$$

$$\left[\left(\frac{y-c}{d-c}\right)^{\frac{\lambda}{d-c}} P_\mu^x(u)\right]_{y=c} = \varphi_1'(x) + \frac{\mu}{(x-a)(b-x)}\varphi_1(x),$$

$$\left[\left(\frac{y-c}{d-c}\right)^{\frac{\lambda}{d-c}} P_\mu^x(u)\right]_{y=d} = \varphi_2'(x) + \frac{\mu}{(x-a)(b-x)}\varphi_2(x)$$

$$\left[\left(\frac{y-c}{d-y}\right)^{\frac{\lambda}{d-c}} u(x,y)\right]_{y=d} = \varphi_2(x), \quad \left[\left(\frac{x-a}{b-x}\right)^{\frac{\mu}{b-a}} P^y_\lambda(u)\right]_{x=b} = \psi_2(y),$$

$$\left[\left(\frac{x-a}{b-x}\right)^{\frac{\mu}{b-a}} (P^y_\lambda)^2 u\right]_{x=a} = \psi^1_1(y) + \frac{\lambda\psi_1(y)}{(y-c)(d-y)},$$

$$\left[\left(\frac{x-a}{b-x}\right)^{\frac{\mu}{b-a}} (P^y_\lambda)^2 U\right]_{x=b} = \psi^1_2(y) + \frac{\lambda\psi_2(y)}{(y-c)(d-y)},$$

where $P^x_\mu \equiv \frac{\partial}{\partial x} + \frac{\mu}{(x-a)(b-x)}$, $P^y_\lambda \equiv \frac{\partial}{\partial y} + \frac{\mu}{(y-c)(d-y)}$.

Integral representation (2) and higher mentioned conditions the solution of the equation (2), give possibility for equation (1) place and investigate the following boundary value problems.

Problem R_1. Find the solution of equation (1) in class $C^2_{xy}(D)$ with one of the following boundary conditions:

1) $$\left[\left(\frac{y-c}{d-y}\right)^{\frac{\lambda}{d-c}} U(x,y)\right]_{y=d} = A_1(x), \quad \left[\left(\frac{x-a}{b-x}\right)^{\frac{\mu}{b-a}} P^y_\lambda\right]_{x=b} = B_1(y);$$

2) $$\left[\left(\frac{y-c}{d-y}\right)^{\frac{\lambda}{d-c}} U(x,y)\right]_{y=c} = A_2(x), \quad \left[\left(\frac{x-a}{b-x}\right)^{\frac{\mu}{b-a}} P^y_\lambda\right]_{x=a} = B_2(y);$$

3) $$\left[\left(\frac{y-c}{d-y}\right)^{\frac{\lambda}{d-c}} P^x_\mu(u)\right]_{y=c} = A_3(x), \quad \left[\left(\frac{x-a}{b-x}\right)^{\frac{\mu}{b-a}} P^y_\lambda\right]_{x=a} = B_3(y);$$

4) $$\left[\left(\frac{y-c}{d-y}\right)^{\frac{\lambda}{d-c}} P^x_\mu(u)\right]_{y=d} = A_4(x), \quad \left[\left(\frac{x-a}{b-x}\right)^{\frac{\mu}{b-a}} P^y_\lambda\right]_{x=b} = B_4(y);$$

where $A_j(x)$, $B_j(y)$, $j = 1,2,3,4-$ are given functions Γ_1 and Γ_2.

Problem R_2. Find the solution of equation (1) in class $C^2_{xy}(D)$ with the following boundary conditions:

1) $$\left[\left(\frac{x-a}{b-x}\right)^{\frac{\mu}{b-a}} (P^y_\lambda)^2(u)\right]_{x=a} = B_5(y), \quad \left[\left(\frac{y-c}{d-y}\right)^{\frac{\lambda}{d-c}} P^x_\mu(u)\right]_{y=c} = A_5(x) \qquad (R^1_2)$$

2) $$\left[\left(\frac{y-c}{d-y}\right)^{\frac{\lambda}{d-c}} P^x_\mu(u)\right]_{y=d} = A_6(y), \quad \left[\left(\frac{x-a}{b-x}\right)^{\frac{\mu}{b-a}} (P^y_\lambda)^2(u)\right]_{x=b} = B_6(y) \qquad (R^2_2)$$

The solutions of problems R_1 and R_2 can be found using the integral representation (2) and properties mentioned above.

For example, the solution of problem R_2^1 reduce to solution of the following linear differential equations with two boundary singular points:

$$\psi_1^1(y) + \frac{\lambda}{(y-c)(d-y)}\psi_1(y) = B_5(y), \tag{15}$$

$$\varphi_1^1(x) + \frac{\mu}{(x-a)(b-x)}\varphi_1(x) = A_5(x). \tag{16}$$

By solving differential equations (15) and (16) according to [3], finding

$$\varphi_1(x) = \begin{cases} M_1^\mu, & a < x \le x_0 \\ M_2^\mu, & x_0 \le x < b \end{cases} \tag{17}$$

where $M_1^\mu[C_1, A_5(x)] = \left(\frac{b-x}{x-a}\right)^{\frac{\mu}{b-a}} [C_1 + T_{a,\mu}^x[A_5(x)]]$, $M_2^\mu[C_2, A_5(x)] = \left(\frac{b-x}{x-a}\right)^{\frac{\mu}{b-a}} [C_2 - T_{b,\mu}^x[A_5(x)]]$, C_1 is an arbitrary constant, C_2 with C_1 connected in the following way

$$C_2 = C_1 + \int_a^b \left(\frac{t-a}{b-t}\right)^{\frac{\mu}{b-a}} \frac{A_5(t)dt}{(t-a)(b-t)}, \tag{18}$$

$$\psi_1(y) = \begin{cases} N_1^\lambda[C_3, B_5(y)], & c < y \le y_0 \\ N_2^\lambda[C_4, B_5(y)], & y_0 \le y < d \end{cases} \tag{19}$$

where $N_1^\lambda[C_3, B_5(y)] = \left(\frac{d-s}{s-c}\right)^{\frac{\lambda}{d-c}} [C_3 + T_{c,\lambda}^y[B_5(y)]]$, $N_2^\lambda[C_4, B_5(y)] = \left(\frac{d-s}{s-c}\right)^{\frac{\lambda}{d-c}} [C_4 - T_{d,\lambda}^y[B_5(y)]]$, C_3 is an arbitrary constant, C_4 with C_3 connected in the following way

$$C_4 = C_3 + \int_c^d \left(\frac{s-c}{d-s}\right)^{\frac{\lambda}{d-c}} \frac{B_5(s)ds}{(s-c)(d-s)} \tag{20}$$

Theorem 0.4. *Let in equation (1), $\lambda, \mu, c, f(x,y)$ satisfy the conditions of Theorem 0.1, and in problem (R_2^1), let $A_5(x) \in C(\Gamma_1)$, $A_5(b) = 0$ with the asymptotic behavior $A_5(x) = O[(b = x)^{\gamma_5}]$, $\gamma_5 > \frac{\mu}{b-a}$, $B_5(y) \in C(\overline{\Gamma}_2)$, $B_5(d) = 0$ with the asymptotic behavior $B_5(y) = O[(d-y)^{\gamma_6}]$, $\gamma_6 > \frac{\lambda}{d-c}$. Then problem ($R_2^1$) is always solvable and its solution is given by formulas (2), (17), (19), (3), (4), (5), (18), (20), where C_1^0, C_1, C_3 are arbitrary constants.*

Remark 0.4. Results similar to Theorem 0.2 are obtained about the solvability of problems R_1 and R_2^2.

References

1. M. Holodnyok, A. Klich, M. Kubichek, M. Marec, Methods of analysis of dynamical models, Moscow, Mir, 1991, 365.
2. N. Rajabov, Integral representation and boundary value problem for certain differential equation with singular lines or singular surfaces, Part 4, TSU publisher, Dushanbe 1985, 147.
3. N. Rajabov, An introduction to the theory of partial differential equations with super-singular coefficients, Tehran 1997, 230.
4. N. Rajabov, Introduction to ordinary differential equations with singular and super-singular coefficients, Dushanbe 1998, 160.
5. N. Rajabov, System of linear integral equations of Volterra type with singular and super-singular kernels. Ill-posed and non-classical problems of mathematical physics and analysis, Proc. Intern. Cons. Samarqand, Uzbekistan, September 11-12, 2000, Kluwer, Utrecht-Boston, 103-124.
6. N. Rajabov, About one Volterra type integral equation, Russian Acad. Sci. Doklady, 2002, v. 383, No. 3, 314-317.
7. N. Rajabov, Volterra type integral equation with interior fixed singular and super-singular point in kernel, proceedings international conference "Differential Equation and its Supplement", Samara 2002, 279-283.
8. N. Rajabov, About one class of Volterra type linear integral equations with interior fixed singular or super-singular point, Topics in analysis and applications, Kluwer, Dordrecht, Boston, London, 2004, 317-326.
9. L. N. Rajabov, About one class hyperbolic equation with singular lines, Bulletin National University, No. 5 (31), 2002, 44-51.
10. N. Rajabov, Volterra type Integral equation with boundary and interior singular and super-singular kernels its application, Dushanbe, 2007, 222.

ASYMPTOTIC STUDY OF AN ANISOTROPIC PERIODIC ROTATING MHD SYSTEM

R. SELMI

Department of Mathematics and Informatics,
Faculty of Sciences, University of Gabes,
Gabes 6000, Tunisia
E-mail: ridha.selmi@isi.rnu.tn

Asymptotic behavior of three-dimensional anisotropic periodic rotating magnetohydrodynamic system is investigated as the Rossby number goes to zero. Since the system is mixed, the divergence free condition is crucial in the proofs. The main tools are the spectral properties of the penalization operator, the energy method, Schochet's method and product laws in anisotropic Sobolev spaces. A filtering process is used to handle the singular perturbation, and elementary Fourier analysis to deal with the nonlinear part.

Keywords: Anisotropic MHD system; asymptotic behavior; limit system; divergence free condition; singular perturbation.

1. Introduction

We consider the following mixed periodic incompressible and perturbed magnetohydrodynamic system, denoted by (MHD^{ε}):

$$\begin{cases} \partial_t u - \nu\Delta_h u + u\cdot\nabla u - \operatorname{curl} b\times b + \dfrac{1}{\varepsilon}\operatorname{curl} b\times e_2 + \dfrac{1}{\varepsilon} u\times e_3 = -\nabla p \\ \partial_t b - \eta\Delta_h b + u\cdot\nabla b - b\cdot\nabla u + \dfrac{1}{\varepsilon}\operatorname{curl}\,(u\times e_2) = 0 \\ \operatorname{div} u = 0, \quad \operatorname{div} b = 0 \\ (u,b)|_{t=0} = (u_0,b_0), \end{cases}$$

where the velocity field u, the induced magnetic perturbation b and the pressure p are unknown functions of time $t\in\mathbb{R}^+$ and space variable $x=(x_1,x_2,x_3)\in\mathbb{T}^3$, e_2 and e_3 are the cartesian coordinate system, ν and η designate respectively the dynamic viscosity and the magnetic diffusivity, ε is the Rossby number destined to tend to zero and $\Delta_h=\partial_1^2+\partial_2^2$. About physical motivation, see [1] and references therein. Applying $\mathbb{P}$, the $L^2(\mathbb{T}^3)$-orthogonal projection onto divergence free vector fields, to the first equation of (MHD^ε), one sees that $U := (u, b)$ satisfies

$$(S^\varepsilon) \quad \begin{cases} \partial_t U + Q(U,U) + a_2(D)U + L^\varepsilon(U) = 0 \quad \text{in} \quad \mathbb{R}^+ \times \mathbb{T}^3 \\ \operatorname{div} u = 0, \quad \operatorname{div} b = 0 \quad \text{in} \quad \mathbb{R}^+ \times \mathbb{T}^3 \\ U|_{t=0} = U_0 \quad \text{in} \quad \mathbb{T}^3, \end{cases}$$

where after simplification, the quadratic term Q, the viscous term $a_2(D)$ and the linear perturbation L^ε are given by

$$\begin{aligned} Q(U,U) &= \begin{pmatrix} \mathbb{P}(u \cdot \nabla u) - \mathbb{P}(b \cdot \nabla b) \\ u \cdot \nabla b - b \cdot \nabla u \end{pmatrix} \\ a_2(D)U &= (-\nu \Delta_h u, -\eta \Delta_h b) \\ L^\varepsilon(U) &= \frac{1}{\varepsilon} L(U) = \frac{1}{\varepsilon} \begin{pmatrix} \partial_2 b + \mathbb{P}(u \times e_3) \\ \partial_2 u \end{pmatrix}. \end{aligned}$$

As in [5], for $(s, s') \in \mathbb{R}^2$ the anisotropic Sobolev space $H^{s,s'}(\mathbb{T}^3)$ is the set of tempered distributions f satisfying, for any $k = (k', k_3) = (k_1, k_2, k_3) \in \mathbb{R}^3$,

$$\|f\|_{H^{s,s'}} := \|(1 + |k'|^2)^{\frac{s}{2}} (1 + |k_3|^2)^{\frac{s'}{2}} \mathcal{F}(f)\|_{L^2} < \infty.$$

Following ideas in our previous work [1], we prove theorem below. Note that product law given by Theorem 1.4 in reference [6] still apply in the periodic case since that if any integral converges in$+\infty$ its corresponding series do.

Theorem 1.1. *Let $s > \frac{1}{2}$ be a real number and $U_0 = (u_0, b_0) \in H^{0,s}(\mathbb{T}^3)$, such that $\operatorname{div} u_0 = 0$ and $\operatorname{div} b_0 = 0$. There exists a positive time T such that for all $\varepsilon > 0$, there exists a unique solution U^ε of (S^ε), satisfying $U^\varepsilon \in C^0_T(H^{0,s}(\mathbb{T}^3))$ and $\nabla_h U^\varepsilon \in L^2_T(H^{0,s}(\mathbb{T}^3))$. Moreover, U^ε fulfils*

$$\|U^\varepsilon(t,.)\|^2_{H^{0,s}(\mathbb{T}^3)} + \min(\nu, \eta) \int_0^t \|\nabla_h U^\varepsilon(\tau,.)\|^2_{H^{0,s}(\mathbb{T}^3)} d\tau \leq \|U_0\|^2_{H^{0,s}(\mathbb{T}^3)}. \quad (1)$$

Furthermore, if a constant c exists, such that $\|U_0\|_{H^{0,s}(\mathbb{T}^3)} \leq c\, min(\nu, \eta)$, then the solution is global.

The aim of this work is to focus on the asymptotic behavior of the unique solution, as the Rossby number ε goes to zero. However, since $(\partial_t U^\varepsilon)$ is not bounded with respect to ε, then the classical methods, based on taking directly such limit in the system, no longer work. L^ε being skew symmetric, we overcome this difficulty by using the method introduced in [10]. This consists in filtering the singular system by the associated group $\mathcal{L}(t)$, to obtain a filtered one, where the penalized perturbation disappears. Thus, it will be possible to look, in the sense of distribution, for the limit system satisfied by the limit V of the filtered solution $V^\varepsilon := \mathcal{L}(-\frac{t}{\varepsilon})U^\varepsilon$. Namely, we establish the following convergence result:

Theorem 1.2. *Let $s > \frac{1}{2}$ and $U_0 = (u_0, b_0) \in H^s(\mathbb{T}^3)$, such that $div\, u_0 = div\, b_0 = 0$. Let $U^\varepsilon = (u^\varepsilon, b^\varepsilon)$ be the family of solutions of (S^ε) given by Theorem 1. Then, for all $s' < s$, the family $V^\varepsilon = (v^\varepsilon, c^\varepsilon)$ converges strongly in $C^0_T(H^{0,s'})$ to the solution V of system (LS) defined by*

$$(LS) \quad \begin{cases} \partial_t V + Q^0(V,V) + a_2(D)V = 0 \quad in \quad [0,T] \times \mathbb{T}^3 \\ div\, v = div\, c = 0 \quad in \quad [0,T] \times \mathbb{T}^3 \\ V|_{t=0} = U_0 = (u_0, b_0) \quad in \quad \mathbb{T}^3 \end{cases}$$

where we set, in the sense of distributions, $Q^0(V,V) := \lim_{\varepsilon \to 0} Q^\varepsilon(V,V)$.

The proof uses the divergence free condition, anisotropic product law [6] and properties of $\mathbb{P}$. In fact, $\partial_3 U_3 = -\mathrm{div}_h\, U_h$ implies that U is more regular with respect to x_3 then expected. We motion here references [7,8], where it was proved global in time existence result of anisotropic Navier-Stokes equation, when ε is sufficiently small but for initial data associated with horizontal mean equal to zero in $H^{0,s}$. Such condition is purely technic and dictated by the use of some Gagliardo-Nirenberg inequality needed in the framework of Littlewood-Paley computation for the critical periodic case $s = \frac{1}{2}$. Here, since $s > \frac{1}{2}$, we can extend existence and uniqueness results proved in [1] for the case $\mathbb{R}^3$ without any additional physically meaningless condition on the initial data. Moreover, although it is local in time, our convergence result presents the advantage to use classical techniques, such as commutator inequalities, to handle the nonlinear part in an elegent and rapid manner in opposition to tedious computation involved by the anisotropic littelwood-Paley theory used in [8].

2. Convergence result

2.1. *Study of the limit system*

Following ideas in [9], we prove the lemma below.

Lemma 2.1. *The "wave equation" given by the following system*

$$\begin{cases} \partial_t U + L(U) = 0 \quad in \quad \mathbb{R} \times \mathbb{T}^3 \\ div\, u_0 = div\, b_0 = 0 \quad in \quad \mathbb{R} \times \mathbb{T}^3 \\ U(0) = U_0 \quad in \quad \mathbb{T}^3 \end{cases} \tag{2}$$

has a global solution $U(t) = \mathcal{L}(t)U_0$, such that for all $(s,s') \in \mathbb{R}^2$ and U_0 in $H^{s,s'}(\mathbb{T}^3)$, $\quad \|\mathcal{L}(t)U_0\|_{H^{s,s'}(\mathbb{T}^3)} = \|U_0\|_{H^{s,s'}(\mathbb{T}^3)} = \|{}^t\mathcal{L}(t)U_0\|_{H^{s,s'}(\mathbb{T}^3)}$.

Proof. System (2) is equivalent to

$$\begin{cases} \partial_t u + \mathbb{P}(u \times e_3) + \partial_2 b = 0 \quad \text{in} \quad \mathbb{R} \times \mathbb{T}^3 \\ \partial_t b + \partial_2 u = 0 \quad \text{in} \quad \mathbb{R} \times \mathbb{T}^3 \\ \operatorname{div} u_0 = \operatorname{div} b_0 = 0 \quad \text{in} \quad \mathbb{R} \times \mathbb{T}^3 \\ U(0) = U_0 \quad \text{in} \quad \mathbb{T}^3. \end{cases} \tag{3}$$

Applying "curl" to the first equation, then by Fourier Transform we get

$$\begin{cases} \partial_t iM(k)\hat{u}(t,k) + ik_3\hat{u}(t,k) - k_2 M(k)\hat{b}(t,k) = 0 \quad \text{in} \quad \mathbb{R} \times \mathbb{T}^3 \\ \partial_t \hat{b}(t,k) + ik_2\hat{u}(t,k) \quad \text{in} \quad \mathbb{R} \times \mathbb{T}^3 \\ k \cdot \hat{u}_0 = k \cdot \hat{b}_0 = 0 \quad \text{in} \quad \mathbb{R} \times \mathbb{T}^3 \\ \hat{U}(0) = \hat{U}_0 \quad \text{in} \quad \mathbb{T}^3. \end{cases} \tag{4}$$

The eigenvalues of $M(k)$ are $\pm i|k|$, 0 and the corresponding eigenvectors are $\rho(k)^{\pm}$ and $\dfrac{k}{|k|}$ which is not divergence free. Vectors $\rho(k)^{\pm}$ are given by

$$\rho(k)^{+} = \begin{cases} \left(1, \, -i\,\frac{k_3}{|k_3|}, \, 0\right) & \text{if} \quad k_1^2 + k_2^2 = 0, \\ \dfrac{\left(\alpha(k), \, \beta(k), \, \gamma(k)\right)}{\left(|\alpha(k)|^2 + |\beta(k)|^2 + |\gamma(k)|^2\right)^{\frac{1}{2}}} & \text{if} \quad k_1^2 + k_2^2 \neq 0, \end{cases}$$

$$\left(\alpha(k), \, \beta(k), \, \gamma(k)\right) = \left(-\,k_1k_3 - i\,k_2\,|k|, \, -k_2k_3 + i\,k_1|k|, \, k_1^2 + k_2^2\right),$$

and $\rho(k)^{-} = \overline{\rho(k)^{+}}$. In this eigenbase, u and b can be rewritten such that

$$\begin{aligned} \hat{u}(t,k) &= \hat{u}^{+}(t,k)\rho(k)^{+} + \hat{u}^{-}(t,k)\rho(k)^{-} \\ \hat{b}(t,k) &= \hat{b}^{+}(t,k)\rho(k)^{+} + \hat{b}^{-}(t,k)\rho(k)^{-}. \end{aligned}$$

In term of variables $\hat{u}^{+}, \hat{u}^{-}, \hat{b}^{+}$ and $\hat{b}^{-}$, system (4) is summarized in

$$\begin{cases} \partial_t \hat{U}^{\pm}(t,k) + A^{+}(k)\hat{U}^{+}(t,k) = 0 \quad \text{in} \quad \mathbb{R} \times \mathbb{T}^3 \\ k \cdot \hat{u} = k \cdot \hat{b} = 0 \quad \text{in} \quad \mathbb{R} \times \mathbb{T}^3, \end{cases} \tag{5}$$

where $\hat{U}^{\pm} = (\hat{u}^{\pm}, \hat{b}^{\pm})$ and

$$A^{\pm}(k) = \begin{pmatrix} \mp i\dfrac{k_3}{|k|} & ik_2 \\ ik_2 & 0 \end{pmatrix}.$$

Let $A = \operatorname{diag}(A^{+}, A^{-})$, its eigenvalues satisfy $\omega_3 = \overline{\omega_2}$, $\omega_4 = \overline{\omega_1}$ and

$$\begin{aligned} \omega_1(k) &= \frac{i}{2|k|}(-k_3 + \sqrt{k_3^2 + 4k_2^2|k|^2}) \\ \omega_2(k) &= \frac{i}{2|k|}(-k_3 - \sqrt{k_3^2 + 4k_2^2|k|^2}). \end{aligned}$$

If we denote by ϱ^j, $1 \le j \le 4$, the corresponding eigenvectors, then

$$\mathcal{F}(U)(t,k) = \sum_{j\in\{1,2,3,4\}} \exp\big(\omega_j(k)t\big) < \mathcal{F}(U)(0,k)|\varrho^j(k) > \varrho^j(k).$$

□

Define V^ε by $V^\varepsilon(t,x) := \mathcal{L}\Big(-\frac{t}{\varepsilon}\Big)U^\varepsilon(t,x) = (v^\varepsilon, c^\varepsilon)$. Then,

$$\mathcal{F}(V^\varepsilon)(t,k) = \sum_{j\in\{1,2,3,4\}} \exp\big(\omega_j(k)t\big) < \mathcal{F}(V^\varepsilon)(0,k)|\varrho^j(k) > \varrho^j(k) \qquad (6)$$

and V^ε satisfies the following system

$$(\tilde{\mathcal{S}}^\varepsilon) \qquad \begin{cases} \partial_t V^\varepsilon + Q^\varepsilon(V^\varepsilon, V^\varepsilon) + a_2(D)V^\varepsilon = 0 \quad \text{in} \quad \mathbb{R}\times\mathbb{T}^3 \\ \operatorname{div} v^\varepsilon = \operatorname{div} c^\varepsilon = 0 \quad \text{in} \quad \mathbb{R}\times\mathbb{T}^3 \\ V^\varepsilon|_{t=0} = U_0 = (u_0, b_0) \quad \text{in} \quad \mathbb{T}^3 \end{cases}$$

where the "filtered" quadratic form Q^ε is given by

$$Q^\varepsilon(V^\varepsilon, V^\varepsilon) = \mathcal{L}\Big(-\frac{t}{\varepsilon}\Big)\mathbb{P}Q^\varepsilon\Big(\mathcal{L}\Big(\frac{t}{\varepsilon}\Big)V^\varepsilon, \mathcal{L}\Big(\frac{t}{\varepsilon}\Big)V^\varepsilon\Big).$$

In Fourier variables $2\mathcal{F}\Big(Q^\varepsilon(V_1^\varepsilon, V_2^\varepsilon)\Big)(n) = (a_{11}, a_{21})$, where

$$\begin{aligned} a_{11} = &\sum_{1\le j,j',j''\le 4} \sum_{k+m=n} e^{-i\frac{t}{\varepsilon}\omega^{j,j',j''}_{k,m,n}} P(n) < v^{j'}(k) m v^{j''}(m)|\varrho^j(n) > \varrho^j(n) \\ - &\sum_{1\le j,j',j''\le 4} \sum_{k+m=n} e^{-i\frac{t}{\varepsilon}\omega^{j,j',j''}_{k,m,n}} P(n) < c^{j'}(k) m c^{j''}(m)|\varrho^j(n) > \varrho^j(n) \\ a_{21} = &\sum_{1\le j,j',j''\le 4} \sum_{k+m=n} e^{-i\frac{t}{\varepsilon}\omega^{j,j',j''}_{k,m,n}} < v^{j'}(k) m c^{j''}(m)|\varrho^j(n) > \varrho^j(n) \\ - &\sum_{1\le j,j',j''\le 4} \sum_{k+m=n} e^{-i\frac{t}{\varepsilon}\omega^{j,j',j''}_{k,m,n}} < c^{j'}(k) m v^{j''}(m)|\varrho^j(n) > \varrho^j(n). \end{aligned}$$

In the expression above, $\omega^{j,j',j''}_{k,m,n} := \omega^{j'}(k) + \omega^{j''}(m) - \omega^j(n)$,

$$P(n) := \mathrm{Id} - \frac{1}{n_1^2+n_2^2+n_3^2}\begin{pmatrix} n_1^2 & n_1n_2 & n_1n_3 \\ n_2n_1 & n_2^2 & n_2n_3 \\ n_3n_1 & n_3n_2 & n_3^2 \end{pmatrix}$$

and for any vector X, $x^j :=< \mathcal{F}(X)(n)|\varrho^j(n) > \varrho^j(n)$. The filtering procedure presents the advantage that the singular perturbation disappears. So, contrarily to $\partial_t U^\varepsilon$ in system $(\mathcal{S}^\varepsilon)$, $\partial_t V^\varepsilon$ in system $(\tilde{\mathcal{S}}^\varepsilon)$ is already bonded in $L^\infty([0,T], H^{-N}(\mathbb{T}^3))$, for some sufficiently large integer N. Consequently, classical compactness argument leads to $U^\varepsilon \to U$ in $C([0,T], H^{s,s'}(\mathbb{T}^3))$, for all $s < 0$ and $s' < \frac{1}{2}$. As in [4], one shoes that in $\mathcal{D}'$ the stationary phase theorem implies that $\lim_{\varepsilon\to 0} Q^\varepsilon(V^\varepsilon, V^\varepsilon) = Q^0(V,V)$. So, when ε goes to 0, one obtains formally the following limit system:

$$(LS) \quad \begin{cases} \partial_t V + Q^0(V,V) + a_2(D)V = 0 \quad \text{in} \quad [0,T] \times \mathbb{T}^3 \\ \operatorname{div} v = \operatorname{div} c = 0 \quad \text{in} \quad [0,T] \times \mathbb{T}^3 \\ V|_{t=0} = U_0 = (u_0, b_0) \quad \text{in} \quad \mathbb{T}^3. \end{cases}$$

Theorem 2.1. *Under assumptions of Theorem 1.1, there exists $T > 0$ and a unique $V \in L^\infty_T(H^{0,s}(\mathbb{T}^3)) \cap L^2_T(H^{1,s}(\mathbb{T}^3))$ solution of (LS) satisfying*

$$\begin{aligned} \|V(t,.)\|^2_{H^{0,s}(\mathbb{T}^3)} &+ \nu \int_0^t \|\nabla_h v(\tau,.)\|^2_{H^{0,s}(\mathbb{T}^3)} d\tau \\ &+ \eta \int_0^t \|\nabla_h c(\tau,.)\|^2_{H^{0,s}(\mathbb{T}^3)} d\tau \le \|U_0\|^2_{H^{0,s}(\mathbb{T}^3)}. \end{aligned} \tag{7}$$

Furthermore, If a constant c exists, such that $\|U_0\|_{H^{0,s}(\mathbb{T}^3)} \le c\, min(\nu,\eta)$, then the solution is global.

The proof follows the lines of the one of Theorem 1.1.

Remark 2.1. Note that the time T appearing in Theorem 2.1 can be different from the one of Theorem 1.1. Here we are not dealing with the possible relation between the life span of the solution of the system (S^ε) and the life span of the solution of the corresponding limit system (LS). But since we are proving a local in time convergence result, any uniform time of existence of both systems, denoted also T, is suitable for our purpose. We mention that the method used in [7] can be followed to prove global well posedeness of (LS) and to deduce the same for (MHD^ε), as ε is small enough.

2.2. *Proof of Theorem 1.2*

Let $W^\varepsilon = V^\varepsilon - V = (v^\varepsilon - v, c^\varepsilon - c) = (W^\varepsilon_1, W^\varepsilon_2)$. Then, W^ε satisfies

$$\begin{cases} \partial_t W^\varepsilon + Q^\varepsilon(W^\varepsilon, W^\varepsilon + 2V) + a_2(D)W^\varepsilon = R^\varepsilon_{osc} \\ \operatorname{div} W^\varepsilon_1 = \operatorname{div} W^\varepsilon_2 = 0 \\ W^\varepsilon|_{t=0} = (0,0), \end{cases} \tag{8}$$

where $R^\varepsilon_{osc} = Q^0(V,V) - Q^\varepsilon(V,V)$. The right hand side of system (8) is an oscillating term which converges weakly to zero but not strongly. The method we use to deal with R^ε_{osc} is inspired from ideas introduced in [10]. It consists to divide R^ε_{osc} into high frequency term $R^{\varepsilon,N}_{osc}$ and low frequency term $R^\varepsilon_{osc,N}$, defined respectively, for any arbitrary cut-off integer $N \geq 1$, by $R^{\varepsilon,N}_{osc} = R^\varepsilon_{osc} - R^\varepsilon_{osc,N}$ and

$$\mathcal{F}\Big(R^\varepsilon_{osc,N}\Big)(t,n) = \mathbf{1}_{\{|n|\leq N\}}\mathcal{F}(R^\varepsilon_{osc;|k|,|m|\leq N})(t,n),$$

where $R^\varepsilon_{osc;|k|,|m|\leq N}$ is the part of R^ε_{osc} for which k and m are such that $|k|,|m| \leq N$. To absorb the low frequency term, we adopt the change of function: $\varphi^\varepsilon_N = W^\varepsilon + \varepsilon\tilde{R}^\varepsilon_{osc,N}$, where $\tilde{R}^\varepsilon_{osc,N}$ is obtained by dividing $\exp(-i\frac{t}{\varepsilon}\omega^{j,j',j''}_{k,m,n})$ by $i\omega^{j,j',j''}_{k,m,n}$ in the expression $R^\varepsilon_{osc,N}$. Function φ^ε_N satisfies:

$$\partial_t\varphi^\varepsilon_N + Q^\varepsilon(\varphi^\varepsilon_N, \varphi^\varepsilon_N - 2\varepsilon\tilde{R}^\varepsilon_{osc,N} + 2V) + a_2(D)\varphi^\varepsilon_N = R^{\varepsilon,N}_{osc} + \varepsilon r^\varepsilon_{osc,N}, \quad (9)$$

where $\varepsilon r^\varepsilon_{osc,N} = -Q^\varepsilon(\varepsilon\tilde{R}^\varepsilon_{osc,N}, \varepsilon\tilde{R}^\varepsilon_{osc,N} - 2V) - \varepsilon a_2(D)(\tilde{R}^\varepsilon_{osc,N}) + \varepsilon\tilde{R}^{\varepsilon,t}_{osc,N}$ and $\tilde{R}^{\varepsilon,t}_{osc,N}$ is the expression of $\tilde{R}^\varepsilon_{osc,N}$ where we have changed, the nonlinear factor by its time derivative. Note that equation (9) presents the advantage that the low frequency terms have disappeared up to an ε. For any real number x, denote by $\langle x\rangle$ the quantity $(1+|x|^2)^{\frac{1}{2}}$ and by Λ_3 the operator defined by $\Lambda_3 = (1+\partial_3^2)^{\frac{1}{2}}$, that is the operator of multiplication by $\langle k_3\rangle$ in the frequency space. Clearly, for all real numbers s and s', Λ_3 is an isometry from $H^{s,s'}$ to $H^{s,s'-1}$.

Taking the scalar product in $H^{0,-\frac{1}{2}}(\mathbb{T}^3)$, equation (9) gives

$$\begin{aligned}\|\varphi^\varepsilon_N\|^2_{0,-\frac{1}{2}} &+ 2\nu\int_0^t \|\nabla_h\varphi^\varepsilon_{N,1}\|^2_{0,-\frac{1}{2}}d\tau + 2\eta\int_0^t \|\nabla_h\varphi^\varepsilon_{N,2}\|^2_{0,-\frac{1}{2}}d\tau \\ &= -2\int_0^t\int_{\mathbb{T}^3} Q^\varepsilon(\varphi^\varepsilon_N, \varphi^\varepsilon_N - 2\varepsilon\tilde{R}^\varepsilon_{osc,N} + 2V)\Lambda_3^{-1}\varphi^\varepsilon_N dx d\tau \quad (10)\\ &\quad -2\int_0^t\int_{\mathbb{T}^3}(R^{\varepsilon,N}_{osc} + \varepsilon r^\varepsilon_{osc,N})\Lambda_3^{-1}\varphi^\varepsilon_N dx d\tau.\end{aligned}$$

Let $\displaystyle\int_{\mathbb{T}^3} Q^\varepsilon(\varphi^\varepsilon_N, \varphi^\varepsilon_N - 2\varepsilon\tilde{R}^\varepsilon_{osc,N} + 2V)\Lambda_3^{-1}\varphi^\varepsilon_N dx d\tau = \sum_{i=1}^4 I_i$, where

$$\begin{aligned}
I_1 &= \int_{\mathbb{T}^3}\mathcal{L}(\frac{-t}{\varepsilon})\mathbb{P}\Big[\Big(\mathcal{L}(\frac{t}{\varepsilon})(\varphi^\varepsilon_N - 2\varepsilon\tilde{R}^\varepsilon_{osc,N} + 2V)\Big)_h\nabla_h\Big(\mathcal{L}(\frac{t}{\varepsilon})\varphi^\varepsilon_N\Big)\Big]\Lambda_3^{-1}\varphi^\varepsilon_N dx\\
I_2 &= \int_{\mathbb{T}^3}\mathcal{L}(\frac{-t}{\varepsilon})\mathbb{P}\Big[\Big(\mathcal{L}(\frac{t}{\varepsilon})(\varphi^\varepsilon_N - 2\varepsilon\tilde{R}^\varepsilon_{osc,N} + 2V)\Big)_3\partial_3\Big(\mathcal{L}(\frac{t}{\varepsilon})\varphi^\varepsilon_N\Big)\Big]\Lambda_3^{-1}\varphi^\varepsilon_N dx\\
I_3 &= \int_{\mathbb{T}^3}\mathcal{L}(\frac{-t}{\varepsilon})\mathbb{P}\Big[\Big(\mathcal{L}(\frac{t}{\varepsilon})\varphi^\varepsilon_N\Big)_h\nabla_h\Big(\mathcal{L}(\frac{t}{\varepsilon})(\varphi^\varepsilon_N - 2\varepsilon\tilde{R}^\varepsilon_{osc,N} + 2V)\Big)\Big]\Lambda_3^{-1}\varphi^\varepsilon_N dx\\
I_4 &= \int_{\mathbb{T}^3}\mathcal{L}(\frac{-t}{\varepsilon})\mathbb{P}\Big[\Big(\mathcal{L}(\frac{t}{\varepsilon})\varphi^\varepsilon_N\Big)_3\partial_3\Big(\mathcal{L}(\frac{t}{\varepsilon})(\varphi^\varepsilon_N - 2\varepsilon\tilde{R}^\varepsilon_{osc,N} + 2V)\Big)\Big]\Lambda_3^{-1}\varphi^\varepsilon_N dx.
\end{aligned}$$

Ideas in [1] lead to

$$\begin{aligned}
|I_1| &\leq \|\varphi_N^\varepsilon - 2\varepsilon\tilde{R}^\varepsilon_{osc,N} + 2V\|_{\frac{1}{2},s}\|\varphi_N^\varepsilon\|^{\frac{1}{2}}_{0,-\frac{1}{2}}\|\varphi_N^\varepsilon\|^{\frac{3}{2}}_{1,-\frac{1}{2}}\\
|I_3| &\leq C\|(\varphi_N^\varepsilon - 2\varepsilon\tilde{R}^\varepsilon_{osc,N} + 2V)\|_{\frac{1}{2},s}\|\varphi_N^\varepsilon\|^{\frac{1}{2}}_{0,-\frac{1}{2}}\|\varphi_N^\varepsilon\|^{\frac{3}{2}}_{1,-\frac{1}{2}}\\
|I_4| &\leq C\|\varphi_N^\varepsilon - 2\varepsilon\tilde{R}^\varepsilon_{osc,N} + 2V\|_{\frac{1}{2},s}\\
&\quad\times\big(\|\varphi_N^\varepsilon\|^{\frac{3}{2}}_{0,-\frac{1}{2}}\|\varphi_N^\varepsilon\|^{\frac{1}{2}}_{1,-\frac{1}{2}} + \|\varphi_N^\varepsilon\|^{\frac{1}{2}}_{0,-\frac{1}{2}}\|\varphi_N^\varepsilon\|^{\frac{3}{2}}_{1,-\frac{1}{2}}\big).
\end{aligned}$$

To estimate I_2, one uses Parseval's formula and property of Λ_3 to obtain

$$\begin{aligned}
I_2 = (2\pi)^{-3}\sum_{k\in\mathbb{Z}}\mathcal{F}\Big(\mathcal{L}(\frac{-t}{\varepsilon})\mathbb{P}\big[(\mathcal{L}(\frac{t}{\varepsilon})(\varphi_N^\varepsilon - 2\varepsilon\tilde{R}^\varepsilon_{osc,N} + 2V))_3\\
\times\,\partial_3(\mathcal{L}(\tfrac{t}{\varepsilon})\varphi_N^\varepsilon)\big]\Big)(k)\frac{1}{\langle k_3\rangle}\mathcal{F}\Big(\varphi_N^\varepsilon\Big)(-k).
\end{aligned}$$

Since one has $\mathcal{F}(\mathbb{P}u) = \mathcal{F}(u) - <\mathcal{F}(u), \frac{k}{|k|}> \frac{k}{|k|}$, then by (6), it comes that

$$\begin{aligned}
I_2 = C\sum_{j,j',j''}\sum_{n\in\mathbb{Z}}\sum_{k\in\mathbb{Z}}\exp(\omega^{j,j',j''}_{k-n,n,k}\frac{t}{\varepsilon})\mathcal{F}(\varphi_N^\varepsilon)(-k)\frac{n_3}{\langle k_3\rangle}\\
\Big\{<(\varphi_N^\varepsilon - 2\varepsilon\tilde{R}^\varepsilon_{osc,N} + 2V)^{j'}(k-n)(\varphi_N^\varepsilon)^{j''}(n)|\varrho^j(k)>\varrho^j(k)-\\
<<(\varphi_N^\varepsilon - 2\varepsilon\tilde{R}^\varepsilon_{osc,N} + 2V)^{j'}(k-n)(\varphi_N^\varepsilon)^{j''}(n)|\varrho^j(k)>\varrho^j(k),\\
\frac{k}{|k|}>\frac{k}{|k|}\Big\}.
\end{aligned}$$

Using the change of variable $(k, n) \leftrightarrow (-n, -k)$, we obtain

$$\begin{aligned}
I_2 = C\sum_{j,j',j''}\sum_{n\in\mathbb{Z}}\sum_{k\in\mathbb{Z}}\exp(\omega^{j,j',j''}_{k-n,n,k}\frac{t}{\varepsilon})\mathcal{F}(\varphi_N^\varepsilon)(-k)\Big(\frac{n_3}{\langle k_3\rangle} - \frac{k_3}{\langle n_3\rangle}\Big)\\
\Big\{<(\varphi_N^\varepsilon - 2\varepsilon\tilde{R}^\varepsilon_{osc,N} + 2V)^{j'}(k-n)(\varphi_N^\varepsilon)^{j''}(n)|\varrho^j(k)>\varrho^j(k)-\\
<<(\varphi_N^\varepsilon - 2\varepsilon\tilde{R}^\varepsilon_{osc,N} + 2V)^{j'}(k-n)(\varphi_N^\varepsilon)^{j''}(n)|\varrho^j(k)>\varrho^j(k),\\
\frac{k}{|k|}>\frac{k}{|k|}\Big\}.
\end{aligned}$$

We recall the following elementary equality due to [6]:

$$\Big|\frac{n_3}{\langle k_3\rangle} - \frac{k_3}{\langle n_3\rangle}\Big| = \Big|\frac{n_3 - k_3}{\langle k_3\rangle} + \frac{(n_3-k_3)k_3(n_3+k_3)}{\langle n_3\rangle\langle k_3\rangle(\langle n_3\rangle + \langle k_3\rangle)}\Big| \leq |n_3 - k_3|\Big(\frac{1}{\langle n_3\rangle} + \frac{1}{\langle k_3\rangle}\Big).$$

Reapply change of variables $(k, n) \leftrightarrow (-n, -k)$, this inequality leads to

$$\begin{aligned}
|I_2| \leq C\sum_{n\in\mathbb{Z}}\sum_{k\in\mathbb{Z}}\frac{|n_3 - k_3|}{\langle k_3\rangle}|\mathcal{F}((\varphi_N^\varepsilon - 2\varepsilon\tilde{R}^\varepsilon_{osc,N} + 2V)_3)(k-n)||\mathcal{F}(\varphi_N^\varepsilon)(n)|\\
|\mathcal{F}(\varphi_N^\varepsilon)(-k)|.
\end{aligned}$$

As W, V and φ_N^ε are divergence free, then

$$|I_2| \le C \sum_{n\in\mathbb{Z}} \sum_{k\in\mathbb{Z}} \sum_{j=1}^{2} \frac{|n_j - k_j||\mathcal{F}((\varphi_N^\varepsilon - 2\varepsilon \tilde{R}_{osc,N}^\varepsilon + 2V)_j)|}{\langle k_3 \rangle} |\mathcal{F}(\varphi_N^\varepsilon)(n)| \\ |\mathcal{F}(\varphi_N^\varepsilon)(-k)|.$$

Let V' such that $\mathcal{F}(V_j') = |\mathcal{F}(V_j)|$. So, $\|V_j'\|_{s,s'} = \|V_j\|_{s,s'}$. Denote $|D_j|$ the operator of multiplication by $|\xi_j|$. Use Parseval's formula to obtain

$$|I_2| \le C \sum_{k\in\mathbb{Z}} \sum_{j=1}^{2} |D_j|(\varphi_N^\varepsilon - 2\varepsilon \tilde{R}_{osc,N}^\varepsilon + 2V)_j' \varphi'^\varepsilon_N \Lambda_3 \varphi'^\varepsilon_N.$$

Since $|D_j|V_j'$ and $\partial_j V_j'$ has the same $H^{s,s'}$ norm, then as for I_3, one has

$$|I_2| \le C \|\varphi_N^\varepsilon - 2\varepsilon \tilde{R}_{osc,N}^\varepsilon + 2V\|_{\frac{1}{2},s} \|\varphi_N^\varepsilon\|_{0,-\frac{1}{2}}^{\frac{1}{2}} \|\varphi_N^\varepsilon\|_{1,-\frac{1}{2}}^{\frac{3}{2}}.$$

Using estimations of $|I_i|$, $1 \le i \le 4$ and $ab \le \frac{1}{4}a^4 + \frac{3}{4}a^{\frac{4}{3}}$, (10) becomes

$$\begin{aligned} &\|\varphi_N^\varepsilon\|_{0,-\frac{1}{2}}^2 + 2\nu \int_0^t \|\nabla_h \varphi_{N,1}^\varepsilon\|_{0,-\frac{1}{2}}^2 d\tau + 2\eta \int_0^t \|\nabla_h \varphi_{N,2}^\varepsilon\|_{0,-\frac{1}{2}}^2 d\tau \\ &\le 2\min(\nu,\eta) \int_0^t \|\varphi_N^\varepsilon\|_{1,-\frac{1}{2}}^2 d\tau + C \int_0^t \|R_{osc}^{\varepsilon,N} + \varepsilon r_{osc,N}^\varepsilon\|_{-1,-\frac{1}{2}}^2 d\tau \qquad (11) \\ &\quad + C \int_0^t \|\varphi_N^\varepsilon\|_{0,-\frac{1}{2}}^2 (\|\varphi_N^\varepsilon - 2\varepsilon \tilde{R}_{osc,N}^\varepsilon + 2V\|_{\frac{1}{2},s}^4 + 1) d\tau. \end{aligned}$$

About the low-frequency terms, we have the following lemma:

Lemma 2.2. *A constant $C_N(T) = C(N,T)$ exists, such that*

$$\|r_{osc,N}^\varepsilon\|_{L_T^2(H^{-1,-\frac{1}{2}})} \le C_N(T)$$
$$\|\tilde{R}_{osc,N}^\varepsilon\|_{L_T^\infty(H^{0,s}) \cap L_T^2(H^{1,s})} \le C_N(T).$$

To prove this lemma, recall that all considered functions are truncated in low frequencies. Moreover, $\omega_{k,m,n}^{j,j',j''} \neq 0$ implies that $\frac{1}{|\omega_{k,m,n}^{j,j',j''}|} \le C(N)$. Since $\varphi_N^\varepsilon = V^\varepsilon - V + R_{osc}^{\varepsilon,N}$, equation (11) leads to

$$\begin{aligned} &\|\varphi_N^\varepsilon\|_{0,-\frac{1}{2}}^2 + 2\nu \int_0^t \|\nabla_h \varphi_{N,1}^\varepsilon\|_{0,-\frac{1}{2}}^2 d\tau + 2\eta \int_0^t \|\nabla_h \varphi_{N,2}^\varepsilon\|_{0,-\frac{1}{2}}^2 d\tau \\ &\le 2\min(\nu,\eta) \int_0^t \|\varphi_N^\varepsilon\|_{1,-\frac{1}{2}}^2 d\tau + C \int_0^t \|R_{osc}^{\varepsilon,N} + \varepsilon r_{osc,N}^\varepsilon\|_{0,-\frac{1}{2}}^2 d\tau \\ &\quad + C \int_0^t \|\varphi_N^\varepsilon\|_{0,-\frac{1}{2}}^2 (\|V^\varepsilon\|_{\frac{1}{2},s}^4 + \|V^\varepsilon\|_{\frac{1}{2},s}^4 + \varepsilon^4 C(N) + 1) d\tau. \end{aligned}$$

It follows that

$$\|\varphi_N^\varepsilon\|_{0,-\frac{1}{2}}^2 \le C \|R_{osc}^{\varepsilon,N}\|_{L_T^2(H^{0,-\frac{1}{2}})}^2 + \varepsilon^2 C_N(T) + C \int_0^t \|\varphi_N^\varepsilon\|_{0,-\frac{1}{2}}^2 h(\tau) d\tau, \qquad (12)$$

where $h(\tau) = \|V^\varepsilon\|^4_{\frac{1}{2},s} + \|V^\varepsilon\|^4_{\frac{1}{2},s} + \varepsilon^4 C(N) + 1$. An interpolation gives

$$\int_0^T \|V^\varepsilon\|^4_{\frac{1}{2},s}(\tau)d\tau \leq \|V^\varepsilon\|^2_{L^\infty_T(H^{0,s})}\|V^\varepsilon\|^2_{L^2_T(H^{1,s})}.$$

The same holds for V. Thus, h belongs to L^1_T. Gronwall's Lemma implies

$$\begin{aligned}\|\varphi^\varepsilon_N\|^2_{0,-\frac{1}{2}} &\leq \left(C\|R^{\varepsilon,N}_{osc}\|^2_{L^2_T(H^{0,-\frac{1}{2}})} + \varepsilon^2 C_N(T)\right)\\ &\times \exp\Big[(C+\varepsilon^4 C(N))T + C\|V^\varepsilon\|_{L^\infty_T(H^{0,s})}\|V^\varepsilon\|^2_{L^2_T(H^{1,s})}\\ &+ C\|V\|_{L^\infty_T(H^{0,s})}\|V\|^2_{L^2_T(H^{1,s})}\Big].\end{aligned} \tag{13}$$

About the high-frequency term $R^{\varepsilon,N}_{osc}$, we introduce the following lemma:

Lemma 2.3. *For any function $f \in \mathcal{C}^0_T(H^{0,s})$ with $s \in \mathbb{R}$, the high frequency term $f^N = \mathcal{F}^{-1}\Big(\mathbf{1}_{[N,+\infty[}\mathcal{F}(f)\Big)$ goes to zero in $\mathcal{C}^0_T(H^{0,s})$, when $N \to \infty$.*

Its proof is based on ideas from [2] and [3]; since $\|f^N(t)\|^2_{H^s} = \sum_{|k|\geq N} |k|^{2s}|\mathcal{F}(f)(t,k)|^2$, Dini's Theorem implies that $\|f^N(t)\|^2_{H^s} \to 0$ uniformly in t. This lemma implies, in a straightforward way, the following:

Proposition 2.1. *$R^{\varepsilon,N}_{osc} \to 0$ in $\mathcal{C}^0_T(H^{0,s}) \cap L^2_T(H^{0,-\frac{1}{2}})$, uniformly in ε, as $N \to \infty$.*

Using Proposition 2.1, letting $\varepsilon \to 0$ and $N \to +\infty$, we get $W^\varepsilon \to 0$ in $\mathcal{C}^0([0,T], H^{0,-\frac{1}{2}}(\mathbb{T}^3))$. An interpolation argument completes the proof.

Remark 2.2. Depending on which is the minimum of ν and η, either $W^\varepsilon_1 \to 0$ in $L^2([0,T], H^{1,-\frac{1}{2}}(\mathbb{T}^3))$ or $W^\varepsilon_2 \to 0$ in $L^2([0,T], H^{1,-\frac{1}{2}}(\mathbb{T}^3))$.

References

1. J. Benameur and R. Selmi: *Study of Anisotropic MHD System in Anisotropic Sobolev Spaces*, Annales Mathématique de la Faculté des Sciences de Toulouse, to appear.
2. I. Gallagher: *Applications of Schochet's Methods to Parabolic Equation*, Journal de Mathématiques Pures et Appliquées **77**, 989-1054, 1998.
3. I. Gallagher: *Asymptotics of the Solutions of Hyperbolic Equations with a Skew-Symmetric Perturbation*, Journal of Differential Equations **150**, 363-384, 1998.
4. E. Grenier: *Oscillatory Perturbations of the Navier-Stokes Equations*, Journal de Mathématiques Pures et Appliquées **76**, 477-498, 1997.
5. D. Iftimie: *Resolution of the Navier-Stokes Equations in Anisotropic Spaces*, Revista Matemática Iberoamericana **15**, 1-36, 1999.

6. D. Iftimie: *A Uniqueness Result for the Navier-Stokes Equations with Vanishing Vertical Viscosity,* SIAM Journal of Mathematical Analysis **33**, 1483-1493, 2002.
7. M. Paicu: *Etude Asymptotique pour les Fluides Anisotropes en Rotation Rapide dans le Cas Periodique*, Journal de Mathématiques Pures et Appliquées **83**, 163-242, 2004.
8. M. Paicu: *Equation Periodique de Navier-Stokes sans Viscosité dans une Direction*, Communications in Partial Differential Equations **30**, 1107-1140, 2005.
9. R. Selmi: *Convergence Results for MHD System*, International Journal of Mathematics and Mathematical Sciences **2006**, Article 28704, 19 pp., 2006.
10. S. Schochet: *Fast Singular Limits of Hyperbolic PDEs*, Journal of Differential Equations **114**, 476-512, 1994.

HILBERT-TYPE BOUNDARY VALUE PROBLEM FOR POLYANALYTIC FUNCTIONS *

YUFENG WANG

School of Mathematics and Statistics, Wuhan University,
Wuhan 430072, P. R. China
E-mail: wh_yfwang@hotmail.com

JINPING WANG

Department of Mathematics, Ningbo University,
Ningbo 315211, P. R. China

We discuss Hilbert-type boundary value problems of polyanalytic functions with the same factors. Using the decomposition of polyanalytic functions and the classical regularization method, we obtain explicit expressions of solutions and conditions of solvability.

Keywords: Polyanalytic function, Hilbert-type boundary value problem, order, poly-Schwarz operator.

1. Introduction

Three kinds of basic boundary value problems (BVPs) of complex equations of order one in complex analysis have been investigated in details by H. Begehr, and the explicit expression of solution and the condition of solvability have been obtained [5]. Those BVPs include the Schwartz problem, the Dirichlet problem and the Newmann problem. By Cauchy-Pompeiu formula and the method of iteration, BVPs of two-order complex equations, including the Poisson equation and the nonhomogeneous Bitsadze equation, have been systematically studied, and the explicit expression of solution and the condition of solvability have also been obtained. In addition, the so-called mixed BVPs of the Poisson equation and the nonhomogeneous

*This first author is supported by Tianyuan Fund of Mathematics (10626039), NNSF of China (10471107) and RFDP of Higher Eduction of China (20060486001). The second author is supported by Natural Science Foundation of Zhejiang Province (Y606093) and Professor Foundation of Ningbo University.

Bitsadze equation have been solved, which are the Dirichlet-Newmann type problem, the Newmann-Dirichlet type problem, and so on [4].

Further, BVPs of the model complex equation of higher order, i.e. polyanalytic equation and polyharmonic equation, have also been investigated, and, by the iteration, the explicit expression of solution and the condition of solvability have been obtained [1,5,7,14].

Because of the similar properties of polyanalytic function with analytic function, Riemann and Hilbert type BVPs of polyanalytic function have been considered, and the explicit expression of solution and the condition of solvability have also been obtained [2,8,12,13].

In this article we will discuss Hilbert boundary value problems of polyanalytic function with the same factors, which is different from Hilbert type BVPs discussed in [13]. Using the regularization method, we finally obtain explicit expressions of solutions and conditions of solvability.

Let $H_n(G)$ denote the set of polyanalytic functions on the open set G as in [8]. In the sequel we need the following decomposition.

Proposition 1.1. *Let $\mathbb{D}$ be the unit disc. Then*

$$H_n(\mathbb{D}\backslash\{0\})=H_1(\mathbb{D}\backslash\{0\})\oplus(\bar{z}+z)H_1(\mathbb{D}\backslash\{0\})\oplus\cdots\oplus(\bar{z}+z)^{n-1}H_1(\mathbb{D}\backslash\{0\}), \quad (1)$$

where $(\bar{z}+z)^j H_1(\mathbb{D}\setminus\{0\}) = \{(\bar{z}+z)^j h(z) : h \in H_1(\mathbb{D}\setminus\{0\})\}$ *for* $j = 0, 1, \cdots, n-1$.

By Proposition 1.1, if $V \in H_n(\mathbb{D}\setminus\{0\})$ then

$$V(z)=V_0(z)+(\bar{z}+z)V_1(z)+\cdots+(\bar{z}+z)^{n-1}f_{n-1}(z) \text{ with } V_j\in H_1(\mathbb{D}\backslash\{0\}) \quad (2)$$

for $j = 0, 1, \cdots, n-1$.

2. Preliminary Results

Let $D(z_0, r) = \{z \in \mathbb{C} : |z - z_0| < r\}$ denote the neighborhood of z_0 on the complex plane $\mathbb{C}$, and $\overset{\circ}{D}(z_0, r) = D(z_0, r) \setminus \{z_0\}$ the deleted neighborhood. Similarly to [2] the growth order of a finite point for polyanalytic functions is introduced as follows.

Definition 2.1. Suppose $f \in H_n(\overset{\circ}{D}(z_0, r))$. If there is an integer m such that $+\infty > \limsup_{z\to z_0} |(z-z_0)^m f(z)| > 0$, then we say f possesses order m at z_0, denoted by $\mathrm{Ord}(f, z_0) = m$. If $\limsup_{z\to z_0} |(z-z_0)^m f(z)| = +\infty$ for an arbitrary integer m, then we also call f possesses order $+\infty$ at z_0, denoted by $\mathrm{Ord}(f, z_0) = +\infty$. We assume $\mathrm{Ord}(f, z_0) = -\infty$ if and only if $f \equiv 0$.

Let $f \in H_n(\overset{\circ}{D}(z_0, r))$, and one has [8]

$$f(z) = f_0(z) + (\bar{z} - \bar{z}_0)f_1(z) + \cdots + (\bar{z} - \bar{z}_0)^{n-1}f_{n-1}(z) \quad (3)$$

where $f_j \in H_1(\overset{\circ}{D}(z_0, r))$, $j = 0, 1, \cdots, n-1$. Observe

$$f_k(z) = \sum_{j=-\infty}^{+\infty} a_{k,j}(z - z_0)^j \text{ with } a_{k,j} \in \mathbb{C}$$

for $k = 0, 1, \cdots, n-1$, and hence

$$f(z) = \sum_{j=-\infty}^{+\infty} p_j(z, \bar{z})(z - z_0)^j \quad (4)$$

with

$$p_j(z, \bar{z}) = \sum_{k=0}^{n-1} a_{k,j-k}(\bar{z} - \bar{z}_0)^k (z - z_0)^{-k}. \quad (5)$$

Similar to [2] we have the following conclusions.

Theorem 2.1. *Suppose* $f \in H_n(\overset{\circ}{D}(z_0, r))$. $\mathrm{Ord}(f, z_0) = m$ *if and only if*

$$f(z) = \sum_{j=-m}^{+\infty} p_j(z, \bar{z})(z - z_0)^j \text{ with } \sum_{k=0}^{n-1} |a_{k,j-k}| \neq 0. \quad (6)$$

In this case the point z_0 *is called a pole of* f.

Corollary 2.1. *Suppose* $f \in H_n(\overset{\circ}{D}(z_0, r))$. *Then* $\mathrm{Ord}(f, z_0) = \max\{\mathrm{Ord}(f_k, z_0) + k,\ k = 0, 1, \cdots, n-1\}$, *where* f_k *is determined by (3).*

Corollary 2.2. *Suppose* $V \in H_n(\mathbb{D}\backslash\{0\})$. *Then* $\mathrm{Ord}(V, 0) = \max\{\mathrm{Ord}(V_k, 0) + k,\ k = 0, 1, \cdots, n-1\}$, *where* V_k *is determined by (2).*

3. Hilbert BVP of Polyanalytic Functions

In this section we investigate the following Hilbert type BVP: Find a $V(z) \in H_n(\mathbb{D})$ such that $\frac{\partial^k V}{\partial \bar{z}^k}$, $k = 0, 1, \cdots, n-1$ are continuous on $\overline{\mathbb{D}}$, satisfying the Hilbert type boundary conditions

$$\mathrm{Re}\left\{[a(t) + ib(t)]\left[\frac{\partial^j V}{\partial \bar{z}^j}\right]^+ (t)\right\} = \gamma_j(t),\ t \in \partial\mathbb{D},\ j = 0, 1, \cdots, n-1, \quad (7)$$

where $a(t), b(t), \gamma_j(t) \in C(\partial\mathbb{D})$, $j = 0, 1, 2, \cdots, n-1$ and $a^2(t) + b^2(t) = 1$, $t \in \partial\mathbb{D}$. When $a(t) = 1, b(t) = 0$ the problem (7) is just the Schwarz problem for polyanalytic functions [1].

Let

$$X(z) = iz^{\kappa} \exp\{S[\alpha](z)\}, \tag{8}$$

where S is the Schwartz operator defined as

$$S[\alpha](z) = \frac{1}{2\pi i} \int_{\partial\mathbb{D}} \alpha(t) \frac{t+z}{t-z} \frac{dt}{t} \quad \text{with} \quad \alpha(t) = \arg\{t^{-\kappa}[a(t) - ib(t)]\} \tag{9}$$

and the index κ is given by

$$\kappa = \frac{1}{2\pi} \{\arg[a(t) - ib(t)]\}_{\partial\mathbb{D}}.$$

Then X has the following properties [8].

(1) $\operatorname{Re}\{[a(t) + ib(t)]X^+(t)\} = 0, \ t \in \partial\mathbb{D}$.
(2) $X(z)$ is analytic in $\mathbb{D}\backslash\{0\}$ and continuous up to $\partial\mathbb{D}$.
(3) $X^+(z) \neq 0, \ z \in \overline{\mathbb{D}}\backslash\{0\}, X^+(t) \in C(\partial\mathbb{D})$.
(4) $X(z)$ possesses order $-\kappa$ at $z = 0$, that is, $\lim_{z\to 0} z^{-\kappa} X(z) = iS[\alpha](0)$.

Our first problem is to find a $V \in H_n(\mathbb{D}\backslash\{0\}) \cap C^{n-1}(\overline{\mathbb{D}}\backslash\{0\})$ such that

$$\begin{cases} \operatorname{Re}\left\{\left[\dfrac{\partial^j V}{\partial \bar{z}^j}\right]^+(t)\right\} = 0, \, t \in \partial\mathbb{D}, \, j = 0, 1, \cdots, n-1, \\ \operatorname{Ord}(V, 0) \leq m. \end{cases} \tag{10}$$

This is the simplest Hilbert BVP.

Introduce the class of symmetric polynomials

$$S\Pi_k = \left\{ q_k(z) = \sum_{j=-k}^{k} c_j z^j : \operatorname{Re} c_0 = 0, c_j = -\bar{c}_{-j}, \ j = 1, 2, \cdots, k \right\}$$

as in [13]. If $k < 0$, we assume $S\Pi_k = \{0\}$.

Theorem 3.1. *The simplest Hilbert BVP (10) is solvable and the set of solutions is* $S\Pi_m \oplus (\bar{z}+z)S\Pi_{m-1} \oplus \cdots \oplus (\bar{z}+z)^{n-1} S\Pi_{m-n+1}$ *where*

$$(\bar{z}+z)^j S\Pi_{m-j} = \{(\bar{z}+z)^j q_{m-j} : q_{m-j} \in S\Pi_{m-j}\}, \ j = 0, 1, \cdots, n-1.$$

Proof. Let $V(z) = V_0(z) + (\bar{z}+z)V_1(z) + \cdots + (\bar{z}+z)^{n-1}V_{n-1}(z)$ with $V_j \in H_1(\mathbb{D} \setminus \{0\})$ for $j = 0, 1, \cdots, n-1$. By Corollary 2.2, $\operatorname{Ord}(V, 0) \leq m$ is equivalent to

$$\operatorname{Ord}(V_j, 0) \leq m - j, \ j = 0, 1, \cdots, n-1. \tag{11}$$

On the other hand, the boundary conditions in (10) are equivalent to

$$\mathrm{Re}V_j^+(t) = 0,\ t \in \partial\mathbb{D}\ j = 0, 1, \cdots, n-1. \tag{12}$$

By [8] we know (11) and (12) imply $V_j \in S\Pi_{m-j},\ j = 0, 1, \cdots, n-1.$ □

Next we consider the homogeneous Hilbert type BVP of (7) as follows

$$\mathrm{Re}\left\{[a(t) + ib(t)]\left[\frac{\partial^j V}{\partial \bar{z}^j}\right]^+(t)\right\} = 0,\ t \in \partial\mathbb{D},\ j = 0, 1, \cdots, n-1. \tag{13}$$

For convenience we introduce the symbol

$$\mathfrak{S}_m = S\Pi_m \oplus (\bar{z} + z)S\Pi_{m-1} \oplus \cdots \oplus (\bar{z} + z)^{n-1} S\Pi_{m-n+1}. \tag{14}$$

Theorem 3.2. *The homogeneous Hilbert type BVP (13) is solvable and its solution is*

$$V(z) = iX(z)q_\kappa(z), \tag{15}$$

where X is defined by (8) and $q_\kappa \in \mathfrak{S}_\kappa$.

Proof. By Property 1, 2 and 3 of X defined by (8), the boundary conditions (13) are equivalent to

$$\mathrm{Re}\left\{\left[\frac{\partial^j}{\partial \bar{z}^j}\left(\frac{V}{iX}\right)\right]^+(t)\right\} = 0,\ t \in \partial\mathbb{D},\ j = 0, 1, \cdots, n-1. \tag{16}$$

By Property 4 of X we know $\mathrm{Ord}(\frac{V}{iX}, 0) \leq \mathrm{Ord}(\frac{1}{X}, 0) = \kappa$. And hence Theorem 3.1 implies

$$\frac{V}{iX} = q_\kappa \in \mathfrak{S}_\kappa,$$

which leads to (15). □

Introduce the poly-Schwarz operator

$$S[\gamma_0, \cdots, \gamma_{n-1}](z) = \sum_{k=0}^{n-1} \frac{(-1)^k}{k!} \frac{1}{2\pi i} \int_{\partial\mathbb{D}} \gamma_k(t)(t - z + \overline{t - z})^k \frac{t+z}{t-z} \frac{dt}{t} \tag{17}$$

for $z \in \mathbb{D}$, where the kernel functions $\gamma_k \in C(\partial\mathbb{D})$, $k = 0, 1, \cdots, n-1$. According to Theorem 3.4 in [1],

$$\mathrm{Re}\left\{\frac{\partial^j}{\partial \bar{z}^j} S[\gamma_0, \cdots, \gamma_{n-1}]\right\}^+(t) = \gamma_j(t),\ t \in \partial\mathbb{D},\ j = 0, 1, \cdots, n-1. \tag{18}$$

Now the boundary conditions (7) are equivalent to

$$\operatorname{Re}\left\{\left[\frac{\partial^j}{\partial \bar{z}^j}\left(\frac{V}{iX}\right)\right]^+(t)\right\} = \frac{\gamma_j(t)}{i[a(t)+ib(t)]X^+(t)}, \quad t \in \partial\mathbb{D}, \tag{19}$$

where $j = 0, 1, \cdots, n-1$. And hence

$$\widetilde{V}(z) = iX(z) \cdot S\left[\frac{\gamma_0}{i(a+ib)X^+}, \cdots, \frac{\gamma_{n-1}}{i(a+ib)X^+}\right](z), \ z \in \mathbb{D} \tag{20}$$

is the special solution of BVP (7). Therefore (19) and (20) give

$$\operatorname{Re}\left\{\left[\frac{\partial^j}{\partial \bar{z}^j}\left(\frac{V-\widetilde{V}}{iX}\right)\right]^+(t)\right\} = 0, \ t \in \partial\mathbb{D}, \ j = 0, 1, \cdots, n-1. \tag{21}$$

When $\kappa \geq 0$, by Theorem 3.2, the solution of BVP (7) may be written as

$$V(z) = \widetilde{V}(z) + iX(z)q_\kappa(z), \ q_\kappa \in \mathfrak{S}_\kappa. \tag{22}$$

When $\kappa < 0$, if the Hilbert type problem (7) is solvable then Theorem 3.2 imply that its solution may be represented as (20) or say

$$V(z) = \sum_{k=0}^{n-1} \frac{(-1)^k}{k!} \frac{X(z)}{2\pi i} \int_{\partial\mathbb{D}} \frac{\gamma_k(t)(t-z+\overline{t-z})^k}{[a(t)+ib(t)]X^+(t)} \frac{t+z}{t-z} \frac{dt}{t} \tag{23}$$

for $z \in \mathbb{D}$. Obviously $V(z)$ defined by (23) is the solution of (7) if and only if $\lim_{z\to 0} V(z)$ exists. If $\lim_{z\to 0} V(z)$ exists, then one has

$$\operatorname{Ord}(V_j, 0) \leq -\kappa - j, \ j = 0, 1, \cdots, n-1 \tag{24}$$

with

$$V_j(z) = \sum_{k=j}^{n-1} \frac{(-1)^k C_k^j}{k!} \cdot \frac{X(z)}{2\pi i} \int_{\partial\mathbb{D}} \frac{\gamma_k(t)(\bar{t}+t)^{k-j}}{[a(t)+ib(t)]X^+(t)} \frac{t+z}{t-z} \frac{dt}{t} \tag{25}$$

for $j = 0, 1, \cdots, n-1$. Observe, at the neighborhood of $z = 0$,

$$\frac{t+z}{t-z} = 1 + 2\sum_{\ell=1}^{+\infty}\left(\frac{z}{t}\right)^\ell = 1 + 2\sum_{\ell=1}^{-\kappa-j-1}\left(\frac{z}{t}\right)^\ell + \frac{z^{-\kappa-j}t^{\kappa+j+1}}{t-z},$$

one has

$$\sum_{k=j}^{n-1} \frac{(-1)^k C_k^j}{k!} \cdot \frac{1}{2\pi i} \int_{\partial\mathbb{D}} \frac{\gamma_k(t)(\bar{t}+t)^{k-j}}{[a(t)+ib(t)]X^+(t)} \frac{dt}{t^{\ell+1}} = 0 \tag{26}$$

for $\ell = 0, 1, \cdots, -\kappa - j - 1$ and $j = 0, 1, \cdots, n-1$. Under the conditions (26), (23) may be rewritten as

$$V(z) = \sum_{k=1}^{n-1} (\bar{z}+z)^j (-1)^j V_j(z) \tag{27}$$

with

$$
\begin{aligned}
V_j(z) =&z^{-j}\sum_{k=j}^{n-1}\frac{(-1)^k C_k^j}{k!}\cdot\frac{\exp\{iS[\alpha](z)\}}{2\pi i}\\
&\int_{\partial\mathbb{D}}\frac{\gamma_k(t)(\bar{t}+t)^{k-j}}{[a(t)+ib(t)]\exp\{iS^+[\alpha](t)\}}\frac{t^j dt}{t-z},
\end{aligned}
\tag{28}
$$

for $j = 0, 1, \cdots, n-1$.

Let $z = re^{i\theta}$, $\theta \in [0, 2\pi]$. By (27) and (28), one gets

$$
\lim_{r\to 0} V(z) = \sum_{j=1}^{n-1} c_j(1+e^{-2i\theta})^j \tag{29}
$$

with

$$
c_j = (-1)^j\sum_{k=j}^{n-1}\frac{(-1)^k C_k^j}{k!}\frac{\exp\{iS[\alpha](0)\}}{2\pi i}\int_{\partial\mathbb{D}}\frac{\gamma_k(t)(\bar{t}+t)^{k-j}t^{j-1}dt}{[a(t)+ib(t)]\exp\{iS^+[\alpha](t)\}} \tag{30}
$$

for $j = 0, 1, \cdots, n-1$.

Lemma 3.1. *Let c_j $(j = 0, 1, \cdots, n-1)$ and c be complex numbers. Then $\sum\limits_{k=j}^{n-1} c_j(1+e^{-2i\theta})^j \equiv c$ for all $\theta \in [0, 2\pi]$ if and only if $c_0 = c$, $c_j = 0$ for $j = 0, 1, \cdots, n \quad 1$.*

Proof. If $\sum_{k=1}^{n-1} c_j(1+e^{-2i\theta})^j \equiv c$ for $\forall\theta \in [0, 2\pi]$, let $\theta = \frac{\pi}{2}$, then $c_0 = c$, and hence $\sum_{k=1}^{n-1} c_j(1+e^{-2i\theta})^j \equiv 0$, which implies $\sum_{k=1}^{n-1} c_j(1+e^{-2i\theta})^{j-1} \equiv 0$. Therefore $c_1 = 0$ similarly. If $c_1 = \cdots = c_k = 0$ then $\sum_{j=k+1}^{n-1} c_j(1+e^{-2i\theta})^j \equiv 0, \forall\theta$ which leads to $\sum_{j=k+1}^{n-1} c_j(1+e^{-2i\theta})^{j-k-1} \equiv 0, \forall\theta$. Hencc $c_{k+1} = 0$. By the principle of induction, we know $c_j = 0$ $(j = 0, 1, \cdots, n-1)$. The converse is obvious. □

By Lemma 3.1, if $V(z)$ is the solution of BVP (7) then $c_j = 0$ $(j = 0, 1, \cdots, n-1)$ or say

$$
\sum_{k=j}^{n-1}\frac{(-1)^k C_k^j}{k!}\cdot\frac{1}{2\pi i}\int_{\partial\mathbb{D}}\frac{\gamma_k(t)(\bar{t}+t)^{k-j}t^{j-1}}{[a(t)+ib(t)]\exp\{iS^+[\alpha](t)\}}dt = 0 \tag{31}
$$

for $j = 0, 1, \cdots, n-1$. Combining (26) and (31) we have

$$
\sum_{k=j}^{n-1}\frac{(-1)^k C_k^j}{k!}\cdot\frac{1}{2\pi i}\int_{\partial\mathbb{D}}\frac{\gamma_k(t)(\bar{t}+t)^{k-j}}{[a(t)+ib(t)]X^+(t)}\frac{dt}{t^{\ell+1}} = 0 \tag{32}
$$

for $\ell = 0, 1, \cdots, -\kappa - j$ and $j = 0, 1, \cdots, n-1$. Conversely, when (32) is fulfilled, BVP (7) is solvable and its solution is (23).

In general we obtain the following result.

Theorem 3.3. *The Hilbert-type BVP (7) is solvable if and only if the conditions (32) are fulfilled, and its solution may be written as (22).*

References

1. H. Begehr and D. Schmersau, The Schwarz problem for polyanalytic functions, ZAA, 2005, **24**(2), 341–351.
2. J. Du and Y. Wang, On boundary value problem of polyanalytic function on the real axis, Complex Variables, 2003, **48**(6), 527–542.
3. M. B. Balk, Polyanalytic Functions, Akademie Verlag, Berlin, 1991.
4. H. Begehr, Complex Analytic Methods for Partial Differential Equations: An Introductory Text, World Scientific, Singapore, 1994.
5. H. Begehr, Boundary value problems in complex analysis I. Bol. Asoc. Mat. Venez., 2005, **12**, 65–85; II. Bol. Asoc. Mat. Venez,. 2005, **12**, 217–250.
6. H. Begehr and G. N. Hile, A hierarchy of integral operators, Rocky Mountain J. Math., 1997, **27**, 669–706.
7. H. Begehr and A. Kumar, Boundary value problem for inhomogeneous polyanalytic equation, Analysis, 2005, **25**, 55–71.
8. J. Du and Y. Wang, Riemann boundary value problem of polyanalytic function and metaanalytic function on the closed curve, Complex Variables, 2005, **50**, 521–533.
9. B. F. Fatulaev. The main Haseman type boundary value problem for metaanalytic function in the case of circular domains, Math. Model. Anal., 2001, **6**, 68–76.
10. J. Lu, Boundary Value Problems for Analytic Functions, World Scientific, Singapore, 1993.
11. N. I. Muskhelishvili, Singular Integral Equations, 2nd ed., Noordhoff, Groningen, 1968.
12. Y. Wang and J. Du, On Riemann boundary value problem of polyanalytic function on the real axis, Acta. Math. Sci., 2004, **24B**, 663–671.
13. Y. Wang and J. Du, Hilbert boundary value problem of polyanalytic function on the unit circumference, Complex Var. Elliptic Equ., 2006, **51**, 923–943.
14. H. Begehr, J. Du and Y. Wang, A Dirichlet problem for polyharmonic functions, Ann. Mat. Pura Appl., to appear.

EXISTENCE AND UNIQUENESS OF THE SOLUTION OF A FREE BOUNDARY PROBLEM FOR A PARABOLIC COMPLEX EQUATION

YONGZHI XU

Department of Mathematics, University of Louisville, Louisville, KY 40292
E-mail: ysxu0001@louisville.edu

We present a proof of the existence and uniqueness of the solution of a free boundary problem for a nonlinear parabolic complex equation of second order in a simply connected region in the complex plane with Dirichlet boundary condition. A more general problem is discussed in [15], but the uniqueness is not proved. The problem is originally motivated by a free boundary problem model of breast cancer. In this paper we convert the free boundary problem to a sequence of boundary value problems of nonlinear parabolic complex equations of second order. Then we use iteration method to construct its solution and prove its existence and uniqueness.

Keywords: Free boundary problem, complex parabolic equation, existence and uniqueness.

1. The free boundary problem

In [15] we considered a free boundary value problem for a parabolic complex equation. Let D_t be a simply connected bounded domain in the $z = x + iy$ plane C with the boundary Γ_t for each $t \geq 0$. $D_0 = \{z||z| < 1\}$ with boundary $\Gamma_0 = \{z||z| = 1\}$. Denote $D = \{(z,t)|z \in D_t, 0 < t \leq T\}$ for some $T > 0$. $\partial D = \partial D^1 \cup \partial D^2$ is the parabolic boundary of D, where ∂D^1 is the bottom of D and ∂D^2 is the side boundary of D, respectively.

Let $\zeta = \varphi(z,t)$ be the bijective comformal map that maps the bounded simply connected region $\overline{D_t}$ onto the closed unit disk $\overline{D_0} = \{\zeta||\zeta| \leq 1\}$ such that $\varphi(0,t) = 0$, $\varphi'(0,t) = 1$, and $z = \psi(\zeta,t)$ is its inverse function. Then $\Gamma_t = \{z|z = \psi(e^{i\theta},t), 0 \leq \theta \leq 2\pi\}$.

Let $u = u(z,t)$ be a function definded in D, and satisfy the nonlinear parabolic complex equation and initial and free boundary value conditions

$$u_t - A_0 u_{z\bar{z}} + Re[A_1 u_z] + A_2 u = A_3, \tag{1}$$

$$u(z,0) = g(z),\ z \in D_0, \tag{2}$$

$$a_1(z,t)\frac{\partial u}{\partial \nu} + a_2(z,t)u = a_3(z,t),\ z \in \Gamma_t,\ t > 0, \tag{3}$$

and the free boundary Γ_t satisfies the condition

$$\frac{\partial Re\{\psi(e^{i\theta},t)\}}{\partial t} = Re\left[\int_{C_t} \mu(u(z,t),t) - \hat{u})dz\right],\ t > 0, \tag{4}$$

where $C_t = \{z|z = \rho\psi(e^{i\theta},t), 0 \le \rho \le 1\}$ is the line segment connecting $z = 0$ and $z = \psi(e^{i\theta},t)$. Since $\Gamma_0 = \{z||z| = 1\}$, we have

$$\psi(\zeta,0) = \zeta.$$

(3) may be written in the complex form

$$2Re[a_1(z,t)\overline{\lambda(z,t)}u_z] + a_2(z,t)u = a_3(z,t),\ z \in \Gamma_t,\ t > 0. \tag{5}$$

Here $z = x + iy$, $A_0 = A_0(z,t,u,u_z,u_{z\bar{z}})$, $A_1 = A_1(z,t,u,u_z)$, $A_2 = A_2(z,t,u)$, and $A_3 = A_3(z,t)$; μ and $\hat{u}$ are known real constants; ν is the unit vector at every point on Γ_t, and $g(z)$, $a_j(z,t)$ $(j = 1,2,3)$, $\lambda(z,t) = \cos(\nu,x) - i\cos(\nu,y)$ are known functions.

2. Preliminary results

2.1. *Initial mixed boundary value problem*

In [10] the following conditions are introduced:

(a) $A_0(z,t,u,u_z,u_{z\bar{z}})$, $A_1(z,t,u,u_z)$, $A_2(z,t,u)$, and $A_3(z,t)$, are measurable for any continuously differentiable function $u(z,t)$ and measurable functions $u_{z\bar{z}} \in L^2(D)$ and satisfy the conditions

$$0 < \delta \le A_0 \le \delta^{-1},\ |A_j| \le k_0,\ j = 1,2,\ L_p[A_3,\bar{D}] \le k_1,\ p > 4. \tag{6}$$

(b) The above functions with respect to $u \in \mathbf{R}$, $u_z \in \mathbf{C}$ are continuous for almost every point $(z,t) \in D$ and $u_{z\bar{z}} \in \mathbf{R}$.

(c) Let $F(z,t,u,u_z,u_{z\bar{z}}) = A_0u_{z\bar{z}} - Re[A_1u_z] - A_2u - A_3$. For almost every point $(z,t) \in D$ and $u \in \mathbf{R}$, $u_z \in \mathbf{C}$, $V \in \mathbf{R}$, there is

$$F(z,t,u,u_z,V^1) - F(z,t,u,u_z,V^2) = \tilde{A}_0(V^1 - V^2), \tag{7}$$

$$\delta < \tilde{A}_0 \le \delta^{-1}. \tag{8}$$

(d) For any $u \in \mathbf{R}, u_z \in \mathbf{C}$, $V \in \mathbf{R}$, there is

$$\begin{aligned} &F(z,t,u^1,u_z^1,V) - F(z,t,u^2,u_z^2,V) \\ &= Re[\tilde{A}_1(u^1 - u^2)_z] + \tilde{A}_2(u^1 - u^2) - (u^1 - u^2)_t \text{ on } \bar{D}, \end{aligned} \tag{9}$$

where $\tilde{A}_j$ $(j = 1, 2)$ satisfy

$$|\tilde{A}_j| \leq k_0, \text{ in } \mathbf{D}.$$

Here $\delta > 0$, $q_0 < 1$, $q_1 > 0$, k_0, k_1 are nonnegative constants. Following Wen [10], the conditions (a)-(c) are called condition C, and the conditions (a)-(d) the condition C'.

We also assume that $g(z)$, $a_j(z,t)$ $(j = 1, 2, 3)$, $\lambda(z,t) = \cos(\nu, x) - i\cos(\nu, y)$ satisfy the conditions

$$C^2_\alpha[g, D_0] \leq k_2, \ a_1(z,0)\frac{\partial g}{\partial \nu} + a_2(z,0)g = a_3(z,0), \ z \in \Gamma_0, \tag{10}$$

$$C^{2,1}_{\alpha,\alpha/2}[\eta, \partial D^2] \leq k_0, \ \eta = (\lambda, a_1, a_2), \ C^{2,1}_{\alpha,\alpha/2}[a_3, \partial D^2] \leq k_2, \tag{11}$$

$$a_j(z,t) \geq 0, \ j = 1, 2, \ a_1 + a_2 \geq 1, \ \cos(\nu, n) \geq \eta > 0 \text{ on } \partial D^2. \tag{12}$$

If $u = u(z,t)$ is a function definded in the unit cylinder $D_0 \times [0, T]$ and satisfies the nonlinear parabolic complex equation and mixed boundary value conditions

$$u_t - A_0 u_{z\bar{z}} + Re[A_1 u_z] + A_2 u = A_3, \text{ in } \partial D_0 \times [0, T], \tag{13}$$

$$u(z, 0) = g(z), \ z \in D_0, \tag{14}$$

$$2Re[a_1(z,t)\overline{\lambda(z,t)}u_z] + a_2(z,t)u = a_3(z,t), \ z \in \partial D_0 \times [0, T]. \tag{15}$$

We call this problem M. This is a special case of the complex parabolic equation considered by Wen in [10]. More precisely in this paper we need only the results in a simple connected domain, and for the case that $Re\{Qu_{zz}\} = 0$, while [10] presented results in a multiple-connected region with the term $Re\{Qu_{zz}\}$. The conditions C and C' presented here are stronger than that in [10]. The following results are special cases of that in [10].

Lemma 2.1. *Let equation (1) satisfy condition C. Then any continuous solution $u(z,t)$ of problem M for (1) satisfies the estimate*

$$C[u, \overline{D}] \leq M_1 = M_1(\delta, \alpha, k, p, D), \tag{16}$$

where $k = k(k_0, k_1, k_2)$, M_1 is a nonnegative constant only dependent on δ, α, k, p, D.

Lemma 2.2. *Under the same condition as in Theorem 2.1, any solution $u(z,t)$ of problem M for (1) satisfies the estimates*

$$C^{1,0}_{\beta,\beta/2}[u, \overline{D^*}] \leq M_2, \ \|u\|_{W^{2,1}_2(D^*)} \leq M_3, \tag{17}$$

where $D^* = \overline{D} \cap \{\cap_{(z^*,t^*)\in\partial D^2}[|z-z^*|^2+|t-t^*|] \geq \epsilon > 0\}$, ϵ *is a small positive number, and* $k=(k_0,k_1,k_2)$, $M_j = M_j(\delta,\alpha,k,p,D)$, $(j=2,3)$ *are nonnegative constants.*

Lemma 2.3. *Under condition* C, *problem* M *for (1) has a solution* $u(z,t)$.

Lemma 2.4. *Under condition* C', *the solution* $u(z,t)$ *of problem* M *for (1) is unique.*

2.2. *The equivalent coupled boundary value problem*

In [15] we proved that the free boundary value problem (1) (2) (4) (5) is equivalent to an initial-boundary value problem of coupled nonlinear differential and integral equations

$$v_t - A_0|\varphi_z|^2 v_{\zeta\bar{\zeta}} + Re[(A_1\varphi_z + \varphi_t)v_\zeta] + A_2 v = A_3, \tag{18}$$

$$v(\zeta,0) = g(\psi(\zeta,0)),\ \zeta \in D_0, \tag{19}$$

$$2Re[a_1(\psi(\zeta,t),t)\overline{\lambda(\psi(\zeta,t),t)\varphi_z v_\zeta}] + a_2(\psi(\zeta,t),t)v = a_3(\psi(\zeta,t),t),\quad \zeta\in\Gamma_0,\ t>0, \tag{20}$$

$$Re[\psi(e^{i\theta},t)] = Re[\psi(e^{i\theta},0)]e^{\int_0^t\int_0^1 \mu[v(\rho e^{i\theta},\tau)-\hat{u}]d\rho d\tau},\ t>0,\ 0\leq\theta\leq 2\pi, \tag{21}$$

where $v(\zeta,t) = u(\psi(\zeta,t),t)$, $\zeta\in D_0$, $t\geq 0$.

The folowing lemmas are proved in [15].

Lemma 2.5. *Under condition* C', *and if* $\psi(e^{i\theta},0)=e^{i\theta}$, *then*

(i) There is $M_1>0$ *such that* $0\leq u(z,t)\leq M_1$, *for* $(z,t)\in D$.

(ii) $e^{-\mu\tilde{u}T} \leq |Re[\psi(e^{i\theta},t)]| \leq e^{\mu(M_1-\tilde{u})T}$, $0\leq\theta\leq 2\pi$, $0\leq t\leq T$.

(iii) $-\mu\tilde{u}|Re[\psi(e^{i\theta},t)]| \leq \left|\frac{\partial Re[\psi(e^{i\theta},t)]}{\partial t}\right| \leq \mu(M_1-\tilde{u})|Re[\psi(e^{i\theta},t)]|$, $0\leq\theta\leq 2\pi$, $0\leq t\leq T$.

Lemma 2.6. *For given* $Re[\psi(e^{i\theta},t)]$ *and* $\psi(0,t)=0$ *for* $0\leq\theta\leq 2\pi$ *and* $t\geq 0$, $\psi(\zeta,t)$ *is determined uniquely in the unit disk* D_0,

$$\psi(\zeta,t) = \frac{1}{2\pi}\int_0^{2\pi} Re[\psi(e^{i\theta},t)]\frac{e^{i\theta}+\zeta}{e^{i\theta}-\zeta}d\theta,\ \text{for } \zeta\in D_0,\ t\geq 0. \tag{22}$$

Moreover, if

$$\psi(e^{i\theta_0},t) = PV\frac{1}{2\pi}\int_0^{2\pi} Re[\psi(e^{i\theta},t)]\frac{e^{i\theta}+e^{i\theta_0}}{e^{i\theta}-e^{i\theta_0}}d\theta,\ \text{for } 0\leq\theta_0\leq 2\pi,\ t\geq 0, \tag{23}$$

where PV *stands for Cauchy principal value, then* $\lim_{\zeta\to e^{i\theta_0}}\psi(\zeta,t) = \psi(e^{i\theta_0},t)$, *and for some constant* $T>0$, $M=M(T)>0$,

$$|\psi(e^{i\theta_0},t)| \leq M(T)|Re[\psi(e^{i\theta_0},t)]|,\ \text{for } 0\leq t\leq T. \tag{24}$$

Lemma 2.7. *Let the mixed problem (1), (2) and (4) satisfy the condition C' and the coefficients in problem (18)-(20) are corresponding coefficients of the Mixed problem (1), (2) and (4). If $\psi(\zeta, t)$ is known, then the Mixed problem (18)-(20) has a unique solution.*

2.3. *The iterated boundary value problem*

We use an iterative algorithm to construct a solution of the (18)-(21). Let $\psi^0(\zeta, t) = \zeta$ for $\zeta \in D_0 = \{\zeta||\zeta| < 1\}$. v^{n+1}, $(n = 0, 1, 2, ...)$ is the solution of the mixed problem on $D_0 \times [0, T]$

$$v_t^{n+1} - A_0^n|\varphi_z^n|^2 v_{\zeta\bar{\zeta}}^{n+1} + Re[(A_1^n\varphi_z^n + \varphi_t^n)v_\zeta^{n+1}] + A_2^n v^{n+1} = A_3^n, \tag{25}$$

$$v^{n+1}(\zeta, 0) = g(\psi^n(\zeta, 0)),\ \zeta \in D_0, \tag{26}$$

$$\begin{aligned} 2Re[a_1(\psi^n(\zeta,t),t)\overline{\lambda(\psi^n(\zeta,t),t)}\varphi_z^n v_\zeta^{n+1}] + a_2(\psi^n(\zeta,t),t)v^{n+1} \\ = a_3(\psi^n(\zeta,t),t), \quad \zeta \in \Gamma_0,\ t > 0, \end{aligned} \tag{27}$$

where $A_0^n = A_0(z, t, u^n, u_z^n, u_{z\bar{z}}^n)$, $A_1^n = A_1(z, t, u^n, u_z^n)$, $A_2^n = A_2(z, t, u^n)$, and $A_3^n = A_3(z, t)$; $\psi^n(\zeta, t)$, $(n = 1, 2, ...)$ are analytic functions in the unit disk for $t \geq 0$, satisfying the boundary condition

$$\begin{aligned} Re[\psi^n(e^{i\theta}, t)] = Re[\psi^n(e^{i\theta}, 0)]e^{\int_0^t \int_0^1 \mu[v^n(\rho e^{i\theta}, \tau) - \hat{u}]d\rho d\tau}, \\ t > 0,\ 0 \leq \theta \leq 2\pi,\ n = 1, 2, 3, \cdots. \end{aligned} \tag{28}$$

In view of Lemmas 2.1–2.4, we have the following.

Theorem 2.1. *Under the condition C', for each interger $n \geq 0$, there exists a unique solution (v^{n+1}, ψ^{n+1}) to the problem (25)-(28). Moreover, we have the estimates*

$$C[v^n, \overline{D}] \leq M_1 = M_1(\delta, \alpha, k, p, D), \tag{29}$$

where $k = k(k_0, k_1, k_2)$, M_1 is a nonnegative constant only dependent on δ, α, k, p, D.

$$C^{1,0}_{\alpha,\alpha/2}[v^n, \overline{D^*}] \leq M_2,\ \|v^n\|_{W_2^{2,1}(D^*)} \leq M_3, \tag{30}$$

where $D^* = \overline{D} \cap \{\cap_{(z^*,t^*)\in\partial D^2}[|z - z^*|^2 + |t - t^*|] \geq \epsilon > 0\}$, ϵ is a small positive number, and $k = (k_0, k_1, k_2)$, $M_j = M_j(\delta, \alpha, k, p, D)$, $(j = 2, 3)$ are nonnegative constants.

3. Existence and uniqueness

In this section we outline the proof of the existence and uniquenesse of the solution of an equivalent coupled boundary value problem for the case that $a_1(z,t) = 0$, $a_2(z,t) = 1$ and $a_3(z,t) = a_3 = constant$. The proof is inspired by [9, Chapter 6]. The general mixed boundary condition case is still an open problem.

First we show that the sequence $\{v^n, \psi^n\}$ converges to a limit $\{v, \psi\}$, where $v \in C^{1/2-\delta}\left([0,T], C^1(D)\right)$ for $\delta > 0$, and $\psi \in C^{3/2-\delta}([0,T])$ for $\delta > 0$.

In view of Sobolev embedding theorem, Theorem 2.1 and Lemmas 2.5, 2.6 imply the following.

Theorem 3.1. *Under the assumptions of Theorem 2.1, we have*
(i) $\{v^n : n = 0, 1, ...\}$ *is compact in* $C^{1/2-\delta}\left([0,T], C^1(D)\right)$, *for* $\delta > 0$;
(ii) $\{\psi^n : n = 0, 1, ...\}$ *is compact in* $C^{3/2-\delta}([0,T])$, *for* $\delta > 0$.

Therefore $\{v^n : n = 0, 1, ...\}$ and $\{\psi^n : n = 0, 1, ...\}$ have limit points v and ψ.

We need to show that v, s are unique and satisfy the free boundary problem. If (v_1^n, ψ_1^n) and (v_2^n, ψ_2^n) are two sequences, then for $i = 1, 2$,

$$v_{i,t}^{n+1} - A_{i,0}^n|\varphi_{i,z}^n|^2 v_{i,\zeta\bar{\zeta}}^{n+1} + Re[(A_{i,1}^n\varphi_{i,z}^n + \varphi_{i,t}^n)v_{i,\zeta}^{n+1}] + A_{i,2}^n v_i^{n+1} = A_{i,3}^n, \quad (31)$$

$$v_i^{n+1}(\zeta, 0) = g(\psi_i^n(\zeta, 0)),\ \zeta \in D_0, \quad (32)$$

$$v_i^{n+1} = a_3,\ \zeta \in \Gamma_0,\ t > 0, \quad (33)$$

$$Re[\psi_i^n(e^{i\theta}, t)] = Re[\psi_i^n(e^{i\theta}, 0)]e^{\int_0^t\int_0^1 \mu[v_i^n(\rho e^{i\theta},\tau) - \hat{u}]d\rho d\tau},$$
$$t > 0,\ 0 \le \theta \le 2\pi,\ n = 1, 2, 3, \cdots. \quad (34)$$

Let $v^{n+1} = v_1^{n+1} - v_2^{n+1}$ and $\psi^{n+1} = \psi_1^{n+1} - \psi_2^{n+1}$. Then

$$v_t^{n+1} - A_{1,0}^n|\varphi_{1,z}^n|^2 v_{\zeta\bar{\zeta}}^{n+1} + Re[(A_{1,1}^n\varphi_{1,z}^n + \varphi_{1,t}^n)v_{\zeta}^{n+1}] + A_{1,2}^n v^{n+1} = F^n, \quad (35)$$

$$v^{n+1}(\zeta, 0) = 0,\ \zeta \in D_0, \quad (36)$$

$$v^{n+1} = 0,\ \zeta \in \Gamma_0,\ t > 0, \quad (37)$$

$$Re[\psi^n(e^{i\theta}, t)] = Re[\psi^n(e^{i\theta}, 0)]e^{\int_0^t\int_0^1 \mu[v_1^n(\rho e^{i\theta},\tau) - \hat{u}]d\rho d\tau}$$
$$\cdot\left[1 - e^{-\int_0^t\int_0^1 \mu[v^n(\rho e^{i\theta},\tau)]d\rho d\tau}\right] \quad t > 0,\ 0 \le \theta \le 2\pi,\ n = 1, 2, 3, \cdots, \quad (38)$$

where $F^n = [A_{2,3}^n - A_{1,3}^n] + \left[A_{2,0}^n|\varphi_{2,z}^n|^2 - A_{1,0}^n|\varphi_{1,z}^n|^2\right] v_{2,\zeta\bar{\zeta}}^{n+1}$

$$+Re\left\{[(A_{2,1}^n\varphi_{2,z}^n + \varphi_{2,t}^n) - (A_{1,1}^n\varphi_{1,z}^n + \varphi_{1,t}^n)]v_{2,\zeta}^{n+1}\right\} + \left[A_{2,2}^n - A_{1,2}^n\right] v_2^{n+1}. \quad (39)$$

In view of Lemma 2.5 (i) for v_i^{n+1}, we know that there is a constant $C > 0$, such that $\|v_i^{n+1}\|_\infty \le C$ and $\|v^{n+1}\|_\infty = \|v_1^{n+1} - v_2^{n+1}\|_\infty \le C$. (We will use C for a general constant that may take different values without mention each time.) From (38) and (24),

$$|\psi^{n+1}(\zeta,t)| \le Ct\|v^{n+1}\|_{L^\infty(D\times[0,t])},\quad 0 \le t \le T. \tag{40}$$

It follows, in view of condition C', that

$$\begin{aligned}|F^n| &\le |A_{2,3}^n - A_{1,3}^n| + \left|A_{2,0}^n|\varphi_{2,z}^n|^2 - A_{1,0}^n|\varphi_{1,z}^n|^2\right| \|v_2^{n+1}\|_{H^2}\\ &+|Re[(A_{2,1}^n\varphi_{2,z}^n+\varphi_{2,t}^n)-(A_{1,1}^n\varphi_{1,z}^n+\varphi_{1,t}^n)]|\|v_2^{n+1}\|_{H^1} + |A_{2,2}^n-A_{1,2}^n|\|v_2^{n+1}\|_{L^\infty}\\ &\le |A_{2,3}^n - A_{1,3}^n| + \left|A_{2,0}^n|\varphi_{2,z}^n|^2 - A_{1,0}^n|\varphi_{1,z}^n|^2\right| \|v_2^{n+1}\|_{H^2}\\ &+|(A_{2,1}^n\varphi_{2,z}^n + \varphi_{2,t}^n) - (A_{1,1}^n\varphi_{1,z}^n + \varphi_{1,t}^n)|\|v_2^{n+1}\|_{H^1} + |A_{2,2}^n - A_{1,2}^n|\|v_2^{n+1}\|_{L^\infty}\\ &\le C_1|\psi(\zeta,t)| \le Ct\|v^{n+1}\|_{L^\infty}.\end{aligned} \tag{41}$$

Note that v^{n+1}, as a solution of the linear parabolic equation (35), has integral representation

$$\begin{aligned}v^{n+1}(\cdot,t) = &-\int_0^t \left[e^{\alpha(t,\tau)L}\left(Re[(A_{1,1}^n\varphi_{1,z}^n + \varphi_{1,t}^n)\right) + \partial_x e_N^{\alpha(t,\tau)L}\left(A_{1,2}^n\right)\right]\\ &v^{n+1}(\cdot,\tau)d\tau + \int_0^t e^{\alpha(t,\tau)L}F^{n+1}d\tau,\end{aligned} \tag{42}$$

where $e^{\alpha(t,\tau)L}$ is the solution operator of the heat operator

$$Lv^{n+1} = v_t^{n+1} - A_{1,0}^n|\varphi_{1,z}^n|^2 v_{\zeta\bar\zeta}^{n+1},$$

with vanishing Dirichlet boundary condition, and $e_N^{\alpha(t,\tau)}$ is the solution operator for the vanishing Neumann boundary condition. So

$$\begin{aligned}\|v^{n+1}(t)\|_{L^\infty(D)} \le M_1 &\int_0^t \tau^{-\frac{r}{2}}\|v^{n+1}\|_{L^\infty(D\times[0,t])}d\tau\\ &+ M_2\int_0^t \tau^{-\frac{r}{2}}\tau\|v^{n+1}\|_{L^\infty(D\times[0,t])}d\tau\end{aligned} \tag{43}$$

For $r < 2$ and $t > 0$ small enough,

$$\|v^{n+1}\|_{L^\infty(D\times[0,t])} \le Ct^{\frac{2-r}{2}}\|v^{n+1}\|_{L^\infty(D\times[0,t])} < \|v^{n+1}\|_{L^\infty(D\times[0,t])}.$$

Since the estimates (40)-(43) are true for $0 \le t \le T$, we can repeat the above argument and conclude that $v_1(\cdot,t) = v_2(\cdot,t)$ in $[0,T]$. This finishes the uniqueness proof.

It must be that $v^{n+1}(\cdot,t) \to 0$ and $\psi^{n+1}(t) \to 0$ in $[0,t]$. Therefore, the limit points v, ψ are unique and satisfy the equivalent coupled boundary value problem (18)-(21). It proves the local existence. But the a priori estimates in Lemma 2.5 hold for $[0, T]$, so by repeating the process we obtain the global existence.

References

1. H. Begehr and R. Gilbert, *Transformations, Transmutations and Kernel Functions*, Longman Scientific & Technical, New York, 1992.
2. H. M. Byrne and M. A. J. Chaplain, Growth of nonnecrotic tumours in the presence and absence of inhibitors, *Math. Biosciences*, 130 (1995), 151-181.
3. J. R. Cannon and J. van der Hoek, Diffusion subject to the specification of mass, *J. Math. Anal. Appl.*, 115, no. 2 (1986), 517-529.
4. Y. S. Choi and K.-Y. Chan, A parabolic equation with nonlocal boundary conditions arising from electrochemistry, *Nonlinear Anal. TMA*, 18, no. 4 (1992), 317-331.
5. A. Friedman and F. Reitich, Analysis of a mathematical model for the growth of tumours, *J. Math. Biol.*, 38 (1999), 262-284.
6. R. Gilbert, *Function Theoretic Methods in Partial Differential Equations*, Academic Press, New York, 1969.
7. R. Gilbert and Y. Xu, A function theory for thin elastic shells, *J. Elasticity*, 22 (1989), 81-93.
8. H. P. Greenspan, Models for the growth of a solid tumour by diffusion, *Studies in Appl. Math.*, 52 (1972), 317-340.
9. M. Taylor, *Partial Differential Equations III: Nonlinear Equations*, Springer, New York, 1996.
10. G. Wen, Initial-mixed boundary value problems for nonlinear parabolic complex equations of second order with measurable coefficients, *Acta Scientiarum Naturalium, Universitatis Pekinensis*, 31, no. 5 (1995), 511-519.
11. G. Wen, *Linear and Nonlinear Elliptic Complex Equations*, Shanghai Science and Technology Publishers, 1986.
12. G. Wen and H. Begehr, *Boundary value problems for elliptic equations and systems*, Longman Scientific & Technical, New York, Wiley, 1990
13. Y. Xu, Dirichlet problem for a class of second order nonlinear elliptic systems, *Complex Variables*, 15, no. 4 (1990), 241-258.
14. Y. Xu, A free boundary problem model of ductal carcinoma in situ, *Discrete and Continous dynamical systems, Series B*, 4, no. 1 (2004), 337-348.
15. Y. Xu, A free boundary problem of parabolic complex equation, *Complex Var. Elliptic Equ.*, 51, Nos. 8-11, (2006), 945-951.
16. N. I. Yurchuk, A mixed problem with an integral condition for some parabolic equations, *Differentsial'nye Uravneniya*, 22, no. 12 (1986), 2117-2126.

FIRST-ORDER DIFFERENTIAL OPERATORS ASSOCIATED TO THE CAUCHY-RIEMANN OPERATOR OF CLIFFORD ANALYSIS

U. YÜKSEL

Department of Mathematics, Atılım University,
06836 İncek, Ankara, Turkey
E-mail: uyuksel@atilim.edu.tr

Consider initial value problems of type

$$\frac{\partial u(t,x)}{\partial t} = \mathcal{L}u(t,x), \quad u(0,x) = u_0(x), \tag{1}$$

where $\mathcal{L}$ is a linear first-order differential operator and the desired solution is sought in a function space defined as the kernel of a linear differential operator $\mathcal{D}$. Mainly two assumptions are required for such initial value problems to be solvable. Firstly, the operators have to be associated, i.e., $\mathcal{L}$ must ransform solutions of $\mathcal{D}u = 0$ again into solutions of this equation. Secondly, an interior estimate

$$\|\mathcal{L}u\|_{\Omega'} \leq \frac{\text{constant}}{\delta} \|u\|_{\Omega}$$

must be true, where Ω' is a subdomain of Ω having the positive distance δ from the boundary of Ω, and $\|\cdot\|$ denotes the norm of a suitably chosen function space (with respcct to Ω' and Ω).

This paper formulates sufficient conditions on the real-valued coefficients of the operator $\mathcal{L}$ having the form

$$\mathcal{L}u := \sum_{A,B,i} C_{B,i}^{(A)} \frac{\partial u_B}{\partial x_i} e_A, \tag{2}$$

where $u = \sum_B u_B e_B$ is a Clifford-algebra-valued function with real-valued components u_B under which $\mathcal{L}$ turns out be associated to the Cauchy-Riemann operator $\mathcal{D}$ of Clifford analysis, i.e., $\mathcal{L}$ transforms monogenic functions into monogenic functions again. The operator $\mathcal{L}$ is constructed in the special case of $n = 2$ using this criterion. See [2] for the construction of $\mathcal{L}$ for $n = 1$. In [2] the initial value problem (1) is also solved in Banach spaces with L_p-norm.

Keywords: Monogenic Functions; Initial Value Problems; Cauchy-Kovalevsky Theorem; Interior Estimates; Associated Operators.

1. A Criterion for the Construction of the Operator $\mathcal{L}$

Let $\mathcal{A}$ be the Clifford algebra generated by $e_0, e_1, \ldots, e_n$, where $e_0 = 1$, $e_j^2 = -1$, $e_ie_j + e_je_i = -2\delta_{ij}$ for $i, j = 1, ..., n$, and Ω be a bounded domain in $\mathbb{R}^{n+1}$ whose points are denoted by $x = (x_0, x_1, ..., x_n)$. A Clifford-algebra-valued function $u = \sum\limits_A u_A e_A$ satisfying the differential equation $\mathcal{D}u = 0$ is called a (left-)monogenic function, where $\mathcal{D} = \frac{\partial}{\partial x_0} + \sum\limits_{j=1}^{n} e_j \frac{\partial}{\partial x_j}$ is the Cauchy-Riemann operator of Clifford analysis (see [1]).

Define the constants $k_{j,A,B}$ for $j = 1, ..., n$ and arbitrary A and B by the relation

$$e_j e_A = \sum_B k_{j,A,B} e_B. \tag{3}$$

Then the following statement is true provided the coefficients of the operator $\mathcal{L}$ are continuously differentiable.

Theorem 1.1. *$\mathcal{L}$ is associated to $\mathcal{D}$ if the coefficients satisfy the systems*

$$\begin{aligned}&\sum_D \left[k_{l,F,D}\left(C_{D,j}^{(A)} - \sum_B k_{j,D,B} C_{B,0}^{(A)}\right)\right]\\ &-\sum_E \left[k_{l,E,A}\left(C_{F,j}^{(E)} - \sum_B k_{j,F,B} C_{B,0}^{(E)}\right)\right] = 0\end{aligned} \tag{4}$$

for all A, $F \in S$ and $j, l = 1, 2, ..., n$ with $j \le l$,

$$\begin{aligned}&\frac{\partial}{\partial x_0}\left(C_{D,j}^{(A)} - \sum_B k_{j,D,B} C_{B,0}^{(A)}\right)\\ &+\sum_E \sum_{m=1}^{n} \left[k_{m,E,A} \frac{\partial}{\partial x_m}\left(C_{D,j}^{(E)} - \sum_B k_{j,D,B} C_{B,0}^{(E)}\right)\right] = 0\end{aligned} \tag{5}$$

for all A, $D \in S$ and $j = 1, 2, ..., n$.

Proof. See [2]. □

Remark 1.1. In Theorem 1.1 we assume that all coefficients of $\mathcal{L}$ are continuously differentiable. However, there are also operators $\mathcal{L}$ associated to $\mathcal{D}$ with a number of continuous coefficients (cf. [2]).

2. Construction of the Operator $\mathcal{L}$

2.1. *The special case of n = 2*

2.1.1. *A system of linear algebraic equations*

If $n = 2$, then we have 4 basis elements $1, e_1, e_2, e_{12}$ and, therefore, $2 \cdot 4 = 8$ equations (3) in the Clifford algebra for the $3 \cdot 4 \cdot 4 = 48$ coefficients $C_{B,i}^{(A)}$. These equations lead to

$$k_{1,0,1} = -k_{1,1,0} = k_{1,2,12} = -k_{1,12,2} = 1,$$

while all other $k_{1AB} = 0$. Analogously

$$k_{2,0,2} = -k_{2,1,12} = -k_{2,2,0} = k_{2,12,1} = 1,$$

and $k_{2AB} = 0$ otherwise. Substituting these values into (4), one obtains a system of 48 linear algebraic equations. 32 of the 48 equations coincide pairwise and therefore only the following $48 - 16 = 32$ equations remain:

$$\begin{aligned}
C_{0,0}^{(0)} + C_{1,1}^{(0)} + C_{0,1}^{(1)} - C_{1,0}^{(1)} &= 0\\
C_{1,2}^{(0)} + C_{2,1}^{(0)} + C_{0,2}^{(1)} - C_{2,0}^{(1)} + C_{0,1}^{(2)} - C_{1,0}^{(2)} &= 0\\
C_{0,0}^{(0)} + C_{2,2}^{(0)} + C_{0,2}^{(2)} - C_{2,0}^{(2)} &= 0\\
C_{0,1}^{(0)} - C_{1,0}^{(0)} - C_{0,0}^{(1)} - C_{1,1}^{(1)} &= 0\\
C_{0,2}^{(0)} + C_{12,1}^{(0)} - C_{1,2}^{(1)} - C_{12,0}^{(1)} - C_{0,0}^{(2)} - C_{1,1}^{(2)} &= 0\\
C_{12,2}^{(0)} - C_{1,0}^{(0)} - C_{1,2}^{(2)} - C_{12,0}^{(2)} &= 0\\
C_{2,0}^{(0)} + C_{12,1}^{(0)} + C_{2,1}^{(1)} - C_{12,0}^{(1)} &= 0\\
C_{0,1}^{(0)} - C_{12,2}^{(0)} - C_{0,0}^{(1)} - C_{2,2}^{(1)} - C_{2,1}^{(2)} + C_{12,0}^{(2)} &= 0\\
C_{0,2}^{(0)} - C_{2,0}^{(0)} - C_{0,0}^{(2)} - C_{2,2}^{(2)} &= 0\\
C_{2,1}^{(0)} - C_{12,0}^{(0)} - C_{2,0}^{(1)} - C_{12,1}^{(1)} &= 0\\
C_{1,1}^{(0)} - C_{2,2}^{(0)} + C_{12,2}^{(1)} - C_{1,0}^{(1)} + C_{2,0}^{(2)} + C_{12,1}^{(2)} &= 0\\
C_{1,2}^{(0)} + C_{12,0}^{(0)} + C_{12,2}^{(2)} - C_{1,0}^{(2)} &= 0\\
-C_{1,2}^{(1)} - C_{2,1}^{(1)} + C_{0,2}^{(0)} - C_{2,0}^{(0)} + C_{0,1}^{(12)} - C_{1,0}^{(12)} &= 0\\
-C_{0,0}^{(1)} - C_{2,2}^{(1)} + C_{0,2}^{(12)} - C_{2,0}^{(12)} &= 0\\
C_{0,2}^{(1)} + C_{12,1}^{(1)} + C_{1,2}^{(0)} + C_{12,0}^{(0)} + C_{0,0}^{(12)} + C_{1,1}^{(12)} &= 0\\
C_{12,2}^{(1)} - C_{1,0}^{(1)} + C_{1,2}^{(12)} + C_{12,0}^{(12)} &= 0\\
C_{0,1}^{(1)} - C_{12,2}^{(1)} + C_{0,0}^{(0)} + C_{2,2}^{(0)} + C_{2,1}^{(12)} - C_{12,0}^{(12)} &= 0\\
C_{0,2}^{(1)} - C_{2,0}^{(1)} + C_{0,0}^{(12)} + C_{2,2}^{(12)} &= 0\\
-C_{1,1}^{(1)} + C_{2,2}^{(1)} + C_{12,2}^{(0)} - C_{1,0}^{(0)} + C_{2,0}^{(12)} + C_{12,1}^{(12)} &= 0\\
-C_{1,2}^{(1)} - C_{12,0}^{(1)} + C_{12,2}^{(12)} - C_{1,0}^{(12)} &= 0\\
C_{0,0}^{(2)} + C_{1,1}^{(2)} + C_{0,1}^{(12)} - C_{1,0}^{(12)} &= 0\\
C_{1,2}^{(2)} + C_{2,1}^{(2)} + C_{0,2}^{(12)} - C_{2,0}^{(12)} - C_{0,1}^{(0)} + C_{1,0}^{(0)} &= 0
\end{aligned} \qquad (6)$$

$$
\begin{aligned}
C^{(2)}_{0,1} - C^{(2)}_{1,0} - C^{(12)}_{0,0} - C^{(12)}_{1,1} &= 0\\
C^{(2)}_{0,2} + C^{(2)}_{12,1} - C^{(12)}_{1,2} - C^{(12)}_{12,0} + C^{(0)}_{0,0} + C^{(0)}_{1,1} &= 0\\
C^{(2)}_{2,0} + C^{(2)}_{12,1} + C^{(12)}_{2,1} - C^{(12)}_{12,0} &= 0\\
C^{(2)}_{0,1} - C^{(2)}_{12,2} - C^{(12)}_{0,0} - C^{(12)}_{2,2} + C^{(0)}_{2,1} - C^{(0)}_{12,0} &= 0\\
C^{(2)}_{2,1} - C^{(2)}_{12,0} - C^{(12)}_{2,0} - C^{(12)}_{12,1} &= 0\\
-C^{(2)}_{1,1} + C^{(2)}_{2,2} - C^{(12)}_{12,2} + C^{(12)}_{1,0} + C^{(0)}_{2,0} + C^{(0)}_{12,1} &= 0\\
-C^{(12)}_{1,2} - C^{(12)}_{2,1} + C^{(2)}_{0,2} - C^{(2)}_{2,0} - C^{(1)}_{0,1} + C^{(1)}_{1,0} &= 0\\
C^{(12)}_{0,2} + C^{(12)}_{12,1} + C^{(2)}_{1,2} + C^{(2)}_{12,0} - C^{(1)}_{0,0} - C^{(1)}_{1,1} &= 0\\
C^{(12)}_{0,1} - C^{(12)}_{12,2} + C^{(2)}_{0,0} + C^{(2)}_{2,2} - C^{(1)}_{2,1} + C^{(1)}_{12,0} &= 0\\
-C^{(12)}_{1,1} + C^{(12)}_{2,2} + C^{(2)}_{12,2} - C^{(2)}_{1,0} - C^{(1)}_{2,0} - C^{(1)}_{12,1} &= 0
\end{aligned}
\tag{7}
$$

This system can be written in matrix form as $M \cdot K = 0$, where the 48×1 column matrix K is

$$
\begin{aligned}
K = \Big(&C^{(0)}_{0,0}\ C^{(1)}_{0,0}\ C^{(2)}_{0,0}\ C^{(12)}_{0,0}\ C^{(0)}_{0,1}\ C^{(1)}_{0,1}\ C^{(2)}_{0,1}\ C^{(12)}_{0,1}\\
&C^{(0)}_{0,2}\ C^{(1)}_{0,2}\ C^{(2)}_{0,2}\ C^{(12)}_{0,2}\ C^{(0)}_{1,0}\ C^{(1)}_{1,0}\ C^{(2)}_{1,0}\ C^{(12)}_{1,0}\\
&C^{(0)}_{1,1}\ C^{(1)}_{1,1}\ C^{(2)}_{1,1}\ C^{(12)}_{1,1}\ C^{(0)}_{1,2}\ C^{(1)}_{1,2}\ C^{(2)}_{1,2}\ C^{(12)}_{1,2}\\
&C^{(0)}_{2,0}\ C^{(1)}_{2,0}\ C^{(2)}_{2,0}\ C^{(12)}_{2,0}\ C^{(0)}_{2,1}\ C^{(1)}_{2,1}\ C^{(2)}_{2,1}\ C^{(12)}_{2,1}\\
&C^{(0)}_{2,2}\ C^{(1)}_{2,2}\ C^{(2)}_{2,2}\ C^{(12)}_{2,2}\ C^{(0)}_{12,0}\ C^{(1)}_{12,0}\ C^{(2)}_{12,0}\ C^{(12)}_{12,0}\\
&C^{(0)}_{12,1}\ C^{(1)}_{12,1}\ C^{(2)}_{12,1}\ C^{(12)}_{12,1}\ C^{(0)}_{12,2}\ C^{(1)}_{12,2}\ C^{(2)}_{12,2}\ C^{(12)}_{12,2}\Big)^T
\end{aligned}
$$

and the 32×48 coefficient matrix M is given in the Appendix. It is easy to show using Mathematica that the rank of the matrix M is 20 and the solution of the system (6)-(7) can be written as

$$
\begin{aligned}
C^{(0)}_{0,0} &= -p_6 + p_9 - p_{16} - p_{17} - p_{26} & C^{(0)}_{2,0} &= -p_8 + p_{11} - p_{19}\\
C^{(1)}_{0,0} &= p_5 + p_{10} - p_{15} - p_{18} + p_{25} & C^{(1)}_{2,0} &= -p_7 - p_{12} + p_{20}\\
C^{(2)}_{0,0} &= p_8 - p_{11} - p_{14} + p_{19} + p_{28} & C^{(2)}_{2,0} &= -p_6 + p_9 - p_{17}\\
C^{(12)}_{0,0} &= -p_7 - p_{12} - p_{13} + p_{20} - p_{27} & C^{(12)}_{2,0} &= -p_5 - p_{10} + p_{18}\\
C^{(0)}_{0,1} &= p_4 + p_5 + p_{25} & C^{(0)}_{2,1} &= p_{20}\\
C^{(1)}_{0,1} &= p_3 + p_6 + p_{26} & C^{(1)}_{2,1} &= p_{19}\\
C^{(2)}_{0,1} &= p_2 - p_7 - p_{27} & C^{(2)}_{2,1} &= p_{18}\\
C^{(12)}_{0,1} &= p_1 - p_8 - p_{28} & C^{(12)}_{2,1} &= p_{17}\\
C^{(0)}_{0,2} &= p_{28} & C^{(0)}_{2,2} &= p_{16}\\
C^{(1)}_{0,2} &= p_{27} & C^{(1)}_{2,2} &= p_{15}\\
C^{(2)}_{0,2} &= p_{26} & C^{(2)}_{2,2} &= p_{14}\\
C^{(12)}_{0,2} &= p_{25} & C^{(12)}_{2,2} &= p_{13}
\end{aligned}
\tag{8}
$$

$$
\begin{array}{ll}
C_{1,0}^{(0)} = p_4 - p_{10} - p_{22} & C_{12,0}^{(0)} = p_{12} \\
C_{1,0}^{(1)} = p_3 + p_9 + p_{21} & C_{12,0}^{(1)} = p_{11} \\
C_{1,0}^{(2)} = p_2 + p_{12} + p_{24} & C_{12,0}^{(2)} = p_{10} \\
C_{1,0}^{(12)} = p_1 - p_{11} - p_{23} & C_{12,0}^{(12)} = p_9 \\
C_{1,1}^{(0)} = p_{16} + p_{17} + p_{21} & C_{12,1}^{(0)} = p_8 \\
C_{1,1}^{(1)} = p_{15} + p_{18} + p_{22} & C_{12,1}^{(1)} = p_7 \\
C_{1,1}^{(2)} = p_{14} - p_{19} - p_{23} & C_{12,1}^{(2)} = p_6 \\
C_{1,1}^{(12)} = p_{13} - p_{20} - p_{24} & C_{12,1}^{(12)} = p_5 \\
C_{1,2}^{(0)} = p_{24} & C_{12,2}^{(0)} = p_4 \\
C_{1,2}^{(1)} = p_{23} & C_{12,2}^{(1)} = p_3 \\
C_{1,2}^{(2)} = p_{22} & C_{12,2}^{(2)} = p_2 \\
C_{1,2}^{(12)} = p_{21} & C_{12,2}^{(12)} = p_1
\end{array}
\tag{9}
$$

where the $p_k = p_k(t, x)$, $k = 1, 2, ..., 28$, are real-valued functions of t and x (see the Appendix).

2.1.2. *A system of partial differential equations*

Analogously, substituting the values of $k_{j,A,B}$ into (5), one obtains the system of 32 linear first order partial differential equations

$$
\begin{array}{ll}
\partial_{x_0} K_{0,1}^{(0)} - \partial_{x_1} K_{0,1}^{(1)} - \partial_{x_2} K_{0,1}^{(2)} = 0 & \partial_{x_0} K_{0,1}^{(1)} + \partial_{x_1} K_{0,1}^{(0)} + \partial_{x_2} K_{0,1}^{(12)} = 0 \\
\partial_{x_0} K_{0,2}^{(0)} - \partial_{x_1} K_{0,2}^{(1)} - \partial_{x_2} K_{0,2}^{(2)} = 0 & \partial_{x_0} K_{0,2}^{(1)} + \partial_{x_1} K_{0,2}^{(0)} + \partial_{x_2} K_{0,2}^{(12)} = 0 \\
\partial_{x_0} K_{1,1}^{(0)} - \partial_{x_1} K_{1,1}^{(1)} - \partial_{x_2} K_{1,1}^{(2)} = 0 & \partial_{x_0} K_{1,1}^{(1)} + \partial_{x_1} K_{1,1}^{(0)} + \partial_{x_2} K_{1,1}^{(12)} = 0 \\
\partial_{x_0} K_{1,2}^{(0)} - \partial_{x_1} K_{1,2}^{(1)} - \partial_{x_2} K_{1,2}^{(2)} = 0 & \partial_{x_0} K_{1,2}^{(1)} + \partial_{x_1} K_{1,2}^{(0)} + \partial_{x_2} K_{1,2}^{(12)} = 0 \\
\partial_{x_0} K_{2,1}^{(0)} - \partial_{x_1} K_{2,1}^{(1)} - \partial_{x_2} K_{2,1}^{(2)} = 0 & \partial_{x_0} K_{2,1}^{(1)} + \partial_{x_1} K_{2,1}^{(0)} + \partial_{x_2} K_{2,1}^{(12)} = 0 \\
\partial_{x_0} K_{2,2}^{(0)} - \partial_{x_1} K_{2,2}^{(1)} - \partial_{x_2} K_{2,2}^{(2)} = 0 & \partial_{x_0} K_{2,2}^{(1)} + \partial_{x_1} K_{2,2}^{(0)} + \partial_{x_2} K_{2,2}^{(12)} = 0 \\
\partial_{x_0} K_{12,1}^{(0)} - \partial_{x_1} K_{12,1}^{(1)} - \partial_{x_2} K_{12,1}^{(2)} = 0 & \partial_{x_0} K_{12,1}^{(1)} + \partial_{x_1} K_{12,1}^{(0)} + \partial_{x_2} K_{12,1}^{(12)} = 0 \\
\partial_{x_0} K_{12,2}^{(0)} - \partial_{x_1} K_{12,2}^{(1)} - \partial_{x_2} K_{12,2}^{(2)} = 0 & \partial_{x_0} K_{12,2}^{(1)} + \partial_{x_1} K_{12,2}^{(0)} + \partial_{x_2} K_{12,2}^{(12)} = 0 \\
\partial_{x_0} K_{0,1}^{(2)} - \partial_{x_1} K_{0,1}^{(12)} + \partial_{x_2} K_{0,1}^{(0)} = 0 & \partial_{x_0} K_{0,1}^{(12)} + \partial_{x_1} K_{0,1}^{(2)} - \partial_{x_2} K_{0,1}^{(1)} = 0 \\
\partial_{x_0} K_{0,2}^{(2)} - \partial_{x_1} K_{0,2}^{(12)} + \partial_{x_2} K_{0,2}^{(0)} = 0 & \partial_{x_0} K_{0,2}^{(12)} + \partial_{x_1} K_{0,2}^{(2)} - \partial_{x_2} K_{0,2}^{(1)} = 0 \\
\partial_{x_0} K_{1,1}^{(2)} - \partial_{x_1} K_{1,1}^{(12)} + \partial_{x_2} K_{1,1}^{(0)} = 0 & \partial_{x_0} K_{1,1}^{(12)} + \partial_{x_1} K_{1,1}^{(2)} - \partial_{x_2} K_{1,1}^{(1)} = 0 \\
\partial_{x_0} K_{1,2}^{(2)} - \partial_{x_1} K_{1,2}^{(12)} + \partial_{x_2} K_{1,2}^{(0)} = 0 & \partial_{x_0} K_{1,2}^{(12)} + \partial_{x_1} K_{1,2}^{(2)} - \partial_{x_2} K_{1,2}^{(1)} = 0 \\
\partial_{x_0} K_{2,1}^{(2)} - \partial_{x_1} K_{2,1}^{(12)} + \partial_{x_2} K_{2,1}^{(0)} = 0 & \partial_{x_0} K_{2,1}^{(12)} + \partial_{x_1} K_{2,1}^{(2)} - \partial_{x_2} K_{2,1}^{(1)} = 0 \\
\partial_{x_0} K_{2,2}^{(2)} - \partial_{x_1} K_{2,2}^{(12)} + \partial_{x_2} K_{2,2}^{(0)} = 0 & \partial_{x_0} K_{2,2}^{(12)} + \partial_{x_1} K_{2,2}^{(2)} - \partial_{x_2} K_{2,2}^{(1)} = 0 \\
\partial_{x_0} K_{12,1}^{(2)} - \partial_{x_1} K_{12,1}^{(12)} + \partial_{x_2} K_{12,1}^{(0)} = 0 & \partial_{x_0} K_{12,1}^{(12)} + \partial_{x_1} K_{12,1}^{(2)} - \partial_{x_2} K_{12,1}^{(1)} = 0 \\
\partial_{x_0} K_{12,2}^{(2)} - \partial_{x_1} K_{12,2}^{(12)} + \partial_{x_2} K_{12,2}^{(0)} = 0 & \partial_{x_0} K_{12,2}^{(12)} + \partial_{x_1} K_{12,2}^{(2)} - \partial_{x_2} K_{12,2}^{(1)} = 0,
\end{array}
\tag{10}
$$

where

$$\begin{array}{ll} K_{0,1}^{(A)} := C_{0,1}^{(A)} - C_{1,0}^{(A)} & K_{2,1}^{(A)} := C_{2,1}^{(A)} - C_{12,0}^{(A)} \\ K_{0,2}^{(A)} := C_{0,2}^{(A)} - C_{2,0}^{(A)} & K_{2,2}^{(A)} := C_{0,0}^{(A)} + C_{2,2}^{(A)} \\ K_{1,1}^{(A)} := C_{0,0}^{(A)} + C_{1,1}^{(A)} & K_{12,1}^{(A)} := C_{2,0}^{(A)} + C_{12,1}^{(A)} \\ K_{1,2}^{(A)} := C_{1,2}^{(A)} + C_{12,0}^{(A)} & K_{12,2}^{(A)} := C_{12,2}^{(A)} - C_{1,0}^{(A)} \end{array} \tag{11}$$

Taking into account (11) and (8), (9), the system (10) can be written in terms of the p_k as follows:

$$\begin{aligned}
\partial_{x_0}(p_5+p_{10}+p_{22}+p_{25})-\partial_{x_1}(p_6-p_9-p_{21}+p_{26})-\partial_{x_2}(-p_7-p_{12}-p_{24}-p_{27}) &= 0\\
\partial_{x_0}(p_8-p_{11}+p_{19}+p_{28})-\partial_{x_1}(p_7+p_{12}-p_{20}+p_{27})-\partial_{x_2}(p_6-p_9+p_{17}+p_{26}) &= 0\\
\partial_{x_0}(-p_6+p_9+p_{21}-p_{26})-\partial_{x_1}(p_5+p_{10}+p_{22}+p_{25})-\partial_{x_2}(p_8-p_{11}-p_{23}+p_{28}) &= 0\\
\partial_{x_0}(p_{12}+p_{24})-\partial_{x_1}(p_{11}+p_{23})-\partial_{x_2}(p_{10}+p_{22}) &= 0\\
\partial_{x_0}(-p_{12}+p_{20})-\partial_{x_1}(-p_{11}+p_{19})-\partial_{x_2}(-p_{10}+p_{18}) &= 0\\
\partial_{x_0}(-p_6+p_9-p_{17}-p_{26})-\partial_{x_1}(p_5+p_{10}-p_{18}+p_{25})-\partial_{x_2}(p_8-p_{11}+p_{19}+p_{28}) &= 0\\
\partial_{x_0}(p_{11}-p_{19})_{x_0}-\partial_{x_1}(-p_{12}+p_{20})-\partial_{x_2}(p_9-p_{17}) &= 0\\
\partial_{x_0}(p_{10}+p_{22})_{x_0}-\partial_{x_1}(-p_9-p_{21})-\partial_{x_2}(-p_{12}-p_{24}) &= 0\\
\partial_{x_0}(p_6-p_9-p_{21}+p_{26})+\partial_{x_1}(p_5+p_{10}+p_{22}+p_{25})+\partial_{x_2}(-p_8+p_{11}+p_{23}-p_{28}) &= 0\\
\partial_{x_0}(p_7+p_{12}-p_{20}+p_{27})_{x_0}+\partial_{x_1}(p_8-p_{11}+p_{19}+p_{28})+\partial_{x_2}(p_5+p_{10}-p_{18}+p_{25}) &= 0\\
\partial_{x_0}(p_5+p_{10}+p_{22}+p_{25})_{x_0}+\partial_{x_1}(-p_6+p_9+p_{21}-p_{26})+\partial_{x_2}(-p_7-p_{12}-p_{24}-p_{27}) &= 0\\
\partial_{x_0}(p_{11}+p_{23})_{x_0}+\partial_{x_1}(p_{12}+p_{24})+\partial_{x_2}(p_9+p_{21}) &= 0\\
\partial_{x_0}(-p_{11}+p_{19})_{x_0}+\partial_{x_1}(-p_{12}+p_{20})+\partial_{x_2}(-p_9+p_{17}) &= 0\\
\partial_{x_0}(p_5+p_{10}-p_{18}+p_{25})_{x_0}+\partial_{x_1}(-p_6+p_9-p_{17}-p_{26})+\partial_{x_2}(-p_7-p_{12}+p_{20}-p_{27}) &= 0\\
\partial_{x_0}(-p_{12}+p_{20})_{x_0}+\partial_{x_1}(p_{11}-p_{19})+\partial_{x_2}(-p_{10}+p_{18}) &= 0\\
\partial_{x_0}(-p_9-p_{21})_{x_0}+\partial_{x_1}(p_{10}+p_{22})+\partial_{x_2}(p_{11}+p_{23}) &= 0\\
\partial_{x_0}(-p_7-p_{12}-p_{24}-p_{27})-\partial_{x_1}(-p_8+p_{11}+p_{23}-p_{28})+\partial_{x_2}(p_5+p_{10}+p_{22}+p_{25}) &= 0\\
\partial_{x_0}(p_6-p_9+p_{17}+p_{26})-\partial_{x_1}(p_5+p_{10}-p_{18}+p_{25})+\partial_{x_2}(p_8-p_{11}+p_{19}+p_{28}) &= 0\\
\partial_{x_0}(p_8-p_{11}-p_{23}+p_{28})-\partial_{x_1}(-p_7-p_{12}-p_{24}-p_{27})+\partial_{x_2}(-p_6+p_9+p_{21}-p_{26}) &= 0\\
\partial_{x_0}(p_{10}+p_{22})-\partial_{x_1}(p_9+p_{21})+\partial_{x_2}(p_{12}+p_{24}) &= 0\\
\partial_{x_0}(-p_{10}+p_{18})-\partial_{x_1}(-p_9+p_{17})+\partial_{x_2}(-p_{12}+p_{20}) &= 0\\
\partial_{x_0}(p_8-p_{11}+p_{19}+p_{28})-\partial_{x_1}(-p_7-p_{12}+p_{20}-p_{27})+\partial_{x_2}(-p_6+p_9-p_{17}-p_{26}) &= 0\\
\partial_{x_0}(p_9-p_{17})-\partial_{x_1}(-p_{10}+p_{18})+\partial_{x_2}(p_{11}-p_{19}) &= 0\\
\partial_{x_0}(-p_{12}-p_{24})-\partial_{x_1}(p_{11}+p_{23})+\partial_{x_2}(p_{10}+p_{22}) &= 0\\
\partial_{x_0}(-p_8+p_{11}+p_{23}-p_{28})+\partial_{x_1}(-p_7-p_{12}-p_{24}-p_{27})-\partial_{x_2}(p_6-p_9-p_{21}+p_{26}) &= 0\\
\partial_{x_0}(p_5+p_{10}-p_{18}+p_{25})+\partial_{x_1}(p_6-p_9+p_{17}+p_{26})-\partial_{x_2}(p_7+p_{12}-p_{20}+p_{27}) &= 0\\
\partial_{x_0}(-p_7-p_{12}-p_{24}-p_{27})+\partial_{x_1}(p_8-p_{11}-p_{23}+p_{28})-\partial_{x_2}(p_5+p_{10}+p_{22}+p_{25}) &= 0\\
\partial_{x_0}(p_9+p_{21})+\partial_{x_1}(p_{10}+p_{22})-\partial_{x_2}(p_{11}+p_{23}) &= 0
\end{aligned} \tag{12}$$

$$\begin{aligned}\partial_{x_0}(-p_9+p_{17})+\partial_{x_1}(-p_{10}+p_{18})-\partial_{x_2}(-p_{11}+p_{19}) &= 0\\ \partial_{x_0}(-p_7-p_{12}+p_{20}-p_{27})+\partial_{x_1}(p_8-p_{11}+p_{19}+p_{28})-\partial_{x_2}(p_5+p_{10}-p_{18}+p_{25}) &= 0\\ \partial_{x_0}(-p_{10}+p_{18})+\partial_{x_1}(p_9-p_{17})-\partial_{x_2}(-p_{12}+p_{20}) &= 0\\ \partial_{x_0}(p_{11}+p_{23})+\partial_{x_1}(-p_{12}-p_{24})-\partial_{x_2}(-p_9-p_{21}) &= 0\end{aligned} \tag{13}$$

2.1.3. *Solution of system (12)-(13) in special cases depending on p_k*

Case 1. Let

$$p_k = \begin{cases}\text{continuously differentiable functions,} & \text{if } 1\le k\le 4 \text{ or } 13\le k\le 16\\ \text{constants,} & \text{if } 9\le k\le 12 \text{ or } 17\le k\le 28\end{cases}$$

Then it can easily be obtained from the system (12)-(13) that p_k, $5\le k\le 8$, are constants.

Case 2. Assume that

$$p_k = \begin{cases}\text{continuously differentiable functions,} & \text{if } 1\le k\le 4 \text{ or } 13\le k\le 16\\ \text{constants,} & \text{if } 17\le k\le 24.\end{cases}$$

Then it follows from the system (12)-(13) that p_k, $9\le k\le 12$, are constants, p_k, $5\le k\le 8$, and p_k, $25\le k\le 28$ are continuously differentiable functions such that $p_{25+k}+p_{5+k}$, $0\le k\le 3$, are constants.

Considering the above cases and taking into account (8) and (9) yield the following.

Lemma 2.1. *Assume that $C^{(A)}_{2,2}$ and $C^{(A)}_{12,2}$ are continuously differentiable functions, and $C^{(A)}_{0,2}$, $C^{(A)}_{1,2}$, $C^{(A)}_{2,1}$, $C^{(A)}_{12,0}$, and $C^{(A)}_{12,1}$ are constants. Assume further that $C^{(A)}_{0,0}$, $C^{(A)}_{0,1}$, $C^{(A)}_{1,0}$, $C^{(A)}_{1,1}$, and $C^{(A)}_{2,0}$ are given as*

$$\begin{aligned}C^{(0)}_{0,0} &= -C^{(2)}_{12,1}+C^{(12)}_{12,0}-C^{(0)}_{2,2}-C^{(12)}_{2,1}-C^{(2)}_{0,2} & C^{(0)}_{0,1} &= C^{(0)}_{12,2}+C^{(12)}_{12,1}+C^{(12)}_{0,2}\\ C^{(1)}_{0,0} &= C^{(12)}_{12,1}+C^{(2)}_{12,0}-C^{(1)}_{2,2}-C^{(2)}_{2,1}+C^{(12)}_{0,2} & C^{(1)}_{0,1} &= C^{(1)}_{12,2}+C^{(2)}_{12,1}+C^{(2)}_{0,2}\\ C^{(2)}_{0,0} &= C^{(0)}_{12,1}-C^{(1)}_{12,0}-C^{(2)}_{2,2}+C^{(1)}_{2,1}+C^{(0)}_{0,2} & C^{(2)}_{0,1} &= C^{(2)}_{12,2}-C^{(1)}_{12,1}-C^{(1)}_{0,2}\\ C^{(12)}_{0,0} &= -C^{(1)}_{12,1}-C^{(0)}_{12,0}-C^{(12)}_{2,2}+C^{(0)}_{2,1}-C^{(1)}_{0,2} & C^{(12)}_{0,1} &= C^{(12)}_{12,2}-C^{(0)}_{12,1}-C^{(0)}_{0,2}\end{aligned}$$

$$\begin{aligned}C^{(0)}_{1,0} &= C^{(0)}_{12,2}-C^{(2)}_{12,0}-C^{(2)}_{1,2} & C^{(0)}_{1,1} &= C^{(0)}_{2,2}+C^{(12)}_{2,1}+C^{(12)}_{1,2}\\ C^{(1)}_{1,0} &= C^{(1)}_{12,2}+C^{(12)}_{12,0}+C^{(12)}_{1,2} & C^{(1)}_{1,1} &= C^{(1)}_{2,2}+C^{(2)}_{2,1}+C^{(2)}_{1,2}\\ C^{(2)}_{1,0} &= C^{(2)}_{12,2}+C^{(0)}_{12,0}+C^{(0)}_{1,2} & C^{(2)}_{1,1} &= C^{(2)}_{2,2}-C^{(1)}_{2,1}-C^{(1)}_{1,2}\\ C^{(12)}_{1,0} &= C^{(12)}_{12,2}-C^{(1)}_{12,0}-C^{(1)}_{1,2} & C^{(12)}_{1,1} &= C^{(12)}_{2,2}-C^{(0)}_{2,1}-C^{(0)}_{1,2}\end{aligned}$$

$$\begin{aligned} C^{(0)}_{2,0} &= -C^{(0)}_{12,1} + C^{(1)}_{12,0} - C^{(1)}_{2,1} \\ C^{(1)}_{2,0} &= -C^{(1)}_{12,1} - C^{(0)}_{12,0} + C^{(0)}_{2,1} \\ C^{(2)}_{2,0} &= -C^{(2)}_{12,1} + C^{(12)}_{12,0} - C^{(12)}_{2,1} \\ C^{(12)}_{2,0} &= C^{(2)}_{2,1} - C^{(12)}_{12,1} - C^{(2)}_{12,0}. \end{aligned}$$

Then $\mathcal{L}$ is associated to $\mathcal{D}$.

Lemma 2.2. *Let $C^{(A)}_{2,2}$ and $C^{(A)}_{12,2}$ be continuously differentiable functions, and $C^{(A)}_{1,2}$, $C^{(A)}_{2,1}$, and $C^{(A)}_{12,0}$ be constants. Suppose that $C^{(A)}_{0,2}$ and $C^{(A)}_{12,1}$ are continuously differentiable functions such that their sum (for each A) is a constant. Suppose further that $C^{(A)}_{0,0}$, $C^{(A)}_{0,1}$, $C^{(A)}_{1,0}$, $C^{(A)}_{1,1}$, and $C^{(A)}_{2,0}$ are given as in Lemma 2.1. Then $\mathcal{L}$ is associated to $\mathcal{D}$.*

3. Appendix

We give the Mathematica code used to solve the system (6)-(7).

```
ClearAll;
M=Table[0,{32},{48}];
M[[1,1]]=1;M[[1,6]]=1;M[[1,14]]=-1;M[[1,17]]=1;M[[2,7]]=1;
M[[2,10]]=1;M[[2,15]]=-1;M[[2,21]]=1;M[[2,26]]=-1;M[[2,29]]=1;
M[[3,1]]=1;M[[3,11]]=1;M[[3,27]]=-1;M[[3,33]]=1;M[[4,2]]=-1;
M[[4,5]]=1;M[[4,13]]=-1;M[[4,18]]=-1;M[[5,3]]=-1;M[[5,9]]=1;
M[[5,19]]=-1;M[[5,22]]=-1;M[[5,38]]=-1;M[[5,41]]=1;M[[6,13]]=-1;
M[[6,23]]=-1;M[[6,39]]=-1;M[[6,45]]=1;M[[7,25]]=1;M[[7,30]]=1;
M[[7,38]]=-1;M[[7,41]]=1;M[[8,2]]=-1;M[[8,5]]=1;M[[8,31]]=-1;
M[[8,34]]=-1;M[[8,39]]=1;M[[8,45]]=-1;M[[9,3]]=-1;M[[9,9]]=1;
M[[9,25]]=-1;M[[9,35]]=-1;M[[10,29]]=1;M[[10,37]]=-1;M[[10,26]]=-1;
M[[10,42]]=-1;M[[11,14]]=-1;M[[11,17]]=1;M[[11,27]]=1;
M[[11,33]]=-1;M[[11,43]]=1;M[[11,46]]=1;M[[12,15]]=-1;
M[[12,21]]=1;M[[12,37]]=1;M[[12,47]]=1;M[[13,8]]=1;M[[13,9]]=1;
M[[13,16]]=-1;M[[13,22]]=-1;M[[13,25]]=-1;M[[13,30]]=-1;
M[[14,2]]=-1;M[[14,28]]=-1;M[[14,34]]=-1;M[[15,4]]=1;M[[15,10]]=1;
M[[15,20]]=1;M[[15,21]]=1;M[[15,37]]=1;M[[15,42]]=1;M[[16,14]]=-1;
M[[16,24]]=1;M[[16,40]]=1;M[[16,46]]=1;M[[17,1]]=1;M[[17,6]]=1;
M[[17,32]]=1;M[[17,33]]=1;M[[17,40]]=-1;M[[17,46]]=-1;M[[18,4]]=1;
M[[18,10]]=1;M[[18,26]]=-1;M[[18,36]]=1;M[[19,13]]=-1;M[[19,18]]=-1;
M[[19,28]]=1;M[[19,34]]=1;M[[19,44]]=1;M[[19,45]]=1;M[[20,16]]=-1;
M[[20,22]]=-1;M[[20,38]]=-1;M[[20,48]]=1;M[[21,3]]=1;M[[21,8]]=1;
M[[21,16]]=-1;M[[21,19]]=1;M[[22,5]]=-1;M[[22,12]]=1;M[[22,13]]=1;
M[[22,23]]=1;M[[22,28]]=-1;M[[22,31]]=1;M[[23,4]]=-1;M[[23,7]]=1;
M[[23,15]]=-1;M[[23,20]]=-1;M[[24,1]]=1;M[[24,11]]=1;M[[24,17]]=1;
M[[24,24]]=-1;M[[24,40]]=-1;M[[24,43]]=1;M[[25,27]]=1;M[[25,32]]=1;
M[[25,40]]=-1;M[[25,43]]=1;M[[26,4]]=-1;M[[26,7]]=1;M[[26,29]]=1;
M[[26,36]]=-1;M[[26,37]]=-1;M[[26,47]]=-1;M[[27,28]]=-1;M[[27,31]]=1;
```

```
M[[27,39]]=-1;M[[27,44]]=-1;M[[28,16]]=1;M[[28,19]]=-1;M[[28,25]]=1;
M[[28,35]]=1;M[[28,41]]=1;M[[28,48]]=-1;M[[29,6]]=-1;M[[29,11]]=1;
M[[29,14]]=1;M[[29,24]]=-1;M[[29,27]]=-1;M[[29,32]]=-1;M[[30,2]]=-1;
M[[30,12]]=1;M[[30,18]]=-1;M[[30,23]]=1;M[[30,39]]=1;M[[30,44]]=1;
M[[31,3]]=1;M[[31,8]]=1;M[[31,30]]=-1;M[[31,35]]=1;M[[31,38]]=1;
M[[31,48]]=-1;M[[32,15]]=-1;M[[32,20]]=-1;M[[32,26]]=-1;M[[32,36]]=1;
M[[32,42]]=-1;M[[32,47]]=1;
Print["Coefficients Matrix M = ", MatrixForm[M]]
Print["Rank of M = ", MatrixRank[M]]
V = NullSpace[M];
Print["MatrixForm[V] = Vectors of NullSpace[M] in Matrix Form = "]
MatrixForm[V]
L1 = Length[V]
Print["Number of Vectors in NullSpace[M] = ", L1]
L2 = Length[V[[1]]]
Print["Number of Components of a Vector in NullSpace[M] = ", L2]
For[i = 1, i <= L1, i++, Print["V(", i, ") = ", V[[i]]]]
For[i = 1, i <= L1, i++, Print["M*V(", i, ") = ", M.V[[i]]]]
Q = Sum[Array[P, L1][[i]]*V[[i]], {i, 1, L1}]
M.Q
Print["Solution of the System ="]
For[i = 1, i <= L2, i++, Print["Q(", i, ") = ", Q[[i]]]]
```

References

1. F. Brackx, R. Delanghe, F. Sommen, *Clifford Analysis*, Pitman, London, 1982.
2. U. Yüksel, *Solution of Initial Value Problems with Monogenic Initial Functions in Banach Spaces with L_p-Norm*, submitted.

SINGULAR INTEGRAL EQUATION ON THE REAL AXIS WITH SOLUTION HAVING SINGULARITY OF HIGHER ORDER AT INFINITE POINT

Shouguo Zhong

School of Mathematics and Statistics, Wuhan University, Wuhan 430072, P. R. China

By discussing the properties of Cauchy-type integrals and Cauchy principal value integrals on the real axis in which kernel densities have singularities of higher order at ∞, we transform the singular integral equations on X with solutions having singularities of higher order at ∞ into ones along a closed contour with solutions having singularities of higher order. For the former we obtain the solutions and solvability conditions of the characteristic equations as well as the extended Noether theorem of complete equations by solving the latter.

Keywords: Singular integral equation; solutions having singularities of higher order; real axis; ∞ point; class$\hat{H}_n^*$.

1. Preliminaries

The solution class of singular integral equations in classical theory is restricted to H or H^* [1], and generalized to the class with singularities of higher order along a curve in [2-6]. In this article by discussing the properties of the Cauchy type integrals and the Cauchy principal value integrals in which kernel densities have singularity of higher order at ∞ on the real axis(including ∞) we transform the singular integral equations on X with solution having singularity of higher order at ∞ into the ones along a circumference with solutions having singularity of higher order at $-i$(but the order is increased by one than the order at ∞), and for the former obtain the solutions and the solvable conditions of the characteristic equations as well as the extended Noether theorem of complete equations from the latter.

In order to get the concise unified forms of the solutions and the conditions of solvability for the above characteristic equations along a circle for further discussion, we abandon the method in [4] (the results of which are in the complicated forms) but use a different method.

Definition 1.1. If $f(x)$ satisfies $(x+i)^{n+1}[(x+i)^{n-1}f(x)]^{(n)} \in \widehat{H}(X)$ (see $\widehat{H}$ in [1]), where n is a nonnegative integer, we denote $f(x) \in \widehat{H}^n(X) \equiv \widehat{H}^n$.

Definition 1.2. If $x \in X' = X \backslash \{\infty\}$, $f(x) = (x+i)^n \tilde{f}(x)$ (n is the same as above) and $\tilde{f} \in \widehat{H}^n(X)$, then we call that $f(x)$ have singularity of order n at $x = \infty$ and denote $f(x) \in \widehat{H}_n^*(X) \equiv \widehat{H}_n^*$. (If $n = 0$, then $f(x) = \tilde{f}(x)$, i.e., $f(x) \in \widehat{H}_0^* = \widehat{H}$.)

Definition 1.3. If $f(x) \in \widehat{H}(X)$ and $f(\infty) = 0$, denote $f(x) \in \widehat{H}_0(X) \equiv \widehat{H}_0$.

Lemma 1.1. *Assume $f(x) \in \widehat{H}^n(X)$, under $x + i = -1/(t+i)$ and let $F(t) = f(x), t \in \Gamma : |t + (i/2)| = 1/2$, then we have derivative formulas*

$$F^{(k)}(t) = (x+i)^{k+1}[(x+i)^{k-1}f(x)]^{(k)}, \quad k = 0, 1, 2, \cdots, n. \tag{1}$$

$$f^{(k)}(x) = (t+i)^{k+1}[(t+i)^{k-1}F(t)]^{(k)}, \quad k = 0, 1, 2, \cdots, n. \tag{2}$$

So by (1)(2), we have $F^{(n)}(t) \in H(\Gamma) \iff f(x) \in \widehat{H}^n(X)$, and for $k \geq 1$, $f^{(k)}(x)$ has a zero of order $k+1$ at ∞, i.e., $f^{(k)}(x) \in \widehat{H}_0$ for $k \geq 1$.

Proof. Use induction [8]. □

Lemma 1.2. *Assume $f(x) = (x+i)^n \tilde{f}(x) \in \widehat{H}_n^*$. Then*

$$F(z) = \frac{1}{2\pi i} \int_{-\infty}^{+\infty} \frac{f(\xi)}{\xi - z} d\xi, \quad z \bar{\in} X \tag{3}$$

exists, the Plemelj formula is established

$$F^{\pm}(x) = \pm\frac{1}{2} f(x) + \frac{1}{2\pi i} \int_{-\infty}^{+\infty} \frac{f(\xi)}{\xi - x} d\xi, \quad x \in X' \tag{4}$$

and for $x \in X, F(x) \in \widehat{H}_0(X)$ and $F(x)$ has a zero of order less than 1 ar $x = \infty$.

Proof. Denote $\tilde{F}(\tau) = \tilde{f}(\xi)$ and make transformation $\tau + i = -1/(\xi + i)$, $w + i = -1/(z+i)$, (3) may be written as

$$F(z) = \frac{(-1)^n (w+i)}{2\pi i} \int_{\Gamma} \frac{\tilde{F}(\tau) d\tau}{(\tau+i)^{n+1}(\tau - w)}, \quad w \notin \Gamma. \tag{5}$$

By Definition 1.1 and Lemma 1.1 we get $\widetilde{F}^{(n)}(\tau) \in H(\Gamma)$, so the right of (5) exists under the meaning of singular integral of higher order at $-i$ [1], that is to say, (3) exists. Similarly $F(x)$ exists when $x \in X'$. Letting

$Z^+(Z^-)$ be the upper (lower) half-plane, $z \in Z^+(Z^-), z \to x \in X'$, in other words, now $w \in D^+ : |w + (i/2)| < 1/2(D^- : \mathbb{C}\setminus\overline{D^+}), w \to t \in \Gamma' = \Gamma\setminus\{i\}$, applying Plemelj formula (3.6) in [7] to (5), then we can obtain (4). By Corollary 2 of Lemma 3 in [3], when $x \in X'$, i.e., $t \in \Gamma'$, $I(t) = (2\pi i)^{-1}\int_\Gamma \tilde{F}(\tau)(\tau+i)^{-(n+1)}(\tau-t)^{-1}d\tau \ \in H^*$ near $-i$, thus $F(x) = (-1)^n(t+i)I(t)$, the right side of $F(x)$ has a zero of order less than 1 at $-i$ and belongs to $H(\Gamma)$, thereby, $F(x) \in \widehat{H}_0(X)$ and $F(x)$ has a zero of order less than 1 at ∞. □

2. Solving Characteristic Equation in $\widehat{H}_n^*$

Consider the characteristic equation with solutions in $\widehat{H}_n^*$

$$a(x)\varphi(x) + \frac{b(x)}{\pi i}\int_{-\infty}^{+\infty}\frac{\varphi(\xi)}{\xi - x}d\xi = f(x), \quad x \in X' \tag{6}$$

where $\varphi(x) = (x+i)^n\tilde{\varphi}(x) \in \widehat{H}_n^*$, $f(x) = (x+i)^n\tilde{f}(x) \in \widehat{H}_n^*$. $a^2(x)-b^2(x) \neq 0 (x \in X), a(\infty) \neq 0,\ a(x), b(x) \in \widehat{H}^n(X)$. Making transformation

$$\tau + i = -1/(\xi + i), \quad t + i = -1/(x+i) \tag{7}$$

(6) is reduced to the characteristic equation with solutions having singularity of order $n+1$ at $-i$ on Γ as follows

$$A(t)\Phi(t) + \frac{B(t)}{\pi i}\int_\Gamma \frac{\Phi(\tau)}{\tau - t}d\tau = F(t), \quad t \in \Gamma' \tag{8}$$

where $\Phi(t) = \tilde{\Phi}(t)/(t+i)^{n+1}, \tilde{\Phi}(t) = \tilde{\varphi}(x)$; $F(t) = \tilde{F}(t)/(t+i)^{n+1}, \tilde{F}(t) = \tilde{f}(x)$. $A(t) = a(x), B(t) = b(x)$ and $\tilde{\Phi}^{(n)}(t), \tilde{F}^{(n)}(t), A^{(n)}(t), B^{(n)}(t) \in H(\Gamma)$ by Lemma 1.1. As the above mentioned reason, now we abandon the results in [4] and use a different method to solve (8).

By Corollary 2 of Lemma 3 in [3], the integral term in (8) belong to H^* near $-i$, moreover $A(-i) = a(\infty) \neq 0$, thus, if (8) has solution, then

$$\tilde{\Phi}^{(r)}(-i) = \left[\tilde{F}(t)/A(t)\right]^{(r)}_{t=-i}, \qquad r = 0, 1, 2, \cdots, n \tag{9}$$

Make interpolation polynomial $T(t)$ of degree n such that $T^{(r)}(-i) = \tilde{\Phi}^{(r)}(-i)$, $r = 0, 1, 2, \cdots, n$. Let

$$\omega(t) = \Phi(t) - T(t)/(t+i)^{n+1} = [\tilde{\Phi}(t) - T(t)]/(t+i)^{n+1} \tag{10}$$

By Lemma 3 in [3], $\omega(t) \in H^*$ near $-i$. Substituting (10) into (8), then (8) is written as

$$A(t)\omega(t) + \frac{B(t)}{\pi i}\int_\Gamma \frac{\omega(\tau)d\tau}{\tau - t} = F(t) - A(t)\frac{T(t)}{(t+i)^{n+1}} \equiv \mathbf{F}(t) \tag{11}$$

(By the extended residue theorem [1], $\int_\Gamma T(\tau)/[(\tau+i)^{n+1}(\tau-t)]d\tau = 0$.) By Lemma 3 in [3], $\mathbf{F}(t) \in H^*$ near $-i$. Thus (11) is a characteristic equation with solutions in the class $H^*(\Gamma)$. By [1], when $\kappa \geq 0$, (11) is always solvable with the general solution

$$\omega(t) = \mathbf{K}^*\mathbf{F} - B^*(t)Z(t)P_{\kappa-1}(t) \tag{12}$$

where

$$\kappa = \operatorname{Ind}_\Gamma \left\{\frac{A(t)-B(t)}{A(t)+B(t)}\right\} = \operatorname{Ind}_X \left\{\frac{a(x)-b(x)}{a(x)+b(x)}\right\},$$

$P_{\kappa-1}(t)$ is an arbitrary polynomial of degree $\kappa - 1$,

$$\mathbf{K}^*\mathbf{F} = A^*(t)\mathbf{F}(t) - B^*(t)Z(t)\frac{1}{\pi i}\int_\Gamma \frac{\mathbf{F}(\tau)}{Z(\tau)}\frac{1}{\tau-t}d\tau,$$

$A^*(t) = A(t)/[A^2(t)-B^2(t)]$, $B^*(t) = B(t)/[A^2(t)-B^2(t)]$, $Z(t) = [A(t)+B(t)]X^+(t) = [A(t)-B(t)]X^-(t)$, $X(w)$ is a canonical function of (11) for $w \notin \Gamma$ [1]. By (10)-(12), when $\kappa \geq 0$ the general solution of (8) is

$$\begin{aligned}\Phi(t) = \mathbf{K}^*F(t) &- \frac{B^*(t)B(t)T(t)}{(t+i)^{n+1}} - B^*(t)Z(t)P_{\kappa-1}(t)\\ &+\frac{B^*(t)Z(t)}{\pi i}\int_\Gamma \frac{A(\tau)T(\tau)d\tau}{Z(\tau)(\tau+i)^{n+1}(\tau-t)}.\end{aligned} \tag{13}$$

Using

$$\frac{1}{X^+(t)} + \frac{1}{X^-(t)} = \frac{2A(t)}{Z(t)}, \qquad \frac{1}{X^+(t)} - \frac{1}{X^-(t)} = \frac{2B(t)}{Z(t)} \tag{14}$$

and the extended residue theorem [1], then

$$\begin{aligned}&\frac{1}{\pi i}\int_\Gamma \frac{A(\tau)T(\tau)}{Z(\tau)(\tau+i)^{n+1}(\tau-t)}d\tau\\ &= \frac{1}{2\pi i}\int_{\Gamma+}\frac{T(\tau)}{X^+(\tau)(\tau+i)^{n+1}(\tau-t)}d\tau - \frac{1}{2\pi i}\int_{\Gamma-}\frac{T(\tau)}{X^-(\tau)(\tau+i)^{n+1}(\tau-t)}d\tau\\ &= \frac{1}{2}\frac{T(t)}{X^+(t)(t+i)^{n+1}} + \frac{1}{2}\frac{1}{n!}\left[\frac{\partial^n}{\partial\tau^n}\frac{T(\tau)}{X^+(\tau)(\tau-t)}\right]_{\tau=-i}\\ &\quad -\left\{\frac{1}{2}\frac{T(t)}{X^-(t)(t+i)^{n+1}}\right.\\ &\quad \left. + \frac{1}{2}\frac{1}{n!}\left[\frac{\partial^n}{\partial\tau^n}\frac{T(\tau)}{X^+(\tau)(\tau-t)}\right]_{\tau=-i} - PP\left[\frac{T(t)}{X(t)(t+i)^{n+1}}\,,\ \infty\right]\right\}\\ &= \frac{B(t)T(t)}{Z(t)(t+i)^{n+1}} + \frac{1}{n!}\left[\frac{\partial^n}{\partial\tau^n}\frac{B(\tau)T(\tau)}{Z(\tau)(\tau-t)}\right]_{\tau=-i} + PP\left[\frac{T(t)}{X(t)(t+i)^{n+1}}\,,\ \infty\right]\end{aligned} \tag{15}$$

where $PP[\cdots]$ in (15) is the principal part of $T(t)/[X(t)(t+i)^{n+1}]$ at ∞. It is a polynomial of degree $\kappa-1$ and can be merged into $P_{\kappa-1}$ in (12). The term after merging is still denoted as $P_{\kappa-1}(t)$ or $P_{\kappa-1}(-(t+i))$ for convenience to simplify latter. Substituting (15) into (13), we get

$$\Phi(t) = \mathbf{K}^* F - B^*(t)Z(t) \times \left(\sum_{k=0}^{n} \frac{1}{k!}\left[\frac{B(\tau)\tilde{F}(\tau)}{Z(\tau)A(\tau)}\right]^{(k)}_{\tau=-i} \frac{1}{(t+i)^{n+1-k}} + P_{\kappa-1}(-(t+i))\right). \tag{16}$$

Applying inverse transformation of (7) into (16) we obtain the general solution of (6) as

$$\varphi(x) = \mathbf{K}^* f - b^*(x)z(x)\left\{\sum_{k=0}^{n} \frac{(-1)^k}{k!}\left[(x+i)^{k+1}\left[(x+i)^{k-1}\frac{b(x)\tilde{f}(x)}{z(x)a(x)}\right]^{(k)}\right]_{x=\infty} \times (x+i)^{n-k} + \frac{C_1}{x+i} + \frac{C_2}{(x+i)^2} + \cdots + \frac{C_\kappa}{(x+i)^\kappa}\right\}, \tag{17}$$

where

$$\mathbf{K}^* f = a^*(x)f(x) - b^*(x)z(x)\frac{1}{\pi i}\int_{-\infty}^{\infty} \frac{f(\xi)}{z(\xi)(\xi - x)}d\xi,$$

$a^*(x) = a(x)/[a^2(x)+b^2(x)]$, $b^*(x) = b(x)/[a^2(x)+b^2(x)]$, $z(x) = [a(x)+b(x)]Y^+(x) = [a(x)-b(x)]Y^-(x)$, $Y(z) = e^{\Gamma(z)}$ for $z \in Z^+$, $Y(z) = [(z+i)/(z-i)]^{-\kappa}e^{\Gamma(z)}$ for $z \in Z^-$, and

$$\Gamma(z) = \frac{1}{2\pi i}\int_{-\infty}^{\infty} \log\left\{\left[\frac{z+i}{z-i}\right]^{-\kappa} \frac{a(x)-b(x)}{a(x)+b(x)}\right\}\frac{1}{x-Z}dx$$

for $z \notin X$.

When $\kappa < 0$, by [1], (11) has the unique solution (12) in which $P_{\kappa-1} \equiv 0$ (similarly, when $\kappa \geq 0$, (6) has the unique solution (17) in which $\sum_{j=0}^{k} c_j/(x+i)^j$ disappears) if and only if

$$\frac{1}{\pi i}\int_\Gamma \frac{\mathbf{F}(t)t^j}{Z(t)}dt = 0, \quad j = 0, 1, \cdots, -\kappa-1 \tag{18}$$

or

$$\frac{1}{\pi i}\int_\Gamma \frac{\mathbf{F}(t)(t+i)^j}{z(t)}dt = 0, \ j = 0, 1, \cdots, -\kappa-1. \tag{19}$$

By (11), (14) and the extended residue theorem [1], then (19) is equivalent to

$$\frac{1}{\pi i}\int_\Gamma \frac{F(t)(t+i)^j}{Z(t)}dt = \frac{1}{n!}\left[\frac{B(t)F(t)(t+i)^j}{A(t)z(t)}\right]^{(n)}_{t=-i}, \ j = 0, 1, \cdots, -\kappa-1. \tag{20}$$

Using (7) to (20) we get the following conditions of solvability of (6) when $\kappa < 0$

$$\frac{1}{\pi i}\int_{-\infty}^{+\infty}\frac{f(x)}{z(x)(x+i)^l}dx=\frac{(-1)^{n+1}}{n!}\left[(x+i)^{n+1}\left[(x+i)^{n-l}\frac{b(x)\tilde{f}(x)}{z(x)a(x)}\right]^{(n)}\right]_{x=\infty} \tag{21}$$

for $l = 1, 2, \cdots, -\kappa$. When $n \geq \kappa$ or $n \leq -\kappa$, by (20)(7), the piecewise representation of (21) can be written as follows respectively

$$\begin{aligned}&\frac{1}{\pi i}\int_{-\infty}^{+\infty}\frac{f(x)}{z(x)(x+i)^l}dx\\&=\frac{(-1)^{(n-l)}}{(n-l-1)!}\left[(x+i)^{n+2-l}\left[(x+i)^{n-l}\frac{b(x)\tilde{f}(x)}{z(x)a(x)}\right]^{(n-l+1)}\right]_{x=\infty}\end{aligned} \tag{22}$$

for $l = 1, 2, \cdots, -\kappa$, or

$$\begin{aligned}&\frac{1}{\pi i}\int_{-\infty}^{+\infty}\frac{f(x)}{z(x)(x+i)^l}dx=\\&\frac{(-1)^{n-l}}{(n-l+1)!}\left[(x+i)^{n+2-l}\left[(x+i)^{n-l}\frac{b(x)\tilde{f}(x)}{z(x)a(x)}\right]^{(n-l+1)}\right]_{x=\infty}\end{aligned} \tag{23}$$

for $1 \leq l \leq n+1$, and it is equal to 0 for $n+2 \leq l \leq -\kappa$.

Theorem 2.1. *When $\kappa \geq 0$, the characteristic equation (6) in $\widehat{H}_n^*$ is always solvable and its general solution containing κ arbitrary constants is (17). When $\kappa < 0$, (6) has the unique solution (17) (in which $\sum_{j=1}^{\kappa} C_j/x+i^j$ disappears) if and only if $-\kappa$ conditions (21) (or (22)) for $n \geq \kappa$, or (23) for $n \leq -\kappa$ hold.*

Special case. When $n = 0$, that is, $\varphi(x), f(x) \in \widehat{H}$, then (17) and (23) are respectively rewritten as

$$\varphi(x) = \mathbf{K}^* f - b^*(x)Z(x)\left[\frac{b(\infty)f(\infty)}{a(\infty)\sqrt{a^2(\infty)-b^2(\infty)}}+\frac{C_1}{x+i}+\cdots+\frac{C_\kappa}{(x+i)^\kappa}\right], \tag{24}$$

$$\frac{1}{\pi i}\int_{-\infty}^{+\infty}\frac{f(x)}{Z(x)(x+i)^l}dx=\begin{cases}\dfrac{-b(\infty)f(\infty)}{a(\infty)\sqrt{a^2(\infty)-b^2(\infty)}}, & l=1,\\ 0, & l=2,\ldots,-\kappa.\end{cases} \tag{25}$$

In particular, when $f \in \widehat{H}_0$, the term $b(\infty)f(\infty)/[a(\infty)\sqrt{a^2(\infty)-b^2(\infty)}]$ in (24)(25) vanishes.

3. Extended Noether Theorem for Complete Equation in $\widehat{H}_n^*$

For the complete equation with solutions in $\widehat{H}_n^*$

$$(\mathbf{K}\varphi)(x) = a(x)\varphi(x) + \frac{1}{\pi i}\int_{-\infty}^{+\infty} \frac{k(x,\xi)\varphi(\xi)}{\xi - x} d\xi = f(x), \quad x \in X', \tag{26}$$

where $\varphi(x), f(x) \in \widehat{H}_n^*$. $a(x), k(x,\xi) \in \widehat{H}^n, b(x) = k(x,x), a^2(x) - b^2(x) \neq 0 (x \in X), a(\infty) \neq 0$, we have the following theorem.

Theorem 3.1. *(Extended Noether Theorem)*

1° *The class of solutions of the homogeneous equation* $(\boldsymbol{K}\varphi)(x) = 0$, $(\boldsymbol{K}'\varphi)(x) = 0$ *does not expand, that is, the number of linearly independent solutions in* $\widehat{H}_n^* \equiv \widehat{H}_0$ *is finite.*

2° $l - l' = \kappa$, *where* l, l' *are the number of linearly independent solutions of* $(\boldsymbol{K}\varphi)(x) = 0, (\boldsymbol{K}'\varphi)(x) = 0$ *in* $\widehat{H}_0$ *respectively.*

3° (26) *is solvable if and only if*

$$\frac{1}{\pi i}\int_{-\infty}^{+\infty} f(x)\psi_j(x)dx$$
$$= \frac{(-1)^{(n+1)}}{n!}\left[(x+i)^{n+1}\left[(x+i)^n \frac{b(x)\tilde{f}(x)}{a(x)}\psi_j(x)\right]^{(n)}\right]_{x=\infty} \tag{27}$$

for $j = 1, 2, \ldots, l'$, *where* $\{\psi_j(x)\}_1^{l'}$ *is a complete system of solutions of* $(\boldsymbol{K}'\varphi)(x) = 0$ *in* $\widehat{H}_0$.

Proof. 1° Under (7), $(\mathbf{K}\varphi)(x) = 0$ is converted into

$$(\mathbf{K}\Phi)(t) = A(t)\Phi(t) + \frac{1}{\pi i}\int_\Gamma \frac{K(t,\tau)\Phi(\tau)}{\tau - t} d\tau = f(t), \quad t \in \Gamma' \tag{28}$$

where $\Phi(t) = \tilde{\Phi}(t)/(t+i)^{n+1}, \tilde{\Phi}(t) = \tilde{\varphi}(x), K(t,\tau) = k(x,\xi)$. By [3], the class of solution of (28) does not expand, i.e., $\widehat{H}_{n+1}^*(\Gamma) \equiv \widehat{H}^*(\Gamma)$ [3]. So $\Phi(t)$ can be written as

$$\Phi(t) = \Phi_*(t)/(t+i)^{\alpha+i\beta}, \quad 0 \leq \alpha < 1, \quad \Phi_*(t) \in H(\Gamma) \tag{29}$$

and (29) may further be rewritten as

$$\Phi(t) = \Phi_{**}(t)/(t+i), \quad \Phi_{**}(t) \in H(\Gamma) \text{ and } \Phi_{**}(-i) = 0. \tag{30}$$

Under (7) denote $\Phi_{**}(t) = \varphi_{**}(x)$ then $\varphi_{**}(x) \in \widehat{H}(X)$ and we easily prove $(\mathbf{K}\varphi_{**})(x) = 0$. Because $(\mathbf{K}\Phi)(t) = 0$ only has finite linearly independent solutions in $H_*(\Gamma)$, $(\mathbf{K}\varphi)(x) = 0$ has the same number of linearly independent solutions in $\widehat{H}_0(X)$. For $(\mathbf{K}'\varphi)(x) = 0$, the conclusions can similarly be proved.

2° Let $(\mathbf{K}'\Phi)(t) = 0$ be the equation corresponding $(\mathbf{K}'\psi)(t)$ under (7). On account of 1°, $l(l')$ is also the number of solutions for $(\mathbf{K}\Phi)(t) = 0\,((\mathbf{K}'\Psi)(t) = 0)$. By [3], $l - l' = \kappa$.

3° Under (7), (26) becomes $(\mathbf{K}\Phi)(t) = F(t)$, where $\Phi(t)$, $K(t,\tau)$ are the same as above, but $F(t) = \tilde{F}(t)/(t+i)^{n+1}, \tilde{F}(t) = \tilde{f}(x)$. By [3], $(\mathbf{K}\Phi)(t) = F(t)$ is solvable if and only if

$$\frac{1}{\pi i}\int_\Gamma F(t)\Psi_j^*(t)dt = \frac{1}{n!}\left[\frac{B(t)\tilde{F}(t)\Psi_j^*(t)}{A(t)}\right]^{(n)}_{t=-i}, \quad j = 1, 2, \cdots, l' \qquad (31)$$

where $\{\Psi_j^*(t)\}_1^{l'}$ is a complete system of solutions of $(\mathbf{K}'\Psi)(t) = 0$ in $H^*(\Gamma)$. Similarly to (30), we write $\Psi_j^*(t)$ as $\Psi_j^*(t) = \Psi_j(t)/(t+i)$, $\Psi_j(t) \in H(\Gamma)$ and $\Psi_j(-i) = 0$. Let $\{\Psi_j(t)\}_1^{l'} = \{\psi_j(x)\}_1^{l'}$; then $\{\psi_j(x)\}_1^{l'}$ is a complete system of solutions of $(\mathbf{K}'\psi)(x) = 0$ in $\widehat{H}_0(X)$. Applying (7) to (31) we have (27). □

References

1. J. Lu, Boundary Value Problems for Analytic Functions, World Scientific, Singapore, 1993.
2. J. Lu, On Integral Equations along Curves the Solutions of Which Have Singularities of Order One, J. Wuhan Univ. (Nat. Sci. Ed.), 1964(1), 1-13 (in Chinese).
3. S. Zhong, Generalized Noether Theorem for Singular Integral Equation in Solutions with Singularities of Higher Order, Chinese Ann. Math., 1998, **19A**(3), 361-366 (in Chinese).
4. S. Zhong, X. Zhao, Characteristic Singular Integral Equation in Solutions with Singularities of Higher Order (Part1), J. Wuhan Univ. (Nat. Sci. Ed.), 1997, **43**(5), 553-559 (in Chinese).
5. S. Zhong, X. Zhao, Characteristic Singular Integral Equation in Solutions with Singularities of Higher Order (Part2), J. Wuhan Univ. (Nat. Sci. Ed.), 1998, **44**(1), 5-8 (in Chinese).
6. S. Zhong, Singular Integral Equations along an Open Arc with Solutions Having Singularities of Higher Order, Acta Math. Scientia, 2005, **25B**(2), 193-200.
7. J. Lu, Singular Integrals of Higher Order and Their Applications to Solving Singular Integral Equations, Sci. and Tech. Wuhan Univ., 1977(2), 106-122 (in Chinese).
8. S. Zhong, Several Types of Formulas to Interchanging Order of Integration for Singular Integrals of Higher Order on the Real Axis, J. Math. (PRC), 2000, **20**(3), 355-358.

Session 06

Dispersive Equations

SESSION EDITORS

V. Georgiev	Università di Pisa, Pisa, Italy
M. Reissig	Technische Universität, Freiberg, Germany

ENERGY ESTIMATES FOR WAVE MODELS WITH TIME-DEPENDENT COEFFICIENTS

C. BÖHME and M. REISSIG

Fakultät für Mathematik und Informatik
TU Bergakademie Freiberg, Germany

This article contains statements for energy estimates of wave models with time-dependent coefficients. The main point is to form a scale-invariant model case of a Klein-Gordon equation.

Keywords: Wave equation, Klein-Gordon equation, energy estimates

1. Introduction

Statements about energy in Cauchy problems of the classical wave equation

$$u_{tt} - \Delta u = 0,\ u(0,x) = u_0(x),\ u_t(0,x) = u_1(x), \tag{1}$$

and the classical Klein Gordon equation

$$u_{tt} - \Delta u + m^2 u = 0,\ u(0,x) = u_0(x),\ u_t(0,x) = u_1(x), \tag{2}$$

form the setting of our investigations. Here, $(t,x) \in [0,\infty) \times \mathbb{R}^n$, $\Delta = \Delta_x$, the data (u_0, u_1) belongs to $H^1(\mathbb{R}^n) \times L^2(\mathbb{R}^n)$, and m^2 is a positive constant. We define the energy of the solution to the wave equation (1) by

$$E_{\mathrm{W}}(u)(t) := \frac{1}{2}\Big(\|u_t(t,\cdot)\|^2 + \|\nabla_x u(t,\cdot)\|^2 \Big), \tag{3}$$

consisting of a kinetic and an elastic energy term. Here, $\|\cdot\|$ denotes the usual L^2-norm in $\mathbb{R}^n$. The energy of the solution to the Klein-Gordon problem (2) is given by

$$E_{\mathrm{KG}}(u)(t) := \frac{1}{2}\Big(\|u_t(t,\cdot)\|^2 + \|\nabla_x u(t,\cdot)\|^2 + m^2\|u(t,\cdot)\|^2 \Big)$$

with an additional potential energy term. Both problems yield *energy conservation*, that is, $E_{\mathrm{W}}(u)(t) = E_{\mathrm{W}}(u)(0)$ and $E_{\mathrm{KG}}(u)(t) = E_{\mathrm{KG}}(u)(0)$, respectively, hold for all times $t \in [0,\infty)$.

In the following sections we will concentrate on wave equations and Klein-Gordon equations with time-dependent coefficients. Because of the appearance of those coefficients, we can not expect energy conservation. However, we are interested in similar energy estimates for suitably defined energies such that we can exclude blow-up as well as decay behavior of the energy for large times.

2. Scale invariant Klein-Gordon equation

As a model case for Klein-Gordon equations with time-dependent mass let us consider the Cauchy problem

$$u_{tt} - \Delta u + \frac{m^2}{(1+t)^2} u = 0,\ u(0,x) = u_0(x),\ u_t(0,x) = u_1(x), \tag{4}$$

where m^2 is a positive constant. The choice of this scale invariant model problem is motivated by studying the damped wave equation with time-dependent dissipation, see [5]. There, the dissipation term $m/(1+t)$ describes a critical case.

Let us return to problem (4) in order to show the following theorem.

Theorem 2.1. *Consider the Cauchy problem (4) with $u_0 \in H^1$, $u_1 \in L^2$. Then*

$$E_{\mathrm{KG}}(u)(t) \leq C E_{\mathrm{KG}}(u)(0)$$

holds for all times t and a positive constant C. The energy $E_{\mathrm{KG}}(u)(t)$ is defined by

$$E_{\mathrm{KG}}(u)(t) := \frac{1}{2}\Big(\|u_t(t,\cdot)\|^2 + \|\nabla_x u(t,\cdot)\|^2 + p(t)^2 \|u(t,\cdot)\|^2 \Big), \tag{5}$$

where

$$p(t) = \begin{cases} \frac{1}{(1+t)^{1/2}}, & m^2 > \frac{1}{4}, \\ \frac{1}{(1+t)^{1/2}(1+\ln(1+t))}, & m^2 = \frac{1}{4}, \\ \frac{1}{(1+t)^{(1+\sqrt{1-4m^2})/2}}, & m^2 \in (0, \frac{1}{4}). \end{cases}$$

The ideas of the proof are based on considerations of the scale invariant model case of the Klein-Gordon equation with singular potential, see [2].

Proof of Theorem 2.1. First, we transform the Cauchy problem (4) with partial Fourier transformation $\hat{u}(t,\xi) = F_{x\to\xi}(u(t,x))(\xi)$ in $\mathcal{S}'$ into

$$\hat{u}_{tt} + |\xi|^2 \hat{u} + \frac{m^2}{(1+t)^2}\hat{u} = 0,\ \hat{u}(0,\xi) = \hat{u}_0(\xi), \hat{u}_t(0,\xi) = \hat{u}_1(\xi).$$

Due to [2], we apply a second transformation with $\hat{u} = \tau^\rho \hat{v}$, $\tau = (1+t)|\xi|$ and $2\rho = 1 + \sqrt{1-4m^2}$ and get $\tau\hat{v}_{\tau\tau} + 2\rho\hat{v}_\tau + \tau\hat{v} = 0$. The third transformation with $z = 2i\tau$ and $\hat{w}(z) = e^{i\tau}\hat{v}(\tau)$ yields

$$z\hat{w}_{zz} + (2\rho - z)\hat{w}_z - \rho\hat{w} = 0. \tag{6}$$

This problem describes a *confluent hypergeometric equation.* We will obtain different fundamental solutions depending on m^2. First, we divide the phase space $[0,\infty) \times \mathbb{R}^n_\xi$ into three zones.

Definition 2.1. For a given constant $N > 0$, the zones $Z_1(N)$, $Z_2(N)$ and $Z_3(N)$ are defined by

$$\begin{aligned} Z_1(N) &:= \{(t,\xi) \in [0,\infty) \times \mathbb{R}^n : (1+t)|\xi| \le N\}, \\ Z_2(N) &:= \{(t,\xi) \in [0,\infty) \times \mathbb{R}^n : |\xi| \le N \le (1+t)|\xi|\}, \\ Z_3(N) &:= \{(t,\xi) \in [0,\infty) \times \mathbb{R}^n : |\xi| \ge N\}. \end{aligned}$$

The function t_ξ is defined by $t_\xi := \frac{N}{|\xi|} - 1$; it separates the zones $Z_1(N)$ and $Z_2(N)$.

2.1. *The case* $m^2 \ne \frac{1}{4}$

Due to [1], $\hat{w}_1(z) = \Phi(\alpha,\beta;z)$ and $\hat{w}_2(z) = z^{1-\beta}\Phi(1+\alpha-\beta, 2-\beta; z)$ form the fundamental system of solutions to the confluent hypergeometric equation $z\hat{w}_{zz} + (\beta - z)\hat{w}_z - \alpha\hat{w} = 0$. The following lemma gives us tools for investigating $\Phi = \Phi(\alpha,\beta;z)$ (see [1]).

Lemma 2.1. *(i)* Φ *is an entire function in* z.
(ii) $\Phi(\alpha,\beta;z) = 1$ *for* $|z| \to 0$.
(iii) $|\Phi(\alpha,\beta;z)| \le C$ *for* $|z| \le N$.
(iv) $|\Phi(\alpha,\beta;z)| \le C_{\alpha\beta}|z|^{\max(\mathrm{Re}(\alpha-\beta),-\mathrm{Re}\alpha)}$ *for large* $|z|$ *with* $0 < \arg z < \pi$.
(v) $(\beta-\alpha)\Phi(\alpha-1,\beta;z) = (\beta-\alpha-z)\Phi(\alpha,\beta;z) + zd_z\Phi(\alpha,\beta;z)$.

The fundamental solutions $\hat{w}_1(z) = \Phi(\rho,2\rho;z)$ and $\hat{w}_2(z) = z^{1-2\rho}\Phi(1-\rho, 2-2\rho; z)$ solve (6). After backward transformation, the solution $\hat{u}(t,\xi)$ can be written as $\hat{u}(t,\xi) = c_1(\xi)\tilde{u}_1(t,\xi) + c_2(\xi)\tilde{u}_2(t,\xi)$, where

$$\begin{aligned} \tilde{u}_1(t,\xi) &= ((1+t)|\xi|)^\rho\, e^{-i(1+t)|\xi|}\, \Phi(\rho, 2\rho; 2i(1+t)|\xi|), \\ \tilde{u}_2(t,\xi) &= ((1+t)|\xi|)^\rho\, e^{-i(1+t)|\xi|}\, (2i(1+t)|\xi|)^{1-2\rho} \\ &\quad \times \Phi(1-\rho, 2-2\rho; 2i(1+t)|\xi|). \end{aligned}$$

The application of (v) from Lemma 2.1 yields the first derivatives given by

$$\tilde{u}_{1,t}(t,\xi) = ((1+t)|\xi|)^{\rho-1}|\xi|e^{-i(1+t)|\xi|}$$

$$\times\Big(\rho\Phi(\rho-1,2\rho;2\mathrm{i}(1+t)|\xi|)+\mathrm{i}(1+t)|\xi|\Phi(\rho,2\rho;2\mathrm{i}(1+t)|\xi|)\Big);$$

$$\begin{aligned}\tilde{u}_{2,t}(t,\xi) &= ((1+t)|\xi|)^{\rho-1}\,|\xi|\,(2\mathrm{i}(1+t)|\xi|)^{(1-2\rho)}\,e^{-\mathrm{i}(1+t)|\xi|}\\ &\quad\times\Big((1-\rho)\Phi(-\rho,2-2\rho;2\mathrm{i}(1+t)|\xi|)\\ &\qquad+\mathrm{i}(1+t)|\xi|\Phi(1-\rho,2-2\rho;2\mathrm{i}(1+t)|\xi|)\Big).\end{aligned}$$

In general, we are interested in the solution in the form $\hat{u}(t,\xi) = E_1(t,s,\xi)\hat{u}(s,\xi) + E_2(t,s,\xi)\hat{u}_t(s,\xi)$, where

$$E_1(t,s,\xi) = \frac{\tilde{u}_{2,t}(s,\xi)\tilde{u}_1(t,\xi)-\tilde{u}_{1,t}(s,\xi)\tilde{u}_2(t,\xi)}{\tilde{u}_{2,t}(s,\xi)\tilde{u}_1(s,\xi)-\tilde{u}_{1,t}(s,\xi)\tilde{u}_2(s,\xi)},$$

$$E_2(t,s,\xi) = \frac{\tilde{u}_1(s,\xi)\tilde{u}_2(t,\xi)-\tilde{u}_2(s,\xi)\tilde{u}_1(t,\xi)}{\tilde{u}_{2,t}(s,\xi)\tilde{u}_1(s,\xi)-\tilde{u}_{1,t}(s,\xi)\tilde{u}_2(s,\xi)}$$

with $s,t \geq 0$. The first derivative is given by $\hat{u}_t(t,\xi) = E_{1,t}(t,s,\xi)\hat{u}(s,\xi) + E_{2,t}(t,s,\xi)\hat{u}_t(s,\xi)$ with

$$E_{1,t}(t,s,\xi) = \frac{\tilde{u}_{2,t}(s,\xi)\tilde{u}_{1,t}(t,\xi)-\tilde{u}_{1,t}(s,\xi)\tilde{u}_{2,t}(t,\xi)}{\tilde{u}_{2,t}(s,\xi)\tilde{u}_1(s,\xi)-\tilde{u}_{1,t}(s,\xi)\tilde{u}_2(s,\xi)},$$

$$E_{2,t}(t,s,\xi) = \frac{\tilde{u}_1(s,\xi)\tilde{u}_{2,t}(t,\xi)-\tilde{u}_2(s,\xi)\tilde{u}_{1,t}(t,\xi)}{\tilde{u}_{2,t}(s,\xi)\tilde{u}_1(s,\xi)-\tilde{u}_{1,t}(s,\xi)\tilde{u}_2(s,\xi)}.$$

First, we determine the denominator $\tilde{u}_{2,t}(s,\xi)\tilde{u}_1(s,\xi) - \tilde{u}_{1,t}(s,\xi)\tilde{u}_2(s,\xi)$. For this, we write the solutions of the fundamental system as $\tilde{u}_1(t,\xi) = f(t,\xi)\hat{w}_1(2\mathrm{i}(1+t)|\xi|)$ and $\tilde{u}_2(t,\xi) = f(t,\xi)\hat{w}_2(2\mathrm{i}(1+t)|\xi|)$, where $f(t,\xi) = ((1+t)|\xi|)^{\rho}e^{-\mathrm{i}(1+t)|\xi|}$ and $\tilde{u}_1$ and $\tilde{u}_2$ as above. The Wronskian of the fundamental solutions is $\hat{w}_1(z)\hat{w}_{2,z}(z) - \hat{w}_{1,z}(z)\hat{w}_2(z) = (1-2\rho)z^{-2\rho}e^{z}$. Thus

$$\begin{aligned}&\tilde{u}_{2,t}(t,\xi)\tilde{u}_1(t,\xi)-\tilde{u}_{1,t}(t,\xi)\tilde{u}_2(t,\xi)\\ &= f\hat{w}_1(f_t\hat{w}_2+f\hat{w}_{2,t})-f\hat{w}_2(f_t\hat{w}_1+f\hat{w}_{1,t})\\ &= f^2(\hat{w}_1\hat{w}_{2,t}-\hat{w}_{1,t}\hat{w}_2) \;=\; f^2 2\mathrm{i}|\xi|(\hat{w}_1\hat{w}_{2,z}-\hat{w}_{1,z}\hat{w}_2)\\ &= ((1+t)|\xi|)^{2\rho}e^{-2\mathrm{i}(1+t)|\xi|}2\mathrm{i}|\xi|(1-2\rho)(2\mathrm{i}(1+t)|\xi|)^{-2\rho}e^{2\mathrm{i}(1+t)|\xi|}\\ &= (2\mathrm{i})^{1-2\rho}(1-2\rho)|\xi|.\end{aligned}$$

Now, we handle the calculations of the multipliers $E_1, E_2, E_{1,t}$ and $E_{2,t}$ in the zones by applying Lemma 2.1. We notice that ρ is a real number with $\rho \in (\frac{1}{2},1)$ for $m^2 < \frac{1}{4}$. Then $1-\rho < \rho$ holds. In the case of $m^2 > \frac{1}{4}$, the constant $\rho = \frac{1}{2} + \frac{\mathrm{i}}{2}\sqrt{4m^2-1}$ is complex. Thus $\mathrm{Re}(1-\rho) = \mathrm{Re}\rho = \frac{1}{2}$.

Zone $Z_1(N)$: We obtain

$$|E_1(t,0,\xi)|,\ |E_2(t,0,\xi)| \lesssim (1+t)^{\mathrm{Re}\rho},$$

$$|E_{1,t}(t,0,\xi)|,\ |E_{2,t}(t,0,\xi)| \lesssim (1+t)^{\mathrm{Re}\rho-1}.$$

Thus, we deduce for the energy terms

$$\begin{aligned}
|\hat{u}_t(t,\xi)| &\lesssim |\hat{u}_0(\xi)| + |\hat{u}_1(\xi)|, \\
|\xi||\hat{u}(t,\xi)| &\lesssim |\hat{u}_0(\xi)| + |\hat{u}_1(\xi)| \quad \text{and} \\
|\hat{u}(t,\xi)| &\lesssim (|\hat{u}_0(\xi)| + |\hat{u}_1(\xi)|) \begin{cases} (1+t)^{1/2}, & m^2 > \frac{1}{4}, \\ (1+t)^{\rho}, & m^2 \in (0, \frac{1}{4}). \end{cases}
\end{aligned}$$

Zone $Z_2(N)$: Calculating the multipliers yields

$$\begin{aligned}
|E_1(t,t_\xi,\xi)| &\lesssim ((1+t_\xi)|\xi|)^{\mathrm{Re}\rho-1}, \quad |E_2(t,t_\xi,\xi)| \lesssim ((1+t_\xi)|\xi|)^{\mathrm{Re}\rho}|\xi|^{-1}, \\
|E_{1,t}(t,t_\xi,\xi)| &\lesssim ((1+t_\xi)|\xi|)^{\mathrm{Re}\rho-1}|\xi|, \quad |E_{2,t}(t,t_\xi,\xi)| \lesssim ((1+t_\xi)|\xi|)^{\mathrm{Re}\rho}.
\end{aligned}$$

We use the estimates for $|\hat{u}(t_\xi,\xi)|$ and $|\hat{u}_t(t_\xi,\xi)|$ determined in zone $Z_1(N)$

$$\begin{aligned}
|\hat{u}(t_\xi,\xi)| &\lesssim (1+t_\xi)^{\mathrm{Re}\rho}(|\hat{u}_0(\xi)| + |\hat{u}_1(\xi)|), \\
|\hat{u}_t(t_\xi,\xi)| &\lesssim (1+t_\xi)^{\mathrm{Re}\rho-1}(|\hat{u}_0(\xi)| + |\hat{u}_1(\xi)|) \lesssim (|\hat{u}_0(\xi)| + |\hat{u}_1(\xi)|).
\end{aligned}$$

This leads us to the estimates of the energy terms in zone $Z_2(N)$:

$$\begin{aligned}
|\hat{u}_t(t,\xi)| &\lesssim |\hat{u}_0(\xi)| + |\hat{u}_1(\xi)|, \\
|\xi||\hat{u}(t,\xi)| &\lesssim |\hat{u}_0(\xi)| + |\hat{u}_1(\xi)| \text{ and} \\
|\hat{u}(t,\xi)| &\lesssim |\hat{u}(t_\xi,\xi)| + ((1+t_\xi)|\xi|)^{\mathrm{Re}\rho}|\xi|^{-1}|\hat{u}_t(t_\xi,\xi)| \\
&\lesssim (|\hat{u}_0(\xi)| + |\hat{u}_1(\xi)|) \begin{cases} (1+t)^{1/2}, & m^2 > \frac{1}{4}, \\ (1+t)^{\rho}, & m^2 \in (0, \frac{1}{4}). \end{cases}
\end{aligned}$$

Zone $Z_3(N)$: Finally, in the zone $Z_3(N)$ we obtain

$$\begin{aligned}
|E_1(t,0,\xi)| &\lesssim C, \quad |E_2(t,0,\xi)| \lesssim |\xi|^{-1}, \\
|E_{1,t}(t,0,\xi)| &\lesssim |\xi|, \quad |E_{2,t}(t,0,\xi)| \lesssim C.
\end{aligned}$$

Consequently,

$$\begin{aligned}
|\hat{u}_t(t,\xi)| &\lesssim |\xi||\hat{u}_0(\xi)| + |\hat{u}_1(\xi)|, \\
|\xi||\hat{u}(t,\xi)| &\lesssim |\xi||\hat{u}_0(\xi)| + |\hat{u}_1(\xi)| \quad \text{and} \\
|\hat{u}(t,\xi)| &\lesssim |\hat{u}_0(\xi)| + |\hat{u}_1(\xi)|.
\end{aligned}$$

Summarizing we have shown for all (t,ξ) in the case $m^2 \neq 1/4$ that

$$E_{\mathrm{KG}}(u)(t) = \frac{1}{2}\Big(\|u_t(t,\cdot)\|^2 + \|\nabla_x u(t,\cdot)\|^2 + p(t)^2\|u(t,\cdot)\|^2 \Big)$$

$$\lesssim \frac{1}{2}\Big(\|u_1(\cdot)\|^2 + \|\nabla_x u_0(\cdot)\|^2 + \|u_0(\cdot)\|^2\Big) = CE_{\mathrm{KG}}(u)(0),$$

where $p(t) = \begin{cases} \frac{1}{(1+t)^{1/2}}, & m^2 > \frac{1}{4}, \\ \frac{1}{(1+t)^{(1+\sqrt{1-4m^2})/2}}, & m^2 \in (0, \frac{1}{4}). \end{cases}$

2.2. *The case* $m^2 = \frac{1}{4}$

We consider the transformed problem (6) with $\rho = 1/2$, the *logarithmic case*. For this problem the fundamental system of solutions is given by $\hat{w}_1(z) = \Psi(1/2, 1; z)$ and $\hat{w}_2(z) = e^z\Psi(1/2, 1; -z)$. In general, we need some properties of the function $\Psi = \Psi(\alpha, \beta; z)$ represented in the following lemma (see [1]).

Lemma 2.2. *(i)* Ψ *is as a single-valued function holomorphic in* $\mathbb{C} \setminus \{0\}$.
(ii) $\Psi(\alpha, 1; z) = -\frac{1}{\Gamma(\alpha)}(\ln z - \psi(\alpha) + 2\gamma) + O(|z \ln z|) \sim \ln z$ *for small* z, *where* $\psi(\alpha) = \frac{\Gamma'(\alpha)}{\Gamma(\alpha)}$ *and* γ *is Euler's constant.*
(iii) $|\Psi(\alpha, \beta; z)| \leq C_{\alpha\beta}|z|^{-\mathrm{Re}\alpha}$ *for large* $|z|$ *with* $-(3/2)\pi < \arg z < (3/2)\pi$.
(iv) $\Psi(\alpha - 1, \beta; z) = (\alpha - \beta + z)\,\Psi(\alpha, \beta; z) - z\, d_z\Psi(\alpha, \beta; z)$.

Transforming back the fundamental solutions yields

$$\tilde{u}_1(t, \xi) = ((1+t)|\xi|)^{\frac{1}{2}} e^{-\mathrm{i}(1+t)|\xi|}\Psi(\frac{1}{2}, 1; 2\mathrm{i}(1+t)|\xi|),$$
$$\tilde{u}_2(t, \xi) = ((1+t)|\xi|)^{\frac{1}{2}} e^{\mathrm{i}(1+t)|\xi|}\Psi(\frac{1}{2}, 1; -2\mathrm{i}(1+t)|\xi|)$$

and their first derivatives

$$\begin{aligned}\tilde{u}_{1,t}(t, \xi) &= ((1+t)|\xi|)^{-\frac{1}{2}}|\xi| e^{-\mathrm{i}(1+t)|\xi|} \\ &\quad\times\Big(-\Psi(-\frac{1}{2}, 1; 2\mathrm{i}(1+t)|\xi|) + i(1+t)|\xi|\Psi(\frac{1}{2}, 1; 2\mathrm{i}(1+t)|\xi|)\Big); \\ \tilde{u}_{2,t}(t, \xi) &= ((1+t)|\xi|)^{-\frac{1}{2}}|\xi| e^{\mathrm{i}(1+t)|\xi|} \\ &\quad\times\Big(-\Psi(-\frac{1}{2}, 1; -2\mathrm{i}(1+t)|\xi|) - i(1+t)|\xi|\Psi(\frac{1}{2}, 1; -2\mathrm{i}(1+t)|\xi|)\Big).\end{aligned}$$

We want to estimate the solution $\hat{u}(t, \xi) = E_1(t, s, \xi)\hat{u}(s, \xi) + E_2(t, s, \xi)\hat{u}_t(s, \xi)$ and the first derivative $\hat{u}_t(t, \xi) = E_{1,t}(t, s, \xi)\hat{u}(s, \xi) + E_{2,t}(t, s, \xi)\hat{u}_t(s, \xi)$, $s, t \geq 0$.

First, we have to determine the denominator $\tilde{u}_{2,t}(s, \xi)\tilde{u}_1(s, \xi) - \tilde{u}_{1,t}(s, \xi)\tilde{u}_2(s, \xi)$. Thus we set $\tilde{u}_1(t, \xi) = f(t, \xi)\hat{w}_1(2\mathrm{i}(1+t)|\xi|)$ and $\tilde{u}_2(t, \xi) =$

$f(t,\xi)\hat{w}_2(2\mathrm{i}(1+t)|\xi|)$, where $f(t,\xi) = ((1+t)|\xi|)^{(1/2)}e^{-\mathrm{i}(1+t)|\xi|}$. We need the Wronskian given by

$$\hat{w}_1(z)\hat{w}_{2,z}(z) - \hat{w}_2(z)\hat{w}_{1,z}(z) = e^{\pi\mathrm{i}\frac{1}{2}}z^{-1}e^z.$$

Furthermore, we have

$$\begin{aligned}
&\tilde{u}_{2,t}(t,\xi)\tilde{u}_1(t,\xi) - \tilde{u}_{1,t}(t,\xi)\tilde{u}_2(t,\xi)\\
&= f\hat{w}_1(f_t\hat{w}_2 + f\hat{w}_{2,t}) - f\hat{w}_2(f_t\hat{w}_1 + f\hat{w}_{1,t})\\
&= f^2(\hat{w}_1\hat{w}_{2,t} - \hat{w}_{1,t}\hat{w}_2) \;=\; f^2 2\mathrm{i}|\xi|(\hat{w}_1\hat{w}_{2,z} - \hat{w}_{1,z}\hat{w}_2)\\
&= f^2(2\mathrm{i}|\xi|)e^{\epsilon\pi\mathrm{i}\frac{1}{2}}(2\mathrm{i}(1+t)|\xi|)^{-1}e^{2\mathrm{i}(1+t)|\xi|)} \;=\; e^{\pi\mathrm{i}\frac{1}{2}}|\xi| \;=\; \mathrm{i}|\xi|.
\end{aligned}$$

We want to investigate the multipliers in the three zones $Z_1(N)$, $Z_2(N)$ and $Z_3(N)$ given in Definition 2.1. The estimates are based on Lemma 2.2.

Zone $Z_1(N)$: We start with $E_1(t,0,\xi)$ by applying Lemma 2.2(ii):

$$\begin{aligned}
|E_1(t,0,\xi)| \lesssim |\xi|^{-1}|\xi|^{\frac{1}{2}}((1+t)|\xi|)^{\frac{1}{2}} \times \big|&\ln(2\mathrm{i}(1+t)|\xi|)\,\ln(-2\mathrm{i}|\xi|)\,e^{-\mathrm{i}t|\xi|}\\
&- \ln(-2\mathrm{i}(1+t)|\xi|)\,\ln(2\mathrm{i}|\xi|)\,e^{\mathrm{i}t|\xi|}\big|.
\end{aligned}$$

The natural logarithm of a complex number $z = ae^{\mathrm{i}b}$ is given by $\ln z = \ln a + \mathrm{i}b$. Hence, we obtain

$$\ln(2\mathrm{i}(1+t)|\xi|)\,\ln(-2\mathrm{i}|\xi|) = \ln(2(1+t)|\xi|)\,\ln(2|\xi|) + \frac{\pi^2}{4} - \mathrm{i}\frac{\pi}{2}\ln(1+t),$$

$$\ln(-2\mathrm{i}(1+t)|\xi|)\,\ln(2\mathrm{i}|\xi|) = \ln(2(1+t)|\xi|)\,\ln(2|\xi|) + \frac{\pi^2}{4} + \mathrm{i}\frac{\pi}{2}\ln(1+t).$$

We have

$$\begin{aligned}
&\left|\ln(2\mathrm{i}(1+t)|\xi|)\,\ln(-2\mathrm{i}|\xi|)\,e^{-\mathrm{i}t|\xi|} - \ln(-2\mathrm{i}(1+t)|\xi|)\,\ln(2\mathrm{i}|\xi|)\,e^{\mathrm{i}t|\xi|}\right|\\
&= \Bigg|\left(\ln(2(1+t)|\xi|)\,\ln(2|\xi|) + \frac{\pi^2}{4}\right)\left(e^{-\mathrm{i}t|\xi|} - e^{\mathrm{i}t|\xi|}\right)\\
&\quad - \mathrm{i}\frac{\pi}{2}\ln(1+t)\left(e^{-\mathrm{i}t|\xi|} + e^{\mathrm{i}t|\xi|}\right)\Bigg|\\
&= \left|\left(\ln(2(1+t)|\xi|)\,\ln(2|\xi|) + \frac{\pi^2}{4}\right)2\sin(-t|\xi|) - \pi\ln(1+t)\cos(-t|\xi|)\right|\\
&\leq \left|\left(\ln(2(1+t)|\xi|)\,\ln(2|\xi|) + \frac{\pi^2}{4}\right)\sin(-t|\xi|) + \pi\ln(1+t)\right|\\
&\lesssim 1 + \ln(1+t).
\end{aligned}$$

We continue the computations for E_1 and obtain in the same way for E_2, $E_{1,t}$ and $E_{2,t}$:

$$|E_1(t,0,\xi)|, |E_2(t,0,\xi)| \lesssim (1+t)^{\frac{1}{2}}(1+\ln(1+t)) \quad \text{and}$$

$$|E_{1,t}(t,0,\xi)|, |E_{2,t}(t,0,\xi)| \lesssim (1+t)^{-\frac{1}{2}}(1+\ln(1+t)).$$

Finally, we get in the zone $Z_1(N)$ for all times t:

$$\begin{aligned}|\hat{u}_t(t,\xi)| &\lesssim |\hat{u}_0(\xi)| + |\hat{u}_1(\xi)|,\\ |\xi||\hat{u}(t,\xi)| &\lesssim |\hat{u}_0(\xi)| + |\hat{u}_1(\xi)| \quad \text{and}\\ |\hat{u}(t,\xi)| &\lesssim (1+t)^{\frac{1}{2}}(1+\ln(1+t))\,(|\hat{u}_0(\xi)| + |\hat{u}_1(\xi)|).\end{aligned}$$

Zone $Z_2(N)$: Similar to the considerations for $m^2 \neq 1/4$ we estimate the energy terms in $Z_2(N)$ by

$$\begin{aligned}|\hat{u}_t(t,\xi)| &\lesssim |\hat{u}_0(\xi)| + |\hat{u}_1(\xi)|,\\ |\xi||\hat{u}(t,\xi)| &\lesssim |\hat{u}_0(\xi)| + |\hat{u}_1(\xi)| \quad \text{and}\\ |\hat{u}(t,\xi)| &\lesssim (1+t)^{\frac{1}{2}}(1+\ln(1+t))(|\hat{u}_0(\xi)| + |\hat{u}_1(\xi)|).\end{aligned}$$

Zone $Z_3(N)$: In the third zone $Z_3(N)$ we obtain for $m^2 = 1/4$ exactly the same results as for $m^2 \neq 1/4$.

Altogether, we can estimate the energy in the case $m^2 = 1/4$ by

$$\begin{aligned}&E_{\mathrm{KG}}(u)(t)\\ &= \frac{1}{2}\Big(\|u_t(t,\cdot)\|^2 + \|\nabla_x u(t,\cdot)\|^2 + \frac{1}{(1+t)(1+\ln(1+t))^2}\|u(t,\cdot)\|^2\Big)\\ &\lesssim \frac{1}{2}\Big(\|u_1(\cdot)\|^2 + \|\nabla_x u_0(\cdot)\|^2 + \|u_0(\cdot)\|^2\Big) = CE_{\mathrm{KG}}(u)(0).\end{aligned}$$

Conclusively, the statements of Theorem 2.1 are shown. □

Remark 2.1. For the moment we have energy estimate only from above by the result of Theorem 2.1. Further investigations should bring lower bounds for the energy $E_{\mathrm{KG}}(u)(t)$.

3. Generalized energy conservation for wave equations

In contrast to the energy estimates of the scale-invariant Klein-Gordon equation, suitably defined energies of wave equations with time-dependent speed of propagation can be estimated from below by a constant depending on the initial energy.

Thus, we introduce the notion *generalized energy conservation*. That is, the estimates

$$C^{-1}E(u)(0) \leq E(u)(t) \leq CE(u)(0)$$

hold for a suitable energy E for a Cauchy problem with a suitable positive constant C.

Let us investigate the energy

$$E_W(u)(t) := \frac{1}{2}\Big(\|u_t(t,x)\|^2 + a(t)^2\|\nabla_x u(t,x)\|^2\Big) \tag{7}$$

for the Cauchy problem of the wave equation

$$u_{tt} - a(t)^2 \triangle u = 0,\ u(0,x) = u_0(x), u_t(0,x) = u_1(x). \tag{8}$$

The application of WKB representations leads to the first generalized energy conservation result given in [4].

Theorem 3.1. *(Reissig-Smith, [4]) Consider the Cauchy problem (8) with* $u_0 \in H^1$, $u_1 \in L^2$. *If* $a = a(t)$ *satisfies*

$$\begin{cases} a(t) \in C^2(\mathbb{R}_+), \\ 0 < a_1 \le a(t) \le a_2 < \infty \text{ for all } t, \\ \left|a^{(k)}(t)\right| \le C_k \left(\frac{1}{1+t}\right)^k, \ k = 1, 2, \end{cases} \tag{9}$$

then generalized energy conservation holds referring to the energy $E_W(u)(t)$ *given in (7) for all times* t *and a positive constant* C.

We can show that a higher regularity in (9) has no influence on the proof of Theorem 3.1. However, in [3] it is proved that higher regularity is useful under an additional stabilization condition on the propagation speed a and, consequently, changed conditions on the derivatives of a up to order $M \ge 2$. We summarize this result in the following theorem.

Theorem 3.2. *(Hirosawa, [3]) Consider the Cauchy problem (8) with* $u_0 \in H^1$, $u_1 \in L^2$. *If* $a = a(t)$ *satisfies*

$$\begin{cases} a(t) \in C^M(\mathbb{R}_+),\ M \ge 2, \\ 0 < a_1 \le a(t) \le a_2 < \infty \text{ for all } t, \\ \int_0^t |a(s) - a_\infty| ds \le C t^p \text{ for all } t \ge 1 \text{ and } p \in [0,1), \\ \left|a^{(k)}(t)\right| \le C_k \left(\frac{1}{1+t}\right)^{qk} \text{ for } k = 1, ..., M \text{ and } q < 1, \end{cases}$$

and $q \ge p + \frac{1-p}{M}$, *then generalized energy conservation holds referring to the energy* $E_W(u)(t)$ *given in (7) for all times* t *and a positive constant* C.

The question of generalized energy conservation for wave equations with low regularity propagation speed appears. We can give the following statement.

Theorem 3.3. *Consider the Cauchy problem (8) with* $u_0 \in H^1$, $u_1 \in L^2$. *If* $a = a(t)$ *satisfies*

$$\begin{cases} 0 < a_0 \le a(t) \le a_1 < \infty \text{ for all } t, \\ a'(t) \in L^1(\mathbb{R}_+), \end{cases}$$

then generalized energy conservation holds referring to the energy $E_W(u)(t)$ *given in (7) for all times* t *and a positive constant* C.

Sketch of the proof. First, we rewrite the Fourier-transformed Cauchy problem by the Liouville transformation for a damped wave equation with time-dependent dissipation term $b = b(t) \in L^1(\mathbb{R}_+)$. Due to [6], we have a scattering result to this problem. That means, the energy to the damped wave problem behaves asymptotically like the energy (3) of the classical wave equation (1). This implies the statements. □

Remark 3.1. We are also interested in statements about generalized energy conservation for Klein-Gordon equations with more general time-dependent mass $m = m(t)$ under certain conditions of the coefficient. Moreover, additional time-depending speed of propagation will play an essential role.

References

1. H. Bateman and A. Erdelyi, *Higher Transcendental Functions, vol. 1*, McGraw-Hill, New York, 1953.
2. D. Del Santo, T. Kinoshita and M. Reissig, *Rendiconti dell'Istituto di Matematica dell'Universita di Trieste* **39** (2007), 1–35.
3. F. Hirosawa, *Mathematische Annalen* **339** (2007), 819–838.
4. M. Reissig and J. Smith, *Hokkaido Mathematical Journal* **34** (2005), 541–586.
5. J. Wirth, *Mathematical Methods in the Applied Sciences* **27** (2004), 101–124.
6. J. Wirth, *Advances in Differential Equations* **12** (2007), 1115–1133.

THE CAUCHY PROBLEM FOR A HYPERBOLIC OPERATOR WITH LOG-ZYGMUND COEFFICIENTS

Daniele Del Santo

Dipartimento di Matematica e Informatica, Università di Trieste
Via A. Valerio 12/1, 34127 Trieste, Italy
E-mail: delsanto@units.it

We use the Littlewood-Paley dyadic decomposition to obtain a well-posedness result for a hyperbolic equation with coefficients log-Zygmund-continuous in t and C^∞ in x.

Keywords: Hyperbolic operators, Cauchy problem, energy estimates, Littlewood-Paley decomposition.

1. Introduction

Consider the operator

$$L = \partial_t^2 - \partial_x(a(t,x)\partial_x), \tag{1}$$

and suppose that L is strictly hyperbolic with bounded coefficients, i. e.

$$\Lambda_0 \geq a(t,x) \geq \lambda_0 > 0 \tag{2}$$

for all $(t,x) \in [0,T] \times \mathbb{R}$. If the coefficient a is Lipschitz–continuous with respect to t, uniformly with respect to x, and $\mathcal{C}^\infty$, bounded with bounded derivatives with respect to x, uniformly with respect to t, then the Cauchy problem for (1) is well–posed in $\mathcal{H}^\infty(\mathbb{R}) = \bigcap_s \mathcal{H}^s(\mathbb{R})$. This conclusion can be deduced from an energy estimate "without loss of derivatives" i.e. from an estimate of the type

$$\begin{aligned} &\sup_{0\leq t\leq T} \{\|u(t,\cdot)\|_{\mathcal{H}^{m+1}} + \|\partial_t u(t,\cdot)\|_{\mathcal{H}^m}\} \\ &\qquad \leq C_m(\|u(0,\cdot)\|_{\mathcal{H}^{m+1}} + \|\partial_t u(0,\cdot)\|_{\mathcal{H}^m} + \int_0^T \|Lu(t,\cdot)\|_{\mathcal{H}^m}\, dt), \end{aligned} \tag{3}$$

which is valid for all $u \in \mathcal{C}^2([0,T],\mathcal{H}^\infty(\mathbb{R}))$.

If the coefficient a is not Lipschitz–continuous with respect to t then the estimate (3) is no more true in general; nevertheless the $\mathcal{H}^\infty$–well–posedness may be recovered from an energy estimate "with loss of derivatives" (see e. g. the estimate (8) below) and it may be proved under weaker regularity assumption on a.

A result of this type was obtained in the well–known paper of Colombini, De Giorgi and Spagnolo [3], where the case of the coefficient a depending only on t is considered. The regularity condition under which the $\mathcal{H}^\infty$–well–posedness is proved is the following one: there exists $C > 0$ such that

$$\int_0^{T-\varepsilon} |a(t+\varepsilon) - a(t)|\, dt \leq C\varepsilon \log(\frac{1}{\varepsilon} + 1) \tag{4}$$

for all $\varepsilon \in (0, T]$. The energy estimate is obtained by using the Fourier transform with respect to x of the equation together with an approximation of the coefficient which is different in different zones of the phase space (the so called approximate energy technique, see [4]).

The case of the coefficient a depending on t and x was considered by Colombini and Lerner in [6], where the regularity condition was: there exists $C > 0$ such that

$$\sup_{(t,x)\in[0,T-\varepsilon]\times\mathbb{R}} |a(t+\varepsilon, x) - a(t,x)|\, dt \leq C\varepsilon \log(\frac{1}{\varepsilon} + 1) \tag{5}$$

for all $\varepsilon \in (0, T]$. In this situation the use of the Littlewood-Paley dyadic decomposition replaces the Fourier transform with respect to x, while the approximate energy technique remains a crucial point.

Recently, in [10], Tarama considered again the case of the coefficient a depending only on t, proving the $\mathcal{H}^\infty$–well–posedness under the condition: there exists $C > 0$ such that

$$\int_\varepsilon^{T-\varepsilon} |a(t+\varepsilon) + a(t-\varepsilon) - 2a(t)|\, dt \leq C\varepsilon \log(\frac{1}{\varepsilon} + 1) \tag{6}$$

for all $\varepsilon \in (0, T/2]$. The improvement with respect to [3] is obtained introducing a new type of approximate energy which involves the second derivatives of the approximating coefficient (see section 3.2 below).

In the present communication the case of the coefficient depending on t and x, under a regularity condition inspired by (6) and (5), will be studied. The dyadic decomposition technique will be applied as in [6] (see also [2] and [7]) together with Tarama's approximate energy.

Acknowledgments

I want to thank J.–M. Bony, F. Colombini, G. Métivier and M. Reissig for the useful discussions on the topics of this communication and the anonymous referee for the corrections and the improvements suggested.

2. Results

Consider the operator (1) under the condition (2). Suppose that there exists $C > 0$ such that

$$\sup_{(t,x)\in[\varepsilon,T-\varepsilon]\times\mathbb{R}} |a(t+\varepsilon,x)+a(t-\varepsilon,x)-2a(t,x)| \le C\varepsilon\log(\frac{1}{\varepsilon}+1) \tag{7}$$

for all $\varepsilon \in (0,T/2]$. Suppose moreover that for all $t \in [0,T]$ the function $x \mapsto a(t,x)$ is $\mathcal{C}^\infty(\mathbb{R})$ and for all $\alpha \in \mathbb{N}$ the functions $(t,x) \mapsto \partial_x^\alpha a(t,x)$ are $\mathcal{L}^\infty([0,T]\times\mathbb{R})$ (we resume all these requirements on the regularity of the coefficient a saying that a is log–Zygmund–continuous in t, uniformly with respect to x, and $\mathcal{B}^\infty$ in x, uniformly with respect to t).

The main result of this communication is the following one.

Theorem 2.1. *The Cauchy problem for (1) is well–posed in* $\mathcal{H}^\infty$.

This result is a consequence of the energy estimate "with loss of derivatives" contained in the following statement.

Theorem 2.2. *There exists* $\beta > 0$ *and, for all* $m \in \mathbb{R}$, *there exists* $C_m > 0$ *such that*

$$\begin{aligned}&\sup_{0\le t\le T}\{\|u(t,\cdot)\|_{\mathcal{H}^{m+1-\beta t}} + \|\partial_t u(t,\cdot)\|_{\mathcal{H}^{m-\beta t}}\}\\ &\quad\le C_m(\|u(0,\cdot)\|_{\mathcal{H}^{m+1}} + \|\partial_t u(0,\cdot)\|_{\mathcal{H}^m} + \int_0^T \|Lu(t,\cdot)\|_{\mathcal{H}^{m-\beta t}}\,dt)\end{aligned} \tag{8}$$

for all $u \in C^2([0,T],\mathcal{H}^\infty(\mathbb{R}))$.

3. Proof of Theorem 2.2

3.1. *Dyadic decomposition*

We collect here some well–known facts on the Littlewood-Paley decomposition, referring to [1] and [6] for the details. Let $\varphi_0 \in C_0^\infty(\mathbb{R}_\xi)$, $0 \le \varphi_0(\xi) \le 1$, $\varphi_0(\xi) = 1$ if $|\xi| \le 1$, $\varphi_0(\xi) = 0$ if $|\xi| \ge 2$, φ_0 even and φ_0 decreasing on $[0,+\infty)$. We set $\varphi(\xi) = \varphi_0(\xi/2) - \varphi_0(\xi)$ and, if ν is an integer greater or equal than 1, $\varphi_\nu(\xi) = \varphi(2^{-\nu}\xi)$. Let w be a tempered distribution in $\mathcal{H}^{-\infty}(\mathbb{R})$; we define

$$w_\nu(x) = \varphi_\nu(D_x)w(x) = \frac{1}{2\pi}\int e^{ix\xi}\varphi_\nu(\xi)\hat{w}(\xi)\,d\xi = \frac{1}{2\pi}\int \hat{\varphi}_\nu(y)w(x-y)\,dy.$$

For all ν, w_ν is an entire analytic function belonging to $\mathcal{L}^2$ and for all $m \in \mathbb{R}$ there exists $K_m > 0$ such that

$$\frac{1}{K_m}\sum_{\nu=0}^\infty \|w_\nu\|_{\mathcal{L}^2}^2 2^{2m\nu} \le \|w\|_{\mathcal{H}^m}^2 \le K_m \sum_{\nu=0}^\infty \|w_\nu\|_{\mathcal{L}^2}^2 2^{2m\nu}.$$

Moreover we have

$$2^{\nu-1}\|w_\nu\|_{\mathcal{L}^2} \leq \|\partial_x w_\nu\|_{\mathcal{L}^2} \leq 2^{\nu+1}\|w_\nu\|_{\mathcal{L}^2}, \tag{9}$$

where the inequality on the right–hand side holds for all ν while the other one holds only for all $\nu \geq 1$.

3.2. *Approximation of the coefficient*

We extend the value of a to $\mathbb{R}^2$ setting $a(t,x) = a(0,x)$ if $t < 0$ and $a(t,x) = a(T,x)$ if $t > T$. The condition (7) remains valid with the supremum evaluated on $\mathbb{R}^2$. We set

$$a_\varepsilon(t,x) = \int_{\mathbb{R}} \rho_\varepsilon(t-s)a(s,x)\,ds,$$

where $\rho_\varepsilon(s) = \frac{1}{\varepsilon}\rho(\frac{s}{\varepsilon})$ with $\rho \in C^\infty(\mathbb{R})$, $0 \leq \rho \leq 1$, $\operatorname{supp}\rho \subseteq [-1,1]$ and $\int \rho(s)\,ds = 1$. We obtain that there exists $C > 0$ such that

$$\sup_{(t,x)\in\mathbb{R}^2} |a_\varepsilon(t,x) - a(t,x)| \leq C\varepsilon \log(\frac{1}{\varepsilon}+1) \tag{10}$$

and

$$\sup_{(t,x)\in\mathbb{R}^2} |\partial_t^2 a_\varepsilon(t,x)| + |\partial_t a_\varepsilon(t,x)|^2 \leq C\frac{1}{\varepsilon}\log(\frac{1}{\varepsilon}+1), \tag{11}$$

for all $\varepsilon > 0$. In particular (10) is obtained from (7) remarking that

$$a_\varepsilon(t,x) - a(t,x) = \frac{1}{2}\int_{|s|\leq\varepsilon} \rho_\varepsilon(s)(a(t-s,x) + a(t+s,x) - 2a(t,x))ds,$$

where we have used the fact that ρ is an even function. Similarly

$$\partial_t^2 a_\varepsilon(t,x) = \frac{1}{2}\int_{|s|\leq\varepsilon} \rho_\varepsilon''(s)(a(t-s,x) + a(t+s,x) - 2a(t,x))ds,$$

and the estimate for $|\partial_t^2 a_\varepsilon(t,x)|$ follows as before using (7), while that one for $|\partial_t a_\varepsilon(t,x)|^2$ is a consequence of a standard argument similar to that one used to prove the Glaeser inequality (see [8] or [9]) using the bound on a_ε and $\partial_t^2 a_\varepsilon$.

Finally, since $0 < \lambda_0 \leq a_\varepsilon(t,x), a(t,x) \leq \Lambda_0$, if we set $\sqrt{a}(t,x) = (a(t,x))^{1/2}$ and $\sqrt{a_\varepsilon}(t,x) = (a_\varepsilon(t,x))^{1/2}$, it is easy to deduce from (11) that there exists $C > 0$ such that

$$\sup_{(t,x)\in\mathbb{R}^2} |\partial_t^2 \sqrt{a_\varepsilon}(t,x)| + |\partial_t \sqrt{a_\varepsilon}(t,x)|^2 \leq C\frac{1}{\varepsilon}\log(\frac{1}{\varepsilon}+1), \tag{12}$$

for all $\varepsilon > 0$.

3.3. *Approximate energy of the ν-component*

Let $u(t,x)$ be a function in $C^2([0,T],\mathcal{H}^\infty(\mathbb{R}^n))$. We set $u_\nu(t,x) = \varphi_\nu(D)u(t,x)$. We obtain $\partial_t^2 u_\nu = \partial_x(a(t,x)\partial_x u_\nu) + \partial_x([\varphi_\nu, a]\partial_x u) + (Lu)_\nu$, where $[\varphi_\nu, a]$ is the commutator between $\varphi_\nu(D_x)$ and a. We introduce the approximate energy of u_ν setting

$$e_{\nu,\varepsilon}(t) = \int_{\mathbb{R}} \frac{1}{\sqrt{a_\varepsilon}} |\partial_t u_\nu + \frac{\partial_t\sqrt{a_\varepsilon}}{2\sqrt{a_\varepsilon}} u_\nu|^2 + \sqrt{a_\varepsilon}|\partial_x u_\nu|^2 + |u_\nu|^2\,dx.$$

We have

$$\begin{aligned}
\frac{d}{dt}e_{\nu,\varepsilon}(t) &= \int \frac{2}{\sqrt{a_\varepsilon}}(\partial_t(\frac{\partial_t\sqrt{a_\varepsilon}}{2\sqrt{a_\varepsilon}}) - (\frac{\partial_t\sqrt{a_\varepsilon}}{2\sqrt{a_\varepsilon}})^2)\mathrm{Re}\,(u_\nu\cdot\overline{(\partial_t u_\nu + \frac{\partial_t\sqrt{a_\varepsilon}}{2\sqrt{a_\varepsilon}} u_\nu)})\,dx \\
&+ \int \frac{2}{\sqrt{a_\varepsilon}}\mathrm{Re}\,(\partial_t^2 u_\nu\cdot\overline{(\partial_t u_\nu + \frac{\partial_t\sqrt{a_\varepsilon}}{2\sqrt{a_\varepsilon}} u_\nu)})\,dx \\
&+ \int \partial_t\sqrt{a_\varepsilon}|\partial_x u_\nu|^2 + 2\sqrt{a_\varepsilon}\mathrm{Re}\,(\partial_x u_\nu\cdot\overline{\partial_x\partial_t u_\nu})\,dx \\
&+ \int 2\mathrm{Re}\,(u_\nu\cdot\overline{(\partial_t u_\nu + \frac{\partial_t\sqrt{a_\varepsilon}}{2\sqrt{a_\varepsilon}} u_\nu)})\,dx \\
&- \int 2\mathrm{Re}\,(u_\nu\cdot\overline{\frac{\partial_t\sqrt{a_\varepsilon}}{2\sqrt{a_\varepsilon}} u_\nu})\,dx.
\end{aligned} \tag{13}$$

From (9) and (12) we deduce that

$$\begin{aligned}
\int \frac{2}{\sqrt{a_\varepsilon}}(\partial_t(\frac{\partial_t\sqrt{a_\varepsilon}}{2\sqrt{a_\varepsilon}}) - (\frac{\partial_t\sqrt{a_\varepsilon}}{2\sqrt{a_\varepsilon}})^2)\mathrm{Re}\,(u_\nu\cdot\overline{(\partial_t u_\nu + \frac{\partial_t\sqrt{a_\varepsilon}}{2\sqrt{a_\varepsilon}} u_\nu)})\,dx \\
\le C\frac{1}{\varepsilon}\log(\frac{1}{\varepsilon}+1)\,2^{-\nu}e_{\nu,\varepsilon}(t),
\end{aligned}$$

where $C > 0$ does not depend on u or ν. Next we have

$$\begin{aligned}
A_1 &:= \int \frac{2}{\sqrt{a_\varepsilon}}\mathrm{Re}\,(\partial_t^2 u_\nu\cdot\overline{(\partial_t u_\nu + \frac{\partial_t\sqrt{a_\varepsilon}}{2\sqrt{a_\varepsilon}} u_\nu)})\,dx \\
&= \int \frac{2}{\sqrt{a_\varepsilon}}\,a\,\mathrm{Re}\,(\partial_x^2 u_\nu\cdot\overline{(\partial_t u_\nu + \frac{\partial_t\sqrt{a_\varepsilon}}{2\sqrt{a_\varepsilon}} u_\nu)})\,dx \\
&+ \int \frac{2}{\sqrt{a_\varepsilon}}\,\partial_x a\,\mathrm{Re}\,(\partial_x u_\nu\cdot\overline{(\partial_t u_\nu + \frac{\partial_t\sqrt{a_\varepsilon}}{2\sqrt{a_\varepsilon}} u_\nu)})\,dx \\
&+ \int \frac{2}{\sqrt{a_\varepsilon}}\,\mathrm{Re}\,((\partial_x([\varphi_\nu, a]\partial_x u) + (Lu)_\nu)\cdot\overline{(\partial_t u_\nu + \frac{\partial_t\sqrt{a_\varepsilon}}{2\sqrt{a_\varepsilon}} u_\nu)})\,dx,
\end{aligned}$$

and

$$\begin{aligned}A_2 &:= \int \partial_t\sqrt{a_\varepsilon}|\partial_x u_\nu|^2 + 2\sqrt{a_\varepsilon}\mathrm{Re}\,(\partial_x u_\nu \cdot \overline{\partial_x\partial_t u_\nu})\,dx \\ &= -\int \frac{2}{\sqrt{a_\varepsilon}}\, a_\varepsilon\,\mathrm{Re}\,(\partial_x^2 u_\nu \cdot \overline{(\partial_t u_\nu + \frac{\partial_t\sqrt{a_\varepsilon}}{2\sqrt{a_\varepsilon}}u_\nu)})\,dx \\ &\quad - \int 2\,\partial_x\sqrt{a_\varepsilon}\,\mathrm{Re}\,(\partial_x u_\nu \cdot \overline{(\partial_t u_\nu + \frac{\partial_t\sqrt{a_\varepsilon}}{2\sqrt{a_\varepsilon}}u_\nu)})\,dx \\ &\quad + \int (\frac{\partial_x\sqrt{a_\varepsilon}\partial_t\sqrt{a_\varepsilon}}{\sqrt{a_\varepsilon}} - \partial_x\partial_t\sqrt{a_\varepsilon})\,\mathrm{Re}\,(\partial_x u_\nu \cdot \overline{u_\nu})\,dx.\end{aligned}$$

Since, from (9) and (10), there exists $C > 0$ not depending on u or ν such that

$$\begin{aligned}&\int \frac{2}{\sqrt{a_\varepsilon}}(a - a_\varepsilon)\,\mathrm{Re}\,(\partial_x^2 u_\nu \cdot \overline{(\partial_t u_\nu + \frac{\partial_t\sqrt{a_\varepsilon}}{2\sqrt{a_\varepsilon}}u_\nu)})\,dx \\ &\qquad \le C\varepsilon\log(\frac{1}{\varepsilon}+1)\,2^\nu\,\|\partial_x u_\nu\|_{\mathcal{L}^2}\|\partial_t u_\nu + \frac{\partial_t\sqrt{a_\varepsilon}}{2\sqrt{a_\varepsilon}}u_\nu\|_{\mathcal{L}^2},\end{aligned}$$

and remarking that $\sup_{(t,x)\in\mathbb{R}^2}|\frac{\partial_x\sqrt{a_\varepsilon}\partial_t\sqrt{a_\varepsilon}}{\sqrt{a_\varepsilon}}| + |\partial_x\partial_t\sqrt{a_\varepsilon}| \le C/\varepsilon$, we obtain

$$\begin{aligned}|A_1 + A_2| &\le C(\varepsilon(1+\log(\frac{1}{\varepsilon}+1))\,2^\nu + 1 + \frac{2^{-\nu}}{\varepsilon})e_{\nu,\varepsilon}(t) \\ &\quad + \int \frac{2}{\sqrt{a_\varepsilon}}\,\mathrm{Re}\,((\partial_x([\varphi_\nu, a]\partial_x u) + (Lu)_\nu)\cdot \overline{(\partial_t u_\nu + \frac{\partial_t\sqrt{a_\varepsilon}}{2\sqrt{a_\varepsilon}}u_\nu)})\,dx.\end{aligned}$$

Similarly, we have

$$\int 2\mathrm{Re}\,(u_\nu \cdot \overline{(\partial_t u_\nu + \frac{\partial_t\sqrt{a_\varepsilon}}{2\sqrt{a_\varepsilon}}u_\nu)})\,dx \le Ce_{\nu,\varepsilon}(t),$$

and

$$\int 2\mathrm{Re}\,(u_\nu \cdot \overline{\frac{\partial_t\sqrt{a_\varepsilon}}{2\sqrt{a_\varepsilon}}u_\nu})\,dx \le C\,\frac{2^{-\nu}}{\varepsilon}\,e_{\nu,\varepsilon}(t).$$

We choose $\varepsilon = 2^{-\nu}$ and we set $e_{\nu,2^{-\nu}}(t) = e_\nu(t)$, $a_{2^{-\nu}} = \alpha_\nu$. As a conclusion we deduce that there exists $C_0 > 0$ not depending on u or ν such that

$$\begin{aligned}\frac{d}{dt}e_\nu(t) &\le C_0(\nu+1)e_\nu(t) \\ &\quad + \int \frac{2}{\sqrt{\alpha_\nu}}\mathrm{Re}\,(\partial_x([\varphi_\nu, a]\partial_x u)\cdot \overline{(\partial_t u_\nu + \frac{\partial_t\sqrt{\alpha_\nu}}{2\sqrt{\alpha_\nu}}u_\nu)})\,dx \\ &\quad + \int \frac{2}{\sqrt{\alpha_\nu}}\mathrm{Re}\,((Lu)_\nu \cdot \overline{(\partial_t u_\nu + \frac{\partial_t\sqrt{\alpha_\nu}}{2\sqrt{\alpha_\nu}}u_\nu)})\,dx.\end{aligned} \tag{14}$$

3.4. *Estimate of the total energy*

We define

$$E_m(t) = \sum_{\nu=0}^{\infty} e^{-2\beta^*(\nu+1)t} 2^{2\nu m} e_\nu(t), \tag{15}$$

with β^* to be fixed. It is possible to prove that there exists C_1, $C_2 > 0$ such that

$$E_m(0) \leq C_1(\|\partial_t u(0,\cdot)\|^2_{\mathcal{H}^m} + \|u(0,\cdot)\|^2_{\mathcal{H}^{m+1}}) \tag{16}$$

and

$$E_m(t) \geq C_2(\|\partial_t u(t,\cdot)\|^2_{\mathcal{H}^{m-\beta t}} + \|u(t,\cdot)\|^2_{\mathcal{H}^{m+1-\beta t}}), \tag{17}$$

where $\beta = \beta^*(\log 2)^{-1}$. Choosing $\beta^* = C_0/2$, from (14) we deduce

$$\begin{aligned} &\frac{d}{dt}E_m(t) \leq \\ &\sum_{\nu=0}^{\infty} e^{-2\beta^*(\nu+1)t} 2^{2\nu m} \int \frac{2}{\sqrt{\alpha_\nu}} \mathrm{Re}\,(\partial_x([\varphi_\nu, a]\partial_x u)\cdot\overline{(\partial_t u_\nu + \frac{\partial_t\sqrt{\alpha_\nu}}{2\sqrt{\alpha_\nu}} u_\nu)})\, dx \\ &\quad + \sum_{\nu=0}^{\infty} e^{-2\beta^*(\nu+1)t} 2^{2\nu m} \int \frac{2}{\sqrt{\alpha_\nu}} \mathrm{Re}\,((Lu)_\nu\cdot\overline{(\partial_t u_\nu + \frac{\partial_t\sqrt{\alpha_\nu}}{2\sqrt{\alpha_\nu}} u_\nu)})\, dx. \end{aligned} \tag{18}$$

It is not difficult to show that there exists $C_3 > 0$ such that

$$\begin{aligned} \sum_{\nu=0}^{\infty} e^{-2\beta^*(\nu+1)t} 2^{2\nu m} \int \frac{2}{\sqrt{\alpha_\nu}} \mathrm{Re}\,((Lu)_\nu\cdot\overline{(\partial_t u_\nu + \frac{\partial_t\sqrt{\alpha_\nu}}{2\sqrt{\alpha_\nu}} u_\nu)})\, dx \\ \leq C_3(E_m(t))^{1/2}\|Lu(t,\cdot)\|_{\mathcal{H}^{m-\beta t}}. \end{aligned} \tag{19}$$

3.5. *Estimate of the commutator term*

The estimate of the first term in the right–hand side part of (18) is more delicate. We refer to [6, Lemmas 4.4 and 4.5] for the details, giving here only a sketch of the argument. First of all it is possible to prove that for all $N \in \mathbb{N}$ there exists $C_N > 0$ such that

$$\|[\varphi_\nu, a]\varphi_\mu\|_{L(\mathcal{L}^2,\mathcal{L}^2)} \leq \begin{cases} C_0 2^{-\nu} & \text{if } |\mu-\nu| \leq 2, \\ C_N 2^{-N\max\{\nu,\mu\}} & \text{if } |\mu-\nu| \geq 3, \end{cases} \tag{20}$$

where $\|\cdot\|_{L(\mathcal{L}^2,\mathcal{L}^2)}$ denotes the norm as operator from $\mathcal{L}^2$ to $\mathcal{L}^2$. Setting $\varphi_{-1} \equiv 0$ we have that

$$[\varphi_\nu, a]\partial_x u = \sum_{\mu=0}^{+\infty}([\varphi_\nu, a](\varphi_{\mu-1} + \varphi_\mu + \varphi_{\mu+1}))\partial_x u_\mu,$$

so that

$$\int \frac{2}{\sqrt{\alpha_\nu}} \mathrm{Re}\,(\partial_x([\varphi_\nu, a]\partial_x u)\cdot\overline{(\partial_t u_\nu + \frac{\partial_t\sqrt{\alpha_\nu}}{2\sqrt{\alpha_\nu}} u_\nu)})\, dx$$

$$= -\sum_{\mu=0}^{\infty} \int \mathrm{Re}\,(([\varphi_\nu, a](\varphi_{\mu-1} + \varphi_\mu + \varphi_{\mu+1})\partial_x u_\mu)$$

$$\cdot\partial_x(\frac{2}{\sqrt{\alpha_\nu}}\overline{(\partial_t u_\nu + \frac{\partial_t\sqrt{\alpha_\nu}}{2\sqrt{\alpha_\nu}} u_\nu)})\, dx,$$

and, consequently,

$$\int \frac{2}{\sqrt{\alpha_\nu}} \mathrm{Re}\,(\partial_x([\varphi_\nu, a]\partial_x u)\cdot\overline{(\partial_t u_\nu + \frac{\partial_t\sqrt{\alpha_\nu}}{2\sqrt{\alpha_\nu}} u_\nu)})\, dx$$

$$\leq 3\sum_{\mu=0}^{\infty} \|([\varphi_\nu, a]\varphi_\mu)\partial_x u_\mu\|_{\mathcal{L}^2} \|\partial_x(\frac{2}{\sqrt{\alpha_\nu}}\overline{(\partial_t u_\nu + \frac{\partial_t\sqrt{\alpha_\nu}}{2\sqrt{\alpha_\nu}} u_\nu)})\|_{\mathcal{L}^2}.$$

It is possible to show that

$$\|\partial_x(\frac{2}{\sqrt{\alpha_\nu}}\overline{(\partial_t u_\nu + \frac{\partial_t\sqrt{\alpha_\nu}}{2\sqrt{\alpha_\nu}} u_\nu)})\|_{\mathcal{L}^2} \leq C\, 2^\nu\, e_\nu^{1/2}(t).$$

Finally the estimate (20) together with Schur's lemma give that there exists $C_4 > 0$ such that

$$\sum_{\nu=0}^{\infty} e^{-2\beta^*(\nu+1)t} 2^{2\nu m} \int \frac{2}{\sqrt{\alpha_\nu}} \mathrm{Re}\,(\partial_x([\varphi_\nu, a]\partial_x u)\cdot\overline{(\partial_t u_\nu + \frac{\partial_t\sqrt{\alpha_\nu}}{2\sqrt{\alpha_\nu}} u_\nu)})\, dx \leq C_4 E_m(t). \tag{21}$$

3.6. *End of the proof*

From (18), (19) and (21) we obtain

$$\frac{d}{dt} E_m(t) \leq C_3 (E_m(t))^{1/2} \|Lu(t,\cdot)\|_{\mathcal{H}^{m-\beta t}} + C_4 E_m(t),$$

and (8) easy follows from (16) and (17).

4. Remarks and Problems

Using the technique sketched in this communication it is possible to prove some local in time energy estimates in Sobolev spaces with "loss of derivatives" for operator having limited regularity in the x variable, e. g. if the coefficient a is C^α with $\alpha > 1$.

Some problems on this topic remain open.

- What to do with hyperbolic operators depending on more than one space variable?
- What to do with systems?
- What to do with a regularity condition like (6) (i. e. an integral condition) but with the coefficient a depending also on x?
- What to do with an isotropic (i. e. involving both t and x) log–Zygmund regularity condition on a?

All these questions will be considered in a forthcoming paper [5].

References

1. J.-M. Bony, *Calcul symbolique et propagation des singularités pour les équations aux dérivées partielles non linéaires*, Ann. Sci. École Norm. Sup. (4) **14** (1981), no. 2, 209–246.
2. M. Cicognani, D. Del Santo and M. Reissig, *A dyadic decomposition approach to a finitely degenerate hyperbolic problem*, Ann. Univ. Ferrara, Sez. VII, Sci. Mat. **52** (2006), 281–289.
3. F. Colombini, E. De Giorgi and S. Spagnolo, *Sur les équations hyperboliques avec des coefficients qui ne dépendent que du temps*, Ann. Scuola Norm. Sup. Pisa Cl. Sci. (4) **6** (1979), no. 3, 511–559.
4. F. Colombini and D. Del Santo, *Strictly hyperbolic operators and approximate energies*, in "Analysis and applications–ISAAC 2001, Proceedings of the 3rd international congress, Berlin, Germany, August 20–25, 2001", pp. 253–277, H. G. W. Begehr ed., Kluwer Academic Publishers, Dordrecht 2003.
5. F. Colombini, D. Del Santo and G. Métivier, in preparation.
6. F. Colombini and N. Lerner, *Hyperbolic operators with non-Lipschitz coefficients*, Duke Math. J. **77** (1995), no. 3, 657–698.
7. D. Del Santo, T. Kinoshita and M. Reissig, *Energy estimates for strictly hyperbolic equations with low regularity in coefficients*, Differential Integral Equations **20** (2007), no. 8, 879–900.
8. G. Glaeser, *Racine carrée d'une fonction différentiable*, Ann. Inst. Fourier **13** (1963), 203–210.
9. T. Nishitani and S. Spagnolo, *An extension of Glaeser inequality and its applications*, Osaka J. Math. **41** (2004), no. 1, 145–157.
10. S. Tarama, *Energy estimate for wave equations with coefficients in some Besov type class*, preprint, 2007.

AN OPTIMAL TRANSPORTATION METRIC FOR TWO NONLINEAR PDE'S

M. Fonte

Radon Institute for Computational and Applied Mathematics
Austrian Academy of Sciences,
Linz, A-4040, Austria
E-mail: massimo.fonte@ricam.oeaw.ac.at

An optimal transportation technique for nonlinear hyperbolic equations is discussed. Based on a Wasserstein-like metric in $H^1(\mathbb{R})$, a distance functional in defined, which yields continuous dependence with respect to the initial conditions of the energy conservative solutions of the Camassa-Holm equation.

Keywords: Optimal transportation; Camassa-Holm equation.

1. Introduction

In this paper there are summarized some global stability results and new techniques for nonlinear PDE equations, obtained by the author in collaboration with A. Bressan, partly contained in [1,2]. It is discussed the existence of a global solution after wave breaking for the Cauchy problem of the *Camassa-Holm* equation

$$u_t + 2\kappa u_x - u_{xxt} + 3uu_x = 2u_x u_{xx} + uu_{xxx},$$

and the construction of a continuous semigroup of solutions in $H^1(\mathbb{R})$. This equation arises from a higher order level of approximation of the asymptotic expansion of the Euler's equation for a shallow water wave model. Here we do not enter into details of the interpretation of such an equation, for the physical motivations we refer to [3–6].

The main part of the work is focused on the stability issue. It is described the construction of a metric on $H^1(\mathbb{R})$, obtained by an optimal transportation argument, for which continuous dependence w.r.t. the initial data is obtained. For a complete and detailed discussion on this problem we suggest [1,2,7]. Finally, an example on the optimal transportation technique for the *Hunter-Saxton* equation is discussed.

In the following we focus our attention on the Camassa-Holm equation with parameter $\kappa = 0$.

1.1. *Non-local formulation of the Camassa-Holm equation*

The Cauchy problem of the Camassa-Holm equation can be written as a scalar conservation law with an additional integro-differential term:

$$u_t + (\frac{1}{2}u^2)_x + P_x = 0\,, \qquad P(x) \doteq \frac{1}{2}e^{-|x|} * \left(u^2 + \frac{u_x^2}{2}\right) \tag{1}$$

and with the initial data specified as

$$u(0,x) = u_0(x), \qquad u_0 \in H^1(\mathbb{R}). \tag{2}$$

Earlier results on the existence and uniqueness of solutions can be found in [8,9]. One can regard (1) as an evolution equation on a space of absolutely continuous functions with derivatives $u_x \in L^2$. Suppose that u is a smooth function, then differentiating (1) w.r.t. x and multiplying by u, u_x one obtain the two conservation laws with source term

$$\left(\frac{u^2}{2}\right)_t + \left(\frac{u^3}{3} + u\,P\right)_x = u_x\,P, \qquad \left(\frac{u_x^2}{2}\right)_t + \left(\frac{uu_x^2}{2} - \frac{u^3}{3}\right)_x = -u_x\,P\,.$$

As a consequence, for regular solutions the total energy

$$E(t) \doteq \int_{\mathbb{R}} \left[u^2(t,x) + u_x^2(t,x)\right] dx \equiv E_0$$

remains constant in time.

As in the case of scalar conservation laws, by the strong nonlinearity of the equations, solutions with smooth initial data can lose regularity in finite time. For the Camassa-Holm equation (1), however, the uniform bound on $\|u_x\|_{L^2}$ guarantees that only the L^∞ norm of the gradient can blow up, while the solution u itself remains Hölder continuous at all times.

In order to construct global in time solutions, several techniques have been developed. A first approach was introduced in [5,10]. Recently, two main methods have been developed. On one hand, one can add a small diffusion term in the right hand side of (1), and recover solutions of the original equations as a vanishing viscosity limit [11,12]. An alternative technique, developed in [13], relies on a new set of independent and dependent variables, specifically designed with the aim of "resolving" all singularities. In terms of these new variables, the solution to the Cauchy problem becomes regular for all times, and can be obtained as the unique fixed point of a contractive transformation.

2. Multi-peakon solutions

In the present paper, we consider yet another approach, related to the solitary solutions of the Camassa-Holm equation. As a starting point we consider all *multi-peakon* solutions, of the form

$$u(t,x) = \sum_{i=1}^{N} p_i(t) e^{-|x-q_i(t)|}. \tag{3}$$

These are obtained by solving the system of O.D.E's

$$\begin{cases} \dot{q}_i = \sum_j p_j \, e^{-|q_i - q_j|}, \\ \dot{p}_i = \sum_{j \neq i} p_i p_j \, \operatorname{sign}(q_i - q_j) \, e^{-|q_i - q_j|}. \end{cases} \tag{4}$$

It is well known [14] that this system can be written in Hamiltonian form:

$$\begin{cases} \dot{q}_i = \dfrac{\partial}{\partial p_i} H(p,q), \\ \dot{p}_i = -\dfrac{\partial}{\partial q_i} H(p,q), \end{cases} \qquad H(p,q) \doteq \frac{1}{2} \sum_{i,j} p_i p_j e^{-|q_i - q_j|}.$$

If all the coefficients p_i are initially positive, then they remain positive and bounded for all times. The solution $u = u(t,x)$ is thus uniformly Lipschitz continuous. We stress, however, that we are not making any assumption about the signs of the p_i. In a typical situation, two peakons can cross at a finite time τ. As $t \to \tau-$ their strengths p_i, p_j and center positions q_i, q_j will satisfy

$$p_i(t) \to +\infty, \qquad p_j(t) \to -\infty, \qquad p_i(t) + p_j(t) \to \bar{p}, \tag{5}$$

$$q_i(t) \to \bar{q}, \quad q_j(t) \to \bar{q}, \qquad q_i(t) < q_j(t) \ \text{ for } t < \tau, \tag{6}$$

for some $\bar{p}, \bar{q} \in \mathbb{R}$. Moreover, $\|u_x(t)\|_{L^\infty} \to \infty$. In this case, we will show that there exists a unique way to extend the multi-peakon solution beyond the interaction time, so that the total energy is conserved.

Having constructed a set of "multi-peakon solutions", our main goal is to show that these solutions form a continuous semigroup, whose domain is dense in $H^1(\mathbb{R})$. Taking the unique continuous extension, we thus obtain a continuous semigroup of solutions of (1), defined on the entire space H^1.

One easily checks that the flow map $\Phi_t : u(0) \mapsto u(t)$ cannot be continuous either as a map from H^1 into itself, or from L^2 into itself. Distances defined in terms of convex norms perform well in connection with linear problems, but occasionally fail when nonlinear features become dominant (see [7, Example 2]).

3. An optimal transportation metric

In the present setting, we construct a new distance $J(u,v)$ between functions $u,v \in H^1$, defined by a problem of optimal transportation. Call $\mathcal{F}$ the family of all strictly increasing absolutely continuous maps $\psi : \mathbb{R} \mapsto \mathbb{R}$ which have an absolutely continuous inverse, and consider the measurable function

$$\varphi(x) \doteq \sup \left\{ \theta \in [0,1] \ \text{ s.t. } \ \theta \cdot \left(1 + u_x^2(x)\right) \leq \left(1 + v_x^2(\psi(x))\right) \psi'(x) \right\},$$

the metric is defined as $J(u,v) \doteq \min_{\psi \in \mathcal{F}} J^\psi$, where

$$J^\psi(u,v) \doteq \int d\Big(\big(x,\, u(x), 2\arctan u_x(x)\big), \big(\psi(x), v(\psi(x)), 2\arctan v_x(\psi(x))\big)\Big) \\ \varphi(x)\left(1+u_x^2(x)\right) dx \\ + \int \Big| \left(1+u_x^2(x)\right) - \left(1+v_x^2(\psi(x))\right) \psi'(x) \Big|\, dx\,, \tag{7}$$

where d is the standard distance in the space $\mathbb{R}^2 \times \mathbb{T}$, $\mathbb{T}$ being the unit circle $[0,2\pi]$ with 0 and 2π identified.

Roughly speaking, $J(u,v)$ is the minimum cost in order to transport the mass distribution with density $1+u_x^2$ located on the graph of u onto the mass distribution with density $1+v_x^2$ located on the graph of v, and the function φ represents the percentage of mass actually transported. With this definition of distance, the main result (Theorem 3.2) shows that

$$\left| \frac{d}{dt} J\big(u(t),\, v(t)\big) \right| \ \leq \ C\, J\big(u(t),\, v(t)\big)$$

for some constant C and any couple of multi-peakon solutions u, v. The distance functional J thus provides the ideal tool to measure continuous dependence on the initial data for solutions to the Camassa-Holm equation. For a multi-peakon solution, as long as all coefficients p_i remain bounded, the solution to the system of ODE's (4) is clearly unique. For each time t, call μ_t the measure having density $u^2(t) + u_x^2(t)$ w.r.t. Lebesgue measure. Consider a time τ where a positive and a negative peakon collide, according to (5)-(6). As $t \to \tau-$, we have the weak convergence $\mu_t \rightharpoonup \mu_\tau$ for some positive measure μ_τ which typically contains a Dirac mass at the point $\bar q$. By energy conservation, we thus have

$$\int \left[u^2(\tau,x) + u_x^2(\tau,x)\right] dx + \mu_\tau\big(\{\bar q\}\big) = \lim_{t\to\tau-} \int \left[u^2(\tau,x) + u_x^2(\tau,x)\right] dx = E_0.$$

There are now two natural ways to prolong the multi-peakon solution beyond time τ: either a conservative solution, such that $E(t) = E_0$ for $t > \tau$, or a dissipative solution, where all the energy concentrated at the point $\bar{q}$ is lost. In this case, the dissipative solution is obtained by simply replacing the two peakons p_i, p_j with one single peakon of strength $\bar{p}$, located at $x = \bar{q}$. On the other hand, the conservative solution contains two peakons emerging from the point $\bar{q}$. As $t \to \tau+$, their strengths and positions satisfy again (5), while (6) is replaced by

$$q_i(t) \to \bar{q}\,, \quad q_j(t) \to \bar{q}\,, \qquad q_i(t) > q_j(t) \ \text{ for } \ t > \tau\,. \tag{8}$$

The vanishing viscosity approach in [11,12] singles out the dissipative solutions. On the other hand, the coordinate transformation approach, based on optimal transport metrics, appear to be well suited for the study of both conservative and dissipative solutions.

The following results are indeed proved (see [1,2]).

Theorem 3.1. *For each initial data $\bar{u} \in H^1$, there exists a solution $u(\cdot)$ of the Cauchy problem (1), (2). Namely, the map $t \mapsto u(t)$ is Lipschitz continuous from $\mathbb{R}$ into L^2, satisfies (2) at time $t = 0$, and the identity*

$$\frac{d}{dt}u = -uu_x - P_x \tag{9}$$

is satisfied as an equality between elements in L^2 at a.e. time $t \in \mathbb{R}$. This same map $t \mapsto u(t)$ is continuously differentiable from $\mathbb{R}$ into L^p and satisfies (9) at a.e. time $t \in \mathbb{R}$, for all $p \in [1, 2[$. The above solution is conservative in the sense that, for a.e. $t \in \mathbb{R}$,

$$E(t) = E_0 \doteq \int \left[\bar{u}^2(x) + \bar{u}_x^2(x)\right] dx\,. \tag{10}$$

Theorem 3.2. *Conservative solutions to (1) can be constructed so that they constitute a continuous flow Φ. Namely, there exists a distance functional J on H^1 such that*

$$\frac{1}{C}\,\|u - v\|_{L^1} \le J(u, v)\, C\, \|u - v\|_{H^1} \tag{11}$$

for all $u, v \in H^1$ and some constant C uniformly valid on bounded sets of H^1. Moreover, for any two solutions $u(t) = \Phi_t\bar{u}$, $v(t) = \Phi_t\bar{v}$ of (1), the map $t \mapsto J\big(u(t), v(t)\big)$ satisfies

$$J\big(u(t), \bar{u}\big) \le C_1\,|t|\,, \tag{12}$$

$$J\big(u(t), v(t)\big) \le J(\bar{u}, \bar{v})\, e^{C_2|t|} \tag{13}$$

for a.e. $t \in \mathbb{R}$ and constants C_1, C_2, uniformly valid as u, v range on bounded sets of H^1.

Somewhat surprisingly, all the properties stated in Theorem 3.1 are still not strong enough to single out a unique solution. To achieve uniqueness, an additional condition is needed.

Theorem 3.3. *Conservative solutions $t \mapsto u(t)$ of (1) can be constructed with the following additional property: For each $t \in \mathbb{R}$, call μ_t the absolutely continuous measure having density $u^2 + u_x^2$ w.r.t. Lebesgue measure. Then, by possibly redefining μ_t on a set of times of measure zero, the map $t \mapsto \mu_t$ is continuous w.r.t. the topology of weak convergence of measures. It provides a measure-valued solution to the conservation law*

$$w_t + (uw)_x = (u^3 - 2uP^u)_x .$$

The solution of the Cauchy problem (1), (2) satisfying the properties stated in Theorem 3.1 and this additional condition is unique.

In the next section, we shall see how the *optimal transportation technique* can be useful for nonlinear equations like the Camassa-Holm equation. The key point is to consider transportation maps which yield Lipschitz continuity for the metric in a space of Radon measures. Since now, we consider the *Hunter-Saxton* [15] equation, and we give a brief heuristic idea for how this technique is involved in.

4. The Hunter-Saxton equation

The Hunter-Saxton equation describes the propagation of waves in a massive vector field of a nematic liquid crystal. Since the physical interpretation and its derivation are beyond to the description of this work, we refer to [7,15]. It can be written in a non-local formulation as a conservation law with a source term:

$$u_t + \left(\frac{u^2}{2}\right)_x = \frac{1}{4}\left(\int_{-\infty}^{x} - \int_{x}^{+\infty}\right) u_x^2(t,y)\,dy \doteq Q^u(t,x), \tag{14}$$

where $t \geq 0$ is the time variable, $x \in \mathbb{R}$ is the space variable in a reference frame, and $u(t,x) \in \mathbb{R}$ is related to the orientation of the liquid crystal molecules in the position x at time t. Results of existence of global dissipative solutions have been proved in [7] by using a method of characteristics, whereas in [16] conservative solutions are estabilished by using a geometric approach, obtaining then periodic solution in a manifold isometric to an open set of an infinite-dimensional L^2 sphere. Here we want to discuss on the stability of these solutions w.r.t. the initial conditions. To our purposes, Hunter-Saxton equation has nothing but the non-local high order part of the Camassa-Holm's source term. Adaptation is thus a matter of calculation.

Suppose that there exists a smooth solution to (14). To the Hunter-Saxton equation we can associate the following two conservation laws

$$(u_x)_t + (uu_x)_x = \frac{u_x^2}{2}, \tag{15}$$

$$(u_x^2)_t + (uu_x^2)_x = 0, \tag{16}$$

which are obtained by computing the derivative of the equation (14) w.r.t. the x variable and then multiplying it by u_x to achieve the second one.

A further conservation law is satisfied by the source term $Q^u(t,x)$. Since for smooth solutions (16) yields the conservation of the energy

$$E_0 \equiv E(t) \doteq \int_{\mathbb{R}} u_x^2(t,x)\, dx$$

the function Q^u can be expressed in the following way

$$Q(t,x) = -\frac{E_0}{4} + \frac{1}{2}\int_{-\infty}^{x} u_x^2(t,y)\, dy.$$

By deriving w.r.t. t we obtain

$$Q_t + uQ_x = 0. \tag{17}$$

The function Q^u is thus constant along the characteristic curves $\xi_u(t,y)$ defined by

$$\frac{\partial}{\partial t}\xi_u(t,y) = u(t,\xi_u(t,y)), \qquad \xi_u(0,y) = y. \tag{18}$$

Let us remark that the previous equation holds for all the time t in which u is a classical solution. It can be seen by the method of characteristics that if $u_0 \not\equiv 0$ is a smooth initial data and for some x_0 we have $u_{0x}(x_0) < 0$, along its outgoing characteristic the gradient blows up in finite time. Since the quantity $E(t)$ remains bounded also at the time of blow up, we can think that a finite amount of energy will be concentrated at the point of blow-up. In [7] the authors focus their attention on solutions that *dissipate* this quantity of energy. As far as the conservative solution is concerned, equation (16) will be satisfied in sense of measures, i.e. it means that thinking at a measure μ^t with absolutely continuous part which satisfies $d\mu^t = u_x^2(t,\cdot)d\mathcal{L}$ (here with $\mathcal{L}$ we indicate the Lebesgue measure), it satisfies

$$\partial_t \mu_t + \partial_x(u\mu_t) = 0 \qquad \text{in } \mathcal{D}'.$$

To find a *conservative solution* to the Cauchy problem (14) with smooth initial condition and finite energy u_0 means then to find a couple $(u(t), \mu_t)$

which satisfies the following system of conservation laws: let μ_0 be the absolutely continuous measure w.r.t. Lebesgue measure defined by $d\mu_0 = {u_0}_x^2 d\mathcal{L}$, then

$$\begin{cases} \partial_t u + \partial_x \left(\dfrac{u^2}{2}\right) = -\dfrac{1}{4}\mu^t(\mathbb{R}) + \dfrac{1}{2}\mu^t(]-\infty, x]) & u(0,x) = \bar{u}(x), \\ \partial_t \mu^t + \partial_x(u\mu^t) = 0 & \mu^t|_{t=0} = \mu_0. \end{cases} \tag{19}$$

Due to the nonlinearity of the problem, as shown in [7, Example 2] for the dissipative solution of the Hunter-Saxton equation, we can aspect that the usual "strong" distance stemming from convex norm is not useful in order to construct a continuous semigroup of solutions. Here we give a sketch of the construction of a metric which yields continuity of solution with respect to the initial data.

4.1. *A transportation map*

Let u_0 and v_0 be two initial data whose associated measures μ_0^1 and μ_0^2 have the same total mass $\mu_0^1(\mathbb{R}) = \mu_0^2(\mathbb{R})$. Suppose that such initial data are not constant in any interval of $\mathbb{R}$, so that the functions Q^{u_0} and Q^{v_0} are absolutely continuous and increasing.

We can thus define a continuous map Ψ_0 for which at every $x \in \mathbb{R}$ it associates the unique point $\Psi_0(x)$ such that

$$Q^{u_0}(x) = Q^{v_0}(\Psi_0(x)). \tag{20}$$

Let us remark that the Ψ_0 is an increasing function, in fact since Q^{u_0} and Q^{v_0} are increasing, then $x < y$ implies

$$Q^{v_0}(\Psi_0(x)) = Q^{u_0}(x) < Q^{u_0}(y) = Q^{v_0}(\Psi_0(y)),$$

so $\Psi_0(x) < \Psi_0(y)$ holds.

Now we want to see how the map Ψ_0 evolves in time. Since by (17) Q^u, Q^v are conserved along the characteristic curves (18), the equalities

$$Q^u(t, \xi_u(t,x)) = Q^{u_0}(x), \qquad Q^v(t, \xi_v(t,x)) = Q^{v_0}(x) \qquad \text{for all } x \in \mathbb{R}$$

yield the definition of the *transportation map*

$$\Psi_t(\xi_u(t,x)) \doteq \xi_v(t, \Psi_0(x)). \tag{21}$$

4.2. *The stability of the nonlinear system of ODE*

As in [13, Section 3], we compute a change of variables in order to obtain a system of ODE with Lipschitz vector field. Let us suppose that the initial data u_0 is in $H^1(\mathbb{R})$. Let us set

$$\omega \doteq 2\arctan(u_x),$$

then ω belongs to the unit circle $\mathbb{T}$. Computing the derivative of ω along the characteristics, having in mind the equation (15) for u_x we have

$$\frac{d}{dt}\omega(t,\xi(t,y)) = 2\frac{(u_x)_t+u(u_x)_x}{1+u_x^2} = \frac{-u_x^2}{1+u_x^2} = \frac{-\tan^2(\omega/2)}{1+\tan^2(\omega/2)} = -\sin^2(\omega/2).$$

It is thus natural to consider the new unknowns

$$\mathbf{X}^u = \mathbf{X}^u(t,y) \doteq \begin{pmatrix} \xi_u(t,y) \\ u(t,\xi_u(t,y)) \\ \omega_u(t,\xi_u(t,y)) \end{pmatrix}$$

which take values in the space $\mathbb{R}^2 \times \mathbb{T}$, and write the corresponding Cauchy problem

$$\begin{cases} \frac{d}{dt}\mathbf{X}^u(t,y) = \begin{pmatrix} u(t,\xi_u(t,y)) \\ Q^u(t,\xi_u(t,y)) \\ -\sin^2(\omega_u(t,\xi_u(t,y))/2) \end{pmatrix} = f(\mathbf{X}^u(t,y)), \\ \mathbf{X}^u(0,y) = \begin{pmatrix} y \\ u_0(y) \\ 2\arctan(u_{0x}(y)) \end{pmatrix}. \end{cases} \tag{22}$$

We remark that by definition, Q^u is far from to be Lipschitz continuous, then we cannot suppose in the previous system of ODE that the function f is a Lipschitz vector field. Hence, to overcome this lack of Lipschitz continuity, we shall make use of the function Ψ_t introduced in 4.1. Let $\mathbf{X}^u$ and $\mathbf{X}^v$ be two solution of the Cauchy problem (22), corresponding to the initial data u_0, v_0 respectively. We allow ourselves to make an abuse of notation by defining the map Ψ_t in the following way:

$$\Psi_t(\mathbf{X}^u(t,y)) \doteq \mathbf{X}^v(t,\Psi_0(y)).$$

We gain a sort of Lipschitz continuity for the function f if we restrict it on the manifold located by Ψ_t. Let us compute the difference of the vector field f evaluated in the points $\mathbf{X}^u(t,y)$ and $\Psi_t(\mathbf{X}^u(t,y))$. It holds

$$|f(\mathbf{X}^u(t,y))-f(\Psi_t(\mathbf{X}^u(t,y)))| = \left| \begin{pmatrix} u(t,\xi_u(t,y)) - v(t,\xi_v(t,\Psi_0(y))) \\ Q^u(t,\xi_u(t,y)) - Q^v(t,\xi_v(t,\Psi_0(y))) \\ \sin^2(\frac{\omega_v(t,\xi_v(\Psi_0))}{2}) - \sin^2(\frac{\omega_u(t,\xi_u)}{2}) \end{pmatrix} \right|.$$

Since by definition of the map Ψ_0 and by the equation (17) we have the identity

$$Q^u(t,\xi_u(t,y)) - Q^v(t,\xi_v(t,\Psi_0(y))) \equiv 0,$$

we deduce the estimate

$$|f(\mathbf{X}^u(t,y)) - f(\Psi_t(\mathbf{X}^u(t,y)))| \leq |\mathbf{X}^u(t,y) - \Psi_t(\mathbf{X}^u(t,y))|$$

for the vector field f. From the previous inequality we can prove that the difference of the two solutions $X^u(t,y)$ and $X^v(t,\Psi_0(y))$ can be estimated by the initial data. In fact, Gronwall Lemma applied to the inequality

$$\begin{aligned}|\mathbf{X}^u(t,y) - \mathbf{X}^v(t,\Psi_0(y))| &\leq |\mathbf{X}^u(0,y) - \mathbf{X}^v(0,\Psi_0(y))| \\ &\quad + \int_0^t |f(\mathbf{X}^u(s,y)) - f(\mathbf{X}^v(s,\Psi_0(y)))|\, ds\end{aligned}$$

yields the estimate $|\mathbf{X}^u(t,y) - \mathbf{X}^v(t,\Psi_0(y))| \leq e^t|\mathbf{X}^{u_0}(y) - \mathbf{X}^{u_0}(\Psi_0(y))|$.

The previous inequality suggests how to introduce the new distance J in order to obtain a stability result for solutions of the Hunter-Saxton equation. The function Ψ_0 can be regarded as a transportation map which transports the measure μ^u into the measure μ^v. The distance will be thus a sort of Wasserstein [17] distance between measures.

References

1. A. Bressan and M. Fonte, *Methods Appl. Anal.* **12**, 191 (2005).
2. M. Fonte, Analysis of singular solutions for two nonlinear wave equations, Ph.D. thesis, International School of Advanced Studies, 2005.
3. R. Camassa and D. Holm, *Phys. Rev. Lett.* **71**, 1661 (1993).
4. A. Constantin and H. McKean, *Comm. Pure Appl. Math.* **52**, 949 (1999).
5. A. Constantin and L. Molinet, *Comm. Math. Phys.* **211**, 45 (2000).
6. R. Johnson, *J. Fluid Mech.* **455**, 63 (2002).
7. A. Bressan and A. Constantin, *SIAM J. Math. Anal.* **37**, 996 (2005).
8. Z. Xin and P. Zhang, *Comm. Pure Appl. Math.* **53**, 1411 (2000).
9. Z. Xin and P. Zhang, *Comm. Partial Differential Equations* **27**, 1815 (2001).
10. A. Constantin and J. Escher, *Indiana Univ. Math. J.* **47**, 1527 (1998).
11. G. Coclite, H. Holden and K. Karlsen, *Discrete Contin. Dyn. Syst.* **13**, 659 (2005).
12. G. Coclite, H. Holden and K. Karlsen, *SIAM J. Math. Anal.* **37**, 1044 (2005).
13. A. Bressan and A. Constantin, *Arch. Rat. Mech. Anal.* **183**, 215 (2007).
14. R. Camassa, D. Holm and J. Hyman, *Adv. Appl. Mech.* **31**, 1 (1994).
15. J. Hunter and R. Saxton, *SIAM J. Appl. Math.* **51**, 1498 (1991).
16. J. Lenells, *Discrete Contin. Dyn. Syst.* **18**, 643 (2007).
17. C. Villani, *Topics in Optimal Transportation*, American Mathematical Society, Providence, RI, 2003.

GENERALIZED ENERGY CONSERVATION FOR WAVE EQUATIONS WITH TIME-DEPENDING COEFFICIENTS UNDER STABILIZATION PROPERTIES

F. HIROSAWA

Department of Mathematical Sciences, Yamaguchi University
Yamaguchi 753-8512, Japan
E-mail: hirosawa@yamaguchi-u.ac.jp

We present an overview of recent research about energy estimates for wave equations with time-depending coefficients taking account of the C^m properties of the coefficients with stabilization properties, which are described by some integrals of the coefficients. By introducing these properties simultaneously, more precise analysis is possible.

Keywords: Wave equation, generalized energy conservation, stabilization.

1. Introduction

Let's consider the asymptotic behavior of the energy in the Cauchy problem

$$\begin{aligned} &\left(\partial_t^2 - a(t)^2\Delta\right) u = 0, \quad (t,x) \in (0,\infty) \times \mathbb{R}^n, \\ &u(0,x) = u_0(x), \quad \partial_t u(0,x) = u_1(x), \quad x \in \mathbb{R}^n, \end{aligned} \tag{1}$$

where $a(t) > 0$ and $a \in C^1([0,\infty))$. Here the energy functional $E(t)$ of (1) is given by

$$E(t) = \frac{1}{2}\left(a(t)^2 \|\nabla u(t,\cdot)\|^2 + \|\partial_t u(t,\cdot)\|^2\right), \tag{2}$$

where $\|\cdot\|$ denotes the usual L^2 norm in $\mathbb{R}^n$. Evidently, the constant propagation speed $a(t) \equiv const.$ concludes the energy conservation $E(t) = E(0)$ for energy solutions for any $t > 0$, but it does not hold in general for variable coefficients. Here we introduce the following property as a generalization of the energy conservation law:

$$E(t) \simeq E(0), \tag{3}$$

which is called GEC, *generalized energy conservation*, where we denote $f(t) \simeq g(t)$ with two positive functions f and g if there exists a positive constant C such that $C^{-1}g(t) \le f(t) \le Cg(t)$. Moreover we denote $f(t) \lesssim g(t)$ if the estimate $f(t) \le Cg(t)$ holds.

We also consider the L^2 well-posedness of the backward Cauchy problem

$$\begin{aligned} &\left(\partial_t^2 - a(t)^2\Delta\right) u = 0, \quad (t,x) \in [0,T) \times \mathbb{R}^n, \\ &u(T,x) = u_0(x), \quad \partial_t u(T,x) = u_1(x), \quad x \in \mathbb{R}^n \end{aligned} \tag{4}$$

with a non-Lipschitz continuous coefficient $a(t)$ at $t = 0$ satisfying $a(t) > 0$ and $a \in C^1((0,T]) \cap L^\infty((0,T))$, where T is a positive small constant. Here we identify the L^2 well-posedness for (4) with the generalized energy conservation on $[0,T]$:

$$E(t) \simeq E(T). \tag{5}$$

Evidently we have the following theorem for (1) and (4):

Theorem 1.1. *Let $a \in C^1((0,\infty))$ satisfy $a(t) \simeq 1$. Then GEC is valid if the L^1 properties $a' \in L^1((0,\infty))$ for (1) and $a' \in L^1((0,T))$ for (4) hold.*

Noting $E'(t) \ge 0$ for $a'(t) > 0$ and $E'(t) \le 0$ for $a'(t) < 0$, the energy can oscillate by effects of oscillating coefficients. However, such effects cannot be taken account by the L^1 properties of $a'(t)$ in Theorem 1.1, thus the assumptions of $a'(t)$ should be relaxed. Indeed, GEC is valid for the following coefficients, though $a'(t)$ does not satisfy the L^1 properties:

$$a(t) = \begin{cases} 2 + \cos\left((\log(1+t))^\gamma\right) & \text{for (1)} \\ 2 + \cos\left((\log t^{-1})^\gamma\right) & \text{for (4)} \end{cases} \tag{6}$$

with $\gamma \le 1$; on the other hand, GEC does not hold in general for $\gamma > 1$ (see [4,6,12]). Actually, GEC is valid, though we generalize $2 + \cos(\cdot)$ in (6) to any positive and uniformly C^2 function $p(\cdot)$.

The main subject in the hereafter is to set appropriate conditions to the coefficients for GEC. As might be expected, they should include a C^2 regularity condition of the coefficients, moreover, the critical number $\gamma = 1$ in the examples (6) must be derived from them.

2. C^2 property of the coefficients

2.1. *Wave equations with variable propagation speeds*

Let us introduce the following theorem, which is an improvement of Theorem 1.1 taking account of C^2 regularity of the coefficients; we shall call C^m regularity of the coefficients and corresponding order of the derivatives by the C^m *property* of the coefficients.

Theorem 2.1 ([5] (cf. [12,16])). *Let $a \in C^2((0,\infty))$ satisfy $a(t) \simeq 1$. If $|a'(t)| \lesssim t^{-1}$ and $|a''(t)| \lesssim t^{-2}$ as $t \to \infty$ for (1) and $t \to +0$ for (4), then the same conclusions as in Theorem 1.1 are valid.*

Sketch of the proof for (1). We suppose that t is large without loss of generality. By partial Fourier transformation with respect to x, the equation of (1) is reduced to

$$v'' + a(t)^2|\xi|^2 v = 0, \tag{7}$$

where v', v'', $\cdots$, $v^{(k)}$ denote the derivatives of v with respect to t. Then the proof is concluded by a uniform estimate of the microlocal energy

$$\mathcal{E}(t,\xi) = \frac{1}{2}\left(a(t)^2|\xi|^2|v(t,\xi)|^2 + |v'(t,\xi)|^2\right).$$

The key of the proof is to estimate $\mathcal{E}(t,\xi)$ by different ways in the two zones $Z_\Psi = \{(t,\xi)\,;\, t \le t_\xi\}$ and $Z_H = \{(t,\xi)\,;\, t > t_\xi\}$, which are called the pseudo-differential zone and the hyperbolic zone, respectively, where t_ξ is determined by

$$t_\xi|\xi| = N \tag{8}$$

with a large constant N. In Z_Ψ, the oscillation of $a(t)$ has only a small effect, thus $\mathcal{E}(t,\xi)$ is estimated as a small perturbation of the constant coefficient case. On the other hand, the oscillations of $a(t)$ are crucial in Z_H, hence more precise analysis taking into account the C^2 property of $a(t)$ may lead to a contribution for a better estimate. For this reason, we set

$$\tilde{a}(t,\xi) = \chi\left(t/t_\xi\right) a_0 + \left(1 - \chi\left(t/t_\xi\right)\right) a(t)$$

as an approximation of $a(t)$ in the respective zones, where a_0 is a constant, $\chi \in C^\infty(\mathbb{R})$ is a monotone decreasing function satisfying $\chi(\tau) \equiv 1$ for $\tau \le 1$ and $\chi(\tau) \equiv 0$ for $\tau \ge 2$. Here we introduce the symbol class in Z_H by

$$\begin{aligned} S\{m,p;k\} = \Big\{ & f(t,\xi) \in C^k([t_\xi,\infty); H^{-\infty}(\mathbb{R}^n))\,; \\ & |f^{(j)}(t,\xi)| \lesssim t^{-(p+j)}|\xi|^m,\ 0 \le j \le k \Big\}. \end{aligned} \tag{9}$$

Then we verify that for any $l > 0$, $f \in S\{m,p;k\}$ and $g \in S\{\tilde{m},\tilde{p};\tilde{k}\}$ it follows that $f^{(j)} \in S\{m,p-j;k-j\}$, $fg \in S\{m+\tilde{m},p+\tilde{p};\min\{k,\tilde{k}\}\}$ and $f \in S\{m+l,p-l;k\}$. Moreover, we have $\tilde{a}, 1/\tilde{a} \in S\{0,0;2\}$.

Let $V = {}^t(\tilde{a}|\xi|v, -iv')$ be a solution of $V' = AV$. By the constant matrix M_0 determined by $M_0^{-1}V = V_1 := {}^t(v' + i\tilde{a}|\xi|v, v' - i\tilde{a}|\xi|v)$, we have

$$V_1' = (\Phi_1 + B_1)\,V_1, \tag{10}$$

where

$$\Phi_1 = \begin{pmatrix} \varphi_{1+} & 0 \\ 0 & \varphi_{1-} \end{pmatrix}, \quad B_1 = \begin{pmatrix} 0 & b_{1-} \\ b_{1+} & 0 \end{pmatrix}, \tag{11}$$

$\varphi_{1\pm} = (\tilde{a}' \pm i|\xi|(a^2 + \tilde{a}^2))/(2\tilde{a}) \in S\{1,0;1\}$ and $b_{1\pm} = (-\tilde{a}' \pm i(a^2 - \tilde{a}^2)|\xi|)/(2\tilde{a}) \in S\{0,1;1\}$; thus the matrix M_0 performs as a diagonalizer of A modulo $S\{0,1;1\}$ by grace the C^1 property of $a(t)$. Actually, Theorem 1.1 is trivially concluded from (10) since $\mathrm{Re}\,\varphi_{1\pm}, b_{1\pm} \in L^1((0,\infty))$, but $\mathrm{Re}\,\varphi_{1\pm}, b_{1\pm} \in S\{0,1;1\}$ are insufficient to conclude Theorem 2.1. Thus the second step of diagonalization procedure by the matrix

$$M_1 = I - \frac{(\tilde{a}')^2}{4\tilde{a}^4|\xi|^2}\begin{pmatrix} 0 & -1 \\ 1 & 0 \end{pmatrix}$$

is required. Indeed, due to the C^2 property of $a(t)$ (10) is reduced to

$$V_2' = (\Phi_1 + R_2)\, V_2 \tag{12}$$

for large N by the transformation $V_2 = M_1^{-1}V_1$, where R_2 satisfies $R_2 \in S\{-1,2;0\}$ and $|R_2| \lesssim |a - a_0||\xi| \lesssim |\xi|$ in Z_Ψ. Here we denote $|R_2| = \max_{1\le j,k\le 2}\{|(R_2)_{jk}|\}$. By carrying out the transformation $W_2 = \Psi_1^{-1}V_2$ with $\Psi_1 = \mathrm{diag}(\exp(-\int_0^t \varphi_{1+}(s,\xi)ds), \exp(-\int_0^t \varphi_{1-}(s,\xi)ds))$, we have $W_2' = \Psi_1^{-1}R_2\Psi_1 W_2$. Consequently, noting $|W_2(t,\xi)|^2 \simeq \mathcal{E}(t,\xi)$, $|\Psi_1^{-1}R_2\Psi_1| \simeq |R_2|$ and

$$\int_0^\infty |R_2(s,\xi)|ds \lesssim t_\xi|\xi| + |\xi|^{-1}\int_{t_\xi}^\infty s^{-2}ds \lesssim 1,$$

we obtain $\mathcal{E}(t,\xi) \simeq \mathcal{E}(0,\xi)$ uniformly with respect to (t,ξ). □

2.2. *Increasing or degenerating coefficients*

Let's generalize Theorem 2.1 to equations with increasing or degenerating coefficients. We assume that $a(t)$ is represented by the following product of increasing or degenerating function $\lambda(t)$ and oscillating function $\omega(t)$:

$$a(t) = \lambda(t)\omega(t), \tag{13}$$

where $\omega(t) \simeq 1$, $\lambda'(t) \ge 0$, $\lim_{t\to\infty}\lambda(t) = \infty$ for (1) and $\lim_{t\to+0}\lambda(t) = 0$ for (4), respectively. Let us introduce the following conditions:

$$\left|\lambda^{(k)}(t)\right| \lesssim \lambda(t)\left(\frac{\lambda(t)}{\Lambda(t)}\right)^k \tag{14}$$

and

$$\left|\omega^{(k)}(t)\right| \lesssim \Xi(t)^k \tag{15}$$

for $k = 1, 2$ as $t \to \infty$ for (1) and $t \to +0$ for (4), where $\Lambda(t) = \int_0^t \lambda(s)ds$ and $\Xi(t)$ is a strictly decreasing function satisfying $\lim_{t\to\infty} \Xi(t) = 0$ and $\lim_{t\to+0} \Xi(t) = \infty$. Then we have the following theorem:

Theorem 2.2 ([6] (cf. [13,15,16])). *Let $\lambda, \omega \in C^2((0,\infty))$. Assume that $\lambda'(t) \le \lambda(t)^2/\Lambda(t)$, (14) for $k = 2$ and (15) for $k = 1, 2$ with*

$$\Xi(t) = \frac{\lambda(t)}{\Lambda(t)}. \tag{16}$$

Then the following weighted generalized energy conservation is valid for (1):

$$\lambda(t)E(0) \simeq E(t). \tag{17}$$

If $t\lambda(t) \le \Lambda(t)$, then the following estimate is valid for (4):

$$E(t) \lesssim E(T). \tag{18}$$

The estimate of $\mathcal{E}(t,\xi)$ in Z_H is an analogy of the proof of Theorem 2.1 after defining t_ξ by $\Lambda(t_\xi)|\xi| = N$. The estimate from above in Z_Ψ is derived by $t\lambda(t) \le \Lambda(t)$; on the other hand, the degeneration of $a(t)$ brings a crucial problem for the estimate from below.

Remark 2.1. The estimate (18) does not ensure the usual meaning of the L^2 well-posedness for (4); indeed, the boundedness of $u(0,x)$ in $\dot{H}^1$ is not concluded by (18). Though the C^∞ well-posedness is valid without the assumption $t\lambda(t) \lesssim \Lambda(t)$ (cf. [15,16]).

2.3. *Dissipative wave equations*

By Liouville transformation, (1) can be reduced to the following Cauchy problem for dissipative wave equation:

$$\begin{aligned} &\left(\partial_t^2 - \Delta + b(t)\partial_t\right) u = 0, \quad (t,x) \in (0,\infty) \times \mathbb{R}^n, \\ &u(0,x) = u_0(x), \quad \partial_t u(0,x) = u_1(x), \quad x \in \mathbb{R}^n. \end{aligned} \tag{19}$$

In particular, if $a(t)$ in the original equation is tending to infinity as $t \to \infty$, then the reduced coefficient $b(t)$, which is represented by $a(\cdot)$ and $a'(\cdot)$, has a positive sign. Moreover, if $\int_0^\infty b(s)ds = \infty$, then the estimate (17) is reduced to a decay estimate of the energy for (19). Indeed, we have the following theorem:

Theorem 2.3 ([14]). *Let $b \in C^1([0,\infty))$ satisfy $b(t) \ge 0$, $tb(t) < 1$ and $|b'(t)| \lesssim t^{-2}$ as $t \to \infty$. Then the following decay estimate holds for (19):*

$$E(t) \lesssim \exp\left(-\int_0^t b(s)ds\right)\left(E(0) + \|u_0(\cdot)\|^2\right). \tag{20}$$

Remark 2.2. The C^1 property of $b(t)$ corresponds to the C^2 property of $a(t)$ for (1), hence this property is crucial for the estimate of the microenergy in the hyperbolic zone.

3. C^m and stabilization properties of the coefficients

It will be a natural expectation that some properties of C^m regularity of the coefficients with $m \geq 3$ contributes to some improvements of the previous results due to the C^2 property. Actually, such an expectation is not realized if we restrict ourselves to the order of γ in the example (6); indeed, Theorem 2.1 has already given the optimal order $\gamma = 1$ due to the C^2 property of the coefficients. However, if we do not persist in the examples (6), then we can derive some improvements from the C^m properties of the coefficients by taking into account a new property of the coefficients, which will be introduced as the *stabilization property.*

3.1. *Stabilization property for strictly hyperbolic equations*

Let $a(t)$ satisfy $a(t) \simeq 1$. We define $\Theta(t) = \int_0^t |a(s) - a_0| ds$ with a positive constant a_0. Here we introduce the *stabilization property* as a non-trivial condition for $\Theta(t)$ by

$$t^{-1}\Theta(t) \to 0 \tag{21}$$

as $t \to \infty$ for (1) and $t \to 0$ for (4). Here we note that the constant a_0 is uniquely determined since (21) is valid. The stabilization property (21) is introduced in [2,8] as an essential condition to the coefficients for the estimate in lower frequency part of the phase space to derive a benefit of the C^m property of the coefficients. Indeed, Theorem 2.1 is improved by taking account of the C^m property as follows:

Theorem 3.1 ([2,8]). *Let $a \in C^m((0,\infty))$ for $m \geq 2$ satisfy $a(t) \simeq 1$. If $\Theta(t) \simeq t^p$ with $0 \leq p < 1$ for (1) and $p > 1$ for (4), and (15) holds for $k = 1, \cdots, m$ with $a(t) = \omega(t)$ and $\Xi(t) = t^{-p+(1-p)/m}$, then the same conclusions as in Theorem 2.1 are valid.*

The restriction to $a'(t)$ is weaker as m larger, and optimality of the order t^{-p} for $a'(t)$ as $m \to \infty$ is proved in [2,8]. The restriction to the order of $a'(t)$ in Theorem 3.1 is weaker than as in Theorem 2.1 since $m \neq 1$ and $p \neq 1$; thus we have an improvement of Theorem 2.1 under assuming both properties C^m with $m \geq 2$ and stabilization property simultaneously.

Consequently, we have nothing new from this theorem for the examples (6), because the stabilization property does not hold for any $\gamma > 0$.

Let us briefly introduce the crucial points in the proof of Theorem 3.1 in contrast to the proof of Theorem 2.1. We define t_ξ by $\Theta(t_\xi)|\xi| = N$. Then the renewal hyperbolic zone is included in the previous one since the stabilization property (21) holds. Noting $\varphi_{1+} = \overline{\varphi_{1-}} =: \varphi_1$ and $b_{1+} = \overline{b_{1-}} := b_1$, the diagonalizer M_1 is represented by

$$M_1 = I + \frac{1}{\varphi_1 - \overline{\varphi_1}} \begin{pmatrix} 0 & -\overline{b_1} \\ b_1 & 0 \end{pmatrix}.$$

Let us define Φ_j and B_j inductively by $\Phi_j = \operatorname{diag}((A_j)_{11}, (A_j)_{22})$ and $B_j = A_j - \Phi_j$, where

$$A_j = M_{j-1}^{-1}\left(\Phi_{j-1} + B_{j-1}\right) M_{j-1} - M_{j-1}^{-1}(M_{j-1})'$$

for $j = 2, \cdots, m$. Then we have the following lemma:

Lemma 3.1 ([7,8]). *For any $j = 1, \cdots, m$ we have the following:*
(i) M_{j-1} are uniformly invertible and $|M_{j-1}| \lesssim 1$.
(ii) The entries of Φ_j and B_j are represented by $(\Phi_j)_{11} = \overline{(\Phi_j)_{22}} =: \varphi_j$ and $(B_j)_{21} = \overline{(B_j)_{12}} \in S\{-j+1, j, m-j\}$, thus M_{j-1} is a diagonalizer modulo $S\{-j+1, j, m-j\}$, where the symbol class is defined after replacing $t^{-(p+j)}$ by $(t^{1/m}\Theta(t)^{1-1/m})^{-(p+j)}$ of (9).
(iii) The estimate $|\int_0^\infty \operatorname{Re}\varphi_j(s,\xi)ds| \lesssim 1$ holds uniformly with respect to ξ.

Let us define $\Psi_m = \operatorname{diag}(\exp(-\int_0^t \varphi_m(s,\xi)ds), \exp(-\int_0^t \overline{\varphi_m}(s,\xi)ds))$. Then $W_m = \Psi_m^{-1} M_m^{-1} \cdots M_1^{-1} V_1$ is a solution of $W_m' = Q_m W_m$, where $Q_m \in S\{-m+1, m, 0\}$ by Lemma 3.1. Therefore, we have

$$\int_0^\infty |Q_m(s,\xi)|ds \lesssim \Theta(t_\xi)|\xi| + |\xi|^{-m+1} \int_{t_\xi}^\infty \left(s^{-\frac{1}{m}}\Theta(s)^{-1+\frac{1}{m}}\right)^m ds \lesssim 1,$$

it follows that $\mathcal{E}(t,\xi) \simeq |W_m(t,\xi)|^2 \simeq |W_m(0,\xi)|^2 \simeq \mathcal{E}(0,\xi)$ uniformly with respect to (t,ξ). Thus Theorem 3.1 is proved.

3.2. *Stabilization property for increasing or degenerating coefficients*

Let $a(t)$ be represented by (13). Then some conclusions of Theorem 2.2 can be improved by the C^m property and the stabilization property $\Theta(t)/\Lambda(t) \to 0$ with $\Theta(t) = \int_0^t \lambda(s)|\omega(s) - \omega_0|ds$.

Theorem 3.2 ([11]). *Let $\lambda, \omega \in C^m((0,\infty))$ and $\lambda(0) > 0$. Assume that $\lambda'(t) \le \lambda(t)^2/\Lambda(t)$, (14) for $k = 2, \cdots, m$ and*

$$(15) \text{ with } \Xi(t) = \frac{\lambda(t)}{\Theta(t)} \left(\frac{\Theta(t)}{\Lambda(t)} \right)^{\frac{1}{m}} \text{ for } k = 1, \cdots, m. \tag{22}$$

Then the estimate $E(t) \lesssim \lambda(t) E(0)$ is valid for (1).

The key of the proof is to separate the phase space by three zones, which are determined by $\Theta(t_{\xi,1})|\xi| = N$ and $\Lambda(t_{2,\xi})|\xi| = N$. Then the C^m and the stabilization property perform in the zones of high and middle frequency parts, respectively (see [9,11]).

Remark 3.1. C^∞ and the Gevrey well-posedness of (4) with $\lambda(0) = 0$ are considered by the C^2 property in [15,16] and by the L^1 property of $a'(t)$ in [3], respectively. Both results are improved by taking account of the C^m and the stabilization property in [9].

3.3. *Stabilization property for dissipative wave equations*

Let us suppose that the coefficient $b(t)$ of (19) is represented by $b(t) = b_0(1+t)^{-1} + \sigma(t)$ with $\sigma(t) \in C^m([0,\infty))$ and a constant $b_0 \in (0,1)$. Then the stabilization property is given by (21) with $\Theta(t) = \int_0^t | \int_0^\tau \sigma(s)ds - \omega_0 | d\tau$, where the constant ω_0 is uniquely determined. Then we have the following theorem:

Theorem 3.3 ([10]). *Let $\sigma \in C^m([0,\infty))$ for $m \ge 1$. Assume that* (21), *$\sup_t | \int_0^t \sigma(s)ds| < \infty$ and*

$$\left| \sigma^{(k)}(t) \right| \lesssim \left(\frac{1}{\Theta(t)} \left(\frac{\Theta(t)}{t} \right)^{\frac{1}{m+1}} \right)^{k-1} \text{ for } k = 0, \cdots, m \tag{23}$$

and for any large t. Then (20) *holds for* (19).

4. Concluding remarks and future tasks

The refined diagonalization procedure with the C^m property of the coefficients makes it possible to derive more precise estimates of the energy under assuming suitable stabilization properties. Here the stabilization property describes an error made from the oscillating behavior of the coefficient. Precisely, it is represented by an integral of difference between a monotone model function and oscillating coefficient. Here we underline that the integral is not in the sense of L^1 but a Riemann integral, therefore, we can derive some benefits of oscillation of the coefficient. Moreover, the error is not from a constant but a monotone function. This means that Theorem 3.2 and 3.3 can be understood as perturbations of results for equations with monotone coefficients.

Thus, our argument is based on the properties for the wave equations with monotone coefficients, therefore, farther developments of the research to such equations are essential. Indeed, the conclusions of Theorem 3.2 and 3.3 may be not sharp because the estimates are only from the above. Concretely, we propose the following problems after clearing up the problems for non-oscillating cases $\omega(t) = \sigma(t) \equiv \omega_0$:

(A) Estimates in both sides of $E(t)$ for Theorem 3.2 (without the assumption $\lambda'(t) \leq \lambda(t)^2/\Lambda(t)$ if it is possible).
(B) Estimates in both sides of $E(t)$ for Theorem 3.3 (without the assumption $b_0 \in (0,1)$ if it is possible).
(C) Generalized energy conservation for (4) under the assumptions (14) and (15) with (22).

The problems (A) and (B) are very close. All the problems above may require introducing some modified energy functionals, which are defined appropriately to each equation.

The equation of (19) can be reduced to the equation of (1) by Liouville transformation; thus Theorem 3.2 and 3.3 are essentially the same. Analogously, the Klein-Gordon equation with variable mass:

$$\left(\partial_t^2 - \Delta + m(t)^2\right) u = 0 \tag{24}$$

can be reduced to the equation of (1). Recently some generalized energy conservation for (24) with suitable monotone functions $m(t)$ is considered in [1]. Thus the base to consider the following problem is already prepared:

(D) GEC for the Cauchy problem of the Klein-Gordon equation (24) with oscillating mass $m(t)$ taking account of the C^m and stabilization property of $m(t)$.

References

1. C. Böhme, Generalized energy conservation for wave models with time-dependent coefficients, Diploma Thesis, TU Bergakademie Freiberg, 2008.
2. M. Cicognani and F. Hirosawa, On the Gevrey well-posedness for strictly hyperbolic Cauchy problems under the influence of the regularity of the coefficients, to appear in *Math. Scand.*

3. F. Colombini, D. Del Santo and T. Kinoshita, On weakly hyperbolic operators with non-regular coefficients and finite order degeneration, *J. Math. Anal. Appl.* **282**, 410–420 (2003).
4. F. Hirosawa, On the Cauchy problem for second order strictly hyperbolic equations with non-regular coefficients, *Math. Nachr.* **256**, 28–47 (2003).
5. F. Hirosawa, Loss of regularity for the solutions to hyperbolic equations with non-regular coefficients -an application to Kirchhoff equation-, *Math. Methods Appl. Sci.* **26**, 783–799 (2003).
6. F. Hirosawa, Loss of regularity for second order hyperbolic equations with singular coefficients, *Osaka J. Math.* **42**, 767–790 (2005).
7. F. Hirosawa, Global solvability for Kirchhoff equation in special classes of non-analytic functions, *J. Differential Equations* **230**, 49–70 (2006).
8. F. Hirosawa, On the asymptotic behavior of the energy for the wave equations with time depending coefficients, *Math. Ann.* **339**, 819–839 (2007).
9. F. Hirosawa, On second order weakly hyperbolic equations with oscillating coefficients and regularity loss of the solutions, preprint.
10. F. Hirosawa and J. Wirth, C^m-theory of damped wave equations with stabilisation, to appear in *J. Math. Anal. Appl.*
11. F. Hirosawa and J. Wirth, Generalised energy conservation law for the wave equations with variable propagation speed, preprint.
12. M. Reissig and J. Smith, L^p-L^q estimate for wave equation with bounded time dependent coefficient, *Hokkaido Math. J.* **34**, 541–586 (2005).
13. M. Reissig and K. Yagdjian, L_p-L_q decay estimates for the solutions of strictly hyperbolic equations of second order with increasing in time coefficients, *Math. Nachr.* **214**, 71–104 (2000).
14. J. Wirth, Wave equations with time-dependent dissipation. I. Non-effective dissipation , *J. Differential Equations* **222**, 487–514 (2006).
15. K. Yagdjian, The Cauchy problem for hyperbolic operators. Multiple characteristics, micro-local approach, Math. Topics, Akademie-Verlag, Berlin, 1997.
16. T. Yamazaki, Unique existence of evolution equations of hyperbolic type with countably many singular of degenerate points, *J. Differential Equations* **77**, 38–72 (1989).

A UNIFIED TREATMENT OF MODELS OF THEORMOELASTICITY, DECAY AND DIFFUSION PHENOMENA

K. JACHMANN and M. REISSIG

Fakultät für Mathematik und Informatik
TU Bergakademie Freiberg, Germany

We present a unified approach for studying qualitative properties of solutions to Cauchy problems for several thermoelasticity models, i.e., classical (or type-1), type-2, type-3 and second sound models, in one space dimension. The approach is based on a diagonalization procedure in phase space for an auxiliary problem.

Keywords: Thermoelasticity; decay estimates; diffusion phenomena.

1. Introduction

1.1. *Background*

Systems of thermoelasticity describe the elastic and thermal behavior of elastic and heat-conducting media.

The linear classical (or type-1) model of thermoelasticity for homogeneous media is in the absence of external body forces and external heat supply given by

$$\begin{cases} u_{tt} - \alpha u_{xx} + \gamma_1 \theta_x = 0, \\ \theta_t - \kappa \theta_{xx} + \gamma_2 u_{tx} = 0, \end{cases} \tag{1}$$

where $u = u(t,x) \in \mathbb{R}$ and $\theta = \theta(t,x) \in \mathbb{R}$, $(t,x) \in \mathbb{R}_{>0} \times \mathbb{R}$, denote the elastic displacement and the temperature difference to some equilibrium, respectively. For the constant coefficients we have the positiveness conditions $\alpha, \kappa, \gamma_1, \gamma_2 > 0$.

The derivation of the classical thermoelasticity model is based on Fourier's law of heat conduction, which implies that the heat equation for the coupled theory is a parabolic one and gives rise to the unphysical property of an infinite propagation speed of thermal disturbances. Replacing Fourier's law by Cattaneo's equation or even by a heat-flux equation of Jeffreys type leads to alternative models.

Examples are thermoelasticity systems with second sound (cf. [1])

$$\begin{cases} u_{tt} - \alpha u_{xx} + \gamma_1 \theta_x = 0, \\ \theta_t + q_x + \gamma_2 u_{tx} = 0, \\ \tau q_t + q + \kappa \theta_x = 0, \end{cases} \tag{2}$$

where $\tau > 0$ denotes the relaxation time, type-2 models (cf. [3])

$$\begin{cases} u_{tt} - \alpha u_{xx} + \gamma_1 \theta_x = 0, \\ \theta_{tt} - \kappa \theta_{xx} + \gamma_2 u_{ttx} = 0 \end{cases} \tag{3}$$

and the more dissipative type-3 models of thermoelasticity (cf. [2]), $\delta > 0$,

$$\begin{cases} u_{tt} - \alpha u_{xx} + \gamma_1 \theta_x = 0, \\ \theta_{tt} - \kappa \theta_{xx} - \delta \theta_{txx} + \gamma_2 u_{ttx} = 0. \end{cases} \tag{4}$$

We are not only interested in studying the Cauchy problems for (1-4), but also in studying the corresponding ones with additional lower order terms in the first equation. One of these is a dissipation term, which can be motivated physically by taking the external force negatively proportional to the velocity.

A lot of work on qualitative properties of solutions to the Cauchy problem of the classical system (1) has been done (cf. [4] and references within), and recently there have been some results on decay estimates for the alternative models (2-4) as well (cf. [5,7]). However, a rather new method of diagonalization in phase space (cf. [6]) can be applied to a more general problem including all above problems at once. Thus we are not only able to quickly reproduce all known results but to study further questions in an efficient way as well. Such questions could be ones related to well-posedness, L^p-L^q decay estimates, diffusion phenomena, propagation of singularities and the like.

1.2. *The problem*

Subjects of our studies are Cauchy problems of the form

$$\begin{cases} U_t + A_0 U + A_1 U_x - A_2 U_{xx} = 0, \\ U(0,x) = U_0(x) \end{cases} \tag{5}$$

for d-dimensional unknowns $U = U(t,x)$ with $(t,x) \in \mathbb{R}_{\geq 0} \times \mathbb{R}$ and A_i being complex and constant d-by-d matrices.

Example 1.1. We consider the classical thermoelasticity model with an additional dissipation term ($m > 0$), that is

$$\begin{cases} u_{tt} - \alpha u_{xx} + \gamma_1 \theta_x + m u_t = 0, \\ \theta_t - \kappa\, \theta_{xx} + \gamma_2 u_{tx} = 0. \end{cases} \tag{6}$$

With the use of $U = (u_+, u_-, \theta)^T$, $u_\pm = u_t \pm \sqrt{\alpha}\, u_x$, we equivalently transform the above system into

$$U_t + A_0 U + A_1 U_x - A_2 U_{xx} = 0 \tag{7}$$

with the matrices

$$A_0 = \begin{pmatrix} \frac{m}{2} & \frac{m}{2} & 0 \\ \frac{m}{2} & \frac{m}{2} & 0 \\ 0 & 0 & 0 \end{pmatrix}, A_1 = \begin{pmatrix} -\sqrt{\alpha} & 0 & \gamma_1 \\ 0 & \sqrt{\alpha} & \gamma_1 \\ \frac{\gamma_2}{2} & \frac{\gamma_2}{2} & 0 \end{pmatrix} \quad \text{and} \quad A_2 = \operatorname{diag}(0, 0, \kappa).$$

Applying partial Fourier transformation we obtain for $V = F_{x\to\xi}(U)$ the initial value problem

$$\begin{cases} V_t + (A_0 + i\xi A_1 + \xi^2 A_2)V = 0, \\ V(0,\xi) = F(U_0)(\xi). \end{cases} \tag{8}$$

In Section 2 we will diagonalize the principal part of the system in (8) in different regions of the phase space correspondingly and derive solution representations for proving results on L^p-L^q decay estimates and some concerning diffusion phenomena in Section 3. We will complete our considerations by discussing some applications in Section 4.

2. Diagonalization in phase space

2.1. *Diagonalization for small frequencies*

In the following we will describe a procedure, which allows us, under certain assumptions, to diagonalize the principal part of the system in (8) in $Z_{int}(\sigma) = \{|\xi| \le \sigma \ll 1\}$.

The matrix A_0 dominates the coefficient matrix $A(\xi) = A_0 + i\xi A_1 + \xi^2 A_2$, and we thus start the procedure with an

Assumption: A_0 is diagonalizable.

We will gradually diagonalize the system in (8), starting off with

Step 0: Diagonalization modulo $O(\xi)$-terms

The vector function $\tilde{V}^{(0)} = \tilde{L}^{(0)} V$, with $\tilde{L}^{(0)}$ ($\tilde{R}^{(0)}$) being the matrix of left (right) eigenvectors to the (in distinct groups of equal numbers ordered) eigenvalues

$$\lambda_{0,1} = \ldots = \lambda_{0,k_1}, \quad \ldots \quad, \lambda_{0,k_{b_1-1}+1} = \ldots = \lambda_{0,k_{b_1}=d}, \tag{9}$$

$\lambda_{0,k_m} \neq \lambda_{0,k_n}$ for all $m \neq n$, of A_0 with $\tilde{L}^{(0)}\tilde{R}^{(0)} = I$, satisfies

$$\tilde{V}_t^{(0)} + (\Lambda_0 + i\xi \tilde{A}_1^{(0)} + \xi^2 \tilde{A}_2^{(0)})\tilde{V}^{(0)} = 0, \tag{10}$$

where $\Lambda_0 = \text{diag}(\lambda_{0,1}, \ldots, \lambda_{0,d})$ and $\tilde{A}_i^{(0)} = \tilde{L}^{(0)} A_i \tilde{R}^{(0)}$.

Step 1: Diagonalization modulo $O(\xi^2)$-terms

The next step of the procedure consists of two substeps.

Substep 1:
We introduce $V^{(1)} = (I + i\xi K_{(1)})\tilde{V}^{(0)}$, where $K_{(1)} = (k_{ij})_{i,j=1}^d \in \mathbb{C}^{d\times d}$ with $k_{ii} = 0$ and non-diagonal entries to be determined later, which satisfies

$$V_t^{(1)} + \left(\Lambda_0 + i\xi A_1^{(1)} + \xi^2 A_2^{(1)} + A_3^{(1)}\right) V^{(1)} = 0 \tag{11}$$

with

$$A_1^{(1)} = [K_{(1)}, \Lambda_0] + \tilde{A}_1^{(0)} \tag{12}$$

and matrices $A_2^{(1)}$ and $A_3^{(1)} = O(\xi^3)$ (i.e. its elements have the behavior $O(\xi^3)$ for $\xi \to 0$) for which we can write down explicit formulas (apart from some remainder $O(\xi^l)$ in case of $A_3^{(1)}$). Here we have used that $(I + i\xi K_{(1)})$ is invertible for $|\xi| \leq \sigma \ll 1$ and that we can write $\tilde{V}^{(0)} = (I + i\xi K_{(1)})^{-1} V^{(1)} = (I - i\xi K_{(1)}(I + i\xi K_{(1)})^{-1})V^{(1)}$.

For the commutator $[K_{(1)}, \Lambda_0] = K_{(1)}\Lambda_0 - \Lambda_0 K_{(1)}$ we have

$$[K_{(1)}, \Lambda_0] = \begin{pmatrix} 0 & k_{12}(\lambda_{0,2} - \lambda_{0,1}) & \ldots & k_{1d}(\lambda_{0,d} - \lambda_{0,1}) \\ k_{21}(\lambda_{0,1} - \lambda_{0,2}) & 0 & \ldots & k_{2d}(\lambda_{0,d} - \lambda_{0,2}) \\ \vdots & \vdots & \ddots & \vdots \\ k_{d1}(\lambda_{0,1} - \lambda_{0,d}) & k_{d2}(\lambda_{0,2} - \lambda_{0,d}) & \ldots & 0 \end{pmatrix}, \tag{13}$$

and we have ordered the eigenvalues of A_0 in (9) in distinct groups of equal numbers. Hence, we can choose $K_{(1)}$ such that $A_1^{(1)}$ is block-diagonal, i.e.

$$A_1^{(1)} = [K_{(1)}, \Lambda_0] + \tilde{A}_1^{(0)} = \text{diag}(B_1^{(1)}, B_2^{(1)}, \ldots, B_{b_1}^{(1)}). \tag{14}$$

Note that if all eigenvalues of A_0 are distinct, then $A_1^{(1)}$ is diagonal if we choose $K_{(1)}$ appropriately. In general this is not the case, and we therefore go on by formulating another

Assumption: $A_1^{(1)}$ is diagonalizable.

Substep 2:
There are matrices $\tilde{L}^{(1)}$ and $\tilde{R}^{(1)}$ of left and right eigenvectors of $A_1^{(1)}$ with

$$\tilde{L}^{(1)}\tilde{R}^{(1)} = I \quad \text{and} \tag{15}$$

$$\tilde{L}^{(1)}\Lambda_0\tilde{R}^{(1)} = \Lambda_0\tilde{L}^{(1)}\tilde{R}^{(1)} = \Lambda_0 \tag{16}$$

so that the eigenvalues of $B_k^{(1)}$ are arranged in distinct groups of equal numbers for each k. In (16) we have used that the entries in Λ_0 in the blocks corresponding to $B_k^{(1)}$ are always equal. The vector $\tilde{V}^{(1)} = \tilde{L}^{(1)}V^{(1)}$ satisfies

$$\tilde{V}_t^{(1)} + \left(\Lambda_0 + i\xi\Lambda_1 + \xi^2\tilde{A}_2^{(1)} + \tilde{A}_3^{(1)}\right)\tilde{V}^{(1)} = 0 \tag{17}$$

with $\Lambda_1 = \operatorname{diag}(\lambda_{1,1},\ldots,\lambda_{1,d})$ and $\tilde{A}_i^{(1)} = \tilde{L}^{(1)}A_i^{(1)}\tilde{R}^{(1)}$.

For step n of the procedure we can, with $a_n = i$ for odd and $a_n = 1$ for even n, we write down a scheme:

Step n: Diagonalization modulo $O(\xi^{n+1})$-terms in $n+1$ substeps

Substep 1: $V^{\left((n-1)\frac{1}{n}\right)} = (I + a_n\xi^n K_{\left((n-1)\frac{1}{n}\right)})\tilde{V}^{(n-1)}$ generates a coefficient matrix with block-diagonal part $\Lambda_0 + i\xi\Lambda_1 + \ldots + a_{n-1}\xi^{n-1}\Lambda_{n-1} + a_n\xi^n A_n^{\left((n-1)\frac{1}{n}\right)}$.

⋮

Substep n: $V^{(n)} = (I + i\xi K_{(n)})V^{\left((n-1)\frac{n-1}{n}\right)}$ generates a coefficient matrix with block-diagonal part $\Lambda_0 + i\xi\Lambda_1 + \ldots + a_{n-1}\xi^{n-1}\Lambda_{n-1} + a_n\xi^n A_n^{(n)}$, $A_n^{(n)} = \operatorname{diag}(B_1^{(n)},\ldots,B_{b_n}^{(n)})$.

Assumption: $A_n^{(n)}$ is diagonalizable ($A_0^{(0)} := A_0$).

Substep $n+1$: $\tilde{V}^{(n)} = \tilde{L}^{(n)}V^{(n)}$ ($V^{(0)} := V$) generates a coefficient matrix with diagonal part $\Lambda_0 + i\xi\Lambda_1 + \ldots + a_n\xi^n\Lambda_n$,
$\Lambda_n = \operatorname{diag}(\lambda_1^{B_1^{(n)}},\ldots,\lambda_{m_1}^{B_1^{(n)}},\ldots,\lambda_1^{B_{b_n}^{(n)}},\ldots,\lambda_{m_{b_n}}^{B_{b_n}^{(n)}})$
$= \operatorname{diag}(\lambda_{n,1},\ldots,\lambda_{n,d})$, $\lambda_j^{B_k^{(n)}}$ ordered in distinct groups of equal numbers for each k.

Lemma 2.1. *In substep k of step n, $n \geq 1$, $1 \leq k \leq n+1$, we do not alter the diagonal part of the coefficient matrix when transforming the system for $V^{\left((n-1)\frac{k-1}{n}\right)}$ ($\tilde{V}^{(n-1)}$ for $k = 1$) into the one for $V^{\left((n-1)\frac{k}{n}\right)}$ ($\tilde{V}^{(n)}$ for $k = n+1$).*

Proof. The proof follows from the choice of the matrices involved in the transformations and the proposed ordering of the eigenvalues $\lambda_{i,j}$, $0 \leq i \leq n-1$, $1 \leq j \leq d$. □

Reconsidering the procedure we should with $A_0^{(0)} := A_0$ at first introduce the assumptions

($\mathbf{A_n}$) The matrices $A_k^{(k)}$ are diagonalizable for $k = 0, 1, \ldots, n$.

Lemma 2.2. *Suppose that the assumption (A_n) holds. Then the characteristic roots $\mu_j = \mu_j(\xi)$ of the coefficient matrix $A(\xi) = A_0 + i\xi A_1 + \xi^2 A_2$ of (8) behave for $|\xi| \leq \sigma \ll 1$ with (in general) complex numbers $\lambda_{k,j}$ as*

$$\mu_j(\xi) = \lambda_{0,j} + i\lambda_{1,j}\,\xi + \lambda_{2,j}\,\xi^2 + \ldots + a_n\lambda_{n,j}\,\xi^n + O(\xi^{n+1}).$$

Assume that ($\mathrm{A_n}$) holds. In generalization of the observation that everything works out especially fine, when all eigenvalues of A_0 are distinct, we introduce the assumptions

($\mathbf{B_n}$) $\forall i, j \in \{1, \ldots, d\}$ with $i \neq j$ $\exists k \in \{0, 1, \ldots, n\} : \lambda_{k,i} \neq \lambda_{k,j}$.

Note that the assumption ($\mathrm{B_n}$) is not satisfied for any n if and only if there are two identical eigenvalues of $A(\xi)$ for small frequencies.

We introduce a number c, assuming that ($\mathrm{A_n}$) and ($\mathrm{B_n}$) hold, by

$$c = \max_{i<j} c_{ij} \quad \text{and} \quad c_{ij} = \min\{k \in \{0, 1, \ldots, n\} : \lambda_{k,i} \neq \lambda_{k,j}\}. \tag{18}$$

Note that c equals the minimal number n for which the assumption ($\mathrm{B_n}$) holds. We obtain with a final diagonalization step (after a careful analysis of the eigenvectors of the coefficient matrix at the end of step n)

Proposition 2.1. *Suppose that the assumptions (A_n) and (B_n) hold for some n. Then the solution of (8) has in $Z_{int}(\sigma)$ the representation*

$$V(t,\xi) = T_{int}^{-1}(\xi)\,\mathrm{diag}(e^{-\mu_1(\xi)t}, \ldots, e^{-\mu_d(\xi)t})\,T_{int}(\xi)\hat{V}_0(\xi),$$

where

$$\begin{aligned} T_{int}(\xi) = (I + O(\xi^{n+1-c}))\,\tilde{L}^{(n)}(I + i\xi K_{(n)}) \cdot \ldots \\ \cdot (I + a_n\xi^n K_{((n-1)\frac{1}{n})}) \cdot \ldots \cdot \tilde{L}^{(1)}(I + i\xi K_{(1)})\tilde{L}^{(0)}. \end{aligned}$$

In the next two sections we will only sketch what should be done to obtain sufficiently nice solution representations for the regions of large and bounded frequencies with a positive distance to zero.

2.2. *Diagonalization for large frequencies*

We diagonalize the principal part of (8) in $Z_{ext}(N) = \{|\xi| \geq N \gg 1\}$. However, for $\xi \in Z_{ext}(N)$ we can rewrite (8) in the form

$$\begin{cases} \partial_t W + (A_2 + i\eta\, A_1 + \eta^2 A_0)W = 0, \\ W(0,\eta) = \eta^{-2} F(U_0)(1/\eta) \end{cases} \tag{19}$$

with $\eta = 1/\xi$, $\hat{\partial}_t := \eta^2\partial_t$ and $W = W(t,\eta) = \eta^{-2}V(t,1/\eta)$.

We can apply the same procedure as in Section 2.1 for small frequencies, and analogous assumptions imply analogous results.

2.3. *Diagonalization for bounded frequencies*

We analyze the problem (8) in $Z_{mid}(\sigma, N) = \{\sigma \leq |\xi| \leq N\}$.

For the practical applications that we are discussing here it is sufficient to work with an assumption

$$\textbf{(C)} \quad \exists C > 0\, \forall \xi \in Z_{mid}(\sigma,N)\, \forall \mu(\xi) \in \mathrm{spec}(A(\xi)) : \mathrm{Re}\,\mu(\xi) \geq C > 0.$$

With (C) and the compactness of $Z_{mid}(\sigma, N)$ we can prove

Proposition 2.2. *The solution V to (8) satisfies in $Z_{mid}(\sigma, N)$*

$$|V(t,\xi)| \lesssim e^{-ct}|\hat{U}_0(\xi)|,$$

where c is a positive constant, provided (C) holds.

Proof. The proof is a simple generalization of the one in [5], Prop. 3.3. □

Remark 2.1. The assumption (C) is sufficient to work with for the practical applications that we are discussing here. There are of course practical useful situations in which the eigenvalues may become imaginary. Those have to be discussed separately.

3. Results for solutions of (5)

3.1. *L^p-L^q decay estimates*

We will restrict to the region of small frequencies, due to the fact that this is for all practical applications that we are discussing here the decay-determining one, i.e., we will study decay estimates for solutions to the Cauchy problem (5) subjected to an operator $\varphi_{int}(D)$ with a smooth symbol $\varphi_{int}(\xi)$ supported in $Z_{int}(\sigma)$, $\varphi_{int}(\xi) = 1$ for $|\xi| \leq \sigma/2$.

We assume $U_0 \in S$, that there exists a number n so that $(\mathrm{A_n})$ and $(\mathrm{B_n})$ hold and

$$\forall \xi \in Z_{int}(\sigma)\, \forall \mu(\xi) \in \mathrm{spec}(A(\xi)) : \mathrm{Re}\,\mu(\xi) \geq 0. \tag{20}$$

For each eigenvalue $\mu_j = \mu_j(\xi)$ we fix with the help of its asymptotic expansion from Lemma 2.2 a decay-determining number $n_{s,j}$ by $n_{s,j} = \min(p_j, q_j)$, where p_j and q_j shall be defined by the following:

(i) We set $p_j = \infty$ if μ_j is purely imaginary in $Z_{int}(\sigma)$.
If the latter is not the case, then we choose p_j to be the minimal one of all numbers $m \in 2\mathbb{N}_0$ with $\operatorname{Re}\lambda_{m,j} > 0$.

(ii) We set $q_j = \infty$, unless there exists a natural number $m \geq 2$ for which $\operatorname{Im}(a_m \lambda_{m,j}) \neq 0$. In that case we choose q_j to be the minimal one of all such numbers m.

Finally, we set $n_s := \max_{j=1,\dots,d} n_{s,j}$.

Theorem 3.1. *Suppose that the above assumptions hold true. Then we have the following L^p-L^q decay estimates for solutions U of (5):*

$$\|\varphi_{int}(D)U\|_{L^q} \lesssim \begin{cases} e^{-ct}\|U_0\|_{L^p}, & n_s = 0, \\ (1+t)^{-\frac{1}{n_s}\left(\frac{1}{p}-\frac{1}{q}\right)}\|U_0\|_{L^p}, & 2 \leq n_s < \infty, \\ \|U_0\|_{L^p}, & n_s = \infty. \end{cases}$$

Here $1 \leq p \leq 2$, $\frac{1}{p} + \frac{1}{q} = 1$, *and c is a positive constant.*

Sketch of the proof. We use the representation for the solution of (8) from Proposition 2.1, derive L^2-L^2 estimates, L^1-L^∞ estimates and interpolate them via the Riesz-Thorin interpolation theorem. For the derivation of L^1-L^∞ estimates we have to make different considerations for different cases for $n_{s,j}$. Main tool for deriving results in the cases of finite $n_{s,j}$ and $q_j < p_j$, i.e. when the imaginary part of the eigenvalue μ_j gives an and determines the decay, is the Lemma of van der Corput.

3.2. *Diffusion phenomena*

Included in our considerations are hyperbolic-parabolic coupled systems, like the classical thermoelasticity model (1). It is certainly of interest whether we can prove an underlying parabolic structure for large times t (from the viewpoint of decay estimates) in such a case. Observations of this type are referred to as diffusion phenomena.

It is natural to prove diffusion phenomena for the decay-determining region of the phase space. We will therefore restrict our considerations here again to the region of small frequencies $Z_{int}(\sigma)$.

Suppose that there exists a number n so that ($\mathrm{A_n}$) and ($\mathrm{B_n}$) hold, that all eigenvalues $\mu_j = \mu_j(\xi)$ (we are looking for an underlying parabolic structure) have a positive real part in $Z_{int}(\sigma)\backslash\{0\}$ and neglect the cases in which we obtain an exponential decay, i.e. in which the number n_s from Section 3.1 vanishes.

We define with the help of the finite numbers p_j from Section 3.1 and $m_s := \max_{j=1,\dots,d} p_j$, $n_s \le m_s$, a reference system:

$$\begin{cases} W_t + \sum_{k=0}^{m_s} a_k M_k \partial_x^k W = 0, \\ W(0,x) = \tilde{L}^{(c)} \cdot \ldots \cdot \tilde{L}^{(0)} U_0(x). \end{cases} \tag{21}$$

Here $a_k = (-1)^{\lfloor k/2 \rfloor}$, $M_k = \mathrm{diag}(m_1^{(k)}, \dots, m_d^{(k)})$ with $m_j^{(k)} = 0$ if $k > p_j$ and $m_j^{(k)} = \lambda_{k,j}$ otherwise, the matrices $\tilde{L}^{(i)}$ are from Section 2.1, and c is the number from (18).

With a similar proof to the one for Theorem 3.1 we can show

Theorem 3.2. *Suppose that the above assumptions hold true. Then we have for the solution U of (5), dual values q, $1 \le p \le 2$ and the to $\tilde{L}^{(i)}$ inverse matrices $\tilde{R}^{(i)}$ from Section 2.1 the estimate*

$$\left\| \varphi_{int}(D) \left(U - \tilde{R}^{(0)} \cdot \ldots \cdot \tilde{R}^{(c)}\, W \right) \right\|_{L^q} \lesssim (1+t)^{-\frac{1}{n_s}\left(\frac{1}{p}-\frac{1}{q}\right) - \frac{1}{m_s}} \left\| U_0 \right\|_{L^p}.$$

Taking the estimates from Theorem 3.1 for $2 \le n_s < \infty$ and

$$\|\varphi_{int}(D) W\|_{L^q} \lesssim (1+t)^{-\frac{1}{n_s}\left(\frac{1}{p}-\frac{1}{q}\right)} \|U_0\|_{L^p}$$

into consideration, and thus that the difference, as stated in Theorem 3.2, is decaying faster, we can say that the asymptotic profile (at least from the point of view of decay estimates) of (5) is given by solutions to (21).

4. Applications

Let us start by considering the Cauchy problem for classical thermoelasticity with an additional dissipation term, i.e. (6) with data $u(0,x) = u_0$, $u_t(0,x) = u_1$ and $\theta(0,x) = \theta_0$.

The procedure works fine in all regions of the phase space. In particular, for small frequencies we obtain, (A_2) and (B_2) are satisfied, for the eigenvalues of $A(\xi) = A_0 + i\xi A_1 + \xi^2 A_2$ with positive and distinct numbers $a_\mp$ the asymptotic behavior

$$\mu_{1,2}(\xi) = a_\mp\, \xi^2 + O(\xi^3), \qquad \mu_3(\xi) = m - \frac{\alpha + \gamma_1\gamma_2}{m}\, \xi^2 + O(\xi^3).$$

This together with the fact that we obtain exponential decay from the regions of bounded and large frequencies yields:

Theorem 4.1. *We assume $u_0, u_1, \theta_0 \in S$. Then we have*

$$\|(u_t, u_x, \theta)\|_{L^q} \lesssim (1+t)^{-\frac{1}{2}\left(\frac{1}{p}-\frac{1}{q}\right)} \|(\langle D\rangle\, u_0, u_1, \theta_0)\|_{L^{p,r_p}}.$$

Here $1 < p \leq 2$, $\frac{1}{p} + \frac{1}{q} = 1$, $r_p = \frac{1}{p} - \frac{1}{q}$ *and* $L^{p,r_p} = \langle D \rangle^{-r_p} L^p$ *denotes the generalized Sobolev space.*

When considering the Cauchy problem for classical thermoelasticity, i.e. (1) with data $u(0,x) = u_0$, $u_t(0,x) = u_1$ and $\theta(0,x) = \theta_0$, then we obtain the same result as in Theorem 4.1 (which is a known fact, cf. [4]). In particular, for the regions of large and bounded frequencies we obtain exponential decay, and for small frequencies (A_1) and (B_1) are satisfied, and we have for the eigenvalues of $A(\xi) = i\xi A_1 + \xi^2 A_2$ (A_i from Example 1.1) the asymptotic behavior

$$\mu_{1,3}(\xi) = \mp i\sqrt{\alpha + \gamma_1\gamma_2}\,\xi + \frac{\kappa\gamma_1\gamma_2}{2(\alpha + \gamma_1\gamma_2)}\,\xi^2 + O(\xi^3),$$

$$\mu_2(\xi) = \frac{\alpha\kappa}{\alpha + \gamma_1\gamma_2}\,\xi^2 + O(\xi^3).$$

For a result on diffusion phenomena we define the reference system:

$$\begin{cases} W_t + M_1 W_x - M_2 W_{xx} = 0, \\ W(0,x) = \tilde{L}^{(1)} U_0(x), \end{cases} \tag{22}$$

with some matrix $\tilde{L}^{(1)} \in \mathbb{R}^{3\times 3}$ coming from the diagonalization procedure, $M_1 = \sqrt{\alpha + \gamma_1\gamma_2}\,\mathrm{diag}\,(-1, 0, 1)$ and $M_2 = \frac{\kappa}{2(\alpha+\gamma_1\gamma_2)}\mathrm{diag}\,(\gamma_1\gamma_2, 2\alpha, \gamma_1\gamma_2)$.

Theorem 4.2. *Assume* $u_0, u_1, \theta_0 \in S$. *Then we obtain for the solution* U *to the initial problem of (7) with* $A_0 \equiv 0$ *and* $\tilde{R}^{(1)}$ *being the inverse of* $\tilde{L}^{(1)}$.

$$\left\| U - \tilde{R}^{(1)} W \right\|_{L^q} \lesssim (1+t)^{-\frac{1}{2}\left(\frac{1}{p}-\frac{1}{q}\right)-\frac{1}{2}} \|(\langle D \rangle u_0, u_1, \theta_0)\|_{L^{p,r_p}}$$

for dual values $q, 1 < p \leq 2$ *and* $r_p = \frac{1}{p} - \frac{1}{q}$.

The asymptotic profile to the Cauchy problem for (1) (from the viewpoint of decay estimates) is thus a parabolic one.

The procedure can not only be applied to all problems discussed in Section 1.1, but also - with some minor changes - to all Cauchy problems for the systems (1-4) with an additional mass term in the first equation.

It can moreover be applied to all above mentioned problems in the 3D case for isotropic media by noting that after applying the Helmholtz decomposition and partial Fourier transformation we arrive at systems of the type

$$V_t + (A_0 + i|\xi|A_1 + |\xi|^2 A_2)V = 0, \tag{23}$$

where the two significant changes to (8) are that $\xi \in \mathbb{R}$ is replaced by $|\xi|$ for $\xi \in \mathbb{R}^3$ and that the matrices $A_i \in \mathbb{C}^{d\times d}$ depend smoothly on $\xi/|\xi|$, i.e. live on the unit sphere $\mathbb{S}^2$.

References

1. D. S. Chandrasekharaiah, Hyperbolic thermoelasticity: a review of recent literature, *Appl. Mech. Rev.* **51** (1998), 705–729.
2. A. E. Green and P. M. Naghdi, On undamped heat waves in an elastic solid, *J. Thermal Stresses* **15** (1992), 253–264.
3. A. E. Green and P. M. Naghdi, Thermoelasticity without energy dissipation, *J. Elasticity* **31** (1993), 189–208.
4. S. Jiang and R. Racke, *Evolution Equations in Thermoelasticity*, Chapman & Hall/CRC, Boca Raton, 2000.
5. M. Reissig and Y. G. Wang, Cauchy problems for linear thermoelastic systems of type III in one space variable, *Math. Methods Appl. Sci.* **11** (2005), 1359–1381.
6. Y. G. Wang, A new approach to study hyperbolic-parabolic coupled systems, *Banach Center Publications* **60** (2003), 227–236.
7. L. Yang and Y. G. Wang, L^p-L^q Decay Estimates for The Cauchy Problem of Linear Thermoelastic Systems with Second Sound in One Space Variable, *Quart. Appl. Math.* **64** (2006), 1–15.

SOME RESULTS ON SPECTRAL ANALYSIS OF NONSELFADJOINT PERTURBATIONS FOR SCHRÖDINGER AND WAVE EQUATIONS

MITSUTERU KADOWAKI *

Department of Mechanical Engineering, Ehime University,
Matsuyama, Ehime, 790-8577, Japan
E-mail: mkadowaki@eng.ehime-u.ac.jp

HIDEO NAKAZAWA *

Department of Mathematics, Chiba Institute of Technology,
Narashino, Chiba, 275-0023, Japan
E-mail: nakazawa.hideo@it-chiba.ac.jp

KAZUO WATANABE

Department of Mathematics, Gakushuin University,
Toshima, Tokyo, 171-8588, Japan
E-mail: kazuo.watanabe@gakushuin.ac.jp

Some results presented in [5], [7], and [8] associated with non-selfadjoint perturbation of Schrödinger and wave equations are summarized and a relation between the spectral structure and the asymptotic behavior of solutions are reviewed. Moreover, the issue of the degree of pole of resolvent is discussed.

Keywords: Nonselfadjoint; spectral structure; Parseval equality; asymptotic behavior; spectral singularity.

1. Introduction—motivation and aim

We shall consider the asymptotic behavior of the solutions of Schrödinger and wave equations with some non-selfadjoint perturbations. Abstract form of our problem is given by

$$i\frac{du(t)}{dt} = (H_0 + V)u(t) \ \big(t \in I \ (\subseteq \mathbb{R})\big), \quad u(0) = u_0 \in X \ (\subseteq \mathcal{H}). \qquad (1)$$

*The first and second author are supported by Grant–in–Aid for Scientific Research (No. 19540189 and No. 17740082), Ministry of Education, Culture, Sports, Science and Technology, Japan.

In the above, H_0 denotes a self-adjoint unperturbed operator with domain $\mathcal{D}(H_0)$ in a Hilbert space $\mathcal{H}$. We consider the two cases

$$(S): \quad H_0 = -\Delta \quad \text{with} \quad \mathcal{D}(H_0) = H^2(\mathbb{R}^N) \quad \text{in} \quad \mathcal{H} = L^2(\mathbb{R}^N)$$

$$(W): \quad H_0 = i\begin{pmatrix} 0 & 1 \\ \Delta & 0 \end{pmatrix} \quad \text{with} \quad \mathcal{D}(H_0) \subseteq H^2(\mathbb{R}^N) \times H^1(\mathbb{R}^N)$$
$$\text{in} \quad \mathcal{H} = \dot{H}^1(\mathbb{R}^N) \times L^2(\mathbb{R}^N)$$

Here H^m represents the usual Sobolev space of order m, and $\dot{H}^1(\mathbb{R}^N)$ denotes the completion of $C_0^\infty(\mathbb{R}^N)$ with respect to the norm $||f|| = \{\int_{\mathbb{R}^N} |\nabla f(x)|^2 dx\}^{1/2}$ for $f \in C_0^\infty(\mathbb{R}^N)$. (S) and (W) denotes the Schrödinger and wave case, respectively. The operator V represents a non-selfadjoint perturbation described below. The set X contains initial data and is given by $X = \mathcal{D}(H_0)$ or $\mathcal{D}(H_0) \subset X \subset \mathcal{H}$.

Example 1.1. [Ikebe (1960): (S) with real potentials] Ikebe treated Schrödinger equation with real short–range potentials [3]. We shall take in (1) as $\mathcal{H} = L^2(\mathbb{R}^3)$, $H_0 = -\Delta$, $X = \mathcal{D}(H_0) = H^2(\mathbb{R}^3)$, $I = \mathbb{R}$, and V is assumed to be a multiplication operator by a real–valued function $V(x)$ satisfying "very short–range" conditions $|V(x)| \le C(1+|x|)^{-2}$ for any $x \in \mathbb{R}^3$. Then the perturbed operator $H = H_0 + V$ becomes self-adjoint and its spectral structure is given by the following: the continuous spectrum $\sigma_c(H)$ fills the non-negative real axis $\sigma_c(H) = [0, \infty)$, and there exist at most finitely many negative eigenvalues $\sigma_p(H) = \{\lambda_j < 0 \ (j = 1, 2, \cdots, m \text{ with } m < \infty)\}$. The spectral decomposition for H is given by

$$H = \sum_{j=1}^{m} \lambda_j P_j + \int_{\sigma_c(H)} \lambda dE(\lambda)$$

and the solution $u(t) = e^{-itH}u(0) \in \mathcal{H}$ of (1) is represented by

$$u(t) = \sum_{j=1}^{m} e^{-it\lambda_j} P_j u_0 + \int_{\sigma_c(H)} e^{-it\lambda} dE(\lambda) u_0.$$

Moreover the inverse wave[(1)] and the scattering operator exist in $\mathcal{H}$:

$$W_\pm = \operatorname*{s-lim}_{t\to\pm\infty} e^{itH_0} e^{-itH}, \quad S = (W_+)^{-1} W_-.$$

[(1)] In the context of dissipative scattering which is the scope of our problem, the inverse wave operator is more useful than the usual one. See, for example, [14].

Thus we find the solution which becomes asymptotically free:

$$||u(t) - e^{-itH_0}W_\pm u_0||_{\mathcal{H}} \to 0 \quad (t \to \pm\infty)^{(2)}.$$

Indicating the usual spectral representation of H_0 by $\mathcal{F}_0(\lambda)$ (e.g., [15]), we can also construct the spectral representation of H by $\mathcal{F}(\lambda) = \mathcal{F}_0(\lambda)W_+$. So, the generalized Parseval formula (cf., [18]) has the form

$$(f,g)_{\mathcal{H}} = \sum_{j=1}^{m} (P_j f, g)_{\mathcal{H}} + \int_{\sigma_c(H)} (\mathcal{F}(\lambda)f, \mathcal{F}(\lambda)g)_{L^2(S^2)}\, d\lambda^{(3)},$$

where P_j denotes the projection operator onto eigenspace associated with the eigenvalue λ_j. Note that the usual one is given by

$$(f,g) = \int_{\sigma_c(H_0)} (\mathcal{F}_0(\lambda)f, \mathcal{F}_0(\lambda)g)_{L^2(S^2)}\, d\lambda.$$

Example 1.2 (Mochizuki (1967-68): (S) with complex potentials). *Mochizuki [12,13] treated the similar problem under complex–valued function $V(x)$ satisfying $\operatorname{Im} V(x) \le 0$ (so H becomes maximal dissipative$^{(4)}$) and proved the existence of the local wave and the scattering operator. Moreover, the representation of the scattering operator and the uniqueness of the inverse scattering problem at high energy is obtained. However, as for the decaying mode, i.e., $\lim_{t\to+\infty} ||u(t)||_{\mathcal{H}} = 0$, the existence of such mode was not mentioned.*

Example 1.3 ((W) with dissipations). *Consider the wave equation with dissipations. We shall take in* (1) *as $\mathcal{H} = H^1(\mathbb{R}^N) \times L^2(\mathbb{R}^N)$, $I = [0,\infty)$, $H_0 = i\begin{pmatrix} 0 & 1 \\ \Delta & 0 \end{pmatrix}$, $\mathcal{D}(H_0) = \Big\{ (f = (f_1, f_2) \in \mathcal{H} \,|\, \Delta f_1 \in L^2(\mathbb{R}^N),\ f_2 \in H^1(\mathbb{R}^N) \Big\}$, and $V = i\begin{pmatrix} 0 & 0 \\ 0 & -b(x) \end{pmatrix}$ with $b(x) > 0$. The energy norm in $\mathcal{H}$ is given by $||u(t)||_{\mathcal{H}} = \left|\left| \begin{pmatrix} w(t) \\ w_t(t) \end{pmatrix} \right|\right|_{\mathcal{H}} = \sqrt{\frac{1}{2}\left(||\nabla w(t)||^2_{L^2} + ||w_t(t)||^2_{L^2} \right)}$, and $w = w(t) = w(x,t)$ satisfies the equation $w_{tt} - \Delta w + b(x) w_t = 0$ in $\mathbb{R}^N \times (0,\infty)$.*

Let us consider the following two cases:

---

(2) $||\cdot||_{\mathcal{H}}$ denotes the norm in $\mathcal{H}$.

(3) $(\cdot,\cdot)_X$ denotes the inner-product in X.

(4) A closed linear operator A is maximal dissipative if and only if $\operatorname{Im}(Af, f)_{\mathcal{H}} \le 0$ for any $f \in \mathcal{D}(A)$ and $\mathcal{R}(H - zE) = \mathcal{H}$ where $\mathcal{R}(H - zE)$ denotes the range of operator $H - zE$ with some z belonging to resolvent set of H. Here E denotes the identity operator.

(a): $0 \leq b(x) \leq b_0(1+|x|)^{-1-\delta}$,
(b): $b_1(1+|x|)^{-1} \leq b(x) \leq b_2$,
for some constants b_0, b_1, b_2 *and* $\delta > 0$. *Then the following results are known:*

(i) Under assumption (a), there exist scattering states ([14] $N \neq 2$, [17] $N \geq 2$): $\lim_{t\to+\infty} ||u(t) - e^{-itH_0}W_+u(0)||_{\mathcal{H}} = 0$.
(ii) Under assumption (b), there exists decaying states ([11], [15]) for suitable initial data: $||u(t)||_{\mathcal{H}} = O(t^{-1})$ $(t \to +\infty)$.
(iii) Under assumption (a) with small b_0 *the spectrum of* H *coincides with the one of* H_0 *if* $N \geq 3$, *and limiting absorption principle also holds for* $N \neq 2$. *([16]).*

These examples suggest the possibility of characterization of asymptotic behavior of solutions by the spectral structure of the generator of H (for other results, see [6] and references therein). So we amass some evidence aiming to develop methodology to treat more general problems.

Before closing this section we shall explain the content of the present paper. In section 2, we briefly introduce our results on Schrödinger equations. One result on spectral singularity is outlined in section 3. Results for the wave equation are mentioned in section 4. About these matter so far, we omit detailed proofs. For them, please refer papers quoted in each section. In the final section 5, we shall argue on the existence of the wave operator for wave equations with dissipation in some layered region. A sketch of the proof is also given.

Acknowledgments

The authors thank to the referee who pointed out our mistakes. They also thank to the editor for his warm and kind response.

2. Schrödinger equations with rank one perturbations

In this section, we shall state some results from [5]. Before it, we should mention the book [1] by Albeverio, Hegh-Krohn and Holden. In it, they studied these types of perturbation systematically. Now let us consider equation (1) with $\mathcal{H} = L^2(\mathbb{R}^1)$, $H_0 = -d^2/dx^2$, $X = \mathcal{D}(H_0) = H^2(\mathbb{R}^1)$, $I = \mathbb{R}$ and $V = V_\alpha = \alpha\,(\cdot,\delta)_{\mathcal{H}}^{(5)}\,\delta$, $\alpha = \alpha_1 + i\alpha_2 \in \mathbb{C}$, $\delta \in \mathcal{H}^{-1}$, where δ

(5) Here, $(\cdot,\cdot)_{\mathcal{H}}$ denotes the dual-coupling between $\mathcal{H}^1$ and $\mathcal{H}^{-1}$. See [1].

denotes the Dirac delta function. Then the operator $H = H_\alpha = H_0 + V_\alpha$ with the domain

$$\mathcal{D}(H_\alpha) = \Big\{ U = u + aH_0(H_0^2+1)^{-1}\delta \,|\, u \in \mathcal{H}^2,\, a \in \mathbb{C},$$
$$(u,\delta) = -a\left(\alpha^{-1} + (\delta, H_0(H_0^2+1)^{-1}\delta)\right)\Big\} \quad (\alpha \not\equiv 0)$$

becomes maximal dissipative for $\alpha_2 < 0$ (self-adjoint for $\alpha_2 \equiv 0$). The action of H_α is defined by $H_\alpha U = H_0 u - a(H_0^2+1)^{-1}\varphi$ for any $U \in \mathcal{D}(H_\alpha)$.

Theorem 2.1 (Spectral structure). *If $\alpha_1 \le 0$ and $\alpha_2 \le 0$, then the spectrum of H_α is given by $\sigma(H_\alpha) = [0,\infty) \cup \{-\alpha^2/4\}$ if $\alpha_1 < 0$, $= [0,\infty)$ if $\alpha_1 = 0$. More precisely, the spectral structure $\sigma(H_\alpha)$ is given by $\sigma_{ess}(H_\alpha) = \sigma_c(H_\alpha) = [0,\infty)$, $\sigma_r(H_\alpha) = \emptyset$, $\sigma_p(H_\alpha) = \sigma_d(H_\alpha) = \{-\alpha^2/4\}$ if $\alpha_1 < 0$, $= \emptyset$ if $\alpha_1 = 0$.*

The projection corresponding to the eigenvalue $-\alpha^2/4$ $(\alpha_1 \ne 0)$ is given by $P_{-\alpha^2/4} f(x) = -\alpha/2 \left(f, e^{(\bar\alpha|\cdot|)/2}\right)_{\mathcal{H}} e^{(\alpha|x|)/2}$. Similarly to the self-adjoint case, we can prove the existence of the wave operator $W(\alpha) = \text{s-}\lim_{t\to+\infty} e^{itH_0}e^{-itH_\alpha}$ under the assumptions $\alpha_1 \le 0$ and $\alpha_2 < 0$.

Theorem 2.2 (Generalized Parseval formula, cf. Pavlov (1966)). (1) *Assume $\alpha_1 \ne 0$. For any f, $g \in \mathcal{H} \cap L^1(\mathbb{R})$ and for any $\alpha \in \{\alpha = \alpha_1 + i\alpha_2; \alpha_1 < 0\}$, it holds*

$$(\mathcal{F}_\alpha f, \mathcal{F}_{\bar\alpha} g)_{\mathcal{H}} = (f,g)_{\mathcal{H}} + \alpha/2 (f, e^{(\bar\alpha|\cdot|)/2})_{\mathcal{H}} (e^{(\alpha|\cdot|)/2}, g)_{\mathcal{H}},$$

where $\mathcal{F}_\alpha = \mathcal{F}_0 W(\alpha)$ is the generalized Fourier transform ($\mathcal{F}_0$ is the usual one).

(2) *Assume $\alpha_1 = 0$. Then we have*

$$\lim_{\varepsilon\to 0} (\mathcal{F}_{i\alpha_2} f, \chi_\varepsilon \mathcal{F}_{-i\alpha_2} g)_{\mathcal{H}}$$
$$= (f,g)_{\mathcal{H}} + (i\alpha_2)/4 \left(\int_{\mathbb{R}} e^{(i\alpha_2|x|)/2} f(x)dx\right)\left(\int_{\mathbb{R}} e^{(i\alpha_2|y|)/2} f(y)dy\right),$$

where χ_ε is the characteristic function of $\{k \in \mathbb{R}\,;\, a \le ||k| + \alpha_2/2|\}$.

Theorem 2.3. *The kernel of the wave operator is given by $Ker W(\alpha) = \mathcal{R}(P_\alpha)$ if $\alpha_1 < 0$, $\alpha_2 < 0$, $= \{0\}$ if $\alpha_1 = 0$, $\alpha_2 < 0$.*

Corollary 2.1. (1) *(Decaying mode)* $u_0 \in \mathcal{R}(P_\alpha)$ *if and only if* $\lim_{t\to+\infty} ||u(t)||_{\mathcal{H}} = 0$. (2) *(Scattering mode)* $(E - P_\alpha)u_0 \ne 0$ *if and only if* $\lim_{t\to+\infty} ||u(t) - e^{-itH_0}W(\alpha)u_0||_{\mathcal{H}} = 0$.

3. Spectral singularity for Schrödinger equations with general rank one perturbations

Why can we argue as above? Because the spectral singularity is the same as in the self-adjoint case. To explain it, we consider (1) with $\mathcal{H} = L^2(\mathbb{R}^1)$, $H_0 = -d^2/dx^2$, $X = \mathcal{D}(H_0) = H^2(\mathbb{R}^1)$, and $V = V_\alpha = \alpha\,(\cdot,\varphi)_{\mathcal{H}}\varphi$ with $\alpha = \alpha_1 + i\alpha_2 \in \mathbb{C}_-$ where $\varphi(x) = \chi_{[0,1]}(x)$ is the characteristic function of $[0,1]$ and $\mathbb{C}_-$ denotes the complex lower half plain. Since $(H_0 - zE)^{-1}u(x) = -(2i\sqrt{z})^{-1}\int_{\mathbb{R}} e^{i\sqrt{z}|x-y|}u(y)dy$ for $\operatorname{Im} z \neq 0$, we have $\left((H_0 - zE)^{-1}\varphi,\varphi\right)_{L^2} = \xi^{-2} - \xi^{-3}(e^{\xi} - 1)$, where we put $\xi = i\sqrt{z}$ and for $\operatorname{Im} z > 0$ we take $\operatorname{Im}\sqrt{z} > 0$. Note that the perturbed resolvent is expressed by $(H - zE)^{-1}u = (H_0 - zE)^{-1}u - \alpha\Gamma(z)^{-1}\left((H_0 - zE)^{-1}u,\varphi\right)$, where $\Gamma(z) = 1 + \alpha\left((H_0 - zE)^{-1}\varphi,\varphi\right)$. Then we have the following two lemmas by explicit calculation [7].

Lemma 3.1. *It holds that* $\left((H_0 + i\varepsilon E)^{-1}\varphi,\varphi\right) = -i\varepsilon^{-1} - c\varepsilon^{-3/2}(e^{-c\sqrt{\varepsilon}} - 1)$ *with* $c = e^{(i\pi)/4}$. *Moreover,* $\lim_{\varepsilon\downarrow 0}\left((H_0 + i\varepsilon E)^{-1}\varphi,\varphi\right) = 0$ *holds.*

Lemma 3.2. *It holds that*

$$\left((H_0 - (\lambda - i0)E)^{-1}\varphi,\varphi\right) = -\lambda^{-1} + \lambda^{-3/2}\sin\sqrt{\lambda} + i\lambda^{-3/4}(\cos\sqrt{\lambda} - 1).$$

Furthermore, $\lim_{\lambda\downarrow 0}\left((H_0 - (\lambda - i0)E)^{-1}\varphi,\varphi\right) = 0$.

From these lemmas we see that $\Gamma(0) \neq 0$. Hence $z = 0$ is not so singular for the resolvent of H. Next we consider the degree of the zeros of $\Gamma(\lambda - i0)$. For simplicity, we put $\mu = \sqrt{\lambda}$.

Theorem 3.1. *There is no* $\mu > 0$ *fulfilling* $\Gamma(\mu^2 - i0) = \Gamma'(\mu^2 - i0) = 0$.

4. Wave equation with rank one dissipations

In this section, we introduce some results [8] on wave equations with dissipative perturbations of rank one. Consider equation (1) with $\mathcal{H} = \dot{H}^1(\mathbb{R}) \times L^2(\mathbb{R})$, $H = H_0 + V$ with $H_0 = i\begin{pmatrix} 0 & 1 \\ d^2/dx^2 & 0 \end{pmatrix}$, $V = i\begin{pmatrix} 0 & 0 \\ 0 & -(\cdot,\varphi)_{L^2}\varphi \end{pmatrix}$ and $\mathcal{D}(H_0) = \mathcal{D}(H) = \left\{ f = (f_1, f_2) \in \mathcal{H} \,|\, \partial_x^2 f_1 \in L^2(\mathbb{R}),\ f_2 \in H^1(\mathbb{R}) \right\}$, where $\varphi \in L^{2,s}(\mathbb{R}) = \left\{u(x) \,|\, \int_{\mathbb{R}}(1 + |x|^2)^s|u(x)|^2dx < \infty\right\}$ $(s > 1/2)$. In this case, $u(t) = \begin{pmatrix} w(t) \\ w_t(t) \end{pmatrix} \in \mathcal{H}$ and $w = w(t) = w(x,t)$ satisfies the equation $w_{tt} - \Delta w + (w_t,\varphi)\varphi = 0$.

Theorem 4.1 (cf. Mochizuki (1976)). (1) $\sigma_p(H) \cap \mathbb{R} = \emptyset$. (2) *The wave operator* $W = s\text{-}\lim_{t\to+\infty} e^{itH_0}e^{-itH}$ *exists in* $\mathcal{H}$ *as a non-trivial operator.*

This implies that the solution $u(t) = e^{-itH}u(0)$ does not have the bound mode. Here, we call the bound mode is neither the scattering mode nor the decaying mode.

Assumption 4.1 (Assumption on φ). (A1) *$\varphi \in L^{2,s+1}(\mathbb{R})$, $s > 1/2$.* **(A2)** *$\Phi(\lambda) \le \Phi(\mu)$ $(0 \le \mu \le \lambda)$, where $\Phi(\lambda) = |\hat{\varphi}(\lambda)|^2 + |\hat{\varphi}(-\lambda)|^2$.*

(To deal with $\Phi(-\lambda)$ for $\lambda < 0$, we extend $\Phi(\lambda)$ as a function on $\mathbb{R}$.)

Notation 4.1. $\Gamma(z) = 1 - iz(r_0(z)\varphi, \varphi)_0$, $\Gamma(\lambda \pm i0) = 1 - i\lambda(r_0(\lambda \pm i0)\varphi, \varphi)_0$, where $r_0(z) = (-\partial^2/\partial x^2 - z^2E)^{-1}$ and $z \in \mathbb{C} \setminus [0, \infty)$, $(\cdot, \cdot)_0$ is inner product in L^2 or dual coupling between $L^{2,s}$ and $L^{2,-s}$ for some $s > 1/2$, $R_0(z) = (H_0 - zE)^{-1}$, $R(z) = (H - zE)^{-1}$, $v_\varphi(z) = (ir_0(z)\varphi, zr_0(z)\varphi)$.

We find that for $z \in \mathbb{C} \setminus \mathbb{R}$ and $f = (f_1, f_2) \in \mathcal{H}$, $R_0(z)f = (r_0(z)(zf_1 + if_2), i\partial_x r_0(z)\partial_x f_1 + zr_0(z)f_2)$. Similar to this, it holds that for any $z \in \mathbb{C} \setminus \mathbb{R}$ satisfying $\Gamma(z) \neq 0$, $R(z)f = R_0(z)f + \frac{i(f, v_\varphi(\bar{z}))}{\Gamma(z)} v_\varphi(z)$.

Theorem 4.2 (Spectral structure). *It holds that $\sigma(H) \cap \mathbb{C}_- = \emptyset$ if $\left|\int_{\mathbb{R}} \varphi(x)dx\right| \le \sqrt{2}$, $= \{i\kappa_0\}$ for some $\kappa_0 < 0$ if $\left|\int_{\mathbb{R}} \varphi(x)dx\right| > \sqrt{2}$. Moreover $i\kappa_0$ is an eigenvalue and its multiplicity is one.*

Put $\varphi_0 = (0, \varphi)$. We define three operators $\mathfrak{F}_0$, $\mathfrak{F}$ and $\mathfrak{G}$ by

$$(\mathfrak{F}_0 f)(\lambda) = \begin{cases} 2^{-1/2}\left(\lambda\hat{f}_1(\lambda) + i\hat{f}_2(\lambda), \lambda\hat{f}_1(-\lambda) + i\hat{f}_2(-\lambda)\right), & \text{if } \lambda > 0, \\ 2^{-1/2}\left(-\lambda\hat{f}_1(-\lambda) - i\hat{f}_2(-\lambda), -\lambda\hat{f}_1(\lambda) - i\hat{f}_2(\lambda)\right), & \text{if } \lambda < 0, \end{cases}$$

$$(\mathfrak{F}f)(\lambda) = (\mathfrak{F}_0 f)(\lambda) + i(f, v_\varphi(\lambda - i0))(\Gamma(\lambda + i0))^{-1}(\mathfrak{F}_0\varphi_0)(\lambda),$$

$$(\mathfrak{G}f)(\lambda) = (\mathfrak{F}_0 u)(\lambda) - i(f, v_\varphi(\lambda - i0))(\overline{\Gamma(\lambda - i0)})^{-1}(\mathfrak{F}_0\varphi_0)(\lambda).$$

Then we see that they become the spectral representation of H_0, H and H^*, respectively.

Theorem 4.3 (Generalized Parseval formula). *Assume (A1) and (A2).*

(1) *If $\Gamma(-i0) \neq 0$, then it holds*

$$(f, g)_{\mathcal{H}} = \begin{cases} \int_{\mathbb{R}} ((\mathfrak{F}f)(\lambda), (\mathfrak{G}g)(\lambda))_{\mathbb{C}^2}\, d\lambda, & \text{if } \Gamma(-i0) > 0, \\ \int_{\mathbb{R}} ((\mathfrak{F}f)(\lambda), (\mathfrak{G}g)(\lambda))_{\mathbb{C}^2}\, d\lambda + (Pf, g)_{\mathcal{H}}, & \text{if } \Gamma(-i0) < 0 \end{cases}$$

for any f, $g \in \mathcal{H}$, where P is the projection onto the eigenspace associated with the eigenvalue $i\kappa_0$ of H.

(2) *If* $\Gamma(-i0) = 0$, *then it holds that for any* $f \in \mathcal{H}$, $g \in \mathcal{E} \equiv \{g \in S(\mathbb{R}) \times S(\mathbb{R}) : (v_\varphi(-i0), g) = 0\}$, *that*

$$(f,g)_{\mathcal{H}} = \int_{\mathbb{R}} ((\mathfrak{F}f)(\lambda), (\mathfrak{G}g)(\lambda))_{\mathbb{C}^2}\, d\lambda.$$

Remark 4.1. A related result can be seen in Gilliam–Schulenberger [2].

Corollary 4.1. *It holds that* $Ker W = \{0\}$ *if* $\left|\int_{\mathbb{R}} \varphi(x)dx\right| \le \sqrt{2}$, $= \mathcal{R}(P)$ *if* $\left|\int_{\mathbb{R}} \varphi(x)dx\right| > \sqrt{2}$.

5. Existence of wave operator for dissipative wave equations in layered region

In the final section, we shall consider the following problems: Fix $N \in \mathbb{N}$ and set $\Omega = \{(x,y) : x \in \mathbb{R}^N, 0 < y < \pi\}$. Consider the following equations for $w = w(x,y,t)$ on $\overline{\Omega}(= \{(x,y) : x \in \mathbb{R}^N, 0 \le y \le \pi\}) \times [0,\infty)$:

$$\begin{cases} w_{tt} - \Delta w + b(x,y)w_t = 0, & (x,y,t) \in \Omega \times (0,\infty), \\ w(x,0,t) = w(x,\pi,t) = 0, & (x,t) \in \mathbb{R}^N \times (0,\infty), \\ w(x,y,0) = w_0(x,y),\ w_t(x,y,0) = w_1(x,y), & (x,y) \in \Omega, \end{cases}$$

where $\Delta = \sum_{j=1}^N \partial^2/\partial x_j^2 + \partial^2/\partial y^2$. At this time, we can obtain only the existence of scattering states if the function $b(x,y)$ is measurable and satisfies $0 \le b(x,y) \le b_3(1+|x|^2)^{-\theta/2}$ ($b_3 > 0$, $\theta > 2$ for $N = 2$, $\theta \ge 2$ for $N \ge 3$). In order to prove this, we prepare with an abstract result from [14] and [4] that originated in [9]:

Theorem 5.1. *Let* $\mathcal{H}$ *be a Hilbert space,* H_0 *a self-adjoint operator in* $\mathcal{H}$, *and* V *a bounded, non-negative self-adjoint operator. Moreover assume the following three conditions: (a1):* $\sigma(H_0) = \sigma_{ac}(H_0) = (-\infty,-m] \cup [m,\infty)$ *or* $[m,\infty)$ *for some* $m \ge 0$; *(a2):* V *is compact; (a3): for any* α, β *satisfying* $m < \alpha < \beta$, *there exist* $C = C_{\alpha,\beta}$ *and* $\eta = \eta_{\alpha,\beta} > 0$ *such that*

$$\sup_{\alpha \le |\operatorname{Re} z| \le \beta, 0 < |\operatorname{Im} z| < \eta} ||\sqrt{V}\left((H_0 - z)^{-1} - (H_0 - \bar{z})^{-1}\right)\sqrt{V}||^{(6)}_{\mathcal{L}(\mathcal{H})} \le C. \quad (2)$$

Then it holds for the perturbed dissipative operator $H = H_0 - iV$ *that*

(1) $\sigma_p(H) \cap \mathbb{R} \ne \emptyset$, *i.e.,* H *has no real eigenvalues.*

(2) *The wave operator* $W = s\text{-}\lim_{t\to+\infty} e^{itH_0}e^{-itH}$ *exists and* $W \not\equiv 0$ *in* $\mathcal{H}$.

---

(6) $||\cdot||_{\mathcal{L}(X)}$ denotes the operator norm in X.

In order to apply Theorem 5.1 above, we put $u(t) = \begin{pmatrix} w(x,y,t) \\ w_t(x,y,t) \end{pmatrix}$, $H_0 = i\begin{pmatrix} 0 & 1 \\ \Delta & 0 \end{pmatrix}$, $V = \begin{pmatrix} 0 & 0 \\ 0 & b(x,y) \end{pmatrix}$, $\mathcal{D}(H) = \mathcal{D}(H_0) = \{f = (f_1, f_2) \in \mathcal{H} \,|\, \Delta f_1 \in L^2(\Omega),\ f_2 \in H^1(\Omega),\ f_1(x,y) = 0\,(x \in \partial\Omega)\}$ where the Hilbert space $\mathcal{H}$ is defined by $\mathcal{H} = \{f \,|\, \|f\| < \infty\}$ with $\|f\| = \sqrt{(f,f)_{\mathcal{H}}}$, $(f,g)_{\mathcal{H}} = \int_0^\pi \int_{\mathbb{R}^N} \left\{\nabla f_1(x,y) \cdot \overline{\nabla g_1(x,y)} + f_2(x,y)\overline{g_2(x,y)}\right\} dxdy$ with $\nabla = (\partial/\partial x_1, \cdots \partial/\partial x_N, \partial/\partial y)$, $f = (f_1, f_2)$, $g = (g_1, g_2)$.

Proposition 5.1. *Assume $\theta > 2$ for $N = 2$ or $\theta \geq 2$ for $N \geq 3$. Then the operators H_0 and V defined above satisfy (a1), (a2) and (a3) in Theorem 5.1 with $m = 1$.*

In order to show this, we define a self-adjoint operator in $L^2(\Omega)$ by $L_0 = -\Delta$ with the domain $\mathcal{D}(L_0) = \{u \in H_0^1(\Omega) \,|\, \Delta u \in L^2(\Omega)\}$.

Lemma 5.1. *Under the same assumptions as in Proposition 5.1, for any $\beta > \alpha > 1$, there exist $\tilde{C} = \tilde{C}_{\alpha,\beta}$ and $\eta = \eta_{\alpha,\beta} > 0$ such that*

$$\sup_{\alpha \leq |\operatorname{Re} z| \leq \beta,\, 0 < |\operatorname{Im} z| < \eta} \|X_{\theta/2}\left((R_0(z) - R_0(\bar{z})\right) X_{\theta/2}\|_{\mathcal{L}(L^2(\Omega))} \leq \tilde{C}, \tag{3}$$

where $X_s = (1 + |x|^2)^{-s/2}$, $R_0(z) = \left(L_0 - z^2 E\right)^{-1}$.

Proof. Using Fourier series, we have for $z \in \mathbb{C} \setminus \mathbb{R}$ and $f \in C_0^\infty(\Omega)$, $(R_0(z)f)(x,y) = 2\pi^{-1}\sum_{n=1}^\infty \sin ny \int_0^\pi \sin ny' (r_n(z)f)(x,y')dy'$, where $r_n(z) = \left(-\Delta_x - (z^2 - n^2)E\right)^{-1}$ and $\Delta_x = \sum_{j=1}^N \partial^2/\partial x_j^2$. Put $\rho(z) = R_0(z) - R_0(\bar{z})$ and $z = \lambda + ik$. We fix $k = 1, 2, 3, \ldots$. Then

$$\begin{aligned}&(\rho(\lambda)f)(x,y) \\ &= \frac{2}{\pi}\sum_{n=1}^{k} \sin ny \int_0^\pi \sin ny' \Big(\big(r_n(\lambda + i0) - r_n(\lambda - i0)\big) f\Big)(x,y')dy',\end{aligned} \tag{4}$$

$$\lim_{k \downarrow 0} \left\| X_s\Big(\rho(z) - \rho(\lambda)\Big)X_s \right\|_{\mathcal{L}(L^2(\Omega))} = 0 \tag{5}$$

for $k < |\lambda| < k+1$ and $s > 1/2$. We prove

$$\|X_{\theta/2}\rho(\lambda)X_{\theta/2}\|_{\mathcal{L}(L^2(\Omega))} = O(1), \quad (|\lambda| \to k + 0). \tag{6}$$

The resolvent estimate (3) follows from (5) and (6). Let $s > 1/2$. Then we have by well-known resolvent estimates (e.g., Kuroda [10]) for $-\Delta_x$,

$$\|X_s r_n(\lambda \pm i0) X_s\|_{\mathcal{L}(L^2(\Omega))} = O(1), \quad (|\lambda| \to k + 0) \tag{7}$$

for $n = 1, 2, 3 \cdots, k-1$. Using (4), (7) and

$$||X_{\theta/2} r_k(\lambda \pm i0) X_{\theta/2}||_{\mathcal{L}(L^2(\Omega))} = O(1), \quad (|\lambda| \to k+0), \tag{8}$$

where $\theta \geq 2$ and $N \geq 3$ (e.g., [4]), we have (6) with $\theta \geq 2$ and $N \geq 3$. We shall prove (6) for $\theta > 2$ and $N = 2$. For $g \in C_0^\infty(\mathbb{R}^2)$, we find

$$\Big((r_k(\lambda+i0) - r_k(\lambda-i0))g \Big)(x) = \int_{\mathbb{R}^2} \left(\frac{i}{2} + E_0(\sqrt{\lambda^2-k^2})(x', x) \right) g(x')dx',$$

where E_0 satisfies $|E_0(\sqrt{\lambda^2 - k^2})(x', x)| \leq C(\lambda^2 - k^2)^{\varepsilon/2}|x' - x|^\varepsilon$ for any $\varepsilon > 0$ (Schlag [19]). Hence, we have for $\theta > 2$,

$$||X_{\theta/2} \left(r_k(\lambda + i0) - r_k(\lambda - i0)\right) X_{\theta/2}||_{\mathcal{L}(L^2(\Omega))} = O(1)$$

as $|\lambda| \to k + 0$. Thus (3), (7) and (8) imply (6) for $\theta > 2$ and $N = 2$. □

Proof of Proposition 5.1. Note that for any $f = (f_1, f_2) \in \mathcal{H}$,

$$\sqrt{V} \left\{ (H_0 - z)^{-1} - (H_0 - \overline{z})^{-1} \right\} \sqrt{V} f = \left(0, \sqrt{b} \left\{ z\rho(z) + \operatorname{Im} z R_0(\overline{z}) \right\} \sqrt{b} f_2 \right).$$

By (3) there exist positive constants C_1 and C_2 independent of z such that

$$||\sqrt{b}\rho(z)\sqrt{b}||_{\mathcal{L}(L^2(\Omega))} \leq C||X_{\theta/2}\rho(z)X_{\theta/2}||_{\mathcal{L}(L^2(\Omega))},$$
$$||\sqrt{b}R_0(\overline{z})\sqrt{b}||_{\mathcal{L}(L^2(\Omega))} \leq C||R_0(\overline{z})X_{\theta/2}||_{\mathcal{L}(L^2(\Omega))}.$$

Thus, we have the desired result if we note

$$\left(X_{\theta/2}\rho(z)X_{\theta/2}f_2, \, f_2 \right)_{L^2(\Omega)} = 2 \operatorname{Re} z \operatorname{Im} z ||R_0(\overline{z})X_{\theta/2}f_2||_{L^2(\Omega)}.$$ □

References

1. S. F. G. Albeverio, R. Hegh-Krohn and H. Holden, *Solvable models in quantum mechanics*, Springer, Berlin, AMS, 1988/2005.
2. D. S. Gilliam and J. R. Schulenberger, Spectral analysis of a dissipative problem in electrodynamics: The Sommerfeld problem, *Acta Appl. Math.*, **6**, pp. 63–94 (1986).
3. T. Ikebe, Eigenfunction expansions associated with the Schrödinger operators and their applications to scattering theory, *Arch. Rational Mech. Anal.*, **5**, pp. 1–34 (1960).
4. M. Kadowaki, Resolvent estimates and scattering states for dissipative systems, *Publ. RIMS Kyoto Univ.*, **38**, pp. 191–209 (2002).
5. M. Kadowaki, H. Nakazawa and K. Watanabe, On the asymptotics of solutions for some Schrödinger equations with dissipative perturbations of rank one, *Hiroshima. Math. J.*, **34**, pp. 345–369 (2004).

6. M. Kadowaki, H. Nakazawa and K.Watanabe, Non-selfadjoint perturbation of Schrödinger and wave equations—Can one completely determine the behavior of solutions by only spectra?, *Advanced studies in Pure Math.*, *Asymptotic Analysis and Singularities—Hyperbolic and dispersive PDEs and fluid mechanics*, Mathematical Society of Japan, Tokyo, **47-1**, pp. 137–157 (2007).
7. M. Kadowaki, H. Nakazawa and K.Watanabe, On the rank one dissipative operator and the Parseval formula, *submitted.*
8. M. Kadowaki, H. Nakazawa and K.Watanabe, Parseval formula for wave equations with dissipative term of rank one, *to appear.*
9. T. Kato, Wave operators and similarity for some non–self adjoint operators, *Math. Ann.* **162**, pp. 258–279 (1966).
10. S. T. Kuroda, *An Introduction to Scattering Theory*, Lecture Note Series N° 51, Matematisk Institut, Aarhus University, 1980.
11. A. Matsumura, Energy decay of solutions of dissipative wave equations, *Proc. Japan Acad.*, **53**, pp. 232–236 (1977).
12. K. Mochizuki, Eigenfunction expansions associated with the Schrödinger operator with a complex potential and scattering inverse problem, *Proc. Japan Acad.*, **43**, pp. 638–643 (1967).
13. K. Mochizuki, Eigenfunction expansions associated with the Schrödinger operator with a complex potential and the scattering theory, *Publ. RIMS Kyoto Univ.*, **4**, pp. 419–466 (1968).
14. K. Mochizuki, Scattering theory for wave equations with dissipative terms, *Publ. RIMS Kyoto Univ.*, **12**, pp. 383–390 (1976).
15. K. Mochizuki, *Scattering theory for wave equations* (Japanese), Kinokuniya, Tokyo, 1984.
16. H. Nakazawa, Principle of limiting absorption for the non-selfadjoint Schrödinger operator with energy dependent potential, *Tokyo. J. Math.*, **23**, pp. 510–536 (2000).
17. H. Nakazawa, On wave equations with dissipations, *Proccedings of* 4th *International conference "Analytical Methods of Analysis and Differential Equations" (AMADE-2006), vol. 3, Differential Equations*, pp. 102–110 (2006).
18. B. S. Pavlov, On a non-selfadjoint Schrödinger operator (Russian), *Prob. Math. Phys. Spectral Theory and Wave Processes, Izdat. Leningrad Univ., Leningrad*, **1**, pp. 102–132 (1966).
19. W. Schlag, Dispersive estimates for Schrödinger operators in dimension two, *Comm. Math. Phys.*, **257**, pp. 87–117 (2005).

ON SCATTERING FOR EVOLUTION EQUATIONS WITH TIME-DEPENDENT SMALL PERTURBATIONS

K. MOCHIZUKI

Department of Mathematics. Chuo University
Kasuga, Bunkyo-ku, Tokyo 112-8551, Japan
E-mail: mochizuk@math.chuo-u.ac.jp

We first summarize an abstract scattering theory for time-dependent small perturbations. Our results give an extension of Kato's results for time-independent smooth perturbations. In the latter half, as examples, we treat the Schrödinger equation with time-dependent complex potential and the wave equation with time-dependent coefficients.

Keywords: Scattering, time-dependent perturbation, Schrödinger equation, wave equation.

1. An abstract result for time-dependent small perturbations

Let $\mathcal{H}$ be a Hilbert space with inner product $(\cdot,\cdot)$ and norm $\|\cdot\|$. We consider in $\mathcal{H}$ the evolution equation

$$i\partial_t u - \Lambda_0 u - V(t)u = 0, \quad t \in \mathbf{R}, \tag{1}$$

where i denotes the imaginary unit, ∂_t the partial derivative with respect to the time variable, Λ_0 is a selfadjoint operator in $\mathcal{H}$ with dense domain $\mathcal{D}(\Lambda_0)$, and $V(t)$ is a Λ_0-compact operator which depends continuously on $t \in \mathbf{R}$.

We choose the initial condition at $t = 0$,

$$u(x,0) = f(x) \in \mathcal{H}, \tag{2}$$

and restrict ourselves to solutions in $\mathcal{H}$. Let $U_0(t) = e^{-it\Lambda_0}$ be the unitary group in $\mathcal{H}$ which represents the solution of the free equation $i\partial_t u_0 - \Lambda_0 u_0 = 0$. Then the perturbed problem (1), (2) reduces to the integral equation

$$u(t) = U_0(t)f - i\int_0^t U_0(t-\tau)V(\tau)u(\tau)d\tau. \tag{3}$$

In the following we restrict ourselves to the class of perturbations with which the following two properties are satisfied.

(I) For given $f \in \mathcal{H}$, equation (3) has a unique solution $u(t) \in C(\mathbf{R}; \mathcal{H})$. We denote by $U(t,s) \in \mathcal{B}(\mathcal{H})$ the evolution operator which maps solutions at time s to those at time t:

$$u(t) = U(t,s)u(s).$$

The unique existence of solutions of (3) implies that for each fixed s and t, $U(t,s)$ defines a bijection on $\mathcal{H}$.

(II) Moreover, the inequality

$$|(V(t)u,v)| \le \eta(t)\|u\|\|v\| + A(t,u)^{1/2}B(t,v)^{1/2} \tag{4}$$

holds for u, $v \in \mathcal{H}$, where $\eta(t)$ is a nonnegative L^1-function of $t \in \mathbf{R}$ and $A(t,u)$, $B(t,v)$ are positive definite quadratic forms satisfying

$$\left|\int_0^{\pm\infty} A(t, U(t,s)f)dt\right| \le C_0\|f\|^2, \tag{5}$$

$$\left|\int_0^{\pm\infty} B(t, U_0(t)g)dt\right| \le C_1\|g\|^2 \tag{6}$$

for some positive constants C_0, C_1 independent of $s \in \mathbf{R}_\pm$ and f, $g \in \mathcal{H}$.

(II) may be called a time-dependent version of the Kato's smooth perturbations (cf. [3,4]). Smallness of $V(t)$ is implicitly required in (5).

With these properties, we obtain the following theorem.

Theorem 1.1. *(i)* $\{U(t,s) : t,s \in \mathbf{R}\}$ *is a family of uniformly bounded operators.*

(ii) For every $s \in \mathbf{R}_\pm$, *there exits the strong limit*

$$Z^\pm(s) = s - \lim_{t\to\pm\infty} U_0(-t+s)U(t,s).$$

(iii) The operator $Z^\pm = Z^\pm(0)$ *satisfies*

$$w - \lim_{s\to\pm\infty} Z^\pm U(0,s)U_0(s) = I \quad (\textit{weak limit}).$$

(iv) If C_0, C_1 *in* (II) *can be chosen to satisfy* $C_0C_1 < 1$, *then* $Z^\pm : \mathcal{H} \longrightarrow \mathcal{H}$ *is a bijection on* $\mathcal{H}$. *Moreover, the scattering operator* $S = Z^+(Z^-)^{-1}$ *is also a bijection.*

Proof. We put $u(t,s) = U(t,s)f$, $u_0(t,s) = U_0(t-s)f_0$. Then we have

$$(u(t,s), u_0(t,s)) = (f, f_0) - i\int_s^t (V(\tau)u(\tau,s), u_0(\tau,s))d\tau$$

from (3). It then follows from (4) of (II) that for any σ, $t \in \mathbf{R}_\pm$,

$$|(u(t,s),u_0(t,s))-(u(\sigma,s),u_0(\sigma,s))| \le \left|\int_\sigma^t \eta(\tau)\|u(\tau,s)\|\,\|u_0(\tau,s)\|d\tau\right| + \left|\int_\sigma^t A(\tau,u(\tau,s))d\tau\right|^{1/2}\left|\int_\sigma^t B(\tau,u_0(\tau,s))d\tau\right|^{1/2}. \quad (7)$$

All the assertions of the theorem are verified from this inequality.

(*i*) We put $\sigma = s$ in (7). Then by (5), (6) of (II)

$$|(u(t,s),u_0(t,s))-(f,f_0)| \le \left|\int_s^t \eta(\tau)\|u(\tau)\|\,\|u_0(\tau)\|d\tau\right| + \sqrt{C_0C_1}\|f\|\,\|f_0\|.$$

Since $U_0(t-s)$ is unitary, it follows that

$$\|u(t,s)\| \le (1+\sqrt{C_0C_1})\|f\| + \int_s^t \eta(\tau)\|u(\tau,s)\|d\tau.$$

The requirement $\eta(t) \in L^1(\mathbf{R})$ and the Gronwall inequality prove the assertion.

(*ii*) Noting (*i*), we have from (7)

$$|(u(t,s),u_0(t,s))-(u(\sigma,s),u_0(\sigma,s))| \le \left\{C_2\|f\|\left|\int_\sigma^t \eta(\tau)d\tau\right| + \left|\int_\sigma^t A(\tau,u(\tau,s))d\tau\right|^{1/2}\sqrt{C_1}\right\}\|f_0\|.$$

Here, for any fixed $s \in \mathbf{R}_\pm$, $\{\cdots\} \to 0$ as $\sigma, t \to \pm\infty$. Thus $U_0(s-t)U(t,s)$ converges strongly in $\mathcal{H}$ as $t \to \pm\infty$.

(*iii*) Let $\sigma = s$ and $t \to \pm\infty$ in (7). Then noting (*i*), we have

$$|(Z^\pm(s)f,f_0)-(f,f_0)| \le \|f\|\left\{C_2\left|\int_s^{\pm\infty}\eta(\tau)d\tau\right|\|f_0\| + \sqrt{C_0}\left|\int_s^{\pm\infty} B(\tau,u_0(\tau,s))d\tau\right|^{1/2}\right\}. \quad (8)$$

Choose here $f = U_0(s)g$ and $f_0 = U_0(s)g_0$. Then

$$|(\{U_0(-s)Z^\pm(s)U_0(s)-I\}g,g_0)| \le \|U_0(s)g\|\left\{C_2\left|\int_s^{\pm\infty}\eta(\tau)d\tau\right|\|U_0(s)g_0\| + \sqrt{C_0}\left|\int_s^{\infty} B(\tau,U_0(\tau)g_0)d\tau\right|^{1/2}\right\}.$$

Since g and g_0 are arbitrary, this implies that as $s \to \pm\infty$,

$$Z^\pm U(0,s)U_0(s) = U_0(-s)Z^\pm(s)U_0(s) \to I$$

weakly in $\mathcal{H}$.

(*iv*) Note that (8) implies

$$|(\{Z^{\pm}(s) - I\}f, f_0)| \le \left\{\left|\int_s^{\pm\infty} \eta(\tau)d\tau\right| C_2 + \sqrt{C_0C_1}\right\}\|f\|\|f_0\|.$$

Since $C_0C_1 < 1$, we can choose $\pm s > 0$ sufficiently large to satisfy

$$\left|\int_s^{\pm\infty} \eta(\tau)d\tau\right| C_2 + \sqrt{C_0C_1} < 1.$$

Thus, $\|Z_{\pm}(s) - I\|_{\mathcal{B}(\mathcal{H})} < 1$ and $Z^{\pm}(s)$ gives a bijection on $\mathcal{H}$. The same property of $Z^{\pm}$ then easily follows. □

2. Schrödinger equations with time dependent complex potentials

We consider the Schrödinger equation

$$i\partial_t u + \Delta u - V(x,t)u = 0 \tag{9}$$

in $(x,t) \in \mathbf{R}^n \times \mathbf{R}$, where Δ is the n dimensional Laplacian and $V(x,t)$ is a complex potential which is bounded and continuous in $\mathbf{R}^n \times \mathbf{R}$.

The Cauchy problem for (9) is reduced to the evolution equation (1) if we choose $\mathcal{H} = L^2(\mathbf{R}^n)$, $\Lambda_0 = -\Delta$ with $\mathcal{D}(\Lambda_0) = H^2$ and $V(t) = V(x,t)$. Properties (I) and (II) are verified based on space-time L^p-L^q estimates of solutions $u(t)$. For this aim, we restrict ourselves to potentials like

(*A*1) $V(x,t) \in L^\nu(\mathbf{R}; L^q)$, where

$$0 \le \frac{1}{q} \le \frac{2}{n} \quad \text{and} \quad \frac{1}{\nu} = 1 - \frac{n}{2q}.$$

$V(x,t)$ also satisfies the smallness condition

$$C_3\|V\|_{L^\infty(\mathbf{R}_\pm;L^{n/2})} < 1 \quad \text{when} \quad \nu = \infty, \tag{10}$$

where $C_3 > 0$ is given later in Lemma 2.2.

A typical example of potentials is

$$V(x,t) = c(1+|x|)^{-\alpha}(1+|t|)^{-\beta}, \tag{11}$$

where $c \in \mathbf{C}$ and α, $\beta \ge 0$ with

$$\frac{\alpha}{2} + \beta > 1 \quad \text{and} \quad |c| \text{ is small if } \beta = 0. \tag{12}$$

Note that potential (11) satisfies (*A*1) if we choose

$$\frac{1}{q} = 0 \ \text{ when } \alpha = 0, \quad \frac{1}{q} = \frac{2}{n} \ \text{ when } \beta = 0$$

and

$$\frac{\max\{0, 2(1-\beta)\}}{n} < \frac{1}{q} < \frac{\min\{\alpha, 2\}}{n}$$

when $\alpha,\ \beta > 0$.

The above potential is considered by Yafaev [13] when c is real and $\beta > 0$ (see also [1,2,6] and [14], where more general real potentials are studied).

Lemma 2.1. *Let $2 \leq p \leq \infty$ and put $\frac{1}{p'} = 1 - \frac{1}{p}$. Then*

$$\|U_0(t)f_0\|_{L^p} \leq (4\pi|t|)^{n/p-n/2}\|f_0\|_{L^{p'}}.$$

Lemma 2.2. *Let $n \geq 3$ and*

$$\frac{n-2}{2n} \leq \frac{1}{p} \leq \frac{1}{2} \quad \text{and} \quad \frac{1}{r} = \frac{n}{2}\left(\frac{1}{2} - \frac{1}{p}\right).$$

Then there exists $C_3 > 0$ such that

$$\left\|\int_0^t U_0(t-\tau)h(\tau)d\tau\right\|_{L^r(\mathbf{R}_\pm;L^p)} \leq C_3\|h\|_{L^{r'}(\mathbf{R}_\pm;L^{p'})}.$$

Lemma 2.1 is classical. Lemma 2.2 is a direct consequence of Lemma 2.1 if $\frac{1}{p} > \frac{n-2}{2n}$. At the end point $\frac{1}{p} = \frac{n-2}{2n}$, it is due to Keel-Tao [5]. As a corollary of this lemma we have the following.

Proposition 2.1. *Let n, p and r be as in Lemma 2.2. Then*

(*i*) *For any $t \in \mathbf{R}_\pm$,*

$$\left\|\int_0^t U_0(-\tau)h(\tau)d\tau\right\| \leq \sqrt{2C_3}\|h\|_{L^{r'}(\mathbf{R}_\pm;L^{p'})}.$$

(*ii*) *For $f_0 \in L^2$, we have $U_0(t)f_0 \in L^r(\mathbf{R}_\pm; L^p)$ and*

$$\|U_0(\cdot)f_0\|_{L^r(\mathbf{R}_\pm;L^p)} \leq \sqrt{2C_3}\|f_0\|.$$

Similar estimates are obtained for perturbed solutions if we require $(A1)$. For $1 \leq \gamma,\ \mu \leq \infty$ and $\pm s \geq 0$, we put

$$Y^{\gamma,\mu}_{\pm,s} = L^\gamma(\mathbf{R}_{\pm,s}; L^\mu),$$

where $\mathbf{R}_{+,s} = [s, \infty)$ for $s \geq 0$ and $\mathbf{R}_{-,s} = (-\infty, s]$ for $s \leq 0$. The space $Y^{\gamma,\mu}_{\pm,0} = L^\gamma(\mathbf{R}_\pm; L^\mu)$ is already used in this section. By $(A1)$ we have $V(x,t) \in Y^{\nu,q}_{\pm,s}$ for any $\pm s \geq 0$. Moreover, as we see from (10), there exist $\pm s \geq 0$ such that

$$C_3\|V\|_{Y^{\nu,q}_{\pm,s}} < 1. \tag{13}$$

In the following we fix such an s, and choose the pair (p, r) related to (q, ν) as

$$\frac{1}{p} = \frac{1}{2}\left(1 - \frac{1}{q}\right) \text{ and } \frac{1}{r} = \frac{1}{2}\left(1 - \frac{1}{\nu}\right). \tag{14}$$

The condition for (q, ν) in $(A1)$ is equivalent to that for (p, r) in Lemma 2.2.

Proposition 2.2. *([10, Theorem 3]) Let $n \geq 3$ and assume $(A1)$. Then the following hold:*

(i) For each $f \in L^2$, the integral equation

$$u(t) = U_0(t - s)f + i \int_s^t U_0(t - \tau)V(\tau)u(\tau)d\tau$$

has a unique solution $u(t) \in Y_{\pm,s}^{r,p}$.

(ii) This solution belongs to $C(\mathbf{R}_{\pm,s}; L^2)$. Moreover, we have

$$\|u\|_{Y_{\pm,s}^{r,p}} \leq \frac{\sqrt{2C_3}}{1 - C_3\|V\|_{Y_{\pm,s}^{\nu,q}}}\|f\| \tag{15}$$

and

$$\left\| \int_s^t U_0(-\tau)V(\tau)u(\tau)d\tau \right\| \leq \frac{2C_3\|V\|_{Y_{\pm,s}^{\nu,q}}}{1 - C_3\|V\|_{Y_{\pm,s}^{\nu,q}}}\|f\|. \tag{16}$$

Now we are ready to verify properties (I), (II) of Section 1. (I) is obtained in Proposition 2.2. (II) is verified as follows. By $(A1)$ and Proposition 2.1 (ii),

$$\left| \int_\sigma^t \int |V||u_0|^2 dx d\tau \right|^{1/2} \leq \|V\|_{Y_{\pm,s}^{\nu,q}}^{1/2} \|u_0\|_{Y_{\pm,s}^{2\nu',2q'}} \leq \sqrt{2C_3}\|V\|_{Y_{\pm,s}^{\nu,q}}^{1/2}\|f_0\|, \tag{17}$$

where we have used the equalities

$$\frac{1}{2q'} = \frac{1}{2}\left(1 - \frac{1}{q}\right) = \frac{1}{p}, \quad \frac{1}{2\nu'} = \frac{1}{r}.$$

On the other hand, by $(A1)$ and (15) of Proposition 2.2 we similarly have

$$\left| \int_\sigma^t \int |V||u|^2 dx d\tau \right|^{1/2} \leq \frac{\sqrt{2C_5}\|V\|_{Y_{\pm,s}^{\nu,q}}^{1/2}}{1 - C_5\|V\|_{Y_{\pm,s}^{\nu,q}}}\|f\|. \tag{18}$$

So we can choose $\eta(t) \equiv 0$ and $A(t)v = B(t)v = |V(x, t)|^{1/2}v$ in (5).

3. Wave equations with time dependent coefficients

Let Ω be an exterior domain in $\mathbf{R}^n$ ($n \geq 3$) with smooth boundary $\partial\Omega$ which is star-shaped with respect to the origin 0. We consider in Ω the wave equation

$$\partial_t^2 w - \Delta w + b_0(x,t)\partial_t w + \sum_{j=1}^{n} b_j(x,t)\partial_j w + c(x,t)w = 0 \tag{19}$$

with Dirichlet boundary condition

$$w(x,t) = 0, \quad (x,t) \in \partial\Omega \times \mathbf{R}, \tag{20}$$

where $b_j(x,t)$ ($j = 0, 1, \cdots, n$) and $c(x,t)$ are real-valued bounded continuous functions.

Let H_D^1 denote the completion of $C_0^\infty(\Omega)$ with respect to the Dirichlet norm

$$\|u\|_D = \left(\int |\nabla u|^2 dx\right)^{1/2}$$

for $u \in C_0^\infty(\Omega)$, and let L^2 be the usual L^2 space in Ω. We define $\mathcal{H}_E = H_D^1 \times L^2$. Then $\mathcal{H}_E$ forms a Hilbert space with the energy norm

$$\|f\|_E = \frac{1}{\sqrt{2}} \left\{\|f_1\|_D^2 + \|f_2\|^2\right\}^{1/2}, \; f = \{f_1, f_2\}.$$

As an evolution equation in $\mathcal{H}_E$, the problem (19), (20) is rewritten in the form

$$i\partial_t \vec{w} = \Lambda_0 \vec{w} + V(t)w, \quad \vec{w} = (w, w_t)^T, \tag{21}$$

where

$$\Lambda_0 = i \begin{pmatrix} 0 & 1 \\ \Delta & 0 \end{pmatrix} \quad \text{and} \quad V(t) = -i \begin{pmatrix} 0 & 0 \\ b(\cdot,t)\cdot\nabla + c(\cdot,t) & b_0(\cdot,t) \end{pmatrix}$$

with $b(x,t)\cdot\nabla = \sum_{j=1}^{n} b_j(x,t)\partial_j$.

The operator Λ_0 becomes selfadjoint in $\mathcal{H}_E$ if the domain is defined by $\mathcal{D}(\Lambda_0) = H_D^2 \times \{H_D^1 \cap L^2\}$, where H_D^2 is the set of functions $f_1 \in H_D^1$ such that $\Delta f_1 \in L^2$. Thus Λ_0 generates a unitary group $\{U_0(t) = e^{-it\Lambda_0}; t \in \mathbf{R}\}$ on $\mathcal{H}_E$, and by use of this $U_0(t)$, problem (21) is reduced to the integral equation

$$\vec{w}(t) = U_0(t)f + \int_0^t U_0(t-s)V(s)\vec{w}(s)ds. \tag{22}$$

Let $\eta = \eta(t)$ be a positive L^1-function of $t \in \mathbf{R}$, and let $\xi = \xi(r)$ be a smooth positive L^1-function of $r > 0$ which also satisfies

$$\xi'(r) \le 0, \qquad \xi'(r)^2 \le 2\xi(r)\xi''(r). \tag{23}$$

Note that the functions $\xi(r) = (1+r)^{-1-\delta}$ or $\xi(r) = (1+r)^{-1}\{\log(e+r)\}^{-1-\delta}$ satisfy (23).

With these functions, we require the following conditions on the coefficients:

$$(A2) \qquad \begin{cases} |b_j(x,t)| \le \epsilon\xi(r) + \eta(t) \quad (j = 0, 1, \cdots, n), \\ |c(x,t)| \le \{\epsilon\xi(r) + \eta(t)\}\dfrac{n-2}{2r}, \end{cases}$$

where ϵ is a small positive constant (see [7,8,11,12] for related conditions).

A typical example of coefficients is

$$\begin{cases} |b_j(x,t)| \le b_{j0}(1+r)^{-\alpha_j}(1+|t|)^{-\beta_j} \quad (j = 0, 1, \cdots, n) \\ |c(x,t)| \le c_0 r^{-1}(1+r)^{-\tilde{\alpha}}(1+|t|)^{-\tilde{\beta}}, \end{cases} \tag{24}$$

where b_{j0}, c_0 are positive constants, and α_j, β_j, $\tilde{\alpha}$, $\tilde{\beta} \ge 0$ satisfy

$$\alpha_j + \beta_j > 1, \quad \tilde{\alpha} + \tilde{\beta} > 1.$$

The constant b_{j0} (or c_0) should be chosen sufficiently small if β_j (or $\tilde{\beta}$) $= 0$.

In order to verify properties (I), (II) to hold we shall make use of space-time weighted energy estimates for solutions $u(t)$.

First we multiply by w_t both sides of (19) and integrate by parts over $\Omega \times [s,t]$. Then noting the boundary condition $w_{0t}|_{\partial\Omega} = 0$, we obtain

$$\|\vec{w}(t)\|_E^2 - \|\vec{w}(s)\|_E^2 + \int_s^t \int P_1 \, dx d\tau = 0, \tag{25}$$

where

$$P_1 = \{b_0 w_t + b \cdot \nabla w + cw\} w_t.$$

Next, we multiply by $\psi\left(w_r + \dfrac{n-1}{2r}w\right)$, $\psi(r) = \displaystyle\int_0^r \xi(\sigma)d\sigma$, both sides of (21), where $w_r = \tilde{x} \cdot \nabla w$ with $\tilde{x} = x/|x|$, and integrate over $\Omega \times [s,t]$ $(s < t)$. Then noting the star-shapedness of $\partial\Omega$, we have

$$\int X(t)dx - \int X(s)dx + \int_s^t \int \{Z + P_2\} dx d\tau \le 0, \tag{26}$$

where

$$X = \psi\left(w_t w_r + \frac{n-1}{2r} w_t w\right),$$

$$Z = \left(\frac{1}{r}\psi - \psi'\right)\left\{|\nabla w|^2 - w_r^2 + \frac{(n-1)(n-3)}{4r^2}w^2\right\}$$
$$+ \frac{1}{2}\psi'(w_t^2 + |\nabla w|^2) - \psi''\frac{n-1}{4r}w^2,$$
$$P_2 = \psi\{b_0 w_t + b \cdot \nabla w + cw\}\left(w_r + \frac{n-1}{2r}w\right).$$

By the use of (25) and (26), the following two propositions hold true.

Proposition 3.1. *([9, Theorem 2]) Let $\xi(r)$, $r > 0$, be a positive L^1-function satisfying* (23)*. Then for any $s < t$ we have*

$$\frac{1}{2}\int_s^t \int \left\{\xi(|\nabla w_0|^2 + w_{0t}^2) - \xi'\frac{n-1}{2r}w_0^2\right\}dxd\tau \le C_4\|\vec{w}_0(0)\|_E^2,$$

where $C_4 = C_4(\xi) = \dfrac{4n-6}{n-2}\|\xi\|_{L^1}$.

Proposition 3.2. *([9, Theorem 3]) Assume (A2) with $\epsilon < \dfrac{2}{9C_4}$. Then*

(*i*) *There exists $C_5 = C_5(\eta) > 0$ such that for any $t \in \mathbf{R}$, we have*

$$\|\vec{w}(t)\|_E^2 \le C_5\|\vec{w}(0)\|_E^2.$$

(*ii*) *There exists $C_6 = C_6(\epsilon, \xi, \eta) > 0$ such that for any $s < t$ we have*

$$\frac{1}{2}\int_s^t \int_\Omega \left\{\xi(|\nabla w|^2 + w_t^2) - \xi'\frac{n-1}{2r}w^2\right\}dxd\tau \le C_6\|\vec{w}(0)\|_E^2.$$

These propositions prove property (II). In fact, since

$$|(V(t)u, v)_E| \le \int \{\epsilon\xi(r) + \eta(t)\}\left\{|\nabla u_1| + \frac{n-2}{2r}|u_1| + |u_2|\right\}|v_2|dx,$$

choosing

$$A(t,u) = \int \epsilon\xi(r)\left\{|\nabla u_1|^2 + \left(\frac{n-2}{2r}|u_1|^2\right)^2 + |u_2|^2\right\}dx,$$
$$B(t,v) = 3\int \xi(r)|v_2|^2 dx$$

in (4), we obtain inequalities (5) and (6) from Propositions 3.2 (*ii*) and 3.1, respectively. Property (I) is also verified by the use of the successive approximation based on these propositions.

References

1. J. Howland, Stationary scattering theory for time dependent Hamiltonians, Math. Ann. **207** (1974), 315-335.
2. A. Jensen, Results in $L^p(\mathbf{R}^d)$ for the Schrödinger equation with time-dependent potential, Math. Ann. **299** (1994), 117-125.
3. T. Kato, Wave operators and similarity for some nonselfadjoint operators, Math. Ann. **162** (1966), 258-279.
4. T. Kato and K. Yajima, Some examples of smooth operators and the associated smoothing effect, Rev. Math. Phys. **1** (1989), 481-496.
5. M. Keel and R. Tao, Endpoint Strichartz estimates, Amer. J. Math. **120** (1998), 955-980.
6. H. Kitada and K. Yajima, A scattering theory for time-dependent long-range potentials, Duke Math. J. **49** (1982), 341-376
7. K. Mochizuki, Scattering theory for wave equations with dissipative terms, Publ. Res. Inst. Math. Sci. **12** (1976), 383-390.
8. K. Mochizuki, Scattering theory for wave equations, Kinokuniya Shoten, 1984 (in Japanese).
9. K. Mochizuki, On scattering for wave equations with time dependent coefficients, Tsukuba J. Math. **31** (2007), 327-342.
10. K. Mochizuki and T. Motai, On decay-nondecay and scattering for Schrödinger equations with time dependent complex potentials, Publ. Res. Inst. Math. Sci. **43** (2007), 1183-1197.
11. K. Mochizuki and H. Nakazawa, Energy decay and asymptotic behavior of solutions to the wave equation with linear dissipation, Publ. Res. Inst. Math. Sci. **32** (1996), 401-414.
12. J. Wirth, Scattering and modified scattering for abstract wave equations with time-dependent dissipation, Adv. Differential Equations, **12** (2007), 1115-1133.
13. D. R. Yafaev, On the violation of the unitarity in time dependent potential scattering, Soviet Math. Dokl., **19** (1973), 1517-1521.
14. K. Yajima, Scattering theory for Schrödinger equations with potentials periodic in time, J. Math. Soc. Japan **29** (1977), 729-743.

COMPARISON OF ESTIMATES FOR DISPERSIVE EQUATIONS

M. RUZHANSKY

Department of Mathematics, Imperial College London
London, United Kingdom
E-mail: m.ruzhansky@imperial.ac.uk

M. SUGIMOTO

Department of Mathematics, Osaka University
Osaka, Japan
E-mail: sugimoto@math.sci.osaka-u.ac.jp

This paper describes a new comparison principle that can be used for the comparison of space-time estimates for dispersive equations. In particular, results are applied to the global smoothing estimates for several classes of dispersive partial differential equations.

Keywords: Comparison principle; dispersive equations; smoothing estimates; weighted estimates.

1. Introduction

In this note we will present a new comparison principle that allows one to compare certain estimates for dispersive equations of different types based on expressions involving their symbols. In particular, a question is that if we have a certain estimate for one equation, whether we can derive a corresponding estimate for another equation. This question is of interest on its own, and it has several applications. Proofs of the statements of this paper can be found in the authors' paper [14].

The main application of this technique that we have in mind is for the global smoothing estimates for dispersive equations. These smoothing estimates are essentially global space-time estimates in weighted Sobolev spaces over L^2, see for example [2,3,5,7,16,17]. There is a known method in the microlocal analysis on how to transform one equation into another, namely the canonical transforms realised in the form of Fourier integral operators [6]. In the global setting one needs to develop global weighted estimates in L^2 for the corresponding classes of Fourier integral operators in order to apply them to the smoothing estimates. Such global estimates for Fourier integral operators have been established by the authors [12] and have been applied to derive new smoothing estimates for Schrödinger equations [12,13]. These techniques allow one to reduce the analysis of dispersive equations to normal forms in one and two dimensions [11]. The comparison principles introduced below allow one to further relate estimates in normal forms, thus establishing comprehensive relations between smoothing estimates for dispersive equations with constant coefficient [14].

In this note, we denote $x = (x_1, \ldots, x_n)$, $\xi = (\xi_1, \ldots, \xi_n)$, and $D_x = (D_1, D_2 \ldots, D_n)$, where D_j denotes $D_{x_j} = \frac{1}{i}\frac{\partial}{\partial x_j}$, for all $j = 1, 2, \ldots, n$, and $i = \sqrt{-1}$.

2. Comparison principles

Let $u(t,x) = e^{itf(D_x)}\varphi(x)$ and $v(t,x) = e^{itg(D_x)}\varphi(x)$ be solutions to the following evolution equations, where $t \in \mathbb{R}$ and $x \in \mathbb{R}^n$:

$$\begin{cases} (i\partial_t + f(D_x))\, u(t,x) = 0, \\ \qquad\qquad u(0,x) = \varphi(x), \end{cases} \tag{1}$$

and

$$\begin{cases} (i\partial_t + g(D_x))\, v(t,x) = 0, \\ \qquad\qquad v(0,x) = \varphi(x). \end{cases} \tag{2}$$

First we state the following result relating several norms involving propagators for equations (1) and (2):

Theorem 2.1. *Let $f \in C^1(\mathbb{R}^n)$ be a real-valued function such that, for almost all $\xi' = (\xi_2, \ldots, \xi_n) \in \mathbb{R}^{n-1}$, $f(\xi_1, \xi')$ is strictly monotone in ξ_1 on the support of a measurable function σ on $\mathbb{R}^n$. Then we have*

$$\left\| \sigma(D_x) e^{itf(D_x)} \varphi(x_1, x') \right\|^2_{L^2(\mathbb{R}_t \times \mathbb{R}^{n-1}_{x'})} = (2\pi)^{-n} \int_{\mathbb{R}^n} |\widehat{\varphi}(\xi)|^2 \frac{|\sigma(\xi)|^2}{|\partial f / \partial \xi_1(\xi)|}\, d\xi$$

for all $x_1 \in \mathbb{R}$, where $x' = (x_2, \ldots, x_n) \in \mathbb{R}^{n-1}$.

The following comparison principle is a straightforward consequence of Theorem 2.1:

Corollary 2.1. *Let $f, g \in C^1(\mathbb{R}^n)$ be real-valued functions such that, for almost all $\xi' = (\xi_2, \ldots, \xi_n) \in \mathbb{R}^{n-1}$, $f(\xi_1, \xi')$ and $g(\xi_1, \xi')$ are strictly* *monotone in* ξ_1 *on the support of a measurable function* χ *on* $\mathbb{R}^n$. *Let* $\sigma, \tau \in C^0(\mathbb{R}^n)$ *be such that, for some* $A > 0$, *we have*

$$\frac{|\sigma(\xi)|}{|\partial_{\xi_1} f(\xi)|^{1/2}} \leq A \frac{|\tau(\xi)|}{|\partial_{\xi_1} g(\xi)|^{1/2}} \tag{3}$$

for all $\xi \in \operatorname{supp} \chi$ *satisfying* $D_1 f(\xi) \neq 0$ *and* $D_1 g(\xi) \neq 0$. *Then we have*

$$\left\| \chi(D_x)\sigma(D_x)e^{itf(D_x)}\varphi(x_1, x') \right\|_{L^2(\mathbb{R}_t \times \mathbb{R}^{n-1}_{x'})} \leq A \| \chi(D_x)\tau(D_x)e^{itg(D_x)}\varphi(\widetilde{x}_1, x') \|_{L^2(\mathbb{R}_t \times \mathbb{R}^{n-1}_{x'})} \tag{4}$$

for all $x_1, \widetilde{x}_1 \in \mathbb{R}$, *where* $x' = (x_2, \ldots, x_n) \in \mathbb{R}^{n-1}$. *Consequently, for any measurable function* w *on* $\mathbb{R}$ *we have*

$$\left\| w(x_1)\chi(D_x)\sigma(D_x)e^{itf(D_x)}\varphi(x) \right\|_{L^2(\mathbb{R}_t \times \mathbb{R}^n_x)} \leq A \| w(x_1)\chi(D_x)\tau(D_x)e^{itg(D_x)}\varphi(x) \|_{L^2(\mathbb{R}_t \times \mathbb{R}^n_x)}. \tag{5}$$

Moreover, if $\chi \in C^0(\mathbb{R}^n)$ *and* $w \neq 0$ *on a set of* $\mathbb{R}$ *with positive measure, the converse is true, namely, if we have estimate* (4) *for all* φ, *for some* $x_1, \widetilde{x}_1 \in \mathbb{R}$, *or if we have estimate* (5) *for all* φ, *and the norms are finite, then we also have inequality* (3).

We remark that inequality (5) in Corollary 2.1 gives the comparison between different weighted estimates. The reason to introduce a cut-off function χ into the estimates is that the relation between symbols may be different for different regions of the frequencies ξ (for example this is the case for the relativistic Schrödinger and for the Klein-Gordon equations, which are of order two for large frequencies and of order zero for small frequencies). In such case we can use this comparison principle to relate estimates for the corresponding ranges of frequencies, thus yielding more refined results, since then we have freedom to choose different σ for different types of behaviour of f'. The assumption $\sigma, \tau \in C^0(\mathbb{R}^n)$ that was made in Corollary 2.1 is for the clarity of the exposition and can clearly be relaxed.

In the case $n = 1$, we neglect $x' = (x_2, \ldots, x_n)$ in a natural way and just write $x = x_1$, $\xi = \xi_1$, and $D_x = D_1$. Similarly in the case $n = 2$, we use the notation $(x, y) = (x_1, x_2)$, $(\xi, \eta) = (\xi_1, \xi_2)$, and $(D_x, D_y) = (D_1, D_2)$. In both cases, we write $\widetilde{x} = \widetilde{x}_1$ in the notation of Corollary 2.1. Then we have the following corollaries in lower dimensions:

Corollary 2.2. *Suppose* $n = 1$. *Let* $f, g \in C^1(\mathbb{R})$ *be real-valued and strictly monotone on the support of a measurable function* χ *on* $\mathbb{R}$. *Let* $\sigma, \tau \in C^0(\mathbb{R})$

be such that, for some $A > 0$, *we have*

$$\frac{|\sigma(\xi)|}{|f'(\xi)|^{1/2}} \leq A \frac{|\tau(\xi)|}{|g'(\xi)|^{1/2}} \tag{6}$$

for all $\xi \in \operatorname{supp} \chi$ *satisfying* $f'(\xi) \neq 0$ *and* $g'(\xi) \neq 0$. *Then we have*

$$\|\chi(D_x)\sigma(D_x)e^{itf(D_x)}\varphi(x)\|_{L^2(\mathbb{R}_t)} \leq A\|\chi(D_x)\tau(D_x)e^{itg(D_x)}\varphi(\widetilde{x})\|_{L^2(\mathbb{R}_t)} \tag{7}$$

for all $x, \widetilde{x} \in \mathbb{R}$. *Consequently, for general* $n \geq 1$ *and for any measurable function* w *on* $\mathbb{R}^n$, *we have*

$$\|w(x)\chi(D_j)\sigma(D_j)e^{itf(D_j)}\varphi(x)\|_{L^2(\mathbb{R}_t \times \mathbb{R}^n_x)} \leq A\|w(x)\chi(D_j)\tau(D_j)e^{itg(D_j)}\varphi(x)\|_{L^2(\mathbb{R}_t \times \mathbb{R}^n_x)}, \tag{8}$$

where $j = 1, 2, \ldots, n$. *Moreover, if* $\chi \in C^0(\mathbb{R})$ *and* $w \neq 0$ *on a set of* $\mathbb{R}^n$ *with positive measure, the converse is true, namely, if we have estimate* (7) *for all* φ, *for some* $x, \widetilde{x} \in \mathbb{R}$, *or if we have estimate* (7) *for all* φ, *and the norms are finite, then we also have inequality* (6).

We have the following comparison principle in two dimensions:

Corollary 2.3. *Suppose* $n = 2$. *Let* $f, g \in C^1(\mathbb{R}^2)$ *be real-valued functions such that, for almost all* $\eta \in \mathbb{R}$, $f(\xi, \eta)$ *and* $g(\xi, \eta)$ *are strictly monotone in* ξ *on the support of a measurable function* χ *on* $\mathbb{R}^2$. *Let* $\sigma, \tau \in C^0(\mathbb{R}^2)$ *be such that, for some* $A > 0$, *we have*

$$\frac{|\sigma(\xi, \eta)|}{|\partial f/\partial\xi(\xi, \eta)|^{1/2}} \leq A \frac{|\tau(\xi, \eta)|}{|\partial g/\partial\xi(\xi, \eta)|^{1/2}} \tag{9}$$

for all $(\xi, \eta) \in \operatorname{supp} \chi$ *satisfying* $\partial f/\partial\xi(\xi, \eta) \neq 0$ *and* $\partial g/\partial\xi(\xi, \eta) \neq 0$. *Then we have*

$$\left\|\chi(D_x, D_y)\sigma(D_x, D_y)e^{itf(D_x, D_y)}\varphi(x, y)\right\|_{L^2(\mathbb{R}_t \times \mathbb{R}_y)} \leq A\|\chi(D_x, D_y)\tau(D_x, D_y)e^{itg(D_x, D_y)}\varphi(\widetilde{x}, y)\|_{L^2(\mathbb{R}_t \times \mathbb{R}_y)} \tag{10}$$

for all $x, \widetilde{x} \in \mathbb{R}$. *Consequently, for general* $n \geq 2$ *and for any measurable function* w *on* $\mathbb{R}^{n-1}$ *we have*

$$\|w(\check{x}_k)\chi(D_j, D_k)\sigma(D_j, D_k)e^{itf(D_j, D_k)}\varphi(x)\|_{L^2(\mathbb{R}_t \times \mathbb{R}^n_x)} \leq A\|w(\check{x}_k)\chi(D_j, D_k)\tau(D_j, D_k)e^{itg(D_j, D_k)}\varphi(x)\|_{L^2(\mathbb{R}_t \times \mathbb{R}^n_x)}, \tag{11}$$

where $j \neq k$ *and* $\check{x}_k = (x_1, \ldots, x_{k-1}, x_{k+1}, \ldots, x_n)$. *Moreover, if* $\chi \in C^0(\mathbb{R}^2)$ *and* $w \neq 0$ *on a set of* $\mathbb{R}^{n-1}$ *with positive measure, the converse is true, namely, if we have estimate* (10) *for all* φ, *for some* $x, \widetilde{x} \in \mathbb{R}$, *or if we have estimate* (10) *for all* φ, *and the norms are finite, then we also have inequality* (9).

By the same argument as used in the proof of Theorem 2.1 and Corollary 2.1, we have a comparison result in the radially symmetric case. Below, we denote the set of the positive real numbers $(0, \infty)$ by $\mathbb{R}_+$.

Theorem 2.2. *Let* $f, g \in C^1(\mathbb{R}_+)$ *be real-valued and strictly monotone on the support of a measurable function* χ *on* $\mathbb{R}_+$. *Let* $\sigma, \tau \in C^0(\mathbb{R}_+)$ *be such that, for some* $A > 0$, *we have*

$$\frac{|\sigma(\rho)|}{|f'(\rho)|^{1/2}} \leq A \frac{|\tau(\rho)|}{|g'(\rho)|^{1/2}} \tag{12}$$

for all $\rho \in \operatorname{supp} \chi$ *satisfying* $f'(\rho) \neq 0$ *and* $g'(\rho) \neq 0$. *Then we have*

$$\begin{aligned} &\|\chi(|D_x|)\sigma(|D_x|)e^{itf(|D_x|)}\varphi(x)\|_{L^2(\mathbb{R}_t)} \\ &\qquad \leq A\|\chi(|D_x|)\tau(|D_x|)e^{itg(|D_x|)}\varphi(x)\|_{L^2(\mathbb{R}_t)} \end{aligned} \tag{13}$$

for all $x \in \mathbb{R}^n$. *Consequently, for any measurable function* w *on* $\mathbb{R}^n$, *we have*

$$\begin{aligned} &\|w(x)\chi(|D_x|)\sigma(|D_x|)e^{itf(|D_x|)}\varphi(x)\|_{L^2(\mathbb{R}_t \times \mathbb{R}^n_x)} \\ &\qquad \leq A\|w(x)\chi(|D_x|)\tau(|D_x|)e^{itg(|D_x|)}\varphi(x)\|_{L^2(\mathbb{R}_t \times \mathbb{R}^n_x)}. \end{aligned} \tag{14}$$

Moreover, if $\chi \in C^0(\mathbb{R}_+)$ *and* $w \neq 0$ *on a set of* $\mathbb{R}^n$ *with positive measure, the converse is true, namely, if we have estimate* (13) *for all* φ, *for some* $x \in \mathbb{R}^n$, *or if we have estimate* (14) *for all* φ, *and the norms are finite, then we also have inequality* (12).

Theorem 2.2 provides an analytic alternative to computations for certain estimates in the radially symmetric case done with the help of special functions [18].

These comparison principles can be extended to provide the relation between Strichartz type norms, and the details and the meaning of the corresponding estimates can be found in authors' paper [14]. Here we just give one corollary:

Corollary 2.4. *Let functions* f, g, σ, τ *be as in Theorem 2.2 and satisfy relation* (12). *Let* $0 < p \leq \infty$. *Then, for any measurable function* w *on* $\mathbb{R}^n$, *we have the estimate*

$$\|w(x)\chi(|D_x|)\sigma(|D_x|)e^{itf(|D_x|)}\varphi(x)\|_{L^p(\mathbb{R}^n_x, L^2(\mathbb{R}_t))}$$

$$\leq A\|w(x)\chi(|D_x|)\tau(|D_x|)e^{itg(|D_x|)}\varphi(x)\|_{L^p(\mathbb{R}^n_x,L^2(\mathbb{R}_t))}. \quad (15)$$

From this, it follows, for example, that for all $0 < p \leq \infty$, the quantities $\|e^{it\sqrt{-\Delta}}\varphi\|_{L^p(\mathbb{R}^n_x,L^2(\mathbb{R}_t))}$, $\||D_x|^{1/2}e^{-it\Delta}\varphi\|_{L^p(\mathbb{R}^n_x,L^2(\mathbb{R}_t))}$, and $\||D_x|e^{it(-\Delta)^{3/2}}\varphi\|_{L^p(\mathbb{R}^n_x,L^2(\mathbb{R}_t))}$ for the propagators of the wave, Schrödinger, and KdV-type equations are equivalent.

3. Some applications

Let us now give some examples of the use of these comparison principles. If both sides in expression (3) in Corollary 2.1 are equivalent, we can use the comparison in two directions, from which it follows that norms on both sides in (4) are equivalent. The same is true for Corollaries 2.2, 2.3 and Theorem 2.2. In particular, we can conclude that many smoothing estimates for the Schrödinger type equations of different orders are equivalent to each other. Indeed, applying Corollary 2.2 in two directions, we immediately obtain that for $n = 1$ and $l, m > 0$, we have

$$\left\||D_x|^{(m-1)/2}e^{it|D_x|^m}\varphi(x)\right\|_{L^2(\mathbb{R}_t)} = \sqrt{\frac{l}{m}}\left\||D_x|^{(l-1)/2}e^{it|D_x|^l}\varphi(x)\right\|_{L^2(\mathbb{R}_t)} \quad (16)$$

for every $x \in \mathbb{R}$, assuming that $\operatorname{supp}\widehat{\varphi} \subset [0, +\infty)$ or $(-\infty, 0]$. On the other hand, still in the case $n = 1$, we have easily

$$\left\|e^{itD_x}\varphi(x)\right\|_{L^2(\mathbb{R}_t)} = \|\varphi\|_{L^2(\mathbb{R}_x)} \quad \text{for all } x \in \mathbb{R}, \quad (17)$$

which is a straightforward consequence of the fact $e^{itD_x}\varphi(x) = \varphi(x+t)$. These observations yield the following result:

Theorem 3.1. *Suppose $n = 1$ and $m > 0$. Then we have*

$$\left\||D_x|^{(m-1)/2}e^{it|D_x|^m}\varphi(x)\right\|_{L^2(\mathbb{R}_t)} \leq C\|\varphi\|_{L^2(\mathbb{R}_x)} \quad (18)$$

for all $x \in \mathbb{R}$. Suppose $n = 2$ and $m > 0$. Then we have

$$\left\||D_y|^{(m-1)/2}e^{itD_x|D_y|^{m-1}}\varphi(x,y)\right\|_{L^2(\mathbb{R}_t\times\mathbb{R}_y)} \leq C\|\varphi\|_{L^2(\mathbb{R}^2_{x,y})} \quad (19)$$

for all $x \in \mathbb{R}$. Each estimate above is equivalent to itself with $m = 1$ which is a direct consequence of equality (17).

Estimates (18) and (19) in Theorem 3.1 in the special case $m = 2$ were shown by Kenig, Ponce and Vega [9, p. 56] and by Linares and Ponce [10, p. 528], respectively. Theorem 3.1 shows that these results, together with their generalisation to other orders m, are in fact just corollaries of the elementary one dimensional fact $e^{itD_x}\varphi(x) = \varphi(x+t)$ once we apply the comparison principle.

By using the comparison principle in the radially symmetric and higher dimensional cases, we have also another type of equivalence of smoothing estimates, which can be found in authors' paper [14]. Let us give one example:

Theorem 3.2. *For $m > 0$ (and any α, β) we have the following relations (in the first equality the left and the right hand sides are finite for the same values of α, β at the same time):*

$$\left\| |x|^{\beta-1}|D_x|^{\beta} e^{it|D_x|^2}\varphi \right\|_{L^2(\mathbb{R}_t\times\mathbb{R}^n_x)} = \sqrt{\frac{m}{2}} \left\| |x|^{\beta-1}|D_x|^{m/2+\beta-1} e^{it|D_x|^m}\varphi \right\|_{L^2(\mathbb{R}_t\times\mathbb{R}^n_x)},$$

and

$$\begin{aligned} \left\| \langle x\rangle^{\alpha-m/2}|D_x|^{\alpha} e^{it|D_x|^m}\varphi(x) \right\|_{L^2(\mathbb{R}_t\times\mathbb{R}^n_x)} &\le \left\| |x|^{\alpha-m/2}|D_x|^{\alpha} e^{it|D_x|^m}\varphi(x) \right\|_{L^2(\mathbb{R}_t\times\mathbb{R}^n_x)} \\ &\le \sup_{\lambda>0} \left\| \langle x\rangle^{\alpha-m/2}|D_x|^{\alpha} e^{it|D_x|^m}\varphi_\lambda(x) \right\|_{L^2(\mathbb{R}_t\times\mathbb{R}^n_x)}, \end{aligned}$$

where $\varphi_\lambda(x) = \lambda^{n/2}\varphi(\lambda x)$, and we take $\alpha \le m/2$ in the last estimate. The operator norms of operators $\langle x\rangle^{\alpha-m/2}|D_x|^{\alpha}e^{it|D_x|^m}$ and $|x|^{\alpha-m/2}|D_x|^{\alpha}e^{it|D_x|^m}$ as mappings from $L^2(\mathbb{R}^n)$ to $L^2(\mathbb{R}_t\times\mathbb{R}^n_x)$ are equal.

As a nice consequence, for $n \ge 3$ and $m > 0$ we can conclude also the estimate

$$\left\| |x|^{-1}|D_x|^{m/2-1} e^{it|D_x|^m}\varphi(x) \right\|_{L^2(\mathbb{R}_t\times\mathbb{R}^n_x)} \le \sqrt{\frac{2\pi}{m(n-2)}}\|\varphi\|_{L^2(\mathbb{R}^n_x)}, \tag{20}$$

where the constant $\sqrt{\frac{2\pi}{m(n-2)}}$ is sharp. This follows from the first equality in Theorem 3.2 with $\beta = 0$ and the best constant in the case $m = 2$ (as shown by Simon [15] as a consequence of constants in Kato's theory).

As a consequence of Theorem 3.2, we have the following:

Corollary 3.1. *Suppose $m > 0$ and $(m-n)/2 < \alpha < (m-1)/2$. Then we have*

$$\left\| |x|^{\alpha-m/2}|D_x|^{\alpha} e^{it|D_x|^m}\varphi(x) \right\|_{L^2(\mathbb{R}_t\times\mathbb{R}^n_x)} \le C\|\varphi\|_{L^2(\mathbb{R}^n_x)}. \tag{21}$$

Suppose $m > 0$ and $(m-n+1)/2 < \alpha < (m-1)/2$. Then we have

$$\left\| |x|^{\alpha-m/2} |D'|^{\alpha} e^{it(|D_1|^m - |D'|^m)} \varphi(x) \right\|_{L^2(\mathbb{R}_t \times \mathbb{R}^n_x)} \leq C \|\varphi\|_{L^2(\mathbb{R}^n_x)}, \tag{22}$$

where $D' = (D_2, \ldots, D_n)$.

Estimate (21) is known in the case $m = 2$ as the Kato–Yajima estimate [8]. The application of the comparison principles also yields some refinements for other equations, for example for the relativistic Schrödinger equation [4], Klein-Gordon equations and wave equations [1]. We refer to [14] for further details.

References

1. M. Ben-Artzi, *Regularity and smoothing for some equations of evolution*, Nonlinear partial differential equations and their applications. College de France Seminar, Vol. XI (Paris, 1989–1991), 1–12, Pitman Res. Notes Math. Ser., 299, Longman Sci. Tech., Harlow, 1994.
2. M. Ben-Artzi and A. Devinatz, *Local smoothing and convergence properties of Schrödinger type equations*, J. Funct. Anal. **101** (1991), 231–254.
3. M. Ben-Artzi and S. Klainerman, *Decay and regularity for the Schrödinger equation*, J. Analyse Math. **58** (1992), 25–37.
4. M. Ben-Artzi and J. Nemirovsky, *Remarks on relativistic Schrödinger operators and their extensions*, Ann. Inst. H. Poincare Phys. Theor. **67** (1997), 29–39.
5. P. Constantin and J. C. Saut, *Local smoothing properties of dispersive equations*, J. Amer. Math. Soc. **1** (1988), 413–439.
6. J. J. Duistermaat and L. Hörmander, *Fourier integral operators. II*, Acta Math. **128** (1972), 183–269.
7. T. Kato, *On the Cauchy problem for the (generalized) Korteweg-de Vries equation*, Studies in applied mathematics, 93–128, Adv. Math. Suppl. Stud., 8, Academic Press, New York, 1983.
8. T. Kato and K. Yajima, *Some examples of smooth operators and the associated smoothing effect*, Rev. Math. Phys. **1** (1989), 481–496.
9. C. E. Kenig, G. Ponce and L. Vega, *Oscillatory integrals and regularity of dispersive equations*, Indiana Univ. Math. J. **40** (1991), 33–69.
10. F. Linares and G. Ponce, *On the Davey-Stewartson systems*, Ann. Inst. H. Poincaré Anal. Non Linéaire **10** (1993), 523–548.
11. M. Ruzhansky and M. Sugimoto, *A new proof of global smoothing estimates for dispersive equations*, Advances in pseudo-differential operators, 65–75, Oper. Theory Adv. Appl., **155**, Birkhäuser, Basel, 2004.
12. M. Ruzhansky and M. Sugimoto, *Global L^2-boundedness theorems for a class of Fourier integral operators*, Comm. Partial Differential Equations **31** (2006), 547–569.
13. M. Ruzhansky and M. Sugimoto, *A smoothing property of Schrödinger equations in the critical case*, Math. Ann. **335** (2006), 645-673.

14. M. Ruzhansky and M. Sugimoto, *Smoothing properties of evolution equations via canonical transforms and comparison principle*, arXiv:math/0612274v4.
15. B. Simon, *Best constants in some operator smoothness estimates*, J. Funct. Anal. **107** (1992), 66–71.
16. P. Sjölin, *Regularity of solutions to the Schrödinger equation*, Duke Math. J. **55** (1987), 699–715.
17. L. Vega, *Schrödinger equations: Pointwise convergence to the initial data*, Proc. Amer. Math. Soc. **102** (1988), 874–878.
18. B. G. Walther, *Regularity, decay, and best constants for dispersive equations*, J. Funct. Anal. **189** (2002), 325–335.

DISPERSIVE ESTIMATES IN ANISOTROPIC THERMOELASTICITY

J. WIRTH*

Department of Mathematics, Imperial College
London SW7 2AZ, UK

We present a unified treatment of the type-1 model of anisotropic thermoelasticity in two space dimensions and derive dispersive estimates for its solutions. The approach is based on diagonalising the full symbol of a corresponding first order system and representing solutions in terms of Fourier integrals with complex phases.

Keywords: Thermoelasticity, a-priori estimates, anisotropic media.

1. The thermo-elastic system

1.1. *Introduction*

The system of thermo-elasticity couples the elastic behaviour of a medium with propagation of heat. There are different ways to model this coupling, see e.g. [1]; we are concerned with the so-called type-1 model based on Fourier's law of heat conduction. It reads as

$$U_{tt} + A(\mathrm{D})U + \gamma\nabla\theta = 0, \tag{1a}$$

$$\theta_t - \kappa\Delta\theta + \gamma\nabla\cdot U_t = 0, \tag{1b}$$

$$U(0,\cdot) = U_1, \qquad U_t(0,\cdot) = U_2, \qquad \theta(0,\cdot) = \theta_0, \tag{1c}$$

where $U(t,\cdot) : \mathbb{R}^2 \to \mathbb{R}^2$ denotes the elastic displacement and $\theta(t,\cdot) : \mathbb{R}^2 \to \mathbb{R}$ the temperature difference to some equilibrium. The thermal conductivity κ and the thermo-elastic coupling γ are assumed to be positive, the elastic operator $A(\mathrm{D})$ is (in general) a self-adjoint and positive 2×2 system of (pseudo-) differential operators.

While for the particular case of *isotropic media*, given by

$$A(\mathrm{D}) = -\mu\Delta - (\lambda+\mu)\nabla\otimes\nabla \tag{2}$$

*Work supported by the Royal Society, Grant CIJN 44364, and EPSRC, EP/E062873/1.

with Lamé constants λ and μ, decay results are well-known in arbitrary space dimension, the treatment for anisotropic media is much harder to tackle in general and only very few results are available for particular cases in two dimensions, [2,3].

We recall the main ideas to investigate the isotropic problem: If we use the Helmholtz decomposition of vector fields (which works in all space dimensions larger than one) and split the displacement into potential and solenoidal part the problem decouples. The solenoidal part satisfies a wave equation with propagation speed $\sqrt{\mu}$, while the potential part satisfies a simpler thermo-elastic system (which is in fact the system of one-dimensional thermo-elasticity). The latter one can be studied in detail, the first one satisfies well known estimates. Decay rates are determined by the wave part (in dimensions larger than or equal to two). The result of this reasoning can be collected in the following statement.

Theorem 1.1. *Let $U(t,x)$ and $\theta(t,x)$ be the solutions to* (1) *for an isotropic medium* (2) *with Schwarz data $U_1, U_2, \theta_0 \in \mathcal{S}(\mathbb{R}^n)$. Then the a-priori estimate*

$$||(U_t, \sqrt{A(\mathrm{D})}U, \theta)||_{L^\infty} \leq C_r(1+t)^{-\delta_n}||(\langle \mathrm{D}\rangle U_0, U_1, \theta_0)||_{H^{1,r}} \tag{3}$$

holds true for $r > n$ and with $\delta_1 = 1/2$ and $\delta_n = (n-1)/2$, $n \geq 2$.

If we analyse this reasoning we see that in the isotropic case solutions are decomposed into hyperbolic parts (solving free wave equations) and parabolic parts (solving a somewhat simpler parabolic type thermo-elastic system). Such a decomposition does not exist in the case of anisotropic media, however we can do it *micro-locally* in certain directions. Our main strategy is to find the hyperbolic components in the solutions, which might determine decay rates, and to show that all other parts behave more nicely.

1.2. *Reformulation of the problem and basic assumptions*

We apply a partial Fourier transform to (1) and write it as system of ordinary differential equations parameterised by the frequency variable $\xi \in \mathbb{R}^2$. This gives

$$\hat{U}_{tt} + A(\xi)\hat{U} + \mathrm{i}\gamma\xi\hat{\theta} = 0, \tag{4a}$$

$$\hat{\theta}_t + \kappa|\xi|^2\hat{\theta} + \mathrm{i}\gamma\xi \cdot \hat{U}_t = 0, \tag{4b}$$

$$\hat{U}(0,\cdot) = \hat{U}_1, \qquad \hat{U}_t(0,\cdot) = \hat{U}_2, \qquad \hat{\theta}(0,\cdot) = \hat{\theta}_0, \tag{4c}$$

where $A(\mathrm{D})$ is replaced by its symbol $A(\xi)$. For the following we assume that

(A1) the symbol is 2-homogeneous, $A(\xi) = |\xi|^2 A(\eta)$, $\eta = \xi/|\xi|$ and depends real-analytically on $\eta \in \mathbb{S}^1$;
(A2) it is symmetric and positive definit, $A(\eta) = A^*(\eta) > 0$;
(A3) its eigenvalues are distinct, $\operatorname{spec} A(\eta) = \{\omega_1^2(\eta), \omega_2^2(\eta)\}$.

We denote the eigenvalues in ascending order, $\omega_1(\eta) < \omega_2(\eta)$. These assumptions imply that they are real-analytic in $\eta \in \mathbb{S}^1$ and we can find real analytic normalised eigenvectors $r_j(\eta) : \mathbb{S}^1 \to \mathbb{S}^1$ corresponding to them. Furthermore, we denote by $M(\eta) = (r_1(\eta)|r_2(\eta))$ the unitary diagonaliser of $A(\eta)$, $\mathcal{D}(\xi) = \operatorname{diag}(\omega_1(\xi), \omega_2(\xi))$ the square root of the corresponding diagonal matrix and $a_j(\xi) = \xi \cdot r_j(\eta)$ auxiliary *coupling functions.*

We are now in a position to formulate (4) as equivalent system of first order. Let for this $V = ((\mathrm{D}_t + \mathcal{D}(\xi))M^*(\eta)\hat{U}, (\mathrm{D}_t - \mathcal{D}(\xi))M^*(\eta)\hat{U}, \hat{\theta})^T$. Then

$$\mathrm{D}_t V = B(\xi) V \tag{5}$$

with

$$B(\xi) = \begin{pmatrix} \omega_1(\xi) & & & & \mathrm{i}\gamma a_1(\xi) \\ & \omega_2(\xi) & & & \mathrm{i}\gamma a_2(\xi) \\ & & -\omega_1(\xi) & & \mathrm{i}\gamma a_1(\xi) \\ & & & -\omega_2(\xi) & \mathrm{i}\gamma a_2(\xi) \\ -\frac{\mathrm{i}\gamma}{2} a_1(\xi) & -\frac{\mathrm{i}\gamma}{2} a_2(\xi) & -\frac{\mathrm{i}\gamma}{2} a_1(\xi) & -\frac{\mathrm{i}\gamma}{2} a_2(\xi) & -\mathrm{i}\kappa|\xi|^2 \end{pmatrix}, \tag{6}$$

which splits into the sum of the 1-homogeneous part $B_1(\xi)$ and the 2-homogeneous part $B_2(\xi)$. Note, that $B_2(\xi)$ has only the lower right corner entry and the functions $a_j(\xi) = |\xi| a_j(\eta)$ indeed describe the coupling between the B_1 and B_2. Of particular importance are directions $\eta \in \mathbb{S}^1$, where one of the coupling functions vanishes.

Definition 1.1. We call a direction $\bar{\eta} \in \mathbb{S}^1$ *hyperbolic* with respect to the eigenvalue $\omega_j^2(\eta)$ if the corresponding coupling function $a_j(\eta)$ vanishes in $\eta = \bar{\eta}$. We denote all non-hyperbolic directions as *parabolic.*

Assumption (A3) essentially guarantees that the upper 4×4 block has distinct eigenvalues. We need one more (slightly technical looking) assumption for our approach, which guarantees that B_1 has distinct eigenvalues. We assume

(A4) $\gamma^2 \neq \omega_2^2(\eta) - \omega_1^2(\eta)$ for all hyperbolic directions w.r.t. $\omega_2^2(\eta)$.

1.3. *Aims*

Assume for the moment that the matrix $B(\xi)$ has only simple eigenvalues, i.e. $\#\mathrm{spec}\, B(\xi) = 5$. Then we can represent the solution $V(t,\xi)$ to (5) as

$$V(t,\xi) = \sum_{\nu(\xi)\in\mathrm{spec}\,B(\xi)} \mathrm{e}^{\mathrm{i}t\nu(\xi)} P_{\nu(\xi)} V_0 \tag{7}$$

with eigenprojections $P_{\nu(\xi)}$ to the eigenvalues ν. Applying the inverse Fourier transform yields expressions for $U(t,x)$ and $\theta(t,x)$ in terms of Fourier integrals. We will see that the eigenvalues $\nu(\xi)$ are in general complex and at least the interesting ones are always separated from the others.

Our main aims are twofold. On the one hand we want to prove properties of these eigenvalues $\nu(\xi)$, especially bounds on their imaginary parts together with asymptotic expansions near interesting ξ (i.e. near all points where real eigenvalues may occur). Results of this type are discussed in Section 2. On the other hand, we want to apply the above given representation of solutions to derive dispersive estimates. This is based on multiplier theorems and tools from harmonic analysis related to stationary phase method and results are given in Section 3. Finally in Section 4 we provide some examples. Full details of all proofs can be found in [4,5].

2. Spectral properties of $B(\xi)$

2.1. *General properties*

The special structure of matrix $B(\xi)$ allows to consider its characteristic polynomial and to derive certain properties from it. We will use that under assumptions (A1) and (A2) the following statement holds true:

Lemma 2.1.

(1) The matrix $B(\xi)$, $\xi \neq 0$, has real eigenvalues if and only if the direction $\eta = \xi/|\xi|$ is hyperbolic.

(2) Let $\bar\eta$ be an isolated hyperbolic direction w.r.to $\omega_j^2(\eta)$. Then for the corresponding 'hyperbolic' eigenvalue of $B(\xi)$ and each fixed $|\xi|$ the limits

$$\lim_{\eta\to\bar\eta} \frac{\mathrm{Im}\,\nu(\xi)}{a_j^2(\eta)} > 0 \tag{8}$$

exist and are positive.

The first statement explains the notion of hyperbolic directions. Let's start looking at the case of isotropic media again. The matrix $A(\eta) = \mu \mathrm{I} + (\lambda+\mu)\eta\otimes\eta$ has the eigenvalues μ and $\lambda+\mu$ with corresponding eigenvectors $\eta^\perp$ and η. Thus the coupling functions are given by $a_1(\eta) \equiv 0$ and $a_2(\eta) \equiv 1$. All directions are hyperbolic and the system decouples, as already mentioned in the introduction.

In general we have only a finite number of hyperbolic directions (due to (A1) the coupling functions are real-analytic on the circle) and there the matrix $B(\xi)$ has two real eigenvalues. In the neighbourhood of these directions we can decompose our vector V into a hyperbolic component and a parabolic one. For the latter one we will see that the imaginary part of the corresponding eigenvalues satisfies certain bounds and essential decay properties come from the hyperbolic component.

2.2. *Asymptotic expansion for small frequencies*

We decompose the matrix $B(\xi)$ into homogeneous components, $B(\xi) = B_1(\xi) + B_2(\xi)$. If ξ is small, $B_1(\xi)$ determines the spectrum of $B(\xi)$ up to terms of order $\mathcal{O}(|\xi|^2)$, provided that the eigenvalues of $B_1(\xi)$ are distinct. Assumption (A4) implies exactly that:

Lemma 2.2. *Assume (A1) and (A2). Then (A4) is equivalent to* $\#\operatorname{spec} B_1(\xi) = 5$. *Furthermore,* $\operatorname{spec} B_1(\xi) = \{0, \pm\tilde{\nu}_1(\xi), \pm\tilde{\nu}_2(\xi)\}$ *with* $\omega_1(\xi) \le \tilde{\nu}_1(\xi) \le \omega_2(\xi) \le \tilde{\nu}_2(\xi)$.

The proof of this statement follows essentially from the characteristic equation of $B_1(\xi)$, which can be written as

$$\tilde{\nu}(\xi) = 0 \qquad \text{or} \qquad \frac{1}{\gamma^2} = \sum_{j=1,2} \frac{a_j^2(\xi)}{\tilde{\nu}_j^2(\xi) - \omega_j^2(\xi)}. \tag{9}$$

If we assume (A4) we can apply a standard diagonalisation scheme to obtain full asymptotic expansions of the eigenvalues as $\xi \to 0$.

Lemma 2.3. *(Reissig-Wirth, [4]) Assume (A1), (A2) and (A4). Then the eigenvalues of the matrix $B(\xi)$ have full asymptotic expansions as $\xi \to 0$. The first terms of these expansions are given by*

$$\nu_0(\xi) = \mathrm{i}\kappa|\xi|^2 b_0(\eta) + \mathcal{O}(|\xi|^3), \tag{10a}$$

$$\nu_j^\pm(\xi) = \pm|\xi|\tilde{\nu}_j(\eta) + \mathrm{i}\kappa|\xi|^2 b_j(\eta) + \mathcal{O}(|\xi|^3), \tag{10b}$$

where the functions $b_j \in C^\infty(\mathbb{S}^1)$ are given by

$$b_0(\eta) = \left(1 + \frac{\gamma^2 a_1^2(\eta)}{\varkappa_1(\eta)} + \frac{\gamma^2 a_2^2(\eta)}{\omega_2^2(\eta)}\right)^{-1} > 0 \tag{11}$$

and

$$b_j(\eta) = \left(1+\gamma^2 a_1^2(\eta)\frac{\tilde\nu_j^2+\omega_1^2(\eta)}{(\tilde\nu_j^2-\omega_1^2(\eta))^2}+\gamma^2 a_2^2(\eta)\frac{\tilde\nu_j^2+\omega_2^2(\eta)}{(\tilde\nu_j^2-\omega_2^2(\eta))^2}\right)^{-1} \geq 0. \qquad (12)$$

Furthermore, $b_j(\eta) > 0$ *if* η *is parabolic and* $b_{j_0}(\eta) = 0$ *if* η *is hyperbolic with respect to the eigenvalue* $\omega_{j_0}^2$.

One particular consequence of this statement is of interest. In combination with Lemma 2.1 (1) and the continuity of $\operatorname{spec} B(\xi)$ it follows that no eigenvalue is allowed to have a negative imaginary part.

Corollary 2.1. *All eigenvalues of* $B(\xi)$ *have non-negative imaginary part.*

Therefore, we see that there exists a constant C such that for *all* t the a-priori estimate

$$||V(t,\cdot)||_2 \leq C||V_0||_2 \qquad (13)$$

is valid. The appearing constant C depends on the norms of the eigenprojections.

2.3. *Asymptotic expansion for large frequencies*

For large $|\xi|$ we employ a two-step diagonalisation procedure, generalising that of [6]. This is necessary since $B_2(\xi)$ is the main part and this matrix has only two distinct eigenvalues. We can diagonalise in a first step to 4,1-block-diagonal structure. It is essential that (A3) guarantees that the 1-homogeneous component of the upper 4×4 block has distinct eigenvalues and can therefore be diagonalised in a second step. This in turn implies full asymptotic expansions.

Lemma 2.4. *(Reissig-Wirth, [4]) Assume (A1) – (A3). Then the eigenvalues of the matrix* $B(\xi)$ *have full asymptotic expansions as* $\xi \to \infty$. *The first terms of these expansions are given by*

$$\nu_0(\xi) = \mathrm{i}\kappa|\xi|^2 - \frac{\mathrm{i}\gamma^2}{\kappa} + \mathcal{O}(|\xi|^{-1}), \qquad (14a)$$

$$\nu_j^{\pm}(\xi) = \pm|\xi|\omega_j(\eta) + \frac{\mathrm{i}\gamma^2}{2\kappa}a_j^2(\eta) + \mathcal{O}(|\xi|^{-1}). \qquad (14b)$$

As consequence we obtain that for $|\xi| \geq c > 0$ and away from hyperbolic directions the eigenvalues are uniformly separated from the real axis.

Corollary 2.2. *Let* $\eta = \xi/|\xi|$ *be parabolic. Then the eigenvalues of* $B(\xi)$ *satisfy* $\operatorname{Im}\nu(\xi) \geq C_\eta > 0$ *for all* $|\xi| \geq c > 0$.

3. Dispersive estimates of solutions

We will give only the main ideas of the proof and cite the results. From Corollary 2.2 it follows that large parabolic frequencies (outside a conic neighbour of hyperbolic directions) lead to exponential decay. Since we expect only polynomial rates, we can concentrate on the remaining directions. For small parabolic directions the imaginary part of the eigenvalues behaves essentially like $|\xi|^2$ and therefore estimates can be done via Hölder inequality and boundedness of Fourier transform

$$||\mathrm{e}^{\mathrm{i}t\nu(\mathrm{D})}f||_{L^\infty} \le ||\mathrm{e}^{-t\mathrm{Im}\,\nu(\xi)}\langle\xi\rangle^{-r}||_{L^1}||f||_{H^{1,r}} \le C(1+t)^{-1/2}||f||_{H^{1,r}}, \quad (15)$$

like for the heat equation.

Near hyperbolic directions we have two different influences. On the one hand the imaginary part vanishes like $a_j^2(\eta) \sim \varphi^{2\ell}$ such that integration over a conic neighbourhood yields from

$$\int_0^c \int_{-\epsilon}^{\epsilon} \mathrm{e}^{-t|\xi|^2\varphi^{2\ell}} \mathrm{d}\varphi|\xi|\mathrm{d}|\xi| \sim t^{-\frac{1}{2\ell}} \quad (16a)$$

$$\int_c^\infty \int_{-\epsilon}^{\epsilon} \mathrm{e}^{-t\varphi^{2\ell}} \mathrm{d}\varphi|\xi|^{1-r}\mathrm{d}|\xi| \sim t^{-\frac{1}{2\ell}}, \qquad r > 2 \quad (16b)$$

as decay rate $t^{1/(2\ell)}$ where ℓ denotes the vanishing order of the coupling function in the hyperbolic direction. On the other hand, the real part of the eigenvalue leads to oscillations in the representation of solutions as Fourier integrals. These oscillations may be studied by means of stationary phase method and may also lead to a decay rate. However, only one of these two strategies can be used. The above mentioned one works for all vanishing orders ℓ, the stationary phase estimate presented below is only possible if the vanishing order ℓ stands in some relation to the contact order $\bar\gamma$ of the Fresnel curve

$$S = \{\, \xi \in \mathbb{R}^2 \;:\; 1 \in \operatorname{spec} A(\xi) \,\} = \bigcup_j \{\, \omega_j^{-1}(\eta)\eta \;:\; \eta \in \mathbb{S}^1 \,\} \quad (17)$$

to its tangent in the corresponding direction η and for the corresponding sheet.

Lemma 3.1. *(Reissig-Wirth, [4]) Let $\bar\eta$ be hyperbolic with respect to the eigenvalue $\omega_{j_0}^2(\eta)$ and denote $\bar\gamma_{j_0}(\bar\eta)$ the contact order between the curve $\omega_{j_0}^{-1}(\eta)\eta$ and its tangent in $\bar\eta$. Let further $\ell_{j_0}(\bar\eta)$ be the vanishing order of $a_{j_0}(\eta)$ in $\bar\eta$. If $\bar\gamma_{j_0}(\bar\eta) < 2\ell_{j_0}(\bar\eta)$, then the estimate*

$$||\mathrm{e}^{\mathrm{i}t\nu_{j_0}^{\pm}(\mathrm{D})}f||_{L^\infty} \le C(1+t)^{-\frac{1}{\bar\gamma}}||f||_{H^{1,r}} \quad (18)$$

is valid for any $r > 2$.

The method of proof is an application of van der Corput Lemma, while the restriction on the contact order stems from the fact that we use the imaginary part of $\nu^{\pm}_{j_0}(\xi)$ as part of the amplitude. The following statement collects our main results on dispersive estimates.

Theorem 3.1. *(Reissig-Wirth, [4]) Let $\bar\eta \in \mathbb{S}^1$ be a fixed direction, $\chi \in C_0^\infty(\mathbb{S}^1)$ a cutoff function for a small neighbourhood of $\bar\eta$ and $\chi(\xi) = \chi(\xi/|\xi|)$. Then the following micro-local dispersive estimates hold true:*

(1) If $\bar\eta$ is hyperbolic with respect to $\omega_j^2(\bar\eta)$ and $2\ell_j(\bar\eta) \le \bar\gamma_j(\bar\eta)$, then for any $r > 2$ we find a constant C such that

$$\|\chi(\mathrm{D})(U_t, \sqrt{A(\mathrm{D})}U, \theta)(t,\cdot)\|_\infty \le C(1+t)^{-\frac{1}{2\ell_j(\bar\eta)}}(\|\chi(\mathrm{D})(U_t, \sqrt{A(\mathrm{D})}U, \theta)(0,\cdot)\|_{H^{1,r}}). \quad (19)$$

(2) If $\bar\eta$ is hyperbolic with respect to $\omega_j^2(\bar\eta)$ and $2\ell_j(\eta) > \bar\gamma_j(\bar\eta)$, then for any $r > 2$ we find a constant C such that

$$\|\chi(\mathrm{D})(U_t, \sqrt{A(\mathrm{D})}U, \theta)(t,\cdot)\|_\infty \le C(1+t)^{-\frac{1}{\bar\gamma_j(\bar\eta)}}(\|\chi(\mathrm{D})(U_t, \sqrt{A(\mathrm{D})}U, \theta)(0,\cdot)\|_{H^{1,r}}). \quad (20)$$

(3) If $\bar\eta$ is not hyperbolic, then for any $r > 2$ we find a constant C such that

$$\|\chi(\mathrm{D})(U_t, \sqrt{A(\mathrm{D})}U, \theta)(t,\cdot)\|_\infty \le C(1+t)^{-1}(\|\chi(\mathrm{D})(U_t, \sqrt{A(\mathrm{D})}U, \theta)(0,\cdot)\|_{H^{1,r}}). \quad (21)$$

4. Examples

We will give some concrete applications of our general results. For this we consider the class of elastic operators defined by $A(\xi) = \mathbb{D}(\xi)\mathcal{S}\mathbb{D}^*(\xi)$ with

$$\mathbb{D}(\xi) = \begin{pmatrix} \xi_1 & \\ & \xi_2 \\ \xi_2 & \xi_1 \end{pmatrix} \quad \text{and} \quad \mathcal{S} = \begin{pmatrix} \tau_1 & \lambda & \sigma_1 \\ \lambda & \tau_2 & \sigma_2 \\ \sigma_1 & \sigma_2 & \mu \end{pmatrix}. \quad (22)$$

For physical reasons one assumes that $\mathcal{S}$ is positive. The choice $\tau_1 = \tau_2 = \lambda + 2\mu$ and $\sigma_1 = \sigma_2 = 0$ leads to isotropic media. We discussed them in the introduction. We will concentrate on anisotropic examples here. They as well as the following pictures are taken from [5]. Depicted are the Fresnel curves for the media together with the coupling cunctions (as curves $(2 + a_j(\eta)\eta$ for $\eta \in \mathbb{S}^1$).

Cubic media are defined by the conditions $\tau_1 = \tau_2$ and $\sigma_1 = \sigma_2 = 0$. For cubic media we obtain eight hyperbolic directions corresponding to the symmetry axes of a square. In all these directions the coupling functions vanish to first order and therefore

Corollary 4.1. *Cubic media have the dispersive decay rate* $t^{-1/2}$.

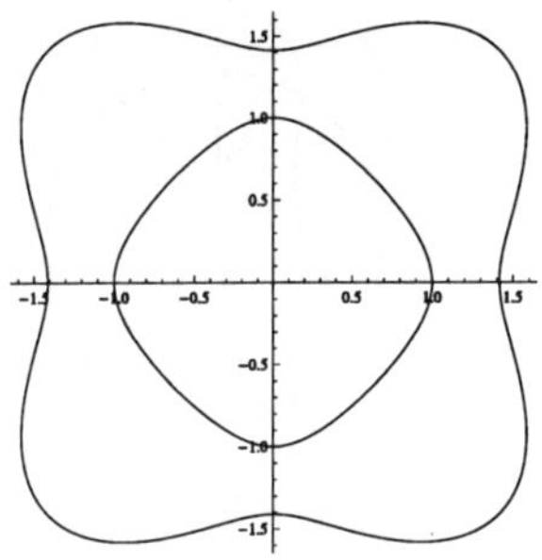

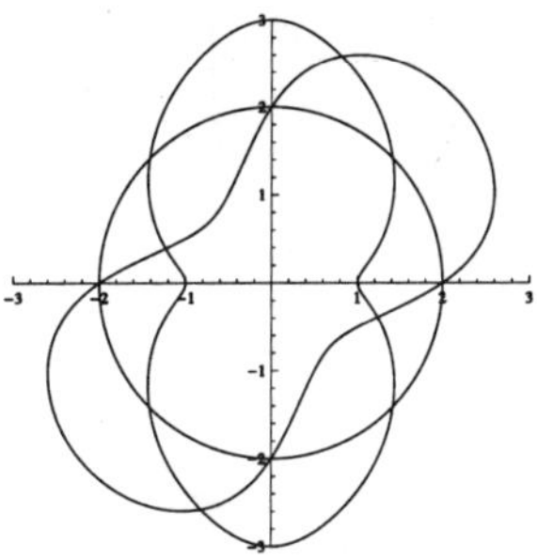

Rhombic media satisfy $\sigma_1 = \sigma_2$. They may have either four or eight hyperbolic directions. Except for the case where $(\lambda+2\mu-\tau_1)(\lambda+2\mu-\tau_2) = 0$ the coupling functions vanish to first order and the dispersive decay rates are $t^{-1/2}$. Only in the latter case, where one coupling function vanishes to third order, we have to check the contact orders for the Fresnel curve in the corresponding direction. We assume $\tau_1 = \lambda + 2\mu$. Following [5] we obtain the contact order 2 unless $\tau_2 = \lambda + \mu$, where the contact order is 4.

Corollary 4.2. *Rhombic media have the dispersive decay rate* $t^{-1/2}$ *except for the case* $\tau_1 = \lambda + 2\mu$, $\tau_2 = \lambda + \mu$ *(or the values interchanged), where the dispersive decay rate is* $t^{-1/4}$.

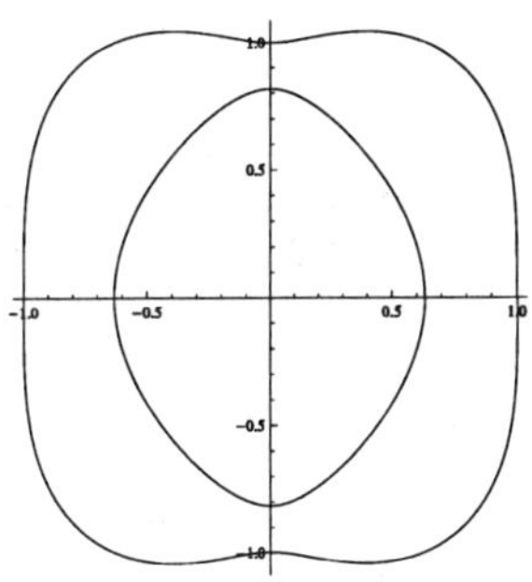

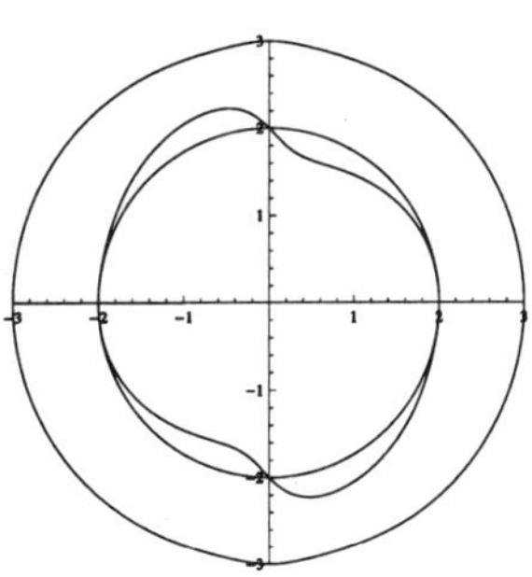

A special anisotropic medium. We chose $\tau_1 = \tau_2 = \lambda + \mu$, $\sigma_1 = 0$ and $\sigma_2 = \mu$. In this case the medium has six hyperbolic directions (forming a regular hexagon), for two of them the coupling function vanishes to second

order. A close examination of the eigenvalues of $A(\xi)$ in these directions shows that the Fresnel curve has an inflection point such that the order of contact is 3. Therefore

Corollary 4.3. *This medium has the dispersive decay rate* $t^{-1/3}$.

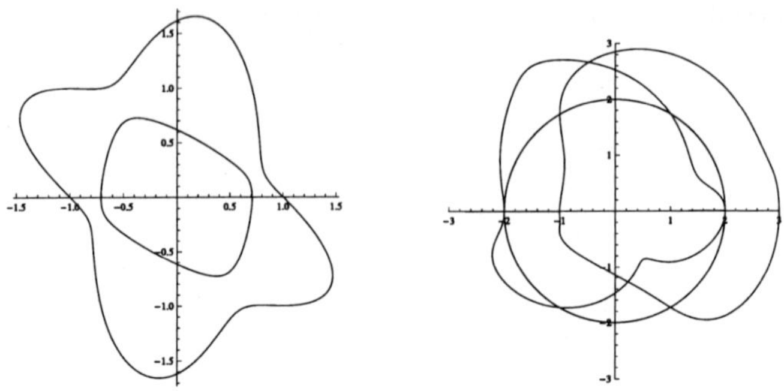

5. Concluding remarks

We will conclude this note with some short remarks on problems occurring for higher dimensions. Unlike for isotropic media, where all dimensions larger than one can be treated similarly, the complexity of the anisotropic problem increases with dimension. While in two dimensions assumption (A3) is not very restrictive and excludes only some particular choices of parameters, starting from three dimensions all interesting cases of anisotropic media have directions where (A3) is violated. Away from these multiplicities our strategy applies and yields asymptotic expansions of eigenvalues (of a similar form) and micro-localised dispersive estimates. The hard problem to tackle is the study of the neighbourhood of degenerate directions.

References

1. A. E. Green and P. M. Naghdi, A re-examination of the basic postulates of thermomechanics, *Proc. R. Soc. London Ser. A* **432**, 171–194 (1991).
2. J. Borkenstein, L^p–L^q-Abschätzungen der linearen Thermoelastizitätsgleichungen für kubische Medien im $\mathbb{R}^2$, Diplomarbeit, Universität Bonn (1993).
3. M. S. Doll, Zur Dynamik (magneto-)thermoelastischer Systeme im $\mathbb{R}^2$, Dissertation, Universität Konstanz (2004).
4. M. Reissig and J. Wirth, Anisotropic thermo-elasiticity in 2D – Part I: A unified treatment, *Asymptotic Anal.* (to appear).
5. J. Wirth, Anisotropic thermo-elasiticity in 2D – Part II: Applications, *Asymptotic Anal.* (to appear).
6. Y.-G. Wang, Microlocal analysis in nonlinear thermoelasticity, *Nonlinear Anal.* **54**, 683–705 (2003).

Session 11

Oscillation of Functional Differential and Difference Equations

SESSION EDITORS

L. Berezansky Ben-Gurion University of the Negev, Be'er-Sheva, Israel
A. Zafer Middle East Technical University, Ankara, Turkey

OSCILLATION CRITERIA FOR A CLASS OF THIRD-ORDER NONLINEAR DELAY DIFFERENTIAL EQUATIONS

M. F. AKTAŞ† and A. TİRYAKİ*

Department of Mathematics, Gazi University,
Faculty of Arts and Sciences, Teknikokullar,
Ankara, 06500, Turkey
† *E-mail: mfahri@gazi.edu.tr*
* *E-mail: tiryaki@gazi.edu.tr*

Using a Riccati-type transformation and the integral averaging technique, some new oscillation criteria for third-order delay differential equations of the form

$$\left(r_2(t)\left(r_1(t)y'\right)'\right)' + p(t)y' + q(t)f(y(g(t))) = 0$$

are established. The results obtained essentially improve earlier results [11].

Keywords: Oscillation; Third order; Delay differential equation; Asymptotic behavior; Riccati type transformation and integral averaging.

1. Introduction

In this paper we consider nonlinear third-order functional differential equations of the form

$$\left(r_2(t)\left(r_1(t)y'\right)'\right)' + p(t)y' + q(t)f(y(g(t))) = 0,\ t \geq T_0, \tag{1}$$

where $T_0 > 0$ is a fixed real number; and r_1, r_2, p, $q \in C[T_0, \infty)$ such that $r_1 > 0$, $r_2 > 0$, $p(t) \geq 0$, $q(t) > 0$, $g \in C^1[T_0, \infty)$ satisfies $g'(t) \geq 0$, $g(t) \leq t$, and $g(t) \to \infty$ as $t \to \infty$ and $f \in C(R, R)$ satisfies $f(u)/u \geq K > 0$ for $u \neq 0$.

We consider only those solutions of Eq. (1) which are defined and nontrivial for all sufficiently large t. Such a solution is called oscillatory if it has arbitrarily large zeros, otherwise it is called nonoscillatory.

Note that if y is a solution of Eq. (1), then $-y$ is a solution of

$$\left(r_2(t)\left(r_1(t)y'\right)'\right)' + p(t)y' + q(t)f^*(y(g(t))) = 0,$$

where $f^*(y) = -f(-y)$ and $yf^*(y) > 0$ for all $y \neq 0$. Since f^* and f are of the same class, we may restrict our attention only to a positive solution of Eq. (1) whenever a nonoscillatory solution of Eq. (1) is concerned.

In the paper, we improve the results obtained in Tiryaki and Aktaş [11]. For example, we see that their results are not applicable to Euler equation

$$y''' + \frac{p_0}{t^2}y' + \frac{q_0}{t^3}y = 0.$$

By using a Riccati type transformation and the integral averaging technique, we establish some new sufficient conditions which insure that every solution of Eq. (1) is oscillatory or converges to zero. For some recent results on these topics we refer the reader to references [1–5,8,9,12,13] and the references cited therein.

The paper is organized as follows. In next section, we shall present some lemmas which are useful in the proof of our main results. In the last section, we establish sufficient conditions and also conditions of Philos-type for oscillation of Eq. (1) and some examples are considered to illustrate the importance of the results.

2. Some Preliminary Lemmas

For the sake of brevity, we define

$$L_0 y(t) = y(t), \quad L_i y(t) = r_i(t)(L_{i-1}y(t))', \quad i = 1, 2,$$
$$L_3 y(t) = (L_2 y(t))' \text{ for } t \in [T_0, \infty).$$

So Eq. (1) can be written as

$$L_3 y(t) + p(t) y' + q(t) f(y(g(t))) = 0.$$

Define the functions

$$R_1(t,s) = \int_s^t \frac{du}{r_1(u)}, \quad R_2(t,s) = \int_s^t \frac{du}{r_2(u)}$$

for $T_0 \le s \le t < \infty$.

Throughout this paper, we assume that

$$R_1(t, T_0) \to \infty \text{ as } t \to \infty \tag{2}$$

and

$$R_2(t, T_0) \to \infty \text{ as } t \to \infty. \tag{3}$$

Lemma 2.1. *Suppose that*

$$(r_2(t)z')' + \frac{p(t)}{r_1(t)}z = 0 \tag{4}$$

is nonoscillatory. If y is a nonoscillatory solution of Eq. (1), then there exists a $T_1 \geq T_0$ such that either $y(t) L_1 y(t) > 0$ or $y(t) L_1 y(t) < 0$ for all $t \geq T_1$.

Lemma 2.2. *Let the assumption (3) holds. If y be a nonoscillatory solution of Eq. (1) which satisfies $y(t) L_1 y(t) \geq 0$ for all large t, then there exists a $T_1 \geq T_0$ such that*

$$L_0 y(t) L_k y(t) > 0, \; k = 0,1,2; \qquad L_0 y(t) L_3 y(t) \leq 0 \tag{5}$$

for all $t \geq T_1$.

A nonoscillatory solution y of Eq. (1) is said to have property V_2 if it satisfies the inequalities (5).

Lemma 2.3. *Assume that there exists a continuous function $g_*(t)$ defined on $[T_0, \infty)$ such that*

$$g_*(t) \leq g(t) \;\text{ for all } t \geq T_0 \text{ and that } \lim_{t\to\infty} g_*(t) = \infty. \tag{6}$$

If y is a solution of Eq. (1) with property V_2, then there exists a $T_1 \geq T_0$ such that $L_1 y(g(t)) \geq R_2(g(t), g_*(t)) L_2 y(t)$ for all $t \geq T_1$.

The proofs of above lemmas proceed along the lines of that of Tiryaki and Aktaş [11, Lemmas 1-3] and hence are omitted.

Lemma 2.4. *Let ρ_2 be a sufficiently smooth positive function defined on $[T_0, \infty)$ and set*

$$\varphi(t) = r_1(t) \left(r_2(t) \rho_2'(t)\right)' + \rho_2(t) p(t).$$

Suppose that there exists a $T_1 \geq T_0$ such that

$$\rho_2'(t) \geq 0, \qquad \varphi(t) \geq 0, \tag{7}$$

$$\int_{T_1}^{\infty} \left(K\rho_2(s) q(s) - \varphi'(s)\right) ds = \infty, \tag{8}$$

where $K\rho_2(t) q(t) - \varphi'(t) \geq 0$ for all $t \in [T_1, \infty)$ and not identically zero in any subinterval of $[T_1, \infty)$. If (2) holds and y is a nonoscillatory solution of Eq. (1) which satisfies $y(t) L_1 y(t) \leq 0$ for all $t \geq T_1$, then $\lim_{t\to\infty} y(t) = 0$.

Proof. Let y be a nonoscillatory solution of Eq. (1). Without loss of generality, we may assume that $y(t) > 0$ and $y(g(t)) > 0$ for $t \geq t_0$ for some t_0 sufficiently large. Then $L_1 y(t)$ becomes negative for all $t \geq t_1$ for some $t_1 \geq t_0$. Let $\lim_{t\to\infty} y(t) = \lambda \geq 0$. Assume that $\lambda \neq 0$.

Multiplying (1) through by $\rho_2(t)$ and integrating from t_1 to t, we obtain

$$\rho_2(t)\,L_2y(t) = C_1 + \rho_2'(t)\,r_2(t)\,L_1y(t) - \varphi(t)\,y(t) - \int_{t_1}^{t} y(s)\left(\frac{f(y(g(s)))}{y(s)}\rho_2(s)\,q(s) - \varphi'(s)\right)ds,$$

where C_1 is a constant. It follows in view of (7) that

$$\rho_2(t)\,L_2y(t) \le C_1 - \lambda\int_{t_1}^{t}(K\rho_2(s)\,q(s) - \varphi'(s))\,ds. \tag{9}$$

Employing (8) we see from (9) that $L_2y(t)$ must take on negative values for t sufficiently large. By using (2) we see that $y(t)$ must be eventually negative, a contradiction. Hence $\lambda = 0$. □

3. Main Results

Theorem 3.1. *Assume that (2)-(3) and (6)-(8) hold, and that Eq. (4) is nonoscillatory. If there exists an eventually positive function $\rho_1(t) \in C^1[T_0,\infty)$ such that*

$$\limsup_{t\to\infty}\int_{T}^{t}\left[K\rho_1(s)\,q(s) - \frac{r_1(g(s))}{r_1^2(s)}\frac{(\rho_1'(s)\,r_1(s) - \rho_1(s)\,p(s)\,R_2(g(s),g_*(s)))^2}{4\rho_1(s)\,R_2(g(s),g_*(s))\,g'(s)}\right]ds = \infty \tag{10}$$

for every $T > T_0$, then every solution $y(t)$ of Eq. (1) is either oscillatory or satisfies $y(t)\to 0$ as $t\to\infty$.

Proof. Let y be a nonoscillatory solution of Eq. (1). Without loss of generality, we may assume that $y(t) > 0$ and $y(g(t)) > 0$ for $t \ge t_0 \ge T_0$. From Lemma 2.1 it follows that $L_1y(t) > 0$ or $L_1y(t) < 0$ for $t \ge t_1 \ge t_0$. If $L_1y(t) > 0$ for $t \ge t_1$, then y has property V_2 for large t from Lemma 2.2. We define

$$\omega(t) = \rho_1(t)\,\frac{L_2y(t)}{y(g(t))} \tag{11}$$

for $t \ge t_1$. In view of Eq. (1) by employing Lemma 2.3 (increasing the size of t_1 if necessary) we have

$$\omega'(t) \le -K\rho_1(t)\,q(t) - \left[A(t)\,\omega^2(t) - \omega(t)\,B(t)\right], \tag{12}$$

where

$$A(t) = \frac{R_2(g(t),g_*(t))\,g'(t)}{r_1(g(t))\,\rho_1(t)}, \qquad B(t) = \frac{\rho_1'(t)}{\rho_1(t)} - p(t)\,\frac{R_2(g(t),g_*(t))}{r_1(t)}.$$

Completing the square in (12) we obtain

$$\omega'(t) \leq -K\rho_1(t)\,q(t) + \frac{B^2(t)}{4A(t)}. \tag{13}$$

Integrating (13) from t_1 to t leads to

$$\int_{t_1}^{t} \left(K\rho_1(s)\,q(s) - \frac{B^2(s)}{4A(s)} \right) ds \leq \omega(t_1)$$

a contradiction with (10).

Let $y(t) > 0$, $L_1 y(t) < 0$, $t \geq t_1$. By Lemma 2.4 we have $\lim_{t\to\infty} y(t) = 0$. The proof is complete.

Example 3.1. Consider the third order differential equation

$$y'''(t) + \frac{p_0}{t^\alpha} y'(t) + \frac{q_0}{t^\beta} y(\lambda t) = 0,\ t \geq 1,\ 0 < \lambda \leq 1, \tag{14}$$

where p_0, q_0, α, and β are some constants. If we choose $\rho_1(t) = \rho_2(t) = t^2$, and $g_*(t) = \frac{\lambda t}{2}$, from Theorem 3.1, every solution $y(t)$ of Eq. (14) is either oscillatory or satisfies $y(t) \to 0$ as $t \to \infty$ for $0 \leq p_0 \leq \frac{1}{4}$, $q_0 > 0$, $\alpha \geq 2$, and $\beta < 3$. On the other hand, if $\alpha = 2$ and $\beta = 3$ we have the same conclusion provided that

$$q_0 > \frac{(4 - \lambda p_0)^2}{8\lambda^2}.$$

In fact, if we take $\lambda = 1$, $p_0 = \frac{1}{4}$ and $q_0 = \frac{25}{4}$, $y_1(t) = \frac{1}{t}$, $y_2(t) = t^2 \cos\left(\frac{3}{2}\ln t\right)$, and $y_3(t) = t^2 \sin\left(\frac{3}{2}\ln t\right)$ are solutions of Euler Eq. (14).

Note that Eq. (14) is a Euler equation, and the results of Tiryaki and Aktaş [11] are not applicable to it.

Example 3.2. Consider the third order delay differential equation

$$y'''(t) + \frac{1}{t^3} y'(t) + \frac{1}{t^3} y(t-1)(4 - \cos y(t-1)) = 0,\ t \geq 2. \tag{15}$$

If we choose $\rho_1(t) = \rho_2(t) = t^2$, and $g_*(t) = \frac{t}{2} - 1$, from Theorem 3.1, every solution y of (15) is either oscillatory or satisfied $y(t) \to 0$ as $t \to \infty$.

Remark 3.1. When $\varphi'(t) \leq 0$, we can take $\int^\infty \rho_2(s)\,q(s)\,ds = \infty$ to replace (8).

We will establish new oscillation criteria for Eq. (1) by using the integral averaging technique due to Philos [7]. We need the class of functions Re. Let

$$D_0 = \{(t,s) : t > s \geq T_0\} \quad \text{and} \quad D = \{(t,s) : t \geq s \geq T_0\}.$$

A function $H \in C(D, \mathbb{R})$ is said to belongs to the class Re if

(*i*) $H(t,t) = 0$ for all $t \geq T_0$; $H(t,s) > 0$ for all $(t,s) \in D_0$;

(*ii*) $H(t,s)$ has a continuous and nonpositive partial derivative on D_0 with respect to the second variable, and for a positive continuous function $h(t,s)$

$$-\frac{\partial H(t,s)}{\partial s} = h(t,s)\sqrt{H(t,s)} \text{ for all } (t,s) \in D_0.$$

For the choice $H(t,s) = (t-s)^n, \ (n \geq 1)$, the Philos-type conditions reduce to the Kamenev-type ones. Other choices of H include

$$\left(\ln\frac{t}{s}\right)^n, \quad \left(\sqrt{t}-\sqrt{s}\right)^n, \quad \left(e^{t-s}-e^{s-t}\right)^n.$$

Theorem 3.2. *Assume that (2)-(3) and (6)-(8) hold, and that Eq. (4) is nonoscillatory. If there exist an eventually positive function $\rho_1(t) \in C^1[T_0,\infty)$, and a function $H \in \mathrm{Re}$ such that*

$$\limsup_{t\to\infty} \frac{1}{H(t,T)} \int_T^t \left[K\rho_1(s) H(t,s) q(s) - \frac{Q^2(t,s)}{4A(s)}\right] ds = \infty \qquad (16)$$

for every $T > T_0$, where

$$A(t) = \frac{R_2(g(t), g_*(t))\, g'(t)}{r_1(g(t))\,\rho_1(t)}, \quad \textit{and}$$

$$Q(t,s) = h(t,s) - \sqrt{H(t,s)}\left(\frac{\rho_1'(s)}{\rho_1(s)} - p(s)\frac{R_2(g(s), g_*(s))}{r_1(s)}\right),$$

then every solution $y(t)$ of Eq. (1) is either oscillatory or satisfies $y(t) \to 0$ as $t \to \infty$.

Corollary 3.1. *Assume that (2)-(3) and (6)-(8) hold, and that Eq. (4) is nonoscillatory. If there exist an eventually positive function $\rho_1(t) \in C^1[T_0,\infty)$, and a function $H \in \mathrm{Re}$ such that for every $T > T_0$,*

$$\limsup_{t\to\infty} \frac{1}{H(t,T)} \int_T^t \rho_1(s) H(t,s) q(s)\, ds = \infty,$$

$$\limsup_{t\to\infty} \frac{1}{H(t,T)} \int_T^t \frac{Q^2(t,s)}{A(s)} ds < \infty,$$

then every solution y of Eq. (1) is oscillatory or satisfies $\lim_{t\to\infty} y(t) = 0$.

Example 3.3. Consider the third-order differential equation

$$\left(\frac{1}{t}y'\right)'' + \frac{1}{4t^3}y' + \frac{1}{t^2}y\left(1+\frac{1}{1+y^2}\right) = 0,\ t \geq 1. \tag{17}$$

If we choose $\rho_1(t) = \rho_2(t) = t$, $H(t,s) = (t-s)^2$, and $g_*(t) = \frac{t}{2}$, from Theorem 3.2, every solution y of (17) is either oscillatory or satisfied $y(t) \to 0$ as $t \to \infty$.

Example 3.4. Consider the third-order delay differential equation

$$\left(\frac{1}{t}\left(\frac{1}{t^4}y'(t)\right)'\right)' + \frac{1}{t^9}y'(t) + \frac{1}{t^2}y\left(\sqrt{t}\right)e^{y(\sqrt{t})} = 0,\ t > 1, \tag{18}$$

where $f(u) = ue^u$ with $K = 1$. Equation $\left(\frac{1}{t}z'\right)' + \frac{1}{t^5}z = 0$ is nonoscillatory (see Parhi [6] and Swanson [10, p. 71]. If we choose $\rho_1(t) = \rho_2(t) = t$, $H(t,s) = (t-s)^2$, and $g_*(t) = \sqrt{\frac{t}{2}}$, from Theorem 3.2, every solution y of (18) is either oscillatory or satisfied $y(t) \to 0$ as $t \to \infty$.

The following two results provide alternative oscillation criteria.

Theorem 3.3. *Let all the assumptions, except (16), of Theorem 3.2 hold. Further, for every $T > T_0$, let*

$$0 < \inf_{s \geq T}\left[\liminf_{t\to\infty}\frac{H(t,s)}{H(t,T)}\right] \leq \infty,$$

and

$$\limsup_{t\to\infty}\frac{1}{H(t,T)}\int_T^t \frac{Q^2(t,s)}{A(s)}ds < \infty\,. \tag{19}$$

Let also $\Psi \in C([T_0,\infty),\mathbb{R})$ be such that

$$\int^\infty \Psi_+^2(s)A(s)\,ds = \infty,$$

where $\Psi_+(s) = \max\{\Psi(t), 0\}$ and

$$\limsup_{t\to\infty}\frac{1}{H(t,T)}\int_T^t\left[K\rho_1(s)H(t,s)q(s) - \frac{Q^2(t,s)}{4A(s)}\right]ds \geq \sup_{t\geq T}\Psi(t)\,.$$

Then every solution $y(t)$ of Eq. (1) is either oscillatory or satisfies $y(t) \to 0$ as $t \to \infty$.

Theorem 3.4. *Let all the assumptions of Theorem 3.3 hold, except condition (19), which is replaced by*

$$\limsup_{t\to\infty} \frac{1}{H(t,T)} \int_T^t H(t,s)\,\rho_1(s)\,q(s)\,ds < \infty$$

for every $T > T_0$. Then, every solution $y(t)$ of Eq. (1) is either oscillatory or satisfies $y(t) \to 0$ as $t \to \infty$.

The proofs of above Theorems proceed along the lines of those of Tiryaki and Aktaş [11] and hence are omitted.

References

1. R. P. Agarwal, S. R. Grace and D. O'Regan, *Oscillation Theory for Difference and Functional Differential Equations*, Kluwer Acad. Publ., Dordrecht, 2000.
2. M. Cecchi and M. Marini, *On the oscillatory behavior of a third order nonlinear differential equation,* Nonlinear Anal. 15 (1990), 141–153.
3. T. Kusano and H. Onose, *Oscillation theorems for delay equations of arbitrary order,* Hiroshima Math. J. 2 (1972), 263–270.
4. N. Parhi and P. Das, *Oscillatory and asymptotic behavior of a class of nonlinear functional differential equations of third order,* Bull. Calcutta Math. Soc. 86 (1994), 253–266.
5. N. Parhi and P. Das, *Oscillatory and asymptotic behaviour of a class of nonlinear differential equations of third order,* Acta Math. Sci. 18 (1998), 95–106.
6. N. Parhi and S. K. Nayak, *Nonoscillation of second-order nonhomogeneous differential equations,* J. Math. Anal. Appl. 102 (1984), 62–74.
7. C. G. Philos, *Oscillation theorems for linear differential equation of second order,* Arch. Math. 53 (1989), 482–492.
8. S. H. Saker, *Oscillation criteria of third-order nonlinear delay differential equations,* Math. Slovaca 56 (2006), 433–450.
9. A. Škerlik, *An integral condition of oscillation for equation $y''' + p(t)y' + q(t)y = 0$ with nonnegative coefficients,* Arch. Math. 31 (1995), 155–161.
10. C. A. Swanson, *Comparison and Oscillation Theory of Linear Differential Equations*, Academic Press, New York, 1968.
11. A. Tiryaki and M. F. Aktaş, *Oscillation criteria of a certain class of third order nonlinear delay differential equations with damping,* J. Math. Anal. Appl. 325 (2007), 54–68.
12. A. Tiryaki and S. Yaman, *Asymptotic behaviour of a class of nonlinear functional differential equations of third order,* Appl. Math. Lett. 14 (2001), 327–332.
13. A. Tiryaki and S. Yaman, *Oscillatory behaviour of a class of nonlinear differential equations of third order,* Acta Math. Sci. B 21 (2001), 182–188.

A CRITERION FOR EXISTENCE OF POSITIVE SOLUTIONS OF THE EQUATION $\dot{y}(t) = -p(t)y^{\alpha}(t-r)$

J. DIBLÍK

Department of Mathematics, Faculty of Electrical Engineering and Communication, Brno University of Technology, Technická 8, Brno, 616 00, Czech Republic
E-mail: diblik@feec.vutbr.cz

M. RŮŽIČKOVÁ

Department of Applied Mathematics, Faculty of Science, Žilina University, Hurbanova 15, Žilina, 010 26, Slovak Republic
E-mail: miroslava.ruzickova@fpv.uniza.sk

Z. SVOBODA

Department of Mathematics, Faculty of Electrical Engineering and Communication, Brno University of Technology, Technická 8, Brno, 616 00, Czech Republic
E-mail: svobodaz@feec.vutbr.cz

We consider the first-order differential equation containing delay

$$\dot{y}(t) = -p(t)y^{\alpha}(t-r)$$

with $p\colon (t_0, \infty) \to (0, \infty)$, $\alpha \in \mathbb{R}$, $\alpha \geq 1$ and $r \in \mathbb{R}$, $r > 0$. A criterion for the existence of positive solutions (for $t \to \infty$) is derived. Connections between the nonlinear case ($\alpha \neq 1$) and the linear case ($\alpha = 1$) are discussed.

Keywords: Positive solution; delay; positivity criterion.

1. Preliminaries

Let $C([a,b], \mathbb{R}^n)$ where $a, b \in \mathbb{R}, a < b, \mathbb{R} = (-\infty, +\infty)$ be the Banach space of continuous functions mapping the interval $[a,b]$ into $\mathbb{R}^n$. In the case $a = -r < 0, b = 0$, we shall denote this space as C, that is, $C = C([-r, 0], \mathbb{R}^n)$. Let

$$\dot{y}(t) = f(t, y_t) \tag{1}$$

be a system of retarded functional differential equations where $f\colon \Omega \mapsto \mathbb{R}^n$ is a quasibounded continuous mapping satisfying a local Lipschitz condition with respect to the second argument and Ω is an open subset in $\mathbb{R} \times C$.

If $\sigma \in \mathbb{R}, A \geq 0$ and $y \in C([\sigma-r, \sigma+A], \mathbb{R}^n)$, then, for each $t \in [\sigma, \sigma+A]$, we define $y_t \in C$ by $y_t(\theta) = y(t+\theta), \theta \in [-r, 0]$. If necessary, we shall assume that the derivatives in the system (1) are right-sided.

In accordance with [14], a function $y(t)$ is said to be a *solution of system* (1) *on* $[\sigma - r, \sigma + A)$ if there are $\sigma \in \mathbb{R}$ and $A > 0$ such that $y \in C([\sigma - r, \sigma+A), \mathbb{R}^n)$, $(t, y_t) \in \Omega$ and $y(t)$ satisfies the system (1) for $t \in [\sigma, \sigma+A)$. For given $\sigma \in \mathbb{R}$, $\varphi \in C$, we say that $y(\sigma, \varphi)$ is a solution of the system (1) through $(\sigma, \varphi) \in \Omega$ if there is an $A > 0$ such that $y(\sigma, \varphi)$ is a solution of the system (1) on $[\sigma - r, \sigma + A)$ and $y_\sigma(\sigma, \varphi) = \varphi$. In view of the above conditions, each element $(\sigma, \varphi) \in \Omega$ determines a unique solution $y(\sigma, \varphi)$ of the system (1) through $(\sigma, \varphi) \in \Omega$ on its maximal interval of existence $I_{\sigma,\varphi} = [\sigma, a), \sigma < a \leq \infty$ which depends continuously on the initial data [14]. A solution $y(\sigma, \varphi)$ of the system (1) is said to be *positive* if

$$y_i(\sigma, \varphi) > 0 \tag{2}$$

on $[\sigma - r, \sigma] \cup I_{\sigma,\varphi}$ for each $i = 1, 2, \ldots, n$. A nontrivial solution $y(\sigma, \varphi)$ of (1) is said to be *oscillatory* on $I_{\sigma,\varphi}$ (subject to $I_{\sigma,\varphi} = [\sigma, \infty)$) if at least one inequality (2) does not hold on any subinterval $[\sigma_1, \infty) \subset [\sigma, \infty)$, $\sigma_1 \geq \sigma$.

2. The Aim of the Paper

Let us consider the following scalar equation with a single delay of the form

$$\dot{y}(t) = -p(t)y^\alpha(t - r) \tag{3}$$

where $p\colon (t_0, \infty) \to (0, \infty)$, $t_0 \in \mathbb{R}$ is a continuous function, $\alpha \in \mathbb{R}$, $\alpha \geq 1$ and $r \in \mathbb{R}$, $r > 0$. Let $t^* \in \mathbb{R}$, $t^* > t_0 + r$. We will derive a new necessary and sufficient criterion for the existence of positive solutions of retarded differential equation (3) for $t \to \infty$. Namely, we prove the following

Theorem 2.1. *For the existence of a positive solution $y = y(t)$ on $[t^* - r, \infty)$ of the equation* (3), *the existence is necessary and sufficient of a positive constant k and a locally integrable function $\lambda\colon [t^* - r, \infty) \to \mathbb{R}$ continuous on $[t^* - r, t^*) \cup [t^*, \infty)$ and satisfying the integral inequality*

$$\lambda(t) \geq k^{\alpha-1} \exp\left[\alpha \int_{t-r}^{t} p(s)\lambda(s)\,\mathrm{d}s + (1 - \alpha)\int_{t^*}^{t} p(s)\lambda(s)\,\mathrm{d}s\right] \tag{4}$$

where $t \in [t^*, \infty)$.

In recent literature, one can find numerous results concerning positive solutions (or their absence) of retarded differential equations. We mention e.g., papers by L. Berezansky and E. Braverman [2], L. Berezansky, Yu. Domshlak and E. Braverman [3], J. Diblík [4,5], J. Diblík and N. Koksch [6], J. Diblík and Z. Svoboda [7,8] Y. Domshlak and I.P. Stavroulakis [9], Á. Elbert and I.P. Stavroulakis [10] and books by R.P. Agarwal, M. Bohner and Wan-Tong Li [1], L.H. Erbe, Q. Kong and B.G. Zhang [11], K. Gopalsamy [12], and I. Györi and G. Ladas [13].

If $\alpha = 1$, then (3) turns into a linear equation. The necessary and sufficient condition for the existence of positive solutions for (3) with $\alpha = 1$ is well known. We recall it here. From Theorem 5 in [8] (compare with [1,4,11] as well) it follows:

Theorem 2.2. *For the existence of a positive solution $y = y(t)$ on $[t^* - r, \infty)$ of the equation* (3) *with $\alpha = 1$ the existence is necessary and sufficient of a locally integrable function $\lambda\colon [t^* - r, \infty) \to \mathbb{R}$ continuous on $[t^* - r, t^*) \cup [t^*, \infty)$ and satisfying the integral inequality*

$$\lambda(t) \geq \exp\left[\int_{t-r}^{t} p(s)\lambda(s)\mathrm{d}s\right] \tag{5}$$

for $t \geq t^*$.

We see that (4) reduces to (5) just for $\alpha = 1$. An interesting fact in this connection is discussed in the paper. It is surprising that the inequality (5) with the function λ having almost the same properties is sufficient for the existence of a positive solution of (3). At the same time, an inequality similar to (5) is necessary for the existence of a positive solution of (3) if $\alpha \in (0, 1)$.

3. Auxiliary Results

In [8], problems are considered on positive solutions of what is called p-type retarded functional differential equations. In the proof of Theorem 2.1, we will employ two results (Lemma 3.1 and Lemma 3.2 below) created by adapting of Lemma 1 and Lemma 2 from [8] for the case of retarded functional differential equations (i.e., in the sense described in the Section 1 (Preliminaries)).

Let us introduce the vectors

$$\rho(t) = (\rho_1(t), \rho_2(t), \ldots, \rho_n(t))^T\colon [t^* - r, \infty) \to \mathbb{R}^n$$
$$\delta(t) = (\delta_1(t), \delta_2(t), \ldots, \delta_n(t))^T\colon [t^* - r, \infty) \to \mathbb{R}^n$$

continuous on $[t^* - r, \infty)$ and continuously differentiable on $[t^*, \infty)$, such that $\rho_i(t) < \delta_i(t)$, $i = 1, 2, \ldots, n$. In the sequel, such type of inequalities between coordinates of vectors will be abbreviated to $\rho(t) \ll \delta(t)$ etc. Let us, moreover, define the set

$$\omega := \{(t, y) : t \geq t^* - r,\ \rho(t) \ll y \ll \delta(t)\}$$

and let $\Omega = (t_0, \infty) \times C$ be put in the next.

Lemma 3.1. *Assume the following:*
(i) For all $j = 1, 2, \ldots, p$ and all $\pi \in C$ for which $(t + \theta, \pi(\theta)) \in \omega$, $\theta \in [-r, 0)$,

$$\delta_j'(t) < f_j(t, \pi) \ \text{ when } \ \pi_j(0) = \delta_j(t) \tag{6}$$

and

$$\rho_j'(t) > f_j(t, \pi) \ \text{ when } \ \pi_j(0) = \rho_j(t). \tag{7}$$

(ii) For all $j = p + 1, p + 2, \ldots, n$ and all $\pi \in C$ for which $(t + \theta, \pi(\theta)) \in \omega$, $\theta \in [-r, 0)$,

$$\rho_j'(t) < f_j(t, \pi) \ \text{ when } \ \pi_j(0) = \rho_j(t)$$

and

$$\delta_j'(t) > f_j(t, \pi) \ \text{ when } \ \pi_j(0) = \delta_j(t).$$

Then there exists an uncountable set $\mathcal{Y}$ of solutions of the system (1) *on $[t^* - r, \infty)$ such that, for each $y(t) \in \mathcal{Y}$,*

$$\rho(t) \ll y(t) \ll \delta(t), \quad t \in [t^* - r, \infty) \tag{8}$$

holds.

Remark 3.1. The number p in Theorem 3.1 and in the sequel can be equal, apart from other values, to *zero* or to n, too. In the first case, the condition (i) is omitted and, in the second one, the condition (ii) is omitted. By analogy, the equality $p = 0$ or $p = n$ is understood in further investigations. Note, moreover, that, in the case $\rho(t) \geq 0$, we deal, as it follows from (8), with positive solutions.

Definition 3.1. We say that a continuous functional $g \colon \Omega \to \mathbb{R}^n$ is *i-strongly decreasing* (or *i-strongly increasing*), $i \in \{1, 2, \ldots, n\}$ if, for each $(t, \varphi) \in \Omega$ and $(t, \psi) \in \Omega$ such that

$$\varphi(\theta) \ll \psi(\theta) \ \text{ where } \theta \in [-r, 0) \ \text{ and } \ \varphi_i(0) = \psi_i(0),$$

the inequality

$$g_i(t,\varphi) > g_i(t,\psi) \quad (\text{or } g_i(t,\varphi) < g_i(t,\psi))$$

holds.

Lemma 3.2. *Let the functional $f\colon \Omega \to \mathbb{R}^n$ be continuous, locally Lipschitzian with respect to the second argument, quasibounded and, moreover,*

- (*i*) *f be i-strongly decreasing if $i = 1,\dots,p$ and i-strongly increasing if $i = p+1,\dots,n$.*
- (*ii*) *$f_i(t,0) \le 0$ for $i = 1,\dots,p$ and $f_i(t,0) \ge 0$ for $i = p+1,\dots,n$ if $(t,0) \in \Omega$.*

If the system (1) *has a positive solution $y = y(t)$ on $[t^* - r,\infty)$, then it also has a positive solution $y = y^*(t)$ on $[t^*,\infty)$ that is* continuously differentiable *on $[t^* - r, t^*)$.*

Proof of Theorem 2.1. NECESSITY. Let y be a positive solution of (3) on $[t^* - r,\infty)$. We set $f(t,\pi) := -p(t)\pi^\alpha(-r)$ where $(t,\pi) \in \Omega$. Then f is a strongly decreasing functional and $f(t,0) = 0$. It is easy to verify that all (remaining) assumptions of Lemma 3.2 (with $i = p = 1$, i.e., we omit the condition (*ii*)) hold. Following this conclusion, there exists a positive solution $y = y^*(t)$ of (3) on $[t^* - r,\infty)$ that is continuously differentiable not only on $[t^*,\infty)$ but on on $[t^* - r, t^*)$, too. We define

$$L(t) := \begin{cases} -\dfrac{y^{*\prime}(t)}{y^*(t)} & \text{if } t \in [t^* - r, t^*) \cup (t^*,\infty), \\ -\dfrac{y^{*\prime}(t^*+0)}{y^*(t^*)} & \text{if } t = t^*. \end{cases}$$

It is easy to verify that, on $[t^* - r,\infty)$,

$$y^*(t) \equiv y^*(t^*) \exp\left[-\int_{t^*}^{t} L(s)\mathrm{d}s\right]. \tag{9}$$

We substitute the representation (9) into (3). On $[t^*,\infty)$, we get

$$-y^*(t^*)L(t)\exp\left[-\int_{t^*}^{t} L(s)\mathrm{d}s\right] \equiv -p(t)(y^*(t^*))^\alpha \exp\left[-\alpha\int_{t^*}^{t-r} L(s)\mathrm{d}s\right]$$

or, equivalently,

$$\lambda(t) \equiv k^{\alpha-1}\exp\left[\alpha\int_{t-r}^{t} p(t)\lambda(s)\mathrm{d}s + (1-\alpha)\int_{t^*}^{t} p(t)\lambda(s)\mathrm{d}s\right] \tag{10}$$

where $k := y^*(t^*) > 0$ and $L(t) := p(t)\lambda(t)$. It means that there are such positive k and a locally integrable function $\lambda\colon [t^* - r, \infty) \to \mathbb{R}$ continuous on $[t^* - r, t^*) \cup [t^*, \infty)$ that (4) holds on $[t^*, \infty)$.

SUFFICIENCY. This part of the proof uses Lemma 3.1. In its formulation we put $j = 1$, $p = n = 1$ (i.e., we omit condition (*ii*) in Lemma 3.1), $\pi \equiv \pi_1(t)$, $f(t,\pi) \equiv f_1(t,\pi) := -p(t)\pi(-r)$ and we define continuous on $[t^* - r, \infty)$ and continuously differentiable on $[t^*, \infty)$ functions $\rho(t) \equiv \rho_1(t) := 0$,

$$\delta(t) \equiv \delta_1(t) := k \exp\left[-\int_{t^*}^{t} p(s)\lambda(s)\,\mathrm{d}s\right]$$

where the constant k and the function λ, mentioned in Theorem 2.1, are used. In this case, the set ω turns into

$$\omega := \left\{(t,y) : t \geq t^* - r,\ \rho(t) = 0 < y < k \exp\left[-\int_{t^*}^{t} p(s)\lambda(s)\,\mathrm{d}s\right] = \delta(t)\right\}.$$

We will verify inequality (6). Assuming $\pi_j(0) = \delta_j(t)$ we get for $t \geq t^*$:

$$\begin{aligned} f(t,\pi) - \delta'(t) &= -\,p(t)\pi^{\alpha}(-r) - \delta'(t) \qquad (11)\\ &= -\,p(t)\pi^{\alpha}(-r) + kp(t)\lambda(t)\exp\left[-\int_{t^*}^{t} p(s)\lambda(s)\,\mathrm{d}s\right]. \end{aligned}$$

Since, because of the assumptions of Lemma 3.1,

$$(t+\theta, \pi(\theta)) \in \omega,\ \theta \in [-r, 0),$$

we have

$$-\pi(-r) > -\delta(t-r) = -k\exp\left[-\int_{t^*}^{t-r} p(s)\lambda(s)\,\mathrm{d}s\right].$$

In this case we can continue to estimate the difference (11):

$$\begin{aligned} f(t,\pi) - \delta'(t) &> -\,p(t)\delta^{\alpha}(t-r) + kp(t)\lambda(t)\exp\left[-\int_{t^*}^{t} p(s)\lambda(s)\,\mathrm{d}s\right]\\ &= -\,k^{\alpha}p(t)\exp\left[-\alpha\int_{t^*}^{t-r} p(s)\lambda(s))\,\mathrm{d}s\right]\\ &\quad + kp(t)\lambda(t)\exp\left[-\int_{t^*}^{t} p(s)\lambda(s)\,\mathrm{d}s\right] \end{aligned}$$

and

$$\begin{aligned} f(t,\pi) - \delta'(t) &> kp(t)\exp\left[-\int_{t^*}^{t} p(s)\lambda(s)\,\mathrm{d}s\right]\\ &\times\left[\lambda(t) - k^{\alpha-1}\exp\left[\alpha\int_{t-r}^{t} p(s)\lambda(s))\,\mathrm{d}s + (1-\alpha)\int_{t^*}^{t} p(s)\lambda(s)\,\mathrm{d}s\right]\right]. \end{aligned}$$

Taking into account assumption (4), we finalize: $f(t,\pi) - \delta'(t) > 0$ for $t \in [t^*, \infty)$ and inequality (6) holds. Finally, we verify inequality (7). Due to the assumptions of Lemma 3.1, inclusion

$$(t+\theta, \pi(\theta)) \in \omega,\ \theta \in [-r, 0)$$

holds and, consequently, we have $\pi(-r) > \rho(t-r) = 0$. Therefore

$$\rho_i'(t) - f(t,\pi) = p(t)\pi(-r) > 0$$

for all $t \geq t^*$. All conditions of Lemma 3.1 are satisfied. This conclusion ends the proof of this part and Theorem 2.1 is proved. □

4. Concluding Remarks

As mentioned before, Theorem 2.1 generalizes Theorem 2.2 into the nonlinear equation (3). Now we show that the inequality for the linear case (5) with the function λ positive not only on interval $[t^*, \infty)$ (as it follows from (5)), but on interval $[t^* - r, t^*)$, too, gives a sufficient condition for the existence of a positive solution for nonlinear equation (3). This statement is formulated as:

Theorem 4.1. *For the existence of a positive solution $y = y(t)$ on $[t^* - r, \infty)$ of the equation* (3) *the existence is sufficient of a locally integrable function λ: $[t^* - r, \infty) \to (0, \infty)$ continuous on $[t^* - r, t^*) \cup [t^*, \infty)$ and satisfying the integral inequality*

$$\lambda(t) \geq \exp\left[\int_{t-r}^{t} p(s)\lambda(s)\mathrm{d}s\right] \tag{12}$$

for $t \geq t^*$.

Proof. We will trace the "sufficient" part of the proof of Theorem 2.1 omitting some technical details. We define continuous on $[t^* - r, \infty)$ and continuously differentiable on $[t^*, \infty)$ functions $\rho(t) := 0$ and

$$\delta(t) := \exp\left[-\int_{t^*-r}^{t} p(s)\lambda(s)\,\mathrm{d}s\right]$$

where the function λ, mentioned in Theorem 4.1 is used. In this case,

$$\omega := \left\{(t,y) : t \geq t^* - r,\ \rho(t) = 0 < y < \exp\left[-\int_{t^*-r}^{t} p(s)\lambda(s)\,\mathrm{d}s\right] = \delta(t)\right\}.$$

We will verify inequality (6). Assuming $\pi_j(0) = \delta_j(t)$ we get for $t \geq t^*$:

$$f(t,\pi) - \delta'(t) = -\,p(t)\pi^{\alpha}(-r) - \delta'(t) \tag{13}$$

$$= - p(t)\pi^{\alpha}(-r) + p(t)\lambda(t)\exp\left[-\int_{t^*-r}^{t} p(s)\lambda(s)\,\mathrm{d}s\right].$$

Since

$$-\pi(-r) > -\delta(t-r) = -\exp\left[-\int_{t^*-r}^{t-r} p(s)\lambda(s)\,\mathrm{d}s\right],$$

we can continue to estimate the difference (13):

$$\begin{aligned} f(t,\pi) - \delta'(t) > & - p(t)\delta^{\alpha}(t-r) + p(t)\lambda(t)\exp\left[-\int_{t^*-r}^{t} p(s)\lambda(s)\,\mathrm{d}s\right] \\ = & - p(t)\exp\left[-\alpha\int_{t^*-r}^{t-r} p(s)\lambda(s))\,\mathrm{d}s\right] \\ & + p(t)\lambda(t)\exp\left[-\int_{t^*-r}^{t} p(s)\lambda(s)\,\mathrm{d}s\right] \\ = & p(t)\exp\left[-\int_{t^*-r}^{t} p(s)\lambda(s)\,\mathrm{d}s\right] \\ & \times\left[\lambda(t) - \exp\left[\int_{t-r}^{t} p(s)\lambda(s)\,\mathrm{d}s\right.\right. \\ & \left.\left. +(1-\alpha)\int_{t^*-r}^{t-r} p(s)\lambda(s)\,\mathrm{d}s\right]\right]. \end{aligned}$$

Using (12), we get

$$\begin{aligned} f(t,\pi) - \delta'(t) > & p(t)\exp\left[-\int_{t^*-r}^{t-r} p(s)\lambda(s)\,\mathrm{d}s\right] \\ & \times\left[1 - \exp\left[(1-\alpha)\int_{t^*-r}^{t-r} p(s)\lambda(s)\,\mathrm{d}s\right]\right]. \end{aligned}$$

Since

$$1-\alpha \le 0 \quad \text{and due to the positivity} \quad \int_{t^*-r}^{t-r} p(s)\lambda(s)\ \mathrm{d}s \ge 0$$

of λ, for $t \ge t^*$, we have

$$1 - \exp\left[(1-\alpha)\int_{t^*-r}^{t-r} p(s)\lambda(s)\,\mathrm{d}s\right] \ge 0.$$

Consequently, $f(t,\pi) - \delta'(t) > 0$ for $t \in [t^*, \infty)$ and (6) holds. Inequality (7) can be verified in a way similar to the proof of Theorem 2.1. □

Now we show that an inequality, similar to inequality (5) is necessary for the existence of a positive solution of (3) with $\alpha \in (0,1)$.

Theorem 4.2. *Let $y = y(t)$ be a positive solution of (3) with $\alpha \in (0,1)$ on $[t^* - r, \infty)$. Then there exists a positive constant k and a locally integrable function $\lambda\colon [t^* - r, \infty) \to \mathbb{R}$ continuous on $[t^* - r, t^*) \cup [t^*, \infty)$ satisfying the integral inequality*

$$\lambda(t) \ge k^{\alpha-1} \exp\left[\alpha \int_{t-r}^{t} p(t)\lambda(s)\mathrm{d}s\right] \tag{14}$$

for $t \ge t^*$.

Proof. We will trace the "necessary" part of the proof of Theorem 2.1 omitting some technical details again. Proceeding as indicated in this part of the proof, we get the identity (10). Since

$$(1-\alpha)\int_{t^*}^{t} p(t)\lambda(s)\mathrm{d}s \ge 0, \quad t \ge t^*$$

we get

$$\begin{aligned}\lambda(t) &\ge k^{\alpha-1} \exp\left[\alpha \int_{t-r}^{t} p(t)\lambda(s)\mathrm{d}s + (1-\alpha)\int_{t^*}^{t} p(t)\lambda(s)\mathrm{d}s\right]\\ &\ge k^{\alpha-1} \exp\left[\alpha \int_{t-r}^{t} p(t)\lambda(s)\mathrm{d}s\right]\end{aligned}$$

and (14) holds. □

In connection with the above investigation, an interesting open problem arises. We formulate it as a conjecture.

Conjecture 4.1. *For the existence of a positive solution $y = y(t)$ on $[t^* - r, \infty)$ of the equation (3) the existence is necessary and sufficient of a locally integrable function $\lambda\colon [t^* - r, \infty) \to \mathbb{R}$ continuous on $[t^* - r, t^*) \cup [t^*, \infty)$ and satisfying the integral inequality*

$$\lambda(t) \ge \exp\left[\int_{t-r}^{t} p(s)\lambda(s)\mathrm{d}s\right]$$

for $t \ge t^*$.

Acknowledgment

This research was supported by the Grant 1/3238/06 of the Grant Agency of Slovak Republic (VEGA), by the Grant 201/08/0469 of Czech Grant Agency (Prague), and by the Council of Czech Government MSM 0021630503 and MSM 0021630529.

References

1. R. P. Agarwal, M. Bohner, W.-T. Li, *Nonoscillation and Oscillation: Theory for Functional Differential Equations,* Marcel Dekker, Inc., 2004.
2. L. Berezansky, E. Braverman, Positive solutions for a scalar differential equation with several delays, *Appl. Math. Lett.,* to appear.
3. L. Berezansky, Y. Domshlak, E. Braverman, On oscillation properties of delay differential equations with positive and negative coefficients, *J. Math. Anal. Appl.* **274** (2002), 81–101.
4. J. Diblík, A criterion for existence of positive solutions of systems of retarded functional differential equations, *Nonlinear Anal.* **38** (1999), 327–339.
5. J. Diblík, Positive and oscillating solutions of differential equations with delay in critical case, *J. Comput. Appl. Math.* **88** (1998), 185–202.
6. J. Diblík, N. Koksch, Positive solutions of the equation $\dot{x}(t) = -c(t)x(t-\tau)$ in the critical case, *J. Math. Anal. Appl.* **250** (2000), 635-659.
7. J. Diblík, Z. Svoboda, Positive solutions of retarded functional differential equations, *Nonlinear Anal.* **63** (2005), e813–e821.
8. J. Diblík, Z. Svoboda, Positive solutions of p-type retarded functional differential equations, *Nonlinear Anal.* **64** (2006), 1831–1848.
9. Y. Domshlak, I. P. Stavroulakis, Oscillation of first-order delay differential equations in a critical case, *Appl. Anal.* **61** (1996), 359–371.
10. Á. Elbert, I. P. Stavroulakis, Oscillation and non-oscillation criteria for delay differential equations, *Proc. Amer. Math. Soc.* **123** (1995), 1503–1510.
11. L. H. Erbe, Q. Kong, B. G. Zhang, *Oscillation Theory for Functional Differential Equations,* Marcel Dekker, Inc., 1995.
12. K. Gopalsamy, *Stability and Oscillations in Delay Differential Equations of Population Dynamics,* Kluwer Academic Publishers, 1992.
13. I. Györi, G. Ladas, *Oscillation Theory of Delay Differential Equations,* Clarendon Press, 1991.
14. J. K. Hale, S. M. V. Lunel, *Introduction to Functional Differential Equations,* Springer-Verlag, 1993.

DISCRETENESS OF THE SPECTRUM OF A NONSELFADJOINT SECOND-ORDER DIFFERENCE OPERATOR *

EBRU ERGÜN

Department of Physics, Ankara University, 06100 Tandoğan, Ankara, Turkey
E-mail: eergun@science.ankara.edu.tr

GUSEIN S. GUSEINOV

Department of Mathematics, Atılım University, 06836 İncek, Ankara, Turkey
E-mail: guseinov@atilim.edu.tr

We investigate a non-Hermitian second-order linear difference operator and establish simple conditions for the discreteness of its spectrum.

1. Introduction

In [1] the following problem was introduced and partly investigated in the Hilbert space $L^2(-\infty,\infty)$:

$$-y''(x)+q(x)y(x)=\lambda\rho(x)y(x),\quad x\in(-\infty,0)\cup(0,\infty), \tag{1}$$

$$y(0^-)=y(0^+),\quad y'(0^-)=e^{2i\delta}y'(0^+), \tag{2}$$

where $y(x)$ is a desired solution, $q(x)$ is a real-valued function, λ is a spectral parameter, $\delta\in[0,\frac{\pi}{2})$, i is the imaginary unit $i=\sqrt{-1}$, and

$$\rho(x)=\begin{cases} e^{2i\delta} & \text{if} \quad x<0,\\ e^{-2i\delta} & \text{if} \quad x>0.\end{cases} \tag{3}$$

The main distinguishing features of problem (1), (2) are that it involves a complex-valued coefficient function $\rho(x)$ of the form (3) and that transition conditions of the form (2) are presented which also involve a complex coefficient. Such a problem is non-Hermitian with respect to the usual inner product of space $L^2(-\infty,\infty)$. Note that non-Hermitian Hamiltonians and

*This work is supported by Grant 106 T549 from the Scientific and Technological Research Council of Turkey (TÜBİTAK).

complex extension of quantum mechanics have recently received a lot of attention [2].

Let $\mathbb{Z}$ denote the set of all integers. For any $l, m \in \mathbb{Z}$ with $l \leq m$, $[l, m]$ will denote the *discrete interval* being the set $\{l, l+1, \ldots, m\}$. Semi-infinite intervals of the form $(-\infty, l]$ and $[l, \infty)$ will denote the discrete sets $\{\ldots, l-2, l-1, l\}$ and $\{l, l+1, l+2, \ldots\}$, respectively. Throughout the paper all intervals will be discrete intervals. Let us set

$$\mathbb{Z}_0 = \mathbb{Z} \setminus \{0, 1\} = \{\ldots, -2, -1\} \cup \{2, 3, \ldots\} = (-\infty, -1] \cup [2, \infty). \tag{4}$$

In this paper we offer a discrete version of problem (1), (2) to be

$$-\Delta^2 y_{n-1} + q_n y_n = \lambda \rho_n y_n, \quad n \in \mathbb{Z}_0, \tag{5}$$

$$y_{-1} = y_1, \quad \Delta y_{-1} = e^{2i\delta} \Delta y_1, \tag{6}$$

where $y = (y_n)_{n \in \mathbb{Z}}$ is a desired solution, Δ is the forward difference operator defined by $\Delta y_n = y_{n+1} - y_n$ so that $\Delta^2 y_{n-1} = y_{n+1} - 2y_n + y_{n-1}$, the coefficients q_n are real numbers given for $n \in \mathbb{Z}_0$, $\delta \in [0, \frac{\pi}{2})$, and ρ_n are given for $n \in \mathbb{Z}_0$ by

$$\rho_n = \begin{cases} e^{2i\delta} & \text{if} \quad n \leq -1, \\ e^{-2i\delta} & \text{if} \quad n \geq 2. \end{cases} \tag{7}$$

For a long time past, mathematicians and physicians are interested in discrete equations (difference equations) [3,4] . Discrete equations arise when differential equations are solved approximately by discretization. On the other hand they often arise independently as mathematical models of many practical events. Discrete equations can easily been algorithmized to solve them on computers. There are only a few works concerning discrete non-Hermitian quantum systems [5–7].

The main result of the present paper is that if

$$q_n \geq c > 0 \quad \text{for} \quad n \in \mathbb{Z}_0 \quad \text{and} \quad \lim_{|n| \to \infty} q_n = \infty,$$

then the spectrum of problem (5), (6) is discrete.

In order to introduce the concept of spectrum for problem (5), (6), define the Hilbert space l_0^2 of complex sequences $y = (y_n)_{n \in \mathbb{Z}_0}$ such that $\sum_{n \in \mathbb{Z}_0} |y_n|^2 < \infty$ with the inner product $\langle y, z \rangle = \sum_{n \in \mathbb{Z}_0} y_n \overline{z}_n$ and norm $\|y\| = \sqrt{\langle y, y \rangle}$, where the bar over a complex number indicates the complex conjugate.

Now we try to rewrite problem (5), (6) in the form of an equivalent operator equation in l_0^2. Introduce the linear set

$$\mathcal{D} = \left\{ y = (y_n)_{n \in \mathbb{Z}_0} \in l_0^2 : (q_n y_n)_{n \in \mathbb{Z}_0} \in l_0^2 \right\}.$$

Taking this set as the domain of L, $L : \mathcal{D} \subset l_0^2 \to l_0^2$ is defined by

$$(Ly)_n = -\Delta^2 y_{n-1} + q_n y_n \quad \text{for} \quad n \in \mathbb{Z}_0,$$

where y_0 and y_1 are defined from the equations

$$y_{-1} = y_1, \quad \Delta y_{-1} = e^{2i\delta} \Delta y_1.$$

Note that y_0 and y_1 are needed when we evaluate $(Ly)_n$ for $n = -1$ and $n = 2$, respectively.

Next, define the operator $M : l_0^2 \to l_0^2$ by $(My)_n = \rho_n y_n$ for $n \in \mathbb{Z}_0$, where ρ_n is given by (7). Since $|\rho_n| = 1$, M is a unitary operator.

Therefore problem (5), (6) can be written as $Ly = \lambda My$, $y \in \mathcal{D}$, or $M^{-1}Ly = \lambda y$, $y \in \mathcal{D}$. This motivates to introduce the following definition.

Definition 1.1. By the spectrum of problem (5), (6) is meant the spectrum of the operator $A = M^{-1}L$ with the domain $\mathcal{D}$ in the space l_0^2.

Remember that (see [8]) if A is a linear operator with a domain dense in a Banach space, then a complex number λ is called a *regular point* of the operator A if the inverse $(A - \lambda I)^{-1}$ exists and represents a bounded operator defined on the whole space. The set of all nonregular points λ is called the *spectrum* of the operator A. Obviously the eigenvalues λ of the operator A belong to its spectrum, since the operator $(A - \lambda I)^{-1}$ does not exist for such points (the operator $A - \lambda I$ is not one-to-one). The set of all eigenvalues is called the *point spectrum* of the operator. The spectrum of the operator A is said to be *discrete* if it consists of a denumerable (i.e., at most countable) set of eigenvalues with no finite point of accumulation.

A linear operator acting in a Banach space and defined on the whole space is called *completely continuous* if it maps bounded sets into relatively compact sets. Any completely continuous operator is bounded and hence its spectrum is a compact subset of the complex plane. As is well known, every nonzero point of the spectrum of a completely continuous operator is an eigenvalue of finite multiplicity (that is, to each eigenvalue there correspond only a finite number of linearly independent eigenvectors); the set of eigenvalues is at most countable and can have only one accumulation point $\lambda = 0$. It follows that if a linear operator A with a domain dense in a Banach space is invertible and its inverse A^{-1} is completely continuous, then the spectrum of A is discrete.

In this paper we show that the operator L is invertible and its inverse L^{-1} is a completely continuous operator. This implies that the operator $A = M^{-1}L$ is invertible and $A^{-1} = L^{-1}M$ is a completely continuous operator. Hence the spectrum of the operator A is discrete.

The paper is organized as follows. In Section 2 we present general properties of solutions of equations of type (5), (6). In Section 3 we show that the operator L is invertible and we describe the structure of the inverse operator L^{-1}. Finally, in Section 4 we show that the operator L^{-1} is completely continuous.

2. Second-Order Linear Difference Equations with Impulse

Consider the second order linear homogeneous difference equation with impulse

$$-\Delta^2 y_{n-1} + p_n y_n = 0, \quad n \in \mathbb{Z}_0 = \mathbb{Z} \setminus \{0, 1\} = (-\infty, -1] \cup [2, \infty), \tag{8}$$

$$y_{-1} = d_1 y_1, \quad \Delta y_{-1} = d_2 \Delta y_1, \tag{9}$$

where $y = (y_n)$ with $n \in \mathbb{Z}$ is a desired solution, the coefficients p_n are complex numbers given for $n \in \mathbb{Z}_0$; d_1, d_2 presented in the "impulse conditions" in (9) are nonzero complex numbers.

Using the definition of Δ-derivative we can rewrite problem (8), (9) in the form

$$-y_{n-1} + \widetilde{p}_n y_n - y_{n+1} = 0, \quad n \in (-\infty, -1] \cup [2, \infty), \tag{10}$$

$$y_{-1} = d_1 y_1, \quad y_0 - y_{-1} = d_2(y_2 - y_1), \tag{11}$$

where

$$\widetilde{p}_n = p_n + 2, \quad n \in (-\infty, -1] \cup [2, \infty).$$

Theorem 2.1. *Let n_0 be a fixed point in $\mathbb{Z}$ and c_0, c_1 be given complex numbers. Then problem (8), (9) has a unique solution (y_n), $n \in \mathbb{Z}$, such that*

$$y_{n_0} = c_0, \quad \Delta y_{n_0} = c_1, \quad \textit{that is,} \quad y_{n_0} = c_0, \quad y_{n_0+1} = c_0 + c_1 = c_1'. \tag{12}$$

Proof. First assume that $n_0 \in (-\infty, -1]$. Solving equation (10) on $(-\infty, -1]$ under the initial conditions (12), we find y_n uniquely for $n \in (-\infty, 0]$. Then we find y_1 and y_2 uniquely from (11) and then we solve equation (10) uniquely on $[2, \infty)$.

Let now $n_0 \in [1, \infty)$. Solving equation (10) on $[2, \infty)$ subject to the initial conditions (12), we find y_n uniquely for $n \in [1, \infty)$. Then we find y_{-1} and y_0 uniquely from (11) and then we solve equation (10) uniquely on $(-\infty, -1]$.

Finally, if $n_0 = 0$, then we find y_0 and y_1 uniquely from the initial condiations (12) with $n_0 = 0$. Then we find y_{-1} and y_2 from the impulse conditions (11). Next, solving equation (10) at first on $(-\infty,-1]$ we find y_n uniquely for $n \in (-\infty,-2]$ and then solving (10) on $[2,\infty)$ we find y_n uniquely for $n \in [3,\infty)$. □

Definition 2.1. For two sequences $y=(y_n)$ and $z=(z_n)$ with $n \in \mathbb{Z}$, we define their Wronskian (or Casoratian) by

$$\begin{aligned} W_n(y,z) &= y_n \Delta z_n - (\Delta y_n) z_n \\ &= y_n z_{n+1} - y_{n+1} z_n, \quad n \in \mathbb{Z}. \end{aligned}$$

Theorem 2.2. *The Wronskian of any two solutions y and z of problem (8), (9) is constant on each of the intervals* $(-\infty,-1]$ *and* $[1,\infty)$:

$$W_n(y,z) = \begin{cases} \omega^-, & n \in (-\infty,-1], \\ \omega^+, & n \in [1,\infty). \end{cases} \tag{13}$$

In addition,

$$\omega^- = d_1 d_2 \omega^+ \tag{14}$$

and

$$W_0(y,z) = -d_2 \omega^+. \tag{15}$$

Proof. Suppose that $y=(y_n)$ and $z=(z_n)$, where $n \in \mathbb{Z}$, are solutions of (8), (9). Let us compute the Δ -derivative of $W_n(y,z)$. Using the product rule for Δ-derivative

$$\begin{aligned} \Delta(f_n y_n) &= (\Delta f_n) y_n + f_{n+1} \Delta g_n \\ &= f_n \Delta g_n + (\Delta f_n) g_{n+1}, \end{aligned}$$

we find that

$$\begin{aligned} \Delta W_n(y,z) &= y_{n+1} \Delta^2 z_n - (\Delta^2 y_n) z_{n+1} \\ &= y_{n+1} p_{n+1} z_{n+1} - p_{n+1} y_{n+1} z_{n+1} = 0 \end{aligned}$$

for $n \in (-\infty,-2] \cup [1,\infty)$. The latter implies that $W_n(y,z)$ is constant on $(-\infty,-1]$ and on $[1,\infty)$. Thus we have (13), where ω^- and ω^+ are some constants (depending on the solutions y and z).

Next using (13) and the impulse conditions in (9) for y_n and z_n, we get

$$\begin{aligned} \omega^- &= W_{-1}(y,z) = y_{-1} \Delta z_{-1} - (\Delta y_{-1}) z_{-1} \\ &= d_1 d_2 [y_1 \Delta z_1 - (\Delta y_1) z_1 = d_1 d_2 W_1(y,z) \\ &= d_1 d_2 \omega^+ \end{aligned}$$

so that (14) is established.

Finally, from the impulse conditions in (11) we find that $y_0 = (d_1 - d_2)y_1 + d_2 y_2$. Substituting this expression for y_0 and z_0 into

$$W_0(y,z) = y_0 z_1 - y_1 z_0,$$

we get

$$W_0(y,z) = -d_2 W_1(y,z) = -d_2 \omega^+.$$

Therefore (15) is also proved. □

Corollary 2.1. *If y and z are two solutions of (8), (9), then either $W_n(y,z) = 0$ for all $n \in \mathbb{Z}$ or $W_n(y,z) \neq 0$ for all $n \in \mathbb{Z}$.*

By using Theorem 2.1, the following two theorems can be proved in exactly the same way when equation (8) does not include any impulse conditions [3].

Theorem 2.3. *Any two solutions of (8), (9) are linearly independent if and only if their Wronskian is not zero.*

Theorem 2.4. *Problem (8), (9) has two linearly independent solutions and every solution of (8), (9) is a linear combination of these solutions.*

We say that $y = (y_n)$ and $z = (z_n)$, where $n \in \mathbb{Z}$, form a *fundamental set* of solutions for (8), (9) provided that they are solutions of (8), (9) and their Wronskian is not zero.

Let us consider the nonhomogeneous equation

$$-\Delta^2 y_{n-1} + p_n y_n = h_n, \quad n \in (-\infty, -1] \cup [2, \infty), \tag{16}$$

with the impulse conditions

$$y_{-1} = d_1 y_1, \quad \Delta y_{-1} = d_2 \Delta y_1, \tag{17}$$

where h_n is a complex sequence defined for $n \in (-\infty, -1] \cup [2, \infty)$. We will extend h_n to the values $n = 0$ and $n = 1$ by setting

$$h_0 = h_1 = 0. \tag{18}$$

Theorem 2.5. *Suppose that $u = (u_n)$ and $v = (v_n)$ form a fundamental set of solutions for the homogeneous problem (8), (9). Then a general solution of the corresponding nonhomogeneous problem (16), (17) is given by*

$$y_n = c_1 u_n + c_2 v_n + x_n, \quad n \in \mathbb{Z},$$

where c_1, c_2 *are arbitrary constants and*

$$x_n = \begin{cases} -\sum_{s=n}^{0} \frac{u_n v_s - u_s v_n}{W_s(u,v)} h_s, & n \le 0, \\ \sum_{s=1}^{n} \frac{u_n v_s - u_s v_n}{W_s(u,v)} h_s, & n \ge 1. \end{cases} \tag{19}$$

Proof. Taking into account (18) it is not difficult to verify that the sequence x_n defined by (19) is a particular solution of (16), (17), namely, x_n satisfies equation (16) and the conditions

$$x_{-1} = \Delta x_{-1} = 0, \quad x_1 = \Delta x_1 = 0.$$

This implies that the statement of the theorem is true. □

3. The Inverse Operator L^{-1}

Let L be the operator defined above in Introduction. In this section we describe the structure of the inverse operator L^{-1}. The following theorem is crucial for doing this.

Theorem 3.1. *Let*

$$q_n \ge c > 0 \quad for \quad n \in \mathbb{Z}_0 = \mathbf{Z} \setminus \{0, 1\}. \tag{20}$$

Then the equation

$$-\Delta^2 y_{n-1} + q_n y_n = 0, \quad n \in \mathbb{Z}_0, \tag{21}$$

subject to the conditions

$$y_{-1} = y_1, \quad \Delta y_{-1} = e^{2i\delta} \Delta y_1, \tag{22}$$

has two linearly independent solutions $\chi = (\chi_n)_{n \in \mathbb{Z}}$ *and* $\psi = (\psi_n)_{n \in \mathbb{Z}}$ *such that*

$$\sum_{-\infty}^{n=0} |\chi_n|^2 < \infty \quad and \quad \sum_{n=0}^{\infty} |\psi_n|^2 < \infty. \tag{23}$$

Note that $\ker L = \{y \in \mathcal{D} : Ly = 0\}$ consists only of the zero element. Indeed, if $y \in \mathcal{D}$ and $Ly = 0$, then y satisfies (21), (22). Since χ and ψ form a fundamental set of solutions of (21), (22), we can write

$$y_n = C_1 \chi_n + C_2 \psi_n, \quad n \in \mathbb{Z},$$

with some constants C_1 and C_2. Hence

$$W_n(y, \psi) = C_1 W_n(\chi, \psi) + C_2 W_n(\psi, \psi), \quad n \in \mathbb{Z}. \tag{24}$$

Next, by $y \in l_0^2$ and (23) we have $y_n \to 0$ as $|n| \to \infty$ and $\psi_n \to 0$ as $n \to \infty$. Besides $W_n(\psi, \psi) = 0$ for all n and $W_n(\chi, \psi)$ is equal to a nonzero constant for $n \geq 1$. Therefore taking the limit in (24) as $n \to \infty$ we get that $C_1 = 0$. It can similarly be shown that $C_2 = 0$. Thus $y = 0$.

It follows that the inverse operator L^{-1} exists. To present its explicit form we introduce the discrete Green function

$$G_{nk} = \frac{1}{W_k(\psi, \chi)} \begin{cases} \chi_k \psi_n & \text{if} \quad k \leq n, \\ \chi_n \psi_k & \text{if} \quad k \geq n, \end{cases}$$

where k and n are in $\mathbb{Z}$.

Theorem 3.2. *The inverse operator L^{-1} is defined on the whole space l_0^2 and for every $f \in l_0^2$*

$$(L^{-1}f)_n = \sum_{k \in \mathbb{Z}_0} G_{nk} f_k, \quad n \in \mathbb{Z}_0. \tag{25}$$

Besides,

$$\|L^{-1}f\| \leq \frac{1}{c \cos \delta} \|f\| \quad \text{for all} \quad f \in l_0^2, \tag{26}$$

where c is a constant from condition (20) and δ is from (22), $\|\cdot\|$ denotes the norm of space l_0^2.

4. Complete Continuity of the Operator L^{-1}

In this section we will show that the operator L^{-1} is completely continuous, that is, it is continuous and maps bounded sets into relatively compact sets.

Theorem 4.1. *Let*

$$q_n \geq c > 0 \quad \text{for} \quad n \in \mathbb{Z}_0 \tag{27}$$

and

$$\lim_{|n| \to \infty} q_n = \infty. \tag{28}$$

Then the operator L^{-1} is completely continuous.

Proof. The operator L^{-1} is continuous by virtue of (26) that holds under the condition (27). In order to show that L^{-1} maps bounded sets into relatively compact sets consider any bounded set S in l_0^2, $S = \{f \in l_0^2 : \|f\| \leq d\}$, and prove that $L^{-1}(S) = Y$ is relatively compact in l_0^2. To this end, we use the following known criterion for relative compactness in l_0^2: *A set $Y \subset l_0^2$ is relatively compact if and only if Y is bounded* *and for every* $\varepsilon > 0$ *there exists a positive integer* n_0 *(depending only on* ε*) such that*

$$\sum_{|n|>n_0} |y_n|^2 \le \varepsilon \quad \textit{for all} \quad y \in Y.$$

Take an arbitrary $f \in S$ and set $L^{-1}f = y$. Then $Ly = f$ or explicitly

$$-\Delta^2 y_{n-1} + q_n y_n = f_n, \quad n \in \mathbb{Z}_0, \tag{29}$$

where y_0 and y_1 are defined from the equations

$$y_{-1} = y_1, \quad \Delta y_{-1} = e^{2i\delta} \Delta y_1. \tag{30}$$

Note that y_0 and y_1 are needed when we write out equation (29) for $n = -1$ and $n = 2$, respectively.

It follows from equations (29), (30) that for any integers $\alpha < 0$ and $\beta > 1$,

$$\operatorname{Re}\left(-\sigma_n \overline{y}_{n+1} \Delta y_n\right)\Big|_{\alpha-1}^{\beta} + (\cos\delta)\left(\sum_{n=\alpha}^{-1} + \sum_{n=2}^{\beta}\right)\left(|\Delta y_n|^2 + q_n |y_n|^2\right)$$
$$= \operatorname{Re}\left(\sum_{n=\alpha}^{-1} + \sum_{n=2}^{\beta}\right)\left(\sigma_n f_n \overline{y}_n\right), \tag{31}$$

where $\sigma_n = \exp(-i\delta)$ if $n \le -1$ and $\sigma_n = \exp(i\delta)$ if $n \ge 0$.

Since $f, y \in l_0^2$ and $|\sigma_n| = 1$, the sum on the right-hand side of (31) converges as $\alpha \to -\infty$, $\beta \to \infty$. Also from $y \in l_0^2$ it follows that $y_n \to 0$ as $|n| \to \infty$. Therefore

$$\left(-\sigma_n \overline{y}_{n+1} \Delta y_n\right)\Big|_{\alpha-1}^{\beta} \to 0 \quad \text{as} \quad \alpha \to -\infty, \ \beta \to \infty.$$

Consequently, we arrive at the equality

$$(\cos\delta) \sum_{n\in\mathbb{Z}_0} \left(|\Delta y_n|^2 + q_n |y_n|^2\right) = \operatorname{Re} \sum_{n\in\mathbb{Z}_0} \sigma_n f_n \overline{y}_n.$$

Hence

$$(\cos\delta) \sum_{n\in\mathbb{Z}_0} q_n |y_n|^2 \le \operatorname{Re} \sum_{n\in\mathbb{Z}_0} \sigma_n f_n \overline{y}_n \tag{32}$$

and therefore using (27) and $\|f\| \le d$ we get

$$\|y\| \le \frac{d}{c\cos\delta} \quad \text{for all} \quad y \in Y. \tag{33}$$

This means that the set $Y = L^{-1}(S)$ is bounded.

From (32) we also have, using (33),

$$\sum_{n\in\mathbb{Z}_0} q_n |y_n|^2 \le \frac{d^2}{c\cos^2\delta} \quad \text{for all} \quad y \in Y. \tag{34}$$

Take now an arbitrary $\varepsilon > 0$. By condition (28) we can choose a positive integer n_0 such that

$$q_n \ge \frac{d^2}{\varepsilon c\cos^2\delta} \quad \text{for} \quad |n| > n_0.$$

Then we get from (34) that

$$\sum_{|n|>n_0} |y_n|^2 \le \varepsilon \quad \text{for all} \quad y \in Y.$$

Thus the completely continuity of the operator L^{-1} is proved. □

Corollary 4.1. *The operator $A = M^{-1}L$ is invertible and its inverse $A^{-1} = L^{-1}M$ is a completely continuous operator. Therefore the spectrum of the operator A is discrete.*

References

1. A. Mostafazadeh, Pseudo-Hermitian description of PT-symmetric systems defined on a complex contour, *J. Phys. A* **38** (2005), 3213–3234.
2. C. M. Bender, Making sense of non-Hermitian Hamiltonians, *Rep. Progr. Phys.* **70** (2007), 947–1018.
3. W. Kelley and A. C. Peterson, *Difference Equations: An Introduction with Applications*, Academic Press, New York, 1991.
4. G. Teshl, *Jacobi Operators and Completely Integrable Nonlinear Lattices*, Math. Surveys Monogr., Vol. 72, Amer. Math. Soc., Providence, R.I., 2000.
5. C. M. Bender, P. N. Meisinger and N. Wang, Finite dimensional PT-symmetric Hamiltonians, *J. Phys. A* **36** (2003), 6791–6797.
6. S. Weigret, How to test for digonalizability: The discretized PT-invariant square-well potential, *Czech. J. Phys.* **55** (2005), 1183–1186.
7. M. Znojil, Matching method and exact solvability of discrete PT-symmetric square wells, *J. Phys. A* **39** (2006), 10247–10261.
8. M. A. Naimark, *Linear Differential Operators, Part II*, Ungar, London, 1968.

EVENTUALLY POSITIVE SOLUTIONS OF SECOND-ORDER SUPERLINEAR DYNAMIC EQUATIONS

R. MERT† and A. ZAFER*

Department of Mathematics, Middle East Technical University,
Ankara, 06531, Turkey
† *E-mail: raziye@metu.edu.tr*
* *E-mail: zafer@metu.edu.tr*

A time scale $\mathbb{T}$ is an arbitrary nonempty closed subset of the real numbers $\mathbb{R}$. We study the existence and nonexistence of positive solutions of a class of second-order superlinear equations defined on $\mathbb{T}$.

Keywords: Time Scale; Delay Dynamic Equation; Oscillation; Eventually Positive/Negative Solution.

1. Introduction

Consider the forced superlinear second order delay dynamic equation

$$x^{\Delta\Delta}(t) + q(t)\,|x(\tau(t))|^{\alpha-1}\,x(\tau(t)) = g(t), \quad t \in [t_0, \infty)_{\mathbb{T}}, \tag{1}$$

where $\mathbb{T}$ is a time scale, $t_0 \in \mathbb{T}$ and $\alpha > 1$ are fixed, and $q, g, \tau \in \mathbb{C}_{rd}([t_0, \infty)_{\mathbb{T}})$. It is assumed that τ is increasing, $\tau(t) \leq t$, $\tau(t) \in \mathbb{T}$ for all $t \in \mathbb{T}$, and $\lim_{t\to\infty} \tau(t) = \infty$. By $[t_0, \infty)_{\mathbb{T}}$ we mean a time scale interval; see the next section for some basics about a time scale.

A. H. Nasr in [7] has studied (1) extensively when $\mathbb{T} = \mathbb{R}$ generalizing the well-known theorems of Atkinson ($\alpha > 1$) and Belohorec ($0 < \alpha < 1$) from unforced equations to forced ones.

If $\mathbb{T} = \mathbb{Z}$, then (1) becomes the superlinear forced delay difference equation

$$\Delta^2 x(n) + q(n)\,|x(\tau(n))|^{\alpha-1}\,x(\tau(n)) = g(n), \quad n \in \{n_0, n_0+1, \ldots\}, \tag{2}$$

which may be considered as a discrete analogue of (1). We should note that (2) has not been sufficiently investigated when compared (1). Moreover, other special cases such as $\mathbb{T} = \overline{q^{\mathbb{Z}}}$, $\mathbb{T} = h\mathbb{Z}$, $\mathbb{T} = \mathbb{P}_{a,b}$ (see [1] for definitions and other special types of time scales) lead to different equations with almost no work so far.

Definition 1.1. A nontrivial solution $x(t)$ of (1) is said to be oscillatory if and only if $x(t)$ is neither eventually positive nor eventually negative; i.e., if and only if given any $t_1 \in [t_0, \infty)_{\mathbb{T}}$ there exists $t_2 \in [t_1, \infty)_{\mathbb{T}}$ such that $x(t_2)x(\sigma(t_2)) \leq 0$.

2. Preliminaries

A *time scale* $\mathbb{T}$ is an arbitrary nonempty closed subset of the real numbers $\mathbb{R}$. The most well-known examples are $\mathbb{T} = \mathbb{R}$ and $\mathbb{T} = \mathbb{Z}$. For details, see the monographs [1,2]. On a time scale $\mathbb{T}$, the forward and backward jump operators are defined by

$$\sigma(t) = \inf\{s \in \mathbb{T} : s > t\} \quad \text{and} \quad \rho(t) = \sup\{s \in \mathbb{T} : s < t\},$$

where $\inf \emptyset := \sup \mathbb{T}$ and $\sup \emptyset := \inf \mathbb{T}$. A point $t \in \mathbb{T}$, $t > \inf \mathbb{T}$, is said to be left-dense if $\rho(t) = t$, right-dense if $t < \sup \mathbb{T}$ and $\sigma(t) = t$, left-scattered if $\rho(t) < t$ and right-scattered if $\sigma(t) > t$. The graininess function μ for a time scale $\mathbb{T}$ is defined by $\mu(t) := \sigma(t) - t$ and for any function $f : \mathbb{T} \to \mathbb{R}$ the notation $f^\sigma(t)$ denotes $f(\sigma(t))$. For $a, b \in \mathbb{T}$ with $a \leq b$ the closed interval $[a, b]_{\mathbb{T}}$ is defined to be the set $\{t \in \mathbb{T} : a \leq t \leq b\}$. Other types of intervals are defined similarly. The set $\mathbb{T}^\kappa := \mathbb{T} \backslash \{m\}$ if $\mathbb{T}$ has a left-scattered maximum m, otherwise $\mathbb{T}^\kappa := \mathbb{T}$.

Definition 2.1. A function $f : \mathbb{T} \to \mathbb{R}$ is called Δ-*differentiable* at a point $t \in \mathbb{T}^\kappa$ if there exists a real number $f^\Delta(t)$ with the property that for any given $\varepsilon > 0$ there is a neighborhood U of t such that

$$|f(\sigma(t)) - f(s) - f^\Delta(t)\,(\sigma(t) - s)| \leq \varepsilon|\sigma(t) - s| \quad \text{for all} \quad s \in U.$$

The number $f^\Delta(t)$ is called the delta (or Hilger) derivative of f at t.

It follows that $f^\Delta(t) = f'(t)$ when $\mathbb{T} = \mathbb{R}$, while $f^\Delta(t) = \Delta f(t) := f(t+1) - f(t)$ if $\mathbb{T} = \mathbb{Z}$. In fact,

$$f^\Delta(t) = \lim_{s \to t} \frac{f(s) - f(t)}{s - t} \quad \text{if } \mu(t) = 0$$

and

$$f^\Delta(t) = \frac{f(\sigma(t)) - f(t)}{\sigma(t) - t} \quad \text{if } \mu(t) > 0.$$

It is easy to see that if f and g are differentiable then

$$(fg)^{\Delta} = f^{\Delta}g + f^{\sigma}g^{\Delta} = fg^{\Delta} + f^{\Delta}g^{\sigma}$$

and

$$\left(\frac{1}{g}\right)^{\Delta} = \frac{-g^{\Delta}}{gg^{\sigma}} \quad (\text{where } gg^{\sigma} \neq 0).$$

A simple useful formula is

$$f^{\sigma} = f + \mu f^{\Delta}. \tag{3}$$

Definition 2.2. A function $f : \mathbb{T} \to \mathbb{R}$ is called rd-continuous, if it is continuous at every right-dense point and if the left-sided limit exists (finite) at every left-dense point.

The set of all such rd-continuous functions defined on a set $\mathbb{D} \subset \mathbb{T}$ is denoted by $\mathbb{C}_{rd}(\mathbb{D})$. As usual the space of functions that are differentiable with rd-continuous derivatives is denoted by $\mathbb{C}^1_{rd}(\mathbb{D})$. It is noteworthy to mention that every rd-continuous function f has an antiderivative F. As in the continuous case, a function F is called an antiderivative of f on $\mathbb{T}$ if $F^{\Delta}(t) = f(t)$ holds for all $t \in \mathbb{T}^{\kappa}$.

The usual chain rule from calculus is no longer valid on an arbitrary time scale (see [1, p. 31]). One form of the following extended chain rule, due to S. Keller [5] and generalized to measure chains by C. Pötzche [6], is as follows.

Theorem 2.1 (1, Theorem 1.90). *Let $f : \mathbb{R} \to \mathbb{R}$ be continuously differentiable and suppose $g : \mathbb{T} \to \mathbb{R}$ is delta differentiable. Then $f \circ g : \mathbb{T} \to \mathbb{R}$ is delta differentiable and the formula*

$$(f \circ g)^{\Delta}(t) = \left\{\int_0^1 f'(g(t) + h\mu(t)g^{\Delta}(t))\, dh\right\} g^{\Delta}(t) \tag{4}$$

holds.

The substitution rule and the Mean Value Theorem on an arbitrary time scale are given as follows.

Theorem 2.2 (1, Theorem 1.98). *Assume $\nu : \mathbb{T} \to \mathbb{R}$ is strictly increasing and $\widetilde{\mathbb{T}} := \nu(\mathbb{T})$ is a time scale. If $f : \mathbb{T} \to \mathbb{R}$ is an rd-continuous function and ν is delta differentiable with rd-continuous derivative , then for $a, b \in \mathbb{T}$,*

$$\int_a^b f(t)\nu^{\Delta}(t)\, \Delta t = \int_{\nu(a)}^{\nu(b)} (f \circ \nu^{-1})(s)\, \widetilde{\Delta} s. \tag{5}$$

Theorem 2.3 (4, Theorem A.2). *Let f be a continuous function on $[a,b]_{\mathbb{T}}$ which is delta differentiable on $[a,b)_{\mathbb{T}}$. Then there exit $\xi, \tau \in [a,b)_{\mathbb{T}}$ such that*

$$f^{\Delta}(\xi) \leq \frac{f(b)-f(a)}{b-a} \leq f^{\Delta}(\tau). \tag{6}$$

3. Positive solutions

Our first theorem is concerned with the existence of an eventually positive solution of (1). Next we give conditions under which there is neither an eventually positive solution nor an eventually negative solution. The combination of the two theorems gives a necessary and sufficient condition for oscillation of solutions of (1).

Theorem 3.1. *Assume that*

(i) $q(t)$ is nonnegative;
(ii) there exists a bounded function $h \in \mathbb{C}^2_{rd}([t_0,\infty)_{\mathbb{T}})$ such that $h^{\Delta\Delta}(t) = g(t)$.

If

$$\int^{\infty} \sigma(s)q(s)\,\Delta s < \infty, \tag{7}$$

then (1) *has an eventually positive solution.*

Proof. In view of (3.1), let $T \in \mathbb{T}$ be sufficiently large so that

$$\int_T^{\infty} \sigma(s)q(s)\,\Delta s \leq \frac{1}{2}\min\left\{\frac{1}{(2M+1)^{\alpha}}, \frac{1/\alpha}{(2M+1)^{\alpha-1}}\right\}, \tag{8}$$

where $M = \sup\{|h(t)| : t \in [t_0,\infty)_{\mathbb{T}}\}$.

Consider the Banach space B_T of all bounded rd-continuous functions $x(t)$, $t \in [T,\infty)_{\mathbb{T}}$ endowed with the norm

$$\|x\| = \sup_{t\in[T,\infty)_{\mathbb{T}}} |x(t)|\,.$$

Let

$$X := \left\{x \in B_T : \frac{1}{2} \leq x(t) \leq 2M+1\right\}.$$

Clearly X is a closed and bounded subset of B_T. Pick a point $T^* \in \mathbb{T}$ in such a way that $T^* > T$ and $\tau^{-1}(T^*) \in \mathbb{T}$. Define a map $A : X \to B_T$ by

$$(Ax)(t) = \begin{cases} (M+1) + h(t) - \int_t^\infty (\sigma(s) - t)q(s)x^\alpha(\tau(s))\,\Delta s, \\ \hfill t \in [\tau^{-1}(T^*), \infty)_\mathbb{T} \\ (M+1) + h(t) - \int_{\tau^{-1}(T^*)}^\infty (\sigma(s) - t)q(s)x^\alpha(\tau(s))\,\Delta s, \\ \hfill t \in [T, \tau^{-1}(T^*))_\mathbb{T}. \end{cases}$$

A maps X into itself. Let $x \in X$. It is clear that $(Ax)(t)$ is rd-continuous (in fact continuous) and $(Ax)(t) \le 2M + 1$. Moreover, in view of (8) we get

$$\begin{aligned} (Ax)(t) &\ge 1 - \int_{\tau^{-1}(T^*)}^\infty \sigma(s)q(s)x^\alpha(\tau(s))\,\Delta s \\ &\ge 1 - (2M+1)^\alpha \int_T^\infty \sigma(s)q(s)\,\Delta s \ge \frac{1}{2}. \end{aligned}$$

A is a contraction mapping. For this, let $x, y \in X$. Then

$$\begin{aligned} |(Ax)(t) - (Ay)(t)| &\le \int_{\tau^{-1}(T^*)}^\infty (\sigma(s) - t)q(s)\,|x^\alpha(\tau(s)) - y^\alpha(\tau(s))|\,\Delta s \\ &\le \int_{\tau^{-1}(T^*)}^\infty \sigma(s)q(s)\,|x^\alpha(\tau(s)) - y^\alpha(\tau(s))|\,\Delta s. \end{aligned}$$

Applying the mean value theorem (continuous version), we see that

$$\|Ax - Ay\| \le \alpha\,\|x - y\| \int_{\tau^{-1}(T^*)}^\infty \sigma(s)q(s)z^{\alpha-1}(\tau(s))\,\Delta s,$$

where $z(\tau(s))$ lies between $x(\tau(s))$ and $y(\tau(s))$. It follows that

$$\|Ax - Ay\| \le \alpha\,\|x - y\|\,(2M+1)^{\alpha-1} \int_T^\infty \sigma(s)q(s)\,\Delta s,$$

and hence by (8),

$$\|Ax - Ay\| \le \frac{1}{2}\,\|x - y\|.$$

Employing the Contraction Mapping Theorem, we conclude that A has a unique fixed point $x \in X$. It is easy to verify that this fixed point is a solution of (1). □

Remark 3.1. If $q(t)$ is nonpositive and (3.1) holds with q replaced by $-q$, then (1) has an eventually negative solution.

Theorem 3.2. *Let $g(t) \equiv 0$, $q(t) \geq 0$, $q(t) \not\equiv 0$ eventually, and $\sigma(t) = O(\tau(t))$ as $t \to \infty$. If*

$$\int^{\infty} \sigma(s)q(s)\,\Delta s = \infty, \tag{9}$$

then (1) has no eventually positive/negative solution.

Proof. Suppose to the contrary that there is an eventually positive solution $x(t)$ of (1). If $x(t)$ is eventually negative, we write $y = -x$ and get back to the previous case.

Choose t_1 sufficiently large so that $x(t) > 0$ and $x(\tau(t)) > 0$ for $t \geq t_1 > t_0$. From (1) we see that $x^{\Delta\Delta}(t) \leq 0$ for $t \geq t_1$. Consequently for some $t_2 \geq t_1$ we have

$$x(t) > 0, \quad x^{\Delta}(t) > 0, \quad x^{\Delta\Delta}(t) \leq 0, \quad t \geq t_2.$$

Define the function $r(t)$ by the Riccati substitution

$$r(t) := -t\frac{x^{\Delta}(t)}{x^{\alpha}(t)}, \quad t \geq t_2.$$

Then $r(t)$ satisfies

$$r^{\Delta}(t) = -\frac{x^{\Delta}(t)}{x^{\alpha}(t)} - \sigma(t)\left[\frac{x^{\Delta}(t)}{x^{\alpha}(t)}\right]^{\Delta}, \quad t \geq t_2.$$

In view of (1), we have

$$r^{\Delta}(t) = -\frac{x^{\Delta}(t)}{x^{\alpha}(t)} + \sigma(t)q(t)\frac{x^{\alpha}(\tau(t))}{x^{\alpha}(\sigma(t))} + \sigma(t)\frac{x^{\Delta}(t)(x^{\alpha})^{\Delta}(t)}{x^{\alpha}(t)x^{\alpha}(\sigma(t))}, \quad t \geq t_2. \tag{10}$$

On the other hand, by the chain rule (4) (with $f(t) = t^{\alpha}, g(t) = x(t)$) we have

$$\begin{aligned}(x^{\alpha})^{\Delta}(t) &= \left\{\int_0^1 \alpha(x(t) + h\mu(t)x^{\Delta}(t))^{\alpha-1}\,dh\right\}x^{\Delta}(t)\\ &\geq \alpha x^{\alpha-1}(t)x^{\Delta}(t), \quad t \geq t_2,\end{aligned} \tag{11}$$

where we have used $x(t) > 0$ and $x^{\Delta}(t) > 0$.

Using (11) in (10) we obtain

$$\begin{aligned}r^{\Delta}(t) &\geq -\frac{x^{\Delta}(t)}{x^{\alpha}(t)} + \sigma(t)q(t)\frac{x^{\alpha}(\tau(t))}{x^{\alpha}(\sigma(t))} + \frac{\alpha\sigma(t)}{t^2}\frac{x^{2\alpha-1}(t)}{x^{\alpha}(\sigma(t))}r^2(t)\\ &\geq -\frac{x^{\Delta}(t)}{x^{\alpha}(t)} + \sigma(t)q(t)\frac{x^{\alpha}(\tau(t))}{x^{\alpha}(\sigma(t))} + \frac{c}{t}\frac{x^{\alpha}(t)}{x^{\alpha}(\sigma(t))}r^2(t), \quad t \geq t_2,\end{aligned} \tag{12}$$

where $c = \alpha x^{\alpha-1}(t_2)$.

Let $s \in \mathbb{T}, s \geq t_2$ be a fixed point. By the mean value theorem (6) applied to $x(t)$ on $[s,t]_{\mathbb{T}}, t > s$, we have

$$\frac{x^{\Delta}(t)}{x(t)} < \frac{1}{t-s}, \tag{13}$$

since $x(t) > 0, x^{\Delta\Delta} \leq 0$. From (13) it follows that

$$x(t) - (t-s)x^{\Delta}(t) > 0, \quad t \geq s. \tag{14}$$

Dividing both sides of (14) by $x(t)x(\sigma(t))$, we see that

$$\left(\frac{t-s}{x(t)}\right)^{\Delta} > 0, \quad t \geq s. \tag{15}$$

Integrating (15) from $\tau(t)$ to $\sigma(t)$, for sufficiently large t we get

$$\frac{x(\tau(t))}{x(\sigma(t))} \geq \frac{\tau(t)-s}{\sigma(t)-s} \tag{16}$$

and hence

$$\frac{x(\tau(t))}{x(\sigma(t))} \geq K, \quad t \geq t_3 \tag{17}$$

for sufficiently large $t_3 \geq t_2$ and some constant $K > 0$ since $\sigma(t) = O(\tau(t))$ as $t \to \infty$.

Similar computations yield that for sufficiently large t_3

$$\frac{x(t))}{\tau(\sigma(t))} \geq L, \quad t \geq t_3 \tag{18}$$

for some constant $L > 0$.

Using (17) and (18) in (12) we see that

$$r^{\Delta}(t) \geq -\frac{x^{\Delta}(t)}{x^{\alpha}(t)} + \beta\sigma(t)q(t) + \gamma\frac{r^2(t)}{t}, \quad t \geq t_3$$

for some constants $\beta, \gamma > 0$. Integrating from t_3 to t we obtain

$$r(t) \geq r(t_3) - \int_{t_3}^{t} \frac{x^{\Delta}(s)}{x^{\alpha}(s)}\,\Delta s + \beta\int_{t_3}^{t} \sigma(s)q(s)\,\Delta s + \gamma\int_{t_3}^{t} \frac{r^2(s)}{s}\,\Delta s. \tag{19}$$

To prove the convergence of the integral

$$\int^{\infty} \frac{x^{\Delta}(s)}{x^{\alpha}(s)}\,\Delta s,$$

we consider the following two cases seperately

Case 1: $\lim_{t\to\infty} x(t) < \infty$.
Case 2: $\lim_{t\to\infty} x(t) = \infty$.

First consider Case 1. In this case, it follows from

$$0 < \frac{x^{\Delta}(t)}{x^{\alpha}(t)} \le \frac{x^{\Delta}(t)}{x^{\alpha}(t_3)}, \quad t \ge t_3$$

and the comparison test.

Next consider Case 2. Clearly, $x([t_3, \infty)_{\mathbb{T}})$ is also a time scale unbounded from above. Let $\widetilde{\mathbb{T}} := x([t_3, \infty)_{\mathbb{T}})$ and $\widetilde{\sigma}$ denote the forward jump operator on $\widetilde{\mathbb{T}}$. Then, by the substitution (5) we have

$$\int_{t_3}^{\infty} \frac{x^{\Delta}(s)}{x^{\alpha}(s)} \Delta s = \int_{x(t_3)}^{\infty} \frac{\widetilde{\Delta} u}{u^{\alpha}},$$

where the integral on the right hand side is convergent. To prove this we will show that $\widetilde{\sigma}(x(t)) = O(x(t))$ as $t \to \infty$ (for details see [3, Corollary 5.14]). We have two possibilities to consider: (i) $x(t)$ is right dense and (ii) $x(t)$ is right scattered. In case (i), $\widetilde{\sigma}(x(t)) = x(t)$ by definition. Let (ii) hold. We see that t is also right scattered and hence $\widetilde{\sigma}(x(t)) = x(\sigma(t))$. Then using (3), (13) and the fact that $\sigma(t) = O(t)$ as $t \to \infty$, it is easy to see that $x(\sigma(t)) = O(x(t))$ as $t \to \infty$. Thus we get the desired result.

From (19) , we see that there exists $t_4 > t_3$ such that

$$r(t) \ge \gamma \int_{t_3}^{t} \frac{r^2(s)}{s} \Delta s, \quad t \ge t_4. \tag{20}$$

Setting

$$z(t) := \gamma \int_{t_3}^{t} \frac{r^2(s)}{s} \Delta s$$

we get

$$\frac{z^{\Delta}(t)}{z^2(t)} \ge \frac{\gamma}{t}, \quad \text{where} \quad z^{\Delta}(t) = \gamma \frac{r^2(t)}{t}.$$

Integrating from t_4 to t, we obtain

$$\int_{t_4}^{t} \frac{z^{\Delta}(s)}{z^2(s)} \Delta s \ge \gamma \int_{t_4}^{t} \frac{\Delta s}{s},$$

which leads to a contradiction as $t \to \infty$. Since we already know that

$$\int^{\infty} \frac{\Delta s}{s} = \infty$$

(see [3, Theorem 5.11]), to get a contradiction we will show that

$$\int^{\infty} \frac{z^{\Delta}(s)}{z^2(s)} \Delta s < \infty.$$

For this, we repeat the previous steps for $z(t)$ instead of $x(t)$. As in the previous case, it can easily be shown that the integral is convergent if $\lim_{t\to\infty} z(t) < \infty$. And when $\lim_{t\to\infty} z(t) = \infty$, we have $z([t_4,\infty)_{\mathbb{T}})$ is a time scale unbounded from above. Let $\widehat{\mathbb{T}} := z([t_4,\infty)_{\mathbb{T}})$ and $\widehat{\sigma}$ denotes the forward jump operator on $\widehat{\mathbb{T}}$. To prove the convergence, we will show that $\widehat{\sigma}(z(t)) = O(z^{\nu}(t))$ as $t \to \infty$ is satisfied for any $\nu \in [1,2)$ (see [3, Corollary 5.14]). If $z(t)$ is right-dense, then $\widehat{\sigma}(z(t)) = z(t)$. Also $\widehat{\sigma}(z(t)) = z(\sigma(t))$ when $z(t)$ is right-scattered. In this case, from (13) we have

$$\begin{aligned}\widehat{\sigma}(z(t)) = z(\sigma(t)) &= z(t) + \gamma\mu(t)\frac{r^2(t)}{t} \\ &= z(t) + \gamma t\mu(t)\left(\frac{x^{\Delta}(t)}{x^{\alpha}(t)}\right)^2 \\ &\le z(t) + \gamma t\mu(t)\frac{1}{(t-s)^2 x^{2\alpha-2}(t)},\end{aligned}$$

where the last expression is $O(z^{\nu}(t))$ as $t \to \infty$. □

The following theorem is a direct consequence of the above results.

Theorem 3.3. *Let $g(t) \equiv 0$, $q(t) \ge 0$, $q(t) \not\equiv 0$ eventually, and $\sigma(t) = O(\tau(t))$ as $t \to \infty$. Then every solution of* (1) *is oscillatory if and only if*

$$\int^{\infty} \sigma(s)q(s)\,\Delta s = \infty.$$

By considering special time scales $\mathbb{T} = \mathbb{R}$, $\mathbb{T} = \mathbb{N}$, and $\mathbb{T} = q^{\mathbb{N}}$ we easily obtain the following oscillation criteria.

Corollary 3.1. *Let $q(t) \ge 0$, $q(t) \not\equiv 0$ eventually. Then every solution of the delay equation*

$$x''(t) + q(t)\,|x(t-r)|^{\alpha-1}\,x(t-r) = 0, \quad (r > 0), \quad t \in [t_0,\infty)_{\mathbb{R}},$$

is oscillatory if and only if

$$\int^{\infty} \sigma(s)q(s)\,ds = \infty.$$

Corollary 3.2. *Let $q(n) \ge 0$, $q(n) \not\equiv 0$ eventually. Then every solution of the delay difference equation*

$$\Delta^2 x(n) + q(n)\,|x(n-k)|^{\alpha-1}\,x(n-k) = 0, \quad (k > 0), \quad n \in \{n_0, n_0+1, \ldots\},$$

is oscillatory if and only if

$$\sum_{n=n_0}^{\infty} (n+1)q(n) = \infty.$$

Corollary 3.3. *Let $Q(t) \geq 0$, $Q(t) \not\equiv 0$ eventually. Then every solution of the delay q-differnce equation*

$$\Delta_q x(t) + Q(t)\,|x(t/q)|^{\alpha-1}\,x(t/q) = 0, \quad (q > 1), \quad t \in [t_0, \infty)_{q^{\mathbb{N}}},$$

where

$$\Delta_q x(t) = \frac{x(qt) - x(t)}{(q-1)t},$$

is oscillatory if and only if

$$\sum^{\infty} q^n Q(q^n) = \infty.$$

References

1. M. Bohner and A. Peterson, *Dynamic Equations on Time Scales*, Boston, Birkhauser, 2001.
2. B. Kaymakçalan, V. Lakshmikantham, and S. Sivasundaram, *Dynamic Systems on Measure Chains*, Dordrecht, Kluwer, 1996.
3. M. Bohner and G. Guseinov, *Dynamic Systems and Applications* **12**, 45 (2003).
4. G. Guseinov, *J. Math. Anal. Appl.* **285**, 107 (2003).
5. S. Keller, Asymptotisches Verhalten Invarianter Faserbündel bei Diskretisierung und Mittelwertbildung im Rahmen der Analysis auf Zeitskalen, Ph.D. thesis, Universität Augsburg, Augsburg, Germany, 1999.
6. C. Pötzche, *J. Comput. Appl. Math.* **141**, 249 (2002).
7. A. H. Nasr, *J. Math. Anal. App.* **212**, 51 (1997).

OSCILLATION CRITERIA FOR SECOND-ORDER NONLINEAR IMPULSIVE DIFFERENTIAL EQUATIONS

A. ÖZBEKLER

Department of Mathematics, Atılım University
İncek, Ankara, 06836, Turkey
E-mail: abdullah@atilim.edu.tr

We establish sufficient conditions for oscillation of second-order nonlinear impulsive differential equations by using the integral averaging technique. In particular, we extend Philos-type oscillation criteria to impulsive differential equations.

Keywords: Oscillation, Impulse, Nonlinear, Second-order, Integral averaging.

1. Introduction

In this study, we are concerned with the problem of oscillation of second-order nonlinear impulsive differential equation of the form

$$\begin{array}{ll} (r(t)\psi(x)\varphi_\alpha(x'))' + p(t)\psi(x)\varphi_\alpha(x') + q(t)f(x) = 0, & t \neq \theta_i; \\ \Delta(r\psi(x)\varphi_\alpha(x')) + q_i f(x) = 0 & t = \theta_i, \end{array} \tag{1}$$

where $\varphi_\alpha(s) := |s|^{\alpha-1}s$ and $\Delta z(t) := z(t^+) - z(t^-)$, $z(t^\pm) = \lim_{\tau\to t^\pm} z(\tau)$. We assume without further mention that

(i) $\alpha > 0$ is a constant;
(ii) $\{\theta_i\}$ is a strictly increasing unbounded sequence of real numbers; $\{q_i\}$ is a real sequence;
(iii) $r, p, q \in \mathrm{PLC}\,[t_0, \infty) := \{u : [t_0, \infty) \to \mathbb{R}$ is continuous on each interval (θ_i, θ_{i+1}), $u(\theta_i^\pm)$ exist, $u(\theta_i) = u(\theta_i^-)$ for $i \in \mathbb{N}\}$; $r(t) > 0$;
(iv) $\psi \in C(\mathbb{R})$, $f \in C^1(\mathbb{R})$ with $\psi(s) > 0$, $sf(s) > 0$ for $s \neq 0$ and

$$\frac{f'(s)}{(\psi(s)|f(s)|^{\alpha-1})^{1/\alpha}} \geq K_\alpha \tag{2}$$

is satisfied; $K_\alpha > 0$ is a constant.

By a solution of (1), we mean a continuous function x defined on $[t_0, \infty)$ such that $x', (r\psi(x)\varphi_\alpha(x'))' \in \text{PLC}\,[t_0, \infty)$ and $x(t)$ satisfies (1) for $t \geq t_0$. Existence of such solutions can be proved in a similar manner performed for equations without impulse effect [1]. As usual, a solution of (1) is called oscillatory if it has arbitrarily large zeros. The equation is called oscillatory if every solution is oscillatory.

Equation (1) without impulse effect can be considered as a natural generalizations of the linear, selfadjoint, half-linear or nonlinear equations

$$x'' + q(t)x = 0, \tag{3}$$

$$(r(t)x')' + q(t)x = 0, \tag{4}$$

$$(r(t)\varphi_\alpha(x'))' + q(t)\varphi_\alpha(x) = 0, \tag{5}$$

$$(r(t)\psi(x)\varphi_\alpha(x'))' + q(t)f(x) = 0, \tag{6}$$

respectively, which have been the object of intensive studies in recent years. During the last two decades there has been a great deal work on oscillatory behavior of solutions for equations (3)-(6) [2–21].

In this study, by using the generalized Riccati technique and the averaging method, we obtain two oscillation criteria for equation (1). As far as averaging method is considered; the first general result in the literature was given by Philos [16] in 1989. He proved two oscillation criteria [16, Theorems 1 and 2] for equation (3) which were considered as extension of Kamenev's criteria [4] and improvement of the result of Yan [20].

Later, the results of Philos have been extended by Li [22], Grace [23] and Manojlovic [14] by use of Riccati technique and the averaging method. In 2000, Manojlovic [15] developed oscillation theory for a general case of the equation (6) in which $f(s)$ is not necessarily of the form $\varphi_\alpha(s)$ and $\psi(s) \neq 1$, without any restriction on the sign of $q(t)$.

In the last decade an intensive investigation of the impulsive differential equations is observed [24–27]. Although numerous aspects of qualitative theory were studied, the first investigation of oscillatory properties of these equations was worked by Gopalsamy and Zhang [27] in 1989. Afterwards, many authors have studied the oscillatory properties of various classes of impulsive differential equations; the results obtained were presented in [26] by Bainov and Simeonov. In 1996, Bainov, Domshlak and Simeonov [24] (see also [26, p. 30]) studied the second order linear impulsive differential equations of the form

$$\begin{aligned} &x''(t) + q(t)x(t) = 0, \; t \neq \theta_i; \\ &\Delta x' + q_i x(t) = 0, \qquad t = \theta_i, \end{aligned} \tag{7}$$

and they have given conditions for oscillation and nonoscillation for equation (7), see [24, Theorem 2].

In present work, we are interested in extending Philos type oscillation criteria [16] to equation (1), and we give some analogous results existing in the literature; namely [16, Theorems 1 and 2] and [24, Theorem 2]; also cf. [14,15,21].

2. Main Results

In order to prove our results we use the following well-known inequality which is due to Hardy, Littlewood and Pólya [28].

Lemma 2.1. *If A, B are nonnegative numbers, then*

$$A^\gamma - \gamma A B^{\gamma-1} + (\gamma-1)B^\gamma \geq 0, \quad \gamma > 1, \tag{8}$$

and the equality holds if and only if $A = B$.

The following theorem is one of the main result of this study.

Theorem 2.1. *Let $D_0 = \{(t,s) : t > s > t_0\}$ and $D = \{(t,s) : t \geq s \geq t_0\}$. Assume $H(t,s) \in C^1(D : (0,\infty))$, $h(t,s) \in C(D_0, \mathbb{R})$ satisfy the conditions*

(i) $H(t,t) = 0$ *for* $t \geq t_0$ *and* $H(t,s) > 0$ *on* D_0*;*
(ii) H *has a continuous and nonpositive partial derivative on* D_0 *with respect to the second variable;*
(iii) $-\dfrac{\partial H}{\partial s}(t,s) + \dfrac{p(s)}{r(s)}H(t,s) - h(t,s)H^{\alpha/(\alpha+1)}(t,s), \quad (t,s) \subset D_0.$

$$\text{If} \quad \limsup_{t\to\infty} \frac{1}{H(t,t_0)} \left\{ \int_{t_0}^{t} \left[H(t,s)q(s) - \Gamma_\alpha r(s)|h(t,s)|^{\alpha+1} \right] ds + \sum_{t_0 \leq \theta_i < t} H(t,\theta_i) q_i \right\} = \infty, \tag{9}$$

where $\Gamma_\alpha = (\alpha+1)^{-(\alpha+1)}\alpha^\alpha K_\alpha^{-\alpha}$, then equation (1) is oscillatory.

Proof. Let $x(t)$ be a nonoscillatory solution of equation (1). We assume that $x(t) \neq 0$ on $[T,\infty)$ for some sufficiently large $T \geq t_0$. Define

$$w(t) := \frac{r(t)\psi(x)\varphi_\alpha(x')}{f(x)}, \qquad t \geq T. \tag{10}$$

Differentiating (10) and making use of (1) and (2), we obtain

$$w'(t) \leq -\frac{p(t)}{r(t)}w(t) - q(t) - K_\alpha \frac{|w(t)|^{1+1/\alpha}}{r^{1/\alpha}(t)}, \qquad t \neq \theta_i; \tag{11}$$

$$\Delta w(t) = -q_i, \qquad t = \theta_i. \tag{12}$$

Multiplying (11), with t replaced by s, by $H(t,s)$ and integrating from T to t, we have

$$\int_T^t H(t,s)q(s)ds \leq -\int_T^t \frac{p(s)}{r(s)} H(t,s)w(s)\,ds$$
$$-K_\alpha \int_T^t H(t,s)\frac{|w(s)|^{1+1/\alpha}}{r^{1/\alpha}(s)}\,ds - \int_T^t H(t,s)w'(s)\,ds. \tag{13}$$

Integration by parts and using (12), the last integral on the right hand side of inequality (13) becomes

$$\int_T^t H(t,s)w'(s)ds = \int_T^t \left[\frac{\partial}{\partial s}\{H(t,s)w(s)\} - w(s)\frac{\partial H}{\partial s}(t,s)\right] ds,$$
$$= -H(t,T)w(T) + \sum_{T\leq\theta_i<t} H(t,\theta_i)q_i - \int_T^t w(s)\frac{\partial H}{\partial s}(t,s)ds. \tag{14}$$

Using (13) and (14), we obtain

$$\int_T^t H(t,s)q(s)\,ds + \sum_{T\leq\theta_i<t} H(t,\theta_i)q_i \leq H(t,T)w(T)$$
$$-\int_T^t \left[K_\alpha\, H(t,s)\,\frac{|w(s)|^{1+1/\alpha}}{r^{1/\alpha}(s)} - |h(t,s)|\,H^{\alpha/(\alpha+1)}(t,s)\,|w(s)|\right] ds. \tag{15}$$

Using inequality (8) with $\gamma = 1 + 1/\alpha$

$$A = \frac{(K_\alpha\, H)^{\alpha/(\alpha+1)}|w|}{r^{1/(\alpha+1)}} \quad \text{and} \quad B = \left(\frac{\alpha}{\alpha+1}\right)^\alpha \left(\frac{r}{K_\alpha^\alpha}\right)^{\alpha/(\alpha+1)} |\,h|^\alpha,$$

we obtain

$$K_\alpha\, H(t,s)\,\frac{|w(s)|^{1+1/\alpha}}{r^{1/\alpha}(s)} - |h(t,s)|\,H^{\alpha/(\alpha+1)}(t,s)\,|w(s)|$$
$$\geq -\Gamma_\alpha\, r(s)\,|\,h(t,s)|^{\alpha+1}. \tag{16}$$

From (15) and (16), we obtain

$$\int_T^t \left[H(t,s)q(s) - \Gamma_\alpha r(s)\,|h(t,s)|^{\alpha+1}\right] ds + \sum_{T\leq\theta_i<t} H(t,\theta_i)q_i$$
$$\leq H(t,T)\,w(T) \tag{17}$$
$$\leq H(t,T)\,|w(T)| \leq H(t,t_0)\,|w(T)| \tag{18}$$

for all $t > T \geq t_0$. In the above inequality we choose $T = T_0$, then we have

$$\int_{t_0}^{t} \left[H(t,s)q(s) - \Gamma_\alpha r(s) \left| h(t,s) \right|^{\alpha+1} \right] ds + \sum_{t_0 \leq \theta_i < t} H(t,\theta_i) q_i$$
$$\leq H(t,t_0) \left\{ \int_{t_0}^{T_0} |q(s)|\, ds + \sum_{t_0 \leq \theta_i < T_0} |q_i| \right\} + H(t,t_0) \left| w(T_0) \right|.$$

It follows that

$$\limsup_{t\to\infty} \frac{1}{H(t,t_0)} \left\{ \int_{t_0}^{t} \left[H(t,s)q(s) - \Gamma_\alpha r(s) \left| h(t,s) \right|^{\alpha+1} \right] ds \right.$$
$$\left. + \sum_{t_0 \leq \theta_i < t} H(t,\theta_i) q_i \right\} \leq \int_{t_0}^{T_0} |q(s)|\, ds + \sum_{t_0 \leq \theta_i < T_0} |q_i| + |w(T_0)| < \infty,$$

which contradicts with (9). This completes the proof. □

As a consequence of the Theorem 2.1, we have the following corollary.

Corollary 2.1. *Let condition (9) in Theorem 2.1 be replaced by*

$$\limsup_{t\to\infty} \frac{1}{H(t,t_0)} \left\{ \int_{t_0}^{t} H(t,s)q(s)\, ds + \sum_{t_0 \leq \theta_i < t} H(t,\theta_i) q_i \right\} = \infty$$

and

$$\limsup_{t\to\infty} \frac{1}{H(t,t_0)} \int_{t_0}^{t} r(s) \left| h(t,s) \right|^{\alpha+1} ds < \infty.$$

Then equation (1) is oscillatory.

Notice that in the special case of half-linear equation (5), for $\psi(s) \equiv 1$ and $f(s) = \varphi_\alpha(s)$, the condition (2) is satisfied with $K_\alpha = \alpha$.

Our second result is the following Theorem which can be considered as an extension of [14, Theorem 2].

Theorem 2.2. *Let the functions H and h be defined as in Theorem 2.1. Moreover, suppose that*

$$0 < \inf_{s \geq t_0} \left\{ \liminf_{t\to\infty} \frac{H(t,s)}{H(t,t_0)} \right\} \leq \infty \tag{19}$$

and

$$\limsup_{t\to\infty} \frac{1}{H(t,t_0)} \int_{t_0}^{t} r(s) \left| h(t,s) \right|^{\alpha+1} ds < \infty. \tag{20}$$

If there exists a function $A \in C([t_0,\infty);\mathbb{R})$ such that

$$\limsup_{t\to\infty}\int_{t_0}^{t}\frac{A_+^{1+1/\alpha}(s)}{r^{1/\alpha}(s)}\,ds=\infty, \tag{21}$$

and for every $T \geq t_0$

$$\limsup_{t\to\infty}\frac{1}{H(t,T)}\left\{\int_T^t\left[H(t,s)q(s)-\Gamma_\alpha r(s)\,|h(t,s)|^{\alpha+1}\right]ds + \sum_{T\leq\theta_i<t}H(t,\theta_i)q_i\right\}\geq A(T), \tag{22}$$

where $A_+(s)=\max\{A(s),0\}$, then equation (1) is oscillatory.

Proof. Without loss of generality, we may assume that there exists a solution $x(t)$ of equation (1) such that $x(t)\neq 0$ on $[T_0,\infty)$ for some sufficiently large $T_0 \geq t_0$. Define $w(t)$ as in (10). As in the proof of Theorem 2.1, we can obtain (15) and (17). It follows that

$$\limsup_{t\to\infty}\frac{1}{H(t,T)}\left\{\int_T^t\left[H(t,s)q(s)-\Gamma_\alpha r(s)\,|h(t,s)|^{\alpha+1}\right]ds + \sum_{T\leq\theta_i<t}H(t,\theta_i)q_i\right\}\leq w(T)$$

for all $T \geq T_0$. Thus by (22) we have

$$A(T)\leq w(T)\quad\text{for all}\quad T\geq T_0 \tag{23}$$

and

$$\limsup_{t\to\infty}\frac{1}{H(t,T_0)}\left\{\int_{T_0}^t H(t,s)q(s)\,ds+\sum_{T_0\leq\theta_i<t}H(t,\theta_i)q_i\right\}\geq A(T_0). \tag{24}$$

Let

$$F(t):=\frac{1}{H(t,T_0)}\int_{T_0}^t|h(t,s)w(s)|\,H^{\alpha/(\alpha+1)}(t,s)\,ds$$

and

$$G(t):=\frac{K_\alpha}{H(t,T_0)}\int_{T_0}^t H(t,s)\,\frac{|w(s)|^{1+1/\alpha}}{r^{1/\alpha}(s)}\,ds$$

for all $t > T_0$. Then, by (15) and (24), we see that

$$\liminf_{t\to\infty}[G(t)-F(t)]=w(T_0)-\limsup_{t\to\infty}\frac{1}{H(t,T_0)}\left\{\int_{T_0}^t H(t,s)q(s)\,ds\right.$$

$$+\sum_{T_0\le\theta_i<t} H(t,\theta_i)q_i\Big\}$$
$$\le w(T_0)-A(T_0)<\infty. \tag{25}$$

Now, we claim that

$$\int_{T_0}^{\infty}\frac{|w(s)|^{1+1/\alpha}}{r^{1/\alpha}(s)}\,ds<\infty. \tag{26}$$

Suppose to the contrary that

$$\int_{T_0}^{\infty}\frac{|w(s)|^{1+1/\alpha}}{r^{1/\alpha}(s)}\,ds=\infty. \tag{27}$$

By (19), there is a positive constant η satisfying

$$\inf_{s\ge t_0}\left\{\liminf_{t\to\infty}\frac{H(t,s)}{H(t,t_0)}\right\}>\eta>0. \tag{28}$$

On the other hand, by (27), for any positive number μ there exists a $T_1>T_0$ such that

$$\int_{T_0}^{t}\frac{|w(s)|^{1+1/\alpha}}{r^{1/\alpha}(s)}\,ds\ge\frac{\mu}{K_\alpha\eta}\quad\text{for all}\quad t\ge T_1 \tag{29}$$

so for all $t\ge T_1$

$$\begin{aligned}G(t)&=\frac{K_\alpha}{H(t,T_0)}\int_{T_0}^{t}H(t,s)\,d\left[\int_{T_0}^{s}\frac{|w(\tau)|^{1+1/\alpha}}{r^{1/\alpha}(\tau)}\,d\tau\right]\\&=\frac{K_\alpha}{H(t,T_0)}\int_{T_0}^{t}\left[-\frac{\partial H}{\partial s}(t,s)\right]\left[\int_{T_0}^{s}\frac{|w(\tau)|^{1+1/\alpha}}{r^{1/\alpha}(\tau)}\,d\tau\right]ds\\&\ge\frac{K_\alpha}{H(t,T_0)}\int_{T_1}^{t}\left[-\frac{\partial H}{\partial s}(t,s)\right]\left[\int_{T_0}^{s}\frac{|w(\tau)|^{1+1/\alpha}}{r^{1/\alpha}(\tau)}\,d\tau\right]ds\\&\ge\frac{\mu}{K_\alpha\eta}\frac{K_\alpha}{H(t,T_0)}\int_{T_1}^{t}\left[-\frac{\partial H}{\partial s}(t,s)\right]ds=\frac{\mu}{\eta}\frac{H(t,T_1)}{H(t,T_0)}.\end{aligned} \tag{30}$$

From (28) we have

$$\liminf_{t\to\infty}\frac{H(t,T_1)}{H(t,t_0)}>\eta>0, \tag{31}$$

there exists $T_2\ge T_1$ such that $H(t,T_1)/H(t,t_0)\ge\eta$ for all $t\ge T_2$. Therefore by (30), $G(t)\ge\mu$ for all $t\ge T_2$, and since μ is arbitrary constant, we conclude

$$\lim_{t\to\infty}G(t)=\infty. \tag{32}$$

Next, consider a sequence $\{t_n\}_{n=1}^{\infty}$ in (T_0, ∞) with $\lim_{t\to\infty} t_n = \infty$ and such that

$$\lim_{n\to\infty} [G(t_n) - F(t_n)] = \liminf_{t\to\infty} [G(t) - F(t)].$$

In the view of (25), there exists a constant M such that

$$G(t_n) - F(t_n) \le M \text{ for all sufficiently large } n. \tag{33}$$

It follows from (32) that

$$\lim_{n\to\infty} G(t_n) = \infty. \tag{34}$$

This and (33) give

$$\lim_{n\to\infty} F(t_n) = \infty. \tag{35}$$

Then by (33) and (34),

$$\frac{F(t_n)}{G(t_n)} - 1 \ge -\frac{M}{G(t_n)} > -\frac{1}{2} \quad \text{for } n \text{ large enough.}$$

Thus,

$$\frac{F(t_n)}{G(t_n)} > \frac{1}{2} \quad \text{for all } n \text{ large enough.}$$

This and (35) imply that

$$\lim_{n\to\infty} \frac{F^{\alpha+1}(t_n)}{G^{\alpha}(t_n)} = \infty. \tag{36}$$

On the other hand, by Hölder's inequality, we have

$$\begin{aligned} F(t_n) &= \frac{1}{H(t_n, T_0)} \int_{T_0}^{t_n} |h(t_n, s)w(s)|\, H^{\alpha/(\alpha+1)}(t_n, s)\, ds \\ &\le \left\{ \frac{K_\alpha}{H(t_n, T_0)} \int_{T_0}^{t_n} H(t_n, s) \frac{|w(s)|^{1+1/\alpha}}{r^{1/\alpha}(s)}\, ds \right\}^{\alpha/(\alpha+1)} \\ &\quad \times \left\{ \frac{1}{K_\alpha^\alpha\, H(t_n, T_0)} \int_{T_0}^{t_n} r(s)|h(t_n, s)|^{\alpha+1}\, ds \right\}^{1/(\alpha+1)} \\ &\le \frac{G^{\alpha/(\alpha+1)}(t_n)}{K_\alpha^{\alpha/(\alpha+1)}} \left\{ \frac{1}{H(t_n, T_0)} \int_{T_0}^{t_n} r(s)|h(t_n, s)|^{\alpha+1}\, ds \right\}^{1/(\alpha+1)}, \end{aligned}$$

and therefore

$$\frac{F^{\alpha+1}(t_n)}{G^{\alpha}(t_n)} \le \frac{1}{K_\alpha^\alpha\, H(t_n, T_0)} \int_{T_0}^{t_n} r(s)|h(t_n, s)|^{\alpha+1}\, ds$$

$$\leq \frac{1}{K_\alpha^\alpha \, \eta \, H(t_n, t_0)} \int_{t_0}^{t_n} r(s)|h(t_n, s)|^{\alpha+1} \, ds$$

for a large n. It follows from (36) that

$$\lim_{n\to\infty} \frac{1}{H(t_n, t_0)} \int_{t_0}^{t_n} r(s)|h(t_n, s)|^{\alpha+1} \, ds = \infty, \tag{37}$$

that is,

$$\limsup_{t\to\infty} \frac{1}{H(t, t_0)} \int_{t_0}^{t} r(s)|h(t, s)|^{\alpha+1} \, ds = \infty,$$

which contradicts (20). Hence (26) holds. Then, it follows from (23) that

$$\int_{T_0}^{t} \frac{A_+^{1+1/\alpha}(s)}{r^{1/\alpha}(s)} \, ds \leq \int_{T_0}^{\infty} \frac{|w(s)|^{1+1/\alpha}}{r^{1/\alpha}(s)} \, ds < \infty$$

which contradicts (21). This completes the proof of Theorem 2.2. □

Note that if $p(t) \equiv 0$, $f(s) = s$, $q_i \equiv 0$ and $\alpha = 1$, we obtain the results of Philos [16].

References

1. Á. Elbert, *A half-linear differential equation*, Colloq. Math. Soc. Janos Bolyai 30: Qualitative Theory of Differential Equations, Szeged (1979), 153–180.
2. H. L. Hong, *On the oscillatory behaviour of solutions of second order nonlinear differential equations*, Publ. Math. Debrecen **52** (1998), 55–68.
3. H. B. Hsu and C. C. Yeh, *Oscillation theorems for second-order half-linear differential equations*, Appl. Math. Lett. **9** (1996), no. 6, 71–77.
4. I. V. Kamenev, *An integral criterion for oscillation of linear differential equations of second order*, Mat. Zametki **23** (1978), 249–251.
5. T. Kusano and M. Naito, *On the number of zeros of nonoscillatory solutions to half-linear ordinary differential equations involving a parameter*, Trans. Amer. Math. Soc. **354** (2002), no. 12, 4751–4765.
6. T. Kusano and Y. Naito, *Oscillation and nonoscillaton criteria for second order quasilinear differential equations*, Acta Math. Hungar. **76** (1997), no. 1-2, 81–99.
7. T. Kusano and N. Yoshida, *Nonoscillation theorems for a class of quasilinear differential equations of second order*, J. Math. Anal. Appl. **189** (1995), 115–127.
8. H. J. Li and C. C. Yeh, *An integral criterion for oscillation of nonlinear differential equations*, Math. Japon. **41** (1995), no. 1, 185–188.
9. H. J. Li and C. C. Yeh, *Nonoscillation criteria for second order half-linear differential equations*, Appl. Math. Lett. **8** (1995), 63–70.
10. H. J. Li and C. C. Yeh, *Nonoscillation theorems for second order quasilinear differential equations*, Publ. Math. Debrecen **47** (1995), no. 3-4, 271–279.

11. H. J. Li and C. C. Yeh, *Oscillation criteria for nonlinear differential equations*, Houston J. Math. **21** (1995), 801–811.
12. H. J. Li and C. C. Yeh, *Oscillations of half-linear second order differential equations*, Hiroshima Math. J. **25** (1995), 585–594.
13. W. C. Lian, C. C. Yeh, and H. J. Li, *The distance between zeros of an oscillatory solution to a half-linear differential equation*, Comput. Math. Appl. **29** (1995), no. 8, 39–43.
14. J. V. Manojlovic, *Oscillation criteria for second-order half-linear differential equations*, Math. Comput. Modelling **30** (1999), 109–119.
15. J. V. Manojlovic, *Oscillation theorems for nonlinear differential equations of second order*, Electron. J. Qual. Theor. Differ. Equ. **2000**, no. 1, 21 pp.
16. Ch. G. Philos, *Oscillation theorems for linear differential equations of second order*, Arch. Math. **53** (1989), 482–492.
17. J. Wang, *Oscillation and nonoscillation theorems for a class of second order quasilinear functional differential equations*, Hiroshima Math. J. **27** (1997), 449–466.
18. J. Wang, *On second order quasilinear oscillations*, Funkcial. Ekvac. **41** (1998), 25–54.
19. P. J. Y. Wong and R. P. Agarwal, *Oscillatory behaviour of solutions of certain second order nonlinear differential equations*, J. Math. Anal. Appl. **198** (1996), 337–354.
20. J. Yan, *Oscillation theorems for second order linear differential equations with damping*, Proc. Amer. Math. Soc. **98** (1986), no. 2, 276–282.
21. X. Yang, *Oscillation results for second-order half-linear differential equations*, Math. Comput. Modelling **36** (2002), 503–507.
22. H. J. Li, *Oscillation criteria for second order linear differential equations*, J. Math. Anal. Appl. **194** (1995), 217–234.
23. S. R. Grace, *Oscillation theorems for nonlinear differential equations of second order*, J. Math. Anal. Appl. **171** (1992), 220–241.
24. D. D. Bainov, Y. I. Domshlak, and P. S. Simeonov, *Sturmian comparison theory for impulsive differential inequalities and equations*, Arch. Math. (Basel) **67** (1996), 35–49.
25. D. D. Bainov and P. S. Simeonov, *Systems with impulse effect: Stability, theory and applications*, Ellis Horwood, Chichester, 1989.
26. D. D. Bainov and P. S. Simeonov, *Oscillation theory of impulsive differential equations*, International Publications, Orlando, Florida, 1998.
27. K. Gopalsamy and B. G. Zhang, *On delay differential equations with impulses*, J. Math. Anal. Appl. **139** (1989), 110–122.
28. G. H. Hardy, J. E. Littlewood, and G. Pólya, *Inequalities*, Cambridge University Press, Cambridge, 1964.

POSITIVE SOLUTIONS OF A NONLINEAR THREE-POINT BOUNDARY VALUE PROBLEM ON TIME SCALES

S. GÜLŞAN TOPAL

Department of Mathematics, Ege University,
Izmir, 35100, Turkey
E-mail: f.serap.topal@ege.edu.tr

AHMET YANTIR

Department of Mathematics, Atılım University,
Ankara, 06100, Turkey
E-mail: ayantir@atilim.edu.tr

We present sufficient conditions for the existence of at least one or at least three positive solutions to a second-order three-point eigenvalue problem on time scales. The proofs are based on fixed point theorems concerning cones in a Banach space.

Keywords: Multipoint Boundary Value Problems, Positive Solutions, Fixed Point Theorems, Time Scales.

1. Introduction

The theory of dynamic equations on time scales unifies the well-known analogies in the concept of difference equations and differential equations. See [3,4] for basic definitions and general theory of time scales. The existence of positive solutions for nonlinear second order multipoint BVPs has been studied by many authors by using the coincidence degree theory and fixed point theorems in cones (see [1,2,6–12] and the references therein).

The multipoint boundary value problems for dynamic equations on time scales arise in a variety of different areas of applied mathematics, physics and engineering. Recently Sun and Lee [13] and Yaslan [14] obtained some existence results for multipoint boundary value problems on time scales.

In this paper we present the values λ for which the following three point BVP has positive solutions

$$u^{\Delta\nabla}(t) + a(t)u^{\Delta}(t) + b(t)u(t) + \lambda h(t)f(t, u(t)) = 0, \ t \in [0,1] \subset \mathbb{T}, \quad (1)$$

$$u(\rho(0)) = 0, \ \alpha u(\eta) = u(\sigma(1)), \tag{2}$$

where $0 < \eta < 1$ and h, f, a, b are real valued functions and satisfy

(C1) $f \in \mathcal{C}([0,1] \times [0,\infty), [0,\infty))$,
(C2) $h \in \mathcal{C}([0,1], [0,\infty))$ and there exists at least one $t_0 \in [0,1]$ such that $h(t_0) > 0$,
(C3) $a \in \mathcal{C}([0,1], [0,\infty))$, $b \in \mathcal{C}([0,1], (-\infty, 0])$.

This paper is organized as follows: In Section 2 starting with some preliminary lemmas we state the Krasnoselskii and Legget-Williams fixed point theorems. In Section 3 we present the existence results.

2. Preliminaries and Fixed Point Theorems

In this section we state the preliminary information that we need for proving the main results.

Lemma 2.1. *Assume that* (C3)*holds. Let* φ_1 *and* φ_2 *be the solutions of*

$$\varphi_1^{\Delta\nabla}(t) + a(t)\varphi_1^{\Delta}(t) + b(t)\varphi_1(t) = 0, \tag{3}$$

$$\varphi_1(\rho(0)) = 0, \ \varphi_1(\sigma(1)) = 1, \tag{4}$$

and

$$\varphi_2^{\Delta\nabla}(t) + a(t)\varphi_2^{\Delta}(t) + b(t)\varphi_2(t) = 0, \tag{5}$$

$$\varphi_2(\rho(0)) = 1, \ \varphi_2(\sigma(1)) = 0, \tag{6}$$

respectively. Then φ_1 *is strictly increasing and* φ_2 *is strictly decreasing on* $[\rho(0), 1]$. *Furthermore* $\varphi_1(t) \geq 0$ *and* $\varphi_2(t) \geq 0$ *on* $[0,1]$.

Lemma 2.2. *Assume that* (C3) *holds. Then* (3)-(4) *and* (5)-(6) *have unique solutions respectively.*

The proofs of the Lemma 2.1 and Lemma 2.2 can be obtained by generalizing the proofs of Lemma 2.1 and Lemma 2.2 of [12] to time scales.

In the rest of the paper we need the following:

(C4) $0 < \alpha\varphi_1(\eta) < 1$.

In the following lemma we state the integral equation equivalent to the solution of the linear three point BVP corresponding to (1)-(2).

Lemma 2.3. *Assume that* (C3) *and* (C4) *hold. Let* $y \in \mathcal{C}[\rho(0), \sigma(1)]$ *and* $\varphi_1(t)$ *and* $\varphi_2(t)$ *be the solutions of* (3)-(4) *and* (5)-(6). *Then the problem*

$$u^{\Delta\nabla}(t) + a(t)u^{\Delta}(t) + b(t)u(t) + y(t) = 0\,, \quad t \in [0,1] \tag{7}$$

$$u(\rho(0)) = 0, \quad \alpha u(\eta) - u(\sigma(1)) = 0, \tag{8}$$

is equivalent to the integral equation

$$u(t) = \int_{\rho(0)}^{\sigma(1)} G(t,s)y(s)\nabla s + A\varphi_1(t), \tag{9}$$

where

$$A = \frac{\alpha}{1-\alpha\varphi_1(\eta)}\Big(\int_{\rho(0)}^{\sigma(1)} G(\eta_i, s)y(s)\nabla s\Big), \tag{10}$$

$$G(t,s) = \frac{1}{d}\begin{cases} \varphi_1(t)\varphi_2(s), & s \geq t \\ \varphi_1(s)\varphi_2(t), & t \geq s, \end{cases} \tag{11}$$

and $d = \varphi_1^{\Delta}(\rho(0))$.

Proof. First we show that the unique solution of Eqs. (7)-(8) can be represented by Eq. (9). From Lemma 2.1 since

$$\begin{vmatrix} \varphi_1(\rho(0)) & \varphi_2(\rho(0)) \\ \varphi_1^{\Delta}(\rho(0)) & \varphi_2^{\Delta}(\rho(0)) \end{vmatrix} = -\varphi_1^{\Delta}(\rho(0)) = -d \neq 0,$$

we know that the homogenous part of Eq. (7) has two linear independent solutions $\varphi_1(t)$ and $\varphi_2(t)$. Now by method of variation of constants, we can obtain the unique solution of BVP (7)-(8) which can be represented by Eq. (9) where A and G are as in Eqs. (10) and (11) respectively.

Next we check the function defined in Eq. (9) is the solution of the BVP (7)-(8). For this purpose we first show that Eq. (9) satisfies Eq. (7). From the definition of the Green's function (11) we obtain the derivatives u^{Δ} and $u^{\Delta\nabla}$ as follows:

$$u^{\Delta}(t) = \frac{1}{d}\Big(\varphi_2^{\Delta}(t)\int_{\rho(0)}^{t} \varphi_1(s)y(s)\nabla s + \varphi_1^{\Delta}(t)\int_{t}^{\sigma(1)} \varphi_2(s)y(s)\nabla s\Big) + A\varphi_1^{\Delta}(t),$$

$$u^{\Delta\nabla}(t) = \frac{1}{d}\Big(\varphi_2^{\Delta\nabla}(t)\int_{\rho(0)}^{\rho(t)} \varphi_1(s)y(s)\nabla s + \varphi_2^{\Delta}(t)\varphi_1(t)y(t)$$
$$+\varphi_1^{\Delta\nabla}(t)\int_{\rho(t)}^{\sigma(1)} \varphi_2(s)y(s)\nabla s - \varphi_1^{\Delta}(t)\varphi_2(t)y(t)\Big) + A\varphi_1^{\Delta\nabla}(t).$$

Plugging the derivatives into Eq. (7), one can verify that the function defined in Eq. (9) satisfies Eq. (7). Further we obtain that Eq. (8) is satisfied by Eq. (9). The first boundary condition of Eq. (8) follows from Eqs. (9), (10) and (11). Now we check the second boundary condition. Taking

$$G(\sigma(1), s) = \frac{1}{d}\varphi_1(s)\varphi_2(\sigma(1)) = 0$$

into account, we obtain

$$u(\sigma(1)) = \int_{\rho(0)}^{\sigma(1)} G(\sigma(1), s)y(s)\nabla s + A\varphi_1(\sigma(1)) = A. \tag{12}$$

Also Eq. (10) enables us to have

$$\begin{aligned}
\alpha u(\eta) &= \alpha\Big(\int_{\rho(0)}^{\sigma(1)} G(\eta, s)y(s)\nabla s + A\varphi_1(\eta)\Big) \\
&= \alpha \int_{\rho(0)}^{\sigma(1)} G(\eta, s)y(s)\nabla s + \frac{\alpha^2\varphi_1(\eta)}{1-\alpha\varphi_1(\eta)} \int_{\rho(0)}^{\sigma(1)} G(\eta, s)y(s)\nabla s \\
&= \Big(\alpha + \frac{\alpha^2\varphi_1(\eta)}{1-\alpha\varphi_1(\eta)}\Big) \int_{\rho(0)}^{\sigma(1)} G(\eta, s)y(s)\nabla s \\
&= \frac{\alpha}{1-\alpha\varphi_1(\eta)} \int_{\rho(0)}^{\sigma(1)} G(\eta, s)y(s)\nabla s = A. \qquad (13)
\end{aligned}$$

Combining Eqs. (12) and (13) finishes the proof. □

In this study, we consider the Banach space $\mathcal{B}$ of continuous functions defined on $[\rho(0), \sigma(1)]$ with the supremum norm. Now we set

$$q(t) = \min\Big\{\frac{\varphi_1(t)}{||\varphi_1(t)||}, \frac{\varphi_2(t)}{||\varphi_2(t)||}\Big\}. \tag{14}$$

Lemma 2.4. *Assume that* (C3)-(C4) *hold. Let* $y \in C([\rho(0), \sigma(1)], [0, \infty))$. *Then the unique solution of BVP* (7)-(8) *satisfies* $u(t) \geq ||u||q(t)$.

Proof. Let t_0 be the point in $(\rho(0), \sigma(1))$ such that $||u|| = u(t_0)$. Next we verify

$$G(t, s) \geq G(t_0, s)q(t). \tag{15}$$

For this purpose we consider the following four cases:

(i) $t_0, t \leq s$: In this case

$$\frac{G(t,s)}{G(t_0,s)} = \frac{\varphi_1(t)}{\varphi_1(t_0)} \geq \frac{\varphi_1(t)}{||\varphi_1||} \geq \min\{\frac{\varphi_1(t)}{|\varphi_1(t)|}, \frac{\varphi_2(t)}{|\varphi_2(t)|}\} = q(t).$$

(ii) $t_0 \leq s \leq t$: In this case

$$\frac{G(t,s)}{G(t_0,s)} = \frac{\varphi_1(s)\varphi_2(t)}{\varphi_1(t_0)\varphi_2(s)} \geq \frac{\varphi_2(t)}{\varphi_2(s)} \geq \frac{\varphi_2(t)}{||\varphi_2||} \geq \min\{\frac{\varphi_1(t)}{|\varphi_1(t)|}, \frac{\varphi_2(t)}{|\varphi_2(t)|}\} = q(t).$$

Inequality (15) can be verified for the cases **(iii)** $t_0, t \geq s$ and **(iv)** $t \leq s \leq t_0$ similarly. We apply Lemma 2.1 for the second and the fourth cases. It follows from (15) and the fact $1 \geq \varphi_1(t) \geq q(t), \ \forall t \in [\rho(0), \sigma(1)]$ that

$$\begin{aligned} u(t) &= \int_{\rho(0)}^{\sigma(1)} G(t,s)y(s)\nabla s + A\varphi_1(t) \geq q(t) \int_{\rho(0)}^{\sigma(1)} G(t_0,s)y(s)\nabla s + A\varphi_1(t) \\ &\geq q(t)\Big\{ \int_{\rho(0)}^{\sigma(1)} G(t_0,s)y(s)\nabla s + A \Big\} \\ &\geq q(t)\Big\{ \int_{\rho(0)}^{\sigma(1)} G(t_0,s)y(s)\nabla s + A\varphi_1(t_0) \Big\} = q(t)u(t_0) = q(t)||u||. \end{aligned}$$ □

Assume that $\xi := \min\{t \in \mathbb{T} : t > \rho(0)\}, \ w := \max\{t \in \mathbb{T} : t < \sigma(1)\}$ both exist and are included in $[\rho(0), \sigma(1)]$, and also satisfy $\rho(0) < \xi < w < \sigma(1)$. Also assume that $\sigma(w) < \sigma(1)$. Consequently let $\tau \in [\xi, w]$ be defined by

$$\int_{\xi}^{w} G(\tau,s)h(s)\nabla s = \max_{t \in [\xi,w]} \int_{\xi}^{w} G(t,s)h(s)\nabla s.$$

Lemma 2.5. *Assume that* (C3)-(C4) *hold. Let* $y \in \mathcal{C}([\rho(0), \sigma(1)], [0, \infty))$. *Then there exists* $\gamma > 0$ *such that the unique solution of BVP* (7)-(8) *satisfies* $u(t) > \gamma||u||$.

Proof. The choice of $\gamma = \min\{q(t) : t \in [\xi, w]\}$ guaranties that $\gamma > 0$ and $u(t) \geq q(t)||u|| > \gamma||u||, \ \forall t \in [\xi, w]$. □

To make use of the fixed point theorems we consider the cone

$$\mathcal{P} = \{u \in \mathcal{B} : u(t) > 0, t \in [\rho(0), \sigma(1)], \text{ and } \min_{t \in [\xi,w]} u(t) \geq \gamma||u||\} \quad (16)$$

on the Banach space $\mathcal{B}$, and set $\mathcal{P}_r = \{x \in \mathcal{P} : ||x|| < r\}$.

Theorem 2.1. [5] *(Krasnoselskii Fixed Point Theorem) Let* E *be a Banach space, and let* $K \subset E$ *be a cone. Assume* Ω_1 *and* Ω_2 *are open, bounded subsets of* E *with* $0 \in \Omega_1, \overline{\Omega_1} \subset \Omega_2$, *and let* $A : K \cap (\overline{\Omega_2} \setminus \Omega_1) \to K$ *be a completely continuous operator such that either*

(i) $\|Au\| \leq \|u\|$ *for* $u \in K \cap \partial\Omega_1$, $\quad \|Au\| \geq \|u\|$ *for* $u \in K \cap \partial\Omega_2$; *or*
(ii) $\|Au\| \geq \|u\|$ *for* $u \in K \cap \partial\Omega_1$, $\quad \|Au\| \leq \|u\|$ *for* $u \in K \cap \partial\Omega_2$,

hold. Then A *has a fixed point in* $K \cap (\overline{\Omega_2} \setminus \Omega_1)$.

Theorem 2.2. [11] *(Legget-Williams Fixed Point Theorem) Let $\mathcal{P}$ be a cone in a real Banach space E. Set*

$$\mathcal{P}(\psi, a, b) = \{x \in \mathcal{P} : a \leq \psi(x), \|x\| \leq b\}.$$

Suppose $A : \overline{\mathcal{P}_r} \to \overline{\mathcal{P}_r}$ be a completely continuous operator and ψ be a non-negative, continuous, concave functional on $\mathcal{P}$ with $\psi(u) \leq \|u\|$ for all $u \in \overline{\mathcal{P}_r}$. If there exist $0 < p < q < l \leq r$ such that

(i) $\{u \in \mathcal{P}(\psi, q, l) : \psi(u) > q\} \neq \emptyset$ *and* $\psi(Au) > q$ *for all* $u \in \mathcal{P}(\psi, q, l)$,
(ii) $\|Au\| < p$ *for all* $\|u\| \leq p$,
(iii) $\psi(Au) > q$ *for* $u \in \mathcal{P}(\psi, q, r)$ *with* $\|Au\| > l$

hold, then A has at least three positive solutions u_1, u_2 and u_3 in $\overline{\mathcal{P}_r}$ satisfying $\|u_1\| < p$, $\psi(u_2) > q$, $p < \|u_3\|$ with $\psi(u_3) < q$.

3. Main Results

We are concerned with determining values of λ, for which there exist positive solutions of three point BVP (1)-(2). We use Krasnoselskii fixed point theorem and Legget-Williams fixed point theorem to prove the main results. From Lemma 2.3 it is clear that the solutions of (1)-(2) are the fixed points of the operator

$$Tu(t) = \lambda\Big\{\int_{\rho(0)}^{\sigma(1)} G(t,s)h(s)f(s,u(s))\nabla s + A\varphi_1(t)\Big\}. \tag{17}$$

To state the main results we need to define the extended real numbers

$$f_0 = \liminf_{u\to 0+} \min_{t\in[\rho(0),\sigma(1)]} \frac{f(t,u)}{u}, \quad f^0 = \limsup_{u\to 0+} \max_{t\in[\rho(0),\sigma(1)]} \frac{f(t,u)}{u},$$

$$f_\infty = \liminf_{u\to \infty} \min_{t\in[\rho(0),\sigma(1)]} \frac{f(t,u)}{u}, \quad f^\infty = \limsup_{u\to \infty} \max_{t\in[\rho(0),\sigma(1)]} \frac{f(t,u)}{u}.$$

Let K and L be defined by

$$K = \min_{t\in[\xi,w]} \int_{\xi}^{w} G(t,s)h(s)\nabla s, \tag{18}$$

$$L = \max_{t\in[\rho(o),\sigma(1)]} \int_{\rho(0)}^{\sigma(1)} G(t,s)h(s)\nabla s = \int_{\rho(0)}^{\sigma(1)} G(s,s)h(s)\nabla s. \tag{19}$$

Theorem 3.1. *Assume that (C1)-(C4) are satisfied. Then there exists at least one positive solution of BVP (1)-(2) for each λ satisfying either one of the following:*

(a) $\dfrac{1}{\gamma K f_\infty} < \lambda < \dfrac{1}{Lf^0}\Big(\dfrac{1-\alpha\varphi(\eta)}{\alpha+1}\Big)$, **(b)** $\dfrac{1}{\gamma K f_0} < \lambda < \dfrac{1}{Lf^\infty}\Big(\dfrac{1-\alpha\varphi(\eta)}{\alpha+1}\Big)$

Proof. We first claim that $T : \mathcal{P} \to \mathcal{P}$. From the nonnegativity of G and from the assumptions (C2) and (C3), if $u \in \mathcal{P}$ then $Tu(t) \geq 0$ for $t \in [\rho(0), \sigma(1)]$. Next if we choose $u \in \mathcal{P}$, then

$$\begin{aligned}
\min_{t\in[\xi,w]} Tu(t) &= \min_{t\in[\xi,w]} \lambda\Big\{ \int_{\rho(0)}^{\sigma(1)} G(t,s)h(s)f(s,u(s))\nabla s + A\varphi_1(t)\Big\} \\
&\geq \lambda\Big\{ \int_{\rho(0)}^{\sigma(1)} q(t)G(t_0,s)h(s)f(s,u(s))\nabla s + A\varphi_1(t)\Big\} \\
&\geq \lambda q(t)\Big\{ \int_{\rho(0)}^{\sigma(1)} G(t_0,s)h(s)f(s,u(s))\nabla s + A\Big\} \\
&\geq \lambda\gamma\Big\{ \int_{\rho(0)}^{\sigma(1)} G(t_0,s)h(s)f(s,u(s))\nabla s + A\Big\} \\
&\geq \lambda\gamma\Big\{ \int_{\rho(0)}^{\sigma(1)} G(t_0,s)h(s)f(s,u(s))\nabla s + A\varphi_1(t_0)\Big\} = \gamma||Tu||.
\end{aligned}$$

Thus $Tu \in \mathcal{P}$. Also completely continuity of $Tu(t)$ can be obtained easily by analysis methods. Now we seek for the fixed points of $Tu(t)$ which belongs to $\mathcal{P}$. Assume (a) holds. From the second part of the inequality there exist $\epsilon > 0$ such that $\lambda L(f^0 + \epsilon)(1+\alpha)/(1-\alpha\varphi_1(\eta)) \leq 1$. The use of the definition of f^0 guarantees that there exists $r_1 > 0$, sufficiently small such that $f(t,u)/u < f^0 + \epsilon, \ \ \forall u \in [0, r_1]$. It follows that $f(t,u) < (f^0 + \epsilon)u$ for $0 \leq u \leq r_1$ and $t \in [\rho(0), \sigma(1)]$. If $u \in \partial\mathcal{P}_{r_1}$ then

$$\begin{aligned}
Tu(t) &= \lambda\Big\{ \int_{\rho(0)}^{\sigma(1)} G(t,s)h(s)f(s,u(s))\nabla s + A\varphi_1(t)\Big\} \\
&\leq \lambda\Big(\frac{1+\alpha}{1-\alpha\varphi_1(\eta)}\Big) \int_{\rho(0)}^{\sigma(1)} G(s,s)h(s)f(s,u(s))\nabla s \\
&\leq \lambda\Big(\frac{1+\alpha}{1-\alpha\varphi_1(\eta)}\Big)(f^0+\epsilon)||u|| \int_{\rho(0)}^{\sigma(1)} G(s,s)h(s)\nabla s \\
&\leq \lambda\Big(\frac{1+\alpha}{1-\alpha\varphi_1(\eta)}\Big)(f^0+\epsilon)||u||L \leq ||u||.
\end{aligned}$$

Hence if $\Omega_1 = \{u \in \mathcal{P} : ||u|| < r_1\}$ then

$$||Tu|| \leq ||u||, \ \forall u \in \partial\mathcal{P}_{r_1} = \mathcal{P} \cap \partial\Omega_1. \tag{20}$$

Now we use the other part of the inequality in part (a). First we consider the case $f_\infty < \infty$. In this case pick ϵ_1 such that $\lambda\gamma K(f_\infty - \epsilon_1) \geq 1$. The use of the definition of f_∞ guarantees that there exists $r > r_1$, sufficiently large so that $f(t,u)/u > f_\infty - \epsilon_1, \ \ \forall u \geq r$. It follows that $f(t,u) > (f_\infty - \epsilon_1)u$

for $0 \leq u \leq r_1$ and $t \in [\rho(0), \sigma(1)]$. Pick r_2 such that $r_2 \geq r/\gamma > r_1$ and let $\Omega_2 = \{u \in \mathcal{P} : ||u|| < r_2\}$. If $u \in \partial\mathcal{P}_{r_2}$ then Lemma 2.5 ensures that

$$Tu(t) = \lambda\Big\{\int_{\rho(0)}^{\sigma(1)} G(t,s)h(s)f(s,u(s))\nabla s + A\varphi_1(t)\Big\}$$

$$\geq \lambda\int_{\rho(0)}^{\sigma(1)} G(t,s)h(s)f(s,u(s))\nabla s \geq \lambda(f_\infty - \epsilon_1)\gamma||u||\int_{\xi}^{w} G(t,s)h(s)\nabla s$$

$$\geq \lambda(f_\infty - \epsilon_1)\gamma||u||K \geq ||u||. \tag{21}$$

Finally we consider the case $f_\infty = \infty$. In this case the inequality becomes $\lambda > 0$. If we choose M sufficiently large so that

$$\lambda M\gamma\int_{\xi}^{w} G(t,s)h(h)\nabla s \geq 1 \quad (or\ \ \lambda M\gamma K \geq 1),$$

for any $t \in [\rho(0), \sigma(1)]$ then there exists $r > r_1$ so that $f(t,u) > Mu$ for $u \geq r$. Let r_2 be defined as above and let $u \in \partial\mathcal{P}_{r_2}$. Then for all $t \in [\rho(0), \sigma(1)]$ we have

$$Tu(t) \geq \lambda M\int_{\rho(0)}^{\sigma(1)} G(t,s)h(s)u(s)\nabla s \geq \lambda M\gamma||u||\int_{\xi}^{w} G(t,s)h(s)\nabla s$$

$$= \lambda M\gamma K||u|| \geq ||u||. \tag{22}$$

From the inequalities (21) and (22)

$$||Tu|| \geq ||u||,\ \forall u \in \partial\mathcal{P}_{r_2} = \mathcal{P} \cap \partial\Omega_2. \tag{23}$$

Thus Krasnoselskii fixed point theorem implies that Tu has a fixed point in $\mathcal{P} \cap (\overline{\Omega_2} \setminus \Omega_1)$. □

Theorem 3.2. *Let $f(t,u)$ satisfy* (C1). *Assume that there exist two positive constants $r_2 > r_1 > 0$ satisfying the following conditions*

(C5) $f(t,u) \leq \lambda^{-1}Mr_2$ *for* $(t,u) \in [\rho(0), \sigma(1)] \times [0, r_2]$,
(C6) $f(t,u) \geq \lambda^{-1}Nr_1$ *for* $(t,u) \in [\rho(0), \sigma(1)] \times [0, r_1]$,

where

$$M = \frac{1 - \alpha\varphi_1(\eta)}{1+\alpha}\Big(\int_{\rho(0)}^{\sigma(1)} G(s,s)h(s)\nabla s\Big)^{-1}, \tag{24}$$

$$N = \Big(\gamma\int_{\xi}^{w} G(t_0,s)h(s)\nabla s\Big)^{-1}, \tag{25}$$

and $t_0 \in (\rho(0), \sigma(1))$ such that $||u|| = u(t_0)$. Then the problem (1)-(2) *has at least one positive solution u satisfying $r_1 \leq ||u|| \leq r_2$.*

Now we state the sufficient conditions which guarantee the existence of at least three solutions for (1)-(2).

Theorem 3.3. *Let $f(t,u)$ satisfy* (C1) *and there exist constants $0 < r_1 < r_2 < r_3$ such that following assumptions hold:*

(C7) $f(t,u) < \lambda^{-1}Mr_1$ *for all* $(t,u) \in [\rho(0), \sigma(1)] \times [0, r_1]$,
(C8) $f(t,u) \geq \lambda^{-1}Nr_2$ *for all* $(t,u) \in [\xi, w] \times [r_2, r_3]$,
(C9) $f(t,u) \leq \lambda^{-1}Mr_3$ *for all* $(t,u) \in [\rho(0), \sigma(1)] \times [\rho(0), r_3]$.

Then (1)-(2) *has at least three positive solutions u_1, u_2 and u_3 satisfying*

$$||u_1|| < r_1, \quad r_2 < \min_{t\in[\xi,w]} |u_2(t)| \leq r_3, \quad r_1 < ||u_3|| \leq r_3, \quad \min_{t\in[\xi,w]} |u_3(t)| < r_2.$$

Proof. We first define a nonnegative, continuous, concave functional $\psi : \mathcal{P} \to [0, \infty)$, $\psi(u) := \min_{t\in[\xi,w]} |u(t)|$. The cone $\mathcal{P}$ as in Eq. (16), M as in Eq. (24) and N as in Eq. (25). Then $\psi(u) \leq \|u\|$ for all $u \in \mathcal{P}$. If $u \in \overline{\mathcal{P}_{r_3}}$, then $\|u\| \leq r_3$. So by the assumption (C9), we obtain

$$\begin{aligned} Tu &\leq \lambda\left(\frac{1+\alpha}{1-\alpha\varphi_1(\eta)}\right)\int_{\rho(0)}^{\sigma(1)} G(s,s)h(s)f(s,u(s))\nabla s \\ &\leq \lambda\left(\frac{1+\alpha}{1-\alpha\varphi_1(\eta)}\right)\lambda^{-1}Mr_3\int_{\rho(0)}^{\sigma(1)} G(s,s)h(s)\nabla s = r_3. \end{aligned}$$

Hence $T : \overline{\mathcal{P}_{r_3}} \to \overline{\mathcal{P}_{r_3}}$. In the same way, if $u \in \overline{\mathcal{P}_{r_1}}$ assumption (C7) yields $||Tu|| < r_1$. Therefore (ii) of Theorem 2.2 is satisfied. To check the condition (i) of Theorem 2.2 we choose $u(t) = r_3$, $\forall t \in [\rho(0), \sigma(1)]$. It is clear that $u(t) = r_3 \in \mathcal{P}(\varphi, r_2, r_3)$. Consequently, since $\varphi(u) = \varphi(r_3) = r_3 > r_2$ then $\{u \in \mathcal{P}(\varphi, r_2, r_3) : \varphi(u) > r_2\} \neq \emptyset$. Moreover by using assumption (C8) and Lemma 2.5 we obtain

$$\begin{aligned} \varphi(Tu) = \min_{t\in[\xi,w]} |Tu(t)| &\geq \lambda\gamma\int_{\xi}^{w} G(t_0,s)h(s)f(s,u(s))\nabla s \\ &\geq \lambda\gamma\lambda^{-1}Nr_2\int_{\xi}^{w} G(t_0,s)h(s)\nabla s = r_2. \end{aligned}$$

Therefore (i) of Theorem 2.2 holds. Since all the conditions of Theorem 2.2 hold Tu has at least three fixed points u_1, u_2 and u_3, satisfying $||u_1|| < r_1$, $r_2 < \min_{t\in[\xi,w]} |u_2(t)| \leq r_3$, $r_1 < ||u_3|| \leq r_3$ and $\min_{t\in[\xi,w]} |u_3(t)| < r_2$. □

References

1. D. R. Anderson, *J. Difference Equations*, **47**, 1-12 (2004).
2. R. I. Avery and J. Henderson, *Comm. Appl. Nonlinear Anal.*, **8**, 27-36 (2001).
3. M. Bohner and A. Peterson, *Dynamic Equations on time scales, An Introduction with Applications*, Birkhäuser, Boston, 2001.
4. M. Bohner and A. Peterson, *Advances in Dynamic Equations on Time Scales*, Birkhäuser, Boston, 2003.
5. D. Guo and V. Lakshmikantham, *Nonlinear Problems in Abstract Cones*, Academic Press, San Diego, 1988.
6. C. P. Gupta, *J. Math. Anal. Appl.*, **168**, 540-551 (1992).
7. C. P. Gupta, *Appl. Math. Comp.*, **89**, 133-146 (1998).
8. V. A. Il'in and E. I. Moiseev, *Differential Equations*, **23**, 803-810 (1987).
9. W. Jiang and Y. Guo, *J. Math. Anal. Appl.*, **327**, 415-424 (2007).
10. M. A. Krasnoselskii, *SIAM Review*, **8**, 122-123 (1964).
11. R. W. Legget and L. R. Williams, *Indiana Univ. Math. J.*, **28**, 673-688 (1979).
12. R. Ma and H. Wang, *J. Math. Anal. Appl.*, **279**, 216-227 (2003).
13. H.-R. Sun and W.-T. Li, *J. Math. Anal. Appl.*, **299**, 508-524 (2004).
14. İ. Yaslan, *J. Comput. Appl. Math.*, **206**, 888-897 (2007).

Session 13

Reproducing Kernels and Related Topics

SESSION EDITORS

D. Alpay	Ben-Gurion University of the Negev, Be'er-Sheva, Israel
A. Berlinet	Université des Sciences et Techniques de Languedoc, Montpellier, France
S. Saitoh	Gunma University, Kiryu, Japan
D.-X. Zhou	City University of Hong Kong, Kowloon, China

REPRODUCING KERNELS FOR HARMONIC FUNCTIONS ON SOME BALLS

K. FUJITA

Faculty of Culture and Education, Saga University,
Saga, 840-8502, Japan
E-mail: keiko@cc.saga-u.ac.jp

In our previous papers we have proved a theorem on harmonic extension of (complex) harmonic functions on "N_p-balls" by using the "harmonic Bergman kernel". Here we consider the "Cauchy integral representation" and give another proof of the theorem on harmonic extension by using the "Cauchy integral kernel".

Keywords: Reproducing kernel; Harmonic function; Integral representation.

Introduction

We have studied holomorphic functions and (complex) harmonic functions on "N_p-balls" (for the definition see Section 1.1). In [4], we gave the "Cauchy integral representation" for harmonic functions on the N_∞-ball (Lie ball) whose integral is taken over a boundary of the "complex light cone" (Section 2.1). Then we have considered the "harmonic Bergman kernel" for harmonic functions on the N_p-ball. By using the harmonic Bergman kernel, in [2], we proved that harmonic functions on a N_p-ball in $\mathbf{C}^2$ can be continued harmonically to a larger Lie ball, then in [3], we proved that result in general dimension (Section 1.2).

In this paper, we will consider the Cauchy integral representation for harmonic functions on the N_p-ball and will give another proof on harmonic continuation by using the "Cauchy kernel" (Section 2.2).

1. Harmonic functions on the N_p-ball

In this Section, we will introduce our definitions and some notation.

1.1. N_p-norm and N_p-ball

For $z=(z_1,\cdots,z_{n+1}), w=(w_1,\cdots,w_{n+1})\in \mathbf{C}^{n+1}$ we denote by

$$z\cdot w=z_1w_1+\cdots+z_{n+1}w_{n+1},\quad z^2=z\cdot z,\quad \|z\|^2=z\cdot\overline{z}.$$

Consider the function

$$N_p(z)=\left(\frac{\left(\|z\|^2+\sqrt{\|z\|^4-|z^2|^2}\right)^{p/2}+\left(\|z\|^2-\sqrt{\|z\|^4-|z^2|^2}\right)^{p/2}}{2}\right)^{1/p}.$$

Then, $N_p(z)$ is a norm on $\mathbf{C}^{n+1}$ if $p\geq 1$, and N_p is monotone increasing in p; that is, we have

$$N_1(z)\leq N_p(z)\leq N_q(z)\leq L(z),\quad 1\leq p<q,$$

where $L(z)=\lim_{p\to\infty}N_p(z)$ is the Lie norm:

$$L(z)=\sqrt{\|z\|^2+\sqrt{\|z\|^4-|z^2|^2}}.$$

Note that

$$N_1(z)=\sqrt{(\|z\|^2+|z^2|)/2}=L^*(z)=\sup\{|z\cdot w|;L(w)\leq 1\}$$

is the dual Lie norm, $N_2(z)=\|z\|$ is the complex Euclidean norm and

$$2^{-1/p}L(z)\leq N_p(z)\leq L(z). \tag{1}$$

For more details on the N_p-norm, see [5].

For $p\geq 1$, we define the N_p-ball $\tilde{B}_p(r)$ with radius r by

$$\tilde{B}_p(r)=\left\{z\in\mathbf{C}^{n+1};N_p(z)<r\right\},\tilde{B}(r)=\tilde{B}_\infty(r)=\left\{z\in\mathbf{C}^{n+1};L(z)<r\right\}.$$

By the definition, $\tilde{B}_p(r)$ is an open convex subset and balanced in $\mathbf{C}^{n+1}$. In particular, $\tilde{B}_p(r)$ is a domain of holomorphy in $\mathbf{C}^{n+1}$. By (1),

$$\tilde{B}(r)\subset\tilde{B}_p(r)\subset\tilde{B}(2^{1/p}r).$$

We denote the closed N_p-ball with radius r by

$$\tilde{B}_p[r]=\left\{z\in\mathbf{C}^{n+1};N_p(z)\leq r\right\},\quad \tilde{B}[r]=\tilde{B}_\infty(r)=\left\{z\in\mathbf{C}^{n+1};L(z)\leq r\right\}.$$

1.2. *Harmonic functions and harmonic Bergman kernel*

The Lie ball is closely related to the harmonic functions because harmonic functions on a real ball can be continued analytically to the Lie ball with the same radius.

We denote the complex Laplacian by

$$\Delta_z \equiv \left(\frac{\partial^2}{\partial z_1^2} + \cdots + \frac{\partial^2}{\partial z_{n+1}^2}\right).$$

If f satisfies the differential equation $\Delta_z f(z) = 0$, then we call f the (*complex*) *harmonic function*. We denote by $\mathcal{O}(\tilde{B}_p(r))$ the space of holomorphic functions on $\tilde{B}_p(r)$ equipped with the topology of uniform convergence on compact sets. Put

$$\mathcal{O}_\Delta(\tilde{B}_p(r)) = \left\{f \in \mathcal{O}(\tilde{B}_p(r)); \Delta_z f(z) = 0\right\}.$$

In [1], we considered the Bergman kernel on

$$H\mathcal{O}_\Delta(\tilde{B}_p(r)) = \left\{f \in \mathcal{O}_\Delta(\tilde{B}_p(r)); \int_{\tilde{B}_p(r)} |f(w)|^2 dV_{\tilde{B}_p(r)}(w) < \infty\right\},$$

where $dV_{\tilde{B}_p(r)}$ denotes the normalized Lebesgue measure on $\tilde{B}_p(r)$. We denote the Bergman kernel on $H\mathcal{O}_\Delta(\tilde{B}_p(r))$ by $B^\Delta_{p,r}(z,w)$ and call it *harmonic Bergman kernel*; that is, for $f \in H\mathcal{O}_\Delta(\tilde{B}_p(r))$ we have the following integral representation:

$$f(w) = \int_{\tilde{B}_p(r)} f(z)\overline{B^\Delta_{p,r}(z,w)} dV_{\tilde{B}_p(r)}(z), \quad w \in \tilde{B}_p(r).$$

Since $\tilde{B}_p(r)$ is a domain of holomorphy, there is a holomorphic function which can not be continued analytically outside of $\tilde{B}_p(r)$. However harmonic functions on $\tilde{B}_p(r)$, $1 \leq p < \infty$, can be continued harmonically outside of $\tilde{B}_p(r)$. By using the harmonic Bergman kernel we proved the following theorems in [2] and [3].

Theorem 1.1. *Let* $1 \leq p < \infty$ *and* $f \in H\mathcal{O}_\Delta(\tilde{B}_p(r))$. *Define*

$$F(w) = \int_{\tilde{B}_p(r)} f(z)\overline{B^\Delta_{p,r}(z,w)} dV_{\tilde{B}_p(r)}(z).$$

Then $F \in \mathcal{O}_\Delta(\tilde{B}(2^{1/p}r))$ *and* $F(w) = f(w)$ *on* $\tilde{B}_p(r)$.

Theorem 1.2. *Let* $1 \leq p < \infty$. *The restriction mapping* r_1 *gives the following linear topological isomorphism:*

$$r_1 : \mathcal{O}_\Delta(\tilde{B}(2^{1/p}r)) \xrightarrow{\sim} \mathcal{O}_\Delta(\tilde{B}_p(r)).$$

We remark that the inverse mapping of r_1 is given by the mapping

$$f \mapsto F(w) = \int_{\tilde{B}_p(r)} f(\rho z)\overline{B^{\Delta}_{p,r}(z/\rho, w)}dV_{\tilde{B}_p(r)}(z),$$

where $\rho < 1$ is sufficiently close to 1.

In the next section, we will give another proof of this theorem by using the Cauchy integral representation.

2. Cauchy integral representation for harmonic functions

Harmonic functions are also closely related to the complex light cone because harmonic functions can be represented by an integral which is taken over a boundary of the complex light cone.

2.1. *Complex light cone and Cauchy integral representation*

We denote the *complex light cone* by

$$\tilde{M} = \{z \in \mathbf{C}^{n+1}; z^2 = 0\}.$$

Put

$$\tilde{M}(r) = \{z \in \tilde{M};\ L(z) < r\}, \quad \tilde{M}[r] = \{z \in \tilde{M};\ L(z) \leq r\},$$
$$M_r = \{z \in \tilde{M};\ L(z) = r\}.$$

In [4], we proved the following theorems:

Theorem 2.1. *Let $f \in \mathcal{O}(\tilde{M}(r))$. For $\rho < 1$ sufficiently close to 1, we have*

$$f(w) = \int_{M_r} f(\rho z)\overline{K_r(z/\rho, w)}dM_r(z), \quad w \in \tilde{M}(\rho r),$$

where

$$K_r(z, w) = \frac{r^{2(n-1)}(r^2 + 2z \cdot \overline{w})}{(r^2 - 2z \cdot \overline{w})^n}$$

is the Cauchy kernel and dM_r is the normalized invariant measure on M_r. Furthermore, if we define

$$F(w) = \int_{M_r} f(\rho z)\overline{K_r(z/\rho, w)}dM_r(z), \quad w \in \tilde{B}(\rho r), \tag{2}$$

then $F \in \mathcal{O}_{\Delta}(\tilde{B}(\rho r))$.

Theorem 2.2. *The restriction mapping r_0 gives the following linear topological isomorphism:*

$$r_0 : \mathcal{O}_\Delta(\tilde{B}(r)) \xrightarrow{\sim} \mathcal{O}(\tilde{M}(r)). \tag{3}$$

The inverse mapping of (3) is given by the mapping $f \mapsto F$ in (2).

Corollary 2.1. *Let $f \in \mathcal{O}_\Delta(\tilde{B}(r))$. For $\rho < 1$ sufficiently close to 1, we have*

$$f(w) = \int_{M_r} f(\rho z)\overline{K_r(z/\rho, w)}dM_r(z), \quad w \in \tilde{B}(\rho r).$$

2.2. *Another proof on harmonic continuation*

Using the above theorems, we can give another proof of Theorem 1.2. Put

$$\tilde{M}_p(r) = \{z \in \tilde{M};\ N_p(z) < r\}, \quad \tilde{M}_p[r] = \{z \in \tilde{M};\ N_p(z) \le r\},$$
$$M_{p,r} = \{z \in \tilde{M};\ N_p(z) = r\}.$$

For $z \in \tilde{M}$, we have $N_p(z) = 2^{1/2-1/p}\|z\|$ and $L(z) = 2^{1/2}\|z\|$. Thus

$$\tilde{M}_p(r) = \tilde{M}(2^{1/p}r), \quad M_{p,r} = M_{2^{1/p}r}, \quad \mathcal{O}(\tilde{M}(2^{1/p}r)) = \mathcal{O}(\tilde{M}_p(r)).$$

Recall that we have $\tilde{B}(r) \subset \tilde{B}_p(r) \subset \tilde{B}(2^{1/p}r)$ and

$$\mathcal{O}_\Delta(\tilde{B}(r)) \supset \mathcal{O}_\Delta(\tilde{B}_p(r)) \supset \mathcal{O}_\Delta(\tilde{B}(2^{1/p}r)).$$

Lemma 2.1. *Let $f \in \mathcal{O}_\Delta(\tilde{B}_p(r))$. For $\rho < 1$ sufficiently close to 1, we have the following integral representation:*

$$f(w) = \int_{M_{p,r}} f(\rho z)\overline{K_{2^{1/p}r}(z/\rho, w)}dM_{p,r}(z), \quad w \in \tilde{B}_p(\rho r), \tag{4}$$

where $dM_{p,r} = dM_{2^{1/p}r}$.

When $p = \infty$, Lemma 2.1 is nothing but Corollary 2.1.

Proof. We denote by r_2 the restriction mapping

$$r_2 : \mathcal{O}_\Delta(\tilde{B}_p(r)) \longrightarrow \mathcal{O}(\tilde{M}_p(r)).$$

Let $f \in \mathcal{O}_\Delta(\tilde{B}_p(r))$. Since $\tilde{B}_p(r)|_{\tilde{M}} = M_{p,r}$, we have

$$f(w) = \int_{M_{p,r}} f(\rho z)\overline{K_{2^{1/p}r}(z/\rho, w)}dM_{p,r}(z), \quad w \in \tilde{M}_p(\rho r)$$

by Theorem 2.1. For $w \in \tilde{B}(2^{1/p}\rho r)$ define

$$F(w) = \int_{M_{p,r}} f(\rho z)\overline{K_{2^{1/p}r}(z/\rho, w)}dM_{p,r}(z), \tag{5}$$

then $F \in \mathcal{O}_\Delta(\tilde{B}(2^{1/p}\rho r))$ by Theorem 2.1. Since $\tilde{B}_p(r) \subset \tilde{B}(2^{1/p}r)$, $F = f$ on $\tilde{B}_p(r)$ by Corollary 2.1. Thus we have (4). □

Now for p with $1 \le p < \infty$, we will prove that the restriction mappings r_1 and r_2 in the following diagram give the linear topological isomorphisms:

$$\begin{array}{ccc} r_0 : \mathcal{O}_\Delta(\tilde{B}(2^{1/p}r)) & \xrightarrow{\sim} & \mathcal{O}(\tilde{M}(2^{1/p}r)) \\ \downarrow r_1 & & || \\ r_2 : \ \mathcal{O}_\Delta(\tilde{B}_p(r)) & \longrightarrow & \mathcal{O}(\tilde{M}_p(r)) \end{array}$$

Another proof of Theorem 1.2. Let $f \in \mathcal{O}_\Delta(\tilde{B}_p(r))$. We denote by α the mapping $f \mapsto F$ in (5). Noting that $\rho < 1$ in (5) is arbitrarily close to 1 and $M_{p,r} = M_{2^{1/p}r}$, we have $(r_1 \circ \alpha)(f) = f$ for $f \in \mathcal{O}_\Delta(\tilde{B}_p(r))$ and $(\alpha \circ r_1)(F) = F$ for $F \in \mathcal{O}_\Delta(\tilde{B}(2^{1/p}r))$ by Lemma 2.1 and Corollary 2.1. Thus we have

$$r_1 : \mathcal{O}_\Delta(\tilde{B}(2^{1/p}r)) \xrightarrow{\sim} \mathcal{O}_\Delta(\tilde{B}_p(r)).$$ □

Since $r_0 = r_2 \circ r_1$, by Theorems 2.2 and 1.2 we have

Corollary 2.2. *The restriction mapping r_2 gives the following linear topological isomorphism:*

$$r_2 : \mathcal{O}_\Delta(\tilde{B}_p(r)) \xrightarrow{\sim} \mathcal{O}(\tilde{M}_p(r)).$$

When $p = \infty$, Corollary 2.2 is nothing but Theorem 2.2.

3. Cauchy kernel and Szegö kernel

We denote by $C(\tilde{M}_p[r])$ the space of continuous functions on $\tilde{M}_p[r]$ and by $L^2(M_{p,r}, dM_{p,r})$ the space of square integrable functions on $M_{p,r}$ with respect to the normalized invariant measure $dM_{p,r}$ on $M_{p,r}$. Let $H\mathcal{O}(\partial\tilde{M}_p(r))$ be the space of closure in $L^2(M_{p,r}, dM_{p,r})$ of the restriction to $M_{p,r}$ of the elements of $\mathcal{O}(\tilde{M}_p(r)) \cap C(\tilde{M}_p[r])$:

$$H\mathcal{O}(\partial\tilde{M}_p(r)) = \overline{\left\{ f \in \mathcal{O}(\tilde{M}_p(r)) \cap C(\tilde{M}_p[r]); \int_{M_{p,r}} |f(w)|^2 dM_{p,r}(w) < \infty \right\}}.$$

As a corollary of Lemma 2.1, we have the following result.

Corollary 3.1. *The Cauchy kernel $K_{2^{1/p}r}(z, w)$ gives the Szegö kernel on $H\mathcal{O}(\partial\tilde{M}_p(r))$. That is, for $f \in H\mathcal{O}(\partial\tilde{M}_p(r))$ we have*

$$f(w) = \int_{M_{p,r}} f(z)\overline{K_{2^{1/p}r}(z, w)}\, dM_{p,r}(z), \quad w \in \tilde{M}_p(r).$$

References

1. K. Fujita, *Harmonic Bergman kernel for some balls*, Univ. Iagel. Acta Math. **41** (2003), 225–234.
2. K. Fujita, *Topics on the Bergman kernel for some balls*, Proceedings of 5th International ISAAC Congress, to appear.
3. K. Fujita, *Reproducing kernels for holomorphic functions on some balls related to the Lie ball*, Ann. Polon. Math. **91** (2007), 219–234.
4. M. Morimoto and K. Fujita, *Analytic functionals and entire functionals on the complex light cone*, Hiroshima Math. J. **25** (1995), 493-512.
5. M. Morimoto and K. Fujita, *Between Lie norm and dual Lie norm*, Tokyo J. Math. **24** (2001), 499–507.

NUMERICAL REAL INVERSION OF THE LAPLACE TRANSFORM BY USING A HIGH-ACCURACY NUMERICAL METHOD

H. FUJIWARA

Graduate School of Informatics,
Kyoto University, Kyoto, 606-8501, Japan
E-mail: fujiwara@acs.i.kyoto-u.ac.jp

T. MATSUURA

Department of Mechanical System Engineering,
Graduate School of Engineering,
Gunma University, Kiryu, 376-8515, Japan
E-mail: matsuura@me.gunma-u.ac.jp

S. SAITOH

Department of Mathematics, Graduate School of Engineering,
Gunma University, Kiryu, 376-8515, Japan
E-mail: ssaitoh@math.sci.gunma-u.ac.jp & kbdmm360@yahoo.co.jp

Y. SAWANO

Department of Mathematics and Information Sciences
Tokyo Metropolitan University,
Minami-Ohsawa 1-1, Hachioji-shi,
Tokyo, 192-0397, Japan
E-mail: yosihiro@tmu.ac.jp

We introduce our three types of numerical inversion formulas of the Laplace transform based on a Fredholm integral equation of the second kind, the Sinc functions (the Sinc method) and the numerical singular value decomposition of the Laplace transform in a certain reproducing kernel Hilbert space. However, we will need the power of the high-accuracy numerical method with multiple-precision arithmetic for difficult situations. We shall give computational experiments for some very difficult situations for the real inversion.

Keywords: Laplace transform, real inversion, numerical inversion, reproducing kernel, Fredholm integral equation, sinc method, compact operator, singular value decomposition, computational experiment, high-accuracy numerical method, multiple-precision.

1. Introduction

We introduce very natural and numerical real inversion formulas of the Laplace transform

$$(\mathcal{L}F)(p) = f(p) = \int_0^\infty e^{-pt}F(t)dt, \quad p > 0 \tag{1}$$

for functions F in some natural function space. The real inversion is very involved and see the recent related books [2] and [6].

We use the reproducing kernel Hilbert space H_K comprised of absolutely continuous functions F on the positive real line $\mathbf{R}^+$ with finite norms

$$\left\{\int_0^\infty |F'(t)|^2 \frac{1}{t} e^t dt\right\}^{1/2}$$

and satisfying $F(0) = 0$. This Hilbert space admits the reproducing kernel

$$K(t,t') = \int_0^{\min(t,t')} \xi e^{-\xi} d\xi \tag{2}$$

(see [11, pp. 55-56]). Then we see that the linear operator $(\mathcal{L}F)(p)p$ on H_K into $L_2(\mathbf{R}^+, dp) = L_2(\mathbf{R}^+)$ is bounded [12]. Therefore, we obtain the following result.

Proposition 1.1. *([12]) For any $g \in L_2(\mathbf{R}^+)$ and for any $\alpha > 0$, the best approximation $F^*_{\alpha,g}$ in the sense*

$$\inf_{F \in H_K} \left\{\alpha \int_0^\infty |F'(t)|^2 \frac{1}{t} e^t dt + \|(\mathcal{L}F)(p)p - g\|^2_{L_2(\mathbf{R}^+)}\right\}$$

exists uniquely and we obtain the representation

$$F^*_{\alpha,g}(t) = \int_0^\infty g(\xi)\left(\mathcal{L}K_\alpha(\cdot,t)\right)(\xi)\xi d\xi. \tag{3}$$

Here, $K_\alpha(\cdot,t)$ is determined by the functional equation

$$K_\alpha(t,t') = \frac{1}{\alpha}K(t,t') - \frac{1}{\alpha}((\mathcal{L}K_{\alpha,t'})(p)p, (\mathcal{L}K_t)(p)p)_{L_2(\mathbf{R}^+)}, \tag{4}$$

where

$$K_{\alpha,t'} = K_\alpha(\cdot,t') \qquad \textit{and} \qquad K_t = K(\cdot,t).$$

2. By a Conventional Method for the Fredholm Integral Equation

We take the Laplace transform of (4) in t and change the variables t and t' as in

$$(\mathcal{L}K_\alpha(\cdot,t))(\xi) = \frac{1}{\alpha}(\mathcal{L}K(\cdot,t))(\xi) - \frac{1}{\alpha}((\mathcal{L}K_{\alpha,t})(p)p, (\mathcal{L}(\mathcal{L}K_\cdot)(p)p))(\xi))_{L_2(\mathbf{R}^+)}. \qquad (5)$$

Therefore, by setting $(\mathcal{L}K_\alpha(\cdot,t))(\xi)\xi = H_\alpha(\xi,t)$, we obtain the Fredholm integral equation of the second type

$$\alpha H_\alpha(\xi,t) + \int_0^\infty H_\alpha(p,t)\frac{1}{(p+\xi+1)^2}dp = -\frac{e^{-t\xi}e^{-t}}{\xi+1}\left(t+\frac{1}{\xi+1}\right) + \frac{1}{(\xi+1)^2}. \qquad (6)$$

We gave a numerical experiment for the typical example

$$F_0(t) = \begin{cases} -te^{-t} - e^{-t} + 1 & \text{for} \quad 0 \le t \le 1 \\ 1 - 2e^{-1} & \text{for} \quad 1 \le t, \end{cases}$$

whose Laplace transform is

$$(\mathcal{L}F_0)(p) = \frac{1}{p(p+1)^2}\left[1 - (p+2)e^{-(p+1)}\right]. \qquad (7)$$

For fixed t, we calculated the integral (6) over $[0, 50]$ with span 0.01 by the Gauss-Kronrod method. For t, we took the values over $[0, 5]$ with span 0.01. By (3), we calculated the inversion by the Gauss-Kronrod method and we obtained quite good results in [8].

3. By the Sinc Method for the Fredholm Integral Equation

In order to solve the integral equation (6), numerically, we employ the sinc method. See [7] for the basic relations of the sampling theory and the theory of reproducing kernels. We set

$$\frac{1}{\pi(z-\overline{u})}\sin\frac{\pi}{h}(z-\overline{u}) =: \frac{1}{h}\operatorname{sinc}\left(\frac{z-\overline{u}}{h}\right)$$

by the notations in [13]. We shall use the double exponential transform following the idea [14]

$$\xi = \varphi(x) = \exp(\frac{\pi}{2}\sinh x), \quad \varphi'(x) = \frac{\pi}{2}\cosh x \exp(\frac{\pi}{2}\sinh x).$$

Then, we have $H_\alpha(\xi,t) = H_\alpha(\varphi(x),t) = \widetilde{H}_\alpha(x,t)$, and so

$$\widetilde{H}_\alpha(x,t) = \sum_{j=-\infty}^{j=\infty} \widetilde{H}_\alpha(jh,t)\,\mathrm{sinc}(\frac{x}{h}-j)$$

and

$$\alpha\widetilde{H}_\alpha(x,t) + \int_{-\infty}^{\infty} \widetilde{H}_\alpha(z,t)\frac{1}{(\varphi(z)+\varphi(x)+1)^2}\varphi'(z)dz = f(\varphi(x),t). \quad (8)$$

We approximate as

$$\widetilde{H}_\alpha(x,t) \simeq \sum_{n=-N}^{n=N} \widetilde{H}_\alpha(jh,t)\,\mathrm{sinc}(\frac{x}{h}-j).$$

Then, by setting

$$A_{lk} \equiv \frac{1}{(\varphi(kh)+\varphi(lh)+1)^2}\varphi'(kh)h,$$

we obtain the equation

$$\alpha\widetilde{H}_\alpha(lh,t) + \sum_k \widetilde{H}_\alpha(kh,t)A_{lk} = f(\varphi(lh),t) \equiv \tilde{f}(lh,t) \quad (9)$$

and the representation

$$F^*(t) = \int_0^{\infty} g(\xi)H_\alpha(\xi,t)d\xi = \int_{-\infty}^{\infty} g(\varphi(x))H_\alpha(\varphi(x),t)\varphi'(x)dx$$
$$\simeq h\sum_i g(\varphi(ih))\widetilde{H}_\alpha(ih,t)\varphi'(ih).$$

We used $h = 0.05$ and for t, we took the span with 0.01. For the simultaneous equations (9), we took from $l = -200$ to $l = 799$, that is, 1000 equations. We solved such equations for $[0, 4.99]$ with the span 0.01 for t. For the numerical experiments, see [9].

4. By Using a Singular System

The singular value decomposition is applicable not only for reconstruction of solutions, but also for Hilbert scales and noise reduction of measurement data. See, for example, [6]. Though the singular value decomposition has various applications, its concrete treatments are hard both mathematically [10] and numerically [4].

We first recall that for the operator $Lf(p) = p(\mathcal{L}f)(p)$, $f \in H_K \to Lf \in L^2(\mathbf{R}^+)$ is an injective and compact linear operator [5]. Since the operator $Lf(p)$ has the singular system, let $\{\lambda_n\}$ be singular values of the operator L,

$\{v_n\}$ and $\{u_n\}$ be complete orthonormal systems of $\mathcal{N}(L)^{\perp}$ (the orthogonal complement of the null space) and $\overline{\mathcal{R}(L)}$ (the closure of the range space), respectively satisfying $Lv_n = \lambda_n u_n$ and $L^* u_n = \lambda_n v_n$. Then we obtain the following real inversion formula (see [6]).

Proposition 4.1. *We consider the Laplace transform $\mathcal{L}f = F$. If the original function f belongs to H_K, then the real inversion of the Laplace transform $\mathcal{L}^{-1}$ is given by*

$$(\mathcal{L}^{-1}F)(t) = \sum_{n=1}^{\infty} \frac{1}{\lambda_n} \left(\int_0^{\infty} F(p)p\, u_n(p)\, dp \right) v_n(t),$$

in the sense of norm convergence in H_K.

Proposition 4.1 implies that for any F with $F(p)p \in L^2(\mathbf{R}^+)$ and a natural number M, spectral cut-off regularization is given by

$$(\mathcal{L}_M^{-1}F)(t) = \sum_{n=1}^{M} \frac{1}{\lambda_n} \left(\int_0^{\infty} F(p)p\, u_n(p)\, dp \right) v_n(t).$$

It is an approximation of the original function in H_K by a function that belongs to H_K, and for $f \in H_K$ we have the convergence $\lim_{M\to\infty} \mathcal{L}_M^{-1}F = f$ in H_K.

5. Numerical Examples

Now we present a discretization of the singular value decomposition of L. Employing a numerical integration scheme with a discretization parameter N, quadrature points x_i and weights $w_i, 0 \le i \le N$, i.e.,

$$\int_0^{\infty} f(x)dx \approx \sum_{i=0}^{N} f(x_i)w_i,$$

we discretize the eigenvalue problem and obtain a linear eigenvalue problem of a matrix whose (i,j)-entry is, for $0 \le j, k \le N$

$$a_{ij} = \sum_{k=0}^{N} \frac{x_k e^{-x_j x_k}}{(x_k+1)^2} \left\{ 1 - e^{-x_i(x_k+1)} \{x_i(x_k+1)+1\} \right\} w_j w_k.$$

Here we suppose that $\tilde{v}_{n,i}$ corresponds to $v_n(x_i)$ and that they satisfy the following normalization condition implied by $\|u_n : L^2(\mathbf{R}^+)\| = 1$:

$$\sum_{j=0}^{N} |\tilde{u}_{n,j}|^2 w_j = 1,$$

where $\tilde{u}_{n,j}$ corresponds to $u_n(x_j)$ and is given by

$$\tilde{u}_{n,j} = \frac{1}{\tilde{\lambda}_n} x_j \sum_{k=0}^{N} \tilde{v}_{n,k} e^{-x_j x_k} w_k.$$

In analogy with the Nyström method, the discretized singular system $\{\tilde{\lambda}_n, \tilde{v}_{n,j}, \tilde{u}_{n,j}\}$ gives approximations of singular functions as

$$u_n^{(N)}(p) = \frac{1}{\tilde{\lambda}_n} p \sum_{k=0}^{N} \tilde{v}_{n,k} e^{-p x_k} w_k,$$

$$v_n^{(N)}(t) = \frac{1}{\tilde{\lambda}_n} \sum_{k=0}^{N} \frac{\tilde{u}_{n,k}}{(x_k+1)^2} \left\{1 - e^{-t(x_k+1)}\{t(x_k+1)+1\}\right\} w_k.$$

The numerical real inversion formula with spectral cut-off regularization is given by

$$(\mathcal{L}_{M,N}^{-1} F)(t) = \sum_{n=1}^{M} \frac{1}{\tilde{\lambda}_n} \left(\sum_{k=0}^{N} F(x_k) x_k \, \tilde{u}_{n,k} \, w_k \right) v_n^{(N)}(t). \qquad (10)$$

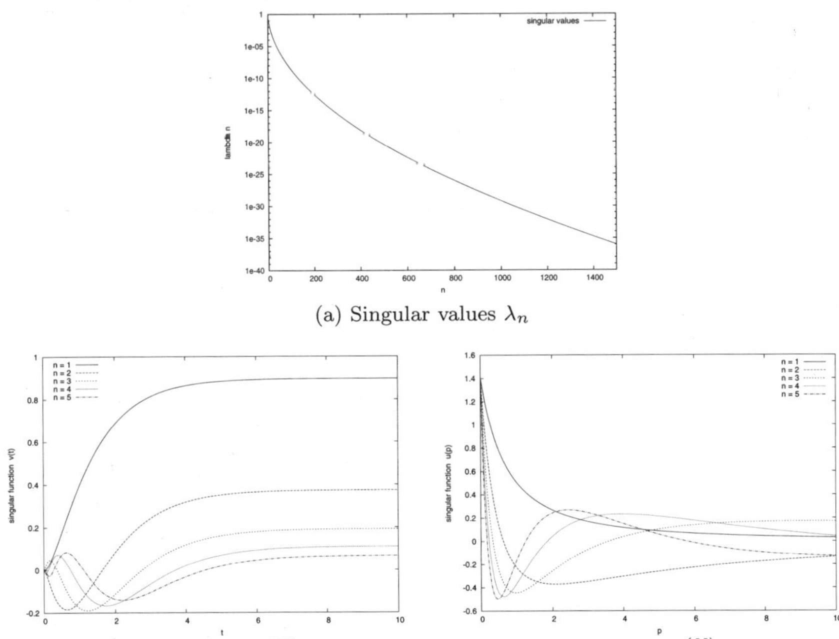

(a) Singular values λ_n

(b) Singular function $v_n^{(N)} \in H_K$

(c) Singular function $u_n^{(N)} \in L^2(\mathbf{R}^+)$

Fig. 1. Numerical singular system for the operator L

Fig. 1 shows numerical singular systems obtained in 400 decimal-digit computation by multiple-precision arithmetic software 'exflib' designed by one of the authors ([3]). In the computation, we adopt the the double exponential rule with the discretization parameter $N = 12000$, and we solve the linear eigenvalue problem by the Householder transformation and the QL method.

Example 1 We consider the function

$$f(t) = \begin{cases} t, & \text{if } 0 \le t < 1; \\ \frac{3}{2} - \frac{1}{2}t, & \text{if } 1 \le t < 3; \\ 0, & \text{if } t \ge 3, \end{cases}$$

whose Laplace transform is

$$F(p) = \frac{1}{2p^2}(2 - 3e^{-p} + e^{-3p}).$$

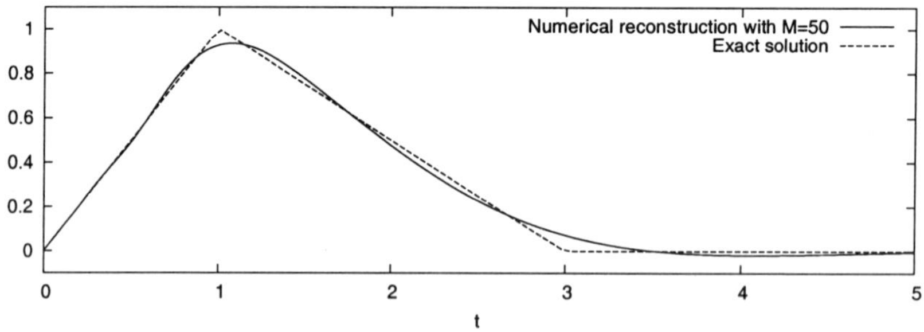

Fig. 2. Numerical real inversion for a continuous function (Example 1)

Fig. 2 shows the numerical real inversion f_M for the spectral number $M = 50$ through the numerical singular value decomposition of L. In this example the original function f belongs to H_K, and the proposed method gives good approximations. See [6], in particular, Fig. 2.2 and Fig. 2.3, where the truncated singular value decomposition for the Laplace transform operator $\mathcal{L}$ appears.

Example 2 The next example is the case when f is a characteristic function of $[\frac{1}{2}, 1] \cup [\frac{3}{2}, \frac{5}{2}] \cup [3, \frac{7}{2}]$. Fig. 3 is the numerical spectral cut-off regularized solutions. The dotted curve is that for $M = 300$, and the solid curve is that for $M = 1500$. We remark that the singular value $\lambda_{300} \approx 3 \times 10^{-16}$ in our numerical computations, and this suggests the $M = 300$ is a limitation in the standard double precision numerical computation. In this case the high-accurate numerical computation is required to large spectral cut-off parameters M.

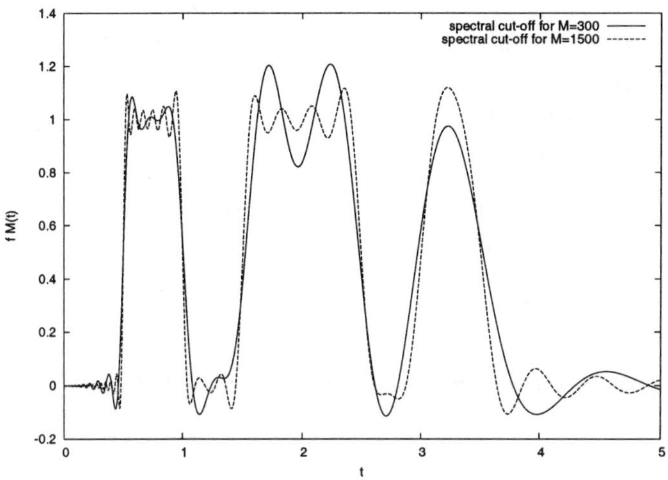

Fig. 3. Numerical real inversion for a characteristic function (Example 2)

Example 3 Finally we consider $F(p) = \exp(-p)$, which is the Laplace transform of the Dirac delta δ_{t-1} in the distribution sense. Our scheme produces Fig. 4 in which the dotted curve and the solid curve show spectral cut-off solutions f_{300} and f_{1500}, respectively. We can realize an effective numerical real inversion by the high-accuracy numerical computation technique and large spectral cut-off parameters.

Acknowledgements

H. Fujiwara is supported in part by the Grant-in-Aid for Young Scientists (B) (No. 17740057). T. Matsuura is supported in part by the Grant-in-Aid for Scientific Research (C)(No.18540110) from the Japan Society for the Promotion Science. S. Saitoh is supported in part by the Grant-in-Aid for Scientific Research (C)(2)(No. 19540164) from the Japan Society for the Promotion of Science. Y. Sawano is supported by Research Fellowships of the Japan Society for the Promotion of Science for Young Scientists.

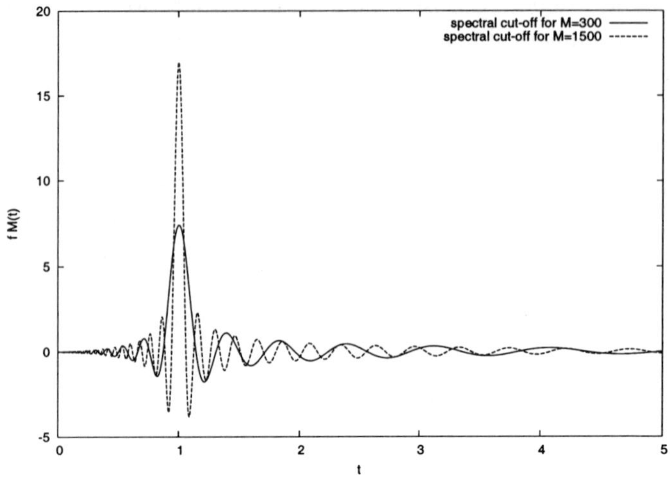

Fig. 4. Numerical real inversion for the Dirac delta (Example 3)

References

1. K. T. Atkinson, *The numerical solution of the eigenvalue problem for compact integral operators*, Trans. Amer. Math. Soc. **129** (1967), 458–465.
2. A. M. Cohen, *Numerical methods for Laplace transform inversion*, Springer, 2007.
3. H. Fujiwara, exflib, a multiple-precision arithmetic software, `http://www-an.acs.i.kyoto-u.ac.jp/~fujiwara/exflib`
4. H. Fujiwara, *High-accurate numerical method for integral equation of the first kind under multiple-precision arithmetic*, Theoret. Appl. Mech. Japan **52** (2003), 192–203.
5. H. Fujiwara, T. Matsuura, S. Saitoh and Y. Sawano, *Real inversion of the Laplace transform in numerical singular value decomposition*, J. Anal. Appl. **6** (2008), 55–68.
6. J. P. Kaipio and E. Somersalo, *Statistical and computational inverse problems*, Springer, 2004.
7. T. Matsuura and S. Saitoh, *Analytical and numerical inversion formulas in the Gaussian convolution by using the Paley-Wiener spaces*, Appl. Anal. **85** (2006), 901–915.
8. T. Matsuura, A. Al-Shuaibi, H. Fujiwara and S. Saitoh, *Numerical real inversion formulas of the Laplace transform by using a Fredholm integral equation of the second kind*, J. Anal. Appl. **5** (2007), 123–136.

9. T. Matsuura, A. Al-Shuaibi, H. Fujiwara, S. Saitoh and M. Sugihara, *Numerical real inversion of the Laplace transform by using the sinc functions*, Far East J. Math. Sci. **27** (2007), 1–14.
10. S. G. Mikhlin and K. L. Smolitskiy, *Approximate methods for solution of differential and integral equations*, American Elsevier, 1967.
11. S. Saitoh, *Integral transforms, reproducing kernels and their applications*, Pitman Res. Notes in Math. Series, **369**, Addison Wesley Longman, 1997.
12. S. Saitoh, *Approximate real inversion formulas of the Laplace transform*, Far East J. Math. Sci. **11** (2003), 53–64.
13. F. Stenger, *Numerical methods based on sinc and analytic functions*, Springer Series in Computational Mathematics, **20**, 1993.
14. H. Takahasi and M. Mori, *Double exponential formulas for numerical integration*, Publ. Res. Inst. Math. Sci. **9** (1974), 511–524.

PRACTICAL INVERSION FORMULAS FOR LINEAR PHYSICAL SYSTEMS

M. YAMADA and S. SAITOIH*

Graduate School of Engineering, Gunma University,
Kiryu 376-8515, Japan
**E-mail: ssaitoh@math.sci.gunma-u.ac.jp & kbdmm360@yahoo.co.jp*

We shall give a new inversion formula for a linear system based on physical experimental data and by using reproducing kernels and Tikhonov regularization. In particular, we will not make any analytical assumption on the linear system, but will use physical experimental data for obtaining an approximate inversion formula for the linear system.

Keywords: Reproducing kernel, linear system, inversion formula, physical system, Tikhonov regularization.

Introduction

Inverse problems in mathematics which are expected to be applied to practical problems will, sometimes, have weak points in the viewpoint that the background theories are not faithful for practical and physical problems. For example, equations are the representations of some ideal models and are not those of faithful models in the real physical world. Sometimes, boundary conditions for the equations are involved in physical units and sometimes their physical realizations and observations are very difficult. Here, we shall give a new inversion formula for a linear system based on physical experimental data and by using reproducing kernels and the Tikhonov regularization. In particular, we will not make any analytical assumption on the linear system, but we use physical experimental data for obtaining an approximate inversion formula for the linear system L. We shall apply the following fundamental theory for this purpose.

Let E be an arbitrary set, and let H_K be a reproducing kernel Hilbert space (RKHS) admitting the reproducing kernel $K(p,q)$ on E. For any Hilbert space $\mathcal{H}$ we consider a bounded linear operator L from H_K into $\mathcal{H}$.

We are generally interested in the best approximate problem

$$\inf_{f \in H_K} \|Lf - \mathbf{d}\|_{\mathcal{H}} \tag{1}$$

for a vector $\mathbf{d}$ in $\mathcal{H}$. However, this extremal problem is involved in the both senses of the existence of the extremal functions in (1) and their representations. See [7] for the details. So, we shall apply its Tikhonov regularization.

For any fixed positive $\alpha > 0$, by introducing the inner product

$$(f, g)_{H_K(L;\alpha)} = \alpha(f, g)_{H_K} + (Lf, Lg)_{\mathcal{H}}, \tag{2}$$

we shall construct the Hilbert space $H_K(L; \alpha)$ comprising functions of H_K. This space, of course, admits a reproducing kernel. Furthermore, we obtain the following fundamental result.

Proposition 0.1. *([8,5,6]) The extremal function $f_{\mathbf{d},\alpha}(p)$ in the Tikhonov regularization*

$$\inf_{f \in H_K} \{\alpha\|f\|^2_{H_K} + \|\mathbf{d} - Lf\|^2_{\mathcal{H}}\} \tag{3}$$

exists uniquely and it is represented in terms of the kernel $K_L(p, q; \alpha)$ as

$$f_{\mathbf{d},\alpha}(p) = (\mathbf{d}, LK_L(\cdot, p; \alpha))_{\mathcal{H}}, \tag{4}$$

where $K_L(p, q; \alpha)$ is the reproducing kernel for the Hilbert space $H_K(L; \alpha)$ and is determined as the unique solution $\tilde{K}(p, q; \alpha)$ of the equation

$$\tilde{K}(p, q; \alpha) + \frac{1}{\alpha}(L\tilde{K}_q, LK_p)_{\mathcal{H}} = \frac{1}{\alpha}K(p, q) \tag{5}$$

with

$$\tilde{K}_q = \tilde{K}(\cdot, q; \alpha) \in H_K \quad \textit{for} \quad q \in E, \tag{6}$$

and

$$K_p = K(\cdot, p) \in H_K \quad \textit{for} \quad p \in E.$$

The reproducing kernel K_L is, of course, represented in the operator form as

$$K_L(\cdot, p; \alpha) = (L^*L + \alpha I)^{-1}K(\cdot, p),$$

where L^* is the adjoint operator of L.

In (4), when $\mathbf{d}$ contains error or noise, we need its error estimate. For this, we can obtain the following general result.

Proposition 0.2. *([4],[5]). In (4), we have the estimate*

$$|f_{\mathbf{d},\alpha}(p)| \leq \frac{1}{\sqrt{\alpha}}\sqrt{K(p,p)}\|\mathbf{d}\|_{\mathcal{H}}.$$

Following Proposition 0.1, we shall look for the approximate inversion $f_{\mathbf{d},\alpha}(p)$ of the linear system

$$Lf = \mathbf{d}$$

for the bounded linear operator L from H_K into $\mathcal{H}$. Here, when α tends to zero, the approximate inversion $f_{\mathbf{d},\alpha}(p)$ tends to the Moore-Penrose generalized inverse of the operator equation in a good way, when it exists, and some detailed behaviour of the convergence is examined. For more details of the convergence rate or the results for noisy data, see ([2],[3]).

1. Approach looking for the inversion

Physically or by computers we can observe only discrete data, so, as a very general algorithm, we shall consider the discrete point data case. In (3), we shall consider the corresponding problem

$$\inf_{f\in H_K}\left\{\alpha\|f\|_{H_K}^2 + \sum_{j=1}^{N}|(Lf)(P_j) - d_j|^2\right\}, \tag{7}$$

for fixed discrete points $\{P_j\}_j$ of the set E and for given values $\mathbf{d} = \{d_j\}_j$; that is, $\mathcal{H}$ is the usual Euclidean space $\mathbf{R}^N$.

In order to use the representation (4), we need $LK_L(\cdot, p; \alpha))$ and it is determined by (5). In (5), we operate L as functions in p and we have

$$\alpha L_p\tilde{K}(p,q;\alpha) + L_p(L\tilde{K}_q, LK_p)_{\mathcal{H}} = L_pK(p,q). \tag{8}$$

Here, when we can take $\alpha = 0$ in the sense of numerical, we can take, of course, $\alpha = 0$ in those arguments.

As stated in Introduction, in many practical problems for the linear system, L is not given analytically and so, here, we shall assume that the system may be used many times, experimentally.

However, in order to use the previous section method, we must realize some physical objects as the N data $\mathbf{d} = \{d_j\}_j$, $N \times N$ values $L_pK(p,q)$ and $N \times N$ values L_pLK_p of real values; that is, f and $\mathbf{d} = \{d_j\}_j$ are numerical representations of some physical objects in the system $Lf = \mathbf{d}$.

Since the reproducing kernel Hilbert space H_K is the function space approximating the solution of the operator equation $Lf = \mathbf{d}$, we can take many simple reproducing kernel Hilbert spaces as in ([7]), however, from the present situation, the reproducing kernel $K(p,q)$ must be realized as the physical object for the present system.

2. Physical viewpoints

We see in our inversion formula (4) that we use a concrete reproducing kernel $K(p,q)$ through (5) but we do not use any Hilbert space structure of the reproducing kernel $K(p,q)$. By the theory of reproducing kernels, for any positive matrix there exists a uniuely determined reproducing kernel Hilbert space ([1,7]).

Furthermore, in our inversion formula (4), we are looking for approximations of the inversion in the function space H_K, so, in general, the space H_K is a sufficient large class of functions in the sense that we can approximate the inverse by the functions in H_K. For example, for any characteristic function on any interval, we can approximate it by the Sobolev Hilbert space of 1 dimension uniformly. This will mean that for the input, we can consider a suitable positive matrix, here, by a suitable positive matrix, we mean that the positive matrix may be realized as the physical data and it will also depend on its physical system.

In connection with these points of view, for example, for the 2 dimensional Sobolev space in ([7]), we shall use the more simple reproducing kernel

$$K(x_1, x_2, y_1, y_2) = \frac{1}{4}\exp(-|x_1 - y_1|)\exp(-|x_2 - y_2|), \tag{9}$$

which is the usual product of the 1 dimensional Sobolev reproducing kernels and its reproducing kernel Hilbert space is the tensor product of the two Hilbert spaces of the one dimensional Sobolev Hilbert space (see [1,7] for this structure).

We shall introduce several simple reproducing kernels on the whole real line space. Note here that for multidimensional spaces, we can consider the products as in (9). Furthermore, the restriction of a reproducing kernel to any subset is again a reproducing kernel. The sum and the usual product of two reproducing kernels on a same set are again reproducing kernels. For these elementary facts, see, ([1,7]). On the whole real space $\mathbf{R}$, the following are reproducing kernels:

(1) Any positive semidefinite matrix.
(2) $\delta(x-y)$ ($\delta(0) = 1$ and $\delta(x) = 0$ for $x \neq 0$).
(3) For any $\alpha > 0$, $\exp(-\alpha|x-y|)$.
(4) $\exp(\alpha xy)$ $\quad(\alpha > 0)$.
(5) $\exp(-\alpha(x-y)^2)$ $\quad(\alpha > 0)$.

(6) $\exp(-|x-y|)(1+|x-y|)$.
(7) $\min(x,y)$.
(8) For any $\alpha > 0$, $\dfrac{\sin(\alpha(x-y))}{x-y}$.

On the half space $\{x > 0\}$, the following are reproducing kernels:
(1) $\dfrac{1}{(x+y)^{2q}} \quad (q \geq \frac{1}{2})$.
(2) $\dfrac{1}{(x^2+y^2)^{2q}} \quad (q \geq \frac{1}{2})$.
(3) $\exp\{\min(x,y)\}$.

On the interval $\{-1 < x < 1\}$, the following are reproducing kernels:
(1) $\dfrac{1}{(1-xy)^{2q}} \quad (q \geq \frac{1}{2})$.
(2) $\log \dfrac{1}{1-xy}$.

Furthermore, note that any reproducing kernel $K(p,q)$ on an arbitrary set E for a separable reproducing kernel Hilbert space is represented in the form, for some functions $\{\varphi_j(p)\}$ on E

$$K(p,q) = \sum_j \varphi_j(p)\overline{\varphi_j(q)},$$

that converges absolutely on $E \times E$. Conversely, any function $K(p,q)$ which is represented in this way for arbitrary complex-valued functions $\{\varphi_j(p)\}$ on E is a reproducing kernel.

3. Exact algorithm

We shall state the exact algorithm looking for the extremal function $f_{\mathbf{d},\alpha}(p)$ in (4) clearly in the setting (7).

1) We set

$$\begin{aligned} X(P,q) &= (L_p\tilde{K}(p,q;\alpha))(P), \\ k(P,q) &= (L_pK(p,q))(P) \end{aligned} \tag{10}$$

and

$$\kappa(P,Q) = (L_qL_pK(p,q))(P,Q). \tag{11}$$

2) As the solution of the regular linear equations (8)

$$\alpha X(P_j,q) + \sum_{j'=1}^{N} X(P_{j'},q)\kappa(P_j,P_{j'}) = k(P_j,q); j = 1,2,...,N, \tag{12}$$

we determine $X(P_j, q)$. Then we obtain the approximate inverse

$$f_{\mathbf{d},\alpha}(p) = \sum_{j=1}^{N} d_j X(P_j, p). \tag{13}$$

Therefore, for some concrete problem for its inversion, we need the experimental data (10) and (11) of the two types in 1) and the procedure 2) is a mathematical problem.

By Proposition 0.2, we note that in (13), the following estimate holds:

$$|f_{\mathbf{d},\alpha}(p)| \leq \frac{1}{\sqrt{\alpha}} \sqrt{K(p,p)} \left(\sum_{j=1}^{N} d_j^2 \right)^{1/2}.$$

The simplest and the most typical case of the above algorithm is that the system L is any type matrix of type m and n (without loss of generality we assume that $n \geq m$), and the positive matrix is the identity matrix of size n.

Acknowledgements

S. Saitoh is supported in part by the Grant-in-Aid for Scientific Research (C)(2)(No. 19540164) from the Japan Society for the Promotion of Science.

References

1. N. Aronszajn, *Theory of reproducing kernels*, Trans. Amer. Math. Soc. **68** (1950), 337–404.
2. H. W. Engl, M. Hanke and A. Neubauer, *Regularization of Inverse Problems* Mathematics and Its Applications, **376**, Kluwer Academic Publishers, 2000.
3. C. W. Groetsch, *Inverse Problems in the Mathematical Sciences*, Vieweg & Sohn Verlagsgesellschaft mbH, Braunschweig/Wiesbaden, 1993.
4. H. Itou and S. Saitoh, *Analytical and numerical solutions of linear singular integral equations*, Int. J. Appl. Stat. **12** (2007), 77–89.
5. T. Matsuura and S. Saitoh, *Analytical and numerical inversion formulas in the Gaussian convolution by using the Paley-Wiener spaces*, Appl. Anal. **85** (2006), 901–915.
6. T. Matsuura, S. Saitoh and D. D. Trong, *Approximate and analytical inversion formulas in heat conduction on multidimensional spaces*, J. Inverse Ill-posed Probl. **13** (2005), 479–493.
7. S. Saitoh, *Integral Transforms, Reproducing Kernels and their Applications*, Pitman Res. Notes in Math. Series, **369**, Addison Wesley Longman, UK, 1997.
8. S. Saitoh, *Approximate real inversion formulas of the Gaussian convolution*, Appl. Anal. **83** (2004), 727–733.

Session 14

Spaces of Differentiable Functions of Several Real Variables and Applications

SESSION EDITORS

V. I. Burenkov   Cardiff University, Cardiff, Wales, UK
S. G. Samko   Universidade do Algarve, Faro, Portugal

WEIGHTED HÖLDER ESTIMATES OF SINGULAR INTEGRALS GENERATED BY A GENERALIZED SHIFT OPERATOR

SADIG K. ABDULLAYEV* and A. A. AKPEROV

Systematic investigation of multi-dimensional singular integrals generated by the generalized shift operator connected with the Laplace-Bessel differential equations started with the works of N. A. Kipriyanov and M. I. Klyuchantsev, where Privalov-type theorems were in particular proved. We study such integrals in the weighted Hölder spaces $H^{\gamma}_{\alpha\beta}$. We obtain sufficient conditions on the parameters α, β, and γ for the boundedness of such singular integral operators in these spaces. Note that we consider the case when the generalized shift is taken with respect to an arbitrary number of variables.

Keywords: Singular integral operator, generalized shift operator, weighted Hölder space.

1. Introduction

Let R_n be the Euclidean space of dimension $n \geq 1$,

$$R^+_{m+k,k} = \{(x_1, ..., x_m, x_{m+1}, ..., x_{m+k}) \in R_{m+k} : x_{m+1} > 0, ..., x_{m+k} > 0\},$$
$$S^+_{m+k,k} = \{x \in R^+_{m+k,k} : |x| = 1\},$$

and let

$$T^s u(x) = c_\nu \int_0^\pi ... \int_0^\pi u\left(x' - s', \sqrt{x^2_{m+1} - 2x_{m+1}s_{m+1}\cos\alpha_1 + s^2_{m+1}}, ...,\right.$$
$$\left.\sqrt{x^2_{m+k} - 2x_{m+k}s_{m+k}\cos\alpha_k + s^2_{m+k}}\right) \prod_{j=1}^{k} \sin^{2\nu_{m+j}-1} \alpha_j \, d\alpha_1 ... d\alpha_k \qquad (1)$$

be the generalized shift operator (GSO) generated by the Laplace-Bessel operator [5–7]

$$\Delta_B = \sum_{i=1}^{m} \frac{\partial^2}{\partial x_i^2} + \sum_{i=m+1}^{m+k} \left(\frac{\partial^2}{\partial x_i^2} + \frac{2\nu_i}{x_i}\frac{\partial}{\partial x_i}\right),$$

*Partially supported by INTAS (Grant 05-1000008-8157).

where $\nu_{m+1} > 0, \ldots, \nu_{m+k} > 0$, $c_\nu = \pi^{-k/2} \prod_{j=1}^{k} \Gamma(\nu_{m+j} + 1/2)\Gamma^{-1}(\nu_{m+j})$ is a normalized constant. We use the notation $x = (x', x_{m+1}, ..., x_{m+k})$, $s = (s', s_{m+1}, ..., s_{m+k})$, $x', s' \in R_m$.

We consider the singular integral

$$Au(x) = \lim_{\varepsilon \to 0+} \int\limits_{\left\{s \in R^+_{m+k,k} : |s| > \varepsilon\right\}} \frac{f(\theta)}{|s|^{m+k+2\nu}} [T^s u(x)] \cdot s_{m+1}^{2\nu_{m+1}} ... s_{m+k}^{2\nu_{m+k}} ds, \quad (2)$$

where $\theta = s/|s|$, $\nu = \nu_{m+1} + ... + \nu_{m+k}$.

2. Preliminaries

The starting point in solving the problem of the existence of singular integral (2) is the easy-to-check representation

$$\begin{aligned} &\int\limits_{\left\{s \in R^+_{m+k,k} : a < |s| < b\right\}} f\left(\frac{s}{|s|}\right) \cdot |s|^{-m-k-2\nu} [T^s u(x)] \cdot s_{m+1}^{2\nu_{m+1}} ... s_{m+k}^{2\nu_{m+k}} ds \\ &\qquad = \frac{c_\nu}{2} \int\limits_{\{y \in R_{m+2k} : a < |\tilde{x} - y| < b\}} f(\tilde{\theta}) \cdot |\tilde{x} - y|^{-m-k-2\nu} \\ &\qquad \times u\left(y', \sqrt{y_{m+1}^2 + \tilde{y}_{m+1}^2}, ..., \sqrt{y_{m+k}^2 + \tilde{y}_{m+k}^2}\right) \\ &\qquad \times |\tilde{y}_{m+1}|^{2\nu_{m+1}-1} \ldots |\tilde{y}_{m+k}|^{2\nu_{m+k}-1} dy, \end{aligned} \quad (3)$$

where $0 < a < b \le \infty$, $x \in R^+_{m+k,k}$, $\tilde{x} = (x', x_{m+1}, ..., x_{m+k}, 0, ..., 0)$, $y = (y', y_{m+1}, ..., y_{m+k}, \tilde{y}_{m+1}, ..., \tilde{y}_{m+k})$, $dy = dy_1 \cdots d\tilde{y}_{m+k}$, $r_{\tilde{x},y} = |\tilde{x} - y|$,

$$\tilde{\theta} = \left(\frac{x' - y'}{r_{\tilde{x},y}}, \frac{\sqrt{(x_{m+1} - y_{m+1})^2 + \tilde{y}_{m+1}^2}}{r_{\tilde{x},y}}, ..., \frac{\sqrt{(x_{m+k} - y_{m+k})^2 + \tilde{y}_{m+k}^2}}{r_{\tilde{x},y}}\right).$$

In the case $u(x) \equiv 1$, we get

$$\begin{aligned} &\int_{S^+_{m+k,k}} f(\theta)\, \theta_{m+1}^{2\nu_{m+1}} ... \theta_{m+k}^{2\nu_{m+k}}\, ds(\theta) \\ &= \frac{c_\nu}{2} \int_{S_{m+2k}} f(\tilde{\theta})\, |\tilde{\theta}_{m+1}|^{2\nu_{m+1}-1} ... \, |\tilde{\theta}_{m+k}|^{2\nu_{m+k}-1} ds\,(\theta), \end{aligned} \quad (4)$$

by passing to polar coordinates, where $S_{m+2k} = \{y \in R_{m+2k} : |y| = 1\}$.

As in the case of the classical singular integrals, the reduction condition

$$\int_{S^+_{m+k,k}} f(\theta) \prod_{j=1}^{k} \theta_{m+j}^{2\nu_{m+j}}\, ds(\theta) = 0 \quad (5)$$

is a necessary condition for the existence of singular integral (2).

Let $\Gamma = \{x \in R_{m+k} : x_{m+1} \dots x_{m+k} = 0\}$ be the boundary of $R^+_{m+k,k}$ and $r_\Gamma(x) = \min_{y\in\Gamma} |x-y|$. Let us take a weight function $\rho(x) = r^\alpha_\Gamma(x)\,(1+|x|)^{\beta-\alpha}$, $x \in R^+_{m+k,k}$, where $r^\alpha_\Gamma(x) = (\min\{x_{m+1},...,x_{m+k}\})^\alpha$, $\alpha > 0$, β is a real number. Introduce the weighted Hölder space $H^\gamma_{\alpha\beta}\left(R^+_{m+k,k}\right)$ [1–4]. Let $\gamma > 0$. By definition, $u \in H^\gamma_{\alpha\beta}\left(R^+_{m+k,k}\right)$ if $\lim\limits_{x\to z\in\Gamma} u(x)\rho(x) = 0$, $\lim\limits_{x\to\infty} u(x)\rho(x) = 0$ and the norm

$$\|u\|_{H^\gamma_{\alpha\beta}} = \sup_{x,y\in R^+_{m+k,k}} \left(|u(x)\rho\,(x) - u\,(y)\,\rho\,(y)| \cdot d^{-\gamma}\,(x,y)\right),$$

is finite, where $d\,(x,y) = |x-y| \cdot [(1+|x|)\cdot(1+|y|)]^{-1}$.

Further "c" is a constant whose exact value is unessential for us: $a(x) \prec b(x)$ means that $a(x) \le c\,b(x)$, where c is independent of x. In the case when $a(x) \prec b(x)$ and $b(x) \prec a(x)$, we write $a(x) \asymp b(x)$. Notice that it is convenient to determine the space $H^\gamma_{\alpha\beta}(R^+_{m+k,k})$ in the terms of inequalities as well.

Lemma 2.1. *If the inequalities*

$$0<\gamma<1,\ \ 0<\alpha-\gamma<1,\ \ 0<\beta+\gamma<m+k. \tag{6}$$

are true, then $u \in H^\gamma_{\alpha\beta}\left(R^+_{m+k,k}\right)$ if and only if

a) $\exists c_1(u) > 0$, $\forall x \in R^+_{m+k,k}$, $|u(x)| < c_1(u)\,r_\Gamma^{\gamma-\alpha}(x)\cdot(1+|x|)^{-\ell}$, where $\ell = 2\gamma+\beta-\alpha$;

b) $\exists c_2(u) > 0$, $\forall x \in R^+_{m+k,k}$, and $|h| \le \frac{r_\Gamma(x)}{8}$

$$|u(x) - u(x+h)| \le c_2(u)\,\rho^{-1}(x)\cdot(1+|x|)^{-2\gamma}\cdot|h|^\gamma.$$

In this case $\|u\| \asymp (\min c_1(u) + \min c_2(u))$.

Introduce the sets

$$\omega_x = \left\{s \in R^+_{m+k,k} : |s-x| < \frac{r_\Gamma\,(x)}{2}\right\},\ \omega'_x = \left\{y \in R_{m+2k} : |\tilde{x}-y| < \frac{r_\Gamma\,(x)}{2}\right\},$$

For further reasonings we need the following.

Corollary 2.1. *If $u \in H^\gamma_{\alpha\beta}\left(R^+_{m+k,k}\right)$, then*

$$a)\ \left|u\left(y', \sqrt{y^2_{m+1}+\tilde{y}^2_{m+1}}, ..., \sqrt{y^2_{m+k}+\tilde{y}^2_{m+k}}\right)\right| \prec \frac{\|u\|_{H^\gamma_{\alpha\beta}}\,(A\,(y,\tilde{y}))^{\gamma-\alpha}}{(1+|y'|+B\,(y,\tilde{y}))^{l}},$$

where $A(y,\tilde{y}) = \min\left\{|y_{m+i}|+|\tilde{y}_{m+i}|,\ i=\overline{1,k}\right\}$ *and* $B(y,\tilde{y}) = \sum\limits_{i=1}^{k}\left(|y_{m+i}|+|\tilde{y}_{m+i}|\right)$.

b) $\forall x\in R^+_{m+k,k}$, $\forall y\in\omega'_x$,

$$\left|u\left(y',\sqrt{y^2_{m+1}+\tilde{y}^2_{m+1}},\ldots,\sqrt{y^2_{m+k}+\tilde{y}^2_{m+k}}\right)-u(x)\right|\prec\frac{\|u\|_{H^\gamma_{\alpha\beta}}d^\gamma(\tilde{x},y)}{\rho(x)}.$$

3. Main result

In this paper we consider a problem on the boundedness of a singular integral operator $A: u\mapsto Au$ in the spaces $H^\gamma_{\alpha\beta}\left(R^+_{m+k,k}\right)$, where $Au(x)$ is determined by formula (2).

Theorem 3.1. *Let* $f(\theta)$ *be bounded and suppose* $\exists C(f)>0$, $\forall\theta_1,\theta_2\in S^+_{m+k,k}$, $|f(\theta_1)-f(\theta_2)|\le C_f|\theta_1-\theta_2|^\delta$, $\delta>0$, *and*

$$\int_{S^+_{m+k,k}} f(\theta)\theta^{2\nu_{m+1}}_{m+1}\ldots\theta^{2\nu_{m+k}}_{m+k}\,ds(\theta)=0.$$

Then

1) for each $u\in H^\gamma_{\alpha\beta}(R^+_{m+k,k})$ *singular integral exists for all* $x\in R^+_{m+k,k}$;

2) if conditions (5) and $\gamma<\delta$ *are fulfilled, then a singular operator* A *is bounded in* $H^\gamma_{\alpha\beta}(R^+_{m+k,k})$.

Proof. Let $x\in R^+_{m+k,k}$. Then allowing for the reduction

$$\int_{S^+_{m+k,k}} f(\theta)\theta^{2\nu_{m+1}}_{m+1}\ldots\theta^{2\nu_{m+k}}_{m+k}\,ds(\theta)=0,$$

we have $|Au(x)|\le c\left(|\tau_1(x;\omega'_x)|+|\tau_2(x;R_{m+2k}\backslash\omega'_x)|\right)$, where

$$\tau_1(x;B)=\int_B\frac{f(\tilde{\theta})}{r^{m+k+2\nu}_{\tilde{x},y}}\left[u\left(y',\sqrt{y^2_{m+1}+\tilde{y}^2_{m+1}},\ldots,\sqrt{y^2_{m+k}+\tilde{y}^2_{m+k}}\right)-u(x)\right]$$
$$\times\left|\tilde{y}_{m+1}\right|^{2\nu_{m+1}-1}\ldots\left|\tilde{y}_{m+k}\right|^{2\nu_{m+k}-1}dy\,;$$

$$\tau_2(x;B)=\int\limits_{(R_{m+2k}\backslash\omega'_x)\cap B}\frac{f(\tilde{\theta})}{r^{m+k+2\nu}_{\tilde{x},y}}u\left(y',\sqrt{y^2_{m+1}+\tilde{y}^2_{m+1}},\ldots,\sqrt{y^2_{m+k}+\tilde{y}^2_{m+k}}\right)$$
$$\times\left|\tilde{y}_{m+1}\right|^{2\nu_{m+1}-1}\ldots\left|\tilde{y}_{m+k}\right|^{2\nu_{m+k}-1}dy\,,$$

and $B\subset\omega'_x$.

Taking into account inequalities $b)$ of Corollary 2.1 we get

$$|\tau_1(x;\omega'_x)|\prec\|f\|\|u\|_{H^\gamma_{\alpha\beta}}\rho(x)^{-1}(1+|x|)^{-2\gamma}$$

$$\times \int_B r_{\tilde{x},y}^{-(m+k+2\nu-\gamma)} |\tilde{y}_{m+1}|^{2\nu_{m+1}-1} \dots |\tilde{y}_{m+k}|^{2\nu_{m+k}-1} \, dy$$

whence

$$|\tau_1(x;\omega_x')| \prec \|f\| \cdot \|u\|_{H^\gamma_{\alpha\beta}} \cdot r_\Gamma^{\gamma-\alpha}(x) \cdot (1+|x|)^{-\ell}. \tag{7}$$

Now, let us estimate from above $\tau_2(x, R_{m+2k} \backslash \omega_x')$. Introduce the sets

$$A_x = \left\{ y \in R_{m+2k} : |\tilde{x} - y| \le \frac{|y|}{2} \right\} \cap (R_{m+2k} \backslash \omega_x')$$

$$B_x = \left\{ y \in R_{m+2k} : \frac{|y|}{2} \le |\tilde{x} - y| \le \frac{3|y|}{2} \right\} \cap (R_{m+2k} \backslash \omega_x'),$$

and $C_x = \{y \in R_{m+2k} : 3|y| \le |\tilde{x} - y|\} \cap (R_{m+2k} \backslash \omega_x')$. Obviously

$$|\tau_2(x, R_{m+2k} \backslash \omega_x')| \prec \|f\| \, \|u\|_{H^\gamma_{\alpha\beta}} (\tau_2(x; A_x) + \tau_2(x; B_x) + \tau_2(x; C_x)). \tag{8}$$

First of all we note that $|\tilde{x} - y| \ge r_\Gamma(x)$ for $y \in R_{m+2k} \backslash \omega_x'$. Now estimate from above the integrals at the right hand side of (7). Note that if $y \in A_x$, then $(1+|x|) \asymp (1+|y|)$ and

$$|\tilde{x} - y| \asymp |x' - y'| + r_\Gamma(x) + \sum_{i=1}^k |x_{m+i} - y_{m+i}| + |\tilde{y}_{m+1}| + \dots + |\tilde{y}_{m+k}|.$$

Then

$$\begin{aligned} \tau_2(x; A_x) &\prec (1+|x|)^{-\ell} \int_0^\infty |\tilde{y}_{m+k}|^{2\nu_{m+k}-1} d\tilde{y}_{m+k} \dots \int_0^\infty |\tilde{y}_{m+1}|^{2\nu_{m+1}-1} d\tilde{y}_{m+1} \\ &\quad \times \int_0^\infty dy_{m+k} \dots \int_0^\infty dy_{m+2} \int_0^\infty (A(y,\tilde{y}))^{\gamma-\alpha} dy_{m+1} \\ &\times \int_{R_m} \left(|z| + r_\Gamma(x) + \sum_{i=1}^k |x_{m+i} - y_{m+i}| + |\tilde{y}_{m+1}| + \dots + |\tilde{y}_{m+k}| \right)^{-m-k-2\nu} dz \\ &\prec r_\Gamma^{\gamma-\alpha}(x) \cdot (1+|x|)^{-\ell}, \end{aligned} \tag{9}$$

where $|z| = |x' - y'|$.

Let us estimate from above $\tau_2(x; B_x)$. Let $|x| \ge 1$ and $y \in B_x$. Then $|\tilde{x} - y| \asymp |y| + |x|$. Put $\mu = (\gamma + \beta) + (k + \gamma - \alpha)$. By (6) $\mu > 0$. Allowing for this fact we get

$$\begin{aligned} \tau_2(x; B_x) \prec &\int_0^\infty |\tilde{y}_{m+k}|^{2\nu_{m+k}-1} d\tilde{y}_{m+k} \dots \int_0^\infty |\tilde{y}_{m+1}|^{2\nu_{m+1}-1} d\tilde{y}_{m+1} \\ &\times \int_0^\infty dy_{m+k} \dots \int_0^\infty dy_{m+2} \int_0^\infty (A(y,\tilde{y}))^{\gamma-\alpha} dy_{m+1} \end{aligned}$$

$$\times \int_{R_m} \left(|y'| + |x| + B(y,\tilde{y})\right)^{-(m+k+\mu)} dy' \prec r_\Gamma^{\gamma-\alpha}(x) \cdot (1+|x|)^{-\ell}.$$

Let $|x| < 1$ and $y \in B_x$. Then for $|y| \geq 1$ $|y| \asymp |y| + |x| + 1$, and also for $|y| < 1$ $(1+|y|) \asymp 1$ and $|\tilde{x} - y| \asymp |y| + |x|$. Therefore

$$\tau_2(x; B_x) \prec \int\limits_{\{y \in R_{m+2k}: |y|<1\}} \frac{|\tilde{y}_{m+k}|^{2\nu_{m+k}-1} \ldots |\tilde{y}_{m+1}|^{2\nu_{m+1}-1} \left(A(y,\tilde{y})\right)^{\gamma-\alpha} dy}{\left(|y'| + r_\Gamma(x) + B(y,\tilde{y})\right)^{m+k+2\nu}}$$

$$+ \int\limits_{\{y \in R_{m+2k}: |y|\geq 1\}} \frac{|\tilde{y}_{m+k}|^{2\nu_{m+k}-1} \ldots |\tilde{y}_{m+1}|^{2\nu_{m+1}-1} \left(A(y,\tilde{y})\right)^{\gamma-\alpha} dy}{\left(|x| + |y| + 1\right)^{m+k+2\nu+\ell}}$$

$$\prec r_\Gamma^{\gamma-\alpha}(x) \cdot (1+|x|)^{-\ell}.$$

The estimate $\tau_2(x; C_x) \prec r_\Gamma^{\gamma-\alpha}(x)\ (1+|x|)^{-\ell}$ is similarly proved. Thus we have proved

$$|\tau_2(x; R_{m+2k} \backslash \omega_x')| \prec \|f\| \cdot \|u\|_{H^\gamma_{\alpha\beta}} \cdot r_\Gamma^{\gamma-\alpha}(x) \cdot (1+|x|)^{-\ell}. \tag{10}$$

So $|A u\ (x)| \leq C_1(u) \cdot r_\Gamma^{\gamma-\alpha}(x) \cdot\ (1+|x|)^{-\ell}$.

On the basis of estimates (7) and (10), the existence of singular integral (2) with density $u \in H^\gamma_{\alpha\beta}\left(R^+_{m+k,\,k}\right)$ at each point is proved by traditional methods.

Put $\omega_1'(x) = \omega_x'(\tilde{x}, 2|h|)$, $\omega_2'(x) = \omega_x'(\widetilde{x+h}, 3|h|)$, $\omega_3'(x) = \omega_x'(\widetilde{x+h}, r_\Gamma(x)/2 - |h|)$. It holds a representation

$$A u(x) - A u(x+h) = \sum_{i=1}^{5} R_i, \tag{11}$$

where

$$R_1 = \left[\int_{\omega_2'} + \int_{\omega_x' \backslash \omega_3'}\right] K(y,\tilde{x})\,(u_1(y) - u(x))\, |\tilde{y}_{m+1}|^{2\nu_{m+1}-1} \ldots |\tilde{y}_{m+k}|^{2\nu_{m+k}-1}\, dy;$$

$$R_2 = -\int_{\omega_2'} K(y,\widetilde{x+h})\,(u_1(y) - u(x+h))\, |\tilde{y}_{m+1}|^{2\nu_{m+1}-1} \ldots |\tilde{y}_{m+k}|^{2\nu_{m+k}-1}\, dy;$$

$$R_3 = -\int_{\omega_x' \backslash \omega_3'} K\left(y, \widetilde{x+h}\right) \cdot u_1(y)\, |\tilde{y}_{m+1}|^{2\nu_{m+1}-1} \ldots |\tilde{y}_{m+k}|^{2\nu_{m+k}-1}\, dy;$$

$$R_4 = \int\limits_{\omega_3' \backslash \omega_2'} \left(K(y,\tilde{x}) - K(y,\widetilde{x+h})\right)(u_1(y) - u(x))$$

$$|\tilde{y}_{m+1}|^{2\nu_{m+1}-1} \ldots |\tilde{y}_{m+k}|^{2\nu_{m+k}-1}\, dy;$$

$$R_5 = \int\limits_{R_{m+2k}\backslash\omega'_x} (K(y,\tilde{x}) - K(y,\widetilde{x+h}))u_1(y)\,|\tilde{y}_{m+1}|^{2\nu_{m+1}-1}\dots|\tilde{y}_{m+k}|^{2\nu_{m+k}-1}\,dy,$$

$$K(y,\tilde{x}) = f(\theta)/r_{\tilde{x},y}^{m+k+2\nu},$$

$$u_1(y) = u\left(y', \sqrt{y_{m+1}^2 + \tilde{y}_{m+1}^2}, \dots, \sqrt{y_{m+k}^2 + \tilde{y}_{m+k}^2}\right).$$

Let $x \in R^+_{m+k,k}$. Then for $x \in R_{m+2k}\backslash\omega'_x$ and $|h| \leq r_\Gamma(x)/8$, it is readily proved that

$$r_{\tilde{x},y} = r_{\widetilde{x+h},\,y}, \tag{12}$$

and

$$|K(y,\tilde{x}) - K(y,\widetilde{x+h})| \prec (c_f + \|f\|)\cdot|h|^\delta \cdot r_{\tilde{x},y}^{-(m+k+2\nu+\delta)}. \tag{13}$$

Using (12), (13) and repeating the above reasoning, we get

$$\begin{aligned}
|R_1| &\prec c_f\,\rho^{-1}(x)\cdot(1+|x|)^{-2\gamma}\cdot\|u\|_{H^\gamma_{\alpha\beta}} \\
&\quad\times\left(\int_{\omega'_2} + \int_{\omega'_x\backslash\omega'_3}\right)\frac{|\tilde{y}_{m+1}|^{2\nu_{m+1}-1}\dots|\tilde{y}_{m+k}|^{2\nu_{m+k}-1}dy}{|\widetilde{x+h}-y|^{m+k+2\nu-\gamma}} \\
&\prec c_f\cdot\|u\|_{H^\gamma_{\alpha\beta}}\cdot\rho^{-1}(x)\cdot(1+|x|)^{-2\gamma}\cdot|h|^\gamma; \\
|R_2| &\prec c_f\cdot\rho^{-1}(x)\cdot(1+|x|)^{-2\gamma}\cdot\|u\|_{H^\gamma_{\alpha\beta}}\cdot|h|^\gamma; \\
|R_3| &\prec c_f\cdot\|u\|_{H^\gamma_{\alpha\beta}}\cdot r_\Gamma^{\gamma-\alpha}(x)\cdot(1+|x|)^{-\ell} \\
&\quad\times\int\limits_{A\backslash\omega'_3}\frac{|\tilde{y}_{m+1}|^{2\nu_{m+1}-1}\dots|\tilde{y}_{m+k}|^{2\nu_{m+k}-1}dy}{|\widetilde{x+h}-y|^{m+k+2\nu}} \\
&\prec c_f\cdot\|u\|_{H^\gamma_{\alpha\beta}}\cdot(1+|x|)^{-2\gamma}\cdot\rho^{-1}(x)\cdot|h|^\gamma, \\
A &= \left\{y\in R_{m+2k} : |\widetilde{x+h}-y| < r_\Gamma(x)/2 + |h|\right\} \supset \omega'_x;
\end{aligned}$$

$$\begin{aligned}
|R_4| &\prec (c_f + \|f\|)\cdot\|u\|_{H^\gamma_{\alpha\beta}}\rho^{-1}(x)\cdot(1+|x|)^{-2\gamma}\cdot|h|^\delta \\
&\quad\times\int\limits_{\omega'_3\backslash\omega'_2}\frac{|\tilde{y}_{m+1}|^{2\nu_{m+1}-1}\dots|\tilde{y}_{m+k}|^{2\nu_{m+k}-1}dy}{|\tilde{x}-y|^{m+k+2\nu+\delta-\gamma}} \\
&\prec (c_f + \|f\|)\cdot\|u\|_{H^\gamma_{\alpha\beta}}\cdot\rho^{-1}(x)\cdot(1+|x|)^{-2\gamma}|h|^\gamma; \\
|R_5| &\prec (c_f + \|f\|)\cdot\|u\|_{H^\gamma_{\alpha\beta}}\cdot|h|^\delta
\end{aligned}$$

$$\times \int\limits_{R_{m+2k} \setminus \omega'_x} \frac{|\tilde{y}_{m+1}|^{2\nu_{m+1}-1} \dots |\tilde{y}_{m+k}|^{2\nu_{m+k}-1} \cdot (A(y,\tilde{y}))^{\gamma-\alpha}\, dy}{r_{\tilde{x},y}^{m+k+2\nu+\delta}\,(1+|y|)^{\ell}}$$

$$\prec (c_f + \|f\|) \cdot \|u\|_{H^{\gamma}_{\alpha\beta}} \cdot \rho^{-1}(x) \cdot (1+|x|)^{-2\gamma} \cdot |h|^{\gamma}.$$

So we get $|Au(x) - Au(x+h)| \leq c_2(u) \cdot \rho^{-1}(x) \cdot (1+|x|)^{-2\gamma} |h|^{\gamma}$. □

References

1. S. K. Abdullayev, *Multidimensional singular integral equations in weighted Hölder spaces.* Institute of Physics of AN of Azerb. SSR, preprint, 1988, No 8, 50 pp (in Russian).
2. S. K. Abdullayev, *Multidimensional singular integral equations in weighted Hölder spaces degenerating on a noncompact set.* Dokl. Akad. Nauk SSSR, 1989, vol. 308, No 6, 1289–1292 (in Russian).
3. S. K. Abdullayev and B. K. Agarzayev, *Weighted Hölder estimates of singular integrals generated by a generalized shift operator.* Transactions of NAS of Azerbaijan Series of Phys.-Techn. and Math. Sci., Math. and Mech., issue XXIV, No 1, Baku, "Elm", 2004.
4. S. K. Abdullayev and A. A. Akperov, *Weighted Hölder estimates of singular integrals generated by a generalized shift operator.* Vestnik Baku State University, ser. fiz-mat. sci., 2007, no. 3, 24–31 (in Russian).
5. N. A. Kipryanov and M. I. Klyuchantsev, *On singular integrals generated by a generalized shift operator. II.* Sibirskii Matem. Zhurn., 1970, vol. XI, No 5, 1061–1083 (in Russian).
6. M. I. Klyuchantsev, *On singular integrals generated by a generalized shift operator. I.* Sibirskii Matem. Zhurn., 1970, vol. XI, No 4, 810–821 (in Russian).
7. B. M. Levitan, *Expansion in series by Bessel functions and Fourier integrals.* Uspekhi Matem. Nauk, vol. 6, No 2 (1951), 102–143 (in Russian).

ON INTERPOLATION PROPERTIES OF GENERALIZED BESOV SPACES

ALEXANDRE ALMEIDA*

Department of Mathematics, University of Aveiro, 3810-193 Aveiro, Portugal
E-mail: jaralmeida@ua.pt

Interpolation properties of Besov spaces with generalized smoothness are studied through wavelet decomposition techniques.

Keywords: Besov spaces; generalized smoothness; real interpolation; function parameter; wavelet decomposition.

1. Introduction

In this note we study real interpolation properties of generalized Besov spaces $B^{\varphi}_{pq}(\mathbb{R}^n)$, where $\varphi \in \mathcal{B}$ (see Section 2.1). We are particularly interested in interpolation with change of p. Even in the classical case $B^s_{pq}(\mathbb{R}^n)$, corresponding to the usual smoothness $\varphi(t) = t^s$, $s \in \mathbb{R}$, it is known that this situation leads to function spaces which are not included in the standard Besov scale. Indeed, we have

$$(B^s_{p_0q}(\mathbb{R}^n), B^s_{p_1q}(\mathbb{R}^n))_{\theta,q} = B^s_{pq(q)}(\mathbb{R}^n), \tag{1}$$

$\frac{1}{p} = \frac{1-\theta}{p_0} + \frac{\theta}{p_1}$, with $1 < p_0 \neq p_1 < \infty$, $1 \leq q < \infty$ and $0 < \theta < 1$ (see [15, Theorem 2.4.1]), where $B^s_{pq(q)}(\mathbb{R}^n)$ is the space obtained as in (2), but replacing the Lebesgue space $L_p(\mathbb{R}^n)$ by the Lorentz space $L_{pq}(\mathbb{R}^n)$.

The counterpart of (1) for spaces $B^{\varphi}_{pq}(\mathbb{R}^n)$ was obtained by Cobos and Fernandez [4] in the Banach case. As far as we know, the quasi-Banach case (when the parameters p, q are allowed to be less than one) was not investigated.

Our paper presents a possible alternative to the approach followed in [4]. We base ourselves on wavelet representations obtained in [1] to con-

*Partially supported by *Unidade de Investigação "Matemática e Aplicações"* (UIMA) of University of Aveiro.

struct a new retraction and a co-retraction, with the advantage of being valid for the full range of the parameters. Although we are not able to give explicit formulas in general, we always have embeddings (from below and from above) for the interpolation spaces (see (19) below). The difficulties in obtaining a full description are related to the behavior of certain interpolation constants involving the function parameter. However, at least in the power case, our statement gives an exact description and, as a consequence, a characterization of Besov type spaces $B^s_{pq(q)}(\mathbb{R}^n)$ is obtained is terms of wavelet coefficients.

Besides the information on the localization of the interpolation spaces, a general aim of this note is to illustrate the usage of wavelet representations in interpolation problems.

This paper is based on a part of the author's PhD thesis [2]. The author would like to thank Prof. A. M. Caetano who suggested the possibility of studying interpolation properties from wavelet decompositions.

2. Function spaces and wavelets

We write $\mathcal{S}(\mathbb{R}^n)$ for the Schwartz class of all infinitely differentiable, rapidly decreasing functions, and $\mathcal{S}'(\mathbb{R}^n)$ for the class of all tempered distributions on $\mathbb{R}^n$. If $f \in \mathcal{S}'(\mathbb{R}^n)$ then Ff (or $\widehat{f}$) stands for the Fourier transform of f, whereas $F^{-1}f$ (or $f^\vee$) denotes the inverse Fourier transform.

2.1. *Function spaces*

$C(\mathbb{R}^n)$ stands for the space of all uniformly continuous bounded functions in $\mathbb{R}^n$ and $C^r(\mathbb{R}^n) = \{f \in C(\mathbb{R}^n) : \; D^\beta f \in C(\mathbb{R}^n), \quad |\beta| \le r\}$, $r \in \mathbb{N}$, with the norm $\|f \,|\, C^r(\mathbb{R}^n)\| := \sum_{|\beta|\le r} \|D^\beta f \,|\, L_\infty(\mathbb{R}^n)\|$.

For $0 < p, q \le \infty$, $L_p(\mathbb{R}^n)$ and $L_{pq}(\mathbb{R}^n)$ denote the well-known Lebesgue and Lorentz spaces, respectively. The Besov spaces $B^s_{pq}(\mathbb{R}^n)$, $s \in \mathbb{R}$, are defined as the collection of all $f \in \mathcal{S}'(\mathbb{R}^n)$ such that

$$\|f \,|\, B^s_{pq}(\mathbb{R}^n)\| := \left(\sum_{j=0}^{\infty} 2^{jsq} \, \|(\varphi_j \widehat{f})^\vee \,|\, L_p(\mathbb{R}^n)\|^q \right)^{1/q} \tag{2}$$

(with the usual modification for "supremum" if $q = \infty$) is finite. They are quasi-Banach spaces not depending on the system $\{\varphi_j\}_{j\in\mathbb{N}_0} \subset \mathcal{S}(\mathbb{R}^n)$ chosen (up to the equivalence of quasi-norms) with the following properties:

$\operatorname{supp} \varphi_0 \subset \{\xi \in \mathbb{R}^n : |\xi| \le 2\}$; $\operatorname{supp} \varphi_j \subset \{\xi \in \mathbb{R}^n : 2^{j-1} \le |\xi| \le 2^{j+1}\}, j \in \mathbb{N}$;

$$\sup_{\xi\in\mathbb{R}^n} |D^\beta \varphi_j(\xi)| \le c\, 2^{-j|\beta|},\ j\in\mathbb{N}_0,\ \beta\in\mathbb{N}_0^n, \quad \text{and} \quad \sum_{j=0}^{\infty}\varphi_j(\xi)=1,\ \xi\in\mathbb{R}^n.$$

Interchanging the quasi-norms in (2) (with respect to p and q), one gets the Triebel-Lizorkin spaces $F^s_{pq}(\mathbb{R}^n)$ (see [14] for further details). It is known that these scales contain some classical spaces as special cases. For instance, $F^s_{p,2}(\mathbb{R}^n) = H^s_p(\mathbb{R}^n)$, $s\in\mathbb{R}$, $1<p<\infty$, are the Bessel potential spaces (which in turn are classical Sobolev spaces if $s\in\mathbb{N}$) and $F^0_{p,2}(\mathbb{R}^n) = h_p(\mathbb{R}^n)$, $0<p<\infty$, are the local Hardy spaces introduced by Goldberg [7] (which, in fact, coincide with $L_p(\mathbb{R}^n)$ if $p>1$).

In this paper we are interested in Besov spaces with generalized smoothness. Roughly speaking, we obtain these spaces by replacing the usual regularity parameter s in (2) by a function with certain properties. Precisely, the space $B^\varphi_{pq}(\mathbb{R}^n)$, $\varphi\in\mathcal{B}$, $0<p,q\le\infty$, is defined as the space of all tempered distributions quasi-normed by

$$\|f\,|\,B^\varphi_{pq}(\mathbb{R}^n)\| := \left(\sum_{j=0}^{\infty}\varphi(2^j)^q\ \|(\varphi_j\widehat{f})^\vee\,|\,L_p(\mathbb{R}^n)\|^q\right)^{1/q} \tag{3}$$

(usual modification if $q=\infty$), where $\mathcal{B}$ denotes the class of all continuous functions $\varphi:(0,\infty)\to(0,\infty)$, such that $\varphi(1)=1$ and $\overline{\varphi}(t) := \sup\{\frac{\varphi(tu)}{\varphi(u)} : u>0\}$ is finite for every $t>0$ (see [10] for some useful properties of this class). Examples of such functions are $\varphi(t) = t^a(1+|\log t|)^b$, with $\overline{\varphi}(t) = t^a(1+|\log t|)^{|b|}$, $a,b\in\mathbb{R}$. Clearly, $B^\varphi_{pq}(\mathbb{R}^n) = B^s_{pq}(\mathbb{R}^n)$ if $\varphi(t)=t^s$.

The spaces $B^\varphi_{pq}(\mathbb{R}^n)$ appear naturally from real interpolation with function parameter between classical Bessel potential spaces [10]. In particular, they cover the spaces $B^{s,\Psi}_{pq}(\mathbb{R}^n)$ introduced by Edmunds and Triebel, where Ψ represents a certain perturbation of the smoothness s (more details may be found in [12]). The study of function spaces with generalized smoothness goes back to the 1970's in connection with some problems of approximation theory investigated by Russian mathematicians [9]. A unified approach to generalized spaces can be found in the recent paper [6], including further references and applications.

Below we shall make use of general weighted sequence spaces as follows. If $\{X_i\}_{i\in I}$ is a countable family of quasi-Banach spaces, $\omega\equiv\{\omega_i\}_{i\in I}$ is a non-negative "sequence" and $0<q\le\infty$, then we will denote by $\ell_q(I,X_i,\omega_i)$ the weighted sequence space of all families $a\equiv\{a_i\}_{i\in I}$, $a_i\in X_i$, $i\in I$, with the quasi-norm

$$\|a\,|\,\ell_q(I,X_i,\omega_i)\| := \left(\sum_{i\in I}(\omega_i\,\|a_i\,|\,X_i\|)^q\right)^{1/q},$$

where $\|a \,|\, \ell_\infty(I, X_i, \omega_i)\|$ must be interpreted as $\sup_{i\in I} \omega_i \, \|a_i \,|\, X_i\|$ if $q = \infty$. We shall omit the index set I if it is clear from the context. When $X_i = X$ for every $i \in I$ then we shall write $\ell_q(I, X, \omega)$, where $\omega \equiv \{\omega_i\}_{i\in I}$. Moreover, in the particular case $X = \mathbb{C}$ we will also omit the "X" writing only $\ell_q(I)$ if $\omega_i = 1$ for all $i \in I$. Putting $\ell_q^\varphi(L_p(\mathbb{R}^n)) := \ell_q(\mathbb{N}_0, L_p(\mathbb{R}^n), \varphi(2^j))$ then we have $\|f \,|\, B^\varphi_{pq}(\mathbb{R}^n)\| = \left\| \{(\varphi_j \widehat{f})^\vee\}_{j\in\mathbb{N}_0} \,|\, \ell_q^\varphi(L_p(\mathbb{R}^n))\right\|$.

Lorentz sequence spaces over $\mathbb{Z}^n$ (the lattice of all points in $\mathbb{R}^n$ with integer components) will be also of interest in the sequel. Given $\varphi \in \mathcal{B}$ and $0 < q \le \infty$, the Lorentz space $\lambda_q(\varphi, \mathbb{Z}^n)$ consists of all bounded complex-valued sequences $a \equiv \{a_m\}_{m\in\mathbb{Z}^n}$ having a finite quasi-norm

$$\|a \,|\, \lambda_q(\varphi, \mathbb{Z}^n)\| := \left(\sum_{k=1}^{\infty} \left[\varphi(k)\, a^*_{k-1}\right]^q \, k^{-1} \right)^{1/q},$$

with the modification for $\sup_{k\in\mathbb{N}} [\varphi(k)\, a^*_{k-1}]$ if $q = \infty$, where $\{a^*_k\}_{k\in\mathbb{N}_0}$ is the decreasing rearrangement of $\{a_m\}_{m\in\mathbb{Z}^n}$, given by

$$a^*_k = \inf\{\delta \ge 0 : \; \#\{m \in \mathbb{Z}^n : \; |a_m| > \delta\} \le k\}, \quad k \in \mathbb{N}_0.$$

If $\varphi(t) = t^{1/p}$ then $\lambda_q(\varphi, \mathbb{Z}^n)$ is the Lorentz sequence space $\ell_{pq}(\mathbb{Z}^n)$, which in turn is the classical space $\ell_q(\mathbb{Z}^n)$ of all q-summable sequences when $p = q$.

2.2. *Wavelet bases*

Let $L_j = L = 2^n - 1$ if $j \in \mathbb{N}$ and $L_0 = 1$. For any $r \in \mathbb{N}$, there are real compactly supported functions

$$\psi_0 \in C^r(\mathbb{R}^n), \quad \psi^l \in C^r(\mathbb{R}^n), \quad l = 1, \ldots, L, \tag{4}$$

with

$$\int_{\mathbb{R}^n} x^\beta \, \psi^l(x)\, dx = 0, \quad \beta \in \mathbb{N}_0^n, \quad |\beta| \le r, \tag{5}$$

such that

$$\{2^{j\frac{n}{2}} \psi^l_{jm} \; : \; j \in \mathbb{N}_0, \; 1 \le l \le L_j, \; m \in \mathbb{Z}^n\} \tag{6}$$

is an orthonormal basis in $L_2(\mathbb{R}^n)$, where

$$\psi^l_{jm}(\cdot) = \begin{cases} \psi_0(\cdot - m) &, \quad j = 0, \; m \in \mathbb{Z}^n, \; l = 1, \\ \psi^l(2^{j-1} \cdot - m), & \; j \in \mathbb{N}, \; m \in \mathbb{Z}^n, 1 \le l \le L. \end{cases} \tag{7}$$

An example of such a system is the *Daubechies wavelet basis* [5]. Further details can be found in [17], [11, Sections 3.8, 3.9] and [18, Chapter 4]. In the sequel Ψ_r, $r \in \mathbb{N}$, will stand for a Daubechies wavelet system $\{\psi^l_{jm}\}_{(l,j,m)\in I}$ with the properties (4)–(7) above, where $I = \{(l,j,m) : j \in \mathbb{N}_0, 1 \le l \le L_j, m \in \mathbb{Z}^n\}$. Below we also consider the notation $I' = \{(l,j) : j \in \mathbb{N}_0, 1 \le l \le L_j\}$.

The properties of this system are sufficiently good to provide unconditional bases in many classical function spaces. It is known that Ψ_r provides an unconditional Schauder basis in the Bessel potential spaces $H^s_p(\mathbb{R}^n)$ if $1 < p < \infty$, $r > |s|$ (in particular in $L_p(\mathbb{R}^n)$ if $1 < p < \infty$), and in the Besov spaces $B^s_{pq}(\mathbb{R}^n)$, with $1 \le p, q < \infty$, $r > |s|$ (see, for instance, [11, Chapter 6]). Based on atomic decompositions, local means and duality theory, Triebel [16] extended these assertions to the entire scales of Besov and Triebel-Lizorkin spaces (including the weighted case, see 8]). In particular, the Daubechies system gives unconditional bases in

$$B^s_{pq}(\mathbb{R}^n) \;\text{ if }\; s \in \mathbb{R},\; 0 < p,q < \infty, \quad r > r(s,p) := \max\left\{ s, \frac{2n}{p} + \frac{n}{2} - s \right\}.$$

Using suitable interpolation techniques, the author showed that Ψ_r also provides an unconditional basis in the generalized Besov spaces $B^\varphi_{pq}(\mathbb{R}^n)$, which is a corollary of the following statement proved in [1].

Theorem 2.1. *Let $\varphi \in \mathcal{B}$, $0 < p < \infty$ and $0 < q \le \infty$. Then there exists $r(\varphi,p)$ such that, for any natural $r > r(\varphi,p)$, there holds: given $f \in \mathcal{S}'(\mathbb{R}^n)$, then $f \in B^\varphi_{pq}(\mathbb{R}^n)$ if and only if it can be represented as*

$$f = \sum_{(l,j,m)\in I} \mu^l_{jm}\, \psi^l_{jm} \quad \textit{with} \quad \mu = \{\mu^l_{jm}\}_{(l,j,m)\in I} \in b^\varphi_{pq} \tag{8}$$

(summability in $\mathcal{S}'(\mathbb{R}^n)$). Moreover, the wavelet coefficients μ^l_{jm} are uniquely determined by

$$\mu^l_{jm} = \mu^l_{jm}(f) := 2^{jn}\langle f, \psi^l_{jm}\rangle, \quad (l,j,m) \in I. \tag{9}$$

Further,

$$\left\| f \,|\, B^\varphi_{pq}(\mathbb{R}^n) \right\| \sim \left\| \mu(f) \,|\, b^\varphi_{pq} \right\|, \quad \mu(f) \equiv \{\mu^l_{jm}(f)\}_{(l,j,m)\in I}. \tag{10}$$

Here b^φ_{pq} denotes the sequence space quasi-normed by

$$\left\| \mu \,|\, b^\varphi_{pq} \right\| := \left(\sum_{(l,j)\in I'} \left(\varphi(2^j)\, 2^{-jn/p} \right)^q \left(\sum_{m\in\mathbb{Z}^n} |\mu^l_{jm}|^p \right)^{q/p} \right)^{1/q} \tag{11}$$

(usual modifications if $p = \infty$ and/or $q = \infty$) and $r(\varphi,p)$ is given by

$$r(\varphi,p) := \max\left(\alpha_{\overline{\varphi}},\, \frac{2n}{p} + \frac{n}{2} - \beta_{\overline{\varphi}} \right), \tag{12}$$

where $\alpha_{\overline{\varphi}}$ and $\beta_{\overline{\varphi}}$ are, respectively, the upper and lower Boyd indices

$$\alpha_{\overline{\varphi}} = \lim_{t\to+\infty} \frac{\log \overline{\varphi}(t)}{\log t} \quad \text{and} \quad \beta_{\overline{\varphi}} = \lim_{t\to 0} \frac{\log \overline{\varphi}(t)}{\log t}.$$

For instance, if $\varphi(t) = t^s$ then $\alpha_{\overline{\varphi}} = \beta_{\overline{\varphi}} = s$ and $b^{\varphi}_{pq} = b^s_{pq}$ is the sequence space considered by Triebel in [16].

Remark 2.1. The previous statement does not make use of duality theory for the spaces $B^{\varphi}_{pq}(\mathbb{R}^n)$. In fact, the symbol $\langle f, \psi^l_{jm}\rangle$ should be interpreted as the sum of two quantities of the type $\langle g, \psi^l_{jm}\rangle$, with $g \in B^s_{pq}(\mathbb{R}^n)$ and $\psi^l_{jm} \in C^r(\mathbb{R}^n)$ (see [1, Remarks 10, 15] for further details).

3. Interpolation with function parameter

3.1. *Preliminaries*

Real interpolation with function parameter is convenient to study interpolation properties of spaces $B^{\varphi}_{pq}(\mathbb{R}^n)$, since we can take the interpolation parameter from the same class $\mathcal{B}$. Recall that the interpolation space $(A_0, A_1)_{\gamma,q}$, $\gamma \in \mathcal{B}$, of compatible quasi-Banach spaces A_0, A_1 is the space of all $f \in A_0 + A_1$ quasi-normed by $\|f\|_{\gamma,q} := \left(\int_0^\infty \left[\gamma(t)^{-1} K(t,f)\right]^q \frac{dt}{t}\right)^{1/q}$ (modification if $q = \infty$), where $K(t,f)$ is the well-known Peetre K-functional (see [3,10] for details). As usual, if $\gamma(t) = t^\theta$ then we simply write $(A_0, A_1)_{\theta,q}$ to denote the corresponding interpolation space.

For $\varphi_0, \varphi_1, \gamma \in \mathcal{B}$, $0 < p, q_0, q_1, q \le \infty$, we have

$$\left(B^{\varphi_0}_{pq_0}(\mathbb{R}^n), B^{\varphi_1}_{pq_1}(\mathbb{R}^n)\right)_{\gamma,q} = B^{\varphi}_{pq}(\mathbb{R}^n), \tag{13}$$

where $0 < \beta_{\overline{\gamma}} \le \alpha_{\overline{\gamma}} < 1$, $\beta_{\overline{\frac{\varphi_0}{\varphi_1}}} > 0$ (or $\alpha_{\overline{\frac{\varphi_0}{\varphi_1}}} < 0$), and $\varphi(t) := \frac{\varphi_0(t)}{\gamma\left(\frac{\varphi_0(t)}{\varphi_1(t)}\right)}$. This statement was proved by Cobos and Fernandez [4] in the Banach case by reducing the problem to the interpolation of appropriate sequence spaces. They observed that $B^{\varphi}_{pq}(\mathbb{R}^n)$ is a retract [3,15] of $\ell^{\varphi}_q(L_p(\mathbb{R}^n))$, where

$$\mathcal{R}\{f_j\}_{j\in\mathbb{N}_0} := \sum_{j=0}^{\infty} F^{-1}(\widetilde{\varphi}_j\, Ff_j), \quad \text{with } \widetilde{\varphi}_j = \sum_{r=-1}^{1} \varphi_{j+r} \quad (\varphi_{-1} \equiv 0), \tag{14}$$

(convergence in $\mathcal{S}'(\mathbb{R}^n)$) is the retraction from $\ell^{\varphi}_q(L_p(\mathbb{R}^n))$ to $B^{\varphi}_{pq}(\mathbb{R}^n)$ and

$$\mathcal{J}f := \{F^{-1}(\varphi_j\, Ff)\}_{j\in\mathbb{N}_0} \tag{15}$$

gives the corresponding co-retraction from $B^{\varphi}_{pq}(\mathbb{R}^n)$ to $\ell^{\varphi}_q(L_p(\mathbb{R}^n))$. It was also observed in [4, Remark 5.4] that the result above remains true in the quasi-Banach case. For this one should replace the space $L_p(\mathbb{R}^n)$ by the Hardy space $h_p(\mathbb{R}^n)$, since the mapping $\mathcal{R}$ above is meaningless if $0 < p < 1$. We refer to [1] for a more detailed description how this question in the general case can be dealt with.

3.2. *Interpolation with change of p*

Real interpolation between spaces $B^{\varphi}_{pq}(\mathbb{R}^n)$ with p fixed produces a Besov space of the same type (cf. formula (13)). However, the change of the parameter p leads to different spaces, which are not included in the scale defined by (3) (cf. (1)). This question was studied in [4, Theorem 5.8] using the same mappings (14), (15). For $1 \le p_0 \ne p_1 < \infty$, $1 \le q < \infty$, $\varphi, \gamma \in \mathcal{B}$, with $0 < \beta_{\overline{\gamma}} \le \alpha_{\overline{\gamma}} < 1$, one has $\left(B^{\varphi}_{p_0q}(\mathbb{R}^n), B^{\varphi}_{p_1q}(\mathbb{R}^n)\right)_{\gamma,q} = B^{\varphi}_{\rho,q}(\mathbb{R}^n)$, with $\rho(t) = t^{\frac{1}{p_0}}\gamma\left(t^{\frac{1}{p_0}-\frac{1}{p_1}}\right)^{-1}$, where $B^{\varphi}_{\rho,q}(\mathbb{R}^n)$ is a certain Besov type space modelled on a generalized Lorentz space instead on the usual L_p-space.

The quasi-Banach case was not considered in [4]. In fact, a direct extension of the Banach case would require some interpolation results for local Hardy spaces which are not available. So, instead of taking $h_p(\mathbb{R}^n)$ in place of $L_p(\mathbb{R}^n)$, we propose a new retraction based on wavelet decompositions obtained in [1]. This will be developed in the sequel.

Let $\Psi_r \equiv \{\psi^l_{jm}\}_{(l,j,m)\in I}$ be a Daubechies wavelet system. If $f \in B^{\varphi}_{p_0q}(\mathbb{R}^n) + B^{\varphi}_{p_1q}(\mathbb{R}^n)$ and $f = f_0 + f_1$, $f_i \in B^{\varphi}_{p_iq}(\mathbb{R}^n)$, $i = 0, 1$ is any decomposition of f, then the quantities $\langle f, \psi^l_{jm}\rangle$, given by

$$\langle f, \psi^l_{jm}\rangle := \langle f_0, \psi^l_{jm}\rangle + \langle f_1, \psi^l_{jm}\rangle, \quad (l, j, m) \in I, \tag{16}$$

with $\langle f_i, \psi^l_{jm}\rangle$ interpreted in the sense of Remark 2.1, are well-defined if we take the natural number r large enough, e.g.

$$r > \max\left(r(\varphi, p_0), r(\varphi, p_1)\right), \tag{17}$$

where $r(\varphi, p_i)$ is defined by (12). Since the quantity $\langle f, \psi^l_{jm}\rangle$ in (16) does not depend on the decomposition of f, we can introduce the following spaces.

Definition 3.1. Let $\varphi \in \mathcal{B}$, $0 < p_0, p_1, q, \nu < \infty$ and $\Upsilon_{p_0,p_1} \equiv \{\rho_j\}_{j\in\mathbb{N}_0}$ be a family of functions (depending on p_0, p_1) "almost" in $\mathcal{B}$ (in the sense that the assumption $\rho_j(1) = 1$ may be dropped). Let also $\Psi_r \equiv \{\psi^l_{jm}\}_{(l,j,m)\in I}$ be a Daubechies wavelet system with r fixed according to (17). We introduce

$$B^{\varphi,(\nu),[\Psi_r]}_{\Upsilon_{p_0,p_1},q}(\mathbb{R}^n) = \left\{f \in B^{\varphi}_{p_0q}(\mathbb{R}^n) + B^{\varphi}_{p_1q}(\mathbb{R}^n) : \ \mu(f) \in b^{\varphi,(\nu)}_{\Upsilon_{p_0,p_1},q}\right\}$$

where in

$$\mu(f) = \{\mu^l_{jm}(f)\}_{(l,j,m)\in I} := \{2^{jn}\langle f, \psi^l_{jm}\rangle\}_{(l,j,m)\in I}, \tag{18}$$

$\langle f, \psi^l_{jm}\rangle$ is given by (16). This space is naturally quasi-normed by

$$\left\| f \,|\, B^{\varphi,(\nu),[\Psi_r]}_{\Upsilon_{p_0,p_1},q}(\mathbb{R}^n)\right\| := \left\| \mu(f) \,|\, b^{\varphi,(\nu)}_{\Upsilon_{p_0,p_1},q}\right\|,$$

where we consider

$$\left\|\mu(f) \,|\, b^{\varphi,(\nu)}_{\Upsilon_{p_0,p_1},q}\right\| := \left(\sum_{(l,j)\in I'} \left[\varphi(2^j)\, 2^{-j\frac{n}{\nu}} \left\| \{\mu^l_{jm}(f)\}_{m\in\mathbb{Z}^n} \,|\, \lambda_q(\rho_j, \mathbb{Z}^n)\right\|\right]^q \right)^{1/q}.$$

Now we are able to formulate the main result.

Theorem 3.1. *Let $\varphi \in \mathcal{B}$, $0 < p_0 \neq p_1 < \infty$, $0 < q < \infty$. Let also $\gamma \in \mathcal{B}$ with $0 < \beta_{\overline{\gamma}} \leq \alpha_{\overline{\gamma}} < 1$. Then*

$$B^{\varphi,(p_0),[\Psi_r]}_{\Upsilon_{p_0,p_1},q}(\mathbb{R}^n) \hookrightarrow (B^{\varphi}_{p_0q}(\mathbb{R}^n), B^{\varphi}_{p_1q}(\mathbb{R}^n))_{\gamma,q} \hookrightarrow B^{\varphi,(p_0),[\Psi_r]}_{\Gamma_{p_0,p_1},q}(\mathbb{R}^n), \qquad (19)$$

where $\Upsilon_{p_0,p_1} \equiv \{A_{\gamma,p_0,p_1}(j)\,\rho_j\}_{j\in\mathbb{N}_0}$, $\Gamma_{p_0,p_1} \equiv \left\{A_{\gamma,p_0,p_1}(j)^{-1}\,\rho_j\right\}_{j\in\mathbb{N}_0}$, *with*

$$\rho_j(t) = \frac{t^{\frac{1}{p_0}}}{\gamma\left(2^{-jn(\frac{1}{p_0}-\frac{1}{p_1})}\, t^{\frac{1}{p_0}-\frac{1}{p_1}}\right)}, \quad j \in \mathbb{N}_0, \qquad (20)$$

and

$$A_{\gamma,p_0,p_1}(j) := \overline{\gamma}\left(2^{-jn(\frac{1}{p_0}-\frac{1}{p_1})}\right)\overline{\gamma}\left(2^{jn(\frac{1}{p_0}-\frac{1}{p_1})}\right), \quad j \in \mathbb{N}_0. \qquad (21)$$

Remark 3.1. We cannot give an explicit formula since the quantities $A_{\gamma,p_0,p_1}(j)$ in (21) cannot be uniformly estimated with respect to j. In general, we only know that $A_{\gamma,p_0,p_1}(j) \geq 1$, which can be easily checked from the properties of the class $\mathcal{B}$. However, at least in the (parameter) case $\gamma(t) = t^\theta$ $(0 < \theta < 1)$, Theorem 3.1 gives an exact interpolation formula (see (22) below), since we have $A_{\gamma,p_0,p_1}(j) \equiv 1$ in this case and, consequently, the corresponding endpoint spaces in (19) coincide.

Corollary 3.1. *Let $\varphi \in \mathcal{B}$, $0 < p_0 \neq p_1 < \infty$, $0 < q < \infty$. Then*

$$(B^{\varphi}_{p_0q}(\mathbb{R}^n), B^{\varphi}_{p_1q}(\mathbb{R}^n))_{\theta,q} = B^{\varphi,(p_0),[\Psi_r]}_{\Gamma_{p_0,p_1},q}(\mathbb{R}^n), \qquad (22)$$

where $\Gamma_{p_0,p_1} \equiv \left\{2^{jn\left(\frac{1}{p_0}-\frac{1}{p}\right)}\, t^{\frac{1}{p}}\right\}_{j\in\mathbb{N}_0}$, *with* $\frac{1}{p} = \frac{1-\theta}{p_0} + \frac{\theta}{p_1}$, $0 < \theta < 1$.

For the classical smoothnes $\varphi(t) = t^s$, formula (22) provides a characterization for the spaces $B^s_{pq(q)}(\mathbb{R}^n)$ considered by Triebel [15] (cf. (1)), since we may write $B^s_{pq(q)}(\mathbb{R}^n) = (B^s_{p_0q}(\mathbb{R}^n), B^s_{p_1q}(\mathbb{R}^n))_{\theta,q}$.

Corollary 3.2. *Let $s \in \mathbb{R}$, $0 < p, q < \infty$. Then $f \in B^s_{pq(q)}(\mathbb{R}^n)$ if and only if $\mu(f) \in b^s_{pq(q)}$, where*

$$\left\| \mu(f) \,|\, b^s_{pq(q)} \right\| := \left(\sum_{(l,j)\in I'} \left[2^{j(s-\frac{n}{p})} \left\| \{\mu^l_{jm}(f)\}_{m\in\mathbb{Z}^n} \,|\, \ell_{pq}(\mathbb{Z}^n) \right\| \right]^q \right)^{1/q} .$$

In particular, $b^s_{pq(p)} = b^s_{pq}$ are the sequence spaces used in [16].

Proof of Theorem 3.1. From Theorem 2.1 we may conclude that $B^{\varphi}_{p_i q}(\mathbb{R}^n)$ is a retract of $b^{\varphi}_{p_i q}$, $i = 0, 1$. In fact,

$$\mathfrak{R}\{\mu^l_{jm}\} = \sum_{(l,j,m)\in I} \mu^l_{jm}\, \psi^l_{jm} \quad \text{(summability in } \mathcal{S}'(\mathbb{R}^n))$$

defines a retraction from $b^{\varphi}_{p_i q}$ into $B^{\varphi}_{p_i q}(\mathbb{R}^n)$, being the associate co-retraction given by $\mathfrak{J} f = \{\mu^l_{jm}(f)\}$ with $\mu^l_{jm}(f) = 2^{jn} \langle f, \psi^l_{jm} \rangle$. Therefore, we have

$$\left\| f \,|\, (B^{\varphi}_{p_0 q}(\mathbb{R}^n), B^{\varphi}_{p_1 q}(\mathbb{R}^n))_{\gamma,q} \right\| \sim \left\| \mathfrak{J} f \,|\, (b^{\varphi}_{p_0 q}, b^{\varphi}_{p_1 q})_{\gamma,q} \right\| . \tag{23}$$

So it remains to interpolate the sequence spaces in the right-hand side of (23). To this end, we observe that $b^{\varphi}_{p_i q}$ may be interpreted as the weighted sequence space $\ell_q\left(\ell_{p_i}(\mathbb{Z}^n), \omega^i\right)$, where $\omega^i = \{\varphi(2^j)\, 2^{-j\frac{n}{p_i}}\}_{(l,j)\in I'}$ (cf. the notation from Section 2.1). From Proposition 3.2 in [13] we obtain

$$\left(\ell_q\left(\ell_{p_0}(\mathbb{Z}^n), \omega^0\right), \ell_q\left(\ell_{p_1}(\mathbb{Z}^n), \omega^1\right)\right)_{\gamma,q} = \ell_q\left((\ell_{p_0}(\mathbb{Z}^n), \ell_{p_1}(\mathbb{Z}^n))_{\gamma_j,q}, \omega^0_j\right),$$

where $\gamma_j(t) = \gamma\left(2^{-jn(\frac{1}{p_0}-\frac{1}{p_1})} t\right)$, $j \in \mathbb{N}_0$. Putting $\widetilde{\gamma}_j = \frac{\gamma_j}{\gamma_j(1)}$ then $\widetilde{\gamma}_j \in \mathcal{B}$ and $(\ell_{p_0}(\mathbb{Z}^n), \ell_{p_1}(\mathbb{Z}^n))_{\gamma_j,q} = \frac{1}{\gamma_j(1)} (\ell_{p_0}(\mathbb{Z}^n), \ell_{p_1}(\mathbb{Z}^n))_{\widetilde{\gamma}_j,q}$, where $\frac{1}{\gamma_j(1)} (\ell_{p_0}(\mathbb{Z}^n), \ell_{p_1}(\mathbb{Z}^n))_{\widetilde{\gamma}_j,q}$ denotes the space $(\ell_{p_0}(\mathbb{Z}^n), \ell_{p_1}(\mathbb{Z}^n))_{\widetilde{\gamma}_j,q}$ equipped with the quasi-norm $\frac{1}{\gamma_j(1)} \|\cdot\|_{\widetilde{\gamma}_j,q}$. But

$$(\ell_{p_0}(\mathbb{Z}^n), \ell_{p_1}(\mathbb{Z}^n))_{\widetilde{\gamma}_j,q} = \lambda_q(\overline{\rho}_j, \mathbb{Z}^n), \tag{24}$$

where $\lambda_q(\overline{\rho}_j, \mathbb{Z}^n)$ are Lorentz sequence spaces and $\overline{\rho}_j := \gamma_j(1)\, \rho_j$, with $\rho_j(t) = \frac{t^{\frac{1}{p_0}}}{\gamma_j\left(t^{\frac{1}{p_0}-\frac{1}{p_1}}\right)}$, $j \in \mathbb{N}_0$ (see [10, Theorem 3]). Formula (24) can be written in the "non-normalized form" as

$$(\ell_{p_0}(\mathbb{Z}^n), \ell_{p_1}(\mathbb{Z}^n))_{\gamma_j,q} = \lambda_q(\rho_j, \mathbb{Z}^n), \quad j \in \mathbb{N}_0.$$

The properties of $\mathcal{B}$ (see [10]) allow us to arrive at the estimates

$$\frac{1}{c}\, (A_{\gamma,p_0,p_1}(j))^{-1} \, \|\cdot \,|\, \lambda_q(\rho_j, \mathbb{Z}^n)\| \le \|\cdot\|_{\gamma_j,q} \le c\, A_{\gamma,p_0,p_1}(j) \, \|\cdot \,|\, \lambda_q(\rho_j, \mathbb{Z}^n)\|,$$

the constants $A_{\gamma,p_0,p_1}(j)$ given by (21) and $c > 1$ independent of j. Thus

$$\lambda_q(A_{\gamma,p_0,p_1}(j)\,\rho_j, \mathbb{Z}^n) \hookrightarrow (\ell_{p_0}(\mathbb{Z}^n), \ell_{p_1}(\mathbb{Z}^n))_{\gamma_j,q} \hookrightarrow \lambda_q(A_{\gamma,p_0,p_1}(j)^{-1}\,\rho_j, \mathbb{Z}^n),$$

where now the "embedding constants" are independent of j. This yields

$$b^{\varphi,(p_0)}_{\Upsilon_{p_0,p_1},q} \hookrightarrow (b^{\varphi}_{p_0 q}, b^{\varphi}_{p_1 q})_{\gamma,q} \hookrightarrow b^{\varphi,(p_0)}_{\Gamma_{p_0,p_1},q},$$

with $\Upsilon_{p_0,p_1} = \{A_{\gamma,p_0,p_1}(j)\,\rho_j\}_{j\in\mathbb{N}_0}$ and $\Gamma_{p_0,p_1} = \{A_{\gamma,p_0,p_1}(j)^{-1}\,\rho_j\}_{j\in\mathbb{N}_0}$, from which we obtain (19) taking into account the equivalence (23). □

References

1. A. Almeida, Wavelet bases in generalized Besov spaces, *J. Math. Anal. Appl.* **304** (2005), 198–211.
2. A. Almeida, *Function Spaces with Generalized Smoothness and Variable Integrability*, PhD Thesis, University of Aveiro, 2005.
3. J. Bergh and J. Löfström, *Interpolation Spaces. An Introduction*, Springer-Verlag, Berlin, 1976.
4. F. Cobos and D. L. Fernandez, Hardy-Sobolev spaces and Besov spaces with a function parameter, *Lecture Notes in Math.*, vol. 1302 (1988), 158–170.
5. I. Daubechies, Orthonormal bases of compactly supported wavelets, *Commun. Pure Appl. Math.* **41** (1988), no. 7, 909–996.
6. W. Farkas and H.-G. Leopold, Characterizations of function spaces of generalized smoothness, *Annali Math. Pura Appl.* **185** (2006), no. 1, 1–62.
7. D. Goldberg, A local version of real Hardy spaces, *Duke Math. J.* **46** (1979), 27–42.
8. D. D. Haroske and H. Triebel, Wavelet bases and entropy numbers in weighted function spaces, *Math. Nachr.* **278** (2005), no. 1-2, 108–132.
9. G. A. Kalyabin and P. I. Lizorkin, Spaces of functions of generalized smoothness, *Math. Nachr.* **133** (1987), 7–32.
10. C. Merucci, Applications of interpolation with a function parameter to Lorentz, Sobolev and Besov spaces, *Lecture Notes in Math.*, vol. 1070 (1984), 183–201.
11. Y. Meyer, *Wavelets and Operators*, Cambridge Univ. Press, Cambridge, 1992.
12. S. Moura, Function spaces of generalized smoothness, *Dissert. Math.* **398** (2001), 1–87.
13. L. E. Persson, *Real Interpolation Between Cross-Sectional L^p-Spaces in Quasi-Banach Bundles*, Technical Report 1, University of Luleå, 1986.
14. H. Triebel, *Theory of Function Spaces*, Birkhäuser, Basel, 1983.
15. H. Triebel, *Interpolation Theory, Function Spaces, Differential Operators*, Johann Ambrosius Barth, Heidelberg, 1995.
16. H. Triebel, A note on wavelet bases in function spaces, *in: Orlicz Centenary Volume, Banach Center Publications*, **64** (2004), 193–206.
17. H. Triebel, *Theory of Function Spaces III*, Birkhäuser, Basel, 2006.
18. P. Wojtaszczyk, *A Mathematical Introduction to Wavelets*, Cambridge Univ. Press, Cambridge, 1997.

PARTIAL HYPOELLIPTICITY OF DIFFERENTIAL OPERATORS

TSEGAYE G. AYELE

Addis Ababa University, Faculty of Science, Addis Ababa, Ethiopia
E-mail: tsegayeg@math.aau.edu.et

WORKU T. BITEW

Temple University, Philadelphia, USA
E-mail:frework@temple.edu

We describe the partial hypoellipticity of differential operators with constant coefficients in terms of fundamental solutions.

Keywords: Convolution, Differential Operator, Hypoellipticity, Fundamental Solution.

1. Introduction

Consider a nonzero linear differential operator $P(D)$ with constant coefficients

$$P(D) = \sum_{\alpha \in K} a_\alpha D^\alpha, \tag{1}$$

where K is a finite set in $\mathbb{N}_0^n$, the space of multi-indices, i.e., $\mathbb{N}_0^n = \mathbb{N}_0 \times \cdots \times \mathbb{N}_0$ (n copies) with $\mathbb{N}_0 = \mathbb{N} \cup \{0\}$. For $\alpha \in \mathbb{N}_0^n$, $|\alpha| = \alpha_1 + \cdots + \alpha_n$, $\max |\alpha| = l$; $a_\alpha \in K$ are constants; $D^\alpha = D_1^{\alpha_1} D_2^{\alpha_2} \cdots D_n^{\alpha_n}$, and $D_j^{\alpha_j} = \frac{1}{i} \frac{\partial^{\alpha_j}}{\partial x_j^{\alpha_j}}$.

Definition 1.1. Let $\Omega \subset \mathbb{R}^n$ be an open set and $u \in D'(\Omega)$. A differential operator (1) is called hypoelliptic if $P(D)u \in C^\infty(\Omega)$ implies $u \in C^\infty(\Omega)$.

For the first time some remarkable results about infinitely differentiability and analyticity of solutions of elliptic differential operators were obtained by Bernstein at the beginning of the 20th century. Later on Petrovskii proved that all classical solutions of equation $P(D)u = f$ with analytic right hand side are analytic if and only if operator $P(D)$ is elliptic.

The conditions for ellipticity and hypoellipticity of operator $P(D)$ in terms of fundamental solutions were described by Hörmander [2]. But the condition for partial hypoellipticity of operator $P(D)$ in terms of fundamental solutions was not established. We look at the case when the solution u of the equation $P(D)u = f$ is conditionally smooth whenever f is a smooth function. That is, we want to study the smoothness of u with respect to some variables.

We will use the expression generalized functions for distributions. Let $\Omega \subset \mathbb{R}^n$ be an open set $m, n \in \mathbb{Z}$ such that $0 \leq m < n$, for $x \in \Omega$ put $x = (x', x'')$ where $x' \in \mathbb{R}^m$ and $x'' \in \mathbb{R}^{n-m}$.

If we denote $\psi_m(x) = \psi(x') \otimes \delta(x'')$, where δ is the Dirac delta function, then $supp(\psi_m) = supp(\psi) \times supp(\delta) = supp(\psi) \times \{0\}$ and later we shall show that the set

$$\Omega_{\psi_m} = \{x : \{x\} - (supp(\psi_m)) \subset \Omega\} = \bigcap_{y \in supp(\psi_m)} (\Omega + y)$$

is open and if $u \in D'(\Omega)$, then the convolution $u * \psi_m$ is defined on the set Ω_{ψ_m} and is a generalized function in $D'(\Omega_{\psi_m})$.

Definition 1.2. Let $\Omega \subset \mathbb{R}^n$ be an open set and $u \in D'(\Omega)$. A differential operator (1) is called partially hypoelliptic with respect to the plane $x'' = 0$, if $P(D)u \in C^\infty(\Omega)$ implies that for any function $\psi \in D(\mathbb{R}^m)$ the convolution $u * \psi_m \in C^\infty(\Omega_{\psi_m})$.

Remark 1.1. If $m = 0$, then $\psi_0 = \delta$ and thus we obtain Definition 1.1.

Definition 1.3. A generalized function $\mathcal{E} \in D'(\mathbb{R}^n)$ is called a fundamental solution for a differential operator (1) if

$$P(D)\mathcal{E} = \delta. \tag{2}$$

The existence of a fundamental solution for an operator (1) in $D'(\mathbb{R}^n)$ was first proved independently by Malgrange (1953) and Ehrenpresis (1954) whereas its existence in the space $S'(\mathbb{R}^n)$ of generalized functions of slow growth (= tempered distributions) was proved by Hörmander (1958).

Equation (2) in the space $S'(\mathbb{R}^n)$ is equivalent to the algebraic equation,

$$P(i\xi)\hat{\mathcal{E}} = (2\pi)^{-\frac{n}{2}}, \tag{3}$$

with respect to the Fourier transform $\mathcal{F}\mathcal{E} = \hat{\mathcal{E}}$. Thus the problem of finding a fundamental solution of slow growth turns out to be a special case of the more general problem of *dividing* a generalized function of slow growth by a polynomial, that is, of the problem of finding a solution $u \in S'(\mathbb{R}^n)$ of the equation

$$P(i\xi)u = f, \tag{4}$$

where P is a nonzero polynomial and f is a specified generalized function in $S'(\mathbb{R}^n)$.

The solvability of the problem of *division* was proved in 1958 independently by Hörmander and Lojasiewicz. As a consequence it was proved that every nonzero linear differential operator with constant coefficients has a fundamental solution of slow growth.

In this paper we prove the following theorem in which a description of the partial hypoellipticity of a differential operator (1) in terms of its fundamental solutions is given.

Theorem 1.1. *For a differential operator (1) to be partially hypoelliptic with respect to the plane $x'' = 0$, it is necessary and sufficient that there exists a fundamental solution $\mathcal{E}$ of (1) such that for any function $\psi_m \in D(\mathbb{R}^n)$ the convolution $\mathcal{E} * \psi_m \in C^\infty(\mathbb{R}^n \backslash supp(\psi_m))$.*

Remark 1.2. From the existence of such fundamental solution it follows that every fundamental solution $\mathcal{E}$ satisfies $\mathcal{E} * \psi_m \in C^\infty(\mathbb{R}^n \backslash supp(\psi_m))$.

Remark 1.3. If $m = 0$, then $\psi_0 = \delta$ and we obtain Hörmander's theorem about description of the hypoellipticity of an operator (1) in terms of fundamental solutions. We will state and prove it later for the sake of completeness.

2. Preliminaries

In this section we will describe the basic tools we need in the proof of our theorem and we prove them for the sake of completeness. For sets A and B we will use the following notations, $A - B = \{x - y : x \in A \text{ and } y \in B\}$ and the distance between A and B, $\rho(A, B) = \inf_{x\in A, y\in B} |x - y|$.

Theorem 2.1. *Let $\Omega \subset \mathbb{R}^n$ be open , $u \in D'(\Omega)$, $\psi \in D(\mathbb{R}^n)$. Then*

$$\Omega_\psi = \{x : \{x\} - supp(\psi) \subset \Omega\} = \bigcap_{y\in supp(\psi)} (\Omega + y),$$

the convolution $u*\psi$ exists on Ω_ψ and $(u*\psi, \varphi)(x) = (u(x), (\psi(y), \varphi(x+y))$ for all $\varphi \in D(\Omega_\psi)$.

Note that for the convolution $u * \psi_m$ to be defined for all $\varphi \in D(\Omega_{\psi_m})$ it is important to show that the set $\Omega_{\psi_m} = \bigcap_{y \in supp(\psi_m)} (\Omega + y)$ is open.

Proposition 2.1. *If $\Omega \subset \mathbb{R}^n$ and $x, y \in \mathbb{R}^n$, then $\rho(x-y, \Omega) = \rho(x, \Omega+y)$.*

Proposition 2.2. *If $\Omega \subset \mathbb{R}^n$ and $x, y, z \in \mathbb{R}^n$, then $|\rho(x, \Omega+y) - \rho(x, \Omega+z)| \leq |y-z|$.*

Proof. $|\rho(x, \Omega+y) - \rho(x, \Omega+z)| = |\rho(x-y, \Omega) - \rho(x-z, \Omega)|$ and $|\rho(x-y, \Omega) - \rho(x-z, \Omega)| \leq |x-y-(x-z)| \leq |y-z|$ by Proposition 2.1. □

Corollary 2.1. *Let $\Omega \subset \mathbb{R}^n$ and $x, y \in \mathbb{R}^n$. Then the function $\rho(x, \Omega+y)$ is continuous in y on $\mathbb{R}^n$.*

Proposition 2.3. *Let $\Omega_\alpha \subset \mathbb{R}^n$, where $\alpha \in A \subset \mathbb{R}$. Then for any $x \in \mathbb{R}^n$ the condition*

$$\rho(x, \bigcup_{\alpha \in A} \Omega_\alpha) > 0 \tag{5}$$

is equivalent to

$$\inf_{\alpha \in A} \rho(x, \Omega_\alpha) > 0\,. \tag{6}$$

Proof. Condition (6) is equivalent to the existence of $c > 0$, such that $\rho(x, \Omega_\alpha) > c$, for all $\alpha \in A$. Then $\mathcal{B}(x, c) \subset \Omega_\alpha^c$, for all $\alpha \in A$, and this means

$$\mathcal{B}(x, c) \subset \bigcap_{\alpha \in A} \Omega_\alpha^c = (\bigcup_{\alpha \in A} \Omega_\alpha)^c \Rightarrow \rho(x, \bigcup_{\alpha \in A} \Omega_\alpha) > 0.$$

Conversely, let condition (5) be satisfied and suppose that condition (6) be not satisfied. If A is a finite set, then there exists $\alpha_0 \in A$ such that $\rho(x, \Omega_{\alpha_0}) = 0$. Hence

$$\rho(x, \bigcup_{\alpha \in A} \Omega_\alpha) \leq \rho(x, \Omega_{\alpha_0}) = 0.$$

That is, $\rho(x, \bigcup_{\alpha \in A} \Omega_\alpha) = 0$. If A is an infinite set, then for each $k \in \mathbb{N}$ there exists $\alpha_k \in A$ such that $\rho(x, \Omega_{\alpha_k}) \leq 1/k$. Consequently,

$$\rho(x, \bigcup_{\alpha \in A} \Omega_\alpha) \leq \rho(x, \Omega_{\alpha_k}) \leq \frac{1}{k}$$

and taking the limit we get $\rho(x, \bigcup_{\alpha \in A} \Omega_\alpha) = 0$. In both cases we arrived at a contradiction to condition (5) and this completes the proof. □

Lemma 2.1. *Let $\Omega \subset \mathbb{R}^n$ be an open set and F compact, then the set $G = \bigcap_{y\in F}(\Omega + y)$ is open.*

Proof. That G is open means that $\rho(x, G^c) > 0$ for all $x \in G$. That is, for all $x \in G$,

$$\rho(x, [\bigcap_{y\in F}(\Omega + y)]^c) = \rho(x, \bigcup_{y\in F}(\Omega + y)^c) = \rho(x, \bigcup_{y\in F}(\Omega^c + y)) > 0.$$

By Proposition 2.3 this is equivalent to $\inf_{y\in F}\rho(x, (\Omega^c + y)) > 0$ for all $x \in G$. Assume to the contrary that there exists $x \in G$ such that $\inf_{y\in F}\rho(x, (\Omega^c + y)) = 0$. This implies that for every $k \in \mathbb{N}$ there exists $y_k \in F$ such that $\rho(x, (\Omega^c + y_k)) \leq 1/k$. We can select a convergent subsequence $\{y_{k_s}\}$ such that $\lim_{s\to\infty} y_{k_s} = y_0$ for $y_0 \in F$. Then by Corollary 2.1, $\rho(x, (\Omega^c + y_0)) = \lim_{s\to\infty}\rho(x, (\Omega^c + y_{k_s})) = 0$. Since $(\Omega^c + y_0)$ is closed by the assumption,

$$x \in (\Omega^c + y_o) \subset \bigcup_{y\in F}(\Omega^c + y) = \bigcup_{y\in F}(\Omega + y)^c = [\bigcap_{y\in F}(\Omega + y)]^c = G^c.$$

This contradicts the fact that $x \in G$. Consequently, $G = \bigcap_{y\in F}(\Omega + y)$ is open. □

From Lemma 2.1 we can conclude that for $\psi \in D(\mathbb{R}^n)$ and open set $\Omega \subset \mathbb{R}^n$ the set $\Omega_\psi = \bigcap_{y\in supp(\psi)}(\Omega + y)$ is open. Next we want to show that $(\psi(y), \varphi(x + y)) \in D(\Omega)$, for all $\psi \in D(\mathbb{R}^n), \varphi \in D(\Omega_\psi)$. To do this we need the following two propositions.

Proposition 2.4. *For $\psi \in D(\mathbb{R}^n)$, $\varphi \in D(\Omega_\psi)$, let*

$$\Lambda(x) = \int_{R^n} \psi(y)\varphi(x + y)dy.$$

Then $supp(\Lambda) \subset supp(\varphi) - supp(\psi)$.

Proof. Let $H = \{x : \text{ for all } y \in supp(\psi), (x + y) \notin supp(\varphi)\}$. Observe that

$$H \subset \{x : \int_{supp(\psi)} \psi(y)\varphi(x + y)dy = 0\}.$$

This implies that $supp(\Lambda) \subset \overline{H^c}$. We know that $x \in H^c \Leftrightarrow$ there exists $y \in supp(\psi)$ such that $x \in (supp(\varphi) - y) \Leftrightarrow x \in supp(\varphi) - supp(\psi)$. Since $H^c = \overline{H^c}$, we have $supp(\Lambda) \subset supp(\varphi) - supp(\psi)$. □

Proposition 2.5. *If $supp(\varphi) \subset \Omega_\psi$, then $supp(\varphi) - supp(\psi) \subset \Omega$.*

Proof. Since

$$supp(\varphi) \subset \bigcap_{y \in supp(\psi)} (\Omega + y) \Leftrightarrow \text{ for all } y \in supp(\psi),$$

$supp(\varphi) \subset \Omega + y$. It follows that there exists $y \in supp(\psi)$ such that $supp(\varphi) - y \subset \Omega \Leftrightarrow supp(\varphi) - supp(\psi) \subset \Omega$. □

Propositions 2.4 and 2.5 imply that $(\psi(y), \varphi(x+y)) \in D(\Omega)$ for all $\psi \in D(R^n)$ and $\varphi \in D(\Omega_\psi)$.

Remark 2.1. i) For any $u \in D'(\Omega)$ and $\psi \in D(\mathbb{R}^n)$ the convolution $u * \psi$ is well defined on Ω_ψ.

ii) Suppose $\Omega \in \mathbb{R}^n$ is an open set, $m, n \in \mathbb{Z}$ are such that $0 \leq m < n$, and for any $\psi \in D(\mathbb{R}^m)$ define $\psi_m(x) = \psi(x') \otimes \delta(x'')$, where $x = (x', x'')$, where $x' \in \mathbb{R}^m, x'' \in \mathbb{R}^{n-m}$. Then $supp(\psi_m) = supp(\psi) \times \{0\}$, and the set $\Omega_{\psi_m} = \bigcap_{y \in supp(\psi_m)} (\Omega + y)$ is open.

Definition 2.1. Let $\Omega \in \mathbb{R}^n$ be open, $u \in D'(\Omega)$ and $\psi \in D(\mathbb{R}^m)$. Then for $\varphi \in D(\Omega_{\psi_m})$, we define

$$(u * \psi_m, \varphi) = (u(x), \int_{supp(\psi)} \psi(y')\varphi(x+y')dy').$$

Proposition 2.6. *Let* $\Lambda(x) = \int_{R^n} \psi(y')\varphi(x+y')dy'$. *Then* $supp(\Lambda) \subset supp(\varphi) - (supp(\psi) \times \{0\})$.

Proof. Define $H = \{x : \text{ for all } y' \in supp(\psi) \times \{0\}, (x+y') \notin supp(\varphi)\}$. Then $H \subset \{x : \int_{R^n} \psi(y')\varphi(x+y')dy' = 0\}$. Hence $supp(\Lambda) \subset H^c$. Now let $x \in H^c$, that is, there exists $y' \in supp(\psi) \times \{0\}$ such that $(x+y') \in supp(\varphi)$. This is equivalent to the existence of $y' \in supp(\psi) \times \{0\} : x \in (supp(\varphi) - y')$. Therefore $x \in supp(\varphi) - supp(\psi) \times \{0\}$. Observe that $supp(\varphi) \subset \Omega_{\psi_m}$ and

$$supp(\varphi) - supp(\psi) \times \{0\} \subset \Omega. \tag{7}$$

This completes the proof. □

The proof of the following theorem is similar to that of Theorem 2.1

Theorem 2.2. *Suppose* $m, n \in \mathbb{Z}$ *are such that* $0 \leq m < n$, $\Omega \subset \mathbb{R}^n$ *is open and* $\psi \in D(\mathbb{R}^m)$. *Then* $(\psi(y'), \varphi(x+y')) \in D(\Omega)$ *for all* $\varphi \in D(\Omega_{\psi_m})$.

Proof. We conclude from (7) and Proposition 2.6 that $(\psi(y'), \varphi(x+y')) \in D(\Omega)$ for all $\psi \in D(R^m)$ and $\varphi \in D(\Omega_{\psi_m})$. □

Remark 2.2. For any $u \in D'(\Omega)$ and $\psi \in D(\mathbb{R}^m)$ the convolution $u * \psi_m$ is well defined on Ω_{ψ_m}.

Lemma 2.2. *Let $\Omega \subset \mathbb{R}^n$ be an open set, $m, n \in \mathbb{Z}$ be such that $0 \leq m < n$ and $\psi \in D(\mathbb{R}^m)$. Let $\varphi, \varphi_k \in D(\Omega_{\psi_m})$ for $k \in \mathbb{N}$ and $\varphi_k \to \varphi$ in $D(\Omega_{\psi_m})$ as $k \to \infty$. Then $(\psi(y'), \varphi_k(x+y')) \to (\psi(y), \varphi(x+y'))$ in $D(\Omega_\psi)$.*

Proof. $\varphi_k \to \varphi$ in $D(\Omega_{\psi_m})$ means $\varphi_k - \varphi \to 0$ in $D(\Omega_{\psi_m})$. Hence

$$(\psi(y'), \varphi_k(x+y')) - (\psi(y'), \varphi(x+y')) = (\psi(y'), (\varphi_k(x+y') - \varphi(x+y)))$$
$$\to (\psi(y'), 0) = 0$$

That is, $(\psi(y'), \varphi_k(x+y')) \to (\psi(y), \varphi(x+y'))$ in $D(\Omega_\psi)$. □

Lemma 2.3. *Let $\Omega \subset \mathbb{R}^n$ be an open set, $m, n \in \mathbb{Z}$ be such that $0 \leq m < n$ and $\psi \in D(\mathbb{R}^m)$. Then $u * \psi_m \in D'(\Omega_{\psi_m})$ for any $u \in D'(\Omega)$.*

Proof. From Remark 2.2 it follows that $u*\psi_m$ is defined on Ω_{ψ_m}, and from Lemma 2.2 it follows that the functional is continuous. Since it is linear on $D(\Omega_{\psi_m})$ we conclude that $u * \psi_m \in D'(\Omega_{\psi_m})$. □

Remark 2.3. If $u \in C^\infty(\Omega)$, then the convolution $u * \psi_m \in C^\infty(\Omega_{\psi_m})$.

3. Description of the partial hypoellipticity in terms of fundamental solutions

It was proved by Petrovskii that all classical solutions of equations $P(D)u = f$ with analytic right hand side are analytic if and only if operator (1) is elliptic. In other words, if $\Omega \subset \mathbb{R}^n$ is an open set, then for any $u \in C^l(\Omega)$ such that $P(D)u$ is an analytic function in Ω, u is an analytic function in Ω if and only if operator (1) is elliptic. In terms of the polynomial P this is equivalent to $\mathcal{P}(\xi) \neq 0$ for all $\xi \neq 0$, where

$$\mathcal{P}(D) = \sum_{\alpha \in K : |\alpha| = l} a_\alpha D^\alpha$$

is the principal part of operator (1).

The corresponding description for hypoelliptic operators was obtained by Hörmander. That is, a differential operator (1) is hypoelliptic if and only if for all multi-indices $\alpha \neq 0$,

$$\frac{P^{(\alpha)}(\xi)}{P(\xi)} \to 0 \quad \text{as} \quad \xi \to \infty.$$

The following is Hörmander's theorem about description of the hypoellipticity in terms of fundamental solutions.

Theorem 3.1. *For an operator (1) to be hypoelliptic it is necessary and sufficient that there exists a fundamental solution $\mathcal{E}$ of (1) such that $\mathcal{E} \in C^\infty(\mathbb{R}^n \backslash \{0\})$.*

Remark 3.1. From the existence of such fundamental solution it follows that any fundamental solution $\mathcal{E} \in C^\infty(\mathbb{R}^n \backslash \{0\})$.

Proof. Necessity: Let $\Omega \in R^n$ be open and $u \in D'(\Omega)$ be a solution of $P(D)u = f$ where $f \in C^\infty(\Omega)$. Since $\mathcal{E}$ is a fundamental solution of operator (2.1), one has $P(D)\mathcal{E} = \delta$ where $\delta \in C^\infty(\mathbb{R}^n \backslash \{0\})$. Consequently by Definition 2.1, $\mathcal{E} \in C^\infty(\mathbb{R}^n \backslash \{0\})$.

Sufficiency: Let $\mathcal{E} \in C^\infty(\mathbb{R}^n \backslash \{0\})$ be a fundamental solution of operator (1) and $u \in D'(\Omega)$ be a solution to the equation $P(D)u = f$ with $f \in C^\infty(\Omega)$. Let Ω' be an arbitrary open set such that $\Omega' \subset\subset \Omega$, and let $\eta \in D(\Omega)$ with $\eta(x) = 1$ for all $x \in \Omega'$. The generalized function ηu is compactly supported in Ω and satisfies the equation $P(D)(\eta u) = \eta f + f_1$ where $\eta f \in D(\Omega)$ and $f_1 \in D'(\mathbb{R}^n)$ such that $supp(f_1) \subset supp(\eta) \backslash \Omega'$. Therefore $\eta u = \delta * \eta u = P(D)\mathcal{E} * \eta u = \mathcal{E} * P(D)\eta u = \mathcal{E} * (\eta f + f_1) = \mathcal{E} * (\eta f) + \mathcal{E} * f_1$. Since $\eta f \in D(\mathbb{R}^n)$, $\mathcal{E} * \eta f \in C^\infty(\mathbb{R}^n)$ and $\mathcal{E} * f_1 \in C^\infty(\mathbb{R}^n \backslash supp(f_1))$, we have $\mathcal{E} * f_1 \in C^\infty(\mathbb{R}^n \backslash (\Omega \backslash \Omega'))$, that is, $\mathcal{E} * f_1 \in C^\infty(\Omega')$. And since $\eta u = u$ in Ω' and $\Omega' \subset\subset \Omega$ is arbitrary, $u \in C^\infty(\Omega)$. □

It is known that operator (1) is partially hypoelliptic with respect to the plane $x'' = 0$ if and only if for all $\alpha \neq 0$

$$\frac{P^{(\alpha)}(\xi', \xi'')}{P(\xi', \xi'')} \to 0 \qquad \text{as } \ \xi'' \to \infty \ \text{ and } \xi' \ \text{ remains bounded.}$$

Lemma 3.1. *Let $G \subset \mathbb{R}^n$ be a closed set, $f \in D'(\mathbb{R}^n) \cap C^\infty(\mathbb{R}^n \backslash G)$ and $g \in \mathcal{E}'(R^n)$. Then $f * g \in C^\infty(\mathbb{R}^n \backslash (\overline{supp(g) + G}))$. Moreover, for all $\gamma > 0$, $(f * g)(x) = (g(y), \eta(y) f(x - y))$ on $\mathbb{R}^n \backslash (\overline{(supp(g)^\gamma) + G})$, where $\eta \in D(\mathbb{R}^n)$, $\eta = 1$ in some neighborhood of $supp(g)$ and $supp(\eta) \subset (supp(g))^\gamma$.*

Proof. Since $f, g \in D'(\mathbb{R}^n)$ and $supp(g)$ is compact, we have $f * g \in D'(\mathbb{R}^n)$ and $(f * g, \varphi) = (f(x)g(y), \eta(y)\varphi(x + y)) = (g(y), (f(x), \eta(y)\varphi(x + y)))$ for all $\varphi \in D(\mathbb{R}^n)$, where η is any function from $D(\mathbb{R}^n)$ such that $\eta = 1$ in some neighborhood of $supp(g)$ and vanishes outside $(supp(g))^\gamma$. One can observe that $supp(g)+G=\bigcup_{y\in G}(supp(g)+y)$ and

$$(supp(g)+G)^c=[\bigcup_{y\in G}(supp(g)+y)]^c=\bigcap_{y\in G}(supp(g)+y)^c=\bigcap_{y\in G}(supp(g)^c+y).$$

Similarly, $((\overline{(supp(g))^\gamma})+G)^c=\bigcap_{y\in G}((\overline{supp(g)})^\gamma)^c+y)$. Now let $supp(\varphi)\subset((\overline{supp(g))^\gamma})+G)^c$. Then $supp(\varphi)\subset\bigcap_{y\in G}((\overline{supp(g))^\gamma})^c+y)$ if and only if $supp(\varphi)-y\subset(\overline{(supp(g))^\gamma})^c$ for all $y\in G$ if and only if $supp(\varphi)-G\subset(\overline{(supp(g))^\gamma})^c$. That is, $(supp(\varphi)-G)\cap(\overline{(supp(g))^\gamma})^c=\varnothing$. Then there exists $\delta_0=\delta_0(\gamma)$ such that, for all δ satisfying $0<\delta\leq\delta_0(\gamma)$, we have $(supp(\varphi)-G^\delta)\cap(\overline{(supp(g))^\gamma})^c=\varnothing$. Equivalently, we can say that $(supp(\varphi)-G^\delta)\subset R^n\backslash\overline{(supp(g))^\gamma}$. Since $supp(\varphi(x+.))=(supp(\varphi)-x)\subset supp(\varphi)-G^\delta$, we have $supp(\eta(.))\cap supp(\varphi(x+.))=\varnothing$. Then for all $x\in G^\delta$ and for all $y\in\mathbb{R}^n$, we have $\eta(y)\varphi(x+y)=0$, That is, $supp(\eta(y)\varphi(.+y))\subset\mathbb{R}^n\backslash G^\delta$. Therefore

$$supp(\eta(y)\varphi(.+y))\subset\mathbb{R}^n\backslash\bigcap_{0<\delta\leq\delta_0(\gamma)}G^\delta=\mathbb{R}^n\backslash G.$$

This implies that $supp(\eta(y)\varphi(.+y))\subset\mathbb{R}^n\backslash G$. Consequently, since $f\in C^\infty(\mathbb{R}^n\backslash G)$,

$$(f(x),\eta(y)\varphi(x+y))=\int_{\mathbb{R}^n}f(x)\eta(y)\varphi(x+y)dx=\int f(x-y)\eta(y)\varphi(x)dx.$$

Note that the function $f(x-y)\eta(y)\varphi(x)\in D(\mathbb{R}^{2n})$ and we assume that if one of the multiples is zero, then the product $f(x-y)\eta(y)\varphi(x)$ is zero even if $f(x-y)$ is undefined, that is $x-y\in G$. Then

$$\begin{aligned}(g(y),\int_{\mathbb{R}^n}f(x-y)\eta(y)\varphi(x)dx)&=\int_{\mathbb{R}^n}(g(y),\eta(y)f(x-y)\varphi(x))dx\\&=\int_{supp(\varphi)}(g(y),\eta(y)f(x-y)\varphi(x))dx\,.\end{aligned}$$

Since for $x\in supp(\varphi)$, $\eta(.)f(x-.)\in D(\mathbb{R}^n)$, the value of $(g(y),\eta(y)f(x-y))$ is defined and $(g(y),\eta(y)(f(x),\varphi(x+y)))=(g(y),\eta(y)f(x-y)),\varphi(x))$. Therefore

$$\begin{aligned}(f*g,\varphi)(x)&=\int_{supp(\varphi)}(g(y),\eta(y)f(x-y))\varphi(x)dx\\&=\int_{\mathbb{R}^n}(g(y),\eta(y)f(x-y))\varphi(x)dx=(g(y),\eta(y)f(x-y)),\varphi(x)),\end{aligned}$$

and we assume that outside $supp(\varphi)$ the function $(g(y),\eta(y)f(x-y))$ is redefined, say, by zero. Since the above equality holds for all functions φ such that $supp(\varphi)\subset\mathbb{R}^n\backslash(\overline{(supp(g))^\gamma}+G)$, we have $(f*g)(x)=$

$(g(y), \eta(y)f(x-y))$. Also for all $x \in \mathbb{R}^n \backslash (\overline{(supp(g))^\gamma} + G)$, the function $\eta(.)f(x-y) \in D(\mathbb{R}^n)$, which implies that $(f * g)(x) = (g(y), \eta(y)f(x-y))$ lies in $C^\infty(\mathbb{R}^n \backslash (supp(g) + G))$, and this completes the proof. □

Remark 3.2. If $\Omega = G^c$, then from Lemma 2.3 and Remark 2.3 we observe that $(f * g)(x) = (g(y), \eta(y)f(x-y)) \in C^\infty(\Omega_g)$, where

$$\Omega_g = \bigcap_{y \in supp(g)} (G^c + y) = (\bigcup_{y \in supp(g)} (G+y))^c$$
$$= (G + supp(g))^c = \mathbb{R}^n \backslash (supp(g) + G).$$

Remark 3.3. If $G = \{0\}$ in Lemma 3.1, then it can be easily shown that $(f*g)(x) = (g(y), \eta(y)f(x-y)) \in C^\infty(\mathbb{R}^n \backslash supp(g))$. See also Vladimirov [1].

Remark 3.4. If $G \subset \mathbb{R}^n$ is a closed set, $f \in D'(\mathbb{R}^n) \cap C^\infty(\mathbb{R}^n \backslash G)$ and $g_1, ... g_n \in \mathcal{E}'(\mathbb{R}^n)$, then $(f * g_1 * ... * g_n)(x)$ exists in $D'(\mathbb{R}^n)$ and we have $(f * g_1 ... * g_n)(x) \in C^\infty(\mathbb{R}^n \backslash (G + supp(g_1) + ... + supp(g_n)))$.

Proof of Theorem 1.1. **Necessity**: Let differential operator (1) be partially hypoelliptic with respect to the plane $x'' = 0$ and $\mathcal{E}$ be its fundamental solution, i.e., $P(D)\mathcal{E} = \delta$. Then using Definition 1.2, and Remarks 2.3, 3.2 and 3.3, we deduce that for any $\psi \in D(\mathbb{R}^m)$ and $\psi_m(x) = \psi(x') \otimes \delta(x'')$, we have $(\mathcal{E} * \psi_m)(x) \in C^\infty(\mathbb{R}^n \backslash supp(\psi_m))$.

Sufficiency: Let $\mathcal{E}$ be a fundamental solution of operator (1) and that $(\mathcal{E} * \psi_m)(x) \in C^\infty(\mathbb{R}^n \backslash supp(\psi_m))$. Moreover, let $u \in D'(\Omega)$ be a solution of the equation $P(D)u = f$ with $f \in C^\infty(\Omega)$. Then $\eta u * \psi_m \in C^\infty(\Omega_{\psi_m})$, where $\eta \in D(\Omega)$, $\eta(x) = 1$ for all $x \in \Omega' \subset\subset \Omega$. Indeed,

$$\eta u * \psi_m = \eta u * (\psi_m * \delta) = \eta u * (\psi_m * P(D)\mathcal{E}) = \eta u * P(D)(\psi_m * \mathcal{E})$$
$$= P(D)(\eta u * (\mathcal{E} * \psi_m)) = P(D)(\eta u) * (\mathcal{E} * \psi_m) = (\eta f + f_1) * (\mathcal{E} * \psi_m),$$

where ηf and f_1 are as in the proof of Theorem 3.1. Since $\eta f \in D(\Omega)$, the convolution $\eta f * (\mathcal{E} * \psi_m) \in C^\infty(R^n)$. By Lemma 3.1,

$$f_1 * (\mathcal{E} * \psi_m) = (\mathcal{E} * \psi_m) * f_1 \in C^\infty(\mathbb{R}^n \backslash (supp(\psi_m) + supp(f_1))$$

and since $\mathbb{R}^n \backslash ((\Omega')^c + supp(\psi_m)) \subset\subset \mathbb{R}^n \backslash (supp(\psi_m) + supp(f_1))$, it follows that $(\mathcal{E} * \psi_m) * f_1 \in C^\infty(\mathbb{R}^n \backslash ((\Omega')^c + supp(\psi_m))$. But

$$(\Omega')^c + supp(\psi_m) = \bigcup_{y \in supp(\psi_m)} ((\Omega')^c + y) = \bigcup (\Omega' + y)^c$$
$$= (\bigcap_{y \in supp(\psi_m)} (\Omega' + y))^c = (\Omega')^c_{\psi_m}.$$

Consequently, $(\mathcal{E} * \psi_m) * f_1 \in C^\infty(\Omega'_{\psi_m})$ and $\eta u * \psi_m \in C^\infty(\Omega'_{\psi_m})$. Since $\Omega' \subset\subset \Omega$ is arbitrary, $\eta u * \psi_m \in C^\infty(\Omega_{\psi_m})$.

Next, let $x_0 \in \Omega_{\psi_m}$. Then there exists $\delta > 0$ such that $B(x_0, \delta) \subset \Omega_{\psi_m}$, $\Omega'_{x0} = B(x_0, \delta) - supp(\psi_m)$ and $\Omega'_{x0} \subset\subset \Omega$. Let $\eta_{x0} \in D(\Omega)$ be such that $\eta_{x0}(x) = 1$ for all $x \in \Omega'_{x0}$. Observe that $\eta_{x0} u * \psi_m = u * \psi_m$ on $B(x_0, \delta)$. That is, $(\eta_{x0} u * \psi_m, \varphi) = (u * \psi_m, \varphi)$ for all $\varphi \in D(B(x_0, \delta)))$. Indeed,

$$\begin{aligned}(\eta_{x0} u * \psi_m, \varphi) &= (\eta_{x_o}(x)u(x), (\psi_m(y), \varphi(x+y)) \\ &= (\eta_{x_o}(x)u(x), ((\psi(y') \otimes \delta(y''), \varphi(x'+y', x''+y'')))) \\ &= (\eta_{x_o}(x)u(x), (\psi(y'), \varphi(x'+y', x''))) \\ &= (u(x), (\eta_{x_o}(x)\psi(y'), \varphi(x+y'))) \\ &= (u(x), \int_{R^m} \eta_{x_o}(x)\psi(y')\varphi(x+y')dy')\end{aligned}$$

since $x = (x', x'') = (x'+y', x'') - (y', 0) \in B(x_0, \delta) - supp(\psi_m)$ for all $y' \in supp(\psi)$. Then for all x with $x + y' \in supp(\varphi) \subset B(x_0, \delta)$, we have $\eta_{x0}(x) = 1$. Consequently

$$(\eta_{x0} u * \psi_m, \varphi) = (u(x), \int_{R^m} \varphi(x+y')\psi(y')dy') = (u * \psi_m, \varphi)$$

for all $\varphi \in D(B(x_0, \delta))$. Hence $u * \psi_m \in C^\infty(\Omega_{\psi_m})$ and we are done. □

Acknowledgment

The authors thank Prof. V. I. Burenkov, Prof. M. L. Goldman and Prof. S. E. Mikhailov for their valuable advice and suggestions.

References

1. V. S. Vladimirov (1979), *Generalized Functions in Mathematical Physics*, Mir Publishers, Moscow.
2. L. Hörmander (1969), *Linear Partial Differential Operators*, New York.
3. T. G. Ayele (1999), *Description of condition of partial hypoellipticity of operators in terms of their fundamental solutions*, All-Union Institute of Scientific and Technical Information, Russian Academy of Sciences, No. 1027-B99 Moscow, 1999.
4. T. G. Ayele (1999), *About smoothness of solutions and conditions of localization for spectral decomposition of operators with constant coefficients*, Ph.D. thesis, Moscow, Peoples' Friendship University of Russia.

SOME EMBEDDINGS AND EQUIVALENT NORMS OF THE $\mathcal{L}^{\lambda,s}_{p,q}$ SPACES

DOUADI DRIHEM

Department of Mathematics L.M.P.A., M'Sila University, P.O. Box 166, M'Sila 28000 Algeria
E-mail: douadidr@yahoo.fr.

Our aim is to give some properties of the $\mathcal{L}^{\lambda,s}_{p,q}$ spaces, especially concerning embeddings and equivalent norms based on maximal functions and local means.

Keywords: Besov spaces, Campanato spaces, Triebel-Lizorkin spaces, $\mathcal{L}^{\lambda,s}_{p,q}$ spaces, local means, maximal functions.

1. Introduction

G. Bourdaud [1] showed that an function $f \in L^2_{\text{loc}}$ belongs to $BMO(\mathbb{R}^n)$ (bounded mean oscillation) if and only if

$$\sup_{B_J} \frac{1}{|B_J|} \sum_{j \geq J} \left\| \mathcal{F}^{-1}(\varphi_j \mathcal{F} f) \mid L^2(B_J) \right\|^2 < \infty, \tag{1}$$

where the supremum is taken over all $J \in \mathbb{Z}$ and all balls B_J of $\mathbb{R}^n$ with radius 2^{-J} and $\{\Psi, \varphi_j\}_{j\in\mathbb{N}}$ is a smooth dyadic resolution of unity in $\mathbb{R}^n$ (see Definition 2.1). This gives that the important space $BMO(\mathbb{R}^n)$ can be described by the Littlewood-Paley decomposition. The idea of G. Bourdaud was used in [7] to get the Littlewood-Paley characterization for Campanato spaces $\dot{\mathcal{L}}^{2,\lambda}$ (which contain as a special case the space $BMO(\mathbb{R}^n)$) and their local versions L^{α}_2, where [7] showed that if $0 \leq \lambda < n+2$, the Campanato space $\dot{\mathcal{L}}^{2,\lambda}$ coincides algebraically and topologically with the space $\dot{\mathcal{L}}^{\lambda,0}_{2,2}(\mathbb{R}^n)$, which can be obtained by replacing in (1), $|B_J|$ by $|B_J|^{\lambda/n}$.

In this work, the focus is to give some properties for the spaces $\mathcal{L}^{\lambda,s}_{p,q}$ $(s \in \mathbb{R}, \lambda \geq 0, 0 < p, q < \infty)$, this new class of function spaces is defined as the set of all tempered distributions f, such that

$$\left\| f \mid \mathcal{L}^{\lambda,s}_{p,q} \right\| = \left(\sup_{B_J} \frac{1}{|B_J|^{\lambda/n}} \sum_{j \geq J^+} 2^{jsq} \left\| \mathcal{F}^{-1}(\varphi_j \mathcal{F} f) \mid L^p(B_J) \right\|^q \right)^{1/q} < \infty,$$

where the supremum is taken over all $J \in \mathbb{Z}$ and all balls B_J of $\mathbb{R}^n$ with radius 2^{-J} and $\{\varphi_j\}_{j\in\mathbb{N}_0}$ is the smooth dyadic resolution of unity in $\mathbb{R}^n$.

In this work we will establish, under appropriate assumptions on the parameters, some embedding theorems between a pair of these spaces, between these spaces and the Besov spaces or the Triebel-Lizorkin spaces. Also we will give a useful characterization of $\mathcal{L}^{\lambda,s}_{p,q}$ spaces based on the so-called local means and maximal functions. Furthermore we will give some equivalent norms of the Local approximation spaces $\mathcal{L}^{\alpha}_2$ for $-n/2 \leq \alpha < 1$. The proof has as a starting point the technique used by H.-Q. Bui, M. Paluszyński and M. Taibleson (see [2,3]) and the simplified version of their papers given by V. S. Rychkov in [9].

2. Definitions and basic properties

As usual, $\mathbb{R}^n$ the n-dimensional real Euclidean space, $\mathbb{N}$ the collection of all natural numbers and $\mathbb{N}_0 = \mathbb{N} \cup \{0\}$. The letter $\mathbb{Z}$ stands for the set of all integer numbers. For $v \in \mathbb{Z}$, we set $v^+ = \max\{v, 0\}$.

By $\mathcal{S}(\mathbb{R}^n)$ we denote the Schwartz space of all complex-valued, infinitely differentiable and rapidly decreasing functions on $\mathbb{R}^n$ and by $\mathcal{S}'(\mathbb{R}^n)$ the dual space of all tempered distributions on $\mathbb{R}^n$. We define the Fourier transform of a function $f \in \mathcal{S}(\mathbb{R}^n)$ by

$$\mathcal{F}(f)(\xi) = \hat{f}(\xi) = (2\pi)^{-n/2} \int_{\mathbb{R}^n} f(x)\, e^{-ix\cdot\xi} dx.$$

Its inverse is denoted by $\mathcal{F}^{-1}f$ or $\overset{\vee}{f}$. Both $\mathcal{F}$ and $\mathcal{F}^{-1}$ are extended to the dual Schwartz space $\mathcal{S}'(\mathbb{R}^n)$ in the usual way.

If $s \in \mathbb{R}, 0 < q \leq \infty$ and $J \in \mathbb{Z}$, then $\ell^s_{q,J}$ is the set of all sequences $\{f_k\}_{k \geq J^+}$ of complex numbers such that

$$\left\| \{f_k\}_{k\geq J^+} \mid \ell^s_{q,J^+} \right\| = \left(\sum_{k\geq J^+} 2^{ksq} |f_k|^q \right)^{1/q} < \infty,$$

with the obvious modification if $q = \infty$. Given two quasi-Banach spaces X and Y, we write $X \hookrightarrow Y$ if $X \subset Y$ and the natural embedding of X in Y is continuous. We write $X \not\hookrightarrow Y$ if there is an f such that $f \in X$ but $f \notin Y$. We shall use c to denote positive constant which may differ at each appearance.

In this section we recall some definitions and some necessary tools.

Definition 2.1. Let Ψ be a function in $\mathcal{S}(\mathbb{R}^n)$ satisfying $0 \leq \Psi(x) \leq 1$, $\Psi(x) = 1$ for $|x| \leq 1$ and $\Psi(x) = 0$ for $|x| \geq \frac{3}{2}$. We put $\varphi_0(x) = \Psi(x)$,

$\varphi_1(x) = \Psi(x/2) - \Psi(x)$ and $\varphi_j(x) = \varphi_1(2^{-j+1}x)$ for $j \geq 2$, then we have $\operatorname{supp}\varphi_j \subset \{x \in \mathbb{R}^n : 2^{j-1} \leq |x| \leq 3 \cdot 2^{j-1}\}$, $\varphi_j(x) = 1$ for $3 \cdot 2^{j-2} \leq |x| \leq 2^j$ and

$$\Psi(x) + \sum_{j \geq 1} \varphi_j(x) = 1 \quad \text{for all } x \in \mathbb{R}^n.$$

The system of functions $\{\varphi_j\}$ is called a smooth dyadic resolution of unity. For any $f \in \mathcal{S}'(\mathbb{R}^n)$, we define the convolution operators Δ_j by $\Delta_j f = \check{\varphi}_j * f$ for $j \in \mathbb{N}$ and $\Delta_0 f = \check{\Psi} * f$. Thus we obtain the Littlewood-Paley decomposition

$$f = \sum_{j \geq 0} \Delta_j f$$

of all $f \in \mathcal{S}'(\mathbb{R}^n)$ (convergence in $\mathcal{S}'(\mathbb{R}^n)$).

Definition 2.2. Let $s \in \mathbb{R}, J \in \mathbb{Z}, 0 < p, q \leq \infty$ and $\Omega \subseteq \mathbb{R}^n$. Then $\ell^s_{q,J^+}(L^p(\Omega))$ is the set of all sequences $\{f_k\}_{k \geq J^+}$ of complex-valued Lebesgue measurable functions such that

$$\left\| \{f_k\}_{k \geq J^+} \mid \ell^s_{q,J^+}(L^p(\Omega)) \right\| = \left(\sum_{k \geq J^+} 2^{ksq} \|f_k \mid L^p(\Omega)\|^q \right)^{1/q} < \infty.$$

with the obvious modification if $p = \infty$ and/or $q = \infty$.

Note that when $J = 0$ and $\Omega = \mathbb{R}^n$ we have $\ell^s_{q,0}(L^p(\mathbb{R}^n)) = \ell^s_q(L^p)$. We now define the spaces $\mathcal{L}^{\lambda,s}_{p,q}$ which will be our main object of study.

Definition 2.3. Let $s \in \mathbb{R}, \lambda \geq 0$ and $0 < p, q < \infty$. The space $\mathcal{L}^{\lambda,s}_{p,q}$ is the collection of all $f \in \mathcal{S}'(\mathbb{R}^n)$ such that

$$\|f \mid \mathcal{L}^{\lambda,s}_{p,q}\| = \left(\sup_{B_J} \frac{1}{|B_J|^{\lambda/n}} \left\| \{\Delta_j f\}_{j \geq J^+} \mid \ell^s_{q,J^+}(L^p(B_J)t) \right\|^q \right)^{1/q} < \infty, \quad (2)$$

where the supremum is taken over all $J \in \mathbb{Z}$ and all balls B_J of $\mathbb{R}^n$ with radius 2^{-J}.

Note that the space $\mathcal{L}^{\lambda,s}_{p,q}$ equipped with the norm (2) is a quasi-Banach space. Now we recall the definition of local approximation spaces.

Definition 2.4. Let $1 \leq p < +\infty, \alpha \geq -\frac{n}{p}$ and $N = \max(-1, [\alpha])$, where $[\alpha]$ is the integer part of α. We say $f \in \mathcal{L}^\alpha_p$ if and only if $f \in L^p_{\text{loc}}(\mathbb{R}^n)$

and for some constant $M = M(f)$, for every cube Q of side length δ, there exists a polynomial P_B of degree $\leq N$ ($P_B = 0$ if $N = -1$) such that

$$\left(\sup_B \frac{1}{|B|^{1+\alpha p/n}} \int_B |f - P_B|^p \, dx\right)^{1/p} < M, \quad \text{if} \quad 0 < \delta < 1,$$

$$\left(\sup_B \frac{1}{|B|} \int_B |f(x)|^p \, dx\right)^{1/p} < M, \quad \text{if} \quad \delta = 1.$$

We denote by $\|f \mid \mathcal{L}_p^\alpha\|$ the infimum of the constants M as above.

The following theorem gives some important properties of the spaces $\mathcal{L}_{2,2}^{\lambda,0}$ that are proved in [7].

Theorem 2.1. *Let $s \in \mathbb{R}$ and $-n/2 \leq \alpha < 1$. Then*

$$\mathcal{L}_{2,2}^{2\alpha+n,0} = \mathcal{L}_2^\alpha \quad \text{and} \quad \mathcal{L}_{2,2}^{2\alpha+n,s} = I^s(\mathcal{L}_2^\alpha).$$

Here $I^s(\mathcal{L}_2^\alpha)$ denotes the image of $\mathcal{L}_2^\alpha$ under I^s (Riesz potential operator).

Remark 2.1. Some import properties of the homogeneous counterpart of the space $\mathcal{L}_{p,q}^{\lambda,s}$ are given in [7].

Now we give a definition of the spaces $B_{p,q}^s$ and $F_{p,q}^s$.

Definition 2.5. (i) Let $s \in \mathbb{R}$ and $0 < p, q \leq \infty$. The Besov space $B_{p,q}^s$ is the collection of all $f \in \mathcal{S}'(\mathbb{R}^n)$ such that

$$\|f \mid B_{p,q}^s\| = \left\|\{\Delta_j f\}_{j\geq 0} \mid \ell_q^s(L^p)\right\| < \infty.$$

(ii) Let $s \in \mathbb{R}, 0 < p < \infty$ and $0 < q \leq \infty$. The Triebel-Lizorkin space $F_{p,q}^s$ is the collection of all $f \in \mathcal{S}'(\mathbb{R}^n)$ such that

$$\|f \mid F_{p,q}^s\| = \left\|\{\Delta_j f\}_{j\geq 0} \mid L^p(\ell_q^s)\right\| < \infty.$$

Here if $s \in \mathbb{R}, 0 < p < \infty$ and $0 < q \leq \infty$, then $L^p(\ell_q^s)$ is the set of all sequences $\{f_k\}_{k\geq 0}$ of complex-valued Lebesgue measurable functions such that

$$\left\|\{f_k\}_{k\geq 0} \mid L^p(\ell_q^s)\right\| = \left\|\left\|\{2^{ks} f_k\}_{k\geq 0} \mid \ell_q\right\|\right\|_p < \infty,$$

with the obvious modification if $q = \infty$. The following lemma is from Young's inequality in $\ell_{q,J+}^0$.

Lemma 2.1. *Let* $0 < a < 1, J \in \mathbb{Z}$ *and* $0 < q \leq \infty$. *Let* $\{\varepsilon_k\}$ *be a sequence of positive real numbers such that* $\left\| \{\varepsilon_k\}_{k \geq J^+} \mid \ell^0_{q,J^+} \right\| = I < \infty$. *The sequence* $\left\{ \eta_k : \eta_k = \sum_{j=k}^{\infty} a^{j-k} \varepsilon_j \right\}$ *is in* ℓ^0_{q,J^+} *with*

$$\left\| \{\eta_k\}_{k \geq J^+} \mid \ell^0_{q,J^+} \right\| \leq c\, I.$$

c depends only on a and q.

3. Embeddings

The following results are generalizations of the results of [6].

Theorem 3.1. *Let* $0 < t < p < \infty$, $s \in \mathbb{R}$, $r - n/t = s - n/p$ *and* $0 < q, q_1 < \infty$. *Then* $\mathcal{L}^{\lambda,r}_{t,q_1} \hookrightarrow \mathcal{L}^{\lambda,s}_{p,q}$ *if and only if* $0 < q_1 \leq q < \infty$.

Theorem 3.2. *Let* $s, r \in \mathbb{R}$, $\lambda \geq 0$, $0 < p, q, q_1 < \infty$ *and* $0 < t \leq \infty$.
(i) Let $0 < t \leq p < \infty$. *Then* $B^{s+\varepsilon}_{t,q} \hookrightarrow \mathcal{L}^{\lambda,s}_{p,q}$ *if and only if* $\varepsilon \geq \lambda/q + n/t - n/p$.
(ii) We have $B^{s}_{p,q_1} \hookrightarrow \mathcal{L}^{0,s}_{p,q}$ *if and only if* $0 < q_1 \leq q < \infty$.
(iii) Let $\lambda > 0$. *Then* $B^{s+\varepsilon}_{p,\infty} \hookrightarrow \mathcal{L}^{\lambda,s}_{p,q}$ *if and only if* $\varepsilon \geq \lambda/q$.
(iv) If $\lambda \geq nq/p$ *and* $\varepsilon > \lambda/q + n/t - n/p$, *we have* $\mathcal{L}^{\lambda,s}_{p,q} \not\hookrightarrow B^{s+\varepsilon}_{t,q}$, $1 < t \leq \infty$.
(v) If $\lambda \geq nq/p$ *and* $r \in \mathbb{R}$, *we have* $\mathcal{L}^{\lambda,s}_{p,q} \not\hookrightarrow B^{r}_{t,q}$, $0 < t \leq 1$.

4. Equivalent quasi-norms

In order to formulate the main result of this paper, let us consider $k_0, k \in \mathcal{S}(\mathbb{R}^n)$ and an integer $S \geq -1$ such that for some $\varepsilon > 0$

$$\left| \hat{k_0}(\xi) \right| > 0 \quad \text{for} \quad |\xi| < 2\varepsilon, \quad \left| \hat{k}(\xi) \right| > 0 \quad \text{for} \quad \frac{\varepsilon}{2} < |\xi| < 2\varepsilon, \qquad (3)$$

and

$$\int_{\mathbb{R}^n} x^\alpha k(x)\, dx = 0 \quad \text{for any} \quad |\alpha| \leq S. \qquad (4)$$

Here (3) are Tauberian conditions, while (4) are moment conditions on k. We recall the notation $k_j(x) = 2^{jn} k(2^j x)$, for $j \geq 1$. For any $a > 0$, $f \in \mathcal{S}'(\mathbb{R}^n)$, $j \in \mathbb{N}_0$ and $x \in \mathbb{R}^n$ we denote (Peetre's maximal functions)

$$k_j^{*,a} f(x) = \sup_{y \in \mathbb{R}^n} \frac{|k_j * f(y)|}{(1 + 2^j |x - y|)^a}, \quad k_{j,J}^{*,a} f(x) = \sup_{y \in B_J} \frac{|k_j * f(y)|}{(1 + 2^j |x - y|)^a}. \qquad (5)$$

Usually $k_j * f$ is called a 'local mean'. We are able now to state the main result of this paper.

Theorem 4.1. *Let $\lambda \geq 0, 0 < p, q < \infty, s < S + 1$ and $a > \max(n/p, \lambda/q)$. Then*

$$\left\| f \mid \mathcal{L}^{\lambda,s}_{p,q} \right\|' = \left(\sup_{B_J} \frac{1}{|B_J|^{\lambda/n}} \left\| \{k_j^{*,a} f\}_{j \geq J^+} \mid \ell^s_{q,J^+} \left(L^p \left(B_J\right)\right) \right\|^q \right)^{1/q} \tag{6}$$

and

$$\left\| f \mid \mathcal{L}^{\lambda,s}_{p,q} \right\|'' = \left(\sup_{B_J} \frac{1}{|B_J|^{\lambda/n}} \left\| \{k_j * f\}_{j \geq J^+} \mid \ell^s_{q,J^+} \left(L^p \left(B_J\right)\right) \right\|^q \right)^{1/q} \tag{7}$$

are equivalent quasi-norms in $\mathcal{L}^{\lambda,s}_{p,q}$.

Theorem 4.2. *Let $\lambda \geq 0, 0 < p, q < \infty, s < S + 1 - \lambda/q$ and $a > n/p$. Then*

$$\left\| f \mid \mathcal{L}^{\lambda,s}_{p,q} \right\|''' = \left(\sup_{B_J} \frac{1}{|B_J|^{\lambda/n}} \left\| \left\{ k_{j,J}^{*,a} f \right\}_{j \geq J^+} \mid \ell^s_{q,J^+} \left(L^p \left(B_J\right)\right) \right\|^q \right)^{1/q} \tag{8}$$

is an equivalent quasi-norm in $\mathcal{L}^{\lambda,s}_{p,q}$.

Remark 4.1. In view of the inequalities

$$k_{j,J}^{*,a} f(x) \leq k_j^{*,a} f(x) \quad \text{and} \quad k_j * f(x) \leq k_{j,J}^{*,a} f(x),$$

for any $f \in \mathcal{S}'(\mathbb{R}^n)$ and any $x \subset B_J$, it follows from Theorem 4.1 that in the case of $a > \max(n/p, \lambda/q)$, (8) is a simple consequence of (6) and (7).

Corollary 4.1. *Let $-n/2 \leq \alpha < 1$ and $a > 0$. Then $\left\| f \mid \mathcal{L}^{2\alpha+n,0}_{2,2} \right\|'$ (in case of $a > \max(n/2, n/2 + \alpha)$), $\left\| f \mid \mathcal{L}^{2\alpha+n,0}_{2,2} \right\|''$ and $\left\| f \mid \mathcal{L}^{2\alpha+n,0}_{2,2} \right\|'''$ (in case of $a > n/2$ and $S > -1 + n/2 + \alpha$) are equivalent norms in $\mathcal{L}^{\alpha}_{2}$.*

The proof of Corollary 4.1 is immediate because $\mathcal{L}^{2\alpha+n,0}_{2,2} = \mathcal{L}^{\alpha}_{2}$ for any $-n/2 \leq \alpha < 1$.

Let $\mu_0, \mu \in \mathcal{S}(\mathbb{R}^n)$ be two positive functions on $\mathbb{R}^n$ such that

$$\mu_0(\xi) = 1 \ \text{if} \ |\xi| \leq 2 \ \text{and} \ \operatorname{supp} \mu_0 \subset \{\xi \in \mathbb{R}^n : |\xi| \leq 4\} \tag{9}$$

and

$$\mu(\xi) = 1 \ \text{if} \ 1/2 \leq |\xi| \leq 2 \ \text{and} \ \operatorname{supp} \mu \subset \{\xi \in \mathbb{R}^n : 1/4 \leq |\xi| \leq 4\}. \tag{10}$$

For any $j \geq 1$ and $\xi \in \mathbb{R}^n$ we define $\mu_j(\xi) = \mu\left(2^{-j}\xi\right)$. The results below will play a key role in the proof of Theorems 4.1 and 4.2.

Theorem 4.3. *Let $s \in \mathbb{R}, \lambda \geq 0, 0 < p, q < \infty$ and $a > \max(n/p, \lambda/q)$. Then*

$$\left\| f \mid \mathcal{L}_{p,q}^{\lambda,s} \right\| = \left(\sup_{B_J} \frac{1}{|B_J|^{\lambda/n}} \left\| \{\Delta_j^{*,a} f\}_{j \geq J^+} \mid \ell_{q,J^+}^s \left(L^p(B_J)\right) \right\|^q \right)^{1/q} \tag{11}$$

is an equivalent quasi-norm in $\mathcal{L}_{p,q}^{\lambda,s}$.

Theorem 4.4. *Let $s \in \mathbb{R}, \lambda \geq 0, 0 < p, q < \infty$ and $a > n/p$. Then*

$$\left\| f \mid \mathcal{L}_{p,q}^{\lambda,s} \right\| = \left(\sup_{B_J} \frac{1}{|B_J|^{\lambda/n}} \left\| \left\{\Delta_{j,J}^{*,a} f\right\}_{j \geq J^+} \mid \ell_{q,J^+}^s \left(L^p(B_J)\right) \right\|^q \right)^{1/q}$$

is an equivalent quasi-norm in $\mathcal{L}_{p,q}^{\lambda,s}$.

Theorem 4.5. *Let $s \in \mathbb{R}, \lambda \geq 0, 0 < p, q < \infty$. Then under the above assumptions on the functions μ_0 and μ,*

$$\left\| f \mid \mathcal{L}_{p,q}^{\lambda,s} \right\|_\mu = \left(\sup_{B_J} \frac{1}{|B_J|^{\lambda/n}} \left\| \left\{\overset{\vee}{\mu}_j * f\right\}_{j \geq J^+} \mid \ell_{q,J^+}^s \left(L^p(B_J)\right) \right\|^q \right)^{1/q}$$

is an equivalent quasi-norm in $\mathcal{L}_{p,q}^{\lambda,s}$.

Proof. We will give only the proof of Theorem 4.1. The idea is from V. S. Rychkov [9].

Step 1. Take any pair of functions φ_0 and $\varphi \in \mathcal{S}(\mathbb{R}^n)$ so that for an $\acute{\varepsilon} > 0$,

$$\left|\overset{\wedge}{\varphi}_0(\xi)\right| > 0 \quad \text{for} \quad |\xi| < 2\acute{\varepsilon}, \quad \left|\overset{\wedge}{\varphi}(\xi)\right| > 0 \quad \text{for} \quad \frac{\acute{\varepsilon}}{2} < |\xi| < 2\acute{\varepsilon},$$

and define for any $a > 0$ the functions $\varphi_j^{*,a} f$ as in (5). Then there is a constant $c > 0$ such that for any $f \in \mathcal{S}'(\mathbb{R}^n)$

$$\left\| f \mid \mathcal{L}_{p,q}^{\lambda,s} \right\|' \leq c \left(\sup_{B_J} \frac{1}{|B_J|^{\lambda/n}} \left\| \{\varphi_j^{*,a} f\}_{j \geq J^+} \mid \ell_{q,J^+}^s \left(L^p(B_J)\right) \right\|^q \right)^{1/q}. \tag{12}$$

To prove (12) we consider the cases $J \geq 1$ (or $J \leq 0$ and $j \geq 1$) and ($J \leq 0$ and $j = 0$) separately. This estimate can be obtained by the same arguments as in [9].

Step 2. We will prove in this step that there is a constant $c > 0$ such that for any $f \in \mathcal{S}'(\mathbb{R}^n)$

$$\left\| f \mid \mathcal{L}_{p,q}^{\lambda,s} \right\|' \leq c \left\| f \mid \mathcal{L}_{p,q}^{\lambda,s} \right\|''. \tag{13}$$

Analogously to (3) find two functions Λ, $\omega \in \mathcal{S}(\mathbb{R}^n)$, so that for an $\varepsilon > 0$, supp $\hat{\Lambda} \subset \{\xi \in \mathbb{R}^n : |\xi| < 2\varepsilon\}$, supp $\hat{\omega} \subset \{\xi \in \mathbb{R}^n : \varepsilon/2 < |\xi| < 2\varepsilon\}$ and for all $f \in \mathcal{S}'(\mathbb{R}^n)$ and $j \in \mathbb{N}_0$

$$k_j * f = \Lambda_j * (k_0)_j * k_j * f + \sum_{m \geq j+1} k_j * \omega_m * k_m * f.$$

By an argument similar to the one in [9], we deduce that for all $f \in \mathcal{S}'(\mathbb{R}^n)$, $x \in B_J$ and $j \in \mathbb{N}$ that

$$\left|k_j^{*,a} f(x)\right|^r \leq c \sum_{m \geq j} 2^{(j-m)Nr+mn} \int_{\mathbb{R}^n} \frac{|k_m * f(z)|^r}{(1 + 2^m |x-z|)^{ar}} dz,$$

where $0 < r < \infty$, $N \in \mathbb{N}$ can be still be taken arbitrarily large and $c > 0$ is independent of j. Together with the corresponding estimate for $k_0^{*,a} f$. Let $J \in \mathbb{Z}$ and B_J a ball of $\mathbb{R}^n$ centered at x_0 with radius 2^{-J}. Then

$$\int_{\mathbb{R}^n} \frac{|k_m * f(z)|^r}{(1 + 2^m |x-z|)^{ar}} dz = \int_{B_{J-1}} (\ldots)\, dz + \sum_{i \geq 0} \int_{B_{J-i-2} \setminus B_{J-i-1}} (\ldots)\, dz. \quad (14)$$

It possible to choose r so that $n \max\left(\frac{1}{a}, \frac{1}{a+n/p-\lambda/q}\right) < r < p$; we make such a choice and fix r for the rest of the proof. Now the function $z \mapsto \frac{1}{(1+|z|)^{ar}}$ is in L^1 and we may use the majorant property $\left(|g|^r * \frac{1}{(1+|\cdot|)^{ar}}\right)(x) \leq cM(|g|^r)(x)$ for the Hardy-Littlewood maximal operator M (see Stein and Weiss [11]). It follows that the right-hand side of (14) is bounded by

$$2^{-mn} M\left(|k_m * f|^r \chi_{B_{J-1}}\right)(x) + \sum_{i \geq 0} 2^{Jn-(iar+mn)} \|k_m * f \mid L^r(B_{J-i-2})\|^r,$$

where we have used $|x-z| \geq |z-x_0| - |x-x_0| \geq 2^{i-J}$ and $2^{(m-J)(n-ar)} \leq 1$. Since the Hardy-Littlewood maximal operator M maps $L^\rho(\mathbb{R}^n)$ into $L^\rho(\mathbb{R}^n)$ for $1 < \rho \leq \infty$ and since $\|k_m * f \mid L^r(B_{J-i-2})\| \leq c\, 2^{(i-J)(n/r-n/p)} \|k_m * f \mid L^p(B_{J-i-2})\|$, we get by Lemma 2.1 and the embedding $\ell^s_{q,(J-i-2)^+}(L^p(B_{J-i-2})) \hookrightarrow \ell^s_{q,J^+}(L^p(B_{J-i-2}))$, for any $i \in \mathbb{N}_0 \cup \{-1\}$, that

$$\begin{aligned} &\left\| \left\{k_j^{*,a} f\right\}_{j \geq J^+} \mid \ell^{s,q}_{J^+}(L^p(B_J)) \right\| \\ &\leq c \left(\sum_{i \geq -1} 2^{i^+(n(1-r/p)-ar)} \left\| \{k_m * f\}_{j \geq (J-i-2)^+} \mid \ell^{s,q}_{J^+}(L^p(B_{J-i-2})) \right\|^r \right)^{1/r} \\ &\leq c\, 2^{\lambda J/q} \left\| \left\{2^{-i(a-n(1/r-1/p)-\lambda/q)}\right\} \mid \ell_r \right\| \left\| f \mid \mathcal{L}^{\lambda,s}_{p,q} \right\|''. \end{aligned}$$

Whence, multiplying through by $2^{-\lambda J/q}$, using $|B_J|^{-\lambda/n} = 2^{\lambda J}$, and taking the sup over all $J \in \mathbb{Z}$ and all balls B_J of $\mathbb{R}^n$ with radius 2^{-J}, we get the estimate (12) by the fact that $\{2^{-i(a-n(1/r-1/p)-\lambda/q)}\} \in \ell_r$.

Step 3. We will prove in this step that for all $f \in \mathcal{S}'(\mathbb{R}^n)$ the following estimates are true:

$$\left\|f \mid \mathcal{L}_{p,q}^{\lambda,s}\right\|' \leq c \left\|f \mid \mathcal{L}_{p,q}^{\lambda,s}\right\| \quad \text{and} \quad \left\|f \mid \mathcal{L}_{p,q}^{\lambda,s}\right\| \leq c \left\|f \mid \mathcal{L}_{p,q}^{\lambda,s}\right\|''. \tag{15}$$

Let μ_0 and $\mu \in \mathcal{S}(\mathbb{R}^n)$ be two positive functions on $\mathbb{R}^n$ satisfying (9) and (10). Let $\overset{\wedge}{\varphi}_0 = \mu_0$ and $\overset{\wedge}{\varphi} = \mu$. The first inequality in (15) is proved by the chain of the estimates

$$\begin{aligned}\left\|f \mid \mathcal{L}_{p,q}^{\lambda,s}\right\|' &\leq c \left(\sup_{B_J} \frac{1}{|B_J|^{\lambda/n}} \left\|\{\varphi_j^{*,a} f\}_{j \geq J^+} \mid \ell_{q,J^+}^s \left(L^p(B_J)\right)\right\|^q\right)^{1/q} \\ &\leq c \left(\sup_{B_J} \frac{1}{|B_J|^{\lambda/n}} \left\|\{\varphi_j * f\}_{j \geq J^+} \mid \ell_{q,J^+}^s \left(L^p(B_J)\right)\right\|^q\right)^{1/q} \\ &\leq c \left\|f \mid \mathcal{L}_{p,q}^{\lambda,s}\right\|,\end{aligned}$$

where the first inequality is (12) (see Step 1), the second inequality is (13) (with φ and φ_0 instead of k and k_0 (see Step 2), and finally the last inequality follows by Theorem 4.5. The proof of the second inequality in (15) is by the chain

$$\begin{aligned}\left\|f \mid \mathcal{L}_{p,q}^{\lambda,s}\right\| &\leq c \left(\sup_{B_J} \frac{1}{|B_J|^{\lambda/n}} \left\|\{\varphi_j^{*,a} f\}_{j \geq J^+} \mid \ell_{q,J^+}^s \left(L^p(B_J)\right)\right\|^q\right)^{1/q} \\ &\leq c \left\|f \mid \mathcal{L}_{p,q}^{\lambda,s}\right\|' \leq c \left\|f \mid \mathcal{L}_{p,q}^{\lambda,s}\right\|'',\end{aligned}$$

where the first inequality is an obvious consequence of (11), the second inequality is (12) (see Step 1) with the roles of k_0 and k, φ_0 and φ, respectively, interchanged, and finally the last inequality is (13) (see Step 2). Hence the theorem is proved. □

References

1. G. Bourdaud, *Analyse fonctionnelle dans l'espace Euclidien*, Publ. Math. Univ. Paris VII, 23 (1987).
2. H. Q. Bui, M. Paluszyński and M. Taibleson, *A maximal characterization of weighted Besov-Lipschitz and Triebel-Lizorkin spaces*, Studia Math. 119 (1996), pp. 219-246.
3. H. Q. Bui, M. Paluszyński and M. Taibleson, *Characterization of the Besov-Lipschitz and Triebel-Lizorkin spaces. The case q < 1*, J. Fourier Anal. Appl. 3, special issue (1997), pp. 837-846.

4. S. Campanato, *Proprietà di Hölderianità di alcune classi di funzioni*, Ann. Scuola Norm. Sup. Pisa 17 (1963), pp. 175-188.
5. S. Campanato, *Proprietà di una famiglia di spazi funzionali*, Ann. Scuola. Norm. Sup. Pisa 18 (1964), pp. 137-160.
6. A. El Baraka, *An embedding theorem for Campanato spaces*, Electron. J. Diff. Eqns. 2002 (2002), N. 66, pp. 1-17.
7. A. El Baraka, *Littlewood-Paley characterization for Campanato spaces*, J. Funct. Spaces Appl. 4 (2) (2006), pp. 193-220.
8. M. Frazier, B. Jawerth and G. Weiss, *Littlewood-Paley theory and the study of function spaces*, CBMS, 79, 1991.
9. V. S. Rychkov, *On a theorem of Bui, Paluszyń ski, and Taibleson*, Proc. Steklov Inst. Math. 227 (1999), pp. 280-292.
10. W. Sickel and H. Triebel, *Hölder inequalities and sharp embeddings in function spaces of $B^s_{p,q}$ and $F^s_{p,q}$ type*, Z. Anal. Anwendungen 14 (1), (1995), pp. 105-140.
11. E. M. Stein and G. Weiss, *Introduction to Fourier analysis on Euclidean spaces*, Princeton Univ. Press, Princeton, NJ, 1971.
12. J.-O. Strömberg and A. Torchinsky, *Weighted Hardy spaces*, Lecture Notes in Math. 1381, Springer, Berlin, 1989.
13. H. Triebel, *Theory of function spaces*, Birkhäuser, Basel, 1983.
14. H. Triebel, *Theory of function spaces II*, Birkhäuser, Basel, 1992.

Session 15

Numerical Functional Analysis

SESSION EDITORS

A. Ashyralyev Fatih University, İstanbul, Turkey

P. E. Sobolevskii Universidade Federal do Ceará, Fortaleza, Brazil

GLOBAL EXPONENTIAL PERIODICITY FOR DISCRETE-TIME HOPFIELD NEURAL NETWORKS WITH FINITE DISTRIBUTED DELAYS AND IMPULSES

H. AKÇA

Mathematical Sciences Department, Faculty of Sciences,
United Arab Emirates University, Al Ain, UAE
E-mail: hakca@uaeu.ac.ae

V. COVACHEV

Department of Mathematics & Statistics, College of Science,
Sultan Qaboos University, Muscat, Sultanate of Oman
& Institute of Mathematics, Bulgarian Academy of Sciences, Sofia, Bulgaria
E-mail: vcovachev@hotmail.com & valery@squ.edu.om

Z. COVACHEVA

Higher College of Technology, Muscat, Sultanate of Oman
& Higher College of Telecommunications and Post, Sofia, Bulgaria
E-mail: zkovacheva@hotmail.com

S. MOHAMAD

Department of Mathematics, Faculty of Science,
Universiti Brunei Darussalam, Brunei Darussalam
E-mail: sannay@fos.ubd.edu.bn

The discrete counterpart of a class of Hopfield neural networks with periodic integral impulsive conditions and finite distributed delays is introduced. The continuation theorem of coincidence degree theory is used to obtain a sufficient condition for the existence of a periodic solution of the discrete system considered. By introducing an appropriate Lyapunov functional, a sufficient condition is obtained for the uniqueness and global exponential stability of the periodic solution.

Keywords: Discrete Hopfield neural networks; Finite distributed delays; Periodic impulses; Global exponential periodicity.

1. Introduction

In the present paper we introduce the discrete counterpart of a class of Hopfield neural networks with periodic integral impulsive conditions and finite distributed delays. We apply the continuation theorem of coincidence degree theory (Gaines and Mawhin [7]) to obtain a sufficient condition for the existence of a periodic solution of the discrete system considered. By introducing an appropriate Lyapunov functional we derive a sufficient condition for the uniqueness and global exponential stability of the periodic solution. For works proving the existence of a periodic solution of differential and difference equations by the coincidence degree theory the reader can see Fan and Agarwal [3], Fan and Wang [4,5], Fan *et al.* [6], Li [9], Li and Kuang [10], Li and Lu [11]. In particular, in Li and Lu [11] the existence of a periodic solution of Hopfield-type neural network with impulses is proved. In Zhou *et al.* [14] one proves the existence of a periodic solution of a discrete-time analogue of a bidirectional associative memory (BAM) neural network with periodic coefficients and finite distributed delays without impulses.

2. Statement of the problem. Main results

We consider a class of Hopfield neural networks with periodic integral impulsive conditions and finite distributed delays, which are formulated in the form of a system of impulsive delay differential equations

$$\frac{dx_i}{dt} = -a_i(t)x_i(t) + \sum_{j=1}^{m} b_{ij}(t)f_j\left(\int_0^{\omega} g_{ij}(s)x_j(t-s)\,ds\right) + I_i(t), \quad t \neq t_k, \tag{1}$$

$$\begin{aligned}\Delta x_i(t_k) &\equiv x_i(t_k+0) - x_i(t_k) \\ &= -\gamma_{ik}x_i(t_k) + \sum_{j=1}^{m} B_{ijk}\Phi_j\left(\int_0^{\omega} c_{ij}(s)x_j(t_k-s)\,ds\right) + \alpha_{ik}, \\ & i = \overline{1,m}, \quad k \in \mathbb{Z},\end{aligned} \tag{2}$$

where m is the number of neurons in the network, $x_i(t)$ is the state of the i-th neuron at time t, $a_i(t) > 0$ is the rate at which the i-th neuron resets its state when isolated from the system, $b_{ij}(t)$ is the synaptic connection weight from the j-th neuron to the i-th one, $f_j(\cdot)$ are signal transmission functions of the j-th neuron, ω is the maximum transmission delay from one neuron to another, $g_{ij}(\cdot)$ and $c_{ij}(\cdot)$ are nonnegative delay kernels, $I_i(t)$ is the external input to the i-th neuron, t_k $(k \in \mathbb{Z})$ are the instants of impulse effect which form a strictly increasing sequence, γ_{ik} ($i = \overline{1,m}$, $k \in \mathbb{Z}$) are positive constants.

We assume that the above system (1), (2) satisfies the following periodicity conditions: $a_i(t)$, $b_{ij}(t)$, $I_i(t)$ are ω-periodic in t; $t_{k+p} = t_k + \omega$, $\gamma_{i,k+p} = \gamma_{ik}$, $B_{ij,k+p} = B_{ijk}$, $\alpha_{i,k+p} = \alpha_{ik}$. Without loss of generality we can assume that

$$0 < t_1 < t_2 < \cdots < t_p < \omega.$$

The Hopfield neural network (1) is similar to the bidirectional associative memory neural network considered in Zhou *et al.* [14].

Combining some ideas of Mohamad and Gopalsamy [13], Akça *et al.* [1], Zhou *et al.* [14] we shall formulate the discrete counterpart of system (1), (2). For a positive integer N we choose the discretization step $h = \omega/N$. For the moment we assume N so large that

$$h < \min_{k=\overline{1,p}} (t_{k+1} - t_k).$$

Then each interval $[nh, (n+1)h]$ contains at most one instant of impulse effect t_k.

For convenience we denote $n = [t/h]$, the greatest integer in t/h, and $n_k = [t_k/h]$. Clearly, we will have $n_{k+p} = n_k + N$ for all $k \in \mathbb{Z}$.

Let $n \in \mathbb{Z}$, $n \neq n_k$. This means that the interval $[nh, (n+1)h]$ contains no instant of impulse effect t_k.

We approximate the integral term in (1) by a sum:

$$\int_0^\omega g_{ij}(s)x_j(t-s)\,ds \approx \sum_{\ell-1}^{N} g_{ij}(\ell h)x_j\left((n-\ell)h\right)\varphi(h),$$

where $\varphi(h) = h + O(h^2)$.

Next we approximate the differential equation (1) on the interval $[nh, (n+1)h]$ by

$$\frac{dx_i}{dt} + a_i(nh)x_i(t) = I_i(nh) + \sum_{j=1}^{m} b_{ij}(nh)f_j\left(\sum_{\ell=1}^{N} g_{ij}(\ell h)x_j\left((n-\ell)h\right)\varphi(h)\right).$$

We multiply both sides of this equation by $\exp\left(a_i(nh)t\right)$ and integrate over the interval $[nh, (n+1)h]$. Thus we obtain

$$x_i\left((n+1)h\right) - x_i(nh) = -\left(1 - e^{-a_i(nh)h}\right)x_i(nh) \tag{3}$$
$$+\frac{1 - e^{-a_i(nh)h}}{a_i(nh)}\left\{I_i(nh) + \sum_{j=1}^{m} b_{ij}(nh)f_j\left(\sum_{\ell=1}^{N} g_{ij}(\ell h)x_j\left((n-\ell)h\right)\varphi(h)\right)\right\}.$$

Henceforth by abuse of notation we write $x_i(n) = x_i(nh)$ and define $\Delta x_i(n) = x_i(n+1) - x_i(n)$ $(i = \overline{1,m},\ n \in \mathbb{Z})$. For convenience we adopt the notation

$$\begin{aligned}
A_i(n) &= 1 - e^{-a_i(nh)h} \qquad (i = \overline{1,m},\ n \in \mathbb{Z} \setminus \{n_k\}_{k\in\mathbb{Z}}),\\
I_i(n) &= \frac{1 - e^{-a_i(nh)h}}{a_i(nh)} I_i(nh) \qquad (i = \overline{1,m},\ n \in \mathbb{Z} \setminus \{n_k\}_{k\in\mathbb{Z}}),\\
b_{ij}(n) &= \frac{1 - e^{-a_i(nh)h}}{a_i(nh)} b_{ij}(nh) \qquad (i,j = \overline{1,m},\ n \in \mathbb{Z} \setminus \{n_k\}_{k\in\mathbb{Z}}),\\
g_{ij}(\ell) &= g_{ij}(\ell h)\varphi(h) \qquad (i,j = \overline{1,m},\ \ell = \overline{1,N}).
\end{aligned}$$

Clearly, we have $0 < A_i(n) < 1$. In particular, if $a_i(t) < \frac{1}{\omega}$, then $A_i(n) < \frac{1}{N}$.

With the above notation equation (3) takes the form

$$\Delta x_i(n) = -A_i(n)x_i(n) + I_i(n) + \sum_{j=1}^{m} b_{ij}(n) f_j\left(\sum_{\ell=1}^{N} g_{ij}(\ell)x_j(n-\ell)\right), \quad (4)$$
$$i = \overline{1,m}, \quad n \neq n_k.$$

Next, for $n = n_k$ the interval $[nh, (n+1)h]$ contains the instant of impulse effect t_k. On this interval we approximate the impulse condition (2) by

$$\Delta x_i(n_k) = -\gamma_{ik}x_i(n_k) + \alpha_{ik} + \sum_{j=1}^{m} B_{ijk}\Phi_j\left(\sum_{\ell=1}^{N} c_{ij}(\ell)x_j(n_k-\ell)\right), \quad (5)$$
$$i = \overline{1,m}, \quad k \in \mathbb{Z},$$

where

$$c_{ij}(\ell) = c_{ij}(\ell h)\varphi(h) \qquad (i,j = \overline{1,m}, \quad \ell = \overline{1,N}).$$

For uniformity of notation we define

$$A_i(n_k) = \gamma_{ik}, \quad I_i(n_k) = \alpha_{ik} \qquad (i = \overline{1,m},\ k \in \mathbb{Z}).$$

Now the difference system (4), (5) can be written in operator form as

$$\Delta x = Hx, \quad (6)$$

where

$$(Hx)_i(n) = -A_i(n)x_i(n) + I_i(n) \quad (7)$$
$$+ \begin{cases} \sum_{j=1}^{m} b_{ij}(n) f_j\left(\sum_{\ell=1}^{N} g_{ij}(\ell)x_j(n-\ell)\right), & n \neq n_k,\\ \sum_{j=1}^{m} B_{ijk}\Phi_j\left(\sum_{\ell=1}^{N} c_{ij}(\ell)x_j(n_k-\ell)\right), & n = n_k. \end{cases}$$

In order to formulate our assumptions, we need some more notation. Let

$$I_N = \{0,\, 1,\, \ldots,\, N-1\},$$

$$\underline{A}_i = \min_{n\in I_N} A_i(n), \quad \overline{A}_i = \sum_{n=0}^{N-1} A_i(n), \qquad i=\overline{1,m}.$$

Now we introduce the following conditions:

H1. $A_i(n+N) = A_i(n)$, $I_i(n+N) = I_i(n)$ for $i=\overline{1,m}$, $n \in \mathbb{Z}$; $n_k \in \mathbb{Z}$ for all $k \in \mathbb{Z}$ and $n_{k+p} = n_k + N$; $b_{ij}(n+N) = b_{ij}(n)$ $(n \neq n_k)$, $B_{ij,k+p} = B_{ijk}$ $(k \in \mathbb{Z})$ for $i,j=\overline{1,m}$.

H2. $\underline{A}_i > 0$, $\overline{A}_i < 1$ for $i=\overline{1,m}$.

H3. The functions $f_j(\cdot)$, $\Phi_j(\cdot)$ $(j=\overline{1,m})$ are bounded on $\mathbb{R}$ and there exist positive constants M_j and L_j such that

$$|f_j(x) - f_j(y)| \le M_j|x-y|, \quad |\Phi_j(x) - \Phi_j(y)| \le L_j|x-y|$$

for all $x, y \in \mathbb{R}$.

H4. $g_{ij}(\ell) \ge 0$, $c_{ij}(\ell) \ge 0$ for $i,j=\overline{1,m}$, $\ell=\overline{1,N}$.

We again introduce some notation:

$$\overline{I}_i = \max_{n\in I_N} |I_i(n)|, \qquad i=\overline{1,m},$$

$$\overline{b}_{ij} = \sup_{n\neq n_k} |b_{ij}(n)|, \quad \overline{B}_{ij} = \max_{k=\overline{1,p}} |B_{ijk}|, \qquad i,j=\overline{1,m}.$$

For an N-periodic sequence $v(n)$ we denote $\tilde{v} = \frac{1}{N}\sum_{n=0}^{N-1} v(n)$; for $i=\overline{1,m}$

$$\rho_i = \overline{I}_i + \frac{1}{N}\sum_{j=1}^{m}\left[(N-p)\overline{b}_{ij}|f_j(0)| + p\overline{B}_{ij}|\Phi_j(0)|\right].$$

Next we denote

$$\mathcal{M}_j = \max\{L_j,\, M_j\}, \qquad j=\overline{1,m},$$

$$G_{ij} = \sum_{\ell=1}^{N} g_{ij}(\ell), \quad C_{ij} = \sum_{\ell=1}^{N} c_{ij}(\ell), i,j=\overline{1,m},$$

$$\mathcal{B}_{ij} = \max\{\overline{b}_{ij},\, \overline{B}_{ij}\}, \quad \mathcal{G}_{ij} = \max\{G_{ij},\, C_{ij}\}, i,j=\overline{1,m}.$$

We introduce the $m\times m$ matrices

$$A = \operatorname{diag}\left(\frac{\underline{A}_i\,(1-\overline{A}_i)}{1+N\underline{A}_i},\ i=\overline{1,m}\right), \qquad B = (\mathcal{B}_{ij}\mathcal{M}_j\mathcal{G}_{ij})_{i,j=1}^{m}.$$

Then we introduce the conditions

H5. $\min\limits_{i=\overline{1,m}} \left(\tilde{A}_i - \mathcal{M}_i \sum\limits_{j=1}^{m} \mathcal{B}_{ji}\mathcal{G}_{ji} \right) > 0.$

H6. $\underline{A}_i > \mathcal{M}_i \sum\limits_{j=1}^{m} \mathcal{B}_{ji}\mathcal{G}_{ji}$ for $i = \overline{1,m}$.

H7. The matrix $A - B$ is an M-matrix (Fiedler [2], Horn and Johnson [8]).

Theorem 2.1. *Suppose that conditions* ***H1–H5, H7*** *hold. Then the equation* (6) *has at least one N-periodic solution.*

Sketch of the proof. We shall prove Theorem 2.1 using Mawhin's continuation theorem (Gaines and Mawhin [7, p. 40]). To state this theorem we need some preliminaries:

Let $\mathbb{X}, \mathbb{Y}$ be real Banach spaces, $L :\ \mathrm{Dom}\, L \subset \mathbb{X} \longrightarrow \mathbb{Y}$ be a linear mapping, and $H : \mathbb{X} \longrightarrow \mathbb{Y}$ be a continuous mapping. The mapping L will be called a Fredholm mapping of index zero if $\dim \mathrm{Ker}\, L = \mathrm{codim}\, \mathrm{Im}\, L < +\infty$ and $\mathrm{Im}\, L$ is closed in $\mathbb{Y}$. If L is a Fredholm mapping of index zero and there exist continuous projectors $P :\ \mathbb{X} \longrightarrow \mathbb{X}$ and $Q :\ \mathbb{Y} \longrightarrow \mathbb{Y}$ such that $\mathrm{Im}\, P = \mathrm{Ker}\, L$, $\mathrm{Ker}\, Q = \mathrm{Im}\, L = \mathrm{Im}\,(I - Q)$, then the mapping $L|_{\mathrm{Dom}\, L \cap \mathrm{Ker}\, P} : (I - P)\mathbb{X} \longrightarrow \mathrm{Im}\, L$ is invertible. We denote the inverse of this mapping by K_P. If Ω is an open bounded subset of $\mathbb{X}$, the mapping H will be called L-compact on $\overline{\Omega}$ if $QH(\overline{\Omega})$ is bounded and $K_P(I - Q)H : \overline{\Omega} \longrightarrow \mathbb{X}$ is compact. Since $\mathrm{Im}\, Q$ is isomorphic to $\mathrm{Ker}\, L$, there exists an isomorphism $J :\ \mathrm{Im}\, Q \longrightarrow \mathrm{Ker}\, L$.

Now Mawhin's continuation theorem can be stated as follows.

Lemma 2.1. *Let L be a Fredholm mapping of index zero, let $\Omega \subset \mathbb{X}$ be an open bounded set and let $H :\ \mathbb{X} \longrightarrow \mathbb{Y}$ be a continuous operator which is L-compact on $\overline{\Omega}$. Assume that the following conditions hold:*

(a) for each $\lambda \in (0, 1)$, $x \in \partial\Omega \cap \mathrm{Dom}\, L$, $Lx \neq \lambda Hx$;

(b) for each $x \in \partial\Omega \cap \mathrm{Ker}\, L$, $QHx \neq 0$;

(c) $\deg\,(JQH,\ \Omega \cap \mathrm{Ker}\, L,\ 0) \neq 0$, *where* $\deg(\cdot)$ *is the Brouwer degree [12]. Then the equation $Lx = Hx$ has at least one solution in $\overline{\Omega} \cap \mathrm{Dom}\, L$.*

Let us choose $\mathbb{X} = \mathbb{Y} = \{x(n) = (x_1(n), x_2(n), \ldots, x_m(n))^T :\ x(n + N) = x(n),\ n \in \mathbb{Z}\}$. If we define $|x_i| = \max\limits_{n \in I_N} |x_i(n)|$, $\|x\| = \sum\limits_{i=1}^{m} |x_i|$, then $\mathbb{X}$ is a Banach space with the norm $\|\cdot\|$. For $x \in \mathbb{X}$, let Hx be defined by (7), $Lx = \Delta x$ and

$$Px = Qx = \tilde{x} = (\tilde{x}_1, \tilde{x}_2, \ldots, \tilde{x}_m)^T.$$

Then Ker $L = \{x \in \mathbb{X} : \ x = h \in \mathbb{R}^m\}$ (vectors with components independent of n), Im $L = \{x \in \mathbb{X} : \ \sum_{n=0}^{N-1} x_i(n) = 0, \ i = \overline{1,m}\}$ is a closed set in $\mathbb{X}$, and codim $L = m$. Thus L is a Fredholm mapping of index zero. It is easy to see that P and Q are continuous projectors and Im $P =$ Ker L, Im $L =$ Ker $Q =$ Im $(I - Q)$, and H is L-compact on $\bar{\Omega}$ for any bounded set $\Omega \subset \mathbb{X}$. Moreover, in condition (c) of Lemma 2.1 the isomorphism J can be taken as the identity operator I.

First for the solutions x of the operator equation $Lx = \lambda Hx$ for $\lambda \in (0, 1)$, that is,

$$\Delta x_i(n) = \lambda (Hx)_i(n), \quad n \in I_N, \ i = \overline{1,m},$$

after lengthy calculations we derive the estimate

$$\frac{\underline{A}_i(1 - \overline{A}_i)}{1 + N\underline{A}_i}|x_i| - \sum_{j=1}^{m} \mathcal{B}_{ij}\mathcal{M}_j\mathcal{G}_{ij}|x_j| \le \rho_i. \tag{8}$$

If we introduce the vectors $|\boldsymbol{x}| = (|x_1|, |x_2|, \ldots, |x_m|)^T$ and $\boldsymbol{\rho} = (\rho_1, \rho_2, \ldots, \rho_m)^T$, then the system of inequalities (8) for $i = \overline{1,m}$ can be written in a matrix form

$$(A - B)|\boldsymbol{x}| \le \boldsymbol{\rho}, \tag{9}$$

where the matrices A and B were introduced in §2. By virtue of condition **H7** the inequality (9) implies

$$|\boldsymbol{x}| \le (A - B)^{-1}\boldsymbol{\rho}.$$

If $(A - B)^{-1}\boldsymbol{\rho} = (C_1^*, C_2^*, \ldots, C_m^*)^T$, this means that the components of each solution of $\Delta x = \lambda Hx$ satisfy $|x_i| \le C_i^*$. If we denote $C^* = \sum_{i=1}^{m} C_i^*$, then each solution of $\Delta x = \lambda Hx$ satisfies $\|x\| \le C^*$.

Now we take $\Omega = \{x \in \mathbb{X} : \ \|x\| < C\}$, where $C > C^*$ will be chosen later. Obviously Ω satisfies condition (a) of Lemma 2.1.

Now let $x \in \partial\Omega \cap \operatorname{Ker} L = \partial\Omega \cap \mathbb{R}^m$, *i.e.*, x is a constant vector in $\mathbb{R}^m$ with $\|x\| = C$. For such x we obtain

$$\|QHx\| \ge \min_{i=\overline{1,m}} \left(\tilde{A}_i - \mathcal{M}_i \sum_{j=1}^{m} \mathcal{B}_{ji}\mathcal{G}_{ji} \right) C - \sum_{i=1}^{m} \rho_i.$$

By condition **H5**

$$\min_{i=\overline{1,m}} \left(\tilde{A}_i - \mathcal{M}_i \sum_{j=1}^{m} \mathcal{B}_{ji}\mathcal{G}_{ji} \right) > 0.$$

Then we can choose $C > C^*$ so large that

$$\min_{i=\overline{1,m}} \left(\tilde{A}_i - \mathcal{M}_i \sum_{j=1}^{m} \mathcal{B}_{ji} \mathcal{G}_{ji} \right) C > \sum_{i=1}^{m} \rho_i.$$

Hence for $x \in \partial\Omega \cap \operatorname{Ker} L$ we have $\|QHx\| > 0$ and $QHx \neq 0$, that is, condition (b) of Lemma 2.1 is satisfied.

To prove (c), we define the mapping $(QH)_\mu : \operatorname{Dom} L \times [0,1] \longrightarrow \mathbb{X}$ by $(QH)_\mu = -\mu\tilde{A} + (1-\mu)QH$, where $\tilde{A}x = (\tilde{A}_1 x_1, \tilde{A}_2 x_2, \ldots, \tilde{A}_m x_m)^T$.

For $x \in \partial\Omega \cap \operatorname{Ker} L$ as above we obtain

$$\|(QH)_\mu x\| \geq \min_{i=\overline{1,m}} \left(\tilde{A}_i - \mathcal{M}_i \sum_{j=1}^{m} \mathcal{B}_{ji} \mathcal{G}_{ji} \right) C - \sum_{i=1}^{m} \rho_i > 0.$$

This means that $(QH)_\mu x \neq 0$ for $x \in \partial\Omega \cap \operatorname{Ker} L$ and $\mu \in [0,1]$. From the homotopy invariance of the Brouwer degree [12] it follows that

$$\deg(QH,\ \Omega \cap \operatorname{Ker} L,\ 0) = \deg(-\tilde{A},\ \Omega \cap \operatorname{Ker} L,\ 0) = (-1)^m \neq 0.$$

According to Lemma 2.1 the equation (6) has at least one N-periodic solution. This completes the proof of Theorem 2.1. □

Theorem 2.2. *Suppose that conditions **H1–H4**, **H6**, **H7** hold. Then the N-periodic solution of* (6) *is unique and globally exponentially stable.*

Sketch of the proof. Let $\boldsymbol{g}_{ij}(\ell) = \max\{g_{ij}(\ell),\ c_{ij}(\ell)\}$, $i,j = \overline{1,m}$, $\ell = \overline{1,N}$. Clearly,

$$\sum_{\ell=1}^{N} \boldsymbol{g}_{ij}(\ell) = \mathcal{G}_{ij}, \qquad i,j = \overline{1,m}.$$

Now we shall use the following assertion.

Lemma 2.2. *Assume that condition **H6** holds. Then there exists $\overline{\lambda} > 1$ such that for any $i = \overline{1,m}$, $n \in I_N$ and $\lambda \in (1, \overline{\lambda}]$ we have*

$$\lambda(1 - A_i(n)) + \mathcal{M}_i \sum_{j=1}^{m} \mathcal{B}_{ji} \sum_{\ell=1}^{N} \lambda^{\ell+1} \boldsymbol{g}_{ji}(\ell) - 1 \leq 0.$$

Now let us suppose that $x^*(n) = (x_1^*(n), x_2^*(n), \ldots, x_m^*(n))^T$ is an N-periodic solution of equation (6), and $x(n) = (x_1(n), x_2(n), \ldots, x_m(n))^T$ is any solution of (6) for $n \geq 0$, defined at least for $n \geq -N$.

From (6) and (7) for $n \in \mathbb{Z}_0^+ = \{n \in \mathbb{Z} :\ n \geq 0\}$, $n \neq n_k$ we derive

$$|x_i(n+1) - x_i^*(n+1)| \leq (1 - A_i(n))|x_i(n) - x_i^*(n)|$$

$$+\sum_{j=1}^{m}\overline{b}_{ij}M_j\sum_{\ell=1}^{N}g_{ij}(\ell)|x_j(n-\ell)-x_j^*(n-\ell)|,$$

while for $n=n_k$ we have

$$|x_i(n_k+1)-x_i^*(n_k+1)|\le(1-A_i(n_k))|x_i(n_k)-x_i^*(n_k)|$$
$$+\sum_{j=1}^{m}\overline{B}_{ij}L_j\sum_{\ell=1}^{N}c_{ij}(\ell)|x_j(n_k-\ell)-x_j^*(n_k-\ell)|.$$

Now we introduce the quantities

$$y_i(n)=\lambda^n|x_i(n)-x_i^*(n)|,\quad \lambda\in(1,\overline{\lambda}],\quad i=\overline{1,m},\quad n\ge -N.$$

Then we have

$$y_i(n+1)\le\lambda(1-A_i(n))y_i(n)+\sum_{j=1}^{m}\mathcal{B}_{ij}\mathcal{M}_j\sum_{\ell=1}^{N}\lambda^{\ell+1}\boldsymbol{g}_{ij}(\ell)y_j(n-\ell),\tag{10}$$
$$\lambda\in(1,\overline{\lambda}],\quad i=\overline{1,m},\quad n\in\mathbb{Z}_0^+.$$

Now we consider a Lyapunov functional $V(n)=V(y_1,y_2,\dots,y_m)(n)$ defined by

$$V(n)=\sum_{i=1}^{m}\left\{y_i(n)+\sum_{j=1}^{m}\mathcal{B}_{ij}\mathcal{M}_j\sum_{\ell=1}^{N}\lambda^{\ell+1}\boldsymbol{g}_{ij}(\ell)\sum_{s=n-\ell}^{n-1}y_j(s)\right\},\quad n\in\mathbb{Z}_0^+.$$

Taking into account (10), we estimate the difference $\Delta V(n)=V(n+1)-V(n)$ for $n\in\mathbb{Z}_0^+$:

$$\Delta V(n)\le\sum_{i=1}^{m}\left\{\lambda(1-A_i(n))+\mathcal{M}_i\sum_{j=1}^{m}\mathcal{B}_{ji}\sum_{\ell=1}^{N}\lambda^{\ell+1}\boldsymbol{g}_{ji}(\ell)-1\right\}y_i(n).$$

By virtue of Lemma 2.2 we have $\Delta V(n)\le 0$ for all $n\in\mathbb{Z}_0^+$, which implies that

$$V(n)\le V(0),\qquad n\in\mathbb{Z}_0^+.\tag{11}$$

On the other hand, we have

$$V(n)\ge\sum_{i=1}^{m}y_i(n)=\sum_{i=1}^{m}\lambda^n|x_i(n)-x_i^*(n)|$$

and

$$V(0)\le\sum_{i=1}^{m}\left\{1+\mathcal{M}_i\sum_{j=1}^{m}\mathcal{B}_{ji}\sum_{\ell=1}^{N}\ell\overline{\lambda}^{\ell+1}\boldsymbol{g}_{ji}(\ell)\right\}\max_{s\in I_{-N}}|x_i(s)-x_i^*(s)|,$$

where $I_{-N} = \{-N, -N+1, \ldots, -1, 0\}$. Here we used that $1 < \lambda \leq \overline{\lambda}$.

Thus from inequality (11) we obtain

$$\sum_{i=1}^{m} |x_i(n) - x_i^*(n)| \leq M\lambda^{-n} \sum_{i=1}^{m} \max_{s \in I_{-N}} |x_i(s) - x_i^*(s)|, \quad n \in \mathbb{Z}_0^+,$$

where

$$M = \max_{i=\overline{1,m}} \left(1 + \mathcal{M}_i \sum_{j=1}^{m} \mathcal{B}_{ji} \sum_{\ell=1}^{N} \ell \overline{\lambda}^{\ell+1} g_{ji}(\ell) \right).$$

This completes the proof of Theorem 2.2. □

References

1. H. Akça, R. Alassar, V. Covachev and Z. Covacheva, *Dyn. Syst. Appl.* **13**, 77 (2004).
2. M. Fiedler, *Special Matrices and Their Applications in Numerical Mathematics*, Martinus Nijhoff, Dordrecht, 1986.
3. M. Fan and S. Agarwal, *Appl. Anal.* **82**, 801 (2002).
4. M. Fan and K. Wang, *Discrete Contin. Dyn. Syst. Ser. B* **4**, 563 (2004).
5. M. Fan and K. Wang, *Math. Comput. Modelling* **35**, 951 (2002).
6. M. Fan, K. Wang and D. Jiang, *Math. Biosci.* **160**, 47 (1999).
7. R. E. Gaines and J. L. Mawhin, *Coincidence Degree and Nonlinear Differential Equations*, Springer-Verlag, Berlin, 1977.
8. R. A. Horn and C. R. Johnson, *Topics in Matrix Analysis*, Cambridge University Press, Cambridge, 1991.
9. Y. K. Li, *Proc. Amer. Math. Soc.* **127**, 1331 (1999).
10. Y. K. Li and Y. Kuang, *J. Math. Anal. Appl.* **255**, 260 (2001).
11. Y. K. Li and L. Lu, *Phys. Lett. A* **333**, 62 (2004).
12. J. W. Milnor, *Topology from the Differentiable Viewpoint*, The University Press of Virginia, Charlottesville, 1969.
13. S. Mohamad and K. Gopalsamy, *Math. Comput. Simul.* **53**, 1 (2000).
14. T. Zhou, Y. Liu, Y. Liu and A. Chen, *Internat. J. Math. Manuscripts* **1**, (2007).

A NOTE ON THE MODIFIED PADÉ DIFFERENCE SCHEMES

A. ASHYRALYEV

Department of Mathematics, Fatih University,
İstanbul, 34500, Turkey
E-mail: aashyr@fatih.edu.tr

The high order of accuracy of the modified Padé difference schemes for approximate solution of the nonlocal boundary value problem for the differential equation

$$v'(t) + Av(t) = f(t)(0 \leq t \leq 1), v(0) = v(\lambda) + \mu, 0 < \lambda \leq 1$$

in an arbitrary Banach space E with a strongly positive operator A are considered. The almost coercive stability and the coercive stability estimates for the solutions of these difference schemes are established.

Keywords: Parabolic equation; Padé difference schemes; Well-posedness.

1. Introduction. Padé difference schemes

It is known that (see, e.g. Dehghan [1], Cannon, Perez-Esteva and Hoek [2], Gordeziani, Natani and Ricci [3], Dautray and Lions [4], Ashyralyev and Özdemir [5] and the references given therein) many applied problems in fluid mechanics, other areas of physics and mathematical biology were formulated into nonlocal mathematical models. However, such problems were not well-investigated in general. The nonlocal boundary value problem

$$v'(t) + Av(t) = f(t) \ (0 \leq t \leq 1), \ v(0) = v(\lambda) + \mu, \ 0 < \lambda \leq 1 \qquad (1)$$

for an abstract parabolic equation in arbitrary Banach space E with the strongly positive operator A is ill-posed in the Banach space $C(E) = C([0,1], E)$ with norm

$$\|\varphi\|_{C(E)} = \max_{0 \leq t \leq 1} ||\varphi(t)||_E.$$

In the papers Ashyralyev, Hanalyev and Sobolevskii [6], Ashyralyev [8], the well-posedness of the nonlocal boundary value problem (1) was established in various Banach spaces under strong positivity of the operator A.

Let $[0,1]_\tau = \{t_k = k\tau, 0 \le k \le N, N\tau = 1\}$ and $2\tau \le \lambda$. We consider the Padé difference schemes of $(p+q)-th$ order of accuracy

$$\frac{u_k - u_{k-1}}{\tau} + \sum_{j=1}^{p} \alpha_j (-A)^j \tau^{j-1} u_k - \sum_{j=1}^{q} \beta_j (-A)^j \tau^{j-1} u_{k-1} = \varphi_k, \quad (2)$$

$$\varphi_k = -\sum_{j=1}^{p} \alpha_j \sum_{i=0}^{j-1} (-A)^{j-i+1} f^{(i)}(t_k) \tau^{j-1}$$

$$+ \sum_{j=1}^{q} \beta_j \sum_{i=0}^{j-1} (-A)^{j-i+1} f^{(i)}(t_{k-1}) \tau^{j-1}, \ 1 \le k \le N,$$

$$u_0 = u_{\frac{\lambda}{\tau}} + \varphi_0 + \mu, \ \frac{\lambda}{\tau} \in Z$$

$$u_0 = \sum_{m=0}^{p+q-1} \frac{(-A)^m}{m!} (\lambda - \left[\frac{\lambda}{\tau}\right] \tau)^m u_{\left[\frac{\lambda}{\tau}\right]} + \varphi_0 + \mu, \ \frac{\lambda}{\tau} \notin Z,$$

$$\varphi_0 = \begin{cases} 0, \ \frac{\lambda}{\tau} \in Z \\ \sum_{m=1}^{p+q-1} \frac{1}{m!} (\lambda - \left[\frac{\lambda}{\tau}\right] \tau)^m \sum_{i=0}^{m-1} (-A)^{m-i+1} f^{(i)}(t_{\left[\frac{\lambda}{\tau}\right]}), \ \frac{\lambda}{\tau} \notin Z, \end{cases}$$

where

$$\begin{cases} \alpha_j = \frac{(p+q-j)!p!(-1)^j}{(p+q)!j!(p-j)!} & \text{for any } j, 1 \le j \le p, \\ \beta_j = \frac{(p+q-j)!q!}{(p+q)!j!(q-j)!} & \text{for any } j, 1 \le j \le q. \end{cases} \quad (3)$$

Let $F_\tau(E)$ be the linear space of mesh functions $\varphi^\tau = \{\varphi_k\}_1^N$ with values in the Banach space E. Next on $F_\tau(E)$, we introduce the Banach spaces $C_\tau(E) = C([0,1]_\tau, E)$, $C_\tau^{\beta,\gamma}(E) = C^{\beta,\gamma}([0,1]_\tau, E)(0 \le \gamma \le \beta < 1)$, $L_{p,\tau}(E) = L_p([0,1]_\tau, E)(1 \le p < \infty)$ with the norms

$$\|\varphi^\tau\|_{C_\tau(E)} = \max_{1 \le k \le N} \|\varphi_k\|_E,$$

$$\|\varphi^\tau\|_{C_\tau^{\beta,\gamma}(E)} = \|\varphi^\tau\|_{C_\tau(E)} + \sup_{1 \le k < k+r \le N} \|\varphi_{k+r} - \varphi_k\|_E \frac{((k+r)\tau)^\gamma}{(r\tau)^\beta},$$

$$\|\varphi^\tau\|_{L_{p,\tau}(E)} = \left(\sum_{K=1}^{N} \|\varphi_k\|_E^p \tau \right)^{1/p}, \ 1 \le p < \infty.$$

The operator A is said to be strongly positive, if its spectrum $\sigma(A)$ lies in the interior of the sector of angle φ, $0 < 2\varphi < \pi$, symmetric with respect to the real axis, and if on the edges of this sector, $S_1(\varphi) = \{\rho e^{i\varphi} : 0 \le \rho \le \infty\}$ and $S_2(\varphi) = \{\rho e^{-i\varphi} : 0 \le \rho \le \infty\}$, and outside of the resolvent $(\lambda - A)^{-1}$

is subject to the bound

$$\left\|(\lambda - A)^{-1}\right\|_{E\to E} \leq \frac{M(\varphi)}{1+|\lambda|} \tag{4}$$

The strong positivity of the operator implies the existence of a bounded operator $R_{q,p}(\tau A) = P_{q,p}(\tau A)Q_{q,p}^{-1}(\tau A)$, defined on the entire space E. Here

$$P_{q,p}(\tau A) = I + \sum_{j=1}^{q} \beta_j(-\tau A)^j, Q_{q,p}(\tau A) = I + \sum_{j=1}^{p} \alpha_j(-\tau A)^j.$$

Note that the Padé difference schemes (2) for $q = p-2, p-1, p$ include difference schemes of arbitrary order of accuracy. Moreover, the corresponding functions $R_{q,p}(z)$, $q = p-2, p-1, p$ are bounded at infinity.

In the paper Ashyralyev [28], the well-posedness of the Padé difference schemes (2) for $q = p-2, p-1$ was established in $C_\tau^{\beta,\gamma}(E)(0 \leq \gamma \leq \beta, 0 < \beta < 1)$, $C_\tau^{\beta,\gamma}(E_{\alpha-\beta})$ $(0 \leq \gamma \leq \beta \leq \alpha < 1)$. The coercive stability in $C_\tau^{\beta,\gamma}(E)(0 \leq \gamma \leq \beta, 0 < \beta < 1)$, $C_\tau^{\beta,\gamma}(E_{\alpha-\beta})$ $(0 \leq \gamma \leq \beta \leq \alpha < 1)$ and $L_{p,\tau}(E)$, $1 \leq p \leq \infty$ spaces for the solution of Crank-Nicholson difference scheme

$$\begin{cases} \frac{u_k - u_{k-1}}{\tau} + \frac{A}{2}(u_k + u_{k-1}) = \varphi_k, \ \varphi_k = f(t_k - \frac{\tau}{2}), t_k = k\tau, 1 \leq k \leq N, \\ u_0 = u_l + \mu, \frac{\lambda}{\tau} \in Z^+, \\ u_0 = (I - l_1 A)\frac{1}{2}(u_l + u_{l+1}) + \mu + l_1\varphi_l, \frac{\lambda}{\tau} \notin Z^+ \end{cases}$$

for the approximate solution of problem (1) can not be established.

In the papers Ashyralyev, Karatay and Sobolevskii [7], Ashyralyev [27], the well-posedness of the modified Crank-Nicholcon difference schemes in $C_\tau^{\beta,\gamma}(E)(0 \leq \gamma \leq \beta, 0 < \beta < 1)$, $C_\tau^{\beta,\gamma}(E_{\alpha-\beta})$ $(0 \leq \gamma \leq \beta \leq \alpha < 1)$, in $L_{p,\tau}(E_{\alpha,p}(E,A))$, $1 \leq p \leq \infty$ spaces are established. Finally, well-posedness of the evolution differential equations have been studied extensively by many researchers (see Ashyralyev [10], Ashiraliev and Sobolevskii [11], Ashyralyev and Sobolevskii [12,13], Guidetti, Karasözen, and Piskarev [14], Ashyralyev, Piskarev and Wei [15], Gavrilyuk and Makarov [16–18], Gordeziani and Avalishvili [19], Shakhmurov [20], Ramos [21], Gulin and Morozova [22], Gulin, Ionkin and Morozova [23], Sapagovas [24,25], Wang Oberguggenberger [26], and the references therein).

In the present paper, the single step modified Padé difference schemes of the high order of accuracy for approximate solutions of this problem are presented. The well-posedness of the modified Padé difference schemes in $L_{p,\tau}(E_{\alpha,p}(E,A))$, $1 \leq p \leq \infty$ spaces are established.

2. Modified Padé difference schemes

Applying the high order approximation formula for $v(0) = v(\lambda) + \mu$ and using the modified Padé difference schemes for the solutions of the initial-value problem for abstract parabolic equation (see Ashyralyev and Sobolevskii [9]), one can present the single step modified Padé difference schemes of $(2p) - th$ order of accuracy

$$
\begin{aligned}
&\frac{u_1 - u_0}{\tau} + \sum_{j=1}^{2p} \alpha_j (-A)^j \tau^{j-1} u_1 = \varphi_1, \qquad (5)\\
&\varphi_1 = -\sum_{j=1}^{2p} \alpha_j \sum_{i=0}^{j-1} (-A)^{j-i+1} f^{(i)}(\tau)\tau^{j-1}, \ \frac{\lambda}{\tau} \notin Z,\\
&\frac{u_k - u_{k-1}}{\tau} + \sum_{j=1}^{p} \alpha_j (-A)^j \tau^{j-1} u_k - \sum_{j=1}^{p-1} \beta_j (-A)^j \tau^{j-1} u_{k-1} = \varphi_k,\\
&\varphi_k = -\sum_{j=1}^{p} \alpha_j \sum_{i=0}^{j-1} (-A)^{j-i+1} f^{(i)}(t_k)\tau^{j-1}\\
&\quad + \sum_{j=1}^{p-1} \beta_j \sum_{i=0}^{j-1} (-A)^{j-i+1} f^{(i)}(t_{k-1})\tau^{j-1},\\
&\quad 2 \le k \le r, \ \frac{\lambda}{\tau} \notin Z; \ 1 \le k \le r, \ \frac{\lambda}{\tau} \in Z,\\
&\frac{u_k - u_{k-1}}{\tau} + \sum_{j=1}^{p} \alpha_j (-A)^j \tau^{j-1} u_k - \sum_{j=1}^{p} \beta_j (-A)^j \tau^{j-1} u_{k-1} = \varphi_k,\\
&\varphi_k = -\sum_{j=1}^{p} \alpha_j \sum_{i=0}^{j-1} (-A)^{j-i+1} f^{(i)}(t_k)\tau^{j-1}\\
&\quad + \sum_{j=1}^{p} \beta_j \sum_{i=0}^{j-1} (-A)^{j-i+1} f^{(i)}(t_{k-1})\tau^{j-1}, \ r+1 \le k \le N,\\
&u_0 = u_{\frac{\lambda}{\tau}} + \varphi_0 + \mu, \ \frac{\lambda}{\tau} \in Z\\
&u_0 = \sum_{m=0}^{2p-1} \frac{(-A)^m}{m!} \left(\lambda - \left[\frac{\lambda}{\tau}\right]\tau\right)^m u_{\left[\frac{\lambda}{\tau}\right]} + \varphi_0 + \mu, \ \frac{\lambda}{\tau} \notin Z,\\
&\varphi_0 = \begin{cases} 0, \ \frac{\lambda}{\tau} \in Z, \\ \sum_{m=1}^{p+q-1} \frac{1}{m!} \left(\lambda - \left[\frac{\lambda}{\tau}\right]\tau\right)^m \sum_{i=0}^{m-1} (-A)^{m-i+1} f^{(i)}\left(t_{\left[\frac{\lambda}{\tau}\right]}\right), \ \frac{\lambda}{\tau} \notin Z. \end{cases}
\end{aligned}
$$

First of all, we will provide the necessary lemmas.

Lemma 2.1. *(Ashyralyev and Sobolevskii [9]) For any $1 \leq k \leq N$, one has the estimates*

$$\begin{cases} \|R_{p,p}^{k}(\tau A)(I+\tau A)^{-1}\|_{E\longrightarrow E} \leq M(\delta), \\ \|k\tau A R_{p,p}^{k}(\tau A)(I+\tau A)^{-2}\|_{E\longrightarrow E} \leq M(\delta). \end{cases} \tag{6}$$

Lemma 2.2. *If $\frac{\lambda}{\tau} \in Z$, then the operator $I - R_{p,p}^{\frac{\lambda}{\tau}-r}(\tau A)R_{p-1,p}^{r}(\tau A)$ has a bounded inverse $T_\tau = (I - R_{p,p}^{\frac{\lambda}{\tau}-r}(\tau A)R_{p-1,p}^{r}(\tau A))^{-1}$ and*

$$\|T_\tau\|_{E\to E} \leq M(\lambda,\delta). \tag{7}$$

Lemma 2.3. *If $\frac{\lambda}{\tau} \notin Z$, then the operator*

$$I - \sum_{m=0}^{2p-1} \frac{(-A)^m}{m!}\left(\lambda - \left[\frac{\lambda}{\tau}\right]\tau\right)^m R_{p,p}^{[\frac{\lambda}{\tau}]-r}(\tau A)R_{p-1,p}^{r-1}R_{0,2p}(\tau A)$$

has a bounded inverse

$$T_\tau = \left(I - \sum_{m=0}^{2p-1} \frac{(-A)^m}{m!}\left(\lambda - \left[\frac{\lambda}{\tau}\right]\tau\right)^m R_{p,p}^{[\frac{\lambda}{\tau}]-r}(\tau A)R_{p-1,p}^{r-1}R_{0,2p}(\tau A)\right)^{-1}$$

and

$$\|T_\tau\|_{E\to E} \leq M(\lambda,\delta). \tag{8}$$

The proofs of Lemmas 2.2 and 2.3 are based on estimates (4) and (6).

Lemma 2.4. *For the solution of problem (2), the following formula holds:*

$$u_k = \begin{cases} R_{0,2p}(\tau A)u_0 + Q_{0,2p}^{-1}(\tau A)\varphi_j\tau, \frac{\lambda}{\tau} \notin Z, k=1, \\ R_{p-1,p}^{k}(\tau A)u_0 + \sum\limits_{j=1}^{k} R_{p-1,p}^{k-j}(\tau A)Q_{p-1,p}^{-1}(\tau A)\varphi_j\tau, \\ k = 2,\cdots,r, \frac{\lambda}{\tau} \notin Z \text{ and } k = 1,\cdots,r, \frac{\lambda}{\tau} \in Z, \\ R_{p,p}^{k}(\tau A)u_0 + \sum\limits_{j=1}^{k} R_{p,p}^{k-j}(\tau A)Q_{p,p}^{-1}(\tau A)\varphi_j\tau, k = r+1,\cdots,N, \\ T_\tau\{\sum\limits_{j=1}^{\frac{\lambda}{\tau}} R_{0,2}^{\frac{\lambda}{\tau}-j}(\tau A)Q_{0,2p}^{-1}(\tau A)\varphi_j\tau + \varphi_0 + \mu\}, \frac{\lambda}{\tau} \in Z, k=0, \\ T_\tau\{\sum_{m=0}^{2p-1} \frac{(-A)^m}{m!}(\lambda - [\frac{\lambda}{\tau}]\tau)^m \\ \times \sum\limits_{j=1}^{[\frac{\lambda}{\tau}]} R_{p,p}^{[\frac{\lambda}{\tau}]-j}(\tau A)Q_{p,p}^{-1}(\tau A)\varphi_j\tau + \varphi_0 + \mu\}, \frac{\lambda}{\tau} \notin Z, k=0. \end{cases} \tag{9}$$

Proof. Let $\frac{\lambda}{\tau} \in Z$. Then, by Ashyralyev and Sobolevskii [9],

$$u_k = \begin{cases} R^k_{p-1,p}(\tau A)u_0 + \sum\limits_{j=1}^{k} R^{k-j}_{p-1,p}(\tau A)Q^{-1}_{p-1,p}(\tau A)\varphi_j\tau, \\ k = 1, \cdots, r, \\ R^{k-r}_{p,p}(\tau A)R^r_{p-1,p}(\tau A)u_0 + \sum\limits_{j=1}^{r} R^{k-r}_{p,p}(\tau A)R^{r-j+1}_{p-1,p}(\tau A)\varphi_j\tau \\ + \sum\limits_{j=r+1}^{k} R^{k-j}_{p,p}(\tau A)R_{p-1,p}(\tau A)\varphi_j\tau, k = r+1, \cdots, N \end{cases} \tag{10}$$

for the solutions of the r-modified Padé difference schemes for the approximate solutions of Cauchy problem

$$u'(t) + Au(t) = f(t)(0 \leq t \leq 1), u(0) = u_0\,. \tag{11}$$

Let $\frac{\lambda}{\tau} \notin Z$. In exactly the same way, one establishes the formula

$$u_k = \begin{cases} R_{0,2p}(\tau A)u_0 + Q^{-1}_{0,2p}(\tau A)\varphi_1\tau, \\ R^{k-1}_{p-1,p}(\tau A)R_{0,2p}(\tau A)u_0 + R^{k-1}_{p-1,p}(\tau A)Q^{-1}_{0,2p}(\tau A)\varphi_1\tau \\ + \sum\limits_{j=2}^{k} R^{k-j-1}_{p-1,p}(\tau A)Q^{-1}_{p-1,p}(\tau A)\varphi_j\tau, k = 2, \cdots, r, \\ R^{k-r}_{p,p}(\tau A)R^{r-1}_{p-1,p}(\tau A)R_{0,2p}(\tau A)u_0 \\ +R^{k-r}_{p,p}(\tau A)R^{r-1}_{p-1,p}(\tau A)Q^{-1}_{0,2p}(\tau A)\varphi_1\tau \\ + \sum\limits_{j=2}^{r} R^{k-r}_{p,p}(\tau A)R^{r-j+1}_{p-1,p}(\tau A)\varphi_j\tau \\ + \sum\limits_{j=r+1}^{k} R^{k-j}_{p,p}(\tau A)R_{p-1,p}(\tau A)\varphi_j\tau, k = r+1, \cdots, N \end{cases} \tag{12}$$

for the solution of the r-modified Padé's difference schemes for the approximate solutions of Cauchy problem (11). Let $r\tau < \lambda$ and $\frac{\lambda}{\tau} \in Z$, then from (10) and the condition

$$u_0 = u_{\lambda/\tau} + \mu \tag{13}$$

it follows that

$$u_0 = R^{\frac{\lambda}{\tau}}_{p-1,p}(\tau A)u_0 + \sum_{j=1}^{\frac{\lambda}{\tau}} R^{\frac{\lambda}{\tau}-j}_{p-1,p}(\tau A)Q^{-1}_{p-1,p}(\tau A)\varphi_j\tau + \mu.$$

By Lemma 2.2, we have that

$$u_0 = T_\tau\Bigg\{\sum_{j=1}^{\frac{\lambda}{\tau}} R^{\frac{\lambda}{\tau}-j}_{p-1,p}(\tau A)Q^{-1}_{p-1,p}(\tau A)\varphi_j\tau + \varphi_0 + \mu\Bigg\}.$$

Let $r\tau > \lambda$ and $\frac{\lambda}{\tau} \in Z$. Then from (10) and (13), it follows that

$$u_0 = R_{p,p}^{\frac{\lambda}{\tau}-r}(\tau A)R_{p-1,p}^{r}(\tau A)u_0 + \sum_{j=1}^{r} R_{p,p}^{\frac{\lambda}{\tau}-r}(\tau A)R_{p-1,p}^{r-j+1}(\tau A)\varphi_j\tau$$

$$+ \sum_{j=r+1}^{\lambda/\tau} R_{p,p}^{\frac{\lambda}{\tau}-j}(\tau A)R_{p-1,p}(\tau A)\varphi_j\tau + \varphi_0 + \mu.$$

By Lemma 2.2, we have that

$$u_0 = T_\tau\Bigg\{\sum_{j=1}^{r} R_{p,p}^{\frac{\lambda}{\tau}-r}(\tau A)R_{p-1,p}^{r-j+1}(\tau A)\varphi_j\tau \qquad (14)$$

$$+ \sum_{j=r+1}^{\frac{\lambda}{\tau}} R_{p,p}^{\frac{\lambda}{\tau}-j}(\tau A)R_{p-1,p}(\tau A)\varphi_j\tau + \varphi_0 + \mu\Bigg\}.$$

Now, let $r\tau < \lambda$ and $\frac{\lambda}{\tau} \notin Z$. Then from formula (14) and the condition

$$u_0 = \sum_{m=0}^{2p-1} \frac{(-A)^m}{m!}(\lambda - \left[\frac{\lambda}{\tau}\right]\tau)^m u_{[\frac{\lambda}{\tau}]} + \varphi_0 + \mu \qquad (15)$$

it follows that

$$u_0 = \sum_{m=0}^{2p-1} \frac{(-A)^m}{m!}(\lambda - \left[\frac{\lambda}{\tau}\right]\tau)^m \Bigg(R_{p-1,p}^{[\frac{\lambda}{\tau}]-1}(\tau A)R_{0,2p}(\tau A)u_0$$

$$+R_{p-1,p}^{[\frac{\lambda}{\tau}]-1}(\tau A)Q_{0,2p}^{-1}(\tau A)\varphi_1\tau+\sum_{j=2}^{[\frac{\lambda}{\tau}]} R_{p-1,p}^{[\frac{\lambda}{\tau}]-j-1}(\tau A)Q_{p-1,p}^{-1}(\tau A)\varphi_j\tau\Bigg)+\varphi_0 + \mu.$$

By Lemma 2.2, we have that

$$u_0 = T_\tau\Bigg\{\sum_{m=0}^{2p-1} \frac{(-A)^m}{m!}(\lambda - \left[\frac{\lambda}{\tau}\right]\tau)^m \Bigg(R_{p-1,p}^{[\frac{\lambda}{\tau}]-1}(\tau A)Q_{0,2p}^{-1}(\tau A)\varphi_1\tau$$

$$+\sum_{j=2}^{[\lambda/\tau]} R_{p-1,p}^{[\frac{\lambda}{\tau}]-j-1}(\tau A)Q_{p-1,p}^{-1}(\tau A)\varphi_j\tau + \varphi_0 + \mu\Bigg)\Bigg\}.$$

Let $r\tau > \lambda$ and $\frac{\lambda}{\tau} \notin Z$. Then from (12) and (15) it follows that

$$u_0 = \sum_{m=0}^{2p-1} \frac{(-A)^m}{m!}(\lambda - \left[\frac{\lambda}{\tau}\right]\tau)^m \times \Bigg(R_{p,p}^{[\frac{\lambda}{\tau}]-r}(\tau A)R_{p-1,p}^{r-1}(\tau A)R_{0,2p}(\tau A)u_0$$

$$+R_{p,p}^{[\frac{\lambda}{\tau}]-r}(\tau A)R_{p-1,p}^{r-1}(\tau A)Q_{0,2p}^{-1}(\tau A)\varphi_1\tau+\sum_{j=2}^{r} R_{p,p}^{[\frac{\lambda}{\tau}]-r}(\tau A)R_{p-1,p}^{r-j+1}(\tau A)\varphi_j\tau$$

$$+\sum_{j=r+1}^{[\frac{\lambda}{\tau}]} R_{p,p}^{[\frac{\lambda}{\tau}]-j}(\tau A)R_{p-1,p}(\tau A)\varphi_j\tau\Big)+\varphi_0+\mu.$$

By Lemma 2.3, we have that

$$u_0=T_\tau\Bigg\{\sum_{m=0}^{2p-1}\frac{(-A)^m}{m!}(\lambda-\left[\frac{\lambda}{\tau}\right]\tau)^m\Big(R_{p,p}^{[\frac{\lambda}{\tau}]-r}(\tau A)R_{p-1,p}^{r-1}(\tau A)Q_{0,2p}^{-1}(\tau A)\varphi_1\tau$$

$$+\sum_{j=2}^{r}R_{p,p}^{[\frac{\lambda}{\tau}]-r}(\tau A)R_{p-1,p}^{r-j+1}(\tau A)\varphi_j\tau$$

$$+\sum_{j=r+1}^{[\lambda/\tau]} R_{p,p}^{[\frac{\lambda}{\tau}]-j}(\tau A)R_{p-1,p}(\tau A)\varphi_j\tau+\varphi_0+\mu\Big)\Bigg\}.$$

This completes the proof of Lemma 2.4. □

In fact, the following inequality

$$\|\{\tau^{-1}(u_k-u_{k-1})\}_1^N\|_{L_{p,\tau}(E)}\le M[\|\varphi^\tau\|_{L_{p,\tau}(E)}+\|A\mu\|_E]$$

does not, generally speaking, hold in an arbitrary Banach space E and for the general strongly positive operator A. Nevertheless, we can establish the almost coercivity and well- posedness of the difference schemes (5) in Banach spaces $L_{p,\tau}(E_{\alpha,p})$, $1\le p\le\infty$.

Theorem 2.1. *Let $1\le p\le\infty$. The solutions of the difference schemes (5) in $L_{p,\tau}(E)$ obey the almost coercivity inequality*

$$\|\{\tau^{-1}(u_k-u_{k-1})\}_1^N\|_{L_{p,\tau}(E)} \tag{16}$$
$$\le M(\delta,\lambda)\Big[\|A(\mu+\varphi_0)\|_E+\min\Big\{\ln\frac{1}{\tau},1+|\ln\|A\|_{E\to E}|\Big\}\|\varphi^\tau\|_{L_{p,\tau}}(E)\Big].$$

Here and in the future $L_{\infty,\tau}(E)=C_\tau(E)$.

Theorem 2.2. *Let $1\le p\le\infty$ and $0<\alpha<1$. The difference schemes (5) is well posed in $L_{p,\tau}(E_{\alpha,p})$ and the following coercivity inequality holds:*

$$\|\{\tau^{-1}(u_k-u_{k-1})\}_1^N\|_{L_{p,\tau}(E_{\alpha,p})} \tag{17}$$
$$\le M(\delta,\lambda)\Big[\|A(\mu+\varphi_0)\|_{E_{\alpha,p}}+\frac{1}{\alpha(1-\alpha)}\|\varphi^\tau\|_{L_{p,\tau}(E_{\alpha,p})}\Big].$$

The proof of estimates (16) and (17) follows the scheme of proof of Ashyralyev [27] and is based on estimates (4), (6), (7), (8) and formula (9).

In practice this abstract result permits to obtain the almost coercivity inequality and the coercive stability estimates for the solutions of difference schemes of the high order of accuracy in the case of the nonlocal boundary value problem for the $2m$-th order multidimensional parabolic equation.

References

1. M. Dehghan, *Math. Prob. Eng.* **2003**, 81 (2003).
2. J. R. Cannon, S. Perez-Esteva and J. van der Hoek, *SIAM J. Num. Anal.* **24**, 499 (1987).
3. N. Gordeziani, P. Natani and P. E. Ricci, *Comp. Math. Appl.* **50**, 1333 (2005).
4. R. Dautray and J. L. Lions, *Analyse Mathematique et Calcul Numerique Pour les Sciences et les Technique,* Masson, Paris, 1988.
5. A. Ashyralyev and Y. Özdemir, *Comp. Math. Appl.* **50**, 1443 (2005).
6. A. Ashyralyev, A. Hanalyev and P. E. Sobolevskii, *AAA* **6**, 53 (2001).
7. A. Ashyralyev, İ. Karatay and P. E. Sobolevskii, *Discrete Dyn. Nat. Soc.* **2**, 273 (2004).
8. A. Ashyralyev, *J. Evol. Equat.* **6**, 1 (2006).
9. A. Ashyralyev and P.E. Sobolevskii, *Well-Posedness of Parabolic Difference Equations,* Birkhäuser Verlag, Basel, Boston, Berlin, 1994.
10. A. Ashyralyev, *Math. Probl. Eng.* **2007**, (2007).
11. A. Ashiraliev and P. E. Sobolevskii, *Dopovidi Akademii Nauk Ukrainskoi RSR Seriya A, Fiziko-Matematichni ta Technichni Nauki* **3**, (1988) (Russian).
12. A. Ashyralyev and P. E. Sobolevskii, *Nonlinear Anal. Theory Methods Appl.* **24,** 257 (2005).
13. A. Ashyralyev, P. E. Sobolevskii, *New Difference Schemes for Partial Differential Equations,* Birkhäuser Verlag, Basel, Boston, Berlin, 2004.
14. D. Guidetti, B. Karasözen, and S. Piskarev, *J. Math. Sci.* **122**, 3013 (2004).
15. A. Ashyralyev, S. Piskarev, S. Wei, *Numer. Func. Anal. Optim.* **23,** 669 (2002).
16. I. P. Gavrilyuk and V. L. Makarov, *Comput. Meth. Appl. Math.* **1**, 333 (2001).
17. I. P. Gavrilyuk and V. L. Makarov, *Math. Comput.* **74**, 555 (2005).
18. I. P. Gavrilyuk and V. L. Makarov, *SIAM J. Numer. Anal.* **43**, 2144 (2005).
19. D. Gordeziani and G. Avalishvili, *Differ. Equ.* **41**, 703 (2005).
20. V. B. Shakhmurov, *J. Math. Anal. Appl.* **292**, 605 (2004).
21. J. I. Ramos, Linearly implicit, *Appl. Math. Comput.* **174**, 1609 (2006).
22. A. V. Gulin and V. A. Morozova, *Differ. Equ.* **39**, 962 (2003) (Russian).
23. A. V. Gulin, N. I. Ionkin and V. A. Morozova, *Differ. Equ.* **37**, 970 (2001) (Russian).
24. M. Sapagovas, *J. Computer Appl. Math.* **88**, 89 (2003).
25. M. Sapagovas, *J. Comp. Appl.Math.* **92**, 77 (2005).
26. Y.G. Wang, M. Oberguggenberger, *J. Math. Anal. Appl.* **233,** 644 (1999).
27. A. Ashyralyev, *Discrete Cont. Dyn. Syst. Series B,* **7**, 29 (2007).
28. A. Ashyralyev, *Abstracts*, International Conference on Modern Analysis and Applications dedicated to the centenary of Mark Krein, Odessa, Ukraine (2007), 17.

NUMERICAL SOLUTION OF A ONE-DIMENSIONAL PARABOLIC INVERSE PROBLEM

A. ASHYRALYEV, A. S. ERDOĞAN* and E. DEMİRCİ

Department of Mathematics, Fatih University,
İstanbul, 34500, Turkey
**E-mail: aserdogan@fatih.edu.tr*

The stable difference schemes for the approximate solutions of the parabolic equation with control parameter

$$\begin{cases} \frac{\partial v(t,x)}{\partial t} = \frac{\partial^2 v(t,x)}{\partial x^2} + p(t)\, v(t,x) + f(t,x), \\ x \in (0,L), \ t \in (0,T], \\ v(0,x) = \varphi(x), \ x \in [0,L], \\ v(t,0) = \alpha(t), \ v(t,L) = \beta(t), \ t \in [0,T], \\ v(t,x^*) = \gamma(t), \ t \in [0,T], \ x^* \in (0,L) \end{cases}$$

are considered. Here $\alpha(t)$, $\beta(t)$, $\gamma(t)$, $f(t,x)$ and $\varphi(x)$ are given sufficiently smooth functions, and $v(t,x)$ and $p(t)$ are unknown functions. Stable difference schemes are constructed. A procedure of modified Gauss elimination method is used for solving these difference schemes. The method is illustrated by numerical examples.

Keywords: Parabolic equation; Difference schemes; Convergence; Inverse problems; Control parameter.

1. Introduction

The parabolic equation with the control parameter is a class of parabolic inverse problems and the equation is non-linear. While determining the solution of these problems, we shall determinate some unknown control parameter. These problems play a very important role in many branches of science and engineering. Some examples were given in temperature overspecification by Dehghan [1], robotics, chemistry (chromatography) by Kimura and Suzuki [2], physics (optical tomography) by Gryazin, Klibanov and Lucas [3].

As a result, considerable efforts have been expanded in formulating numerical solution methods that are both accurate and efficient. Methods of solutions parabolic inverse problems have been studied by researchers

Cannon, Lin and Xu [4], Chao-rong Ye and Zhi-zhong Sun [5] and Dehghan [6]. Our goal in this work is to investigate the first and second order of accuracy difference schemes for approximately solving this problem. A procedure of modified Gauss elimination method is used for solving these difference schemes.

We consider the parabolic equation with control parameter

$$\begin{cases} \frac{\partial u(t,x)}{\partial t} = \frac{\partial^2 u(t,x)}{\partial x^2} + p(t)\, u(t,x) + f(t,x), \\ x \in (0,L), \ t \in (0,T], \\ u(0,x) = \varphi(x), \ x \in [0,L], \\ u(t,0) = \alpha(t), \ u(t,L) = \beta(t), t \in [0,T], \\ u(t,x^*) = \gamma(t), \ t \in [0,T], \ x^* \in (0,L). \end{cases} \tag{1}$$

1.1. *Evaluating the Approximate Value of the Control Parameter $P(t)$*

First, using the parabolic equation, we get

$$u_{txx}(t,x) = u_{xxxx}(t,x) + p(t)\, u_{xx}(t,x) + f_{xx}(t,x), \tag{2}$$

$$u_{txxxx}(t,x) = u_{xxxxxx}(t,x) + p(t)\, u_{xxxx}(t,x) + f_{xxxx}(t,x). \tag{3}$$

From equations (2) and (3), it follows

$$p(t) = \frac{u_t(t,x)}{u(t,x)} - \frac{u_{xx}(t,x)}{u(t,x)} - \frac{f(t,x)}{u(t,x)},$$

$$\begin{aligned} p'(t) = {} & \frac{u_{tt}(t,x)\, u(t,x) - u_t^2(t,x)}{u^2(t,x)} \\ & - \frac{u(t,x)\left(u_{xxxx}(t,x) + p(t)\, u_{xx}(t,x) + f_{xx}(t,x)\right)}{u^2(t,x)} \\ & + \frac{u_{xx}(t,x)\, u_t(t,x) - f_t(t,x)\, u(t,x) + f(t,x)\, u_t(t,x)}{u^2(t,x)}, \end{aligned}$$

$$\begin{aligned} p''(t) = {} & \frac{u_{ttt}(t,x)\, u(t,x) - u_{tt}(t,x)\, u_t(t,x)}{u^2(t,x)} \\ & - \frac{u_t(t,x)\left(u_{xxxx}(t,x) + p(t)\, u_{xx}(t,x) + f_{xx}(t,x)\right)}{u^2(t,x)} \\ & - \frac{u_{xxxxxx}(t,x) + p(t)\, u_{xxxx}(t,x) + f_{xxxx}(t,x) - p_t(t)\, u_{xx}(t,x)}{u(t,x)} \\ & + \frac{p(t)\left(u_{xxxx}(t,x) + p(t)\, u_{xx}(t,x) + f_{xx}(t,x)\right) + f_{txx}(t,x)}{u(t,x)} \\ & + \frac{\left(u_{xxxx}(t,x) + p(t)\, u_{xx}(t,x) + f_{xx}(t,x)\right) u_t(t,x)}{u^2(t,x)} \end{aligned}$$

$$+\frac{u_{xx}(t,x)\,u_{tt}(t,x)-f_{tt}(t,x)\,u(t,x)+f(t,x)\,u_{tt}(t,x)}{u^{2}(t,x)}$$
$$-\frac{2\left(u_{tt}(t,x)\,u(t,x)-u_{t}^{2}(t,x)\right)}{u^{3}(t,x)}$$
$$-\frac{2u(t,x)\left(u_{xxxx}(t,x)+p(t)\,u_{xx}(t,x)+f_{xx}(t,x)\right)}{u^{3}(t,x)}$$
$$-\frac{2\left(-u_{xx}(t,x)\,u_{t}(t,x)+f_{t}(t,x)\,u(t,x)-f(t,x)\,u_{t}(t,x)\right)}{u^{3}(t,x)}.$$

Using these formulas and putting $x=x^{*},\ t=0$, we get

$$p(0)=\frac{\gamma_{t}(0)}{\varphi(x^{*})}-\frac{\varphi_{xx}(x^{*})}{\varphi(x^{*})}-\frac{f(0,x^{*})}{\varphi(x^{*})},$$
$$p'(0)=\frac{\gamma_{tt}(0)\,\varphi(x^{*})-\gamma_{t}^{2}(0)}{\varphi^{2}(x^{*})}$$
$$-\frac{\varphi(x^{*})\left(\varphi_{xxxx}(x^{*})+p(0)\,\varphi_{xx}(x^{*})+f_{xx}(0,x^{*})\right)}{\varphi^{2}(x^{*})}$$
$$+\frac{\varphi_{xx}(x^{*})\gamma_{t}(0)-f_{t}(0,x^{*})\,\varphi(x^{*})+f(0,x^{*})\,\gamma_{t}(0)}{\varphi^{2}(x^{*})},$$
$$p''(0)=\frac{\gamma_{ttt}(0)\,\varphi(x^{*})-\gamma_{tt}(0)\,\gamma_{t}(0)}{\varphi^{2}(x^{*})}$$
$$-\frac{\gamma_{t}(0)\left(\varphi_{xxxx}(x^{*})+p(0)\,\varphi_{xx}(x^{*})+f_{xx}(0,x^{*})\right)}{\varphi^{2}(x^{*})}$$
$$-\frac{\varphi_{xxxxxx}(x^{*})+p(0)\,\varphi_{xxxx}(x^{*})+f_{xxxx}(0,x^{*})-p_{t}(0)\,\varphi_{xx}(x^{*})}{\varphi(x^{*})}$$
$$+\frac{p(0)\left(\varphi_{xxxx}(x^{*})+p(0)\,\varphi_{xx}(x^{*})+f_{xx}(0,x^{*})\right)+f_{txx}(0,x^{*})}{\varphi(x^{*})}$$
$$-\frac{-\left(\varphi_{xxxx}(x^{*})+p(0)\,\varphi_{xx}(x^{*})+f_{xx}(0,x^{*})\right)\gamma_{t}(0)}{\varphi^{2}(x^{*})}$$
$$-\frac{-\varphi_{xx}(x^{*})\gamma_{tt}(0)+f_{tt}(0,x^{*})\,\varphi(x^{*})-f(0,x^{*})\,\gamma_{tt}(0)}{\varphi^{2}(x^{*})}$$
$$-\frac{2\left(\gamma_{tt}(0)\,\varphi(x^{*})-\gamma_{t}^{2}(0)\right)}{\varphi^{3}(x^{*})}$$
$$-\frac{2\varphi(x^{*})\left(\varphi_{xxxx}(x^{*})+p(0)\,\varphi_{xx}(x^{*})+f_{xx}(0,x^{*})\right)}{\varphi^{3}(x^{*})}$$
$$-\frac{2\left(-\varphi_{xx}(x^{*})\gamma_{t}(0)+f_{t}(0,x^{*})\,\varphi(x^{*})-f(0,x^{*})\,\gamma_{t}(0)\right)}{\varphi^{3}(x^{*})}.$$

Second, the approximate formulas for $p(t_k)$, $t_k = k\tau$, $k = 1, 2, \cdots$ will be constructed. For $p(\tau)$, we have

$$p(\tau) = p(0) + p'(0)\,\tau + p''(0)\,\frac{\tau^2}{2} + o\left(\tau^3\right). \tag{4}$$

$p(2\tau)$ can be calculated by using the formula

$$p(2\tau) = -3p(0) + 4p(\tau) - 2\tau p'(0) + o\left(\tau^3\right). \tag{5}$$

Then, $p(t_k)$, $k \geq 3$, can be calculated by using the iteration algorithm

$$\begin{aligned} p(t_k) &= \left(p(t_{k-3}) - 3p(t_{k-2}) + 3p(t_{k-1})\right)\left(1 + \tau^4\right) + o\left(\tau^3\right), \\ t_k &= k\tau,\ k = \overline{3, N},\ N\tau = T. \end{aligned} \tag{6}$$

The proofs of formulas (5) and (6) are based on the method of undetermined coefficients. Neglecting the last small terms, we get following formulas for the approximate value of $p(t)$.

$$\begin{cases} p_0 = p(0), \\ p_1 = p(0) + p'(0)\,\tau + p''(0)\,\frac{\tau^2}{2}, \\ p_2 = -3p(0) + 4p_1 - 2\tau p'(0) \\ p_k = \left(p_{k-3} - 3p_{k-2} + 3p_{k-1}\right)\left(1 + \tau^4\right),\ 3 \leq k \leq N. \end{cases} \tag{7}$$

1.2. *Difference Schemes*

Now, we will consider the stable difference schemes of the first order of accuracy

$$\begin{cases} \frac{u_n^k - u_n^{k-1}}{\tau} - \frac{u_{n+1}^k - 2u_n^k + u_{n-1}^k}{h^2} - p_k u_n^k = \varphi_n^k, \\ \varphi_n^k = f(t_k, x_n),\ t_k = k\tau,\ 1 \leq k \leq N,\ N\tau = T, \\ x_n = nh, 1 \leq n \leq M-1\ , Mh = L, \\ u_n^0 = \varphi(x_n)\,,\ x_n = nh, 0 \leq n \leq M, \\ u_0^k = \alpha(t_k)\,,\ u_M^k = \beta(t_k)\,,\ t_k = k\tau,\ 0 \leq k \leq N \end{cases} \tag{8}$$

and second order of accuracy

$$\begin{cases} \frac{u_n^k - u_n^{k-1}}{\tau} - \frac{u_{n+1}^k - 2u_n^k + u_{n-1}^k}{2h^2} - \frac{u_{n+1}^{k-1} - 2u_n^{k-1} + u_{n-1}^{k-1}}{2h^2} \\ -\frac{p_k u_n^k + p_{k-1} u_n^{k-1}}{2} = \varphi_n^k, \\ \varphi_n^k = f(t_k - \frac{\tau}{2}, x_n), t_k = k\tau,\ 1 \leq k \leq N,\ N\tau = T, \\ x_n = nh, 1 \leq n \leq M-1\ , Mh = L, \\ u_n^0 = \varphi(x_n)\,,\ x_n = nh, 0 \leq n \leq M, \\ u_0^k = \alpha(t_k)\,,\ u_M^k = \beta(t_k)\,,\ t_k = k\tau,\ 0 \leq k \leq N \end{cases} \tag{9}$$

for the numerical solution of problem (1).

2. Numerical Results

For numerical analysis, we consider the problem

$$\begin{cases} \frac{\partial u(t,x)}{\partial t} = \frac{\partial^2 u(t,x)}{\partial x^2} + p(t)\,u(t,x) \\ +(\pi^2 - (t+1)^2)e^{-t^2}\sin \pi x, \ x \in (0,1), \ t \in (0,1], \\ u(0,x) = \sin \pi x, \ x \in [0,1], \\ u(t,0) = 0, \ u(t,1) = 0, \ u(t,\frac{1}{4}) = \frac{\sqrt{2}}{2}e^{-t^2}, \ t \in [0,1]. \end{cases} \tag{10}$$

The exact solution of the given problem is $u(t,x) = e^{-t^2}\sin \pi x$ and for the control parameter $p(t)$ is $1+t^2$.

First, we will calculate the approximate value of the control parameter $p(t)$. In our numerical process, by putting $t=0$ and $x=\frac{1}{4}$, we can obtain

$$p(0) = 1, \ p'(0) = 0, \ p''(0) = 2.$$

Then, by using equation (7), we can calculate all p_k, $k = \overline{0,N}$.

Second, applying the first order of accuracy difference scheme (8), we obtain $(N+1)\times(M+1)$ system of linear equations and we write them in the matrix form

$$AU_{n-1} + BU_n + CU_{n+1} = R\varphi_n, \quad 1 \le n \le M-1, \ U_0 = U_M = \widetilde{0}, \tag{11}$$

where

$$C = A = \begin{bmatrix} 0 & 0 & 0 & \cdots & 0 & 0 & 0 & 0 \\ x & 0 & 0 & \cdots & 0 & 0 & 0 & 0 \\ 0 & x & 0 & \cdots & 0 & 0 & 0 & 0 \\ . & . & . & . & . & . & . & . \\ 0 & 0 & 0 & \cdots & 0 & x & 0 & 0 \\ 0 & 0 & 0 & \cdots & 0 & 0 & x & 0 \end{bmatrix}_{(N+1)\times(N+1)},$$

$$B = \begin{bmatrix} 1 & 0 & 0 & \cdots & 0 & 0 & 0 \\ v & y_1 & 0 & \cdots & 0 & 0 & 0 \\ 0 & v & y_2 & \cdots & 0 & 0 & 0 \\ . & . & . & . & . & . & . \\ 0 & 0 & 0 & \cdots & v & y_{N-1} & 0 \\ 0 & 0 & 0 & \cdots & 0 & v & y_N \end{bmatrix}_{(N+1)\times(N+1)}.$$

Here, $x = -\frac{1}{h^2}$, $v = -\frac{1}{\tau}$, $y_k = \frac{1}{\tau} + \frac{2}{h^2} - p_k$, $1 \le k \le N$,

$$U_s = \begin{bmatrix} u_s^0 \\ \ldots \\ u_s^N \end{bmatrix}_{(N+1)x1} \quad for\ s = n+1, n, n-1, \quad \varphi = \begin{bmatrix} \varphi(t_0) \\ \ldots \\ \varphi(t_N) \end{bmatrix}_{(N+1)\times 1},$$

$$R = \begin{bmatrix} 1\ 0 \cdots 0 \\ 0\ 1 \cdots 0 \\ \cdot\ \cdot\ \cdot\ \cdot \\ 0\ 0 \cdots 1 \end{bmatrix}_{(N+1)\times(N+1)}, \varphi_n = \begin{bmatrix} \sin \pi x_n \\ \varphi_n^1 \\ \cdots \\ \varphi_n^{N-1} \\ \varphi_n^N \end{bmatrix}_{(N+1)\times 1}.$$

So, we have the second order difference equation with respect to n with matrix coefficients. The solution of the system of linear equation can be obtained by using the modified Gauss elimination method. We will seek a solution of matrix equation of the form

$$U_n = \alpha_{n+1} U_{n+1} + \beta_{n+1},\ n = M-1, \cdots, 2, 1. \tag{12}$$

Here,

$$\alpha_1 = \begin{bmatrix} 0\ 0 \cdots 0 \\ 0\ 0 \cdots 0 \\ \cdot\ \cdot\ \cdot\ \cdot \\ 0\ 0 \cdots 0 \end{bmatrix}_{(N+1)\times(N+1)} \quad \text{and } \beta_1 = \begin{bmatrix} 0 \\ 0 \\ \cdot \\ 0 \end{bmatrix}_{(N+1)\times 1}$$

and α_k and β_k are derived from

$$\begin{aligned} \alpha_{n+1} &= -(B + C\alpha_n)^{-1} A, \\ \beta_{n+1} &= (B + C\alpha_n)^{-1} (R\varphi_n - C\beta_n), n = 1, 2, \cdots, M-1. \end{aligned}$$

So, after calculating α_j's and β_j's, U_n can be obtained by using the boundary condition at $x = 1$ as follows

$$\begin{aligned} U_M &= \tilde{0}, \\ U_n &= \alpha_{n+1} U_{n+1} + \beta_{n+1},\ n = M-1, \cdots, 2, 1. \end{aligned}$$

Finally, applying the second order of accuracy difference scheme (9), we have again $(N+1) \times (M+1)$ system of linear equations and we write them in the matrix form (11), where

$$C = A = \begin{bmatrix} 0\ 0\ 0 \cdots 0\ 0\ 0 \\ q\ q\ 0 \cdots 0\ 0\ 0 \\ 0\ q\ q \cdots 0\ 0\ 0 \\ \cdot\ \cdot\ \cdot\ \cdot\ \cdot\ \cdot\ \cdot \\ 0\ 0\ 0 \cdots q\ q\ 0 \\ 0\ 0\ 0 \cdots 0\ q\ q \end{bmatrix}_{(N+1)\times(N+1)},$$

$$B = \begin{bmatrix} 0 & 0 & 0 & \cdots & 0 & 0 & 0 \\ z_1 & g_1 & 0 & \cdots & 0 & 0 & 0 \\ 0 & z_2 & g_2 & \cdots & 0 & 0 & 0 \\ \cdot & \cdot & \cdot & \cdot & \cdot & \cdot & \cdot \\ 0 & 0 & 0 & \cdots & z_{N-1} & g_{N-1} & 0 \\ 0 & 0 & 0 & \cdots & 0 & z_N & g_N \end{bmatrix}_{(N+1)\times(N+1)}.$$

Here, $q = -\frac{1}{2h^2}$, $z_k = -\frac{1}{\tau} + \frac{1}{h^2} - p_{k-1}$, $1 \le k \le N$, $g_k = \frac{1}{\tau} + \frac{1}{h^2} - p_k$, $1 \le k \le N$. In order to get the solution of difference scheme (9), again we use modified Gauss elimination method and MATLAB programs. Now, we will give the results of the numerical analysis. The numerical solutions are recorded for different values of N and M and u_n^k represents the numerical solutions of these difference schemes at (t_k, x_n).

Table 1 gives the error analysis between the exact solution of $p(t)$ and the solutions derived by the numerical process.

Table 1. Error analysis for $p(t)$.

	N=20	N=40	N=80
Max. Error	0.0083	0.0044	0.0022

Table 2 gives the error analysis between the exact solution and the solutions derived by difference schemes. Table 2 is constructed for $N = M = 20$, 40 and 80 respectively. For comparison, the errors are computed by

$$E = \max_{\substack{1\le k\le N \\ 1\le n\le M}} \left| u(t_k, x_n) - u_n^k \right|.$$

Table 2. Error analysis for the exact solution $u(t, x)$.

Method	N=M=20	N=M=40	N=M=80
1st order of accuracy	0.0048	0.0030	0.0016
2nd order of accuracy	0.0026	0.0006	0.0002

To find the control parameter $p(t)$, we use $u(t, x^*) = \gamma(t)$ for $t = 0$, values of its first and second derivatives and smoothness of $\gamma(t)$ in t. Therefore, we will consider the error between $\gamma(t_k)$ and $u^k_{[\frac{x^*}{h}]}$. Table 3 gives the maximum error for $h=\frac{1}{M}$ and $N=M=20$, 40 and 80 respectively.

Table 3. Error analysis between $\gamma(t_k)$ and $u^k_{\left[\frac{x^*}{h}\right]}$

Method	N=M=20	N=M=40	N=M=80
1st order of accuracy	0.0034	0.0021	0.0011
2nd order of accuracy	0.0018	4.5662×10^{-4}	1.1529×10^{-4}

Thus, the second order of accuracy difference scheme is more accurate comparing with the first order of accuracy difference scheme.

Remark 2.1. In (6), we can put $1+\tau^p$ for $p=4,5,6$. Nevertheless without $1+\tau^p$, we can use identity

$$p(t_k)=p(t_{k-3})-3p(t_{k-2})+3p(t_{k-1})+o\left(\tau^3\right), \qquad (13)$$
$$t_k=k\tau, k=\overline{3,N}, N\tau=T.$$

Neglecting last terms, we can obtain the algorithm for p_k. The results are given below.

Table 4. Error analysis for $p(t)$.

	N=20	N=40	N=80
Max. Error	1.2212×10^{-14}	4.7695×10^{-13}	8.3393×10^{-13}

Table 5. Error analysis for the exact solution $u(t,x)$.

Method	N=M=20	N=M=40	N=M=80
1st order of accuracy	0.0049	0.0030	0.0016
2nd order of accuracy	0.0026	0.0006	0.0002

Table 6. Error analysis between $\gamma(t_k)$ and $u^k_{\left[\frac{x^*}{h}\right]}$

Method	N=M=20	N=M=40	N=M=80
1st order of accuracy	0.0034	0.0021	0.0011
2nd order of accuracy	0.0018	4.5333×10^{-4}	1.1327×10^{-4}

As is seen from Table 4, even though the errors are very small, the result is unstable in the approximate solution of the control parameter $p(t)$. Since errors are very small, Tables 5-6 are nearly the same as Tables 2-3. Note that we obtain the multiplication factor $1+\tau^4$ experimentally and use it in (6) for the stable approximate solution of the control parameter $p(t)$.

3. Conclusions

In this study, we developed a new way to obtain the control parameter and also we use first and second order of accuracy difference schemes. For N=M=20, we found the maximum error as 0.0049 while in Chao-rong Ye and Zhi-zhong Sun's work [5], it is 0.0077. In M. Dehghan's article [1] the errors are given for h=1/50 and s=1/4. We reached more accurate results for small M and N numbers.

Difference schemes of high order of accuracy can be investigated by using the operator approach in [7].

References

1. M. Dehghan, *Appl. Numer. Math* **124**, 17 (2001).
2. T. Kimura, T. Suzuki, *SIAM J. Appl. Math.* **53**, 1747 (1993).
3. Y. A. Gryazin, M. V. Klibanov, T. R. Lucas, *Inverse Probl.* **15**, 373 (1999).
4. J. R. Cannon, Y. L. Lin, S. Xu, *Inverse Probl.* **10**, 227 (1994).
5. C.-R. Ye and Z.-Z. Sun, *Appl. Math. Comput.* **188**, 214 (2007).
6. M. Dehghan, *Appl. Math. Comput.* **135**, 491 (2003).
7. A. Ashyralyev, P. E. Sobolevskii, *New Difference Schemes for Partial Differential Equations*, Birkhäuser Verlag, Basel, Boston, Berlin, 2004.

NUMERICAL SOLUTION OF NONLOCAL BOUNDARY VALUE PROBLEMS FOR ELLIPTIC-PARABOLIC EQUATIONS

A. ASHYRALYEV

Department of Mathematics, Fatih University,
İstanbul, 34500, Turkey
E-mail: aashyr@fatih.edu.tr

O. GERÇEK

İstanbul Vocational School, Fatih University,
İstanbul, 34500, Turkey
E-mail: ogercek@fatih.edu.tr

A numerical method is proposed for solving elliptic-parabolic partial differential equations with nonlocal boundary conditions. Stable difference schemes of first and second orders of accuracy are presented. Stability and almost coercive stability of these difference schemes are established. The method is illustrated by numerical examples.

Keywords: Elliptic-parabolic equation; Difference schemes.

1. Introduction

Methods of solving nonlocal boundary value problems for elliptic-parabolic differential equations have been studied by Bazarov and Soltanov [1], Glazatov [2], Ashyralyev and Soltanov [3] and Karatopraklieva [4].

Let Ω be the open unit cube in the n-dimensional Euclidean space $\mathbb{R}^n$ $(0 < x_k < 1,\ 1 \le k \le n)$ with boundary $S, \overline{\Omega} = \Omega \cup S$. In $[-1,1] \times \Omega$, mixed boundary value problem

$$\begin{cases} -u_{tt} - \sum_{r=1}^{n}(a_r(x)u_{x_r})_{x_r} = g(t,x),\ 0 < t < 1,\ x \in \Omega, \\ u_t + \sum_{r=1}^{n}(a_r(x)u_{x_r})_{x_r} = f(t,x),\ -1 < t < 0,\ x \in \Omega, \\ u(t,x) = 0,\ -1 \le t \le 1,\ x \in S, \\ u(-1,x) = u(1,x) + \mu(x),\ x \in \overline{\Omega}, \\ u(0+,x) = u(0-,x),\ u_t(0+,x) = u_t(0-,x),\ x \in \overline{\Omega} \end{cases} \quad (1)$$

for multidimensional mixed elliptic-parabolic equation is considered.The functions $a_r(x) \geqslant a > 0$ $(x \in \Omega)$, $g(t,x)$ $(t \in (0,1),\ x \in \overline{\Omega})$ and $f(t,x)$ $(t \in (-1,0), x \in \overline{\Omega})$ are smooth. Here a is a constant.

In present paper, we are interested in studying stable difference schemes for the numerical solution of the nonlocal boundary value problem for elliptic-parabolic differential equation (1). The first and second order of accuracy difference schemes are presented. The stability and almost coercive stability of these difference schemes are established. A procedure of modified Gauss elimination method is used for solving these difference schemes for the one-dimensional elliptic-parabolic differential equation. The method is illustrated by a numerical example.

2. Difference Schemes

The discretization of problem (1) is carried out in two steps. In the first step, the grid sets

$$\widetilde{\Omega}_h = \{x = x_m = (h_1 m_1, \cdots, h_n m_n),\ m = (m_1, \cdots, m_n)\ ,$$
$$0 \leq m_r \leq N_r,\ h_r N_r = 1,\ r = 1, \cdots, n\},$$
$$\Omega_h = \widetilde{\Omega}_h \cap \Omega,\ S_h = \widetilde{\Omega}_h \cap S$$

are defined. To the differential space operator A by the formula

$$Au = -\sum_{r=1}^{n} (a_r(x) u_{x_r})_{x_r},$$

acting in the space of functions $u(x)$, defined in $\overline{\Omega}$ and satisfying the Dirichlet boundary condition: $u(x) = 0$ in S, we assign the difference operator A_h^x by the formula

$$A_h^x u^h = -\sum_{r=1}^{n} \left(a_r(x) u^h_{\bar{x}_r} \right)_{x_r, j_r} \tag{2}$$

acting in the space of grid functions $u^h(x)$, satisfying the conditions $u^h(x) = 0$ for all $x \in S_h$. With the help of A_h^x, we arrive at the nonlocal boundary value problem

$$\begin{cases} -\frac{d^2 u^h(t,x)}{dt^2} + A_h^x u^h(t,x) = g^h(t,x), \quad 0 \leq t \leq 1, x \in \Omega_h, \\ \frac{du^h(t,x)}{dt} - A_h^x u^h(t,x) = f^h(t,x), \ -1 \leq t \leq 0, x \in \Omega_h, \\ u^h(-1,x) = u^h(1,x) + \mu^h(x), x \in \widetilde{\Omega}_h, \\ u^h(0+,x) = u^h(0-,x),\ \frac{du^h(0+,x)}{dt} = \frac{du^h(0-,x)}{dt}, x \in \widetilde{\Omega}_h \end{cases} \tag{3}$$

for an infinite system of ordinary differential equations.

In the second step, we replace problem (3) by the first order of accuracy difference scheme

$$\begin{cases} -\frac{u_{k+1}^h(x)-2u_k^h(x)+u_{k-1}^h(x)}{\tau^2}+A_h^x u_k^h(x)=g_k^h(x), \\ g_k^h(x)=g^h(t_k,x_n),\ t_k=k\tau,\ 1\le k\le N-1,\ \ N\tau=1, x\in\Omega_h, \\ \frac{u_k^h(x)-u_{k-1}^h(x)}{\tau}-A_h^x u_{k-1}^h(x)=f_k^h(x), \\ f_k^h(x)=f^h(t_{k-1},x_n),\ \ t_{k-1}=(k-1)\tau,\ -N+1\le k\le 0, x\in\Omega_h, \\ u_{-N}^h(x)=u_N^h(x)+\mu^h(x), x\in\widetilde{\Omega}_h, \\ u_1^h(x)-u_0^h(x)=u_0^h(x)-u_{-1}^h(x), x\in\widetilde{\Omega}_h \end{cases} \tag{4}$$

and the second order of accuracy difference scheme

$$\begin{cases} -\frac{u_{k+1}^h(x)-2u_k^h(x)+u_{k-1}^h(x)}{\tau^2}+A_h^x u_k^h(x)=g_k^h(x), \\ g_k^h(x)=g^h(t_k,x_n),\ t_k=k\tau,\ 1\le k\le N-1,\ \ N\tau=1, x\in\Omega_h, \\ \frac{u_k^h(x)-u_{k-1}^h(x)}{\tau}-\frac{A_h^x}{2}\left(u_k^h(x)+u_{k-1}^h(x)\right)=f_k^h(x), \\ f_k^h(x)=f^h(t_{k-\frac12},x_n),\ t_{k-\frac12}=(k-\frac12)\tau,\ -N+1\le k\le 0, x\in\Omega_h, \\ u_{-N}^h(x)=u_N^h(x)+\mu^h(x), x\in\widetilde{\Omega}_h, \\ -u_2^h(x)+4u_1^h(x)-3u_0^h(x)=3u_0^h(x)-4u_{-1}^h(x)+u_{-2}^h(x), x\in\widetilde{\Omega}_h. \end{cases} \tag{5}$$

To formulate the results, one needs to introduce the space $L_{2h}=L_2(\widetilde{\Omega}_h)$ of all the grid functions $\varphi^h(x)=\{\varphi(h_1m_1,\cdots,h_nm_n)\}$ defined on $\widetilde{\Omega}_h$, equipped with the norm

$$\|\varphi^h\|_{L_2(\widetilde{\Omega}_h)}=\left(\sum_{x\in\overline{\Omega}_h}\left|\varphi^h(x)\right|^2 h_1\cdots h_n\right)^{1/2}.$$

Theorem 2.1. *Let τ and $|h|=\sqrt{h_1^2+\cdots+h_n^2}$ be sufficiently small numbers. Then the solutions of difference scheme (4) and (5) satisfy the stability and almost coercivity estimates*

$$\max_{-N\le k\le N-1}\left\|u_k^h\right\|_{L_{2h}}\le C_1\left[\max_{-N+1\le k\le 0}\left\|f_k^h\right\|_{L_{2h}}+\max_{1\le k\le N-1}\left\|g_k^h\right\|_{L_{2h}}+\left\|\mu^h\right\|_{L_{2h}}\right],$$

$$\max_{1\le k\le N-1}\left\|\tau^{-2}\left(u_{k+1}^h-2u_k^h+u_{k-1}^h\right)\right\|_{L_{2h}}+\max_{-N+1\le k\le 0}\left\|\tau^{-1}\left(u_k^h-u_{k-1}^h\right)\right\|_{L_{2h}}$$

$$+\sum_{r=1}^{n}\left\|{u^h}_{x_r\overline{x}_r,\,j_r}\right\|_{L_{2h}}\le C_2\left[\sum_{r=1}^{n}\left\|{\mu^h}_{x_r\overline{x}_r,\,j_r}\right\|_{L_{2h}}+\tau\sum_{r=1}^{n}\left\|\left(f_0^h\right)_{x_r\overline{x}_r,\,j_r}\right\|_{L_{2h}}\right.$$

$$\left.+\ln\frac{1}{\tau+|h|}\left(\max_{-N\le k\le -1}\left\|f_k^h\right\|_{L_{2h}}+\max_{1\le k\le N-1}\left\|g_k^h\right\|_{L_{2h}}\right)\right],$$

where C_1 and C_2 do not depend on $\tau, h, \mu^h(x)$ and $g_k^h(x)$, $1\le k\le N-1$, f_k^h, $-N+1\le k\le 0$.

Proof. The proof of Theorem 2.1 is based on the symmetry properties of the difference operator A_h^x defined by the formula (2) in L_{2h} and the following theorem on the coercivity inequality for the solution of the elliptic difference problem in L_{2h} Sobolevskii [5]. □

Theorem 2.2. *For the solution of the elliptic difference problem*

$$A_h^x u^h(x) = \omega^h(x), \ x \in \Omega_h, \ \ u^h(x) = 0, \ x \in S_h, \tag{6}$$

the following coercivity inequality holds:

$$\sum_{r=1}^{n} \left\| u^h_{\bar{x}_r x_r, j_r} \right\|_{L_{2h}} \leq C_3 \left\| \omega^h \right\|_{L_{2h}}.$$

3. Numerical Results

We have not been able to obtain a sharp estimate for the constants figuring in the stability inequality. Therefore, we will give the following results of numerical experiments of the nonlocal boundary value problem

$$\begin{cases} \frac{\partial u}{\partial t} + \frac{\partial^2 u}{\partial x^2} = (1-t)\sin x, \ \ -1 < t < 0, \ 0 < x < \pi, \\ \frac{\partial^2 u}{\partial t^2} + \frac{\partial^2 u}{\partial x^2} = -t \sin x, \ \ 0 < t < 1, \ 0 < x < \pi, \\ u\left(1, x\right) = u\left(-1, x\right) + 2\sin x, \ 0 \leq x \leq \pi, \\ u(0+, x) = u(0-, x), \ u_t(0+, x) = u_t(0-, x), \ 0 \leq x \leq \pi, \\ u\left(t, 0\right) = u\left(t, \pi\right) = 0, \ -1 \leq t \leq 1 \end{cases} \tag{7}$$

for elliptic-parabolic equation.

In the first step, we apply the first order of accuracy difference scheme (4) to get system of equations in the matrix form

$$\begin{cases} AU_{n+1} + BU_n + CU_{n-1} = D\varphi_n, \ 1 \leq n \leq M-1, \\ U_0 = \widetilde{0}, \ U_M = \widetilde{0}, \end{cases} \tag{8}$$

where A, B, C are $(2N+1) \times (2N+1)$ matrices defined by

$$A = C = \begin{bmatrix} 0\ 0 \cdot\cdot\ 0\ 0\ 0\ 0 \cdot\cdot\ 0\ 0 \\ a\ 0 \cdot\cdot\ 0\ 0\ 0\ 0 \cdot\cdot\ 0\ 0 \\ 0\ a \cdot\cdot\ 0\ 0\ 0\ 0 \cdot\cdot\ 0\ 0 \\ \cdot\cdot\ \cdot\cdot\ \cdot\cdot\ \cdot\cdot\ \cdot\cdot\ \cdot\cdot\ \cdot\cdot\ \cdot\cdot\ \cdot\cdot\ \cdot\cdot \\ 0\ 0 \cdot\cdot\ a\ 0\ 0\ 0 \cdot\cdot\ 0\ 0 \\ 0\ 0 \cdot\cdot\ 0\ 0\ a\ 0 \cdot\cdot\ 0\ 0 \\ 0\ 0 \cdot\cdot\ 0\ 0\ 0\ a \cdot\cdot\ 0\ 0 \\ \cdot\cdot\ \cdot\cdot\ \cdot\cdot\ \cdot\cdot\ \cdot\cdot\ \cdot\cdot\ \cdot\cdot\ \cdot\cdot\ \cdot\cdot\ \cdot\cdot \\ 0\ 0 \cdot\cdot\ 0\ 0\ 0\ 0 \cdot\cdot\ a\ 0 \\ 0\ 0 \cdot\cdot\ 0\ a\ 0\ 0 \cdot\cdot\ 0\ 0 \end{bmatrix}, \quad B = \begin{bmatrix} -1\ 0\ 0\ 0 \cdot\cdot\ 0\ 0\ 0\ 0\ 1 \\ b\ \ c\ 0\ 0 \cdot\cdot\ 0\ 0\ 0\ 0\ 0 \\ 0\ \ b\ c\ 0 \cdot\cdot\ 0\ 0\ 0\ 0\ 0 \\ \cdot\cdot\ \cdot\cdot\ \cdot\cdot\ \cdot\cdot\ \cdot\cdot\ \cdot\cdot\ \cdot\cdot\ \cdot\cdot\ \cdot\cdot\ \cdot\cdot \\ 0\ \ 0 \cdot\cdot\ b\ c\ 0\ 0\ 0 \cdot\cdot\ 0 \\ 0\ \ 0 \cdot\cdot\ 0\ d\ e\ d\ 0 \cdot\cdot\ 0 \\ 0\ \ 0 \cdot\cdot\ 0\ 0\ d\ e\ d \cdot\cdot\ 0 \\ \cdot\cdot\ \cdot\cdot\ \cdot\cdot\ \cdot\cdot\ \cdot\cdot\ \cdot\cdot\ \cdot\cdot\ \cdot\cdot\ \cdot\cdot\ \cdot\cdot \\ 0\ \ 0 \cdot\cdot\ \cdot\cdot\ \cdot\cdot\ \cdot\cdot\ \cdot\cdot\ d\ e\ d \\ 0\ \ 0 \cdot\cdot\ 0\ f\ g\ c\ 0 \cdot\cdot\ 0 \end{bmatrix},$$

and D is $(2N+1)\times(2N+1)$ identity matrix, φ_n, U_s are $(2N+1)\times 1$ column vectors as

$$\varphi_n = \begin{bmatrix} \varphi_n^{-N} \\ \cdots \\ \varphi_n^0 \\ \cdots \\ \varphi_n^N \end{bmatrix}, \quad U_s = \begin{bmatrix} U_s^{-N} \\ \cdots \\ U_s^0 \\ \cdots \\ U_s^N \end{bmatrix} \quad \text{for } s = n \pm 1, n.$$

with

$$\varphi_n^k = \begin{cases} 2\sin x_n, k = -N, \\ (1-t_{k-1})\sin x_n, \quad -N+1 \le k \le 0, \\ -t_k \sin x_n, \quad 0 \le k \le N-1, \\ \sin x_n, \quad k = N. \end{cases}$$

So, we have the second order difference equation (8) with respect to n with matrix coefficients. To solve this difference equation, we have applied a procedure of modified Gauss elimination method. Hence, we obtain a solution of the matrix equation in the form

$$\begin{cases} U_j = \alpha_{j+1} U_{j+1} + \beta_{j+1}, \quad j = M-1, \cdots, 2, 1, \\ U_M = 0, \end{cases}$$

where α_j $(j = 1, \cdots, M)$ are $(2N+1)\times(2N+1)$ square matrices and β_j $(j = 1, \cdots, M)$ are $(2N+1)\times 1$ column matrices defined by

$$\begin{cases} \alpha_{j+1} = -(B + C\alpha_j)^{-1} A, \\ \beta_{j+1} = (B + C\alpha_j)^{-1}(D\varphi_j - C\beta_j), \end{cases}$$

where $j = 1, \cdots, M-1$, α_1 is the $(2N+1)\times(2N+1)$ zero matrix and β_1 is the $(2N+1)\times 1$ zero matrix.

In the second step, we apply the second order difference scheme (5) to get a system of linear equations in the matrix form

$$\begin{cases} AU_{n+1} + BU_n + CU_{n-1} = D\varphi_n, \quad 1 \le n \le M-1, \\ U_0 = \widetilde{0}, \ U_M = \widetilde{0}, \end{cases} \tag{9}$$

where A, B, C are $(2N+1)\times(2N+1)$ matrices defined by

$$A=C=\begin{bmatrix} 0\,0\,0\,0 \cdots \cdots 0\,0\,0\,0 \\ x\,x\,0\,0 \cdots \cdots 0\,0\,0\,0 \\ 0\,x\,x\,0 \cdots \cdots 0\,0\,0\,0 \\ \cdots\cdots\cdots\cdots\cdots\cdots \\ 0\,0 \cdots x\,x\,0 \cdots \cdots 0\,0 \\ 0\,0 \cdots 0\,0\,a \cdots \cdots 0\,0 \\ 0\,0 \cdots 0\,0\,0\,a \cdots 0\,0 \\ \cdots\cdots\cdots\cdots\cdots\cdots \\ 0\,0 \cdots 0\,0\,0\,0 \cdots a\,0 \\ 0\,0\,0\,0 \cdots \cdots 0 \cdots 0\,0 \end{bmatrix}, \quad B=\begin{bmatrix} -1\,0\,0\,0 & \cdots & 0\,0\,0\,0\,1 \\ y\;\;z\,0\,0 & \cdots & 0\,0\,0\,0\,0 \\ 0\;\;y\;z\,0 & \cdots & 0\,0\,0\,0\,0 \\ \cdots\cdots & \cdots & \cdots\cdots \\ 0\;\;0 \cdots y & z & 0\,0\,0 \cdots 0 \\ 0\;\;0 \cdots 0 & d & e\,d\,0 \cdots 0 \\ 0\;\;0 \cdots 0 & 0 & d\,e\,d \cdots 0 \\ \cdots\cdots & \cdots & \cdots\cdots \\ 0\;\;0 \cdots \cdots & \cdots & \cdots \cdots d\,e\,d \\ 0\;\;0 \cdots 1 & -4 & 6\,4\,1 \cdots 0 \end{bmatrix},$$

and D is the $(2N+1)\times(2N+1)$ identity matrix, φ_n, U_s are $(2N+1)\times 1$ column vectors as

$$\varphi_n=\begin{bmatrix} \varphi_n^{-N} \\ \cdots \\ \varphi_n^{0} \\ \cdots \\ \varphi_n^{N} \end{bmatrix}, \quad U_s=\begin{bmatrix} U_s^{-N} \\ \cdots \\ U_s^{0} \\ \cdots \\ U_s^{N} \end{bmatrix} \qquad \text{for } s=n\pm 1, n,$$

with

$$\varphi_n^k=\begin{cases} 2\sin x_n, \ k=-N, \\ (1-(t_k-\frac{\tau}{2}))\sin x_n, \ -N+1\le k\le 0, \\ -t_k\sin x_n, \ 1\le k\le N-1, \\ 0, \ k=N. \end{cases}$$

So, we again have the second order difference equation (9) with respect to n with matrix coefficients. To solve this difference equation, we have applied the same procedure of modified Gauss elimination method.

Finally, to compare the results, we give the errors of the numerical solution for first order of accuracy difference and second order accuracy difference schemes computed by

$$E_M^N=\max_{\substack{-N\le k\le N \\ 1\le n\le M-1}}\left|u(t_k,x_n)-u_n^k\right|,$$

where $u(t_k,x_n)$ and u_n^k represent the exact solution and the numerical solution at (t_k,x_n) respectively. The results are shown in the following table.

Table 1. Comparison of the errors

Method	N=M=20	N=M=30	N=M=60
Difference scheme (4)	0.0468	0.0319	0.0163
Difference scheme (5)	0.00080	0.00036	0.00009

It is observed that, the second order of accuracy difference scheme is more accurate compared to the first order of accuracy difference scheme.

References

1. D. Bazarov and H. Soltanov, *Some Local and Nonlocal Boundary Value Problems for Equations of Mixed and Mixed-Composite Types*, Ylym, Ashgabat, 1995 (Russian).
2. S. N. Glazatov, *Nonlocal Boundary Value Problems for Linear and Nonlinear Equations of Variable Type*, Sobolev Institute of Mathematics SB RAS, Novosibirsk, 1998 (Russian).
3. A. Ashyralyev and H. Soltanov, On elliptic-parabolic equations in a Hilbert space, *Proc. of the IMM and CS of Turkmenistan*, Turkmenistan, 1995 (Russian).
4. M. G. Karatopraklieva, *Differensial'nye Uravneniya*, **27**, 68 (1991) (Russian).
5. P. E. Sobolevskii, *On Difference Methods for the Approximate Solution of Differential Equations*, Voronezh State University Press, Voronezh, 1975 (Russian).

ON ONE DIFFERENCE SCHEME OF SECOND ORDER OF ACCURACY FOR HYPERBOLIC EQUATIONS

A. ASHYRALYEV

Department of Mathematics, Fatih University,
İstanbul, 34500, Turkey
E-mail: aashyr@fatih.edu.tr

M. E. KÖKSAL

Department of Mathematics, Fatih University,
İstanbul, 34500, Turkey
& Department of Mathematics, Gebze Institute of Technology,
Kocaeli, 41400, Turkey
E-mail: mekoksal@fatih.edu.tr

A numerical solution of a mixed problem for hyperbolic equations with dependent coefficients is considered. The difference scheme of the second order of accuracy is presented. The stability estimates for the solution of this difference scheme are constructed. The numerical analysis is given for the wave equation.

Keywords: Hyperbolic equation; Difference schemes; Numerical solution.

1. Introduction. The Difference Scheme

It is known (see, for example, Krein [1], Fattorini [2] and Piskarev [3]) that various initial-boundary value problems for the hyperbolic equations can be reduced to the initial value problem

$$\frac{d^2u(t)}{dt^2} + A(t)u(t) = f(t), \quad 0 < t < T, \quad u(0) = \varphi, \ u'(0) = \psi \tag{1}$$

for differential equations in a Hilbert space H. Here $A(t)$ are the self-adjoint positive definite operators in H with a t-independent domain $D = D(A(t))$.

Of great interest is the study of absolute stable difference schemes of a high order of accuracy for hyperbolic partial differential equations, in which stability was established without any assumptions to respect of the grid steps τ and h. Such type stability inequalities for the solutions of the first order of accuracy difference scheme

$$\begin{cases} \tau^{-2}(u_{k+1} - 2u_k + u_{k-1}) + A_k u_{k+1} = f_k, \ A_k = A(t_k), \ f_k = f(t_k), \\ t_k = k\tau, \ 1 \leq k \leq N-1, \ N\tau = T, \ \tau^{-1}(u_1 - u_0) = \psi, \ u_0 = \varphi \end{cases} \tag{2}$$

for the approximate solution of problem (1) were established for the first time in Sobolevskii and Chebotaryeva [4].

Let Ω be the open unit cube in the n-dimensional Euclidean space $\mathbb{R}^n$ $(0 < x_k < 1, \ 1 \leq k \leq n)$ with boundary S, $\overline{\Omega} = \Omega \cup S$. We consider the initial-boundary value problem for the multidimensional hyperbolic equation

$$\begin{cases} \frac{\partial^2 u(t,x)}{\partial t^2} - \sum\limits_{r=1}^{n} (a_r(t,x)u_{x_r})_{x_r} = f(t,x), \\ x = (x_1, \ldots, x_n) \in \Omega, \ 0 < t < 1, \\ u(0,x) = \varphi(x), \ \frac{\partial u(0,x)}{\partial t} = \psi(x), \ x \in \overline{\Omega}; \ u(t,x) = 0, \ x \in S, \end{cases} \tag{3}$$

where $a_r(t,x)$ $(x \in \Omega)$, $\varphi(x)$, $\psi(x)$ $(x \in \overline{\Omega})$ and $f(t,x)$ $(t \in (0,1), \ x \in \Omega)$ are given smooth functions and $a_r(t,x) \geq a > 0$.

The Hilbert space $L_2(\overline{\Omega})$ of all square integrable functions defined on $\overline{\Omega}$, equipped with the norm

$$\|f\|_{L_2(\overline{\Omega})} = \left\{ \int \cdots \int_{x \in \overline{\Omega}} |f(x)|^2 dx_1 \cdots dx_n \right\}^{\frac{1}{2}}$$

is introduced. Problem (3) has a unique smooth solution $u(t,x)$ for smooth $a_r(t,x)$ and $f(t,x)$ functions. This allows us to reduce the mixed problem (3) to the initial value problem (1) in a Hilbert space $H = L_2(\Omega)$ with a self-adjoint positive definite operators $A(t)$ defined by (3).

The discretization of problem (3) is carried out in two steps. In the first step, the grid sets

$$\overline{\Omega}_h = \{x = x_j = (h_1 j_1, \cdots, h_n j_n), j = (j_1, \cdots, j_n), 0 \leq j_r \leq M_r,$$

$$h_r M_r = 1, \ r = 1, \cdots, n\}, \ \Omega_h = \overline{\Omega} \cap \Omega, \ S_h = \overline{\Omega} \cap S$$

are defined.

The Banach space $L_{2h} = L_2\left(\overline{\Omega}_h\right)$ of the grid functions

$$\varphi^h(x) = \{\varphi(h_1 m_1, \cdots, h_n m_n)\}$$

defined on $\overline{\Omega}_h$, equipped with the norm

$$\|\varphi^h\|_{L_{2h}} = \left(\sum_{x \in \overline{\Omega}_h} |\varphi^h(x)|^2 h_1 \cdots h_n \right)^{\frac{1}{2}}$$

is introduced. To the differential operators $A^x(t)$ generated by the problem (3), we assign the difference operators $A_h^x(t)$ by the formula

$$A_h^x(t)u^h(t,x) = -\sum_{r=1}^{n}(a_r(t,x)u^h_{\bar{x}_r})_{x_r,j_r} \tag{4}$$

acting in the space of grid functions $u^h(t,x)$, satisfying the conditions $u^h(t,x) = 0$ for all $x \in S_h$. It is known that $A_h^x(t)$ are self-adjoint positive definite operators in L_{2h}. With the help of $A_h^x(t)$ we arrive at the initial value problem

$$\begin{cases} \frac{d^2u^h(t,x)}{dt^2} + A_h^x(t)u^h(t,x) = f^h(t,x),\ 0 < t < 1,\ x \in \overline{\Omega}_h, \\ u^h(0,x) = \varphi^h(x),\ \frac{du^h(0,x)}{dt} = \psi^h(x),\ x \in \overline{\Omega}_h \end{cases} \tag{5}$$

for an infinite system of ordinary differential equations.

In the second step, applying (3), one can construct the first order of accuracy in t and the second order of accuracy in the space variables

$$\begin{cases} \frac{1}{\tau^2}(u_{k+1}^h - 2u_k^h + u_{k-1}^h) + A_h^x(t_{k+1})u_{k+1}^h = f_k^h, \\ f_k^h = f^h(t_{k+1},x),\ t_k = k\tau, x \in \Omega_h, 1 \le k \le N-1,\ N\tau = 1, \\ \frac{1}{\tau}(u_1^h - u_0^h) = \psi^h,\ \psi^h = \psi^h(x), \\ u_0^h = \varphi^h,\ \varphi^h = \varphi^h(x),\ x \in \overline{\Omega}_h; u_k^h = 0,\ x \in S_h \end{cases} \tag{6}$$

for approximate solution (3). Abstract theorems given in Sobolevskii and Chebotaryeva [4] permit us to establish the stability of the difference scheme (6).

We are interested in studying the high order of accuracy two-step difference schemes for the approximate solutions of the problem (3). In the present paper, the second order of accuracy difference scheme

$$\begin{cases} \frac{1}{\tau^2}(u_{k+1}^h - 2u_k^h + u_{k-1}^h) + \frac{1}{2}A_h^x(t_k)u_k^h + \frac{1}{4}A_h^x(t_k)u_{k+1}^h \\ +\frac{1}{4}A_h^x(t_k)u_{k-1}^h = f_k^h, f_k^h = f^h(t_k,x), \\ t_k = k\tau,\ x \in \Omega_h,\ 1 \le k \le N-1,\ N\tau = 1, \\ \frac{1}{\tau}(u_1^h - u_0^h) + \tau A_h^x(0)(u_1^h - u_0^h) = \frac{\tau}{2}(f_0^h - A_h^x(0)u_0^h) + \psi^h, \\ f_0^h = f^h(0,x),\ \psi^h = \psi^h(x),\ u_0^h = \varphi^h,\ \varphi^h = \varphi^h(x),\ x \in \overline{\Omega}_h; \\ u_k^h = 0,\ x \in S_h \end{cases} \tag{7}$$

is presented.

Theorem 1.1. *Let τ and $|h|$ be sufficiently small numbers. Then, for the solution of the difference schemes (6) and (7), the following inequality is valid:*

$$\max_{1\leq k\leq N-1}\left\|\frac{u^h_{k+1}-2u^h_k+u^h_{k-1}}{\tau^2}\right\|_{L_{2h}}+\max_{1\leq k\leq N-1}\sum_{r=1}^{n}\left\|\left(\frac{u^h_{k+1}+u^h_k}{2}\right)_{\overline{x}_r x_r,j_r}\right\|_{L_{2h}}$$
$$\leq C_1\left[\max_{1\leq k\leq N-1}\left\|\tau^{-1}\left(f^h_k-f^h_{k-1}\right)\right\|_{L_{2h}}+\left\|f^h_0\right\|_{L_{2h}}\right.$$
$$\left.+\sum_{r=1}^{n}\left\|\psi^h_{\overline{x}_r,j_r}\right\|_{L_{2h}}+\sum_{r=1}^{n}\left\|\varphi^h_{\overline{x}_r x_r,j_r}\right\|_{L_{2h}}\right],$$

where C_1 is independent of τ, h, f^h_k, ψ^h and φ^h.

The proof of Theorem 1.1 is based on the symmetry properties of the operators $A^x_h(t)$ defined by (4) and the following theorem on the coercivity inequality for the solution of the elliptic difference problem in L_{2h}.

Theorem 1.2. *For the solutions of the elliptic difference problem*

$$A^x_h u^h(x)=w^h(x),\ \ x\in\Omega_h, u^h(x)=0,\ \ x\in S_h \tag{8}$$

the following coercivity inequality holds (Sobolevskii [5]):

$$\sum_{r=1}^{n}\left\|u^h_{x_r\overline{x}_r,j_r}\right\|_{L_{2h}}\leq C_2\left\|w^h\right\|_{L_{2h}}.$$

In the following section, applying the difference schemes (6) and (7), numerical methods are proposed for solving a one-dimensional hyperbolic partial differential equation with dependent coefficients.

2. Numerical Results

For numerical analysis the following initial-boundary value problem

$$\begin{cases}\frac{\partial^2 u(t,x)}{\partial t^2}-(t+x)\frac{\partial^2 u(t,x)}{\partial x^2}=(1+t+x)\exp(-t)\sin x,\\ 0<t<1,\ 0<x<\pi,\\ u(0,x)=\sin x,\ u_t(0,x)=-\sin x,\ \ 0\leq x\leq\pi,\\ u(t,0)=u(t,\pi)=0,\ 0\leq t\leq 1\end{cases} \tag{9}$$

is considered.

First, applying the first order of accuracy difference scheme (6), the first order of accuracy difference scheme

$$\begin{cases} \frac{u_n^{k+1}-2u_n^k+u_n^{k-1}}{\tau^2} - g(t_{k+1},x_n)\frac{u_{n+1}^{k+1}-2u_n^{k+1}+u_{n-1}^{k+1}}{h^2} = f(t_k,x_n), \\ t_k = k\tau,\ x_n = nh, 1 \le k \le N-1, 1 \le n \le M-1, \\ u_n^0 = \varphi(x_n),\ x_n = nh,\ 0 \le n \le M, \\ \frac{u_n^1-u_n^0}{\tau} = \psi(x_n),\ x_n = nh,\ 0 \le n \le M, \\ u_0^k = u_M^k = 0, 0 \le k \le N, \\ g(t,x) = t+x,\ f(t,x) = 2\exp(-t)\sin x, \\ \varphi(x) = \sin x,\ \psi(x) = -\sin x \end{cases} \tag{10}$$

for the approximate solutions of the problem (9) is presented. We have $(M+1) \times (M+1)$ system of linear equations (10) and we write them in the matrix form

$$\begin{cases} A^k\, U^{k+1} + BU^k + C\, U^{k-1} = D\vartheta^k, \quad 1 \le k \le N-1, \\ U^0 = \widetilde{\varphi(x_n)}, U^1 = U^0 + \tau\widetilde{\psi(x_n)},\ \ 0 \le n \le M, \end{cases} \tag{11}$$

where A^k, B, C, and D are $(M+1) \times (M+1)$ matrices given by

$$A^k = \begin{bmatrix} 0 & 0 & 0 & \dots & 0 & 0 & 1 \\ a_1^{k+1} & b_1^{k+1} & a_1^{k+1} & \dots & 0 & 0 & 0 \\ 0 & a_2^{k+1} & b_2^{k+1} & \dots & 0 & 0 & 0 \\ \dots & \dots & \dots & \dots & \dots & \dots & \dots \\ 0 & 0 & 0 & \dots & b_{M-2}^{k+1} & a_{M-2}^{k+1} & 0 \\ 0 & 0 & 0 & \dots & a_{M-1}^{k+1} & b_{M-1}^{k+1} & a_{M-1}^{k+1} \\ 1 & 0 & 0 & \dots & 0 & 0 & 0 \end{bmatrix},$$

$$B = \begin{bmatrix} 0 & 0 & 0 & \dots & 0 & 0 & 0 \\ 0 & c & 0 & \dots & 0 & 0 & 0 \\ 0 & 0 & c & \dots & 0 & 0 & 0 \\ \dots & \dots & \dots & \dots & \dots & \dots & \dots \\ 0 & 0 & 0 & \dots & c & 0 & 0 \\ 0 & 0 & 0 & \dots & 0 & c & 0 \\ 0 & 0 & 0 & \dots & 0 & 0 & 0 \end{bmatrix}, \quad C = \begin{bmatrix} 0 & 0 & 0 & \dots & 0 & 0 & 0 \\ 0 & d & 0 & \dots & 0 & 0 & 0 \\ 0 & 0 & d & \dots & 0 & 0 & 0 \\ \dots & \dots & \dots & \dots & \dots & \dots & \dots \\ 0 & 0 & 0 & \dots & d & 0 & 0 \\ 0 & 0 & 0 & \dots & 0 & d & 0 \\ 0 & 0 & 0 & \dots & 0 & 0 & 0 \end{bmatrix},$$

$$D = \begin{bmatrix} 1 & 0 & \dots & 0 \\ 0 & 1 & \dots & 0 \\ \dots & \dots & \dots & \dots \\ 0 & 0 & \dots & 1 \end{bmatrix}, \quad U^s = \begin{bmatrix} U_0^s \\ U_1^s \\ \dots \\ U_M^s \end{bmatrix}_{(M+1)\times(1)}, \text{ where } s = k \pm 1, k.$$

Here

$$a_n^{k+1} = -\frac{t_{k+1}+x_n}{h^2}, \quad b_n^{k+1} = \frac{1}{\tau^2} + \frac{t_{k+1}+x_n}{h^2},$$

$0 \leq k \leq N-1$, $0 \leq n \leq M$, $c = -\frac{2}{\tau^2}$, $d = \frac{1}{\tau^2}$, $e = \frac{1}{\tau}$,

$$\vartheta_k^n = \begin{cases} 0, & n = 0, \\ f(t_k, x_n), & 1 \leq n \leq M-1, \\ 0, & n = M, \end{cases}, \vartheta^k = \begin{bmatrix} \vartheta_0^k \\ \vartheta_1^k \\ \cdots \\ \vartheta_M^k \end{bmatrix}_{(M+1)\times 1},$$

$$\varphi_n = \begin{bmatrix} \varphi_0 \\ \varphi_1 \\ \cdots \\ \varphi_M \end{bmatrix}_{(M+1)\times 1}, \psi_n = \begin{bmatrix} \psi_0 \\ \psi_1 \\ \cdots \\ \psi_M \end{bmatrix}_{(M+1)\times 1}.$$

So, we have the second order difference equation (11) with respect to k with matrix coefficients. To solve this difference equation we apply the following procedure:

$$\begin{cases} U^{k+1} = \left(A^k\right)^{-1} D\vartheta^k - \left(A^k\right)^{-1} BU^k - \left(A^k\right)^{-1} CU^{k-1}, \\ k = 1, 2, \cdots, N-1, \\ U^0 = \widetilde{\sin(x_n)},\ U^1 = (1-\tau)\widetilde{\sin(x_n)}, \qquad 0 \leq n \leq M. \end{cases}$$

Second, applying the second order of accuracy difference scheme (7), the following second order of accuracy difference scheme for the approximate solutions of the problem (9)

$$\begin{cases} \frac{u_n^{k+1} - 2u_n^k + u_n^{k-1}}{\tau^2} - g(t_k, x_n)\left(\frac{u_{n+1}^k - 2u_n^k + u_{n-1}^k}{2h^2} + \frac{u_{n+1}^{k+1} - 2u_n^{k+1} + u_{n-1}^{k+1}}{4h^2}\right. \\ \left. + \frac{u_{n+1}^{k-1} - 2u_n^{k-1} + u_{n-1}^{k-1}}{4h^2}\right) = f(t_k, x_n),\ x_n = nh,\ t_k = k\tau, \\ 1 \leq k \leq N-1, 1 \leq n \leq M-1, \\ u_n^0 = \varphi(x_n), x_n = nh,\ 0 \leq n \leq M, \\ \frac{u_n^1 - u_n^0}{\tau} = \frac{\tau}{2}\left(\frac{u_{n+1}^1 - 2u_n^1 + u_{n-1}^1}{h^2} + f(0, x_n)\right) + \psi(x_n), \\ x_n = nh,\ 1 \leq n \leq M-1, u_0^k = u_M^k = 0,\ 0 \leq k \leq N, \\ g(t,x) = t + x,\ f(t,x) = 2\exp(-t)\sin x, \\ \varphi(x) = \sin x,\ \psi(x) = -\sin x \end{cases} \tag{12}$$

is presented. We have again $(M+1) \times (M+1)$ system of linear equations (12) and we write them in the matrix form

$$\begin{cases} A^k\, U^{k+1} + B^k U^k + C^k\, U^{k-1} = D\vartheta^k, \quad 1 \leq k \leq N-1, \\ U^0 = \widetilde{\varphi(x_n)}, EU^1 = vU^0 + \widetilde{\gamma(x_n)},\ 0 \leq n \leq M, \end{cases} \tag{13}$$

where

$$A^k = \begin{bmatrix} 0 & 0 & 0 & \dots & 0 & 0 & 1 \\ a_1^k & b_1^k & a_1^k & \dots & 0 & 0 & 0 \\ 0 & a_2^k & b_2^k & \dots & 0 & 0 & 0 \\ \dots & \dots & \dots & \dots & \dots & \dots & \dots \\ 0 & 0 & 0 & \dots & b_{M-2}^k & a_{M-2}^k & 0 \\ 0 & 0 & 0 & \dots & a_{M-1}^k & b_{M-1}^k & a_{M-1}^k \\ 1 & 0 & 0 & \dots & 0 & 0 & 0 \end{bmatrix}_{(M+1)\times(M+1)},$$

$$B^k = \begin{bmatrix} 0 & 0 & 0 & \dots & 0 & 0 & 0 \\ c_1^k & d_1^k & c_1^k & \dots & 0 & 0 & 0 \\ 0 & c_2^k & d_2^k & \dots & 0 & 0 & 0 \\ \dots & \dots & \dots & \dots & \dots & \dots & \dots \\ 0 & 0 & 0 & \dots & d_{M-2}^k & c_{M-2}^k & 0 \\ 0 & 0 & 0 & \dots & c_{M-1}^k & d_{M-1}^k & c_{M-1}^k \\ 0 & 0 & 0 & \dots & 0 & 0 & 0 \end{bmatrix}_{(M+1)\times(M+1)},$$

$$C^k = \begin{bmatrix} 0 & 0 & 0 & \dots & 0 & 0 & 0 \\ a_1^k & b_1^k & a_1^k & \dots & 0 & 0 & 0 \\ 0 & a_2^k & b_2^k & \dots & 0 & 0 & 0 \\ \dots & \dots & \dots & \dots & \dots & \dots & \dots \\ 0 & 0 & 0 & \dots & b_{M-2}^k & a_{M-2}^k & 0 \\ 0 & 0 & 0 & \dots & a_{M-1}^k & b_{M-1}^k & a_{M-1}^k \\ 0 & 0 & 0 & \dots & 0 & 0 & 0 \end{bmatrix}_{(M+1)\times(M+1)},$$

D is the $(M+1)\times(M+1)$ identity matrix, and

$$E = \begin{bmatrix} 0 & 0 & 0 & \dots & 0 & 0 & 1 \\ j_1 & p_1 & j_1 & \dots & 0 & 0 & 0 \\ 0 & j_2 & p_2 & \dots & 0 & 0 & 0 \\ \dots & \dots & \dots & \dots & \dots & \dots & \dots \\ 0 & 0 & 0 & \dots & j_{M-2} & p_{M-2} & 0 \\ 0 & 0 & 0 & \dots & j_{M-1} & p_{M-1} & j_{M-1} \\ 1 & 0 & 0 & \dots & 0 & 0 & 0 \end{bmatrix}_{(M+1)\times(M+1)}.$$

Here

$$a_n^k = -\frac{t_k + x_n}{4h^2},\ b_n^k = \frac{1}{\tau^2} + \frac{t_k + x_n}{2h^2},\ c_n^k = -\frac{t_k + x_n}{2h^2},$$
$$d_n^k = -\frac{2}{\tau^2} + \frac{t_k + x_n}{h^2},\ j_n = -\frac{\tau x_n}{2h^2},\ p_n = \frac{1}{\tau^2} + \frac{\tau x_n}{h^2},$$

$\gamma_n = (-1 + \frac{\tau}{2}(1 + x_n))\sin(x_n),\ v = \frac{1}{\tau},\ 0 \le k \le N,\ 0 \le n \le M,$

$$\vartheta_k^n = \begin{cases} 0, & n = 0, \\ f(t_k, x_n), & 1 \le n \le M-1, \\ 0, & n = M, \end{cases} \quad \vartheta^k = \begin{bmatrix} \vartheta_0^k \\ \vartheta_1^k \\ \cdots \\ \vartheta_M^k \end{bmatrix}_{(M+1)\times 1},$$

$$\varphi_n = \begin{bmatrix} \varphi_0 \\ \varphi_1 \\ \cdots \\ \varphi_M \end{bmatrix}_{(M+1)\times 1}, \quad \gamma_n = \begin{bmatrix} \gamma_0 \\ \gamma_1 \\ \cdots \\ \gamma_M \end{bmatrix}_{(M+1)\times 1}.$$

So, we have the second order difference equation (13) with respect to k with matrix coefficients. To solve this difference equation we apply the procedure

$$\begin{cases} U^{k+1} = \left(A^k\right)^{-1} D\vartheta^k - \left(A^k\right)^{-1} B^k U^k - \left(A^k\right)^{-1} C^k U^{k-1}, \\ k = 1, 2, \cdots, N-1, \\ U^0 = \widetilde{\varphi(x_n)},\ U^1 = E^{-1} v U^0 + E^{-1}\widetilde{\gamma(x_n)}, \qquad 0 \le n \le M. \end{cases}$$

Now, the results of the numerical analysis are given. For their comparison, the errors computed by

$$E_0 - \max_{1 \le k \le N-1} \left(\sum_{n=1}^{M-1} \left| u(t_k, x_n) - u_n^k \right|^2 h \right)^{1/2},$$

of the numerical solutions are recorded for different values of $N=M$, where $u(t_k, x_n)$ represents the exact solution and u_n^k represents the numerical solution at (t_k, x_n). The error for the approximate solutions obtained by using the first and second order of accuracy difference schemes are shown in Table 1 for $N = M = 20$, 30, 40 and 50 respectively.

Table 1. Comparison of the errors E_0 for the approximate solutions

Method	N=M=20	N=M=30	N=M=40	N=M=50
Scheme (6)	0.0108	0.0074	0.0056	0.0046
Scheme (7)	0.0027	0.0012	0.0006	0.0004

Thus, the second order of accuracy difference schemes are more accurate when compared with the first order of accuracy difference scheme.

Remark 2.1. In papers by Ashyralyev and Köksal [6,7], two types of second order of accuracy difference schemes generated by $A^{1/2}(t)$ for the numerical solution of the abstract problem (1) are presented. The stability estimates for the solution of these difference schemes and for the first and second order difference derivatives are established. However, for the practical realization of these difference schemes it is necessary to construct an operator $A^{1/2}(t)$. This action is very difficult for a computer. Therefore, in spite of the theoretical results, the role of their applications to the numerical solution of the boundary value problem is not great.

References

1. S. G. Krein, *Linear Differential Equations in a Banach Space*, Nauka, Moscow, 1966 (Russian).
2. H. O. Fattorini, *Second Order Linear Differential Equations in Banach Space*, Elsevier Science Publishers B.V., North-Holland, 1985.
3. S. Piskarev, *Differential Equations in a Banach Space and Their Approximation*, Moscow State University Publishing House, Moscow, 2005 (Russian).
4. P. E. Sobolevskii and L. M. Chebotaryeva, Izv. Vyssh. Uchebn. Zav. Matematika **5**, 103 (1977) (Russian).
5. P. E. Sobolevskii, *Difference Methods for the Approximate Solution of Differential Equations*, Voronezh State University Press, Voronezh, 1975 (Russian).
6. A. Ashyralyev and M. E. Köksal, *Numer. Func. Anal. Opt.* **26**, 739 (2005).
7. A. Ashyralyev and M. E. Köksal, *Discrete Dyn. Nat. Soc.* **2007**, 1 (2007).

ON WELL-POSEDNESS OF ABSTRACT HYPERBOLIC PROBLEMS IN FUNCTION SPACES

A. ASHYRALYEV

Department of Mathematics, Fatih University,
İstanbul, 34500, Turkey
E-mail: aashyr@fatih.edu.tr

M. MARTINEZ[†] and J. PASTOR[‡]

Department of Applied Mathematics, University of Valencia, Valencia, Spain
[†] *E-mail: Martinez@uv.es*
[‡] *E-mail: Vicente.Pastor@uv.es*

S. PISKAREV*

Scientific Research Computer Center, Moscow State University,
Vorobjevy Gory, Moscow 119899, Russia
E-mail: piskarev@gmail.com

This paper is devoted to the analysis of abstract hyperbolic differential equations in $C([0,T];E^{\theta})$, $C^{\alpha}([0,T];E)$ and $L^{p}([0,T];E^{\theta})$ spaces. The presentation is based on C_0-cosine operators theory and a functional analysis approach. For the solutions of second order differential equations, the weak maxima regularity estimates are established.

Keywords: Hyperbolic problem; Banach spaces; Maximal regularity; Weak maximal regularity.

1. Introduction

The importance of coercive (maximal regularity, well-posedness) inequalities is well-known. Ashyralyev, Piskarev and Weis [2], Ashyralyev and Sobolevskii [4,5], Guidetti [12], Keyantuo and Lizama [17], Piskarev [23]. There are a lot of papers which investigated the maximal regularity property for second order differential equations in time: Ashyralyev and

*Research partially supported by Russian Foundation for Basic Research 07-01-00269, 07-01-92104 and by University of Valencia.

Sobolevskii [3], Clement and Guerre-Delabrire [8], Hoppe [14], Keyantuo and Lizama [16]. Recently, Ashyralyev, Cuevas and Piskarev [1], considered the well-posedness of difference schemes for abstract elliptic equations in $L^p([0,T];E)$ spaces. The present paper is devoted to the analysis of abstract hyperbolic differential equations in $C([0,T];E^\theta), C^\alpha([0,T];E)$ and $L^p([0,T];E^\theta)$ spaces.

Let $B(E)$ denote the Banach algebra of all linear bounded operators on a complex Banach space E. The set of all linear closed densely defined operators in E will be denoted by $\mathrm{C}(E)$. In a Banach space E, let us consider the following inhomogeneous Cauchy problem:

$$u'(t) = Au(t) + f(t),\ t \in [0,T], u(0) = u_0, \tag{1}$$

where the operator $A \in \mathrm{C}(E)$ generates a C_0-semigroup and $f(\cdot)$ is some function from $[0,T]$ into E. Problem (1) can be considered in various functional spaces. The most popular situations are the following settings: the well-posedness in $C([0,T];E), C^{\alpha,0}([0,T];E)$, and $L^p([0,T];E)$ spaces (see Ashyralyev and Sobolevskii [4]).

We say that problem (1) is well posed, say in $C([0,T];E)$, if, for any $f(\cdot) \in C([0,T];E)$ and any $u_0 \in D(A)$,

(*i*) the problem (1) is uniquely solvable, i.e., $u(\cdot)$ satisfies the equation and boundary condition (1), $u(\cdot)$ is continuously differentiable on $[0,T]$, $u(t) \in D(A)$ for any $t \in [0,T]$ and $Au(\cdot)$ is continuous on $[0,T]$;

(*ii*) the operator $(f(\cdot), u^0) \to u(\cdot)$ as an operator from $C([0,T];E) \times D(A)$ to $C([0,T];E)$ is continuous.

The coercive well-posedness in the space $\Upsilon([0,T];E)$ for problem (1) means that it is well-posed in the space $\Upsilon([0,T];E)$ and

$$\|u'(\cdot)\|_{\Upsilon([0,T];E)} + \|Au(\cdot)\|_{\Upsilon([0,T];E)} \leq C\left(\|f(\cdot)\|_{\Upsilon([0,T];E)} + \|u^0\|_{\tilde{E}}\right), \tag{2}$$

where $\tilde{E}$ is some subspace of E. For results of the coercive well-posedness see Ashyralyev and Sobolevskii [4], Guidetti, Karasözen and Piskarev [11].

We have to note that if (1) is coercive well-posed in the space $C([0,T];E)$, then (see Eberhardt and Greiner [18]) operator A should be bounded or the space E contains a subspace isomorphic to c_0. It means that in general problem (1) is not coercive well-posed in $C([0,T];E)$ space. However, problem (1) is classically well-posed in $C([0,T];E^\alpha)$, (see Da Prato and Grisvard [9,10], Sobolevskii [24]), where $E^\alpha = (E, D(A))_\alpha$ is a suitable interpolation space. It was proved (see Ashyralyev and Sobolevskii [3] that coercive well-posedness in $C^{\alpha,0}([0,T];E)$ space is equivalent to the condition that A generates analytic C_0 -semigroup.

In the mean time the situation in $L^p([0,T];E)$ space is not complete. One has only an extrapolation theorem and one could get coercive inequality just for the interpolation space instead of E.

The necessary and sufficient conditions for coercive well-posedness of problem (1) in $L^p([0,T];E)$ with E to be a UMD space was obtained in Kalton and Lancien [15], Kunstmann and Weis [21] and Weis [27].

Theorem 1.1. *(Weis [27]) Let A generate a bounded analytic semigroup $\exp(\cdot A)$ on a UMD–space E. Then problem (1) is coercive well-posed in the space $L^p(\mathbb{R}_+;E)$ if and only if one of the sets i), ii) or iii) is R–bounded:*

i) $\{\lambda(\lambda I - A)^{-1} : \lambda \in i\mathbb{R}, \lambda \neq 0\}$;
ii) $\{\exp(tA), tA\exp(tA) : t > 0\}$;
iii) $\{\exp(zA) : |\arg z| \leq \delta\}$.

The case of a second order equation is very different from the first order equation case. In a Banach space E, let us consider the Cauchy problem

$$u''(t) = Au(t) + f(t), \quad t \in [0,T], \quad u(0) = u^0, u'(0) = u^1 \tag{3}$$

with the operator $A \in \mathrm{C}(E)$ generating a C_0-cosine operator function $C(\cdot, A)$. We will write $A \in \mathrm{C}(M,\omega)$ if $\|C(t,A)\| \leq Me^{\omega|t|}$, $t \in \mathbb{R}$.

Definition 1.1. A function $u(\cdot)$ is called a *classical solution* of problem (3) if $u(\cdot)$ is twice continuously differentiable, $u(t) \in D(A)$ for all $t \in [0,T]$, and $u(\cdot)$ satisfies relations (3).

If $f(\cdot) \in C([0,T];E)$ and $u(\cdot)$ is a classical solution of (3), then

$$u(t) = C(t,A)u^0 + S(t,A)u^1 + \int_0^t S(t-s,A)f(s)\,ds, \quad t \in [0,T]. \tag{4}$$

As in the case of C_0-semigroups of operators, the function $u(\cdot)$ given by (4) is not a classical solution in general.

Proposition 1.1. *(Fattorini [13]) Let $A \in C(M,\omega)$, and let either*
(i) $f(\cdot), Af(\cdot) \in C([0,T);E)$ and $f(t) \in D(A)$ for $t \in [0,T]$ or
(ii) $f(\cdot) \in C^1([0,T];E)$.

Then the function $u(\cdot)$ from (4) with $u^0 \in D(A)$ and $u^1 \in E^1$ is a classical solution of problem (3) on $[0,T]$. Here E^1 is Kisynskii space, i.e. E^1 is the space with the norm $\|x\|_{E^1} := \|x\| + \sup_{0<t\leq 1} \|C'(t,A)x\|$.

Definition 1.2. The function $u(\cdot) \in C([0,T);E)$ given by expression (4) is called a *mild solution* of problem (3).

Let us consider the following homogenous uniformly well-posed Cauchy problem:

$$u''(t) = Au(t), \quad t \in I\!R; u(0) = u^0, \quad u'(0) = u^1. \tag{5}$$

Define the matrix operator $\mathrm{A} := \begin{pmatrix} 0 & I \\ A & 0 \end{pmatrix} : E^1 \times E \to E^1 \times E$ acting on an element $(x, y) \in E^1 \times E$ by the formula $\mathrm{A}(x, y) = (y, Ax)$ that is given on the domain of $D(\mathrm{A}) = D(A) \times E^1$. In what follows, an element $(x, y) \in E^1 \times E$ will be written as the vector $\begin{pmatrix} x \\ y \end{pmatrix}$. The operator A generates (Kisynski [19]) the following C_0-groups of operators on the Banach space $E^1 \times E$:

$$\exp(t\mathrm{A}) \begin{pmatrix} x \\ y \end{pmatrix} := \begin{pmatrix} C(t, A) & S(t, A) \\ AS(t, A) & C(t, A) \end{pmatrix} \begin{pmatrix} x \\ y \end{pmatrix}, \quad t \in I\!R.$$

We have to note that the studies of problem (5) by reducing it to an equivalent first order system is sometimes inconvenient, since the space E^1 is defined either through the C_0-cosine operator-valued function $C(\cdot, A)$ or through infinitely many powers of the resolvent. Therefore, certain additional conditions that allow us to reduce problem (5) to a system without use of the space E^1 are of interest.

Let the uniformly well-posed problem (5) have the form

$$u''(t) = \mathcal{B}^2 u(t) \quad t \in I\!R; \quad u(0) = u^0, \quad u'(0) = u^1, \tag{6}$$

where $\mathcal{B} \in \mathrm{C}(E)$.

Definition 1.3. We say that a solution $u(\cdot)$ of problem (6) satisfies Condition (K) if $u'(\cdot) \in C([0, T]; \mathrm{D}(\mathcal{B}))$.

Proposition 1.2. *(Guidetti, Karasözen and Piskarev [11]) Problem* (6) *has a unique solution satisfying Condition* (K) *if and only if the following Cauchy problem is uniformly well posed on the space* $E \times E$*:*

$$\begin{pmatrix} u \\ v \end{pmatrix}' (t) = \begin{pmatrix} 0 & \mathcal{B} \\ \mathcal{B} & 0 \end{pmatrix} \begin{pmatrix} u \\ v \end{pmatrix} (t), \quad t \in I\!R, \quad \begin{pmatrix} u \\ v \end{pmatrix} (0) = \begin{pmatrix} u_0 \\ v_0 \end{pmatrix}. \tag{7}$$

An analog of Condition (K), which allows us to simplify the study of problem (5) by using C_0-semigroups, is the following Condition (F).

Definition 1.4. A C_0-cosine operator-valued function $C(\cdot, A)$ satisfies *Condition* (F) if the following conditions hold:

(i) there exists $\mathcal{B} \in \mathrm{C}(E)$ such that $\mathcal{B}^2 = A$, and $\mathcal{B}$ commutes with any operator from $B(E)$ commuting with A;

(ii) an operator $S(t, A)$ maps E into $D(\mathcal{B})$ for any $t \in I\!R$;
(iii) the function $\mathcal{B}S(t, A)x$ is continuous in $t \in I\!R$ for any fixed $x \in E$.

Proposition 1.3. *(Fattorini [13]) Under Condition (F), for each $t \in I\!R$, we have $\mathcal{B}S(t, A) \in B(E)$ and $D(\mathcal{B}) \subseteq E^1$.*

Proposition 1.4. *(Fattorini [13]) There exist a Banach space E and a C_0-cosine operator-valued function $C(\cdot, A)$ (even uniformly bounded) such that Condition (F) does not hold.*

Proposition 1.5. *(Cioranescu and Keyantuo [7]) Let $A \in C(M, 0)$, and let E be a UMD space. Then Condition (F) holds.*

Theorem 1.2. *(Travis and Webb [26]) Let $\mathcal{A}$ and $\mathcal{B}$ be operators satisfying condition (i) of Definition 1.4, and let $0 \in \rho(\mathcal{B})$. The following conditions are equivalent:*

(i) the C_0-cosine operator-valued function $C(\cdot, A)$ satisfies Condition (F);

(ii) the operator $\mathcal{B}$ generates a C_0-group $\exp(\cdot\mathcal{B})$ on E;

(iii) the operator $\begin{pmatrix} 0 & \mathcal{B} \\ \mathcal{B} & 0 \end{pmatrix}$ with the domain $D(A) \times D(\mathcal{B})$ generates a C_0-group on $E \times E$;

(iv) the operator $\mathbb{A} := \begin{pmatrix} 0 & I \\ A & 0 \end{pmatrix}$ with the domain $D(A) \times D(\mathcal{B})$ generates a C_0-group $\exp(\cdot\mathbb{A})$ on $D(\mathcal{B}) \times E$, where $D(\mathcal{B})$ is the Banach space of elements $D(\mathcal{B})$ endowed with the graph norm;

(v) the embedding $D(\mathcal{B}) \subseteq E^1$ holds;

(vi) $D(\mathcal{B}) = E^1$.

Theorem 1.3. *(Krein [20]) Let the operator $\mathcal{B} = \sqrt{A}$ in problem (3) have a bounded inverse $\mathcal{B}^{-1} \in B(E)$ and be a generator of a C_0-group, and let the function $f(\cdot)$ have one of the following properties:*

(i) $f(\cdot) \in C^1([0, T); E)$;

(ii) $\mathcal{B}f(\cdot) \in C([0, T); E)$.

Then for any $u^0 \in D(A)$ and $u^1 \in D(\mathcal{B})$, there exists a unique classical solution of problem (3) given by formula (4) in the form

$$u(t) = \frac{1}{2}\Big(\exp(t\mathcal{B}) + \exp(-t\mathcal{B})\Big)u^0 + \frac{1}{2}\Big(\exp(t\mathcal{B}) - \exp(-t\mathcal{B})\Big)\mathcal{B}^{-1}u^1 \qquad (8)$$
$$+ \frac{1}{2}\int_0^t \Big(\exp((t-s)\mathcal{B}) + \exp(-(t-s)\mathcal{B})\Big)\mathcal{B}^{-1}f(s)\,ds, \quad t \in [0, T].$$

Denote $(S(\cdot, A) * f)(t) := \int_0^t S(t-s, A)f(s)\,ds, \quad t \in [0, T]$.

Definition 1.5. We say that a C_0-cosine operator function $C(\cdot, A)$ has the *maximal regularity* (*MR-property*) if $S(\cdot, A) * f \in C^2([0,T]; E)$ (or, which is equivalent, $C(\cdot, A) * f \in C([0,T]; \mathrm{D}(A))$ for all $f(\cdot) \in C([0,T]; E)$.

Definition 1.6. Let $\tilde{E}$ be a Banach space being a subspace of the initial space E, and let $\Upsilon([0,T]; E)$ be the Banach space of functions with values in E. Problem (3) is said to be *coercively solvable* in the pair of spaces $(\tilde{E}, \Upsilon([0,T]; E))$ (in other words, the solution $u(\cdot)$ has the *maximal regularity property*) if for any right-hand side $f(\cdot) \in \Upsilon([0,T]; E)$, there exists a classical solution $u(\cdot)$ of the Cauchy problem (3), for each t, the value of the solution $u(t)$ belongs to $\tilde{E}$, and the following coercive inequality holds:

$$\|u''(\cdot)\|_{\Upsilon([0,T];E)} + \|Au(\cdot)\|_{\Upsilon([0,T];E)} \le M\Big[\|f(\cdot)\|_{\Upsilon([0,T];E)} + \|u^0\|_{\tilde{E}} + \|u^1\|_{\tilde{E}}\Big].$$

Theorem 1.4. *(Chyan, Shaw and Piskarev [6]) Let problem* (3) *be coercively solvable in the pair* $(D(A), C([0,T]; E))$. *Then* $A \in B(E)$.

In case of $L^p([0,T]; E)$ space the situation is the same.

Definition 1.7. The problem (3) is said to be *coercively solvable* in $L^p([0,T]; E)$, $1 \le p \le \infty$, if for any $f(\cdot) \in L^p([0,T]; E)$, there exists a unique solution $u(\cdot)$ satisfying the equation almost everywhere such that $u(0) = u^0$, $u'(0) = u^1$, $u''(\cdot), Au(\cdot) \in L^p([0,T]; E)$, and the following coercive inequality holds:

$$\begin{aligned} &\|u''(\cdot)\|_{L^p([0,T];E)} + \|Au(\cdot)\|_{L^p([0,T];E)} \\ &\le M(p)\Big(\|f(\cdot)\|_{L^p([0,T];E)} + \|u^0\|_{D(A)} + \|u^1\|_{E^1}\Big). \end{aligned} \tag{9}$$

Theorem 1.5. *(Palencia and Piskarev [22]) Let problem* (3) *be coercively solvable in* $L^p([0,T]; E)$ *with a certain* $1 \le p \le \infty$. *Then A is bounded.*

2. The weak maximal regularity (WMR) property

As we saw in the Introduction maximal regularity for hyperbolic problem (3) in $C([0,T]; E), L^p([0,T]; E)$ spaces implies boundedness of operator A. So for the case of an unbounded operator A it is very natural to give the following definition in case one would like to consider some kind of maximal regularity.

Definition 2.1. We say that a C_0-cosine operator function $C(\cdot, A)$ has the weak *maximal regularity* (*WMR-property*) or the *maximal regularity with loss (MRL-property)* if $S(\cdot, A) * f \in C^2([0,T]; E)$ (or, which is equivalent,

$C(\cdot,A)*f \in C([0,T];\mathrm{D}(A))$ for all $f(\cdot) \in F([0,T];\tilde{E}) \subset C([0,T];E)$, where $\tilde{E} \subset E$.

Definition 2.2. Let $\check{E}$ and $\hat{E}$ be Banach spaces being subspaces of the original space E, and let $C([0,T];E)$ be the Banach space of continuous functions with values in E. Problem (3) is said to be weakly coercively solvable in the pair of spaces $((\check{E},\hat{E}),C([0,T];E))$ (in other words, the solution $u(\cdot)$ has the weak maximal regularity property) if for any right-hand side $f(\cdot) \in F([0,T];E)$, there exists a classical solution $u(\cdot)$ of the Cauchy problem (3), for each t, the value of the solution $(u(t),u'(t))$ belongs to $\check{E} \times \hat{E}$, and the following weak coercive inequality holds:

$$\|u''(\cdot)\|_{C([0,T];E)}\|+\|Au(\cdot)\|_{C([0,T];E)} \leq M\Big[\|f(\cdot)\|_{F([0,T];E)}+\|u^0\|_{\check{E}}+\|u^1\|_{\hat{E}}\Big].$$

Remark 2.1. Note that in the cases $F([0,T];E) = C([0,T];E)$ and $\check{E} = D(A), \hat{E} = D(\mathcal{B})$ the weak *maximal regularity* (*WMR-property*) is equivalent to the maximal regularity in $C([0,T];E)$.

Theorem 2.1. *Assume that condition (F) is satisfied and* $f(\cdot) \in C^1([0,T];E) \cap C([0,T];D(\mathcal{B})$. *Then problem* (3) *is weakly coercively solvable in the pair* $((D(A),D(\mathcal{B})),C([0,T];E))$ *and*

$$\|u''(\cdot)\|_{C([0,T];E)}\| + \|Au(\cdot)\|_{C([0,T];E)} \leq M\big(\|Au^0\|_E + \|\mathcal{B}u^1\|_E\big) + \min\big(\|f(\cdot)\|_{C^1([0,T];E)}, \|f(\cdot)\|_{C([0,T];D(\mathcal{B}))}\big).$$

3. Maximal regularity in $C([0,T];E^\theta)$, $C^\alpha([0,T];E)$ spaces

As one can see from the Introduction, the open questions on maximal regularity occur in the area of maximal regularity for second order equation in the spaces like $C([0,T];E^\theta), C^\alpha([0,T];E), 0 \leq \alpha < 1, L^p([0,T];E^\theta)$ spaces. In the case of first order in time Cauchy problem (1) is not coercive well-posed in $C([0,T];E)$ space, but it is coercive well-posed in the spaces $C([0,T];E^\theta), C^{\alpha,0}([0,T];E)$, where E^θ is interpolation space. In the case of second order in time Cauchy problem (3) the maximal regularity property is naturally not independent of spaces.

Proposition 3.1. *(Sova [25], Travis and Webb [26]) For all* $t,s \in I\!R$, *we have the relations*

(*i*) $C(t,A) = C(-t,A), \quad S(-t,A) = -S(t,A), \quad S(0,A) = 0;$
(*ii*) $S(t+s,A) + S(t-s,A) = 2S(t,A)C(s,A);$
(*iii*) $S(t+s,A) = S(t,A)C(s,A) + S(s,A)C(t,A);$
(*iv*) $C(t+s,A) - C(t-s,A) = 2AS(t,A)S(s,A);$
(*v*) $C(2t,A) = 2C(t,A)^2 - I, C(t,A)^2 - AS(t,A)^2 = I.$

Theorem 3.1. *Let an operator A be a generator of C_0-cosine operator function $C(t,A)$. Assume that Cauchy problem (3) is coercive well-posed in $C^\alpha([0,T];E)$ space. Then the operator A is bounded.*

Proof. Let us define the sequence of operators

$$L_{\xi_j,t}x = \frac{1}{\xi_j^\alpha}(g(t+\xi_j) - g(t))x, \quad \xi_j \to 0 \text{ as } j \to \infty,$$

where $g(t)x = A\int_0^t S(t-s,A)S(s,A)x$ for any $x \in E$. The function $g(\cdot)$ represents the function $Au(\cdot)$ which we have in (4) in case $f(t) = S(t,A)x$. It is clear that such $f(\cdot) \in C^1([0,T];E) \subset C^\alpha([0,T];E)$. If the problem (3) is coercively well-posed in $C^\alpha([0,T];E)$ space then $g(\cdot) \in C^\alpha([0,T];E)$ and therefore sequence $\{L_{\xi_j,t}x\}_{j=1}^\infty$ bounded for any $x \in E$. By uniform boundedness principle the norms $\|L_{\xi_j,t}\|$ are uniformly bounded as $j \to \infty$. In the mean time $g(t) = S(t,A) - tC(t,A)$ and therefore $L_{\xi_j,t} = \frac{1}{\xi_j^\alpha}\Big(S(t+\xi_j,A) - S(t,A)\Big) - \frac{1}{\xi_j^\alpha}\Big((t+\xi_j)C(t+\xi_j,A) - tC(t,A)\Big)$. Hence $\|C(t+\xi_j,A) - C(t,A)\| \le C\xi_j^\alpha$ for any $t \in (0,T)$ and $\xi_j \to 0$. This implies that the operator A must be bounded (see Piskarev [23]). □

Theorem 3.2. *Let an operator A be a generator of C_0-cosine operator function $C(t,A)$. Assume that Cauchy problem (3) is coercive well-posed in $C([0,T];E^\theta)$ space. Then the operator A is bounded.*

Proof. Consider the function $g(t)x = A\int_0^t S(t-s,A)C(s,A)(-A)^{-\alpha}xds$. For any $x \in E$ the function $f(s) = C(s,A)(-A)^{-\alpha}x$ belongs to $C([0,T];E^\alpha)$ space and therefore well-posedness in $C([0,T];E^\alpha)$ implies that $g(\cdot) \in C([0,T];E^\alpha)$. Then integrating $A\int_0^t S(t-s,A)C(s,A)(-A)^{-\alpha}xds$ by simple calculations which use (*ii*) from Proposition 3.1 one can find that $\|tAS(t,A)\| \le constant$. So for any $t > 0$ one gets by (iv) from Proposition 3.1 that

$$\|C(t+h,A) - C(t-h,A)\| \le \|2AS(t,A)\|\|S(h,A)\| \to 0$$

as $h \to 0$ and therefore A is bounded. □

From Theorem 1.3 the following results follow.

Theorem 3.3. *Assume that condition* (F) *is satisfied and* $f(\cdot) \in C^1([0,T];E^\theta) \cap C([0,T];D(\mathcal{B}^{1+2\theta}))$. *Then the problem* (3) *is weakly coercively solvable in the pair* $((D(A^{1+2\theta}), D(\mathcal{B}^{1+2\theta})), C(([0,T];E^\theta))$ *and*

$$\|u''(\cdot)\|_{C([0,T];E^\theta)} + \|Au(\cdot)\|_{C([0,T];E^\theta)} \le M\big(\|Au^0\|_{E^\theta} + \|\mathcal{B}u^1\|_{E^\theta}\big) + \min\big(\|f(\cdot)\|_{C^1([0,T];E^\theta)}, \|f(\cdot)\|_{C([0,T];D(\mathcal{B}^{1+2\theta}))}\big)$$

Theorem 3.4. *Assume that condition* (F) *is satisfied and*

$$f(\cdot) \in C^{1+\alpha}([0,T];E) \cap C^\alpha([0,T];D(B))$$

Then the problem (3) *is weakly coercively solvable in the pair* $((D(A^{1+\frac{\alpha}{2}}), D(\mathcal{B}^{1+\alpha})), C^\alpha([0,T];E))$ *and*

$$\|u''(\cdot)\|_{C^\alpha([0,T];E)} + \|Au(\cdot)\|_{C^\alpha([0,T];E)} \le M\big(\|A^{1+\frac{\alpha}{2}}u^0\|_E + \|\mathcal{B}^{1+\alpha}u^1\|_E\big) + \min\big(\|f(\cdot)\|_{C^{1+\alpha}([0,T];E)}, \|f(\cdot)\|_{C^\alpha([0,T];D(\mathcal{B}))}\big)$$

Remark 3.1. As was mentioned in Theorem 1.5 the coercive well-posedness of (3) in $L^p([0,T];E)$ spaces holds in general if and only if A is bounded. Even in the case of periodic functions spaces $L^p_{2\pi}(\mathbb{R};E), C^\alpha_{2\pi}(\mathbb{R};E)$ the situation is not changed: in Keyantuo and Lizama [16] it was shown that maximal regularity holds if and only if $\{-k^2 : k \in \mathbb{Z}\} \subset \rho(A), \sup_{k\in\mathbb{Z}} \|k(-k^2I - A)^{-1}\| < \infty$, which is not true even for general bounded C_0-cosine operator function in the Hilbert space E. In the particular case of UMD spaces E maximal regularity of Cauchy problem (3) for second order in time differential equations is defined by the location of the spectrum of operator A, but not by the smoothness of the space E, that is the function $\lambda^2(\lambda^2 I - A)^{-1}$ must be a Fourier multiplier which is not true in general. So we do not a coercive well-posedness of (3) in $L^p([0,T];E^\theta)$ space in general.

Theorem 3.5. *Assume that condition* (F) *is satisfied and* $f(\cdot) \in W^{1,p}([0,T];E^\theta) \cap L^p([0,T];D(\mathcal{B}^{1+2\theta}))$. *Then the problem* (3) *is weakly coercively solvable in the pair* $((D(A^{1+2\theta}), D(\mathcal{B}^{1+2\theta})), L^p([0,T];E^\theta))$ *and*

$$\|u''(\cdot)\|_{L^p([0,T];E^\theta)}\| + \|Au(\cdot)\|_{L^p([0,T];E^\theta)} \le M\big(\|Au^0\|_{E^\theta} + \|\mathcal{B}u^1\|_{E^\theta}\big) + \min\big(\|f(\cdot)\|_{W^{1,p}([0,T];E^\theta)}, \|f(\cdot)\|_{L^p([0,T];D(\mathcal{B}^{1+2\theta}))}\big)$$

References

1. A. Ashyralyev, C. Cuevas and S. Piskarev, *Numer. Func. Anal. Opt.* **29**, 1 (2008).

2. A. Ashyralyev, S. Piskarev and L. Weis, *Numer. Func. Anal. Opt.* **23**, 669 (2002).
3. A. Ashyralyev and P. E. Sobolevskii, *Discrete Dyn. Nat. Soc.* 183 (2005).
4. A. Ashyralyev and P. E. Sobolevskii, *Well-Posedness of Parabolic Difference Equations*, Birkhäuser Verlag, Basel, Boston, Berlin, 1994.
5. A. Ashyralyev and P. E. Sobolevskii, *New Difference Schemes for Partial Differential Equations*, Birkhäuser Verlag, Basel, Boston, Berlin, 2004.
6. D. K. Chyan, S. Y. Shaw and S. Piskarev, *J. London Math. Soc.* **59**, 1023 (1999).
7. I. Cioranescu and V. Keyantuo, *Semigroup Forum,* **63**, 429 (2001).
8. P. Clement and S. Guerre-Delabrire, *Mat. Appl.* **9**, 245 (1999).
9. G. Da Prato and P. Grisvard, *J. Math. Pures Appl.* **54**, 305 (1975).
10. G. Da Prato and P. Grisvard, *C. R. Acad. Sci. Paris* **283**, 709 (1976).
11. D. Guidetti, B. Karasözen and S. Piskarev, *J. Math. Sci.* **122**, 3013 (2004).
12. D. Guidetti, *Numer. Func. Anal. Opt.* **28**, 307 (2007).
13. H. O. Fattorini, *Second Order Linear Differential Equations in Banach Spaces*, Elsevier Science Publishers B.V., North-Holland, 1985.
14. H. W. R. Hoppe, *SIAM J. Numer. Anal.* **19**, 1110 (1982).
15. N. J. Kalton and G. Lancien, *Math. Z.* **235**, 559 (2000).
16. V. Keyantuo and C. Lizama, *Math. Z.* **235**, 489 (2006).
17. V. Keyantuo and C. Lizama, *Studia Math.* **168**, 25 (2005).
18. B. Eberhardt and G. Greiner, *Acta Appl. Math.* **27**, 47 (1992).
19. J. Kisynski, *Studia Math.* **44**, 93 (1972).
20. S. G. Krein, *Linear Differential Equations in Banach Space*, American Mathematical Society, Providence, R.I., 1971.
21. P. C. Kunstmann and L. Weis, *Lecture Notes in Math.*, Springer, Berlin, 2004.
22. C. Palencia and S. Piskarev, *Semigroup Forum* **63**, 127 (2001).
23. S. Piskarev, *Differential equations in Banach space and it's approximation*, Moscow State University Publish House, Moscow 2005 (Russian).
24. P. E. Sobolevskii, *Trudy Nauchn.-Issled. Inst. Mat. Voronezh. Gos. Univ.* 68 (1975) (Russian).
25. M. Sova, *Rozpr. Mat.* **49**, 1 (1966).
26. C. C. Travis and G. F. Webb, *Acta Math. Acad. Sci. Hung.* **32**, 75 (1978).
27. L. Weis, *Math. Ann.* **319**, 735 (2001).

A NOTE ON DIFFERENCE SCHEMES OF SECOND ORDER OF ACCURACY FOR HYPERBOLIC-PARABOLIC EQUATIONS

A. ASHYRALYEV and Y. ÖZDEMİR*

Department of Mathematics, Fatih University,
İstanbul, 34500, Turkey
**E-mail: yozdemir@fatih.edu.tr*

The second order of accuracy difference schemes for approximately solving the multipoint nonlocal boundary value problem for the differential equations

$$\begin{cases} \frac{d^2u(t)}{dt^2} + Au(t) = f(t),\ (0 \le t \le 1),\ \frac{du(t)}{dt} + Au(t) = g(t)\ (-1 \le t \le 0), \\ u(-1) = \sum\limits_{j=1}^{N} \alpha_j u(\mu_j) + \sum\limits_{j=1}^{L} \beta_j u'(\lambda_j) + \varphi,\ \sum\limits_{j=1}^{N} |\alpha_j|,\ \sum\limits_{j=1}^{L} |\beta_j| \le 1,\ 0 < \mu_j, \lambda_j \le 1 \end{cases}$$

in a Hilbert space H with the selfadjoint positive operator A are presented. Stability estimates for the solution of these difference schemes are established. Stability estimates for the solutions of these difference schemes of nonlocal boundary value problems for hyperbolic-parabolic equations are obtained.

Keywords: Hyperbolic-parabolic equation; Nonlocal boundary value problems; Stability.

1. Introduction

Methods of solutions of nonlocal boundary value problems for hyperbolic-parabolic differential equations have been studied extensively by Dzhuraev [1], Karatoprakliev [2], Karatoprakliева [3], Korzyuk and Lemeshevsky [4], Berdyshev and Karimov [5], Salakhitdinov and Urinov [6], Ashyralyev and Özdemir [7].

The nonlocal boundary value problem for differential equation

$$\begin{cases} \frac{d^2u(t)}{dt^2} + Au(t) = f(t), (0 \le t \le 1), \\ \frac{du(t)}{dt} + Au(t) = g(t), (-1 \le t \le 0) \\ u(-1) = \sum_{j=1}^{N} \alpha_j u(\mu_j) + \sum_{j=1}^{L} \beta_j u'(\lambda_j) + \varphi, \\ \sum_{j=1}^{N} |\alpha_j|, \sum_{j=1}^{L} |\beta_j| \le 1, 0 < \mu_j, \lambda_j \le 1 \end{cases} \quad (1)$$

in a Hilbert space H with self-adjoint positive definite operator A is considered.

In the paper Ashyralyev and Özdemir [7], the stability estimates for the solution of the problem (1) are established. In applications, the stability estimates for the solutions of the mixed type boundary value problems for hyperbolic-parabolic equations are obtained.

In the present paper, the second order of accuracy difference schemes for the approximate solution of the boundary value problem (1) are presented. The stability estimates for the solution of these difference schemes and their first and second order difference derivatives are established. The stability estimates for the solutions of these difference schemes of nonlocal boundary value problem for hyperbolic-parabolic equation are obtained.

2. The Difference Schemes. Two Auxiliary Lemmas

Throughout this paper, for simplicity, $\min \mu_j$, $\min \lambda_j > 2\tau$ will be considered. Let us associate the boundary value problem (1) with the corresponding second order of accuracy difference scheme

$$\begin{cases} \tau^{-2}(u_{k+1} - 2u_k + u_{k-1}) + Au_k + \frac{\tau^2}{4}A^2u_{k+1} = f_k, f_k = f(t_k), \\ t_k = k\tau,\ 1 \le k < N, \tau^{-1}(I + \tau^2 A)(u_1 - u_0) = Z_1, \\ Z_1 = \frac{\tau}{2}(f(0) - Au_0) + (g(0) - Au_0), \\ \tau^{-1}(u_k - u_{k-1}) + A(I + \frac{\tau}{2}A)u_k = (I + \frac{\tau}{2}A)g_k, \\ g_k = g(t_k - \frac{\tau}{2}), t_k = k\tau, -(N-1) \le k \le 0, \\ u_{-N} = \sum\limits_{j=1}^{N} \alpha_j (u_{[\mu_j/\tau]} + (\mu_j - [\mu_j/\tau]\tau) \frac{u_{[\mu_j/\tau]} - u_{[\mu_j/\tau]-1}}{\tau}) + \varphi \\ + \sum\limits_{j=1}^{L} \beta_j (\frac{u_{[\lambda_j/\tau]} - u_{[\lambda_j/\tau]-1}}{\tau} + (\lambda_j - [\lambda_j/\tau]\tau + \frac{\tau}{2})(f_{[\lambda_j/\tau]} - Au_{[\lambda_j/\tau]})). \end{cases} \tag{2}$$

First of all, let us give some lemmas that will be needed below.

Lemma 2.1. *The following estimates*

$$\|R(\pm\tau A^{1/2})\|_{H\to H} \le 1,\ \ \|\tau A^{1/2}R(\pm\tau A^{1/2})\|_{H\to H} \le 1, \tag{3}$$

$$\|\tau^2 AR(\pm\tau A^{1/2})\|_{H\to H} \le 1,\ \ (k\tau)^\beta \|A^\beta R^k\|_{H\to H} \le 1, \tag{4}$$

hold, where $k \ge 1, 0 \le \beta \le 1$,

$$R(\pm\tau A^{1/2}) = (I \pm i\tau A^{1/2} - \tfrac{\tau^2}{2}A)^{-1} \text{ and } R = (I + \tau A + \tfrac{\tau^2}{2}A^2)^{-1}.$$

Lemma 2.2. *We denote*

$$B = \textstyle\sum_{j=1}^{N} \alpha_j (1 + (\frac{\mu_j}{\tau} - [\mu_j/\tau]))\{\frac{1}{2}[R^{[\mu_j/\tau]-1}(\tau A^{1/2})(I - \frac{i\tau A^{1/2}}{2})$$

$$+ R^{[\mu_j/\tau]-1}(-\tau A^{1/2})(I + \tfrac{i\tau A^{1/2}}{2})]R^N - \tfrac{1}{2i}A^{-1/2}(I + \tfrac{\tau^4 A^2}{4})$$
$$\times [R^{[\mu_j/\tau]}(-\tau A^{1/2}) - R^{[\mu_j/\tau]}(\tau A^{1/2})] \left(I + \tau^2 A\right)^{-1} A(\frac{\tau}{2} + 1)R^N\})$$
$$- \textstyle\sum_{j=1}^N \alpha_j(\frac{\mu_j}{\tau} - [\mu_j/\tau])\{\frac{1}{2}[R^{[\mu_j/\tau]-2}(\tau A^{1/2})(I - \frac{i\tau A^{1/2}}{2})$$
$$+ R^{[\mu_j/\tau]-2}(-\tau A^{1/2})(I + \tfrac{i\tau A^{1/2}}{2})]R^N$$
$$- \tfrac{1}{2i}A^{-1/2}(I + \tfrac{\tau^4 A^2}{4})[R^{[\mu_j/\tau]-1}(-\tau A^{1/2}) - R^{[\mu_j/\tau]-1}(\tau A^{1/2})]$$
$$\times \left(I + \tau^2 A\right)^{-1} A(\tfrac{\tau}{2} + 1)R^N + \textstyle\sum_{j=1}^L \beta_j \frac{1}{\tau}(I - (\lambda_j - [\lambda_j/\tau]\tau + \frac{\tau}{2})A)$$
$$\times \{\tfrac{1}{2}[R^{[\lambda_j/\tau]-1}(\tau A^{1/2})(I - \tfrac{i\tau A^{1/2}}{2})$$
$$+ R^{[\lambda_j/\tau]-1}(-\tau A^{1/2})(I + \tfrac{i\tau A^{1/2}}{2})]R^N$$
$$- \tfrac{1}{2i}A^{-1/2}(I + \tfrac{\tau^4 A^2}{4})[R^{[\lambda_j/\tau]}((-\tau A^{1/2}) - R^{[\lambda_j/\tau]}((\tau A^{1/2})](I + \tau^2 A)^{-1}$$
$$\times A(\tfrac{\tau}{2} + 1)R^N\} - \textstyle\sum_{j=1}^L \beta_j \frac{1}{\tau}\{\frac{1}{2}[R^{[\lambda_j/\tau]-2}(\tau A^{1/2})(I - \frac{i\tau A^{1/2}}{2})$$
$$+ R^{[\lambda_j/\tau]-2}(-\tau A^{1/2})(I + \tfrac{i\tau A^{1/2}}{2})]R^N - \tfrac{1}{2i}A^{-1/2}(I + \tfrac{\tau^4 A^2}{4})$$
$$\times [R^{[\lambda_j/\tau]-1}(-\tau A^{1/2}) - R^{[\lambda_j/\tau]-1}(\tau A^{1/2})](I + \tau^2 A)^{-1}A(\tfrac{\tau}{2} + 1)R^N.$$

The operator $I - B$ has an inverse $T_\tau = \left(I - B\right)^{-1}$ and the estimate

$$\|T_\tau\|_{H \to H} \le M \tag{5}$$

holds, where M does not depend on τ.

3. The Main Theorem

Theorem 3.1. *Suppose that $\varphi \in D(A), g_0 \in D(A^{1/2})$ and $f_0 \in D(A^{1/2})$. Then, for the solutions of the difference scheme (2), the stability estimates*

$$\max_{-N\le k\le N} \|u_k\|_H \le M_1[\|\varphi\|_H + \|A^{-1/2}f_0\|_H + \|A^{-1/2}g_0\|_H \tag{6}$$
$$+ \max_{1\le k<N} \|A^{-1/2}(f_k - f_{k-1})\tau^{-1}\| + \max_{-N<k\le 0} \|A^{-1/2}(g_k - g_{k-1})\tau^{-1}\|_H],$$

$$\max_{-N\le k\le N} \|A^{1/2}u_k\|_H \le M_1[\|A^{1/2}\varphi\|_H + \|f_0\|_H + \|g_0\|_H \tag{7}$$
$$+ \max_{1\le k\le N-1} \|(f_k - f_{k-1})\tau^{-1}\|_H + \max_{-(N-1)\le k\le 0} \|(g_k - g_{k-1})\tau^{-1}\|_H],$$

$$\max_{1\le k\le N-1} \|(u_{k+1} - 2u_k + u_{k-1})\tau^{-2}\|_H + \max_{-N\le k\le N} \|Au_k\|_H \tag{8}$$
$$+ \max_{-(N-1)\le k\le 0} \|(u_k - u_{k-1})\tau^{-1}\|_H \le M_1[\|A\varphi\|_H + \|A^{1/2}f_0\|_H$$
$$+ \|(f_1 - f_0)\tau^{-1}\|_H + \max_{1\le k\le N-2} \|(f_{k+1} - 2f_k + f_{k-1})\tau^{-2}\|_H + \|A^{1/2}g_0\|_H$$

$$+\|(g_0-g_{-1})\tau^{-1}\|_H+\max_{-(N-1)\le k\le -1}\|(g_{k+1}-2g_k+g_{k-1})\tau^{-2}\|_H]$$

hold, where M_1 *is independent of* $\tau, f_k, 0\le k<N, g_k, -N<k\le 0$ *and* φ.

Proof. Initially, the formula for the solution of the difference scheme (2) will be obtained. It is known that there are unique solutions of the initial value difference problems

$$\begin{cases}\tau^{-2}(u_{k+1}-2u_k+u_{k-1})+Au_k+\frac{\tau^2}{4}A^2u_{k+1}=f_k, f_k=f\left(t_k\right),\\ t_k=k\tau,\ 1\le k<N, u_0=\xi,\ \tau^{-1}(u_1-u_0)=\psi,\end{cases}\tag{9}$$

$$\begin{cases}\tau^{-1}\left(u_k-u_{k-1}\right)+A(I+\frac{\tau}{2}A)u_k=(I+\frac{\tau}{2}A)g_k,\\ g_k=g(t_k-\frac{\tau}{2}), t_k=k\tau, -\left(N-1\right)\le k\le 0,\end{cases}\tag{10}$$

with u_{-N} given, and the following formulas hold:

$$\begin{aligned}u_1&=\xi+\tau\psi, u_k=\tfrac{1}{2}[R^{k-1}(\tau A^{1/2})(I-\tfrac{i\tau A^{1/2}}{2})+R^{k-1}(-\tau A^{1/2})\\ &\times(I+\tfrac{i\tau A^{1/2}}{2})]\xi+\tfrac{1}{2i}A^{-1/2}(I+\tfrac{\tau^4A^2}{4})[R^k(-\tau A^{1/2})-R^k(\tau A^{1/2})]\psi\\ &+\sum_{s=1}^{k-1}\tfrac{\tau}{2i}A^{-1/2}[R^{k-s}(-\tau A^{1/2})-R^{k-s}(\tau A^{1/2})]f_s,\ 2\le k\le N,\end{aligned}\tag{11}$$

$$u_k=R^{N+k}u_{-N}+\tau\sum_{s=-N+1}^{k}(I+\tfrac{\tau}{2}A)R^{k-s+1}g_s,\ -\left(N-1\right)\le k\le 0.\tag{12}$$

Using estimates (11), (12) and the formulas

$$\tau^{-1}(u_1-u_0)=(I+\tau^2A)^{-1}(-(\tfrac{\tau}{2}+1)Au_0+\tfrac{\tau}{2}f(0)+g_0),$$

$u_0=\xi$, $\tau^{-1}(u_1-u_0)=\psi$, we get

$$\xi=u_0=R^Nu_{-N}+\tau\sum_{s=-N+1}^{0}(I+\tfrac{\tau}{2}A)R^{-s+1}g_s,$$

$$\begin{aligned}\psi&=\tau^{-1}(u_1-u_0)=(I+\tau^2A)^{-1}\{-A(\tfrac{\tau}{2}+1)\\ &\times\left[R^Nu_{-N}+\tau\sum_{s=-N+1}^{0}(I+\tfrac{\tau}{2}A)R^{-s+1}g_s\right]+\tfrac{\tau}{2}f(0)+g_0\}.\end{aligned}$$

Therefore,

$$\begin{aligned}u_1&=(I-\tau A(\tfrac{\tau}{2}+I)(I+\tau^2A)^{-1})\Big[R^Nu_{-N}\\ &+\tau\sum_{s=-N+1}^{0}(I+\tfrac{\tau}{2}A)R^{-s+1}g_s\Big]+(I+\tau^2A)^{-1}\{\tfrac{\tau^2}{2}f(0)+\tau g_0\},\end{aligned}\tag{13}$$

$$u_k = \tfrac{1}{2}[R^{k-1}(\tau A^{1/2})(I - \tfrac{i\tau A^{1/2}}{2}) + R^{k-1}(-\tau A^{1/2})(I + \tfrac{i\tau A^{1/2}}{2})]$$

$$\times [R^N u_{-N} + \tau \sum_{s=-N+1}^{0} (I + \tfrac{\tau}{2}A)R^{-s+1}g_s] + \tfrac{1}{2i}A^{-1/2}(I + \tfrac{\tau^4 A^2}{4})$$

$$\times [R^k(-\tau A^{1/2}) - R^k(\tau A^{1/2})](I + \tau^2 A)^{-1}\{-A(\tfrac{\tau}{2} + 1)$$

$$\times [R^N u_{-N} + \tau \sum_{s=-N+1}^{0} (I + \tfrac{\tau}{2}A)R^{-s+1}g_s] + \tfrac{\tau}{2}f(0) + g_0\}$$

$$+ \sum_{s=1}^{k-1} \tfrac{\tau}{2i}A^{-1/2}[R^{k-s}(-\tau A^{1/2}) - R^{k-s}(\tau A^{1/2})]f_s, \ 2 \le k \le N. \quad (14)$$

Then, using the condition

$$u_{-N} = \sum_{j=1}^{N} \alpha_j(u_{[\mu_j/\tau]} + (\mu_j - [\mu_j/\tau]\tau)\tfrac{u_{[\mu_j/\tau]} - u_{[\mu_j/\tau]-1}}{\tau})$$

$$+ \sum_{j=1}^{L} \beta_j(\tfrac{u_{[\lambda_j/\tau]} - u_{[\lambda_j/\tau]-1}}{\tau} + (\lambda_j - [\lambda_j/\tau]\tau - \tfrac{\tau}{2})(f_{[\lambda_j/\tau]} - Au_{[\lambda_j/\tau]})) + \varphi,$$

we obtain the formula

$$u_{-N} = T_\tau[\sum_{j=1}^{N} \alpha_j(1 + (\tfrac{\mu_j}{\tau} - [\mu_j/\tau]))\{\tfrac{1}{2}[R^{[\mu_j/\tau]-1}(\tau A^{1/2})(I \quad \tfrac{i\tau A^{1/2}}{2}) \quad (15)$$

$$+ R^{[\mu_j/\tau]-1}(-\tau A^{1/2})(I + \tfrac{i\tau A^{1/2}}{2})]\tau \sum_{s=-N+1}^{0} (I + \tfrac{\tau}{2}A)R^{-s+1}g_s$$

$$+ \tfrac{1}{2i}A^{-1/2}(I + \tfrac{\tau^4 A^2}{4})[R^{[\mu_j/\tau]}(-\tau A^{1/2}) - R^{[\mu_j/\tau]}(\tau A^{1/2})](I + \tau^2 A)^{-1}$$

$$\times\{-A(\tfrac{\tau}{2} + 1)\tau \sum_{s=-N+1}^{0} (I + \tfrac{\tau}{2}A)R^{-s+1}g_s + \tfrac{\tau}{2}f(0) + g_0\}$$

$$+ \sum_{s=1}^{[\mu_j/\tau]-1} \tfrac{\tau}{2i}A^{-1/2}[R^{[\mu_j/\tau]-s}(-\tau A^{1/2}) - R^{[\mu_j/\tau]-s}(\tau A^{1/2})]f_s\}$$

$$- \sum_{j=1}^{N} \alpha_j(\tfrac{\mu_j}{\tau} - [\mu_j/\tau])\{\tfrac{1}{2}[R^{[\mu_j/\tau]-2}(\tau A^{1/2})(I - \tfrac{i\tau A^{1/2}}{2})$$

$$+R^{[\mu_j/\tau]-2}(-\tau A^{1/2})(I+\tfrac{i\tau A^{1/2}}{2})]\tau\sum_{s=-N+1}^{0}(I+\tfrac{\tau}{2}A)R^{-s+1}g_s$$

$$+\tfrac{1}{2i}A^{-1/2}(I+\tfrac{\tau^4A^2}{4})[R^{[\mu_j/\tau]-1}(-\tau A^{1/2})-R^{[\mu_j/\tau]-1}(\tau A^{1/2})](I+\tau^2A)^{-1}$$

$$\times\{-A(\tfrac{\tau}{2}+1)\tau\sum_{s=-N+1}^{0}(I+\tfrac{\tau}{2}A)R^{-s+1}g_s+\tfrac{\tau}{2}f(0)+g_0\}$$

$$+\sum_{s=1}^{[\mu_j/\tau]-2}\tfrac{\tau}{2i}A^{-1/2}[R^{[\mu_j/\tau]-1-s}(-\tau A^{1/2})$$

$$-R^{[\mu_j/\tau]-1-s}(\tau A^{1/2})]f_s\}+\sum_{j=1}^{L}\beta_j\tfrac{1}{\tau}(I-(\lambda_j-[\lambda_j/\tau]\tau+\tfrac{\tau}{2})A)$$

$$\times\{\tfrac{1}{2}[R^{[\lambda_j/\tau]-1}(\tau A^{1/2})(I-\tfrac{i\tau A^{1/2}}{2})+R^{[\lambda_j/\tau]-1}(-\tau A^{1/2})(I+\tfrac{i\tau A^{1/2}}{2})]$$

$$\times\tau\sum_{s=-N+1}^{0}(I+\tfrac{\tau}{2}A)R^{-s+1}g_s+\tfrac{1}{2i}A^{-1/2}(I+\tfrac{\tau^4A^2}{4})$$

$$\times[R^{[\lambda_j/\tau]}(-\tau A^{1/2})-R^{[\lambda_j/\tau]}(\tau A^{1/2})](I+\tau^2A)^{-1}$$

$$\times\{-A(\tfrac{\tau}{2}+1)\tau\sum_{s=-N+1}^{0}(I+\tfrac{\tau}{2}A)R^{-s+1}g_s+\tfrac{\tau}{2}f(0)+g_0\}$$

$$+\sum_{s=1}^{[\lambda_j/\tau]-1}\tfrac{\tau}{2i}A^{-1/2}[R^{[\lambda_j/\tau]-s}(-\tau A^{1/2})-R^{[\lambda_j/\tau]-s}(\tau A^{1/2})]f_s\}$$

$$-\sum_{j=1}^{L}\beta_j\frac{1}{\tau}\{\tfrac{1}{2}[R^{[\lambda_j/\tau]-2}(\tau A^{1/2})(I-\tfrac{i\tau A^{1/2}}{2})+R^{[\lambda_j/\tau]-2}(-\tau A^{1/2})$$

$$\times(I+\tfrac{i\tau A^{1/2}}{2})]\tau\sum_{s=-N+1}^{0}(I+\tfrac{\tau}{2}A)R^{-s+1}g_s+\tfrac{1}{2i}A^{-1/2}(I+\tfrac{\tau^4A^2}{4})$$

$$\times[R^{[\lambda_j/\tau]-1}(-\tau A^{1/2})-R^{[\lambda_j/\tau]-1}(\tau A^{1/2})](I+\tau^2A)^{-1}\{-A(\tfrac{\tau}{2}+1)$$

$$\times\tau\sum_{s=-N+1}^{0}(I+\tfrac{\tau}{2}A)R^{-s+1}g_s+\tfrac{\tau}{2}f(0)+g_0\}$$

$$+\sum_{s=1}^{[\lambda_j/\tau]-2}\tfrac{\tau}{2i}A^{-1/2}[R^{[\lambda_j/\tau]-1-s}(-\tau A^{1/2})-R^{[\lambda_j/\tau]-1-s}(\tau A^{1/2})]f_s\}$$

$$+\sum_{j=1}^{L}\beta_j(\lambda_j-[\lambda_j/\tau]\tau+\tfrac{\tau}{2})f_{[\lambda_j/\tau]}+\varphi].$$

So, for the solution of the difference scheme (2), formulas (12), (13), (14) and (15) are established. The proof of estimates (6), (7) and (8) follows the scheme of proof of the paper Ashyralyev and Özdemir [7] and is based on estimates (3), (4) and formulas (12), (13), (14) and (15). This completes the proof of Theorem 3.1. □

Note that in a similar manner the same stability estimates of Theorem 3.1 for solutions of the difference scheme of the second order of accuracy

$$\begin{cases} \tau^{-2}(u_{k+1}-2u_k+u_{k-1})+\frac{1}{2}Au_k+\frac{1}{4}A(u_{k+1}+u_{k-1})=f_k, \\ f_k=f(t_k);\ t_k=k\tau,\ 1\le k<N, \tau^{-1}(I+\tau^2 A)(u_1-u_0)=Z_1, \\ Z_1=\frac{\tau}{2}(f(0)-Au_0)+(g(0)-Au_0), \\ \tau^{-1}(u_k-u_{k-1})+A(I+\frac{\tau}{2}A)u_k=(I+\frac{\tau}{2}A)g_k, \\ g_k=g(t_k-\frac{\tau}{2}),\ \ t_k=k\tau,\ -(N-1)\le k\le 0, \\ u_{-N}=\sum\limits_{j=1}^{N}\alpha_j(u_{[\mu_j/\tau]}+(\mu_j-[\mu_j/\tau]\tau)\frac{u_{[\mu_j/\tau]}-u_{[\mu_j/\tau]-1}}{\tau})+\sum\limits_{j=1}^{L}\beta_j \\ \times(\frac{u_{[\lambda_j/\tau]}-u_{[\lambda_j/\tau]-1}}{\tau}+(\lambda_j-[\lambda_j/\tau]\tau-\frac{\tau}{2})(f_{[\lambda_j/\tau]}-Au_{[\lambda_j/\tau]}))+\varphi \end{cases} \tag{16}$$

for numerical solutions of the nonlocal boundary value problem (1) can be obtained.

4. Application

The nonlocal boundary value problem for one-dimensional hyperbolic-parabolic equation

$$\begin{cases} v_{tt}-(a(x)v_x)_x+\delta v=f(t,x), 0<t<1,\ 0<x<1, \\ v_t-(a(x)v_x)_x+\delta v=g(t,x), -1<t<0,\ 0<x<1, \\ v(-1,x)=\sum\limits_{j=1}^{N}\alpha_j v\,(\mu_j,x)+\sum\limits_{j=1}^{L}\beta_j v_t(\lambda_j,x)+\varphi(x), 0\le x\le 1, \\ \sum\limits_{j=1}^{N}|\alpha_j|, \sum\limits_{j=1}^{L}|\beta_j|\le 1, 0<\mu_j,\lambda_j\le 1, \\ v(t,0)=v(t,1), v_x(t,0)=v_x(t,1), -1\le t\le 1, \\ v(0+,x)=v(0-,x), v_t(0+,x)=v_t(0-,x), 0\le x\le 1 \end{cases} \tag{17}$$

is considered.

The discretization of problem (17) is carried out in two steps. In the first step, let us define the grid space

$$[0,1]_h=\{x: x_n=nh, 0\le n\le K, Kh=1\}.$$

We introduce the Hilbert space $L_{2h} = L_2([0,1]_h)$ of the grid functions $\varphi^h(x) = \{\varphi_n\}_{n=1}^{K-1}$ defined on $[0,1]_h$, equipped with the norm

$$\|\varphi^h\|_{L_{2h}} = (\sum_{n=1}^{K-1} |\varphi_n|^2 h)^{1/2}.$$

To the differential operator A generated by the problem (17), we assign the difference operator A_h^x by the formula

$$A_h^x \varphi^h(x) = \{-(a(x)\varphi_{\bar{x}})_{x,n} + \delta\varphi_n\}_1^{K-1}, \tag{18}$$

acting in the space of grid functions $\varphi^h(x)$ satisfying the conditions

$$\varphi_0 = \varphi_K, \varphi_1 - \varphi_0 = \varphi_K - \varphi_{K-1}.$$

It is known that A_h^x is a self-adjoint positive definite operator in L_{2h}. With the help of A_h^x, we arrive at the nonlocal boundary value problem

$$\begin{cases} \frac{d^2v^h(t,x)}{dy^2} + A_h^x v^h(t,x) = f^h(t,x),\ 0 < t < 1,\ x \in [0,1]_h, \\ \frac{dv^h(t,x)}{dy} + A_h^x v^h(t,x) = f^h(t,x), -1 < t < 0,\ x \in [0,1]_h, \\ v^h(-1,x) = \sum\limits_{j=1}^{N} \alpha_j v^h(\mu_j, x) + \sum\limits_{j=1}^{L} \beta_j v_t^h(\lambda_j, x) + \varphi^h(x), x \in [0,1]_h, \\ v^h(0+,x) = v^h(0-,x), \frac{dv^h(0+,x)}{dy} = \frac{dv^h(0-,x)}{dy}, x \in [0,1]_h \end{cases} \tag{19}$$

for an infinite system of ordinary differential equations.

In the second step, problem (19) is replaced by the difference scheme (2)

$$\begin{cases} \frac{u_{k+1}^h(x) - 2u_k^h(x) + u_{k-1}^h(x)}{\tau^2} + A_h^x u_k^h(x) + \frac{\tau^2}{4}(A_h^x)^2 u_{k+1}^h(x) = f_k^h(x), \\ f_k^h(x) = \{f(t_k, x_n)\}_1^{K-1}, x \in [0,1]_h, t_k = k\tau, 1 \le k \le N-1, \\ N\tau = 1;\ \tau^{-1}(I + \tau^2 A_h^x)(u_1^h(x) - u_0^h(x)) = Z_1^h(x), \\ Z_1^h(x) = \frac{\tau}{2}(f_0^h(x) - A_h^x u_0^h(x)) + (g_0^h(x) - A_h^x u_0^h(x)), \\ f_0^h(x) = \{f(0,x_n)\}_1^{K-1}, g_0^h(x) = \{g(0,x_n)\}_1^{K-1}, \\ \frac{u_k^h(x) - u_{k-1}^h(x)}{\tau} + A_h^x(I + \frac{\tau}{2}A_h^x)u_k^h(x) = (I + \frac{\tau}{2}A_h^x)g_k^h(x), x \in [0,1]_h, \\ g_k^h(x) = \{g(t_k - \frac{\tau}{2}, x_n)\}_1^{K-1}, t_k = k\tau, -(N-1) \le k \le 0, \\ u_{-N}^h(x) = \sum\limits_{j=1}^{N} \alpha_j (u_{[\mu_j/\tau]}^h(x) + (\mu_j - [\mu_j/\tau]\tau) \\ \times \frac{u_{[\mu_j/\tau]}^h(x) - u_{[\mu_j/\tau]-1}^h(x)}{\tau}) + \sum\limits_{j=1}^{L} \beta_j (\frac{u_{[\lambda_j/\tau]}^h(x) - u_{[\lambda_j/\tau]-1}^h(x)}{\tau} \\ +(\lambda_j - [\frac{\lambda_j}{\tau}]\tau - \frac{\tau}{2})(f_{[\lambda_j/\tau]}^h(x) - A_h^x u_{[\lambda_j/\tau]}^h(x))) + \varphi^h(x), x \in [0,1]_h. \end{cases} \tag{20}$$

Theorem 4.1. *Let τ and h be sufficiently small numbers. Then the solutions of difference scheme (20) satisfy the stability estimates*

$$\max_{-N \le k \le N} \|u_k^h\|_{L_{2h}} + \max_{-N+1 \le k \le N} \|(u_k^h - u_{k-1}^h)\tau^{-1}\|_{L_{2h}} + \max_{-N \le k \le N} \|(u_k^h)_{x,n}\|_{L_{2h}}$$

$$\leq M_2[\|f_0^h\|_{L_{2h}} + \max_{2\leq k\leq N-1} \|(f_k^h - f_{k-1}^h)\tau^{-1}\|_{L_{2h}}$$

$$+\|g_0^h\|_{L_{2h}} + \max_{-N+1\leq k\leq 0} \|(g_k^h - g_{k-1}^h)\tau^{-1}\|_{L_{2h}} + \|\varphi_{\bar{x},n}^h\|_{L_{2h}}],$$

$$\max_{1\leq k\leq N-1} \|(u_{k+1}^h - 2u_k^h + u_{k-1}^h)\tau^{-2}\|_{L_{2h}}$$

$$+ \max_{-N\leq k\leq N} \|(u_k^h)_{x\bar{x},n}\|_{L_{2h}} + \max_{-N+1\leq k\leq 0} \|(u_k^h - u_{k-1}^h)\tau^{-1}\|_{L_{2h}}$$

$$\leq M_2[\|(f_0^h)_{x,n}\|_{L_{2h}} + \|(f_1^h - f_0^h)\tau^{-1}\|_{L_{2h}}$$

$$+\max_{2\leq k\leq N-1} \|(f_{k+1}^h - 2f_k^h + f_{k-1}^h)\tau^{-2}\|_{L_{2h}} + \|(g_0^h)_{x,n}\|_{L_{2h}} + \|(g_0^h - g_{-1}^h)\tau^{-1}\|_{L_{2h}}$$

$$+ \max_{-N+1\leq k\leq -1} \|(g_{k+1}^h - 2g_k^h + g_{k-1}^h)\tau^{-2}\|_{L_{2h}} + \|\varphi_{x\bar{x},n}^h\|_{L_{2h}}].$$

Here, M_2 is independent of $\tau, h, \varphi^h(x), f_k^h(x)$ and $g_k^h(x)$.

The proof of Theorem 4.1 is based on the abstract Theorem 3.1 and the symmetry properties of the operator A_h^x defined by the formula (18).

Applying the second order of accuracy difference scheme (16), we can construct the second order of accuracy difference schemes with respect to one variable for approximate solutions of the boundary value problems (17). This approach permits us to obtain the stability estimates for the solution of this difference scheme.

References

1. T. D. Dzhuraev, *Boundary Value Problems for Equations of Mixed and Mixed-Composite Types*, Tashkent, FAN, 1979 (Russian).
2. G. D. Karatoprakliev, *Differensial'nye Uravneniya*, **25**, 1355 (1989) (Russian).
3. M. G. Karatopraklieva, *Differensial'nye Uravneniya* **27**, 68 (1991) (Russian).
4. V. I. Korzyuk and S. V. Lemeshevsky, *Math. Mod. Anal.* **6**, 106 (2001).
5. A. S. Berdyshev and E. T. Karimov, *Cent. Eur. J. Math.* **4**, 183 (2006).
6. M. S. Salakhitdinov and A. K. Urinov, *Boundary Value Problems for Equations of Mixed Type with a Spectral Parameter*, Tashkent, FAN, 1997 (Russian).
7. A. Ashyralyev and Y. Özdemir, *Taiwanese J. Math.* **11**, 1075 (2007).

NUMERICAL SOLUTIONS OF BITSADZE-SAMARSKII PROBLEM FOR ELLIPTIC EQUATIONS

A. ASHYRALYEV and E. ÖZTÜRK*

Department of Mathematics, Fatih University,
İstanbul, 34500, Turkey
**E-mail: eozturk@fatih.edu.tr*

A numerical method is proposed for solving Bitsadze-Samarskii-type nonlocal boundary value problem for multidimensional elliptic partial differential equations. The second and fourth orders of accuracy stable difference schemes are presented. The stability and almost coercive stability of these difference schemes are established. The method is illustrated by numerical examples.

Keywords: Bitsadze-Samarskii problem; Elliptic equation; Nonlocal boundary value problems; Difference scheme.

1. Introduction

Methods of solutions of Bitsadze-Samarskii nonlocal boundary value problems for elliptic differential equations have been studied extensively by Krein [1], Gorbachuk [2], Berikelashvili [3,10], Gordeziani [4], Kapanadze [5], Il'in and Moiseev [6], Bitsadze and Samarskii [7], Ashyralyev [8], Soldatov [9], Makarov, Lazurchak and Bandyrskii [11], Kilbas and Repin [12], Gurbanov and Dosiev [13], Sobolevskii [14], Balakishiyev and Dosiev [15].

Let Ω be the unit open cube in R^n $(x = (x_1, \cdots, x_n) : 0 < x_k < 1, 1 \leq k \leq n)$ with boundary S, $\overline{\Omega} = \Omega \cup S$. In $[0,1] \times \Omega$, the Bitsadze-Samarskii type nonlocal boundary value problem for the multidimensional elliptic equation

$$\begin{cases} -u_{tt} - \sum\limits_{r=1}^{n} (a_r(x) u_{x_r})_{x_r} = f(t,x), 0 < t < 1, x = (x_1, \ldots, x_n) \in \Omega, \\ u(0,x) = \varphi(x), u(1,x) = \sum\limits_{j=1}^{J} \alpha_j u(\lambda_j, x) + \psi(x), x \in \overline{\Omega}, \\ \sum\limits_{j=1}^{J} |\alpha_j| \leq 1, 0 < \lambda_1 < \cdots < \lambda_J < 1; u(t,x) = 0, x \in S \end{cases} \tag{1}$$

is considered. Here, $a_r(x), (x \in \Omega), \psi(x), \varphi(x)$ $(x \in \overline{\Omega})$ and $f(t,x)$ $(t \in$

$(0,1),\ x \in \Omega)$ are smooth functions and $a_r(x) \geq a > 0$.

In the present paper, we are interested in studying the stable difference schemes for the numerical solution of the problem (1). The second and fourth orders of accuracy difference schemes are presented. The stability and almost coercive stability of these difference schemes are established. A procedure of modified Gauss elimination method is used for solving these difference schemes for the two-dimensional elliptic differential equation. The method is illustrated by numerical examples.

2. Difference Schemes

The discretization of problem (1) is carried out in two steps. In the first step, let us define the grid sets

$$\overline{\Omega}_h = \{x = x_m = (h_1 m_1, \cdots, h_n m_n), m = (m_1, \cdots, m_n), 0 \leq m_r \leq N_r$$
$$h_r N_r = 1, r = 1, \cdots, n\}, \Omega_h = \overline{\Omega}_h \cap \Omega, S_h = \overline{\Omega}_h \cap S.$$

We introduce the Hilbert space $L_{2h} = L_2(\overline{\Omega}_h)$ of the grid functions $\varphi^h(x) = \{\varphi(h_1 m_1, \cdots, h_n m_n)\}$ defined on $\overline{\Omega}_h$, equipped with the norm

$$\left\|\varphi^h\right\|_{L_{2h}} = \Big(\sum_{x \in \overline{\Omega}_h} \left|\varphi^h(x)\right|^2 h_1 \cdots h_n\Big)^{1/2}.$$

To the differential operator A generated by the problem (1), we assign the difference operator A_h^x, by the formula

$$A_h^x u^h = -\sum_{r=1}^{n} (a_r(x) u^h_{\bar{x}_r})_{x_r, j_r} \tag{2}$$

acting in the space of grid functions $u^h(x)$, satisfying the conditions $u^h(x) = 0$ for all $x \in S_h$. It is known that A_h^x is a self-adjoint positive definite operator in L_{2h}. With the help of A_h^x, we arrive at the nonlocal boundary value problem for an infinite system of ordinary differential equations

$$\begin{cases} -\frac{d^2 u^h(t,x)}{dt^2} + A_h^x u^h(t,x) = f^h(t,x),\ 0 < t < 1,\ x \in \Omega_h, \\ u^h(0,x) = \varphi^h(x);\ u^h(1,x) = \sum_{j=1}^{J} \alpha_j u^h(\lambda_j, x) + \psi^h(x),\ x \in \overline{\Omega}_h. \end{cases} \tag{3}$$

In the second step, the second order of accuracy difference scheme

$$\begin{cases} -\frac{u^h_{k+1}(x) - 2u^h_k(x) + u^h_{k-1}(x)}{\tau^2} + A_h^x u^h_k(x) = \varphi^h_k, \\ \varphi^h_k = f^h(t_k, x), t_k = k\tau, 1 \leq k \leq N-1,\ N\tau = 1, x \in \Omega_h, \\ u^h_0(x) = \varphi^h(x);\ u^h_N(x) = \sum_{j=1}^{J} \alpha_j (u^h_{[\lambda_j/\tau]}(x) \\ +(u^h_{[\lambda_j/\tau]+1}(x) - u^h_{[\lambda_j/\tau]}(x))(\frac{\lambda_j}{\tau} - [\lambda_j/\tau])) + \psi^h(x), x \in \overline{\Omega}_h \end{cases} \tag{4}$$

and the fourth order of accuracy difference scheme

$$\begin{cases} -\frac{u_{k+1}^h(x)-2u_k^h(x)+u_{k-1}^h(x)}{\tau^2} + A_h^x u_k^h(x) + \frac{\tau^2}{12}(A_h^x)^2 u_k^h(x) = \varphi_k^h, \\ \varphi_k^h = f^h(t_k,x) + \frac{\tau^2}{12}(\frac{f^h(t_{k+1},x)-2f^h(t_k,x)+f^h(t_{k-1},x)}{\tau^2} \\ +A_h^x f^h(t_k,x)), t_k = k\tau, 1 \le k \le N-1, x \in \Omega_h \\ u_0^h(x) = \varphi^h(x), x \in \overline{\Omega}_h, \\ u_N^h(x) = \sum\limits_{j=1}^{J} \alpha_j (u_{[\lambda_j/\tau]}^h(x) + (\lambda_j - [\lambda_j/\tau]\tau)((u_{[\lambda_j/\tau]+1}^h(x) \\ -u_{[\lambda_j/\tau]}^h(x))\frac{1}{\tau} - \frac{\tau}{2}(A_h^x u_{[\lambda_j/\tau]}^h(x) - f^h([\lambda_j/\tau]\,\tau,x)) \\ -\frac{\tau^2}{6}(A_h^x(u_{[\lambda_j/\tau]+1}^h(x) - u_{[\lambda_j/\tau]}^h(x))\frac{1}{\tau} \\ -(f^h(([\lambda_j/\tau]+1)\tau,x) - f^h(([\lambda_j/\tau]-1)\tau,x))\frac{1}{2\tau}) \\ +(A_h^x u_{[\lambda_j/\tau]}^h(x) - f^h([\lambda_j/\tau]\,\tau,x))\frac{(\lambda_j-[\lambda_j/\tau]\tau)^2}{2} \\ +(A_h^x(u_{[\lambda_j/\tau]+1}^h(x) - u_{[\lambda_j/\tau]}^h(x))\frac{1}{\tau} - \frac{1}{\tau}\left(f^h(([\lambda_j/\tau]+1)\tau,x)\right. \\ -f^h([\lambda_j/\tau]\,\tau,x)))\frac{(\lambda_j-[\lambda_j/\tau]\tau)^3}{6}) + \psi^h(x), x \in \overline{\Omega}_h \end{cases} \tag{5}$$

are presented. In the formulation of the result would be to denote $f_k^h(x) = f^h(t_k,x)$.

Theorem 2.1. *Let τ and $|h|$ be sufficiently small numbers. Then the solutions of difference scheme (4) satisfy the stability and almost coercive stability estimates*

$$\max_{0\le k\le N} \|u_k^h\|_{L_{2h}} \le M_1[\|\varphi^h\|_{L_{2h}} + \|\psi^h\|_{L_{2h}} + \max_{1\le k\le N-1} \|f_k^h\|_{L_{2h}}],$$

$$\max_{1\le k\le N-1} \|\left(u_{k+1}^h - 2u_k^h + u_{k-1}^h\right)\tau^{-2}\|_{L_{2h}} + \max_{0\le k\le N} \sum_{r=1}^{n} \|\left(u_k^h\right)_{x_r\overline{x_r},\, j_r}\|_{L_{2h}}$$

$$\le M_2[\sum_{r=1}^{n} \|\varphi^h_{x_r\overline{x_r},\, j_r}\|_{L_{2h}} + \sum_{r=1}^{n} \|\psi^h_{x_r\overline{x_r},\, j_r}\|_{L_{2h}} + \ln\frac{1}{\tau+|h|} \max_{1\le k\le N-1} \|f_k^h\|_{L_{2h}}].$$

Here, M_1 does not depend on $\tau, h, \psi^h(x), \varphi^h(x)$ and $f_k^h(x), 1 \le k \le N-1$.

Theorem 2.2. *Let τ and $|h|$ be sufficiently small numbers. Then the solutions of difference scheme (5) satisfy the stability and almost coercive stability estimates*

$$\max_{0\le k\le N} \|u_k^h\|_{L_{2h}} \le M_1[\|\varphi^h\|_{L_{2h}} + \|\psi^h\|_{L_{2h}} + \max_{0\le k\le N} \|f_k^h\|_{L_{2h}}],$$

$$\max_{1\le k\le N-1} \|\left(u_{k+1}^h - 2u_k^h + u_{k-1}^h\right)\tau^{-2}\|_{L_{2h}} + \max_{0\le k\le N} \sum_{r=1}^{n} \|\left(u_k^h\right)_{x_r\overline{x_r},\, j_r}\|_{L_{2h}}$$

$$\le M_2[\sum_{r=1}^{n} \|\varphi^h_{x_r\overline{x_r},\, j_r}\|_{L_{2h}} + \sum_{r=1}^{n} \|\psi^h_{x_r\overline{x_r},\, j_r}\|_{L_{2h}} + \ln\frac{1}{\tau+|h|} \max_{0\le k\le N} \|f_k^h\|_{L_{2h}}].$$

Here, M_2 is independent of $\tau, h, \psi^h(x), \varphi^h(x)$ and $f_k^h, 0 \le k \le N$.

Proof. The proofs of Theorems 2.1-2.2 are based on the symmetry properties of the difference operator A_h^x defined by the formula (2) and on the formulas

$$u_k^h = (I - R^{2N})^{-1}\{(R^k - R^{2N-k})u_0^h + (R^{N-k} - R^{N+k})u_N^h$$
$$-(R^{N-k} - R^{N+k})(I + \tau B)(2I + \tau B)^{-1}B^{-1}\sum_{i=1}^{N-1}(R^{N-i} - R^{N+i})\varphi_i\tau\}$$
$$+(I + \tau B)(2I + \tau B)^{-1}B^{-1}\sum_{i=1}^{N-1}(R^{|k-i|} - R^{k+i})\varphi_i\tau,$$

$$u_N^h = D^{-1}\left(\sum_{j=1}^{J}\alpha_j\left((I - R^{2N})^{-1}\{(R^{[\lambda_j/\tau]} - R^{2N-[\lambda_j/\tau]})u_0^h\right.\right.$$
$$-(R^{N-[\lambda_j/\tau]} - R^{N+[\lambda_j/\tau]})(I + \tau B)(2I + \tau B)^{-1}B^{-1}$$
$$\times\sum_{i=1}^{N-1}(R^{N-i} - R^{N+i})\varphi_i\tau\} + (I + \tau B)(2I + \tau B)^{-1}B^{-1}$$
$$\times(\sum_{i=1}^{[\lambda_j/\tau]}R^{[\lambda_j/\tau]-i}\varphi_i\tau + \sum_{i=[\lambda_j/\tau]+1}^{N-1}R^{i-[\lambda_j/\tau]}\varphi_i\tau - \sum_{i=1}^{N-1}R^{[\lambda_j/\tau]+i}\varphi_i\tau)$$
$$-(\frac{\lambda_j}{\tau} - [\lambda_j/\tau])\left((I - R^{2N})^{-1}\{\tau B(R^{[\lambda_j/\tau]+1} + R^{2N-[\lambda_j/\tau]})u_0^h\right.$$
$$+(R^{N-[\lambda_j/\tau]-1} + R^{N+[\lambda_j/\tau]})(2I + \tau B)^{-1}\sum_{i=1}^{N-1}(R^{N-i} - R^{N+i})\varphi_i\tau^2\}$$
$$+(2I + \tau B)^{-1}(\sum_{i=1}^{[\lambda_j/\tau]}R^{[\lambda_j/\tau]-i}\varphi_i\tau^2 - \sum_{i=[\lambda_j/\tau]+1}^{N-1}R^{i-[\lambda_j/\tau]-1}\varphi_i\tau^2$$
$$\left.\left.\left.-\sum_{i=1}^{N-1}R^{[\lambda_j/\tau]+i}\varphi_i\tau^2)\right)\right) + \psi\right)$$

for difference scheme (4), and on formulas

$$u_N^h = D^{-1}(\sum_{j=1}^{J}\alpha_j\left\{(I - R^{2N})^{-1}\left[(R^{[\lambda_j/\tau]} - R^{2N-[\lambda_j/\tau]}\right.\right.$$
$$+(\lambda_j - [\lambda_j/\tau]\tau)\{-BR^{[\lambda_j/\tau]+1} - BR^{2N-[\lambda_j/\tau]} - \tfrac{\tau}{2}(AR^{[\lambda_j/\tau]}$$
$$-AR^{2N-[\lambda_j/\tau]}) + \tfrac{\tau^2}{6}(ABR^{[\lambda_j/\tau]+1} + ABR^{2N-[\lambda_j/\tau]})\}$$
$$+\tfrac{(\lambda_j-[\lambda_j/\tau]\tau)^2}{2}(AR^{[\lambda_j/\tau]} - AR^{2N-[\lambda_j/\tau]}) - \tfrac{(\lambda_j-[\lambda_j/\tau]\tau)^3}{6}AB$$
$$\times(R^{[\lambda_j/\tau]+1} + R^{2N-[\lambda_j/\tau]}))u_0 + (I + \tau B)(2I + \tau B)^{-1}B^{-1}$$
$$\times\left(R^{N+[\lambda_j/\tau]} - R^{N-[\lambda_j/\tau]} + (\lambda_j - [\lambda_j/\tau]\tau)\right.$$
$$\times[-BR^{N-[\lambda_j/\tau]} - BR^{N+[\lambda_j/\tau]+1} - \tfrac{\tau}{2}(AR^{N+[\lambda_j/\tau]} - AR^{N-[\lambda_j/\tau]})$$

$$
\begin{aligned}
&+\tfrac{\tau^2}{6}(\tau BR^{N-[\lambda_j/\tau]}+\tau BR^{N+[\lambda_j/\tau]+1})]+\tfrac{(\lambda_j-[\lambda_j/\tau]\tau)^2}{2}\\
&\times(AR^{N+[\lambda_j/\tau]}-AR^{N-[\lambda_j/\tau]})-\tfrac{(\lambda_j-[\lambda_j/\tau]\tau)^3}{6}(\tau BR^{N-[\lambda_j/\tau]}\\
&+\tau BR^{N+[\lambda_j/\tau]+1})\Big)\sum_{i=1}^{N-1}(R^{N-i}-R^{N+i})\varphi_i\tau\Big]\\
&+(I+\tau B)(2I+\tau B)^{-1}B^{-1}(\sum_{i=1}^{[\lambda_j/\tau]}R^{[\lambda_j/\tau]-i}\varphi_i\tau+\sum_{i=[\lambda_j/\tau]+1}^{N-1}R^{i-[\lambda_j/\tau]}\varphi_i\tau\\
&-\sum_{i=1}^{N-1}R^{[\lambda_j/\tau]+i}\varphi_i\tau-(\lambda_j-[\lambda_j/\tau]\tau)\tfrac{\tau}{2}(A\sum_{i=1}^{[\lambda_j/\tau]}R^{[\lambda_j/\tau]-i}\varphi_i\tau\\
&+A\sum_{i=[\lambda_j/\tau]+1}^{N-1}R^{i-[\lambda_j/\tau]}\varphi_i\tau-A\sum_{i=1}^{N-1}R^{[\lambda_j/\tau]+i}\varphi_i\tau))+(2I+\tau B)^{-1}\\
&\times(\lambda_j-[\lambda_j/\tau]\tau)\{\sum_{i=1}^{[\lambda_j/\tau]}-R^{[\lambda_j/\tau]-i}\varphi_i\tau+\sum_{i=[\lambda_j/\tau]+1}^{N-1}R^{i-[\lambda_j/\tau]-1}\varphi_i\tau\\
&+\sum_{i=1}^{N-1}R^{[\lambda_j/\tau]+i}\varphi_i\tau\}-(\lambda_j-[\lambda_j/\tau]\tau)\tfrac{\tau}{6}A(2I+\tau B)^{-1}(\sum_{j=1}^{[\lambda_j/\tau]}-R^{[\lambda_j/\tau]-i}\varphi_i\tau^2\\
&+\sum_{i=[\lambda_j/\tau]+1}^{N-1}R^{i-[\lambda_j/\tau]-1}\varphi_i\tau^2+\sum_{i=1}^{N-1}R^{[\lambda_j/\tau]+i}\varphi_i\tau^2)+\tfrac{(\lambda_j-[\lambda_j/\tau]\tau)^2}{2}\\
&\times A(I+\tau B)(2I+\tau B)^{-1}B^{-1}(\sum_{i=1}^{[\lambda_j/\tau]}R^{[\lambda_j/\tau]-i}\varphi_i\tau+\sum_{i=[\lambda_j/\tau]+1}^{N-1}R^{i-[\lambda_j/\tau]}\varphi_i\tau\\
&-\sum_{i=1}^{N-1}R^{[\lambda_j/\tau]+i}\varphi_i\tau)+\tfrac{(\lambda_j-[\lambda_j/\tau]\tau)^3}{6}A(2I+\tau B)^{-1}(\sum_{j=1}^{[\lambda_j/\tau]}-R^{[\lambda_j/\tau]-i}\varphi_i\tau\\
&+\sum_{i=[\lambda_j/\tau]+1}^{N-1}R^{i-[\lambda_j/\tau]-1}\varphi_i\tau+\sum_{i=1}^{N-1}R^{[\lambda_j/\tau]+i}\varphi_i\tau)\\
&+((\lambda_j-[\lambda_j/\tau]\tau)\tfrac{\tau}{2}-\tfrac{(\lambda_j-[\lambda_j/\tau]\tau)^2}{2})f^h([\lambda_j/\tau]\tau,x)+(\lambda_j-[\lambda_j/\tau]\tau)\\
&\times\tfrac{\tau}{12}(f^h(([\lambda_j/\tau]+1)\tau,x)-f^h(([\lambda_j/\tau]-1)\tau,x))\\
&-\tfrac{(\lambda_j-[\lambda_j/\tau]\tau)^3}{6\tau}(f^h(([\lambda_j/\tau]+1)\tau,x)-f^h([\lambda_j/\tau]\tau,x))\Big\}+\psi)
\end{aligned}
$$

for difference scheme (5). Here

$$D = \begin{cases} I - R^{2N} - \sum_{j=1}^{J} \alpha_j \{R^{N-[\lambda_j/\tau]} - R^{N+[\lambda_j/\tau]} - (\frac{\lambda_j}{\tau} - [\lambda_j/\tau])\tau B \\ \times (R^{N-[\lambda_j/\tau]} + R^{N+[\lambda_j/\tau]+1})\} \text{ for (4)}, \\ I - R^{2N} - \sum_{j=1}^{J} \alpha_j \{R^{N-[\lambda_j/\tau]} - R^{N+[\lambda_j/\tau]} + (\lambda_j - [\lambda_j/\tau]\tau) \\ \times \{B(R^{N-[\lambda_j/\tau]} + R^{N+[\lambda_j/\tau]+1}) - \frac{\tau}{2}(AR^{N-[\lambda_j/\tau]} - AR^{N+[\lambda_j/\tau]}) \\ -\frac{\tau^3}{6}(AB(R^{N-[\lambda_j/\tau]} + R^{N+[\lambda_j/\tau]+1}))\} + \frac{(\lambda_j - [\lambda_j/\tau]\tau)^2}{2} \\ \times (AR^{N-[\lambda_k/\tau]} - AR^{N+[\lambda_k/\tau]}) + \frac{(\lambda_j - [\lambda_j/\tau]\tau)^3}{6} \\ \times AB(R^{N-[\lambda_j/\tau]} + R^{N+[\lambda_j/\tau]+1})\} \text{ for (5)}, \end{cases}$$

where $R = (I + \tau B)^{-1}$, $B = \frac{\tau C}{2} + \sqrt{\frac{\tau^2 C^2}{4} + C}$, $A = A_{h,}^{x}$, and $C = A$ for (4), $C = \frac{A}{2} + \frac{A^2}{12}$ for (5). □

3. Numerical Results

We have not been able to obtain a sharp estimate for the constants figuring in the stability inequality. Therefore, the following numerical results of the nonlocal boundary value problem

$$\begin{cases} -\frac{\partial^2 u(t,x)}{\partial t^2} - \frac{\partial^2 u(t,x)}{\partial x^2} + u = -\exp(-t)\sin x, 0 < t < 1, 0 < x < \pi, \\ u(0,x) = \sin x, u(1,x) = u(\frac{1}{2},x) + (\exp(-1) - \exp(-\frac{1}{2}))\sin x, \\ 0 \le x \le \pi;\ u(t,0) = u(t,\pi) = 0, 0 \le t \le 1 \end{cases} \tag{6}$$

for elliptic equation are presented. First, applying the second order of accuracy difference scheme (4), we get system of equations in the matrix form

$$AU_{n+1} + BU_n + CU_{n-1} = D\varphi_n, 1 \le n \le M-1; U_0 = U_M = \tilde{0},$$

where

$$A = \begin{bmatrix} 0\,0\,0 \cdot 0\,0\,0 \\ 0\,a\,0 \cdot 0\,0\,0 \\ 0\,0\,a \cdot 0\,0\,0 \\ \cdot\;\cdot\;\cdot\;\cdot\;\cdot\;\cdot\;\cdot \\ 0\,0\,0 \cdot a\,0\,0 \\ 0\,0\,0 \cdot 0\,a\,0 \\ 0\,0\,0 \cdot 0\,0\,0 \end{bmatrix}_{(N+1)\times(N+1)}, B = \begin{bmatrix} 1\,0\,0 & 0 & \cdot\,0\,0 \\ c\,b\,c & 0 & \cdot\,0\,0 \\ 0\,c\,b & c & \cdot\,0\,0 \\ \cdot\;\cdot\;\cdot & \cdot & \cdot\;\cdot\;\cdot \\ 0\,0\,\cdot & \cdot & \cdot\,0\,0 \\ 0\,0\,\cdot & c & b\,c\,0 \\ 0\,0\,\cdot & -1 & \cdot\,0\,1 \end{bmatrix}_{(N+1)\times(N+1)}$$

and $C = A, D = I$,

$$U_s = \begin{bmatrix} U_s^0 \\ U_s^1 \\ \cdots \\ U_s^N \end{bmatrix}_{(N+1)\times 1}, \varphi_n = \begin{bmatrix} \varphi_n^0 \\ \varphi_n^1 \\ \cdots \\ \varphi_n^N \end{bmatrix}_{(N+1)\times 1} \quad \text{for } s = n \pm 1, n.$$

Here,

$$\begin{array}{l} a = \frac{1}{h^2}, \\ b = -\frac{2}{\tau^2} - \frac{2}{h^2} - 1, \\ c = \frac{1}{\tau^2}, \end{array} \text{ and } \varphi_n^k = \begin{cases} \sin(x_n), k = 0, \\ -\exp(-t_k)\sin(x_n), 1 \le k \le N-1, \\ (\exp(-1) - \exp(-\frac{1}{2}))\sin(x_n), k = N. \end{cases}$$

We seek a solution of the matrix equation in the form

$$U_j = \alpha_{j+1} U_{j+1} + \beta_{j+1}, j = M-1, \cdots, 2, 1, U_M = \tilde{0},$$

where α_j $(j = 1, \cdots, M)$ are $(N+1) \times (N+1)$ square matrices and β_j $(j = 1, \cdots, M)$ are $(N+1) \times 1$ column matrices defined by

$$\alpha_{j+1} = -(B + C\alpha_j)^{-1} A, \quad \beta_{j+1} = (B + C\alpha_j)^{-1}(D\varphi_j - C\beta_j),$$

where $j = 1, \cdots, M-1$. Here, α_1 is $(N+1) \times (N+1)$ zero matrix, β_1 is $(N+1) \times 1$ zero matrix and I is $(N+1) \times (N+1)$ identity matrix.

Second, applying the fourth order of accuracy in one variable difference scheme (5), we get system of linear equations in the matrix form

$$\begin{cases} AU_{n+2} + BU_{n+1} + CU_n + DU_{n-1} + EU_{n-2} = R\varphi_n, \\ 2 \le n \le M-2, U_0 = U_M = \tilde{0}, \\ U_{M-1} = \frac{4}{5}U_{M-2} - \frac{1}{5}U_{M-3}, U_1 = \frac{4}{5}U_2 - \frac{1}{5}U_3, \end{cases}$$

where

$$A = \begin{bmatrix} 0\,0\,0 \cdot 0\,0\,0 \\ 0\,a\,0 \cdot 0\,0\,0 \\ 0\,0\,a \cdot 0\,0\,0 \\ \cdot\;\cdot\;\cdot\;\cdot\;\cdot\;\cdot\;\cdot \\ 0\,0\,0 \cdot a\,0\,0 \\ 0\,0\,0 \cdot 0\,a\,0 \\ 0\,0\,0 \cdot 0\,0\,0 \end{bmatrix}_{(N+1)\times(N+1)}, B = \begin{bmatrix} 0\,0\,0 \cdot 0\,0\,0 \\ 0\,b\,0 \cdot 0\,0\,0 \\ 0\,0\,b \cdot 0\,0\,0 \\ \cdot\;\cdot\;\cdot\;\cdot\;\cdot\;\cdot\;\cdot \\ 0\,0\,0 \cdot b\,0\,0 \\ 0\,0\,0 \cdot 0\,b\,0 \\ 0\,0\,0 \cdot 0\,0\,0 \end{bmatrix}_{(N+1)\times(N+1)}$$

$$C = \begin{bmatrix} 1\,0\,0 & 0 & \cdot\,0\,0 \\ d\,c\,d & 0 & \cdot\,0\,0 \\ 0\,d\,c & 0 & \cdot\,0\,0 \\ \cdot\;\cdot\;\cdot & \cdot & \cdot\;\cdot\;\cdot \\ 0\,0\,\cdot & \cdot & \cdot\,0\,0 \\ 0\,0\,\cdot & d & c\,d\,0 \\ 0\,0\,\cdot & -1 & \cdot\,0\,1 \end{bmatrix}_{(N+1)\times(N+1)}, D = \begin{bmatrix} 0\,0\,0\cdot 0\,0\,0 \\ 0\,e\,0\cdot 0\,0\,0 \\ 0\,0\,e\cdot 0\,0\,0 \\ \cdot\;\cdot\;\cdot\;\cdot\;\cdot\;\cdot\;\cdot \\ 0\,0\,0\cdot e\,0\,0 \\ 0\,0\,0\cdot 0\,e\,0 \\ 0\,0\,0\cdot 0\,0\,0 \end{bmatrix}_{(N+1)\times(N+1)}$$

$$U_s = \begin{bmatrix} U_s^0 \\ U_s^1 \\ \cdot\cdot \\ U_s^N \end{bmatrix}_{(N+1)\times 1}, \varphi_n = \begin{bmatrix} \varphi_n^0 \\ \varphi_n^1 \\ \cdot\cdot \\ \varphi_n^N \end{bmatrix}_{(N+1)\times 1},$$

$s = n \pm 2, n \pm 1, n$, $E = A$, and $R = I$. Here,

$$b = \frac{-12h^2 - 4\tau^2 - 2h^2\tau^2}{12h^4}, \quad c = \frac{12h^4 + 4h^2\tau^2 + 6\tau^2 24h^2 + \tau^2 h^4 - 2\tau^2 h^2}{12h^4} + \frac{2}{\tau^2},$$
$$e = \frac{-12h^2 - 4\tau^2 - 2\tau^2 h^2}{12h^4}, \quad d = -\frac{1}{\tau^2}, \quad a = -\frac{\tau^2}{12h^4}, \quad \text{and}$$

$$\varphi_n^k = \begin{cases} \sin(x_n), k = 0, \\ (1 + \frac{\tau^2}{4})\exp(-t_k)\sin x_n, \ 1 \le k \le N-1, \\ (\exp(-1) - \exp(-\frac{1}{2}))\sin x_n, \ k = N. \end{cases}$$

We seek a solution of the matrix equation in the form

$$U_j = \alpha_{j+1}U_{j+1} + \beta_{j+1}U_{j+2} + \gamma_{j+1}, j = M-3, \cdots, 2, 1,$$

where α_j, β_j $(j = 1, \cdots, M)$ are $(N+1) \times (N+1)$ square matrices and γ_j are $(N+1) \times 1$ column matrices defined by

$$\begin{aligned} \alpha_{j+1} &= -(C + D\alpha_j + E\beta_{j-1} + E\alpha_{j-1}\alpha_j)^{-1}(B + D\beta_j + E\alpha_{j-1}\beta_j), \\ \beta_{j+1} &= -(C + D\alpha_j + E\beta_{j-1} + E\alpha_{j-1}\alpha_j)^{-1}A, \\ \gamma_{j+1} &= ((C + D\alpha_j + E\beta_{j-1} + E\alpha_{j-1}\alpha_j)^{-1} \\ &\quad \times (R\varphi_j - D\gamma_j - E\alpha_{j-1}\gamma_j - E\alpha_{j-1}\gamma_j - E\gamma_{j-1}), \end{aligned}$$

where $j = 2, \cdots, M-2$. Here, α_1 and β_1 are $(N+1)\times(N+1)$ zero matrices, I is $(N+1) \times (N+1)$ identity matrix, $\alpha_2 = \frac{4}{5}I$ and $\beta_2 = -\frac{1}{5}I$, γ_1 and γ_2 are $(N+1) \times 1$ zero matrices and

$$\begin{cases} U_M = \tilde{0}; \ U_{M-1} = [(4I - \alpha_{M-2})\gamma_{M-1} - \gamma_{M-2}]\,P; \\ U_{M-2} = [(\beta_{M-2} + 5I)\,U_{M-1} + \gamma_{M-2}]\,[(4I - \alpha_{M-2})]^{-1}, \end{cases}$$

where $P = [(\beta_{M-2} + 5I) - (4I - \alpha_{M-2})\alpha_{M-1}]^{-1}$.

Now, the results of the numerical analysis are given. The errors of the numerical solutions for different values of M and N are computed by

$$E_M^N = \max_{1\leq k\leq N-1}\left(\sum_{n=1}^{M-1}\left|u(t_k,x_n)-u_n^k\right|^2 h\right)^{1/2}$$

where $u(t_k,x_n)$ represents the exact solution and u_n^k represents the numerical solution at (t_k,x_n). The results are shown in Table 1.

Table 1. Error analysis for the exact solution $u(t,x)$.

Method	N=6, M=450	N=8, M=450	N=10, M=450
Scheme (4)	3.2795e-004	1.8589e-004	1.1921e-004
Scheme (5)	8.8061e-007	6.7656e-007	6.1898e-007

Thus, the fourth order of accuracy in one variable difference scheme (5) is more accurate compared with the second order of accuracy in one variable difference scheme (4).

References

1. S. G. Krein, *Linear Differential Equations in Banach Space*, Nauka, Moscow, 1966 (Russian).
2. V. L. Gorbachuk and M. L. Gorbachuk, *Boundary Value Problems for Differential-Operator Equations*, Naukova Dumka, Kiev, 1984 (Russian).
3. G. Berikelashvili, *Comput. Methods Appl. Math.* **3**, 35 (2003).
4. D. G. Gordeziani, *Abstracts of reports of Inst. Appl. Math.* **2**, 38 (1970).
5. D. V. Kapanadze, *Dif. Equat.* **23**, 1426 (1987).
6. V. A. Il'in and E. I. Moiseev, *Mat. Mod.* **2**, 139 (1990).
7. A. V. Bitsadze and A. A. Samarskii, *Dokl. Akad. Nauk.* **10**, 398 (1969).
8. A. Ashyralyev, *Abstracts, II. Türk Dünyası Matematik Sempozyumu*, 4 (2007).
9. A. P. Soldatov, *Doklady Akademii Nauk.* **410**, 607 (2006) (Russian).
10. G. K. Berikelashvili, *Differential Equations* **39**, 945 (2003).
11. V. L. Makarov, I. I. Lazurchak, and B. I. Bandyrskii, *Cybernetics and Systems Analysis* **39**, 862 (2003) (Russian).
12. A. A. Kilbas and O. A. Repin, *Differential Equations* **39**, 674 (2003).
13. I. A. Gurbanov and A. A. Dosiev, *On the Numerical Solution of Nonlocal Boundary Problems for Quasilinear Elliptic Equations, Approximate Methods for Operator Equations*, Baku State Univ., 1984 (Russian).
14. P. E. Sobolevskii, *On Difference Method for Approximate Solution of Differential Equations*, Voronezh State Univ., Vorenezh, 1975 (Russian).
15. B. B. Balakishiyev and A. A. Dosiev, *Investigation of the Method of Nets for Solving Nonlocal Boundary Value Problems with Singularities, Approximate Methods for Solving Operator Equations*, Baku State Univ., 1991 (Russian).

A NOTE ON PARABOLIC DIFFERENCE EQUATIONS

A. ASHYRALYEV and Y. SÖZEN*

Department of Mathematics, Fatih University,
İstanbul, 34500, Turkey
** E-mail: ysozen@fatih.edu.tr*

We consider the parabolic differential equation

$$u'(t) + \mathbf{A}u(t) = f(t) \quad (-\infty < t < \infty)$$

in a general Banach space $\mathbf{E}$ with the strongly positive operator $\mathbf{A}$. The almost coercivity inequality in $C(\mathbb{R}_\tau, \mathbf{E})$ and the well-posedness in the Hölder space $C^\alpha(\mathbb{R}_T, \mathbf{E})$ are established for the solutions of the Padé difference schemes.

Keywords: Abstract parabolic equation; Almost coercivity inequality; Well-posedness; Padé difference scheme.

1. Introduction

The role of coercivity inequalities (maximal regularity, well-posedness) in the study of boundary-value problems for parabolic and elliptic differential equations is well known (see, e.g. Ladyzhenskaya, Solonnikov and Ural'tseva [1], Vishik, Myshkis and Oleinik [2]).

Main role of the coercivity inequalities for difference problems lies in that they present a special type of stability, which allows the existence of exact, i.e. two-sided estimates for the rate of convergence approximate solutions with respect to the corresponding coercivity norms.

It is quite possible that there are cases where the discretized problem is well posed, although the differential problem is not.

Well-posedness of local and nonlocal boundary value problems for abstract parabolic differential and difference equations in Banach spaces have been studied extensively by many researchers (see, Ladyzhenskaya, Solonnikov and Ural'tseva [1], Vishik, Myshkis and Oleinik [2], Sobolevskii [3], Ashyralyev, Sözen and Sobolevskii [4], Ashyralyev, Sobolevskii [5,9,10], Sobolevskii [6], Ashyralyev, Piskarev and Wei [7], Guidetti, Karasözen and Piskarev, [8] and the references therein).

In this paper, well-posedness of parabolic equations is investigated. The paper is organized as follows. In §2, the parabolic differential equation in the Banach space $\mathbf{E}$ is considered. In §3, Padé difference scheme for parabolic differential equation is studied. The almost coercivity inequality for solutions of this difference scheme is established. §4 presents the well-posedness of this difference scheme in different analogues of Hölder spaces.

2. Well-posedness of the parabolic differential equation

We consider the parabolic differential equation

$$\frac{du(t)}{dt} + \mathbf{A}u(t) = f(t), \ -\infty < t < \infty \tag{1}$$

in an arbitrary Banach space $\mathbf{E}$. Here, $u(t)$ and $f(t)$ are unknown and given abstract functions, defined on $\mathbb{R}$ with values in $\mathbf{E}$; $\mathbf{A}$ is a linear unbounded closed operator acting in $\mathbf{E}$ with dense domain $D(\mathbf{A}) \subset \mathbf{E}$.

We call a function $u(t)$ is called a *solution* of the problem (1) if the following conditions are satisfied:

i. $u(t)$ is continuously differentiable and bounded, its derivative is bounded.
ii. The element $u(t)$ belongs to $D(\mathbf{A})$ for all $t \in \mathbb{R}$ and the function $Au(t)$ is continuous and bounded in $\mathbb{R}$.
iii. $u(t)$ satisfies the equation (1).

From now on, we will refer to a solution of problem (1) defined in this manner as a solution of problem (1) in the Banach space $C(\mathbf{E}) = C(\mathbb{R}, \mathbf{E})$ of all continuously bounded functions $\varphi(t)$ defined on $\mathbb{R}$ with values in $\mathbf{E}$ equipped with the sup-norm

$$\| \varphi \|_{C(\mathbf{E})} = \sup_{-\infty<t<\infty} ||\varphi(t)||_{\mathbf{E}}. \tag{2}$$

The *well-posedness* of (1) in $C(\mathbf{E})$ means if, for the solutions of (1), the following *coercivity inequality*

$$\|u'\|_{C(\mathbf{E})} + \|\mathbf{A}u(t)\|_{C(\mathbf{E})} \leq M_C \|f\|_{C(\mathbf{E})} \tag{3}$$

holds for some $1 \leq M_C < \infty$ that does not depend on $f(t) \in C(E)$.

Throughout the paper, we will assume that the operator $-\mathbf{A}$ generates a semigroup $\exp\{-t\mathbf{A}\}$ $(t \geq 0)$ with exponentially decreasing norm when $t \to +\infty$, namely,

$$\| \mathbf{e}^{-t\mathbf{A}} \|_{\mathbf{E}\to\mathbf{E}} \leq M\mathbf{e}^{-\delta t}. \tag{4}$$

Then from coercivity inequality the estimate

$$\|\mathbf{A}e^{-t\mathbf{A}}\|_{\mathbf{E}\to\mathbf{E}} \leq Mt^{-1}. \tag{5}$$

follows. This is known as the Sobolovskii analyticity of the semigroup $e^{-t\mathbf{A}}$ for $t > 0$ (Ashyralyev and Sobolevskii [5]).

Theorem 2.1. *Let* $\mathbf{A}$ *generate an exponentially decreasing semigroup. Then, if problem (1) is well-posed in the declared sense, the semigroup generated by* $\mathbf{A}$ *is analytic. Futhermore,* $\forall f \in C(\mathbf{E})$*, the solution* u *is given by (Ashyralyev, Sözen and Sobolevskii [4])*

$$u(t) = \int_{-\infty}^{t} e^{-(t-s)\mathbf{A}} f(s)\, ds. \tag{6}$$

Clearly, formula (6) defines solution of problem (1) in $C(\mathbf{E})$, if, for example, $\mathbf{A}f(t) \in C(\mathbf{E})$ or $f'(t) \in C(E)$. It actually turns out that formula (6) defines solution of problem (1) in $C(\mathbf{E})$ under essentially less restriction on smoothness of function $f(t)$. Note also that from (4), (5),

$$\|\mathbf{A}^{\gamma}[e^{-t\mathbf{A}} - e^{-(t+\tau)\mathbf{A}}]\|_{\mathbf{E}\to\mathbf{E}} \leq M\frac{\tau^{\beta}}{t^{\beta+\gamma}} \tag{7}$$

follows for any $0 < t < t+\tau$, $0 \leq \gamma \leq 1$, and $0 \leq \beta \leq 1$.

The well-posedness of problem (1) can be established on the assumption (4), (5) if one considers this problem in the Hölder space $C^{\beta}(\mathbf{E}) = C^{\beta}(\mathbb{R}, \mathbf{E})$, $\beta \subset (0,1)$, of all $\mathbf{E}$-valued abstract functions $\varphi(t)$ defined on $\mathbb{R}$ with the norm

$$\|\varphi\|_{C^{\beta}(\mathbf{E})} = \sup_{-\infty<t<\infty} \|\varphi(t)\|_{\mathbf{E}} + \sup_{-\infty<t<t+\tau<\infty} \frac{\|\varphi(t+\tau) - \varphi(t)\|_{\mathbf{E}}}{\tau^{\beta}}. \tag{8}$$

Theorem 2.2. *(Ashyralyev, Sözen and Sobolevskii [4]) Let* $\mathbf{A}$ *be the infinitesimal generator of an analytic, exponentially decreasing semigroup. Then,* $\forall \beta \in (0,1)$*, problem (1) is well posed in* $C^{\beta}(\mathbf{E})$*. Moreover, if* $f \in C^{\beta}(\mathbf{E})$ *and* u *is the corresponding solution, then*

$$\|u'\|_{C^{\beta}(\mathbf{E})} + \|\mathbf{A}u\|_{C^{\beta}(\mathbf{E})} \leq \frac{M}{\beta(1-\beta)}\|f\|_{C^{\beta}(\mathbf{E})} \tag{9}$$

with $M \in \mathbb{R}^{+}$*, independent of* β *and* f*.*

In the paper Ashyralyev, Sözen and Sobolevskii [4], we also established the well-posedness of the corresponding Rothe difference scheme.

3. Almost coercivity inequality. Padé difference schemes

Let $\mathbb{R}_\tau$ denote the grid space $\{t_k = k\tau; k \in \mathbb{Z}\}$, where $\tau > 0$ is small number. The exact single-step difference scheme of the solutions of (1) is

$$\frac{u_k - u_{k-1}}{\tau} + \frac{(\mathbf{I} - \mathbf{e}^{-\tau\mathbf{A}})u_{k-1}}{\tau} = f_k, \tag{10}$$

where $f_k = \frac{1}{\tau}\int_{t_{k-1}}^{t_k} \mathbf{e}^{-(t_k-s)\mathbf{A}} f(s)ds, \quad k \in \mathbb{Z}$.

Replacing $\mathbf{e}^{-\tau\mathbf{A}}$ with the Padé approximant $\mathfrak{R}_{q,p}(\tau\mathbf{A})$, where $q = p-1, p-2$, and f_k with close element $f_k^{q,p}$ such that $\|f_k - f_k^{q,p}\|_{\mathbf{E}} \le \tau^{q+p}$, we obtain the Padé difference schemes

$$\frac{u_k - u_{k-1}}{\tau} + \mathbf{A}_{q,p}u_{k-1} = f_k^{q,p} \tag{11}$$

for the approximate solutions of the parabolic differential equation (1). Here, $u_k \in D(\mathbf{A})$, $f_k^{q,p} \in \mathbf{E}$ are unknown and given elements, and $\mathbf{A}_{q,p}$ is the operator $\frac{1}{\tau}(\mathbf{I} - \mathfrak{R}_{q,p}(\tau\mathbf{A}))$.

Let $C(\mathbb{R}_\tau, \mathbf{E})$ be the Banach space of all bounded grid functions $v^\tau = \{v_k\}_{k=-\infty}^{\infty}$ defined on the grid space $\mathbb{R}_\tau$ with the sup-norm

$$\|v^\tau\|_{C(\mathbb{R}_\tau,\mathbf{E})} = \sup_{-\infty<k<\infty} \|v_k\|_{\mathbf{E}} \tag{12}$$

and let $\mathfrak{D}_\tau : C(\mathbb{R}_\tau, \mathbf{E}) \to C(\mathbb{R}_\tau, \mathbf{E})$ denote the operator defined by the rule

$$v^\tau = \mathfrak{D}_\tau u^\tau, \; v_k = \frac{u_k - u_{k-1}}{\tau}, \; k \in \mathbb{Z}. \tag{13}$$

Then, equation (11) will be considered as the operator equation

$$\mathfrak{D}_\tau u^\tau + \mathbf{A}_{q,p}u^\tau = f^\tau \tag{14}$$

in $C(\mathbb{R}_\tau, \mathbf{E})$. Here $\mathbf{A}_{q,p}u^\tau = \{\mathbf{A}u_k\}_{k=-\infty}^{\infty}$ and $f^\tau = \{f_k\}_{k=-\infty}^{\infty}$.

From the strong positivity of the operator $\mathbf{A}$, the existence of a bounded operator $\mathfrak{R}_{q,p}(\tau\mathbf{A})$ defined on the entire Banach space $\mathbf{E}$ follows. Therefore, for every f^τ, a unique solution $u^\tau = u^\tau(f^\tau)$ exists for problem (11) and

$$u_k = \sum_{i=-\infty}^{k} \mathfrak{R}_{q,p}^{k-i}(\tau\mathbf{A}) f_i \, \tau, \; k \in \mathbb{Z} \tag{15}$$

holds. If, moreover, the spectral angle $\varphi(\mathbf{A}, \mathbf{E})$ of the operator $\mathbf{A}$ is sufficiently small ($\varphi(\mathbf{A}, \mathbf{E}) < \frac{\pi}{2p}$), then for $q = p-1, p-2$, we have

$$\|\mathfrak{R}_{q,p}^k(\tau\mathbf{A})\|_{\mathbf{E}\to\mathbf{E}} \le M(1+\tau\delta)^{-k}, \; k \ge 1, \tag{16}$$

$$\|k\tau\mathbf{A}_{q,p}\mathfrak{R}_{q,p}^k(\tau\mathbf{A})\|_{\mathbf{E}\to\mathbf{E}} \le M(1+\tau\delta)^{-\frac{k}{2}}, \; k \ge 1, \tag{17}$$

$$\|\mathbf{A}_{q,p}(\tau\mathbf{A})^{-1}\|_{\mathbf{E}\to\mathbf{E}} \leq M. \tag{18}$$

$$\|\mathfrak{R}^{-k}(\tau\mathbf{A})\mathfrak{R}^{k}_{q,p}(\tau\mathbf{A})\|_{\mathbf{E}\to\mathbf{E}} \leq M,\ k \geq 1, \tag{19}$$

where $\mathfrak{R}(\tau\mathbf{A}) = (\mathbf{I} + \frac{\tau\mathbf{A}}{p})^{-1}$.

For the proofs of (16) and (17), we refer the reader to Ashyralyev and Sobolevskii [5].

The problem (14) is said to be *coercively stable* (*well posed*) in $C(\mathbb{R}_\tau, \mathbf{E})$, if we have the coercive stability

$$\|\mathbf{A}_{q,p}u^\tau\|_{C(\mathbb{R}_\tau,\mathbf{E})} \leq M\|f^\tau\|_{C(\mathbb{R}_\tau,\mathbf{E})} \tag{20}$$

where M is independent not only of f^τ but also of τ.

Since the problem (1) in the space $C(\mathbb{R}, \mathbf{E})$ is not well-posed for the general positive operator $\mathbf{A}$ and space $\mathbf{E}$, then the well-posedness of the difference problem (14) in $C(\mathbb{R}_\tau, \mathbf{E})$ norm does not take place uniformly with respect to $\tau > 0$. This means that the *coercivity norm*

$$\| u^\tau \|_{\mathcal{K}_\tau(\mathbf{E})} = \|\mathbf{A}_{q,p}u^\tau\|_{C(\mathbb{R}_\tau,\mathbf{E})} + \|\mathfrak{D}_\tau u^\tau\|_{C(\mathbb{R}_\tau,\mathbf{E})} \tag{21}$$

tends to ∞ as $\tau \to 0^+$. The investigation of the difference problem (14) permits to establish the order of growth of this norm to ∞ as follows.

Theorem 3.1. *For the solution of the Padé difference problem (14), we have the almost coercivity inequality*

$$\| u^\tau \|_{\mathcal{K}_\tau(\mathbf{E})} \leq M(p)\min\left\{\ln\frac{1}{\tau}, 1 + |\ln\| \mathbf{A} \|_{\mathbf{E}\to\mathbf{E}}|\right\} \| f^\tau \|_{C(\mathbb{R}_\tau,\mathbf{E})}. \tag{22}$$

Proof. Using formula (15) and the substitution $m = k - i$, we get

$$\mathbf{A}_{q,p}u_k = \sum_{i=-\infty}^{k} \mathbf{A}_{q,p}\mathfrak{R}^{k-i}_{q,p}(\tau\mathbf{A})f_i\,\tau = \sum_{m=0}^{\infty} \mathbf{A}_{q,p}\mathfrak{R}^{m}_{q,p}(\tau\mathbf{A})f_{k-m}\,\tau$$

$$= \sum_{m=1}^{[\frac{1}{\tau}]} \mathbf{A}_{q,p}\mathfrak{R}^{m}_{q,p}(\tau\mathbf{A})f_{k-m}\,\tau + \sum_{m=[\frac{1}{\tau}]+1}^{\infty} \mathbf{A}_{q,p}\mathfrak{R}^{m}_{q,p}(\tau\mathbf{A})f_{k-m}\,\tau + \mathbf{A}_{q,p}f_k\tau$$

$$= J_1 + J_2 + J_3,$$

where $[\cdot]$ stands for the integer part.

Let us estimate J_2. Using estimates (16) and (17), we obtain

$$\|J_2\|_{\mathbf{E}} \leq \sum_{m=[\frac{1}{\tau}]+1}^{\infty} \|\tau\mathbf{A}_{q,p}\mathfrak{R}^{[\frac{m}{2}]}_{q,p}\mathfrak{R}^{m-[\frac{m}{2}]}_{q,p}\|_{\mathbf{E}\to\mathbf{E}}\|f_{k-m}\|_{\mathbf{E}}$$

$$\leq M\|f^\tau\|_{C(\mathbb{R}_\tau,\mathbf{E})} \sum_{m=[\frac{1}{\tau}]+1}^{\infty} \frac{1}{(1+\tau\delta)^m\, m} \leq M(\delta)\|f^\tau\|_{C(\mathbb{R}_\tau,\mathbf{E})}.$$

Let us estimate J_1. Using estimates (18) and (19), we get

$$\|J_1\|_{\mathbf{E}} \leq \sum_{m=1}^{[\frac{1}{\tau}]} \|\tau \mathbf{A}_{q,p}\mathfrak{R}_{q,p}^m(\tau\mathbf{A})f_{k-m}\|_{\mathbf{E}}$$

$$\leq \sum_{m=1}^{[\frac{1}{\tau}]} \|\tau \mathbf{A}_{q,p}\mathfrak{R}_{q,p}^m(\tau\mathbf{A})\|_{\mathbf{E}\to\mathbf{E}}\|f_{k-m}\|_{\mathbf{E}}$$

$$\leq \sum_{m=1}^{[\frac{1}{\tau}]} \|\tau \mathbf{A}_{q,p}(\tau\mathbf{A})^{-1}\|_{\mathbf{E}\to\mathbf{E}}\|\tau\mathbf{A}\mathfrak{R}^m\|_{\mathbf{E}\to\mathbf{E}}\|\mathfrak{R}^{-m}\mathfrak{R}_{q,p}^m\|_{\mathbf{E}\to\mathbf{E}}\|f_{k-m}\|_{\mathbf{E}}$$

$$\leq M(p)\|f^\tau\|_{C(\mathbb{R}_\tau,\mathbf{E})} \sum_{m=1}^{[\frac{1}{\tau}]} \|\tau\mathbf{A}\mathfrak{R}^m\|_{\mathbf{E}\to\mathbf{E}}.$$

By Ashyralyev and Sobolevskii [5, Theorem 1.2],

$$\sum_{m=1}^{[\frac{1}{\tau}]} \tau\|\mathbf{A}\mathfrak{R}^m(\tau\mathbf{A})\|_{\mathbf{E}\to\mathbf{E}} \leq M(p)\min\left\{\ln\left[\frac{1}{\tau}\right], 1+|\ln\|\,\mathbf{A}\,\|_{\mathbf{E}\to\mathbf{E}}|\right\}.$$

Thus

$$\|J_1\|_{\mathbf{E}} \leq M(p)\min\left\{\ln\left[\frac{1}{\tau}\right], 1+|\ln\|\,\mathbf{A}\,\|_{\mathbf{E}\to\mathbf{E}}|\right\}\|\,f^\tau\,\|_{C(\mathbb{R}_\tau,\mathbf{E})}.$$

Combining the estimates for $\|J_1\|_{\mathbf{E}}$ and $\|J_2\|_{\mathbf{E}}$, we obtain

$$\|\mathbf{A}_{q,p}u_k\|_{\mathbf{E}} \leq M(p)\min\left\{\ln\left[\frac{1}{\tau}\right], 1+|\ln\|\,\mathbf{A}\,\|_{\mathbf{E}\to\mathbf{E}}|\right\}\|\,f^\tau\,\|_{C(\mathbb{R}_\tau,\mathbf{E})}$$

for all k. It follows from this that

$$\|\mathbf{A}_{q,p}u^\tau\|_{C(\mathbb{R}_\tau,\mathbf{E})} \leq M(p)\min\left\{\ln\frac{1}{\tau}, 1+|\ln\|\,\mathbf{A}\,\|_{\mathbf{E}\to\mathbf{E}}|\right\}\|\,f^\tau\,\|_{C(\mathbb{R}_\tau,\mathbf{E})}.$$

By the triangle inequality, this last estimate and equation (14) yield

$$\|\mathfrak{D}_\tau u^\tau\|_{C(\mathbb{R}_\tau,\mathbf{E})} \leq M(p)\min\left\{\ln\frac{1}{\tau}, 1+|\ln\|\,\mathbf{A}\,\|_{\mathbf{E}\to\mathbf{E}}|\right\}\|\,f^\tau\,\|_{C(\mathbb{R}_\tau,\mathbf{E})}.$$

Theorem 3.2 is proved. □

4. Well-posedness of difference scheme

Now, the difference equation (11) is considered as the operator equation (14) in the Banach space $\mathcal{C}^\alpha(\mathbb{R}_\tau,\mathbf{E})$ $(0<\alpha<1)$ of grid functions $\varphi^\tau=\{\varphi_k\}_{k=-\infty}^{\infty}$ with norm

$$\|\varphi^\tau\|_{\mathcal{C}^\alpha(\mathbb{R}_\tau,\mathbf{E})}=\|\varphi^\tau\|_{\mathcal{C}(\mathbb{R}_\tau,\mathbf{E})}+\sup_{-\infty<i<i+r<\infty}\frac{\|\varphi_{i+r}-\varphi_i\|_{\mathbf{E}}}{(r\tau)^\alpha}.$$

The *well-posedness* of (14) in the space $\mathcal{C}^\alpha(\mathbb{R}_\tau,\mathbf{E})$ means that for solutions u^τ of (14) in $\mathcal{C}^\alpha(\mathbb{R}_\tau,\mathbf{E})$ coercive inequality

$$\|\mathfrak{D}_\tau u^\tau\|_{\mathcal{C}^\alpha(\mathbb{R}_\tau,\mathbf{E})}+\|\mathbf{A}_{q,p}u^\tau\|_{\mathcal{C}^\alpha(\mathbb{R}_\tau,\mathbf{E})}\le M_C(\alpha)\|f^\tau\|_{\mathcal{C}^\alpha(\mathbb{R}_\tau,\mathbf{E})}\tag{23}$$

holds for some $1\le M_C(\alpha)<\infty$ that is independent of f^τ and positive small number τ.

Theorem 4.1. *The difference equation (14) is well-possed in Banach space $\mathcal{C}^\alpha(\mathbb{R}_\tau,\mathbf{E})$ $(0<\alpha<1)$ and for its solutions coercivity inequality*

$$\|\mathfrak{D}_\tau u^\tau\|_{\mathcal{C}^\alpha(\mathbb{R}_\tau,\mathbf{E})}+\|\mathbf{A}_{q,p}u^\tau\|_{\mathcal{C}^\alpha(\mathbb{R}_\tau,\mathbf{E})}\le\frac{M(p)}{\alpha(1-\alpha)}\|f^\tau\|_{\mathcal{C}^\alpha(\mathbb{R}_\tau,\mathbf{E})}$$

holds for some $1\le M(p)<\infty$, which does not depend on $f^\tau\in\mathcal{C}^\alpha(\mathbb{R}_\tau,\mathbf{E})$, $\alpha\in(0,1)$ and positive small number τ.

Proof. Let us estimate $\|\mathbf{A}u^\tau\|_{\mathcal{C}(\mathbb{R}_\tau,\mathbf{E})}$. Using formula (15) and the identity

$$\tau\mathbf{A}_{q,p}\mathfrak{R}_{q,p}^{k-i}=\mathfrak{R}_{q,p}^{k-i}-\mathfrak{R}_{q,p}^{k-i+1}\tag{24}$$

we obtain

$$\mathbf{A}_{q,p}u_k=\sum_{i=-\infty}^{k-1}\tau\mathbf{A}_{q,p}\mathfrak{R}_{q,p}^{k-i}(\tau\mathbf{A})(f_i-f_k)+f_k.\tag{25}$$

Estimates (16), (17) and the substitution $j=k-i$ will imply

$$\begin{aligned}\|\mathbf{A}_{q,p}u_k\|_{\mathbf{E}}&\le\sum_{i=-\infty}^{k-1}\|\tau\mathbf{A}_{q,p}\mathfrak{R}_{q,p}^{k-i}(\tau\mathbf{A})\|_{\mathbf{E}\to\mathbf{E}}\|f_i-f_k\|_{\mathbf{E}}+\|f_k\|_{\mathbf{E}}\\&=\sum_{j=1}^{\infty}\|\tau\mathbf{A}_{q,p}\mathfrak{R}_{q,p}^{j}(\tau\mathbf{A})\|_{\mathbf{E}\to\mathbf{E}}\|f_{k-j}-f_k\|_{\mathbf{E}}+\|f_k\|_{\mathbf{E}}\\&\le\|f^\tau\|_{\mathcal{C}^\alpha(\mathbb{R}_\tau,\mathbf{E})}\left(\sum_{j=1}^{\infty}\frac{M\tau^\alpha}{(1+\tau\delta)^{\frac{j}{2}}\,j^{1-\alpha}}+1\right)\end{aligned}$$

$$\leq M\|f^\tau\|_{C^\alpha(\mathbb{R}_\tau,\mathbf{E})}\left(\int_0^\infty \frac{\mathbf{e}^{\frac{-\delta x}{2}}}{x^{1-\alpha}}dx+1\right)\leq \frac{M(\delta)}{\alpha}\|f^\tau\|_{C^\alpha(\mathbb{R}_\tau,\mathbf{E})}$$

for all k. Therefore, it follows that for all k,

$$\|\mathbf{A}_{q,p}u_k\|_{\mathbf{E}}\leq \frac{M(\delta)}{\alpha}\|f^\tau\|_{C^\alpha(\mathbb{R}_\tau,\mathbf{E})}.$$

Hence, we obtain

$$\|\mathbf{A}_{q,p}u^\tau\|_{C(\mathbb{R}_\tau,\mathbf{E})}\leq \frac{M(\delta)}{\alpha}\|f^\tau\|_{C^\alpha(\mathbb{R}_\tau,\mathbf{E})}. \tag{26}$$

Next, the estimate for the α-Hölder norm $H^\alpha\left(\mathbf{A}_{q,p}u^\tau\right)$ of $\mathbf{A}_{q,p}u^\tau$ will be established, where

$$H^\alpha\left(\varphi^\tau\right)=\sup_{-\infty<k<k+r<\infty}\frac{\|\varphi_{k+r}-\varphi_k\|_{\mathbf{E}}}{(r\tau)^\alpha}.$$

From (25), it follows that

$$\begin{aligned}\mathbf{A}_{q,p}u_{k+r}-\mathbf{A}_{q,p}u_k &= \sum_{i=k-r}^{k+r-1}\tau\mathbf{A}_{q,p}\mathfrak{R}_{q,p}^{k+r-i}(\tau\mathbf{A})(f_i-f_{k+r})\\ &\quad -\sum_{i=k-r}^{k-1}\tau\mathbf{A}_{q,p}\mathfrak{R}_{q,p}^{k-i}(\tau\mathbf{A})(f_i-f_k)\\ &\quad +\sum_{i=-\infty}^{k-r-1}\tau\mathbf{A}_{q,p}\left[\mathfrak{R}_{q,p}^{k+r-i}(\tau\mathbf{A})-\mathfrak{R}_{q,p}^{k-i}(\tau\mathbf{A})\right](f_i-f_k)\\ &\quad +\sum_{i=-\infty}^{k-r-1}\tau\mathbf{A}_{q,p}\mathfrak{R}_{q,p}^{k+r-i}(\tau\mathbf{A})(f_k-f_{k+r})+(f_{k+r}-f_k)\\ &= J_1+J_2+J_3+J_4.\end{aligned}$$

Let us estimate $\{J_m\}_{m=1}^4$ for $m=1,\cdots,4$ separately.

For $\|J_1\|_{\mathbf{E}}$, using (16), (17) and the substitution $j=k+r-i$, we get

$$\|J_1\|_{\mathbf{E}}\leq \sum_{i=k-r}^{k+r-1}\|\tau\mathbf{A}_{q,p}\mathfrak{R}^{k+r-i+1}(\tau\mathbf{A})\|_{\mathbf{E}\to\mathbf{E}}\|f_i-f_{k+r}\|_{\mathbf{E}}$$

$$\leq M^2\|f^\tau\|_{C^\alpha(\mathbb{R}_\tau,\mathbf{E})}\sum_{i=k-r}^{k+r-1}\frac{\tau^\alpha}{(k+r-i)^{1-\alpha}}=M^2\|f^\tau\|_{C^\alpha(\mathbb{R}_\tau,\mathbf{E})}\sum_{j=1}^{2r}\frac{\tau^\alpha}{j^{1-\alpha}}.$$

It results in

$$\|J_1\|_{\mathbf{E}}\leq (r\tau)^\alpha\frac{M_1}{\alpha}\|f^\tau\|_{C^\alpha(\mathbb{R}_\tau,\mathbf{E})}.$$

In a similar manner, we can show that

$$\|J_2\|_{\mathbf{E}} \le (r\tau)^{\alpha}\frac{M_2}{\alpha}\|f^{\tau}\|_{C^{\alpha}(\mathbb{R}_{\tau},\mathbf{E})}.$$

For $\|J_3\|_{\mathbf{E}}$, using (16), (17), we obtain

$$\|J_3\|_{\mathbf{E}} \le \sum_{i=-\infty}^{k-r-1} \left\|\tau \mathbf{A}_{q,p}\left[\mathfrak{R}_{q,p}^{k+r-i}(\tau\mathbf{A}) - \mathfrak{R}_{q,p}^{k-i}(\tau\mathbf{A})\right]\right\|_{\mathbf{E}\to\mathbf{E}} \|f_i - f_k\|_{\mathbf{E}}$$

$$\le M_3\|f^{\tau}\|_{C^{\alpha}(\mathbb{R}_{\tau},\mathbf{E})} \sum_{i=-\infty}^{k-r-1} \frac{r\tau^{\alpha}}{(k-i)^{2-\alpha}}.$$

The substitution $j = k - i$ gives

$$\|J_3\|_{\mathbf{E}} \le M_3\|f^{\tau}\|_{C^{\alpha}(\mathbb{R}_{\tau},\mathbf{E})} \sum_{j=r+1}^{\infty} \frac{r\tau^{\alpha}}{j^{2-\alpha}} \le (r\tau)^{\alpha} M_3 \|f^{\tau}\|_{C^{\alpha}(\mathbb{R}_{\tau},\mathbf{E})} \frac{1}{(1-\alpha)}$$

for all k. Thus, we have

$$\|J_3\|_{\mathbf{E}} \le (r\tau)^{\alpha}\frac{M_3}{1-\alpha}\|f^{\tau}\|_{C^{\alpha}(\mathbb{R}_{\tau},\mathbf{E})}.$$

Next, using (16), (17) and (24), we get $J_4 = \mathfrak{R}_{q,p}^{2r+1}(\tau\mathbf{A})(f_k - f_{k+r})$. It follows that

$$\|J_4\|_{\mathbf{E}} \le \left\|\mathfrak{R}_{q,p}^{2r+1}(\tau\mathbf{A})\right\|_{\mathbf{E}\to\mathbf{E}} \|f_k - f_{k+r}\|_{\mathbf{E}} < M(r\tau)^{\alpha}\|f^{\tau}\|_{C^{\alpha}(\mathbb{R}_{\tau},\mathbf{E})}$$

for all k. Then $\|J_4\|_{\mathbf{E}} \le M(r\tau)^{\alpha}\|f^{\tau}\|_{C^{\alpha}(\mathbb{R}_{q,p}\tau,\mathbf{E})}$. Combining all the estimates for $\{J_m\}_{m=1}^{4}$, we get

$$H^{\alpha}(\mathbf{A}_{q,p}u^{\tau}) \le \frac{M_4}{(1-\alpha)\alpha}\|f^{\tau}\|_{C^{\alpha}(\mathbb{R}_{\tau},\mathbf{E})}.$$

Using estimates (26) and (4), we get

$$\|\mathbf{A}_{q,p}u^{\tau}\|_{C^{\alpha}(\mathbb{R}_{\tau},\mathbf{E})} \le \frac{M_4}{(1-\alpha)\alpha}\|f^{\tau}\|_{C^{\alpha}(\mathbb{R}_{\tau},\mathbf{E})}. \tag{27}$$

Then, from the triangle inequality, (14), and (27),

$$\|\mathfrak{D}_{\tau}u^{\tau}\|_{C^{\alpha}(\mathbb{R}_{\tau},\mathbf{E})} \le \frac{M_4}{(1-\alpha)\alpha}\|f^{\tau}\|_{C^{\alpha}(\mathbb{R}_{\tau},\mathbf{E})}$$

follows. This concludes the proof of Theorem 4.1. □

References

1. O. A. Ladyzhenskaya, V. A. Solonnikov and N. N. Ural'tseva, *Linear and Quasilinear Equations of Parabolic Type*, Nauka, Moscow, 1967 (Russian).
2. M. I. Vishik, A. D. Myshkis and O. A. Oleinik, *PDE in: Math. in USSR* **1**, 1917 (1959) (Russian).
3. P. E. Sobolevskii, *Dokl. Akad. Nauk SSSR* **157**, 52 (1964) (Russian).
4. A. Ashyralyev, Y. Sözen and P. E. Sobolevskii, *AAA* **2007**, 1155 (2007).
5. A. Ashyralyev and P. E. Sobolevskii, *Well-Posedness of Parabolic Difference Equations*, Birkhäuser Verlag, Basel, Boston, Berlin, 1994.
6. P. E. Sobolevskii, *Dokl. Acad. Nauk SSSR* **201**, 1063 (1971) (Russian).
7. A. Ashyralyev, S. Piskarev and S. Wei, *Numer. Funct. Anal. Optim.* **23**, 669 (2002).
8. D. Guidetti, B. Karasözen and S. Piskarev, *J. Math. Sci.* **122**, 3013 (2004).
9. A. Ashyralyev and P. E. Sobolevskii, *Nonlin. An.-Theo. Methods Appl.* **24**, 257 (1995).
10. A. Ashyralyev and P. E. Sobolevskii, *New Difference schemes for Partial Differential Equations*, Birkhäuser Verlag, Operator Theory and Appl., Basel, Boston, Berlin, 2004.
11. H. Amann, *Math. Nachr.* **186**, 5 (1997).

SEMIEXPLICIT MULTISYMPLECTIC INTEGRATION OF NONLINEAR SCHRÖDINGER EQUATION

AYHAN AYDIN

Department of Mathematics, Atılım University,
06836, Ankara, Turkey
E-mail: aydin@atilim.edu.tr

BÜLENT KARASÖZEN

Department of Mathematics and Institute of Applied Mathematics,
Middle East Technical University, 06531 Ankara, Turkey
E-mail: bulent@metu.edu.tr

We apply Lobatto IIIA-IIIB-type multisymplectic discretization in space and time to the nonlinear Schrödinger equation. The resulting scheme is semiexplicit in time and therefore more efficient than implicit multisymplectic schemes. Numerical results confirm excellent long-time conservation of the local and global conserved quantities like energy, momentum and norm.

Keywords: nonlinear Schrödinger equation, multisymplectic integrators, Lobatto IIIA-IIIB methods.

1. Introduction

Recently the field of numerical analysis experienced a growing interest in the numerical investigation of ordinary and partial differential equations (pde's) by using geometric integrators which preserve the essential features of the underlying system after discretization. Symplectic integrators for ordinary differential equations (ode's) have been thoroughly analyzed and the concept of symplectic time integration has also been extended to Hamiltonian pde's [7,13]. More recently nonlinear pde's were also studied numerically using multi-symplectic integrators (see [3,13] for an overview).

A multi-symplectic structure of a pde is generated by a pair of skew-symmetric matrices $\mathbf{M}, \mathbf{K} \in \mathbb{R}^{n\times n}$ and a multi-symplectic Hamiltonian $S(z)$ which is a smooth function on $\mathbb{R}^n$:

$$\mathbf{M}z_t + \mathbf{K}z_x = \nabla_z S(z), \qquad z \in \mathbb{R}^n. \tag{1}$$

The Hamiltonian system (1) is multi-symplectic in the sense that associated with **M** and **K** are the two-forms

$$\omega = \frac{1}{2}(dz \wedge \mathbf{M}dz), \qquad \kappa = \frac{1}{2}(dz \wedge \mathbf{K}dz), \tag{2}$$

which define a space-time symplectic structure governed by the multi-symplectic conservation law (MSCL)

$$\partial_t \omega + \partial_x \kappa = 0. \tag{3}$$

This multi-symplecticity generalizes the symplectic conservation law of Hamiltonian ode's.

The multi-symplectic structure gives rise to local conservation laws which can be derived from a multi-symplectic form of Nöther's theorem. The invariance of the Hamiltonian $S(z)$ with respect to space-time shifts implies the following local energy and momentum conservation laws (LECL and LMCL), respectively

$$\begin{array}{l} \text{LECL: } E_t + F_x = 0,\, E = S - \frac{1}{2}z^T\mathbf{K}z_x,\, F = \frac{1}{2}z^T\mathbf{K}z_t \\ \text{LMCL: } I_t + G_x = 0,\; G = S - \frac{1}{2}z^T\mathbf{M}z_t,\, I = \frac{1}{2}z^T\mathbf{M}z_x. \end{array} \tag{4}$$

Additionally, norm preservation follows from the phase symmetry.

Multi-symplectic integrators based on simultaneous space-time discretization of the pde's preserve the MSCL (3) in the discrete sense. In contrast to the symplectic integrators; the multi-symplectic conservation law is not exactly preserved by the multi-symplectic integrators, different multi-symplectic integrators preserve different discretizations of the multi-symplectic conservation law. However, preservation of the multi-symplectic structure by a numerical scheme does not imply the exact preservation of the local conservation laws (4) or of global invariants of the pde's. Due to their local nature the multi-symplectic integrators preserve the conserved quantities locally better than the symplectic integrators. The word "local" here means that the conservation properties do not depend on the specific domain or the boundary conditions of the pde's.

The Runge-Kutta (RK) methods play an important role in the numerical solution of symplectic and multi-symplectic pde's. It was shown [17] that the scalar wave equation can be integrated using concatenation of Gauss-Legendre Runge-Kutta methods in space and time, which leads to a multi-symplectic integrator. It is known that the Preissmann scheme corresponds to the concatenation of mid-point methods in time and space. The main weakness of RK based multi-symplectic integrators is their implicit nature that causes theoretical and practical problems due to the solutions of large nonlinear systems [18,19]. The nonlinear Schrödinger (NLS) equation is an often used example for testing the performance of multi-symplectic integrators like the Preissmann scheme or multi-symplerctic pseudospectral methods [2,4,10–12,17].

The partitioned Runge-Kutta (PRK) methods are another useful class of symplectic integrators which are applied successfully to symplectic ode's yielding explicit methods. The approach for symplectic PRK methods was extended to Hamitonian pde's in multi-symplectic form and the conditions which have to be satisfied formally by the multi-symplectic PRK methods were given [9]. For higher order RK and PRK methods, the conditions for preserving the multi-symplecticity become very complicated and intractable. Different several multi-symplectic RK and PRK methods are considered based on different multi-symplectic formulations of the nonlinear wave equation [15]. Recently several PRK methods were applied to Hamiltonian pde's like nonlinear Schrödinger equation, nonlinear Dirac equation, Boussinesq equation and to the scalar wave equation [1,8,9,14,15].

In this paper, we apply in Section 2 the 2-stage Lobatto IIIA-IIIB multi-symplectic integrator developed in [18,19] to the NLS equation. In Section 3 we present numerical results for the long term conservation of local and global invariants like energy, momentum and norm. The paper ends with some concluding remarks.

2. Multi-symplectic discretization of the nonlinear Schrödinger equation

The focusing one-dimensional nonlinear Schrödinger (NLS) equation

$$i\psi_t + \psi_{xx} + 2|\psi|^2\psi = 0, \tag{5}$$

can be written in a multi-symplectic form by letting $\psi = p + iq$ and introducing the new variables $v = p_x$, $w = q_x$ [17]. Separating into real and imaginary parts, the NLS equation (5) can be written in the multi-symplectic form (1) with

$$z = \begin{pmatrix} p \\ q \\ v \\ w \end{pmatrix}, \quad \mathbf{M} = \begin{pmatrix} 0 & 1 & 0 & 0 \\ -1 & 0 & 0 & 0 \\ 0 & 0 & 0 & 0 \\ 0 & 0 & 0 & 0 \end{pmatrix}, \quad \mathbf{K} = \begin{pmatrix} 0 & 0 & -1 & 0 \\ 0 & 0 & 0 & -1 \\ 1 & 0 & 0 & 0 \\ 0 & 1 & 0 & 0 \end{pmatrix},$$

and Hamiltonian $S = \frac{1}{2}((p^2+q^2)^2 + v^2 + w^2)$.

Implementing the relations (4) for the NLS equation yields the local energy and momentum conservation laws (LECL and LMCL):

$$\begin{aligned} &E_t + F_x = 0,\ E = \tfrac{1}{2}((p^2+q^2)^2 - v^2 - w^2), F = vp_t + wq_t \\ &I_t + G_x = 0,\ I = pw - qv, G = (p^2+q^2)^2 + v^2 + w^2 - (pq_t - p_t q). \end{aligned} \tag{6}$$

Additionally we have a norm conservation law for the NLS equation

$$N_t + M_x = 0, \quad N = \frac{1}{2}(p^2+q^2), \quad M = qv - pw.$$

For periodic boundary conditions, the local conservation laws can be integrated in x to obtain global conservation of energy, momentum and norm defined as

$$\frac{d}{dt}\mathcal{E}(z) = 0, \quad \frac{d}{dt}\mathcal{I}(z) = 0, \quad \frac{d}{dt}\mathcal{N}(z) = 0,$$

where $\mathcal{E}(z) = \int_0^L E(z)dx, \mathcal{I}(z) = \int_0^L \mathcal{I}(z)dx,\ \mathcal{N}(z) = \int_0^L N(z)dx.$

For the space and time discretization of the NLS equation we follow the formulation of PRK methods as multi-symplectic integrators in [18,19]. An s-stage time discretization of the differential equation $z_t = f(z)$ consists of a set of equations coupling the node values z_i to the stage vectors $Z_{i,n}$ is given by

$$\begin{aligned} Z_{i,j}^{(\gamma)} &= z_i^{(\gamma)} + \Delta t \sum_{k=1}^{s} a_{jk}^{(\gamma)} f(Z_{i,j}), \quad i = 1, \cdots, s \\ z_{i+1} &= z_i + \Delta t \sum_{j=1}^{s} b_j^{(\gamma)} f(Z_{i,j}). \end{aligned}$$

Here the vector of dependent variables z is partitioned in two parts, $z^{(\gamma)}$, $\gamma = 1, 2$. The coefficients $b_j^{(\gamma)}$ and $a_{jk}^{(\gamma)}$ are chosen to satisfy certain order conditions. For the case of canonical Hamiltonian system the conditions for symplecticity for a two-partition PRK discretization are [7]

$$-a_{kj}^{(1)} b_k^{(2)} - b_j^{(1)} a_{jk}^{(2)} + b_j^{(1)} b_k^{(2)} = 0, \quad j, k = 1, \ldots s.$$

For a PRK discretization in space, the set of equations is given as

$$\begin{aligned} Z_{i,j}^{(\gamma)} &= z_i^{(\gamma)} + \Delta x \sum_{k=1}^{s} a_{jk}^{(\gamma)} \partial_x Z_{i,j}^{(\gamma)}, \quad i = 1, \cdots, s \\ z_{i+1} &= z_i + \Delta x \sum_{j=1}^{s} b_j^{(\gamma)} \partial_x Z_{i,j}^{(\gamma)} \end{aligned}$$

where the variables $\partial_x Z_{i,j}$ have to satisfy the multi-symplectic conservation law (3)

$$M\partial_t Z_{i,j} + K\partial_x Z_{i,j} = \nabla S(Z_{i,j}).$$

We use here the second-order 2-stage symplectic Lobatto IIIA-IIIB method given by the following RK table:

$$\text{IIIA:}\quad \begin{array}{c|cc} 0 & 0 & 0 \\ 1 & 1/2 & 1/2 \\ \hline & 1/2 & 1/2 \end{array} \qquad\qquad \text{IIIB:}\quad \begin{array}{c|cc} 0 & 1/2 & 0 \\ 1 & 1/2 & 0 \\ \hline & 1/2 & 1/2 \end{array}$$

Grouping the dependent variables as $z^{(1)} = \{p, q\}$ and $z^{(2)} = \{v, w\}$, semi-discretization of NLS equation by the Lobatto IIIA-IIIB discretization results [18,19] in

$$\begin{aligned} \partial_t p_i &= -\frac{q_{i-1} - 2q_i + q_{i+1}}{\Delta x^2} - 2(p_i^2 + q_i^2)q_i, \\ \partial_t q_i &= \frac{p_{i-1} - 2p_i + p_{i+1}}{\Delta x^2} + 2(p_i^2 + q_i^2)p_i. \end{aligned} \tag{7}$$

These equations correspond precisely to second order finite difference approximation of the NLS equation in space. This implies that the 2-stage Lobatto IIIA-IIIB discretisation in space is equivalent to second order finite differences in space applied to such a pde.

Using the same partitioning of the variables in space and time yields implicit integrators [18]. Choosing the partitioning $z^{(3)} = (p, v), z^{(4)} = (q, w)$, it can be seen that the variables v and w do not appear in (7), so that 2-stage Lobatto IIIA-IIIB method in time can be applied to the partitioned ode (7) containing only the variables p and q. Second order Lobatto IIIA-IIIB in time in this form is commonly known as generalized leapfrog which is in general an implicit method. Applied to (7), we obtain the integrator that maps (p_i^n, q_i^n) to (p_i^{n+1}, q_i^{n+1}) in the following way:

$$\begin{aligned} q_i^{n+(1/2)} &= q_i^n + \frac{\Delta t}{2\Delta x^2}(p_{i-1}^n - 2p_i^n + p_{i+1}^n) + \Delta t((q_i^{n+(1/2)})^2 + (p_i^n)^2)p_i^n, \\ p_i^{n+1} &= p_i^n - \frac{\Delta t}{\Delta x^2}(q_{i-1}^{n+(1/2)} - 2q_i^{n+(1/2)} + q_{i+1}^{n+(1/2)}) \\ &\quad - \Delta t((p_i^n)^2 + (p_i^{n+1})^2 + 2(q_i^{n+(1/2)})^2)q_i^{n+(1/2)}, \\ q_i^{n+1} &= q_i^{n+(1/2)} + \frac{\Delta t}{\Delta x^2}(p_{i-1}^{n+1} - 2p_i^{n+1} + p_{i+1}^{n+1}) \\ &\quad + \Delta t((q_i^{n+(1/2)})^2 + (p_i^{n+1})^2)p_i^{n+1}. \end{aligned} \tag{8}$$

In this system the nonlinearities enter as scalar quadratic equations which can be solved explicitly. Applying one Newton iteration step to the first and second equation in (8), one can compute $q_i^{n+(1/2)}$ and p_i^{n+1}, $i = 1, \ldots, N$. The last equation is explicit in p_i^{n+1}, $i = 1, \ldots, N$. In contrast to this, the implicit multi-symplectic methods like the Preismann scheme require several Newton steps in order to find an approximate solution. The computational stencil for the 2-stage Lobatto IIIA-IIIB method in Fig. 1 reveals that it can be interpreted as a method on staggered grids, because the values of the p_i's and q_i's are computed on different nodes.

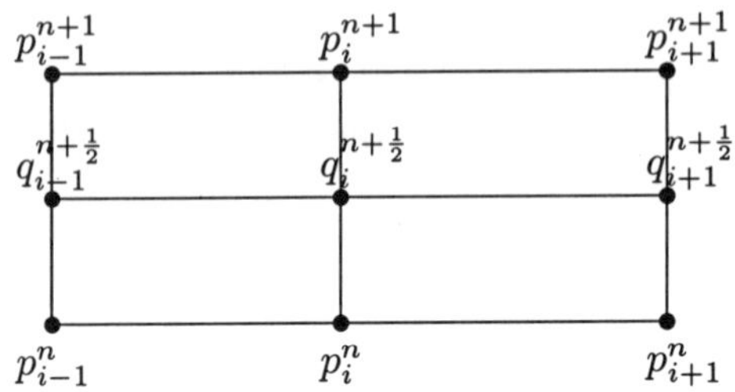

Fig. 1. Stencil for the Lobatto IIIA-IIIB method for the NLS equation

Similarly, the Volterra lattice equation which is an approximation of the Korteweg de Vries (KdV) equation as a Poisson system was discretized in time using the implicit and explicit Euler method [5] (first order Lobatto IIIA-IIIB method). Due to the quadratic terms appearing in the Volterra lattice, linear systems of equations have to be solved at each time step like in (8).

Higher order $s-$stage Lobatto IIIA-IIIB discretization in space and time can be constructed. But for Lobatto IIIA-IIIB methods, space-time discretizations with more than 2-stages lead to implicit integrators [19]. Only discretizations in space with higher order Lobatto IIIA-IIIB methods and applying composition methods in time lead to explicit integrators.

Another important feature of multi-symplectic integrators is the preservation of the linear dispersion relations of the pde's. The dispersion properties of the implicit multi-symplectic integrators like the Gauss-Legendre RK methods and including the Preissman scheme were investigated for the NLS equation and other multi-symplectic pde's [6,15,16]. These multi-symplectic integrators preserve the discrete dispersion relations for the linearized pde's and do not exhibit spurious solutions. Any symmetric and A-stable RK methods like the Lobatto IIIA-IIIB method preserve the discrete dispersion relations too [6]. It was shown that, similarly, a space-time discretization of the Boussinesq equation by the 2-stage Lobatto IIIA-IIIB method preserves the numerical dispersion relation and there are no spurious solutions [1].

3. Discretization of conserved quantities and numerical results

We can use the residual for the local energy conservation law as [8,9]

$$RE=\frac{1}{\Delta t}\sum_{m=1}^{2} b_m^{(1)}\left(E(\mathbf{z}_{i,m}^{n+1})-E(\mathbf{z}_{i,m}^{n})\right)+\frac{1}{\Delta x}\sum_{k=1}^{2} b_k^{(2)}\left(F(\mathbf{z}_{i+1}^{n,k})-F(\mathbf{z}_{i}^{n,k})\right) \quad (9)$$

where $\mathbf{z}_{i,m}^{n+1} \approx z(c_m\Delta x,(n+1)\Delta t)$, $\mathbf{z}_{i,m}^{n} \approx z(c_m\Delta x, n\Delta t)$ and $\mathbf{z}_{i+1}^{n,k} \approx z((i+1)\Delta x, \widetilde{c}_k\Delta t)$, $\mathbf{z}_i^{n,k} \approx z(i\Delta x, \widetilde{c}_k\Delta t)$. The residual in (9) does not mean the local error at the mesh point x_i, t_n, it was derived from the local integration over the rectangular domain $(x_i,t_n),(x_{i+1},t_n),(x_{i+1},t_n),(x_{i+1},t_{n+1})$ [8].

Since $c_1=0, c_2=1$ and the weights $b_i^{(2)}=b_i^{(2)}=1/2,\ i=1,2$ for the Lobatto IIIA-IIIB method, we can write the residual (9) as

$$RE=\frac{E_i^{n+1}-E_i^n+E_{i+1}^{n+1}-E_{i+1}^n}{2\Delta t}+\frac{F_{i+1}^n-F_i^n+F_{i+1}^{n+1}-F_{i+1}^{n+1}}{2\Delta x} \quad (10)$$

where

$$\begin{aligned} E_i^n &= -\frac{1}{2}\left[\left((p_i^n)^2+(q_i^n)^2\right)^2-(v_i^n)^2-(w_i^n)^2\right], \\ F_i^n &= -v_i^n\left(\frac{p_i^{n+1}-p_i^n}{\Delta t}\right)-w_i^n\left(\frac{q_i^{n+1}-q_i^n}{\Delta t}\right). \end{aligned}$$

We can use the residual for the momentum conservation law as [9]

$$RI=\frac{1}{\Delta t}\sum_{m=1}^{2} b_m\left(I(\mathbf{z}_m^{\mathbf{1}})-I(\mathbf{z}_m^{\mathbf{0}})\right)+\frac{1}{\Delta x}\sum_{k=1}^{2}\widetilde{b}_k\left(G(\mathbf{z}_{\mathbf{1}}^k)-G(\mathbf{z}_{\mathbf{0}}^k)\right) \quad (11)$$

which can be rewritten as

$$RI=\frac{I_i^{n+1}-I_i^n+I_{i+1}^{n+1}-I_{i+1}^n}{2\Delta t}+\frac{G_{i+1}^n-G_i^n+G_{i+1}^{n+1}-G_{i+1}^{n+1}}{2\Delta x}, \quad (12)$$

where

$$\begin{aligned} I_i^n &= -\frac{1}{2}\left(p_i^n w_i^n - q_i^n v_i^n\right), \\ G_i^n &= -\frac{1}{2}\Bigg[\left((p_i^n)^2+(p_i^n)^2\right)^2+(v_i^n)^2+(w_i^n)^2-(p_i^n)\left(\frac{q_i^{m+1}-q_i^m}{\Delta t}\right) \\ &\quad +(q_i^n)\left(\frac{p_i^{m+1}-p_i^m}{\Delta t}\right)\Bigg] \end{aligned}$$

The global energy GE and the global momentum GI for both integrators are given by

$$GE = \Delta x \sum_{j=1}^{N} (E_j^n - E^0), \qquad GI = \Delta x \sum_{j=1}^{N} (I_j^n - I^0)$$

where E^0 and I^0 are the initial energy and momentum respectively. The local and global norm conservation laws can be defined analogously.

Time discretization by Lobatto IIIA-IIIB methods does not the preserve phase symmetry [19] and the quadratic invariants appearing in the norm conservation $N = (p^2 + q^2)/2$ can not be preserved exactly. But the Lobatto IIIA-IIIB methods preserve the quadratic invariants like $I = pw - qv$ in the momentum exactly [20].

For a numerical experiment we consider the NLS equation (5) with the initial condition [10] $\psi_0(x) = 0.5(1 + 0.1\cos(\mu x))$, $\mu = 2\pi/L$, $L = 2\sqrt{2}\pi$. We set the spatial mesh-size to $\Delta x = L/64$ and the time step to $\Delta t = 5 \times 10^{-3}$.

In Fig. 2 the numerical results obtained by the Lobatto IIIA-IIIB multi-symplectic integrator are presented over the time domain $[0, 100]$. We see that the surface exhibits a quasi periodic behavior in time. We notice that the spatial period is $2\pi/\mu$ or L; within that length there is only one peak in space. The wave oscillates between the near uniform state and the one hump state. From Fig. 2 we see that the errors in local energy and momentum are concentrated in the region where there are peaks in the solution. The local errors are bounded and do not grow in time. The global conservation laws are preserved very well over the long time integration. Numerical results obtained for the preservation of the locally and globally conserved quantities confirm the theoretically predicted ones. The global momentum is preserved up to the machine accuracy. The errors for the global energy and momentum remain bounded as in the case of the implicit multi-symplectic integration of NLS equation.

4. Conclusions

Multi-symplectic Lobatto IIIA-IIIB space-time discretization of the NLS equation is second order accurate in time and space variables and requires less computation than the Preissman scheme. Local and global energy can be preserved more accurately using semi-discretization in space using a higher order Lobatto IIIA-IIIB pair and symmetric composition in time.

References

1. A. Aydın and B. Karasözen, Symplectic and multi-symplectic Lobatto methods for the "good" Boussinesq equation, submitted.
2. T. J. Bridges and S. Reich, *Phys. Lett. A* **284**, 184 (2001).
3. T. J. Bridges and S. Reich, *J. Phys. A* **39**, 5287 (2006).
4. J. B. Chen, M. Z. Qin and Y. F. Tang, *Comput. Math. Appl.* **43**, 1095 (2002).
5. T. Ergenç and B. Karasözen, *Appl. Numer. Math.* **56**, 879 (2006).
6. J. Frank, B.E. Moore and S. Reich, *SIAM J. Sci. Comput.* **28**, 260 (2006).
7. E. Hairer, C. Lubich and G. Wanner, *Geometric Numerical Integration Structure-Preserving Algorithms for Ordinary Differential Equations*, 2nd ed., Springer, New York, 2005.
8. J. Hong and C. Li, *J. Comput. Phys.* **211**, 448 (2006).
9. J. Hong, H. Liu and G. Sun, *Math. Comp.* **75**, 167 (2006).
10. A. L. Islas and D. A. Karpeev and C. M. Schober, *J. Comput. Phys.* **173**, 116 (2001).
11. A. L. Islas and C. M. Schober, *J. Comput. Phys.* **197**, 585 (2004).
12. A. L. Islas and C. M. Schober, *Math. Comput. Simulation* **69**, 290 (2005).
13. B. Leimkuhler and S. Reich, *Simulating Hamitonian Dynamics*, Cambridge University Press, 2004.
14. Q. Li, Y. Song and Y. Wang, *Appl. Math. Comput,* **177**, 36 (2006).
15. H. Liu and K. Zhang, *IMA J. Numer. Anal.* **26**, 252 (2006).
16. B. E. Moore, *A Modified Equations Approches for Multi-symplectic Integration Methods*, Ph.D. thesis, Department of Mathematics and Statistics, School of Electronics and Physical Sciences, University of Surrey, 2003.
17. S. Reich, *J. Comput. Phys.* **157**, 473 (2000).
18. B. N. Ryland and R. I. McLachlan, *On multi-symplecticity of partitioned Runge-Kutta methods*, technical report, Center for Mathematics and Computer Science (CWI), (2007).
19. B. N. Ryland, R. I. McLachlan and J. Frank, *Int. J. Comput. Math.* **84**, 874 (2007).
20. Y. Sun, *Quadratic invariants and multi-symplecticity of partitioned Runge-Kutta methods for Hamiltonian PDE's*, Institute of Computational Mathematics and Scientific/Engineering Computing, Beijing, China.

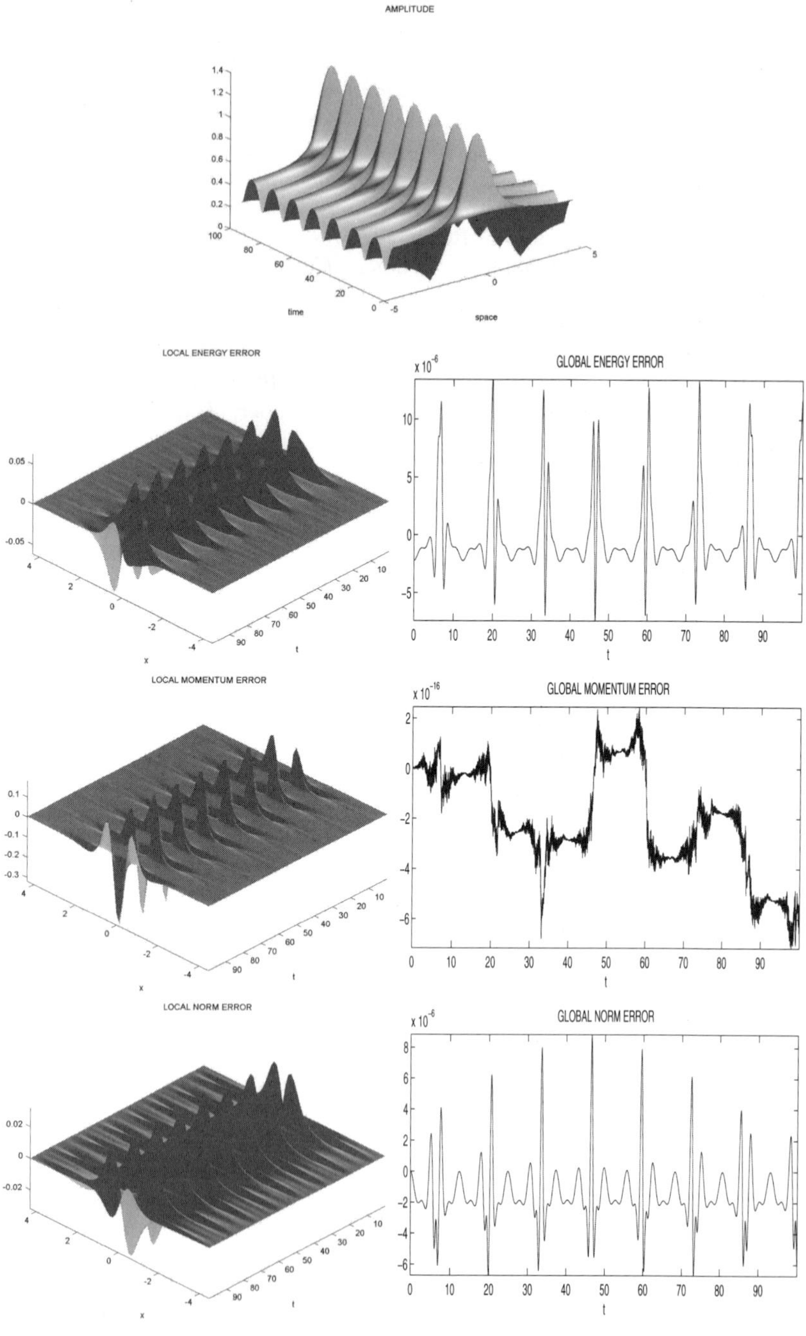

Fig. 2. Lobatto IIIA-IIIB integration of NLS

NONLOCAL PROBLEMS WITH SPECIAL GLUING FOR A PARABOLIC-HYPERBOLIC EQUATION

A. S. BERDYSHEV

Department of Mathematics, Syrdarya University,
Zhetisay, 160500, Kazakhstan
E-mail: berdyshev@mail.ru

N. A. RAKHMATULLAEVA

Institute of Mathematics and Information Technologies,
Uzbek Academy of Sciences,
Tashkent, 100125, Uzbekistan
E-mail: rakhmatullaeva@mail.ru

A nonlocal problem for a parabolic-hyperbolic equation with three lines of type changing is considered. To prove the uniqueness of solution, the method of energy integrals is used. The existence is proved by the method of integral equations.

Keywords: Non-local problem; Parabolic-hyperbolic equation; Energy integral; Integral equation.

1. Introduction

The necessity of study of boundary-value problems for parabolic-hyperbolic type equations was justified from the application side by Gel'fand [1]. He gave an example connected to the movement of gas in a channel surrounded by a porous environment. Inside the channel, the movement of gas was described by the wave equation, outside by the diffusion equation. Also Zolina [2,3] has studied many boundary-value problems, including the famous Tricomi problem for model parabolic-hyperbolic equation and she gives some physical interpretation of these problems. After these works was published a set of works, which are devoted to the study of local and non-local problems for the second order parabolic-hyperbolic equations. A review of the basic results in these directions till 1986 was given in the book by Dzhuraev, Sopuev and Mamazhonov [4]. Note that boundary-value problems for parabolic-hyperbolic type equations with one line of changing type were considered by many authors.

We note the works of Egamberdiev [5], Eleeev [6] and Abdullaev [7] studying local and non-local boundary-value problems for parabolic-hyperbolic type equations with two lines of type changing. But till now, even the basic boundary-value problems for parabolic-hyperbolic equations with three lines of type changing were not investigated.

One of the basic differences of these problems from others consists in the fact that the value of the required function will be given only in hyperbolic parts of the set area.

2. Formulation of the problem

We consider an equation

$$0 = \begin{cases} u_{xx} - u_y - \lambda u, (x,y) \in \Omega_0, \\ u_{xx} - u_{yy} + \mu u, (x,y) \in \Omega_i \ \left(i = \overline{1,3}\right) \end{cases} \tag{1}$$

in a domain $\Omega = \Omega_0 \cup \Omega_i \ \left(i = \overline{1,3}\right)$, where Ω_0 is a domain bounded by segments AB, BC, CD, DA of straight lines $y = 0$, $x = 1$, $y = 1$ and $x = 0$, i.e. a square $\{0 \le x \le 1, \ 0 \le y \le 1\}$; Ω_2 is a characteristic triangle bounded by the segment AD of the axis y and by the characteristics $AK : x + y = 0$, $DK : y - x = 1$ of the equation (1), outgoing from the points A, D and crossing at $K\left(-\frac{1}{2}, \frac{1}{2}\right)$; Ω_3 is a characteristic triangle bounded by the segment BC and by the characteristics $BM : x - y = 1$, $CM : x - 2 = -y$ of the equation (1), outgoing from the points B, C and crossing at $M\left(\frac{3}{2}, \frac{1}{2}\right)$; Ω_1 a characteristic triangle bounded by the segment AB and by the characteristics $AN : x + y = 0$, $BN : x - y = 1$ of the equation (1), outgoing from the points A, B and crossing at $N\left(\frac{1}{2}, -\frac{1}{2}\right)$. In equation (1), λ and μ are given real parameters.

Problem $\mathbf{S^3}$. Find a function $u(x,y)$ from the set of functions

$$W \equiv C\left(\overline{\Omega}\right) \cap C^1(\Omega_1 \cup AB \cup BC \cup AD) \cap C^{2,1}_{x,y}(\Omega_0) \cap C^2(\Omega_i),$$

satisfying the equation (1) in the domains $\Omega_0 \cup \Omega_i$, where $i = \overline{1,3}$, and the following non-local conditions

$$A^{0,\sqrt{\mu}}_{0x}\left[u\left(\theta_1\right)\right] + a_1(x)u(x,0) = b_1(x), \ 0 \le x \le 1, \tag{2}$$

$$A^{0,\sqrt{\mu}}_{0y}\left[u\left(\theta_2\right)\right] + a_2(y)u(0,y) = b_2(y), \ 0 \le y \le 1, \tag{3}$$

$$A^{0,\sqrt{\mu}}_{0y}\left[u\left(\theta_3\right)\right] + a_3(y)u(1,y) = b_3(y), \ 0 \le y \le 1. \tag{4}$$

Moreover, on the lines of type changing the following gluing conditions should be fulfilled:

$$u_y(x,+0) = \alpha u_y(x,-0), \ 0 < x < 1, \tag{5}$$

$$u_x(+0,y) = \beta u_x(-0,y), \ 0 < y < 1, \tag{6}$$

$$u_x(1-0,y) = \gamma u_x(1+0,y), \ 0 < y < 1. \tag{7}$$

Here, $a_i(t)$, $b_i(t)$ $(i = \overline{1,3})$ are given real-valued functions. Moreover

$$\begin{cases} a_1(x) \neq -\frac{1}{2}, a_3(y) \neq -\frac{1}{2}, b_1(0) = b_2(0), \\ a_1(0) = a_2(0), a_2(y) \neq -\frac{1}{2}, \alpha, \beta, \gamma \in \mathbb{R} \setminus \{0\}, \\ \theta_1 = \left(\frac{x}{2}, -\frac{x}{2}\right), 0 < x < 1, \theta_2 = \left(-\frac{y}{2}, \frac{y}{2}\right), \theta_3 = \left(\frac{y+1}{2}, \frac{y-1}{2}\right), 0 < y < 1 \end{cases}$$

and

$$A_{mx}^{n,\sqrt{\mu}}[f(x)] = f(x) - \int_m^x f(t)\left(\frac{t-m}{x-m}\right)^n \frac{\partial}{\partial t} J_0[\sqrt{\mu(x-m)(x-t)}]\, dt,$$

for $m, n = 0, 1$ is an integral operator introduced and studied by Salakhitdinov and Urinov [8].

3. The main results

Theorem 3.1. *Problem S^3 has a unique solution if the following conditions regarding the given functions and parameters are fulfilled:*

$$\lambda \geq 0, \ \alpha > 0, \ \beta > 0, \gamma > 0, \quad a_i(z), b_i(z) \in C^2[0,1], \ i = 1,2,3,$$

$$\frac{1}{1+2a_1(1)} > 0, \ \frac{1}{1+2a_2(1)} > 0, \ \frac{1}{1+2a_3(1)} < 0,$$

$$\left(\frac{1}{1+2a_1(x)}\right)' < 0, \ \left(\frac{1}{1+2a_2(y)}\right)' < 0, \ \left(\frac{1}{1+2a_3(y)}\right)' > 0.$$

Proof. To prove this theorem, we find principal functional relations in the domains Ω_i, $(i = \overline{1,3})$.

$$(1+2a_1(x))\tau_1^-(x) = \int_0^x J_0\left[\sqrt{\mu}(x-t)\right] \nu_1^-(t)\, dt + b_1(x), \ 0 \leq x \leq 1, \tag{8}$$

$$(1+2a_2(y))\tau_2^-(y) = \int_0^y \nu_2^-(t) J_0\left[\sqrt{\mu}(y-t)\right]\, dt + b_2(y), \ 0 \leq y \leq 1, \tag{9}$$

$$(1+2a_3(y))\tau_3^+(y) = \int_0^y \nu_3^+(t) J_0\left[\sqrt{\mu}(y-t)\right]\, dt + b_3(y), \ 0 \leq y \leq 1. \tag{10}$$

In the domain Ω_0, we have the following equality:

$$\iint_{\Omega_0} (u_x^2 + \lambda u^2)\, dx\, dy + \frac{1}{2}\int_0^1 (u^2(x,1) - u^2(x,0))\, dx$$
$$- \int_0^1 \tau_3^-(y)\nu_3^-(y)\, dy + \int_0^1 \tau_2^+(y)\nu_2^+(y)\, dy = 0. \tag{11}$$

Here,

$$\begin{aligned} u(x;+0) &= \tau_1^+(x); \quad u_y(x;+0) = \nu_1^+(x);\\ u(x;-0) &= \tau_1^-(x); \quad u_y(x;-0) = \nu_1^-(x);\\ u(+0;y) &= \tau_2^+(y); \quad u_x(+0;y) = \nu_2^+(y);\\ u(-0;y) &= \tau_2^-(y); \quad u_x(-0;y) = \nu_2^-(y);\\ u(1+0;y) &= \tau_3^+(y); \quad u_x(1+0;y) = \nu_3^+(y);\\ u(1-0;y) &= \tau_3^-(y); \quad u_x(1-0;y) = \nu_3^-(y). \end{aligned}$$

4. The uniqueness of the solution

First of all, we prove that $\tau_1(x)=0$. Passing to the limit in the domain Ω_0 as $y \to +0$ from the equation $u_{xx}-u_y-\lambda u=0$, we obtain

$$(\tau_1^+(x))'' - \nu_1^+(x) - \lambda\tau_1^+(x) = 0.$$

From here, we find $\nu_1^+(x) = (\tau_1^+(x))'' - \lambda\tau_1^+(x)$ and substitute in the integral

$$I_1 = \int_0^1 \tau_1^+(x)\nu_1^+(x)\,dx.$$

Then, we get

$$\begin{aligned} I_1 &= \int_0^1 \tau_1^+(x)\left(\tau_1^+(x)\right)''\,dx - \lambda\int_0^1 \left(\tau_1^+(x)\right)^2\,dx\\ &= \tau_1^+(x)\left(\tau_1^+(x)\right)'\Big|_0^1 - \int_0^1 (\tau_1^{+'}(x))^2\,dx - \lambda\int_0^1 \left(\tau_1^+(x)\right)^2\,dx\\ &= -\int_0^1 (\tau_1^{+'}(x))^2\,dx - \lambda\int_0^1 \left(\tau_1^+(x)\right)^2\,dx. \end{aligned} \tag{12}$$

Considering $a_1(0) \neq -1/2$, from (2) we have $\tau_1^-(0) = \tau_1^+(0) = 0$. Also taking $a_3(0) \neq -1/2$ into account, from (4) we get $\tau_1^-(1) = \tau_1^+(1) = 0$. Assuming $\lambda \geq 0$, from (12) we obtain $I_1 \leq 0$.

Now, considering the gluing condition (5), we get

$$I_1 = \int_0^1 \tau_1^+(x)\nu_1^+(x)\,dx = \alpha\int_0^1 \tau_1^-(x)\nu_1^-(x)\,dx.$$

Instead of $\tau_1(x)$, we substitute its represention by the formula (8) at $b_1(x)=0$ and get

$$I_1 = \alpha\int_0^1 \nu_1^-(x)\left[\frac{1}{1+2a_1(x)}\int_0^x J_0\left[\sqrt{\mu}(x-t)\right]\nu_1^-(t)\,dt\right]dx$$

$$= \alpha \int_0^1 \frac{1}{1+2a_1(x)} \int_0^x J_0\left[\sqrt{\mu}(x-t)\right] \nu_1^-(t)\nu_1^-(x)\,dx\,dt.$$

Using integral representation

$$J_0\left[\sqrt{\mu}(x-t)\right] = \frac{1}{\Gamma\left(\frac{1}{2}\right)\pi} \int_{-1}^1 (1-z^2)^{-1/2} \cos\sqrt{\mu}(x-t)z\,dz,$$

we get

$$I_1 = \frac{\alpha}{\Gamma(1/2)\pi} \int_{-1}^1 \frac{dz}{(1-z^2)^{\frac{1}{2}}} \int_0^1 \frac{dx}{1+2a_1(x)} \int_0^x \nu_1^-(t) \cos\sqrt{\mu}(x-t)z\,dt.$$

We denote

$$M(z,x) = \int_0^x \nu_1^+(t) \cos\sqrt{\mu}tz\,dt \quad \text{and} \quad N(z,x) = \int_0^x \nu_1^+(t) \sin\sqrt{\mu}tz\,dt.$$

Then, we obtain

$$I_1 = \frac{\alpha}{\Gamma(1/2)\pi} \int_{-1}^1 (1-z^2)^{-\frac{1}{2}}\,dz \left[\frac{1}{1+2a_1(1)}(M^2(z,1)+N^2(z,1)) - \int_0^1 \left(\frac{1}{1+2a_1(y)}\right)' [M^2(z,x)+N^2(z,x)]\,dy\right].$$

If $\alpha > 0$, then $\dfrac{1}{1+2a_1(1)} > 0$ and $\left(\dfrac{1}{1+2a_1(x)}\right)' < 0$. This implies $I_1 > 0$. From here $I_1 = 0$ and $\tau_1(x) = 0$.

Now, we prove the inequalities

$$I_2 = \int_0^1 \tau_2^+(y)\nu_2^+(y)\,dy \geq 0, \quad \text{and} \quad I_3 = \int_0^1 \tau_3^-(y)\nu_3^-(y)\,dy \leq 0. \tag{13}$$

First, we prove that $I_2 \geq 0$. Considering (6), (9) and the integral representation of Bessel's function

$$I_2 = \frac{\beta}{2\Gamma\left(\frac{1}{2}\right)\pi} \int_{-1}^1 (1-z^2)^{-\frac{1}{2}}\,dz \left[\frac{1}{2a_2(1)+1}(M^2(z,1)+N^2(z,1))\right] - \int_0^1 \left(\frac{1}{1+2a_2(y)}\right)' [M^2(z,y)+N^2(z,y)]\,dy.$$

If $\beta > 0$, then $\dfrac{1}{2a_2(1)+1} > 0$ and $\left(\dfrac{1}{1+2a_2(y)}\right)' < 0$. This implies $I_2 \geq 0$.

Second, we prove that $I_3 \leq 0$. Instead of τ_3^+, substituting its representation by formula (10) and considering (7), we get

$$I_3 = \int_0^1 \tau_3^-(y)\nu_3^-(y)\,dy = \int_0^1 \frac{\nu_3^-(y)}{1+2a_3(y)} \int_0^y \nu_3^+(t) J_0\left[\sqrt{\mu}(y-t)\right] dt\,dy$$

$$= \gamma \int_0^1 \frac{1}{1+2a_3(y)} \left[\int_0^y \nu_3^-(y)\nu_3^-(t) J_0 \left[\sqrt{\mu}(y-t)\right] dt \right] dy.$$

We use the following integral representation

$$J_0\left[\sqrt{\mu}(y-t)\right] = \frac{1}{\Gamma\left(\frac{1}{2}\right)\pi} \int_{-1}^1 (1-z^2)^{-\frac{1}{2}} \cos \sqrt{\mu}(y-t) z \, dz,$$

then after some transformations, we have

$$I_3 = \frac{\gamma}{2\Gamma\left(\frac{1}{2}\right)\pi} \int_{-1}^1 (1-z^2)^{-\frac{1}{2}} dz \left[\frac{1}{1+2a_3(1)} (M^2(z,1) + N^2(z,1)) \right.$$
$$\left. - \int_0^1 \left(\frac{1}{1+2a_3(y)} \right)' [M^2(z,y) + N^2(z,y)] \, dy. \right]$$

If $\gamma > 0$, then $\dfrac{1}{1+2a_3(1)} < 0$ and $\left(\dfrac{1}{1+2a_3(y)}\right)' > 0$. This implies $I_3 \leq 0$. Taking (14), (15) into account from (5), we get $\tau_3^-(y) = 0, \tau_2^+(y) = 0$.

Finally considering $\tau_1^+(x) = u(x;+0) = 0$, $\tau_2^+(y) = u(+0;y) = 0$, $\tau_3^-(y) = u(-1;y) = 0$, from the solution of the first boundary-value problem for parabolic equations we have $u(x,y) = 0$ in the domain $\overline{\Omega}_0$. As $u(x,y) \in C\left(\overline{\Omega}\right)$ follows that $u(x,y) = 0$ in $\overline{\Omega}$.

5. The existence of the solution

First of all, in the domain Ω_0 passing to the limit as $y \to +0$ in the equation (1) and considering conditions (2), (4) we get

$$\tau_1(x) = r(x) + \int_0^1 G^*(x,t,\lambda)\nu_1^+(t) \, dt. \tag{14}$$

Differentiating (8) and (14) with respect to x once and equating expressions of $\tau'(x)$ after some transformations

$$\nu_1^+(x) - \int_0^1 \nu_1^+(t) K_1^*(x,t,\lambda,\mu) \, dt = k_1(x). \tag{15}$$

Since K_1^* has only a weak singularity, we can write the solution of (15) via a resolvent kernel

$$\nu_1^+(x) = k_1(x) + \int_0^y k_1(t) R_{K_1^*} \, dt,$$

where $R_{K_1^*}$ is a resolvent-kernel of $K_1^*(x,t,\lambda,\mu)$.

Now, differentiating (9), (10) once with respect to y and using the solution of the first BVP for equation (1), we obtain

$$\nu_2^-(y) + \int_0^y E_1(y,y_1)\nu_2^-(y_1)\,dy_1 + \int_0^y E_2(y,y_1)\nu_3^-(y_1)\,dy_1 = E_3(y), \quad (16)$$

$$\nu_3^-(y) + \int_0^y E_4(y,y_1)\nu_2^-(y_1)\,dy_1 + \int_0^y E_5(y,y_1)\nu_3^-(y_1)\,dy_1 = E_6(y). \quad (17)$$

From (17), we find $\nu_3^-(y)$, substitute in (16) and get

$$\nu_2^-(y) + \int_0^y \nu_2^-(t)K_2^*(x,t,\lambda,\mu)\,dt = k_2(y). \quad (18)$$

Hence, we find all needed functions to restore the solution of the problem in the domain Ω_0 as a solution of the first BVP and in the domains Ω_i $(i=\overline{1,3})$ as a solution of the Cauchy problem. □

References

1. I. M. Gel'fand, *UMN* **XIV**, 3 (1959).
2. L. A. Zolina, *Zh. Vychisl. Mat. Mat. Fiz.* **11**, 991 (1968) (Russian).
3. L. A. Zolina, *Izv. Akad. Nauk UzSSR, Ser. Fiz.-Mat. Nauk* **11**, 14 (1967) (Russian).
4. T. D. Dzhuraev, A. Sopuev and M. Mamazhonov, *Boundary Value Problems for The Parabolic-Hyperbolic Type Equations*, Tashkent, FAN, 1986 (Russian).
5. U. Egamberdiev, *Uzbek Math. J.* **3**, 92 (1984).
6. V. A. Eleev and V. N. Lesev, *Vladikavkaz Math. J.* **3**, 4 (2001).
7. A. S. Abdullaev, *On Some Boundary-Value Problems for Mixed Parabolic-Hyperbolic Equations with Two Parallel Lines of Type Changing. Mixed Type Equations and Free Boundary Problems*, Tashkent, FAN, 1987 (Russian).
8. M. S. Salakhitdinov and A. K. Urinov, *Boundary Value Problems for Equations of Mixed Type with a Spectral Parameter*, Tashkent, FAN, 1997 (Russian).

EXACT EMBEDDING THEOREMS FOR TRACES AND EXTENSIONS

M. D. RAMAZANOV*

Russian Acad. Sci., Inst. Math., Ufa. Sci. Ctr.,
Ul Chernyshevskogo 112, Ufa 450000, Russia
E-mail: ramazanovmd@yandex.ru

General exact theorems about traces of functions from Banach spaces on smooth $(n-1)$-dimensional surfaces Γ are established.

1. The Problem

We consider Banach spaces B consisting of complex-valued functions of n real variables, $f : \mathbb{R}^n \to \mathbb{C}$. Let B be a closure of $C_0^\infty(\mathbb{R}^n)$ in some norm $\|*\|_B$ and be embedded in $C(\mathbb{R}^n)$. We suppose that the ball in B is strictly convex and the corresponding sphere is smooth. Our interest is to get general exact theorems about traces of functions from B on smooth $(n-1)$-dimensional surfaces Γ. Namely, it needs to determine the Banach space $B(\Gamma)$ which consists of traces of f belonging to B. The first answer is obviously

$$\|f'\|_{B(\Gamma)} = \inf_{g|_\Gamma = f'} \|g\|_B \,. \tag{1}$$

We want to give a constructive description for this infimum norm. Let Π be the operator giving the restriction of f on Γ. Then $\Pi : B \to B(\Gamma)$ is a bounded operator from B onto all $B(\Gamma)$. Our aim is to conserve this property. Namely, we are seeking to find constructively a space B' on Γ such that the operator $\Pi : B \to B'$ would be onto and bounded map. Let the surface Γ be given as

$$\Gamma = \{x \in \mathbb{R}^n | x = \varphi(y') \equiv \Phi(y',0),\ \Phi\colon y \to x \text{ is a diffeomorphism in } \mathbb{R}^n\}\,.$$

*This research has been done with the support of grant of RFFR 06-01-00597 and Program No. 14 of RAS.

Without loss of generality, we can consider only the case when $\Gamma = \{x|\, x \in \mathbb{R}^n,\ x_n = 0\}$.

In our case the infimum in formula (1) is a minimum and the corresponding minimizer is unique. This implies that there exists an extension operator $S : B(\Gamma) \to B$, $S(f') = \arg\min\limits_{\Pi g = f'} \|g\|_B$ such that $\Pi S(f') = f'$ for all $f' \in B(\Gamma)$. Our second task is to give a formula for the operator S.

2. Solutions of Problems

For a given function $f \in B$ we define an extreme functional l_f as follows: $l_f = \arg\sup\limits_{\|l\|_{B^*}=1} |(l, f)|$ (that is the element of the set for which the supremum is attained) and correspondingly an extreme f for $l \in B^*$ as $f_l = \arg\sup\limits_{\|f\|_B=1} |(l, f)|$. Since the spases B and $B(\Gamma)$ are reflexible, then by Hahn-Banach theorem it follows that the extreme functional l_f exists for any f and correspondingly there exists f_l for any l. Let N_B be the operator $f \to l_f$ restricted to the unit sphere, N_{B^*} is the operator $l \to f_l$ restricted to the unit B^*-sphere. Observe that $N_B \cdot N_{B^*} = I$ is an identical operator. That is, $N_{B^*} = N_B^{-1}$. According to our assumptions the operator N can be expressed as the derivative of the functional $f \to \|f\|_B$.

Theorem 2.1. $N_B(f) = D\,\|f\|_B$, *where D is the Gâteaux derivative.*

Proof. Let us show that the B^* norm of $D\,\|f\|_B$ is equal to one.

$$\|D\,\|f\|_B\|_{B^*} = \sup_{h\in B\backslash 0} [|(D\,\|f\|_B\,, h)|\,/\,\|h\|_B] \equiv \sup_{h\in B\backslash 0} \left| \frac{d}{dt} \frac{\|f+th\|_B}{\|h\|_B} \Big|_{t=0} \right|$$
$$= \sup_{h\in B\backslash 0} \Big| \lim_{t\in\mathbb{R}^1, t\to 0} [(\|f+th\|_B - \|f\|_B)/t\,\|h\|_B]| \le \sup_{h\in B\backslash 0} \overline{\lim_{t\to 0}} \frac{\|th\|_B}{|t|\,\|h\|_B} - 1.$$

If $h = f$, we have $|(D\,\|f\|_B\,, f)/\,\|f\|_B| = \left|\frac{d}{dt}[\|f+tf\|_B\,/\,\|f\|_B]\right| = \left|\lim\limits_{t\to 0}[\|f+tf\|_B - \|f\|_B]/t\,\|f\|_B\right| = \left|\lim\limits_{t\to 0} \frac{|1+t|-1}{t}\right| = 1.$ □

In this proof we have used that the functional $f : m \to |(m, f)|$ ($\|m\|_{B^*} = 1$) achieves its maximum at the point $m = D\,\|f\|$, *i.e.*, $D\,\|f\| = N(f)$.

Now we give a sufficient and necessary condition when pairs (B, B') of spaces are related by the theorem on traces and extensions, *i.e.*, $B' = B(\Gamma)$.

Theorem 2.2. *The B' norm is equal to the infimum norm of $B(\Gamma)$ if and only if*

$$\|l'\|_{B'^*} = \|\delta_n l'\|_{B^*}\,,\ \forall\, l' \in B'^* \tag{2}$$

for any $l' \in B'^*$, *where* $\delta_n l'$ *is a direct product of the distribution* $l'(x_1, \ldots, x_{n-1})$ *and the Dirac function* $\delta_n \equiv \delta(x_n)$.

Proof. Note that

$$\|\delta_n l'\|_{B^*} = \sup_{g \in B \setminus 0} [|(\delta_n l', g)| \,/\, \|g\|_B] = \sup_{g \in B \setminus 0} [|(l', \Pi g)| \,/\, \|g\|_B].$$

We have $\sup\limits_{g \in B \setminus 0} \ldots = \sup\limits_{f' \in B' \setminus 0} \sup\limits_{\Pi g = f'} \ldots$. Hence

$$\begin{aligned} \|\delta_n l'\|_{B^*} &= \sup_{f' \in B' \setminus 0} \sup_{\Pi g = f'} [|(l', f')| \,/\, \|g\|_B] \\ &= \sup_{f' \in B' \setminus 0} [|(l', f')| \,/ \inf_{g, \Pi g = f'} \|g\|_B] = \|l'\|_{B'^*}. \end{aligned}$$

□

Corollary 2.1. *The* B' *norm can be rewritten as*

$$\begin{aligned} \inf_{g|_\Gamma = f|_\Gamma} \|g\|_B \equiv \|f'\|_{B'} &= \sup_{l' \in B' \setminus 0} [|(l', f')| \,/\, \|l'\|_{B'^*}] \\ &= \sup_{l' \in B' \setminus 0} [|(l', f')| \,/\, \|\delta_n l'\|_{B^*}]. \end{aligned}$$

For any pair (B, B') of spaces that are connected by exact embedding theorems we can construct the operator S providing an extension of any function $f' \in B'$ on B.

Theorem 2.3. *Let the pair* (B, B') *satisfy the conditions of Theorem 2.2. Then*

$$S(f') = \|f'\|_{B'} N_{B^*}(\delta_n N_{B'}(f')) = \|f'\|_{B'} D \, \|\delta_n D \, \|f'\|_{B'}\|_{B^*}. \tag{3}$$

Proof. First we suppose that $\|f'\|_{B'} = 1$. By the definition of S, we have

$$S : f' \to S(f') \equiv f, \quad \Pi f = f', \quad \|f\|_B = \|f'\|_{B'} = 1.$$

Hence $f' : m' \to (m', f')$, $\|l'_{f'}\|_{B'^*} = 1 = |(l'_{f'}, f')| = (l'_{f'}, f')$ because $l'_{f'}$ is the extreme functional for f'.

Consider the chain of equalities $(\delta_n l'_{f'}, f) = (l'_{f'}, \Pi f) = (l'_{f'}, f') = 1$. But $\|\delta_n l'_{f'}\|_{B^*} = \|l'_{f'}\|_{B'^*} = 1$ and therefore $\|f\|_B = 1$. Thus f is an extreme function for $\delta_n \cdot l'$, *i.e.*, $f = N_{B^*}(\delta_n l')$. Hence $S(f') \equiv f = N_{B^*}(\delta_n l'_{f'}) = N_{B^*}(\delta_n N'(f'))$ for any f' belonging to the unit sphere in B'. Now we extend the operator S from the unit sphere for any $f' \in B'$, applying the homogeneity property, *i.e.*, $S(\lambda f') = \lambda S(f')$ for any λ. Thus if we take $\lambda = \|f'\|_{B'}$, then we obtain the proof. □

3. Examples

Now we show that there are several Banach spaces B that can lead to one B'-space on Γ. Let $B = H_p^\mu$ be the Hörmander space with norm $\|f\|_{H_p^\mu} = \left(\int_{\mathbb{R}^n} d\xi \left|\tilde{f}(\xi)\mu(\xi)\right|^p\right)^{1/p}$, where $\tilde{f}$ is the Fourier transform of f and $\mu(\xi) \neq 0$ for any ξ. The condition $\int \frac{d\xi}{|\mu(\xi)|^{p'}} < \infty$, where $\frac{1}{p} + \frac{1}{p'} = 1$, leads to an embedding of H_p^μ to C. If $B' = H_p^\nu(\mathbb{R}^{n-1})$, then the condition $\|l'\|_{B'^*} = \|\delta_n l'\|_{B^*}$ becomes the formula $\frac{1}{|\nu(\xi')|} = \left(\int \frac{d\xi n}{|\mu(\xi)|^{p'}}\right)^{1/p'}$. Let $x \in \mathbb{R}^2$, $\nu(\xi_1) = 1 + \mathrm{i}\xi_1$, $p' = 2$. Then $\left|\nu^2(\xi_1)\right| = (1+\xi_1^2)^2$ and we can take $\mu(\xi) = \nu(\xi_1)\,(1+\xi_2^2)^{\mathrm{m}} / \int \frac{dt}{(1+t^2)^{2m}}$ for any $m > 1/4$. Thus several B spaces give one trace space B'.

Let us consider another example:

$$\begin{aligned}
&B = l_1(\mathbb{R}^3) = \{x | x \in \mathbb{R}^3, \|x\|_B = |x_1| + |x_2| + |x_3|\},\\
&B^* = m(\mathbb{R}^3) = \{y | y \in \mathbb{R}^3, \|y\|_{B^*} = \max(|y_1|, |y_2|, |y_3|)\},\\
&\Gamma = \{x | x \in \mathbb{R}^3, x_3 = 0\},\\
&B' = l_1(\mathbb{R}^2) = \{x | x \in \mathbb{R}^2, \|x\|_{B'} = |x_1| + |x_2|\},\\
&B'^* = m(\mathbb{R}^2) = \{y | y \in \mathbb{R}^2, \|y\|_{B'^*} = \max(|y_1|, |y_2|)\}.
\end{aligned}$$

Now the unit ball is not strongly convex and the sphere is not smooth. But condition (2) is fulfilled: $\forall\, l' \in B'^*$ $\|l'\|_{B'^*} = \max(|l_1'|, |l_2'|) = \max(|l_1'|, |l_2'|, 0) = \|\delta_3 l'\|_{B^*}$, where $\delta_3 : \begin{pmatrix} x_1 \\ x_2 \\ x_0 \end{pmatrix} \to \begin{pmatrix} x_1 \\ x_2 \\ 0 \end{pmatrix}$. The operator S is

$$S : l' = (l_1, l_2) \to (l_1, l_2, 0) \tag{4}$$

and this element is unique. It is interesting what happens if the condition (4) is broken. Let us take $f' = \begin{pmatrix} 0 \\ 1 \end{pmatrix}$, $N_{B'}(f') = \begin{pmatrix} \theta \\ 1 \end{pmatrix}$ $\forall\, \theta \in [-1, 1]$,

$$\delta_3 N_{B'}(f') = (\theta, 1, 0),\ N_{B^*}(\delta_3 N_{B'}(f')) = \begin{cases} \begin{pmatrix} 0 \\ 1 \\ 1 \end{pmatrix}, & \text{if } |\theta| \neq 1, \\ \begin{pmatrix} \alpha \operatorname{sgn} \theta \\ 1 - \alpha \\ 0 \end{pmatrix} & \begin{array}{l} \text{with any} \\ \alpha \in [0,1], \\ \text{if } |\theta| = 1. \end{array} \end{cases}$$

We see that the formal application of algorithm (3) gives many answers but only one is carried out (see (4)).

Now let us give an example of an application of the embedding theorems in the anisotropic case. Let $B = W^{4,3}_{2,x_1,x_2}(\mathbb{R}^2)$ and $\Gamma =$

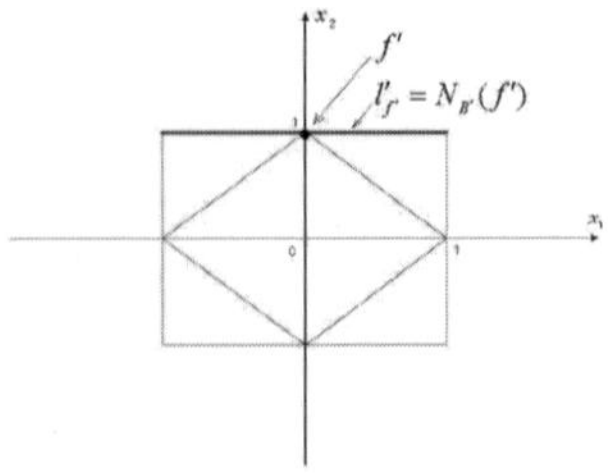

Fig. 1. Spaces B', B'^*

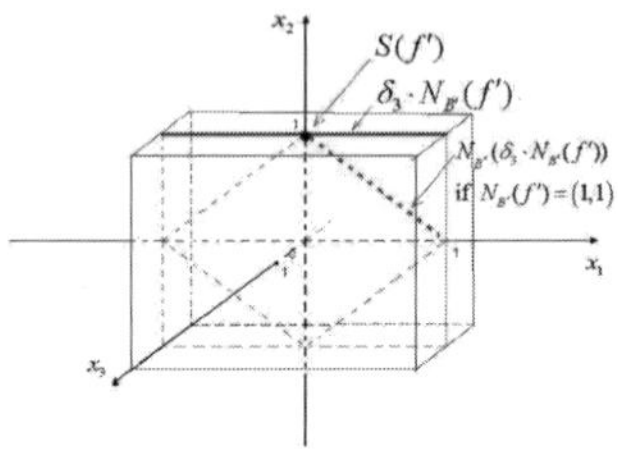

Fig. 2. Spaces B, B^*

$\{x|\ x \in \mathbb{R}^2,\ x_2 = x_1^3/3\}$. In this case we can calculate the norm which is equivalent to the $B'(\Gamma)$ norm. The equivalent norm can be written as

$$\|f'\|^2_{B'} = \int dt\, |f'(t)|^2 + \int dt \int d\xi_1 \frac{\tilde{f}'(\xi_1)\bar{f}'(t)\exp(2\pi i t\xi_1)}{\int d\xi_2\ 1/\left(((\xi_1 - \xi_2 t^2)^2 + 1)^4 + (1+\xi_1^2)^3\right)}.$$

4. Hypotheses

We suppose that the condition $\|l'\|_{B'^*} = \|\delta_n l'\|_{B^*}$ is sufficient for the condition that the map $S : f' \mapsto \|f'\|_{B'} N_{B^*}(\delta_n N_{B'}(f'))$ gives the correct extension from $B'(\Gamma)$ to $B(\mathbb{R}^n)$ in the general case. Furthermore it is possible even when the space B is not reflexive.

We saw that the operator Π acts from $B(\mathbb{R}^n)$ on the factor space of equivalent classes of functions coinciding on Γ and equipped with the infimum norm. We suppose that in the general case (when Γ is an arbitrary set in $\mathbb{R}^n$) the operator of multiplication by δ_n has to be replaced by the action of Π^* which is the conjugate operator to Π.

5. Remarks

First we shall tell about Theorem 1. We thank professor Lech Maligranda for advice to acquaint myself with works of S. Mazur [1,2], G. Köthe [3] and E. J. McShane [4]. In reality these articles and the article of R. C. James [5] and unpublished work of Dr. I. Solomesh (1980) contain very similar calculations and results, especially in works [4,5]. We have got our results independently and published our theorems in 2002 [6]. Our first research results about these problems were published in 1969, 1970 [7–10].

References

1. S. Mazur, *Studia Math.* **4**, 70 (1933).
2. S. Mazur, *Studia Math.* **4**, 128 (1933).
3. G. Köthe, *Math. Ann.* **277**, 489 (1987).
4. E. J. McShane, *Proc. Amer. Math. Soc.* **1**, 402 (1950).
5. R. C. James, *Trans. Amer. Math. Soc.* **61**, 265 (1947).
6. M. D. Ramazanov, *Doklady RAN* **382**, 744 (2002) (in Russian).
7. M. D. Ramazanov, *Doklady RAN* **185**, 1239 (1969) (in Russian).
8. M. D. Ramazanov, *Doklady RAN* **190**, 784 (1970) (in Russian).
9. M. D. Ramazanov, *Doklady RAN* **377**, 158 (2001) (in Russian).
10. M. D. Ramazanov, *Doklady RAN* **143**, 174 (2007) (in Russian),

Session 16

Integrable Systems

SESSION EDITORS

M. Gürses	Bilkent University, Ankara, Turkey
I. T. Habibullin	Bilkent University, Ankara, Turkey

THE CHARACTERISTIC CHERN-TYPE CLASSES AND INTEGRABILITY OF MULTIDIMENSIONAL DIFFERENTIAL SYSTEMS ON RIEMANNIAN MANIFOLDS

A. K. PRYKARPATSKY

Department of Nonlinear Mathematical Analysis,
Institute for Applied Problems of Mechanics and Mathematics of NAS,
Lviv, 79060, Ukraine
& Department of Applied Mathematics,
The AGH University of Science and Technology,
Krakow, 30059, Poland
E-mail: pryk.anat@ua.fm

N. N. BOGOLUBOV, Jr.

V. A. Steklov Mathematical Institute of RAS,
Moscow, 117946 Russian Federation
& Abdus Salam International Center for Theoretical Physics,
Trieste, Italy
E-mail: nikolai_bogolubov@hotmail.com

The differential-geometric properties of generalized de Rham-Hodge complexes naturally related with integrable multidimensional differential systems of Gromov type are analyzed. The geometric structure of Chern-type characteristic classes are studied, special differential invariants of Chern type are constructed. The integrability of multidimensional nonlinear differential systems on Riemannian manifolds is discussed.

Keywords: De Rham-Hodge complexes; Chern characteristic classes; Differential invaraiants; Multidimensioanl systems; Integrability.

1. Introduction: The differential-geometric problem setting

We start with a smooth finite-dimensional Riemannian manifold M and two complex vector bundles over it: the tangent bundle $T(M)$ and a bundle $E(M)$, endowed with some Hermitian scalar structure $<\cdot,\cdot>_E$ on fibers E. The related vector fields on M are denoted as $\mathcal{T}(M)$ and smooth sections of $E(M)$ as $\mathcal{E}(M)$. As usual [29,31,35,40], one can introduce on $\mathcal{E}(M)$ a

Cartan connection Γ by means of a connection mapping

$$d_{\mathrm{A}} : \mathcal{E}(M) \to \mathcal{T}^*(M) \otimes \mathcal{E}(M), \tag{1}$$

which satisfies the following property:

$$d_{\mathrm{A}}(f\alpha + \beta) := df \otimes \alpha + f d_{\mathrm{A}}\alpha + d_{\mathrm{A}}\beta \tag{2}$$

for any smooth function $f \in D(M)$ and $\alpha, \beta \in \mathcal{E}(M)$. Let $\Lambda(M) := \oplus_{p=0}^{\dim M} \Lambda^p(M)$ denote the usual [1,2,29,35,40] Grassmann algebra of differential forms on M. If we define the associated linear bundles $\Lambda^p(M,E) := \Lambda^p(M) \otimes \mathcal{E}(M)$ for $p = \overline{0,m}$, the connection mapping (1) can be naturally extended on $\Lambda^p(M,E)$ as

$$d_{\mathrm{A}} : \Lambda^p(M,E) \to \Lambda^{p+1}(M,E), \tag{3}$$

satisfying the related Leibniz rule

$$d_{\mathrm{A}}(f^{(p)} \wedge \alpha^{(q)}) := df^{(p)} \wedge \alpha^{(q)} + (-1)^p f^{(p)} \wedge d_{\mathrm{A}}\alpha^{(q)} \tag{4}$$

for any $f^{(p)} \in \Lambda^p(M)$ and $\alpha^{(q)} \in \Lambda^q(M,E)$, $q, p = \overline{0,m}$.

The connection operation (3) possesses a very interesting and important property. Its composition $d_{\mathrm{A}}^2 := d_{\mathrm{A}} d_{\mathrm{A}}$ is a linear mapping over smooth functions ring $D(M)$:

$$d_{\mathrm{A}}^2(f\alpha^{(p)} + \beta^{(p)}) = f d_{\mathrm{A}}^2 \alpha^{(p)} + d_{\mathrm{A}}^2 \beta^{(p)}, \tag{5}$$

where $f \in D(M)$ and $\alpha^{(p)}, \beta^{(p)} \in \Lambda^p(M,E)$, $p = \overline{0,m}$, are arbitrary. The resulting linear tensor mapping $\Omega^{(2)} := d_{\mathrm{A}}^2 : \mathcal{E}(M) \to \Lambda^2(M,E)$ is called the curvature tensor and is of great importance for geometrical analysis of integrable multi-dimensional differential systems of M. Gromov type [3], generated by means of some Cartan integrable [6,27,38–40] ideals $I(\alpha) \subset \Lambda(M, End\ E) := \Lambda(M) \otimes End\ \mathcal{E}(M)$ on Riemannian manifolds M. Moreover, one can construct the smooth integral submanifold imbedding mapping $i_\alpha : M_\alpha \to M$ for the ideal $I(\alpha) \subset \Lambda(M, End\ E)$, satisfying the following determining condition: The curvature 2-form $\Omega^{(2)} \in \Lambda^2(M,E)$ reduced on M_α vanishes, that is, $i_\alpha^* \Omega^{(2)} = 0$. This implies also that the related reduced co-chain

$$E \to \Lambda^0(M_\alpha, E) \overset{d_\alpha}{\to} \Lambda^1(M_\alpha, E) \overset{d_\alpha}{\to} \ldots \Lambda^{m_\alpha}(M_\alpha, E) \overset{d_\alpha}{\to} 0 \tag{6}$$

is a de Rham complex, that is $d_\alpha^2 = 0$, where, by definition, $d_\alpha := i_\alpha^* d_{\mathrm{A}}$ and $m_\alpha := \dim M_\alpha$. Since the submanifold $M_\alpha \subset M$ also possesses the induced Riemannian structure $g_\alpha : T(M_\alpha) \times T(M_\alpha) \to \mathbb{R}$, we can construct from (6) a suitably generalized de Rham-Hodge complex of Hilbert spaces $H_\Lambda^p(M_\alpha)$, $p = \overline{0, m_\alpha}$, whose properties, as it was shown before in [5,23,40,41], make it possible to describe the so-called Delsarte-Lions transmutation operators of Volterra type, serving for constructing integrable multi-dimensional differential systems on Riemannian manifolds and finding their of special type exact solutions.

On the other hand, one can consider the generalized chain of modules over M

$$E \to \Lambda^0(M,E) \overset{d_{\mathrm{A}}}{\to} \Lambda^1(M,E) \overset{d_{\mathrm{A}}}{\to} \dots \overset{d_{\mathrm{A}}}{\to} \Lambda^m(M,E) \to 0, \tag{7}$$

which is not, evidently, a de Rham-Hodge complex, but determines such very important [28,31,32,35] geometric objects as the Chern characteristic classes and characters. Their properties and related geometric aspects of the integrability problem of multi-dimensional differential systems of Gromov type [3] on Riemannian manifolds will be treated below.

2. The differential invariants

The connection mapping (3) on M can be equivalently rewritten locally on an open neighborhood $U \subset M$ by means of the expression

$$d_{\mathrm{A}}|_U = d + A^{(1)}, \tag{8}$$

where $A^{(1)} \in \Lambda(U, End\ E)$ is some suitably determined $End\ \mathcal{E}(U)$-valued differential 1-form on $U \subset M$. In local coordinates of a point $u \in U$ we can write

$$A^{(1)} := \sum_{i=1}^{m} A_i(u) du^i, \tag{9}$$

where $A_i(u) \in End\ \mathcal{E}(U)$, $i = \overline{1,m}$. Making use of the representation (8), one can easily obtain the following local expression for the curvature 2-form $\Omega^{(2)}|_U \in \Lambda^2(U) \otimes End\ \mathcal{E}(U)$:

$$\Omega^{(2)}|_U = dA^{(1)} + A^{(1)} \wedge A^{(1)}. \tag{10}$$

The expression (10) is very convenient for local determining the related with the complex (7) local cohomology characteristic Chern type classes

$$ch_j(\mathrm{A})|_U := tr(\frac{i}{2\pi}\Omega^{(2)}|_U)^j \in \Lambda^{2j}(U), \tag{11}$$

where $j \in \mathbb{Z}_+$. The expression (11), owing to the linear vector bundle $E(M)$ properties, can be invariantly extended upon the whole manifold M as correctly defined differential forms on M, thereby determining the characteristic Chern type classes

$$ch_j(\mathrm{A}) = tr(\frac{i}{2\pi}\Omega^{(2)})^j \in \Lambda^{2j}(M) \tag{12}$$

for $j \in \mathbb{Z}_+$, defined already on M. It is a standard [30,31,35] result that classes (12) are closed and do not depend on the choice of a connection mapping (3).

As a consequence of these properties one can define for every linear Hermitian fiber bundle $E(M)$ over M the set of corresponding characteristic Chern type classes

$$ch_j(E, M) := [ch_j(\mathrm{A})] \tag{13}$$

for $j \in \mathbb{Z}_+$ by means of which the Chern character $ch(E, M)$ of this linear fiber bundle $E(M)$ is determined as

$$ch(E, M) := \oplus_{j \in \mathbb{Z}_+} (j!)^{-1} ch_j(E, M) = [tr \exp(\frac{i}{2\pi} \Omega^{(2)})]. \tag{14}$$

The Chern character, as is well known [31,32,35], finds a great deal of applications to modern differential topology and mathematical physics problems.

Concerning applications to strongly integrable multi-dimensional differential systems on Riemannian manifolds, we will consider the Cartan geometric picture, developed before in [6,27,38–40]. Within this picture a studied nonlinear multi-dimensional differential system $\hat{\alpha}$ is represented in the form of a Cartan integrable ideal $I(\alpha) \subset \Lambda(M, End\ E)$ with coefficients from $End\ \mathcal{E}(M)$, where $\mathcal{E}(M)$ is a specially chosen Hermitian vector fiber bundle over some suitably chosen finite-dimensional Riemannian manifold M. The corresponding integral submanifold $M_\alpha \subset M$ of the ideal $I(\alpha)$, in general, is equivalent to the set of independent variables of our multi-dimensional integrable nonlinear differential system.

Note also here, that we call our multi-dimensional differential system strongly integrable, if it allows a suitable connection Γ_λ parametrically dependent on $\lambda \in \mathbb{R}$, whose curvature 2-form $\Omega_\lambda^{(2)} \in \Lambda^2(M, End\ E)$ is vanishing upon the integral submanifold $M_\alpha \subset M$ of the ideal $I(\alpha) \subset \Lambda(M, End\ E)$. The latter condition is, evidently, equivalent to the inclusion

$$\Omega_\lambda^{(2)} \in I(\alpha) \tag{15}$$

for all allowed values of $\lambda \in \mathbb{R}$. If the connection Γ does not depend nontrivially on parameter $\lambda \in \mathbb{R}$, a multi-dimensional differential system is called integrable.

On the other hand, the condition (15) serves [39,40] for finding the corresponding connection mapping (3), if it a priori assumes to exist. The resulting search algorithm details depend 38–40] strongly on the related so-called holonomy group properties of the connection Γ_λ, $\lambda \in \mathbb{R}$, on the principal fiber bundle $P(M, G)$ naturally associated with the bundle $E(M)$, where G is a so-called structure Lie group of the connection Γ_λ, $\lambda \in \mathbb{R}$. Concerning the mentioned algorithm details one can consult further [38–40].

The condition (15) appears to be very important as we can obtain under the conditions $H^{2j}(M,\mathbb{R}) = 0$, $j \in \mathbb{Z}_+$, right away that

$$ch_j(\mathrm{A}) := d\ \chi_j(\mathrm{A}) \tag{16}$$

for some suitably determined global and real differential $(2j - 1)$-form $\chi_j(A) \in \Lambda^{2j-1}(M)$, $j \in \mathbb{Z}_+$, on the manifold M. Moreover, since, evidently, all degrees $(\frac{i}{2\pi}\Omega^{(2)})^j \in I(\alpha)$ for $j \in \mathbb{Z}_+$, from the condition $i_\alpha^* I(\alpha) = 0$, where $i_\alpha : M_\alpha \to M$ is the integral submanifold imbedding mapping, one easily gets that $i_\alpha^* ch_j(A) = 0$, or equivalently

$$d\chi_j(\mathrm{A}) = 0 \tag{17}$$

for all $j \in \mathbb{Z}_+$, giving rise to new differential Chern type invariants on M_α. One can formulate the obtained result as the following theorem.

Theorem 2.1. *If an integrable multi-dimensional nonlinear differential system $\hat{\alpha}$ is equivalent to the Cartan integrable ideal $I(\alpha) \subset \Lambda(M, End\ E)$ on a Riemannian manifold M, satisfying the cohomology conditions $H^{2j}(M,\mathbb{R}) = 0$, $j \in \mathbb{Z}_+$, then it possesses a set of differential Chern type invariants (17) on the suitable integral submanifold $M_\alpha \subset M$ of the ideal $I(\alpha)$. If nontrivial, these differential invariants describe, in particular, a related moduli space of the linear fiber bundle $E(M)$.*

It is useful to mention here that most differential invariants (17) reduce at higher indices $j \in \mathbb{Z}_+$ to identical zero. Indeed, all differential invariants $\chi_j(A)$ for $j \geq [\dim M_\alpha/2]+1$ are identically zero. In particular, for the case of multi-dimensional integrable nonlinear differential systems, for which $\dim M_\alpha = 2$ or 3 one easily gets, that only one differential invariant can exist if any. Moreover, if for such differential systems the corresponding structure Lie groups are special linear ones, for which the Lie algebras are traceless, one easily gets that, on the whole, no invariant exists on M_α.

This result is instructive enough for the mathematical theory of geometrically integrable multi-dimensional nonlinear differential systems possessing suitable connection mappings (3) on $\Lambda(M, E)$ under additional cohomology conditions $H^{2j}(M,\mathbb{R}) = 0$, $j \in \mathbb{Z}_+$. Namely, for these connection mappings to exist on $\Lambda(M, E)$ in the higher-dimensional case $\dim M_\alpha \geq 4$, the nontrivial differential invariants can be proved to exist. The latter gives rise, in particular, to some topological obstacles to be satisfied. Indeed, a nontrivial differential invariant entails right away nontrivial cohomology constraints $H^s(M,\mathbb{R}) \neq 0$ for some $s \in \mathbb{Z}_+$, thereby contradicting the above zero-cohomology conditions at these values $s \in \mathbb{Z}_+$. Otherwise, if a priori $H^{2s}(M,\mathbb{R}) \neq 0$ for some $s \in \mathbb{Z}_+$, the related characteristic Chern type classes $ch_s(E,M)$ for these $s \in \mathbb{Z}_+$ are, in general, strongly nontrivial, thereby defining differential $(2s-1)$-forms $\chi_s(A) \in \Lambda^{2s-1}_{loc}(M)$ only locally, owing to the classical Poincaré lemma. These local differential forms, if reduced upon the integral submanifold $M_\alpha \subset M$, give rise to a set of differential multivalued quasi-invariants $\chi_s(A) \in \Lambda^{2s-1}_{loc}(M_\alpha)$, where $d\,\chi_s(A) = 0$ and whose existence can mean, in particular, that our nonlinear differential system is in some sense ill-posed. We remark here that related differential system structures can be studied also by means of the corresponding de Rham-Hodge complexes (6). These aspects will also be briefly discussed below.

3. The structure of related Delsarte-Lions transmutation operators

Consider the cohomology complex (6) and define on its spaces $\Lambda^p(M_\alpha,E)$, $p \in \mathbb{Z}_+$, where $M_\alpha \subset M$ is the compact integral submanifold of the Riemannian manifold M, the standard Hodge star *-operation

$$^* : \Lambda^p(M_\alpha,E) \to \Lambda^{m_\alpha - p}(M_\alpha,E), \tag{18}$$

where $m_\alpha := \dim M_\alpha$ and for any $\beta \in \Lambda^p(M_\alpha,E)$ the form $^*\beta \in \Lambda^{m_\alpha-p}(M_\alpha,E)$ is such that the following [1,7,35,40] conditions hold:

i) $< \gamma, {}^*\beta >_{(m_\alpha - p)} := << \gamma, {}^*\beta >_E >_{m_\alpha - p} = << \gamma\wedge, \beta >_E, d\mu_{g_\alpha} >_{m_\alpha}$ for any $\gamma \in \Lambda^{m_\alpha-p}(M_\alpha,E)$, where $d\mu_{g_\alpha}$ is the invariant measure on M_α, induced by the dual imbedding mapping $i^*_\alpha : \Lambda(M_\alpha) \to \Lambda(M_\alpha)$ from the corresponding invariant measure $d\mu_g$ on the Riemannian manifold M, endowed with the positive definite Riemannian metrics $g : T(M)\times T(M) \to \mathbb{R}$; the scalar product

$$\begin{aligned} &< \beta_1^{(1)} \wedge \beta_2^{(1)} \wedge ... \wedge \beta_k^{(1)}, \gamma_1^{(1)} \wedge \gamma_2^{(1)} \wedge ... \wedge \gamma_k^{(1)} >_k \\ &:= \det\{< \beta_i^{(1)}, \gamma_j^{(1)} >_1 : i,j = \overline{1,k}\}, \end{aligned} \tag{19}$$

where $< \beta_i^{(1)}, \gamma_j^{(1)} >_1 := < \hat{g}_\alpha^{-1}\beta_i^{(1)}, \hat{g}_\alpha^{-1}\gamma_j^{(1)} >_{g_\alpha}$ for any $\beta_i^{(1)}, \gamma_j^{(1)} \in \Lambda^1(M_\alpha)$, $i,j = \overline{1,k}$, and $\hat{g}_\alpha : T(M_\alpha) \to T^*(M_\alpha)$ is the canonical isomorphism, generated by the corresponding metrics $< \cdot,\cdot >_{g_\alpha}$ on $T(M_\alpha)$;

ii) $(m_\alpha - p)$-dimensional volume $|{}^*\beta|$ of the form $^*\beta \in \Lambda^{m_\alpha-p}(M_\alpha,E)$ equals p -dimensional volume $|\beta|$ of a form $\beta \in \Lambda^p(M_\alpha,E)$;

iii) m_α-dimensional measure $< \beta\wedge, {}^*\beta >_E \geq 0$ at a fixed orientation on M_α.

Owing to the conditions i)-iii) one can endow the spaces $\Lambda^p(M_\alpha, E)$, $p \in \mathbb{Z}_+$, with the natural scalar product

$$(\beta, \gamma) := \int_{M_\alpha} < \beta\wedge, *\gamma >_E = \int_{M_\alpha} < \beta, \gamma >_{(p)} d\mu_{g_\alpha} \tag{20}$$

for any $\beta, \gamma \in \Lambda^p(M_\alpha, E)$. Subject to the scalar product (20) we can naturally construct the corresponding Hilbert space

$$\mathcal{H}_\Lambda(M_\alpha) := \overset{m_\alpha}{\underset{k=0}{\oplus}} \mathcal{H}^k_\Lambda(M_\alpha)$$

well suitable for our further consideration. Notice also here that the Hodge star *-operation satisfies the following easily checkable property: For any $\beta, \gamma \in H^k_\Lambda(M_\alpha)$, $k = \overline{0, m_\alpha}$,

$$< \beta, \gamma >_{(k)} = < {}^*\beta, {}^*\gamma >_{(m_\alpha - k)}, \qquad (\beta, \gamma) = ({}^*\beta, {}^*\gamma), \tag{21}$$

that is, the Hodge operation $^* : H_\Lambda(M_\alpha) \to H_\Lambda(M_\alpha)$ is an isometry and its standard adjoint $(\cdot)'$ with respect to the scalar product (20) operation satisfies the condition $({}^*)' = ({}^*)^{-1}$.

Denote by d'_α the formally adjoint expression to the Cartan type connection mapping $d_\alpha : H_\Lambda(M_\alpha) \to H_\Lambda(M_\alpha)$ in the Hilbert space $H_\Lambda(M_\alpha)$. Here $d_\alpha := i^*_\alpha d_A$, where $d_A{:}H_\Lambda(M) \to H_\Lambda(M)$ is a suitable Cartan connection mapping and $i_\alpha : M_\alpha \to M$ is the corresponding integral submanifold imbedding mapping, associated with a given multi-dimensional nonlinear integrable differential system on the Riemannian manifold M. Making use of these operations d'_α and d_α in $H_\Lambda(M_\alpha)$, one can naturally define [7,34,35] a generalized Laplace-Hodge operator $\Delta_\alpha : H_\Lambda(M_\alpha) \to H_\Lambda(M_\alpha)$ as

$$\Delta_\alpha := d'_\alpha d_\alpha + d_\alpha d'_\alpha. \tag{22}$$

Take a form $\beta \in H_\Lambda(M_\alpha)$ satisfying the equality

$$\Delta_\alpha \beta = 0. \tag{23}$$

Such a form is called harmonic. One can also verify that a harmonic form $\beta \in H_\Lambda(M_\alpha)$ satisfies simultaneously the following two adjoint conditions:

$$d'_\alpha \beta = 0, \quad d_\alpha \beta = 0, \tag{24}$$

easily stemming from (22) and (24).

It is not hard to check that the differential operation

$$d^*_\alpha := {}^*d'_\alpha({}^*)^{-1} \tag{25}$$

in $H_\Lambda(M_\alpha)$ defines also a usual [26–28,30,33] Cartan type connection mapping in $H_\Lambda(M_\alpha)$. The corresponding dual to (6) co-chain

$$E \to \Lambda^0(M_\alpha, E) \overset{d_\alpha^*}{\to} \Lambda^1(M_\alpha, E) \overset{d_\alpha^*}{\to} \dots \overset{d_\alpha^*}{\to} \Lambda^m(M_\alpha, E) \overset{d_\alpha^*}{\to} 0 \tag{26}$$

is, evidently, a de Rham complex too, as the property $d_\alpha^* d_\alpha^* = 0$ holds owing to the definition (25).

Denote further by $H^k_{\Lambda(\alpha)}(M_\alpha)$, $k = \overline{0, m_\alpha}$, the cohomology groups of d_α-closed and by $H^k_{\Lambda(\alpha^*)}(M_\alpha)$, $k = \overline{0, m_\alpha}$, the cohomology of d_α^*-closed differential forms, respectively, and by $H^k_{\Lambda(\alpha^*\alpha)}(M_\alpha)$, $k = \overline{0, m_\alpha}$, the abelian groups of harmonic differential forms from the sub-spaces $H^k_\Lambda(M_\alpha, E)$, $k = \overline{0, m_\alpha}$. Before formulating the next results, define the standard sub-spaces for harmonic forms $H^k_{\Lambda(\alpha^*\alpha)}(M_\alpha)$ and cohomology groups $H^k_{\Lambda(\alpha)}(M_\alpha)$, $H^k_{\Lambda(\alpha^*)}(M_\alpha)$ for $k = \overline{0, m_\alpha}$. Assume also that the Laplace-Hodge operator (22) is elliptic in $H^0_\Lambda(M_\alpha)$. Now by reasonings similar to those in [7,30,34,35,41] one can formulate the following a slightly generalized de Rham-Hodge theorem.

Theorem 3.1. *The groups of harmonic forms $H^k_{\Lambda(\alpha^*\alpha)}(M_\alpha)$, $k = \overline{0, m_\alpha}$, are, respectively, isomorphic to the cohomology groups $(H^k(M_\alpha, \mathbb{R}))^{|\Sigma|}$, $k = \overline{0, m_\alpha}$, where $H^k(M_\alpha, \mathbb{R})$ is the $k-$th cohomology group of the manifold M_α with real coefficients, $\Sigma \subset \mathbb{R}^p$, $p \in \mathbb{Z}_+$, $|\Sigma| := card\ \Sigma$, is a set of suitable "spectral" parameters labeling the linear space of independent d_α^* -closed 0-forms from $H^0_{\Lambda(\alpha)}(M_\alpha)$ and, moreover, the direct sum decompositions*

$$\begin{aligned} &\mathcal{H}^k_{\Lambda(\alpha^*\alpha)}(M_\alpha) \oplus \Delta\mathcal{H}^k_\Lambda(M_\alpha) = \mathcal{H}^k_\Lambda(M_\alpha) \\ &= \mathcal{H}^k_{\Lambda(\alpha^*\alpha)}(M_\alpha) \oplus d_\alpha \mathcal{H}^{k-1}_\Lambda(M_\alpha) \oplus d'_\alpha \mathcal{H}^{k+1}_\Lambda(M_\alpha) \end{aligned} \tag{27}$$

hold for any $k = \overline{0, m_\alpha}$.

Another variant of the statement above was formulated in [10,11,23,41] and reads as the following generalized de Rham-Hodge theorem.

Theorem 3.2. *The generalized cohomology groups $H^k_{\Lambda(\alpha)}(M_\alpha)$, $k = \overline{0, m_\alpha}$, are isomorphic, respectively, to the cohomology groups $(H^k(M_\alpha, \mathbb{R}))^{|\Sigma|}$, $k = \overline{0, m_\alpha}$.*

The next lemma [10,11,22,24,41] is fundamental for the proof of the above isomorphism Theorem 3.2.

Lemma 3.1. *There exist a set of differential $(k+1)$-forms $\mathcal{Z}^{(k+1)}[\varphi^{(0)}(\lambda), d_\alpha \psi^{(k)}] \in \Lambda^{k+1}(M_\alpha, \mathbb{R})$, $k = \overline{0, m_\alpha - 1}$, and a set of k-forms $\mathcal{Z}^{(k)}[\varphi^{(0)}(\lambda), \psi^{(k)}] \in \Lambda^k(M_\alpha, \mathbb{R})$, $k = \overline{0, m_\alpha - 1}$, parametrized by a* *set* $\Sigma \ni \lambda$ *and semilinear in* $(\varphi^{(0)}(\lambda), \psi^{(k)}) \in H_0^* \times H_\Lambda^k(M_\alpha)$, *such that*

$$\mathcal{Z}^{(k+1)}[\varphi^{(0)}(\lambda), d_\alpha \psi^{(k)}] = d\mathcal{Z}^{(k)}[\varphi^{(0)}(\lambda), \psi^{(k)}] \tag{28}$$

for all $k = \overline{0, m_\alpha - 1}$ *and* $\lambda \in \Sigma$.

Based now on Lemma 3.1, one can construct the cohomology group isomorphism claimed in Theorem 3.2 formulated above. Namely, following [10,11,13,41], let us take some singular simplicial [29,30,33–35] complex $K(M_\alpha)$ of the manifold M_α and introduce linear mappings $B_\lambda^{(k)} : H_\Lambda^k(M_\alpha) \to C_k(M_\alpha, \mathbb{R}))$, $k = \overline{0, m_\alpha - 1}$, $\lambda \in \Sigma$, where $C_k(M_\alpha, \mathbb{R})$, $k = \overline{0, m_\alpha - 1}$, are as before free abelian groups over the field $\mathbb{R}$ generated, respectively, by all k-chains of simplexes $S^{(k)} \in C_k(M_\alpha, \mathbb{R})$, $k = \overline{0, m_\alpha - 1}$, from the singular simplicial complex $K(M_\alpha)$, as

$$B_\lambda^{(k)}(\psi^{(k)}) := \sum_{S^{(k)} \in C_k(M_\alpha, \mathbb{R}))} S^{(k)} \int_{S^{(k)}} \mathcal{Z}^{(k)}[\varphi^{(0)}(\lambda), \psi^{(k)}] \tag{29}$$

with $\psi^{(k)} \in H_\Lambda^k(M_\alpha)$, $k = \overline{0, m_\alpha - 1}$. The following theorem [10–12,22,41] based on mappings (29) holds.

Theorem 3.3. *The set of operations (29) parametrized by* $\lambda \in \Sigma$ *realizes the isomorphism of cohomology groups formulated in Theorem 3.2.*

Assume now that the Riemannian compactified manifold $M = M_\alpha \times \mathbb{R}^s$, $\dim M = s + \dim M_\alpha \in \mathbb{Z}_+$, and $E := \mathbb{R}^N$, where $M_\alpha \simeq \overset{m_\alpha}{\underset{j=1}{\times}} M_{\alpha,j}$, $M_{\alpha,j} := [0, T_j) \subset \mathbb{R}_+$, $j = \overline{1, m_\alpha}$, and put

$$d_\alpha = \sum_{j=1}^{m_\alpha} dt_j \wedge A_j(t|\partial), \quad A_j(t|\partial) := \partial/\partial t_j + A_j(t), \tag{30}$$

with $A_j(t)$, $j = \overline{1, m_\alpha}$, being matrices parametrically dependent on $t \in M_\alpha^{m_\alpha}$. It is also assumed that operators $A_j(t|\partial) : H_\Lambda^0(M_\alpha) \to H_\Lambda^0(M_\alpha)$, $j = \overline{1, m_\alpha}$, are commuting with each other.

Take now such a fixed pair $(\varphi^{(0)}(\lambda), \psi^{(0)}(\mu)dt) \in H_0^* \times H_{\Lambda(\alpha)}^{m_\alpha}(M_\alpha)$, parametrized by elements $(\lambda, \mu) \in \Sigma \times \Sigma$, for which owing to both Theorem 3.3 and the Stokes theorem [26,27,30,35] the equality

$$B_\lambda^{(m_\alpha)}(\psi^{(0)}(\mu)dt) = S_{(t)}^{(m_\alpha)} \int_{\partial S_{(t)}^{(m_\alpha)}} \Omega^{(m_\alpha - 1)}[\varphi^{(0)}(\lambda), \psi^{(0)}(\mu)d\tau] \tag{31}$$

holds, where $S_{(t)}^{(m_\alpha)} \in C_{m_\alpha}(M_\alpha, \mathbb{R})$ is some fixed element parametrized by an arbitrarily chosen point $t \in S_{(t)}^{(s)} \subset M_\alpha$. Consider next the integral expressions

$$\Omega_{(t)}(\lambda,\mu):=\int_{\partial S^{(m_\alpha)}_{(t)}}\Omega^{(m_\alpha-1))}[\varphi^{(0)}(\lambda),\psi^{(0)}(\mu)d\tau],$$

$$\Omega_{(t_0)}(\lambda,\mu):=\int_{\partial S^{(m_\alpha)}_{(t_0)}}\Omega^{(m_\alpha-1)}[\varphi^{(0)}(\lambda),\psi^{(0)}(\mu)d\tau], \tag{32}$$

with a point $t_0\in S^{(m_\alpha)}_{(t_0)}\subset M_\alpha$, being taken fixed, $\lambda,\mu\in\Sigma$, and interpret them as the corresponding kernels [15,16] of the integral invertible operators of Hilbert-Schmidt type $\Omega_{(t)},\Omega_{(t_0)}:L_2^{(\rho)}(\Sigma;\mathbb{R})\to L_2^{(\rho)}(\Sigma;\mathbb{R})$, where ρ is some finite Borel measure on the parameter set Σ. In particular, we assumed above that the boundaries $\partial S^{(m_\alpha)}_{(t)}:=\sigma^{(m_\alpha-1)}_{(t)}$ and $\partial S^{(m_\alpha)}_{(t_0)}:=\sigma^{(m_\alpha-1)}_{(t_0)}$ are taken homological to each other as $t\to t_0\in M_\alpha$. Define now the expressions

$$\Omega_\pm:\psi^{(0)}(\eta)\to\tilde{\psi}^{(0)}(\eta) \tag{33}$$

for any $\psi^{(0)}(\eta)\in H^0_\Lambda(M_\alpha)$, $\eta\in\Sigma$, and some $\tilde{\psi}^{(0)}(\eta)\in H^0_\Lambda(M_\alpha)$, where, by definition, for any $\eta\in\Sigma$,

$$\begin{aligned}\tilde{\psi}^{(0)}(\eta)&:=\psi^{(0)}(\eta)\cdot\Omega^{-1}_{(t)}\Omega_{(t_0)}\\&=\int_\Sigma d\rho(\mu)\int_\Sigma d\rho(\xi)\psi^{(0)}(\mu)\Omega^{-1}_{(t)}(\mu,\xi)\Omega_{(t_0)}(\xi,\eta),\end{aligned} \tag{34}$$

motivated by the expression (31). Suppose now that the elements (34) are ones being related to some other Delsarte-Lions transformed cohomology group $H^0_{\Lambda(\tilde{\alpha})}(M_\alpha)$, that is, the condition

$$d_{\tilde{\alpha}}\tilde{\psi}^{(0)}(\eta)=0 \tag{35}$$

holds for $\tilde{\psi}^{(0)}(\eta)\in H^0_{\Lambda(\alpha)}(M_\alpha)$, $\eta\in\Sigma$, and some new connection mapping

$$d_{\tilde{\alpha}}:=\sum_{j=1}^{m_\alpha}dt_j\wedge\tilde{\mathrm{A}}_j(t|\partial), \tag{36}$$

in $H_\Lambda(M_\alpha)$, where $\tilde{A}_j(t;\partial):=\partial/\partial t_j+\tilde{A}_j(t)$, $j=\overline{1,m_\alpha}$, are parametrically dependent on $t\in M_\alpha$.

Put now

$$\tilde{\mathrm{A}}_j:=\Omega_\pm\mathrm{A}_j\Omega_\pm^{-1} \tag{37}$$

for each $j=\overline{1,m_\alpha}$, where $\Omega_\pm:H^0_\Lambda(M_\alpha)\to H^0_\Lambda(M_\alpha)$ are the corresponding Delsarte-Lions transmutation operators [8,9] corresponding to some elements $S_\pm(\sigma^{(m_\alpha-1)}_{(t)},\sigma^{(m_\alpha-1)}_{(t_0)})\in C_{m_\alpha}(M_\alpha,\mathbb{R})$ naturally generated by homological to each other boundaries $\partial S^{(m_\alpha)}_{(t)}=\sigma^{(m_\alpha-1)}_{(t)}$ and $\partial S^{(m_\alpha)}_{(t_0)}=\sigma^{(m_\alpha-1)}_{(t_0)}$.

Since all the differential expressions $A_j : H^0_\Lambda(M_\alpha) \to H^0_\Lambda(M_\alpha)$, $j = \overline{1, m_\alpha}$, were taken commuting, the same property also holds for the transformed operators (37), that is $[\tilde{A}_j, \tilde{A}_k] = 0$, $k, j = \overline{0, m_\alpha}$. The latter is, evidently, equivalent owing to (37), to the general expression

$$d_{\tilde{\alpha}} = \Omega_\pm d_\alpha \Omega_\pm^{-1}. \tag{38}$$

For the conditions (38) and (35) to be satisfied, let us consider the expressions

$$\tilde{B}_\lambda^{(m_\alpha)}(\tilde{\psi}^{(0)}(\eta)dt) = S_{(t)}^{(m_\alpha)} \tilde{\Omega}_{(t)}(\lambda, \eta), \tag{39}$$

based on (31) and related with the corresponding external differentiation (38), where $S_{(t)}^{(m_\alpha)} \in C_{m_\alpha}(M, \mathbb{R})$ and $(\lambda, \eta) \in \Sigma \times \Sigma$. Assume further that there are also defined mappings

$$\Omega_\pm^\circledast : \varphi^{(0)}(\lambda) \to \tilde{\varphi}^{(0)}(\lambda) \tag{40}$$

with $\Omega_\pm^\circledast : H^0_{\Lambda(\alpha)}(M_\alpha) \to H^0_{\Lambda(\alpha)}(M_\alpha)$, being some operators associated (but not necessarily adjoint!) with the corresponding Delsarte-Lions transmutation operators $\Omega_\pm : H^0_\Lambda(M_\alpha) \to H^0_\Lambda(M_\alpha)$ and satisfying the standard relationships $\tilde{A}_j^* := \Omega_\pm^\circledast A_j^* \Omega_\pm^{\circledast,-1}$, $j = \overline{1, m_\alpha}$. The proper Delsarte-Lions type operators $\Omega_\pm : H^0_\Lambda(M_\alpha) \to H^0_\Lambda(M_\alpha)$ are related with two different realizations of the action (34) under the necessary conditions

$$d_{\tilde{\alpha}} \tilde{\psi}^{(0)}(\eta) = 0, \quad d_{\tilde{\alpha}}^* \tilde{\varphi}^{(0)}(\lambda) = 0, \tag{41}$$

needed to be satisfied and meaning, evidently, that the embeddings $\tilde{\varphi}^{(0)}(\lambda) \in H^0_{\Lambda(\tilde{\alpha}^*)}(M_\alpha)$, $\lambda \in \Sigma$, and $\tilde{\psi}^{(0)}(\eta) \in H^0_{\Lambda(\tilde{\alpha})}(M_\alpha)$, $\eta \in \Sigma$, are satisfied. Now we need to formulate a lemma very important for conditions (41) to hold.

Lemma 3.2. *The invariance property*

$$\mathcal{Z}^{(m_\alpha)} = \Omega_{(t_0)} \Omega_{(t)}^{-1} \mathcal{Z}^{(m_\alpha)} \Omega_{(t)}^{-1} \Omega_{(t_0)} \tag{42}$$

holds for any t and $t_0 \in M_\alpha$.

As a result of (42) and the symmetry invariance between cohomology spaces $H^0_{\Lambda(\alpha)}(M_\alpha)$ and $H^0_{\Lambda(\tilde{\alpha})}(M_\alpha)$ one obtains the pair of related mappings

$$\begin{aligned} \psi^{(0)} &= \tilde{\psi}^{(0)} \tilde{\Omega}_{(t)}^{-1} \tilde{\Omega}_{(t_0)}, \quad \varphi^{(0)} = \tilde{\varphi}^{(0)} \tilde{\Omega}_{(t)}^{\circledast,-1} \tilde{\Omega}_{(t_0)}^\circledast, \\ \tilde{\psi}^{(0)} &= \psi^{(0)} \Omega_{(t)}^{-1} \Omega_{(t_0)}, \quad \tilde{\varphi}^{(0)} = \varphi^{(0)} \Omega_{(t)}^{\circledast,-1} \Omega_{(t_0)}^\circledast, \end{aligned} \tag{43}$$

where the integral operator kernels in $L_2^{(\rho)}(\Sigma;\mathbb{R})\otimes L_2^{(\rho)}(\Sigma;\mathbb{R})$ are defined as

$$\begin{aligned}\tilde{\Omega}_{(t)}(\lambda,\mu) &:= \int_{\sigma_{(t)}^{(m_\alpha)}} \tilde{\Omega}^{(m_\alpha-2)}[\tilde{\varphi}^{(0)}(\lambda),\tilde{\psi}^{(0)}(\mu)d\tau], \\ \tilde{\Omega}^{\circledast}_{(t)}(\lambda,\mu) &:= \int_{\sigma_{(t)}^{(m_\alpha)}} \bar{\tilde{\Omega}}^{(m_\alpha-2),\intercal}[\tilde{\varphi}^{(0)}(\lambda),\tilde{\psi}^{(0)}(\mu)d\tau]\end{aligned} \tag{44}$$

for all $(\lambda,\mu)\in\Sigma\times\Sigma$, giving rise to proper Delsarte-Lions transmutation operators ensuring the pure differential nature of the transformed expressions (37).

Note here also that owing to (42) and (43) the operator property

$$\Omega_{(t_0)}\Omega_{(t)}^{-1}\Omega_{(t_0)} + \tilde{\Omega}_{(t_0)}\Omega_{(t)}^{-1}\Omega_{(t_0)} = 0 \tag{45}$$

holds for any t and $t_0 \in M_\alpha$ meaning that $\tilde{\Omega}_{(t)} = -\Omega_{(t_0;)}$.

Similarly, one can now define additional three subspaces

$$\begin{aligned}\mathcal{H}_0 &:= \{\psi^{(0)}(\mu)\in\mathcal{H}^0_{\Lambda(\alpha)}(M_\alpha) : d_\alpha\psi^{(0)}(\mu)=0,\quad \mu\in\Sigma\},\\ \tilde{\mathcal{H}}_0 &:= \{\tilde{\psi}^{(0)}(\mu)\in\mathcal{H}^0_{\Lambda(\tilde{\alpha})}(M_\alpha) : d_{\tilde{\alpha}}\tilde{\psi}^{(0)}(\mu)=0,\quad \mu\in\Sigma\},\\ \tilde{\mathcal{H}}_0^* &:= \{\tilde{\varphi}^{(0)}(\eta)\in\mathcal{H}^0_{\Lambda(\tilde{\alpha})}(M_\alpha) : d^*_{\tilde{\alpha}}\tilde{\varphi}^{(0)}(\eta)=0,\ \eta\in\Sigma\}\end{aligned} \tag{46}$$

closed and dense in $H^0_\Lambda(M_\alpha)$, and construct the actions

$$\Omega_\pm : \psi^{(0)} \mapsto \tilde{\psi}^{(0)} := \psi^{(0)}\Omega_{(t)}^{-1}\Omega_{(t_0)},\quad \Omega^{\circledast}_\pm : \varphi^{(0)} \mapsto \tilde{\varphi}^{(0)} := \varphi^{(0)}\Omega^{\circledast,-1}_{(t)}\Omega^{\circledast}_{(t_0)} \tag{47}$$

on arbitrary but fixed pairs of elements $(\varphi^{(0)}(\lambda),\psi^{(0}(\mu)) \in H_0^*\times H_0$, parametrized by the set Σ, where by definition, one needs that all obtained pairs $(\tilde{\varphi}^{(0)}(\lambda),\tilde{\psi}^{(0)}(\mu))$, $\lambda,\mu\in\Sigma$, belong to $H^0_{\Lambda(\tilde{\alpha})}(M_\alpha)\times H^0_{\Lambda(\tilde{\alpha})}(M_\alpha)$. Note also that the related operator property (45) can be compactly written as

$$\tilde{\Omega}_{(t)} = \tilde{\Omega}_{(t_0)}\Omega_{(t)}^{-1}\Omega_{(t_0)} = -\Omega_{(t_0)}\Omega_{(t)}^{-1}\Omega_{(t_0)}. \tag{48}$$

Construct now from the expressions (47) the operator kernels from the Hilbert space $L_2^{(\rho)}(\Sigma;\mathbb{R})\otimes L_2^{(\rho)}(\Sigma;\mathbb{R})$

$$\begin{aligned}\Omega_{(t)}(\lambda,\mu)-\Omega_{(t_0)}(\lambda,\mu) &= \int_{\partial S_{(t)}^{(m_\alpha)}} \Omega^{(m_\alpha-1)}[\varphi^{(0)}(\lambda),\psi^{(0)}(\mu)d\tau] \\ &\quad - \int_{\partial S_{(t_0)}^{(m_\alpha)}} \Omega^{(m_\alpha-1)}[\varphi^{(0)}(\lambda),\psi^{(0)}(\mu)d\tau] \\ &= \int_{S_{(\pm)}^{(m_\alpha)}(\sigma_{(t)}^{(m_\alpha-1)},\sigma_{(t_0)}^{(m_\alpha-1)})} d\Omega^{(m_\alpha-1)}[\varphi^{(0)}(\lambda),\psi^{(0)}(\mu)d\tau]\end{aligned} \tag{49}$$

$$= \int\limits_{S_{(\pm)}^{(m_\alpha)}(\sigma_{(t)}^{(m_\alpha-1)},\sigma_{(t_0)}^{(m_\alpha-1)})} \mathcal{Z}^{(m_\alpha)}[\varphi^{(0)}(\lambda),\psi^{(0)}(\mu)d\tau],$$

and, similarly,

$$\begin{aligned}
\Omega_{(t)}^{\circledast}(\lambda,\mu)-\Omega_{(t_0)}^{\circledast}(\lambda,\mu) &= \int_{\partial S_{(t)}^{(m_\alpha)}} \bar{\Omega}^{(m_\alpha-1),\intercal}[\varphi^{(0)}(\lambda),\psi^{(0)}(\mu)d\tau] \\
&\quad - \int_{\partial S_{(t_0)}^{(m_\alpha)}} \bar{\Omega}^{(m_\alpha-1),\intercal}[\varphi^{(0)}(\lambda),\psi^{(0)}(\mu)d\tau] \qquad (50)\\
&= \int\limits_{S_{\pm}^{(m_\alpha)}(\sigma_{(t)}^{(m_\alpha-1)},\sigma_{(t_0)}^{(m_\alpha-1)})} d\bar{\Omega}^{(m_\alpha-1),\intercal}[\varphi^{(0)}(\lambda),\psi^{(0)}(\mu)d\tau] \\
&= \int\limits_{S_{\pm}^{(m_\alpha)}(\sigma_{(t)}^{(m_\alpha-1)},\sigma_{(t_0)}^{(m_\alpha-1)})} \mathcal{Z}^{(m_\alpha-1),\intercal}[\varphi^{(0)}(\lambda),\psi^{(0)}(\mu)d\tau],
\end{aligned}$$

where $\lambda,\mu \in \Sigma$, and by definition, m_α -dimensional surfaces $S_{+}^{(m_\alpha)}(\sigma_{(t)}^{(m_\alpha-1)},\sigma_{(t_0)}^{(m_\alpha-1)})$ and $S_{-}^{(m_\alpha)}(\sigma_{(t)}^{(m_\alpha-1)},\sigma_{(t_0)}^{(m_\alpha-1)}) \subset C_{m_\alpha-1}(M_\alpha)$ are spanned smoothly without self-intersection between two homological cycles $\sigma_{(t)}^{(m_\alpha-1)} = \partial S_{(t)}^{(m_\alpha)}$ and $\sigma_{(t_0)}^{(m_\alpha-1)} := \partial S_{(t_0)}^{(m_\alpha)} \in C_{m_\alpha}(M_\alpha,\mathbb{R})$ in such a way that the boundary $\partial(S_{+}^{(m_\alpha)}(\sigma_{(t)}^{(m_\alpha-1)},\sigma_{(t_0)}^{(m_\alpha-1)}) \cup S_{-}^{(m_\alpha)}(\sigma_{(t)}^{(m_\alpha-1)},\sigma_{(t_0)}^{(m_\alpha-1)})) = \varnothing$. Since the integral operator expressions $\Omega_{(t_0)},\Omega_{(t_0)}^{\circledast} : L_2^{(\rho)}(\Sigma;\mathbb{R}) \to L_2^{(\rho)}(\Sigma;\mathbb{R})$ are at a fixed point $t_0 \in M_\alpha$, evidently constant and assumed to be invertible, for extending the actions given (47) on the whole Hilbert space $H_\Lambda^0(M_\alpha)\times H_\Lambda^0(M_\alpha)$ one can apply to them the classical constants variation approach, making use of the expressions (50). As a result, we easily obtain the Delsarte-Lions transmutation integral operator expressions

$$\begin{aligned}
\Omega_{\pm} &= \mathbf{1} - \int_{\Sigma\times\Sigma} d\rho(\xi)d\rho(\eta)\tilde{\psi}(t;\xi)\Omega_{(t_0)}^{-1}(\xi,\eta) \qquad (51)\\
&\quad \times \int\limits_{S_{\pm}^{(m_\alpha)}(\sigma_{(t)}^{(m_\alpha-1)},\sigma_{(t)}^{(m_\alpha-1)})} \mathcal{Z}^{(m_\alpha)}[\varphi^{(0)}(\eta),\cdot],\\
\Omega_{\pm}^{\circledast} &= \mathbf{1} - \int_{\Sigma\times\Sigma} d\rho(\xi)d\rho(\eta)\tilde{\varphi}(t;\eta)\Omega_{(t_0)}^{\circledast,-1}(\xi,\eta)\\
&\quad \times \int\limits_{S_{\pm}^{(m_\alpha)}(\sigma_{(t)}^{(m_\alpha-1)},\sigma_{(t_0)}^{(m_\alpha-1)})} \mathcal{Z}^{(m_\alpha),\intercal}[\cdot,\psi^{(0)}(\xi)d\tau]
\end{aligned}$$

for fixed pairs $(\varphi^{(0)}(\xi), \psi^{(0)}(\eta)) \in H_0^* \times H_0$, $\xi, \eta \in \Sigma$, and $(\tilde{\varphi}^{(0)}(\lambda), \tilde{\psi}^{(0)}(\mu)) \in \tilde{H}_0^* \times \tilde{H}_0$, $\lambda, \mu \in \Sigma$, being bounded invertible integral operators of Volterra type on the whole space $H \times \tilde{H}^*$. Applying the standard arguments, one can show also that transformed sets of operators $\tilde{A}_j := \Omega_\pm A_j \Omega_\pm^{-1}$, $j = \overline{1, m_\alpha}$, and $\tilde{A}_k^* := \Omega_\pm^{\circledast} A_k^* \Omega_\pm^{\circledast,-1}$, $k = \overline{1, m_\alpha}$, prove to be, respectively, purely differential too. Thereby, one can formulate [23,40,41] the following final theorem.

Theorem 3.4. *The expressions (51) are bounded invertible Delsarte-Lions transmutation integral operators of Volterra type onto $H_\Lambda^0(M_\alpha) \times H_\Lambda^0(M_\alpha)$, transforming, respectively, given commuting sets of differential expressions A_j, $j = \overline{1, m_\alpha}$, and their formally adjoint ones A_k^*, $k = \overline{1, m_\alpha}$, into the pure differential sets of expressions $\tilde{A}_j := \Omega_\pm A_j \Omega_\pm^{-1}$, $j = \overline{1, m_\alpha}$, and $\tilde{A}_k^* := \Omega_\pm^{\circledast} A_k^* \Omega_\pm^{\circledast,-1}$, $k = \overline{1, m_\alpha}$. Moreover, the suitably constructed closed subspaces $H_0 \subset H_{\Lambda(\alpha)}^0(M_\alpha)$ and $\tilde{H}_0 \subset H_{\Lambda(\tilde{\alpha})}^0(M_\alpha)$ such that the operators Ω and $\Omega^{\circledast} \in B(H_\Lambda^0(M_\alpha))$ depend strongly on the topological structure of the generalized cohomology groups $H_{\Lambda(\alpha)}^0(M_\alpha)$ and $H_{\Lambda(\tilde{\alpha})}^0(M_\alpha)$, being parametrized by elements $S_\pm^{(m_\alpha)}(\sigma_{(t)}^{(m_\alpha-1)}, \sigma_{(t_0)}^{(m_\alpha-1)}) \in C_{m_\alpha}(M_\alpha, \mathbb{R})$.*

4. Conclusion

The study done above presents some of the recent results devoted to the development of a generalized de Rham-Hodge theory [10,22,30,33–35,40,41] and related differential-geometric aspects of Chern characteristic classes, concerning special differential complexes with Cartan type connections, which give rise to effective analytical tools of studying multi-dimensional integrable nonlinear differential systems of M. Gromov type [3] on Riemannian manifolds. Some results on the structure of the Delsarte-Lions transmutation operators can be easily adapted for constructing effective transformations of Cartan-type connections for multidimensional integrable Davey-Stewartson-type nonlinear differential system on a Riemannian manifold M, vanishing on a three-dimensional integral submanifold $M_\alpha \subset M$. The results obtained can be used for studying a wide class of exact special solutions of such differential systems, having applications [4,14,31,32,35,37,40] at solving some problems of modern differential topology and mathematical physics.

Acknowledgements

The authors are cordially thankful to the Abdus Salam International Center for Theoretical Physics in Trieste, Italy, for the hospitality during their 2007 research scholarships. A.P. also appreciates interesting discussions of the results with Profs. Y. Kondratyev and Y. Kozicki during the Third International Conference on Infinite Dimensional Systems, held June 23–28, 2007 in Kazimierz Dolny, Poland, and to Profs. O. Çelebi, K. Zheltukhin, G. Barsegian, O. Zhdanov and V. Golubeva during the 2007 International ISAAC Congress, held August 13–18, 2007 in Ankara, Turkey. The last but not the least thanks belong to the referee of the article for many insightful and qualified comments which helped in an essential way in improving the exposition.

References

1. R. Abraham, J. Marsden, Foundations of mechanics, Cummings, NY, USA, 1978, 806P.
2. V. I. Arnold, Mathematical methods of classical mechanics, Springer, NY, 1978, 472P.
3. M. Gromov, Partial differential relations, Springer, NY, 1986, 536P.
4. Y. A. Mitropolski, N. N. Bogoliubov, Jr., A. K. Prykarpatsky and V. H. Samoilenko, Integrable dynamical systems: Differential-geometric and spectral aspects, Nauka Dumka, Kiev, 1987, 296P. (in Russian).
5. N. N. Bogoliubov, Y. A. Prykarpatsky, A. M. Samoilenko and A. K. Pryakrpatsky, A generalized de Rham-Hodge theory of multidimensional Delsarte transformations of differential operators and its applications for nonlinear dynamic systems, Physics of Particles and Nuclei, 2005, 36, N1, P.111–121.
6. E. Cartan, Lecons sur invariants integraux, Hermann, Paris, 1971, 260P.
7. S. S. Chern, Complex manifolds, Chicago University Publ., USA, 1956, 240P.
8. J. Delsarte, Sur certaines transformations fonctionelles relative aux equations lineaires aux derives partielles du second ordre, C. R. Acad. Sci. Paris, 1938, 206, P.178–182.
9. J. Delsarte and J. Lions, Transmutations d'operateurs differentielles dans le domain complex, Comment. Math. Helv., 1957, 52, P.113–128.
10. I. V. Skrypnik, Periods of A-closed forms, Proceedings of the USSR Academy of Sciences, 1965, 160, N4, 1965, P.772–773 (in Russian).
11. I. V. Skrypnik, Harmonique fields with singularities, Ukr. Math. Journal, 1965, 17, N4, P.130–133 (in Russian).
12. I. V. Skrypnik, The generalized de Rham theorem, Proc. UkrSSR Acad. Sci., 1965, 1, P.18–19 (in Ukrainian).
13. Y. B. Lopatynski, On harmonic fields on Riemannian manifolds, Ukr. Math. Journal, 1950, 2, P.56–60 (in Russian).
14. L. D. Faddeev and L. A. Takhtadjyan, Hamiltonian approach to soliton theory, Moscow, Nauka, 1986, 528P. (in Russian).

15. Y. M. Berezansky, Eigenfunctions expansions related with selfadjoint operators, Kiev, Nauk. Dumka Publ., 1965, 798P. (in Russian).
16. F. A. Berezin and M. A. Shubin, Schrodinger equation, Moscow, the Moscow University Publisher, 1983, 470P. (in Russian).
17. A. L. Bukhgeim, Volterra equations and inverse problems, Moscow, Nauka, 1983, 208P. (in Russian).
18. V. B. Matveev and M. I. Salle, Darboux-Backlund transformations and applications, NY, Springer, 1993, 245P.
19. L. P. Nizhnik, Inverse scattering problems for hyperbolic equations, Kiev, Nauk. Dumka Publ., 1991, 232P. (in Russian).
20. I. C. Gokhberg and M. G. Krein, Theory of Volterra operators in Hilbert spaces and its applications, Moscow, Nauka, 1967, 506P. (in Russian).
21. Y. V. Mykytiuk, Factorization of Fredholmian operators, Mathematical Studii, Proceedings of Lviv Mathematical Society, 2003, 20, N2, P.185–199 (in Ukrainian).
22. A. M. Samoilenko, A. K. Prykarpatsky and V. G. Samoylenko, The structure of Darboux-type binary transformations and their applications in soliton theory, Ukr. Mat. Journal, 2003, 55, N12, 1704–1723 (in Ukrainian).
23. Y. A. Prykarpatsky, A. M. Samoilenko and A. K. Prykarpatsky, The multidimensional Delsarte transmutation operators, their differential-geometric structure and applications, Part 1, Opuscula Mathematica, 2003, 23, P.71–80.
24. Y. A. Prykarpatsky, A. M. Samoilenko and A. K. Prykarpatsky, The de Rham-Hodge-Skrypnik theory of Delsarte transmutation operators in multidimension and its applications. Reports on Mathem. Physics, 2005, 55, N3, P.351–363.
25. J. Golenia, Y. A. Prykarpatsky, A. M. Samoilenko and A. K. Prykarpatsky, The general differential-geometric structure of multidimensional Delsarte transmutation operators in parametric functional spaces and their applications in soliton theory, Part 2, Opuscula Mathematica, 2004, 24, P.71–83.
26. C. Godbillon, Geometrie differentielle et mechanique analytique, Paris, Hermann, 1969, 188P.
27. Sternberg S., Lectures on differential geometry, Prentice Hall, USA, 1956, 410P.
28. A. S. Mishchenko and A. T. Fomenko, Introduction to differential geometry and topology, Moscow University Publ., 1983, 439P.
29. S. Kobayashi and K. Nomizu, Foundations of differential geometry, John Wiley and Sons, NY, v.1, 1963, 344P.; v.2, 1969, 357P.
30. R. Teleman, Elemente de topologie si varietati diferentiabile, Bucuresti Publ., Romania, 1964, 390P.
31. J. D. Moore, Lectures on Seiberg-Witten invariants, 2nd ed., Springer, 2001, 160P.
32. S. Donaldson, An application of gauge theory to four-dimensional topology, J. Differential Geometry, 1983, 18, P.279–315
33. G. de Rham, Varietes differentielles, Hermann, Paris, 1955, 249P.
34. G. de Rham, Sur la theorie des formes differentielles harmoniques, Ann. Univ. Grenoble, 1946, 22, P.135–152.

35. F. Warner, Foundations of differential manifolds and Lie groups, Academic Press, NY, 1971, 346P.
36. S. P. Novikov, Topology, Institute of Computer Reserch Publ., Moscow, 2002 (in Russian).
37. A. M. Samoilenko and Y. A. Prykarpatsky, Algebraic-analytic aspects of completely integrable dynamical systems and their perturbations, Kyiv, NAS, Inst. Mathem. Publisher, v.41, 2002, 245P. (in Ukrainian).
38. D. L. Blackmore, Y. A. Prykarpatsky and R. V. Samulyak, The integrability of Lie-invariant geometric objects generated by ideals in Garssmann algebras, J. Nonlinear Math. Phys., 1998, 5, P.54–67.
39. A. K. Prykarpatsky and I. V. Mykytiuk, Algebraic integrability of nonlinear dynamical systems on manifolds: Classical and quantum aspects, Kluwer Acad. Publishers, the Netherlands, 1998, 553P.
40. O. Y. Hentosh, M. M. Prytula and A. K. Prykarpatsky, Differential-geometric integrability fundamentals of nonlinear dynamical systems on functional menifolds, 2nd revised ed., Lviv University Publisher, Lviv, Ukraine, 2006, 408P.
41. A. M. Samoilenko, Y. A. Prykarpatsky and A. K. Prykarpatsky, The spectral and differential-geometric aspects of a generalized de Rham-Hodge theory related with Delsarte transmutation operators in multidimension and its applications to spectral and soliton problems, Nonlinear Analysis, 2006, 65, P.395–432
42. V. Zakharov, S. Manakov, S. Novikov, L. Pitaevsky, Theory of solitons: The inverse scattering problem, Moscow, Nauka Publisher, 1980, 320P. (in Russian).
43. V. E. Zakharov, Integrable systems in multidimensional spaces, Lect. Notes in Phys., 1982, 153, P.190–216.

INTRODUCTORY BACKGROUND TO MODERN QUANTUM MATHEMATICS WITH APPLICATION TO NONLINEAR DYNAMICAL SYSTEMS

A. K. PRYKARPATSKY

Department of Nonlinear Mathematical Analysis,
Institute for Applied Problems of Mechanics and Mathematics of NAS,
Lviv, 79060, Ukraine
& Department of Applied Mathematics,
The AGH University of Science and Technology,
Krakow, 30059, Poland
E-mail: pryk.anat@ua.fm

J. GOLENIA

Department of Applied Mathematics,
The AGH University of Science and Technology,
Krakow, 30059, Poland
E-mail: goljols@tlen.pl

N. N. BOGOLUBOV, Jr.

V. A. Steklov Mathematical Institute of RAS,
Moscow, 117946 Russian Federation
& Abdus Salam International Center for Theoretical Physics,
Trieste, Italy
E-mail: nikolai_bogolubov@hotmail.com

U. TANERİ

Department of Applied Mathematics and Computer Science,
Eastern Mediterranean University EMU,
Famagusta, North Cyprus
& Institute of Graduate Studies, Kyrenia American University GAÜ,
Kyrenia, North Cyprus
E-mail: ufuk.taneri@gmail.com

Introductory background to a new mathematical physics discipline—Quantum Mathematics—is discussed and analyzed both from historical and analytical points of view. The magical properties of the second quantization method, invented by Fock in 1934, are demonstrated, and an impressive application to the nonlinear dynamical systems theory is considered.

Keywords: Quantum mathematics; Creation and annihilation operators; Nonlinear dynamical systems;Representation.

1. Introduction

There is a broad and inclusive view of modern mathematical physics by many mathematicians and mathematical physicists. During the last century, modern mathematical physics evolved within at least four components which illustrate [19] the development of the mathematics and quantum physics synergy:

1) the use of ideas from mathematics in shedding new light on the existing principles of quantum physics, either from a conceptual or from a quantitive point of view;

2) the use of ideas from mathematics in discovering new "laws of quantum physics";

3) the use of ideas from quantum physics in shedding new light on existing mathematical structures;

4) the use of ideas from quantum physics in discovering new domains in mathematics.

Each one of these topics plays some role in understanding the modern mathematical physics. However, our success in directions 2) and 4) is certainly more modest than our success in directions 1) and 3). In some cases it is difficult to draw a clear-cut distinction between these two sets. In fact, we are lucky when it is possible to make progress in directions 2) and 4); so much so that when we achieve a major progress, historians like to speak of a revolution. In any case, many of mathematical physicists strive to understand within their research efforts these deep and lofty goals. There are many situations however, when mathematical physicists' research efforts are directed toward one other more mundane aspect:

5) the use of ideas from quantum physics and mathematics to benefit "economic competitiveness".

Here too, one might subdivide this aspect into conceptual understanding on one hand (such as the mathematical model of Black and Sholes for pricing of derivative securities in financial markets) and invention on the other: the creation of new algorithms or materials (e.g. quantum computers) which might revolutionize technology or change our way of life. As in the first four cases, the boundary between these domains is not sharp, and it remains open to views and interpretations. This fifth string can be characterized as "applied" mathematical physics. We will restrict our analysis to the first four strands characterizing modern quantum physics and mathematics aspects; it is believed that that most of the profound applied directions arise after earlier fundamental quantum physics and mathematics progress.

We have passed through an extraordinary 35-year period of development of modern fundamental mathematics and quantum physics. Much of this development has been derived from one subject to understand the other. Not only the concepts from diverse fields have been united: statistical physics, quantum field theory and functional integration; gauge theory and geometry; index theory and knot invariants, etc., but also, new phenomena have been recognized and new areas have emerged whose significance for both mathematics and modern quantum physics is only partially understood: for example, non-commutative geometry, super-analysis, mirror symmetry, new topological invariants of manifolds, and the general notion of geometric quantization.

There is no doubt that, over the past 35 years, the ideas from quantum physics have led to far greater inventions of new mathematics than the ideas from mathematics have in discoveries of laws of quantum physics. Recognition of this underlines the opportunities for future progress in the opposite direction: a new understanding of the quantum nature of the world is certainly our expectation!

Great publicity and recognition has been attached to the progress made in modern geometry, representation theory, and deformation theory due to this interaction. We will not also dwell here on the substantial progress in analysis and probability theory, which unfortunately is more difficult to understand because of its delicate dependence on subtle notions of continuity.

On the other hand, there are deep differences between pure mathematics and modern quantum physics fundamentals. They have evolved from different cultures and they each have a distinctive set of values of their own, suited for their different realms of universality. But both subjects are strongly based on *intuition*, some natural and some acquired, which form our understanding. Quantum physics describes the natural micro-world. Hence, physicists appeal to observation in order to verify the validity of a physical theory. And, although much of mathematics arises from the natural world, mathematics has no analogous testing grounds - mathematicians appeal to their own set of values, namely mathematical proof, to justify validity of a mathematical theory. In mathematical physics, when announcing results of a mathematical nature, it is necessary to claim a theorem when the proof meets the mathematical community standards for a proof; otherwise, it is necessary to make a conjecture with a detailed outline for support. Most of physics, on the other hand, has completely different standards.

There is no question that the interaction between modern mathematics and quantum physics will change radically during this running century. We do hope however, that this evolution will preserve the positive experience of being a mathematician, a pure physicist, or a mathematical physicist, so that it remains attractive to the brightest and gifted young students today and tomorrow.

It is instructive to look at the beginning of the 20th century and trace the way mathematics has been exerting influence on modern and classical quantum physics, and next observe the way the modern quantum physics is nowadays exerting so impressive influence on modern mathematics. With the latter, application of modern quantum mathematics to studying nonlinear dynamical systems in functional spaces will for example be a significant topic of our present work. We will begin with a brief history of quantum mathematics:

The beginning of the 20th *century:*

- P. A. M. Dirac — first realized and used in quantum physics the fact that the commutator operation $D_a : \mathcal{A} \ni b \longrightarrow [a,b] \in \mathcal{A}$, where $a \in \mathcal{A}$ is fixed and $b \in \mathcal{A}$, is a differentiation of an associative algebra $\mathcal{A}$; moreover, he first constructed a spinor matrix realization of the Poincaré symmetry group $\mathcal{P}(1,3)$ and invented the famous Dirac δ-function [8] (1920-1926);
- J. von Neumann — first applied the spectral theory of self-adjoint operators in Hilbert spaces to explain the radiation spectra of atoms and the stability of the related matter [30] (1926);
- V. Fock — first introduced the notion of many-particle Hilbert space, Fock space, and introduced the related creation and annihilation operators acting in it [13] (1932);
- H. Weyl — first understood the fundamental role of the notion of symmetry in physics and developed a physics-oriented group theory; moreover he showed the importance of different representations of classical matrix groups for physics and studied the unitary representations of the Heisenberg-Weyl group related with creation and annihilation operators in Fock space [33] (1931).

The end of the 20th *century:*

New developments are due to

- L. Faddeev with co-workers — quantum inverse spectral theory transform [10] (1978);
- V. Drinfeld, S. Donaldson, E. Witten — quantum groups and algebras, quantum topology, quantum super-analysis [21,22,34] (1982–1994);
- Y. Manin, R. Feynman — quantum information theory [11,12,28] (1980–1986);
- P. Shor, E. Deutsch, L. Grover and others — quantum computer algorithms [18,20,32] (1985–1997).

As one can observe, many exciting and highly important mathematical achievements were strictly motivated by the impressive and deep influence of quantum physics ideas and ways of thinking, leading nowadays to an altogether new scientific field often called quantum mathematics.

Following this quantum mathematical way of thinking, we will demonstrate below that a wide class of strictly nonlinear dynamical systems in functional spaces can be treated as a natural object in specially constructed Fock spaces in which the corresponding evolution flows are completely linearized. Thereby, the powerful machinery of classical mathematical tools can be applied to studying the analytical properties of exact solutions to suitably well posed Cauchy problems.

2. Mathematical preliminaries: Fock space and its realizations

Let Φ be a separable Hilbert space, F be a topological real linear space and $\mathcal{A} := \{A(\varphi) : \varphi \in F\}$ a family of commuting self-adjoint operators in Φ (i.e. these operators commute in the sense of their resolutions of the identity). Consider the Gelfand rigging [2] of the Hilbert space Φ, i.e., a chain

$$\mathcal{D} \subset \Phi_+ \subset \Phi \subset \Phi_- \subset \mathcal{D}' \tag{1}$$

in which Φ_+ and Φ_- are further Hilbert spaces, and the inclusions are dense and continuous, i.e. Φ_+ is topologically (densely and continuously) and quasi-nucleously (the inclusion operator $i : \Phi_+ \longrightarrow \Phi$ is of the Hilbert-Schmidt type) embedded into Φ, Φ_- is the dual of Φ_+ with respect to the scalar product $< .,. >_\Phi$ in Φ, and $\mathcal{D}$ is a separable projective limit of Hilbert spaces, topologically embedded into Φ_+. Then, the following structural theorem [2,3] holds:

Theorem 2.1. *Assume that the family of operators $\mathcal{A}$ satisfies the following conditions:*

a) $\mathcal{D} \subset DomA(\varphi)$, $\varphi \in F$, *and the closure of the operator* $A(\varphi) \uparrow \mathcal{D}$ *coincides with* $A(\varphi)$ *for any* $\varphi \in F$, *that is* $\overline{A(\varphi) \uparrow \mathcal{D}} = A(\varphi)$ *in* Φ;

b) the Range $A(\varphi) \uparrow \mathcal{D} \subset \Phi_+$ *for any* $\varphi \in F$;

c) for every $f \in \mathcal{D}$ *the mapping* $F \ni \varphi \longrightarrow A(\varphi)f \in \Phi_+$ *is linear and continuous;*

d) there exists a strong cyclic (vacuum) vector $|\Omega\rangle \in \bigcap_{\varphi \in F} DomA(\varphi)$, *such that the set of all vectors* $|\Omega\rangle$, $\prod_{j=1}^{n} A(\varphi_j)|\Omega\rangle$, $n \in \mathbb{Z}_+$, *is total in* Φ_+ *(i.e. their linear hull is dense in* Φ_+*).*

Then there exists a probability measure μ *on* $(F', C_\sigma(F'))$, *where* F' *is the dual of* F *and* $C_\sigma(F')$ *is the* $\sigma-$*algebra generated by cylinder sets in* F' *such that, for* $\mu-$*almost every* $\eta \in F'$ *there is a generalized joint eigenvector* $\omega(\eta) \in \Phi_-$ *of the family* $\mathcal{A}$, *corresponding to the joint eigenvalue* $\eta \in F'$, *that is*

$$< \omega(\eta), A(\varphi)f >_\Phi = \eta(\varphi) < \omega(\eta), f >_\Phi \tag{2}$$

with $\eta(\varphi) \in \mathbb{R}$ *denoting the pairing between* F *and* F'.

The mapping

$$\Phi_+ \ni f \longrightarrow < \omega(\eta), f >_\Phi := \hat{f}(\eta) \in \mathbb{C} \tag{3}$$

for any $\eta \in F'$ *can be continuously extended to a unitary surjective operator* $\mathcal{F} : \Phi \longrightarrow L_2^{(\mu)}(F'; \mathbb{C})$, *where*

$$\mathcal{F}\ f(\eta) := \hat{f}(\eta) \tag{4}$$

for any $\eta \in F'$ *is a generalized Fourier transform corresponding to the family* $\mathcal{A}$. *Moreover, the image of the operator* $A(\varphi)$, $\varphi \in F'$, *under the* $\mathcal{F}-$*mapping is the operator of multiplication by the function* $F' \ni \eta \to \eta(\varphi) \in \mathbb{C}$.

We assume additionally that the main Hilbert space Φ possesses the standard Fock space (bose)-structure [4,6,31], that is,

$$\Phi = \oplus_{n \in \mathbb{Z}_+} \Phi_{(s)}^{\otimes n}, \tag{5}$$

where subspaces $\Phi_{(s)}^{\otimes n}$, $n \in \mathbb{Z}_+$, are the symmetrized tensor products of a Hilbert space $\mathcal{H} := L_2(\mathbb{R}^m; \mathbb{C})$. If a vector $g := (g_0, g_1, ..., g_n, ...) \in \Phi$, its norm is

$$\|g\|_\Phi := \left(\sum_{n \in \mathbb{Z}_+} \|g_n\|_n^2 \right)^{1/2}, \tag{6}$$

where $g_n \in \Phi_{(s)}^{\otimes n} \simeq L_{2,(s)}((\mathbb{R}^m)^n; \mathbb{C})$ and $\| \ ... \ \|_n$ is the corresponding norm in $\Phi_{(s)}^{\otimes n}$ for all $n \in \mathbb{Z}_+$. Note here that concerning the rigging structure (1), there holds the corresponding rigging for the Hilbert spaces $\Phi_{(s)}^{\otimes n}$, $n \in \mathbb{Z}_+$, that is,

$$\mathcal{D}_{(s)}^n \subset \Phi_{(s),+}^{\otimes n} \subset \Phi_{(s)}^{\otimes n} \subset \Phi_{(s),-}^{\otimes n} \tag{7}$$

with some suitably chosen dense and separable topological spaces of symmetric functions $\mathcal{D}_{(s)}^n$, $n \in \mathbb{Z}_+$. Concerning expansion (5) we obtain by means of projective and inductive limits [2,3,6] the quasi-nucleous rigging of the Fock space Φ in the form (1):

$$\mathcal{D} \subset \Phi_+ \subset \Phi \subset \Phi_- \subset \mathcal{D}'.$$

Consider now any vector $|(\alpha)_n\rangle \in \Phi_{(s),}^{\otimes n}$ $n \in \mathbb{Z}_+$, which can be written [2,4,26] in the following canonical Dirac ket-form:

$$|(\alpha)_n\rangle := |\alpha_1, \alpha_2, ..., \alpha_n\rangle, \tag{8}$$

where, by definition,

$$|\alpha_1, \alpha_2, ..., \alpha_n\rangle := \frac{1}{\sqrt{n!}} \sum_{\sigma \in S_n} |\alpha_{\sigma(1)}\rangle \otimes |\alpha_{\sigma(2)}\rangle ... |\alpha_{\sigma(n)}\rangle \tag{9}$$

and $|\alpha_j\rangle \in \Phi_{(s)}^{\otimes 1}(\mathbb{R}^m; \mathbb{C}) := \mathcal{H}$ for any fixed $j \in \mathbb{Z}_+$. The corresponding scalar product of base vectors as (9) is given as

$$\begin{aligned} \langle (\beta)_n | (\alpha)_n \rangle &:= \langle \beta_n, \beta_{n-1}, ..., \beta_2, \beta_1 | \alpha_1, \alpha_2, ..., \alpha_{n-1}, \alpha_n \rangle \\ &= \textstyle\sum_{\sigma \in S_n} \langle \beta_1 | \alpha_{\sigma(1)} \rangle ... \langle \beta_n | \alpha_{\sigma(n)} \rangle := per\{\langle \beta_i | \alpha_j \rangle : i, j = \overline{1, n}\}, \end{aligned} \tag{10}$$

where "*per*" denotes the permanent of matrix and $\langle .|. \rangle$ is the corresponding product in the Hilbert space $\mathcal{H}$. Based now on representation (8) one can define an operator $a^+(\alpha) : \Phi_{(s)}^{\otimes n} \longrightarrow \Phi_{(s)}^{\otimes (n+1)}$ for any $|\alpha\rangle \in \mathcal{H}$ by

$$a^+(\alpha) |\alpha_1, \alpha_2, ..., \alpha_n\rangle := |\alpha, \alpha_1, \alpha_2, ..., \alpha_n\rangle, \tag{11}$$

which is called the "creation" operator in the Fock space Φ. Its adjoint operator $a(\beta) := (a^+(\beta))^* : \Phi_{(s)}^{\otimes (n+1)} \longrightarrow \Phi_{(s)}^{\otimes n}$ with respect to the Fock space Φ (5) for any $|\beta\rangle \in \mathcal{H}$, called the "annihilation" operator, acts as

$$a(\beta) |\alpha_1, ..., \alpha_{n+1}\rangle := \sum_{\sigma \in S_n} \langle \beta, \alpha_j \rangle |\alpha_1, ..., \alpha_{j-1}, \hat{\alpha}_j, \alpha_{j+1}, ..., \alpha_{n+1}\rangle, \tag{12}$$

where the "*hat*" over a vector denotes that it should be omitted from the sequence.

It is easy to check that the commutator relationship

$$[a^+(\alpha), a(\beta)] = \langle \alpha, \beta \rangle \tag{13}$$

holds for any vectors $|\alpha\rangle \in \mathcal{H}$ and $|\beta\rangle \in \mathcal{H}$. Expression (13), owing to the rigged structure (1), can be naturally extended to the general case, when vectors $|\alpha\rangle$ and $|\beta\rangle \in \mathcal{H}_-$, conserving its form. In particular, taking $|\alpha\rangle := |\alpha(x)\rangle = \frac{1}{\sqrt{2\pi}} e^{i\langle\lambda,x\rangle} \in \mathcal{H}_- := L_{2,-}(\mathbb{R}^m;\mathbb{C})$ for any $x \in \mathbb{R}^m$, one easily gets from (13) that

$$[a^+(x), a(y)] = \delta(x-y), \tag{14}$$

where we put, by definition, $a^+(x) := a^+(\alpha(x))$ and $a(y) := a(\alpha(y))$ for all $x, y \in \mathbb{R}^m$ and denoted by $\delta(\cdot)$ the classical Dirac delta-function.

The construction above makes it possible to observe easily that there exists a unique vacuum vector $|\Omega\rangle \in \mathcal{H}_+$, such that for any $x \in \mathbb{R}^m$

$$a(x)|\Omega\rangle = 0, \tag{15}$$

and the set of vectors

$$\left(\prod_{j=1}^{n} a^+(x_j)\right)|\Omega\rangle \in \Phi_{(s)}^{\otimes n} \tag{16}$$

is total in $\Phi_{(s)}^{\otimes n}$, that is, their linear integral hull over the dual functional spaces $\hat{\Phi}_{(s)}^{\otimes n}$ is dense in the Hilbert space $\Phi_{(s)}^{\otimes n}$ for every $n \in \mathbb{Z}_+$. This means that for any vector $g \in \Phi$ the representation

$$g = \oplus_{n\in\mathbb{Z}_+} \int_{(\mathbb{R}^m)^n} \hat{g}_n(x_1, ..., x_n) a^+(x_1) a^+(x_2) ... a^+(x_n)|\Omega\rangle \tag{17}$$

holds with the Fourier type coefficients $\hat{g}_n \in \hat{\Phi}_{(s)}^{\otimes n}$ for all $n \in \mathbb{Z}_+$, with $\hat{\Phi}_{(o)}^{\otimes 1} := \mathcal{H} \simeq L_2(\mathbb{R}^m;\mathbb{C})$. The latter is naturally endowed with the Gelfand type quasi-nucleous rigging dual to

$$\mathcal{H}_+ \subset \mathcal{H} \subset \mathcal{H}_-, \tag{18}$$

making it possible to construct a quasi-nucleous rigging of the dual Fock space $\hat{\Phi} := \oplus_{n\in\mathbb{Z}_+} \hat{\Phi}_{(s)}^{\otimes n}$. Thereby, chain (18) generates the dual Fock space quasi-nucleous rigging

$$\hat{\mathcal{D}} \subset \hat{\Phi}_+ \subset \hat{\Phi} \subset \hat{\Phi}_- \subset \hat{\mathcal{D}}' \tag{19}$$

with respect to the central Fock type Hilbert space $\hat{\Phi}$, where $\hat{\mathcal{D}} \simeq \mathcal{D}$, easily following from (1) and (18).

Construct now the self-adjoint operator

$$a^+(x)a(x) := \rho(x) : \Phi \to \Phi, \tag{20}$$

called the density operator at a point $x \in \mathbb{R}^m$, satisfying the commutation properties

$$\begin{aligned} [\rho(x), \rho(y)] &= 0, \\ [\rho(x), a(y)] &= -a(y)\delta(x-y), \\ [\rho(x), a^+(y)] &= a^+(y)\delta(x-y) \end{aligned} \tag{21}$$

for all $y \in \mathbb{R}^m$.

Now, if one is to construct the self-adjoint family $\mathcal{A} := \{\int_{\mathbb{R}^m} \rho(x)\varphi(x)dx : \varphi \in F\}$ of linear operators in the Fock space Φ, where $F := \mathcal{S}(\mathbb{R}^m;\mathbb{R})$ is the Schwartz functional space, one can derive, making use of Theorem 2.1, that there exists the generalized Fourier transform (4), such that

$$\Phi(\mathcal{H}) = L_2^{(\mu)}(\mathcal{S}';\mathbb{C}) \simeq \int_{\mathcal{S}'}^{\oplus} \Phi_\eta d\mu(\eta) \tag{22}$$

for some Hilbert space sets Φ_η, $\eta \in F'$, and a suitable measure μ on $\mathcal{S}'$, with respect to which the corresponding joint eigenvector $\omega(\eta) \in \Phi_+$ for any $\eta \in F'$ generates the Fourier transformed family $\hat{\mathcal{A}} = \{\eta(\varphi) \in \mathbb{R} : \ \varphi \in \mathcal{S}\}$. Moreover, if $\dim \Phi_\eta = 1$ for all $\eta \in F$, the Fourier transformed eigenvector $\hat{\omega}(\eta) := \Omega(\eta) = 1$ for all $\eta \in F'$.

Now we will consider the family of self-adjoint operators $\mathcal{A}$ as generating a unitary family $\mathcal{U} := \{U(\varphi) : \varphi \in F\} = \exp(i\mathcal{A})$, where for any $\rho(\varphi) \in \mathcal{A}$, $\varphi \in F$, the operator

$$U(\varphi) := \exp[i\rho(\varphi)] \tag{23}$$

is unitary, satisfying the abelian commutation condition

$$U(\varphi_1)U(\varphi_2) = U(\varphi_1 + \varphi_2) \tag{24}$$

for any $\varphi_1, \varphi_2 \in F$.

Since, in general, the unitary family $\mathcal{U} = \exp(i\mathcal{A})$ is defined in some Hilbert space Φ, not necessarily being of Fock type, the important problem of describing its Hilbertian cyclic representation spaces arises, within which the factorization

$$\rho(\varphi) = \int_{\mathbb{R}^m} a^+(x)a(x)\varphi(x)dx \tag{25}$$

jointly with relationships (21) hold for any $\varphi \in F$. This problem can be treated using mathematical tools devised both within the representation theory of C^*-algebras [9] and the Gelfand–Vilenkin [14] approach. Below we will describe the main features of the Gelfand–Vilenkin formalism, being much more suitable for the task, providing a reasonably unified framework of constructing the corresponding representations.

Definition 2.1. Let F be a locally convex topological vector space, $F_0 \subset F$ be a finite dimensional subspace of F. The space $F^0 \subseteq F'$ defined by

$$F^0 := \{\xi \in F' : \ \xi|_{F_0} = 0\}, \tag{26}$$

is called the annihilator of F_0.

The quotient space $F'^0 := F'/F^0$ may be identified with $F_0' \subset F'$, the adjoint space of F_0.

Definition 2.2. Let $A \subseteq F'$; then the subset

$$X_{F^0}^{(A)} := \{\xi \in F' : \xi + F^0 \subset A\} \tag{27}$$

is called the cylinder set with base A and generating subspace F^0.

Definition 2.3. Let $n = \dim F_0 = \dim F_0' = \dim F'^0$. One says that a cylinder set $X^{(A)}$ has a Borel base if A is Borel when regarded as a subset of $\mathbb{R}^n$.

The family of cylinder sets with Borel base forms an algebra of sets.

Definition 2.4. The measurable sets in F' are the elements of the $\sigma-$algebra generated by the cylinder sets with Borel base.

Definition 2.5. A cylindrical measure in F' is a real-valued $\sigma-$pre-additive function μ defined on the algebra of cylinder sets with Borel base and satisfying the conditions $0 \leq \mu(X) \leq 1$ for any X, $\mu(F') = 1$ and $\mu\left(\coprod_{j\in\mathbb{Z}_+} X_j\right) = \sum_{j\in\mathbb{Z}_+} \mu(X_j)$, if all sets $X_j \subset F'$, $j \in \mathbb{Z}_+$, have a common generating subspace $F_0 \subset F$.

Definition 2.6. A cylindrical measure μ satisfies the commutativity condition if and only if for any bounded continuous function $\alpha : \mathbb{R}^n \longrightarrow \mathbb{R}$ of $n \in \mathbb{Z}_+$ real variables, the function

$$\alpha[\varphi_1, \varphi_2, ..., \varphi_n] := \int_{F'} \alpha(\eta(\varphi_1), \eta(\varphi_2), ..., \eta(\varphi_n)) d\mu(\eta) \tag{28}$$

is sequentially continuous in $\varphi_j \in F$, $j = \overline{1,m}$. (It is well known [14,15] that in countably normed spaces the properties of sequential and ordinary continuity are equivalent.)

Definition 2.7. A cylindrical measure μ is countably additive if for any cylinder set $X = \coprod_{j\in\mathbb{Z}_+} X_j$, which is the union of countably many mutually disjoints cylinder sets $X_j \subset F'$, $j \in \mathbb{Z}_+$, $\mu(X) = \sum_{j\in\mathbb{Z}_+} \mu(X_j)$.

The following propositions hold.

Proposition 2.1. *A countably additive cylindrical measure μ can be extended to a countably additive measure on the $\sigma-$algebra generated by the cylinder sets with Borel base. Such a measure will also be called a cylindrical measure.*

Proposition 2.2. *Let F be a nuclear space. Then any cylindrical measure μ on F', satisfying the continuity condition, is countably additive.*

Definition 2.8. Let μ be a cylindrical measure in F'. The Fourier transform of μ is the nonlinear functional

$$\mathcal{L}(\varphi) := \int_{F'} \exp[i\eta(\varphi)]d\mu(\eta). \tag{29}$$

Definition 2.9. The nonlinear functional $\mathcal{L} : F \longrightarrow \mathbb{C}$ on F, defined by (29), is called positive definite, if for all $f_j \in F$ and $\lambda_j \in \mathbb{C}$, $j = \overline{1,n}$, the condition

$$\sum_{j,k=1}^{n} \bar{\lambda}_j \mathcal{L}(f_k - f_j)\lambda_k \geq 0 \tag{30}$$

holds for any $n \in \mathbb{Z}_+$.

Proposition 2.3. *The functional $\mathcal{L} : F \longrightarrow \mathbb{C}$ on F , defined by (29), is the Fourier transform of a cylindrical measure on F', if and only if it is positive definite, sequentially continuous and satisfies the condition $\mathcal{L}(0) = 1$.*

Suppose now that we have a continuous unitary representation of the unitary family $\mathcal{U}$ in a Hilbert space Φ with a cyclic vector $|\Omega\rangle \in \Phi$. Then we can put

$$\mathcal{L}(\varphi) := \langle\Omega|U(\varphi)|\Omega\rangle \tag{31}$$

for any $\varphi \in F := \mathcal{S}$, the Schwartz space on $\mathbb{R}^m$, and observe that functional (31) is continuous on F owing to the continuity of the representation. Therefore, this functional is the generalized Fourier transform of a cylindrical measure μ on $\mathcal{S}'$:

$$\langle\Omega|U(\varphi)|\Omega\rangle = \int_{\mathcal{S}'} \exp[i\eta(\varphi)]d\mu(\eta). \tag{32}$$

From the spectral point of view, based on Theorem 2.1, there is an isomorphism between the Hilbert spaces Φ and $L_2^{(\mu)}(\mathcal{S}';\mathbb{C})$, defined by $|\Omega\rangle \longrightarrow \Omega(\eta) = 1$ and $U(\varphi)|\Omega\rangle \longrightarrow \exp[i\eta(\varphi)]$ and next extended by linearity upon the whole Hilbert space Φ.

In the the non-cyclic case there exists a finite or countably infinite family of measures $\{\mu_k : k \in \mathbb{Z}_+\}$ on $\mathcal{S}'$, with $\Phi \simeq \oplus_{k\in\mathbb{Z}_+} L_2^{(\mu_k)}(\mathcal{S}';\mathbb{C})$ and the unitary operator $U(\varphi) : \Phi \longrightarrow \Phi$ for any $\varphi \in \mathcal{S}'$ corresponds in all $L_2^{(\mu_k)}(\mathcal{S}';\mathbb{C})$, $k \in \mathbb{Z}_+$, to $\exp[i\eta(\varphi)]$. This means that there exists a single cylindrical measure μ on $\mathcal{S}'$ and a $\mu-$measurable field of Hilbert spaces Φ_η on $\mathcal{S}'$, such that

$$\Phi \simeq \int_{\mathcal{S}'}^{\oplus} \Phi_\eta d\mu(\eta), \tag{33}$$

with $U(\varphi) : \Phi \longrightarrow \Phi$, corresponding [14] to the operator of multiplication by $\exp[i\eta(\varphi)]$ for any $\varphi \in \mathcal{S}$ and $\eta \in \mathcal{S}'$. Thereby, having constructed the nonlinear functional (29) in an exact analytical form, one can retrieve the representation of the unitary family $\mathcal{U}$ in the corresponding Hilbert space Φ of the Fock type, making use of the suitable factorization (25) as follows: $\Phi = \oplus_{n\in\mathbb{Z}_+}\Phi_n$, where

$$\Phi_n = \underset{f_n \in L_{2,s}((R^m)^n;\mathbb{C})}{span} \left\{ \prod_{j=\overline{1,n}} a^+(x_j)|\Omega\rangle \right\}, \tag{34}$$

for all $n \in \mathbb{Z}_+$. The cyclic vector $|\Omega\rangle \in \Phi$ can be, in particular, obtained as the ground state vector of some unbounded self-adjoint positive definite Hamilton operator $\mathbb{H} : \Phi \longrightarrow \Phi$, commuting with the self-adjoint particles number operator

$$\mathbb{N} := \int_{\mathbb{R}^m} \rho(x)dx, \tag{35}$$

that is $[\mathbb{H}, \mathbb{N}] = 0$. Moreover, the conditions

$$\mathbb{H}|\Omega\rangle = 0 \tag{36}$$

and

$$\inf_{g\in dom\mathbb{H}} \langle g, \mathbb{H}g\rangle = \langle\Omega|\mathbb{H}|\Omega\rangle = 0 \tag{37}$$

hold for the operator $\mathbb{H} : \Phi \longrightarrow \Phi$, where $dom\,\mathbb{H}$ denotes the domain of definition of $\mathbb{H}$.

To find the functional (31), which is called the generating Bogolubov type functional for moment distribution functions

$$F_n(x_1, x_2, ..., x_n) := \langle\Omega| : \rho(x_1)\rho(x_2)...\rho(x_n) : |\Omega\rangle, \tag{38}$$

where $x_j \in \mathbb{R}^m$, $j = \overline{1,n}$, and the normal ordering operation $: \cdot :$ is defined as

$$: \rho(x_1)\rho(x_2)...\rho(x_n) : = \prod_{j=1}^{n} \left(\rho(x_j) - \sum_{k=1}^{j} \delta(x_j - x_k) \right), \tag{39}$$

it is convenient to choose the Hamilton operator $\mathbb{H} : \Phi \longrightarrow \Phi$ in the [5,15,16] algebraic form

$$\mathbb{H} := \frac{1}{2} \int_{\mathbb{R}^m} K^+(x)\rho^{-1}(x)K(x)dx + V(\rho), \tag{40}$$

which is equivalent in the Hilbert space Φ to the positive definite operator expression

$$\mathbb{H} := \frac{1}{2} \int_{\mathbb{R}^m} (K^+(x) - A(x;\rho))\rho^{-1}(x)(K(x) - A(x;\rho))dx, \tag{41}$$

where $A(x;\rho) : \Phi \rightarrow \Phi$, $x \in \mathbb{R}^m$, is some specially chosen linear self-adjoint operator. The "potential" operator $V(\rho) : \Phi \longrightarrow \Phi$ is, in general, a polynomial (or analytical) functional of the density operator $\rho(x) : \Phi \longrightarrow \Phi$ and the operator is given as

$$K(x) := \nabla_x \rho(x)/2 + iJ(x), \tag{42}$$

where the self-adjoint "current" operator $J(x) : \Phi \longrightarrow \Phi$ can be defined (but non-uniquely) from the equality

$$\partial\rho/\partial t = \frac{1}{i}[\mathbb{H}, \rho(x)] = - < \nabla_x \cdot J(x) >, \tag{43}$$

holding for all $x \in \mathbb{R}^m$. Such an operator $J(x) : \Phi \longrightarrow \Phi$, $x \in \mathbb{R}^m$ can exist owing to the commutation condition $[\mathbb{H}, \mathbb{N}] = 0$, giving rise to the continuity relationship (43), if taking into account that supports *supp* ρ of the density operator $\rho(x) : \Phi \longrightarrow \Phi$, $x \in \mathbb{R}^m$, can be chosen arbitrarily owing to the independence of (43) on the potential operator $V(\rho) : \Phi \longrightarrow \Phi$, but its strict dependence on the corresponding representation (33). Denote also that representation (41) holds only under the condition that there exists such a self-adjoint operator $A(x;\rho) : \Phi \longrightarrow \Phi$, $x \in \mathbb{R}^m$, that

$$K(x)|\Omega\rangle = A(x;\rho)|\Omega\rangle \tag{44}$$

for all ground states $|\Omega\rangle \in \Phi$, correspond to suitably chosen potential operators $V(\rho) : \Phi \longrightarrow \Phi$.

The self-adjointness of the operator $A(x;\rho) : \Phi \longrightarrow \Phi$, $x \in \mathbb{R}^m$, can be stated following schemes from works [5,16], under the additional condition of the existence of such a linear anti-unitary mapping $T : \Phi \longrightarrow \Phi$ that the invariance conditions

$$T\rho(x)T^{-1} = \rho(x), \qquad T\, J(x)\, T^{-1} = -J(x), \qquad T|\Omega\rangle = |\Omega\rangle \tag{45}$$

hold for any $x \in \mathbb{R}^m$. Therefore, owing to conditions (45), the expressions

$$K^*(x)|\Omega\rangle = A(x;\rho)|\Omega\rangle = K(x)|\Omega\rangle \tag{46}$$

hold for any $x \in \mathbb{R}^m$, giving rise to the self-adjointness of the operator $A(x;\rho) : \Phi \longrightarrow \Phi$, $x \in \mathbb{R}^m$.

Based now on the construction above one easily deduces from expression (43) that the generating Bogolubov type functional (31) obeys for all $x \in \mathbb{R}^m$ the functional-differential equation

$$[\nabla_x - i\nabla_x\varphi]\,\frac{1}{2i}\frac{\delta\mathcal{L}(\varphi)}{\delta\varphi(x)} = A\left(x;\frac{1}{i}\frac{\delta}{\delta\varphi}\right)\mathcal{L}(\varphi), \tag{47}$$

whose solutions should satisfy the Fourier transform representation (32). In particular, a wide class of special so-called Poissonian white noise type solutions to the functional-differential equation (47) was obtained in [5,16] by means of functional-operator methods in the generalized form

$$\mathcal{L}(\varphi) = \exp\left\{A\left(\frac{1}{i}\frac{\delta}{\delta\varphi}\right)\right\}\exp\left(\bar{\rho}\int_{\mathbb{R}^m}\{\exp[i\varphi(x)]-1\}dx\right), \tag{48}$$

where $\bar{\rho} := \langle\Omega|\rho|\Omega\rangle \in \mathbb{R}_+$ is a Poisson distribution density parameter.

Consider now the case that the basic Fock space $\Phi = \otimes_{j=1}^{s}\Phi^{(j)}$, where $\Phi^{(j)}$, $j = \overline{1,s}$, are Fock spaces corresponding to the different types of independent cyclic vectors $|\Omega_j\rangle \in \Phi^{(j)}$, $j = \overline{1,s}$. This, in particular, means that the suitably constructed creation and annihilation operators $a_j(x), a_k^+(y) : \Phi \longrightarrow \Phi$, $j,k = \overline{1,s}$, satisfy the commutation relations

$$\begin{gathered}[a_j(x), a_k(y)] = 0,\\ [a_j(x), a_k^+(y)] = \delta_{jk}\delta(x-y)\end{gathered} \tag{49}$$

for any $x, y \in \mathbb{R}^m$.

Definition 2.10. A vector $|u\rangle \in \Phi$, $x \in \mathbb{R}^m$, is called coherent with respect to a mapping $u \in L_2(\mathbb{R}^m;\mathbb{R}^s) := M$, if it satisfies the eigenfunction condition

$$a_j(x)|u\rangle = u_j(x)|u\rangle \tag{50}$$

for each $j = \overline{1,s}$ and all $x \in \mathbb{R}^m$.

It is easy to check that the coherent vectors $|u\rangle \in \Phi$ exist. Indeed, the vector expression

$$|u\rangle := \exp\{(u, a^+)\}|\Omega\rangle, \tag{51}$$

where $(.,.)$ is the standard scalar product in the Hilbert space M, satisfies the defining condition (50), and moreover, the norm

$$\|u\|_\Phi := \langle u|u\rangle^{1/2} = \exp(\frac{1}{2}\|u\|^2) < \infty, \tag{52}$$

since $u \in M$ and its norm $\|u\| := (u, u)^{1/2}$ is bounded.

3. The Fock space embedding method, nonlinear dynamical systems and their complete linearization

Consider any function $u \in M := L_2(\mathbb{R}^m; \mathbb{R}^s)$ and observe that the Fock space embedding mapping

$$\xi : M \ni u \longrightarrow |u\rangle \in \Phi, \tag{53}$$

defined by means of the coherent vector expression (51) realizes a smooth isomorphism between Hilbert spaces M and Φ. The inverse mapping $\xi^{-1} : \Phi \longrightarrow M$ is given by the exact expression

$$u(x) = \langle \Omega|a(x)|u\rangle, \tag{54}$$

holding for almost all $x \in \mathbb{R}^m$. Owing to condition (52), one finds from (54) that the corresponding function $u \in M$.

In the Hilbert space M, let's now define a nonlinear dynamical system (which can, in general, be non-autonomous) in partial derivatives

$$du/dt = K[u], \tag{55}$$

where $t \in \mathbb{R}_+$ is the corresponding evolution parameter, $[u] := (t, x; u, u_x, u_{xx}, ..., u_{rx}), r \in \mathbb{Z}_+$, and a mapping $K : M \longrightarrow T(M)$ is Frechet smooth. Assume also that the Cauchy problem

$$u|_{t=+0} = u_0 \tag{56}$$

is solvable for any $u_0 \in M$ in an interval $[0, T) \subset \mathbb{R}^1_+$ for some $T > 0$. Thereby, the smooth evolution mapping is defined

$$T_t : M \ni u_0 \longrightarrow u(t|u_0) \in M, \tag{57}$$

for all $t \in [0, T)$.

It is now natural to consider the commutative diagram

$$\begin{array}{ccc} M & \xrightarrow{\xi} & \Phi \\ T_t \downarrow & & \downarrow \mathbb{T}_t \\ M & \xrightarrow{\xi} & \Phi, \end{array} \tag{58}$$

where the mapping $\mathbb{T}_t : \Phi \longrightarrow \Phi$, $t \in [0, T)$, is defined from the conjugation relationship

$$\xi \circ T_t = \mathbb{T}_t . \circ \xi \tag{59}$$

Now take a coherent vector $|u_0\rangle \in \Phi$, corresponding to $u_0 \in M$, and construct the vector

$$|u\rangle := \mathbb{T}_t \cdot |u_0\rangle \tag{60}$$

for all $t \in [0, T)$. Since vector (60) is, by construction, coherent, that is

$$a_j(x)|u\rangle := u_j(x, t|u_0)|u\rangle \tag{61}$$

for each $j = \overline{1, s}$, $t \in [0, T)$ and almost all $x \in \mathbb{R}^m$, owing to the smoothness of the mapping $\xi : M \longrightarrow \Phi$ with respect to the corresponding norms in the Hilbert spaces M and Φ, we derive that coherent vector (60) is differentiable with respect to the evolution parameter $t \in [0, T)$. Thus, one can easily find [25,26] that

$$\frac{d}{dt}|u\rangle = \hat{K}[a^+, a]|u\rangle, \tag{62}$$

where

$$|u\rangle|_{t=+0} = |u_0\rangle \tag{63}$$

and a mapping $\hat{K}[a^+, a] : \Phi \longrightarrow \Phi$ is defined by the exact analytical expression

$$\hat{K}[a^+, a] := (a^+, K[a]). \tag{64}$$

As a result of the consideration above we obtain the following theorem.

Theorem 3.1. *Any smooth nonlinear dynamical system (55) in Hilbert space $M := L_2(\mathbb{R}^m; \mathbb{R}^s)$ is representable by means of the Fock space embedding isomorphism $\xi : M \longrightarrow \Phi$ in the completely linear form (62).*

We now make some comments concerning the solution to the linear equation (62) under the Cauchy condition (63). Since any vector $|u\rangle \in \Phi$ has the series representation

$$|u\rangle = \bigoplus_{n=\sum_{j=1}^{s} n_j \in \mathbb{Z}_+} \frac{1}{(n_1!\dots n_s!)^{1/2}} \int_{(\mathbb{R}^m)^n} f^{(n)}_{n_1\dots n_s}(x_1^{(1)},\dots,x_{n_1}^{(1)}; \\ x_1^{(2)},\dots,x_{n_2}^{(2)};\dots;x_1^{(s)},\dots,x_{n_s}^{(s)}) \prod_{j=1}^{s} \left(\prod_{k=1}^{n_j} dx_k^{(j)} a_j^{+}(x_k^{(j)})\right) |\Omega\rangle, \tag{65}$$

where for any $n = \sum_{j=1}^{s} n_j \in \mathbb{Z}_+$,

$$f^{(n)}_{n_1\dots n_s} \in \bigotimes_{j=1}^{s} L_{2,s}((\mathbb{R}^m)^{n_j};\mathbb{C}) \simeq L_{2,s}(\mathbb{R}^{mn_1} \times \dots \times \mathbb{R}^{mn_s};\mathbb{C}), \tag{66}$$

and

$$\|u\|_{\Phi}^2 = \sum_{n=\sum_{j=1}^{s} n_j \in \mathbb{Z}_+} \|f^{(n)}_{n_1 n_2\dots n_s}\|_2^2 = \exp(\|u\|^2). \tag{67}$$

By substituting (65) into equation (62), reduces (62) to an infinite recurrent set of linear evolution equations in partial derivatives on coefficient functions (66). The latter can often be solved [25] step by step analytically in exact form, thereby, making it possible to obtain, owing to representation (54), the exact solution $u \in M$ to the Cauchy problem (56) for our nonlinear dynamical system in partial derivatives (55).

Remark 3.1. Concerning some applications of nonlinear dynamical systems like (53) in mathematical physics problems, it is very important to construct their so-called conservation laws or smooth invariant functionals $\gamma : M \longrightarrow \mathbb{R}$ on M. Making use of the quantum mathematics technique described above one can suggest an effective algorithm for constructing these conservation laws in exact form.

Indeed, consider a vector $|\gamma\rangle \in \Phi$, satisfying the linear equation

$$\frac{\partial}{\partial t}|\gamma\rangle + \hat{K}^*[a^+, a]|\gamma\rangle = 0. \tag{68}$$

Then the following result [25] holds.

Proposition 3.1. *The functional*

$$\gamma := \langle u|\gamma\rangle \tag{69}$$

is a conservation law for dynamical system (53), that is,

$$d\gamma/dt|_K = 0 \tag{70}$$

along any orbit of the evolution mapping (57).

4. Conclusion

Within the scope of this work we have described the main mathematical preliminaries and properties of the quantum mathematics techniques suitable for analytical studying of the important linearization problem for a wide class of nonlinear dynamical systems in partial derivatives in Hilbert spaces. This problem was analyzed in much detail using the Gelfand-Vilenkin representation theory [14] of infinite dimensional groups and the Goldin-Menikoff-Sharp theory [15–17] of generating Bogolubov type functionals, classifying these representations. The related problem of constructing Fock type space representations and retrieving their creation-annihilation generating structure still needs a deeper investigation within the approach devised. Here we mention only that some aspects of this problem within the so-called Poissonian white noise analysis were studied in a series of works [1,2,24,27], based on some generalizations of the Delsarte type characters technique. It is also necessary to mention the related results obtained in [23,25,26], devoted to the application of the Fock space embedding method to finding conservation laws and the so called recursion operators for the well known Korteweg-de Vries type nonlinear dynamical systems. We plan to devote our next investigations to concerning some important applications of the methods devised in the work to concrete dynamical systems.

Acknowledgements

Two of the authors (N.B. and A.P.) are cordially thankful to the Abdus Salam International Center for Theoretical Physics in Trieste, Italy, for the hospitality during their 2007 research scholarships. A.P. is also thankful to Profs. Y. Kondratyev and Y. Kozicki for interesting discussions of the results during the Third International Conference on Infinite Dimensional Systems, held June 23–28, 2007 in Kazimierz Dolny, Poland, and to Profs. O. Çelebi, K. Zheltukhin, G. Barsegian, O. Zhdanov and V. Golubeva during the 2007 International ISAAC Congress, held August 13–18, 2007 in Ankara, Turkey. The last but not least thanks belong to the referee of the article for many insightful and qualified comments which helped in an essentialway in improving the exposition.

References

1. S. Albeverio, Y. G. Kondratiev and L. Streit, How to generalize white noise analysis to non-Gaussian measures, Preprint Bi Bo S, Bielefeld, 1992.
2. Y. M. Berezansky, A generalization of white noise analysis by means of theory of hypergroups, Reports on Math. Phys., 38, N.3 (1996), pp. 289–300.

3. Y. M. Berezansky and Y. G. Kondratiev, Spectral methods in infinite dimensional analysis, v.1 and 2, Kluwer, 1995.
4. N. N. Bogolubov and N. N. Bogolubov, Jr., Introduction into quantum statistical mechanics. World Scientific, NJ, 1986.
5. N. N. Bogoliubov, Jr., A. K. Prykarpatsky, Quantum method of generating Bogolubov functionals in statistical physics: Current Lie algebras, their representations and functional equations, Physics of Elementary Particles and Atomic Nucleus, v.17, N.4 (1986), pp. 791–827 (in Russian).
6. F. A. Berezin, The second quantization method, Nauka Publisher, Moscow, 1986 (in Russian).
7. N. N. Bogoliubov, Collected works, v.2, Naukova Dumka, Kiev, 1960 (in Russian).
8. P. A. M. Dirac, The principles of quantum mechanics, Oxford University Press, 1932.
9. J. Dixmier, C^*-algebras, Amsterdam, North-Holland, 1982.
10. L. D. Faddeev, E. K. Sklyanin, Quantum mechanical approach to completely integrable field theories, Proceed. of the USSR Academy of Sciences (DAN), 243 (1978), pp. 1430–1433 (in Russian).
11. R. Feynman, Quantum mechanical computers, Found. Physics, 16 (1986), pp. 507–531.
12. R. Feynman, Simulating physics with computers. Intern. Journal of Theor. Physics, 21 (1982), pp. 467–488.
13. V. A. Fock, Konfigurationsraum und zweite Quantelung, Zeischrift Phys., Bd. 75 (1932), pp. 622–647.
14. I. Gelfand and N. Vilenkin, Generalized functions, 4, Academic Press, New York, 1964.
15. G. A. Goldin, Nonrelativistic current algebras as unitary representations of groups, Journal of Mathem. Physics, 12(3), 1971, pp. 462–487.
16. G. A. Goldin, J. Grodnik, R. T. Powers and D. Sharp, Nonrelativistic current algebra in the N/V limit, J. Math. Phys., 15, (1974), pp. 88–100.
17. G. A. Goldin, R. Menikoff and F. H. Sharp, Diffeomorphism groups, gauge groups, and quantum theory. Phys. Rev. Lett. 51 (1983), pp. 2246–2249.
18. L. K. Grover, Quantum mechanics helps in searching for a needle in a haystack, Phys. Rev. Lett., 79 (1997), pp. 325–328.
19. A. Jaffe and F. Quinn, Theoretical mathematics: Toward a cultural synthesis of mathematics and theoretical physics, Bull. Amer. Math. Soc., 29 (1993), pp. 1–13 Zeischrift Phys., Bd. 75 (1932), pp. 622–647.
20. D. Deutsch, Quantum theory, the Church-Turing principle and the universal quantum computer, Proc. Roy. Soc. London, A400 (1985), pp. 97–117.
21. V. G. Drinfeld, Quantum groups, Proceed. of the Int. Congress of Mathematicians, 1986, pp. 798–820.
22. S. K. Donaldson, An application of gauge theory to four dimensional topology, J. Diff. Geom., 17 (1982), pp. 279–315.
23. K. Kowalski and W.-H. Steeb, Symmetries and first integrals for nonlinear dynamical systems: Hilbert space approach, I and II, Progress of Theoretical Physics, 85, N.4 (1991), pp. 713–722 and 85, N.4 (1991), pp. 975–983.

24. Y. G. Kondratiev, L. Streit, W. Westerkamp and J.-A. Yan, Generalized functions in infinite dimensional analysis, II, AS preprint, 1995.
25. K. Kowalski, Methods of Hilbert spaces in the theory of nonlinear dynamical systems, World Scientific, 1994.
26. K. Kowalski and W.-H. Steeb, Nonlinear dynamical systems and Carleman linearization, World Scientific, 1991.
27. E. W. Lytvynov, A. L. Rebenko and G. V. Shchepaniuk, Wick calculus on spaces of generalized functions compound Poisson white noise, Reports on Math. Phys., 39, N.2 (1997), pp. 219–247.
28. Y. I. Manin, Computable and uncomputable, Moscow, Sov. Radio, 1980 (in Russian).
29. Y. I. Manin, Classical computation, quantum computation and P. Shor's factoraizing algorithm, Seminaire N. Bourbaki, 1998–1999, exp. N.862, p. 375–404.
30. J. von Neumann, Mathematische Grundlagen der Quanten Mechanik, J. Springer, Berlin, 1932.
31. A. K. Prykarpatsky, U. Taneri and N. N. Bogolubov, Jr., Quantum field theory and application to quantum nonlinear optics, World Scientific, NY, 2002.
32. P. W. Shor, Polynomial time algorithms for prime factorization and discrete logarithms on a quantum computer, SIAM Journ. Comput., 26, N.5 (1997), pp. 1484–1509.
33. H. Weyl, The theory of groups and quantum mechanics, Dover, New York, 1931.
34. E. Witten, Nonabelian bosonization in two dimensions, Commun. Mathem. Physics, 92 (1984), pp. 455–472.

Session 17

General Session

SESSION EDITOR

A. Ö. Çelebi Yeditepe University, İstanbul, Turkey

COHEN p-NUCLEAR MULTILINEAR MAPPINGS

DAHMANE ACHOUR

Laboratoire de Mathématiques Pures et Appliquées, Université de M'sila, 28000 M'sila, Algeria,
E-mail: dachourdz@yahoo.fr

We define the concept of Cohen p-nuclear multilinear operators. Some properties and composition theorems are shown.

Keywords: Absolutely p-summing m-linear operator, Cohen strongly p-summing, Cohen p-nuclear multilinear operator.

1. Introduction and Notation

In a series of papers [1,2,5,8,9] the authors have developed the theory of summing multilinear operators, generalizing the nice behavior of linear p-summing operators to the multilinear setting. Among other things, we introduced the concept of multilinear operators of type p-nuclear. In the linear case the concept of Cohen p-nuclear operators $(1 \leq p \leq \infty)$ was introduced by Cohen in [4] and generalized to Cohen (p,q)-nuclear $(1 \leq q \leq \infty)$ by Apiola in [3]. A linear operator T between two Banach spaces X, Y is Cohen p-nuclear for $(1 < p < \infty)$ if there is a positive constant C such that for all $n \in \mathbb{N}$; $x_1, ..., x_n \in X$ and $y_1^*, ..., y_n^* \in Y^*$ we have

$$\left\| (\langle T(x_i), y_i^* \rangle)_{1 \leq i \leq n} \right\|_{l_1^n} \leq C \sup_{x^* \in B_{X^*}} \|(x^*(x_i))\|_{l_p^n} \sup_{y \in B_Y} \|(y_i^*(y))\|_{l_{p^*}^n}.$$

The smallest constant C, which is denoted by $n_p(T)$, such that the above inequality holds, is called the Cohen p-nuclear norm on the space $\mathcal{N}_p(X,Y)$ of all Cohen p-nuclear operators from X into Y which is a Banach space. For $p = 1$ and $p = \infty$ we have $\mathcal{N}_1(X,Y) = \pi_1(X,Y)$ (the Banach space of all 1-summing operators), and $\mathcal{N}_\infty(X,Y) = \mathcal{D}_\infty(X,Y)$ (the Banach space of all strongly ∞-summing operators).

In this paper, we introduce a multilinear generalization of the notion of Cohen p-nuclear linear operators. Some properties and composition theorems are shown.

This paper is organized as follows: In Section 1, we give some basic definitions and properties. We introduce in Section 2, a multilinear version of Cohen p-nuclear operators studied by Cohen for which the resulting vector space is a Banach space. We prove, if T verifies a domination theorem then T is p-nuclear we asked if the statement of this theorem characterizes p-nuclear operators. In Section 3, some composition theorems are obtained.

Now, we fix the notation used in this paper. Let $m \in \mathbb{N}$ and $X_1, ..., X_m, Y$ be Banach spaces over the real numbers, we denote by $\mathcal{L}(X_1, ..., X_m; Y)$ the Banach space of all continuous m-linear mappings from $X_1 \times ... \times X_m$ into Y, under the norm

$$\|T\| = \sup_{x_k \in B_{X_k}} \|T(x_1, ..., x_m)\|$$

where B_{X_k} denotes the closed unit ball of X_k centred at 0. The vector space of all bounded linear operators from X into Y will be noted by $\mathcal{B}(X, Y)$.

Let now X be a Banach space and $1 \leq p \leq \infty$. We denote by $l_p^n(X)$ the space of all sequences $(x_i)_{1 \leq i \leq n}$ in X with the norm

$$\left\| (x_i)_{1 \leq i \leq n} \right\|_{l_p^n(X)} = (\sum_{i=1}^{n} \|x_i\|^p)^{\frac{1}{p}}$$

and by $l_p^{n\ \omega}(X)$ the space of all sequences $(x_i)_{1 \leq i \leq n}$ in X with the norm

$$\|(x_n)\|_{l_p^{n\ \omega}(X)} = \sup_{\|\xi\|_{X^*}=1} (\sum_{i=1}^{n} |\langle x_i, \xi \rangle|^p)^{\frac{1}{p}}$$

where X^* denotes the dual (topological) of X. We know (see [6]) that $l_p(X) = l_p^{\omega}(X)$ for some $1 \leq p < \infty$ iff $dim(X)$ is finite. If $p = \infty$, we have $l_\infty(X) = l_\infty^{\omega}(X)$.

We begin by presenting the several classes of multilinear mappings related to the concept of absolutely summing operators.

• Consider m in $\mathbb{N}$. An m-linear operator $T \in \mathcal{L}(X_1, ..., X_m; Y)$ is Cohen strongly p- summing $(1 < p \leq \infty)$, if and only if there exists a constant $C > 0$ and a Radon probability measure μ on $B_{Y^{**}}$ such that for all $(x^1, ..., x^m) \in X^1 \times ... \times X^m$, we have

$$|\langle T(x^1, ..., x^m), y^* \rangle| \leq C \prod_{j=1}^{m} \|x^j\| \left(\int_{B_{Y^{**}}} |y^*(y^{**})|^{p^*} d\mu \right)^{\frac{1}{p^*}} \qquad (1)$$

Again the class of all Cohen strongly p-summing m-linear operators from $X_1 \times ... \times X_m$ into Y, which is denoted by $\mathcal{D}_p^m(X_1, ..., X_m; Y)$ is a Banach space with the norm $d_p^m(T)$ which is the smallest constant C such that the inequality (1) holds. For $p = 1$, we have $\mathcal{D}_1^m(X_1, ..., X_m, Y) = \mathcal{L}(X_1, ..., X_m; Y)$.

• One will say that an m-linear operator $T \in \mathcal{L}(X_1, ..., X_m; Y)$ is absolutely p-summing $(1 \leq p < \infty)$ if there is a constant $C > 0$ such that for any $n \in \mathbb{N}$ and $(x_i^j)_{1 \leq i \leq n} \subset X_j$ $(1 \leq j \leq m)$, we have

$$(\sum_{i=1}^{n} \|T(x_i^1, ..., x_i^m)\|^p)^{\frac{1}{p}} \leq C \prod_{j=1}^{m} \left\|(x_i^j)_{1 \leq i \leq n}\right\|_{l_p^{n\,\omega}(X_j)}. \tag{2}$$

The space of all absolutely p-summing m-linear mappings from $X_1 \times ... \times X_m$ into Y will be denoted by $\mathcal{L}_p^m(X_1, ..., X_m; Y)$, and the infimum of the C for which the inequality (2) always holds defines a norm, $\|T\|_p$ for $\mathcal{L}_p^m(X_1, ..., X_m; Y)$.

The Cohen strongly p-summing mappings was introduced by the work of Achour-Mezrag [1]. The multilinear mappings of absolutely summing type are motivated by Pietsch in [9] and more recently in [2,7,10].

2. General Properties

We introduce the following generalization of the concept p-nuclear operators defined in [4], also we give some properties concerning this class of operators.

Definition 2.1. An m-linear operator $T : X_1 \times ... \times X_m \longrightarrow Y$ is Cohen p-nuclear $(1 < p < \infty)$, if and only if there is a constant $C > 0$ such that for any $x_1^j, ..., x_n^j \in X_j$, $(j = 1, ..., m)$, and any $y_1^*, ..., y_n^* \in Y^*$, we have

$$\left|\sum_{i=1}^{n} \langle T(x_i^1, ..., x_i^m), y_i^* \rangle\right| \leq C \prod_{j=1}^{m} \left\|(x_i^j)\right\|_{l_p^{n\,\omega}(X_j)} \sup_{y \in B_Y} \|(y_i^*(y))\|_{l_{p^*}^n} \tag{3}$$

Again the class of all Cohen p-nuclear m-linear operators from $X_1 \times ... \times X_m$ into Y, which is denoted by $\mathcal{N}_p^m(X_1, ..., X_m; Y)$ is a Banach space with the norm $n_p^m(T)$ which is the smallest constant C such that the inequality (3) holds. It is clear that, for every $T \in \mathcal{N}_p^m(X_1, ..., X_m; Y)$, T is continuous and $\|T\| \leq n_p^m(T)$.

Example 2.1. Let K be a compact Hausdorff space, let μ be a positive regular Borel measure on K and let $1 \leq p < \infty$. Each $g \in L_p(\mu)$ defines an m-linear multiplication operator $T_g \in \mathcal{L}(^m C(K); L_p(\mu)), T_g(f^1, ..., f^m) = g \cdot f^1 \cdot ... \cdot f^m$. This map is Cohen p-nuclear and $n_p^m(T_g) = \|g\|_{L_p(\mu)}$.

Proposition 2.1. *a) For* $p = 1$, *we have* $\mathcal{N}_1^m(X_1, ..., X_m; Y) = \pi_1^m(X_1, ..., X_m; Y)$.

b) For $p = \infty$, *we have* $\mathcal{N}_\infty^m(X_1, ..., X_m; Y) = \mathcal{D}_\infty^m(X_1, ..., X_m; Y)$.

Proof. (a) Let $T \in \mathcal{N}_1^m(X_1,...,X_m;Y)$, then

$$\begin{aligned}\left|\sum_{i=1}^{n}\left\langle T\left(x_i^1,...,x_i^m\right),y_i^*\right\rangle\right| &\le n_1^m(T)\prod_{j=1}^{m}\left\|\left(x_i^j\right)\right\|_{l_1^{n\,\omega}(X_j)}\sup_{y\in B_Y}\left\|\left(y_i^*(y)\right)\right\|_{l_\infty^n}\\ &\le n_1^m(T)\prod_{j=1}^{m}\left\|\left(x_i^j\right)\right\|_{l_1^{n\,\omega}(X_j)}\left\|y_i^*\right\|\end{aligned}$$

On the other hand, we have

$$\sum_{i=1}^{n}\left\|T\left(x_i^1,...,x_i^m\right)\right\| = \sup\left\{\left|\sum_{i=1}^{n}\left\langle T\left(x_i^1,...,x_i^m\right),y_i^*\right\rangle\right|, \sup\left\|y_i^*\right\|\le 1\right\}.$$

This implies

$$\sum_{i=1}^{n}\left\|T\left(x_i^1,...,x_i^m\right)\right\| \le n_1^m(T)\prod_{j=1}^{m}\left\|\left(x_i^j\right)\right\|_{l_1^{n\,\omega}(X_j)}.$$

Then T is absolutely 1-summing and $\|T\|_1 \le n_1^m(T)$.

Conversely, let T be a absolutely 1-summing m-linear operator, we have

$$\begin{aligned}\left|\sum_{i=1}^{n}\left\langle T\left(x_i^1,...,x_i^m\right),y_i^*\right\rangle\right| &\le \sup_i\left\|y_i^*\right\|\sum_{i=1}^{n}\left\|T\left(x_i^1,...,x_i^m\right)\right\|\\ &\le \|T\|_1\prod_{j=1}^{m}\left\|\left(x_i^j\right)\right\|_{l_1^{n\,\omega}(X_j)}\sup_i\left\|y_i^*\right\|\\ &\le \|T\|_1\prod_{j=1}^{m}\left\|\left(x_i^j\right)\right\|_{l_1^{n\,\omega}(X_j)}\sup_{y\in B_Y}\left\|\left(y_i^*(y)\right)\right\|_{l_\infty^n}.\end{aligned}$$

This T is 1-nuclear m-linear operator and $n_1^m(T) \le \|T\|_1$.

(b) It is obvious, and this completes our proof. □

Proposition 2.2. *Consider $m \in \mathbb{N}$ and $T \in \mathcal{L}(X_1,...,X_m;Y)$. The operator T is Cohen p-nuclear if and only if, for all $n \in \mathbb{N}$ and all $v \in \mathcal{B}(l_p^n, Y^*)$, we have*

$$\left|\sum_{i=1}^{n}\left\langle T\left(x_i^1,...,x_i^m\right),v\left(e_i\right)\right\rangle\right| \le C\prod_{j=1}^{m}\left\|\left(x_i^j\right)\right\|_{l_p^{n\,\omega}(X_j)}\|v\|.$$

Proof. Let $v : l_{p^*} \longrightarrow X$ be a linear operator such that $v(e_i) = x_i$ (namely, $v = \sum_1^\infty e_j \otimes x_j$, e_j denotes the unit vector basis of l_p). We can use $\|v\| = \|(x_n)\|_{l_p^\omega(X)}$, as we wanted to prove. □

In the next theorem we shall present an integral condition which is sufficient to guarantee that an multilinear operator T belong to $\mathcal{N}_p^m(X_1,...,X_m;Y)$.

Theorem 2.1. *Let $T \in \mathcal{L}(X_1, ..., X_m; Y)$ and $1 < p < \infty$. Suppose there exists a constant $C \geq 0$ and positive Radon probability measures $\mu_j \in C(B_{X_j^*})^*$ $(1 \leq j \leq m)$ and $\lambda \in C(B_{Y^{**}})^*$ such that for all $(x^1, ..., x^m) \in X_1 \times .... \times X_m$ and $y^* \in Y^*$, we have*

$$\left|\left\langle T\left(x^1, ..., x^m\right), y^*\right\rangle\right| \leq C \prod_{j=1}^{m} \|x_j\|_{L_p(B_{X_j^*}, \mu_j)} \|y^*\|_{L_{p^*}(B_{Y^{**}}, \lambda)}. \tag{4}$$

Then T is Cohen p-nuclear and $n_p^m(T) \leq C$.

Proof. Let $(x_i^1, ..., x_i^m) \in X_1 \times .... \times X_m$ and $y_i^* \in Y^*$ by (4), then

$$\left|\left\langle T\left(x_i^1, ..., x_i^m\right), y_i^*\right\rangle\right| \leq C \prod_{j=1}^{m} \|x_j\|_{L_p(B_{X_j^*}, \mu_j)} \|y^*\|_{L_{p^*}(B_{Y^{**}}, \lambda)},$$

for all $1 \leq i \leq n$, and

$$\begin{aligned}\left|\sum_{i=1}^{n} \left\langle T\left(x_i^1, ..., x_i^m\right), y_i^*\right\rangle\right| &\leq \sum_{i=1}^{n} \left|\left\langle T\left(x_i^1, ..., x_i^m\right), y_i^*\right\rangle\right| \\ &\leq C \sum_{i=1}^{n} \Big(\prod_{j=1}^{m} \left\|x_i^j\right\|_{L_p(B_{X_j^*}, \mu_j)} \|y_i^*\|_{L_{p^*}(B_{Y^{**}}, \lambda)}\Big).\end{aligned}$$

We can use Hölder's inequality in order to write

$$\begin{aligned}&\left|\sum_{i=1}^{n} \left\langle T\left(x_i^1, ..., x_i^m\right), y_i^*\right\rangle\right| \\ &\leq C\Big(\sum_{i=1}^{n} \prod_{j=1}^{m} \|x_j\|^p_{L_p(B_{X_j^*}, \mu_j)}\Big)^{\frac{1}{p}} \Big(\sum_{i=1}^{n} \|y^*\|_{L_{p^*}(B_{Y^{**}}, \lambda)}{}^{p^*}\Big)^{\frac{1}{p^*}} \\ &\leq C \prod_{j=1}^{m} \Big(\sum_{i=1}^{n} \int_{B_{X_j^*}} |\langle x_i^j, x^{j*}\rangle|^p d\mu_j(x^{j*})\Big)^{\frac{1}{p}} \Big(\sum_{i=1}^{n} \int_{B_{Y^{**}}} |y_i^*(y^{**})|^{p^*} d\lambda(y^{**})\Big)^{\frac{1}{p^*}} \\ &\leq C \prod_{j=1}^{m} \sup_{x_j^* \in B_{X_j^*}} \Big(\sum_{i=1}^{n} |x^{j*}(x_i^j)|^p\Big)^{\frac{1}{p}} \sup_{y \in B_Y} \Big(\sum_{i=1}^{n} |y_i^*(y)|^{p^*}\Big)^{\frac{1}{p^*}}\end{aligned}$$

Therefore T is p-nuclear and $n_p^m(T) \leq C$, as we wanted to prove. □

Question 2.1. Does $T \in \mathcal{L}(X_1, ..., X_m; Y)$ Cohen p-nuclear imply there exist Radon probability measures $\mu_j \in C(B_{X_j^*})^*$ $(1 \leq j \leq m)$ and $\lambda \in C(B_{Y^{**}})^*$ such that for all $(x^1, ..., x^m) \in X_1 \times .... \times X_m$ and $y^* \in Y^*$, we have (4)?

3. Composition Theorems

In this section, we give some composition theorems. At first, it is easy to verify the following result.

Proposition 3.1. *Let $T \in \mathcal{L}(X_1,...,X_m;Y)$, $w \in \mathcal{L}(Y,Z)$ and $u_j \in \mathcal{L}(E_j,X_j)\,(1 \leq j \leq m)$.*

(i) If T is Cohen p-nuclear, then $w \circ T$ is Cohen p-nuclear and $n_p^m(w \circ T) \leq \|w\|\, n_p^m(T)$.

(ii) If T is Cohen p-nuclear, then $T \circ (u_1,...,u_m)$ is Cohen p-nuclear and $n_p^m(T \circ (u_1,...,u_m)) \leq n_p^m(T) \prod_{j=1}^{m} \|u_j\|$.

Theorem 3.1. *Let $m \in \mathbb{N}$ and $X_1,...,X_m$, Y, Z be Banach spaces. Let $u : Y \to Z$ be a strongly p-summing operator and let $S : Y_1 \times ... \times Y_m \longrightarrow Y$ be absolutely p-summing. Then, the operator $T = uS : X_1 \times .... \times X_m \longrightarrow Z$, is p-nuclear and $n_p^m(T) \leq \|S\|_p\, \pi_{p*}(u^*)$.*

Proof. Let $(x_i^j) \subset X_j$ and $(z_i^*) \subset Z^*$. We have

$$\begin{aligned}
\sum_{i=1}^{n} \langle T(x_i^1,...x_i^m), z_i^*\rangle &\leq \sum_{i=1}^{n} |\langle S(x_i^1,...x_i^m), u^*(z_i^*)\rangle| \\
&\leq (\sum_{i=1}^{n} \|S(x_i^1,...,x_i^m)\|^p)^{\frac{1}{p}} (\sum_{i=1}^{n} \|u^*(z_i^*)\|^{p^*})^{\frac{1}{p^*}} \\
&\leq \|S\|_p \prod_{j=1}^{m} \left\|(x_i^j)\right\|_{l_p^{n\,\omega}(X_j)} (\sum_{i=1}^{n} \|u^*(z_i^*)\|^{p^*})^{\frac{1}{p^*}}
\end{aligned}$$

since u^* is p^*-summing. Then

$$\begin{aligned}
&\left|\sum_{i=1}^{n} \langle T(x_i^1,...x_i^m), z_i^*\rangle\right| \\
&\leq \|S\|_p\, \pi_{p*}(u^*) \prod_{j=1}^{m} \left\|(x_i^j)\right\|_{l_p^{n\,\omega}(X_j)} \sup_{z \in B_{Z^{**}}} (\sum_{i=1}^{n} |\langle z_i^*, z^{**}\rangle|^{p^*})^{\frac{1}{p^*}},
\end{aligned}$$

and thus T is p-nuclear and $n_p^m(T) \leq \|S\|_p\, \pi_{p*}(u^*)$. □

Theorem 3.2. *Let $X_1,...,X_m$, $Y_1,...,Y_m$; Z be Banach spaces, with $\dim(Z^*) < \infty$. Let, for every $1 \leq j \leq m$, $v_j : X_j \to Y_j$ be a p-summing operator and let $R : Y_1 \times .... \times Y_m \longrightarrow Z$ be absolutely p-summing. Then the operator $T = R(v_1,...,v_m) : X_1 \times ... \times X_m \longrightarrow Z$ is p-nuclear and*

$$n_p^m(T) \leq \|R\|_p \prod_{j=1}^{m} \pi_p(v_j).$$

Proof. Let $x^1, ..., x^m \in X_1 \times ... \times X_m$ and $z^* \in Z^*$. We can use Hölder's inequality and (2) to obtain

$$\begin{aligned}&\left|\sum_{i=1}^{n} \langle R\left(v_1\left(x_i^1\right), ..., v_m\left(x_i^m\right)\right), z_i^*\rangle\right| \\ &\leq (\sum_{i=1}^{n} \left\|R\left(v_1\left(x_i^1\right), ..., v_m\left(x_i^m\right)\right)\right\|^p)^{\frac{1}{p}} (\sum_{i=1}^{n} \|z_i^*\|^{p^*})^{\frac{1}{p^*}} \\ &\leq \|R\|_p \prod_{j=1}^{m} \left\|(v_j(x_i^j))_{1 \leq i \leq n}\right\|_{l_p^{n,\omega}} (\sum_{i=1}^{n} \|z_i^*\|^{p^*})^{\frac{1}{p^*}}.\end{aligned}$$

On the other hand, $\left\|(v_j(x_i^j))_{1 \leq i \leq n}\right\|_{l_p^{n,\omega}} \leq (\sum_{i=1}^{n} \left\|v_j(x_i^j)\right\|^p)^{\frac{1}{p}}$. Because v_j is p-summing and $\dim(Z^*) < \infty$, we have

$$\begin{aligned}&\sum_{i=1}^{n} \left|\langle R\left(v_1\left(x_i^1\right), ..., v_m\left(x_i^m\right)\right), z_i^*\rangle\right| \\ &\leq \|R\|_p \prod_{j=1}^{m} \pi_p\left(v_j\right) \prod_{j=1}^{m} \left\|(x_i^j)\right\|_{l_p^{n\,\omega}(X_j)} \|(z_i^*)\|_{l_{p^*}^{n\,\omega}(Z^*)},\end{aligned}$$

and hence T is p-nuclear, and $n_p^m(T) \leq \|R\|_p \prod_{j=1}^{m} \pi_p\left(v_j\right)$. □

References

1. D. Achour and L. Mezrag, On the Cohen strongly p-summing multilinear operators, *J. Math. Anal. Appl.* 327 (2007), 550–563.
2. R. Alencar and M. C. Matos, Some classes of multilinear mappings between Banach spaces, *Pub. Dep. An. Mat. Univ. Complut. Madrid* 12 (1989).
3. H. Apiola, Duality between spaces of p-summable sequences, (p, q)-summing operators and characterizations of nuclearity, *Math. Ann.* 219 (1976), 53–64.
4. J. S. Cohen, Absolutely p-summing, p-nuclear operators and their conjugates, *Math. Ann.* 201 (1973), 177–200.
5. V. Dimant, Strongly p-summing multilinear operators, *J. Math. Anal. Appl.* 278 (2003), 182–193.
6. J. Diestel, H. Jarchow, A. Tonge, Absolutely summing operators, *Cambridge University Press*, 1995.
7. S. Geiss, Ideale multilinearer Abbildungen, *Diplomarbeit*, 1984.
8. D. Pérez-Garcìa, I. Villanueva, Multiple summing operators on Banach spaces, *J. Math. Anal. Appl.* 285 (2003), 86–96.
9. A. Pietsch, Ideals of multilinear functionals: Designs of a theory, *Proceedings of the Second International Conference on Operator Algebras, Ideals, and their Applications in Theoretical Physics*, Leipzig, Teubner-Texte, 1983, 185–199.
10. B. Schneider, On absolutely p-summing and related multilinear mappings, *Wissenchaftliche Zeitschrift der Brandenburger Landeshochschule* 35 (1991), 105–117.

NUMERICAL MODELING OF A SYSTEM OF MUTUAL REACTION-DIFFUSION TYPE

MERSAID ARIPOV*

National University of Uzbekistan, Uzbekistan
E-mail: mirsaidaripov@mail.ru

ABDUGAPPAR KHYDAROV

National University of Uzbekistan, Uzbekistan

The methods of numerical modeling of nonlinear processes of mutual diffusion-reaction systems are offered. For modeling of diffusion-reaction systems, a degenerate-type parabolic system equations are used. The properties of weak solutions of initial-boundary value problems for the nonlinear reaction-diffusion systems in the presence of convective transfer, the volumetric absorption or source using a self-similar approach are investigated.

1. Introduction

We consider n components of a gas reacting with each other in the course of R independent reactions at constant temperature. The system of equations of local balance of mass in case of n components $u_i(t,x)$ $(i = \overline{1,n})$ in nonlinear media at the presence of convective transfer may be written in the form [1-2]

$$\frac{\partial u_i}{\partial t} = \nabla \left(u_{3-i}^{s_{3-i}} \nabla u_i\right) - div(v_i(t)u_i) + \varepsilon u_i^{p_i} u_{3-i}^{q_i}, \quad \left(t > 0,\ x \in \mathbb{R}^N;\ i = 1,2\right). \tag{1}$$

The different properties of solutions of system (1) such as blow-up, localization of solution, asymptotical behavior of self similar solution are studied. The numerical modeling aspect of a solution of system (1) is considered in [5]. They notice that some of their results required strong justifications. Below, the method of investigation of initial value problem for (1) based

*Corresponding author.

on the method of nonlinear decomposition [3-4] is offered. This method is valid for $n > 1$. For simplicity we consider the case $n = 2$.

2. Construction of the self-similar system

The system of self-similar equations

$$\frac{1}{2}\varphi_i \xi \frac{df_i}{d\xi} + \xi^{1-N}\frac{d}{d\xi}\left(\xi^{N-1} f_{3-i}^{s_{3-i}} \frac{df_i}{d\xi}\right) + \psi_i\left(f_i^{p_i} f_{3-i}^{q_i} - f_i\right) = 0$$

corresponding to equation (1) is constructed as follows:

$$u_i(t,x) = \bar{u}_i(t)\omega_i(\tau, x),$$

$$\omega_i(\tau, x) = f_i(\xi), \quad |\xi| = \frac{\left(\sum\limits_{i=1}^{N}\left(x_i - \int_0^t v_i(t)dt\right)\right)^{1/2}}{\tau^{1/2}(t)}, \qquad (2)$$

where the functions

$$\bar{u}_i(t) = d_i (T \pm t)^{-\alpha_i}$$

are solutions of the system

$$\frac{d\bar{u}_i}{dt} = \varepsilon \bar{u}_i^{p_i} \bar{u}_{3-i}^{q_i}$$

and

$$\tau = d_1^{s_1} \frac{(T+t)^{1-\alpha_1 s_1}}{\alpha_1 s_1 - 1},$$

where

$$d_i = \left[\alpha_{3-i}^{q_i} \alpha_i^{1-p_{3-i}}\right]^{\frac{1}{p_i}}, \quad \alpha_1 s_2 - 1 > 0, \quad \alpha_1 s_2 = \alpha_2 s_1,$$

$$\varphi_i = d_1^{s_1} d_{3-i}^{-s_{3-i}}, \quad \psi_i = \frac{\alpha_i \varphi_i}{\alpha_i s_i - 1},$$

$$\alpha_i = (1 - p_{3-i} + q_i)p^{-1}, \quad p = q_1 q_2 - (1-p_1)(1-p_2), \qquad i = 1, 2.$$

After substituting

$$f_i(\xi) = \bar{f}_i(\xi) y_i(\eta), \quad \eta = -\ln(a - \xi^2), \quad \bar{f}_i(\xi) = A_i (a - \xi^2)^{\frac{1}{s_i}}$$

in (2), we have

$$\begin{aligned}[y_i' + a_{i1} y_i]' + [a_{i2} + a_{i6} y_{3-i}^{-s_{3-i}} + a_{i3} a_{3-i,1} + a_{i3} y_{3-i}^{-1} y_{3-i}'][y_i' + a_{i1} y_i] \\ + a_{i4} y_i y_{3-i}^{-s_{3-i}} + a_{i5} y_i^{p_i} y_{3-i}^{q_i - s_{3-i}} = 0,\end{aligned} \qquad (3)$$

where

$$a_{i1} = -\frac{1}{s_i},$$

$$a_{i2} = 1 + a_{i1} + \frac{N}{2ae^{\eta} - 2},$$

$$a_{i3} = \tfrac{1}{2}s_{3-i},$$

$$a_{i4} = -\tfrac{1}{4}\psi_i A_{3-i}^{-s_{3-i}}(ae^{\eta} - 1)^{-1},$$

$$a_{i5} = \tfrac{1}{4}\psi_i A_i^{p_i - 1} A_{3-i}^{q_i - s_{3-i}} \frac{e^{-l_i \eta}}{a - e^{-\eta}},$$

$$a_{i6} = \tfrac{1}{4}\varphi_i A_{3-i}^{-s_{3-i}},$$

$$l_i = 1 + \frac{q_i}{s_{3-i}} + \frac{p_i - 1}{s_i}, \quad i = 1, 2,$$

$$b_{i1} = \frac{2 - s_i}{2s_i^2},$$

$$b_{i2} = -\tfrac{1}{4s_i}\varphi_i A_{3-i}^{-s_{3-i}},$$

$$b_{i3} = \tfrac{1}{4a}\psi_i A_i^{p_i - 1} A_{3-i}^{q_i - s_{3-i}}.$$

3. Main results

Theorem 3.1. *For the existence of a solution to system (3) in the form*

$$y_i(\eta) = y_i^0 + o(1), \quad \eta \to +\infty, \quad i = 1, 2, \quad 0 < y_i^0 < +\infty, \tag{4}$$

it is necessary that the numbers y_i^0 $(i = 1, 2)$ be a solution of the following corresponding nonlinear algebraic system:

1. $l_i = 0, \ b_{i1} + b_{i2} z_{3-i}^{-s_{3-i}} + b_{i3} z_i^{p_i - 1} z_{3-i}^{q_i - s_{3-i}} = 0,$
2. $l_1 = 0, \ l_2 > 0, \ b_{11} + b_{12} z_2^{-s_2} + b_{13} z_1^{p_1 - 1} z_2^{q_1 - s_2} = 0, \ b_{21} + b_{22} z_1^{-s_1} = 0,$
3. $l_1 > 0, \ l_2 = 0, \ b_{11} + b_{12} z_2^{-s_2} = 0, \ b_{21} + b_{22} z_1^{-s_1} + b_{23} z_1^{q_2 - s_1} z_2^{p_2 - 1} = 0,$
4. $l_i > 0, \ b_{i1} + b_{i2} z_{3-i}^{-s_{3-i}} = 0.$

4. Conclusion

Primary investigations of the qualitative properties of system (1) allowed us to carry out a numerical experience depending on different values of numerical parameters. To reach this goal, the asymptotical solutions were used as initial approximation. For the linearization of system (1), Newton,

Piker and a special method were used. The results of numerical experiments showed the effectivity of suggested approach. Asymptotes of different type solutions to the system (2) allowed also modeling of mutual reaction-diffusion processes in visualization form with animations which had the possibility of observing the evolution of the studied problem. Blow-up analysis of solutions for different values of parameters was performed too.

References

1. M. Holodnyok, A. Klich, M. Kubichek, M. Marec, Methods of Analysis of Dynamical Models; Moscow, Mir, 1991, 365pp.
2. A. A. Samarskii, V. A. Galaktionov, S. P. Kurduomov, A. P. Mikhajlov, Blow-up in Parabolic Equations; Berlin, New York, Walter de Gruyter, 1995, 535pp.
3. M. Aripov, Approximate Self Similar Approach to Solving of Quasilinear Parabolic Equations: Experimentation, Modeling and Computation in Flow, Turbulence and Combustion; John Wiley and Sons, 1997, vol. 2, 19–25.
4. M. Aripov, Asymptotics of the Solution of the Non-Newton Polytropical Filtration Equation; ZAMM, vol. 80 (2000), suppl. 3, 767–768.
5. S. P. Kurduomov, E. S. Kurkina, O. V. Telkovskaya. Blow-up in Two-Component Media; Mathematical Modeling, 1989, vol. 5.

ON COMPLEMENTED SUBSPACES OF $E_0^{l_2}(a) \times E_\infty^{l_2}(b)$

E. KARAPINAR
İzmir University of Economics,
Departments of Mathematics
İzmir, 35330, Turkey
E-mail: erdalkarapinar@yahoo.com

It is known that a complemented subspace of a finite-type nuclear Köthe space $E_0^{l_2}(a)$ has a basis due to Mityagin [2]. By using the results of Prada [4] and Zakharyuta [6], it is obtained that any stable complemented subspace of $E_0^{l_2}(a) \times E_\infty^{l_2}(b)$ is basic, where a_i or b_i tends to infinity.

Keywords: Complemented Subspaces, Köthe Spaces.

1. Introduction

Let $A = (a_{in})_{i\in I, n\in\mathbb{N}}$ be a matrix of real numbers such that $0 \le a_{in} \le a_{in+1}$. Köthe space $K^{l_p}(A)$ defined by the matrix A is a Fréchet space of all sequences $\xi = (\xi_i)$ such that

$$|\xi|_n := \left(\sum_{i\in I}(a_{in})^p|\xi_i|^p\right)^{1/p} < \infty, \qquad \forall n \in \mathbb{N}$$

with the topology, generated by the system of semi-norms $\{|\xi|_n = \|\xi\|_{l_p} : n \in \mathbb{N}\}$. In the case $X = K^{l_2}(A)$, $Y = K^{l_2}(B)$, where $A = (a_{ip})$, $B = (b_{jp})$, $i, j \in \mathbb{N}$, the spaces $X \times Y$ and $X\hat{\otimes}_{l_2}Y$ are naturally isomorphic, respectively, to the spaces $K(C)$ and $K(D)$, where $c_{k,n}$ equals to $a_{i,n}$ if $k = 2i-1$ and $b_{i,n}$ if $k = 2i$, and also $D = (d_{\nu p})$, $d_{\nu p} = a_{ip}b_{jp}$, $\nu = (i,j) \in \mathbb{N}^2 = \mathbb{N}\times\mathbb{N}$.

Let ω_+ be the set of all sequences with positive terms. Here we consider two important particular cases: For $a, b \in \omega_+$,

$$E_0^{l_2}(a) := K^{l_2}(\exp(-\frac{1}{p}a_i)) \text{ and } E_\infty^{l_2}(b) := K^{l_2}(\exp(qb_i)),$$

whose elements are called finite- and infinite-type power series, respectively. For $a \in \omega_+$, a subspace $E_\nu^{l_2}(\tilde{a})$ of $E_\nu^{l_2}(a)$ is called basic if $\tilde{a} = (a_{i_k})$, where

$\nu = 0, \infty$.

The sequence $a = (a_n)$ will be called stable if a satisfies the condition $\lim_n \sup \frac{a_{2n}}{a_n} < \infty$. This is equivalent to saying that $E_\nu^{l_2}(a)$ isomorphic to $(E_\nu^{l_2}(a))^2$, where $\nu = 0, \infty$.

Let X and Y be locally convex spaces. A linear operator $T : X \to Y$ is called a *near isomorphism* if $T(X)$ is closed in Y and an open map from X onto $T(X)$, $\dim(T) < \infty$, $co\dim(T) < \infty$. The locally convex spaces X and Y are said to be *nearly isomorphic*, denoted by $X \approx Y$ if there is a near isomorphism T from X onto Y. Note that if X and Y are isomorphic, we shall write $X \simeq Y$.

Definition 1.1. Let X and Y be locally convex spaces. A linear operator $T : X \longrightarrow Y$ will be called compact if there exists a neighborhood U in X such that its image $T(U)$ is precompact in Y.

We shall say that an ordered pair of locally convex spaces (X, Y) satisfies condition $\mathcal{K}$, and write $(X, Y) \in \mathcal{K}$, if every linear continuous operator $T : X \longrightarrow Y$ is compact.

Proposition 1.1. *([6]) Let X, Y be locally convex spaces and $(X, Y) \in \mathcal{K}$. Then $(X_0, Y_0) \in \mathcal{K}$ for every subspace X_0 that is topologically complemented in X and any subspace Y_0 of Y.*

Proof. Let $T : X_0 \longrightarrow Y_0$ be an arbitrary linear continuous operator. By the assumption, there exists a subspace X_1 in X such that $X_1 \oplus X_2$. Let $T : X \longrightarrow Y$ be the linear continuous operator $T(x) = \begin{cases} T_0 x \text{ if } x \in X_0, \\ 0 \quad \text{if } x \in X_1. \end{cases}$ Since $(X, Y) \in \mathcal{K}$, this operator is compact and therefore T_0 is compact too. Hence $(X_0, Y_0) \in \mathcal{K}$. □

Proposition 1.2. *([4]) A complemented subspace of $X_1 \times X_2$, where X_1, X_2 are Fréchet spaces such that $(X_1, X_2) \in \mathcal{K}$, is isomorphic to a subspace of $X_1 \times X_2$ of the form $L_1 \times L_2$, where L_1 (respectively L_2) is a complemented subspace of X_1 (respectively X_2).*

Nuclearity of the classical Köthe spaces $K^{l_p}(A)$ is equivalent to the Grothendieck-Pietsch condition [3]: $\forall p \;\; \exists q > p$ and $(\xi_n) \in l_1$ with $a_n^p \leq \xi_n a_n^q, \;\; n \in N$.

Mityagin [2] proved that a complemented subspace of finite type of a nuclear Köthe space $E_0^{l_2}(a)$ has a basis, and asked about the existence of basis for a complemented subspace of infinite type of a nuclear Köthe space $E_\infty^{l_2}(b)$. In this short paper, we observe that any stable complemented subspace of $E_0^{l_2}(a) \times E_\infty^{l_2}(b)$ is basic, where a_i or b_i tends to infinity.

2. Main

Theorem 2.1. *Let L be a complemented subspace of $E_0^{l_2}(a) \times E_\infty^{l_2}(b)$, where $a, b \in \omega_+$. If $L \simeq L^2$ and $a_i \to \infty$ (or $b_i \to \infty$), then L is isomorphic to a basic subspace of $E_0^{l_2}(a) \times E_\infty^{l_2}(b)$ of the form $L_1 \times L_2$, where L_1, respectively L_2, is a complemented subspace of $E_0^{l_2}(a)$, respectively $E_\infty^{l_2}(b)$.*

Proof. Let $X_1 = E_0^{l_2}(\tilde{a})$ and $X_2 = E_\infty^{l_2}(\tilde{a})$. By Proposition 1.2, if L is a complemented subspace of $X_1 \times X_2$, then there exists L_1, L_2 such that L_ν is a complemented subspace of X_ν, where $\nu = 1, 2$.

Due to Mityagin [2], we know that any complemented subspace has a basis for the finite-type power series, i.e, it is of the same type with the original space, $L_1 \simeq E_0^{l_2}(\tilde{a})$, where $\tilde{a} := (a_{i_n})$.

Since $a_i \to \infty$, we have $(X_1, X_2) \in \mathcal{K}$. Combining with Proposition 1.1, we get $(L_1, L_2^2) \in \mathcal{K}$, $(L_1^2, L_2) \in \mathcal{K}$. Let $T : L \to L^2$ be an isomorphism such that $T = [T_{ij}]$ where $T_{ij} : L_j \to L_i^2$. By Douady Lemma [1], $T = \begin{bmatrix} T_{11} & T_{12} \\ 0 & T_{22} \end{bmatrix}$ is a near isomorphism.

In view of the result of Zahariuta [6, Prop. 4], we obtain that $T_{11} : L_1 \to L_1^2$ and $T_{22} : L_2 \to L_2^2$ are near isomorphisms, that is, $L \approx L^2$ if and only if $L_1 \approx L_1^2$ and $L_2 \approx L_2^2$.

By Wagner [5], if $L_2 \approx L_2^2$ then L_2 has a basis. Hence $L_2 = E_\infty^{l_2}(\tilde{b})$, where $\tilde{b} := (b_{j_k})$. □

The following result is due to the fact that $l_2 \hat{\otimes}_{l_2} l_2 = l_2$.

Corollary 2.1. *Let L be a complemented subspace of $E_0^{l_2}(a) \hat{\otimes}_{l_2} E_\infty^{l_2}(b)$, where $a, b \in \omega_+$. If $L \simeq L^2$ and $a_i \to \infty$ (or $b_i \to \infty$), then L is isomorphic to basic subspace of $E_0^{l_2}(a) \hat{\otimes}_{l_2} E_\infty^{l_2}(b)$ of the form $L_1 \hat{\otimes}_{l_2} L_2$, where L_1 (respectively L_2), is a complemented subspace of $E_0^{l_2}(a)$, (respectively $E_\infty^{l_2}(b)$).*

Acknowledgements

The author thanks Prof. Dr. V. Zakharyuta for his helpful discussions and endless support.

References

1. A. Douady, *Un espace de Banach dont le group lineaire n'est pas convexe*, Indag. Math. **25** (1965), 27–29.
2. B. S. Mityagin, *The equivalence of bases in Hilbert scales*, Studia Math. **37** (1970), 111–137 (in Russian).
3. A. Pietsch, *Nukleare lokalkonvexe Räume*, Berlin, 1965.
4. J. Prada, *On idempotent operators on Fréchet spaces*, Arch. Math. **43** (1984), 179–182.
5. M. J. Wagner, *Stable complemented subspaces of (s) have basis*, Seminar Lecture, A-G Funktionalanalysis Düsseldorf/Wuppertal, 1985.
6. V. P. Zahariuta, *On the isomorphism of Cartesian products of locally convex spaces*, Studia Math. **46** (1973), 201–221.

ASYMPTOTIC FORMULAS FOR EIGENVALUES AND EIGENFUNCTIONS OF A NONSELFADJOINT STURM-LIOUVILLE OPERATOR

KHANLAR R. MAMEDOV[†] and HAMZA MENKEN[*]

Mersin University, Faculty of Science and Letters,
Çiftlikköy Campus, 33343, Mersin, Turkey
[†] *E-mail: hanlar@mersin.edu.tr,*
[*] *E-mail: hmenken@mersin.edu.tr*

We consider the nonselfadjoint Sturm-Liouville operator with periodic and antiperiodic boundary conditions. We obtain new accurate asymptotic formulas for the eigenvalues and eigenfunctions of the boundary value problem when the potential $q(x) \in C^{(4)}[0,1]$ is a complex-valued function satisfying the condition $q(0) = q(1)$.

Keywords: Nonselfadjoint Sturm-Liouville operator, periodic and antiperiodic boundary conditions, accurate asymptotic formula.

1. Introduction

We consider the nonselfadjoint differential operator generated by the differential equation

$$\ell(y) \equiv y'' + q(x)y = \lambda y, \ 0 < x < 1, \tag{1}$$

and the boundary conditions

$$y(0) = (-1)^{\theta} y(1), \quad y'(0) = (-1)^{\theta} y'(1), \tag{2}$$

where $\theta = 1, 2$, and $q(x)$ is a complex-valued function in the space $C^{(4)}[0,1]$ satisfying the condition $q(0) = q(1)$. Without loss of generality, we can assume that $\int_0^1 q(x)dx = 0$. Under this condition, we get a new asymptotic form of the solution of the equation (1) with accuracy of the term $O(\lambda^{-7})$. Using this form of the solution, accurate asymptotic estimates for eigenvalues and eigenfunctions of the boundary value problems (1), (2) are obtained, and it is shown that all but finitely many eigenvalues are simple. The main goal of this paper is to obtain the asymptotic formulas for eigenvalues and eigenfunctions. The asymptotic formulas for the eigenfunctions of the boundary problems (1), (2) have important role in obtaining spectral expansion formulas. In [1,2], using the asymptotic formulas in the case of $q(0) \neq q(1)$, the basis property in $L_p(0,1)$ $(1 < p < \infty)$ were proved. We note that eigenvalues and eigenfunction of a differential operator under differing conditions have been investigated by many authors for various purposes. In [3], the relation between the order of accuracy of the asymptotic form of Sturm-Liouville operator and the smoothness of the coefficient $q(x)$ was investigated, and the asymptotic formulas were obtained when $q(x) \in W_2^n(0,1)$ was an real-valued function. The asymptotic formulas for eigenvalues and eigenfunctions of a differential operator order n were given in [4]. The accurate asymptotic estimates of the eigenvalues of the classical Sturm-Liouville boundary value problem were constructed in [5], and in the general case, they were obtained in [6].

First, we investigate the asymptotic estimates of the boundary value problem (1), (2), when $\theta = 1$.

2. Asymptotic formulas for eigenvalues of the antiperiodic problem

Theorem 2.1. *All eigenvalues of the boundary-value problem (1), (2) when $\theta - 1$, beyond a certain point, are simple and form two infinite sequences $\lambda_{k,1}, \lambda_{k,2}$, $k = n_0, n_0 + 1, \cdots$, where n_0 is a positive integer and*

$$\lambda_{k,1} = -\left[(2k+1)\pi\right]^2 + \frac{q'(1) - q'(0) - \int_0^1 q^2(x)dx}{[2(2k+1)\pi]^2} + O(\frac{1}{k^3}), \tag{3}$$

$$\lambda_{k,2} = -\left[(2k+1)\pi\right]^2 - \frac{q'(1) - q'(0) + \int_0^1 q^2(x)dx}{[2(2k+1)\pi]^2} + O(\frac{1}{k^3}). \tag{4}$$

Proof. Consider equation (1) or

$$y'' + q(x)y + \mu^2 y = 0, \tag{5}$$

where $\mu = \sqrt{-\lambda}$ and $\sqrt{r}e^{i\varphi/2}$ for $-\pi < \varphi \leq \pi$. It is well-known (see [3,4]) that the eigenvalues of the boundary problem (1), (2) are asymptotically located in pairs and the relation

$$\mu_{k,j} = \sqrt{-\lambda_{k,j}} = (2k+1)\pi + O\left(\frac{1}{k^{1/2}}\right) \quad (k = n_0, n_0 + 1, \cdots, j = 1, 2),$$

holds for all $\mu_{k,1}, \mu_{k,2} \in Q = \{\mu : \operatorname{Re}\mu \geq 0, |\operatorname{Im}\mu| \leq c_o\}$, where $c_o > 0$. Let $S_0 - ic_o \equiv T$, where $S_0 = \{\mu : 0 \leq \arg\mu \leq \frac{\pi}{2}\}$ and $Q \subset T$. According to

[3,4], for μ in the region T of the complex plane, the equation (5) has two linearly independent solutions $\varphi_1(x,\mu)$, $\varphi_2(x,\mu)$ satisfying the relations

$$\varphi_j(x,\mu) = e^{\mu\omega_j x}\left[\sum_{m=0}^{6}\frac{u_m(x)}{(2\omega_j\mu)^m} + O(\frac{1}{\mu^7})\right], j = 1,2,$$

$$\varphi_j'(x,\mu) = \mu\omega_j e^{\mu\omega_j x}\left[u_0(x) + \sum_{m=1}^{6}\frac{u_m(x) + 2u_{m-1}'(x)}{(2\omega_j\mu)^m} + O(\frac{1}{\mu^7})\right], j = 1,2,$$

where $\omega_1 = -\omega_2 = i$, $u_0(x) \equiv 1$, and

$$u_m(x) = -\int_0^x \ell\left(u_{m-1}(t)\right)dt,\ m = 1,2,3,4,5,6.$$

It follows that

$$\varphi_j(0,\mu) = 1 + O(\frac{1}{\mu^7}),$$

$$\begin{aligned}
\varphi_j(1,\mu) = e^{\mu\omega_j}\Big\{1 &- \frac{1}{(2\omega_j\mu)^3}[q'(1) - q'(0) + \int_0^1 q^2(t)dt] + \frac{1}{(2\omega_j\mu)^4}[q''(1) \\
&-q''(0) + \frac{5}{2}q^2(1) - \frac{3}{2}q^2(0) - q(0)q(1)] - \frac{1}{(2\omega_j\mu)^5}[q'''(1) - q'''(0) \\
&+7q(1)q'(1) - 5q(0)q'(0) - q(0)q'(1) - q(1)q'(0) \\
&+(q(1) - q(0))\int_0^1 q^2(t)dt + 2\int_0^1 q^3(t)dt - \int_0^1 q'^2(t)dt] \\
&+\frac{1}{(2\omega_j\mu)^6}[q^{(4)}(1) - q^{(4)}(0) + 9q(1)q''(1) - 7q(0)q''(0) \\
&-q(0)q''(1) - q(1)q''(0) + \frac{11}{2}q'^2(1) - \frac{9}{2}q'^2(0) - q'(0)q'(1) \\
&+\frac{15}{2}q^3(1) - \frac{7}{2}q^3(0) - \frac{5}{2}q(0)q^2(1) - \frac{3}{2}q(1)q^2(0) \\
&+(q'(1) - q'(0))\int_0^1 q^2(t)dt + \frac{1}{2}\left(\int_0^1 q^2(t)dt\right)^2] + O(\frac{1}{\mu^7})\Big\},
\end{aligned}$$

$$\begin{aligned}
\varphi_j'(0,\mu) = \mu\omega_j\Big\{1 &- \frac{2q(0)}{(2\omega_j\mu)^2} + \frac{2q'(0)}{(2\omega_j\mu)^3} - \frac{1}{(2\omega_j\mu)^4}[2q''(0) + 2q^2(0)] \\
&+\frac{1}{(2\omega_j\mu)^5}[2q'''(0) + 8q(0)q'(0)] - \frac{1}{(2\omega_j\mu)^6}[2q^{(4)}(0) \\
&+10q'^2(0) + 12q(0)q''(0) + 4q^3(0)] + O(\frac{1}{\mu^7})\Big\},
\end{aligned}$$

$$\varphi_j'(1,\mu) = \mu\omega_j e^{\mu\omega_j}\Big\{1 - \frac{q(0)+q(1)}{(2\omega_j\mu)^2} + \frac{1}{(2\omega_j\mu)^3}[q'(1) + q'(0) - \int_0^1 q^2(t)dt]$$

$$-\frac{1}{(2\omega_j\mu)^4}[q''(1)+q''(0)+\frac{3}{2}q^2(1)+\frac{3}{2}q^2(0)-q(0)q(1)]$$
$$+\frac{1}{(2\omega_j\mu)^5}[q'''(1)+q'''(0)+5q(1)q'(1)+5q(0)q'(0)-q(0)q'(1)$$
$$-q(1)q'(0)+(q(1)+q(0))\int_0^1 q^2(t)dt-2\int_0^1 q^3(t)dt+\int_0^1 q'^2(t)dt]$$
$$-\frac{1}{(2\omega_j\mu)^6}[q^{(4)}(1)+q^{(4)}(0)+\frac{13}{2}q'^2(1)+\frac{9}{2}q'^2(0)-q'(0)q'(1)$$
$$+7q(1)q''(1)+7q(0)q''(0)-q(0)q''(1)-q(1)q''(0)+\frac{7}{2}q^3(1)$$
$$+\frac{7}{2}q^3(0)-\frac{3}{2}q(0)q^2(1)-\frac{3}{2}q(1)q^2(0)$$
$$+(q'(1)+q'(0))\int_0^1 q^2(t)dt-\frac{1}{2}\left(\int_0^1 q^2(t)dt\right)^2]+O(\frac{1}{\mu^7})\Big\}.$$

Let us substitute all these expressions into the characteristic determinant

$$\Delta(\mu)=\begin{vmatrix} U_1(\varphi_1) & U_1(\varphi_2) \\ U_2(\varphi_1) & U_2(\varphi_2)\end{vmatrix},$$

where $U_1(y)=y(1)+y(0)$, $U_2(y)=y'(1)+y'(0)$. By elementary transformations, we obtain the relation

$$-\frac{e^{i\mu}}{i\mu}\Delta(\mu)=e^{2i\mu}\Big\{1-\frac{2q(0)}{(2i\mu)^2}-\frac{1}{(2i\mu)^3}\int_0^1 q^2(t)dt-\frac{1}{(2i\mu)^4}[2q''(0)-\frac{1}{2}q^2(1)$$
$$+\frac{3}{2}q^2(0)+q(0)q(1)]-\frac{1}{(2i\mu)^5}[q(1)q'(1)-q(0)q'(1)-q(0)q'(0)$$
$$+q(1)q'(0)-2q(0)\int_0^1 q^2(t)dt+2\int_0^1 q^3(t)dt-\int_0^1 q'^2(t)dt]$$
$$-\frac{1}{(2i\mu)^6}[2q^{(4)}(0)+\frac{1}{2}q'^2(1)+\frac{21}{2}q'^2(0)-q'(0)q'(1)-q(1)q''(1)$$
$$+11q(0)q''(0)+q(0)q''(1)+q(1)q''(0)-2q^3(1)+3q^3(0)$$
$$+3q(0)q^2(1)-\frac{1}{2}\left(\int_0^1 q^2(t)dt\right)^2]+O(\frac{1}{\mu^7})\Big\}$$
$$+2e^{i\mu}\Big\{1-\frac{2q(0)}{(2i\mu)^2}-\frac{1}{(2i\mu)^4}[2q''(0)+2q^2(0)]-\frac{1}{(2i\mu)^6}[2q^{(4)}(0)$$
$$+12q(0)q''(0)+10q'^2(0)+4q^3(0)]+O(\frac{1}{\mu^7})\Big\}$$

$$+\Big\{1-\frac{2q(0)}{(2i\mu)^2}+\frac{1}{(2i\mu)^3}\int_0^1 q^2(t)dt-\frac{1}{(2i\mu)^4}[2q''(0)-\frac{1}{2}q^2(1)$$
$$+\frac{3}{2}q^2(0)+q(0)q(1)]+\frac{1}{(2i\mu)^5}[q(1)q'(1)-q(0)q'(0)-q(0)q'(1)$$
$$+q(1)q'(0)-2q(0)\int_0^1 q^2(t)dt+2\int_0^1 q^3(t)dt-\int_0^1 q'^2(t)dt]$$
$$-\frac{1}{(2i\mu)^6}[2q^{(4)}(0)+\frac{1}{2}q'^2(1)+\frac{21}{2}q'^2(0)-q'(0)q'(1)-q(1)q''(1)$$
$$+11q(0)q''(0)+q(0)q''(1)+q(1)q''(0)-2q^3(1)+3q^3(0)$$
$$+3q(0)q^2(1)-\frac{1}{2}\left(\int_0^1 q^2(t)dt\right)^2]+O(\frac{1}{\mu^7})\Big\}, \tag{6}$$

for $\mu \in T$ sufficiently large in absolute value.

Let $b(\mu)$ be the coefficient of $e^{2i\mu}$ in (6). It can be easily seen that the relation

$$b^{-1}(\mu)=1+\frac{2q(0)}{(2i\mu)^2}+\frac{1}{(2i\mu)^3}\int_0^1 q^2(t)dt+\frac{1}{(2\omega_j\mu)^4}[2q''(0)-\frac{1}{2}q^2(1)$$
$$+\frac{11}{2}q^2(0)+q(0)q(1)]+\frac{1}{(2i\mu)^5}[q(1)q'(1)-q(0)q'(0)$$
$$-q(0)q'(1)+q(1)q'(0)+2q(0)\int_0^1 q^2(t)dt+2\int_0^1 q^3(t)dt$$
$$-\int_0^1 q'^2(t)dt]+\frac{1}{(2i\mu)^6}[2q^{(4)}(0)+\frac{1}{2}q'^2(1)+\frac{21}{2}q'^2(0)$$
$$-q'(0)q'(1)-q(1)q''(1)+19q(0)q''(0)+q(0)q''(1)$$
$$+q(1)q''(0)-2q^3(1)+17q^3(0)+q(0)q^2(1)$$
$$+4q(1)q^2(0)+\left(\int_0^1 q^2(t)dt\right)^2]+O(\frac{1}{\mu^7}) \tag{7}$$

holds for $\mu \in T$ sufficiently large in absolute value.

Thus, for $\mu \in T$ sufficiently large in absolute value, the equation $\Delta(\mu)=0$ is equivalent to the equation

$$-(i\mu)^{-1}b^{-1}(\mu)\Delta(\mu)e^{i\mu}=0. \tag{8}$$

Using (6), (7) and the relation $q(1)=q(0)$, by the equation (8) we obtain

$$e^{i\mu}+1=\frac{q'(1)-q'(0)-\int_0^1 q^2(t)dt}{(2i\mu)^3}+O(\frac{1}{\mu^4}), \tag{9}$$

$$e^{i\mu}+1=-\frac{q'(1)-q'(0)+\int_0^1 q^2(t)dt}{(2i\mu)^3}+O(\frac{1}{\mu^4}). \tag{10}$$

By Rouche's theorem, we have asymptotic expressions for the roots $\mu_{k,1}$ and $\mu_{k,2}$, $k = n_0, n_0+1, \cdots$, of the equations (9) and (10), respectively, where n_0 is a positive integer

$$\mu_{k,1} = -\left[(2k+1)\pi\right]^2 + \frac{q'(1) - q'(0) - \int_0^1 q^2(t)dt}{[2(2k+1)\pi]^2} + O(\frac{1}{k^3}), \tag{11}$$

$$\mu_{k,2} = -\left[(2k+1)\pi\right]^2 - \frac{q'(1) - q'(0) + \int_0^1 q^2(t)dt}{(4k\pi)^2} + O(\frac{1}{k^3}). \tag{12}$$

Note that $\mu_{k,1}$ and $\mu_{k,2}$ are simple roots of the equations (9) and (10), respectively. From the relations (11), (12) and the relations $\lambda_{k,1} = -\mu_{k,1}^2$, $\lambda_{k,2} = -\mu_{k,2}^2$, we obtain the formulas (3)-(4) and observe that these eigenvalues are simple. The theorem is proved. □

3. Asymptotic formulas for eigenfunctions of the antiperiodic problem

Theorem 3.1. *The eigenfunctions of the boundary value problem (1), (2) when $\theta = 1$, corresponding the eigenvalues above, are of the form*

$$y_{k,1}(x) = \sin(2k+1)\pi x + O(\frac{1}{k}), \tag{13}$$

$$y_{k,2}(x) = \cos(2k+1)\pi x + O(\frac{1}{k}). \tag{14}$$

Proof. Let us calculate $U_1(\varphi_1(x,\mu_{k,1}))$ and $U_1(\varphi_2(x,\mu_{k,1}))$. Since

$$e^{i\mu_{k,1}} = -1 + \frac{q'(1) - q'(0) - \int_0^1 q^2(t)dt}{(2i\mu_{k,1})^3} + O(\frac{1}{\mu_{k,1}^4}),$$

we have

$$\begin{aligned} U_1(\varphi_1(x,\mu_{k,1})) &= \varphi_1(1,\mu_{k,1}) + \varphi_1(0,\mu_{k,1}) \\ &= e^{i\mu_{k,1}}\left\{1 - \frac{q'(1) - q'(0) + \int_0^1 q^2(t)dt}{(2i\mu_{k,1})^3} + O(\frac{1}{\mu_{k,1}^4})\right\} \\ &\quad + \left[1 + O(\frac{1}{\mu_{k,1}^4})\right] \\ &= \frac{2\,(q'(1) - q'(0))}{(2i\mu_{k,1})^3} + O(\frac{1}{\mu_{k,1}^4}). \end{aligned} \tag{15}$$

Similarly,

$$U_1(\varphi_2(x,\mu_{k,1})) = \varphi_2(1,\mu_{k,1}) + \varphi_2(0,\mu_{k,1})$$

$$= -\frac{2\,(q'(1) - q'(0))}{(2i\mu_{k,1})^3} + O(\frac{1}{\mu_{k,1}^4}).$$

Since $U_1(\varphi_j(x, \mu_{k,1})) \neq 0$, $j = 1, 2$, we seek the eigenfunction $y_{k,1}(x)$ corresponding to the eigenvalue $\lambda_{k,1}$ in the form

$$y_{k,1}(x) = -\frac{(2i\mu_{k,1})^2}{4\,[q'(1) - q'(0)]} \begin{vmatrix} \varphi_1(x, \mu_{k,1}) & \varphi_2(x, \mu_{k,1}) \\ U_1(\varphi_1(x, \mu_{k,1})) & U_1(\varphi_2(x, \mu_{k,1})) \end{vmatrix}. \tag{16}$$

To normalized the eigenfunction $y_{k,1}(x)$ we assume that $q'(1) - q'(0) \neq 0$. From the equalities

$$\varphi_j(x, \mu_{k,1}) = e^{\mu_{k,1}\omega_j x}\left\{1 + \sum_{v=1}^{3} \frac{u_v(x)}{(2w_j\mu_{k,1})^v} + O(\frac{1}{\mu_{k,1}^4})\right\}, \quad j = 1, 2$$

and the formulas (15), (16) we obtain

$$y_{k,1}(x) = \cos \mu_{k,1} x + O(\frac{1}{\mu_{k,1}}).$$

Therefore, the eigenfunction $y_{k,1}(x)$ satisfies the asymptotic formula (13). In a similar way, we obtain

$$\begin{aligned} U_2(\varphi_1(x, \mu_{k,2})) &= \varphi_1'(1, \mu_{k,2}) + \varphi_1'(0, \mu_{k,2}) \\ &= i\mu_{k,2} e^{i\mu_{k,2}} \left[1 - \frac{q(1) + q(0)}{(2i\mu_{k,2})^2} + \frac{q'(1) + q'(0) - \int_0^1 q^2(t)dt}{(2i\mu_{k,2})^3} \right. \\ &\quad \left. + O(\frac{1}{\mu_{k,2}^4})\right] + i\mu_{k,2}\left[1 - \frac{2q(0)}{(2i\mu_{k,2})^2} + \frac{2q'(0)}{(2i\mu_{k,2})^3} + O(\frac{1}{\mu_{k,1}^4})\right] \\ &= -\frac{q'(1) - q'(0)}{(2i\mu_{k,1})^2} + O(\frac{1}{\mu_{k,1}^3}), \end{aligned}$$

and

$$U_2(\varphi_2(x, \mu_{k,2})) = -\frac{q'(1) - q'(0)}{(2i\mu_{k,1})^2} + O(\frac{1}{\mu_{k,1}^3}).$$

Since $U_2(\varphi_j(x, \mu_{k,1})) \neq 0$, $j = 1, 2$, we can seek the eigenfunction $y_{k,2}(x)$ corresponding to the eigenvalues $\lambda_{k,2}$ in the form

$$y_{k,2}(x) = -\frac{(2i\mu_{k,2})^2}{2i\,[q'(1) - q'(0)]} \begin{vmatrix} \varphi_1(x, \mu_{k,2}) & \varphi_2(x, \mu_{k,2}) \\ U_2(\varphi_1(x, \mu_{k,2})) & U_2(\varphi_2(x, \mu_{k,2})) \end{vmatrix}.$$

Thus, we obtain

$$y_{k,2}(x) = \sin \mu_{k,2} x + O(\frac{1}{\mu_{k,2}}).$$

This completes the proof of the theorem. □

4. Asymptotic formulas for eigenvalues and eigenfunctions of the periodic problem

In a similar way, the following results are obtained for the boundary problem (1), (2) when $\theta = 2$.

Theorem 4.1. *All eigenvalues of the boundary value problem (1), (2) when $\theta = 2$, beyond a certain point, are simple and form two infinite sequence $\lambda_{k,1}$, $\lambda_{k,2}$, $k = n_0, n_0 + 1, \cdots$, where n_0 is a positive integer and*

$$\lambda_{k,1} = -(2k\pi)^2 - \frac{q'(1) - q'(0) + \int_0^1 q^2(t)dt}{(4k\pi)^2} + O(\frac{1}{k^3}),$$

$$\lambda_{k,2} = -(2k\pi)^2 + \frac{q'(1) - q'(0) - \int_0^1 q^2(t)dt}{(4k\pi)^2} + O(\frac{1}{k^3}).$$

The corresponding eigenfunctions are of the form

$$y_{k,1}(x) = \sin 2k\pi x + O(\frac{1}{k}),$$

$$y_{k,2}(x) = \cos 2k\pi x + O(\frac{1}{k}).$$

References

1. N. B. Kerimov and K. R. Mamedov, On the Riesz basis property of the root functions in certain regular boundary value problems, *Math. Notes* **V. 64, N.4**, pp. 483–487 (1998).
2. V. M. Kurbanov, A theorem on equivalent bases for a differential operator, *Dokl. Akad. Nauk* **V. 406, N.1**, pp. 17–20 (2006).
3. V. A Marchenko, *Sturm-Liouville Operators and Applications*, Birkhauser Verlag, 1986.
4. M. A. Naimark, *Linear Differential Operators, Part I*, Frederick Ungar Pub. Co., New York, 1967.
5. V. A. Chernyatin, Higher-order spectral asymptotics for the Sturm-Liouville operator, *Differ. Equ.* **V. 38, N.2**, pp. 206–215 (2002).
6. V. A. Vinokurov and V. A. Sadovnichii, *Izv. RAN. Ser. Mat.* **V. 64, N.2**, pp. 47–108 (2000).

ON THE MULTISUBLINEAR OPERATORS

LAHCÈNE MEZRAG

Laboratoire de Mathmatiques Pures et Appliquées, Université de M'sila, 28000 M'sila, Algeria
E-mail: lmezrag@yahoo.fr

KHALIL SAADI

Laboratoire de Mathématiques Pures et Appliquées, Université de M'sila, 28000 M'sila, Algeria
E-mail: kh_saadi@yahoo.fr

We introduce and study the class of multisublinear operators. We give an approach to a Hahn-Banach theorem for this category like the one given by Ioffe in the linear case.

Keywords: Banach lattices, Cohen strongly p-summing multilinear operators, Hahn-Banach theorem, multisublinear operators.

1. Introduction

Let X be a Banach space and Y be a Banach lattice. Let $T : X \longrightarrow Y$ be a sublinear operator, namely, subadditive and positively homogeneous. We denote by ∇T the set of all linear operators $u : X \longrightarrow Y$ such that $u(x) \leq T(x)$ for all x in X. We know (see [1]) that ∇T is not empty if Y is a complete Banach lattice and $T(x) = \sup \{u(x) : u \in \nabla T\}$, moreover, the supremum is attained. If Y is simply a Banach lattice then ∇T is empty in general (see [12]). In this paper we introduce and study the category of m-sublinear or multisublinear operators and the relation with m-linear or multilinear operators. We try to find a relation between multisublinear and multilinear like the above equality. This is the first original motivation of our work. In the case of sublinear operators, to have $T(x) = \sup \{u(x) : u \in \nabla T\}$, we have used the Hahn-Banach theorem (see [11]): Let X_0 be a subspace of X and $u : X_0 \longrightarrow Y$ be a linear operator such that $u \leq T$, then u extends to a linear operator $\widetilde{u} : X \longrightarrow Y$ (Y is a complete Banach lattice) such that $\widetilde{u} \leq T$. For this, we attempt to give a multilinear version of Hahn-Banach theorem. The celebrated Hahn-Banach theorem for linear functionals has been generalized in many different directions. In particular, for linear and bilinear operators. In our knowledge, the first who has studied this problem is A. Grothendieck in his thesis [7]. He has shown that every bilinear integral form can be extended. In [9] Hayden has posed the following conjecture: Let X, Y be two Banach spaces and let X_0, Y_0 be two subspaces respectively. Consider $u \in \mathcal{B}(X_0, Y_0; \mathbb{R})$, the space of all continuous bilinear forms on $X_0 \times Y_0$ such that $\|u\| \leq C \|x\| \|y\|$ for an absolute constant C. Then, there is $\widetilde{u} \in \mathcal{B}(X, Y; \mathbb{R})$ such that $\widetilde{u}/X_0 \times Y_0 = u$ and $\|\widetilde{u}\| \leq C \|x\| \|y\|$. He proved this conjecture in the case of Hilbert spaces and showed in [10] that it is not characterize Hilbert spaces. The problem of the extension of bilinear and multilinear operators has been studied by many authors. Contrary to the case of continuous linear operators, multilinear bounded forms on Banach spaces do not extend from subspaces to the whole spaces (see [4, p. 10]). A number of papers [2,3,6] among so many others have considered this problem from a subspace X_0 to a larger X by regarding the manner of the embedding of X_0 into X. But there is several positive answers for particular problems. Let $m \in \mathbb{N}$ and $X_1, ..., X_m$ be real or complex Banach spaces and Y be a complete Banach lattice. Let E_j be a subspace of X_j for $1 \leq j \leq m$ and T be a continuous m-sublinear operator from $X_1 \times ... \times X_m$ into Y. If u_0 is a continuous m-linear operator from $E_1 \times ... \times E_m$ into Y such that $u_0 \leq T$. Does thereexist a continuous m-linear operator u from $X_1 \times ... \times X_m$ into Y such that $u/E_1 \times ... \times E_m = u_0$ and $u \leq T$? This question remains partially open, it depends on u. We study this problem for some particular cases. If $T(x^1, ..., x^m) = T_1(x^1)...T_m(x^m)$, where T_j is a symmetrical sublinear operator, then this problem turned out to be true for some particular m-linear operators u. Such operators T cover a large gamut of m-sublinear operators. If $u(x^1, ..., x^m) = u_1(x^1)...u_m(x^m)$ such that $u_j \leq T_j$, then the extension holds. Our approach consists of studying these types of problems.

2. The multisublinear operators

In this section we give some terminology concerning Banach lattices. These spaces are well-known. For more details, the reader can consult the references [13,14]. But for our convenience we recall some facts. We introduce also the class of multisublinear operators. Firstly, we start by recalling the abstract definition of Banach lattices. A real Banach lattice (resp. a real complete Banach lattice) X is an ordered vector space equipped with a lattice (resp. a complete lattice) structure and a Banach space norm satisfying the following compatibility condition:

$$\forall x, y \in X, \quad |x| \leq |y| \Longrightarrow \|x\| \leq \|y\|,$$

where $(|x| = \sup\{x, -x\}$. Note that this implies obviously that for any $x \in X$ the elements x and $|x|$ have the same norm. We denote by $X^+ = \{x \in X : x \geq 0\}$. An element x of X is positive if $x \in X^+$. Let $X_1, ..., X_m$ be Banach lattices then the product space $X_1 \times ... \times X_m$ is also a vector lattice equipped with the natural order

$$x \leq y \Longleftrightarrow x^1 \leq y^1, ..., x^m \leq y^m \quad \text{for all } x, y \in X_1 \times ... \times X_m,$$

$$\sup\{x, y\} = \left(\sup\{x^1, y^1\}, ..., \sup\{x^m, y^m\}\right).$$

We denote by $(X_1 \times ... \times X_m)^+$ the set $\left\{(x^1, ..., x^m) : x^j \in X_j^+\right\}$.

To render this presentation possible, we assume that X is a Banach lattice of equivalent classes of measurable functions of a measure space (Ω, μ) with the usual order, because we use the multiplication in X.

We continue this section by defining the notion of multisublinear operators. It appeared firstly in [5] for bisublinear operators. A mapping T from a Banach space X into a Banach lattice Y is said to be sublinear if for all x, y in X and λ in $\mathbb{R}^+$ we have $T(\lambda x) = \lambda T(x)$ and $T(x+y) \leq T(x)+T(y)$. Consider m in $\mathbb{N}$. Let $X_1, ..., X_m, Y$ be Banach spaces over the real numbers. We will denote by $\mathcal{L}(X_1, ..., X_m; Y)$ the space of all continuous m-linear operators from $X_1 \times ... \times X_m$ into Y.

Definition 2.1. Let $X_1, ..., X_m$ be Banach spaces and Y be a Banach lattice. A mapping T from $X_1 \times ... \times X_m$ into Y which is sublinear in each variable separately is called m-sublinear or multisublinear.

In ther words, T is multisublinear means that it is positively homogeneous and subadditive in every component. Note that the sum of two multisublinear operators is a multisublinear operator and the multiplication by a positive number is also a multisublinear operator.

Let us denote by

$$\mathcal{SL}(X_1, ..., X_m; Y) = \{m\text{-sublinear mappings } T : X_1 \times ... \times X_m \longrightarrow Y\}.$$

If $X_1 = ... = X_m = X$, we write simply $\mathcal{SL}(^m X, Y)$. We equip $\mathcal{SL}(X_1, ..., X_m; Y)$ with the natural order induced by Y

$$T_1 \leq T_2 \Longleftrightarrow T_1(x) \leq T_2(x), \quad \forall x \in X_1 \times ... \times X_m, \tag{1.1}$$

$$\nabla T = \{u \in \mathcal{L}(X_1, ..., X_m; Y) : u \leq T \text{ (i.e., } \forall x \in X_1 \times ... \times X_m,\ u(x) \leq T(x))\}.$$

For a particular T, the set ∇T is not empty by Theorem 2.1 below if Y is a complete Banach lattice. As a consequence, for all $x = (x^1, ..., x^m)$ in $X_1 \times ... \times X_m$ and j in $\{1, ..., m\}$, we have

$$u \in \nabla T \Longleftrightarrow -T(x^1, ..., -x^j, ..., x^m) \leq u(x) \leq T(x). \tag{1.2}$$

Remark 2.1. We do not know if ∇T exists if T is an arbitrary multisublinear operator.

Definition 2.2. We say now that an m-sublinear operator T is:

(a) symmetrical with respect to j if $T(x) = T(x^1, ..., -x^j, ..., x^m)$ for all x in $X_1 \times ... \times X_m$.

(b) positive if $T(x) \geq 0$ for all x in $X_1 \times ... \times X_m$.

The symmetry (for one j) implies the positivity; the converse is false. If T is symmetrical with respect one j, then

$$u \in \nabla T \Longleftrightarrow |u(x)| \leq T(x), \quad \forall x \in X,$$

$$T(x^1, ..., \lambda x^j, ..., x^m) = |\lambda| T(x), \quad \text{for all } \lambda.$$

Let T be an m-sublinear operator from $X_1 \times ... \times X_m$ into a Banach lattice Y. The operator T is continuous, if and only if, there is a positive constant C such that for all $x \in X_1 \times ... \times X_m$, $\|T(x)\| \leq C \|x^1\| ... \|x^m\|$. In this case we say that T is bounded and we put

$$\|T\| = \sup\{\|T(x)\| : \|x^1\|_{B_{X_1}} = 1, ..., \|x^m\|_{B_{X_m}} = 1\}.$$

We will denote by $\mathcal{SB}(X_1, ..., X_m; Y)$ the set of all bounded multisublinear operators from $X_1 \times ... \times X_m$ into Y.

Example 2.1.

1- If u is multilinear then $|u|$ is multisublinear.

2- The operator $T(x^1, ..., x^m) = \|x^1\| ... \|x^m\|$ is multisublinear.

3- Consider $T : l_m \times ... \times l_m \longrightarrow l_1$ defined by $T((x_n^1)_n, ..., (x_n^m)_n) = (|x_n^1 ... x_n^m|)_n$. The operator T is m-sublinear and continuous with $\|T\| \leq 1$. Indeed, by using a general form of Hölder's inequality, we obtain

$$\sum_{n=1}^{\infty} |x_n^1 ... x_n^m| \leq \left(\sum_{n=1}^{\infty} |x_n^1|^m\right)^{\frac{1}{m}} ... \left(\sum_{n=1}^{\infty} |x_n^m|^m\right)^{\frac{1}{m}}.$$

4- Let us now define the notion of m-quasilinear operators. An operator T from $X_1 \times ... \times X_m$ into a Banach lattice Y is called to be m-quasilinear if for all $x = (x^1, ..., x^m)$ in $X_1 \times ... \times X_m$, $y^j \in X_j$ $(1 \leq j \leq m)$ and λ in $\mathbb{R}$, we have

$$\begin{aligned} &\text{(i) } \left|T(x^1, ..., \lambda x^j, ..., x^m)\right| &&= |\lambda| \, |T(x)| \text{ for all } 1 \leq j \leq m, \\ &\text{(ii) } \left|T(x^1, ..., x^j + y^j, ..., x^m)\right| &&\leq |T(x)| + \left|T(x^1, ..., y^j, ..., x^m)\right|. \end{aligned}$$

Note that in general the sum of two m-quasilinear operators is not an m-quasilinear operator but the multiplication by a scalar is m-quasilinear. If we put $\varphi(x) = |T(x)|$ then φ is a symmetrical m-sublinear operator. If T is m-sublinear and symmetrical for one j, then T is m-quasilinear.

Remark 2.2. Let m in $\mathbb{N}$. Let $X_1, ..., X_m, E_1, ..., E_m, Y, Z$ be Banach spaces such that Y, Z are Banach lattices.

(a) Consider T in $\mathcal{SL}(X_1, ..., X_m; Y)$ and u in $\mathcal{L}(Y; Z)$ (assume that u is positive, i.e., $u(y) \geq 0$ if $y \in Y^+$). Then, $u \circ T \in \mathcal{SL}(X_1, ..., X_m; Z)$.

(b) Consider u_j in $\mathcal{L}(E_j, X_j)$ $(1 \leq j \leq m)$ and T in $\mathcal{SL}(X_1, ..., X_m; Y)$. Then, $T \circ (u_1, ..., u_n) \in \mathcal{SL}(E_1, ..., E_m; Y)$.

(c) $\forall T \in \mathcal{SL}(X_1, ..., X_m; Y)$ and $\forall \lambda \in \mathbb{R}$, we have $\lambda T(x) \leq T(x_1, ..., \lambda x_j, ..., x_m)$ for all j.

As an immediate consequence, $T \leq u$ $(u \in \mathcal{L}(X_1, ..., X_m; Y)) \Rightarrow T = u$. Indeed, we have for all x in $X_1 \times ... \times X_m$ that $T \leq u \Longrightarrow T(x) \leq u(x) \Longrightarrow T(x^1, ..., -x^j, ..., x^m) \leq u(x^1, ..., -x^j, ..., x^m) \Longrightarrow$ (by (1.2) $-T(x) \leq T(x^1, ..., -x_j, ..., x^m) \leq -u(x) \Longrightarrow T(x) \geq u(x)$.

3. Extension of Hahn-Banach theorem for the multisublinear operators

The main objective in this section is to study a particular version of Hahn-Banach theorem for this category of operators. We establish under some suppositions the relation between multilinear and multisublinear operators.

The problem of extension of a multilinear operator u defined on a subspace $E_1 \times ... \times E_m$ of $X_1 \times ... \times X_m$ $(E_j \subset X_j)$ such that $u \leq T$ where $T \in \mathcal{SL}(X_1, ..., X_m; Y)$ is not true in general as we have seen above. In this section, we study an extension for the multisublinear of the form

$$T(x^1, ..., x^m) = T_1(x^1) ... T_m(x^m). \tag{2.1}$$

The following theorem generalizes the Hahn-Banach theorem in the case of the multisublinear operators. It seems simple, but its applications are very useful.

Theorem 3.1. *Let $X_1, ..., X_m$ be Banach spaces and let Y be a complete Banach lattice. Consider T in $\mathcal{SL}(X_1, ..., X_m; Y)$ of the form (2.1), where*

$T_j : X_j \to Y$ is a symmetrical bounded sublinear operator. Then, for all x_0 in $X_1 \times ... \times X_m$ there is $u_{x_0} \in \nabla T$ such that $T(x_0) = u_{x_0}(x_0)$, i.e., the supremum is attained, $T(x) = \sup\{u(x) : u \in \nabla T\}$.

Proof. Let $x_0 = (x_0^1, ..., x_0^m)$ be in $X_1 \times ... \times X_m$. Consider $v_{x_0} : G = \mathbb{R}x_0^1 \times ... \times \mathbb{R}x_0^m \to Y$ defined by $v_{x_0}\left(\lambda_1 x_0^1, ..., \lambda_m x_0^m\right) = \lambda_1 T_1\left(x_0^1\right) ... \lambda_m T_m\left(x_0^m\right)$ for all $(\lambda_j)_{1 \leq j \leq m} \subset \mathbb{R}$. The operator v_{x_0} is multilinear on G. We have by Remark 2.2 (c), for any j, that $\lambda_j T_j(x_0^j) \leq T_j(\lambda_j x_0^j)$. Thus, this implies $v_{x_0} \leq T$ on G. For $v_j : \mathbb{R}x_0^j \to Y$ defined by $v_j(\lambda x_0^j) = \lambda_j T_j(x_0^j)$ and $T_j : X_j \to Y$, we have $v_j \leq T_j$ on X_j. By Hahn-Banach theorem (see [11]) applied to sublinear operators, there is a linear extension of v_j noted u_j such that $u_j \leq T_j$ on X_j and $u_j/_{X_j} = v_j$. Since T_j is positive (T_j is symmetrical) for all j we get $|u_j\left(x^j\right)| \leq T_j\left(x^j\right)$ for all $x^j \in X_j$. Hence, we have $|u_1\left(x^1\right) ... u_m\left(x^m\right)| \leq T_1\left(x^1\right) ... T_m\left(x^m\right)$, $\forall x \in X_1 \times ... \times X_m$.

Furthermore if we put $u(x) = u_1\left(x^1\right) ... u_m\left(x^m\right)$, we obtain $|u(x)| \leq T(x)$, $\forall x \in X_1 \times ... \times X_m$. Consequently u is multilinear and $u(x) \leq T(x)$ for all $x \in X_1 \times ... \times X_m$ and hence $u/_G = v_1 ... v_m$. This concludes the proof because $u(x_0) = u(x_0^1, ..., x_0^m) = v_{x_0}(x_0^1, ..., x_0^m) = T_1\left(x_0^1\right) ... T_m\left(x_0^m\right) = T(x_0)$. □

Remark 3.1. We have made some assumptions on T in Theorem 3.1. Without these assumptions, we have no answer in general.

Corollary 3.1. *Let $X_1, ..., X_m$ be Banach spaces and Y be a complete Banach lattice. Let T be in $\mathcal{SB}(X_1, ..., X_m; Y)$ such that $T(x^1, ..., x^m) = T_1(x^1) ... T_m(x^m)$, where $T_j : X_j \to Y$ is a symmetrical bounded sublinear operator. Then, the following properties are equivalent.*

(i) The operator T is bounded.

(ii) For every u in ∇T, u is bounded.

Proof. The first implication is obvious. Concerning the second, we use Banach-Steinhauss theorem for multilinear operators (we are grateful to V. Dimant and I. Villanueva for their help concerning Banach-Steinhauss theorem). As we have seen there is no Hahn-Banach theorem, also no uniform continuity and no open mapping theorem, but there is closed graph theorem and a Banach-Steinhaus theorem. □

Corollary 3.2. *With the same conditions as in the above corollary, we have for all $x = (x^1, ..., x^m)$ in $X_1 \times ... \times X_m$*

$$\begin{array}{llll} (i) & \|T(x)\| & = & \sup_{u\in\nabla T} \|u(x)\|, \\ (ii) & \|T\| & = & \sup_{u\in\nabla T} \|u\|. \end{array}$$

Proof. (i) Let $x = (x^1, ..., x^m)$ be in $X_1 \times ... \times X_m$. There exists $u_x \in \nabla T$ such that $\|T(x)\| = \|u_x(x)\| \leq \sup_{u\in\nabla T} \|u(x)\|$. This is one direction.

For the other direction, we have for $u \in \nabla T$,

$$\begin{array}{rcl} -u(x) & = & u(-x^1, ..., x^m) \leq T(-x^1, ..., x^m) \\ & \leq & T_1(-x^1) ... T_m(x^m) \ (T_1 \text{ is symmetrical}) \\ & \leq & T_1(x^1) ... T_m(x^m) \leq T(x). \end{array}$$

This implies that $|u(x)| \leq T(x)$ and hence $\|u(x)\| \leq \|T(x)\|$. Thus $\sup_{u\in\nabla T} \|u(x)\| \leq \|T(x)\|$.

(ii) This follows immediately from (i). □

Remark 3.2. In general, we have $\|u\| \leq 2\|T\|$. Indeed, $u(x) \leq T(x)$ implies that $-u(x^1, ..., x^m) \leq T(-x^1, ..., x^m)$ and hence

$$\begin{array}{rcl} |u(x)| & \leq & \sup\{T(x), T(-x^1, ..., x^m)\} \\ & \leq & \sup\{|T(x)|, |T(-x^1, ..., x^m)|\} \\ & \leq & |T(x)| + |T(-x^1, ..., x^m)|. \end{array}$$

Corollary 3.3. *Let* $T \in \mathcal{SB}(X_1, ..., X_m; \mathbb{K})$ ($\mathbb{K} = \mathbb{R}$ *or* $\mathbb{C}$) *be given by* $T(x) = \|x^1\| ... \|x^m\|$. *Then*

(i) $\nabla T = \{u \in \mathcal{L}(X_1, ..., X_m; \mathbb{K}) : \|u\| \leq 1\} = B_{\mathcal{L}(X_1,...,X_m;\mathbb{K})}$.

(ii) $\|x^1\| ... \|x^m\| = \sup\{|u(x)| : u \in \mathcal{B}_{\mathcal{L}(X_1,...,X_m;\mathbb{K})}\}$ *and* $\|T\| = \sup\{\|u\| : u \in \mathcal{B}_{\mathcal{L}(X_1,...,X_m;\mathbb{K})}\} = 1$.

Proof. (i) is easy and (ii) is clear from Corollary 3.2. □

Corollary 3.4. *Consider* $x_0 \in X_1 \times ... \times X_m$. *Then there exists* $u \in \mathcal{L}(X_1, ..., X_m; \mathbb{K})$ *such that* $\|u\| = \|x_0^1\| ... \|x_0^m\|$ *and* $u(x_0) = \|x_0^1\|^2 ... \|x_0^m\|^2$.

Proof. Let $1 \leq j \leq m$. We define T_j on X_j by $T_j(x^j) = \|x_0^j\| \|x^j\|$, and u_j on $\mathbb{R}x_0^j$ by $u_j(\lambda_j x_0^j) = \lambda_j \|x_0^j\|^2$. We have $u_j(\lambda_j x_0^j) \leq \|x_0^j\| \|\lambda_j x_0^j\| = T_j(\lambda_j x_0^j)$, $\forall j$, $1 \leq j \leq m$. Then $u_0 = u_1 ... u_m \leq T = T_1 ... T_m$ on $G = \mathbb{R}x_0^1 \times ... \times \mathbb{R}x_0^m$. By Theorem 3.1, there exists $u \in \mathcal{L}(X_1, ..., X_m; \mathbb{K})$ such that $u(x) \leq T(x)$, $\forall x \in X_1 \times ... \times X_m$ and $u/G = u_0$. Hence $\|u\| = \|x_0^1\| ... \|x_0^m\|$ because $\|u_0\| = \|x_0^1\| ... \|x_0^m\|$ and $u(x_0) = u_0(x_0) = \|x_0^1\|^2 ... \|x_0^m\|^2$. □

References

1. D. Achour and L. Mezrag. *Little Grothendieck's theorem for sublinear operators.* J. Math. Anal. Appl. **296** (2004), 541-552.
2. D. Carando. *Extendable polynomials on Banach spaces.* J. Math. Anal. Appl. **223** (1999), 359-372.
3. J. M. F. Castillo, R García and J. A. Jaramillo. *Extension of bilinear forms on Banach spaces.* Proc. Amer. Math. Soc. **129** (2001), 3647-3656.
4. A. Defant, K. Floret. *Tensor Norms and Operator Ideals.* North-Holland, 1993.
5. K. Donner. *Extension of Positive Operators and Korovkin Theorems.* Lecture Notes in Mathematics, **90**.
6. P. Calindo, D. García, M. Maestre and J. Mujica. *Extension of multilinear mappings on Banach spaces.* Studia Math. **108** (1994), 55-76.
7. A. Grothendieck. *Produits tensoriels topologiques et espaces nucleaires.* Memoirs Amer. Math. Soc. **16** (1955).
8. L. Grafakos and T. Tao. *Multilinear interpolation between adjoint operators.* J. Funct. Anal. **199** (2003), 379-385.
9. T. L. Hayden. *A conjecture on the extension of bilinear functionals.* Amer. Math. Monthly **74** (1967), 1108-1109.
10. T. L. Hayden. *The extension of bilinear functionals.* Pacific J. Math. **22** (1967), 99-108.
11. A. Ioffe. *A new proof of the equivalence of the Hahn-Banach extension and the least upper bound properties.* Proc. Amer. Math. Soc. **82** (1981), 385-389.
12. Y. E. Linke. *Linear operators without subdifferentials.* Sibirsk. Mat. Zh. **32** (1991), 219-221.
13. J. Lindenstrauss and L. Tzafriri. *Classical Banach Spaces, I and II.* Springer-Verlag, Berlin, 1996.
14. P. Meyer-Nieberg. *Banach Lattices.* Springer-Verlag, Berlin, Heidelberg, New-York, 1991.

PROBABILISTIC MODULAR SPACES

K. NOUROUZI

Department of Mathematics, K. N. Toosi University of Technology,
P.O. Box 15875-4416, Tehran, Iran
E-mail: nourouzi@kntu.ac.ir

The notion of a probabilistic modular space is given, and then some results in convexity is examined in such spaces.

Keywords: Modular; Probabilistic modular; Convexity.

1. Introduction

On a real linear space X, a real functional ρ is called a modular on X if it satisfies the following conditions:

1. $\rho(x) = 0$ iff $x = 0$,
2. $\rho(x) = \rho(-x)$,
3. $\rho(\alpha x + \beta y) \leq \rho(x) + \rho(y)$, for all $x, y \in X$ and $\alpha, \beta \geq 0$, $\alpha + \beta = 1$.

The theory of modulars was initiated by Nakano [4] and generalized by Musielak and Orlicz [3]. For more details, the reader is referred to [1] and [2]. The main purpose of this note is to introduce the concept of *probabilistic modular spaces*. In fact, following the notion of a probabilistic norm [5], this is carried out, and then we give some results in convexity.

Throughout this note, X denotes a real vector space and Δ stands for the set of all nondecreasing functions $f : \mathbb{R} \to \mathbb{R}_0^+$ satisfying $\inf_{t\in\mathbb{R}} f(t) = 0$ and $\sup_{t\in\mathbb{R}} f(t) = 1$. The symbol $\wedge$ means the minimum.

Definition 1.1. A pair (X, μ) is called a probabilistic modular space (briefly, $\mathcal{P}$-modular space) if μ is a mapping from X into Δ (for $x \in X$, the function $\mu(x)$ is denoted by μ_x, and $\mu_x(t)$ is the value μ_x at $t \in \mathbb{R}$) satisfying the following conditions:

1. $\mu_x(0) = 0$,
2. $\mu_x(t) = 1$ for all $t > 0$ iff $x = 0$,
3. $\mu_{-x}(t) = \mu_x(t)$,
4. $\mu_{\alpha x+\beta y}(s + t) \geq \mu_x(s) \wedge \mu_y(t)$ for all $x, y \in X$, and $\alpha, \beta, s, t \in \mathbb{R}_0^+$

with $\alpha + \beta = 1$.

If we replace 4 by

$$\mu_{\alpha x+\beta y}(s+t) \geq \mu_x(\frac{s}{\alpha}) \wedge \mu_y(\frac{t}{\beta}),$$

for all $x, y \in X$, and $\alpha, \beta, s, t \in \mathbb{R}_0{}^+$, $\alpha + \beta = 1$, we say that (X, μ) is convex. We will also say (X, μ) is β-homogeneous, where $\beta \in (0, 1]$, if

$$\mu_{\alpha x}(t) = \mu_x(\frac{t}{|\alpha|^\beta}),$$

for every $x \in X$, $t > 0$, and $\alpha \in \mathbb{R}$.

Example 1.1. Suppose that ρ is a modular on X. Define

$$\mu_x(t) = \begin{cases} 0, & t \leq 0 \\ \frac{t}{t+\rho(x)}, & r\, t > 0, \end{cases}$$

for all $x \in X$. Then (X, μ) is a $\mathcal{P}$-modular space.

Example 1.2. Suppose that ρ is a modular on X. Define

$$\mu_x(t) = \begin{cases} 0, & t \leq \rho(x) \\ 1, & t > \rho(x), \end{cases}$$

for all $x \in X$. Then (X, μ) is a $\mathcal{P}$-modular space.

Example 1.3. Define $\rho : \mathbb{R} \to \mathbb{R}$ by

$$\rho(x) = \begin{cases} 0, & x = 0 \\ 1, & x \neq 0. \end{cases}$$

Then it is evident that $(\mathbb{R}, \rho)$ is a modular space. If we define μ as

$$\mu_x(t) = \frac{t}{t + \rho(x)},$$

then $(\mathbb{R}, \mu)$ is a non β-homogeneous $\mathcal{P}$-modular space for every $\beta \in (0, 1]$ since $\frac{t}{t+1} \neq \frac{t}{t+2^\beta}$ for all $\beta \in (0, 1]$.

Definition 1.2. Let (X, μ) be a $\mathcal{P}$-modular space.

- A sequence (x_n) in X is said to be μ-convergent to a point $x \in X$ and denoted by $x_n \xrightarrow{\mu} x$ if for every $t > 0$ and $r \in (0, 1)$, there exists a positive integer k such that $\mu_{x_n - x}(t) > 1 - r$ for all $n \geq k$.
- The μ-closure of a subset E of X is denoted by $\overline{E}$ and defined as the set of all $x \in X$ such that there is a sequence (x_n) of elements of E such that $x_n \to x$. The subset E is μ-dense in X if $\overline{E} = X$.

- For $x \in X$, $t > 0$, and $0 < r < 1$, the μ-ball centered at x with radius r is defined by
$$B(x, r, t) = \{y \in X : \mu_{x-y}(t) > 1 - r\}.$$
- An element $x \in E$ is called a μ-interior point of E if there are $r \in (0,1)$ and $t > 0$ such that $B(x, r, t) \subseteq E$. We say that E is μ-open in X if each element of E is a μ-interior point.

According to the notions above, we have the folowing elementary results.

Lemma 1.1. *The μ-limit of any μ-convergent sequence is unique.*

Proof. Let (X, μ) be a $\mathcal{P}$-modular space and (p_n) be a sequence in X. If $p_n \xrightarrow{\mu} p$ and $p_n \xrightarrow{\mu} q$, then
$$\mu_{\frac{1}{2}(p-q)}(t) = \mu_{\frac{1}{2}(p_n-p)+\frac{1}{2}(q-p_n)}(t) \geq \mu_{(p_n-p)}(\frac{t}{2}) \wedge \mu_{(p_n-q)}(\frac{t}{2})$$
for all $t > 0$. Thus
$$\mu_{\frac{1}{2}(p-q)}(t) \geq \lim_{n\to\infty} (\mu_{(p_n-p)}(\frac{t}{2}) \wedge \mu_{(p_n-q)}(\frac{t}{2})) = 1,$$
for all $t > 0$. This implies that $p = q$. □

Lemma 1.2. *The operations + and · in the β-homogeneous $\mathcal{P}$-modular space (X, μ) are continuous.*

Proof. Let $\epsilon > 0$ and $r \in (0,1)$. Suppose $x_n \xrightarrow{\mu} x$, $y_n \xrightarrow{\mu} y$, and $\alpha_n \xrightarrow{\mathbb{R}} \alpha$, where $x, y, x_n, y_n \in X$ and $\alpha, \alpha_n \in \mathbb{R}$, for all n. We may assume $\alpha_n \neq \alpha$ and $\alpha_n \neq 0$ for all n. Then there exists $k \in \mathbb{N}$ such that for every $n \geq k$, we have $|\alpha_n| < |\alpha| + 1$,
$$\mu_{x_n-x}\left(\frac{\epsilon}{2^{2(\beta+1)}|\alpha_n|^\beta}\right) \wedge \mu_{y_n-y}\left(\frac{\epsilon}{2^{(\beta+1)}}\right) > 1 - r,$$
and
$$\mu_x\left(\frac{\epsilon}{2^{2(\beta+1)}|\alpha_n - \alpha|^\beta}\right) > \mu_{x_n-x}\left(\frac{\epsilon}{2^{2(\beta+1)}(|\alpha| + 1)^\beta}\right) > 1 - r.$$
Putting $\gamma = \mu_{(\alpha_n x_n + y_n)-(\alpha x - y)}(\epsilon)$, we obtain
$$\begin{aligned}\gamma &\geq \mu_{x_n-x}\left(\tfrac{\epsilon}{2^{2(\beta+1)}|\alpha_n|^\beta}\right) \wedge \mu_x\left(\tfrac{\epsilon}{2^{2(\beta+1)}|\alpha_n-\alpha|^\beta}\right) \wedge \mu_{y_n-y}\left(\tfrac{\epsilon}{2^{\beta+1}}\right)\\ &> 1 - r,\end{aligned}$$
for all $n \geq k$. This completes the proof. □

2. Convex sets in $\mathcal{P}$-modular spaces

In this section, we aim to give some elementary results in convexity. By a convex set in a vector space X, it is understood a subset of X containing the segment joining every two points of it.

Theorem 2.1. *An open ball in a convex $\mathcal{P}$-modular space (X, μ) is convex.*

Proof. Let $B(x, r, t)$ be an arbitrary open ball in (X, μ). If $y, z \in B(x, r, t)$ and $\lambda \in (0, 1)$, then

$$\begin{aligned} \mu_{x-(\lambda y-(1-\lambda)z)}(t) &= \mu_{\lambda(x-y)+(1-\lambda)(x-z)}(\lambda t + (1-\lambda)t) \\ &\geq \mu_{(x-y)}\left(\tfrac{\lambda t}{\lambda}\right) \wedge \mu_{(x-z)}\left(\tfrac{(1-\lambda)t}{(1-\lambda)}\right) \\ &= \mu_{(x-y)}(t) \wedge \mu_{(x-z)}(t) > 1 - r. \end{aligned}$$

Hence $\lambda y + (1-\lambda)z \in B(x, r, t)$. □

Theorem 2.2. *Suppose that A is a convex subset of a convex β-homogeneous $\mathcal{P}$-modular space (X, μ). Suppose further that $a \in A$, and x is aμ-interior point of A. Then every element of*

$$\mathfrak{T} = \{\lambda x + (1-\lambda)a; \lambda \in (0, 1)\}$$

is a μ-interior point of A.

Proof. If $q \in \mathfrak{T}$, then $q = \lambda x + (1-\lambda)a$ for some $\lambda \in (0, 1)$. There exist $r_0 \in (0, 1)$ and $t_0 > 0$ such that $B(x, r_0, t_0) \subseteq A$. We have

$$B(q, r_0, \lambda^\beta t_0) \subseteq A.$$

Indeed, if $y \in B(q, r_0, \lambda^\beta t_0)$, then $\mu_{\lambda^{-1}(y-q)}(t_0) = \mu_{y-q}(\lambda^\beta t_0) > 1 - r_0$, which implies that $\mu_{x+\lambda^{-1}(y-q)-x}(t_0) > 1 - r_0$. Then

$$u = x + \lambda^{-1}(y - q) \in B(x, r_0, t_0),$$

that is, $y = a + \lambda(u - a) \in A$. □

Theorem 2.3. *Suppose that A is a convex subset of a convex β-homogeneous $\mathcal{P}$-modular space (X, μ). If x is a μ-interior point of A, and y belongs to the μ-closure of A, then all points of*

$$\mathfrak{T} = \{\lambda y + (1-\lambda)x; \lambda \in (0, 1)\},$$

are in the μ-interior of A.

Proof. Let x be an μ-interior point of A, and y a μ-closure point of A. Then there exists a sequence (a_n) of element of A such that $a_n \xrightarrow{\mu} y$. If $z \in \mathfrak{T}$, then $z = (1-\lambda)x + \lambda y$, for some $\lambda \in (0,1)$. Define $z_n = (1-\lambda)x + \lambda a_n$. There exist $r_0 \in (0,1)$ and $t_0 > 0$, such that $B(x, r_0, t_0) \subseteq A$. By the same argument presented in the proof of the preceeding theorem, we obtain $B(z_n, r_0, \lambda^\beta t_0) \subseteq A$, for all n. Since $z_n \xrightarrow{\mu} z$ and $z_n \in B(z_n, r_0, \lambda^\beta t_0)$ for all $n \geq n_0$, we have $z \in B(z_{n_0}, r_0, \lambda^\beta t_0)$, that is, $z \in A$. □

Corollary 2.1. *If the μ-interior of a convex set A is a nonempty, then*
1) the μ-closure of μ-interior of A is equal to $\overline{A}$,
2) A and $\overline{A}$ have the same μ-interior.

Proof. 1) It is clear that $\overline{\mu - intA} \subseteq \overline{A}$. Conversely, let $y \in \overline{A}$ and $x \in \mu - intA$. By Theorem 2.3, $T = \{\lambda x + (1-\lambda)y; \lambda \in (0,1)\} \subseteq \mu - intA$. Now if $\lambda_n \in (0,1)$ with $\lambda_n \to 0$, then $\{\lambda_n x + (1-\lambda_n)\}$ is a sequence in T, which is μ-convergent to y as $\lambda_n \to 0$. Hence $y \in \overline{\mu - intA}$.

2) The inclusion $\mu - intA \subseteq \overline{A}$ is trivial. Conversely, choose $x \in \mu - intA$. If $y \in \mu - int\overline{A}$, then $T \subseteq \mu - intA$. Let $z_\lambda = (1-\lambda)^{-1}(y - \lambda x)$ where $\lambda \in (0,1)$. Then $z_\lambda \xrightarrow{\mu} y$ as $\lambda \to 0$. Thus $z_\lambda \in \overline{A}$ for some λ. Thus $y = \lambda x + (1-\lambda)z_\lambda \in \mu - intA$ and therefore $\mu - int\overline{A} \subseteq intA$. □

Acknowledgments

The author would like to thank the Institute for Studies in Theoretical Physics and Mathematics (IPM), Tehran, Iran, for supporting this research (Grant no. 86470033).

References

1. W. M. Kozlowski, *Modular Function Spaces*, Marcel Dekker, 1988.
2. J. Musielak, *Orlicz Spaces and Modular Spaces*, Lecture Notes in Mathematics, vol. 1034, Springer-Verlag, 1983.
3. J. Musielak and W. Orlicz, On modular spaces, *Studia Math.* **18** (1959), 49–65.
4. H. Nakano, *Modular Semi-Ordered Spaces*, Tokyo, Japan, 1959.
5. B. Schweizer and A. Sklar, *Probabilistic Metric Spaces*, North Holland, New York, Amsterdam, Oxford, 1983.

AN APPLICATION OF EXTREMAL POINTS IN DUAL SPACES

S. REZAPOUR

Department of Mathematics, Azarbaidjan University of Tarbiat Moallem, Azarshahr, Tabriz, Iran
E-mail: sh.rezapour@azaruniv.edu

By using extremal points of dual spaces, we shall improve an incorrect result in the well-known book of Ivan Singer [3].

Keywords: ε-approximation, extremal point.

1. Introduction

Let X be a normed space and W a subspace of X. A point $w_0 \in W$ is said to be an ε-approximation for $x \in X$ if $\|x - w_0\| \leq d(x, W) + \varepsilon$. For $x \in X$, put $P_{W,\varepsilon}(x) = \{y \in W : \|x-y\| \leq d(x, W) + \varepsilon\}$. Note that $P_{W,\varepsilon}(x)$ is a non-empty, closed, bounded and convex subset of X, and when $\varepsilon = 0$, $P_W(x) = P_{W,\varepsilon}(x)$ is the set of all best approximations of x in W. A subspace W is called k-semi-Chebyshev subspace whenever for every $x \in X$, there do not exist more than k linearly independent elements in $P_W(x)$. But $P_W(x)$ may be empty. Throughout, (S, μ) is a positive σ-finite measure space.

There are some works about ε-approximation, for example, [1,2]. Also, there are some mistakes in [3] (see, for example, [4]).

Let $x \in X$ and $r > 0$. We say that a set $A \subseteq X$ supports the cell $S(x, r) = \{y \in X : \|x - y\| \leq r\}$ whenever $d(x, A) = r$. Let $B \subseteq X$ be a closed convex subset. We say that $D \subseteq B$ is extremal if (a) D is closed and convex, (b) $x, y \in B$, $\lambda x + (1 - \lambda)y \in D$ and $0 < \lambda < 1$ imply $x, y \in D$. An extremal subset of B consisting of a single point is called an extremal point of B. We shall denote by $\mathcal{E}(B)$ the set of all extremal points of B. It is known that $\mathcal{E}(C) = \mathcal{E}(B) \bigcap C$, whenever C is an extremal subset of B. Let $x \in X$, $f \in X^*$ and α be a real number. We say that $H = \{y \in X : Re\, f(x - y) \geq \alpha\}$ is extremal whenever $f \in \mathcal{E}(S_{X^*})$.

Here we list some known results which we will need in the main results.

Lemma 1.1 ([2]; Lemma 1.2). *Let W be a closed subspace of a normed linear space X, $x \in X$ and $\varepsilon > 0$ be given. Then $M \subseteq P_{W,\varepsilon}(x)$ if and only if there exists $f \in X^*$ such that $\|f\| = 1$, $f|_W = 0$ and $f(x-y) \geq \|x-y\| - \varepsilon$ for all $y \in M$.*

Lemma 1.2 ([3]; Lemma 1.13; p. 83). *Let $X = L^1(S,\mu)$ and $f \in X^*$. Then, $f \in \mathcal{E}(S_{X^*})$ if and only if there exists $h \in L^\infty(S,\mu)$ such that $|h| = 1$, $\mu -$ a.e. on S and*

$$f(g) = \int_S gh\, d\mu, \quad for\ all\ g \in X. \tag{1}$$

Lemma 1.3 ([3]; Theorem 1.19; p. 85). *Let G be a subspace of the normed space $X = L^1(S,\mu)$, $f \in X$ and $g_0 \in G$. Then, $g_0 \in P_G(f)$ if and only if for each $g \in G$ there exists $f^g \in L^\infty(S,\mu)$ such that $|f^g| = 1$, $\mu -$ a.e. on S,*

$$Re \int_S (g_0 - g) f^g\, d\mu \geq 0, \tag{2}$$

$$Re \int_S (f - g_0) f^g\, d\mu = \int_S |f - g_0|\, d\mu. \tag{3}$$

Finally, we will show that there exists a logical error in the next lemma.

Lemma 1.4 ([3]; Theorem 4.3; p. 133). *Let W be a subspace of $X = L^1(S,\mu)$ and $0 \leq k < \infty$. Then the following are equivalent:*
(1) W is a k-semi-Chebyshev subspace.
(2) There do not exist a measurable subset A of S with $\mu(A) > 0$, a function $f_0 \in L^\infty(S,\mu)$ and $k+1$ linearly independent elements $g_0, g_1, \cdots, g_k$ in W such that

$$\|f_0\|_\infty = 1, \tag{4}$$

$$\int_S f_0 h d\mu = 0, for\ all\ h \in W, \tag{5}$$

$$\sum_{i=0}^{k} |g_i| \neq 0,\ \mu - a.e.\ on\ A, \tag{6}$$

$$f_0 = \pm \operatorname{sign}\ g_i,\ \ \mu - a.e.\ on\ S \backslash ker(g_i),\ (i = 0, 1, \cdots, k), \tag{7}$$

$$g_i = 0,\ \mu - a.e.\ on\ S \backslash A,\ (i = 0, 1, \cdots, k). \tag{8}$$

2. Main Results

Proposition 2.1. *Let W be a subspace of a normed space $(X, \|.\|)$, $x \in X$, $\varepsilon > 0$ be given, and $w_0 \in P_{W,\varepsilon}(x)$. Then, there exists $f \in X^*$ such that*

$f \in \mathcal{E}(S_{X^*})$, $Re\, f(w_0) \geq 0$ and $Re\, f(x - w_0) \geq \|x - w_0\| - \varepsilon$.

Proof. Put $\mathcal{M}^{\varepsilon}_{x,w_0} = \{f \in X^* : \|f\| = 1, Re\, f(x - w_0) \geq \|x - w_0\| - \varepsilon\}$. Note that $\mathcal{M}^{\varepsilon}_{x,w_0}$ is convex and $\sigma(X^*, X)$-compact, where $\sigma(X^*, X)$ is the weak topology. Now by using Lemma 1.1, choose $f_1 \in \mathcal{M}^{\varepsilon}_{x,w_0}$ such that $f_1(w_0) = 0$. Now, define $\psi : X^* \longrightarrow \mathbb{C}$ by $\psi(f) = f(w_0)$. Then, ψ is linear and continuous for $\sigma(X^*, X)$. Thus, $\psi(\mathcal{M}^{\varepsilon}_{x,w_0})$ is convex and compact. Since $\psi(f_1) = f_1(w_0) = 0$, so $0 \in \psi(\mathcal{M}^{\varepsilon}_{x,w_0})$.

Now, suppose that $\Delta = \{z \in \mathbb{C} : Re\, z \geq 0\}$, and $\xi \subseteq \Delta$ be a support line of $\psi(\mathcal{M}^{\varepsilon}_{x,w_0})$ parallel to the boundary of Δ. Then, ξ, whence also Δ, contains an extremal point ξ_0 of $\psi(\mathcal{M}^{\varepsilon}_{x,w_0})$. Thus, by using a virtue of the Krein-Milman theorem, there exists $f_0 \in \mathcal{E}(\mathcal{M}^{\varepsilon}_{x,w_0} \bigcap \psi^{-1}(\xi_0))$. Therefore, $f_0 \in \mathcal{E}(\mathcal{M}^{\varepsilon}_{x,w_0})$, and so $f_0 \in \mathcal{E}(S_{X^*})$. But, $f_0(w_0) = \psi(f_0) = \xi_0 \in \Delta$, hence $Re\, f_0(w_0) \geq 0$. Since $f_0 \in \mathcal{M}^{\varepsilon}_{x,w_0}$, $Re\, f_0(x - w_0) \geq \|x - w_0\| - \varepsilon$. □

Proposition 2.2. *Let W be a subspace of a normed space $(X, \|.\|)$, $x \in X$, $w_0 \in W$, $f \in X^*$ and $\varepsilon > 0$ be given. Then the following are equivalent:*
(i) $f \in \mathcal{E}(\mathcal{M}^{\varepsilon}_{x,w_0})$ and $Re\, f(w_0) \geq 0$.
(ii) $H = \{y \in X : Re\, f(x - y) \geq \|x - w_0\| - \varepsilon\}$ is extremal, supports the cell $S(x, \|x - w_0\| - \varepsilon)$, $Re\, f(w_0) \geq 0$ and $0, w_0 \in H$.

Proof. (i)⇒ (ii). It is clear that H is extremal and $w_0 \in H$. If $y \in H$, then $\|x - y\| \geq Re\, f(x - y) \geq \|x - w_0\| - \varepsilon$. Thus,

$$d(x, H) = inf_{y \in H} \|x - y\| = \|x - w_0\| - \varepsilon,$$

that is, H supports the cell $S(x, \|x - w_0\| - \varepsilon)$. Since $Re\, f(w_0) \geq 0$, $Re\, f(x) \geq Re\, f(x - w_0) \geq \|x - w_0\| - \varepsilon)$. Hence, $0 \in H$.

(ii)⇒ (i). Since H is extremal, $f \in \mathcal{E}(S_{X^*})$. □

Theorem 2.1. *Let W be a subspace of a normed space $(X, \|.\|)$, $x \in X$, $\varepsilon > 0$ be given. Then, $w_0 \in P_{W,\varepsilon}(x)$ if and only if for each $w \in W$, there exists $f^w \in \mathcal{E}(S_{X^*})$ such that*

$$Re\, f^w(w_0 - w) \geq 0 \quad \text{and} \quad Re\, f^w(x - w_0) \geq \|x - w_0\| - \varepsilon.$$

Proof. First suppose that $w_0 \in P_{W,\varepsilon}(x)$ and $w \in W$. Then, $w_0 - w \in P_{W,\varepsilon}(x - w)$, so by Proposition 2.1, there exists $f^w \in \mathcal{E}(S_{X^*})$ such that $Re\, f^w(w_0 - w) \geq 0$ and $Re\, f^w(x - w_0) \geq \|x - w_0\| - \varepsilon$. For the converse part, let $w \in W$ be arbitrary. Then, we have $\|x - w_0\| - \varepsilon \leq Re\, f^w(x - w_0) = Re\, f^w(x - w) + Re\, f^w(w - w_0) \leq Re\, f^w(x - w) \leq \|x - w\|$. Therefore, $w_0 \in P_{W,\varepsilon}(x)$. □

Corollary 2.1. *Let W be a subspace of a normed space $(X, \|.\|)$, $x \in X$, and $\varepsilon > 0$ be given. Then $w_0 \in P_{W,\varepsilon}(x)$ if and only if for each $w \in W$, $H_w = \{y \in X : Re\, f^w(x-y) \geq \|x - w_0\| - \varepsilon\}$ is extremal, supports the cell $S(x, \|x - w_0\| - \varepsilon)$, and $w_0, w \in H$, where f^w is the functional obtained in Theorem 2.1.*

We can obtain the following result from Lemma 1.2 and Theorem 2.1.

Corollary 2.2. *Let G be a subspace of the normed space $X = L^1(S, \mu)$, $f \in X$, $g_0 \in G$ and $\varepsilon > 0$ be given. Then, $g_0 \in P_{G,\varepsilon}(f)$ if and only if for each $g \in G$ there exists $f^g \in L^\infty(S, \mu)$ such that $|f^g| = 1$, $\mu -$ a.e. on S,*

$$Re \int_S (g_0 - g) f^g \, d\mu \geq 0, \tag{9}$$

$$Re \int_S (f - g_0) f^g \, d\mu \geq \int_S |f - g_0| \, d\mu - \varepsilon. \tag{10}$$

Now, we show that there is an error in the Lemma 1.4. In fact, by the second part of the lemma, $\int_S f_0 g_i \, d\mu = 0$ $(i = 0, 1, \cdots, k)$. Hence, $0 = \int_S f_0 g_i \, d\mu = \int_{S \setminus ker(g_i)} f_0 g_i \, d\mu = \pm \int_{S \setminus ker(g_i)} (\text{sign}\, g_i) g_i \, d\mu$. Thus $\int_S |g_i| \, d\mu = 0$, and so $g_i = 0$ $(i = 0, 1, \cdots, k)$. Therefore, there do not exist linearly independent elements in W that satisfy conditions (4)-(6).

This shows that the second statement is always true whereas the first statement may be incorrect. This logical error shows that the result of [3] is incorrect. Now, by using Lemma 1.3, we can improve the result as follows.

Proposition 2.3. *Let G be a subspace of $X = L^1(S, \mu)$ and $0 \leq k < \infty$. Then the following are equivalent:*
(1) G is a k-semi-Chebyshev subspace.
(2) There do not exist $f \in X$, $k+1$ linearly independent elements $g_0, g_1, \cdots, g_k$ in G such that the following statement holds:
For each $g \in G$ and $i = 0, 1, \cdots, k$ there exists $f_i^g \in L^\infty(S, \mu)$ such that

$$|f_i^g| = 1, \quad \mu - a.e. \text{ on } S, \tag{11}$$

$$Re \int_S (g_i - g) f_i^g \, d\mu \geq 0, \ (i = 0, 1, \cdots, k), \tag{12}$$

$$Re \int_S (f - g_i) f_i^g \, d\mu = \int_S |f - g_i| \, d\mu \ (i = 0, 1, \cdots, k). \tag{13}$$

Proof. (1)⇒(2). First suppose that G is a k-semi-Chebyshev subspace. If there exist $f \in X$, $k+1$ linearly independent elements $g_0, g_1, \cdots, g_k$ in G

such that the following statement holds: For each $g \in G$ and $i = 0, 1, \cdots, k$ there exists $f_i^g \in L^\infty(S, \mu)$ such that $|f_i^g| = 1$, $\mu - a.e.$ on S,

$$Re \int_S (g_i - g) f_i^g \, d\mu \geq 0, \ (i = 0, 1, \cdots, k), \tag{14}$$

$$Re \int_S (f - g_i) f_i^g \, d\mu = \int_S |f - g_i| \, d\mu \ (i = 0, 1, \cdots, k). \tag{15}$$

Then we have

$$\begin{aligned}
\int_S |f - g_i| \, d\mu &= Re \int_S (f - g_i) f_i^g \, d\mu \\
&= Re \int_S (f - g) f_i^g \, d\mu + Re \int_S (g - g_i) f_i^g \, d\mu \\
&\leq Re \int_S (f - g) f_i^g \, d\mu \leq | \int_S (f - g) f_i^g \, d\mu| \\
&\leq \int_S |(f - g) f_i^g| \, d\mu = \int_S |(f - g)| \, d\mu
\end{aligned}$$

for all $g \in G$. Thus, $g_i \in P_G(f)$ for all $i = 0, 1, \cdots, k$. Therefore, G is not a k-semi-Chebyshev subspace. This contradiction shows that (2) holds.

(2)$\Rightarrow$(1). If G is not a k-semi-Chebyshev subspace, then there exist $f \in X$ and at least $k+1$ linearly independent elements $g_0, g_1, \cdots, g_k$ in G such that $g_i \in P_G(f)$ for all $i = 0, 1, \cdots, k$. Since $g_i \in P_G(f)$, by Lemma 1.3, for each $g \in G$ and $i = 0, 1, \cdots, k$ there exists $f_i^g \in L^\infty(S, \mu)$ such that $|f_i^g| = 1$, $\mu - a.e.$ on S,

$$Re \int_S (g_i - g) f_i^g \, d\mu \geq 0, \tag{16}$$

$$Re \int_S (f - g_i) f_i^g \, d\mu = \int_S |f - g_i| \, d\mu. \tag{17}$$

This contradiction completes the proof. □

References

1. S. Rezapour, *ε-weakly Chebyshev subspaces of Banach spaces*, Anal. Theory Appl. **19** (2003), 130–135.
2. S. Rezapour, *Weak compactness of the set of ε-extensions*, Bull. Iranian Math. Soc. **30** (2004), 13–20.
3. I. Singer, *Best approximation in normed linear spaces by elements of linear subspaces*, Springer-Verlag, Berlin, 1970.
4. X. Xu, *A result on best proximity pair of two sets*, J. Approx. Theory **54** (1988), 322–325.

ON A PROBLEM FOR A PARABOLIC-HYPERBOLIC EQUATION WITH A NONSMOOTH LINE OF TYPE CHANGING

A. K. URINOV* and I. U. KHAYDAROV

Ferghana State University,
Ferghana, 712000, Uzbekistan
**E-mail: urinovak@mail.ru*

The uniqueness of a boundary value problem for a parabolic-hyperbolic equation with a nonsmooth line of type changing will be proved.

Keywords: Nonlocal problem; parabolic-hyperbolic equation; energy integral; integral equation.

1. Formulation of the problem

Consider the equation

$$0 = L_\lambda u \equiv \begin{cases} u_{xx} - u_y - \lambda_1^2 u, & (x,y) \in \Omega_0, \\ u_{xx} sign\, y + u_{yy} sign\, x - \lambda_2^2\, u\, sign\,(x+y)\,, & (x,y) \in \Omega_j \cup \Omega_j^*, \end{cases}$$

where λ_1, λ_2 are given complex numbers ($j = 1, 2$), and let $\Omega_0 = \{(x,y) : 0 < x < 1, 0 < y < 1\}$, $\Omega_1 = \{(x,y) : -y < x < 1+y, (-1/2) < y < 0\}$, $\Omega_2 = \{(x,y) : -x < y < 1+x, (-1/2) < x < 0\}$, where Ω_1^*, Ω_2^* domains, which are symmetric with respect to straight line $x + y = 0$ to domains Ω_1, Ω_2, respectively. Let $O(0,0)$, $A(1,0)$, $B(0,1)$, $A^*(0,-1)$, $B^*(-1,0)$, $C(1/2,-1/2)$, $D(-1/2,1/2)$, $A_0(1,1)$, $OA(OB^*)$, $OB(OA^*)$, $OC(OD)$, AA_0 are segments of straight lines $y = 0, x = 0, x + y = 0, x = 1$, respectively; $\Omega = \Omega_0 \cup \Omega_1 \cup \Omega_2 \cup \Omega_1^* \cup \Omega_2^* \cup OA \cup OB \cup OC \cup OD$.

Boundary-value problems for equation $L_\lambda u = 0$ in various subdomains of Ω studied by many authors [1,2,3]. In work [4] for the first time an inner-boundary problem for the equation $L_\lambda u = 0$ in the domain Ω was formulated, which was called problem F_1, and also the unique solvability theorem for this problem was stated. In the present work, problem F_1 is investigated in detail.

Problem F_1. Find a function $u(x,y)$ such that

1) u is a regular solution of the equation $L_\lambda u = 0$ in the domains Ω_0, Ω_1, Ω_2, Ω_1^*, Ω_2^*;

2) $u(x,y) \in C(\overline{\Omega}) \cap C^1((\Omega \cup OA^* \cup OB^*) \setminus (OC \cup OD))$;

3) u satisfies

$$u(x,y) = \varphi(y), (x,y) \in \overline{AA_0}; \tag{1}$$

$$\begin{aligned} &u(0,y) = f_1(y), \quad (0,y) \in \overline{A^*O}; \\ &u_x(0,y) - u_x(0,-y) = f_2(y), \quad (0,y) \in A^*O; \end{aligned} \tag{2}$$

$$\begin{aligned} &u(x,0) = g_1(x), \quad (x,0) \in \overline{B^*O}; \\ &u_y(x,0) - u_y(-x,0) = g_2(x), \quad (x,0) \in B^*O. \end{aligned} \tag{3}$$

Here $\varphi(y)$, $f_j(y)$, $g_j(x)$ $(j = 1,2)$ are given functions; moreover $\varphi(y) \in C^1[0,1]$; $f_1(t), g_1(t) \in C[-1,0] \cap C^2(-1,0)$, $f_1(0) = g_1(0)$; $f_2(t), g_2(t) \in C^1(-1,0)$, and they can have singularities of order less than 1 as $t \to 0$ and $t \to -1$.

2. Main functional relations obtained from hyperbolic subdomains

Let $u(x,y)$ be a solution of the problem F_1. Suppose that functions $u_x(x,0), u_x(0,y), u_y(x,0), u_y(0,y)$ are integrable and continuously differentiable in the intervals $(-1,0), (1,0)$. Then function $u(x,y)$ can be found by Riemann's formula as a solution of Cauchy problem for the equation $L_\lambda u = 0$ in the domains Ω_j and Ω_j^* $(j = 1,2)$ [5]:

$$\begin{aligned} u(x,y) &= \frac{1}{2}\left[u(x+y,0) + u(x-y,0)\right] \\ &+ \frac{\lambda_2 y}{2} \int_{x-y}^{x+y} \frac{J_1\left[\lambda_2\sqrt{(x-\xi)^2 - y^2}\right]}{\sqrt{(x-\xi)^2 - y^2}} u(\xi,0)\, d\xi \\ &+ \frac{1}{2} \int_{x-y}^{x+y} J_0\left[\lambda_2\sqrt{(x-\xi)^2 - y^2}\right] \left[u_\eta(\xi,\eta)\big|_{\eta=0}\right] d\xi \end{aligned} \tag{4}$$

for $(x,y) \in \Omega_1 \cup \Omega_2^*$,

$$\begin{aligned} u(x,y) &= \frac{1}{2}\left[u(0,y+x) + u(0,y-x)\right] \\ &+ \frac{\lambda_2 x}{2} \int_{y-x}^{y+x} \frac{J_1\left[\lambda_2\sqrt{(y-\eta)^2 - x^2}\right]}{\sqrt{(y-\eta)^2 - x^2}} u(0,\eta)\, d\eta \\ &+ \frac{1}{2} \int_{y-x}^{y+x} J_0\left[\lambda_2\sqrt{(y-\eta)^2 - x^2}\right] \left[u_\xi(\xi,\eta)\big|_{\xi=0}\right] d\eta \end{aligned} \tag{5}$$

for $(x,y) \in \Omega_2 \cup \Omega_1^*$, where $J_s(x)$ is the Bessel function of the first kind of s-th order. Denote

$$\begin{array}{ll} \tau_1(x) = u(x,0), \quad 0 \le x \le 1; & \nu_1(x) = u_y(x,0), \quad 0 < x < 1; \\ \tau_2^*(y) = u(0,y), \quad -1 \le y \le 0; & \nu_2^*(y) = u_x(0,y), \quad -1 < y < 0. \end{array} \tag{6}$$

From the statement of the problem F_1, it follows that

$$\lim_{x\to -y+0} u(x,y) = \lim_{x\to -y-0} u(x,y), 0 \le x \le 1/2 \tag{7}$$

Substituting (4), (5) into (7) and taking (6) into account, after some evaluations we obtain

$$[\tau_1(x)-\tau_2^*(-x)] + \int_0^x [\tau_1(t)-\tau_2^*(-t)]\frac{x}{t}\frac{\partial}{\partial x} J_0\left[\lambda_2\sqrt{t(t-x)}\right] dt = \gamma(x) \tag{8}$$

for $0 \le x \le 1$, where $\gamma(x) = \int_0^x [\nu_1(t) + \nu_2^*(-t)] J_0\left[\lambda_2\sqrt{t(t-x)}\right] dt$. Transforming equation (8) according to the work [6, p. 59], we find

$$\tau_1(x) - \tau_2^*(-x) = \gamma(x) - \int_0^x \gamma(t)\frac{\partial}{\partial t} J_0\left[\lambda_2\sqrt{x(x-t)}\right], \quad 0 \le x \le 1.$$

Substituting presentation of the function $\gamma(x)$ and taking

$$\int_\xi^x J_0\left[\lambda_2\sqrt{\xi(\xi-t)}\right]\frac{\partial}{\partial t} J_0\left[\lambda_2\sqrt{x(x-t)}\right] dt = J_0\left[\lambda_2\sqrt{\xi(\xi-x)}\right] - J_0[\lambda_2(x-\xi)],$$

into account, we have

$$\tau_1(x) - \tau_2^*(-x) = \int_0^x [\nu_1(t) + \nu_2^*(-t)] J_0[\lambda_2(x-t)]\, dt, \quad 0 \le x \le 1.$$

Taking (2) and (6) into account, from the last equality we get

$$\tau_1(x) = f_1(-x) + \int_0^x [\nu_1(t) + \nu_2(t) + f_2(-t)] J_0[\lambda_2(x-t)]\, dt. \tag{9}$$

Similarly using formulas (4), (5) and condition (3) we obtain

$$\tau_2(x) = g_1(-x) + \int_0^x [\nu_1(t) + \nu_2(t) + g_2(-t)] J_0[\lambda_2(x-t)]\, dt \tag{10}$$

for $0 \le x \le 1$, where $\tau_2(y) = u(0,y)$, $0 \le y \le 1$; $\nu_2(y) = u_x(0,y)$, $0 < y < 1$. From (9) and (10) follows that

$$\tau_2(x) = \tau_1(x) + \Phi_0(x), \quad 0 \le x \le 1, \tag{11}$$

where $\Phi_0(x) = g_1(-x) - f_1(x) + \int_0^x [g_2(-t) - f_2(-t)] J_0[\lambda_2(x-t)]\, dt.$
Further, transforming equation (9) in view of $\nu_1(x) + \nu_2(x)$ we get

$$\nu_1(x) + \nu_2(x) = -f_2(-x) + C_{0x}^{0,\lambda_2}[\tau_1(x) - f_1(-x)], \quad 0 < x < 1, \tag{12}$$

where $C_{0x}^{0,\lambda_2}[p(x)] \equiv p'(x) + \lambda_2^2 \int_0^x \frac{J_1[\lambda_2(x-t)]}{\lambda_2(x-t)} p(t)\, dt.$

Equalities (11) and (12) are main functional relations among $\tau_1(x)$, $\tau_2(x)$, $\nu_1(x)$ and $\nu_2(x)$, obtained from hyperbolic subdomains of Ω.

3. The uniqueness of solution

Lemma 3.1. *If $u(x,y)$ is a solution of homogeneous problem F_1 and $v(x,y) = e^{\delta(x+y)}u(x,y)$, $\forall\delta \in \backslash\{0\}$, then the equality*

$$\iint_{\Omega_0} \left[\left(\mathrm{Re}\,\lambda_1^2 - \delta^2 - \delta\right)|v(x,y)|^2 + |v_x(x,y)|^2\right] dxdy$$

$$+\frac{1}{2}\int_0^1 |v(x,1)|^2\, dx + \int_0^1 |T_1'(x)|^2\, dx + \left(\mathrm{Re}\lambda_1^2 - \delta^2 - \frac{1}{2}\right)\int_0^1 |T_1(x)|^2 dx$$

$$+\mathrm{Re}\int_0^1 e^{2\delta x}\,\bar{\tau}_1(x)\, C_{0x}^{0,\lambda_2}[\tau_1(x)]\, dx \equiv 0 \qquad (13)$$

is valid, where $T_1(x) = v(x,0)$, $0 \le x \le 1$, and $\bar{\tau}_1(x)$ is the conjugate function of $\tau_1(x)$.

Proof. Let $u(x,y)$ be a solution of the problem F_1 at $\varphi(y) \equiv f_j(y) \equiv$ $\equiv g_j(x) \equiv 0$ $(j = 1,2)$. Then equalities

$$u_{xx} - u_y - \lambda_2^2 u = 0, \quad (x,y) \in \Omega_0, \qquad (14)$$

$$\begin{aligned} \tau_2(x) &= \tau_1(x), \qquad 0 \le x \le 1; \\ \nu_2(x) &= -\nu_1(x) + C_{0x}^{0,\lambda_2}[\tau_1(x)], \quad 0 < x < 1 \end{aligned} \qquad (15)$$

are true. Making the substitution $u(x,y) = e^{-\delta(x+y)}v(x,y)$ in (14) and (15), we have

$$v_{xx} - v_y - 2\delta v_x - (\lambda_1^2 - \delta^2 - \delta)v = 0, \quad (x,y) \in \Omega_0. \qquad (16)$$

$$T_1(x) = T_2(x), \quad 0 \le x \le 1, \qquad (17)$$

$$N_1(x) + N_2(x) = 2\delta T_1(x) + e^{\delta x} C_{0x}^{0,\lambda_2}\left[e^{-\delta x}T_1(x)\right], \; 0 < x < 1, \qquad (18)$$

where $T_2(x) = v(0,t)$, $N_1(t) = v_y(t,0)$, $N_2(t) = v_x(0,t)$, $0 \le t \le 1$. Multiplying two sides of (16) by $\bar{v}(x,y)$, which is the conjugate function of $v(x,y)$, we rewrite

$$(\bar{v}v_x)_x - \bar{v}v_y - 2\delta\bar{v}v_x - (\lambda_1^2 - \delta^2 - \delta)|v|^2 = 0, \quad (x,y) \in \Omega_0. \qquad (19)$$

We integrate identity (19) along quadrangle, bounded by straight lines $x = \varepsilon$, $x = 1 - \varepsilon$, $y = h$, $y = 1$, where ε and h are enough small positive numbers. Then, we use Gauss-Ostrogradskij's formula to obtained integral. Selecting real part of the obtained equality, passing to the limit as $h \to 0$, $\varepsilon \to 0$, considering $\varphi(y) \equiv 0$, we get

$$\iint_{\Omega_0} \left[\left(\operatorname{Re} \lambda_1^2 - \delta^2 - \delta \right) |v(x,y)|^2 + |v_x(x,y)|^2 \right] dxdy + \frac{1}{2} \int_0^1 |v(x,1)|^2 dx$$

$$+ \operatorname{Re} \int_0^1 \overline{T_2}(y) N_2(y) dy - \delta \int_0^1 |T_2(x)|^2 dx - \frac{1}{2} \int_0^1 |T_1(x)|^2 dx = 0, \quad (20)$$

where $\bar{T}_2(t)$ is the conjugate function of $T_2(t)$. Further, we integrate (14) by x along the segment $\{(x,y): \; y = h\,, \; \varepsilon < x < 1-\varepsilon\}$. In the obtained equality passing to the limit as $h \to 0$, $\varepsilon \to 0$ and taking the real part, considering $T_1(0) = T_1(1) = 0$, we have

$$\operatorname{Re} \int_0^1 \overline{T_1}(x) N_1(x) dx = -\int_0^1 |T_1'(x)|^2 dx - (\operatorname{Re}\lambda_1^2 - \delta^2 - \delta) \int_0^1 |T_1(x)|^2 dx, \quad (21)$$

Substituting to this integral $N_2(x)$ from (18), considering (17) and $\tau_1(x) = e^{-\delta x} T_1(x)$, we obtain

$$\operatorname{Re} \int_0^1 \bar{T}_2(x) N_2(x)\, dx = Re \int_0^1 e^{2\delta x} \overline{\tau}_1(x)\, C_{0x}^{0,\lambda_2} \left[\tau_1(x) \right] dx$$

$$+2\delta \int_0^1 |T_1(x)|^2 dx - \operatorname{Re} \int_0^1 \overline{T}_1(x)\, N_1(x)\, dx \quad (22)$$

From (20), (21), (22) the validity of the identity (13) follows. □

Lemma 3.2 ([7]). *If* $\delta \le -|\operatorname{Im}\lambda_2|$, *then we have*

$$\operatorname{Re} \int_0^1 e^{2\delta x} \bar{\tau}_1(x)\; C_{0x}^{0,\lambda_2} \left[\tau_1(x) \right]\, dx \ge 0, \quad (23)$$

Theorem 3.1. *If the inequality*

$$\operatorname{Re} \lambda_1^2 \ge (\operatorname{Im}\lambda_2)^2 + 1/2, \quad (24)$$

is valid, then the problem F_1 *can not have more than one solution.*

Proof. Let $u(x,y)$ be a solution of the problem F_1 and $\delta = -|\operatorname{Im} \lambda_2|$. Then according to lemmas 3.1, 3.2, identity (13) and inequality (23) are true. Moreover, by virtue of (24), the first and the forth items of (13) are non-negative. If we consider this fact, from (13) follows, that $v(x,y) \equiv 0$ in $\bar{\Omega}_0$. Then $u(x,y) \equiv 0$ in $\bar{\Omega}_0$. From here also follows $u_y(x,0) \equiv 0$, $0 \le x \le 1$, $u_x(0,y) \equiv 0$, $0 \le y \le 1$. Taking this into account we get $f_2(y) \equiv g_2(x) \equiv 0$. From (2) and (3) follows that $u_y(x,0) \equiv 0$, $-1 \le x \le 0$, $u_x(0,y) \equiv 0$,

$-1 \le y \le 0$. Because of these and using formulas (4), (5), we can conclude that $u(x,y) \equiv 0 \ \ (x,y) \in \bar{\Omega}_j \cup \bar{\Omega}_j^*, \ j = 1,2$. Hence, $u(x,y) \equiv 0$ in $\bar{\Omega}$, which means that the problem F_1 cannot have more than one solution. □

It follows that the homogeneous problem F_1 can have trivial solution if
1) $\lambda_1, \lambda_2 \in R$ and $(\lambda_1,\lambda_2) \in G_1 = \{(\lambda_1,\lambda_2) : |\lambda_1| \ge 1/2, -\infty < \lambda_2 < +\infty\}$;
2) $\lambda_2 = i\lambda_3$; $\lambda_1, \lambda_3 \in R$ and $(\lambda_1,\lambda_3) \in G_2 = \{(\lambda_1,\lambda_3) : \lambda_1^2 - \lambda_3^2 \ge 1/2\}$;
3) $\lambda_1 = \lambda_2 = \lambda \in C$ and $\lambda \in G_3 = \left\{\lambda : (\mathrm{Re}\lambda)^2 - 2(\mathrm{Im}\,\lambda)^2 \ge 1/2\right\}$.

4. The existence of the solution

Let $u(x,y)$ be a solution of the problem F_1. Then equality (14) and relations (11), (12) are true. Passing to the limit as $y \to +0$, from (14) we get the equation $\tau_1''(x) - \lambda_1^2 \tau_1(x) = \nu_1(x)$.

Solution of this equation satisfying boundary conditions $\tau_1(0) = f_2(0)$, $\tau_1(1) = \varphi(0)$ in the segment $0 \le x \le 1$ is represented as

$$\tau_1(x) = \Phi_1(x) + \int_0^1 K(x,t)\,\nu_1(t)\,dt, \tag{25}$$

where

$$\Phi_1(x) = f_1(0) + x[\varphi(0) - f_1(0)] + \lambda_1^2 \int_0^1 K(x,t)\,[f_1(0) + t[\varphi(0) - f_1(0)]]\,dt,$$

$$K(x,t) = \begin{cases} sh\lambda_1(1-t)\,sh\lambda_1 x/(\lambda_1 sh\lambda_1), & 0 \le x \le t, \\ sh\lambda_1 t\ sh\lambda_1(1-x)/(\lambda_1 sh\lambda_1), & t \le x \le 1. \end{cases}$$

Next in the domain Ω_0 consider the first the boundary-value problem for equation (14), satisfying conditions $u(0,y) = \tau_2(y)$, $u(x,0) = \tau_1(x)$, $u(1,y) = \varphi(y)$.

Solution of this problem is given defined by the formula [1]

$$u(x,y) = \int_0^y \tau_2(\eta)\,e^{\lambda_1^2(\eta-y)} G_\xi(x,y;0,\eta)\,d\eta \tag{26}$$

$$-\int_0^y \varphi(\eta)\,e^{\lambda_1^2(\eta-y)} G_\xi(x,y;1,\eta)\,d\eta + \int_0^1 \tau_1(\xi)\,e^{-\lambda_1^2 y} G(x,y;\xi,0)\,d\xi,$$

where $G(x,y;\xi,\eta)$

$$= \frac{1}{2\sqrt{\pi(y-\eta)}} \sum_{n=-\infty}^{+\infty} \left\{ \exp\left[-\frac{(x-\xi+2n)^2}{4(y-\eta)}\right] - \exp\left[-\frac{(x+\xi+2n)^2}{4(y-\eta)}\right] \right\}.$$

Differentiating (26) by x and setting $x = 0$, we have

$$\nu_2(y) = -\int_0^y \left[\tau_2(\eta)\,e^{\lambda_1^2(\eta-y)}\right]_\eta' N(0,y;0,\eta)\,d\eta$$

$$+\int_0^1 \tau_1'(\xi)\, e^{-\lambda_1^2 y} N(0,y;\xi,0)\, d\xi + \Phi_2(y), \quad 0<y<1, \tag{27}$$

where $\Phi_2(y) = \int_0^y \left[\varphi(\eta)\, e^{\lambda_1^2(\eta-y)}\right]_\eta' N(0,y;1,\eta)\, d\eta$ and

$$N(0,y;\xi,\eta) = \frac{1}{[\pi(y-\eta)]^{1/2}} \sum_{n=-\infty}^{+\infty} \exp\left[-(\xi-2n)^2 \Big/ 4(y-\eta)\right].$$

Accepting functions $\tau_1(x)$, $\tau_2(x)$, $\nu_2(x)$ from equations (11), (12), (25), (27), we obtain a Fredholm integral equation of the second kind regarding to function $\nu_1(x)$, which is equivalent (in the sense of solvability in the class of functions wanted) to the problem F_1:

$$\nu_1(y) + \int_0^1 K_2(y,t)\, \nu_1(t)\, dt = \Phi_4(x), \quad 0<x<1, \tag{28}$$

where

$$K_2(x,t) = K_1(x,t) - K_x'(x,t) - \lambda_2^2 \int_0^x \frac{J_1[\lambda_2(x-z)]}{\lambda_2(x-z)} K(z,t)\, dz,$$

$$\Phi_4(x) = -f_2(-x) - \Phi_3(x) + C_{0x}^{0,\lambda_2}[\Phi_1(x) - f_1(-x)],$$

$$K_1(y,t) = -\int_0^1 \left[K(\eta,t)\, e^{\lambda_1^2(\eta-y)}\right]_\eta' N(0,y;0,\eta)\, d\eta$$

$$+\int_0^1 K_\xi'(\xi,t)\, e^{-\lambda_1^2 y} N(0,y;\xi,0)\, d\xi,$$

$$\Phi_3(y) = \Phi_2(y) + \int_0^1 \Phi_1'(\xi)\, e^{-\lambda_1^2 y} N(0,y;\xi,0)\, d\xi$$

$$-\int_0^y \left\{[\Phi_2(\eta) + \Phi_1(\eta)]\, e^{\lambda_1^2(\eta-y)}\right\}_\eta' N(0,y;0,\eta)\, d\eta.$$

The kernel $K_2(x,t)$ is continuous and continuously differentiable with respect to x in $0 \le x < t$, $t < x \le 1$, and $\Phi_3(x) \in C^1(0,1)$. Moreover, it can have a singularity of order less than 1 at $x \to 0$, $x \to 1$. Hence to the equation (28) we can apply Fredholm's alternative. The solvability of (28) follows from the uniqueness of solution of the problem F_1.

After finding the function $\nu_1(x)$ from (28), functions $\tau_1(x)$, $\tau_2(x)$ and $\nu_2(x)$ can be found by formulas (25), (11) and (27), respectively. By virtue of aforementioned we conclude that unique solution exists for problem F_1. In the domain Ω_0 it determines by formula (26), and in the domains Ω_j, Ω_j^* $(j=1,2)$ by formulas (4), (5). Moreover functions $u_x(x,y)|_{A^*O}$ and $u_y(x,y)|_{B^*O}$ can be found from (2) and (3) with the help of $\nu_1(x)$, $\nu_2(x)$.

It is not difficult to be sure that when (24) condition is fulfilled, the problem F_1 is uniquely solvable, even we replace conditions (2), (3) with the conditions

$$u_x(x,y) = f_1(y),\ (x,y) \in A^*O;\ \ u(x,y) + u(x,-y) = f_2(y),\ (x,y) \in \overline{A^*O};$$
$$u_y(x,y) = g_1(x),\ (x,y) \in B^*O;\ \ u(x,y) + u(-x,y) = g_2(x),\ (x,y) \in \overline{B^*O}.$$

Note that in works [8,9] F_1-type problems were investigated for elliptic-hyperbolic-type equations

$$u_{xx} sign\, y + u_{yy} sign\, x - \lambda^2 u\, sign(x,y) = 0,$$
$$u_{xx} sign\, y + u_{yy} sign\, x + \frac{2\beta}{|x|} u_x + \frac{2\beta}{|y|} u_y = 0$$

in the domain Ω, when Ω_0 is quarter disk, and condition (1) replaced with condition $u(x,y)|_{\sigma_0} = \varphi(x,y)$, where $\sigma_0 = \{(x,y) : x^2 + y^2 = 1, x > 0, y > 0\}$, $0 < \beta = const < 1/2$.

References

1. T. D. Dzhuraev, A. Sopuev, M. Mamazhonov, Boundary-value problems for parabolic-hyperbolic equations, Tashkent, Fan, 1986, 220 pp.
2. K. B. Sabytov, To the theory of mixed parabolic-hyperbolic type equations with spectral parameter, Differentsialnije uravnenija, 1988, V. XXIV, 117–126.
3. E. T. Karimov, Boundary-value problems for parabolic-hyperbolic type equations with spectral parameter, Ph.D. Thesis, Tashkent, 2006.
4. A. K. Urinov, I. U. Khaydarov, On a problem for parabolic-hyperbolic type equation with non-smooth line of type change, Abstracts, 6th international ISAAC congress, Ankara, Turkey, 2007, p. 105.
5. A. N. Tikhonov, A. A. Samarskij, Equations of mathematical physics, Moscow, Nauka, 1977, 737 pp.
6. I. N. Vekua, New methods of solving for elliptic equations, Moscow, Gostekhizdat, 1948, 296 pp.
7. M. S. Salakhitdinov, A. K. Urinov, Boundary-value problems for mixed type problems with spectral parameter, Tashkent, Fan, 1997, 168 pp.
8. A. K. Urinov, I. T. Tojiboev, On a uniqueness of solution of non-local problem, Proceedings of international conference "Modern problems of differential equations, theory of operators and space technologies", Almaty, 2006, 120–121.
9. A. K. Urinov, K. T. Karimov, On the unique solvability of a problem for mixed type equation with two singular coefficients, Proceedings of international conference "Modern problems of differential equations, theory of operators and space technologies", Almaty, 2006, 119–120.

Addenda to

Session 06

Dispersive Equations

SESSION EDITORS

V. Georgiev   Università di Pisa, Pisa, Italy
M. Reissig   Technische Universität, Freiberg, Germany

ASYMPTOTIC BEHAVIOR AND REGULARITY FOR NONLINEAR DISSIPATIVE WAVE EQUATIONS IN $\mathbb{R}^n$

G. TODOROVA

University of Tennessee, Knoxville, TN, USA
E-mail: todorova@math.utk.edu

D. UĞURLU

Abant İzzet Baysal University, Bolu, Turkey
E-mail: ugurlu_d@ibu.edu.tr

B. YORDANOV

University of Tennessee, Knoxville, TN, USA
E-mail: yordanov@math.utk.edu

We discuss several results on asymptotic behavior and regularity for wave equations with nonlinear damping $u_{tt} - \Delta u + |u_t|^{m-1}u_t = 0$ in $\mathbb{R}_+ \times \mathbb{R}^n$, where $m > 1$. In particular, we show that $\nabla u \in L^{m+1}(I \times \mathbb{R}^n)$ for all initial data $(u, u_t)|_{t=0} \in H^1(\mathbb{R}^n) \times L^2(\mathbb{R}^n)$ when $m \leq (n+2)/n$ and $I \subset \mathbb{R}_+$ is a compact interval. We also consider global well-posedness in Sobolev spaces $H^k(\mathbb{R}^n) \times H^{k-1}(\mathbb{R}^n)$ for $k > 2$. The strength of nonlinear damping is critical when $m = n/(n-2)$ and $n \geq 3$, which makes classical techniques ineffective. We outline the proof of global well-posedness for $m = 3$ and $n = 3$ (critical) under the additional condition that $(u, u_t)|_{t=0}$ are spherically symmetric. Finally, we present scattering results for spherically symmetric global solutions.

Keywords: Wave equation, nonlinear damping, regularity, asymptotic behavior.

1. Introduction

The asymptotic behavior and regularity of solutions to wave equations with nonlinear damping $(m > 1)$

$$u_{tt} - \Delta u + |u_t|^{m-1}u_t = 0, \quad (t, x) \in \mathbb{R}_+ \times \mathbb{R}^n, \tag{1}$$

are not only challenging questions but also starting points for understanding basic mechanisms in nonlinear hyperbolic equations, such as scattering-damping [18–20] and smoothing-focussing [3,9,14]. A classical result of Lions and Strauss [15] states that (1) is globally well-posed for all initial data

$$u(0,x) = u_0(x), \quad u_t(0,x) = u_1(x), \quad x \in \mathbb{R}^n, \tag{2}$$

with $(u_0, u_1) \in H^k(\mathbb{R}^n) \times H^{k-1}(\mathbb{R}^n)$ and $1 \leq k \leq 2$; see also [9,14,22]. Moreover, if

$$u_0 \in H^2(\mathbb{R}^n), \quad u_1 \in H^1(\mathbb{R}^n) \cap L^{2m}(\mathbb{R}^n),$$

problem (1)–(2) has a unique global solution u with the following properties:

(a) $u \in L^\infty(\mathbb{R}_+, H^2(\mathbb{R}^n)), \quad u_t \in L^\infty(\mathbb{R}_+, H^1(\mathbb{R}^n))$;

(b) $E_1(u,t) = E_1(u,t_0) - \int_{t_0}^t \int |u_s|^{m+1} dx ds, \quad t \geq t_0,$

where $E_k(u,t) = \frac{1}{2} \sum_{|\alpha|=k-1} \int (|\nabla^\alpha u_t|^2 + |\nabla^\alpha \nabla u|^2) dx$;

(c) $E_1(u,t) \leq E_1(u,t_0), \quad E_2(u,t) \leq E_2(u,t_0), \quad t \geq t_0$;

(d) $\mathrm{supp}(u,u_t) \subset \{|x| \leq t + R\}$ if $\mathrm{supp}(u_0,u_1) \subset \{|x| \leq R\}$.

Hence, $E_1(u,t)$ and $E_2(u,t)$ are non-increasing functions of t. A natural question is whether these norms decay to zero or not as $t \to \infty$. The answers are known in several cases described in Section 2. Another interesting problem is suggested by identity (b) which implies $u_t \in L^{m+1}(\mathbb{R}_+ \times \mathbb{R}^n)$ even for data $(u_0, u_1) \in H^1(\mathbb{R}^n) \times L^2(\mathbb{R}^n)$. Section 3 shows that the gain of regularity in ∇u is similar if $m \leq (n+2)/n$: $\nabla u \in L^{m+1}(I \times \mathbb{R}^n)$ for every compact $I \subset \mathbb{R}_+$. The final Section 4 deals with the question $E_k(u,t) < \infty$ at all $t \in \mathbb{R}_+$ for $k > 2$ and arbitrarily large $E_k(u,0)$. It is expected, based on the invariant scaling of (1), that either $m < \infty$ and $n = 1, 2$, or $m \leq n/(n-2)$ and $n \geq 3$ will be a sufficient condition. However, the critical cases $m = n/(n-2)$, $n \geq 3$ remain open as they are difficult to study by current perturbation techniques. This paper outlines authors contribution[25] which covers $m = 3$, $n = 3$, and spherically symmetric data in $H^k(\mathbb{R}^3) \times H^{k-1}(\mathbb{R}^3)$, $k \geq 3$.

2. Asymptotic behavior of energy

The long time behavior of solutions to the Cauchy problem (1)–(2) is not well-understood yet. We should mention that the boundary value problem $x \in \Omega$, where $\Omega \subset \mathbb{R}^n$ is a bounded domain, is quite different. Nakao [21] and Haraux [7] have found polynomial decay rates of $E_1(u,t)$ under the Dirichlet boundary condition $u = 0$ on $\partial\Omega$. The current state of this problem and its generalizations for localized damping and source is presented in [13]. A surprising similarity with the Cauchy problem is that no decay estimates are known for $E_2(u,t)$ and higher-order norms.

There are three types of results about the asymptotic behavior of (1)–(2) with data $(u_0,u_1) \in H^1(\mathbb{R}^n) \times L^2(\mathbb{R}^n)$: Decay estimates for $E_1(u,t)$, lower bounds on $E_1(u,t)$, and $E_1(u-u_+,t) \to 0$ as $t \to \infty$ for a finite-energy solution u_+ of the wave equation, i.e., scattering in the energy space. Clearly, the existence of scattering states implies that the energy approaches a non-zero constant. Below is a short review of significant results.

A sufficient condition for scattering, due to Motai and Mochizuki [20], is $m > 1+2/(n-1)$, $n \geq 2$. In such cases, $E_1(u,t)$ can not decay as $t \to \infty$. The latter result is generalized to certain exterior domains in $n=3$ in [17].

The strongest decay estimate for problem (1)–(2) has long been logarithmic. For $1 < m < 1+2/n$ and $0 < \mu < 2/(m-1)$, Mochizuki and Motai [19] have established

$$E_1(u,t) \leq C_\mu \{\ln(2+t)\}^{-\mu}, \tag{3}$$

where C_μ depends on μ and (u_0,u_1). The gap between conditions for scattering and energy decay means that the asymptotic behavior of $E_1(u,t)$ is unknown if $1+2/n \leq m \leq 1+2/(n-1)$. Estimate (3) is very weak but it may be sharp in the critical cases $n \geq 2$, $m = 1+2/(n-1)$. Kubo [12] has found such modified asymptotic profiles in $n=2$, $m=3$.

The logarithmic decay rate of energy is unlikely to be sharp, however, if $1 < m < 1+2/n$. Then the invariant scaling and nonlinear diffusion phenomenon for (1) strongly suggest polynomial rates depending on n and m. To explain the first argument, we consider a change of variables

$$u(t,x) \mapsto u_\lambda(t,x) = \lambda^{(2-m)/(m-1)} u(\lambda(t-t_0), \lambda(x-x_0)), \tag{4}$$

with $\lambda > 0$ and $(t_0,x_0) \in \mathbb{R} \times \mathbb{R}^n$. Equation (1) is invariant under these transformations. Since $\{u_\lambda(t,x)\}$ resembles a delta family concentrating at (t_0,x_0), as $\lambda \to \infty$, it can be used for testing conjectures about the asymptotic behavior and regularity of u. Changes may occur at critical m where the energy is λ-invariant. We readily calculate

$$E_k(u_\lambda,t) = \lambda^{2/(m-1)+2(k-1)-n} E_k(u, \lambda(t-t_0)) \tag{5}$$

and set $t=1$, $t_0=0$, and $k=1$ to obtain

$$E_1(u,\lambda) = \lambda^{n-2/(m-1)} E_1(u_\lambda, 1).$$

Thus the behavior of $u(\lambda,x)$ as $\lambda \to \infty$ is determined by the relative smoothness of $u_\lambda(1,x)$ compared to $u_\lambda(0,x)$; any non-trivial smoothing estimate of $u_\lambda(1,x)$ will imply a decay estimate of $u(\lambda,x)$. The former type of results is very plausible since the nonlinear damping has a regularizing effect on solutions. Notice that always $E_1(u_\lambda, 1) \to \infty$ as $\lambda \to \infty$, so the energy of u is not expected to decay when $m \geq 1 + 2/n$.

To present the second argument in favor of polynomial decay rates, we assume that (1) exhibits the nonlinear diffusion phenomenon. Here the relevant correspondence is $u_t \approx v$, where v solves the fast diffusion equation

$$(v^m)_t - \Delta v = 0, \quad (t,x) \in \mathbb{R}_+ \times \mathbb{R}^n.$$

Every positive solution v has a polynomial asymptotic profile

$$v(t,x) \sim t^{-n\gamma}(\alpha + \beta|x|^2/t^{2m\gamma})^{-1/(m-1)}, \tag{6}$$

with α, $\beta > 0$, and $\gamma = (n - m(n-2))^{-1}$; see [4] and the references therein. The wave equation with a nonlinear damping (1) is formally transformed into the fast diffusion equation if u_{tt} is neglected, the remaining terms are differentiated with respect to t, and u_t is replaced by v. Such manipulations cannot be justified although they are valid if the damping is linear; see [16].

Theorem 2.1. *Let u be a solution of (1)–(2) with compactly supported $(u_0, u_1) \in H^2(\mathbb{R}^n) \times H^1(\mathbb{R}^n)$. If $n \geq 3$ and $1 < m \leq (n+2)/(n+1)$, then*

$$E_1(u,t) \leq C_a t^{-a}, \quad t \to \infty,$$

where $a > 0$ depends only on m and n, while C_a depends on a and (u_0, u_1).

Unfortunately, the above decay rate [26] is implicit. Here the the lack of control on $\|u\|_{L^2}$ is the main difficulty. This is also an essential difference with the wave equation in a bounded domain or the Klein-Gordon equation $u_{tt} - \Delta u + u + |u_t|^{m-1}u_t = 0$. The proof of Theorem 2.1 uses "parabolic" effects coming from the presence of damping in (1). A crucial idea is to introduce weights similar to the asymptotic profile of fast diffusion (6).

3. Smoothing effects

The regularizing effect of nonlinear damping on solutions of (1)–(2) has already been studied. Examples are constructed of piecewise C^2 data with jump discontinuities which are partially smoothed during the evolution when $m > 1 + 2/(n-1)$; see [9] (radial data) and [14] (general data).

We consider the complementary range $1 < m \leq 1 + 2/n$ and general data $(u_0, u_1) \in H^1(\mathbb{R}^n) \times L^2(\mathbb{R}^n)$. Recall that $u_t \in L^{m+1}(\mathbb{R}_+ \times \mathbb{R}^n)$, so u_t is more regular than u_1 for almost all $t > 0$; see Section 1 (b). The partial smoothing of u_t is a significant difference with the linear wave equation.

A natural question is whether the regularizing effect extends to ∇u. The answer is affirmative, if the nonlinear damping is relatively weak.

Theorem 3.1. *Assume that $1 < m \leq 1 + 2/n$. For every compact interval $I \subset \mathbb{R}_+$, the solution u of (1) satisfies*

$$\|\nabla u\|_{L^{m+1}(I \times R^n)} \leq C \left(E_1^{1/(m+1)}(u,0) + E_1^{m/(m+1)}(u,0) \right),$$

where C depends only on m, n, and I.

We begin with a simple estimate for the wave operator $\Box = \partial_t^2 - \Delta$.

Lemma 3.1. *Assume that $1 < p \leq q < \infty$ and $1/p - 1/q = 1/(n+1)$. There exists a C depending on p, q, and n, such that*

$$\|\nabla u\|_{L^q(R^{n+1})} \leq C\|\partial_t u\|_{L^q(R^{n+1})} + C\|\Box u\|_{L^p(R^{n+1})}$$

for every compactly supported distribution $u \in \mathcal{E}'(\mathbb{R}^{n+1})$ which has finite norms on the right side.

Proof of Theorem 3.1. Let $I \subset J$ be compact intervals in $\mathbb{R}_+$ and $\chi \in C_0^\infty(\mathbb{R}_+)$ be a cut-off function satisfying

$$\chi(t) = \begin{cases} 1, & \text{if } t \in I, \\ 0, & \text{if } t \in \mathbb{R} \setminus J. \end{cases}$$

We denote the first two derivatives of χ by χ' and χ''. Applying Lemma 3.1 to χu, where u is a solution of (1), we have

$$\|\nabla_x(\chi u)\|_{L^q(R^{n+1})} \leq C\|\partial_t(\chi u)\|_{L^q(R^{n+1})} + C\|\Box(\chi u)\|_{L^p(R^{n+1})}.$$

Since $\Box(\chi u) = -\chi|u_t|^{m-1}u_t + 2\chi' u_t + \chi'' u$ and $\partial_t(\chi u) = \chi u_t + \chi' u$, the initial estimate becomes

$$\begin{aligned} \|\chi \nabla_x u\|_{L^q(R^{n+1})} \leq\ & C(\|\chi u_t\|_{L^q(R^{n+1})} + \|\chi' u\|_{L^q(R^{n+1})}) \\ & + C\|\chi|u_t|^m\|_{L^p(R^{n+1})} \\ & + C(\|\chi' u_t\|_{L^p(R^{n+1})} + \|\chi'' u\|_{L^p(R^{n+1})}). \end{aligned} \tag{7}$$

The strongest regularization occurs when

$$q = m + 1, \qquad p = (m+1)(n+1)/(m+n+2), \tag{8}$$

as the nonlinear damping requires $q \leq m + 1$ for $u_t \in L^q(\mathbb{R}_+ \times \mathbb{R}^n)$.

Notice also that $m \leq 1 + 2/n$ implies $p \leq (m+1)/m$. Hence, the most singular term in (7) is estimated by Hölder's inequality:

$$\|\chi|u_t|^m\|_{L^p(R^{n+1})} \leq C\|u_t\|^m_{L^{m+1}(J \times R^n)} \leq C E_1^{m/(m+1)}(u,0). \tag{9}$$

We can assume that $u(t,\cdot)$ is compactly supported, since the estimates are local in time and the equation has a finite speed of propagation.

The other terms in (7) are trivially bounded by $E_1(u,0)$. Recall that the integration takes place on a compact set $\{(t,x) : t \in J,\ |x| \le t+R\}$ and $p \le (m+1)/m < 2$, where p and q are defined in (8).

We begin with the norms involving u_t on the right side of (7):

$$\begin{aligned}\|\chi u_t\|_{L^q(R^{n+1})} + \|\chi' u_t\|_{L^p(R^{n+1})} &\le \|u_t\|_{L^{m+1}(J\times R^n)} + C\|u_t\|_{L^2(J\times R^n)} \\ &\le CE_1^{1/(m+1)}(u,0) + CE_1^{1/2}(u,0). \quad (10)\end{aligned}$$

The Sobolev embedding $H^1(\mathbb{R}^{n+1}) \hookrightarrow L^q(\mathbb{R}^{n+1})$ for $q \le 2+2/n$ implies

$$\begin{aligned}\|\chi' u\|_{L^q(R^{n+1})} &\le C(\|u_t\|_{L^2(J\times R^n)} + \|\nabla u\|_{L^2(J\times R^n)}) \\ &\le CE_1^{1/2}(u,0). \qquad (11)\end{aligned}$$

Finally, we use $H^1(\mathbb{R}^{n+1}) \hookrightarrow L^2(\mathbb{R}^{n+1})$ and $p < 2$:

$$\|\chi'' u\|_{L^p(R^{n+1})} \le \|\nabla u\|_{L^2(J\times R^n)} \le CE_1^{1/2}(u,0). \qquad (12)$$

Substituting estimates (9), (10), (11), and (12) into (7), we obtain

$$\|\nabla_x u\|_{L^{m+1}(I\times R^n)} \le C\left(E_1^{1/2}(u,0) + E_1^{1/(m+1)}(u,0) + E_1^{m/(m+1)}(u,0)\right).$$

Since $1/(m+1) \le 1/2 \le m/(m+1)$, we can drop $E_1^{1/2}(u,0)$ from the last estimate. The proof of Theorem 3.1 is complete. □

Proof of Lemma 3.1. Define the Fourier transform $\mathcal{F}$ as

$$\mathcal{F}u(\tau,\xi) = \int_{R^{n+1}} e^{-i(\tau t+\xi x)} u(t,x)\,dx\,dt, \quad (\tau,\xi) \in \mathbb{R}^{n+1},$$

and denote its inverse by $\mathcal{F}^{-1}$. From $\mathcal{F}(\partial_t u) = i\tau\mathcal{F}u$ and $\mathcal{F}(\nabla u) = i\xi\mathcal{F}u$, we have $\mathcal{F}(\Box u) = (-\tau^2 + |\xi|^2)\mathcal{F}u$. To express ∇u in terms of $\partial_t u$ and $\Box u$, we rewrite this identity as $|\xi|\mathcal{F}u = |\tau|\mathcal{F}u + (|\xi|+|\tau|)^{-1}\mathcal{F}(\Box u)$ and apply $\mathcal{F}^{-1}$ to both sides:

$$\nabla u = \mathcal{F}^{-1}\left(\xi|\xi|^{-1}\mathrm{sign}(\tau)\mathcal{F}(\partial_t u)\right) + i\mathcal{F}^{-1}\left(\xi|\xi|^{-1}(|\xi|+|\tau|)^{-1}\mathcal{F}(\Box u)\right). \quad (13)$$

Here $\mathrm{sign}(\tau) = \tau/|\tau|$ for $\tau \ne 0$. It is well known [5] that $\xi/|\xi|$ and $\mathrm{sign}(\tau)$ are $L^q(\mathbb{R}^{n+1})$ Fourier multipliers, while $(|\xi|+|\tau|)^{-1}$ is an $L^p(\mathbb{R}^{n+1}) - L^q(\mathbb{R}^{n+1})$ Fourier multiplier if $1/p - 1/q = 1/(n+1)$ and $1 < q < \infty$. Using (13),

$$\begin{aligned}\|\nabla u\|_{L^q(R^{n+1})} &\le \|\mathcal{F}^{-1}(\xi|\xi|^{-1}\mathrm{sign}(\tau)\mathcal{F}(\partial_t u))\|_{L^q(R^{n+1})} \\ &\quad + \|\mathcal{F}^{-1}(\xi|\xi|^{-1}(|\xi|+|\tau|)^{-1}\mathcal{F}(\Box u))\|_{L^q(R^{n+1})} \\ &\le C\|\partial_t u\|_{L^q(R^{n+1})} + C\|\Box u\|_{L^p(R^{n+1})}.\end{aligned}$$

□

4. Regularity and scattering for critical damping

The global well-posedness of (1) is a highly nontrivial problem in Sobolev spaces $H^k(\mathbb{R}^n) \times H^{k-1}(\mathbb{R}^n)$ with $k > 2$ since the monotonicity of $|u_t|^{m-1}u_t$ no longer implies *a priori* estimates. Standard techniques treat the nonlinear damping as a perturbation and restrict its growth $m \leq n/(n-2)$ if $n \geq 3$; this is implied by scaling laws (4), (5) and $E_2(u,t) \leq E_2(u,0)$:

$$E_2(u_\lambda, t) = \lambda^{2/(m-1)+2-n} E_2(u, \lambda t) \leq E_2(u, 0), \quad \lambda \to \infty,$$

for $m \geq n/(n-2)$, $n \geq 3$. Hence the existence of concentrating solutions (increasing norms of order $k > 2$) cannot be excluded for large m. There is some possibility to handle the critical case $m = n/(n-2)$ by improved classical techniques similar to the nonlinear wave and Schrodinger equations; see [1,6,23] for the former and [2,24] for the latter. So far the global well-posedness of (1) is established along these lines in a special case: $m = 3$, $n = 3$, and radial data [25]. We present the main ideas and analytic tools.

Our result concerns mainly large data, as the global existence and asymptotic behavior are well-understood for small data; see [8,11,17].

Let us define the Sobolev spaces of spherically symmetric functions

$$H^k_{\mathrm{rad}}(\mathbb{R}^3) = \{u \in H^k(\mathbb{R}^3) : u \text{ is a function of } |x|\},$$

for $k > 1$. These are invariant under the evolution determined by (1)–(2).

We write $Du = (\nabla u, u_t)$ for the space-time derivative of u. The following is partial answer to the question about global well-posedness [25].

Theorem 4.1. *Let* $m = 3$, $n = 3$, *and* $(u_0, u_1) \in H^3_{\mathrm{rad}}(\mathbb{R}^3) \times H^2_{\mathrm{rad}}(\mathbb{R}^3)$. *Then problem (1)–(2) admits a unique global solution* u, *such that*

$$D^\alpha u \in C(\mathbb{R}_+, H^{3-|\alpha|}_{\mathrm{rad}}(\mathbb{R}^3)), \qquad |\alpha| \leq 3.$$

Moreover, u *satisfies the estimates*

$$\sum_{|\alpha|\leq 2} \int_0^t \|D^\alpha u(s)\|^2_{L^\infty}\, ds \leq C_1(u_0, u_1), \quad t \geq 0,$$

$$\sum_{|\alpha|\leq 2} E_1(D^\alpha u, t) \leq C_2(u_0, u_1), \quad t \geq 0,$$

for implicit constants $C_k(u_0, u_1)$, $k = 1, 2$, *which are finite whenever the norm* $\|u_0\|_{H^3} + \|u_1\|_{H^2}$ *is finite.*

Remark 4.1. We rely on the "forbidden" $L^1_t L^2_x - L^2_t L^\infty_x$ Strichartz estimate for the wave equation in $\mathbb{R} \times \mathbb{R}^3$; see [10]. The estimate is valid only for spherically symmetric solutions which explains the condition for spherically symmetric data.

Proof of Theorem 4.1 (outline). Recall a useful fact: for $|\alpha| \le 1$,

$$E_1(D^\alpha u, t) + (2|\alpha|+1)\int_0^t \|u_s D^\alpha u_s\|_{L^2}^2\, ds = E_1(D^\alpha u, 0).$$

Step 1. Let u be a local solution of (1)–(2), such that $E_3(u,t) < \infty$ for $t \in [0, T_*)$. The equation for $D^\alpha u$, $|\alpha| = 2$, involves non-dissipative terms:

$$\Box D^\alpha u + 3u_t^2 D^\alpha u_t + \sum_{\beta+\gamma=\alpha} c_{\beta,\gamma} u_t D^\beta u_t D^\gamma u_t = 0. \tag{14}$$

From $E_2(u,t) \le E_2(u,0)$, the L^2-norms of these terms are bounded by

$$\int_0^t \|D^\alpha u(s)\|_{L^\infty}^2\, ds, \quad |\alpha| \le 2. \tag{15}$$

Step 2. We can handle $L_t^2 L_x^\infty$ norms by means of

$$\left(\int_0^t \|u(s)\|_{L^\infty}^2\, ds\right)^{1/2} \le C E_1^{1/2}(u,0) + C\int_0^t \|\Box u(s)\|_{L^2}\, ds,$$

which is an endpoint Strichartz estimate. If $|\alpha| = 1$, the inequality leads to

$$\left(\int_0^t \|D^\alpha u(s)\|_{L^\infty}^2\, ds\right)^{1/2} \le C E_2^{1/2}(u,0) + C\int_0^t \|u_s^2(s) D^\alpha u_s(s)\|_{L^2}\, ds.$$

It is not trivial to bound the right side by the $L_t^2 L_x^\infty$ norm on the left side. We need a simple "non-concentration" lemma.

Lemma 4.1. *Assume that* $(u_0, u_1) \in H^3(\mathbb{R}^3) \times H^2(\mathbb{R}^3)$ *and* $|\alpha| = 1$. *Let* u *be the solution of problem (1)–(2) extended on a maximal interval* $[0, T_*)$. *For every* $\epsilon > 0$ *there exists* $\delta \in (0, T_0)$ *such that*

$$\sup_{t\in[T_0,T_*)} \int_{t-\delta}^t \left(\|u_s(s)\|_{L^4}^4 + \|u_s(s) D^\alpha u_s(s)\|_{L^2}^2\right) ds < \epsilon.$$

This result follows from $\|u_s(s)\|_{L^4}^4 + \|u_s(s) D^\alpha u_s(s)\|_{L^2}^2 \in L^1([0, T_*))$. Applying Lemma 4.1, we find $C > 0$ and $\delta = \delta(u_0, u_1, \alpha) > 0$, such that

$$\left(\int_0^t \|D^\alpha u(s)\|_{L^\infty}^2\, ds\right)^{1/2} \le C E_2^{1/2}(u,0)(1 + C E_2^{1/2}(u,0))^{1+t/\delta} \tag{16}$$

for all $t \in [0, T_*)$. The estimate for $|\alpha| = 2$ in (15) is similar.

Step 3. To complete the proof of $E_3(u,t) < \infty$, we combine the standard energy estimate from (14) and the $L_t^2 L_x^\infty$-estimate (16) with $|\alpha| = 1, 2$. □

The solutions constructed in Theorem 4.1 are asymptotically free, since the nonlinear term u_t^3 is supercritical for scattering theory in $\mathbb{R}^3$.

Theorem 4.2. *Let* $m = 3$, $n = 3$, *and* $(u_0, u_1) \in H^3_{\mathrm{rad}}(\mathbb{R}^3) \times H^2_{\mathrm{rad}}(\mathbb{R}^3)$. *The global solution* u *of problem (1)–(2) is asymptotically free:*

$$\|D^\alpha u(t) - D^\alpha u_+(t)\|_2 \to 0, \qquad t \to \infty,$$

for $1 \le |\alpha| \le 3$, *where* u_+ *is a solution of* $\Box u_+ = 0$ *with initial data*

$$Du_+(x, 0) \in H^2_{\mathrm{rad}}(\mathbb{R}^3) \times H^1_{\mathrm{rad}}(\mathbb{R}^3).$$

Proof (outline). We use the estimates in Theorem 4.1 and Hölder's inequality to show a sufficient condition for scattering:

$$\int_0^\infty \|D^\alpha u_t^3(t)\|_2 \, dt < \infty, \quad 0 \le |\alpha| \le 2. \qquad \Box$$

References

1. H. Bahouri, P. Gérard, *High frequency approximation of solutions to critical nonlinear wave equations*, Amer. J. Math. 121 (1999), 131–175.
2. J. Bourgain, *Global wellposedness of defocusing critical nonlinear Schrödinger equation in the radial case*, J. Amer. Math. Soc. 12 (1999), 145–171.
3. R. Carles, J. Rauch, *Focusing of Spherical Nonlinear Pulses in* $\mathbb{R}^{1+3}$, Proc. Amer. Math. Soc. 130 (2002), 791–804.
4. J. A. Carrillo, J. L. Vázquez, *Fine asymptotics for fast diffusion equations*, Comm. Partial Differential Equations 28 (2003), 1023–1056.
5. J. Duoandikoetxea, *Fourier Analysis*, transl. and rev. D. Cruz-Uribe, Amer. Math. Soc., Providence, 2000.
6. M. Grillakis, *Regularity and asymptotic behaviour of the wave equation with a critical nonlinearity*, Ann. of Math. 132 (1990), 485–509.
7. A. Haraux, *Comportement à l'infini pour une équation d'ondes non linéaire dissipative*, C. R. Acad. Sci. Paris Sér. A-B 287 (1978), A507–A509.
8. L. Hörmander, *Lectures on Nonlinear Wave Equations*, Math. Appl., vol. 26, Springer, Berlin, 1997.
9. J.-L. Joly, G. Métivier, J. Rauch, *Nonlinear hyperbolic smoothing at a focal point*, Michigan Math. J. 47 (2000), 285–312.
10. S. Klainerman, M. Machedon, *Space-time estimates for null forms and the local existence theorem*, Comm. Pure Appl. Math. 46 (1993), 1221–1268.
11. S. Klainerman, G. Ponce, *Global, small amplitude solutions to nonlinear evolution equations*, Comm. Pure Appl. Math. 63 (1983), 133–141.
12. H. Kubo, *Large-time behavior of solutions to semilinear wave equations with dissipative terms*, preprint, 2006.
13. I. Lasiecka, D. Toundykov, *Energy decay rates for the semilinear wave equation with nonlinear localized damping and source terms—an intrinsic approach*, preprint, 2005.

14. J. Liang, *Nonlinear hyperbolic smoothing at a focal point*, preprint, 2004.
15. J.-L. Lions, W. A. Strauss, *Some non-linear evolution equations*, Bull. Soc. Math. France 93 (1965), 43–96.
16. A. Matsumura, *On the asymptotic behavior of solutions of semi-linear wave equations*, Publ. Res. Inst. Math. Sci. 12 (1976/77), 169–189.
17. T. Matsuyama, *Asymptotics for the nonlinear dissipative wave equation*, Trans. Amer. Math. Soc. 355 (2003), 865–899.
18. K. Mochizuki, *Decay and asymptotics for wave equations with dissipative terms*, Lecture Notes in Phys., vol. 39, Springer, 1975, pp. 486–490.
19. K. Mochizuki, T. Motai, *On energy decay-nondecay problems for wave equations with nonlinear dissipative term in* $\mathbb{R}^N$, J. Math. Soc. Japan 47 (1995), 405–421.
20. T. Motai, K. Mochizuki, *On asymptotic behaviors for wave equation with a nonlinear dissipative term in* $\mathbb{R}^N$, Hokkaido Math. J. 25 (1996), 119–135.
21. M. Nakao, *Asymptotic stability of the bounded or almost periodic solution of the wave equation with nonlinear dissipative term*, J. Math. Anal. Appl. 58 (1977), 336–343.
22. J. Serrin, G. Todorova, E. Vitillaro, *Existence for a nonlinear wave equation with damping and source terms*, Differential Integral Equations 16 (2003), 13–50.
23. J. Shatah, M. Struwe, *Regularity results for nonlinear wave equations*, Ann. of Math. 138 (1993), 503–518.
24. T. Tao, M. Visan, X. Zhang, *Global well-posedness and scattering for the defocusing mass-critical nonlinear Schrödinger equation for radial data in high dimensions*, Duke Math. J. 140 (2007), 165–202.
25. G. Todorova, D. Uğurlu, B. Yordanov, *Regularity and scattering for the wave equation with a critical nonlinear damping*, preprint, 2007.
26. G. Todorova, B. Yordanov, *The energy decay problem for wave equations with nonlinear dissipative terms in* $\mathbb{R}^n$, Indiana Univ. Math. J. 56 (2007), 389–416.

THE BLOW-UP BOUNDARY FOR A SYSTEM OF SEMILINEAR WAVE EQUATIONS

HIROSHI UESAKA

Department of Mathematics, College of Science and Technology, Nihon University,
Tokyo, 101-8308, Chiyodaku Kanda Surugadai 1-8, Japan
E-mail: uesaka@math.cst.nihon-u.ac.jp

We shall show the existence of a blow-up boundary of blow-up solutions for a system of semilinear wave equations. The blow-up boundary is proved to be Lipschitz continuous. It is well-known for a single semilinear wave equation that Caffarelli and Friedman investigated the blow-up boundary of blow-up solutions and proved various remarkable results.

Keywords: Blow-up boundary, system of semilinear wave equations.

1. Introduction

Let u be a real-valued solution for some wave equation in $\mathbf{R}^3 \times [0, \infty)$ and let u blow-up in finite time. The set Γ is defined by

$$\Gamma = \partial\{u = \infty\} \cap \{t > 0\}$$

and it is given by $\Gamma : \ t = \varphi(x)$. Then we call u a *pointwise blow-up solution*. The set of $(x, \varphi(x))$ is called the *blow-up boundary* of u.

Caffarelli and Friedman investigated a blow-up boundary for a single semilinear wave equation and published their results in [2,3]. Later Alinhac summarized a part of their results in his book [1]. They studied the following Cauchy problem for a semilinear wave equation:

$$\begin{cases} \Box u = F(u) = u^p \text{ in } \mathbf{R}^3 \times (0, T), \\ u(x, 0) = f(x), \ \ \partial_t u(x, 0) = g(x) \text{ in } \mathbf{R}^3, \end{cases} \tag{1}$$

where $\Box = \partial_t^2 - \triangle$, T is a positive constant and $p > 1$. Caffarelli and Friedman proved that the blow-up boundary is a Lipschitz continuous surface and further it is a C^1 space-like one.

Let u be a pointwise blow-up solution within $(0, T]$ for some nonlinear wave equation. T is considered in two ways as follows:

1. T is determined from a pointwise blow-up solution given in advance (see [4,5,8]), or,

2. we can manage initial data for any fixed T given in advance so that a solution blows up pointwise within $(0,T]$ (see [1,2]).

We consider the following Cauchy problem for a system in $\mathbf{R}^3 \times [0,T)$ with $0 < T < \infty$ and assign initial data to it, that is, we are interested in

$$\Box_1 u_1 = F_1(u_1,u_2),\ \Box_2 u_2 = F_2(u_1,u_2) \text{ in } \mathbf{R}^3 \times (0,T), \tag{2}$$

where $\Box_k = \partial_t^2 - c_k^2\Delta,\ c_1 \geq c_2 > 0$, and

$$\begin{cases} u_1(x,0) = f_1(x),\ \partial_t u_1(x,0) = g_1(x) \text{ in } \mathbf{R}^3, \\ u_2(x,0) = f_2(x),\ \partial_t u_2(x,0) = g_2(x) \text{ in } \mathbf{R}^3, \end{cases} \tag{3}$$

where $f_k(x) \in C^3(\mathbf{R^3}),\ g_k(x) \in C^2(\mathbf{R^3})$ and additional assumptions are supposed later. The function $F_k = F_k(u_1,u_2)\,(k=1,2)$ is given by

$$F_k(u_1,u_2) = a_k u_1^{p_k} u_2^{q_k} \quad \text{or} \quad a_k u_1^{p_k} + b_k u_2^{q_k},$$

where $p_k, q_k > 1$ and $a_k, b_k > 0$.

Let

$$\begin{aligned} u_{k,0}(x,t) &= \frac{\partial}{\partial t}\left(\frac{t}{4\pi}\int_{|\omega|=1} f_k(x+c_k t\omega)d\omega\right) + \frac{t}{4\pi}\int_{|\omega|=1} g_k(x+c_k t\omega)d\omega, \\ &= \frac{1}{4\pi}\int_{|\omega|=1}\big[f_k(x+c_k t\omega) \\ &\qquad + t\,(c_k\omega\cdot\nabla f_k(x+c_k t\omega) + g_k(x+c_k t\omega)\big]d\omega, \end{aligned} \tag{4}$$

where $k=1,2$ and $x,\omega \in \mathbf{R}^3,\ |\omega|=1$, and $\int_{|\omega|=1}\cdots d\omega$ means the integral over the unit sphere. Then the problem (2)–(3) is converted as usual into the system of integral equations:

$$u_k(x,t) = u_{k,0}(x,t) + \frac{1}{4\pi}\int_0^t (t-s)ds\int_{|\omega|=1} F_k(\vec{u}(x+c_k(t-s)\omega,s))d\omega, \tag{5}$$

where $\vec{u} = (u_1,u_2) \in \mathbf{R}^2$.

Many results concerning (2)–(3) have been proved, for example, existence, regularity, blow-up and other properties of solutions. Here we do not give the proof of the existence of solutions. But we suppose the following.

A.1.1. *The Cauchy problem (2)–(3) has a local unique C^2-solution $\vec{u}$ within $(0,T]$ which satisfies the integral equation (5).*

2. Positive Solutions

Let R be any fixed positive number. We denote a ball of radius R with the center at the origin by $B_R = \{x \in \mathbf{R}^3 \mid |x| \leq R\}$. Let T be any fixed positive constant.

We formulate two assumptions in this section. These are essentially the same as those in [2].

A.2.1. *Let $X = x + c_1 t\omega$, where $x \in B_R$, $0 \leq t \leq T$ and $\omega \in \mathbf{R}^3, |\omega| \leq 1$. We suppose*

(1) $f_k(x) \in C^3(\mathbf{R}^3)$ and $g_k(x) \in C^2(\mathbf{R}^3)$,
(2) $f_k(X) \geq 0$ and $g_k(X) \geq c_k|\nabla f_k(X)|$.

A backward characteristic cone for $\square_k$ is defined and denoted by

$$C_k(x,T) = \{(\xi,\tau) \in \mathbf{R}^3 \times [0,\infty) \mid |x-\xi| \leq c_k(T-\tau)\}.$$

We define $K_{k,R,T} = \cup_{x \in B_R} C_k(x,T)$.

Theorem 2.1. *Suppose **A.1.1** and **A.2.1**. Let $\vec{u}$ be a solution of (2)–(3). Then $u_k(x,t) \geq 0$ for $k = 1,2$ in the domain of existence of solutions $\vec{u}$ contained in $K_{1,R,T}$.*

Theorem 2.1 is essential for proving other results. Not only the result of Theorem 2.1 but also the reasoning to prove the theorem are often applied to get statements which we shall introduce below.

Proof. We only give an outline. The assumption ***A.2.1*** shows that $u_{0,k}(x,t) \geq 0$ for $k = 1,2$ (see (5)). Assume either u_1 or u_2 becomes negative at some point $(x,t) \in K_{1,R,T}$. We take the backward characteristic cone $C_k(x,t)$ and use the finite speed of propagation of solutions. By means of Keller's argument [5,7,8] we are led to a contradiction if $u_k < 0$. □

We remark that $u_k(x,t) > 0$ holds if $f_k(X) > 0$ in ***A.2.1.*** Moreover, we suppose the following.

A.2.2. *Let $k = 1,2$. Let each λ_k be any fixed positive number. Put $X = x + c_1 t\omega$, $x \in B_R$, $0 \leq t \leq T$. Suppose*

(1) $f_k(x) \in C^4(B_{R+T})$ and $g_k(x) \in C^3(B_{R+T})$,
(2) $g_k(X) \geq \lambda_k|\nabla f_k(X)|$,
(3) $c_k^2 \Delta f_k(X) + F_k(\vec{f}(X)) \geq (c_k + \lambda_k)|\nabla g_k(X)| + \lambda_k c_k |\nabla^2 f_k(X)|$,

where $|\nabla^2 f(x)| = \sqrt{\sum_{i,j=1}^3 |\frac{\partial^2}{\partial x_i \partial x_j} f(x)|^2}$.

Theorem 2.2. *Suppose that the assumptions* ***A.1.1*** *and* ***A.2.1-2*** *are satisfied. Let* $\vec{u} = (u_1, u_2)$ *be a solution of (2)–(3). Then*

$$\partial_t u_k(x,t) \geq \lambda_k |\nabla u_k(x,t)| \text{ in } K_{1,R,T}. \tag{6}$$

Proof. We define $L_k(x,t) = \partial_t u_k(x,t) + \lambda_k \mathbf{e} \cdot \nabla u_k(x,t)$, where $\mathbf{e}$ is any unit vector of $\mathbf{R}^3$. Then L_k satisfies

$$\begin{cases} \Box_k L_k(x,t) = \partial_1 F_k(\vec{u}) L_1(x,t) + \partial_2 F_k(\vec{u}) L_2(x,t), \\ L_k(x,0) \quad = g_k(x) + \lambda_k \mathbf{e} \cdot \nabla f_k(x), \\ \partial_t L_k(x,0) = c_k^2 \Delta f_k(x) + F_k(\vec{f}(X)) + \lambda_k \mathbf{e} \cdot \nabla g_k(x), \end{cases}$$

where $\partial_l F_k(s_1, s_2) = \partial_{s_l} F_k(s_1, s_2)$ for $k, l = 1, 2$.

According to the same reasoning as in the proof of Theorem 2.1, the assumption **A.2.2** assures that the above equation has a positive solution L_k. The desired result follows from $L_k \geq 0$. □

3. Blow-up

Let $\vec{u} = (u_1, u_2)$ be a solution of (2)–(3) with positive components. It is called *a pointwise blow-up solution* in B_R if there exists a positive constant $T_k(x) < \infty$ depending on $x \in B_R$ such that

$$\lim_{t \nearrow T_k(x)} u_k(x,t) = \infty \, (k = 1, 2).$$

We shall show that there exists a sufficient condition for (2)–(3) to have a pointwise blow-up solution in B_R.

Lemma 3.1 (Keller [5]). *Let* $f(s) = As^p + B$, *where* $A > 0$, $B \geq 0$ *and* $p > 1$. *Let* v *be the solution of the ordinary differential equation*

$$\begin{cases} v''(t) = f(v(t)) \; (t > 0, \, ' = \frac{d}{dt}), \\ v(0) = \alpha, \; v'(0) = \beta, \end{cases} \tag{7}$$

where α, $\beta \geq 0$. *Then there exists some finite positive* T *such that*

$$\lim_{t \nearrow T} v(t) = \infty. \tag{8}$$

Suppose the following.

A.3.1. $f_k(X) \geq \alpha$ *and* $g_k(X) \geq \beta + c_k |\nabla f_k(X)|$ *for* $X = x + c_1 t\omega$, $x \in B_R$, $0 \leq t \leq T$.

Then we have the following result.

Theorem 3.1. *Let* $\vec{u} = (u_1, u_2)$ *be the* C^2*-solution of (2)–(3). Let* T *be given in Lemma 3.1. Moreover, let us suppose that* ***A.3.1*** *holds. If* $\vec{u}$ *keeps* *to be a solution from the class C^2 till u_k attains ∞, then there exists $T_k(x)$ depending on $x \in B_R$ such that $0 < T_k(x) \leq T$ and*

$$\lim_{t \nearrow T_k(x)} u_k(x,t) = \infty \text{ for } x \in B_R.$$

Proof. We first put $f(v) = F_1(v,\alpha)$ in (7). Then Lemma 3.1 holds. Let $w_1(x,t) = u_1(x,t) - v(t)$. We remark that $F_1(u_1,u_2) \geq F_1(u_1,\alpha)$ for $t \geq 0$ because u_k is monotone increasing in t. By the mean value theorem

$$F_1(u_1,u_2) - F_1(v,\alpha) \geq F_1(u_1,\alpha) - F_1(v,\alpha) = \partial_1 F_1(u_1',\alpha)w_1, \qquad (9)$$

where u_1' is a point between u_1 and v, and $\partial_1 F_1(u_1',\alpha) > 0$ when $u_1', \alpha > 0$.

We make the same computation for u_2. Then we have

$$\begin{cases} \Box_k w_k = F_k(\vec{u}) - F_k(\vec{v}) \geq \partial_k F_k(\vec{u}')w_k, \\ w_k(x,0) = f_k(x) - \alpha, \ \partial_t w_k(x,0) = g_k(x) - \beta, \end{cases}$$

where $\vec{u}' = (u_1',\alpha)$ or (α,u_2'), and $\vec{v} = (v,\alpha)$ or (α,v).

According to the same argument as in the proof to Theorem 2.1, we can get $u_k(x,t) \geq v(t)$ for $x \in B_R$, $t \leq T$. By Lemma 3.1, there exists $T_k(x)$ such that $0 < T_k(x) \leq T$ and $\lim_{t \nearrow T_k(x)} u_k(x,t) = \infty$. Thus the proof is completed. □

4. Blow-up Boundary

Let c_1 and c_2 be the propagation speeds and $c_1 \geq c_2$. We shall prove that a blow-up boundary actually exists over D_R and it is Lipschitz continuous with Lipschitz constant c_1^{-1}. First we define the *influence domain* (see [1,6]).

Definition 4.1. Let κ be a positive constant. The open set $\mathcal{F}_\kappa$ is called an influence domain if it satisfies

(1) $\mathcal{F}_\kappa \subset \mathbf{R}^3 \times [0,\infty)$,
(2) $C_\kappa(x,t) \subset \mathcal{F}_\kappa$ for $(x,t) \in \mathcal{F}_\kappa$,

where $C_\kappa(x,t) = \{(x',t') \in \mathbf{R}^3 \times [0,\infty) \mid |x-x'| \leq \kappa(t-t')\}$.

Lemma 4.1. *Let $\mathcal{F}_\kappa$ be an influence domain and let $\psi(x) := \sup\{t,\ (x,t) \in \mathcal{F}_\kappa\}$. Then the following is true: either $\psi(x) \equiv \infty$ or*

$$|\psi(x) - \psi(y)| \leq \frac{1}{\kappa}|x-y|.$$

The proof is easy, so we omit it (see [1,6]).

We assume the following.

A.4.1. *Put* $\lambda_1 = \lambda_2 = c_1$ *in* ***A.2.2*** *and in Theorem 2.2.*

Let $\vec{u} = (u_1, u_2)$ be a solution of (2)–(3) satisfying the assumptions ***A.1.1***, ***A.2.1-2*** and ***A.4.1***. We suppose that there exists a positive constant T such that $\vec{u}(x,t)$ does not exist as a C^2-solution for $t > T$. Let x be any fixed point in B_R. Then $u_k(x,t)$ is monotone increasing in t. Moreover, we let $\{u_{k,n}\}$, $k = 1, 2$ to be a sequence of successively approximating solutions to u_k in $K_{1,R,T}$. Here $u_{k,0}$ is equal to (4), and the approximating solutions $u_{k,n}$ satisfy for $n \geq 1$ in $K_{1,R,T}$

$$\begin{cases} \Box u_{k,n} = F_k(\vec{u}_{n-1}), \\ u_{k,n}(x,0) = f_k(x) \text{ and } \partial_t u_{k,n}(x,0) = g_k(x), \end{cases} \tag{10}$$

where $\vec{u}_n = (u_{1,n}, u_{2,n})$. Using (5), $\vec{u}_n$ is defined as an integral of $\vec{u}_{n-1}$. Then $u_{k,n}$ has the following properties in $K_{1,R,T}$:

(1) $u_{k,1} \leq u_{k,2} \leq \cdots \leq u_{k,n} \leq \cdots$,
(2) $\partial_t u_{k,n}(x,t) \geq c_1 |\nabla u_{k,n}(x,t)|$,
(3) $u_{k,n}(x,t)$ is monotone increasing in t.

We can prove these properties easily by the expression of the integral form and by the same argument as those for proving the properties for u_k.

Let $x \in B_R$. We define a subset $\Omega_k \subset K_{k,R,T}$ by

$$\Omega_k = \{(x,t) \mid u_k(x,t) = \lim_{n\to\infty} u_{k,n} \text{ exists and is of } C^2\}.$$

We set $\Omega = \Omega_1 \cap \Omega_2$. We define $\varphi_k(x)$ for $x \in B_R$ by $\varphi_k(x) = \sup\{t \mid (t,x) \in \Omega_k\}$, and $\varphi(x)$ by $\varphi(x) = \min(\varphi_1(x), \varphi_2(x))$. We shall show that Ω is the influence domain, $\varphi_1(x) = \varphi_2(x)$, and $\lim_{t\nearrow\varphi(x)} u_k(x,t) = \infty$.

Theorem 4.1. *Let $\vec{u}$ be the solution stated in the above paragraphs. Then $\vec{u}$ is a blow-up solution of class C^2. The following holds:*

(1) $\varphi(x) = \varphi_1(x) = \varphi_2(x)$, *whence* $\Omega = \Omega_1 = \Omega_2$,
(2) $|\varphi(x) - \varphi(y)| \leq c_1^{-1}|x - y|$ *for* $x, y \in B_R$,
(3) $\lim_{t\nearrow\varphi(x)} u_k(x,t) = \infty$ *for* $k = 1, 2$.

Proof. Let x be any point in B_R. We shall prove $\lim_{t\nearrow\varphi(x)} u_k(x,t) = \infty$ for $k = 1, 2$. We take any $t < \varphi(x)$ and fix it. We set $m = (x,t)$. We define the backward characteristic cone $C_k(m)$ by

$$C_k(m) = \{\, (x',t') \mid |x - x'| \leq c_k(t - t'), t' \geq 0 \,\}.$$

We suppose that $\lim_{\tau\nearrow\varphi(x)} u_k(x,\tau) = A < \infty$. Let $m' = (x',t')$ be any point $\in C_1(m)$. Then $0 \leq u_k(m') \leq A$ holds for $k = 1, 2$. We shall show this inequality. By $[mm']$ we denote the segment connecting m and m'. Let p be any point on $[mm']$, so $p = m + s(m - m')$ with $0 \leq s \leq 1$. By the Taylor formula and (2) of the above property we have

$$\begin{aligned} u_{k,n}(m) &= u_{k,n}(m') + \int_0^1 \{(t-t')\partial_t u_{k,n}(p) + (x-x')\cdot \nabla u_{k,n}(p)\}\, ds \\ &\geq u_{k,n}(m'), \end{aligned} \tag{11}$$

because $\partial_t u_{k,n}(p) \geq c_1|\nabla u_{k,n}(p)| > \frac{|x-x'|}{t-t'}|\nabla u_{k,n}(p)|$. Thus we have shown $0 \leq u_k(m') \leq A$.

We compare the first and second order derivatives of $u_{k,n}$ with $W = M\exp(Bt)$ and its derivative. By the same argument as in Lemma 2.3 from [2] or in the proof of Theorem 4.1 of Chapter 3 from [1], if we choose M and B suitably large we can show that the absolute value of these derivatives are less than some positive constant C, which depends on A, B and M, and is independent of n.

By an argument similar to that in Satz 1 from [4] we can show that $u_{k,n}$ converges in $C_k(m)$ to u_k as $n \to \infty$ and u_k is of class C^2 in $C_k(m)$. Thus both u_1 and u_2 exist together with their derivatives up to second order at $(x, \varphi(x))$. This is contradiction. There is a possibility that either u_1 or u_2 does not go to ∞ at $x \in B_R$ as $t \nearrow \varphi(x)$. But by means of the construction of the solution such a possibility is excluded. Then $\lim_{t\nearrow\varphi(x)} u_k(x,t) = \infty$ for $k = 1, 2$. Consequently, we have $\varphi_1 = \varphi_2$, $\Omega_1 = \Omega_2$ and $C_1(m) \subset \Omega$. Then Ω is the influence domain and $\varphi(x)$ is Lipschitz continuous with Lipschitz constant c_1^{-1} by Lemma 4.1. Thus the proof is complete. □

5. Blow-up Rates

In this section we study the order of infinity of $u_k(x,t)$ when $t \to \varphi(x)$. We cannot yet prove such results for different propagation speeds and for product type nonlinearity. For this reason we suppose the following.

A.5.1. (1) $c_1 = c_2 = 1$,
(2) $F_k(\vec{u}) = a_k u_1^{p_k} + b_k u_2^{q_k}$, *where* $a_k, b_k > 0$ *and* $p_k, q_k > 1$.

We put $G_k(u_1) = a_k u_1^{p_k}$ and $H_k(u_2) = b_k u_2^{q_k}$. Let $\vec{u}$ be the pointwise blow-up solution. The function J_k is defined in Ω by

$$J_k(x,t) = \partial_t^2 u_k(x,t) - F_k(\vec{u}) + m_k\partial_t u_k(x,t) = \Delta u_k(x,t) + m_k\partial_t u_k(x,t),$$

where m_k is some positive large constant given afterward, and it satisfies

$$\begin{cases} \Box J_k(x,t) &= G_k'(u_1)J_1(x,t) + H_k'(u_2)J_2(x,t) \\ &\quad + G_k''(u_1)|\nabla u_1|^2 + H_k''(u_2)|\nabla u_2|^2, \\ J_k(x,0) &= \triangle f_k(x) + m_k g_k, \\ \partial_t J_k(x,0) &= \triangle g_k + m_k(\triangle f_k + F_k(\vec{f})). \end{cases}$$

By the same reasoning as in the proof of positivity of solutions (Theorem 2.1) we get the following lemma.

Lemma 5.1. *Suppose $f_k(x) > 0$ and that the assumptions **A.1.1**, **A.2.1-2** and **A.5.1** are satisfied. Let m_k be sufficiently large such that $\partial_t J_k(X,0) \geq |\nabla J_k(X,0)|$ holds. Then we have $J_k(x,t) \geq 0$ in Ω.*

Theorem 5.1. *Let $\vec{u}$ be the solution stated in Lemma 3.1. Then there exists a suitable positive constant C_k such that*

$$u_k(x,t) \leq \frac{C_k}{(\varphi(x)-t)^{p_k'}} \quad in\ \Omega, \quad where \quad p_k' = \frac{2}{p_k - 1}. \tag{12}$$

Proof. Let $k = 1$. We have $\partial_t^2 u_1 + m_1 \partial_t u_1 \geq a_1 u_1^{p_1} + b_1 u_2^{q_1} \geq a_1 u_1^{p_1}$. By multiplying this inequality by $\partial_t u_1 e^{2m_1 t}$ and by integrating it from t_0 to t, where $t_0 < t < \varphi(x)$, we have

$$(\partial_t u_1(\cdot,t))^2 \geq C_1 \int_{t_0}^{t} \partial_s (u_1(\cdot,s)^{p_1+1}) ds + C_2 \geq C_3 u_1(\cdot,t)^{p_1+1}.$$

We choose a suitable constant C and we have

$$\partial_t u_1(x,t) \geq C u_1(x,t)^{\frac{p_1+1}{2}}, \ C = \sqrt{C_3}. \tag{13}$$

We integrate both sides of (13) from 0 to t and get (12) for $k=1$. We can carry out the same calculation for u_2 and get (12) for $k=2$. □

Remark 5.1. We do not have an estimate for u_1 and u_2 from below, but at least we have

$$u_1(x,t) + u_2(x,t) \geq \frac{C}{(\varphi(x)-t)^{r'}},$$

where $r' = \frac{2}{r-1}$, $r = \max\{p_1, p_2, q_1, q_2\}$.

Remark 5.2. We can consider the same problem in $\mathbf{R}^d \times [0,T)$ for $d = 1, 2$.

References

1. S. Alinhac, *Blowup for Nonlinear Hyperbolic Equations*, Birkhäuser, 1995.
2. L. A. Caffarelli and A. Friedman, The blow-up boundary for nonlinear wave equations, *Trans. Amer. Math. Soc.* **297** (1986), 223–241.
3. L. A. Caffarelli and A. Friedman, Differentiability of the blow-up curve for one dimensional nonlinear wave equations, *Arch. Ration. Mech. Anal.* **91** (1985), 83–98.
4. K. Jörgens, Das Anfangswertproblem im Grossen für eine Klasse nichtlinearer Wellengleichungen, *Math. Z.* **77** (1961), 295–308.
5. J. B. Keller, On solutions of nonlinear wave equations, *Comm. Pure Appl. Math.* **10** (1957), 523–530.
6. H. Lindblad, Blow-up for solutions of $\Box u = |u|^p$ with small initial data, *Comm. Partial Differential Equations* **15** (1990), 757–821.
7. H. Uesaka, Oscillation or nonoscillation property for semilinear wave equations, *J. Comput. Appl. Math.* **164/165** (2004), 723–730.
8. H. Uesaka, Non-negative solutions of the Cauchy problem for semilinear wave equations and non-existence of global non-negative solutions, to appear, *Advances in Analysis* (Proceedings of 5th ISAAC Congress 2005), World Scientific.

AUTHOR INDEX